Spring Technical Meeting of the Eastern States Section of the Combustion Institute 2018

AA004495

State College, Pennsylvania, USA
4 - 7 March 2018

ISBN: 978-1-5108-6185-5

Printed from e-media with permission by:

Curran Associates, Inc.
57 Morehouse Lane
Red Hook, NY 12571

Some format issues inherent in the e-media version may also appear in this print version.

Copyright© (2018) by The Combustion Institute
All rights reserved.

Printed by Curran Associates, Inc. (2018)

For permission requests, please contact The Combustion Institute at the address below.

The Combustion Institute
5001 Baum Boulevard
Suite 644
Pittsburgh, PA 15213-1851
USA

Phone: (412) 687-1366
Fax: (412) 687-0340

Office@CombustionInstitute.org

Additional copies of this publication are available from:

Curran Associates, Inc.
57 Morehouse Lane
Red Hook, NY 12571 USA
Phone: 845-758-0400
Fax: 845-758-2633
Email: curran@proceedings.com
Web: www.proceedings.com

TABLE OF CONTENTS

OXYGENATES

SOOTING TENDENCIES

SOOTING FLAME SIMULATIONS

SOOT

DROPLETS

SPRAY FLAMES

NEW DEVICES

COAL AND BIOMASS

REACTION KINETICS IV

REACTION KINETICS V

REACTION KINETICS VI

NOX KINETICS

FIRE - INSTRUMENTATION

POOL FIRES

SOLID PROPELLANTS AND EXPLOSIVES

REFRIGERANTS

TURBULENT FLAMES

TURBULENT COMBUSTION MODELING

Spring Technical Meeting
Eastern States Section of the Combustion Institute
March 4-7, 2018
State College, Pennsylvania

Evaluated Rate Constants for *i*-Butane + H and CH_3: Shock Tube Experiments with Bayesian Model Optimization

Laura A. Mertens,[1,] Ifthikar A. Awan[1] and Jeffrey A. Manion[1]*
[1]Chemical Sciences Division, National Institute of Standards and Technology, Gaithersburg, MD, 20899-8320
[*]Corresponding Author Email: laura.mertens@nist.gov

Abstract: The reactions of *i*-butane with CH_3 and H were investigated with shock tube experiments (870 K to 1130 K and 140 kPa to 360 kPa). Propene and *i*-butene, measured with GC/FID and MS, were quantified as characteristic of H-abstraction from the primary and tertiary carbons, respectively. A comprehensive Cantera[1] kinetics model based on JetSurF 2.0[2] was optimized to these experimental measurements and literature data – including early experiments at low temperatures, measurements of the total rate constant, and measurement from ethane for the rate constant for primary carbons – using the Method of Uncertainty Minimization using Polynomial Chaos Expansions (MUM-PCE) – pioneered by David Sheen and Hai Wang.[3] For both H and CH_3, the optimization increased the rate H-abstraction from the tertiary carbon relative to the primary carbon. We combined our primary and tertiary rate constants with previous results from our group on *n*-butane[4] to get site-specific rate constants for the reaction of H and CH_3 with a generic primary, secondary and tertiary carbon.
***Key Words**: kinetics, pyrolysis, combustion, modeling, H-abstraction*

1. Introduction

Combustion chemistry is driven by chain reactions propagated by free radicals, like H and methyl (CH_3).[5] H and CH_3 radicals readily react with hydrocarbons, typically abstracting a H to form H_2 or CH_4, respectively. The rate of this reaction is highly dependent on the structure of the carbon center from which the H is abstracted. Due to the relative strengths of their C-H bonds, tertiary carbons react faster with H and CH_3 than secondary carbons, which react faster than primary carbons. The global structure and the mass of the alkane do not strongly affect the rate of hydrogen abstraction, so rate constants for a large hydrocarbon fuel are often calculated from the number of its primary, secondary and tertiary carbons.[6]

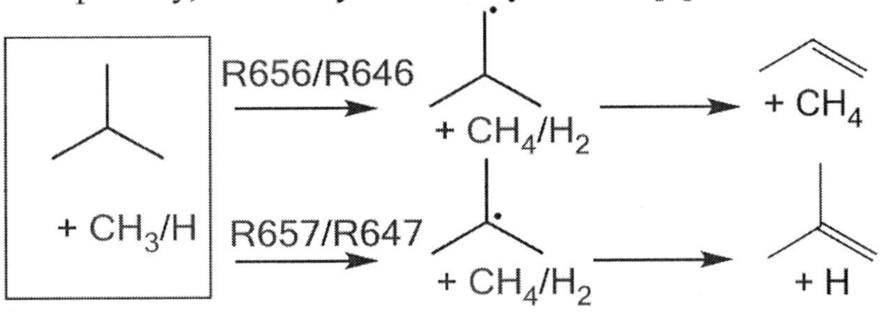

Figure 1. Reactions of CH_3 (red, reactions R656 and R657) and H (blue, reactions R646 and R647) with *i*-butane.

Here, we found relative rate constants for the reaction of *i*-butane with H and CH_3 radicals with shock tube experiments. The shock tube data was combined with literature data to find to find rates for abstraction at both the primary and tertiary carbons using MUM-PCE. Combining primary and tertiary rate constants from this work with previous work on *n*-butane,[4] we propose self-consistent rate constants for H-abstraction from primary, secondary and tertiary carbon centers by H and CH_3.

2. Methods

All experiments were performed in the NIST shock tube reactor which has been described elsewhere.[4] H and CH_3 radicals were produced from pyrolysis of C_2H_5I and *di-tert*-butylperoxide (*dt*BPO), respectively. After the H or CH_3 reacted with an excess of *i*-butane, the products – especially propene and *i*-butene, produced from H-abstraction from the primary carbon and secondary carbons, respectively– were measured with gas chromatography with flame ionization and mass spectrometric detection. The ratio of [propene]/[*i*-butene] approximately equals the branching ratio of k_{656}/k_{657} for experiments with *dt*BPO and of k_{646}/k_{647} for experiments with C_2H_5I. An excess of toluene was added to scavenge free radicals. The shock temperature and pressure were (868 to 1131) K and (130 to 360) kPa.

Table 1. Rate Constants for the Main Reactions of the Prior Model. Parameters are for $k = A \times T^n \times \exp(-E_a/RT)$. *A* is in units of mol, s, cm. E_a/R is in units of K.

Reaction number	Reaction		$\log_{10}A$	n	E_a/R	Reference
645	$CH_3 + iC_3H_7 \leftrightarrow iC_4H_{10}$	k_∞	15.15	-0.68	0	JetSurF 2.0[2]
		k_0	61.62	-13.33	1964.3	
646	$H + iC_4H_{10} \leftrightarrow H_2 + iC_4H_9$	k	6.257	2.54	3400.	Tsang and JetSurF 2.0[2, 7]
647	$H + iC_4H_{10} \leftrightarrow H_2 + tC_4H_9$	k	5.780	2.4	1300.	
656	$CH_3 + iC_4H_{10} \leftrightarrow CH_4 + iC_4H_9$	k	0.132	3.65	3600.	
657	$CH_3 + iC_4H_{10} \leftrightarrow CH_4 + tC_4H_9$	k	-0.044	3.46	2314	
673	Toluene + H $\leftrightarrow$ Benzyl + H_2	k	14.17	0	4500.	Sheen *et al.*[8]
674	Toluene + H $\leftrightarrow$ Benzene + CH_3	k	6.230	2.2	2000.	Sheen *et al.*[8]
676	Toluene + CH_3 $\leftrightarrow$ Benzyl + CH_4	k	11.50	0	4780.6	JetSurF 2.0[2, 9]
805	Benzyl + CH_3 $\leftrightarrow$ Ethylbenzene	k	13.08	0	111.2	Brand *et al.*[10]

We created a chemical model with the Cantera software package[1] based off the JetSurF 2.0 chemical mechanism.[2] This mechanism was modified by David Sheen to include compounds commonly used for shock tube experiments at NIST[11]'[4, 8] and C_2H_5I decomposition from Bentz *et al.* Rate constants for the decomposition of *i*-butyl radicals were taken from experiments in our laboratory that measured that *i*-C_4H_9 is dominated by C-C scission. Under 1050 K, the kinetics model predicts that >96 % of propene is from R646 and R656 – abstraction from the primary carbon. For the entire temperature range of the experiments, ≥97 % of *i*-butane is from abstraction from the tertiary carbon, R647 and R657.

The Cantera kinetics was optimized to the experimental data using Method of Uncertainty Minimization using Polynomial Chaos Expansions (MUM-PCE) with the mumpce 0.1 software package published by David Sheen.[3, 12] First we performed a sensitivity analysis to find what reactions – called "active parameters" – are most important to predicting the outcome of the shock experiment (Table 1). With Bayes theorem, MUM-PCE then optimized the active parameters' rate constants to better match experimental data from the shock tube and relevant literature data.[13] The literature data included measurements of raw experimental data, total rate constants ($k_p + k_s$), relative rate constants, and primary rate constants from ethane.

3. Results and Discussion

Product distributions are shown for a typical trial on Figure 2 for all species with abundances of 1 % or more of acetone. Besides acetone, which is produced from *dt*PBO decomposition, the two most abundant species are methane and ethane, both from CH_3 radical chemistry: methane from R656 and R657 and ethane from CH_3 self-recombination. As the main products of CH_3 reaction with *i*-butane, propene and *i*-butene were also detected in significant amounts. We also measured aromatic compounds from the reactions of CH_3 with toluene, including ethylbenzene (R805), benzene (R674, H reactant produced from R657) and bibenzyl. All products except acetone increase with temperature (Figure 2), and this increase accelerates above 1050 K. This correlates with the decomposition of *i*-butane (R645), which starts to decompose at 1050 K, producing additional CH_3 radicals.

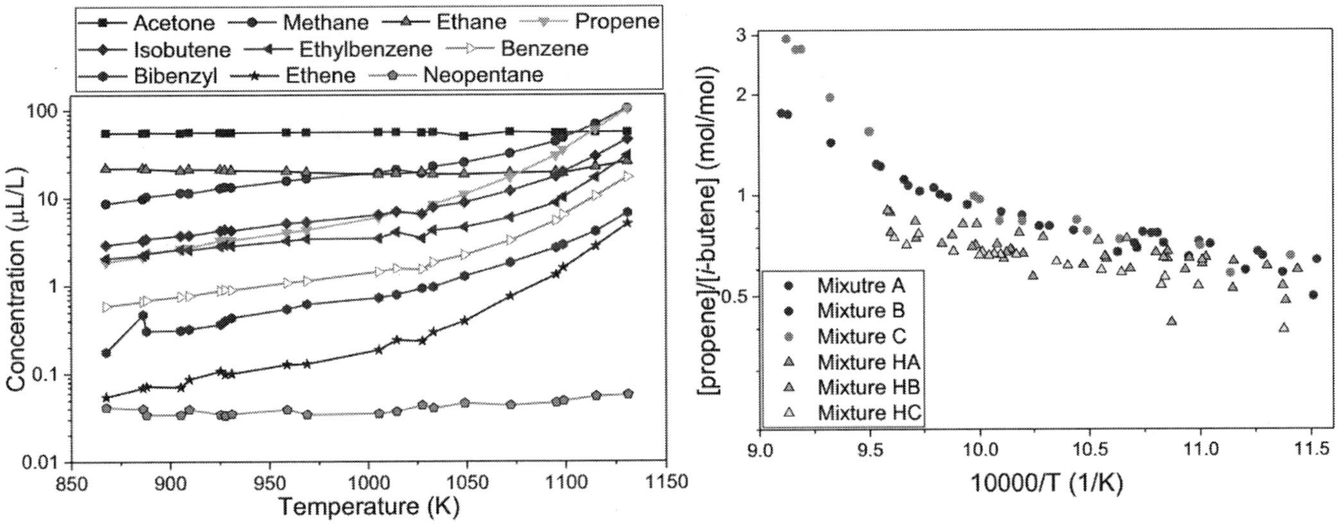

Figure 2. Left: Products a mixture with *dt*BPO. Right: [propene]/[i-butene] for all mixtures. Mixtures A, B and C have *dt*BPO and mixtures HA, HB and HC

As with mixtures *dt*BPO, mixtures with C_2H_5I had significant amounts of *i*-butene and propene from R646 and R647, respectively. Since CH_3 radicals were produced as co-products with propene from R646, we observed methane at concentrations about equal to those of propene. Ethane concentration, due to the lower CH_3 concentrations for mixtures with C_2H_5I – about an order of magnitude less than methane concentrations. We, again, measured significant amounts of benzene, ethylbenzene, and bibenzyl from toluene chemistry.

Assuming no side chemistry, the ratio of [propene]/[*i*-butene] is equal to the branching ratio of k_{646}/k_{647} for H-dominated mixtures with C_2H_5I and k_{656}/k_{657} CH_3-dominated mixtures with *dt*BPO. Figure 1 shows that for both H and CH_3 the ratio is about equal to 1 at temperatures between 868 and 1052K. This ratio increases with temperature, as tertiary abstraction is more favorable at lower temperatures due to the lower C-H bond strengths (and therefor lower activation energy for reaction R646 and R647) for tertiary carbons.[7] The branching ratio is higher for CH_3 than H, meaning that H is more selective for the tertiary carbon than CH_3.

The data from these shock tube experiments (including measurements of the ratio, propene, ethene, benzene, methane and ethane). We used MUM-PCE to find the sensitivity of the prior model – or more specifically, the model's predictions for all experiments used for the optimization – to changing each of its 1487 rate constants. This analysis lead to the choice of nine rate constants to optimize – which we call active parameters (Table 1). For most of these nine active parameters (Table 1), we only optimized the A-factor (which is the same as multiplying the rate constant by a temperature- and pressure-independent constant). For reactions R646, R647, R656, and R657, we optimized both the A-factor and the activation energy, E_a, as these reactions are the primary focus of this work and are the most well-constrained by our experiment and the chosen literature data.

Table 2. Posterior Active Parameters with Factorial Uncertainties (f_A for A and f_{Ea} for E_a). A is in units of mol, s, cm. E_a/R is in units of K.

Reaction number		$\log_{10}A$	f_A	n	E_a/R	f_{Ea}
645	k_∞	15.57	1.32	-0.68	0	-
	k_0	61.62	-	-13.33	1964.3	-
646	k	6.17	1.55	2.54	3390	1.09
647	k	5.88	1.45	2.4	1409	1.18
656	k	0.233	1.67	3.65	3819	1.14
657	k	0.191	1.41	3.46	2362	1.08
673	k	14.23	1.42	0	4500	-
674	k	6.15	1.35	2.2	2000	-
676	k	11.54	1.45	0	4780.6	-
805	k	12.83	1.61	0	111.2	-

We first optimized all active parameters on Table 1 to only our data We updated the kinetics model with the optimized rate constants for R645, R673, R673, R676 and R805, and reoptimized R646, R647, R656 and R657 to (1) the literature data and (2) our rate constants for R646, R647, R656 and R657 found from the first optimization. This two-step optimization was done so that our data would have approximately equal weight to the other experiments.

The posterior model also well describes our experimental data for most products and most experimental conditions, to within 35 %. The optimization decreased this relative rate constant to match the experimental ratios. From 950 -1200 K the ratio of k_{656}/k_{657} was lowered by (61 to 62) %, mostly due to a 72 % increase in the A-factor for CH_3 reaction with the tertiary carbon of *i*-butane (R657). The optimization also increased the relative amount of tertiary abstraction for H reaction with *i*-butane (26 to 28) % from 950 K to 1100 K), by increasing the A-factor of R647 by 25 % and decreasing the A-factor of R646 by 18 %.

The relative rate constants for the posterior model are (again, on a per-H basis):

$$\text{H} + i\text{-butane: } k_p/k_t = 10^{(-0.660 \pm 0.250)} \times T^{0.14} \times \exp(-(1982 \pm 397)/T)$$

$$\text{CH}_3 + i\text{-butane: } k_p/k_t = 10^{(-0.912 \pm 0.268)} \times T^{0.19} \times \exp(-(1457 \pm 567)/T)$$

and are applicable to temperatures from (270 to 1327) K. All uncertainties are 2σ.

By combining k_p and k_t from this work with k_s/k_p (k_s denotes the rate constant for H-abstraction from a secondary carbon) from previous work by Manion *et al.*[4] on *n*-butane, we provide a full set of self-consistent rate constants for H and CH_3 H-abstraction from primary, secondary and tertiary carbons. For both CH_3 and H reaction with a generic hydrocarbon, we use H-abstraction rate constants for primary carbons from this work, and find the secondary rate constant by multiplying the relative rate constants by Manion *et al.*[4] for k_s/k_p by our value of k_p. While Manion *et al.* did find rate constants for k_p with MUM-PCE for both H and CH_3, the values of k_p from this work are based on a larger data set.

We found the following rate constants for H reaction with a generic hydrocarbon on a per-H basis, (270 to 1327) K:

$$k_p = 10^{(5.22 \pm 0.19)} \times n^{2.54} \times \exp(-(3391 \pm 305)\ \mathrm{K} / T)\ \mathrm{cm^3\ mol^{-1}\ s^{-1}}$$
$$k_s = 10^{(5.64 \pm 0.20)} \times n^{2.4} \times \exp(-(2145 \pm 336)\ \mathrm{K} / T)\ \mathrm{cm^3\ mol^{-1}\ s^{-1}}$$
$$k_t = 10^{(5.88 \pm 0.16)} \times n^{2.4} \times \exp(-(1409 \pm 253)\ \mathrm{K} / T)\ \mathrm{cm^3\ mol^{-1}\ s^{-1}}$$

These values are plotted on Figure 3 and compared to evaluated rate constants by Tsang.[7, 14] The rate constants are only slightly updated from the prior model.

We found the following rate constants for CH_3 reaction with a generic hydrocarbon on a per-H basis, (270 to 1327) K:

$$k_p = 10^{(-0.722 \pm 0.223)} \times n^{3.65} \times \exp(-(3819 \pm 535)\ \mathrm{K} / T)\ \mathrm{cm^3\ mol^{-1}\ s^{-1}}$$
$$k_s = 10^{(0.095 \pm 0.231)} \times n^{3.46} \times \exp(-(3193 \pm 547)\ \mathrm{K} / T)\ \mathrm{cm^3\ mol^{-1}\ s^{-1}}$$
$$k_t = 10^{(0.191 \pm 0.149)} \times n^{3.46} \times \exp(-(2362 \pm 189)\ \mathrm{K} / T)\ \mathrm{cm^3\ mol^{-1}\ s^{-1}}$$

These values are plotted and compared to evaluated rate constants from Tsang.[7, 15] While the primary rate constant agrees well with Tsang, our rate constants for abstraction from a secondary carbon and tertiary carbons are slightly lower and slightly higher, respectively, than Tsang's rate constants.

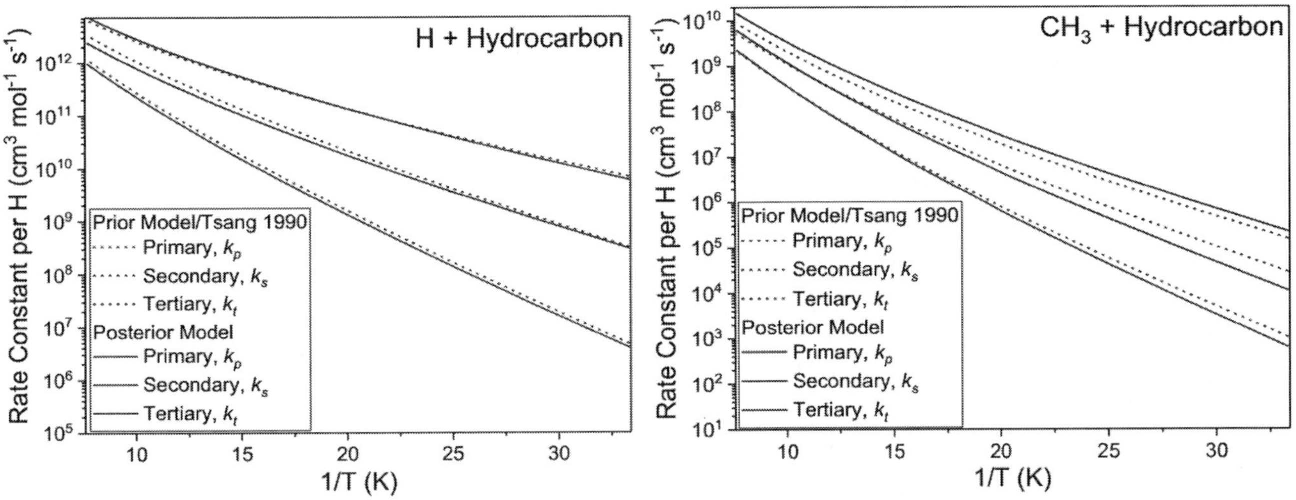

Figure 3. Our evaluated rate constants for H and CH_3 + a generic primary, secondary and tertiary carbon on a per-H basis compared to previous values by Tsang.[7, 15]

As expected due to the relative bond strengths, the activation energy for H-abstraction decreases from primary to secondary to tertiary for reaction with both H and CH_3. The A-factors increase from primary to secondary to tertiary, which ensures that the rate constants will not cross if they are extrapolated to higher temperatures. The rate constant for H and CH_3 reaction with any hydrocarbon can be found with the following formulate:

$$k_{total} = k_p \times N_{Hp} + k_s \times N_{Hs} + k_t \times N_{Ht}$$

where N_{Hp}, N_{Hs}, and N_{Ht} are the number of H's attached to primary, secondary and tertiary carbons, respectively.

4. Conclusions

We successfully used the MUM-PCE software package[12] to provide evaluated rate constants for H and CH_3 reaction with *i*-butane. The MUM-PCE method optimized a Cantera kinetics model to the experimental data we found using the NIST shock tube as well as to a range of literature data. The resulting rate constants were combined with previous work from our laboratory on *n*-butane[4] to give self-consistent rate constants for H and CH_3 abstraction of a generic H on primary, secondary and tertiary carbons.

9. Acknowledgments
David Sheen for providing technical support for Cantera and MUM-PCE. The National Academy of Science's NRC Research Associateship Program for financial support (LAM).

10. References

[1] D.G. Goodwin, H.K. Moffat, and R.L. Speth, *Cantera.* 2017.
[2] Wang, H., et al., *JetSurF version 2.0.* 2010.
[3] D.A. Sheen and H. Wang, The method of uncertainty quantification and minimization using polynomial chaos expansions, Combust. Flame 158 (2011) 2358-2374.
[4] J.A. Manion, D.A. Sheen, and I.A. Awan, Evaluated Kinetics of the Reactions of H and CH3 with *n*-Alkanes: Experiments with *n*-Butane and a Combustion Model Reaction Network Analysis, J. Phys. Chem. A 119 (2015) 7637-7658.
[5] H.J. Curran, Rate constant estimation for C-1 to C-4 alkyl and alkoxyl radical decomposition, International Journal of Chemical Kinetics 38 (2006) 250-275.
[6] N. Cohen, Are reaction rate coefficients additive? Revised transition state theory calculations for OH + alkane reactions, International Journal of Chemical Kinetics 23 (1991) 397-417.
[7] W. Tsang, Chemical Kinetic Data Base for Combustion Chemistry Part 4. Isobutane, Journal of Physical and Chemical Reference Data 19 (1990) 1-68.
[8] D.A. Sheen, C.M. Rosado-Reyes, and W. Tsang, Kinetics of H atom attack on unsaturated hydrocarbons using spectral uncertainty propagation and minimization techniques. Proceedings of the Combustion Institute, 34 (2013) 527-536.
[9] J. A. Kerr and M. J. Parsonage, Evaluated Kinetic Data on Gas Phase Hydrogen Transfer Reactions of Methyl Radicals, Berichte der Bunsengesellschaft für physikalische Chemie, 80 (1976) 825-825.
[10] U. Brand, *et al.*, Carbon-carbon and carbon-hydrogen bond splits of laser-excited aromatic molecules. 1. Specific and thermally averaged rate constants, J. Phys. Chem. 94 (1990) 6305-6316.
[11] D.A. Sheen. Personal correspondence.
[12] D.A. Sheen. mumpce 0.1 documentation, 2017 April 6, Available from: https://davidasheen.github.io/mumpce_py/#
[13] J. A. Manion, *et al.*, NIST Chemical Kinetics Database, NIST Standard Reference Database 17, Version 7.0 (Web Version), Release 1.6.8, Data version 2015.12, National Institute of Standards and Technology, Gaithersburg, Maryland, 20899-8320. http://kinetics.nist.gov/
[14] W. Tsang, Chemical kinetic data-base for combution chemistry. 3. Propane. Journal of Physical and Chemical Reference Data, 17 (1988) 887-952.
[15] W. Tsang, Chemical Kinetic Data Base for Combustion Chemistry 5. Propene. Journal of Physical and Chemical Reference Data, 20 (1991) 221-273

Sub Topic: Reaction Kinetics

Spring Technical Meeting
Eastern States Section of the Combustion Institute
March 4-7, 2018
State College, Pennsylvania

Kinetics of H Atom Addition to Cyclopentene

Jeffrey A. Manion and Iftikhar A. Awan*

[1]Chemical Sciences Division
National Institute of Standards and Technology
Gaithersburg, MD, 20899-8320
**Corresponding author: jeffrey.manion@nist.gov*

Abstract: To provide benchmark information needed to develop kinetic models of the combustion and pyrolysis of hydrocarbon ring structures, we have used the single pulse shock tube technique to study the kinetics of H atom addition to cyclopentene at 863 K to 1167 K and pressures of 160 kPa to 370 kPa. Addition of H to the pi bond leads to the cyclopentyl radical, which rapidly decomposes to ethene and allyl radical. Rate constants for the overall process were determined relative to a reference reaction via post-shock GC/FID/MS monitoring of products. A Transition-State-Theory/Rice-Ramsberger-Kassel-Marcus (TST/RRKM) model has been applied in conjunction with evaluated literature data to convert the primary measurements to a high pressure limiting rate expression for H addition. Results are compared with related systems. Near 1000 K, our data require a minimum value of 1.5 for branching between beta C-C and C-H bond scission in cyclopentyl radicals to maintain established trends in H addition rates. This minimum is consistent with a branching value of about 3 that we determined in previous experiments, but conflicts with much smaller values derived by current computations and those used in recent kinetics models to describe jet-stirred reactor studies of cyclopentane combustion.

Keywords: Kinetics, Fuels, Cyclopentyl, Shock Tube, RRKM

1. Introduction

Cyclopentene (CPE) is a combustion intermediate and component of some fuels. It is the smallest hydrocarbon having an unsaturated C5 ring and is a prototypical species useful for predicting the behavior of larger analogs. Such ring structures may be present in liquid fuels or may arise from the short chain radicals and olefinic species formed during the breakup of larger fuels. The reaction of these species with small radicals leads to polycyclic aromatic hydrocarbons (PAH's), finally terminating with particulate soot,[1] which is an environmental and regulatory concern. Our interest is related to the development of reliable kinetic models for the combustion of cyclic fuels and their conversion to PAH. Benchmark information on the base ring structure is expected to provide useful information needed to reliably extend the models to larger systems

We report herein the use the shock tube technique to determine rate constants for the addition of H atoms to cyclopentene at temperatures near 1000 K. We also use a TST/RRKM model to explore the relationship of this process to the subsequent decomposition of the cyclopentyl adduct radical. Addition of H to cyclopentene forms the cyclopentyl radical, which at higher temperatures undergoes rapid ring opening and decomposition to ethene and allyl radical.[2-5] Ring opening competes with the re-ejection of a hydrogen atom, the kinetics of which is related to H addition through detailed balance. The overall decomposition is:

cyclopentene (CPE) + H $\rightarrow$ ethene + allyl (R1)

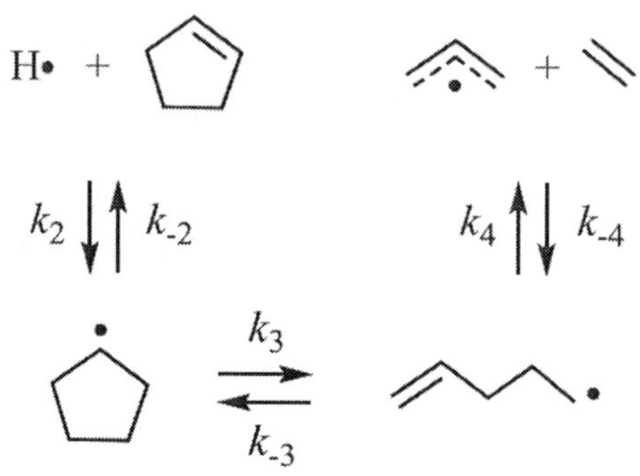

Scheme 1: Decomposition of cyclopentene (CPE) induced by addition of H atoms.

As depicted in Scheme 1, The process consists of three elementary reactions, R2 to R4:

cyclopentene (CPE) + H ⇄ cyclopentyl (R2)

cyclopentyl ⇄ pent-4-en-1-yl (R3)

pent-4-en-1-yl ⇄ ethene + allyl (R4)

Although the mechanism is established, there is conflicting information about the various rate constants, and it is the balance of the various elementary reactions that determines the overall behavior under conditions of interest.

In experimental work reported in 2011 we measured the branching between cyclopentene and ethene products in the decomposition of cyclopentyl radical under conditions similar to the present experiments and developed a TST/RRKM model of R2 to R4.[2] Our measured ratio for CC/CH beta scission in cyclopentyl (k_3/k_{-2}) was about an order of magnitude larger than values derived by computation.[5] In the intervening years a number of relevant studies have been reported. Pertinent to k_{-3} (Scheme 1), Wang *et al.*[6] in 2015 derived structure activity rate estimation rules for ring closure in C_4 to C_8 alkenyl radicals. Herbinet *et al.*[1] studied and modeled cyclopentene pyrolysis and formation of the first aromatic ring in a jet-stirred reactor (JSR) in 2016. Finally, in 2017 Al Rashidi *et al.*[7] reported a study of cyclopentane combustion in a JSR and developed a detailed kinetic model to describe the results. They found unusual inhibition behavior and their model showed that the relative and absolute rate constants for cyclopentyl radical decomposition were critical parameters. They were unable to model their results using the rate constants and k_3/k_{-2} branching ratios we reported in 2011.

In the present work, we use thermal precursors to create H atoms in the presence of a large excess of cyclopentene and an additional reference compound. The stable ethene product of R1 is quantitated in post-shock GC analyses and compared with the amount of product formed by attack of H on the reference species This directly yields relative rate constants, which are then converted to absolute values of R1 based on the reference rate. We combine our data with computations and various literature data to create a detailed kinetic model that relates the results to the elementary reactions of Scheme 1. Our data and analysis lead to reliable high pressure rate expression for k_2 and an independent method of estimating k_3/k_{-2}.

2. Methods / Experimental

Experiments are carried out in a heated single pulse shock tube having an ≈ 500 μs reaction time and described in earlier publications.[8] In the current studies, reactants are highly diluted in a bath gas of argon and we use small quantities of a precursor [≈50 μL/L (ppm)] to create a limited number of H atoms in the presence of much larger quantities (≈10,000 μL/L) of cyclopentene and 1,3,5-trimethylbenzene (135TMB). The latter compound serves as both a rate reference for H atom reactions and a radical scavenger. A Hewlett Packard 6890 N gas chromatograph equipped with FID and MS detection is used to measure the post-shock products. Including systematic errors, analytical uncertainties (2σ) for the main products are estimated as 6 %.

Generation of H atoms. We have used three different methods to thermally produce H atoms. On the ≈ 500 μs time scale of our experiments, thermolysis of hexamethylethane (HME) is an effective source of H at temperatures >1000 K:

hexamethylethane → 2 tert-butyl (R5)

tert-butyl → isobutene + H (R6)

To probe lower temperatures, a few experiments were conducted with mixtures containing small amounts (≈25 μL/L) of tert-butyl peroxide (tBPO) and a large amount of H_2 (≈20 %), which leads to H atoms via the following sequence:

tert-butyl peroxide → 2 $(CH_3)_3C$-O → 2 acetone + 2 CH_3 (R7)

$CH_3 + H_2 \rightarrow CH_4 + H$ (R8)

Although useful as a confirmatory technique, we found this method to be of limited value due to the loss of methyl radicals to side reactions. The decomposition of 2-iodopropane proved to be a more effective source of H atoms at temperatures < 1000 K. 2-Iodopropane undergoes both molecular elimination of HI and fission of the weak C-I bond. The latter process generates an isopropyl radical that rapidly ejects H:

2-iodopropane → propene + HI (R9)

2-iodopropane → i-C_3H_7 + I (R10)

i-C_3H_7 → propene + H (R11)

We determine rate constants for the reaction of H with CPE relative to known values for the reaction of H with 135TMB. Attack of H on 135TMB leads either to H_2 and dimethylbenzyl radical (DMB) or displacement of methyl to give m-xylene:

H + 135TMB → H_2 + dimethylbenzyl (DMB) (R12)

H + 135TMB → CH_3 + m-xylene (R13)

Near 1000 K the branching is about 2:1 in favor of abstraction.[9] DMB does not readily propagate radical chains and inhibits secondary chemistry through recombination reactions. Formation of m-xylene through R13 is a highly specific rate reference for reactions of H atoms. For R13 we use the rate parameters of Tsang et al.[9], k(H + 135TMB → m-xylene + CH_3) = 6.7×10^{13} exp(−3255/T) $cm^3\ mol^{-1}\ s^{-1}$. At our temperatures the uncertainty in k_{13} is about a factor of 1.5.[9] The existing database of rate constants measured relative to R13 provides valuable comparisons and a consistent scaling of absolute values.

3. Results and Discussion

Matheu et al.[10] used a mechanism generating computercode to identify 70 possible product channels in the H + cyclopentene addition reaction, but concluded that the reactions of Scheme 1 are the only important unimolecular processes. Our experimental product data support this

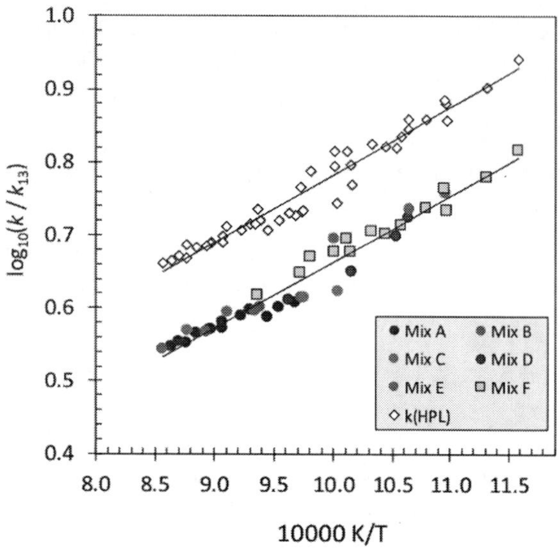

Figure 1. Experimental rate constant ratios for k_1/k_{13} (filled markers) and the corresponding high pressure limit values for $k_{2(\infty)}/k_{13}$ (unfilled markers) derived after applying an experimentally based correction for branching in the decomposition of the cyclopentyl radical (see text).

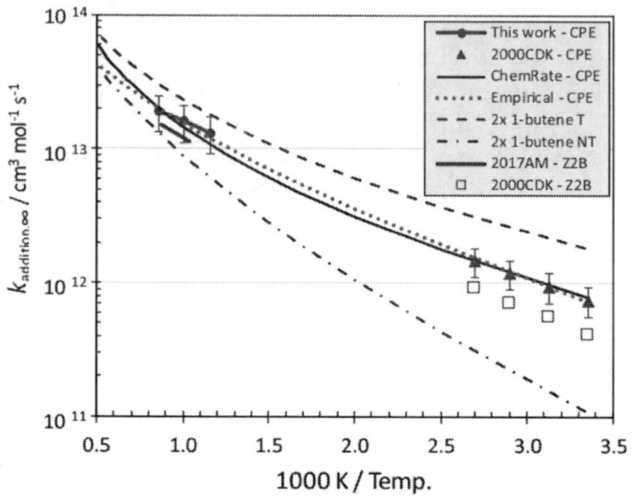

Figure 2. High pressure rate constants for addition of H to CPE and selected alkenes.; Z2B = (Z)-2-butene; T = terminal site; NT = nonterminal site. Dotted and solid lines indicate respective empirical and TST fits to the CPE data. Indicated uncertainties are 2σ. Data for 1-butene are multiplied by 2 to normalize the number of sites relative to CPE. Data sources: CPE[11]; 1-butene[8]; Z2B[11, 12].

conclusion. Thus, ethene formation is directly related to the overall reaction CPE + H → cyclopentyl → ethene + allyl (R1). The rate of this reaction relative to R13 is given by the molar yields of ethene and *m*-xylene normalized by the ratio of the reactants, k_1/k_{13} = [ethene][135TMB]/[*m*-xylene][CPE]. These data are plotted in Figure 1 and give:

$$\frac{k_1}{k_{13}} = 10^{-0.196\pm0.062}\exp[(1995 \pm 60)\ \mathrm{K}/T);\ 863\ \mathrm{K}\ \text{to}\ 1167\ \mathrm{K}$$

The given uncertainties are 2σ and represent precision only. Including systematic errors, the overall uncertainty (2σ) in the relative rate is estimated as about ± 15 %. Using k_{13} we derive:

$$k_1 = 4.27 \times 10^{13} \exp(-1260\ \mathrm{K}/T)\ \mathrm{cm^3\ mol^{-1}s^{-1}};\ 863\ \mathrm{K}\ \text{to}\ 1167\ \mathrm{K}$$

The uncertainty (2σ) in the absolute rate constant is about a factor of 1.5 and is due mainly to the uncertainty in the rate of the reference reaction. Because addition of H to cyclopentene, R2, is partially reversible, k_1 represents the minimum possible value of k_2. Assuming all H additions yield either ethene or cyclopentene as a stable olefin, then $k_2(\infty) = k_1(\gamma+1)/\gamma$, where γ is the molar product ratio, [ethene]/[CPE], in the decomposition of the cyclopentyl radicals formed by R2.

As reported in 2011,[9] we have previously measured values of γ by creating cyclopentyl radicals under dilute conditions and directly observing the ethene/CPE product ratio. G3MP2B3 computations were used to compute energies and other properties and an RRKM master equation analysis of the system depicted in Scheme 1 was carried out with ChemRate software[13]. The initial theoretical model was minimally tuned to match available data. Presently, we have slightly updated the 2011 model to better match additional literature information and then used it to derive values of γ for the chemically activated cyclopentyl radicals formed in the current experiment. Computed γ values are about 3 and little changed from the thermal values under our conditions. They lead to values of $k_{2(\infty)}$ that are about 30 % larger than k_1 (Figure 1). For $k_{2(\infty)}$, we find:

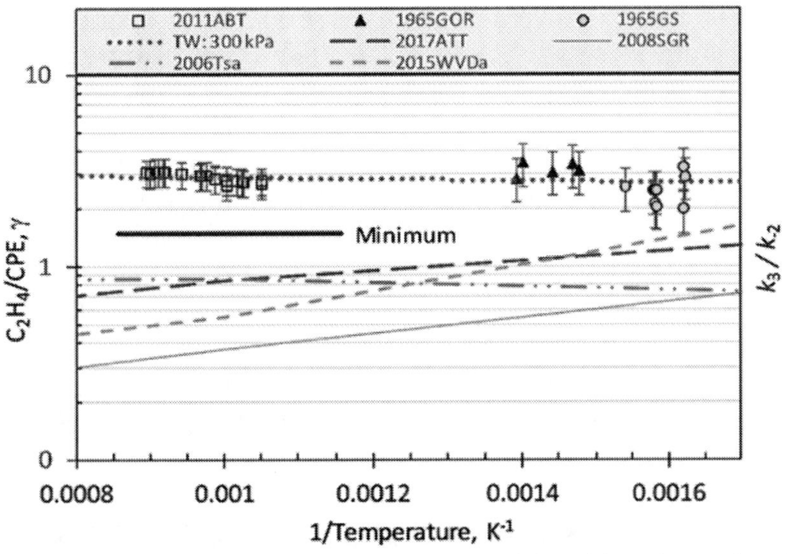

Figure 3. Branching in cyclopentyl decomposition. The symbols and dotted line show ethene/cyclopentene product ratios (γ, left axis) from experiment and our model at 300 kPa, respectively; the other lines are k_3/k_{-2} values (right axis) at the high pressure limit, which are approximately equivalent. The indicated minimum is that required by our current experiments to maintain consistency in H atom addition rates. Product ratios of 1965GOR[3] 1965GS[4] are plotted only at their higher temperatures where they relate to k_3/k_{-2}. Values from 1965GS are corrected for cyclopentene formed by disproportionation of cyclopentyl using $k_{dis}/k_{rec} = 1$ as reported in that work. Data sources: TW, this work; 2011ABT[2]; 2006Tsa[14]; 2017ATT[7]; 2015WVDa[6]; 2008SGR[5].

$$k_{2(\infty)} = 5.37 \times 10^{13} \exp(-1213\ \mathrm{K}/T)\ \mathrm{cm^3\ mol^{-1}s^{-1}};\ 863\ \mathrm{K\ to}\ 1167\ \mathrm{K}$$

Our TST/RRKM model has been tuned to fit the present data as well as those of Clarke *et al.* at 298 K to 370 K, and results in $k_{2(\infty)} = 9.09\mathrm{x}10^7 T^{1.783}\exp(-324\ \mathrm{K}/T)$ between 298 K and 2000 K. Comparisons with pertinent data are presented in Figure 2. Data near 1000 K are from this laboratory and are all derived relative to k_{13}. The addition rate of H to cyclopentene lies between those for terminal and nonterminal addition of H to 1-butene and is slightly faster than the analogous reaction with (Z)-2-butene, a noncyclic analog.

We have derived $k_{2(\infty)}$ using γ values of about 3. At our temperatures, values of γ less than 1.1 would require, on a per site basis, that the high pressure rate constant for H atom addition to cyclopentene exceed that for terminal addition of H to 1-butene. Such a result would be strongly at odds with known trends in rates of H atom addition to olefins. To maintain self-consistency in H atom addition rates, the reasonable lower limit of γ is about 1.5. This derived minimum is consistent with our measured γ values (about 3), but is obtained in an independent manner. It is at odds, however, with γ values derived from computations and used in current kinetics models of cyclopentane combustion, all of which are in the range of 0.3 to 0.8 near 1000 K.

Figure 3 summarizes experimental values of γ and theoretical values for the closely related branching ratio k_3/k_{-2}. The two are not strictly identical because of the slight reversibility of ring opening (about 5%), but all current models agree that the difference is minimal. Note that extrapolation of the experimental γ values to high pressure limits would slightly increase differences with the k_3/k_{-2} values. The experimental product ratios exhibit only a weak temperature dependence between 600 K and 1100 K. These values are derived from GC measurements having high precision, so significant errors would have to be systematic and related to the mechanism assumed in the data analyses. Both Gordon[3] and Gunning and Stock[4] conclude that in their studies cyclopentene is formed at lower temperatures primarily by disproportionation reactions of cyclopentyl. The data from these works plotted in Figure 3 are from their higher temperatures where C-H bond scission is believed to be the dominant source of cyclopentene, conditions where ethene/cyclopentene ratios should be closely related to k_3/k_{-2}. Notice that contributions from disproportionation would tend to make the derived ratios too small and would further increase the discrepancy with the values from theory or used in the modeling work.

In their experimental and modeling study of cyclopentane combustion in a JSR, Al Rashidi *et al.*[7] showed that the relative and absolute rate constants for cyclopentyl radical decomposition were critical parameters in their detailed kinetic model. They were unable, however, to reproduce the observed reactivity and inhibition behavior without using branching ratios much smaller than we reported in 2011.[2] Nonetheless, the present data and analysis appears to independently confirm the 2011 results. The reasons for this apparent disagreement remain to be explained.

4. Conclusions

Shock tube methods have been used to investigate the addition of H atoms to cyclopentene at temperatures of (863 to 1167) K and pressures of (160 to 370) kPa. Addition of H to the double bond leads to a cyclopentyl radical that rapidly ring opens and decomposes to ethene and allyl radical. We have determined rate constants for the overall process and related these to the high pressure limiting rate constant for H atom addition. The data allow the relative and absolute rates of CC and CH β-scission in cyclopentyl radical to be deduced. Near 1000 K, ratios of CC to CH scission smaller than about 1.5 are shown to result in inconsistencies in known relative rates of H atom addition to olefins. The present data are consistent with branching values near 3 as determined in our previous work, but do not agree with reported much smaller values derived from theoretical studies or those seemingly required in detailed kinetic models of cyclopentane combustion to explain the observed global reactivity behavior.

6. References

[1] O. Herbinet, A. Rodriguez, B. Husson, F. Battin-Leclerc, Z.D. Wang, Z.J. Cheng, F. Qi, J. Phys. Chem. A 120 (2016) 668-682.
[2] I.A. Awan, D.R. Burgess, Jr., W. Tsang, J.A. Manion, Proc. Combust. Inst. 33 (2011) 341-349.
[3] A.S. Gordon, Can. J. Chem. 43 (1965) 570-581.
[4] H.E. Gunning, R.L. Stock, Can. J. Chem.-Rev. Can. Chim. 42 (1964) 357-370.
[5] B. Sirjean, P.A. Glaude, M.F. Ruiz-Lopez, R. Fournet, J. Phys. Chem. A 112 (2008) 11598-11610.
[6] K. Wang, S.M. Villano, A.M. Dean, J. Phys. Chem. A 119 (2015) 7205-7221.
[7] M.J. Al Rashidi, S. Thion, C. Togbe, G. Dayma, M. Mehl, P. Dagaut, W.J. Pitz, J. Zador, S.M. Sarathy, Proc. Combust. Inst. 36 (2017) 469-477.
[8] J.A. Manion, I.A. Awan, J. Phys. Chem. A 119 (2015) 429–441.
[9] W. Tsang, J.P. Cui, J.A. Walker, Single Pulse Shock Tube Studies on the Reactions of Hydrogen Atoms with Unsaturated Compounds, Proceedings of the 17th International Symposium on Shock Waves & Shock Tubes, Bethlehem, PA, AIP Conference Proceedings 208, American Institute of Physics, Melville, NY, 1990, pp. 63 - 73, DOI: 10.1063/1061.39497.
[10] D.M. Matheu, W.H. Green, J.M. Grenda, Int. J. Chem. Kinet. 35 (2003) 95-119.
[11] J.S. Clarke, N.M. Donahue, J.H. Kroll, H.A. Rypkema, J.G. Anderson, J. Phys. Chem. A 104 (2000) 5254-5264.
[12] I.A. Awan, J.A. Manion, unpublished work, 2018.
[13] V. Mokrushin, V. Bedanov, W. Tsang, M. Zachariah, V.D. Knyazev, W.S. McGivern, ChemRate, National Institute of Standards and Technology, Gaithersburg, Maryland, 1996-2011.
[14] W. Tsang, J. Phys. Chem. A 110 (2006) 8501-8509.

Spring Technical Meeting
Eastern States Section of the Combustion Institute
March 4-7, 2018
State College, Pennsylvania

Intramolecular Phenyl Migration in Hexylbenzene Radicals: An Ab initio/TST Study

*Sarah Khanniche**, *William H. Green*

Department of Chemical Engineering, Massachusetts Institute of Technology, 77 Massachusetts Avenue Cambridge, MA, 02139, USA
**Corresponding Author Email: sarahkha@mit.edu*

Abstract: This study provides an insight into the phenyl shift of hexylbenzene radicals by means of quantum chemical calculations. A two-step mechanism was considered with a ring closure by *exo*-cycloaddition, followed by a ring opening by β-scission. Each channel involves spirocyclic transition states (TS1rad*x* and TS2rad*x*) connected by a cyclic intermediate (Ringrad*x*). Barrier heights were computed at the CBS-QB3 level. Canonical transition state theory was applied over the temperature range of 500-2000 K and the results were fitted to the modified Arrhenius form.
Keywords: *Electronic Structure Calculations, Potential Energy Surfaces, Arrhenius Parameters*

1. Introduction

Pyrolysis of alkyl-aromatic linkages is important in a wide range of important systems, from conversion of biomass and coal to liquid and gaseous fuels and to coking of heavy oils. It is also thought to be important in formation of coke deposits on hot surfaces, and in depolymerization of waste materials. Developing detailed mechanisms for pyrolysis of heavy hydrocarbons is challenging. In this context, quantum chemistry computations can prove efficient to obtain parameters required to build reliable models. Typical pyrolysis models include dissociation, βscission, addition, H-abstraction, and isomerization reactions. A key isomerization reaction class is phenyl migration (PM). [1], [2] In this work, we focus on investigating the PM of alkyl aromatic compounds with longer alkyl chain starting with hexylbenzene (HB) as a model compound [3]. Figure 1 illustrates the reaction mechanism associated with phenyl shift channels. Briefly, HB decomposes to hexylbenzene radicals (HBrad) by hydrogen abstraction reactions. Two types of radicals are considered that are here called HBrad*x* with $x = 2$ and $x = 4$, where x corresponds to the location of the carbon radical relative to the phenyl group. Once formed, each HBrad*x* undergoes a ring closure by *exo*-cycloaddition, followed by a ring opening by CCβ-scission. [1], [2] The bicyclic intermediates that result from the exo-cycloaddition of HBrad*x* at the ipso position are here called Ringrad*x* while *x*-HBrad correspond to HB isomers.

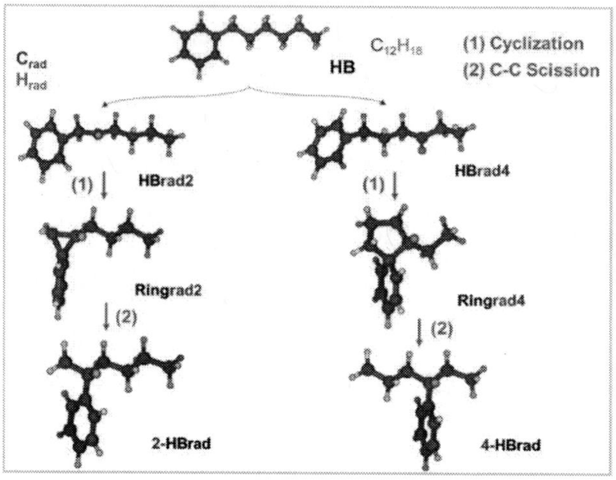

Figure 1: Schematic illustration of the reaction mechanism of phenyl migration reactions of hexylbenzene (HB). The color code is: C, H, Crad, and Hrad for HBrad*x*.

2. Computational Methods

Geometry optimizations for the lowest energy conformer of the reactants, products, intermediated and transition states (TS) were performed at the CBS-QB3 high-level theory [4]. First-order saddle points were characterized by one imaginary frequency and intrinsic reaction coordinate (IRC) calculations were performed to ensure connection between the TSs and the desired reactants and products. Torsions around CC single bonds not in rings were treated as hindered rotors. Relaxed potential energy scans at the B3LYP/CBSB7 level were carried out with a 10° stepsize. Each potential was then fitted by a Fourier series up to 16 terms [5]. Electronic structure calculations were performed using the Gaussian 03 and 09 suites of programs [6], [7]. Kinetic parameters were computed with the open-source CanTherm software package using the rigid-rotor harmonic oscillator approximation with 1-D hindered rotor corrections as implemented in Cantherm [8]. High pressure limit (HPL) rate coefficients were calculated using conventional Transition State Theory [9], [10] with a one-dimensional Eckart tunneling correction [11]. The HPL rate constants were fitted to the modified Arrhenius equations.

3. Results and Discussion

Potential energy surfaces (PES) for the phenyl shift reactions are represented in Figures 2 and 3. For each stationary point, zero-point corrected Energies (ZPE) computed at the CBS-QB3 level are plotted relative to the appropriate HBrad*x*. TS1rad*x* and TS2rad*x* on the PES are connected by Ringrad*x* intermediates. For the *exo*-cyclic intermediate to be formed, the HB radicals must overcome a barrier corresponding to the transition structure TS1rad*x*. The barrier height for TS1rad2 is predicted to be 13.5 kcal mol^{-1}. A lowest energetic barrier is observed for TS1rad4 (9.9 kcal mol^{-1}) with the formation of the five-membered cycle. Ringrad4 intermediate appears to be highly stabilized (*i.e.* 5 kcal mol^{-1} below HBrad4), if one compares to the three-membered cyclic structure, Ringrad2, which is located 10.5 kcal mol^{-1} above HBrad4. TS2rad*x* characterizes the energetic barrier of the β dissociation reaction. According to our calculations, TS2rad*x* is lying slightly above TS1rad*x*. The gap between TS1rad2 and TS2rad2 is only 0.8 kcal mol^{-1} and increases up to 1.8 kcal mol^{-1} for $x = 4$.

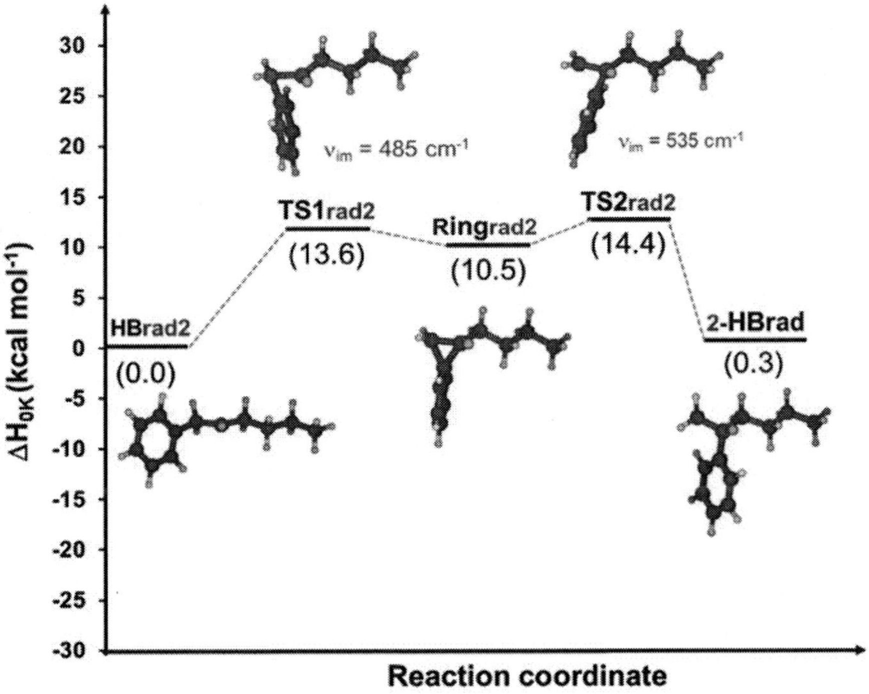

Figure 2: Potential energy surface of HBrad2 1,2- PM reaction at 0 K computed at the CBS-QB3 level of theory.

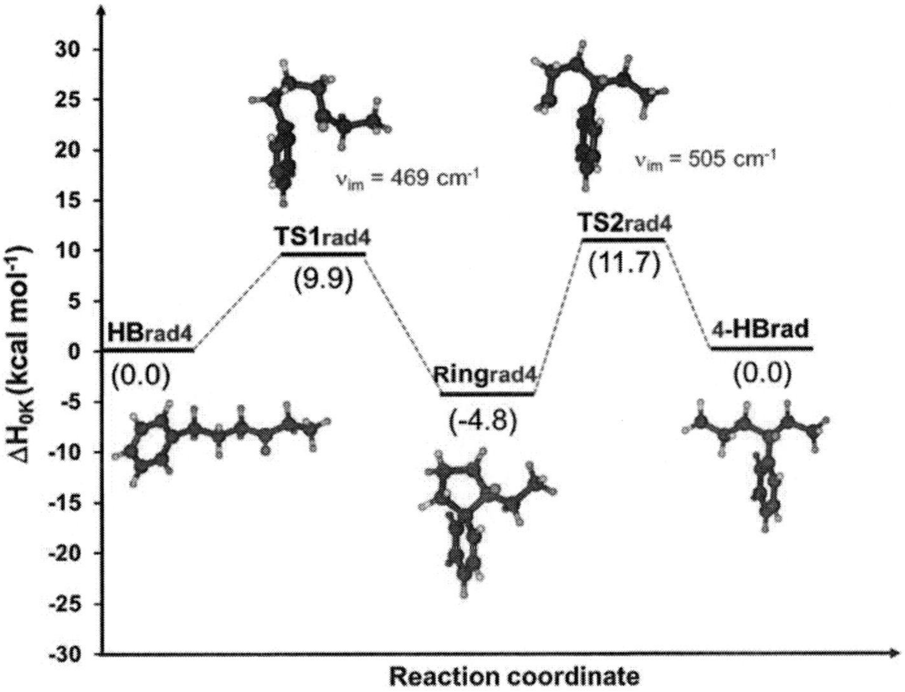

Figure 3: Potential energy surface of HBrad4 1,4- PM reaction at 0 K computed at the CBS-QB3 level of theory.

Arrhenius parameters for $k1radx$ and $k2radx$ fitted over $500 \leq T$ (K) ≤ 2000 are collected in Table 1. $k1radx$ and $k2radx$ correspond to the HPL rate coefficients of the HBrad$x$ → Ringrad$x$ and Ringrad$x$ → x-HB channels, respectively, for $x = 2$ and $x = 4$. The activation energy is higher by 3.7 kcal mol^{-1} for the three- membered ring transition state than for the five- membered cycle TS, *i.e.* Ea(TS1rad2) > Ea(TS1rad4). Indeed, increasing the size of the TS ring reduces the activation energy of the cycloaddition reactions as highlighted previously [5]. Meanwhile, A(TS1rad2) > A(TS1rad4) with $A_{cycloaddition} < A_{\beta scission}$. This can be explained by the fact that pre-exponential factors tend to diminish when reducing the length of the aliphatic chain and thus the number of hindered rotors within the TS structure.

Table 1: Pre-exponential factors A in s^{-1}, activation energies, E_a in kcal mol^{-1} and temperature coefficient, n obtained over the temperature range 500-2000 K.

Rate constants	Reaction	A	n	E_a
$k1$rad2	HBrad2 → Ringrad2	1.46×10^7	1.30	13.15
$k2$rad2	Ringrad2 → 2-HB	4.17×10^{13}	0.38	4.74
$k1$rad4	HBrad4 → Ringrad4	9.06×10^2	2.15	7.55
$k2$rad4	Ringrad4 → 4-HB	5.35×10^{12}	0.51	17.30

4. Conclusions

This work reports electronic structure calculations at the CBSQB3 level of phenyl migration reactions of hexylbenzene radicals. Two radicals were considered that differ from each other by the position of the radical along the C6 chain. The title reaction was predicted to take place by two consecutive steps characterized by two transition structures (TS1rad$x$ and TS2rad$x$) that were connected by a bicyclic Ringrad$x$ intermediate. The first step corresponds to the formation of the C–C bond by *exo*-cycloaddition of HB radicals and the second results in a ring opening by β-scission. The PES show that the formation of the five-membered ring leads to the lowest energetic barrier.

5. References

[1] V. V. Kislov, A. M. Mebel, J. Aguilera-Iparraguirre, and W. H. Green, Reaction of Phenyl Radical with Propylene as a Possible Source of Indene and Other Polycyclic Aromatic Hydrocarbons: An Ab Initio/RRKM-ME Study, J. Phys. Chem. A 116 (2012) 4176–4191.
[2] Z. Wang, L. Zhang, and F. Zhang, Kinetics of Homoallylic/Homobenzylic Rearrangement Reactions under Combustion Conditions, J. Phys. Chem. A 118 (2014) 6741–6748.
[3] A. G. Carr, C. A. Class, L. Lai, Y. Kida, T. Monrose, and W. H. Green, Supercritical water treatment of crude oil and hexylbenzene: An experimental and mechanistic study on alkylbenzene decomposition, Energy Fuels 29 (2015) 5290–5302.
[4] J. A. Montgomery, M. J. Frisch, J. W. Ochterski, and G. A. Petersson, A complete basis set model chemistry. VI. Use of density functional geometries and frequencies, J. Chem. Phys. 110, (1999) 2822–2827.

[5] S. Sharma, S. Raman, and W. H. Green, Intramolecular hydrogen migration in alkylperoxy and hydroperoxyalkylperoxy radicals: Accurate treatment of hindered rotors, J. Phys. Chem. A 114 (2010) 5689–5701.
[6] M.J. Frisch, G.W. Trucks, H.B. Schlegel, G.E. Scuseria, M.A. Robb, J.R. Cheeseman, J. Montgomery, J.A. Montgomery, T. Vreven, K.N. Kudin, J.C. Burant, J.M. Millam, S.S. Iyengar, J. Tomasi, V. Barone, B. Mennucci, M. Cossi, G. Scalmani, N. Rega, G.A. Petersson, H. Nakatsuji, M. Hada, M. Ehara, K. Toyota, R. Fukuda, J. Hasegawa, M. Ishida, T. Nakajima, Y. Honda, O. Kitao, H. Nakai, M. Klene, X. Li, J.E. Knox, H.P. Hratchian, J.B. Cross, V. Bakken, C. Adamo, J. Jaramillo, R. Gomperts, R.E. Stratmann, O. Yazyev, A.J. Austin, R. Cammi, C. Pomelli, J.W. Ochterski, P.Y. Ayala, K. Morokuma, G.A. Voth, P. Salvador, J.J. Dannenberg, V.G. Zakrzewski, S. Dapprich, A.D. Daniels, M.C. Strain, O. Farkas, D.K. Malick, A.D. Rabuck, K. Raghavachari, J.B. Foresman, J.V. Ortiz, Q. Cui, A.G. Baboul, S. Clifford, J. Cioslowski, B.B. Stefanov, G. Liu, A. Liashenko, P. Piskorz, I. Komaromi, R.L. Martin, D.J. Fox, T. Keith, M.A. Al-Laham, C.Y. Peng, A. Nanayakkara, M. Challacombe, P.M.W. Gill, B. Johnson, W. Chen, M.W. Wong, C. Gonzalez, J.A. Pople, Gaussian 03, Revision C.02, 2004.
[7] M. J. Frisch, G. W. Trucks, H. B. Schlegel, G. E. Scuseria, M. A. Robb, J. R. Cheeseman, G. Scalmani, V. Barone, B. Mennucci, G. A. Petersson, H. Nakatsuji, M. Caricato, X. Li, H. P. Hratchian, A. F. Izmaylov, J. Bloino, G. Zheng, J. L. Sonnenberg, M. Hada, M. Ehara, K. Toyota, R. Fukuda, J. Hasegawa, M. Ishida, T. Nakajima, Y. Honda, O. Kitao, H. Nakai, T. Vreven, J. A. Montgomery, Jr., J. E. Peralta, F. Ogliaro, M. Bearpark, J. J. Heyd, E. Brothers, K. N. Kudin, V. N. Staroverov, T. Keith, R. Kobayashi, J. Normand, K. Raghavachari, A. Rendell, J. C. Burant, S. S. Iyengar, J. Tomasi, M. Cossi, N. Rega, J. M. Millam, M. Klene, J. E. Knox, J. B. Cross, V. Bakken, C. Adamo, J. Jaramillo, R. Gomperts, R. E. Stratmann, O. Yazyev, A. J. Austin, R. Cammi, C. Pomelli, J. W. Ochterski, R. L. Martin, K. Morokuma, V. G. Zakrzewski, G. A. Voth, P. Salvador, J. J. Dannenberg, S. Dapprich, A. D. Daniels, O. Farkas, J. B. Foresman, J. V. Ortiz, J. Cioslowski, and D. J. Fox, *Gaussian 09, Revision D.01*. Gaussian, Inc., Wallingford CT, 2013.
[8] J.W. Allen, S. Sharma, M.R. Harper, W.H. Green, CanTherm. <http://github.com/GreenGroup/RMG-Py>.
[9] H. Eyring, The activated complex in chemical reactions, J. Chem. Phys. 3 (1935) 107–115.
[10] H. D. Beckey, H. S. Johnston, Gas phase reaction rate theory. The Ronald Press Company, New York, 1966.
[11] C. Eckart, The penetration of a Potential Barrier by Electrons, Phys Rev 35 (1930) 1303–1309.

Spring Technical Meeting
Eastern States Section of the Combustion Institute
March 4-7, 2018
State College, Pennsylvania

A fully-automated kinetic mechanism lumping algorithm

Lara Backer[1], *Perrine Pepiot*[1]

[1]*Mechanical and Aerospace Engineering, Cornell University, Ithaca, NY*

This work introduces a new reaction lumping algorithm which reduces the fuel pyrolysis pathways contained in a detailed mechanism to a representative set of global 'lumped' reactions, based on the assumption that fuel cracks rapidly at high temperatures. As a novel addition to prior lumping strategies, fuel pyrolysis reactions in the detailed mechanism are automatically identified based on species structural information. Lumped reaction parameters are computed using least squares fitting of species production and consumption rates exclusively due to the fuel pyrolysis reactions in the detailed mechanism. These rates are pre-computed in reference simulations run using the detailed mechanism. As a result, this reduction strategy is fully self-contained, depending only on the provided mechanism without any expert knowledge or optimization. The lumping methodology is presented here, along with an application to an n-dodecane mechanism. Quality of the procedure is assessed by comparing flame speeds and ignition delay times generated with the detailed and lumped mechanisms at a range of initial conditions. ***Keywords: Kinetic mechanism reduction, reaction lumping, high-temperature fuel cracking, n-dodecane***

1 Introduction

As detailed kinetic mechanisms describe increasingly complex and often multi-component fuels, mechanism reduction strategies targeting reduction of large hydrocarbon reaction kinetics are essential in decreasing computational expense. Lumping as a reduction technique is particularly suitable for simplifying complex fuel kinetic mechanisms because of the abundance of large hydrocarbon intermediate species and reactions which contain them. This has previously been accomplished by grouping functionally similar intermediates [1, 2], by grouping fast and rate-limiting reactions [3], and by grouping both species and reactions based on prior knowledge of reaction classifications [4, 5]. One significant drawback to prior lumping methods is the lack of a fully autonomous approach. Additionally, available mechanisms often do not have clearly defined rate rules or specified fuel pyrolysis sub-mechanisms, making reaction lumping impossible without significant manual effort.

At high temperatures, it is well known that fuels undergo rapid pyrolysis to form species fragments with a small number of carbon atoms, typically between 1 and 4. These fragments then oxidize, which is a rate-limiting process [3]. Gao *et al.* [6] performed a bifurcation analysis on reactions in a *n*-dodecane mechanism to discover that ignition and extinction phenomena are generally insensitive to the details of fuel cracking at temperatures above 1100K and pressures up to 30atm.

Pyrolysis reactions containing the fuel or intermediates with more than 4 carbon atoms may therefore be decoupled from the mechanism and simplified with minimal impact on key combustion phenomena, as long as the evolution of small species is reasonably well described.

In this work, a new automatic high temperature lumping methodology is proposed based on this assumption that fuel rapidly pyrolyzes. Using a dataset generated from simulations using the detailed mechanism, the pyrolysis reaction pathways are automatically lumped to form a minimal set of reactions sufficient to describe fuel consumption and formation of species with up to 4 carbon atoms. The lumping algorithm is applied to the high temperature n-dodecane detailed mechanism developed in [7]. The resulting lumped mechanism is validated against the detailed mechanism, with experimental data provided for reference.

2 Lumping algorithm

2.1 Overview

We consider the high temperature oxidation of a large alkane molecule, referred to as the fuel F in the following sections. The set of elementary reactions describing this process can be separated into two subsets: the fuel pyrolysis reaction set $\mathbb{R}_p$, describing the fuel breakdown without any direct oxidation, and the remainder of the reactions, called the core reactions and denoted by $\mathbb{R}_c$.

The proposed lumping process replaces $\mathbb{R}_p$ of cardinality $n\mathbb{R}_p$ in the detailed mechanism with a much smaller set of global lumped reactions $\mathbb{R}_\ell$ with cardinality $n\mathbb{R}_\ell$, which, when added to the set of core reactions $\mathbb{R}_c$, form the lumped mechanism. Because H-abstraction and β-scissions are favored over oxidation at high temperature conditions [8], the reactions included in $\mathbb{R}_\ell$ are assumed to be of the form:

$$R + F \rightarrow \nu_1 P_1 + \nu_2 P_2 + ..., \tag{1}$$

where F is the fuel molecule, R is an optional radical species, and ν_i and P_i are the stoichiometric coefficients and names of the lumped reaction products, respectively, the number of which has yet to be determined.

In the lumped reactions, the fuel F directly decomposes into small molecular weight species P_i, bypassing large hydrocarbon intermediates in the detailed mechanism which rapidly decompose along with the fuel. This set of intermediate pyrolysis species $\mathbb{S}_p$ can then be removed from the detailed kinetic mechanism, leaving the set of species $\mathbb{S}_\ell$ in the lumped mechanism.

The lumped reaction products, stoichiometric coefficients, reactants, and reaction rates are chosen so that the production and consumption of species $\mathbb{S}_\ell$ remaining in the lumped mechanism arising from reactions in $\mathbb{R}_\ell$ are as close as possible to the production and consumption rates of species in $\mathbb{S}_\ell$ arising from reactions in $\mathbb{R}_p$. This matching of rates is made possible by a database generated from reference simulations using the detailed mechanism prior to defining the lumped reactions.

2.2 Selecting the set of reactions to lump, $\mathbb{R}_p$

The set of fuel pyrolysis reactions $\mathbb{R}_p$ contains the set of reactions describing the breakdown of the fuel to small hydrocarbon fragments at high temperatures. $\mathbb{R}_p$ is built by traversing all reactions contained in the detailed mechanism. Reactions which form intermediate fuel pyrolysis species without any C-C bond reconfiguration between the fuel and the intermediate species, as determined using transition state analysis [9], are identified as fuel pyrolysis reactions as follows.

If F or a species in $\mathbb{S}_p$ is a reactant in a reaction r contained in the detailed mechanism, and if F or the species in $\mathbb{S}_p$ directly forms a product with more than 4 carbon atoms, r is tagged as a fuel pyrolysis reaction and that product is added to $\mathbb{S}_p$. $\mathbb{S}_p$ and $\mathbb{R}_p$ are initialized as empty sets; the identification of reactions and species in these sets is recursive and stops only when nothing else can be added to $\mathbb{S}_p$ or $\mathbb{R}_p$. All other species and reactions are grouped into the core species and reaction sets, $\mathbb{S}_c$ and $\mathbb{R}_c$.

This procedure identifies the set of fuel pyrolysis reactions and ensures that the only reactions to lump are those associated with the breakdown of fuel F.

2.3 Reference simulations

As described earlier, lumped reaction rates, R, ν_i, and P_i in Eq. 1 are defined based on a dataset populated with time-dependent data recorded from N 0D isochor simulations spanning a range of initial conditions using the detailed mechanism. A dataset entry contains parameters $\{T, C_i, \dot{C}_{-,S_i}, \dot{C}_{+,S_i}\}$. In order, these values represent the temperature, the concentration, and the consumption and production rates of a given species S_i from pyrolysis reactions included in $\mathbb{R}_p$. The latter are computed using the equations:

$$\dot{C}_{-,i}(t) = \sum_{j=1}^{n\mathbb{R}_p} \nu_{j,i}^{R} \omega_j(t) \tag{2}$$

$$\dot{C}_{+,i}(t) = \sum_{j=1}^{n\mathbb{R}_p} \nu_{j,i}^{P} \omega_j(t) \tag{3}$$

ν^R denotes the reactant's stoichiometric coefficients, ν^P the product's stoichiometric coefficients, and ω the reaction rate.

2.4 Selecting lumped reaction reactants and products

To identify the optional radical reactants R involved in the lumped reactions, we first define a cutoff α_-:

$$\max_{n \in N} \left(\frac{\int_{t=0}^{t_{\max,n}} \dot{C}_{-,R}(t)dt}{\int_{t=0}^{t_{\max,n}} \dot{C}_{-,F}(t)dt} \right) \geq \alpha_- \tag{4}$$

If R is included in $\mathbb{S}_\ell$ and has a normalized consumption rate satisfying Eq. 4, an H-abstraction reaction is added to the list of lumped reactions $\mathbb{R}_\ell$ of the form $R + F \rightarrow RH + ...$

Similarly, to identify potential products P_i, we define a cutoff α_+:

$$\frac{1}{N}\sum_{n\in N}\left(\frac{\int_{t=0}^{t_{\max,n}} \dot{C}_{+,S_i}(t)dt}{\max\limits_{s\in\mathbb{S}_c}\int_{t=0}^{t_{\max,n}} \dot{C}_{+,s}(t)dt}\right) \geq \alpha_+ \tag{5}$$

If the normalized production term for a species S_i in $\mathbb{S}_\ell$ satisfies Eq. 5, it is added to all lumped reactions as a product P_i.

2.5 Computing lumped reaction rate coefficients

The temperature-dependent rate coefficients k for the lumped reactions are computed to ensure that the consumption of the fuel F and selected reactants R in the lumped reactions match the consumption of F and R in the simulation dataset.

The consumption of R is used to compute the rate coefficient for the corresponding abstraction reaction

$$\dot{C}_{-,R} = k_R(T_b)C_F C_R \tag{6}$$

and the consumption of F is used to compute the rate coefficient for the decomposition reaction, where nR is the number of optional lumped reactants.

$$\dot{C}_{-,F} = \max(0, \dot{C}_{-,F} - \sum_{j=1}^{nR}\dot{C}_{-,j}) = k_F(T_b)C_F \tag{7}$$

To compute k as a function of temperature, entries in the simulation dataset are partitioned based on temperature into a user-defined number of bins n_b. An average bin temperature T_b is computed from all entries in a bin. For each lumped reaction and each bin, the overdetermined system in Eq. 6 or 7 is solved with least squares to yield $k(T_b)$.

For each lumped reaction, constant Arrhenius parameters A and E are obtained by fitting $k(T_b)$ and T_b over $b = 1, n_b$:

$$\ln(A) = \ln(k(T_b)) + \frac{E}{R_c T_b} \tag{8}$$

where A is the Arrhenius pre-exponential factor, E is the activation energy E, and R_c is the ideal gas constant.

2.6 Computing lumped reaction product stoichiometric coefficients

Stoichiometric coefficients of lumped reaction products are computed from the core species production rates in the simulation dataset. For each unique P_i, ν_{j,P_i} is computed for each lumped reaction j to minimize the residual r_{P_i}:

$$r_{P_i} = ||\sum_{j=1}^{n\mathbb{R}_\ell}\nu_{j,P_i}\omega_j - \dot{C}_{+,P_i}|| \tag{9}$$

Conservation of elements is enforced by employing linear programming to compute the minimum change to each $\nu \geq 0$ that results in a balance of each element in each reaction.

3 Results: Application to High-T n-dodecane Mechanism

The algorithm was applied to lump n-dodecane pyrolysis reactions in a high temperature mechanism containing 180 species and 1960 reactions, developed by Narayanaswamy *et al.* [7].

Reference simulations were run with initial conditions $\phi = 1$, $P = 1$, 5, and 20 atm, and $T = 1100 - 1600$K in 10K increments. Using cutoffs $\alpha_- = 0.2$ and $\alpha_+ = 0.48$, 4 lumped reactions were identified with 9 unique products. Lumped reaction parameters are shown in Table 1.

Reaction			A	E
n-$C_{12}H_{26}$ + H	$\rightarrow$	0.38395 H_2 + 1.0715 p-C_4H_9 + 0.4013 C_2H_5 + 0.9975 C_2H_4 + 0.8795 C_3H_6 + 0.7593 n-C_3H_7	2.55×10^{15}	34.4
n-$C_{12}H_{26}$ + OH	$\rightarrow$	H_2O + 0.4873 n-C_3H_7 + 1.28205 C_2H_4 + 1.9744 C_3H_6 + 0.5127 p-C_4H_9	2.1×10^{14}	10.7
n-$C_{12}H_{26}$ + HO_2	$\rightarrow$	H_2O_2 + 0.8148 p-C_4H_9 + 0.1852 n-C_3H_7 + 1.25925 C_2H_4 +1.8889 C_3H_6	1.4×10^{15}	92.4
n-$C_{12}H_{26}$	$\rightarrow$	1.0589 p-C_4H_9 + 0.701 n-C_3H_7 + 0.24 C_2H_5 + 0.9157 C_3H_6 + 1.21715 C_2H_4 + 5e-5 H_2	2.1×10^{16}	312.8

Table 1: Lumped fuel cracking reactions. Arrhenius parameters are in cm, s, mol and kJ.

The final lumped mechanism contains 172 fewer reactions and 26 fewer species, and is validated against the detailed mechanism and experimental data for ignition delays and flame speeds in Fig. 1 and 2. The experimental definition of ignition delay time was used to define simulation ignition delay time. Good agreement is found between the lumped and detailed mechanisms for key combustion features, particularly when compared to experimental uncertainty.

4 Conclusions

A novel systematic lumping process for high-temperature combustion was developed and applied to a high-temperature n-dodecane kinetic mechanism. The resulting lumped mechanism was validated against the detailed mechanism for a range of conditions. This method is notable for several reasons: (1) fuel cracking pathways are targeted in the reduction, (2) this method is automated and requires minimal user input, and (3) no optimization or *ad-hoc* tuning is required. In the future, the lumping approach presented here can be extended to represent the decomposition of a fuel blend as a single lumped fuel.

Acknowledgments

This research was funded by the National Science Foundation Graduate Research Fellowship under Grant No. DGE-1144153.

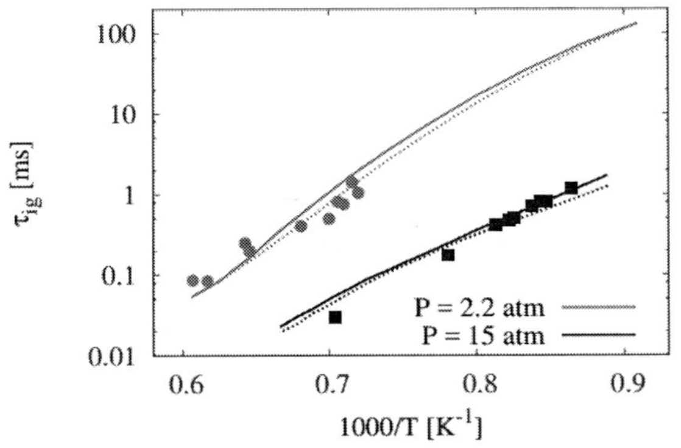

Figure 1: Constant volume ignition delays for detailed (solid lines) and lumped (dashed lines) mechanisms at high temperatures. Shock tube experiments (symbols) are by Davidson *et al.* [10] at P = 2.2 atm for *n*-dodecane in Argon at X=400ppm and ϕ = 1.0, and by Vasu *et al.* [11] at P=15atm for *n*-dodecane in Argon at X=1000ppm and ϕ = 0.5 with an uncertainty of 10%.

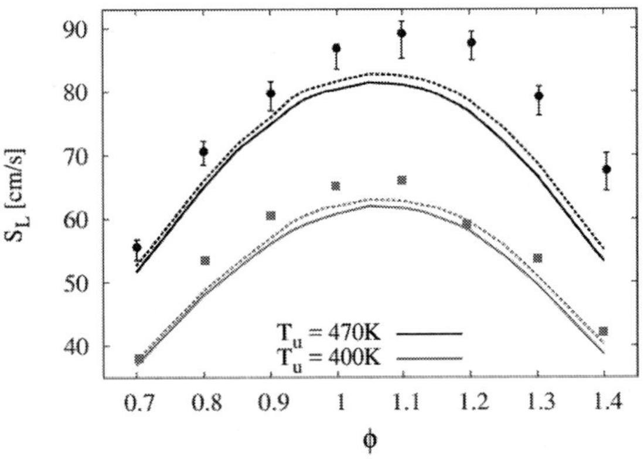

Figure 2: Laminar flame speeds at P=1atm for detailed mechanism (solid lines), lumped mechanism (dashed lines) and experiments (symbols) by Hui and Sung [12] at unburnt flame temperature T_u=400K, Kumar and Sung [13] at T_u=470K

References

[1] Syed Sayeed Ahmed, Fabian Mauß, Gladys Moréac, and Thomas Zeuch. *Physical Chemistry Chemical Physics*, 9 (2007) 1107–1126.

[2] P Pepiot-Desjardins and H Pitsch. *Combustion Theory and Modelling*, 12 (2008) 1089–1108.

[3] Xiaoqing You, Fokion N Egolfopoulos, and Hai Wang. *Proceedings of the Combustion Institute*, 32 (2009) 403–410.

[4] E Ranzi, T Faravelli, P Gaffuri, and A Sogaro. *Combustion and Flame*, 102 (1995) 179–192.

[5] Alessandro Stagni, Alberto Cuoci, Alessio Frassoldati, Tiziano Faravelli, and Eliseo Ranzi. *Industrial & Engineering Chemistry Research*, 53 (2013) 9004–9016.

[6] Yang Gao, Ruiqin Shan, Sgouria Lyra, Cong Li, Hai Wang, Jacqueline H Chen, and Tianfeng Lu. *Combustion and Flame*, 163 (2016) 437–446.

[7] Krithika Narayanaswamy, Perrine Pepiot, and Heinz Pitsch. *Combustion and Flame*, 161 (2014) 866 – 884.

[8] Tomasz Malewicki and Kenneth Brezinsky. *Proceedings of the Combustion Institute*, 34 (2013) 361–368.

[9] C Laurent, CJ Frewin, and P Pepiot. *Combustion and Flame*, 173 (2016) 387–401.

[10] DF Davidson, Z Hong, GL Pilla, A Farooq, RD Cook, and RK Hanson. *Proceedings of the Combustion Institute*, 33 (2011) 151–157.

[11] S.S. Vasu, D.F. Davidson, Z. Hong, V. Vasudevan, and R.K. Hanson. *Proceedings of the Combustion Institute*, 32 (2009) 173 – 180.

[12] Xin Hui and Chih-Jen Sung. *Fuel*, 109 (2013) 191–200.

[13] Kamal Kumar and Chih-Jen Sung. *Combustion and Flame*, 151 (2007) 209–224.

Spring Technical Meeting
Eastern States Section of the Combustion Institute
March 4–7, 2018
State College, Pennsylvania

Introducing a workflow for improving kinetic models: case study using butanol

Sai Krishna Sirumalla[1], *Morgan A. Mayer*[2], *Kyle E. Niemeyer*[2], *and Richard H. West*[1,*]

[1]*Department of Chemical Engineering, Northeastern University, Boston, MA 02115, USA*
[2]*School of Mechanical, Industrial, and Manufacturing Engineering, Oregon State University, Corvallis, OR 97331, USA*
[*]*Corresponding author:* r.west@northeastern.edu

Abstract: We demonstrate a new workflow that uses a combination of experimental data, data-science methods, and critical appraisal, to bring detailed kinetic models of combustion chemistry closer to the "truth." Starting with a prior model for butanol combustion, we identify over 1600 conflicting parameters in other published kinetic models, rank them according to their impact on ignition delay times, then resolve the most important discrepancies by performing literature evaluations. The updated model is tested against a range of experimental data.
***Keywords:* Butanol, Chemical kinetic models, Model comparison, Uncertainty**

1. Introduction

When designing a combustion device or modeling a flame there is real utility in a model that optimally reproduces trusted experimental data. The equations and parameters comprising such a model do not need to be physically meaningful for the model to be useful, and thus two models can coexist with different parameter values without either being incorrect.

However, in the paradigm of detailed kinetic modeling of combustion, the aim is for every reaction in the model to describe a real reaction, and its kinetic parameters to describe the actual chemical physics of that reaction. In this ideal situation the same parameters would work equally well to describe the reaction in another model of another fuel and at different conditions. When this transfer fails (e.g., when a reaction rate needs to be fast in one model and slow in another, to get both models to work well), it suggests that either one or more of the models are structurally incorrect, perhaps missing some reaction pathways, or that some physics is not being modeled correctly (such as for roaming reactions [1, 2], non-thermalized product distributions [3], non-equilibrium effects [4], etc.), or the errors are masking other errors in different parameters.

In this paradigm, once we include all the important reaction pathways, understand how to model all the physics, and get all the parameters sufficiently close to their true values, our models will be truly predictive. And as long as we head in that direction, we feel we are making progress.

A great benefit of this approach, besides the intellectual satisfaction of approaching the truth, is that the models are predictive rather than interpolative, both for new fuels and for conditions that have not been measured experimentally.

If each parameter is physically meaningful on its own, then it has a "true" value independent of the model it is in. Therefore whenever there is a disagreement between two models, one of them must be closer to the truth. Despite this, significant discrepancies exist between parameters in detailed kinetic models being published to this day. For example, in the supplementary data of the 2016 Combustion Symposium proceedings, 15% of the 2,600 reactions we have identified in two or more models disagree by over one order of magnitude at 1000 K, and some by as much as 31 orders of magnitude; for 4% of the species found across several models, the published values for standard enthalpies of formation varied by more than 50 kJ/mol [5].

Despite efforts to collate kinetic expressions such as NIST[6] and PrIMe[7], the vast majority of kinetic modeling data exists only in scattered CHEMKIN input files. Due to the constraints of the CHEMKIN input format (designed to fit on a 1960's era 80-column punch-card), the majority of these files use ambiguous nicknames and abbreviations for the species identifiers, making direct comparison between models a challenge.

1.1 Objectives

Our vision is a unified database of thermochemical and kinetic parameters which should work in all models, a database of the models, a database of all experimental data, and tools to easily work with these databases.

In a first step we developed a tool based on the open-source Reaction Mechanism Generator (RMG) [8] to help identify molecular structures of species in detailed kinetic models [9]. Using this, dozens of recent kinetic models were imported into a common database while identifying and labeling tens of thousands of chemical reaction rates.

We have also developed a simple and clear data format, ChemKED [10], to describe experimental conditions and measurements, along with tools to validate the data syntax, and perform simulations using a provided kinetic model [11]. Using this we have collected data from a range of experimental studies.

Butanol was chosen to conduct a case study on detecting and determining the importance of discrepancies in the literature [5]. Taking the well-validated LLNL model for butanol isomers by Sarathy et al. [12] as a base model, we found $\sim$1,200 reaction rates and $\sim$400 thermochemistry expressions that were in conflict with several other models in the literature, i.e., at least two other recently published kinetic models had used different parameters.

Using a genetic algorithm to maximize and minimize ignition delay times we demonstrated that a wide latitude (e.g., an order of magnitude in ignition delay time) was possible by cherry-picking from these published parameters. We also showed an example where merely trying to bring a model closer to the experimental data would lead one to pick a parameter that is probably further from the truth. We propose that a better workflow would be to first rank the parameters by importance then independently appraise them to select the "most truthful" values. In this work we demonstrate that proposed workflow in full.

2. Methodology

2.1 Workflow

The steps involved in our workflow are:

1. Collate experimental data from the literature, converting to ChemKED format, including uncertainties when available.
2. Collate detailed kinetic models from the literature, identifying the molecular structures represented by the species names, to allow comparison between models.
3. Select a base model (in this case butanol), and from it generate alternative models by substituting parameters found in other models, one substitution per model.
4. Run each of the alternative models (in this case ~1,600) against all experimental data (only shock tubes for now), using PyTeCK, to generate overall "error metric" for each alternative.
5. Rank parameter substitutions according to highest impact (positive *or* negative) on the error metric.
6. For the *N* most impactful parameters, follow the paper trail to determine the original source of the values, and decide which are most likely to be more accurate.
7. Construct a new model using all the selected substitutions.
8. (optional) Return to step 3 using the new model as the base and removing from consideration the discrepancies already resolved.
9. Evaluate the final model against all available experimental data, using PyTeCK.

2.2 Experimental data

We extended the ignition delay data set used in our prior study [5] to include the measurements of Bec et al. [13] and Zhu et al. [14] taken using the constrained-reaction volume filling strategy. In addition, we incorporated the laminar flame speed measurements for the four butanol isomers of Veloo and Egolfopoulos [15] in our model performance analysis. These data cover equivalence ratios of 0.7–1.5 for butanol/air at an unburned temperature of 434 K and 1 atm.

2.3 Model evaluation and validation

Building on our previous work, we used the Python-based tool PyTeCK [11, 16] to determine an error metric for the baseline and improved model across the full dataset for butanol isomer autoignition. PyTeCK uses ChemKED files [10] to create Cantera [17] simulations at the specified experimental conditions, and then calculates ignition delay for each point. We have extended our analysis to include laminar flame speed measurements; Cantera [17] `FreeFlame` objects simulate a one-dimensional freely propagating flat flame based on experimental conditions. Each simulation uses a 50 cm domain, with automatic grid refinement criteria of 0.1 for the maximum difference in slope and curvature and 3.0 for the maximum ratio in spacing around a point (i.e., `set_refine_criteria(ratio=3.0, slope=0.1, curve=0.1)`.

Using the simulated and experimental values for ignition delay time and laminar flame speed, we determine a performance metric for a given model, adopted from Olm et al. [18, 19]. The overall error of a model is quantified with the average error E over the N datasets given by

$$E = \frac{1}{N}\sum_{i=1}^{N} E_i\,, \tag{1}$$

where E_i is the average error for the ith dataset

$$E_i = \frac{1}{N_i}\sum_{j=1}^{N_i}\left(\frac{Y_{ij}^{\text{exp}} - Y_{ij}^{\text{sim}}}{\sigma(Y_{ij}^{\text{exp}})}\right)^2\,, \tag{2}$$

$Y_{ij} = y_{ij}$ if $\sigma(y_{ij}^{\text{exp}}) \approx$ constant or $Y_{ij} = \log(y_{ij})$ if $\sigma(\log y_{ij}^{\text{exp}}) \approx$ constant, which apply to laminar flame speeds and ignition delay time measurements, respectively [18, 19]; N_i is the number of data points in set i, and $y_{ij}^{\text{exp}}/y_{ij}^{\text{sim}}$ represents the experimental and simulated value of the ith data point in the jth data set. σ is the standard deviation, and the log function indicates the natural logarithm.

The standard deviation of an experimental dataset, $\sigma(Y_{ij}^{\text{exp}})$ is determined by fitting a spline to the experimentally measured values with respect to the variable changing in the dataset (typically temperature for ignition delay measurements or equivalence ratio for laminar flame speeds), following the approach used by Olm et al. [18, 19]. This was implemented via the `interpolation.UnivariateSpline` function [20] of `SciPy` [21]. A minimum value of $\sigma = 0.10$ was established, used to override any small calculated values. This approach reduces the importance of experimental data sets with scatter, but does not attempt to identify or resolve systematic errors or inconsistencies between experimental data sets.

2.4 Model curation

Building on our previous work [5] which ran all ~1600 alternatives, a new model is constructed by substituting the parameters that have the highest impact on the PyTeCK error metric. In this case, a paper trail is carried out to find the sources of the parameters. The paper trail is started using the comments in CHEMKIN files. Most models do not contain references for their thermochemistry data, but by exhaustively searching for matching polynomial coefficients one can usually track down or guess the original source. For kinetics, the original source may be traced from a paper trail, given that model authors often include brief comments usually indicating secondary sources for values. For the discrepancies found to have the largest impact on model's ability to reproduce experimental ignition delay times of n-butanol [5], we tracked the sources and critically appraised them, choosing a preferred value in each case.

2.5 Ignition delay times

High temperature shock tube ignition delay simulations were run for all four butanol isomers with both models using Cantera [17] at 42 bar, and $\phi = 1.0$ in 4:96 O_2:N_2 mixture in the temperature range of 600 K to 1600 K, reflecting the conditions used to validate the original model [12]. Figure 1 shows that for all the four isomers there's only a marginal difference in ignition delay times, below five percent, except for tertiary butanol at high temperatures which reaches 15%.

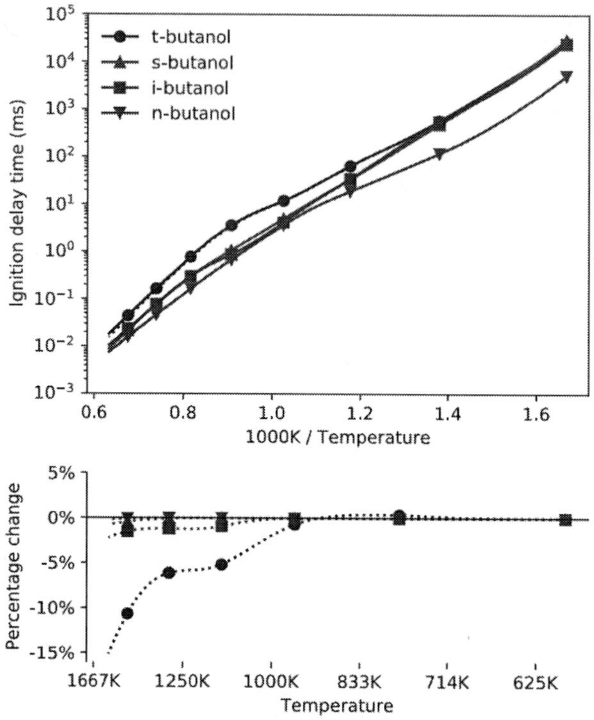

Figure 1: Ignition delay times from both models for all four isomers with the percentage changes

2.6 Laminar flame speeds

Atmospheric premixed laminar flame speeds (S_u°) were calculated for n-butanol at 353 K for different equivalence ratios (ϕ) using Cantera [17] and the solutions were converged on approximately 200 grid points `set_refine_criteria(ratio=3.0, slope=0.1, curve=0.1)`. Figure 2 shows that the modifications made to the model change the flame speeds by less than 1%, which is less than the numerical uncertainty due to grid convergence.

2.7 Limitations and discussion

It is worth remembering that a lack of discrepancy in the literature does not necessarily mean a lack of uncertainty. Often an uncertain estimate is universally adopted—i.e., the present work, aiming to detect and correct discrepancies, is not a substitute for uncertainty quantification studies.

The current workflow will not detect missing pathways in the base model: Reactions in the base model with very slow rates will be detected, but zero-rate reactions (i.e., missing reactions) will not be detected. This could be addressed using software such as Reaction Mechanism Generator (RMG) [8] to propose missing pathways.

Although laminar flame speeds were used in the evaluation of the final model, the selection of important parameters was done only using ignition delay times to avoid the expense of converging 1,600 flame calculations. An approach using local sensitivity analysis or a perturbation method could reduce the computational cost and it feasible to screen for discrepancies important to flame speeds. A similar analysis for speciation data, either from ignition or steady-state experiments,

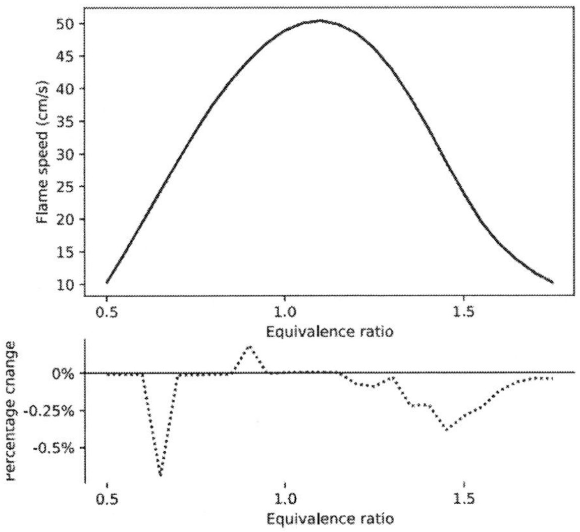

Figure 2: Laminar flame speeds of n-butanol from both models at 353 K and 1 atm with the percentage changes

would be interesting.

Presently, only butanol has been analyzed. As described in the introduction, physically meaningful parameters should be consistent across all models, and this workflow (including the automated tools to run multiple validations) will become especially powerful as the data set is expanded.

For many discrepancies, we agreed with the LLNL Butanol model; although the few changes we made have only a small impact on this model, the alternative parameters that we rejected came from other recent models, and updating those may have a larger impact.

3. Conclusions

We demonstrate new tools and a workflow that uses data science techniques to find discrepancies among published kinetic models, and screen them for importance using experimental validation data. Following the paper trail and evaluating the sources of the conflicting parameters allows the conflicts to be resolved, and the models updated. In most cases when resolving a conflict we agreed with the original model authors, so the modifications made to the butanol model are minor. This is confirmed by our ignition delay and laminar flame speed simulations with the old and new models. However, our aim was not just to make yet another butanol model, but to develop and demonstrate a workflow and procedure that reduces uncertainty caused by discrepancies. We expect that updating the other models (which contained the conflicting parameters) may have more significant effects, and envision a procedure to do this automatically, eventually converging upon a self-consistent set of detailed kinetic models, as well as experimental validation data.

Acknowledgments

This material is based upon work supported by the National Science Foundation under Grant nos. 1403171, 1605568, and 1535065; the Women and Minorities in Engineering program at Oregon State University; and the Department of Chemical Engineering at Northeastern University.

References

[1] S. J. Klippenstein, Y. Georgievskii, and L. B. Harding, Statistical theory for the kinetics and dynamics of roaming reactions. J. Phys. Chem. A 115 (2011) 14370–14381. DOI: 10.1021/jp208347j.

[2] R. H. West and C. F. Goldsmith, *The Impact of Roaming Radicals on the Combustion Properties of Transportation Fuels*, arXiv:1708.01004 [physics.chem-ph] Combust. Flame (submitted), 2017.

[3] M. P. Burke, C. F. Goldsmith, Y. Georgievskii, and S. J. Klippenstein, Towards a quantitative understanding of the role of non-Boltzmann reactant distributions in low temperature oxidation, Proc. Combust. Inst. 35 (2015) 205–213. DOI: 10.1016/j.proci.2014.05.118.

[4] N. J. Labbe, R. Sivaramakrishnan, C. F. Goldsmith, Y. Georgievskii, J. A. Miller, and S. J. Klippenstein, Ramifications of including non-equilibrium effects for HCO in flame chemistry, Proc. Combust. Inst. 36 (2017) 525–532. DOI: 10.1016/j.proci.2016.06.038.

[5] S. K. Sirumalla, M. A. Mayer, K. E. Niemeyer, and R. H. West, Assessing impacts of discrepancies in model parameters on autoignition model performance: a case study using butanol, Combust. Flame 190 (2018) 284–292. DOI: 10.1016/j.combustflame.2017.11.018.

[6] J. A. Manion, R. E. Huie, R. D. Levin, D. R. Burgess Jr, V. L. Orkin, W. Tsang, W. S. McGivern, J. W. Hudgens, V. D. Knyazev, D. B. Atkinson, E. Chai, A. M. Tereza, C. Y. Lin, T. C. Allison, W. G. Mallard, F. Westley, J. T. Herron, R. F. Hampson, and D. H. Frizzell, *NIST Chemical Kinetics Database, NIST Standard Reference Database 17, Version 7.0 (Web Version)*, NIST, Gaithersburg, MD, http://kinetics.nist.gov/, 2017.

[7] M. Frenklach, Transforming data into knowledge—Process Informatics for combustion chemistry, Proceedings of the Combustion Institute 31 (2007) 125–140. DOI: 10.1016/j.proci.2006.08.121.

[8] C. W. Gao, J. W. Allen, W. H. Green, and R. H. West, Reaction Mechanism Generator: Automatic construction of chemical kinetic mechanisms, Comput. Phys. Commun. 203 (2016) 212–225. DOI: 10.1016/j.cpc.2016.02.013.

[9] R. H. West, *RMG-Py importer branch*, commit 9d9ed1cb1440e405939c066a6ba34746c1ae5f4b, 2014, URL: https://github.com/rwest/RMG-Py/tree/importer/examples/chemkinImporter/USC_Mech_ii.

[10] B. W. Weber and K. E. Niemeyer, ChemKED: a human- and machine-readable data standard for chemical kinetics experiments, Int. J. Chem. Kinet. (2018), DOI: 10.1002/kin.21142.

[11] K. E. Niemeyer, PyTeCK: a Python-based automatic testing package for chemical kinetic models, Proceedings of the 15th Python in Science Conference (SciPy 2016) (2016), pp. 82–89, URL: `http://conference.scipy.org/proceedings/scipy2016/kyle_niemeyer.html`.

[12] S. M. Sarathy, S. Vranckx, K. Yasunaga, M. Mehl, P. Oßwald, W. K. Metcalfe, C. K. Westbrook, W. J. Pitz, K. Kohse-Höinghaus, R. X. Fernandes, and H. J. Curran, A comprehensive chemical kinetic combustion model for the four butanol isomers, Combust. Flame 159 (2012) 2028–2055. DOI: 10.1016/j.combustflame.2011.12.017.

[13] I. L. R. Bec, Y. Zhu, D. F. Davidson, and R. K. Hanson, Shock Tube Measurements of Ignition Delay Times for the Butanol Isomers Using the Constrained-Reaction-Volume Strategy, International Journal of Chemical Kinetics 46 (2014) 433–442. DOI: 10.1002/kin.20859.

[14] Y. Zhu, D. F. Davidson, and R. K. Hanson, 1-Butanol ignition delay times at low temperatures: An application of the constrained-reaction-volume strategy, Combust. Flame 161 (2014) 634–643. DOI: 10.1016/j.combustflame.2013.06.028.

[15] P. S. Veloo and F. N. Egolfopoulos, Flame propagation of butanol isomers/air mixtures, Proc. Combust. Inst. 33 (2011) 987–993.

[16] K. E. Niemeyer, *PyTeCK version 0.2.1*, https://doi.org/10.5281/zenodo.546270, 2017, DOI: 10.5281/zenodo.546270.

[17] D. G. Goodwin, H. K. Moffat, and R. L. Speth, *Cantera: An Object-oriented Software Toolkit for Chemical Kinetics, Thermodynamics, and Transport Processes*, http://www.cantera.org, Version 2.3.0, 2017, DOI: 10.5281/zenodo.170284.

[18] C. Olm, I. G. Zsély, R. Pálvölgyi, T. Varga, T. Nagy, H. J. Curran, and T. Turányi, Comparison of the performance of several recent hydrogen combustion mechanisms, Combust. Flame 161 (2014) 2219–2234. DOI: 10.1016/j.combustflame.2014.03.006.

[19] C. Olm, I. G. Zsély, T. Varga, H. J. Curran, and T. Turányi, Comparison of the performance of several recent syngas combustion mechanisms, Combust. Flame 162 (2015) 1793–1812. DOI: 10.1016/j.combustflame.2014.12.001.

[20] S. van der Walt, S. C. Colbert, and G. Varoquaux, The NumPy Array: A Structure for Efficient Numerical Computation, Comput. Sci. Eng. 13 (2011) 22–30. DOI: 10.1109/MCSE.2011.37.

[21] E. Jones, T. Oliphant, P. Peterson, and et al., *SciPy: Open source scientific tools for Python*, http://www.scipy.org/, 2001–.

Spring Technical Meeting
Eastern States Section of the Combustion Institute
March 4-7, 2018
State College, Pennsylvania

An Analytical Jacobian Generator for Reduced Chemical Kinetic Models Involving Quasi-Steady-State Assumptions

Pushan Sharma[1], Himanshu Goyal[2], Perrine Pepiot[1]

[1]*Sibley School of Mechanical and Aerospace Engineering, Cornell University, Ithaca, NY*
[2]*Robert F. Smith School of Chemical and Biomolecular Engineering, Cornell University, Ithaca, NY*

Recognizing the role of chemistry in the development of novel, more efficient engine technologies, numerical simulations of practical combustion systems are employing increasingly complex chemical kinetic models. A major challenge in such simulations is the cost of evaluating the chemical source terms, which often dominates the overall CPU cost, especially when using implicit numerical methods. In this case, a significant contributor to the chemical source term CPU cost is the evaluation of the jacobian matrix by the stiff ODE solver employed. This contribution can however be significantly decreased if an analytical formulation is explicitly provided to the solver. While automatic analytical jacobian code generators have been developed in the literature, in this work, we extend the concept to handle the case of reduced kinetic models that involve quasi-steady state assumptions. The algorithm includes smart ordering of the quasi-steady-state variables to minimize the number of operations needed to invert the QSS algebraic system, and the automatic generation of the routines needed to evaluate the analytical jacobian corresponding to the non-QSS species. The algorithm is validated and its performance assessed using a reduced n-dodecane oxidation model in the context of partially stirred reactor simulations.

1 Introduction

Detailed numerical simulations of complex combustion phenomena, e.g., engine technologies, require accurate and comprehensive chemical kinetic models. However, detailed modeling using a complex kinetic mechanism is still quite a daunting task. A wide range of techniques aiming at minimizing the computational demand have been developed and successfully applied in past literature, including skeletal reduction [1–3] or chemical lumping [4, 5]. The focus of this work is on the introduction of quasi steady state (QSS) assumptions, which convert the differential equations corresponding to a carefully selected species, also called QSS species into computationally more efficient algebraic expressions, thus reducing the number of ODEs that have to be solved and typically reducing significantly the stiffness of the remaining equations [6, 7]. Apart from chemistry reduction, different mathematical techniques were also developed focusing algorithm speed-up, e.g., more efficient solvers to integrate the governing ordinary differential equations. Due to the inherent stiffness of the ODE system, chemical kinetic models use robust solvers using backward differentiation methods which require repeated jacobian matrix evaluations [8–10]. The computational cost of evaluating an approximate jacobian scales quadratically with number of species (n_s^2), whereas the factorization of the matrix scales as n_s^3, where n_s is the number of species [11, 12]. For a medium to large size mechanism, jacobian matrix factorization is the most expensive step [11]. An analytical jacobian matrix formulation rather than the finite difference approximation reduces the computational cost significantly.

Motivated by the potential reduction in computational cost, a few analytical jacobian generator package have been developed in last few years. Safta et al. [13] developed TChem toolkit to calculate analytical jacobian using an interface, not allowing customized source code for different kinetic models. Youssefi [14] and Bisetti [11] also developed analytical jacobian matrix source code to reduce both computational cost and numerical error. The latter used species concentrations rather than mass fraction, which helped increasing the sparsity of the analytical jacobian matrix. Perini et al [12] used analytical jacobian for constant volume reactor which again helps the matrix more sparse as the off-diagonal elements would only be non-zero because of direct species interactions and fall off or pressure dependent reactions. Perini et al [12] reported a 80% faster result than a finite difference approach of jacobian matrix formulation. Unfortunately, most of the combustion CFD tools are built on a constant pressure assumption, thus reducing sparsity of the matrix. Recently GPU or hybrid CPU/GPU [15] have been used rather than pure CPU to speedup the computational time. Dijkmans et al. [16] took a GPU based analytical jacobian approach to accelerate the integration algorithm even further. Niemeyer et al. [17] developed pyJac, an open source analytical jacobian tool optimized for SIMD processors.

In this work, the concept of analytical jacobian formulation is extended further for reduced kinetic mechanisms involving QSS species. This package calculates QSS variables using their corresponding algebraic expressions and automatically develops the analytical jacobian matrix formulation for non-QSS species. The next section discusses the formulation of the analytical jacobian. A reduced n-dodecane oxidation model is used to validate and asses the performance of the current algorithm in a partially stirred reactor simulation.

2 Theory and Formulation

The present study considers the time evolution of reactive gaseous mixtures using a modular code in Fortran in a partially stirred reactor environment whose solution is achieved through the integration of a system of ODE. Here, the required jacobian matrix is derived analytically using quasi steady state (QSS) approximation. The variables in the jacobian matrix are taken to be the non-QSS species mass fractions (Y_i) and the average system temperature (T). The QSS species mass fractions are calculated using their corresponding algebraic expressions, thus reducing the computational cost significantly. However, the QSS species concentrations not only depend on non-QSS species, but also on other QSS species (only linear coupling is considered). Numerically optimized decoupled, explicit expressions for each QSS species have been achieved by Gauss pivoting [18].

The $(N_{nq} + 1) \times (N_{nq} + 1)$ jacobian matrix, $\mathcal{J}$, is filled with partial derivatives such as (N_{nq} = number of non-QSS species) $\mathcal{J}_{i,j} = \frac{\partial f_i}{\partial \phi_j}$ where ϕ corresponds to the variables (mass fraction and temperature) and

$$f = \frac{\partial \phi}{\partial t} = \left\{ \frac{\partial Y_1}{\partial t}, \frac{\partial Y_2}{\partial t}, ..., \frac{\partial Y_{N_{nq}}}{\partial t}, \frac{\partial T}{\partial t} \right\} \tag{1}$$

where

$$\frac{\partial Y_i}{\partial t} = \frac{W_i}{\rho}\dot{\omega}_i \quad \text{and} \quad \frac{\partial T}{\partial t} = \frac{-1}{\rho c_p}\sum_{i=1}^{N_{nq}} h_i W_i \dot{\omega}_i \quad \text{for} \quad i = 1 \text{ to } N_{nq} \tag{2}$$

2.1 Derivatives with respect to Mass Fraction

The mass fraction derivatives, $\frac{\partial \dot{Y}_i}{\partial Y_j}$, where both i and j vary from 1 to N_{nq}, can be simplified by rewriting it in terms of concentration derivatives $\left(\frac{\partial \dot{C}_i}{\partial C_j}\right)$. From Eq. 2

$$\mathcal{J}_{i,j} = \frac{\partial \dot{Y}_i}{\partial Y_j} = \left(\frac{W_i}{\rho}\frac{\partial \dot{C}_i}{\partial C_j} - \dot{Y}_i \frac{W_j}{\rho} \right) \frac{\partial C_i}{\partial Y_j} \tag{3}$$

where

$$\frac{\partial \dot{C}_i}{\partial C_j} = \sum_{k=1}^{N_R} \left(\nu_{i,k} \frac{\partial \omega_k}{\partial C_j} \right) \tag{4}$$

and

$$\frac{\partial C_i}{\partial Y_j} = -C_i \frac{W}{W_j} + \delta_{ij}\frac{\rho}{W_i} \tag{5}$$

In the above equations, $\nu_{i,k} = \nu^{''}_{i,k} - \nu^{'}_{i,k}$. $\nu^{''}$ and $\nu^{'}$ are the stoichiometric coefficients of products and reactants in a reaction respectively, W is the mixture molecular weight, W_i is the molecular weight of the i-th species, ρ is the mixture density, N_R is the total number of reactions, δ_{ij} is the Kronecker Delta function, and ω_k is the reaction rate of k-th reaction. The derivative of the reaction rate with respect to concentration in Eq. 4 will be different for different types of reactions. For a simple reaction containing only non-QSS species, it will only depend on the reactant concentrations and rate coefficient of forward ($k_{f,k}$) and backward ($k_{b,k}$) reactions.

$$\frac{\partial \omega_k}{\partial C_j} = k_{f,k}\nu^{'}_{j,k} C_j^{\nu^{'}_{j,k}-1} \prod_{i \neq j} C_i^{\nu^{'}_{i,k}} - k_{b,k}\nu^{''}_{j,k} C_j^{\nu^{''}_{j,k}-1} \prod_{i \neq j} C_i^{\nu^{''}_{i,k}} \tag{6}$$

However, for a reaction containing a QSS species (Q_k is the QSS species in the k-th reaction), an additional term corresponding to the derivatives of the QSS species with respect to other non-QSS species is required as shown in Eq. 7.

$$\frac{\partial \omega_k}{\partial C_j} = k_{f,k}\left(\nu'_{j,k}C_j^{\nu'_{j,k}-1}\prod_{i\neq j} C_i^{\nu'_{i,k}} + \prod_{i\neq Q_k} C_i^{\nu'_{i,k}}\frac{\partial Q_k}{\partial C_j}\right) - k_{b,k}\left(\nu''_{j,k}C_j^{\nu''_{j,k}-1}\prod_{i\neq j} C_i^{\nu''_{i,k}} + \prod_{i\neq Q_k} C_i^{\nu''_{i,k}}\frac{\partial Q_k}{\partial C_j}\right) \tag{7}$$

It is important to notice that in Eq. 6, for a particular reaction k, $\frac{\partial \omega_k}{\partial C_j}$ is nonzero only when i and j are the reactants. However, in Eq. 7, $\frac{\partial \omega_k}{\partial C_j}$ is nonzero when i corresponds to the reactants, but j corresponds to all non-QSS species that the concentration of Q_k depends on, which includes, but is not limited to the reactants only. The rate coefficients ($k_{f,k}$ and $k_{b,k}$) are calculated using Arrhenius equation. The effects of third-body and pressure dependent reactions are also considered in this formulation of this analytical jacobian, although not discussed here in detail.

On the other hand, the temperature derivatives with respect to species mass fraction can be directly derived from Eq. 2.

$$\mathcal{J}_{N_{nq}+1,j} = \frac{\partial \dot{T}}{\partial Y_j} = \frac{-1}{c_p}\left(c_{p,j}\dot{T} + \sum_{j=1}^{nq} h_i \frac{\partial \dot{Y}_i}{\partial Y_j}\right) \tag{8}$$

where c_p and h are the specific heat and the specific enthalpy respectively.

2.2 Derivatives with respect to Temperature

The species mass fraction derivatives with temperature, $\frac{\partial \dot{Y}_i}{\partial T}$ is also evaluated in terms of concentrations as shown in Eq. 9.

$$\mathcal{J}_{i,N_{nq}+1} = \frac{\partial \dot{Y}_i}{\partial T} = \frac{W_i}{\rho}\left(\frac{\partial \dot{C}_i}{\partial T} + \frac{\dot{C}_i}{T}\right) \tag{9}$$

The analytical expression of temperature derivative with respect to temperature is

$$\mathcal{J}_{N_{nq}+1,N_{nq}+1} = \frac{\partial \dot{T}}{\partial T} = \frac{-1}{c_p}\left(\sum_{i=1}^{N_{nq}} \frac{\partial c_{p,i}}{\partial T} Y_i \dot{T} + \sum_{i=1}^{N_{nq}} c_{p,i}\dot{Y}_i + \sum_{i=1}^{N_{nq}} h_i \frac{\partial \dot{Y}_i}{\partial T}\right) \tag{10}$$

3 Results and Discussion

In order to evaluate the accuracy and the performance of this analytical jacobian matrix with QSS species approximation, we consider a skeletal n-dodecane mechanism with 54 species and 269 reactions, developed by Yao et al. [19]. This skeletal mechanism was further reduced to a mechanism with 47 species and 198 reactions using YARC, an automatic reduction software described in Pepiot [18]. The reduced mechanism is validated against experimental results and the skeletal mechanism itself for laminar flames and auto-ignition, as shown in Fig. 1. In the present work, the results using analytical jacobian is compared to the numerical approximation based on finite differences with and without QSS approximations. For simplicity, it is made sure that the algebraic expressions corresponding to each QSS species are linear. With this constraint, total 19 QSS species are identified. So, the final reduced n-dodecane oxidation model consists of 28 non-QSS and 19 QSS species.

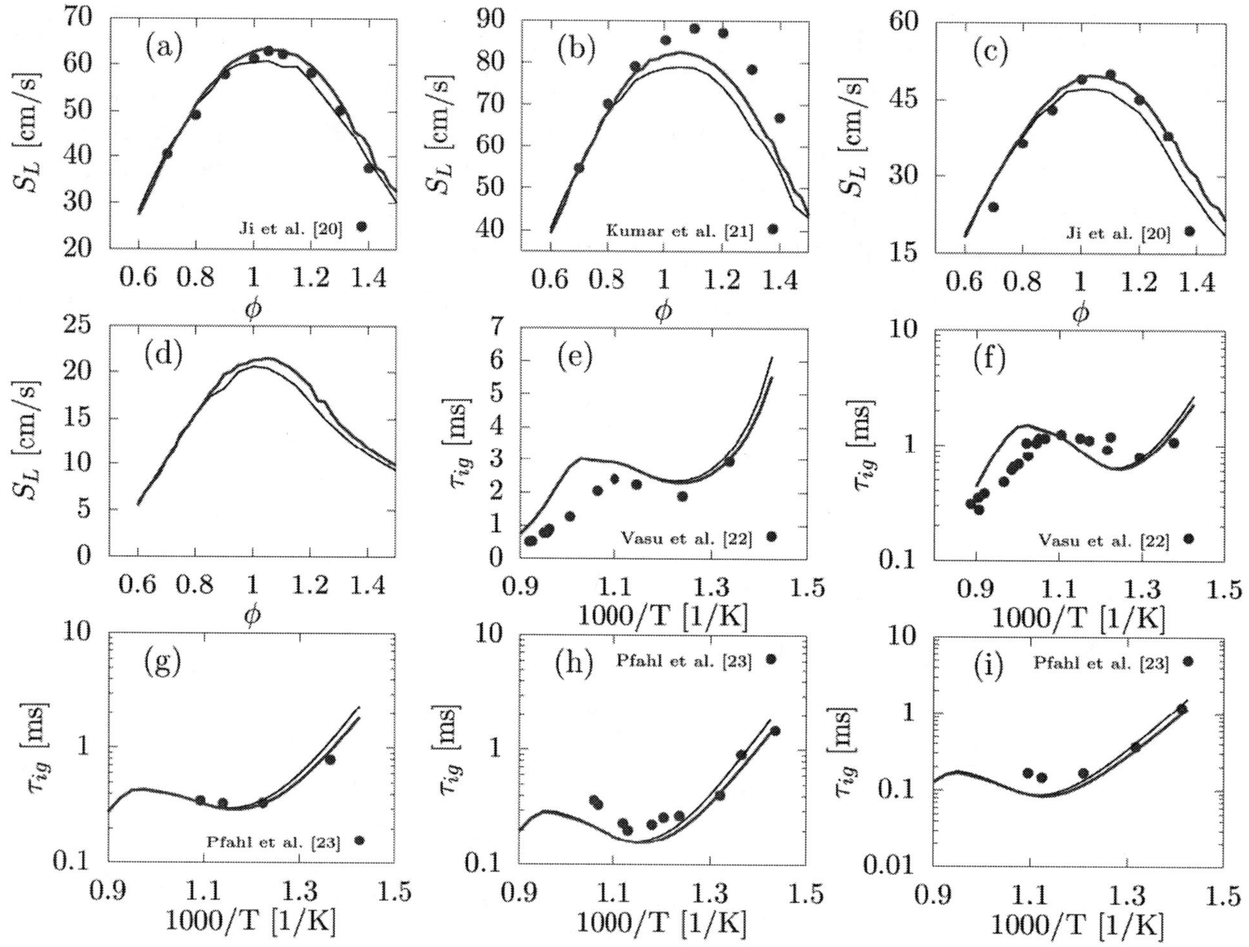

Figure 1: Comparison of Laminar flame speed (S_L) and ignition delay time (τ_{ig}) between experimental data [20–23] (blue symbols), simulation using skeletal mechanism [19] (black line) and reduced mechanism (red line) at different conditions: (a) $p = 1$ bar, $T = 400$K, (b) $p = 1$ bar, $T = 470$K, (c) $p = 3$ bar, $T = 400$K, (d) $p = 35$ bar, $T = 400$K, (e) $p = 20$ bar, $\phi = 0.5$, (f) $p = 20$ bar, $\phi = 1$, (g) $p = 50$ bar, $\phi = 0.67$, (h) $p = 50$ bar, $\phi = 1$, (i) $p = 50$ bar, $\phi = 2$.

To examine the performance of the analytical jacobian algorithm, the CPU time is evaluated in a partially stirred reactor (PaSR) environment. In a PaSR, the flow field is represented by a number of reactive particles carrying their own species composition and temperature. This is an adiabatic and statistically spatially homogeneous reactor, although not perfectly mixed at the molecular level. The composition of each particle evolves in time due to inflow/outflow/pairing, mixing and reaction. A characteristic residence time corresponds to the inflow/outflow rate inside the reactor. Each particle is assigned a partner and the pairing of these two particles, specified by a pairing time, provides a local mixing process inside the reactor. Although PaSR is not a mathematical representation of the real system, the simplicity of this model permits to explore the effect of reduced

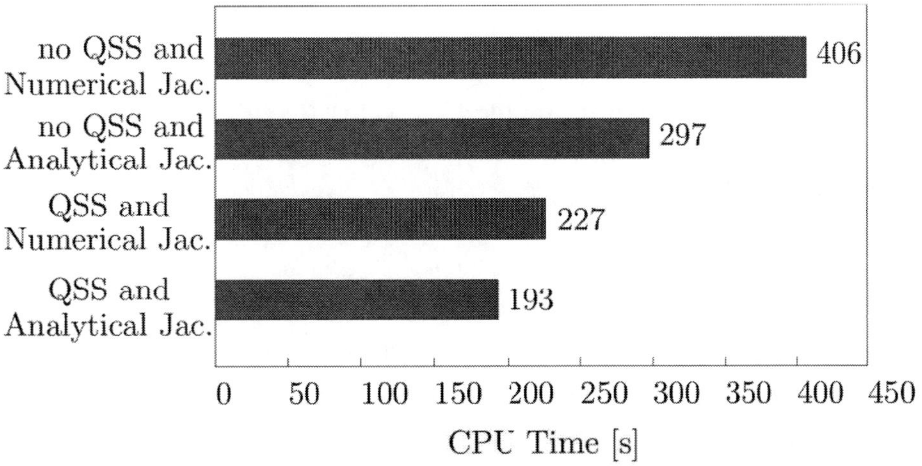

Figure 2: Comparison of CPU time in 4 different situations: (1) no QSS and numerical jacobian, (2) no QSS and analytical jacobian, (3) 19 QSS species and numerical jacobian, (4) 19 QSS species and analytical jacobian.

chemistry models without a significant computational cost [7, 24]. The PaSR parameters chosen for the simulations are: total 100 gas particles, residence time of 0.01s, pairing and mixing times of 0.001s, total 3 inflow streams with mass flow rates of 0.05, 0.893 and 0.057kg/s and at 2271K, 300K and 300K respectively. The first inflow contains mostly N_2, CO_2 and trace amounts of H_2O, CO, O_2, OH, H_2, O, H, HO_2, H_2O_2, HCO and CH_2O. The second and third inflow compositions are air and pure n-dodecane respectively. As expected, the results using numerical and analytical jacobian exactly coincides with each other for temperature and all the individual species. However, the use of analytical jacobian is proved to be quite efficient in terms of CPU cost. Figure 2 compares CPU time required under aforementioned PaSR conditions in 4 different combinations: (1) no QSS and numerical jacobian, (2) no QSS and analytical jacobian, (3) with 19 QSS species and numerical jacobian and (4) with 19 QSS species and analytical jacobian. Almost a 15% reduction in CPU time using analytical jacobian (case 4) than numerical jacobian (case 3) is observed under QSS approximation. It must be noted that chemical source term integration takes majority of the time, thus reduction in overall CPU time is dominated by this term that involves the analytical jacobian.

4 Conclusions

In the present work, an analytical jacobian matrix evaluation approach with QSS species approximation for constant pressure chemical kinetics has been developed. This package automatically develops the analytical jacobian for non-QSS species and then calculates the QSS species concentrations using their corresponding linear algebraic expressions, thus reducing the computational cost significantly. The results using this analytical jacobian is also validated against the solution obtained by numerical jacobian. All the simulations were performed with a n-dodecane reduced oxidation model in a partially stirred reactor environment and it turned out that the plots corresponding to analytical and numerical jacobians match exactly and coincides with each other. The analytical jacobian with QSS approximation also performs almost 15% faster than the numerical jacobian, thus reducing the overall CPU cost of numerical simulations involving complex chemical kinetic models. However, it should be noted that no optimization of the jacobian in terms of number of operations performed has been done yet.

Acknowledgements

This material is based upon work supported by the U.S. Department of Energy, Office of Energy Efficiency and Renewable Energy, Co-Optimization of Engines and Fuels (Co-Optima) initiative under Award Number DE-EE0007978.

References

[1] Tianfeng Lu and Chung K Law. *Progress in Energy and Combustion Science*, 35 (2009) 192–215.

[2] Tianfeng Lu and Chung K Law. *Combustion and Flame*, 144 (2006) 24–36.

[3] Perrine Pepiot-Desjardins and Heinz Pitsch. *Combustion and Flame*, 154 (2008) 67–81.

[4] Tianfeng Lu and Chung K Law. *Combustion and flame*, 148 (2007) 117–126.

[5] Perrine Pepiot-Desjardins and Heinz Pitsch. *Combustion Theory and Modelling*, 12 (2008) 1089–1108.

[6] SH Lam and DA Goussis. *International Journal of Chemical Kinetics*, 26 (1994) 461–486.

[7] Himanshu Goyal and Perrine Pepiot. *Energy & Fuels*, 31 (2017) 12120–12132.

[8] CF Curtiss and Joseph O Hirschfelder. *Proceedings of the National Academy of Sciences*, 38 (1952) 235–243.

[9] Peter N Brown, George D Byrne, and Alan C Hindmarsh. *SIAM journal on scientific and statistical computing*, 10 (1989) 1038–1051.

[10] Alan C Hindmarsh, Peter N Brown, Keith E Grant, Steven L Lee, Radu Serban, Dan E Shumaker, and Carol S Woodward. *ACM Transactions on Mathematical Sofware (TOMS)*, 31 (2005) 363–396.

[11] Fabrizio Bisetti. *Combustion Theory and Modelling*, 16 (2012) 387–418.

[12] Federico Perini, Emanuele Galligani, and Rolf D Reitz. *Energy & Fuels*, 26 (2012) 4804–4822.

[13] Cosmin Safta, Habib N Najm, and Omar Knio. *Sandia Report, SAND2011-3282*, (2011) journal.

[14] Mehrnaz Rouhi Youssefi. Development of analytic tools for computational flame diagnostics. Master's thesis, University of Connecticut, 2011.

[15] Yu Shi, William H Green, Hsi-Wu Wong, and Oluwayemisi O Oluwole. *Combustion and Flame*, 158 (2011) 836–847.

[16] Thomas Dijkmans, Carl M Schietekat, Kevin M Van Geem, and Guy B Marin. *Computers & Chemical Engineering*, 71 (2014) 521–531.

[17] Kyle E Niemeyer, Nicholas J Curtis, and Chih-Jen Sung. *Computer Physics Communications*, 215 (2017) 188–203.

[18] Perrine Pepiot. *Automatic strategies to model transportation fuel surrogates*. PhD thesis, Stanford University, 2008.

[19] Tong Yao, Yuanjiang Pei, Bei-Jing Zhong, Sibendu Som, Tianfeng Lu, and Kai Hong Luo. *Fuel*, 191 (2017) 339–349.

[20] Chunsheng Ji, Enoch Dames, Yang L Wang, Hai Wang, and Fokion N Egolfopoulos. *Combustion and Flame*, 157 (2010) 277–287.

[21] Kamal Kumar and Chih-Jen Sung. *Combustion and Flame*, 151 (2007) 209–224.

[22] Subith S Vasu, David F Davidson, Z Hong, V Vasudevan, and Ronald K Hanson. *Proceedings of the Combustion Institute*, 32 (2009) 173–180.

[23] U Pfahl, K Fieweger, and G Adomeit. Self-ignition of diesel-relevant hydrocarbon-air mixtures under engine conditions. In *Symposium (International) on Combustion*, Vol. 26, pages 781–789. Elsevier, 1996.

[24] Youwen Liang, Stephen B Pope, and Perrine Pepiot. *Combustion and Flame*, 162 (2015) 3236–3253.

Spring Technical Meeting
Eastern States Section of the Combustion Institute
March 4-7, 2018
State College, Pennsylvania

ChemKED for profile-resolved data: A discussion of some salient data standard features

Francis M. Haas[1,], C. Franklin Goldsmith[2], Michael P. Burke[3], Bryan W. Weber[4], and Kyle E. Niemeyer[5]*

[1]Mechanical Engineering Department, Rowan University, Glassboro, NJ 08028
[2]School of Engineering, Brown University, Providence, RI 02912
[3]Department of Mechanical Engineering, Columbia University, New York, NY 10027
[4]Department of Mechanical Engineering, University of Connecticut, Storrs, CT 06269
[5]School of Mechanical, Industrial, and Manufacturing Engineering, Oregon State University, Corvallis, OR 97331
[*]Corresponding Author Email: haas@rowan.edu

Abstract: Development of the open, human- and machine-readable ChemKED data standard for describing common observables used in fundamental combustion kinetics experiments has presently progressed to robustly describe ignition delay times and related supporting measurements (e.g., RCM volume histories). These single-valued "global" measurements offer some insight into combustion kinetics; however, additional understanding can be determined from interpretation of "profile" measurements typical of those from jet stirred reactors, flow reactors, species-resolving shock tubes, and burner-stabilized flames. Despite the utility of such measurements, the combustion community presently lacks a widely adopted data standard to facilitate archiving, sharing, and automated manipulation of profile data. Unlike global observables, profile measurements generally resolve the (co)evolution of one or more observables (e.g., species mole fractions and/or temperature) as a function of a continuous and independent spatial, temporal, or thermal coordinate intrinsic to a single chemical kinetics experiment. The numerous profile-resolving experiment types, their intrinsic experimental variables, and (multiple) observables each pose a potential challenge for transforming the existing global-type ChemKED data structure into one capable of handling profile-resolved data. Toward stimulating further discussion of a broadly acceptable data standard for profile measurements, the present work offers several concrete examples of these challenges and proposes an initial approach toward a generally satisfactory solution.

Keywords:* *Chemical Kinetics, Experimental Data, Data Standard, Species Profiles

1. Introduction

Experimental measurements of "fundamental" gas phase combustion phenomena have been in widespread use for much of the last century [1] as a means gaining insight into the chemistry of fuel pyrolysis/oxidation and pollutant formation/destruction. Measured observables such as laminar burning velocity, ignition delay time, and flow reactor species evolution profiles, among others, are ubiquitous inputs for development and validation of chemical kinetic models, constraint of elementary reaction rate coefficients, and exploration of distinct mechanistic features in reacting systems. Despite – or perhaps as a consequence of – the utility, abundance, and historical accumulation of these data, a broadly accepted standard for their archiving, sharing, and automated manipulation remains lacking. One distinct consequence of this reality is that the tools of data science cannot easily be deployed to extract further value from the existing experimental database scattered across $O(10^6)$ pages of literature relevant to the combustion discipline. The absence of a widely adopted, *community-based infrastructure* for data management poses a barrier to progress in the combustion kinetics community [2, 3] and all but assures costly and unnecessary redundancy in acquisition and utilization of experimental measurements.

The contemporary work of Weber & Niemeyer [3] critically reviews several existing initiatives to address community data management, including PrIMe [2, 4], ReSpecTh [5, 6], and CloudFlame [7]. It concludes that there remains a need for an *open, decentralized* combustion kinetics data management framework, and then addresses this need by offering ChemKED, an open-source, YAML-based [8], human- and machine-readable data standard developed to manage fundamental (combustion) kinetics experiments. The present ChemKED v0.3.0 format detailed in [3] includes a functional data standard for documenting autoignition experiments in shock tubes and rapid compression machines. However, the kinetics discipline is informed by a significantly more diverse array of experiments, and envisioned updates to ChemKED will include handling of other experiments measuring global observables (e.g., laminar burning velocity or extinction strain rate) as well as experiments that measure profile-resolved observables (e.g., flow reactor species profile evolution or shock tube laser schlieren histories).

This work intends to stimulate discussion around developing a ChemKED data standard for profile-resolved experiments as a distinct, albeit diverse, class of experiments sharing the common feature of one or more (co)evolving observables (e.g., species mole fraction(s) or temperature) varying as a function of one continuous and independent intrinsic experimental coordinate (e.g., spatial dimension, time, or temperature). Results of these experiments are typically considered to be discretized series samplings of strongly correlated, continuously varying observables responding to the intrinsic experimental coordinate. This functional definition, while imperfect, differentiates these types of experiments from global experiments, in which individual measurements are (at most) weakly correlated and do not depend on an intrinsic, continuously varying coordinate dimension. Global observables are generally characterized by point-like dimensionality.

The distinction between global and profile-resolved experiments is not entirely unambiguous. For example, measurements from end-sampling flow reactor experiments typical of the ones described by Herrmann et al. [9] and Guo et al. [10] are typically treated in the literature as profile data resolved against a nominal reactor temperature as the intrinsic experimental coordinate. However, each nominal reactor temperature corresponds to a different axial temperature profile along the reactor, and may also correspond to changes in system conditions that challenge assumptions regarding (de)coupling of reactions and heat transfer with the flow field and reactor wall. Either or both of these factors may be important for kinetic insight and accurate experimental modeling [11]. Instead of treating such experiments as profile-resolved, perhaps such experiments would be better treated in ChemKED as a set of related global measurements, each subject to its own respective temperature profile. This would be similar to a series of related RCM autoignition experiments, each subject to its own inferred volume history, or to the *outlet concentration measurement* experiment type specified by ReSpecTh [6].

Throughout this paper, several points underlying to-be-resolved issues such as the preceding one related to ChemKED keys [3] `kind` and `experiment-type` will be bolded and set apart from the body text (as below) to indicate that they merit further discussion toward developing ChemKED for profile-resolved data.

- **What distinction should be made between global and profile data, and what implication does this have for organization of the corresponding `datapoints`? Is ambiguity tolerable, or is strong formalism required for future revisions to the ChemKED schema?**

Discussion points developed in the present work will be neither comprehensive nor conclusive. Nevertheless, they can serve as a starting point to arouse discussions that further the development of ChemKED, particularly for the management and profitable use of profile-resolved experiments.

2. "Meta" data Considerations

Expansion of ChemKED to include profile-resolved data from common kinetics experiments necessitates expansion of values associated with the `apparatus kind` and `experiment-type` keys. The `kind` key indicates the type of apparatus used to perform the archived experiment, and in v0.3.0, can assume only the values of `shock tube` and `rapid compression machine`. Both of these devices can be used to acquire profile-resolved data; however, `stirred reactor`, `flow reactor`, and `burner-stablized flame` are among other apparatuses commonly used to acquire profile measurements. These certainly merit inclusion in future versions of the ChemKED standard; nevertheless, other canonical apparatuses like `static reactor` and `counterflow flame` also inform combustion kinetics with profile data, suggesting prioritization of `apparatus kind` inclusion will be necessary to thoroughly develop each according to its respective experimental nuances evident in the literature.

- **What priority, based on either perceived importance to the community or user interest, do different `kinds` of experimental apparatuses have for inclusion into later versions of ChemKED?**

The current ChemKED v0.3.0 supports `ignition delay` as the only value for the `experiment-type` key. Profile-resolved data to be included in future revisions of the standard will require one or more new values be added for this key. Notwithstanding the previously discussed ambiguity of certain measurements as either global or profile-resolved, the value "`profile measurement`" (or an alternative string of words of similar generality) for the `experiment-type` key appears to address a large subset of profile-resolving `apparatus kinds` and extends to experiments that merge results from dissimilar diagnostics (e.g., FTIR and GC) onto the same intrinsic experimental coordinate. Moreover, it subsumes the inferred natural language value human readers would ascribe to strings such as "`species profile`." Nevertheless, this generality would frustrate manual and automated searching and manipulation of ChemKED records. There also exists an opposite extreme in which values associated with `experiment-type` are narrowly defining and related to the most direct observable evidence of experimental speciation measurements (e.g., `gas chromatograms` or frequency-resolved `laser absorption profiles`). Despite the advantage of clarity and indexing of the ChemKED archive, use of overly specific values for `experiment-type` may lead to a burgeoning list of values for this key and hazard confusion or inadvertent misuse among users. Such specificity can likely be included in `datapoints` or as explanatory notes in the metadata. A middle ground would define a limited list of values for experiment-type that describes the majority of profile-resolving experiments, thereby achieving balance between generality and specificity.

- **What balance between generality and specificity is appropriate for the profile-resolving experiment-types? Is development of a list of values for this key driven by the same considerations of importance to the community/user interest as new kinds of apparatus?**

Finally, thorough implementation of new apparatus kinds and experiment-types ideally involves internal consistency. For example, addition of stirred reactor and flow reactor to the available kinds of apparatus opens up the combination of these apparatuses with ignition delay as their experiment-type.

3. Datapoints Considerations

While ChemKED v0.3.0 exclusively supports autoignition delay time experiments, several features of the existing schema can transfer directly for use with many profile-resolved measurements. These features include, for example, keys for nominal reaction conditions (e.g., temperature and pressure) and the ability to assign them as common properties, as well as reactant composition and species handling.

The largest difference between autoignition and profile-resolved data appears in the datapoints portion of the ChemKED record. For autoignition data, each individual Datapoint collapses into a point representation even if it includes its own volume-history profile. For profile-resolved data, a working draft standard uses a comma separated value table stored in a separate .csv file to describe the variation of one target observable with the intrinsic experimental coordinate (e.g., a single species mole fraction evolution as a function of time). A full multi-species profile involving N dependent observables would use N associated two- or four-column .csv files to specify datapoints for the experiment without or with associated observable and intrinsic coordinate uncertainty, respectively. If N is a large value, this approach may be massively redundant with respect to the intrinsic experimental coordinate. Assuming instead, as is often the case, that profile sampling for each dependent observable is simultaneous with respect to the intrinsic profile coordinate, an alternative realization for profile data would combine the N observables into a single $N+1$ column .csv file. Inclusion of profile-resolved uncertainty could be accommodated by a similar (up to) $2(N+1)$ column .csv file, with keys in the associated ChemKED file indicating absence/presence and type (relative or absolute) of assigned uncertainty. More flexibly, an (up to) $4N$ column .csv file would accommodate asynchronous multi-observable profile experiments with optional treatment of uncertainties. Provided that consideration of non-ideal facility effects and assumptions about initial/boundary conditions for profile-resolving experiments can be deferred to later versions of ChemKED, then the manner in which profile data is stored and manipulated is likely the single most important decision for the initial release of a ChemKED data standard version with a functional approach to profile-resolved data archiving.

- **How should profile data interface with the central ChemKED file? Are certain treatments of associated .csv profile data storage files to be preferred for ease of human interface or compatibility with past/contemporary means of profile data storage? What complexities are foreseen for implementation of the alternative .csv realizations presented here (or otherwise).**

4. Concluding Remarks

The preceding discussion advances several important considerations for evolving the present ChemKED v0.3.0 data standard beyond its description of global autoignition experiments in shock tube and rapid compression machine facilities to include other kinetics experiments of the profile-resolving type. These considerations regarding (1) the difference between global and profile-resolved experiments, (2) priorities for introduction of new `kind`s of experimental apparatuses, (3) a balance between generality and specificity in the definition of new `experiment-type`s, and (4) file structure alternatives for separate profile data storage are neither exhaustive nor conclusive as bears on revision to ChemKED v0.3.0. Nevertheless, they offer leaping off points for further discussion and progress toward a ChemKED featuring a functional standard for profile-resolved experiments.

5. References

[1] B. Lewis, G. von Elbe, Combustion, flames, and explosions of gases, Cambridge University Press, 1938.

[2] M. Frenklach, Proceedings of the Combustion Institute, 31 (2007) 125-140. https://doi.org/10.1016/j.proci.2006.08.121

[3] B.W. Weber, K.E. Niemeyer, International Journal of Chemical Kinetics, in press. (2018). https://doi.org/10.1002/kin.21142

[4] X. You, A. Packard, M. Frenklach, International Journal of Chemical Kinetics, 44 (2012) 101-116. https://doi.org.10.1002/kin.20627

[5] T. Varga, T. Turányi, E. Czinki, T. Furtenbacher, A.G. Császár, ReSpecTh: a joint reaction kinetics, spectroscopy, and thermochemistry information system, 7th European Combustion Meeting (2015), paper P1-04.

[6] T. Varga, C. Olm, Á. Busai, I.G. Zsély, ReSpecTh kinetics data format specification v2.0, http://respecth.chem.elte.hu/, July 28, 2017.

[7] G.L. Goteng, N. Nettyam, S.M. Sarathy, CloudFlame: Cyberinfrastructure for Combustion Research, 2013 International Conference on Information Science and Cloud Computing Companion (2013), 294-299.

[8] O. Ben-Kiki, C. Evans, I. döt Net, YAML Ain't Markup Language (YAML™) Version 1.2, http://www.yaml.org/spec/1.2/spec.html, 2009.

[9] F. Herrmann, P. Oßwald, K. Kohse-Höinghaus, Proceedings of the Combustion Institute, 34 (2013) 771-778. https://doi.org/10.1016/j.proci.2012.06.136

[10] H. Guo, W. Sun, F.M. Haas, T. Farouk, F.L. Dryer, Y. Ju, Proceedings of the Combustion Institute, 34 (2013) 573-581. https://doi.org/10.1016/j.proci.2012.05.056

[11] F.L. Dryer, F.M. Haas, J. Santner, T.I. Farouk, M. Chaos, Progress in Energy and Combustion Science, 44 (2014) 19-39. https://doi.org/10.1016/j.pecs.2014.04.002

Spring Technical Meeting
Eastern States Section of the Combustion Institute
March 4-7, 2018
State College, Pennsylvania

Shock Tube and Kinetic Modelling Study of Organo Phosphorus Compounds used as Chemical Weapon Simulants

*Sneha Neupane[a], Frank Barnes[a], Samuel Barak[a], Erik Ninnemann[a], Zachary Loparo[a], Artëm E. Masunov[b], Subith S. Vasu[a]**

[a]Center for Advanced Turbomachinery and Energy Research, Mechanical and Aerospace Engineering, University of Central Florida, Orlando, FL 32816
[b]Nano Science Technology Center, Department of Chemistry, Department of Physics, University of Central Florida, Orlando, FL 32816
***Corresponding Email:** subith@ucf.edu

Abstract: Incineration is an effective way of destruction of chemical warfare agent. Increased threat of chemical warfare stockpile has renewed interest in their combustion chemistry. However, due to high toxicity of chemical warfare agents, experiments to study their combustion chemistry are carried out using organo phosphorous simulants. In this study kinetics study of oxidation of tri-ethyl phosphate (TEP) was carried out in a shock tube. Laser absorption spectroscopy was used to measured intermediate species (carbon monoxide) time histories. TEP is a simulant of Sarin-GB, a very toxic nerve agent which is used as a weapon of mass destruction. The intermediate CO time history data during TEP oxidation reported in this study will be useful in validation of kinetic models of organo phosphorous compounds and chemical warfare agents such as Sarin-GB.

Keywords: Organo phosphorous compounds, simulants, tri-ethyl phosphate, chemical kinetics

1. Introduction

Simulants are used in incineration studies, whose chemical structure and combustion rates would be similar to those of chemical warfare (CW) agents. Previous experimental work on these compounds were mainly limited to flow reactors (at temperatures lower than 1000K and at slow heating rates) [1-3] and flame studies (as dopants) [4-9], where typically, final products speciation analysis is carried out. There are major gaps in our knowledge of the high temperature decomposition pathways leading to final products and combustion chemistry of CW agents and simulants that are crucial to characterize their burning process, which will help to design better strategies to counter weapons of mass destruction (WMD). This gap can be partially filled by reasonable theoretical estimates, but benchmark experiments are needed to validate these estimations. Thus, in the current study, pyrolysis and oxidation of a chemical warfare simulant was carried out in a shock tube (which provides very fast heating rates) along with mid-IR laser absorption spectroscopy techniques for in-situ measurements of CO time history. Shock tube provides an ideal tool to investigate high-temperature chemical kinetics. The current study will reveal crucial knowledge related to pathways and reactions of simulants which will provide greater and much needed fundamental understanding for counter-WMD operations. Figure 1 shows the chemical structure of Sarin-GB and its simulant, Tri-ethyl phosphate (TEP)

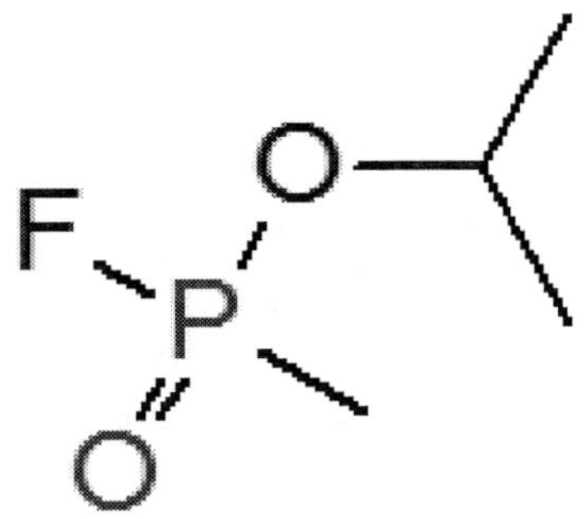

Sarin, GB

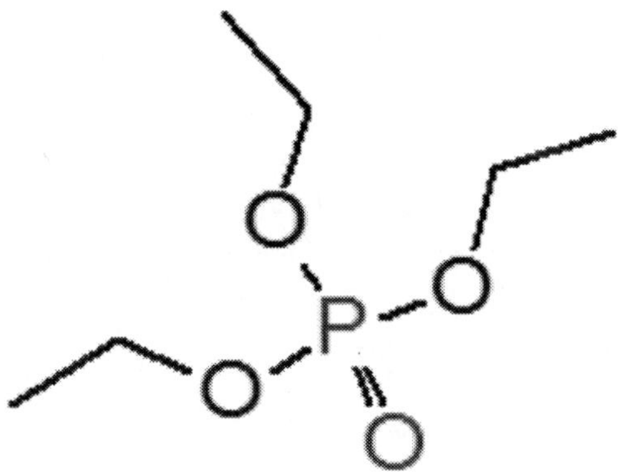

Tri- ethyl Phosphate (TEP)

Figure 1 Molecular structure of Sarin-GB and its simulant Tri-ethyl phosphate (TEP)

2. Methods / Experimental

2.1 Shock Tube Facility

Experiments were performed in the reflected shock region of high purity, heated, double-diaphragm, stainless steel shock tube at UCF shock tube facility (details in [10-15]). The driven section of the shock tube was heated to T_1= 373 K using eight sections of a custom made heating jacket and four benchtop PID temperature controllers. Before each experiment, the driven section was evacuated to < 5.3 Pa using a rotatory vane pump (Agilent DS102) before filling with 3333-8666 Pa of test mixtures (pre-shock pressure, P_1). Incident shock velocity was measured using five piezoelectric pressure transducers and extrapolated to obtain the incident shock velocity at the end wall. 5mm diaphragm was used in current work to get the desired P_5 (pressure behind reflected shock wave) of ~1.3 atm (131, 722 Pa). Conditions behind the reflected shock wave (P_5 and T_5

with uncertainties 1.8% and 2%, respectively) were calculated based on the measured velocity, P_1, T_1, and one-dimensional normal shock equations. The shock tube configuration and use of driver inserts allowed to obtain a constant P_5 behind the reflected shock region for test times of ~ 2.5 ms (P_5 was also monitored using Kistler type 603B1 sensor).

2.2 Fuel/oxidizer Mixture Preparation

Research grade TEP (> 99% pure; Acros Organics) and gases (O_2, Ar, > 99.999% purity; nexAir) were used to prepare test gas mixture in a high purity, 33 L, Teflon coated, stainless steel, heated mixing facility (heated to 423 K) designed specifically to accommodate low vapor pressure OPC compounds. The temperature of mixing manifold and transfer line to the shock tube was maintained at 392 $\pm$ 5 K. Before injecting liquid TEP, the tank was evacuated to P<2.67 Pa with a rotary vane pumps (Agilent DS102). Partial pressures were measured with 100 Torr- and 10,000 Torr full scale range baratrons (MKS Baratron E27D and 628D). The mixtures were let to mix for at least 3 hours before the first shock tube experiments using magnetically driven stirrer (to ensure homogeneity of the mixtures).

3. Results and Discussion

A quantum cascade laser (QD4580CM1) operating at 4580.4 nm was employed to probe the R(10) transition line for CO. A 50/50 beam splitter was used to split the laser light into two beams: transmitted (I_{tr}) and reference (I_{ref} to monitor laser power fluctuation). Transmitted beam was directed through the test section via two sapphire windows and a narrow band pass filter (Thorlabs model #FB4500-500) was used to filter out emissions from hot gasses and was collected in a thermoelectrically cooled MCT (HgCdTe) Vigo Systems PVI-2TE-5.0 photovoltaic detector. The CO mole fraction was obtained from the Beer–Lambert law:

$$\alpha = -\ln(\frac{I_{tr}}{I_{ref}})_v = \sigma(v,T,P)\frac{P_{tot}}{RT}\chi L \quad (1)$$

where α_v is absorbance, σ [cm^2/molecule] is absorption cross section, P [atm] is pressure, and T [K] is temperature, L [cm] is path length, and χ is the mole fraction of the absorbing species. The CO cross section (σ [cm^2/molecule]) at T_5 and P_5 were taken from the HITRAN database [16]. Major pyrolysis products during TEP pyrolysis (C_2H_4, C_2H_5OH, CH_4, C_3H_6, C_2H_2) identified using the LLNL's kinetic model showed no absorption feature near 4.6 μm allowing for interference free detection of CO (uncertainty $\pm$2.6%) at the chosen wavelength (Figure 2).

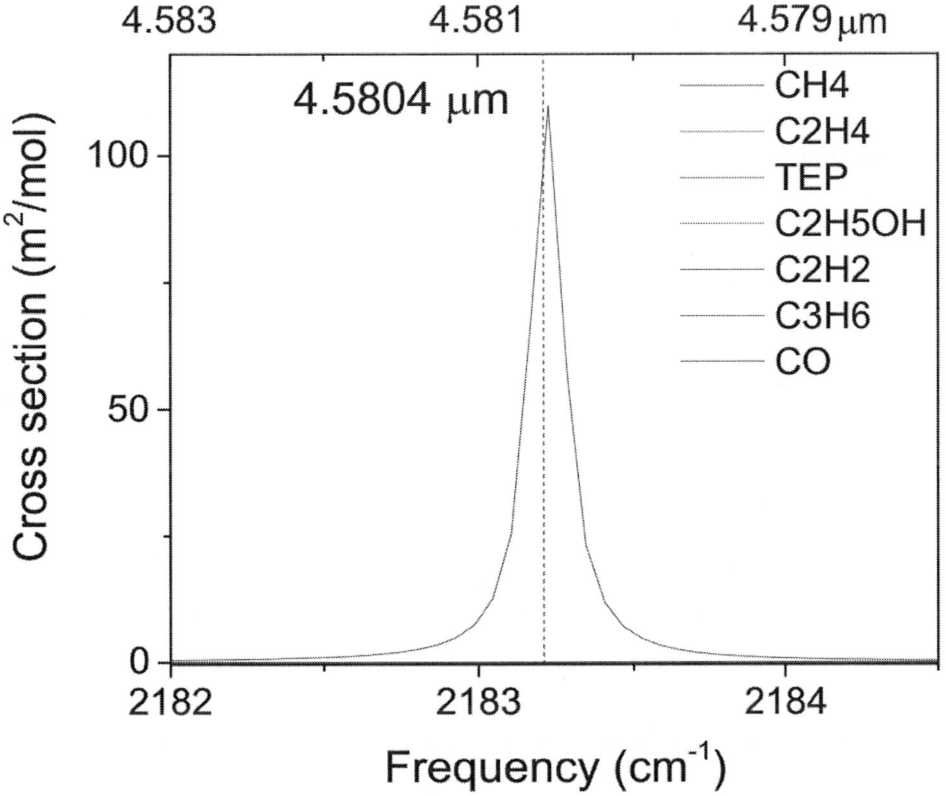

Figure 2 Cross section (at STP) of CO and possible interfering species during TEP pyrolysis near 4.6 μm showing no absorption feature from other combustion products at the chosen wavelength [17]

Figures 3 shows the measured CO time history yield during oxidation of TEP at 1195 K and 1.493 atm as compared to model prediction using LLNL's OPC incineration mechanism in Chemkin Pro software [18]. The time scale of predicted CO formation was slower than the experiment. Constant pressure model was used for all simulations in Chemkin Pro.

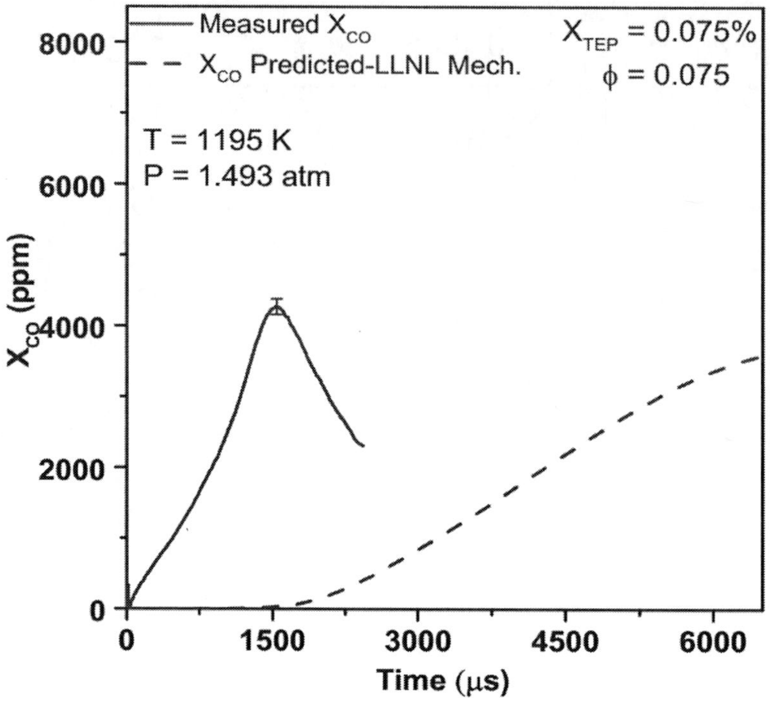

Figure 3 Measure and predicted CO time history during TEP oxidation

4. Conclusions

Oxidation of TEP – a simulant for sarin-GB was carried out at incineration conditions (temperatures = 1195 K and pressure ~1.4 atm). Intermediate CO concentrations were measured using direct laser absorption spectroscopy with a quantum cascade laser operating at 4.5804 nm. Measured CO time histories were compared with model predictions using LLNL's detailed OPC incineration mechanism and the model did not adequately predict the CO yield at the conditions tested in this study. Further investigation into kinetic parameters of TEP decomposition reactions is required to improve the model.

5. Acknowledgements

The project was sponsored by the Department of the Defense, Defense Threat Reduction Agency (grant number: HDTRA1-16-1-0009). The content of the information does not necessarily reflect the position or the policy of the federal government, and no official endorsement should be inferred.

6. References

[1] E. Zegers, E. Fisher, Gas-phase pyrolysis of diethyl methylphosphonate, Combustion science and technology 116 (1996) 69-89.
[2] E.J.P. Zegers, E.M. Fisher, Gas-Phase Pyrolysis of Diisopropyl Methylphosphonate, Combustion and Flame 115 (1998) 230-240.
[3] E. Zegers, E. Fisher, Pyrolysis of triethyl phosphate, Combustion science and technology 138 (1998) 85-103.
[4] D.A. Knyazkov, A.G. Shmakov, O.P. Korobeinichev, Application of molecular beam mass spectrometry in studying the structure of a diffusive counterflow flame of CH4/N2 and O2/N2 doped with trimethylphosphate, Combustion and Flame 151 (2007) 37-45.

[5] O.P. Korobeinichev, V.M. Shvartsberg, A.G. Shmakov, The chemistry of combustion of organophosphorus compounds, Russian Chemical Reviews 76 (2007) 1094.
[6] O. Korobeinichev, A. Chernov, T. Bolshova, The chemistry of the destruction of organophosphorus compounds in flames—IV: destruction of DIMP in a flame of H 2+ O 2+ Ar, Combustion and flame 123 (2000) 412-420.
[7] O.P. Korobeinichev, S.B. Ilyin, V.M. Shvartsberg, A.A. Chernov, The destruction chemistry of organophosphorus compounds in flames—I: Quantitative determination of final phosphorus-containing species in hydrogen-oxygen flames, Combustion and flame 118 (1999) 718-726.
[8] D. Knyazkov, S. Yakimov, O. Korobeinichev, A. Shmakov, Effect of trimethylphosphate additives on the flammability concentration limits of premixed methane-air mixtures, Combustion, Explosion, and Shock Waves 44 (2008) 9-17.
[9] O.P. Korobeinichev, V.M. Shvartsberg, A.G. Shmakov, T.A. Bolshova, T.M. Jayaweera, C.F. Melius, W.J. Pitz, C.K. Westbrook, H. Curran, Flame inhibition by phosphorus-containing compounds in lean and rich propane flames, Proceedings of the Combustion Institute 30 (2005) 2353-2360.
[10] E. Ninnemann, B. Koroglu, O. Pryor, S. Barak, L. Nash, Z. Loparo, J. Sosa, K. Ahmed, S. Vasu, New insights into the shock tube ignition of H2/O2 at low to moderate temperatures using high-speed end-wall imaging, Combustion and Flame 187 (2018) 11-21.
[11] B. Koroglu, S. Neupane, O. Pryor, R.E. Peale, S.S. Vasu, High temperature infrared absorption cross sections of methane near 3.4 μm in Ar and CO2 mixtures, Journal of Quantitative Spectroscopy and Radiative Transfer 206 (2018) 36-45.
[12] O.M. Pryor, S. Barak, B. Koroglu, E. Ninnemann, S.S. Vasu, Measurements and interpretation of shock tube ignition delay times in highly CO2 diluted mixtures using multiple diagnostics, Combustion and Flame 180 (2017) 63-76.
[13] Z.E. Loparo, J.G. Lopez, S. Neupane, W.P. Partridge, K. Vodopyanov, S.S. Vasu, Fuel-rich n-heptane oxidation: A shock tube and laser absorption study, Combustion and Flame 185 (2017) 220-233.
[14] G. Barari, O. Pryor, B. Koroglu, J. Lopez, L. Nash, S.M. Sarathy, S.S. Vasu, High Temperature Shock Tube Experiments and Kinetic Modeling Study of Diisopropyl Ketone Ignition and Pyrolysis, Combustion and Flame 177 (2017) 207-218.
[15] B. Koroglu, O. Pryor, J. Lopez, L. Nash, S.S. Vasu, Shock tube ignition delay times and methane time-histories measurements during excess CO2 diluted oxy-methane combustion, Combustion and flame 164 (2016) 152-163.
[16] I.E. Gordon, L.S. Rothman, C. Hill, R.V. Kochanov, Y. Tan, P.F. Bernath, M. Birk, V. Boudon, A. Campargue, K.V. Chance, B.J. Drouin, J.M. Flaud, R.R. Gamache, J.T. Hodges, D. Jacquemart, V.I. Perevalov, A. Perrin, K.P. Shine, M.A.H. Smith, J. Tennyson, G.C. Toon, H. Tran, V.G. Tyuterev, A. Barbe, A.G. Császár, V.M. Devi, T. Furtenbacher, J.J. Harrison, J.M. Hartmann, A. Jolly, T.J. Johnson, T. Karman, I. Kleiner, A.A. Kyuberis, J. Loos, O.M. Lyulin, S.T. Massie, S.N. Mikhailenko, N. Moazzen-Ahmadi, H.S.P. Müller, O.V. Naumenko, A.V. Nikitin, O.L. Polyansky, M. Rey, M. Rotger, S.W. Sharpe, K. Sung, E. Starikova, S.A. Tashkun, J.V. Auwera, G. Wagner, J. Wilzewski, P. Wcisło, S. Yu, E.J. Zak, The HITRAN2016 molecular spectroscopic database, Journal of Quantitative Spectroscopy and Radiative Transfer 203 (2017) 3-69.
[17] S.W. Sharpe, T.J. Johnson, R.L. Sams, P.M. Chu, G.C. Rhoderick, P.A. Johnson, Gas-phase databases for quantitative infrared spectroscopy, Applied spectroscopy 58 (2004) 1452-1461.
[18] ANSYS Chemkin-Pro, http://www.ansys.com/products/fluids/ansys-chemkin-pro, San Diego, CA., 2017.

Spring Technical Meeting
Eastern States Section of the Combustion Institute
March 4-7, 2018
State College, Pennsylvania

Carbon addition lowers initiation and iodine release temperatures from iodine oxide-based biocidal energetic materials

*Tao Wu, Xizheng Wang, Scott Holdren, Michael R. Zachariah**

Department of Chemistry and Biochemistry, University of Maryland College Park, MD 20742, USA
**Corresponding Author Email: mrz@umd.edu*

Abstract: Iodine oxides are of significant interest as strong oxidizers in energetic formulation that also operate as iodine release agents for neutralization of spores using strong thermal pulses. In this paper incorporation of carbon black (CB) as the main fuel or additive into iodine oxides-based energetic materials are shown to lower both initiation and iodine release temperatures compared to those of Al/iodine oxides and Ta/iodine oxides thermites. Those lowering effects were triggered by a condensed phase CB-iodine oxides reaction explored by high heating rate time-resolved temperature-jump time-of-flight mass spectrometry and low heating rate thermogravimetric analysis/differential scanning calorimetry results. We observe that other carbon allotropes, such as carbon nanotubes and functionalized graphene sheets, also feature the similar effect as CB. Fourier transform infrared spectroscopy analysis shows that the presence of CB lowered the bond energy of iodine oxide on the surface to trigger the CB-iodine oxide initiation at a relatively lower temperature.
Keywords: Iodine/Iodic Biocide, Carbon, Thermites, Ignition, Energetic

1. Introduction [12pt]

High efficiency neutralization of biological warfare agents has become a major research focus within the United States, in addition to many other countries, due to the increased threat of bioterrorism [1]. Preliminary laboratory studies have suggested that an ideal neutralization process should contain not only a high thermal but also long-lasting biocidal agent release [2]. The main problem with conventional energetic materials is low neutralization efficiency since a thermal neutralization mechanism is dominant in this case [2]. Therefore, it has been proposed that simultaneously delivering a rapid thermal pulse with a remnant biocidal agent would prolong the exposure time and improve the inactivation process [3]. Halogens-based energetic materials show the most promise because of their excellent biocidal properties [4] when compared to silver-containing energetic materials. Among all halogens, iodine stands out owing to its strong neutralization effect.

Aluminum has a passivating alumina shell that stabilizes the material at ambient conditions. However, once the aluminum core is heated near the melting temperature, obtaining sufficient mobility, it can diffuse through the oxide shell and interact with any available oxidizer ultimately leading to initiation [5]. However, iodine oxide releases O_2 at a relatively low temperature (~450 C) which is well below the Al melting point, where reaction is most advantaged [6]. To explore the fuel melting aspect of initiation, we have also used tantalum as a fuel since it has a much higher melting point (~3000 °C) than Al (660 °C), and thus would not melt at the point of initiation. However, the immobile tantalum core will not react with iodine oxide until some cracks appeared in the Ta_2O_5 shell upon heating (higher than 500 °C) due to the amorphous to crystalline phase change of the oxide shell [7]. Those cracks serve as a low resistance pathway for the oxygen to react with the tantalum core [7].

In this work, we employed carbon as a fuel to better interpret the initiation process of iodine oxide-based energetic materials since it has a high melting point (3550 °C) and no passivating shell. Therefore, there will be direct contact between the fuel and oxidizer enabling further understanding of the initiation process. Results obtained from both high and slow heating rate methods indicate that carbon black (CB) and other carbon allotropes can lower the iodine release temperature from iodine oxides such as I_2O_5, HIO_3 and HI_3O_8, via a condensed phase reaction mechanism. More importantly, we also found that CB as an additive with only 3 wt% can also significantly lower the initiation and iodine release temperatures for both Al/iodine oxides and Ta/iodine oxides thermites, indicating the potential use of CB in biocidal applications.

2. Methods / Experimental

2.1. Materials:

The aluminum nanopowders (Al) (Alex, ~80 nm) were purchased from Novacentrix. The active Al was 81 % by mass, determined by thermogravimetric analysis. Nano-tantalum powders (<50 nm, Ta) were purchased from Global Advanced Metals. Nano-sized carbon black (~50 nm, CB) was obtained from Cabot Corporation. Multi-wall carbon nanotubes and iodine pentoxide (I_2O_5) purchased from Sigma-Aldrich were directly used as received. Functionalized graphene sheets (FGS) with C:O molar ratio of 15 and 60 were obtained from Aksay et.al [8] and labeled as FGS_{15} and FGS_{60}, respectively. All the other chemicals were of analytical grade and used as purchased without further treatment.

2.2. Preparation of thermites:

Aluminum, tantalum or CB was stoichiometrically mixed with I_2O_5, HIO_3 and HI_3O_8, respectively, in hexane followed by 30 minutes of sonication and left uncapped at room temperature to dry.

2.3. Time-resolved temperature-jump time-of-flight mass spectrometer (T-Jump/TOFMS) measurement and high-speed imaging

The decomposition of each oxide listed above was investigated using T-Jump/TOFMS [2]. Typically, a ~1 cm long platinum wire (76 um in width) with a thin coating of oxidizer sample was rapidly joule-heated to about 1200 °C by a 3 ms pulse at a heating rate of ~10^5 °C/s. The current and voltage signals were recorded, and the temporal temperature of the wire was determined using the Callendar–Van Dusen equation. Spectra were collected at 0.1 ms intervals. The detailed experimental set-up has been previously described [2].

2.4 Thermogravimetric analysis/differential scanning calorimetry (TGA/DSC) measurement

TGA/DSC were conducted using an SDT Q600 (TA instruments). The measurement was performed under a 100 mL min^{-1} argon flow with ~ 1.0 mg samples placed into an alumina pan and heated from room temperature up to 550 °C (or higher temperature) at a rate of 5 °C/min.

2.5 Fourier Transform Infrared Spectroscopy (FTIR) analysis

Attenuated total reflection (ATR) FTIR spectra of I_2O_5 samples with and without CB were collected using a Nicolet iS-50R spectrometer equipped with a room temperature deuterated triglycine sulfate (DTGS) detector FTIR spectroscopy. A Thermo Scientific Smart iTX accesory was installed to collect the ATR spectra shown here at 4 cm^{-1}resolution and averaged over 25 scans.

3. Results and Discussion

To investigate the performance of iodine oxides/iodic acids as oxidizers in thermite systems, we have previously employed Al and Ta as fuels. However, both aluminum and tantalum have an oxide shell, which to some extent restricted our understanding on how those thermites react during initiation [9], and thus motivated the use of carbon. Commercial iodine pentoxide was first examined because it is the most extensively studied and can be obtained easily [10].

T-Jump/TOFMS was employed to study the initial events of bare oxidizers or thermites under rapid heating [11]. To understand the relationship between iodine release and initiation, we plotted iodine release temperatures vs initiation temperatures in Fig. 1A. The dotted red line represents the iodine/O_2 release temperatures (~450 °C) from neat I_2O_5 obtained through a high heating method. Note that, the y-axis in Fig. 1A is the iodine/O_2 release temperature of the fuel/I_2O_5 mixture.

Different initiation temperatures were obtained for thermites with different fuels including Al, Ta and CB. We have previously argued that the initiation of Al-based and Ta-based thermites is dominated by the melting of aluminum and the phase change of the Ta_2O_5 shell, respectively [9]. Results in Fig. 1A shows that the initiation temperature is correlated with the iodine release temperature for all cases except for Al/I_2O_5 in which its initiation was controlled by the mobilization of the aluminum core. When aluminum is employed as the fuel, iodine release is actually delayed to a significantly higher temperature compared with neat I_2O_5 (550 vs. 450 °C). This delay might be caused by the interaction between the alumina shell and I_2O_5, which to some extent stabilizes I_2O_5 and therefore delays its decomposition. Similar effects of halogen interactions have been reported in previous studies [12].

As to CB, the initiation temperature of CB/I_2O_5 is ~330 °C (determined from release of CO_2 product since there is virtually no optical signature of initiation), and more than 100 °C lower than the iodine release temperature from the neat I_2O_5, which implies a condensed phase reaction mechanism may be dominant here. Since CB/I_2O_5 ignition was initiated at a very low temperature, 3 wt% CB was introduced into both Al/I_2O_5 and Ta/I_2O_5 systems to investigate the effect of CB as an additive.

Fig. 1A also shows that addition of just 3 wt% CB into Al/I_2O_5 system, not only lowers the iodine release temperature but also the initiation temperature. The incorporated CB offsets the influence of aluminum on the iodine release from I_2O_5 and makes it behave like that of the bare oxidizer. As to why initiation seems not limited by the melting point of Al: the condensed phase reaction of CB/I_2O_5 is initiated first and presumably the energy generated kicks the temperature above the point of making the Al core mobile. Thus, addition of CB lowers the initiation temperature of an aluminum-based thermite.

As a comparison, $CB_{3wt\%}$/I_2O_5 (non-stoichiometric) shows a lower initiation and iodine release temperatures than $CB_{3wt\%}$/Al/I_2O_5, which indicates the presence of aluminum inhibits the condensed phase reaction of CB/I_2O_5 and therefore delays the iodine release from I_2O_5. On the other hand, with only 3wt% addition of CB, the iodine release temperature of I_2O_5 was brought lower by about 50 °C compared to the neat I_2O_5, which indicates that the condensed phase CB/I_2O_5 reaction occurred before the I_2O_5 decomposition. Furthermore, the addition of 3 wt% CB into Ta/I_2O_5 system also bring both iodine release and initiation to lower temperatures, indicating the general effect of CB additive on different fuel/I_2O_5 systems.

Slow heating rate TGA in argon is shown in Fig. 1B. Neat I_2O_5 shows one minor weight loss at around 210 °C that corresponds to the dehydration of HI_3O_8 and another major weight loss at around 370 °C that results from the decomposition of I_2O_5. When CB was mixed with I_2O_5 (red curve in Fig. 1B) the onset decomposition temperature is 140 °C lower than that of neat I_2O_5. Even with only 3wt% of CB, the TGA curve of I_2O_5 follows the same kinetics as CB/I_2O_5 at temperatures below 250 °C. However, as temperature increases (250<T<300 °C), the reaction effectively stops until it reaches the neat I_2O_5 condition resuming decomposition at a rate the same as neat I_2O_5. In fact, the actual weight loss of $CB_{3wt\%}/I_2O_5$ at 300 °C, ~35%, is very close the theoretical weight loss ~37% based on the reaction of 3 wt% CB and I_2O_5 (accounted for the 80% hydration HI_3O_8 specie). The unchanged TGA result of Al/I_2O_5 compared with that of I_2O_5 indicates that aluminum does not pose any significant influence on the decomposition of I_2O_5 at low heating rates, which is quite different from the T-jump/TOFMS experiment in which the presence of aluminum delays the iodine release of I_2O_5. Moreover, for $CB_{3wt\%}/Al/I_2O_5$, the decomposition started at a temperature very close to that of CB/I_2O_5, thus aluminum does not influence the CB/I_2O_5 reaction at a low heating rate and again differs from the T-jump/TOFMS experiment.

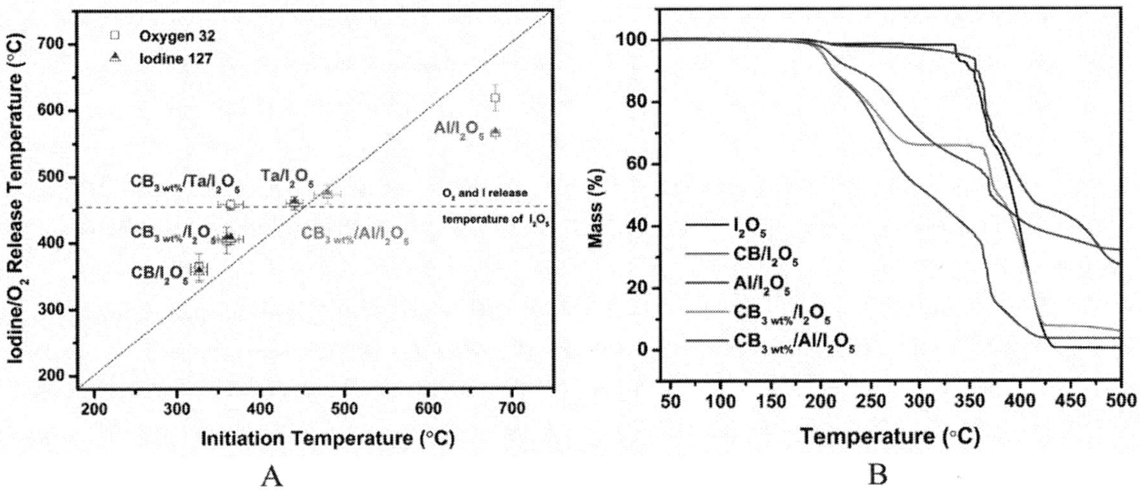

Fig. 1. (A) Release temperatures of iodine/oxygen vs. initiation temperatures of different thermites; (B) TGA results of neat I_2O_5, CB/I_2O_5, Al/I_2O_5 and $CB_{3\ wt\%}/Al/I_2O_5$ thermites.

Additional carbon allotropes including CNT and FGS were evaluated by stoichiometrically mixing with I_2O_5. Fig. 2A shows the relation between iodine release temperatures and initiation temperature of those energetic composites employing carbon allotropes as the fuel from T-Jump mass spectrometry. The horizontal dotted red line represents the oxygen and iodine release temperatures of neat I_2O_5, and the dotted blue line would refer to the perfect correlation between iodine release and the onset reaction temperature. For all samples, the initiation temperatures are either lower to or the same as the corresponding iodine release temperatures and the iodine release temperature is lower than bare I_2O_5. All four stoichiometric thermites employing FGS_{15}, FGS_{60}, CNT or CB as the fuel initiated almost at the same temperature ~310 °C, implying that the different properties among those carbon materials [13] appear to have no significant effect on initiation. We also note that above a threshold level of carbon the ignition temperature is not affected. So, 3% and 8% of CB addition have essentially the same ignition temperature.

We now turn to mechanistic effects. Since we conclude that surface reaction is clearly playing a significant role, FTIR attenuated total reflection (ATR, surface) was employed to deduce any structure changes just prior to reaction [13]. FTIR-ATR spectra of bare I_2O_5, $CB_{3wt\%}/I_2O_5$ and stoichiometric CB/I_2O_5 at room temperature is shown in Fig. 2B. The spectra show a steady red-shift (too lower wavenumbers) with increasing CB content. For clarity, the stoichiometric CB/I_2O_5 is labeled as $CB_{8wt\%}/I_2O_5$. The redshifts indicate weaker iodine-oxygen bonds due to the presence of CB. In addition, no new peak is detected with CB-added samples which suggests no new carbon-iodine or carbon-oxide bonds are formed between CB and iodine oxide. Keeping in mind that the ATR is a surface sensitive technique while the KBr pellet method is a bulk technique we can conclude that the surface of I_2O_5 particles even under room temperature conditions is affected by CB in a manner to enable a lower decomposition temperature.

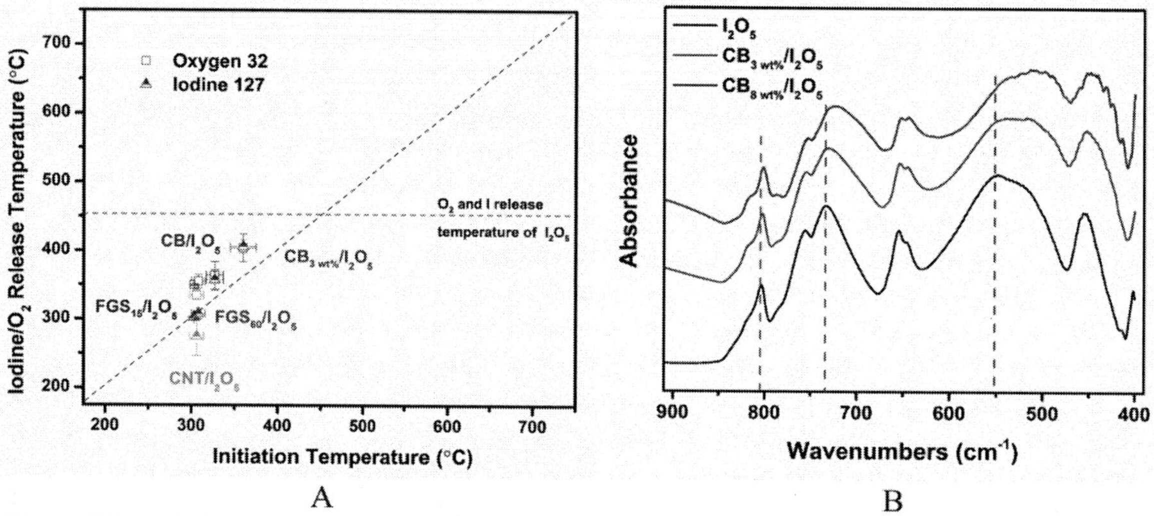

Fig. 2. The relation between iodine/oxygen release temperatures and initiation temperatures of thermites employing carbon allotropes as the fuel; (B) ATR-FTIR spectra of bare I_2O_5, $CB_{3wt\%}/I_2O_5$ and $CB_{8wt\%}/I_2O_5$ at room temperature showing I-O bond red-shift upon addition of CB.

4. Conclusions

In this paper, we employed carbon (CB, CNT and FGS) as both a fuel and as an additive to aluminum and tantalum fuels to study the influence of a fuel without nascent oxide shell on initiation of iodine oxide-based thermites. We found that the initiation temperature of CB/I_2O_5 is much lower than the initiation temperature of Al/I_2O_5 and Ta/I_2O_5. When CB is used as an additive to other fuels we observe a decrease in the overall initiation temperature and a lowering of the iodine release temperature to below that of the nascent oxidizer. Other carbon allotropes showed similar results. FTIR analysis indicated at CB lowered the iodine-oxygen bond energy of I_2O_5 on the surface and thus triggered this condensed phase reaction. These results indicate that CB can be employed as an additive to any metal fuel as a means to lower both the initiation and iodine release temperature.

5. Acknowledgements

This research was funded by the Defense Threat Reduction Agency.

6. References

[1] K.T. Sullivan, C.W. Wu, N.W. Piekiel, K. Gaskell, M.R. Zachariah, Synthesis and reactivity of nano-Ag2O as an oxidizer for energetic systems yielding antimicrobial products. Combust. Flame 160 (2013) 438.

[2] K. Sullivan, N. Pickiel, S. Chowdhury, C.W. Wu, C. Johnson, M.R. Zachariah, Ignition and Combustion Characteristics of Nanoscale Al/AgIO3: A Potential Energetic Biocidal System, Combust. Sci. Technol. 183 (2011) 205.

[3] H. Wang, G. Jian, W. Zhou, J. DeLisio, V. Lee, M-R. Zachariah, Metal Iodate-based Energetic Composites and their Combustion and Biocidal Performance, ACS Appl. Mater. Interfaces **7** (2015) 17363.

[4] Y. Aly, S. Zhang, M. Schoenitz, V.K. Hoffmann, E.L. Dreizin, M. Yermakov, R. Indugula, S.A. Grinshpun, Iodine-containing Aluminum-based Fuels for Inactivation of Bioaerosols. Combust. Flame 161 (2014) 303.

[5] G. Jian, N.W. Piekiel, M.R. Zachariah, Time-Resolved Mass Spectrometry of Nano-Al and Nano-Al/CuO Thermite under Rapid Heating: A Mechanistic Study, J Phys Chem C 116 (2012) 26881-26887.

[6] G.Q. Jian, S. Chowdhury, K. Sullivan, M.R. Zachariah, Nanothermite reactions: Is gas phase oxygen generation from the oxygen carrier an essential prerequisite to ignition? Combust Flame 160 (2013) 432-437.

[7] J.B. DeLisio, G.C. Egan, S.-C. Liou, W.-A. Chiou, M.R. Zachariah, Probing the Oxidation Mechanism of Ta Nanoparticles via In-Situ and Ex-Situ Ultra-Fast Heating TEM/STEM, Microscopy and Microanalysis 22 (2016) 780-781.

[8] C. Punckt, M.A. Pope, Y.F.M. Liu, I.A. Aksay, Structure-Dependent Electrochemistry of Reduced Graphene Oxide Monolayers, J Electrochem Soc 163 (2016) H491-H498.

[9] T. Wu, X. Wang, J.B. DeLisio, H. Wang, M.R. Zachariah, Performance of iodine oxides/iodic acids as oxidizers in thermite systems, Combust Flame In press (2017).

[10] D.K. Smith, K. Hill, M.L. Pantoya, J.S. Parkey, M. Kesmez, Reactive characterization of anhydrous iodine (v) oxide (I2O5) with aluminum: amorphous versus crystalline microstructures, Thermochim Acta 641 (2016) 55-62.

[11] X.Z. Wang, W.B. Zhou, J.B. DeLisio, G.C. Egan, M.R. Zachariah, Doped delta-bismuth oxides to investigate oxygen ion transport as a metric for condensed phase thermite ignition, Phys Chem Chem Phys 19 (2017) 12749-12758.

[12] J.B. DeLisio, X.L. Hu, T. Wu, G.C. Egan, G. Young, M.R. Zachariah, Probing the Reaction Mechanism of Aluminum/Poly(vinylidene fluoride) Composites, J Phys Chem B 120 (2016) 5534-5542.

[13] T. Wu, M.X. Chen, L. Zhang, X.Y. Xu, Y. Liu, J. Yan, W. Wang, J.P. Gao, Three-dimensional graphene-based aerogels prepared by a self-assembly process and its excellent catalytic and absorbing performance, J Mater Chem A 1(26) (2013) 7612-7621.

[14] P.M. Sherwood, J.J. Turner, Vibrational Spectra of Compounds in Iodine Pentoxide-Water System and Sodium Iodate, Spectrochim Acta a-M A 26 (1970) 1975-&.

Spring Technical Meeting
Eastern States Section of the Combustion Institute
March 4-7, 2018
State College, Pennsylvania

Thermochemistry and Kinetic Analysis on the Reactions and Mechanisms of $CH_3SCH_2CH_3$ with Oxygen and OH Radicals

Guanghui Song[1], Joseph W Bozzelli[1,]*

[1]Department of Chemical, Biological and Pharmaceutical Engineering, New Jersey Institute of Technology, University Heights, Newark, NJ 07102 USA
[*]Corresponding Author Email: joseph.w.bozzelli@njit.edu

Abstract: The oxidation of Methyl ethyl sulfide ($CH_3SCH_2CH_3$, methylthioethane) under atmospheric and combustion conditions is initiated by OH radicals. The Methyl ethyl sulfide (MES) radicals generated after loss of a H atom will react with O_2 to form peroxy radicals. The subsequent reaction mechanisms of MES peroxy radicals are studied under B3LYP/6-31G(d, p), B3LYP/6-31+G(2d, p), CBS-QB3 and M062X/6-311+G(2d, p) levels. Transition states and kinetic coefficients of intramolecular hydrogen atom transfer, sulfuric alkoxy ring formation, formation of aldehydes and molecular elimination channels are characterized to evaluate the important reaction paths and the kinetics. Thermochemical properties of reactants, transition states and products are obtained at the CBS-QB3 level for calculation of the high pressure limit kinetic parameters. Kinetic parameters are determined versus pressure and temperature based on the Multifrequency Quantum Rice Ramsperger Kassel (QRRK) analysis with master equation analysis for falloff. The major reaction paths under atmospheric conditions are ring formation and formation of aldehydes.
Keywords: methyl ethyl sulfide, partial oxidation, transition state theory, Arrhenius equation

1. Introduction

Methyl ethyl sulfide ($CH_3SCH_2CH_3$, methylthioethane, MES) is a widely used chemical agent but exerts significantly negative effects to the atmosphere and the environment by generating sulfuric oxides and smog [1] There is limited information about the kinetics of sulfuric hydrocarbon partial oxidation in the atmosphere.[2, 3] In this study, the important reaction steps involved in the initial stages of MES partial oxidation are analyzed, and the kinetic parameters are determined as a function of temperature and pressure. In addition to the models of partial oxidation, the thermochemical and kinetic data under varied pressures and temperatures are applicable to the modeling initial reactions in combustion of MES's.[4]

2. Reaction pathways

According to the recent studies by Mansilla *et al.*[5] and Cao *et al.*[6], the partial oxidation of MES in the atmosphere in the absence of NO_x or Cl radicals begins from the abstraction of one hydrogen atom from each MES molecule by one hydroxyl radical.[5, 7-9] In this paper, the symbol ("•" or "j") represents the radical site located on the preceding carbon or oxygen atom.

$CH_3SCH_2CH_3 + HO\bullet \rightarrow CH_2\bullet SCH_2CH_3 + H_2O$
$CH_3SCH_2CH_3 + HO\bullet \rightarrow CH_3SCH\bullet CH_3 + H_2O$
$CH_3SCH_2CH_3 + HO\bullet \rightarrow CH_3SCH_2CH_2\bullet + H_2O$

The radicals formed after loss of one hydrogen atom from the different carbon sites will react with the oxygen ($^3O_2$) to form the alkyl sulfide peroxy radicals[6, 7]:

$CH_2{\bullet}SCH_2CH_3 + O_2 \rightarrow CH_2(OO{\bullet})SCH_2CH_3$ ΔHrxn -26.4 kcal mol^{-1}
$CH_3SCH{\bullet}CH_3 + O_2 \rightarrow CH_3SCH(OO{\bullet})CH_3$ ΔHrxn -31.1 kcal mol^{-1}
$CH_3SCH_2CH_2{\bullet} + O_2 \rightarrow CH_3SCH_2CH_2OO{\bullet}$ ΔHrxn -34.1 kcal mol^{-1}

This paper focuses on the reactions of the chemically activated $CH_2(OO{\bullet})SCH_2CH_3^*$ adduct formed from the $CH_2{\bullet}SCH_2CH_3$ MES radical reaction with O_2.

The reactions investigated in this paper are separated into the following classes: [4, 6, 7, 10, 11]
Reverse reaction back to $CH_2{\bullet}SCH_2CH_3$ radical plus O_2;
Intramolecular hydrogen transfer from three carbon sites to the radical peroxy oxygen;
Intramolecular H-shift from the ipso carbon to the peroxy oxygen to form aldehyde plus OH;
Cyclic ether ring formation, four and five member thiol rings plus OH radical ;
Intramolecular addition of peroxy oxygen radical to sulfur with subsequent simultaneous weaker O—O and C—S bond cleavages, leading to the formations of C=O and S=O double bonds.

3. Calculation Methods

Thermochemical properties of the parent molecules and their radicals were calculated using several density functional theory and composite *ab initio* computation chemistry methods in the Gaussian 09 program suite.[12-14], with the use of isodesmic reactions for standard enthalpies of formation. The composite CBS-QB3 calculation method has been used for the thermochemical properties of each reactant, product and transition state, in order to achieve improvements on the accuracy over the DFT methods.[15-19] Contributions to the entropy and heat capacity of each molecule and radical from translation, vibrations and external rotation were determined using the Statistical Mechanics for Heat Capacity and Entropy, C_p and S (SMCPS) program.[11, 20] The SMCPS program utilizes the rigid-harmonic oscillator approximation from the optimized structures obtained in the CBS-QB3 calculation.[21, 22] The scaling factor of 0.99 for zero-point vibration energy was used.[21-24].

High-pressure limit reaction rate constants (k_∞) were determined based on the canonical transition state theory (CTST). The values of k_∞ were fit to the three-parameter form of the Arrhenius equation to yield the parameters of A, n, and E_a over the temperature range of 200-3000K. [4, 11, 24, 25] Temperature and pressure-dependent rate constants are calculated using the multichannel, multifrequency Quantum Rice-Ramsperger-Kassel (QRRK) analysis for $k(E)$ based on a master equation analysis for falloff and stabilization as implemented in the CHEMASTER code.[26]

Variational transition state theory (VTST) analysis was performed for the association of O_2 to the radical sites of methyl ethyl sulfide and their reverse reactions. The calculations on optimized structure, energy and vibration frequencies at 0.05 Å steps under the M06-2X/6-311+G(2d,p) level. [26].

4. Results and Analysis

4.1 Variational Transition State Theory (VTST) Analysis

The Potential energy surface for the steps in the CCSC—OO• bond length is shown in Figure 1. CBS QB3 calculations with work reactions place the reaction energy are 27. kcal mol^{-1} The M06-2X/6-311+G(2d,p) VTST calculations on the R-OO bond dissociation show no saddle point barrier and place the TS geometry with the CCSC—OO• bond length in the region of 1.95 to 2.25 Å for the dissociation.

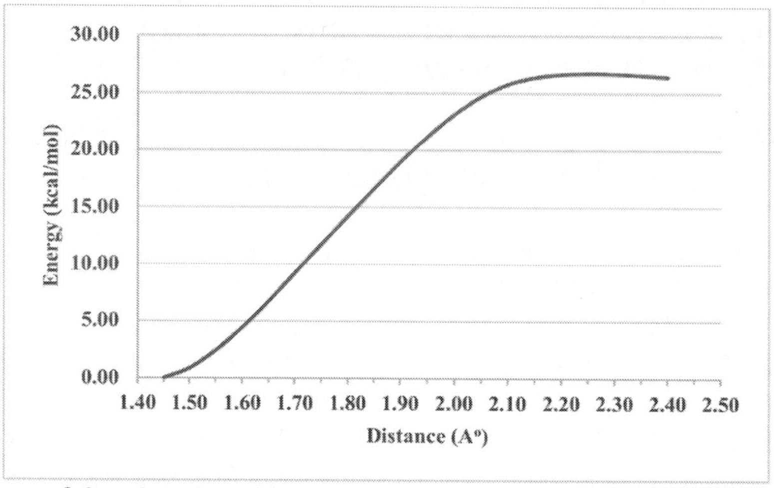

Figure 1: Scan of C—OO• bond length for VTST calculation of •$OOCH_2SCH_2CH_3$.

4.3 Thermochemical Kinetic Analysis for •$OOCH_2SCH_2CH_3$ Reactions

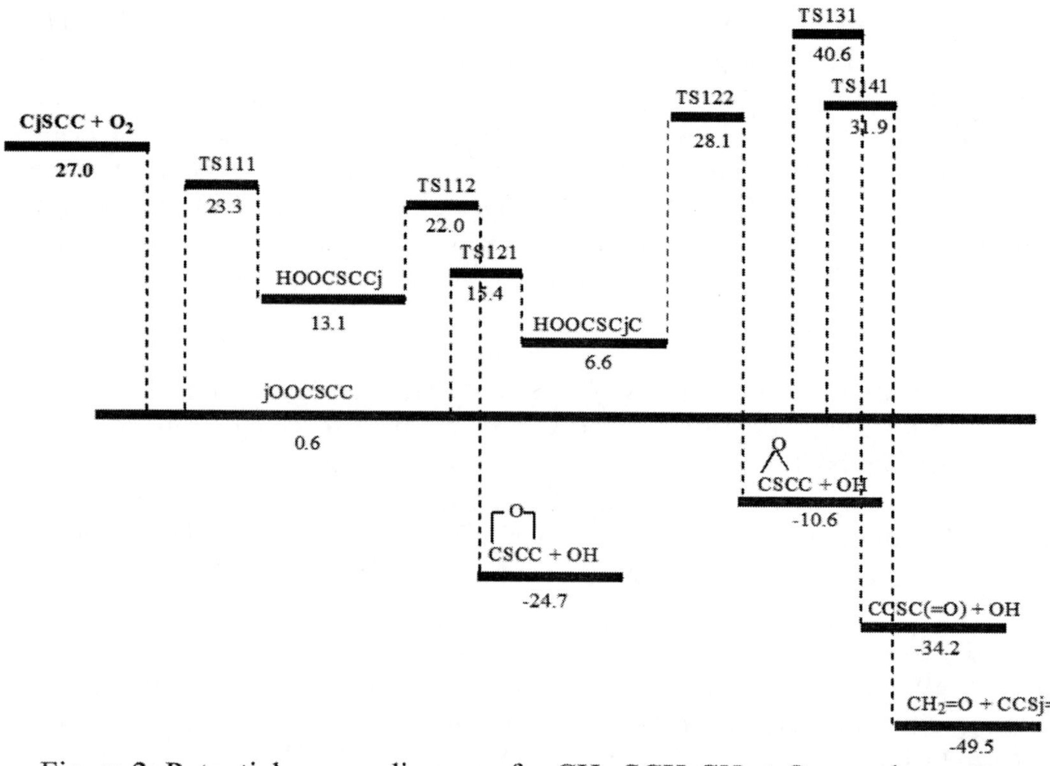

Figure 2: Potential energy diagrams for CH_2•SCH_2CH_3 + O_2 reaction system.

TS111 in Figure 2 illustrates the intramolecular transfer of a H atom from the primary methyl carbon on the ethyl group of CH_2(OO•)SCH_2CH_3 to the peroxy radical site, through a six member ring TST structure, resulting in formation of CH_2(OOH)SCH_2CH_2•. This reaction has a barrier of 22.7 kcal mol^{-1}. The methyl radical on this newly formed CH_2(OOH)SCH_2CH_2• intermediate can undergo substitution into the carbon oxygen of hydroperoxide to form a cyclic five member ring 'thiol – ether' with elimination of an OH radical over a barrier of only 7 kcal mol^{-1}. The $CH_2(OOH)SCH_2CH_2$• radical can also undergo beta scission reaction to form a thiol radical $CH_2(OOH)S$• + ethylene (CH_2=CH_2). The thiol radical can further react via a beta scission reaction to CH2=S + HO_2.

TS121 in Figure 2 illustrates the intramolecular transfer of a H atom from the secondary methyl carbon on the ethyl group of $CH_2(OO$•$)SCH_2CH_3$ to the peroxy radical site, through a five member ring TST structure, resulting in formation of $CH_2(OOH)SCH$•CH_3. This 5 member ring formation TS121 has a 10 kcal mol^{-1} lower barrier than that of the 6 member ring TS111 at only 14 kcal mol^{-1}. The lower barrier is a result of the sulfur atom bonding to the secondary carbon and formation of partial resonance in the sulfur--carbon radical system. Following this hydrogen shift, a barrier of 21.5 kcal/mol obtained for a 4 member ring, thiol–ether, Cy-(CSCO)-C formed from HOOCSC•C. This barrier is higher than that required for the five member Cy(CSCCO) ring formation from HOOCSCC• at only 9 kcal/mol.

Dissociation of •OOCSCC into CH_2=O and CCS•=O TS141 occurs over a barrier of 32.1 kcal mol^{-1} via peroxy radical addition to the sulfur atom, double bond formation of this peroxy oxygen to the sulfur and cleavage of the RO—O-.SCC bond forming an alkoxy radical. This occurs while the alkoxy undergoes beta scission to cleave the primary methyl carbon from the sulfur with the overall barrier of 32 kcal mol^{-1}.

The CCSC(=O) + OH radical TS131, aldehyde plus hydroxyl, formation has a barrier of 40 kcal/mol, which is the highest among all the reactions of •OOCSCC.

Dissociation of •OOCSCC into CH_2=O and CCS•=O occurs via intramolecular H atom transfer from the ipso carbon, carbon bonded to the peroxy group, to start to form a hydroperoxyde and a radical on this ipso carbon. This radical formation then initiates π bond formation between the carbon radical and the bonded oxygen and cleavage of the OH radical. This barrier is 40 kcal mol^{-1}, which is well above the chemical activation energy.

While the CCSC(=O) and CH2(=O) aldehyde formation channels are both among the most exothermic reactions; they also have higher barriers.

The chain branching reaction of $CH_2(OO$•$)SCH_2CH_3$ to $CH_2(O$•$)SCH_2CH_3$ + O• atom is endothermic by 59 kcal mol^{-1} from the stabilized peroxy radical and 31 kcal mol^{-1} above initially formed peroxy radical.

The pressure- and temperature- dependent rate constants are calculated using the multichannel and multifrequency quantum Rice-Ramsperger-Kassel (QRRK) analysis in the Chemaster code.[11] Calculated chemical activation rate constants k(E) versus vs. 1000 /T K at pressure of 1 atm leading to the different products are illustrated in Figure 3; and in figure 4 vs. Pressure at 298K.

Stabilization to the peroxy radical is important at room temperature and reaction to CH2=O + CCS•(=O) , CyCOCCS + OH, and Cy(COCS)-• at higher temperatures.

A high temperature reduces CjSCC activation by O_2 but is good for •$OOCH_2SCH_2CH_3$ dissociation and 4- or 5- member ring formation that is accompanied by OH generation. A high pressure helps with •$OOCH_2SCH_2CH_3$ generation but reduces the rates of all the subsequent reaction steps including the isomerization of •$OOCH_2SCH_2CH_3$. The isomerization of •$OOCH_2SCH_2CH_3$ needs an optimized temperature to proceed. A temperature of 600~800K should be the best temperature for the CH_2•$SCH_2CH_3 + O_2$ reaction system under P=1 atm.

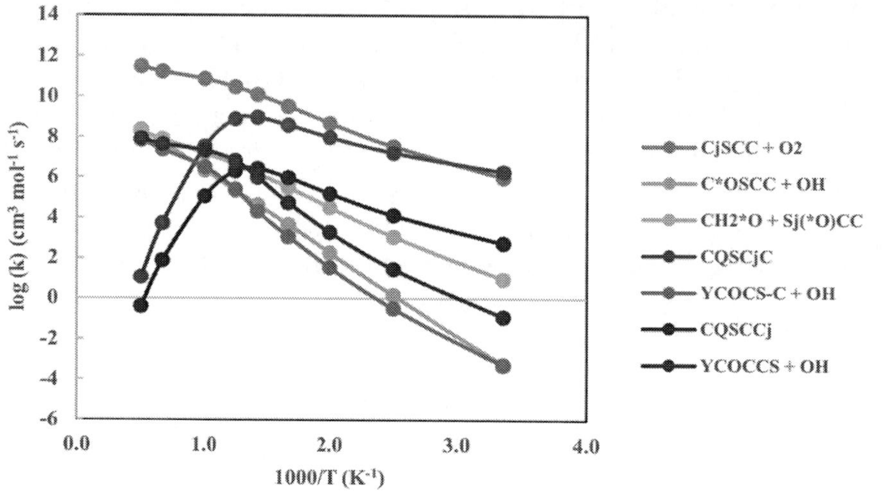

Figure 3 k vs. temperature for •$OOCH_2SCH_2CH_3$ under P=1 atm.

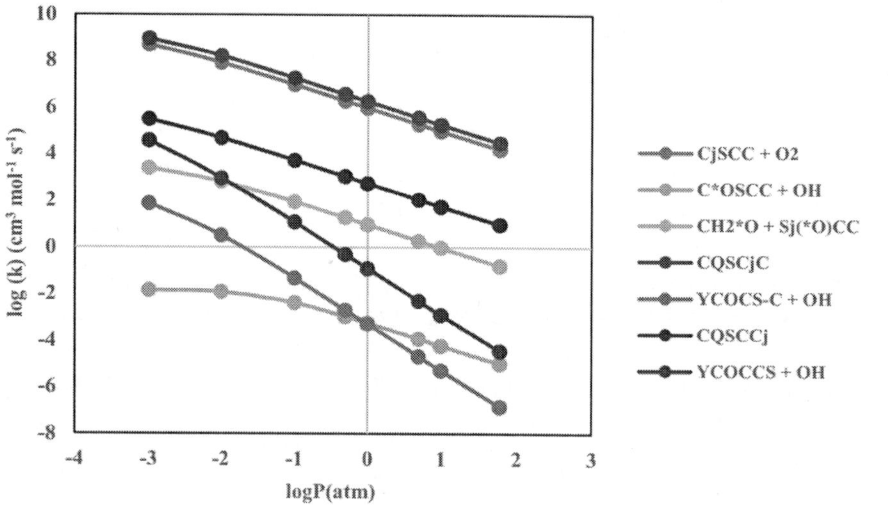

Figure 4 k(E) vs. pressure for •$OOCH_2SCH_2CH_3$ under T= 298 K.

5. Summary

Thermochemical and kinetic studies of methyl ethyl sulfide partial oxidation initiated by OH radical are presented. The CH_2•$SCH_2CH_3 + O_2$ -> •$OOCH_2SCH_2CH_3$* reaction system is illustrated as an example. The presence of sulfur reduces the chemical activation energy for $^3O_2$ association to carbon atoms adjacent to the sulfur moiety. The sulfur also presents new paths in formation of sulfur – oxygen bond systems. Theoretical investigations to MES partial oxidation reaction paths should be helpful in understanding the atmospheric chemistry and combustion of sulfides.

6. Acknowledgements

We acknowledge the NJIT Academic and Research Computing Systems resources and staff, for computational facilities and assistance on the "Kong HPC" computer system.

7. References

[1] S. Pillai and J. W. Bozzelli, "Computational study on structures, thermochemical properties, and bond energies of disulfide oxygen (S-S-O)-bridged CH 3SSOH and CH 3SS(=O)H and radicals," *Journal of Physical Organic Chemistry,* vol. 25, no. 6, pp. 475-485, 2012.
[2] M. B. Williams, P. Campuzano-Jost, A. J. Pounds, and A. J. Hynes, "Experimental and theoretical studies of the reaction of the OH radical with alkyl sulfides: 2. Kinetics and mechanism of the OH initiated oxidation of methylethyl and diethyl sulfides; observations of a two channel oxidation mechanism," *Physical Chemistry Chemical Physics,* vol. 9, no. 31, pp. 4370-4382, 2007.
[3] M. B. Williams, P. Campuzano-Jost, A. J. Hynes, and A. J. Pounds, "Experimental and theoretical studies of the reaction of the OH radical with alkyl sulfides: 3. Kinetics and mechanism of the OH initiated oxidation of dimethyl, dipropyl, and dibutyl sulfides: reactivity trends in the alkyl sulfides and development of a predictive expression for the reaction of OH with DMS," *Journal of Physical Chemistry A,* vol. 113, no. 24, pp. 6697-6709, 2009.
[4] F. Jin, R. Asatryan, and J. W. Bozzelli, "Thermodynamic and kinetic analysis on the reaction of dimethyl sulfide radical with oxygen," (in English), *International Journal of Quantum Chemistry,* vol. 112, no. 8, pp. 1945-1958, 2012.
[5] G. Oksdath-Mansilla, A. B. Peñéñory, I. Barnes, P. Wiesen, and M. A. Teruel, "Photodegradation of (CH 3CH 2) 2S and CH 3CH 2SCH 3 initiated by OH radicals at atmospheric pressure. Product yields and mechanism in NOx free air," *Atmospheric Environment,* vol. 55, pp. 263-270, 2012.
[6] J. Cao, W. Wang, Y. Zhang, T. Zhang, J. Lv, and C. Li, "Computational study on the reaction of CH3SCH2CH3 with OH radical: Mechanism and enthalpy of formation," (in English), *Theoretical Chemistry Accounts,* vol. 129, no. 6, pp. 771-780, 2011.
[7] R. Atkinson and J. Arey, "Atmospheric Degradation of Volatile Organic Compounds," *Chemical Reviews,* vol. 103, no. 12, pp. 4605-4638, 2003.
[8] G. Oksdath-Mansilla, A. B. Peñéñory, M. Albu, I. Barnes, P. Wiesen, and M. A. Teruel, "FTIR relative kinetic study of the reactions of CH3CH2SCH2CH3 and CH3CH2SCH3 with OH radicals and Cl atoms at atmospheric pressure," *Chemical Physics Letters,* vol. 477, no. 1–3, pp. 22-27, 2009.
[9] A. J. Hynes, P. H. Wine, and D. H. Semmes, "Kinetics and mechanism of hydroxyl reactions with organic sulfides," *The Journal of Physical Chemistry,* vol. 90, no. 17, pp. 4148-4156, 1986.
[10] L. Zhu, J. W. Bozzelli, and L. M. Kardos, "Thermochemical properties, Δ fH o(298), S o(298), and C p o(298), (T), for n-butyl and n-pentyl hydroperoxides and the alkyl and peroxy radicals, transition states, and kinetics for intramolecular hydrogen shift reactions of the peroxy radicals," *Journal of Physical Chemistry A,* vol. 111, no. 28, pp. 6361-6377, 2007.
[11] C. Y. Sheng, J. W. Bozzelli, A. M. Dean, and A. Y. Chang, "Detailed Kinetics and Thermochemistry of C2H5 + O2: Reaction Kinetics of the Chemically-Activated and Stabilized CH3CH2OO• Adduct," *The Journal of Physical Chemistry A,* vol. 106, no. 32, pp. 7276-7293, 2002.
[12] J. A. Pople, M. Head-Gordon, D. J. Fox, K. Raghavachari, and L. A. Curtiss, "Gaussian-1 theory: A general procedure for prediction of molecular energies," *The Journal of Chemical Physics,* vol. 90, no. 10, pp. 5622-5629, 1989.
[13] J. A. Pople, "Nobel lecture: Quantum chemical models," *Reviews of Modern Physics,* vol. 71, no. 5, pp. 1267-1274, 1999.
[14] M. J. Frisch *et al.*, "Gaussian 03, Revision C.02," ed, 2003.

[15] J. A. Montgomery Jr, M. J. Frisch, J. W. Ochterski, and G. A. Petersson, "A complete basis set model chemistry. VII. Use of the minimum population localization method," *Journal of Chemical Physics,* Article vol. 112, no. 15, pp. 6532-6542, 2000.
[16] J. A. Montgomery Jr, M. J. Frisch, J. W. Ochterski, and G. A. Petersson, "A complete basis set model chemistry. VI. Use of density functional geometries and frequencies," *Journal of Chemical Physics,* vol. 110, no. 2-12, pp. 2822-2827, 1999.
[17] K. B. Wiberg and G. A. Petersson, "A Computational Study of RXHn X-H Bond Dissociation Enthalpies," *Journal of Physical Chemistry A,* vol. 118, no. 12, pp. 2353-2359, Mar 27 2014.
[18] K. B. Wiberg, G. B. Ellison, J. M. McBride, and G. A. Petersson, "Substituent Effects on O-H Bond Dissociation Enthalpies: A Computational Study," *Journal of Physical Chemistry A,* vol. 117, no. 1, pp. 213-218, Jan 10 2013.
[19] V. G. Kiselev, N. P. Gritsan, V. E. Zarko, P. I. Kalmykov, and V. A. Shandakov, "Multilevel quantum chemical calculation of the enthalpy of formation of 1,2,5 oxadiazolo 3,4-e 1,2,3,4 -tetrazine-4,6-Di-N-Dioxide," *Combustion Explosion and Shock Waves,* vol. 43, no. 5, pp. 562-566, Sep-Oct 2007.
[20] K. B. Wiberg, "Ab Initio Molecular Orbital Theory," *Journal of Computational Chemistry,* vol. 7, no. 3, pp. 379-379, 1986.
[21] I. K. Ortega *et al.*, "From quantum chemical formation free energies to evaporation rates," *Atmospheric Chemistry and Physics,* vol. 12, no. 1, pp. 225-235, 2012.
[22] R. Casasnovas, J. Frau, J. Ortega-Castro, A. Salvà, J. Donoso, and F. Muñoz, "Simplification of the CBS-QB3 method for predicting gas-phase deprotonation free energies," *International Journal of Quantum Chemistry,* vol. 110, no. 2, pp. 323-330, 2010.
[23] A. G. Vandeputte, M. K. Sabbe, M. F. Reyniers, and G. B. Marin, "Modeling the Gas-Phase Thermochemistry of Organosulfur Compounds," *Chemistry-a European Journal,* vol. 17, no. 27, pp. 7656-7673, Jun 2011.
[24] C. A. Class, J. Aguilera-Iparraguirre, and W. H. Green, "A kinetic and thermochemical database for organic sulfur and oxygen compounds," *Physical Chemistry Chemical Physics,* vol. 17, no. 20, pp. 13625-13639, 2015.
[25] S. Yommee and J. W. Bozzelli, "Cyclopentadienone Oxidation Reaction Kinetics and Thermochemistry for the Alcohols, Hydroperoxides, and Vinylic, Alkoxy, and Alkylperoxy Radicals," *Journal of Physical Chemistry A,* vol. 120, no. 3, pp. 433-451, 2016.
[26] S. Snitsiriwat and J. W. Bozzelli, "Thermochemistry, reaction paths, and kinetics on the tert -isooctane radical reaction with O2," *Journal of Physical Chemistry A,* vol. 118, no. 26, pp. 4631-4646, 2014.

Spring Technical Meeting
Eastern States Section of the Combustion Institute
March 4-7, 2018
State College, Pennsylvania

Ion-molecule reactions relevant to the detection of atmospheric oxidized mercury by chemical ionization mass spectrometry

Francisco J. Guzman[1], *Matthew Cooper*[2], *John Antley*[2], *Joseph Bozzelli*[2], *and Alexei Khalizov*[2,*]

[1]*Department of Chemical Engineering, New Jersey Institute of Technology, 323 Dr Martin Luther King Jr Blvd, Newark, NJ 07102, USA*
[2] *Department of Chemistry & Environmental Science, New Jersey Institute of Technology, 323 Dr Martin Luther King Jr Blvd, Newark, NJ 07102, USA*
[*]*Corresponding Author Email: khalizov@njit.edu*

Abstract: Mercury is a persistent environmental pollutant arising from anthropogenic and natural sources. Gaseous oxidized mercury (GOM) is significantly more water soluble than elemental Hg and should be the main form responsible for deposition unto environmental sinks, however, the speciation of GOM is poorly understood due to its relatively low concentrations and labile nature. We investigated ion-molecule chemistry of gas-phase $HgCl_2$ with SF_6^-, CO_3^- , and $CO_2{\bullet}O_2^-$ reagent ions by ion-drift chemical ionization mass spectrometry. Additionally, DFT and ab-initio computations coupled with rate constant calculations based on average dipole orientation (ADO) theory were used to evaluate the product stability, elucidate reaction pathways and corroborate experimental results. We found that the reaction of $HgCl_2$ with SF_6^-, CO_3^-, and $CO_2{\bullet}O_2^-$ forms HgCl2•F$^-$, $HgCl_2{\bullet}CO_3^-$, $HgCl_2O^-$, and $HgCl_2O_2^-$ product ions. These reactions are efficient in converting gaseous oxidized mercury to molecular ions, although competing reactions with O_3, NO_x, and H_2O may limit their application in atmospheric applications. The calculated stability of the complexes is in the 11 to 57 kcal-mol-1range.
Keywords: *detection of oxidized mercury, reaction thermochemistry, and ADO theory*

1. Introduction

Mercury (Hg) is pervasive environmental pollutant [1], which exists in the environment predominately in three major forms - gaseous elemental mercury ($Hg_{(g)}^0$, GEM), gaseous oxidized mercury ($Hg_{(g)}^{2+}$, GOM), and particulate bound mercury ($Hg_{(p)}$, PBM) - and can be transported in the environment by numerous physical mechanisms [2]. The small concentrations of mercury make its detection and speciation a challenge. In-stack continuous emissions measurements from a coal fuel power plant reported 67 to 72 ppb NO_x, 319 to 566 ppb SO_2, and 0.35 to 0.41 ppb of total gaseous mercury (TGM = GEM + GOM) [3]. The atmospheric concentrations are many orders of magnitude lower. Estimates of global Hg emissions are ~5200 Mg per year, where 70% of the yearly emissions are from natural sources, which include re-emissions from previously deposited natural and anthropogenic sources, and 30% are new anthropogenic emissions [4]. Clearly, even if all new anthropogenic emissions of mercury are ceased immediately, there is a major mercury re-circulation problem that must be studied. High fidelity measurements of TGM including GOM speciation are thus of critical importance.

TGM in the environment can be measured using direct sampling techniques such as proton transfer mass spectrometry (PTRMS) and ion-drift chemical ionization mass spectrometry (ID-CIMS) [5]. In direct sampling, the user controls the volume of incoming sample allowing for real time measurements and speciation of the sample. For the detection and speciation of atmospheric GOM, the available sampling techniques must have very low detection limits, short sampling times, and avoid transformation of the measured species by either thermal decomposition, transformations in the condensed phase, or surface-catalyzed reactions with the chamber walls.

PTRMS is soft ionization technique where H_3O^+ ions are used to transfer a proton to the target trace gas, typically organic compounds [6]. A computational study by Dibble, Zelie, and Jiao has investigated the applicability of PTRMS toward the detection of GOM. The author's found PTRMS unsuitable for detection of $HgBr_2$ and $HgCl_2$ due to the high PA of water. The study also analyzed the protonation of BrHgY species (Y = NO_2, ClO, BrO, HOO) with H_3O^+ and showed the stable products formed where predominately $BrHgYH^+$ [7]. However, in order to achieve the required limit of detection for atmospheric GOM, the drift tube of PTRMS must operate near atmospheric pressure. This would require excessively high voltages, on the order of 50 kV across the drift tube in order create sufficiently strong electric field to break up the $H_3O^+(H_2O)_{n=1,2,..}$ water clusters.

ID-CIMS is similar to PTRMS, but can employ a broad range of positive or negative reagent ions [8]. In ID-CIMS the reaction mechanism could be either charge transfer, ion transfer, or a combination of the two.

A reagent ion appropriate for GOM detection must meet the following criteria: (1) ΔH_{rxn} between GOM and R should be exothermic, (2) clustering of the reagent ion with water vapor should be small, (3) reactivity with other atmospheric trace gases such as O_3 and NO_2 should be low. Reagent ions SF_6^-, $CO_2{\bullet}O_2^-$, and CO_3^- were selected because of their use in prior studies for the detection of trace atmospheric gases. SF_6^- was used for the detection of $ClNO_3$, O_3, NO_2, N_2O_5, and SO_2 [9, 10]. Huey, Hanson, and Howard showed that most of the reactions with SF_6^- produced multiple products indicative of a complex transformation pathways [9].

SF_6^- reacts via charge and ion transfer mechanisms, depending on the trace gas it reacted with. However, the fast reaction with O_3 could limit the usefulness SF_6^- as a reagent ion in atmospheric applications [9]. Reagent ion CO_3^- was used for the detection of SO_2, H_2SO_4, and HNO_3 using CIMS [11-13]. The $CO2{\bullet}O2^-$ complex was used for HNO_3 and HCO_2H detection using an ion-molecule reaction mass spectrometry system [13, 14].

This paper is organized as follows. We present results of our electron affinity (EA) and ion affinity (IA) calculations. This is followed by a discussion of the experimental products ions formed with an elucidation of the reaction pathways based on our DFT and *ab-initio* calculations and a kinetic discussion based on average dipole orientation (ADO) theory. For this work, we used commercially available $HgCl_2$ as our GOM species.

2. Methods / Experimental

Ab-initio and Density Functional Theory (DFT) calculations were carried out using Revision C.01 of the Gaussian 09 program [15]. To account for relativistic effects of the molecules studied, small core energy consistent effective-core potentials (ECP) of the Stuttgart/Köln type where used for Hg [16]. The ECP used for Hg is the ECP60MDF which has the inner 60 electrons replaced by the potential and accounts for relativistic effects. The aug-cc-pVTZ (AVTZ) basis set is used for all of the atoms in the molecules in this study (C, Cl, F, Hg, O and S) the AVTZ basis sets without a pseudo-potential were used [17, 18]. It is implied that the AVTZ basis set for Hg include the pseudo-potential. DFT methods used in this study are M06-2X and PBE0. CCSD(T) is a hybrid truncated coupled cluster method with single and doublet excitations (CCSD) and a non-iterative triplets [19] In our calculations, structures were optimized at the DFT level of theory, followed by a single point CCSD(T)/AVTZ//DFT/AVTZ calculation.

In experiments, GOM detection was performed using ID-CIMS. Neutral GOM molecules were reacted with appropriate reagent ions in a drift tube to form well-defined product ions, which were detected by a quadrupole mass spectrometer (Extrel) equipped with a low-noise counting electron multiplier. Reagent ions were generated by passing SF_6/N_2 and CO_2/O_2 mixtures through corona discharge (-1000 V, total gas pressure in the drift tube 1.4-2.0 Torr). $HgCl_2$ vapor was introduced in a flow of He from the headspace of a glass container containing solid $HgCl_2$ on fiberglass. The concentration of $HgCl_2$ in the drift tube was around 2×10^{10} molecule cm^{-3}.

3. Results and Discussion

Figure 1 shows the structures of GOM product species formed. For the GOM species all the molecules were developed for this study with the exception of adjusting the initial Hg-Cl bond length as per the work of Shepler, Balabanov, and Peterson [20].

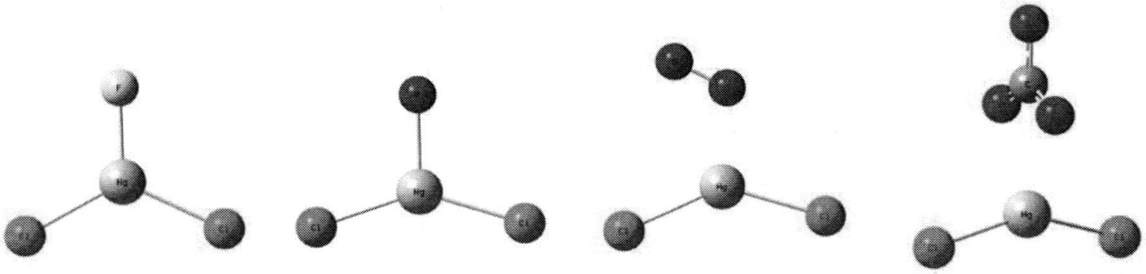

Figure 1: Structures of GOM product ions from reactions listed in Tables 2, 3, and 4. All structures calculated at the DFT M06-2X/AVTZ level of theory. From left to right; $HgCl_2{\bullet}F^-$, $HgCl_2{\bullet}O^-$, $HgCl_2{\bullet}O_2^-$, and $HgCl_2{\bullet}CO_3^-$

Figure 2 shows the measured mass spectra of the reagent and product ions generated between the reaction of SF_6 and $HgCl_2^-$ and between $CO_2{\bullet}O_2^-$ and $HgCl_2$. Figure 2a shows the reagent ion SF_6^- corresponding to a peak at 146 amu. SF_5^- is being formed as a reagent ion by-product at a lower intensity compared to SF_6^-, indicated by the peak at 127 amu. Figure 2b shows the $Hg_{(g)}^{2+}$ product ions formed, $HgCl_2^-$ and $HgCl_2{\bullet}F^-$. $HgCl_2^-$ located between 266 and 271 amu and $HgCl_2{\bullet}F^-$ is between 285 and 295 amu. $HgCl_2{\bullet}F^-$ has a peak intensity of 5.9 kHz and 2.2 kHz for $HgCl_2^-$. Figure 2c shows the reagent ions CO_3^- and $CO_2{\bullet}O_2^-$, identified by the peaks at 60 and 76 amu, respectively. Figure 2d shows the $Hg_{(g)}^{2+}$ product ions formed where $HgCl_2^-$ is between 266 and 271 amu, $HgCl_2{\bullet}O^-$ is between 282 and 298 amu, $HgCl_2{\bullet}O_2^-$ is between 298 and 310 amu, and $HgCl_2 \bullet [CO_2 \bullet O_2^-]$ between 326 and 336 amu. In descending order, the peak intensties for the $Hg_{(g)}^{2+}$ product ions are 4.2 kHz for $HgCl_2 \bullet CO_3^-$, 2.2 kHz for $HgCl_2^-$, 1.4 kHz for $HgCl_2 \bullet O^-$, and 0.6 kHz for $HgCl_2 \bullet O_2^-$.

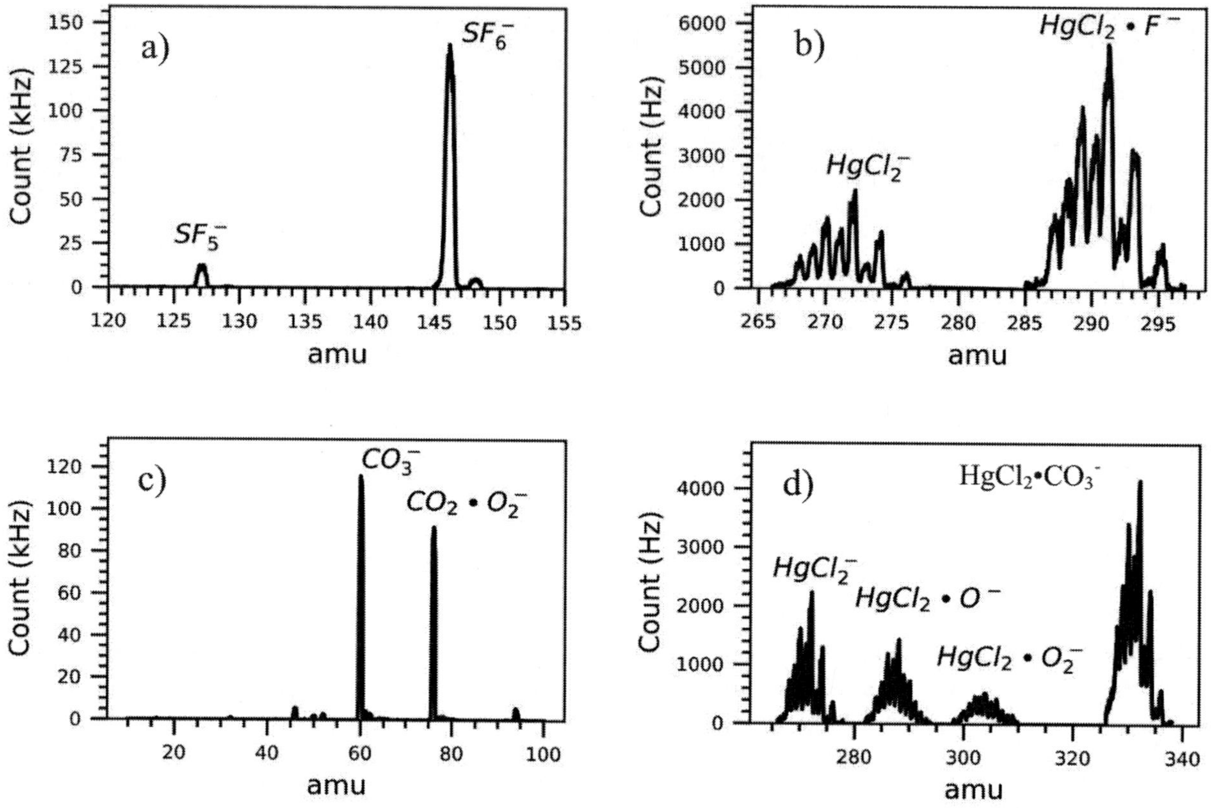

Figure 2: Mass spectra of the reagent ions generated from SF_6 (a) and corresponding product ions produced in reactions with $HgCl_2$ (b) . Mass spectra of reagent ions generated from CO_3^- (c) and corresponding product ions produced in reactions with $HgCl_2$ (d)

Table 1 shows the electron affinities (EA) and ion affinities (IA) for the reagent ions, $HgCl_2$, and selected product species in units of eV. Both EA and IA are calculated as $-\Delta H_{rxn}$. For the EA calculations, both DFT methods are within 0.03 to 0.25 eV of each other. Neither DFT method is consistently higher or lower than each other. Compared to DFT, CCSD(T) results are lower in energy. For the IA calculations, the M06-2X energy is typically higher than PBE0 with values ranging between 0.09 to 0.19 eV. Compared to DFT, CCSD(T) IA are again lower in energy.

Table 1: Electron (EA) and Ion (IA) affinities for reagent ions, $HgCl_2$ in units of eV.

Reaction	**M06-2X/AVTZ**	**PBE0/AVTZ**	**CCSD(T)/AVTZ// M06-2X/AVTZ**
	EA/IA = - ΔH_{rxn} (298 K) (eV)		
Electron Affinities			
$SF_6 \rightarrow SF_6^-$	1.47	1.63	1.23
$SF_5 \rightarrow SF_5^-$	4.24	3.99	4.12
$O_2 \rightarrow O_2^-$	2.01	2.04	1.70

$O_3 \rightarrow O_3^-$	2.74	2.47	
$CO_3 \rightarrow CO_3^-$	4.10	4.27	
$CO_2 \cdot O_2 \rightarrow CO_2 \cdot O_2^-$	3.16	3.19	2.72
$HgCl_2 \rightarrow HgCl_2^-$	1.84	1.65	1.60
Ion Affinities			
$SF_6 + F^- \rightarrow SF_7^-$	0.34	0.24	0.27
$SF_5 + F^- \rightarrow SF_6^-$	2.59	2.50	2.29
$CO_2 + O_2^- \rightarrow CO_2 \cdot O_2^-$	1.15	1.15	1.02
$HgCl_2 + F^- \rightarrow HgCl_2 \cdot F^-$	2.51	2.32	2.33
$HgCl_2 + O_2^- \rightarrow HgCl_2 \cdot O_2^-$	1.58	1.43	1.49

Table 2 shows the enthalpies of reaction forming $HgCl_2F^-$ starting with reagent ion SF_6^-. We categorize the reactions into two distinct sets: (1) reactions with neutral $HgCl_2$ and (2) reactions with anion $HgCl_2^-$ formed from neutral $HgCl_2$. $HgCl_2$ is a closed shell singlet molecule and $HgCl_2^-$ is an open shell doublet molecule (a radical - anion). Product ion $HgCl_2 \cdot F^-$ is a closed shell singlet. For the SF_n molecules the spin multiplicities are as follows: SF_7^-, SF_6, SF_2, and SF_4 are closed shell singlets; SF_6^-, SF_5 and SF_3 are open shell doublets. For the reaction set of neural $HgCl_2$, the most exothermic reactions is $HgCl_2 + SF_7^- \rightarrow HgCl_2 \cdot F^- + SF_6$ followed by the reaction $HgCl_2 + SF_6^- \rightarrow HgCl_2^- + SF_6$. We note that the measurements show only trace quantities of SF_7^-. The reaction $HgCl_2 + SF_6^- \rightarrow HgCl_2 \cdot F^- + SF_5$ is slightly endothermic according DFT calculations but slightly exothermic according to the CCSD(T) calculations. For the reaction set with anion $HgCl_2^-$ the reaction $HgCl_2^- + SF_5 \rightarrow HgCl_2 \cdot F^- + SF_4$ is the most exothermic. The k_{ADO} rate constant is calculated using the polarizabilities and dipole moments of the neutral species from our M06-2X/AVTZ calculations. At a constant temperature k_{ADO} is not a strong function of the ion, but rather the polarizability and dipole moment of the neutral species.

Table 2: Reaction enthalpies in kcal/mol and k_{ADO} in cm^3/s of the formation of $HgCl_2F^-$ starting with reagent ion SF_6^- calculated at three different theory levels. k_{ADO} is calculated using M06-2x/AVTZ.

Reaction	**M06-2X/AVTZ**		**PBE0/AVTZ**	**CCSD(T)/AVTZ// M06-2X/AVTZ**
	k_{ADO} (10^{-10} cm^3/s)	**ΔH_{rxn} (298 K) (kcal/mol)**		
Reactions with neutral $HgCl_2$				
$HgCl_2 + SF_7^- \rightarrow HgCl_2{\cdot}F^- + SF_6$	5.50	-50.05	-48.11	-47.54
$HgCl_2 + SF_6^- \rightarrow HgCl_2{\cdot}F^- + SF_5$	5.72	1.85	4.12	-0.85
$HgCl_2 + SF_6^- \rightarrow HgCl_2^- + SF_6$	5.72	-8.55	-0.46	-8.46
Reactions with anion $HgCl_2^-$				
$HgCl_2^- + SF_6 \rightarrow HgCl_2{\cdot}F^- + SF_5$	5.21	10.40	4.58	7.61
$HgCl_2^- + SF_5 \rightarrow HgCl_2{\cdot}F^- + SF_4$	5.83	-54.39	-50.56	-57.80
$HgCl_2^- + SF_3 \rightarrow HgCl_2{\cdot}F^- + SF_2$	7.57	-38.41	-31.99	

Table 3 shows the enthalpies of reaction for the formation of $HgCl_2{\cdot}O^-$, $HgCl_2{\cdot}O_2^-$, $HgCl_2{\cdot}CO_3^-$, $HgCl_2{\cdot}[CO_2{\cdot}O_2^-]$ and $HgCl_2 \cdot CO_2^-$ from reactions with reagent ions O_2^-, $CO_2 \cdot O_2^-$ and CO_3^-. The results are presented by ascending reagent ion size with reactions with only one product ion presented first. All of the reactions are exothermic, with the most exothermic reaction being $HgCl_2 + O_2^- \rightarrow HgCl_2{\bullet}O_2^-$ and the least exothermic being $HgCl_2 + CO_2{\bullet}O_2^- \rightarrow HgCl_2{\bullet}O_2^- + CO_2$. The spin multiplicities for the reagent ions and Carbon/Oxygen species are: (1) closed shell singlet: O^- and CO_2; (2) open shell doublet: O_2^-, CO_3^-, $HgCl_2{\bullet}O^-$, $HgCl_2{\bullet}O_2^-$, and $HgCl_2{\bullet}CO_3^-$.

Table 3: Reaction enthalpies in kcal/mol and k_{ADO} in cm^3/s of the formation of $HgCl_2{\bullet}O^-$, $HgCl_2{\bullet}O_2^-$, and $HgCl_2{\bullet}CO_3^-$ from reagent ions O_2^-, CO_3^-, and $CO_2{\bullet}O_2^-$ calculated at three different theory levels. k_{ADO} is calculated using M06-2x/AVTZ.

Reaction	**M06-2X/AVTZ**		**PBE0/AVTZ**	**CCSD(T)/AVTZ// M06-2X/AVTZ**
	k_{ADO} (10^{-10} cm^3/s)	**ΔH_{rxn} (298 K) (kcal/mol)**		
$HgCl_2 + O^- \rightarrow HgCl_2{\bullet}O^-$	14.3	-46.63	-64.75	-39.77
$HgCl_2 + O_2^- \rightarrow HgCl_2{\bullet}O_2^-$	10.4	-36.39	-33.07	-34.29
$HgCl_2 + CO_3^- \rightarrow HgCl_2{\bullet}CO_3^-$	7.95	-31.23	-24.79	-29.43
$HgCl_2 + CO_2{\bullet}O_2^- \rightarrow HgCl_2{\bullet}O_2^- + CO_2$	7.23	-9.83	-6.62	-10.86

Based on our computational results, we propose a two-step mechanism for the formation of $HgCl_2F^-$ from SF_6^-. First, a charge transfer reaction occurs where ionized SF_6^- gives an electron to $HgCl_2$, forming $HgCl_2^-$. This is supported by the reaction $HgCl_2 + SF_6^- \rightarrow HgCl_2^- + SF_6$ being exothermic and the EA of $HgCl_2$ being larger than that of SF_6. SF_6^- prefers to donate an electron to $HgCl_2$ as this results in a lower energy state. The SF_6 F^- IA is also much less than that of $HgCl_2$. A common feature of the exothermic reactions in Table 1 is that the SF_n product formed is in the singlet state. GOM products could be either in the singlet or doublet state, so we argue that the SF_n reactions are mediated by the product state of SF_n.

Once $HgCl_2^-$ is formed, it reacts with a neutral SF_n (n = 5 or 3) species forming the detected $HgCl_2{\bullet}F^-$. It is likely that the dominant pathway is the reaction with SF_5 rather than with SF_3 as SF_5 is detected in significantly higher quantities. Another potential pathway for $HgCl_2{\bullet}F^-$ is directly from the reaction of SF_7^- and $HgCl_2$. Although this reaction is thermodynamically feasible, we could not measure significant quantities SF_7^- and hence ruled out this pathway.

For reactions between $HgCl_2$ and $CO_2{\bullet}O_2^-$ and $HgCl_2$ and CO_3^- and $CO_2{\bullet}O_2^-$, mass spectra show the formation of $HgCl_2{\bullet}O^-$, $HgCl_2{\bullet}O_2^-$, and $HgCl_2{\bullet}CO_3^-$. As seen in Table 3, all of the reactions are exothermic and have GOM products in either the singlet or doublet state. From our thermodynamic results, we suggest the reactions 2 thru 4 in Table 3 as the mechanism due to the abundance of CO_3^- and $CO_2{\bullet}O_2^-$. Between $HgCl_2{\bullet}O^-$ and $HgCl_2{\bullet}O_2^-$, k_{ADO} is in agreement with the measured mass spectra where the rate constant and measured spectra is higher for $HgCl_2{\bullet}O^-$. However, k_{ADO} is lower for $HgCl_2{\bullet}CO_3^-$ which has the highest count of the formed C/O/Hg species.

4. Conclusions

GOM detection and speciation has been achieved using ID-CIMS. A potential pathway forming $HgCl_2{\bullet}F^-$ starting from reagent ion SF_6^- has been thermodynamically identified. First, a charge transfer reaction between $HgCl_2$ and SF_6^- forms $HgCl_2^-$. $HgCl_2^-$ then reacts with SF_5, forming $HgCl_2{\bullet}F^-$. The $HgCl_2{\bullet}O^-$, $HgCl_2{\bullet}O_2^-$, and $HgCl_2{\bullet}CO_3^-$ GOM product ions have been observed and a potential mechanism identified. $HgCl_2{\bullet}CO_3^-$ can form directly from reagent ions CO_3^-. $HgCl_2{\bullet}O^-$ can be formed but requires a source of O^-. $HgCl_2{\bullet}O_2^-$ can form O_2^- or the $CO_2{\bullet}O_2^-$ reagent ion.

5. Acknowledgements

FJG would like to acknowledge the ARCS staff at NJIT for assistance in setting up the calculations.

6. References

[1] A. Carpi, Mercury from Combustion Sources: A Review of the Chemical Species Emitted and Their Transport in the Atmosphere, Water, Air, and Soil Pollution 98 (1997) 241-254.
[2] W.H. Schroeder, J. Munthe, Atmospheric Mercury - An Overview, Atmospheric Environment 32 (1998) 809-822.
[3] M.S. Landis, J.V. Ryan, A.F. ter Schure, D. Laudal, Behavior of mercury emissions from a commercial coal-fired power plant: the relationship between stack speciation and near-field plume measurements, Environ Sci Technol 48 (2014) 13540-13548.
[4] N. Pirrone, S. Cinnirella, X. Feng, R.B. Finkelman, H.R. Friedli, J. Leaner, R. Mason, A.B. Mukherjee, G.B. Stracher, D.G. Streets, K. Telmer, Global mercury emissions to the atmosphere from anthropogenic and natural sources, Atmospheric Chemistry and Physics 10 (2010) 5951-5964.
[5] M.S. Gustin, H.M. Amos, J. Huang, M.B. Miller, K. Heidecorn, Measuring and modeling mercury in the atmosphere: a critical review, Atmospheric Chemistry and Physics 15 (2015) 5697-5713.
[6] J. de Gouw, C. Warneke, Measurements of volatile organic compounds in the earths atmosphere using proton-transfer-reaction mass spectrometry, Mass Spectrometry Reviews 26 (2007) 223-257.
[7] T.S. Dibble, M.J. Zelie, Y. Jiao, Quantum chemistry guide to PTRMS studies of as-yet undetected products of the bromine-atom initiated oxidation of gaseous elemental mercury, J Phys Chem A 118 (2014) 7847-7854.
[8] E.C. Fortner, J. Zhao, R.Y. Zhang, Development of ion drift-chemical ionization mass spectrometry, Analytical Chemistry 76 (2004) 5436-5440.
[9] L.G. Huey, D.R. Hanson, C.J. Howard, Reactions of SF6- and I- with Atmospheric Trace Gases, The Journal of Physical Chemistry 99 (1995) 5001-5008.
[10] A.A. Viggiano, A.I. Fernandez, J. Troe, Ion-molecule kinetics at 15-700 Torr, Phys. Chem. Chem. Phys. 7 (2005) 1533-1539.
[11] D. Salcedo, P.W. Villalta, V. Varutbangkul, J.C. Wormhoudt, R.C. Miake-Lye, D.R. Worsnop, J.O. Ballenthin, W.F. Thorn, A.A. Viggiano, T.M. Miller, R.C. Flagan, J.H. Seinfeld, Effect of relative humidity on the detection of sulfur dioxide and sulfuric acid using a chemical ionization mass spectrometer, International Journal of Mass Spectrometry 231 (2004) 17-30.
[12] E. Arijs, A. Barassin, E. Kopp, C. Amelynck, V. Catoire, H.P. Fink, C. Guimbaud, U. Jenzer, D. Labonnette, W. Luithardt, E. Neefs, D. Nevejans, N. Schoon, A.M. Van Bavel, Stratospheric chemical ionization mass spectrometry: nitric acid detection by different ion molecule reaction schemes, International Journal of Mass Spectrometry 181 (1998) 99-111.
[13] J. Viidanoja, T. Reiner, F. Arnold, Laboratory investigations of negative ion molecule reactions of formic and acetic acids: implications for atmospheric measurements by ion-molecule reaction mass spectrometry, International Journal of Mass Spectrometry 181 (1998) 31-41.

[14] C. Amelynck, D. Fussen, E. Arijs, Reactions of nitric acid with di- and trichloride ions, di- and tri-iodide ions and with CO−4 in the gas phase, International Journal of Mass Spectrometry and Ion Processes 133 (1994) 13-28.
[15] M.J. Frisch, G.W. Trucks, H.B. Schlegel, G.E. Scuseria, M.A. Robb, J.R. Cheeseman, G. Scalmani, V. Barone, G.A. Petersson, H. Nakatsuji, X. Li, M. Caricato, A.V. Marenich, J. Bloino, B.G. Janesko, R. Gomperts, B. Mennucci, H.P. Hratchian, J.V. Ortiz, A.F. Izmaylov, J.L. Sonnenberg, Williams, F. Ding, F. Lipparini, F. Egidi, J. Goings, B. Peng, A. Petrone, T. Henderson, D. Ranasinghe, V.G. Zakrzewski, J. Gao, N. Rega, G. Zheng, W. Liang, M. Hada, M. Ehara, K. Toyota, R. Fukuda, J. Hasegawa, M. Ishida, T. Nakajima, Y. Honda, O. Kitao, H. Nakai, T. Vreven, K. Throssell, J.A. Montgomery Jr., J.E. Peralta, F. Ogliaro, M.J. Bearpark, J.J. Heyd, E.N. Brothers, K.N. Kudin, V.N. Staroverov, T.A. Keith, R. Kobayashi, J. Normand, K. Raghavachari, A.P. Rendell, J.C. Burant, S.S. Iyengar, J. Tomasi, M. Cossi, J.M. Millam, M. Klene, C. Adamo, R. Cammi, J.W. Ochterski, R.L. Martin, K. Morokuma, O. Farkas, J.B. Foresman, D.J. Fox, Gaussian 09, Revision C.01, Wallingford, CT, 2009.
[16] D. Figgen, G. Rauhut, M. Dolg, H. Stoll, Energy-consistent pseudopotentials for group 11 and 12 atoms: adjustment to multi-configuration Dirac–Hartree–Fock data, Chemical Physics 311 (2005) 227-244.
[17] T.H. Dunning, Gaussian basis sets for use in correlated molecular calculations. I. The atoms boron through neon and hydrogen, The Journal of Chemical Physics 90 (1989) 1007-1023.
[18] D.E. Woon, T.H. Dunning, Gaussian basis sets for use in correlated molecular calculations. III. The atoms aluminum through argon, The Journal of Chemical Physics 98 (1993) 1358-1371.
[19] K. Raghavachari, G.W. Trucks, J.A. Pople, M. Head-Gordon, A Fifth-Order Pertubation Comparision of Electron Correlation Theories, Chemical Physical Letters 157 (1989) 479-483.
[20] B.C. Shepler, N.B. Balabanov, K.A. Peterson, Ab Initio Thermochemistry Involving Heavy Atoms: An Investigation of the Reactions Hg+ IX (X = I, Br, Cl, O), J. Phys. Chem. A 109 (2005).

Spring Technical Meeting
Eastern States Section of the Combustion Institute
March 4-7, 2018
State College, Pennsylvania

Double Luminous Zones in Inverse Laminar Jet Diffusion Flames

Z. Wang[1], *P.B. Sunderland*[1,*], *R.L. Axelbaum*[2],
[1]*Department of Fire Protection Engineering, University of Maryland College Park, College Park, Maryland, 20742 USA*
[2]*Department of Energy, Environmental & Chemical Engineering, Washington University in St. Louis, St. Louis 63130, USA*
[*]*Corresponding author email: pbs@umd.edu*

Abstract: The canonical nonpremixed hydrocarbon flame has a thin blue luminous zone, often surrounded by soot. However, previous experimental and computational work indicates that double luminous zones are possible. This work examines double luminous zones in inverse gas jet diffusion flames. Over 100 inverse flames were observed using methane and propane as fuels, and nitrogen, argon, and helium as diluents. The double luminous zones are visible with the naked eye and in color images for various burner diameters, fuels and inerts, dilution levels, and flame lengths. The inner zone is blue, thick, and teardrop shaped, and the outer zone is blue-green, much thinner, and does not appear to be closed on the flame axis. Double luminous zones were visible in every IDF studied. NDFs were considered at otherwise similar conditions, and none exhibited a double luminous zone. Thin-filament pyrometry and intentional partial premixing were performed. The results indicate that the inner flame is stoichiometric and the outer zone is a rich premixed flame. Several possible mechanisms are considered. These include reactant leakage at the flame base, reactant leakage across the stoichiometric contour, misalignment between the peaks in temperature and reaction rates, and soot luminosity.

Keywords: reactant leakage; thin-filament pyrometry; triple flames

1. Introduction

The canonical nonpremixed hydrocarbon flame has a thin blue luminous zone, often surrounded by soot. However, experimental and computational work indicates that double luminous zones are possible.

Saito et al. [1] observed two blue zones in small CH_4 normal gas jet diffusion flames (NDFs). Weinberg and co-workers [2,3] imaged such zones in similar flames. Gülder and co-workers [4] observed double reaction zones of CH_4 NDFs from 1 – 100 atm. These studies found the inner and outer zones involved CH_4 and CO/H_2 consumption, respectively. Sung et al. [5] studied spherical CH_4-doped H_2 inverse diffusion flames and observed double luminous zones separated by up to 8 mm. Toward the fuel there was a green zone associated with CH_4 consumption and toward the oxidizer there was a blue zone associated with H_2 consumption.

Diffusion flames with two reaction fronts have been predicted for halogens and cool flames [6], but experimental evidence is lacking.

On close inspection, many published color images of inverse gas-jet diffusion flames (IDFs) reveal double luminous zones [7–10], although they are close together and easily missed – few studies mentioned them [7]. They are evident for various fuels (CH_4, C_2H_4, C_2H_6, C_3H_8, and C_4H_{10}), in normal gravity and microgravity, and in coflowing or quiescent ambients (and lifted flames). The double zones are not evident in some cases [11-13], perhaps owing to high soot loading or low image quality. This IDF phenomenon, which seems counterintuitive based on classical diffusion flame theory, is examined here.

2. Experimental

Diffusion flames with monocomponent hydrocarbon fuels were studied using a co-flow burner. The outer tube had a diameter of 101.6 mm and contained a ceramic honeycomb to ensure flow uniformity. The inner tube was stainless steel with an inside diameter of 3 mm, a wall thickness of 1.5 mm, and a discharge height above the honeycomb of 5 mm. A similar burner with 15 mm inner tube diameter was also used. The burner was fitted with a 155 mm long chimney whose top was sealed by aluminum foil with a 15 mm round hole on axis. For most tests a secondary flame was maintained above this hole. The NDFs had fuel in the inner tube, surrounded by oxidizer, whereas for IDFs these were reversed.

The burner was connected to flows of gaseous methane, propane, oxygen, and nitrogen. Flow rates were controlled with metering valves and measured with calibrated rotameters. Over 100 IDFs were observed, with methane and propane as fuels, various burner diameters and materials, dilution levels, and flow rates, and stoichiometric mixture fractions of 0.064 – 0.78. For a few tests, qualitative thin-filament pyrometry was performed using a horizontal 14 μm SiC fiber.

Flame images were recorded with a Nikon D100 digital camera and saved as nef files. To avoid gamma corrections, the conversion to tif files was performed using Dcraw [14]. The red, green, and blue (RGB) intensities were extracted with ImageJ and averaged to find grayscale intensity. For some images a 10 nm bandpass filter (at 430 nm) was used. For some an intensified ultraviolet camera (Xybion ISG-250-GQU-3) with a 10 nm bandpass filter (at 310 nm) was used.

3. Results and Discussion

Figure 1a shows a representative IDF. Visible here is a double luminous zone, where the inner zone is blue, thick, and teardrop shaped, and the outer zone is blue-green, much thinner, and does not appear to be closed on the flame axis. The locations of peak intensity for the inner and outer zones are typically separated by 0.7 mm at the half height of the inner zone. Also visible are yellow soot emissions.

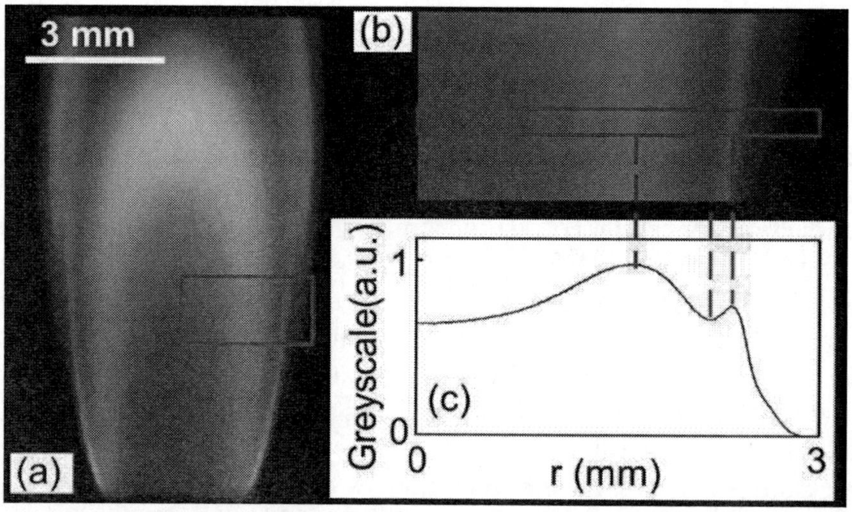

Fig. 1. The double luminous zone structure in an IDF. (a) Photograph of an IDF with an O_2 mole fraction of 0.77 and O_2/N_2 flow rate of 7.83 mg/s in the inner tube, and a CH_4 mole fraction of 0.30 and CH_4/N_2 flow rate of 42.1 mg/s in the outer tube. (b) A 3X zoom view of the red-bordered region in Fig. 1a. (c) A plot of grayscale intensity for the red-bordered region in Fig. 1b.

Figure 1b shows a 3X zoomed in image of the red-outlined region in Fig. 1a. This confirms that between the inner and outer zones is a relatively dark region. Figure 1c is a plot of greyscale intensity (averaged across the rectangle width) versus radius for the red-outlined region in Fig. 1b. The plot shows two distinct peaks, corresponding to the inner and outer zones.

Double luminous zones were visible in every IDF studied. NDFs were considered at otherwise similar conditions, and none exhibited a double luminous zone.

Partial premixing was performed to identify where premixed flames could be expected, as shown in Fig. 2. When O_2 was added to the ambient gas of the Fig. 2a flame, the outer luminous zone became more prominent (see Fig. 2b). When, in addition, CH_4 was added to the oxidizer stream, a third blue zone appeared inside the others (see Fig. 2c) that had the shape of a premixed Bunsen burner flame. Figures 2b and 2c are examples of double and triple flames [15,16]. These tests indicate that the double luminous zone structure of IDFs involves an inner zone that is near stoichiometric and an outer zone that is rich premixed.

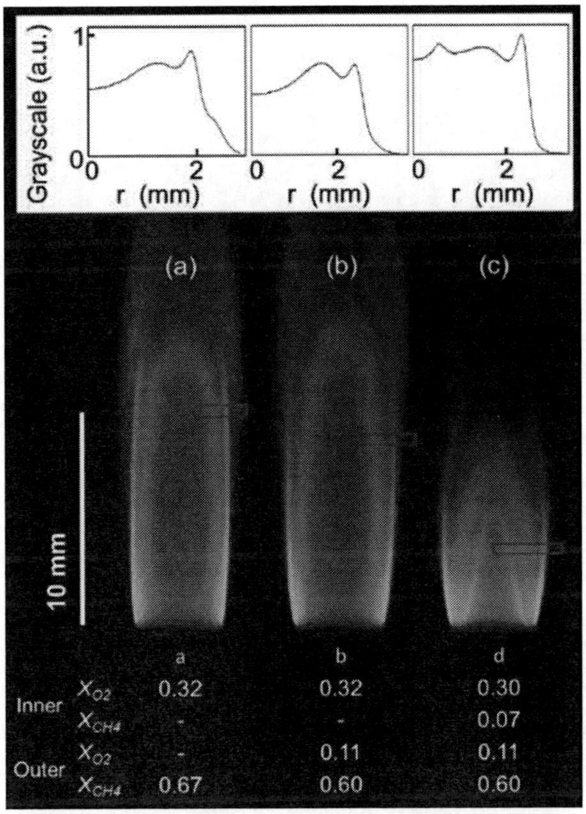

Fig. 2. Effects of partial premixing. Red-bordered regions indicate where the luminous zones are most distinct, and grayscales are plotted for these regions. (a) An image of an IDF with an O_2 flow rate of 6.06 mg/s and a CH_4 flow rate of 52 mg/s. (b) An image with O_2 added to the ambient gas. (c) An image with O_2 added to the ambient gas and CH_4 added to the burner gas.

This finding is supported by qualitative thin-filament pyrometry, which found the inner luminous zone to be hotter than the outer zone. Previous studies have found, using OH planar laser-induced fluorescence, that what appears to be the inner luminous zone of IDFs is nearly stoichiometric [17,18]. Although Refs. [7-10] used stainless burners, we observed double luminous zones with brass and quartz burners, suggesting that these luminous zones are not affected by burner material.

All the flame zones visible in Figs. 2a-2c have visible emissions at 430 nm. Emissions at 310 nm were detected in all luminous zones except the outermost zones in Figs. 2a-2c.

Double luminous zones in IDFs involving diluents other than N_2 are also studied. Figure 3 illustrates two representative IDF images with Ar and He as diluents. Both IDFs have clear double luminous zones.

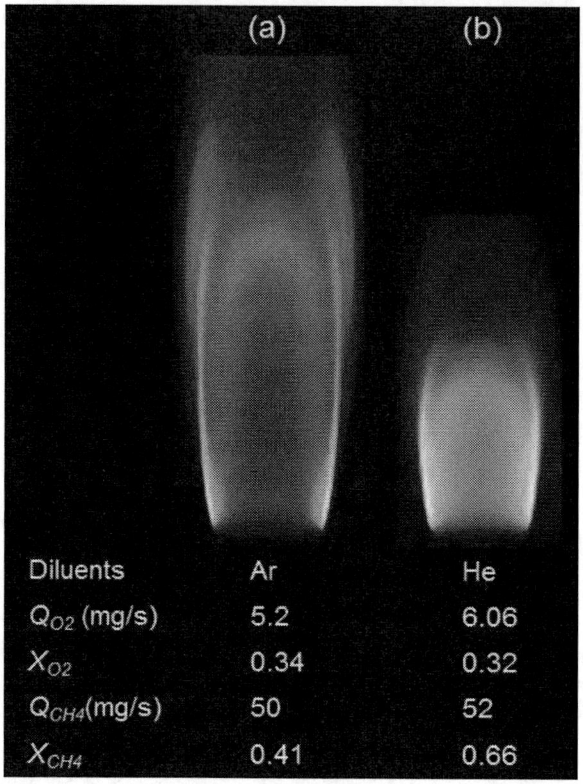

Fig. 3. Representative methane IDF images with (a) Ar and (b) He and as diluents. The corresponding O_2 flow rates (Q_{O2}), O_2 mole fractions (X_{O2}), CH_4 flow rates (Q_{CH4}), and CH_4 mole fractions (X_{CH4}) are also listed.

The mechanism(s) responsible for IDF double luminous zones is unknown, but several possibilities are considered here.

Significant reactant leakage at the flame base is possible [1-3]. Modeling predicts O_2 mole fractions of 0.04 – 0.05 on the fuel side of IDFs [19]. Gradients in concentration and velocity at the base are steeper than those in typical NDFs. Furthermore IDF flame anchor regions are generally narrower than the burner [20,21], in contrast to NDF behavior. If sufficient oxidizer leaks, a rich premixed flame could result, and Figs. 2b and 2c suggest this would have the shape of the outer luminous zone. This mechanism does not exist in spherical microgravity flames, where double luminous zones have not been observed [22, 23].

Reactant leakage across the stoichiometric contour is possible. Such leakage can be either oxygen [24] or fuel [5]. For the present IDFs, the outer and inner blue zones may correspond to the consumption of CH4 and CO/H_2, respectively.

The peaks in temperature and concentrations of products and radicals are never in perfect alignment in diffusion flames [21, 25, 26]. If these peaks are sufficiently separated, this could give rise to double luminous zones.

Soot luminosity is responsible for the orange and yellow emissions from IDFs The close proximity of soot luminosity to the outer luminous zone (see Fig. 1) suggests that the zone could be associated with soot precursors. However, the zone's color and presence inside the soot zone (Fig. 2a) is unexpected behavior of soot precursors. Furthermore the zone appears even in flames with no orange or yellow emissions.

Conclusions

In this work, double luminous zones in IDFs were studied. The inner flames were brighter with a complete teardrop shape, while the outer flames were open at the top. The inner flame was found to stoichiometric and the outer zone was a rich premixed flame. Several possible mechanisms are considered. These include reactant leakage at the flame base, reactant leakage across the stoichiometric contour, misalignment between the peaks in temperature and reaction rates, and soot luminosity.

Acknowledgements

This research was funded by NASA's International Space Station Research Program.

References

[1] K. Saito, F.A. Williams, A.S. Gordon, Structure of laminar coflow methane air diffusion flames, J. Heat Transfer 108 (1986) 640–648.

[2] F. Weinberg, F. Carleton, Ionization and chemiluminescence during the progressive aeration of methane flames, Combust. Flame 156 (2009) 2276–2284.

[3] F. Weinberg, F. Carleton, R. Houdmont, D. Dunn-Rankin, S. Karnani, Syngas formation in methane flames and carbon monoxide release during quenching, Combust. Flame 158 (2011) 273–280.

[4] P.H. Joo, M.R.J. Charest, C.P.T. Groth, Ö.L. Gülder, Comparison of structures of laminar methane–oxygen and methane–air diffusion flames from atmospheric to 60 atm, Combust. Flame 160 (2013) 1990–1998.

[5] C.J. Sung, D.L. Zhu, C.K. Law, On micro-buoyancy spherical diffusion flames and a double luminous zone structure of the hydrogen/methane flame, Symp. Int. Combust. 27 (1998) 2559–2566.

[6] V. Nayagam, D.L. Dietrich, F.A. Williams, A Burke-Schumann analysis of dual-flame structure supported by a burning droplet, Int. Commun. Heat Mass 87 (2017) 84-89.

[7] P.B. Sunderland, S.S. Krishnan, J.P. Gore, Effects of oxygen enhancement and gravity on normal and inverse laminar jet diffusion flames, Combust. Flame 136 (2004) 254–256.

[8] C.J. Unrau, R.L. Axelbaum, P. Biswas, P. Fraundorf, Synthesis of single-walled carbon nanotubes in oxy-fuel inverse diffusion flames with online diagnostics, Proc. Combust. Inst. 31 II (2007) 1865–1872.

[9] B.M. Kumfer, S.A. Skeen, R.L. Axelbaum, Soot inception limits in laminar diffusion flames with application to oxy-fuel combustion, Combust. Flame 154 (2008) 546–556.

[10] B. Wang, Q. Eri, T. Li, R. Duan, Experimental investigation on the effect of oxygen enhancement on radiation distribution in inverse diffusion flames, ASME 2016 International Mechanical Engineering Congress and Exposition, (2016).

[11] C.R. Shaddix, T.C. Williams, Measurements of the velocity field in laminar ethylene inverse jet diffusion flames, Combust. Flame 156 (2009) 942-945.

[12] F. Xu, X. Liu, D.T. Stephen, Synthesis of carbon nanotubes on metal alloy substrates with voltage bias in methane inverse diffusion flames, Carbon 44 (2006) 570-577.

[13] B. Stelzner, F. Hunger, A. Laugwitz, M. Gräbner, S. Voss, K. Uebel, M. Schurz, R. Schimpke, S. Weise, S. Krzack, D. Trimis, Development of an inverse diffusion partial oxidation flame and model burner contributing to the development of 3rd generation coal gasifiers. Fuel Process. Technol. 110 (2013.) 33-45.

[14] H. Guo, J.A. Castillo, P.B. Sunderland, Digital camera measurements of soot temperature and soot volume fraction in axisymmetric flames. Appl. Opt. 52 (2013) 8040-8047.

[15] F. Bisetti, S.M. Sarathy, M. Toma, S.H. Chung, Stabilization and structure of n-heptane tribrachial flames in axisymmetric laminar jets, Proc. Combust. Inst. 35 (2015) 1023–1032.

[16] M.K. Kim, S.H. Chung, H.H. Kim, Effect of AC electric fields on the stabilization of premixed bunsen flames, Proc. Combust. Inst. 33 (2011) 1137–1144.

[17] M.A. Mikofski, T.C. Williams, C.R. Shaddix, L.G. Blevins, Flame height measurement of laminar inverse diffusion flames, Combust. Flame 146 (2006) 63–72.

[18] F. Escudero, A. Fuentes, R. Demarco, J.L. Consalvi, F. Liu, J.C. Elicer-Cortés, C. Fernandez-Pello, Effects of oxygen index on soot production and temperature in an ethylene inverse diffusion flame, Exp. Therm Fluid Sci. 73 (2016) 101–108.

[19] K.T. Wu, R.H. Essenhigh, Mapping and structure of inverse diffusion flames of methane. Symp. Int. Combust. 20 (1984) 1925-1932.

[20] S.S. Krishnan, J.M. Abshire, P.B. Sunderland, Z.-G. Yuan, J.P. Gore, Analytical predictions of shapes of laminar diffusion flames in microgravity and earth gravity, Combust Theory Model. 12 (2008) 605-620.

[21] B. Stelzner, F. Hunger, S. Voss, J. Keller, C. Hasse, D. Trimis, Experimental and numerical study of rich inverse diffusion flame structure, Prog. Energy Combust. Sci. 34 (2013) 1045-1055.

[22] P.B. Sunderland, R.L. Axelbaum, D.L. Urban, B.H. Chao, S. Liu, Effects of structure and hydrodynamics on the sooting behavior of spherical microgravity diffusion flames, Combust. Flame 132 (2003) 25–33.

[23] P.B. Sunderland, D.L. Urban, D.P. Stocker, B.-H. Chao, R.L. Axelbaum, Sooting limits of microgravity spherical diffusion flames in oxygen-enriched air and diluted fuel, Combust. Sci. Technol. 176 (2004) 2143-2164.

[24] F.A. Williams, Progress in knowledge of flamelet structure and extinction, Prog. Energy Combust. Sci. 26 (2000) 657–682.

[25] C.S. Panoutsos, Y. Hardalupas, A.M.K.P. Taylor, Numerical evaluation of equivalence ratio measurement using OH∗ and CH∗ chemiluminescence in premixed and non-premixed methane–air flames. Combust. Flame 156 (2009) 273-291.

[26] S.A. Skeen, G. Yablonsky, R.L. Axelbaum, Characteristics of non-premixed oxygen-enhanced combustion: I. The presence of appreciable oxygen at the location of maximum temperature, Combust. Flame 156 (2009) 2145–2152.

Spring Technical Meeting
Eastern States Section of the Combustion Institute
March 4-7, 2018
State College, Pennsylvania

Low-temperature multistage diffusion flames

Omar R. Yehia[1,], Christopher B. Reuter[1], Yiguang Ju[1]*

[1]*Department of Mechanical and Aerospace Engineering, Princeton University, Princeton, NJ 08544, USA*
[*]*Corresponding Author Email: oyehia@princeton.edu*

Abstract: We demonstrate the existence of steady low-temperature multistage diffusion flames in the counterflow configuration and examine their character as well as that of premixed flames. The character of both diffusion and premixed multistage flames is such that intermediates are produced and consumed in separate oxidation layers. For hydrocarbon fuels in thermodynamic conditions of practical interest, there exist three distinct temperature-dependent classes of reactions that dictate three different oxidation stages, namely low-temperature peroxy chemistry, intermediate-temperature chemistry, and high-temperature small radical chemistry. For ethers of practical interest, such as diethyl and dibutyl ether, the intermediate-temperature stage is pronounced owing to the additional contribution of the intermediates produced from the scission of the fuel radical, along with the intermediates produced from the decomposition of ketohydroperoxide. For both non-premixed and premixed flames, multistage flames exist for intermediate Damköhler numbers, limited by the classical high-temperature hot flame in the limit of large Damköhler number and a single-stage low-temperature cool flame in the limit of small Damköhler number approaching that of extinction. An abrupt transition between the two-stage flame and single-stage hot flame occurs, whereas a smooth transition between both flames occur for rich premixed flames, for the conditions considered.
Keywords: Cool flame; low-temperature chemistry, multistage diffusion flame, counterflow.

1. Introduction

The study of the chemical and asymptotic [1] structure of flames has profoundly influenced the current understanding of flames. Such studies have revelated that different reaction layers, in which different sets of chemistries occur, are defining characteristics of all real flames. Further, since all real fuels of practical interest are governed by complex kinetics, detailed studies of the role of finite-rate kinetics on flame structure and dynamics has become indispensable in the effort to design next-generation advanced engines. The design of advanced compression ignition engines has necessitated a fundamental understanding of the low-temperature reactivity of real fuels. Such low-temperature reactivity has been shown to manifest in low-temperature cool flames and to significantly influence engine combustion dynamics [2]. Specifically, studies of premixed hydrocarbon flames at low temperatures (300K – 900K) have revealed that, owing to the characteristically slow chain-branching sequence of hydrocarbons at low temperatures (peroxy chemistry), multiple distinct oxidation layers can exist and can support multiple co-existent stages of heat release in premixtures. While such premixed character has been known to exist [3], both the chemistry and structure of such flames were not well-understood. Recently, owing to advances in optical diagnostics and in the understanding of low-temperature chemistry [4], such configurations have received renewed interest [5] and have revealed much regarding the role of different temperature-dependent classes of reactions on the existence of such flames, although the conclusions derived from such studies have been limited as the kinetic-transport coupling, which is a fundamental characteristic of all flames, is obscured due to the significant flame-boundary coupling interactions.

Consequently, there remain a variety of outstanding fundamental questions regarding the nature of such flames including the chemical character and dynamics of multistage premixed flames, and whether non-premixed flames of such character can even exist. The latter question is motivated by an abundance of recent studies regarding non-premixed low-temperature flames at both terrestrial [6–8] and microgravity conditions [9,10], none of which have revealed such phenomena. Therefore, it is the purpose of this work to demonstrate the existence of low-temperature multistage diffusion flames at terrestrial and atmospheric conditions and examine their character in relation to that of multistage premixed flames. In premixed flames, the temperature profile dictates the existence of each stage and its coupling with the other stages. For diffusion flames, owing to the initially unmixed reactants, an additional complexity arises. Specifically, partial fuel oxidation in the low-temperature stage, which is characteristically slow, produces intermediates and a considerable amount of reactant leakage through the first stage. For intermediate Damköhler numbers, the intermediates can then be oxidized in the second stage and repopulate the radical pool that, in turn, promoting secondary fuel consumption. Finally, for sufficiently large Damköhler number, the flame temperature becomes high enough such that small radical and hydrogen peroxide chemistry become sufficiently fast to result in the classical high-temperature flame.

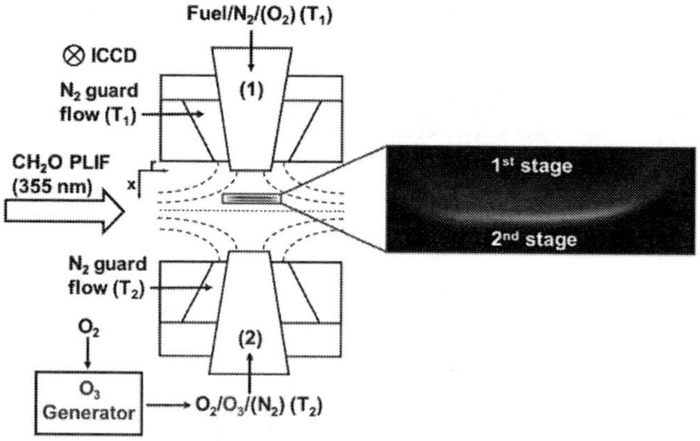

Figure 1: Experimental apparatus and chemiluminescence (f/4.5 lens with six-second exposure) image of low-temperature multistage diffusion flame.

2. Experimental and computational methods

The experimental apparatus includes a counterflow burner, as depicted in Fig. 1. A detailed account of the experimental setup and optical diagnostics is provided elsewhere [11] and will only be briefly reviewed here. The upper nozzle, augmented with a vaporization chamber, is heated to a fixed temperature T_1 (550K in this work), through which a mixture of fuel/nitrogen (non-premixed) or fuel/oxygen mixtures (premixed) are introduced. The mixture is then ejected through the nozzle to impinge upon another stream (stream 2), which is maintained at ambient conditions. The second stream is an oxidizer (O_2/O_3) in the non-premixed cases, and an inert (nitrogen) in the premixed cases. The burner separation is maintained at 2.25 cm throughout the study. Optical diagnostics are performed using unfiltered chemiluminescence measurements and filtered planar-laser induced fluorescence of the formaldehyde molecule, both of which are captured onto an intensified CCD. The laser-induced fluorescence is performed by targeting the 4_0^1 absorption band in the $\tilde{A}^1A_2 \leftarrow \tilde{X}^1A_1$ system near 355 nm using the output of a pulsed Nd:YAG laser (Quantel, Q-smart 850) at 5 Hz and ~200 mJ/pulse with a 200-μm sheet along the nozzle centerline. The fluorescence signal was then filtered through a 390-nm long-pass optical filter. Finally, intrusive thermometry was performed using a 254-μm K-type thermocouple along the nozzle centerline. We consider diethyl (DEE) and dibutyl (DBE) ethers as fuels, which are promising biofuels due to their high cetane numbers and energy densities. Pure and ozone-doped oxygen are used as oxidizers.

Axisymmetric flow in the counterflow configuration is modeled by transforming and discretizing the steady conservation equations of mass, momentum, energy, and species using OPPDIF [12] in CHEMKIN. The resulting one-dimensional equations are closed with an appropriate equation of state and a kinetic model for the chemical source term [13,14]. Experimental and computational conditions are compared using the hydrodynamic strain rate [15], defined on the oxidizer side, which is a function of inlet conditions only.

3. Results and discussion

3.1. Multistage premixed flames

We begin by first considering the different types of flame structures that are dictated by mixture stoichiometry and flame temperature. Starting from a stoichiometric condition and slightly in excess ($\phi \approx 2.5$), the flame temperatures are high enough such that small radical chemistry and hydrogen-oxygen branching completely dominate the flame chemistry, and a single-stage of heat release is observed. Such a flame is the classical high-temperature flame in which most of the reactants are converted to the carbon dioxide and water. As the mixture is made richer ($\phi \approx 4.0$), the flame temperature drops dramatically, and multiple distinct oxidation stages occur, with distinct peaks in heat release (not shown) and temperature (Fig. 2). Further increase in fuel concentration reduces the flame temperature until the high-temperature (third stage) ceases ($\phi \approx 19$) after which the intermediate-temperature stage (second stage) further ceases ($\phi \approx 27$). The first stage, supported by peroxy chemistry, persists for even larger equivalence ratios, where the extinction condition is dictated by the boundary temperatures. Interestingly, the analysis suggests that a single-stage low-temperature cool flame is, in fact, a near limit structure, and that the dominant mode of low-temperature burning is multistage in character (two-stage or three-stage).

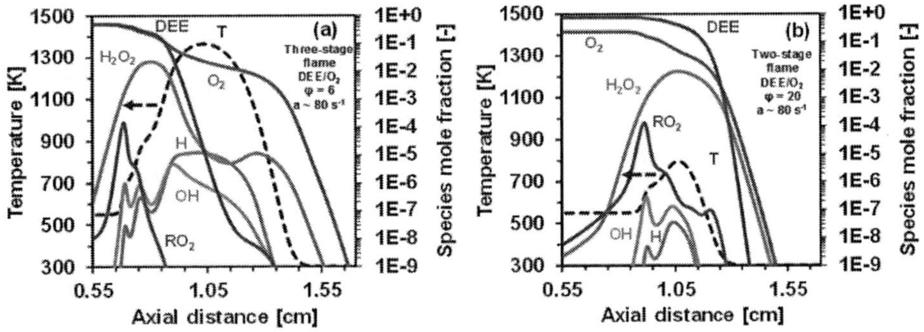

Figure 2: Computed DEE/O_2 premixed flame structure of (a) three-stage flame, (b) two-stage flame.

By examining the flame structure and the rate of heat and hydroxyl production, the different classes of reactions that support each stage become clear, as shown in Figs. 2 – 3. Briefly, the first stage is supported predominately by the low-temperature peroxy branching sequence, in which the fuel undergoes H-abstraction to form the fuel radical that is then oxidized to form the peroxy radical. The peroxy radical then isomerizes and undergoes secondary oxidation to eventually produce ketohydroperoxide and a hydroxyl radical. The ketohydroperoxide further decomposes to liberate a second hydroxyl radical and a smaller intermediate. The decomposition of ketohydroperoxide produces the second radical needed for the peroxy branching sequence but also produces the intermediates needed for oxidation in the second stage. The oxidation of such intermediates replenishes the radical pool farther from the fuel, which acts as a radical sink. In the case of ethers, a significant contribution to both the first and second oxidation stages is also from the scission of the fuel radical. Specifically, owing to the weak carbon-oxygen bond, the fuel radical can readily undergo scission to produce intermediates that can be oxidized in the second stage as shown in Fig. 3, which we refer to as the side-branch chemistry.

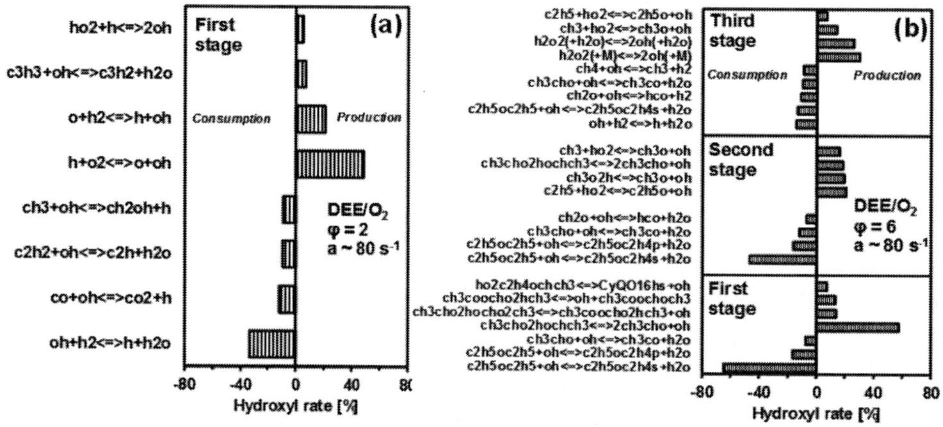

Figure 3: Computed hydroxyl rate contributions in (a) single-stage high-temperature flame, (b) three stage flame.

Of particular importance is the role of hydrogen peroxide in resulting in chain-branching in the third oxidation stage. From Figs. 2 – 3, the decomposition of hydrogen peroxide, which is produced primarily from the recombination of the hydroperoxyl radical, only supports the third (high-temperature) stage. The abrupt and significant consumption decomposition of hydrogen peroxide only occurs in the third stage, where the local temperature is sufficiently high (in excess of 1000K). That is, in the first and second stages, where the maximum temperature is at most around ~900K, the decomposition of hydrogen peroxide is sufficiently slow to prevent autocatalytic oxidation of the fuel. Instead, the radical pool is primarily replenished from intermediate oxidation and fuel peroxy chemistry from the first and the second stages.

3.2. Multistage diffusion flames

Having established the chemical character of multistage premixed flames, we now demonstrate, perhaps for the first time, that such chemical character can also manifest in non-premixed flames. Specifically, multistage diffusion flames are shown to exist for Damköhler numbers far away from that of extinction. However, for sufficiently large Damköhler numbers, an abrupt ignition to a hot diffusion flame occurs, unlike the continuous transition in the case of initially premixed reactants. As previously described, the case of initially unmixed reactants adds additional complexity to the flame structure, as all three stages are coupled both through flame temperature and through the migration of reactants across the different reaction stages. For the same boundary conditions mentioned previously (atmospheric pressure, T_I = 550K), multistage diffusion flames of DBE were established in the counterflow configuration for the first time.

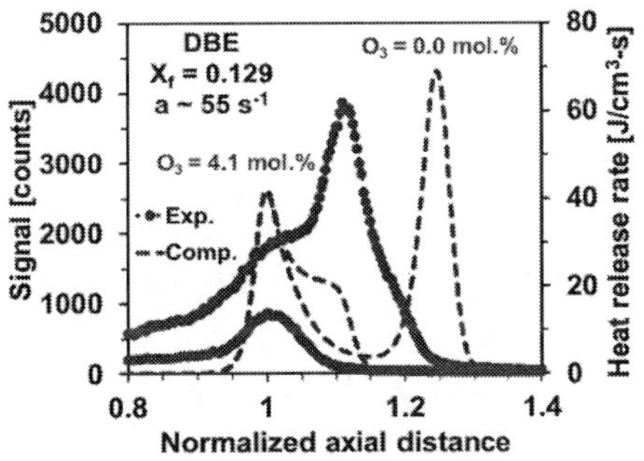

Figure 4: Measured chemiluminescence signal and computed heat release rate profiles for low-temperature multistage diffusion flames of DBE.

Chemiluminescence along the burner centerline along with computed values of the heat release rate are shown in Fig. 4. We note that both computations reveal a multistage character of flames with and without ozone doping. However, direct chemiluminescence imaging of the flames, without the use of ozone, will not show two distinct stages since the chemiluminescence signal is almost exclusively electronically-excited formaldehyde. Whereas, in the case of ozone-doped flames, the radical concentration is much higher in the second stage (more than an order of magnitude), such that chemiluminescence of other species, including excited hydroxyl, is also visible. Further, it is also clear that ozone-doping does not influence the oxidation in the first stage, as its heat release rate is identical to the case without ozone. However, ozone-doping dramatically promotes both the heat release rate and the hydroxyl radical pool (not shown) as it significantly accelerates the consumption of the second-stage intermediates.

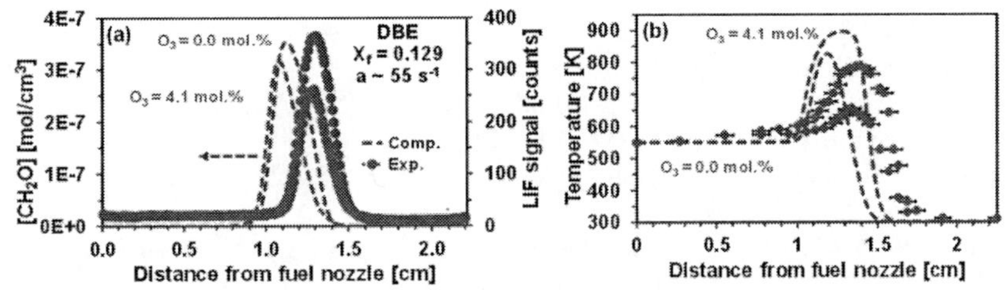

Figure 5: (a) Formaldehyde laser-induced fluorescence signal and (b) temperature profiles of DBE/O_2 flames with and without ozone doping.

To further examine the structure of such flames, we examine the formaldehyde structure and the temperature of such flames using laser-induced fluorescence and thermocouple thermometry, as shown in Figure 5. It is clear that ozone-doping indeed promotes intermediate oxidation, as is evident by the reduction in both the formaldehyde maximum concentration and the formaldehyde profile width. Specifically, ozone-doping accelerates the formation of the formyl radical and subsequently carbon monoxide by consuming formaldehyde produced in the first stage.

Finally, we conclude by relating the current findings to configurations of practical interest. The multistage character of non-premixed and premixed flames is particularly relevant important in flows in highly turbulent or microgravity environments, where long residence times are possible. For both non-premixed and premixed cases, the multistage behavior described likely controls the transition from the low-temperature and intermediate-temperature burning modes to the classical high-temperature hot flame.

4. Conclusions

We experimentally demonstrate the existence of low-temperature multistage diffusion flames and examine the structure and character of premixed multistage flames of diethyl and dibutyl ether. Multistage diffusion flames have been established, for the first time, in the counterflow configuration and were interrogated using thermometry, chemiluminescence imaging, laser-induced fluorescence. Further, for the non-premixed configuration, ozone-doping of the oxidizer side promotes the oxidation of the intermediates in the second stage. For thermodynamic conditions of practical interest, three distinct temperature-dependent classes of reactions dictate the existence of three distinct oxidation zones in multistage flames. The first stage is governed by low-temperature peroxy chemistry, the second by intermediate-temperature chemistry, and the third by conventional high-temperature small radical chemistry. For both premixed and non-premixed configurations, multistage flames can exist for intermediate Damköhler numbers, bounded by the classical single-stage high-temperature flame in the limit of large Damköhler number, and a single-stage low-temperature cool flame in the limit of low Damköhler number.

5. Acknowledgements

The authors acknowledge financial support from the NSF grant CBET-1507358, NASA ISS Post-Graduate award NNX15AB67G, and NASA microgravity grant NNX16AK07G. ORY is grateful for support by the Daniel and Florence Guggenheim Foundation fellowship at Princeton University. CBR is supported by the NDSEG fellowship program of the U.S. Department of Defense.

6. References

[1] J.D. Buckmaster, et al., Proc. Combust. Inst. 30 (2005) 1–19.
[2] Y. Ju, W. Sun, M.P. Burke, X. Gou, Z. Chen, Proc. Combust. Inst. 33 (2011) 1245–1251.
[3] W.G. Agnew, J.T. Agnew, Symp. Combust. 10 (1965) 123–138.
[4] J. Zádor, C.A. Taatjes, R.X. Fernandes, Prog. Energy Combust. Sci. 37 (2011) 371–421.
[5] S. Kikui et al., Proc. Combust. Inst. 35 (2015) 3405–3412.
[6] S.H. Won, B. Jiang, P. Diévart, C.H. Sohn, Y. Ju, Proc. Combust. Inst. 35 (2015) 881–888.
[7] S. Deng, D. Han, C.K. Law, Combust. Flame 176 (2017) 143–150.
[8] V.R. Katta, W.M. Roquemore, Proc. Combust. Inst. 0 (2016) 1–8.
[9] V. Nayagam, D.L. Dietrich, P. V. Ferkul, M.C. Hicks, F.A. Williams, Combust. Flame 159 (2012) 3583–3588.
[10] T.I. Farouk, D.L. Dietrich, F.E. Alam, F.L. Dryer, Proc. Combust. Inst. 0 (2016) 1–8.
[11] J.K. Lefkowitz, S.H. Won, Y. Fenard, Y. Ju, Proc. Combust. Inst. 34 (2013) 813–820.
[12] A.E. Lutz, R.J. Kee, J.F. Grcar, F.M. Rupley, OPPDIF: A Fortran Program for Computing Opposed-Flow Diffusion Flames, Albuquerque, NM, and Livermore, CA (United States), 1997.
[13] Z. Tang, L. Zhang, X. Chen, G. Tang, Energy & Fuels 31 (2017) 2803–2813.
[14] L. Cai et al., Combust. Flame 161 (2014) 798–809.
[15] K. Seshadri, F.A. Williams, Int. J. Heat Mass Transf. 21 (1978) 251–253.

Sub Topic: Reaction Kinetics

Spring Technical Meeting
Eastern States Section of the Combustion Institute
March 4–7, 2018
State College, Pennsylvania

Multi-modal counterflow flames under autoignitive conditions

Temistocle Grenga[1,], Jonathan F. MacArt[1], and Michael E. Mueller[1]*

[1]*Department of Mechanical and Aerospace Engineering, Princeton University, Princeton, 08544 NJ, USA*

[*]*Corresponding author:* tgrenga@princeton.edu

Abstract: In practical systems, combustion occurs in multi-model regimes rather than in asymptotic limits of nonpremixed flames, premixed flames, or autoignition as commonly assumed in turbulent combustion models. In canonical configurations, multi-modal combustion is critical in the stabilization of lifted jet flames. At low temperatures, stabilization occurs kinetmatically through a "triple" flame with regions of both premixed and nonpremixed combustion. At high temperatures, autoignition is activated, and the role of autoignition versus premixed flame propagation for flame stabilization depends on the residence times and flow speed. While detailed simulations of laminar lifted jet flames are computationally tractable, for turbulent lifted jet flames, the lift-off heights require very large computational domains, and simulations of such flames with DNS becomes extremely expensive. Therefore, in work, the counterflow configuration is investigate as a more compact alternative. A series of detailed simulations of DME/air counter flames at elevated pressure are performed spanning a range of boundary conditions in both streams. The counterflow configuration is shown to exhibit multi-reaction zone structures, that is double and triple "flame" structures, analogous to lifted jet flames but only if the fuel and oxidizer streams are changes from pure components to partially premixed components. In other words, fuel/air boundary conditions gives only one reaction zone, but, under certain conditions, rich/lean boundary conditions can give up to three reaction zones.
Keywords: ***Multi-modal combustion, Counterflow, DME, G-Scheme***

1. Introduction

Multi-modal regimes are how combustion phenomena manifest themselves in practical systems. The asymptotic regimes, nonpremixed and premixed combustion, has been largely, and quite successfully, used as simplified models in numerical calculation. However, to respond to the current request of accuracy and reliability of predictive numerical models it is necessary to deeply investigate the multi-modal combustion phenomena and create more sophisticated models that can be used in nowadays super-computers.

Multi-modal combustion regimes, in the form of two-dimensional tribrachial structures, has been observed in nonpremixed laminar lifted flames at nonautoignitive conditions [1] where the local flame propagation speed is balanced by the flow speed. At elevated pressures and temperatures, autoignition phenomena may change the thermal and chemical structure of these tribrachial structures, as well as the stabilization mechanism. This is compounded by the fact that many fuels exhibit complex autoignition behavior due to the competition between low- and high-temperature chemistry [2]. A computational investigation by Krisman *et al.* [3] revealed that multibrachial thermal structures were present in nonpremixed DME/air flames at 40 atm and 700-1500 K, in which nonpremixed combustion, premixed combustion, low-temperature ignition, and high-temperature ignition all occur to varying degree. Deng *et al.* investigated the negative temperature coefficient regime at low temperature experimentally for nonpremixed flames in the counterflow configuration [4].They further found that the presence of these modes varied with velocity and temperature in laminar nonpremixed coflow flames of DME/air [5]. Numerical simulations of multi-modal combustion featuring autoignition in turbulent regimes is extremely computationally expensive even at relatively low Reynolds number [6], and this strongly limits the range of parameters that can be explored. Therefore, a more compact counterflow configuration is considered.

DME oxidation has been extensively studied both experimentally and computationally to investigate ignition delay time and emissions under practical engine conditions [7]. Low temperature oxidation was studied over temperature and pressure ranges of 550-850 K and 12-18 atm, respectively, by Curran *et al.* [8], and high temperature oxidation was studied at atmospheric pressure by Fischer *et al.* [9]. Zheng *et al.* [10] studied ignition in counterflow flames of nitrogen diluted DME and heated air at pressures from 1.3 to 3 atm.

In the present study, a series of partially premixed DME/air laminar counterflow flames at elevated pressure are computationally investigated. The simulations are performed for several mixtures. The aim is to investigate multi-modal combustion regimes in the counterflow configuration in comparison with the coflow configuration, specifically, quantifying the dominant chemical pathways using the G-Scheme to differentiate between premixed flame propagation and autoignition. The compactness of the counterflow geometry will allow a study of the multi-modal combustion regimes, for turbulent cases, with a computational cost at least one order of magnitude lower than the coflow configuration.

2. DNS configurations

The coflow case used for comparison is the same configuration investigated by Deng *et al.* [11]. A central axisymmetric DME stream at 300 K is surrounded by a air coflow stream at 900 K, both streams are at 30 atm. The fuel nozzle diameter is 0.8 mm, while the coflow is 3.9 mm, they are separated by a no-slip wall with thickness 0.04 mm.

The counterflow geometry presents one inlet for each side with height H =0.8 mm and length $2H$, while the combustion chamber is $3H$ length and $5H$ high. A DME skeletal mechanism of 39 species and 175 reactions [12] in several configurations, including purely nonpremixed and premixed flames, which are not discussed here. In Table 1, the temperature and mixtures compositions are reported for the cases analyzed in this work.

Additional numerical simulations of DME/air counterflow flames have been conducted using FlameMaster [13] in a 1D domain with the same configurations used for the multi-dimensional simulations and reported in Table 1.

3. Numerical methods

The equations for conservation of mass, momentum, energy, and species are solved in the low Mach number limit using NGA, a semi-implicit, iterative algorithm of Desjardins *et al.* [14]. The species and temperature equations are evolved monolithically in time using a linearly implicit, approximately factorized scheme with the exact chemical Jacobian [15].

Case	$x = 0.0$ cm T (K)	$x = 0.0$ cm Composition	$x = 5.6$ cm T (K)	$x = 5.6$ cm Composition
FA	300.0	$Y_{DME} = 0.250$, $Y_{O_2} = 0.000$, $Y_{N_2} = 0.750$	900.0	$Y_{DME} = 0.000$, $Y_{O_2} = 0.232$, $Y_{N_2} = 0.768$
RA	460.8	$Y_{DME} = 0.183$, $Y_{O_2} = 0.190$, $Y_{N_2} = 0.627$	900.0	$Y_{DME} = 0.000$, $Y_{O_2} = 0.232$, $Y_{N_2} = 0.768$

Table 1: Description of the test cases for DME/air counterflow flames at 30 atm, rich flame ($\phi = 2.0$), air and DME diluted with nitrogen.

4. The G-Scheme analysis

The analysis of multi-scale physics, such as combustion chemistry, is a challenging task because of the simultaneous contribution of a wide range of time scales to the system dynamics. However, it is typical that the dynamics can develop very slow and very fast time scales separated by a range of active time scales. The G-Scheme is a time accurate explicit solver [16] that exploits all opportunities for on-line reduction of stiff systems, which arise when the gaps in the fast/active and slow/active time scales become sufficiently large. In the G-Scheme, it is assumed that the dynamics are (locally) decomposed into active ($\mathbb{A}$), slow ($\mathbb{H}$), fast ($\mathbb{T}$), and, when applicable, invariant subspaces ($\mathbb{E}$).

The characterization of the local tangent space structure of the subspaces can be of great significance in the analysis of the dynamics with the aim of achieving a low-dimensional description. The *G-Scheme Participation Index* (GPI) [17] can be used to identify the reactions most contributing to the local dynamics. The *CSP Participation Index* [18] provides a measure of the contribution of the k-th reaction to the i-th mode at the state $\mathbf{x}$ normalized with respect to the contribution of all reactions at the state $\mathbf{x}$:

$$P_k^i(\mathbf{x}) = \frac{|C_k^i(\mathbf{x})r^k(\mathbf{x})|}{\sum_{k'=1}^{2N_r} \left|C_{k'}^i(\mathbf{x})r^{k'}(\mathbf{x})\right|}, \qquad C_k^i = \mathbf{b}^i(\mathbf{x}) \cdot \mathbf{S}_k, \qquad 0 \leq P_k^i(\mathbf{x}) \leq 1,$$

where $\mathbf{b}^j$ are contravariant basis vectors, $\mathbf{S}_k$ and r^k are the stoichiometric vector and the rate associated with the k-th reaction, and k ranges from 1 to the number of reversible reactions, N_r. The GPI provides a measure of the contribution that the k-th reaction contributes to the subspace, normalized with respect to the contributions of all reactions to that subspace and is defined as:

$$\mathscr{P}_k^s = \sum_{i=N_{s,b}}^{N_{s,e}} w_i^s P_k^i,$$

where s identifies any of the subspaces $\mathbb{A}$, $\mathbb{H}$, and $\mathbb{T}$, $N_{s,b}$ and $N_{s,e}$ identify the first and last mode, respectively, of the subspace, and w_i^s is the corresponding weight for the subspace. The associated weights are given by

$$w_i^{\mathbb{A}} := \frac{(g^i)^2}{\sum_{j=N_{\mathbb{A},b}}^{N_{\mathbb{A},e}} (g^j)^2} \qquad \text{and} \qquad w_i^{\mathbb{H}/\mathbb{T}} := \frac{1}{N_{\mathbb{H}/\mathbb{T},e} - N_{\mathbb{H}/\mathbb{T},b} + 1}.$$

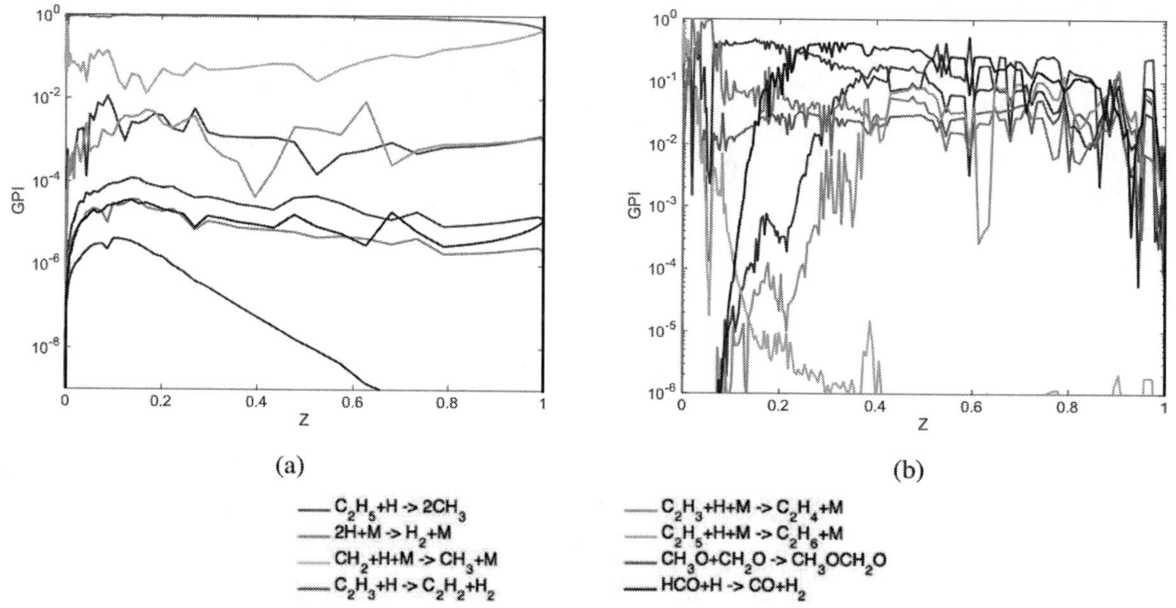

Figure 1: GPI values along the mixture fraction for (a) the lifted tribrachial flame, and (b) the counterflow geometry case FA.

The weight $w_i^{\mathbb{A}}$ of each active mode involves the mode amplitude g^i squared: this definition gives more weight to energy-carrying modes. For the fast and slow subspaces, the mode amplitudes cannot be used as weights since, by definition, the fast mode amplitudes should be zero (or nearly so), and the slow mode amplitudes should be nearly zero.

5. Results and conclusions

Figure 1 shows the GPI values in the Active subspace for both configuration studied. The data in the coflow case has been collected along the curve determined by the gradient of the mixture fraction, while for the counterflow case the data has been obtained from the 1D simulation described above which correspond to the symmetry plane in the multidimensional simulation. The same set of reactions constitute the skeleton of the chemical path for both case. However, the production/consumption of intermediate species is constant in the coflow configuration (Fig. 1(a)), while in the counterflow case the relative importance of reactions changes with Z. This latter behaviour is similar to the results obtained in the analysis of the GPI along a curve having a constant mixture fraction value in the coflow configuration. The reason is the different orientation of the flame in the flame in the two configuration with respect to the gradient of the mixture fraction. In the coflow, the flame is normal to the mixture fraction gradient so the diffusive effects and history effects can be seen only along a curve having a constant mixture fraction value.

Figure 1 shows the GPI values in the Active subspace for counterflow configuration in the case FA and RA as listed in Table 1. The rich mixture inlet on the left side of the domain (Fig. 2(b)) is highlighted by the high values of the GPI for the reactions involving the hydrocarbons and hydrogen (yellow and red lines). The presence of the premixed flame at $x = 1.5$ cm makes the grow of the GPI for the other reactions. Similar behaviour has been found in the coflow configuration

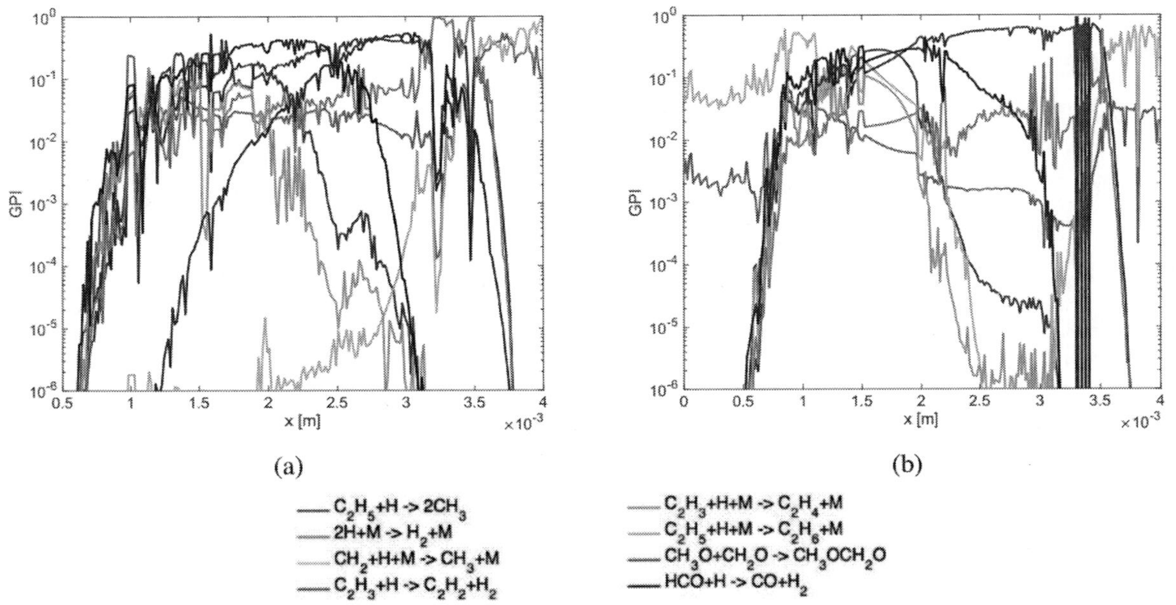

Figure 2: GPI values along the mixture fraction gradient for counterflow geometry (a) case FA, and (b) case RA.

along curves having the a mixture fraction larger than the stoichiometric value.

Mixing between air and fuel (or lean and rich mixtures) occurs in the direction of mixture fraction gradient in the counterflow configuration, and in the normal direction in the coflow case. The analysis with the GPI proves that the chemistry paths in the two configuration are similar in the mixing direction. This suggests that a reduced order manifold has to consider the different orientation of the mixture fraction gradient in the two configurations.

6. Acknowledgements

The authors gratefully acknowledge valuable support in the form of computational time on the TIGRESS high performance computer center at Princeton University, which is jointly supported by the Princeton Institute for Computational Science and Engineering (PICSciE) and the Princeton University Office of Information Technology's Research Computing department.

References

[1] J. Buckmaster, Edge-flames, Progress in Energy and Combustion Science 28 (2002) 435–475.

[2] C. K. Law and P. Zhao, NTC-affected ignition in nonpremixed counterflow, Combustion and Flame 159 (2012) 1044–1054.

[3] A. Krisman, E. R. Hawkes, M. Talei, A. Bhagatwala, and J. H. Chen, Polybrachial structures in dimethyl ether edge-flames at negative temperature coefficient conditions, Proceedings of the Combustion Institute 35 (2015) 999–1006.

[4] S. Deng, P. Zhao, D. Zhu, and C. K. Law, NTC-affected ignition and low-temperature flames in nonpremixed DME/air counterflow, Combustion and Flame 161 (2014) 1993–1997.

[5] S. Deng, P. Zhao, M. E. Mueller, and C. K. Law, Autoignition-affected stabilization of laminar nonpremixed DME/air coflow flames, Combustion and Flame 162 (2015) 3437–3445.

[6] Y. Minamoto and J. H. Chen, {DNS} of a turbulent lifted {DME} jet flame, Combustion and Flame 169 (2016) 38–50.

[7] H. Yamada, M. Yoshii, and A. Tezaki, Chemical mechanistic analysis of additive effects in homogeneous charge compression ignition of dimethyl ether, Proceedings of the Combustion Institute 30 (2005) 2773–2780.

[8] H. Curran, S. Fischer, and F. Dryer, The reaction kinetics of dimethyl ether. II: Low-temperature oxidation in flow reactors, International Journal of Chemical Kinetics 32 (2000) 741–759.

[9] S. Fischer, F. Dryer, and H. Curran, The reaction kinetics of dimethyl ether. I: High-temperature pyrolysis and oxidation in flow reactors, International Journal of Chemical Kinetics 32 (2000) 713–740.

[10] X. Zheng, T. Lu, C. Law, C. Westbrook, and H. Curran, Experimental and computational study of nonpremixed ignition of dimethyl ether in counterflow, Proceedings of the Combustion Institute 30 (2005) 1101–1109.

[11] S. Deng, P. Zhao, M. E. Mueller, and C. K. Law, Stabilization of laminar nonpremixed DME/air coflow flames at elevated temperatures and pressures, Combustion and Flame 162 (2015) 4471–4478.

[12] A. Bhagatwala, Z. Luo, H. Shen, J. A. Sutton, T. Lu, and J. H. Chen, Numerical and experimental investigation of turbulent DME jet flames, Proceedings of the Combustion Institute 35 (2015) 1157–1166.

[13] H. Pitsch, *FlameMaster, A C++ computer program for 0D combustion and 1D laminar flame calculations.*

[14] O. Desjardins, G. Blanquart, G. Balarac, and H. Pitsch, High order conservative finite difference scheme for variable density low Mach number turbulent flows, Journal of Computational Physics 227 (2008) 7125–7195.

[15] J. F. MacArt and M. E. Mueller, Semi-implicit iterative mehtods for low Mach number turbulent reacting flow: Operator splitting versus approximation factorization, Journal of Computational Physics 326 (2016) 569–595.

[16] M. Valorani and S. Paolucci, The G-Scheme: A framework for multi-scale adaptive model reduction, Journal of Computational Physics 228 (2009) 4665–4701.

[17] T. Grenga, S. Paolucci, and M. Valorani, A G-Scheme based analysis methodology of combustion processes, Combustion and Flame 189C (2018) 275–287.

[18] D. Goussis and S. Lam, A Study of Homogeneous Methanol Oxidation Kinetic Using CSP, Proc. Comb. Inst. 24 (1992) 113–120.

Spring Technical Meeting
Eastern States Section of the Combustion Institute
March 4-7, 2018
State College, Pennsylvania

Transient Interactions between a Premixed Double Flame and a Vortex

Christopher B. Reuter[1,], Viswanath R. Katta[2], Omar R. Yehia[1], Yiguang Ju[1]*

[1]Department of Mechanical and Aerospace Engineering
Princeton University, Princeton, NJ 08544, USA
[2]Innovative Scientific Solutions, Inc.,
7610 McEwen Road, Dayton, OH 45440, USA
[*]Corresponding Author Email: cbreuter@princeton.edu

The interaction between a laminar flame and a vortex is an important study for understanding the fundamentals of turbulent combustion. In the past, however, flame-vortex interactions have been investigated only for high-temperature flames. In this study, the impact of a vortex on a premixed double flame, which consists of a coupled cool flame and a hot flame, is examined experimentally and computationally using dimethyl ether/oxygen/ozone mixtures. The double flame is first shown to occur near the extinction limit of the hot flame. The transient interactions between double flames and impinging vortices are then investigated experimentally using a micro jet and numerically in two-dimensional transient modeling. It is seen that the vortex can extinguish the near-limit hot flame locally, resulting in a lone cool flame. It is found that there can be up to three different transient flame structures coexisting: an extinguished flame hole, a cool flame, and a double flame. Moreover, flame curvature is shown to play an important role in determining whether the vortex weakens or strengthens the cool flame and double flame.
Keywords: Flame-vortex interaction, counterflow premixed flame, double flame

1. Introduction

Flame-vortex interactions can be considered the building blocks of turbulent combustion theory [1]. Turbulence itself consists of many vortices operating on different length scales and time scales, and turbulent combustion is often modeled by imposing thin laminar flamelets onto a turbulent flow field [2, 3]. Thus, fundamental interactions between a laminar flame and a single vortex have received the attention of many studies [4-7]. These investigations have provided important insights into the role of flame stretch [5], local extinction and re-ignition [6], and preferential diffusion [7]. Nevertheless, all of these studies have been limited to high-temperature flames and, for the most part, simple fuels such as hydrogen and methane.

Many real transportation fuels, however, possess complex chemical behavior that varies significantly at high and low temperatures [8]. This chemical complexity results in the two-stage ignition behavior characteristic of jet and diesel fuels. While the second-stage high-temperature ignition chemistry centers upon a well-known set of reactions involving small radicals, the first-stage ignition chemistry is a much less understood process. Under real engine conditions, low-temperature chemistry is responsible for many important phenomena such as ignition-assisted flame propagation [9], cool flame waves [10], and multi-stage flames [11, 12]. Therefore, it is important to examine how these more complicated flame structures interact with a vortex.

This study examines the vortex-flame interactions of a particularly unique and near-limit flame structure—a cool premixed flame and a hot premixed flame coupled together as a double flame. The steady-state double flame is first investigated in a 1-D laminar counterflow setup. The differences between traditional hot flames, double flames, and cool flames are noted. The impact of an impinging vortex on the double flame structure is then studied. Flame-vortex interactions are explored through experiments and 2-D transient numerical simulations.

2. Experiment

The experimental platform for this study, shown is Fig. 1, is centered upon a counterflow burner coupled to an ozone generator [13-15]. In order to investigate flame-vortex interactions, a micro-nozzle design based upon that of Takagi [7, 16] is integrated into the burner. The separation distance between the two main nozzles is held at 22.5 mm. The upper main nozzle issues pure N_2 and is heated to 550 K through PID control. The lower main nozzle expels a mixture of oxygen, ozone, and dimethyl ether (DME). The ozone is produced in situ by passing pure oxygen through a non-equilibrium plasma ozone generator (Ozone Solutions, TG-20).

CH_2O planar laser-induced fluorescence (PLIF) is employed to visualize the instantaneous cool flame structure. A Nd:YAG laser (Quantel, Q-smart 850) emits a beam (~250 mJ/pulse) at a wavelength of 355 nm, which is expanded into a 200-µm-thick vertical sheet and aligned with the centerline of the burner. An ICCD camera (Princeton Instruments, PI-MAX 4) operating at a gate width of 250 ns captures the fluorescence with the aid of a low-pass filter (390 nm).

The transient vortices in the experiment are generated by abruptly expelling nitrogen from a stainless steel micro-nozzle positioned along the centerline of the burner. The micro-nozzle, which protrudes 4 mm beyond the upper main nozzle exit, has an inner diameter of 0.41 mm and an outer diameter of 0.51 mm. The axial

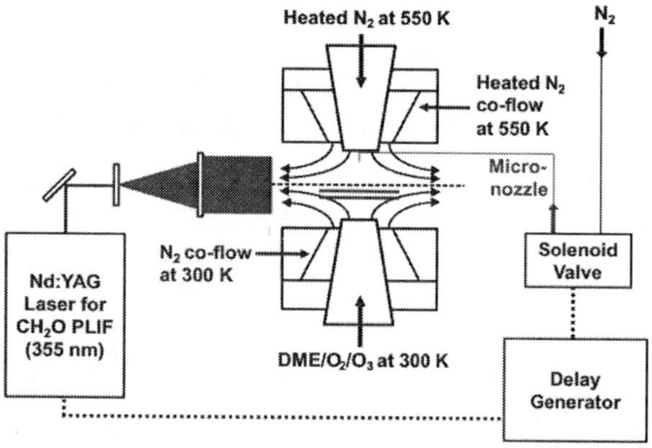

Fig. 1. Schematic of the experimental counterflow burner used to establish stable cool flames and investigate the interaction between a double flame and a vortex.

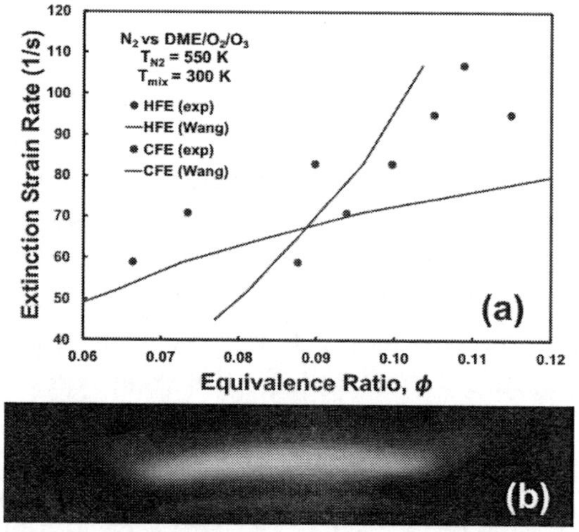

Fig. 2. (a) Measured and computed extinction limits for DME cool flames and hot flames. (b) A near-limit double flame at $a = 59\ s^{-1}$ and $\phi = 0.087$.

velocity of the vortex is approximated by mass conservation. Similar to [16], a delay generator controls the timing between the laser pulse and the solenoid valve feeding the micro-nozzle.

3. Numerical modeling

A time-dependent axisymmetric model known as UNICORN (UNsteady Ignition and COmbustion using ReactioNs) [17, 18] is used for the numerical simulation of flame-vortex interactions. The model solves the Navier-Stokes, species-conservation, and energy-conservation equations written in the cylindrical-coordinate (z-r) system. The kinetic model developed by Wang et al. [19] is used for describing DME combustion. The ozone submodel of Ombrello [20] is also incorporated. Details of the solution algorithm and the boundary conditions used are described in [21] and [22].

4. Results and discussion

4.1 Steady-state near-limit double flames

A double premixed flame structure occurs when the low-temperature reaction zone begins to propagate noticeably faster than the high-temperature reaction zone [12]. This can especially happen for cases in which the cool flame extinction limit exceeds the hot flame extinction limit [14]. Figure 2a shows this phenomenon occurring for premixed DME/O_2/O_3 flames at strain rates below 90 s^{-1}. In these cases, as the fuel concentration decreases, the hot flame weakens. As a result, two distinct flame fronts become apparent—a faster cool flame in the front and a slower near-limit hot flame trailing it. Figure 2b shows a visible image of a typical double flame at a = 59 s^{-1}. The double flame eventually extinguishes to form a lone cool flame.

Figure 3 shows the differences in the experimentally measured steady-state flame structures of a hot flame, double flame, and cool flame. The hot flame in Fig. 3a at $\phi = 0.095$ exhibits a typical flame structure—the chemiluminescence signal is monotonic, and the highest concentrations of CH_2O occur in the preheat zone. When the equivalence ratio is lowered to $\phi = 0.090$ in Fig. 3b, a double flame structure is apparent. The chemiluminescence signal now peaks twice. The width of the CH_2O profile also broadens noticeably, as the CH_2O produced in the low-temperature reaction zone must wait for the arrival of the high-temperature reaction zone to be consumed. The cool flame at $\phi = 0.0875$ in Fig. 3c is the result of the high-temperature reaction zone extinguishing. There is a monotonic and weak chemiluminescence signal, and the CH_2O concentration only decreases noticeably with the arrival of the stagnation plane.

Comparative numerical calculations of these three steady-state flame structures are seen in Fig. 4. Although the exact equivalence ratios where the transitions occur are slightly different, the computed CH_2O profiles qualitatively agree with the experiments. The hot flame's CH_2O profile is clearly the thinnest, and the cool flame's highlights the role of CH_2O as not simply an intermediate but one of the primary products of low-temperature oxidation. The double flame shows an obvious two-stage fuel consumption behavior indicative of two distinct reaction zones.

4.2 Transient double flames

The remainder of this study concerns the transient response of double flames to flow perturbations in the form of impinging micro jets. Figure 5 depicts the evolution of a double flame interacting with a micro jet issued from the inert side at a velocity of approximately 13.6 m/s. In Fig. 5a, a steady-state double flame can be seen at a strain rate of a = 75 s^{-1} and an

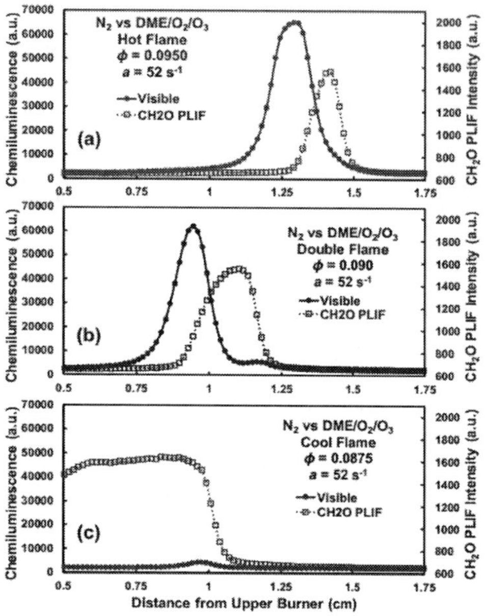

Fig. 3. Experimental measurements of centerline chemiluminescence and CH_2O PLIF for a (a) hot flame, (b) double flame, and (c) cool flame.

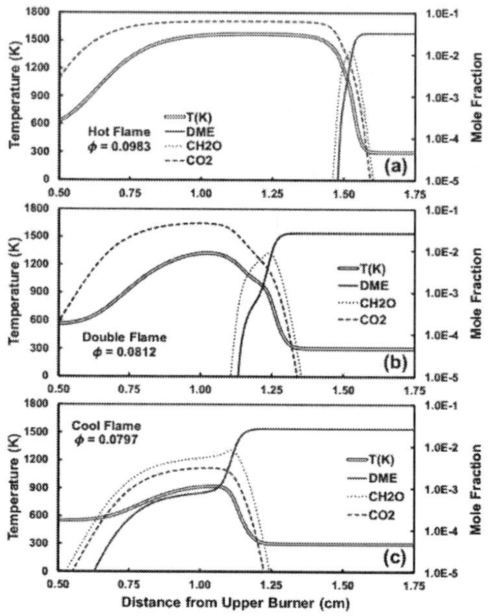

Fig. 4. Computed steady-state flame structure for a premixed (a) hot flame, (b) double flame, and (c) cool flame at a strain rate of $a = 52$ s^{-1}.

equivalence ratio of $\phi = 0.094$. 10 ms later in Fig. 5b, the double flame encounters the micro jet. It is soon clear in Fig. 5c that there are three different flame structures present: a post-extinction flame hole in the center, a cool flame surrounding the flame hole (the thickened CH_2O layer), and a double flame (thinner CH_2O layer) outside the cool flame. The hot flame within the double flame structure continues to extinguish radially outward for the next several ms until it is completely vanished in Fig. 5d. The cool flame finally experiences complete extinction (Fig. 5e) approximately 100 ms after first encountering the micro jet. The above results show that the vortex-double flame interaction can lead to different and interesting transient flame structures.

The experimental conditions from Fig. 5 are simulated in Fig. 6 using the 2-D UNICORN code. Temperature (black lines), OH concentration (white lines), CH_2O concentration (color plot on the left side), and heat release rate (color plot on the right side) are shown at four instants of interaction. In Fig. 6a, the vortex issued from the 0.4-mm-diameter tube has grown to ~3.2 mm as it approaches the post-flame zone. After a few milliseconds, the vortex has grown further (Fig. 6b) and has started extinguishing the hot flame. As expected, the extinction of the hot

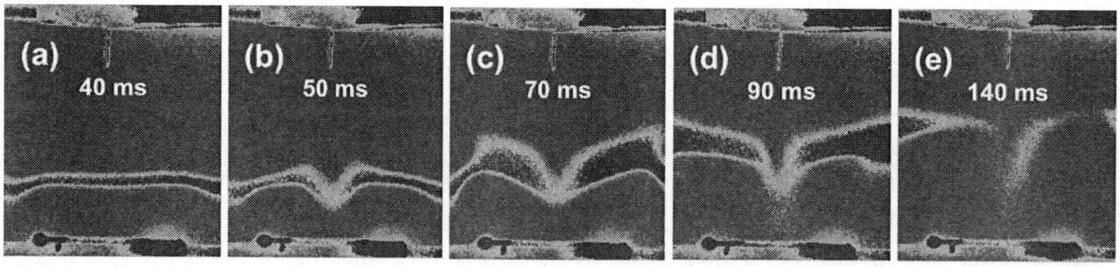

Fig. 5. CH_2O PLIF sequence showing the impact of a vortex on a counterflow double flame.

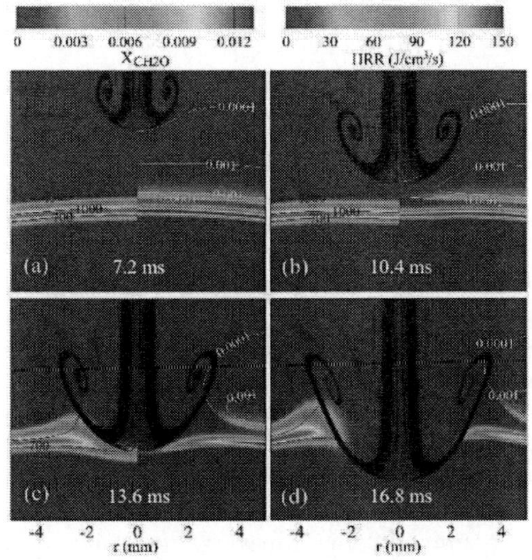

Fig. 6. 2-D simulation showing an impinging vortex (originating on the inert side) interacting with a counterflow double flame.

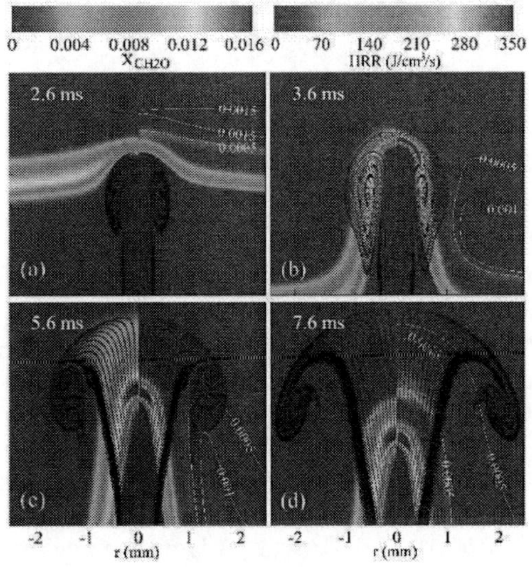

Fig. 7. 2-D simulation showing an impinging vortex (originating on the premixture side) interacting with a counterflow double flame.

flame leads to a broadening of the CH_2O distribution (left side of Fig. 6c). The predicted broadening of the CH_2O distribution and increase in separation between the cool and hot flames near the edges of the vortex mirrors the experiment, although the sequence proceeds more quickly in the computation. The vortex has extinguished both the cool flame and the hot flame at the centerline in Fig. 6d.

Due to the difficulties in experimentally creating a DME/O_2/O_3 micro jet, a second simulation is performed in which the micro jet is generated on the premixture side. The velocity of the vortex is maintained at 13.6 m/s. As seen in Figs. 7a and 7b, the hot flame is extinguished first even when the vortex originates on the premixture side. The CH_2O concentration in the cool flame initially decreases (Fig. 7a), but further interaction with the vortex (Fig. 7b) strengthens the flame. Interestingly, after initially extinguishing, the hot flame propagates back towards the centerline and eventually re-ignites completely (Fig. 7d). The double flame then approaches a steady-state structure with the flame curvature enhancing both the cool flame and the hot flame. Thus, it appears that previous understandings of flame stretch and preferential diffusion, which were derived for hot flames, can also be applied to the aerodynamics of premixed cool flames.

5. Conclusion

Stable premixed double flames and their interactions with transient vortices have been examined in a counterflow burner experimentally and computationally. The double flame structure occurs for near-limit hot flames when the hot flame extinction limit is lower than the cool flame extinction limit. Examination of vortex-double flame interactions shows that the weaker hot flame always extinguishes first, independent of the vortex velocity or location. It is shown that there can be three different co-existing transient flame structures: an extinction flame hole, a cool flame, and a double flame. Moreover, it is found through numerical simulations that curvature effects on cool flames and double flames resemble those of classic hot flames; i.e, negative curvature enhances the cool flame of a $Le > 1$ mixture.

6. Acknowledgement

The authors recognize the support of NSF grant CBET-1507358, NASA ISS Post-Graduate award NNX15AB67G, and NASA microgravity grant NNX16AK07G. CBR is supported by the NDSEG fellowship program of the U.S. Department of Defense. ORY is grateful for support from the Daniel and Florence Guggenheim Foundation Fellowship at Princeton University.

7. References

[1] V.R. Katta, C.D. Carter, G.J. Fiechtner, W.M. Roquemore, J.R. Gord, J.C. Rolon, Interaction of a vortex with a flat flame formed between opposing jets of hydrogen and air, Symp. (Int.) Combust. 27 (1998) 587-594.
[2] F.A. Williams, A review of some theoretical considerations of turbulent flame structure, Analytical and Numerical Methods for Investigation of Flow Fields with Chemical Reactions, Especially Related to Combustion, AGARD Conference Proceedings 164, 1975.
[3] N. Peters, Turbulent Combustion, Cambridge University Press, Cambridge, UK, 2000.
[4] J. Jarosinski, J.H.S. Lee, R. Knystautas, Interaction of a vortex ring and a laminar flame, Symp. (Int.) Combust. 22 (1989) 505-514.
[5] W.L. Roberts, J.F. Driscoll, A laminar vortex interacting with a premixed flame: Measured formation of pockets of reactants, Combust. Flame 87 (1991) 245-256.
[6] J.C. Rolon, F. Aguerre, S. Candel, Experiments on the interaction between a vortex and a strained diffusion flame, Combust. Flame 100 (1995) 422-429.
[7] T. Takagi, Y. Yoshikawa, K. Yoshida, M. Komiyama, S. Kinoshita, Studies on strained non-premixed flames affected by flame curvature and preferential diffusion, Symp. (Int.) Combust. 26 (1996) 1103-1110.
[8] F.L. Dryer, Chemical kinetic and combustion characteristics of transportation fuels, Proc. Combust. Inst. 35 (2015) 117-144.
[9] S.H. Won, B. Windom, B. Jiang, Y. Ju, The role of low temperature fuel chemistry on turbulent flame propagation, Combust. Flame 161 (2014) 475-483.
[10] R.N. Dahms, G.A. Paczko, S.A. Skeen, L.M. Pickett, Understanding the ignition mechanism of high-pressure spray flames, Proc. Combust. Inst. 36 (2017) 2615-2623.
[11] Y. Ju, W. Sun, M.P. Burke, X. Gou, Z. Chen, Multi-timescale modeling of ignition and flame regimes of n-heptane-air mixtures near spark assisted homogeneous charge compression ignition conditions, Proc. Combust. Inst. 33 (2011) 1245-1251.
[12] Y. Ju, C.B. Reuter, S.H. Won, Numerical simulations of premixed cool flames of dimethyl ether/oxygen mixtures, Combust. Flame 162 (2015) 3580-3588.
[13] S.H. Won, B. Jiang, P. Diévart, C.H. Sohn, Y. Ju, Self-sustaining n-heptane cool diffusion flames activated by ozone, Proc. Combust. Inst. 35 (2015) 881-888.
[14] C.B. Reuter, S.H. Won, Y. Ju, Experimental study of the dynamics and structure of self-sustaining premixed cool flames using a counterflow burner, Combust. Flame 166 (2016) 125-132.
[15] C.B. Reuter, M. Lee, S.H. Won, Y. Ju, Study of the low-temperature reactivity of large n-alkanes through cool diffusion flame extinction, Combust. Flame 179 (2017) 23-32.
[16] K. Yoshida, T. Takagi, Transient local extinction and reignition behavior of diffusion flames affected by flame curvature and preferential diffusion, Symp. (Int.) Combust. 27 (1998) 685-692.
[17] V. Katta, L. Goss, W. Roquemore, Numerical investigations of transitional H 2/N 2 jet diffusion flames, AIAA J. 32 (1994) 84-94.
[18] V.R. Katta, W.M. Roquemore, Calculation of multidimensional flames using large chemical kinetics, AIAA J. 46 (2008) 1640.
[19] Z. Wang, X. Zhang, L. Xing, L. Zhang, F. Herrmann, K. Moshammer, F. Qi, K. Kohse-Höinghaus, Experimental and kinetic modeling study of the low- and intermediate-temperature oxidation of dimethyl ether, Combust. Flame 162 (2015) 1113-1125.
[20] T. Ombrello, S.H. Won, Y. Ju, S. Williams, Flame propagation enhancement by plasma excitation of oxygen. Part I: Effects of O3, Combust. Flame 157 (2010) 1906-1915.
[21] V.R. Katta, W.M. Roquemore, Response of hydrogen-air opposing-jet diffusion flame to different types of perturbations, Proc. Combust. Inst. 28 (2000) 2055-2062.
[22] V.R. Katta, W.M. Roquemore, J. Gord, Examination of laminar-flamelet concept using vortex/flame interactions, Proc. Combust. Inst. 32 (2009) 1019-1026.

Sub Topic: Laminar Flames

Spring Technical Meeting
Eastern States Section of the Combustion Institute
March 4-7, 2018
State College, Pennsylvania

Laminar Flame Propagation in Supercritical Hydrogen/Air and Methane/Air Mixtures

Wenkai Liang, Weiyu Li and Chung K. Law

Department of Mechanical and Aerospace Engineering, Princeton University, Princeton NJ 08544, USA

Abstract: The laminar propagation of hydrogen/air and methane/air flames in supercritical conditions was computationally simulated for the planar configuration, incorporating descriptions of supercritical thermodynamics and transport as well as high-pressure chemical kinetics. The inaccuracies associated with the use of the ideal gas assumptions for various components of the supercritical description were systematically assessed with progressively more complete formulation. Specifically, for hydrogen/air flames, the laminar flame speeds at high pressures were found to increase with the use of the non-ideal equation of state (EoS), and is mainly due to the density modification of the initial mixture. Including the thermodynamic properties of enthalpy and heat capacity reduces the flame speed because of the correspondingly reduced adiabatic flame temperature. Diffusive transport properties were found to have small effect because of the inherent insensitivity of the laminar burning rate to their variations. For methane/air flames, the use of recently reported high-pressure chemical kinetics considerably affects the laminar flame speed, even for the same flame temperature.
Keywords: Laminar Flame Speed; Hydrogen; Methane; Supercritical Combustion.

1. Introduction

Combustion under high-pressure conditions holds the potential for improved thermodynamic efficiency, enhanced power generation, and reduced emission of some pollutants. It underlies the technology of internal combustion engines such as the automotive engine, the gas turbine, and the rocket engine [1-3], for which the operating pressure can reach as high as hundreds of atm [3]. At such high pressures, the fluid can be locally or even globally at the supercritical state, and as such is subjected to considerable real-fluid effects. Recognizing that studies of high-pressure combustion have frequently adopted ideal-gas approximations to describe the thermodynamics, transport and chemical kinetics components of the problem, it behoove us to assess the extent of inaccuracy embedded in such approximations.

Most previous investigations on supercritical and transcritical combustion have involved *diffusion* flames [4-8]. Specifically, Ribert et al. [4] numerically investigated the hydrogen/oxygen counterflow diffusion flames of general fluids by incorporating subcritical and supercritical thermodynamic and transport properties, while Pons et al. [5] studied the effects of mass transfer in transcritical non-premixed counterflow and identified the sharp density gradients under such conditions. Furthermore, the scaling laws of flame thickness, heat release rate and species profiles in hydrogen and hydrocarbon counterflow diffusion flames were assessed by numerical simulations in [6, 7]. Recently, Juanós and Sirignano [8] evaluated different levels of assumptions of ideal and dilute gases for the high-pressure real fluid effects in methane/air and methane/water-air diffusion flames for methane hydrate applications. Only limited studies, however, have been conducted on *premixed* flames at supercritical conditions [9, 10].

The current work numerically investigates the real-fluids effect on flame propagating at supercritical conditions. In the following, we shall first present the models of supercritical fluid properties and the numerical method. Then the planar laminar flame speeds and burning fluxes of hydrogen/air and methane/air mixtures are assessed and the inaccuracies of these parameters due to the non-ideal thermodynamic and transport properties and high-pressure chemical kinetics are evaluated and discussed.

2. Numerical Models

The planar flame simulation is performed by the PREMIX package [11] from the CHEMKIN code. For hydrogen, the recent kinetic model by Burke et al. [12] is used; while for methane, the GRI 3.0 mechanism [13], the USC 2.0 mechanism [14], the HP mechanism [15] and the recently updated high-pressure mechanism by Hashemi et al. [16] are adopted.

To account for the non-ideality of EoS and enthalpy formulation, we use the Soave-Redlich-Kwong (SRK) EoS [17]. For the (diffusive) transport properties, the real fluid models for both the thermal conductivity and mass diffusivity are incorporated. The thermal conductivity is evaluated from the correlation of Ely and Hanley [18]. For mass diffusion, the binary diffusion coefficients are corrected for high-pressure conditions by the approach of Takahashi [19]. Furthermore, the non-ideal diffusion flux formulation in [10] based on the Hirschfelder-Curtiss approximation is used and the Soret diffusion term [20] is included as well. The predictions of density, heat capacity, thermal conductivity and mass diffusivity agree well with the experimental data, as demonstrated in the literature [5-8, 10].

In order to identify the inaccuracies due to each of the updated models, five different cases with progressively more real-fluid properties were designed. Specifically, Case 1 is the complete ideal-gas description serving as the starting, reference case for comparison. Cases 2-5 incorporate four real-fluid corrections: Case 2 replaces the ideal-gas law by the SRK EoS; Case 3 further incorporates the non-ideal enthalpy formulation in the energy equation; Case 4 adds the high-pressure thermal conductivity correction; and Case 5 adds the high-pressure mass diffusion model. Effects of high-pressure chemistry are then implemented.

3. Results and Discussion

We start by analyzing the prototypical, adiabatic planar flames of hydrogen/air and methane/air mixtures propagating in the doubly infinite domain. Figure 1 shows the critical points (T_c and P_c) for these two mixtures calculated by using the SRK EoS as the inflection point in the P-v diagram. It is seen that the critical points of the hydrogen/air mixture are lower than those of methane, hence indicating that higher pressures are needed to induce supercritical effects for methane. In addition, the critical temperatures for both mixtures are always below the room temperature, so high critical pressure is the primary limitation in reaching the supercritical state. Moreover, for hydrogen/air, the critical point decreases with increasing equivalence ratio, while for methane/air the trend is opposite. This is due to the relatively high critical point of methane (T_c = 190 K and P_c = 46 atm) and relatively low critical point of hydrogen (T_c = 33 K and P_c = 12 atm), as compared to nitrogen (T_c = 126 K and P_c = 33 atm) and oxygen (T_c = 154 K and P_c = 50 atm).

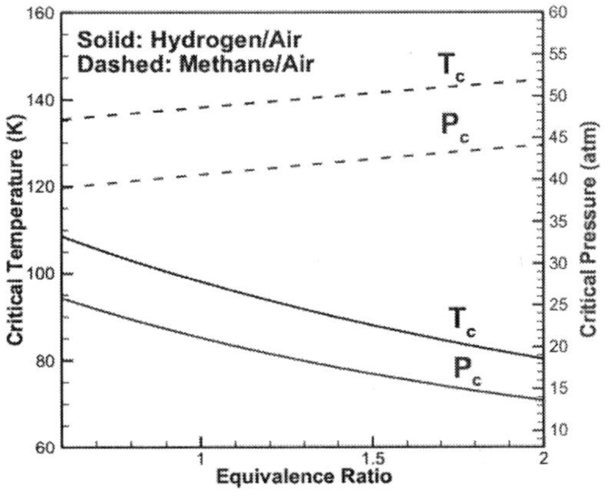

Figure 1: Critical temperature and pressure for hydrogen/air and methane/air mixtures at various equivalence ratios.

Figure 2 then shows the laminar flame speed of hydrogen/air mixtures at 100 atm with different assumptions (Cases 1-5). In addition, to quantify the error induced by using the temperature form of the energy equation, Case 3* is the case in which the thermodynamic properties are updated with the supercritical models but with the (incorrect) temperature equation instead of the proper enthalpy equation.

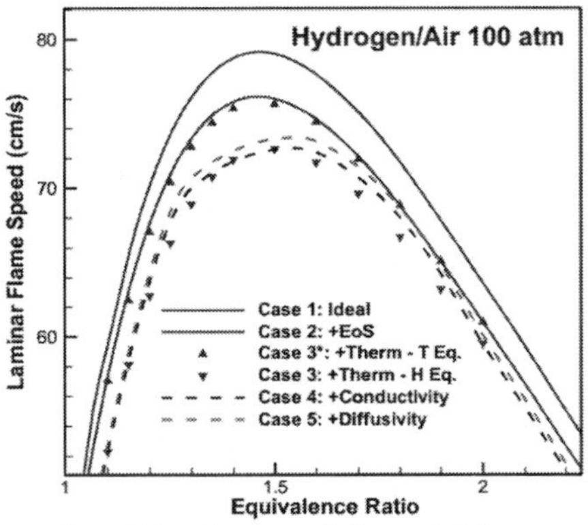

Figure 2: Laminar flame speeds of hydrogen/air at 100 atm with different assumptions.

To quantify the effect of EoS, by comparing Cases 1 and 2 we find that using the SRK EoS results in higher flame speeds than those of the ideal gas. Since the density of hydrogen is much smaller than that of air, its mixture density decreases with increasing equivalence ratio. The results then show that, by incorporating the SRK EoS, the density becomes lower than that of the ideal gas because the mixture behaves as a real fluid rather than an ideal gas at such high pressures. Further we note that the eigenvalue of the one-dimensional planar flame is the burning flux (i.e., the burning rate), $f = \rho S_L$, rather than the laminar flame speed S_L, where ρ is the unburned density. It is then seen that the burning fluxes of Cases 1 and 2 are almost identical. This therefore indicates that the updated EoS does not change the eigenvalue of the flame propagation, and as such the increase of the laminar flame speed observed in Fig. 2 is primarily due to the decrease in the upstream density, while the burning flux is minimally affected.

We next assess the effects of thermodynamic properties and the format of the energy equation by comparing Cases 2-3. Figure 2 shows that the laminar flame speed progressively decreases from Case 2 to Case 3. While this real-fluid thermodynamics effect is opposite to that of the EoS, at stoichiometric condition the laminar flame speed of Case 3 is still lower than Case

1 by around 6 cm/s and Case 2 by around 10 cm/s, which is beyond the uncertainty of most flame speed measurements. The cause for this reduction is the corresponding reduction of the adiabatic flame temperature, and through it the global reaction rate, as shown for Cases 2-3 in Fig. 3. To further elucidate this effect, we note that the adiabatic flame temperature can be approximately expressed as $T_{ad} = \Delta H/C_p$, where ΔH is the difference of enthalpy between the reactants and products and C_p is the heat capacity; the effect due to product dissociation is relatively small at such high pressures and as such not included in the assessment. For real fluids, ΔH decreases and C_p slightly increases; consequently both effects lead to the reduced T_{ad} as shown in Fig. 3 by comparing Case 2 and Case 3*. Furthermore, by using the enthalpy form of the equation, we note that since the mixing rule for enthalpy at the supercritical state is no longer valid such that $h < \sum_i Y_i h_i$. This effect further reduces ΔH and hence leads to even smaller T_{ad}, as shown in Fig. 3 by comparing Case 3* and Case 3.

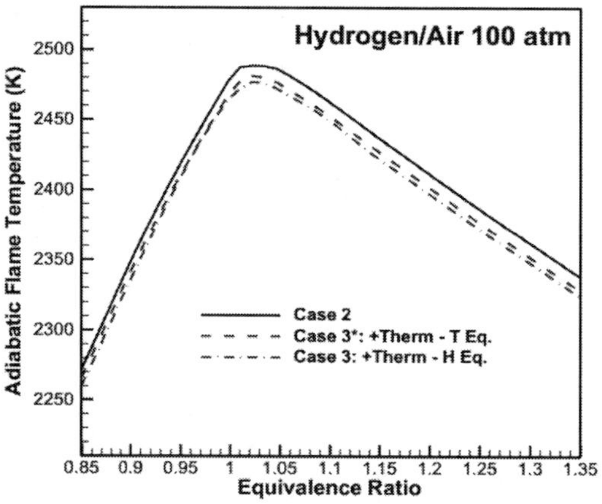

Figure 3: Adiabatic flame temperature of hydrogen/air at 100 atm for Cases 2-3.

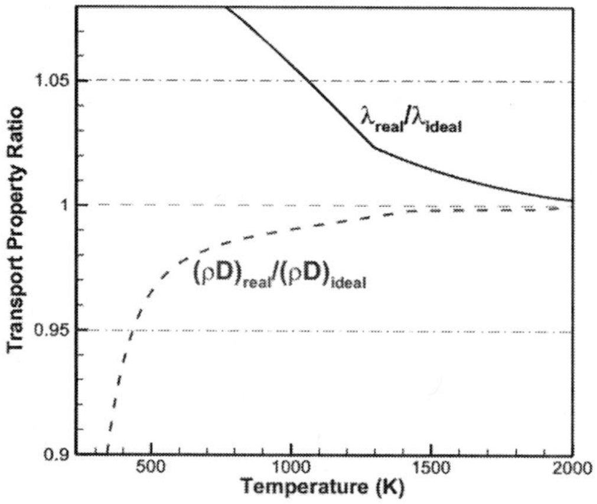

Figure 4: Transport property ratios of the stoichiometric hydrogen/air flame at 100 atm.

We next examine the effects of transport properties on the laminar flame speeds. Case 3 to Case 5 in Fig. 2 shows that there is little effect on the flame speed when either the heat conductivity or the mass diffusivity is modified with the real-fluid models. In order to understand this finding, we plot the ratios for these two transport properties, namely thermal conductivity, λ, and the density-weighted mass diffusivity, ρD, of the ideal gas and real fluids in Fig. 4 with the 5%

deviation demonstrated. It is seen that the real-fluid models only have notable influence on the transport properties at relatively low temperatures, which is the condition when the fluid approach the liquid side of the phase diagram.

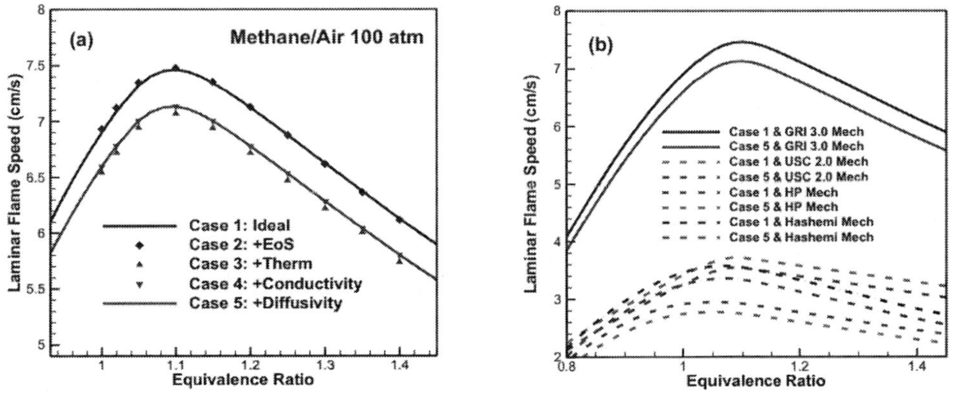

Figure 5: Laminar flame speeds of methane/air at 100 atm with (a) different assumptions and (b) different kinetic models.

Figure 5(a) shows the laminar flame speeds of methane/air at 100 atm with different thermodynamic assumptions, by using the GRI 3.0 mechanism. It is seen that different from hydrogen/air mixtures, the update of EoS (Cases 1 and 2) has little effect on the flame speed, which implies that the density change of methane/air is not as large as that for hydrogen/air. This is reasonable in that, as shown in Fig. 2, the critical point of methane/air is much higher than that of hydrogen/air, so there is less real-gas compression for the methane/air mixtures. For the effects of thermodynamics and transport (Cases 2-5), the findings are similar to the hydrogen case, which shows the updated energy equation reduces the flame speed while transport has very small influence.

To assess the effect of chemical kinetic models adopted, GRI 3.0 mechanism [16], USC 2.0 mechanism [17], HP mechanism [18] and the recently developed high-pressure methane mechanism of Hashemi et al. [19] were used to compare with each other, as shown in Fig. 5(b). The HP and Hashemi mechanisms are designed to include more accurate pressure-dependent rates and have been validated against high-pressure experimental data up to 200 atm. The results show that predictions using the USC 2.0, HP and Hashemi mechanisms, with the flame temperature being either that of Case 1 for the ideal gas or that of Case 5 for the real fluid, are much lower than those using the GRI 3.0 mechanism. This finding indicates that the accuracy of the kinetic mechanisms used perhaps has the largest sensitivity for such high-pressure applications than the updates of EoS, thermodynamics and transport.

4. Conclusions

The planar hydrogen/air and methane/air flame propagation at supercritical conditions have been simulated, with descriptions of real-gas EoS, thermodynamics, transport and chemistry incorporated into the numerical simulations at progressively more complete levels of implementation. For hydrogen/air mixtures, it is found to increase due to the non-ideal equation of state, which is mainly caused by the density modification of the initial mixture. Including the thermodynamic description would reduce the laminar flame speed because of the decreased adiabatic flame temperature through the real-fluid model. Furthermore, the enthalpy format of the energy equation leads to lower laminar flame speed compared to the temperature format of the energy equation. This is because the mixing rule at supercritical state further reduces the mixture enthalpy hence decreases the flame temperature. Transport models, however, are found to have minimal effect on the laminar flame speed even at very high pressures. For methane/air flames, use of the recently-developed high-pressure kinetics mechanism leads to substantially reduced flame speed at supercritical states; the reason is kinetic in nature in terms of revised rate parameters instead of the different flame temperatures resulting from different supercritical thermodynamic descriptions.

5. Acknowledgements

This research was sponsored by the US Air Force Office of Scientific Research under the technical monitoring of Dr. Mitat Birkan. WL was in addition supported by a Maeder Fellowship and a Harari Fellowship of Princeton University.

References

[1] A. H. Lefebvre, Gas turbine combustion, CRC press, 1998.
[2] R. D. Reitz, Combust. Flame, vol. 160, no. 1, pp. 1–8, 2013.
[3] G. P. Sutton and O. Biblarz, Rocket propulsion elements, Wiley & Sons, 1966.
[4] G. Ribert, N. Zong, V. Yang, L. Pons, N. Darabiha, and S. Candel, Combust. Flame, vol. 154, no. 3, pp. 319–330, 2008.
[5] L. Pons, N. Darabiha, S. Candel, G. Ribert, and V. Yang, Combust. Theory Model., vol. 13, no. 1, pp. 57–81, 2009.
[6] H. F. Huo, X. J. Wang, and V. Yang, Combust. Flame, vol. 161, no. 12, pp. 3040–3050, 2014.
[7] X. Wang, H. Huo, and V. Yang, Combust. Sci. Technol., vol. 187, no. 1–2, pp. 60–82, 2014.
[8] A. Jordà Juanós and W. A. Sirignano, Combust. Flame, vol. 181, pp. 54–70, 2017.
[9] S. Candel, M. Juniper, G. Singla, P. Scouflaire, and C. Rolon, Combust. Sci. Technol., vol. 178, no. 1–3, pp. 161–192, 2006.
[10] V. Giovangigli, L. Matuszewski, and F. Dupoirieux, Combust. Theory Model., vol. 15, no. 2, pp. 141-182, 2011.
[11] R.J. Kee, J.F. Grcar, M.D. Smooke MD, J.A. Miller, E. Meeks, PREMIX: a Fortran program for modeling steady laminar one-dimensional premixed flames. Sandia National Laboratories Report. 1985(SAND85-8249).
[12] M. P. Burke, M. Chaos, Y. Ju, F. L. Dryer, and S. J. Klippenstein, Int. J. Chem. Kinet., vol. 44, no. 7, pp. 444–474, 2012.
[13] G.P. Smith, D.M. Golden, M. Frenklach, N.W. Moriarty, B. Eiteneer, M. Goldenberg, C.T. Bowman, R.K. Hanson, S. Song, W.C. Gardiner, V.V. Lissianski. GRI 3.0 Mechanism. Gas Research Institute (http://www. me. berkeley. edu/gri_mech). 1999.
[14] H. Wang, X. You, A.V. Joshi, S.G. Davis, A. Laskin, F. Egolfopoulos and C.K. Law, USC Mech Version II. High-Temperature Combustion Reaction Model of H2/CO/C1-C4 Compounds. http://ignis.usc.edu/USC_Mech_II.htm, 2007.
[15] H. Zhao, J. Fu, F.M. Haas, Y. Ju, Combustion and Flame, 183 (2017) 253–260.
[16] H. Hashemi, J. M. Christensen, S. Gersen, H. Levinsky, S. J. Klippenstein, and P. Glarborg, Combust. Flame, vol. 172, pp. 349–364, 2016.
[17] G. Soave, Chemical Engineering Science. 1972 Jun 1;27(6):1197-203.
[18] J.F. Ely, H.J. Hanley, Industrial & Engineering Chemistry Fundamentals. 1983 Feb;22(1):90-7.
[19] S. Takahashi, Journal of Chemical Engineering of Japan. 1975 Jan 31;7(6):417-20.
[20] W. Liang, Z. Chen, F. Yang, and H. Zhang, Proc. Combust. Inst., vol. 34, no. 1, pp. 695–702, 2013.

The 2018 (41st) Spring Technical Meeting
Organized by the Eastern States Section of the Combustion Institute (ESSCI)
and hosted by the Pennsylvania State University
State College, PA, March 4-7, 2018

Various Regimes of Premixed Flame Propagation in Obstructed Channels with both Extremes Open

Abdulafeez Adebiyi[1], Elizabeth Ridgeway[1], Rawan Alkandari[1], Amanda Cathreno[1], Damir Valiev[2] and V'yacheslav Akkerman[1,*]

[1]*Computational Fluid Dynamics and Applied Multi-Physics Center*
Center for Innovation in Gas Research and Utilization (CIGRU)
Center for Alternative Fuels, Engines and Emissions (CAFEE)
Department of Mechanical and Aerospace Engineering, West Virginia University
Morgantown, WV 26506-6106, USA
[2]*Center for Combustion Energy, Tsinghua University, Beijing 100084, China*

[*]*Corresponding Author Email: Vyacheslav.Akkerman@mail.wvu.edu*

Abstract: Premixed flames are found to accelerate extremely fast and they even may trigger a detonation when propagating in a semi-open obstructed channel (one extreme of the channel is closed; the flame is ignited at the closed extreme and moves towards the open one). However, industrial and laboratory pipes oftentimes have both extremes open, with a flame ignited at one of them as considered in this work. Specifically, flame propagation through a comb-shaped array of obstacles, in-built in a channel with both extremes open, is studied by means of the computational simulations of the reacting flow equations with fully-compressible hydrodynamics, transport properties and Arrhenius chemical kinetics. The parametric study includes various blockage ratios, spacing between the obstacles, and thermal expansion ratios. Depending on these parameters, we have found either flame oscillations, or acceleration, or their sequence. All the features observed differ conceptually from fast flame acceleration in semi-open channels. Such a difference is caused by a fact that while the entire flame-generated jet-flow is pushed towards a single exit in semi-open channels, in channels with two extremes open, this jet-flow is distributed between the upstream and downstream flows, thereby moderating flame propagation. The flame oscillations are nonlinear in all cases studied. The oscillation period grows with the blockage ratio but decreases with the thermal expansion ratio. The present results also support the recent experiments, modelling and theory of flames in obstructed channels with both ends open, which all yielded steady or quasi-steady flame propagation prior to the onset of flame acceleration. Indeed, the oscillations can be treated as the fluctuations around quasi-steady solution, followed by acceleration.

Keywords: **premixed flame; obstructed channel; flame oscillation; flame acceleration; numerical simulation.**

1. Introduction

A premixed flame propagating in a narrow channel may experience various behaviors including oscillations [1], acceleration [2,3] or a sequence of these trends [4]. In channels with smooth (unobstructed) walls, these effects are mostly caused by mechanistic boundary conditions such as nonslip walls. Specifically, flames use to accelerate in a "semi-open" channel (that with one end open), with nonslip, adiabatic walls, when propagating from the closed extreme to the open one, because the entire new volume generated in the burning process is pushed towards the singe exit in that case [2]. In contrast, when both channel extremes are open, the newly-generated volume is distributed between the upstream and downstream flows, towards both ends, thus moderating the effect. Consequently, the flame may oscillate around a regime of "quasi-steady" propagation instead of acceleration [1]. Other effects such as turbulence, thermal boundary conditions and non-equidiffusivity and may also influence both the accelerative and oscillating trends [5–7].

It is nevertheless noted that flame acceleration (FA) in channels with adiabatic, nonslip walls is actually relatively moderate; moreover, the acceleration rate decreases with the channel width. This conceptually differs from the case of obstructed conduits, where a flame may accelerate enormously [8]. In fact, there was a general belief that turbulence is the major driver of FA in such an obstructed configuration [8-12] until Bychkov *et al.* [13,14] have revealed a shockless, conceptually-laminar mechanism of ultrafast FA in semi-open channels or tubes with a comb-shaped array of the tightly-packed obstacles. Here, FA is driven by delayed combustion in the pockets between the obstacles, and it is equally strong for narrow or wide pipes, with turbulence playing only a supplementary role.

However, industrial and laboratory applications of obstructed conduits oftentimes have both ends open, with ignition occurring at one of them. Therefore, there is a critical need to scrutinize premixed flame propagation (FP) in open obstructed pipes and channels such as illustrated in Fig. 1. In particular, will the difference between the open and semi-open obstructed conduits be as paramount as that for unobstructed ones? The latter question is the focus of the present work. Specifically, FP through a comb-shaped array of obstacles in channels of various blockage ratios and widths, with both extremes open, is investigated by means of the two-dimensional (2D) computational simulations of the combustion equations with fully-compressible hydrodynamics and Arrhenius chemistry. The present work follows the modelling [15], the experiments and the in-situ modelling [16], and the analytical formulation [17] that all yielded steady FP prior to the onset of FA in such an "open-open" obstructed conduit. In the present study, in the narrow channels, we have found regular oscillations of the burning rate, which can actually be treated as fluctuations around a quasi-steady solution, being thereby in logical agreement with the works [15–17], but opposing ultrafast FA in semi-open obstructed channels [13,14]. Moreover, these results qualitatively reproduce the situation of *unobstructed* channels, with FA driven by wall friction in semi-open channels [2] and the flame oscillations (FOs) in the open-open ones [1]. In contrast, in wider channels, a transition from FOs to FA is found, which also agrees with [15-17].

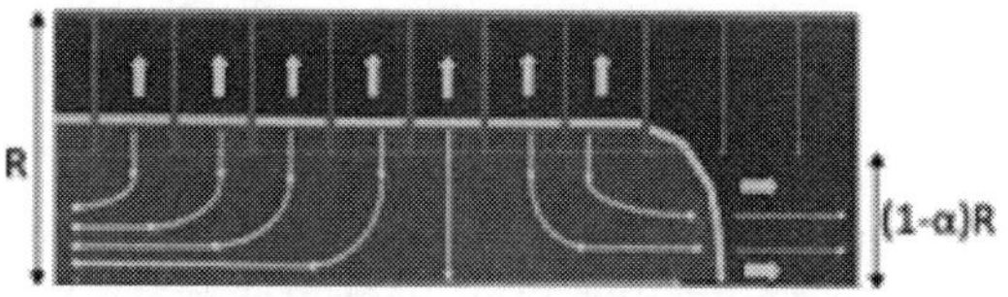

Figure 1: A schematic of flame propagation in an obstructed channel with both extremes open.

2. Formulation

We have performed computational simulations of the 2D reacting flow equations, including fully compressible hydrodynamics, transport properties (heat conduction, diffusion and viscosity), and chemical kinetics represented by a single irreversible Arrhenius reaction of the first order, with the activation energy E_a and the constant of time dimension τ_R. More details of the numerical scheme and the solver can be found, for instance, in [1–4,14]. The basic equations read:

$$\frac{\partial}{\partial t}\rho+\frac{\partial}{\partial x_i}(\rho u_i)=0, \qquad \frac{\partial}{\partial t}(\rho u_i)+\frac{\partial}{\partial x_j}(\rho u_i u_j+\delta_{i,j}P)-\gamma_{i,j}=0, \tag{1}$$

$$\frac{\partial}{\partial t}\left(\rho\varepsilon+\frac{1}{2}\rho u_i u_j\right)+\frac{\partial}{\partial x_i}\left(\rho u_i h+\frac{1}{2}\rho u_i u_j u_j+q_i-u_j\gamma_{i,j}\right)=0, \tag{2}$$

$$\frac{\partial}{\partial t}(\rho Y)+\frac{\partial}{\partial x_i}\left(\rho u_i Y-\frac{\zeta}{Sc}\frac{\partial Y}{\partial x_i}\right)=-\frac{\rho Y}{\tau_R}\exp\left(-E_a / R_p T\right), \tag{3}$$

where Y is a progress variable representing the mass fraction of the fuel mixture; $\varepsilon = QY + C_V T$ and $h = QY + C_P T$ the specific internal energy and enthalpy, respectively; $Q = C_P T_f(\Theta - 1)$ is the energy release in the reaction, with the specific heats at constant volume, C_V, and pressure, C_P, the initial fuel temperature, $T_f = 300K$, pressure, $P_f = 1\,\text{bar}$, and density, $\rho_f = 1.16\,\text{kg/m}^3$, and the thermal expansion ratio $\Theta = T_b / T_f = \rho_f / \rho_b$ in the range $\Theta = 5 \sim 10$. To avoid any impact of diffusional-thermal instability, here we consider equidiffusive burning. For this reason, we take $\Pr = Sc = 1$ such that the Lewis number, $Le \equiv Sc/\Pr = 1$. The unstretched laminar flame speed is taken as large as $S_L = 3.47$ m/s, which resembles typical methane-air combustion. The speed of sound in this fuel mixture is 100 times larger, $c_s = 347$ m/s, such that the hydrodynamics is almost incompressible at the initial stage of burning. Both the fuel mixture and burnt matter are assumed to be ideal gases of constant molecular weight, $M = 2.9$ kg/kmol, such that the equation of state reads $P = \rho R_P T / M$, with the universal gas constant $R_P = 8.31\,\text{kJ/(kmol}\cdot\text{K)}$. The stress tensor $\gamma_{i,j}$ and the energy diffusion vector q_i are given by

$$\gamma_{i,j} = \zeta\left(\frac{\partial u_i}{\partial x}+\frac{\partial u_j}{\partial x_i}-\frac{2}{3}\frac{\partial u_k}{\partial x_k}\delta_{i,j}\right), \qquad q_i = -\zeta\left(\frac{C_P}{\Pr}\frac{\partial T}{\partial x_i}+\frac{Q}{\text{Sc}}\frac{\partial Y}{\partial x_i}\right), \tag{4}$$

where $\zeta \equiv \rho\nu$ is the dynamic viscosity, being $\zeta_f = 1.7\times10^{-5}\,kg/(\text{m}\cdot\text{s})$ in the fuel mixture. Then the thermal flame thickness can be defined, conventionally, as $L_f = \zeta_f / \Pr \rho_f S_L = 4.22\times10^{-6}\,\text{m}$.

A flame propagates in a long 2D channel of width $2R$, with both extremes open, where the fraction $2R\alpha$ is blocked by the obstacles with a spacing (the distance between two neighboring obstacles) Δz. In this respect, the major dimensionless quantities of this study are the blockage ratio α, the scaled spacing $\Delta z/R$, the flame Reynolds number $\text{Re} = RS_L/\nu = R/\Pr L_f = R/L_f$, and the instantaneous scaled burning rate U_w/S_L. Similar to [1,2], the latter quantity is evaluated in this work by means of the instantaneous scaled surface area (length in 2D) of a corrugated flame front, $U_w(t)/S_L \approx S_f(t)/2R$. To be specific, the following configurational parameters are employed: Re = 12, 14, 36 and 48; α = 1/3, 1/2, and 2/3; $\Delta z/R$ = 1/4, 1/2, and 1; and $\Theta = 5{\sim}10$.

The walls of the channel and obstacles are adiabatic, $\mathbf{n}\cdot\nabla T = 0$, and free-slip, $\mathbf{n}\cdot\mathbf{u} = 0$, where $\mathbf{n}$ is the normal vector at the wall/obstacle surface. The absorbing (non-reflecting) boundary conditions are employed at both opened extremes to prevent the reflection of the sound and weak shocks. Initially, the parameters at the right (unburnt) end are $\rho = \rho_f$, $T = T_f$, $u_z = -S_L$, while those at the left (burnt) end are $\rho = \rho_f/\Theta$, $T = \Theta T_f$, $u_z = -\Theta S_L$. The initial flame structure was imitated by the classical Zeldovich-Frank-Kamenetsky (ZFK) solution for a planar flame front [19]. Such a planar ZFK flame was initiated at a distance $50\,L_f$ from the left (burnt) extreme.

3. Results and Discussion

We have investigated premixed FP and the flame-generated flow velocity fields in obstructed channels with both ends open. An initially planar ZFK flame (not shown in the figures) acquires promptly a curved shape. It is observed that the length and curvature of the flame front vary as the flame spreads. Specifically, the variations of the scaled burning rate, U_w / S_L, shows that the flame front oscillates in a periodic manner. As a result, we have analyzed the flame dynamics and morphology and the evolution of the velocity field during such an oscillation.

The color temperature snapshots of Fig. 2 show the flame front evolution for $\alpha = 1/3$, Fig. 2a, $\alpha = 1/2$, Fig. 2b, and $\alpha = 2/3$, Fig. 2c, respectively. In all three cases we have $\Delta z = R/4$ and $\Theta = 8$ such that the temperature varies from 300 K in the fresh fuel mixture (blue) till 2400 K in the burnt gas (red). The flame segment in a free (unobstructed) part of the channel generally acquires a concave shape, with formation of a cusp, directed towards the burnt gas at the centerline. The other flame segments enter the pocket between the obstacles.

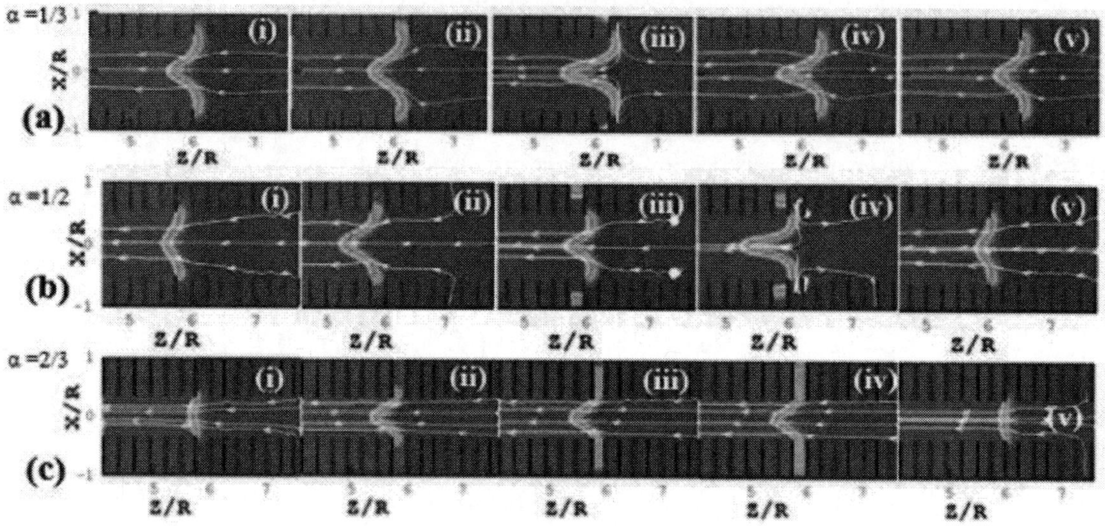

Figure 2: Consecutive color temperature snapshots (from T = 300 K, blue, in the fuel mixture till T = 2400K, red, in the burnt matter) and the flow streamlines (white) for propagation of a flame with $\Theta = 8$ in a channel with the obstacle spacing $\Delta z = R/4$ and various blockage ratios: $\alpha = 1/3$ (a), 1/2 (b), and 2/3 (c).

At the initial stage of each FO, the flame front is almost quiescent as shown by the snapshots (i)–(ii) in all the figures. The duration of such a stationary stage increases with α, whereas the concavity of the flame front is smaller for larger α. During this stage, a flame starts consuming the fuel mixture in the pocket; the burning gas expands, and a newly-generated gas volume bends a flame segment in the pocket, making it convex, while that near the centerline remains concave. Consequently, the cumulative surface area (length) of the flame front grows, thereby promoting the total burning rate and yielding FA at this stage. However, as combustion proceeds in a pocket, the flame eventually enters the next pocket and its front gets broken by the obstacle. Such a breaking of the flame front terminates FA, so the flame decelerates, as shown by snapshots (iii)–(iv) in Figs. 2(a-c) until the shape (v) of Figs. 2(a,b) is acquired. The acceleration-deceleration scenario described above repeats on a regular basis thus yielding flame oscillations.

To quantify these oscillations, we have plotted the time evolution of the scaled instantaneous burning rate, U_w / S_L, for various blockage ratios, α, spacing, Δz, and the thermal expansion ratios Θ. In particular, Fig. 3 scrutinizes the impact of the blockage ratio on the FOs by means of comparing the cases of $\alpha = 1/3$, 1/2, and 2/3 (Figs. 3a – 3b, respectively), with fixed $\Delta z = R/4$ and $\Theta = 8$ in all three cases. It is seen that the FOs are nonlinear, and this nonlinearity is stronger for larger α. In the $\alpha = 1/3$ channel, a secondary peak appears for each FO. Overall, there are three basic quantities characterizing the oscillations of the burning rate, namely: the oscillation period τ_p, the oscillation amplitude ΔU, which slowly dampens with time in Fig. 3, and the average burning rate during an oscillation $\langle U_w \rangle$. According to Fig. 3, τ_p increases with α such that the largest period, with the most profound nonlinearity, is seen for $\alpha = 2/3$. The latter effect can be attributed to the fact that larger blockage ratios correspond to deeper pockets and thereby longer duration of a near-quiescent stage observed.

The role of the spacing between the obstacles Δz is described by Fig. 4, where the cases of $\Delta z = R/4$, Fig. 4a, $\Delta z = R/2$, Fig. 4b, and $\Delta z = R$, Fig. 4c are compared for the fixed $\alpha = 1/3$ and $\Theta = 8$ in all there cases. The effect of spacing is found to be noticeable. In particular, one can see that both the oscillation period and nonlinearity get promoted with the increase in $\Delta z/R$.

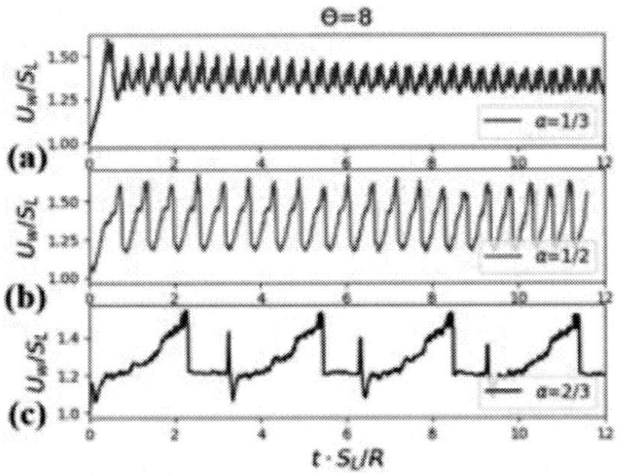

Figure 3: The scaled burning rate U_w/S_L vs the scaled time tS_L/R for $\Theta = 8$, $\Delta z = R/4$ and various $\alpha = 1/3$ (a), 1/2 (b) and 2/3 (c).

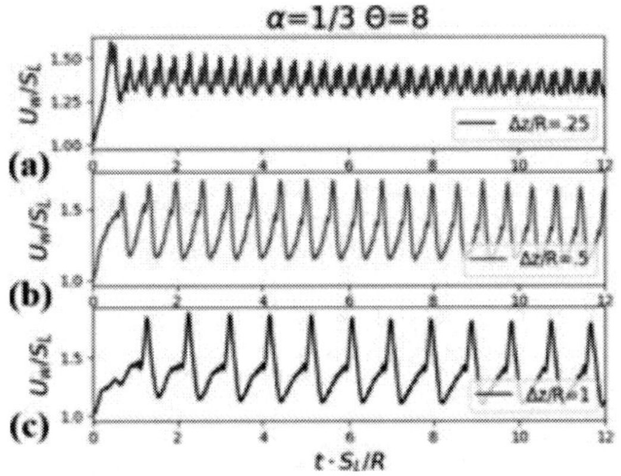

Figure 4: The scaled burning rate U_w/S_L vs the scaled time tS_L/R for $\alpha = 1/3$, $\Theta = 8$, and various $\Delta z = R/4$ (a), R/2 (b) and R (c).

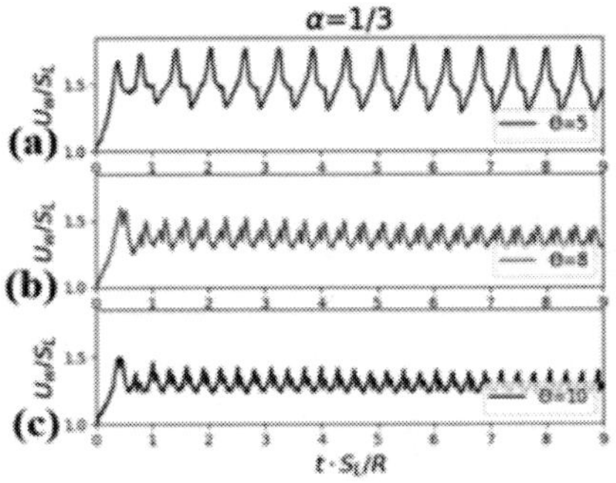

Figure 5: The scaled burning rate U_w/S_L vs the scaled time tS_L/R for $\alpha = 1/3$, $\Delta z = R/4$ and various $\Theta = 5$ (a), 8 (b) and 10 (c).

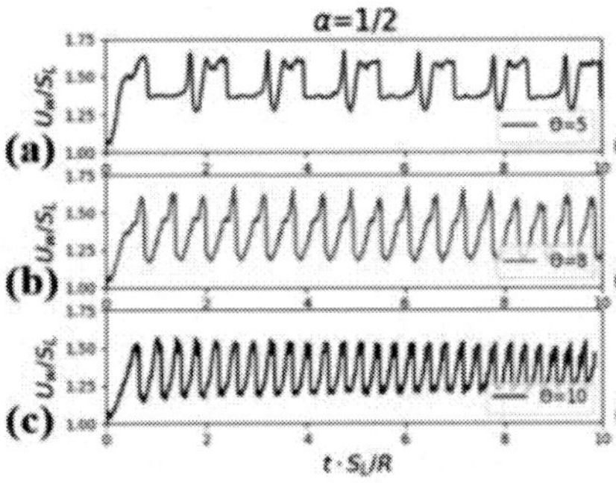

Figure 6: The scaled burning rate U_w/S_L vs the scaled time tS_L/R for $\alpha = 1/2$, $\Delta z = R/4$ and various $\Theta = 5$ (a), 8 (b) and 10 (c).

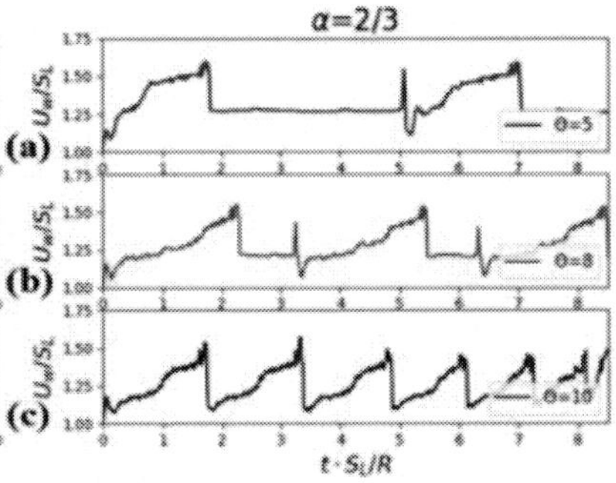

Figure 7: The scaled burning rate U_w/S_L vs the scaled time tS_L/R for $\alpha = 2/3$, $\Delta z = R/4$ and various $\Theta = 5$ (a), 8 (b) and 10 (c).

Finally, we have studied the impact of thermal expansion of the burning matter on the FOs by varying the expansion ratios in the range $\Theta = 5\sim10$. Specifically, Fig. 5 compares the evolutions of the burning rate in the cases of $\Theta = 5$, Fig. 5a, $\Theta = 8$, Fig. 5b, and $\Theta = 10$, Fig. 5c, with the same $\Delta z = R/4$ and $\alpha = 1/3$ in all three cases. It is clearly seen that the oscillation period reduces with Θ, and the same nonlinearity does. Indeed, instead of the regular and smooth oscillations for $\Theta = 10$ in Fig. 5c, we can see strong nonlinearity and irregularity for $\Theta = 5$ in Fig. 5a.

Figures 6 and 7 are the counterparts of Fig. 5 for $\alpha = 1/2$ and $2/3$, respectively. In fact, all the three figures yield the same conclusion: the reduction in the expansion factor Θ promotes both nonlinearity and the oscillation period τ_p. At the same time, nonlinearity itself is much stronger in Figs. 6 and 7 as compared to Fig. 5. The latter is obviously due to the larger values of α.

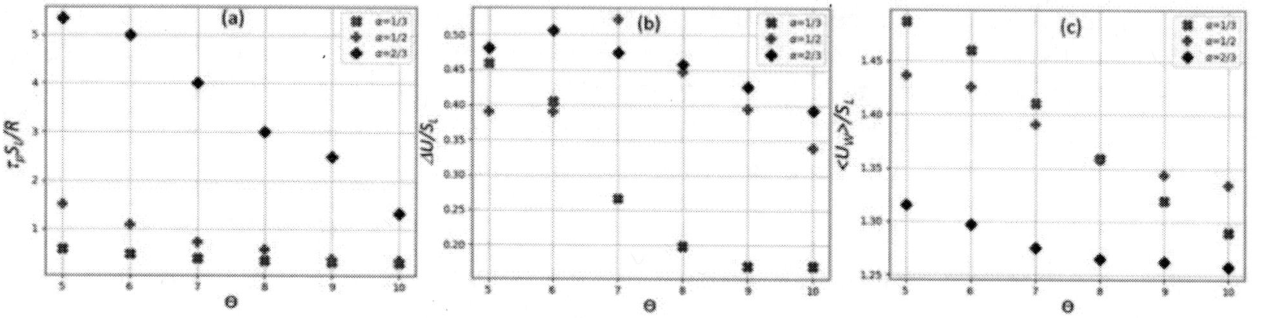

Figure 8: The scaled oscillation period $\tau_p S_L/R$ (a), the scaled oscillation amplitude $\Delta U/S_L$ (b) and the scaled average burning rate during an oscillation $\langle U_w \rangle / S_L$ (c) vs the thermal expansion ratio Θ for the obstacles spacing $\Delta z = R/4$ and various blockage ratios $\alpha = 1/3, 1/2, 2/3$ in each figure.

To quantify the qualitative conclusions of Figs. 5–7, we also plotted the scaled oscillation period, $\tau_p S_L/R$, the scaled oscillation amplitude, $\Delta U/S_L$ as well as the scaled average burning rate, $\langle U_w \rangle / S_L$, versus the thermal expansion ratio Θ, see Figs. 8a – 8c, respectively. Specifically, Fig. 8a shows that τ_p reduces with Θ for any blockage ratio, but the effect is minor for $\alpha = 1/3$ and it is quite moderate for $\alpha = 1/2$. In contrast, the reduction in τ_p is drastic for $\alpha = 2/3$. The average burning rate $\langle U_w \rangle$, Fig. 8c, also reduces with Θ for any α, but here the situation is qualitatively opposite to Fig. 8c: the effect appears minor for a large blockage ratio, $\alpha = 2/3$, but it is notable for smaller α. As for the oscillation amplitude ΔU, Fig. 8b, it reduces with Θ when the blockage ratio is moderate, $\alpha = 1/3$, while for larger α, the Θ-dependence of ΔU is non-monotonic.

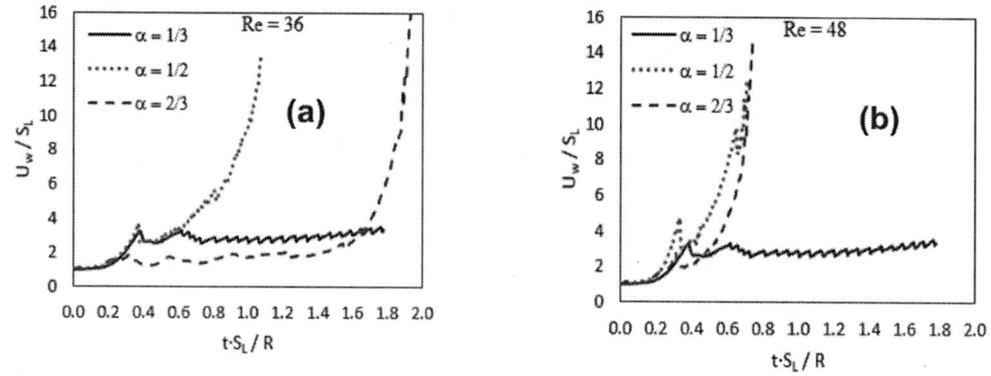

Figure 9: The scaled burning rate U_w/S_L vs the scaled time $t\,S_L/R$ for $R/L_f = 36$ (a) and 48 (b), with $\Theta = 8$, $\Delta z = R/4$ and various $\alpha = 1/3$ (solid, black), 1/2 (dotted, red) and 2/3 (dashed, blue) in each figure.

While all FOs discussed above have been obtained in relatively narrow channels, $Re = R/L_f = 12\sim24$, we next employ wider channels with both extremes open, with Re = 36, Fig. 10a, and 48, Fig. 10b, considering $\alpha = 1/3$ (solid), 1/2 (dotted) and 2/3 (dashed) in both cases. It is seen that the initial FOs are eventually replaced by sudden and prompt FA for $\alpha = 1/2$ and $2/3$, while this is not the case for a relatively small blockage ratio, $\alpha = 1/3$. Presumably, a transition from FOs to FA requires more time for narrower channels and smaller α such that it is observed only in some of the simulation runs reported here. Obviously, such a FOs-FA transition of Fig. 9 resembles that of Fig. 10 (summarizing the simulations [15,16], experiments [16] and theory [17]), where quasi-steady FP is followed by sudden and strong FA. Moreover, even the FOs differ from FA, they can be treated as the fluctuations around steady FP. In this respect, such a quasi-steady regime of FP resembles initial steady FP of [15-17]; compare Fig. 11 with the initial stages at Fig. 10. In addition, being conceptually different from ultrafast FA in semi-open obstructed channels [13], the FOs revealed in the present work actually resemble the FOs found in *unobstructed* channels with both extremes open [1]. Consequently, the present study generally agrees with [1,15-17].

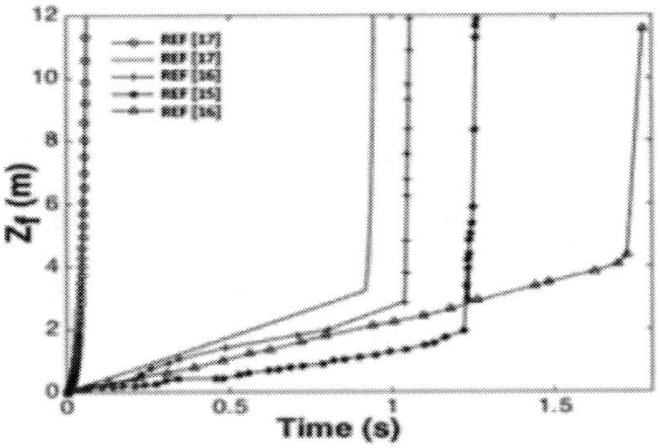

Figure 10: Simulations [15,16], experiments [16] and theory [17]: Steady FP at the initial stage of burning in obstructed pipes with both extremes open/vented {extracted from [17]}

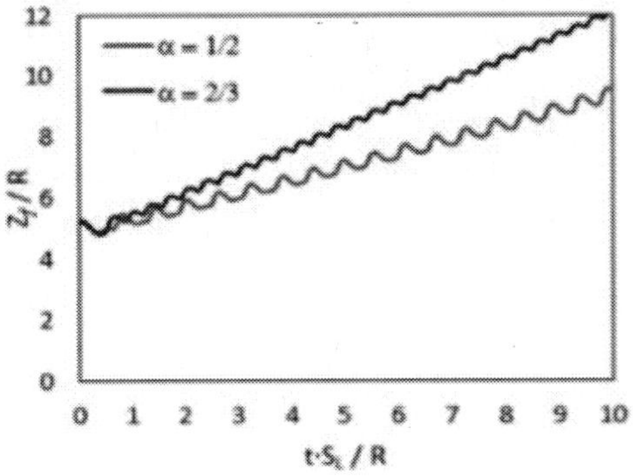

Figure 11: Present work: Quasi-steady FP (FOs) in a channel of half-width $R = 12\ L_f$ with both extremes open.

4. Conclusions

The present study scrutinizes premixed FP in obstructed channels with both extremes open by means of the computational simulations. Specifically, the channels of half-widths $R/L_f = 12, 24, 36, 48$; the blockages ratios $\alpha = 1/3, 1/2, 2/3$; and the spacing between the obstacles $\Delta z/R = 1/4, 1/2, 1$ are considered, with the thermal expansion ratios in the range $5 \leq \Theta \leq 10$. In narrow channels, $R \leq 24\ L_f$, the FOs are found as illustrated by the color temperature snapshots of Fig. 2, whereas the time evolution of the burning rate is quantified in Figs. 3-7. The oscillations of the burning rate are described by the oscillation period, τ_p, amplitude, ΔU, and the average burning rate $\langle U_w \rangle$. The latter three quantities are presented versus the thermal expansion ratio Θ in Fig. 8.

It is shown that Θ and α influence the FOs both qualitatively and quantitatively. In particular, the oscillation period grows with α but decreases with Θ. While these FOs differ conceptually from ultrafast FA in semi-open obstructed channels [13,14], they resemble the FOs found in unobstructed channels with both ends open [1]. Moreover, treated as fluctuations around steady FP, the oscillations agree with the experiments [16], modelling [15,16] and theory [17] of flames in channels with both extremes open or vented in terms that all of the studies [15-17] identified a steady stage of FP prior to fast FA. Moreover, a transition from the FOs to FA is found in wider channels, $R/L_f \geq 36$, for $\alpha \geq 1/2$, thus yielding extra agreement with the previous studies [15-17].

5. Acknowledgements

This research was funded by the NSF through the CAREER Award No. 1554254 (V.A.).

6. References

[1] V. Akkerman, V. Bychkov, A. Petchenko, L.-E. Eriksson, Combust. Flame 145 (2006) 675.
[2] V. Bychkov, A. Petchenko, V. Akkerman, L.-E. Eriksson, Phys. Rev. E 72 (2005) 046307.
[3] V. Akkerman, V. Bychkov, A. Petchenko, L.-E. Eriksson, Combust. Flame 145 (2006) 206.
[4] V. Akkerman, C.K. Law, V. Bychkov, L.-E. Eriksson, Phys. Fluids 22 (2010) 053606.
[5] E.S. Oran, V.N. Gamezo, Combust. Flame 148 (2007) 4.
[6] G. Pizza, C.E. Frouzakis, J. Mantzaras, A.G. Tomboulides, K. Boulouchos, Combust. Flame 152 (2008) 433.
[7] C. Cui, M. Matalon, T.L. Jackson, AIAA Journal 43 (2005) 1284.
[8] G. Roy, S. Frolov, A. Borisov, D. Netzer, Prog. Energy Combust. Sci. 30 (2004) 545.
[9] G. Ciccarelli, C. Fowler, M. Bardon, Shock Waves 14 (2005) 161.
[10] M. Kuznetsov, V. Alekseev, I. Matsukov, S. Dorofeev, Shock Waves 14 (2005) 205.
[11] G. Ciccarelli, S. Dorofeev, Prog. Energy Combust. Sci. 34 (2008) 499.
[12] V. Gamezo, T. Ogawa, E. Oran, Combust. Flame 155 (2008) 302.
[13] V. Bychkov, D. Valiev, L.-E. Eriksson, Phys. Rev. Lett. 101 (2008) 164501.
[14] D. Valiev, V. Bychkov, V. Akkerman, C.K. Law, L.-E. Eriksson, Combust. Flame 157 (2010) 1012.
[15] P. Middha, O.R. Hansen, Proc. Saf. Prog. 27 (2008) 192.
[16] J. Yanez, M. Kuznetsov, V. Bykov, 24th ICDERS, Taipei, Taiwan, July 28–Aug 2, 2013, paper #164.
[17] V. Bychkov, J. Sadek, V. Akkerman, Phys. Review E 95 (2017) 013111.
[18] A. Petchenko, V. Bychkov, V. Akkerman, L.E. Eriksson, Phys. Rev. Lett. 97 (2006) 164501.
[19] I. Zeldovich, G.I. Barenblatt, V. Librovich, G. Makhviladze, Mathematical Theory of Combustion and Explosions, .Consultants Bureau, New York, 1985.

The 2018 (41st) Spring Technical Meeting
Organized by the Eastern States Section of the Combustion Institute (ESSCI)
and hosted by the Pennsylvania State University
State College, PA, March 4-7, 2018

Computational Simulations of Nonequidiffusive Premixed Flames in Obstructed Pipes

Abdulafeez Adebiyi[1], Gbolahan Idowu[1], Damir Valiev[2] and V'yacheslav Akkerman[1,*]

[1] *Computational Fluid Dynamics and Applied Multi-Physics Center*
Center for Innovation in Gas Research and Utilization (CIGRU)
Center for Alternative Fuels, Engines and Emissions (CAFEE)
Department of Mechanical and Aerospace Engineering, West Virginia University
Morgantown, WV 26506-6106, USA
[2] *Center for Combustion Energy, Tsinghua University, Beijing 100084, China*
[*] *Corresponding Author Email: Vyacheslav.Akkerman@mail.wvu.edu*

Abstract: The impact of the Lewis number, Le, on the dynamics and morphology of a premixed flame front, spreading through a toothbrush-like array of obstacles in a semi-open channel, is studied by means of the computational simulations of the reacting flow equations with fully-compressible hydrodynamics and Arrhenius chemical kinetics. The computational approach employs a cell-centered, finite-volume numerical scheme, which is of the 2nd-order accuracy in time, 4th-order in space for the convective terms, and of the 2nd-order in space for the diffusive terms. The channels of blockage ratios 0.33~0.67 are considered, with the Lewis numbers in the range $0.2 \le Le \le 2.0$ employed. It is shown that the Lewis number influences the flame evolution substantially. Specifically, flame acceleration weakens for $Le>1$ (inherent to fuel-lean hydrogen or fuel-rich hydrocarbon burning), presumably, due to a thickening of the flame front. In contrast, $Le<1$ flames (such as that of rich hydrogen or lean hydrocarbon) acquire an extra strong folding of the front and thereby accelerate even much faster. The later effect can be devoted to the onset of the diffusional-thermal combustion instability.

Keywords: **premixed combustion, nonequidiffusive flames, Lewis number, obstructed channels.**

1. Introduction

Flame acceleration (FA) and deflagration-to-detonation transition (DDT) are usually considered as the fire safety demands, but they are also expected to improve various advanced combustion technologies such as pulse- and rotation-detonation engines [1] or micro-combustors [2]. Among geometries associated with FA and DDT, obstructed pipes provide the fastest regime of burning [3]. While flame propagation through obstacles is oftentimes associated with turbulence, shocks [4], or hydraulic resistance [5], Bychkov *et al.* [6-9] identified a conceptually laminar, shockless mechanism of ultrafast FA in semi-open channels or tubes equipped with a toothbrush-like array of obstacles. This mechanism is illustrated in Fig. 1, and it is devoted to a powerful jet-flow along the channel centerline, generated by a cumulative effect of delayed combustion in the pockets between the obstacles. According to the analytical formulation [6], substantiated by the computational simulations [6,7,9], a flame accelerates in a two-dimensional (2D) channel as

$$\frac{U_{tip}}{S_L} = \Theta \exp\left[\frac{(\Theta-1)}{(1-\alpha)}\frac{S_L}{R}t\right] = \Theta \exp(\sigma\tau), \qquad \sigma = \frac{(\Theta-1)}{(1-\alpha)}, \tag{1}$$

where $U_{tip} \equiv dZ_{tip} / dt$ is the flame tip velocity in the laboratory reference frame, S_L the planar flame speed, $\Theta \equiv \rho_u / \rho_b$ the thermal expansion in the burning process, R the half-width of the channel, α the blockage ratio, and $\tau = S_L t / R$ the scaled time. This FA is extremely powerful indeed: say, for typical $\Theta = 8$ and $\alpha = 1/2$, the scaled exponential acceleration rate of Eq. (1) is $\sigma = (\Theta - 1)/(1-\alpha) = 14$! It is noted that σ drastically depends on α ; it grows with α and Θ, thereby promoting FA, but it does not depend on R. To some extent, this makes the Bychkov acceleration mechanism scale-invariant (Reynolds-independent) and, thereby, relevant to various scales, including micro-combustors, industrial conduits as well as mining and subway tunnels.

The theory and modelling [6] adopted a number of simplifications, including the conventional approach of equidiffusive combustion, when the Lewis number, defined as the thermal-to-mass diffusivities ratio is unity, $Le = 1$. However, $Le \neq 1$ oftentimes in the practical reality, which leads to the interplay between the internal and global flame structures such as the diffusional-thermal instability. The impact of Le on FA has been found substantial in *unobstructed* channels [10-12]. In particular, Bilgili *et al.* [12] has shown that FA in unobstructed channels is drastically promoted in a $Le < 1$ premixtures, due to enormous extra elongation ("channeling") of the flame front. In contrast, $Le > 1$ combustion leads to a flame "thickening", thereby moderating FA.

May we expect the same or similar impacts of Le in an obstructed channel? Answering this question constitutes the focus of the present work. Initially, we anticipated a similar effect for $Le > 1$ flames, but not in the case of $Le < 1$, because the obstacles were expected to prevent the flame channeling and, thereby, drastic elongation of the flame surface area, accompanied by FA. In contrast, we have found that the role of Le is significant in *both* cases, with moderation of acceleration for $Le > 1$ flames and strong promotion of FA in the case of $Le < 1$ combustion. In fact, the impact of the Lewis number is found to be as strong as that of the blockage ratio α.

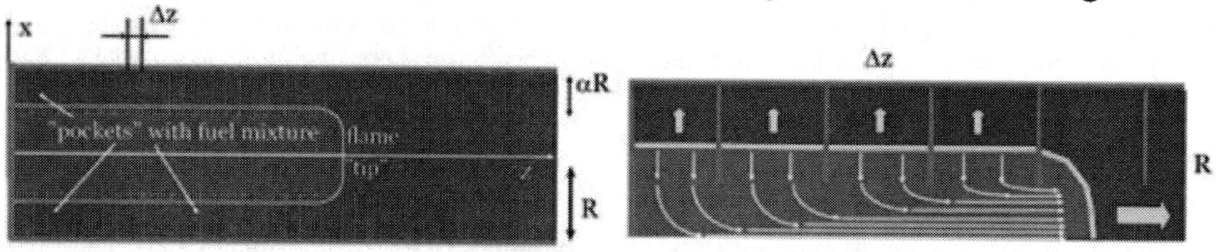

Figure 1: An illustration of the Bychkov mechanism of flame acceleration in an obstructed channel.

2. Numerical Method

We performed the computational simulations of the following 2D, Cartesian set of equations:

$$\frac{\partial}{\partial t}\rho + \frac{\partial}{\partial x_i}(\rho u_i) = 0, \qquad \frac{\partial}{\partial t}(\rho u_i) + \frac{\partial}{\partial x_j}(\rho u_i u_j + \delta_{i,j} P) - \gamma_{i,j} = 0, \tag{2}$$

$$\frac{\partial}{\partial t}\left(\rho\varepsilon + \frac{1}{2}\rho u_i u_j\right) + \frac{\partial}{\partial x_i}\left(\rho u_i h + \frac{1}{2}\rho u_i u_j u_j + q_i - u_j \gamma_{i,j}\right) = 0, \tag{3}$$

$$\frac{\partial}{\partial t}(\rho Y) + \frac{\partial}{\partial x_i}\left(\rho u_i Y - \frac{\zeta}{Sc}\frac{\partial Y}{\partial x_i}\right) = -\frac{\rho Y}{\tau_R}\exp(-E_a / R_p T), \tag{4}$$

where Y is the mass fraction of the fuel mixture; $\varepsilon = QY + C_V T$ and $h = QY + C_P T$ the specific internal energy and enthalpy, respectively; $Q = C_P T_f (\Theta - 1)$ the energy release in the reaction, with the specific heats at constant volume, C_V, and pressure, C_P, the initial fuel temperature,

$T_f = 300K$, pressure, $P_f = 1\,\text{bar}$, and density, $\rho_f = 1.16\,\text{kg/m}^3$. The stress tensor $\gamma_{i,j}$ and the energy diffusion vector q_i read

$$\gamma_{i,j} = \zeta\left(\frac{\partial u_i}{\partial x_j} + \frac{\partial u_j}{\partial x_i} - \frac{2}{3}\frac{\partial u_k}{\partial x_k}\delta_{i,j}\right), \qquad q_i = -\zeta\left(\frac{C_p}{\mathrm{Pr}}\frac{\partial T}{\partial x_i} + \frac{Q}{\mathrm{Sc}}\frac{\partial Y}{\partial x_i}\right), \qquad (5)$$

where $\zeta \equiv \rho\nu$ is the dynamic viscosity, being $\zeta_f = 1.7\times10^{-5}\,kg/(\text{m}\cdot\text{s})$ in the fuel mixture. Then the thermal flame thickness can be defined, conventionally, as $L_f = \zeta_f / \mathrm{Pr}\,\rho_f S_L = 4.22\times10^{-5}\,\text{m}$. As a result, Eqs. (2) – (5) include fully-compressible hydrodynamics, transport properties (heat conduction, diffusion and viscosity), and combustion imitated by a single irreversible Arrhenius reaction of the first order, with the activation energy E_a and the constant of time dimension τ_R. More details of the numerical scheme and the solver can be found, for instance, in Refs. [6,7,9].

Similar to Fig. 1, the flame propagates in a long 2D channel of width $2R$, where the fraction $2R\alpha$ is blocked by the obstacles with spacing (the distance between two neighboring obstacles) Δz. The channel width is described by the Reynolds number associated with flame propagation, $\mathrm{Re} \equiv RS_L/\nu = R/\mathrm{Pr}\,L_f = R/L_f$. In the present work, we used $\Delta z/R = 1/4$, $\mathrm{Re} = 24, 36, 48$, and $\alpha = 1/3; 1/2; 2/3$. The walls of the channel and obstacles are adiabatic, $\mathbf{n}\cdot\nabla T = 0$, and free-slip, $\mathbf{n}\cdot\mathbf{u} = 0$, where $\mathbf{n}$ is the normal vector at the wall. The absorbing (non-reflecting) boundary conditions are employed at the open end to prevent the reflection of the sound waves and weak shocks. The left end of the free part of the channel is closed while the other end, on the right, is open, with the boundary conditions $\rho = \rho_f$, $P = P_f$, $u_z = 0$ adopted. The initial flame structure is imitated by the classical Zeldovich-Frank-Kamenetsky (ZFK) solution for a hemispherical flame front [7], of initial radius $5.1L_f$, ignited at the centerline, and the closed end of the channel.

Both the fuel mixture and burnt matter are assumed to be ideal gases of constant molecular weight, $M = 29$ kg/kmol, such that the equation of state is $P = \rho R_P T/M$, with the universal gas constant $R_P = 8.31\,\text{kJ}/(\text{kmol}\cdot\text{K})$. The chemical properties of the fuel mixture are chosen to model methane-air burning with a Lewis number in the range $0.2 \le Le \le 2.0$, thermal expansion $\Theta = 8$, and the planar flame speed $S_L = 34.7$ cm/s. The initial speed of sound in this fuel mixture is 10^3 times larger, $c_0 = 347$ m/s, such that hydrodynamics is almost incompressible at the initial stage of burning. In this respect, in addition to the standard scaling of Eq. (1), U_{tip}/S_L, we also scaled U_{tip} by the local, instantaneous speed of sound, $c_{tip} = \sqrt{(c_p/c_v)\times(R_p/M)\times T_{tip}}$. Such a flame tip Mach number $M_{tip}(t) \equiv U_{tip}/c_{tip}$ appears a good measure of the current level of compressibility as well as of the instantaneous stage of the DDT process, being $M_{tip} << 1$ at the initial, quasi-isobaric stage of burning and approaching an order of unity by the time of detonation triggering.

3. Results and Discussion

We have performed extensive computational simulations of premixed FA in the configuration described above, and for various Le, α and Re. Because of the limit of space, we present only a single combination of color temperature snapshots, but, presumably, the most illustrative. Figure 2 (a-i) shows the flame shapes and positions attained at various Le, $Le = 0.2$, 1.0, 2.0, and α, $\alpha = 1/3; 1/2; 2/3$ but at the *same time instant*, $\tau = 0.075$, and for the same $\mathrm{Re} = 24$ in all nine simulation runs. The respective $M_{tip}|_{\tau=0.075}$ are also depicted. The flames are represented by the color temperature snapshots, from 300 K in the fuel (blue) till 2400 K in the burnt matter (red). It is seen that the role of the Lewis number is paramount and as strong as that of the blockage ratio. Indeed, when both the effect of large α and nonequidiffusivity (with $Le < 1$) work together, as in Fig. 2c for $\alpha = 2/3$ and $Le = 0.2$, then the flame front is drastically fold; it has propagated through $17R$ and is, presumably, about to trigger detonation as the flame tip Mach number is as high as $M_{tip} = 0.6$. Consequently, FA is enormous in this case. In contrast, a $Le > 1$ flame in a channel with small blockage ratio accelerates very slow, as observed in Fig. 2g for $\alpha = 1/3$ and $Le = 2.0$. In other cases of Fig. 2, the effects of α and Le compete such that we observe almost equivalent flames and M_{tip} in the pairs 2d and 2h; 2a and 2e; an even 2b and 2i, respectively.

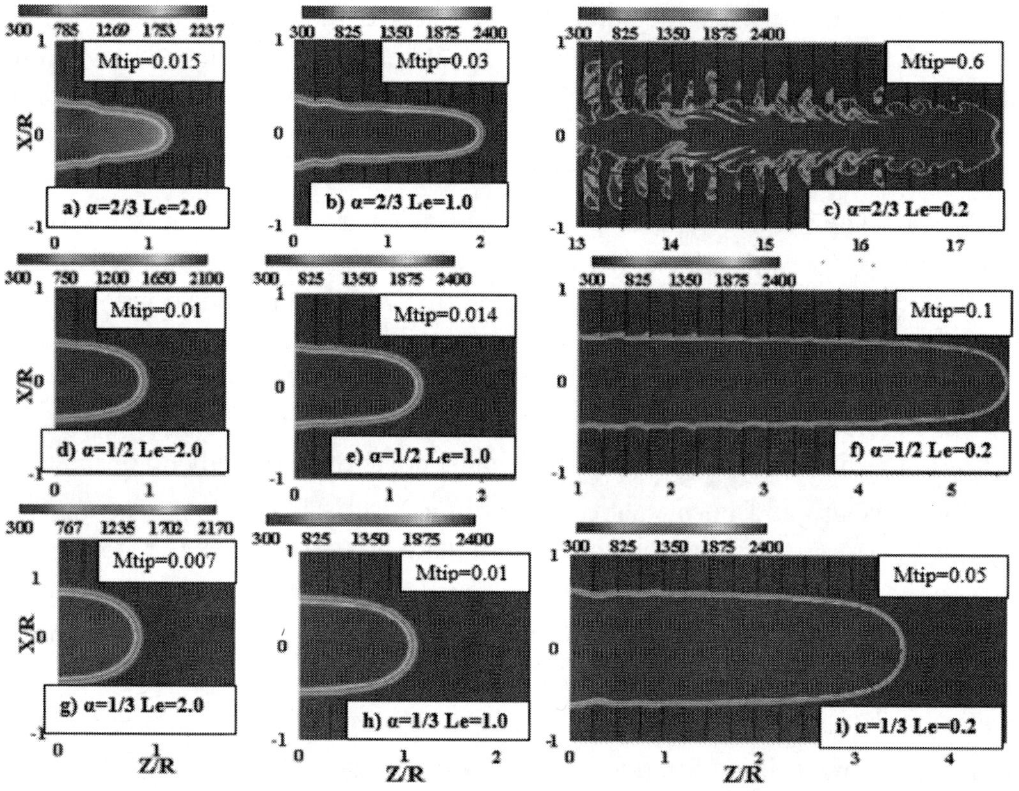

Figure 2: Color temperature snapshots [from 300 K in the fuel (blue) till 2400 K in the burnt matter (red)] taken at the same scaled time $\tau = tS_L/R = 0.075$ for Re = 24 and various Le = 0.2, 1.0, 2.0 and α = 1/3; 1/2; 2/3.

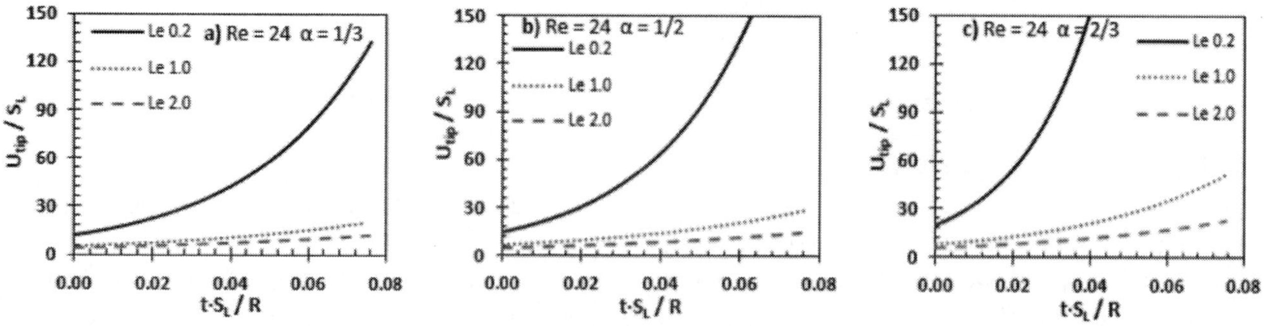

Figure 3: The scaled flame tip U_{tip}/S_L vs the scaled time τ for Re = 24 and α = 1/3 (a), 1/2 (b), and 2/3 (c).

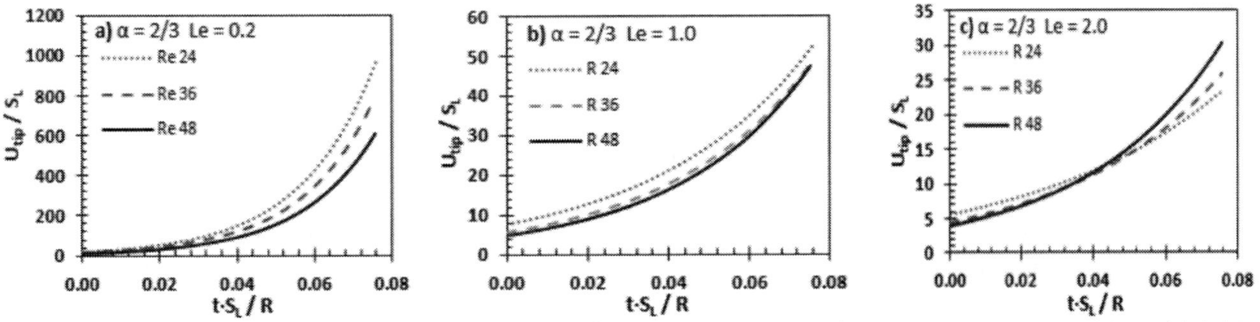

Figure 4: The scaled flame tip velocity U_{tip}/S_L vs the scaled time τ for α = 2/3 and *Le* = 0.2 (a), 1.0 (b), and 2.0 (c).

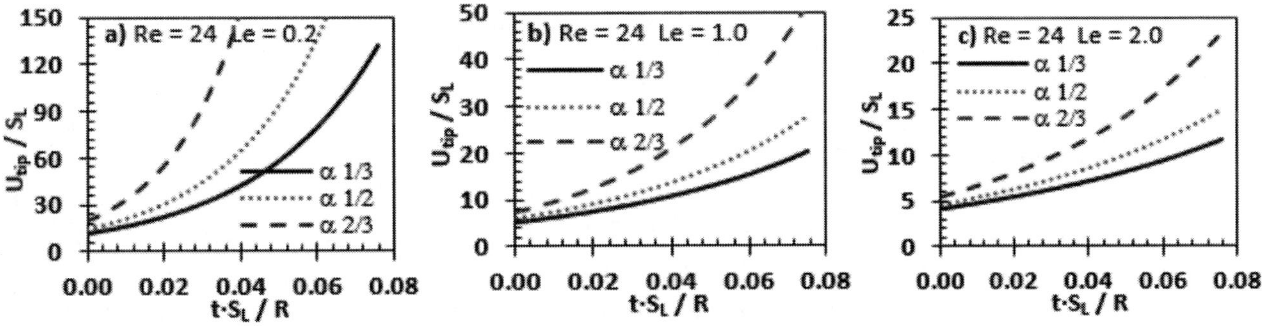

Figure 5: The scaled flame tip velocity U_{tip}/S_L vs the scaled time τ for Re = 24 and *Le* = 0.2 (a), 1.0 (b), and 2.0 (c).

To quantify the impact of *Le*, in Fig. 3 we present the time evolution of the scaled flame tip velocity, U_{tip} / S_L, for several blockage ratios, α = 1/3 (3a), 1/2 (3b), and 2/3 (3c), with various *Le*, *Le* = 0.2, 1.0, 2.0 in each plot. It is seen that the effect of *Le* is very strong, especially for the *Le* < 1 flames. Indeed, in all three Figs. 3(a-c), *Le* = 0.2 promotes U_{tip} almost by an order of magnitude as compared to the equidiffusive case, *Le* = 1. The effect of Le > 1 is substantially weaker, but *Le* = 2.0, nevertheless, noticeably moderates FA as compared to the *Le* = 1 cases.

Figure 4 scrutinizes the role of the channel width for various *Le* and α. It is seen that the impact of Re is minor as all the curves for Re = 24, 36 and 48 go very close in all three Figs. 4 (a-c). This actually supports the Bychkov formulation [6] predicting Re-independent FA, Eq. (1). On the other hand, Fig. 4 shows a very intriguing result: even the Re-dependence is quite weak, the impact of *Le* may change it, up to the opposite one. Indeed, FA weakens with Re for $Le \leq 1$ flames, Fig. 4(a, b), but it is promoted with Re in the *Le* > 1 case, Fig. 4c. In this light, we can potentially look for a certain threshold *Le* that would correspond to the change of the trend and thus provide the complete Re-independence.

Figure 5 shows the role of the blockage ratio α. It is seen that α-dependence is significant and much stronger than Re-dependence for all *Le* considered. At the same time, the impact of *Le* on α-dependence is less than that on Re-dependence: α-dependence does not change sign due to *Le*, but there is a noticeably quantitative effect such that α-dependence is stronger for *Le* < 1 flames.

Finally, we have studied all the acceleration trends observed, and when acceleration exhibited an exponential trend, then the exponential acceleration rate was calculated. The result is plotted versus *Le* in Fig. 6, for Re = 24, 36 and 48 in Figs. 6 (a-c), respectively, with α = 1/3, 1/2 and 2/3 in each figure. As expected, the acceleration rate σ appears largest for non-equidiffusive cases of *Le* < 1. More specifically, the exponential acceleration trend is seen for *Le* = 0.2 only for low α, α = 1/3, 1/2. The absence of this trend for α = 1/2 is attributed to a strong competition between the *Le* and α effects in that scenario.

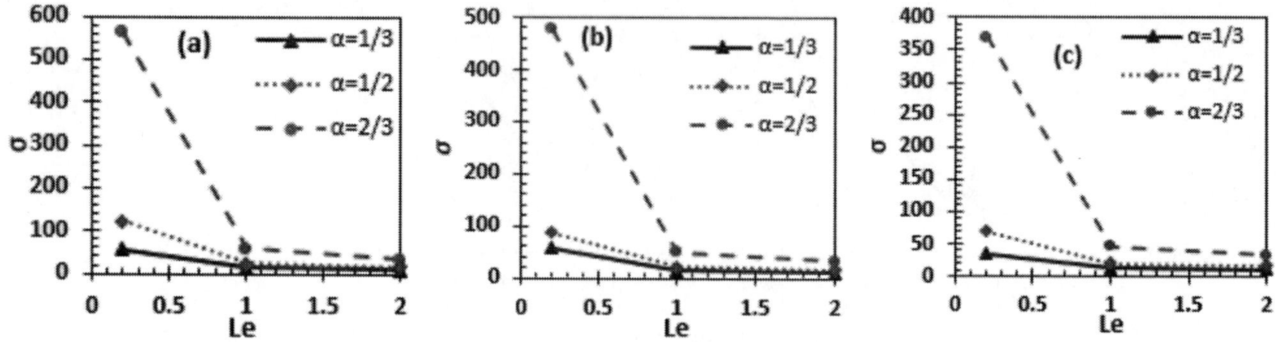

Figure 6: The exponential acceleration rate σ vs the Lewis number *Le* for Re = 24 (a), 36 (b), and 48 (c), with α = 1/3, 1/2, and 2/3 in each figrure.

4. Conclusions

In this work, a thorough investigation of non-equidiffusive FA in obstructed channels has been performed, by means of computational simulations, and a profound impact of *Le* on FA has been identified. This effect is compared to that of α and it has been found to be as strong. The non-equidiffusive scenarios involve non-unity *Le*, such that in the case of $Le < 1$, promotion of flame acceleration is discovered. It is found that *Le* influences α-dependence. In contrast, moderation of FA has been observed for $Le > 1$ flames. In addition, a unique trend is noticed for the *Le*-impact on Re. Indeed, the Lewis number may change Re-dependence of FA to an opposite trend.

Finally, we aim to unify the analysis. Similar to Bychkov *et al.* [6], the combination $\sigma Z_f/\Theta R$ is plotted versus $\sigma t S_L/R$ in Fig. 7. Here, different colors represent different Lewis numbers (blue, orange and grey for $Le = 0.2$, 1.0 and 2.0, respectively), with three blockage ratios for each case, namely $\alpha = 1/3$ (solid), 1/2 (dashed) and 1/3 (dotted). While Ref. [6] was limited to equidiffusive flames, $Le = 1$, with all the data collapsing into a single curve; here we have various *Le*, and the set of curves associated with different *Le* (i.e. different colors) differ. However, for each given *Le*, the plots (of the same color) collapse into a single curve, thereby fitting the Bychkov model.

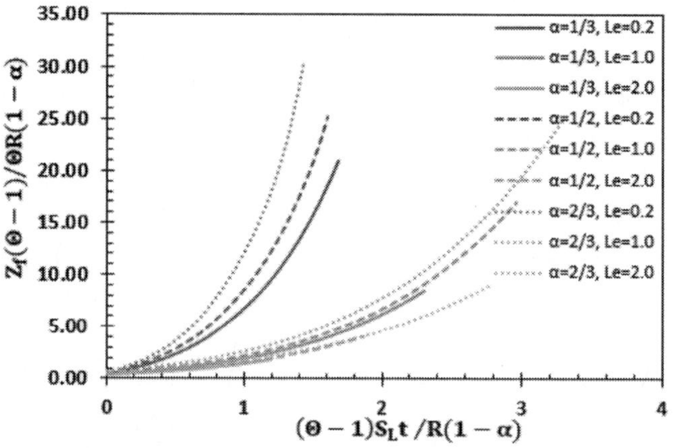

Figure 7: Scaled flame tip position $\sigma Z_f/\Theta R$ vs the scaled time $\sigma t S_L/R$ for various $Le = 0.2$ (blue), 1.0 (orange) and 2.0 (grey), with three different $\alpha = 1/3$ (solid), ½ (dashed) and 2/3 (dotted) for each given *Le*.

5. Acknowledgements

This research was funded by the NSF through the CAREER Award No. 1554254 (V.A.).

6. References

[1] G. Roy, S. Frolov, A. Borisov, D. Netzer, Prog. Energy Combust. Sci. 30 (2004) 545.
[2] Y. Ju, K. Maruta, Prog. Energy Combust. Sci. 37 (2011) 669.
[3] E.S. Oran, V.N. Gamezo, Combust. Flame 148 (2007) 4.
[4] G. Ciccarelli, S. Dorofeev, Prog. Energy Combust. Sci. 34 (2008) 499.
[5] I. Brailovsky, L. Kagan, G. Sivashinsky, Phil. Trans. R. Soc. A 370 (2012) 625.
[6] V. Bychkov, D. Valiev, L.-E. Eriksson, Phys. Rev. Lett. 101 (2008) 164501.
[7] D. Valiev, V. Bychkov, V. Akkerman, C.K. Law, L.-E. Eriksson, Combust. Flame 157 (2010) 1012.
[8] V. Bychkov, V. Akkerman, D. Valiev, C.K. Law, Combust. Flame 157 (2010) 2008.
[9] O. Ugarte, V. Bychkov, J. Sadek, D. Valiev, V. Akkerman, Phys. Fluids 28 (2016) 093602.
[10] J.B. Bell, R.K. Cheng, M.S. Day, I.G. Shepherd, Proc. Combust. Inst. 31 (2007) 1309.
[11] C. Cui, M. Matalon, T.L. Jackson, AIAA Journal 43 (2005) 1284.
[12] S. Bilgili, D. Valiev, V. Bychkov, V. Akkerman, 9th U.S. National Combustion Meeting, May 17-20, 2015, paper #3E08, pp. 1-10.

Sub Topic: Fire

2018 Spring Technical Meeting
Eastern States Section of the Combustion Institute
Hosted by Pennsylvania State University
March 04-07, 2018

Transient Ellipsoidal Flames in Microgravity

A. Markan[1,*], *H. R. Baum*[1], *P. B. Sunderland*[1], *J. G. Quintiere*[1], *J. L. de Ris*[2]

[1]*Fire Protection Engineering, University of Maryland, College Park, MD*
[2]*retired, FM Global, Norwood, MA*
[*]*Corresponding Author Email: akshit.markan@gmail.com*

Abstract: The Burning Rate Emulator (BRE) is a porous gas-fueled burner that examines the burning of condensed phase fuels. Previous BRE experiments conducted in NASA Glenn's 5-s Zero Gravity Research Facility show the evolution of an approximately ellipsoidal flame moving away from the burner. Steady flames were not achieved during the five second duration. Hence, a transient combustion model formulated in oblate ellipsoidal coordinates is developed to analyze the behavior of microgravity BRE flames. The model is axially symmetric and considers the burning of gaseous fuel leaving the surface of a porous ellipsoidal particle in microgravity. A composite solution is generated as a product of an exact steady state solution and an asymptotic transient solution that becomes exact far from the burner. The BRE burner geometry is idealized as an axially symmetric porous disc. The microgravity results are compared to the transient model to help predict the flame behavior beyond the duration of the test.
Keywords: combustion modelling, laminar flames, microgravity

1. Introduction

A porous burner has been recently developed to emulate the combustion of condensed phase fuels [1-3]. Experiments using the burner were conducted in NASA Glenn's 5-s microgravity facility in order to provide a material flammability test aboard spacecraft. The burner flames appeared to grow over the duration of the experiment while transforming from a near ellipsoidal shape to a circular shape far from the burner. The current study proposes a transient mathematical model for the flame structure over such a circular burner in microgravity conditions. The analysis considers a 1D spherical conduction problem with heated flow as the starting point. The conduction problem is related to the combustion problem using a model based on the burning of small particles formulated in ellipsoidal coordinates [4,5].

2. Spherical conduction problem with heated flow

A porous sphere is considered with heated fluid at temperature T_s flowing out of the surface with ambient temperature T_∞. The sphere has a radius R and the environment has a specific heat c_p, density ρ and thermal conductivity k which are assumed to be constant. The desired quantity is the transient temperature T as a function of the radial distance r and the fluid flow rate $\dot{M}$ at the surface. The energy equation governs the temperature profile of the fluid surrounding the sphere [6].

$$\frac{\partial \theta}{\partial \tau} = \frac{\partial^2 \theta}{\partial y^2} + \frac{(2y-b)}{y^2}\frac{\partial \theta}{\partial y} \tag{1}$$

Here, the non-dimensional terms are:

$$\theta = \frac{T-T_\infty}{T_s-T_\infty}, \qquad y = \frac{r}{R}, \qquad b = \frac{\dot{M}c_p}{4\pi kR}, \qquad \tau = \frac{kt}{\rho c_p R^2}. \tag{2}$$

A composite transient solution to the conduction problem is presented below as a combination of the exact steady state solution and the asymptotic transient solution in the far field. The steady state solution for the energy equation is given as

$$\theta = \frac{e^{-b/y}-1}{e^{-b}-1}. \tag{3}$$

The effect of fluid flow becomes negligible far from the sphere and hence, the transient solution [7] in the far field can be expressed as

$$\theta = \frac{1}{y}erfc\left(\frac{y-1}{2\sqrt{\tau}}\right), \tag{4}$$

where $(1/y)$ represents the steady state solution for the conduction problem without fluid flow. The composite solution is attained by replacing $(1/y)$ with steady state solution given in Eq. (3).

$$\theta = \left(\frac{e^{-b/y}-1}{e^{-b}-1}\right) erfc\left(\frac{y-1}{2\sqrt{\tau}}\right) \tag{5}$$

To verify the accuracy of the composite solution, Eq. (5) is substituted in the PDE given by Eq. (1). The error in the PDE is determined with time for different values of y (r/R). It is considered that a sphere of diameter 25 mm has a fluid flowing through it at a surface mass flux of $\dot{m}_0'' = 3.582$ g/m^2-s, which corresponds to the mass flow rate used during the 5-s drop experiments [3]. The PDE error is shown in Figure 1. It can be seen that the error is less than 1% for all radial positions and at all times. Thus, the logic behind the composite thermal solution can be used as a benchmark in further analysis involving combustion.

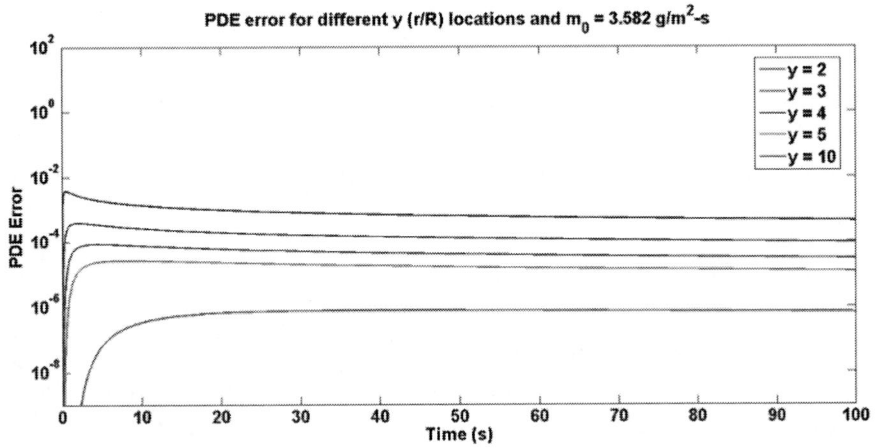

Figure 1: PDE error for the composite solution to the spherical conduction problem with flow

3. Ellipsoidal combustion model

The burning of fuel leaving the surface of an ellipsoidal body is studied in a microgravity environment. The model considers fuel is injected through the surface at a constant mass flow rate $\dot{M}$. The schematic and geometry of the ellipsoidal body is as shown in Figure 2.

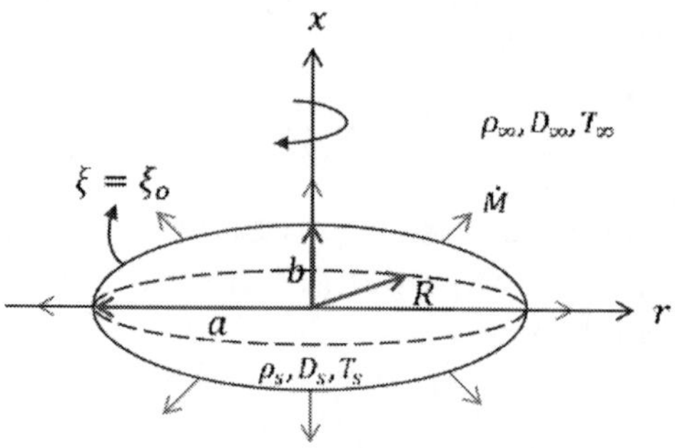

Figure 2: Schematic and geometry of the ellipsoidal body

The ellipsoid body is at a constant temperature T_s which is the vaporization temperature of the fuel being injected. This temperature T_s corresponds to the density ρ_s and diffusivity D_s. Initially at time $t = 0$, the environment surrounding the burner is at a constant temperature T_∞, density ρ_∞ and diffusivity D_∞. The flame which is initially at the surface starts spreading outwards while the density ρ, velocity u and temperature T vary in space and time. (x, r) is a cylindrical coordinate system in which the ellipsoidal burner is defined. The burner is symmetric about the x axis and has a semi-major axis a and a semi-minor axis b. Also, since the burner is symmetric, the semi-major axis can be written as $a = R$. The problem can be simplified by converting to oblate ellipsoidal coordinate system given by (ξ, η). The system aligns with the cylindrical coordinate system such that the surface of the burner is defined by $\xi = \xi_o$. The two coordinate systems can be related using a quantity c as shown below:

$$\xi = \frac{1}{\sqrt{2}c}\left[x^2 + r^2 - c^2 + \sqrt{(x^2 + r^2 - c^2)^2 + (2xc)^2}\right]^{1/2}, \qquad \eta = \frac{x}{c\xi} \tag{6}$$

The quantity c can be related to the semi-major and semi-minor axis as follows:

$$a = c\sqrt{1 + {\xi_o}^2}, \qquad b = c\xi_o \tag{7}$$

3.1. Conservation laws

The temperature T and oxygen mass fraction Y_O are combined into a mixture fraction variable Z [4] which can be written as

$$Z = \frac{c_p(T - T_\infty) + \Delta h_c(Y_O - Y_{O,\infty})/S}{c_p(T_s - T_\infty) - \Delta h_c Y_{O,\infty}/S}. \tag{8}$$

Here, $Y_{O,\infty}$ is the ambient oxygen mass fraction, Δh_c is the heat of combustion per mass of fuel, c_p is the specific heat of the gas and S is the stoichiometric which denotes the mass of oxygen consumed per unit mass of fuel. It is assumed that the Lewis number $Le \equiv k/\rho c_p D = 1$. The model assumes that the entire combustion process can be represented the evolution of the mixture fraction. The energy and species conservation equations are combined into a single mixture fraction equation. Thus, the mass conservation and mixture fraction equations in ellipsoidal coordinates [5] take the form:

$$\frac{\partial \rho}{\partial t} + \frac{1}{c^2(\xi^2+\eta^2)}\frac{\partial}{\partial \xi}\left[\rho(\xi^2 + 1)\frac{\partial \phi}{\partial \xi}\right] = 0 \tag{9a}$$

$$\frac{\partial(\rho Z)}{\partial t} + \frac{1}{c^2(\xi^2+\eta^2)}\left\{\frac{\partial}{\partial \xi}\left[\rho Z(\xi^2 + 1)\frac{\partial \phi}{\partial \xi}\right] - \frac{\partial}{\partial \xi}\left[\rho D(\xi^2 + 1)\frac{\partial Z}{\partial \xi}\right]\right\} = 0 \tag{9b}$$

Here, ϕ is the velocity potential such that $\vec{u} = \partial\phi/\partial\xi$. The boundary conditions for this problem are that the mixture fraction $Z = 1$ on the surface of the burner which is defined by $\xi = \xi_o$ and $Z = 0$ in the far field where $\xi \to \infty$. Also, the velocity potential ϕ takes a constant value along the burner surface due to no-slip condition and in the far field $\phi \to 0$.

3.2. Composite solution

A composite solution is presented for the ellipsoidal combustion problem similar to the thermal solution for the spherical conduction problem in the previous section. To get a steady state solution, we can assume that the quantities of interest do not vary in the time domain. ρD can be assumed to be a function of Z. For the sake of simplicity, ρD is taken as a constant. The continuity and energy equations are integrated and solved by applying the boundary conditions for the mixture fraction on the burner surface and in the far field. This gives a steady state solution for the mixture fraction.

$$Z = \frac{1-\exp\left[b\left(\arctan(\xi)-\frac{\pi}{2}\right)\right]}{1-\exp\left[b\left(\arctan(\xi_o)-\frac{\pi}{2}\right)\right]}, \qquad b = \frac{\dot{M}}{4\pi c\rho D}. \tag{10}$$

To get the far field transient solution, we can assume that velocity is really small ($\vec{u} \to 0$). Also, since $\xi \to \infty$ in the far-field and $-1 \le \eta \le 1$, it is reasonable to assume that $\xi^2 + 1 \approx \xi^2$ and $\xi^2 + \eta^2 \approx \xi^2$. Hence, the mixture fraction equation (9b) reduces to the form as shown below.

$$\frac{\partial Z}{\partial \tau} = \frac{1}{\xi^2}\frac{\partial}{\partial \xi}\left(\xi^2 \frac{\partial Z}{\partial \xi}\right) \tag{11}$$

Here, τ is the dimensionless time given by $\tau = D_\infty t/c^2$. This form of partial differential equation can be readily solved and the analytical solution is given as:

$$Z = \frac{1}{\xi}\operatorname{erfc}\left(\frac{\xi-\xi_o}{2\sqrt{\tau}}\right) \tag{12}$$

The above expression is the transient solution for the mixture fraction Z in the far-field. If we look at steady state for equation (8), i.e., $\tau \to \infty$, we obtain that Z varies with $1/\xi$. This shows that the error function part of the solution is the transient component whereas the $1/\xi$ indicates the steady component. Hence, to get the composite solution in the entire domain, we replace $1/\xi$ in equation (12) with the exact steady state solution derived in equation (10). Hence, the final solution for the mixture fraction can be written as,

$$Z = \left\{\frac{1-\exp\left[b\left(\arctan(\xi)-\frac{\pi}{2}\right)\right]}{1-\exp\left[b\left(\arctan(\xi_o)-\frac{\pi}{2}\right)\right]}\right\}\operatorname{erfc}\left(\frac{\xi-\xi_o}{2\sqrt{\tau}}\right), \qquad b = \frac{\dot{M}}{4\pi c\rho D} \tag{13}$$

A piecewise linear state relation given by equation (8) expresses the mixture fraction in terms of the temperature and oxygen mass fraction. Further state relations are chosen to connect the fuel mass fraction and products mass fraction to the mixture fraction as shown below.

$$Z = \frac{SY_F-(Y_O-Y_{O,\infty})}{SY_{Fb}+Y_{O,\infty}} \tag{14a}$$

$$Z = \frac{(1+S)Y_F+Y_P}{(1+S)Y_{Fb}+Y_{Pb}} \tag{14b}$$

The quantities Y_{Fb} and Y_{Pb} denote the fuel and product mass fractions at the burner surface respectively. The state relations emerge from the assumptions of Fick's Law of diffusion and equal diffusivity for all species, a plausible approximation if nitrogen is the dominant molecular species. These relations satisfy the mixture fraction equation and the boundary conditions.

The reaction taking place around the burner is assumed to be infinitely fast such that the oxidizer and fuel cannot coexist. The fuel and oxygen are separated by the infinitesimally thin flame sheet which is located on a curve given by,

$$Z = Z_{fl} = \frac{Y_{O,\infty}}{SY_{Fb}+Y_{O,\infty}} \tag{15}$$

Since the thin flame sheet separates the fuel and the oxygen, the reaction zone can be divided into two domains, i.e., the oxidizer side and the fuel side. On the fuel side of the flame where $Z \geq Z_{fl}$, $Y_O = 0$ and equation (14a) determines Y_F. Similarly, on the oxidizer side of the flame where $Z \leq Z_{fl}$, $Y_F = 0$ and equation (14a) determines Y_O. The product mass fraction is given by equation (14b) and the temperature can be found using equation (8).

4. Prediction of 5 s Drop Tests

The results of the 5-s BRE microgravity experiments are utilized to validate the transient mathematical model. During these experiments, fuel at a constant flow rate is passed through the burner surface. This is in accordance with the transient model. The burner geometry can be idealized as a flat circular disc with fuel flowing out from one side so as to apply the ellipsoidal model. Two particular tests are selected for this purpose with parameters shown in Table 1. The measurements during the test include the heat flux and the temperature at two locations on the burner surface, one at the center and the other at an offset radius R^* (R^* = 8.25 mm for the 25 mm burner, R^* = 16 mm for the 50 mm burner). The flame shape is recorded using analog video. The surface temperature during the 5 second test does not change by much and hence, it can be taken as a constant input for the mathematical model.

Table 1. Tests to study the mathematical model

D (mm)	Fuel	X_{O_2}	p (atm)	$\dot{m}''$ (g/m^2-s)	T_s (°C)
25	C_2H_4	0.30	0.5	3.53	34.5
50	C_2H_4	0.26	0.81	3.47	89.0

It is important to account for the fact that the experiment runs for only about 5 s and a steady flame is not achieved during the test. Hence, the transient model would provide a prediction of the flame shape and heat flux beyond the duration of the test. The composite transient solution presented in section 3 is utilized to provide a prediction of the flame shape and the heat flux for the 5 s drop tests. The ellipsoidal flame location ξ_{fl} is obtained by substituting the mixture fraction value Z_{fl} at the flame in the composite solution. The heat flux at the surface is obtained by differentiating the composite solution with the ellipsoidal coordinate ξ. Figure 3 shows the predicted heat flux and flame shape for the tests listed in Table 1. The predicted flame shape is denoted by dotted lines superimposed over the flame images. The graph shows the predicted heat flux at the surface along with the experimental heat flux at the two locations.

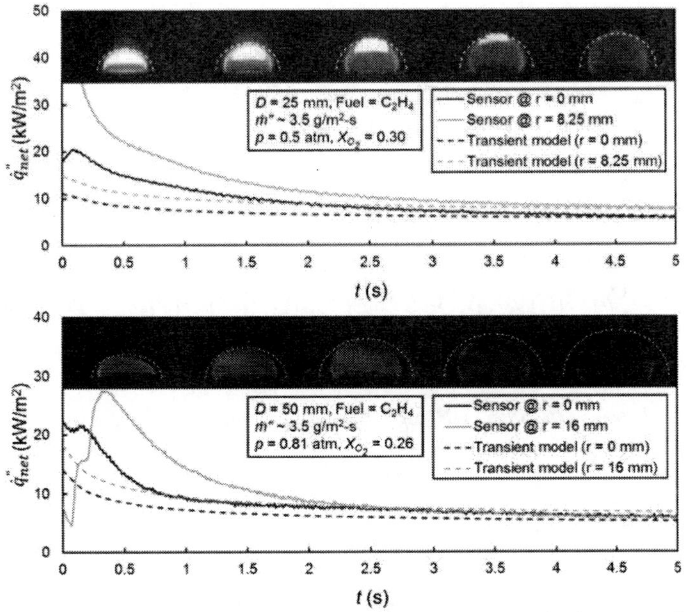

Figure 3: Comparison of 5-s microgravity tests with the transient model

5. Conclusions

A transient mathematical model for the BRE microgravity flame is presented. It is based on a model that predicts the burning of small firebrands and employs oblate ellipsoidal coordinates. A composite solution is generated by combining an exact steady state solution and a far-field asymptotic transient solution. The model accurately predicts the flame shape and heat flux for the microgravity BRE experiments beyond the 5-s duration. There is good agreement between the predicted heat flux and the experimental heat flux after the initial two seconds. The future work involves the inclusion of the possibility of radiative extinction in the model.

Acknowledgements

This research was funded by NASA's International Space Station Research Program, with D.P. Stocker serving as contract monitor.

References

[1] Y. Zhang, M. Kim, H. Guo, P.B. Sunderland, J.G. Quintiere, J. de Ris, D. Stocker, Combustion and Flame, 162 (2015) 3449-3455.

[2] Y. Zhang, M. Kim, P.B. Sunderland, J.G. Quintiere, J. de Ris, Experimental Thermal and Fluid Science, 73 (2016) 87-93.

[3] A. Markan, P.B. Sunderland, J.G. Quintiere, J.L. de Ris, D.P. Stocker, H.R. Baum, Combust. Flame, 2018 (accepted).

[4] H. Baum and A. Atreya, Fire Safety Science 11, 2014.

[5] W. Magnus, F. Oberhettinger, Formulas and Theorems for the Functions of Mathematical Physics, Chelsea, New York. p. 149, 1943.

[6] S. Goldstein, Lectures on Fluid Mechanics, Interscience, New York, p. 89, 1960.

[7] H. S. Carslaw, J. C. Jaeger, Conduction of Heat in Solids, Oxford University Press, 2nd ed, London, 1959, p. 215.

Spring Technical Meeting
Eastern States Section of the Combustion Institute
March 4-7, 2018
State College, Pennsylvania

Shock tube studies of methyl propanoate and methane kinetics

Shirin Jouzdani[1], Deshawn M.Coomb[1], Xuan Zheng[1], Ben Akhi-Kumgeh[1,]*

[1]Department of Mechanical and Aerospace Engineering
Syracuse University, Syracuse NY 13244
[*]Corresponding Author Email: bakihkum@syr.edu

Abstract:
The biodiesel surrogate, methyl propanoate, is more reactive than methane. Mixtures of the two can be used to control the combustion process in various combustion systems. This paper presents a shock tube study of the influence of chemical interactions resulting from mixing the two fuels on observable combustion properties, such as global chemical time scales and species time histories. Experiments are carried out at pressures of about 4 atm and 7 atm and temperatures in the range of 1000 K-1500 K. By means of direct laser absorption, CO evolution is measured. The CO absorbance is used to determine a pyrolysis time by means of which the effects of pressure and temperature on pyrolysis are probed. The evidence of chemical interactions during ignition is observed through a reduction of methane ignition delay times resulting from the inclusion of methyl propanoate. The ignition delay times of the mixtures equally show the increased resistance to ignition of methyl propanoate in the presence of methane. With respect to CO formation during methyl propanoate pyrolysis, the presence of methane is observed not to significantly influence the pyrolysis time, showing limited radical withdrawal by methane during the propanoate pyrolysis. Time histories of CO is also obtained for further model validation.
Methyl propanoate, methane ignition, fuel blend kinetics, pyrolysis, laser absorption spectroscopy

1. Introduction

Natural gas and biodiesel are of interest to modern combustion systems. Natural gas is a less pollutant fuel. Biodiesel is seen as a viable bio-derived fuel that can contribute to reducing environmental footprints of combustion and because of its renewability. Biodiesel can be used in most of today's diesel engines by blending with diesel fuel. They are composed of long-chain fatty acid esters. Their chemical kinetics have been approached by studying smaller methyl esters. Coniglio et al. [1] reviewed the experimental and kinetic modeling work related to biodiesel and its surrogates. In order to develop detailed chemical kinetic mechanisms for long-chain esters many studies have focused on the short chain esters.

Methyl propanoate (MP) is a small chain methyl esters which has been used to understand biodiesel combustion. Experimental data on its pyrolysis and laminar flames are more extensive than autoignition studies. Zhao et al. [2] studied the pyrolysis of MP in a laminar flow reactor at low pressures, and developed a kinetic mechanism for MP pyrolysis at low pressure. Farooq et al. [3] measured species time-histories in pyrolysis of MP and ethyl propanoate in a shock tube at temperatures between 1250- 1750 K and proposed a kinetic mechanism. Felsmann et al. [4] reported experimental measurements made of laminar low-pressure flames used to develop its kinetic mechanism. Ning et al. [5] conducted pyrolysis measurements of MP behind reflected shock waves at temperatures between 1292 - 1551 K at 1.6 atm, as well as chemical kinetic modeling to refine the chemical kinetic model of Felsmann et al. [4]. Furthermore, Wang et al. [6] reported an experimental study of several small methyl and ethyl esters, including MP, in laminar flames. The experimental data were compared with predictions from a number of kinetic models from literature which showed that improvements in mechanisms were needed.

Experimental studies on the autoignition of methyl propanoate are limited. Akih-Kumgeh et al. [7] measured autoignition delay times of C1-C4 methyl esters, investigating the structure reactivity trends. Zhang et al. [8] measured ignition delay times of MP in a shock tube a various conditions. A kinetic model was developed and validated against the ignition delay and pyrolysis data of Zhao et al. [2]. Recently, Kumar et al. [9] studied the ignition of MP at high pressure and low-to-intermediate temperature conditions in a rapid compression machine. Comparisons of the reactivity of MP to methyl ethanoate and methyl butanoate were made. There are no fundamental studies which explore the chemical influence of MP on the combustion of small alkanes.

Natural gas is being pursued as an alternative fuel in several engine designs. Engines using natural gas can either be dedicated to pure natural gas, or use a bi-fuel strategy. Wei and P. Geng [10] reviewed the use of natural gas in diesel engines. The composition of natural gas can vary due to several factors, but the main component is methane. As a result of the high methane content, natural gas suffers from poor ignition characteristics and requires a pilot fuel to be ignited. Variations of the natural gas composition can affect its autoignition characteristics [11]. Fuel composition can also affect the combustion and emission characteristics of engines [12]. Biodiesel and natural gas have been tested in several engine configurations, Yang et al. [13] completed as study of the effects of methane addition to biodiesel on the performance of diesel engine. The result showed that ignition delay increase with increasing the methane percentage and both CO and soot emissions reduced. Hiremath et al. [14] conducted a comparative analysis dual fuel (DF) and homogeneous compression ignition (HCCI) engines with methyl ester and compressed natural gas. The analysis showed that the HCCI mode with natural gas induction was more efficient than DF mode with both diesel and biodiesel fuels, in terms of efficiency and emissions. Also the natural gas HCCI engine operated more efficiently with diesel fuel compared to biodiesel fuel.

Here, we present an experimental study on the ignition and pyrolysis of methyl propanoate (MP) and methane behind reflected shock waves, using laser absorption methods to measure time-histories CO. The aim in to discover the effect of chemical interactions of the two fuels on observable combustion properties which affect combustion system performance.

2. Methods / Experimental

The investigations are carried out in a shock tube with an inner diameter of 10 cm, a driven section of 6.0 m, and a driver section of 3.0 m. Test mixtures are prepared in a 150-liter mixing tank using the manometric technique, with a subsequent mixing time of at least 12 hours. Research grade samples of MP (Sigma-Aldrich, 99.9 + %) with methane, oxygen, and argon, helium (Airgas, >99.999 %) are used in mixture preparation and shock generation.

Ignition is detected based on CH chemiluminescence signals recorded by a photodiode, equipped with a 430 ± 10 nm narrow band filter on the side wall. The ignition delay time is determined by extrapolating the maximum slope of the CH emission signal to the baseline photodiode signal. Simulations of ignition delay times, fuel and CO concentration profiles are carried out using the

CANTERA software package, assuming the chemical system behind reflected shock waves as a constant volume reactor.

CO time-histories are measured near 4.6 μm using a mid-infrared quantum-cascade laser (QCL) on the basis of which a characteristic pyrolysis time scale is defined. A pyrolysis time is defined based on a concentration correlation function in which the rate of formation of a major species, j, is used to represent the fuel being depleted:

$$f_{cc} \propto X_j \frac{dX_j}{dt} \tag{1}$$

Further details about the pyrolysis time can be found in our previous work [15]. In this work the pyrolysis time has been obtained in fuels with MP by focusing on absorption measurement of CO, because of its strong mid IR absorption bands without significant interference.

The pyrolysis time scale allows us to appreciate the temperature sensitivity of pyrolysis over a wide temperature range. For further constraining kinetic models, quantitative CO concentration can be obtained from its absorbance using the standard Beer-Lambert's Law:

$$\ln\left(\frac{I_0}{I}\right) = \alpha L = \sigma(\lambda, P, T)NL \tag{2}$$

Where I and I_0 are the transmitted and incident intensities, respectively, while α is the absorption coefficient with units of inverse length. Also, αL, is the absorbance, σ - the absorption cross-section, N is the fuel concentration and L is the shock tube inner diameter.

3. Results and Discussion

The results are presented below, starting with ignition delay times, followed by pyrolysis times and selected CO time histories. The experiments are compared to prediction of models by Zhao et al. [2] and Zhang et al. [8].

Ignition delay times for stoichiometric fuel/O_2/Ar mixtures are measured. The relative ignition behavior of 3% MP and 3% CH_4 are studied as shown in Fig. 1. It is observed that MP is much more reactive than CH_4. Further, blends of equal proportion of both fuels as well as 80% CH_4 with 20% MP are studied. The results in Fig. 1 show that the ignition delay times of CH_4 are about 8 times longer than of those of 80/20 blend at 1377 K and about 18 times longer that of those of 50/50 blend at 1360 K. The observed reactivity trends can be understood as resulting from the very shorts bonds of CH_4 which are resistant to reaction initiation and propagation. The highly reactive MP is connected to the presence of weaker C-H bonds and more reactive radicals that promote reactions once the fuel is attached. With addition of a bit of MP, many radicals are generated. Their reactions with CH_4 are marked by lower activation barriers.

The observed ignition delay times are also compared with predictions obtained using two chemical kinetic models from the literature [2, 8]. It is observed that the model by Zhang el al. [8] predicts the measured ignition delay times better than the model by Zhao et al. [2]. Both models capture the reactivity trend showing the significant effect of MP on methane ignition.

The global kinetics of the pyrolysis of MP and CH_4 can be assessed using the pyrolysis time [15]. The effect of fuel concentration on pyrolysis times of MP and its blend with equal proportion of CH_4 is shown in Fig. 2. The pyrolysis time is observed to be fairly insensitive to MP concentration and the presence of methane, although there is a very small effect in the case of adding methane at low temperatures. It is possible that a very weak effect of fuel concentration also exists but it cannot be clearly discerned from the concentration doubling used here.

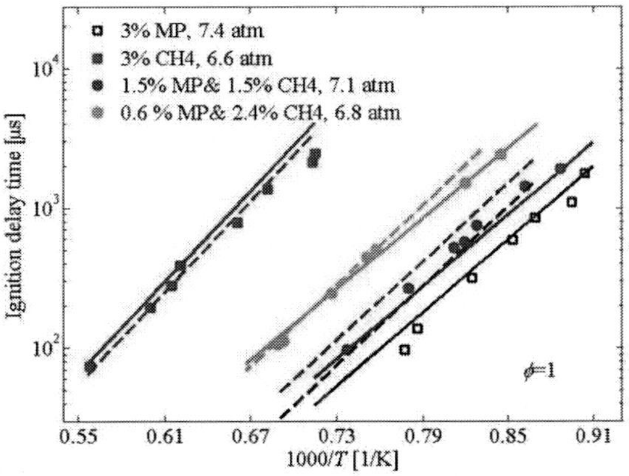

Figure 1: Experimental and model predictions of ignition delay times of MP, CH_4 and blend of both. Solid lines: model by Zhang et al. [8] Dashed lines: model by Zhao et al. [2]

Figure. 2a also shows the dependence of MP and CH_4 pyrolysis time scales on temperature. An Arrhenius-type dependence with a global activation energy between 50 to 60 kcal/mol is observed. This is in contrast to global activation energies of ignition process which is about 40 kcal/mol in this study.

The observed pyrolysis times are also compared with model predictions in Fig. 2b. Shown are results for mixtures of 3% MP and a mixture of MP and CH_4 in argon at average pressure 4 atm. It is observed that model by Zhao et al. [2] more closely captures the pyrolysis times at 3% MP while showing greater deviations for the blend mixture. The model by Zhang et al. [8] predicts much shorter pyrolysis times than measured. This is in contrast with the agreement observed for ignition, where the model by Zhang et al. [8] is better.

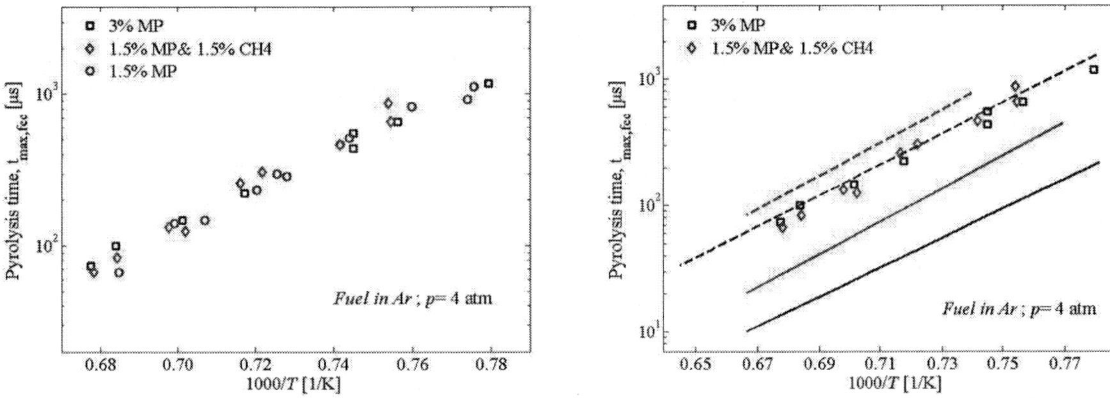

(a) (b)

Figure 2: a) Pyrolysis time scales of MP and its equal blend with CH_4. b) Comparison of pyrolysis time scales with simulations .Solid lines: model by Zhang et al. [8] Dashed lines: model by Zhao et al. [2]

Selected experiments are processed to obtain CO time histories for further comparison with model predictions. The absorption cross-sections are obtained from the HITRAN corresponding to the R9 CO absorption band (2179.8 cm−1). CO production is faster for 3% MP, 1.5% MP and the blend of 1.5% MP with 1.5% CH_4, respectively as shown in Fig. 3a. In every experiment shown here, CO formation is initially very fast, but after a short while the CO production becomes very slow. The measured profiles are also compared with model predictions in Fig. 3b. It is found that both models under-predict the initial rate of CO formation but the model by Zhang et al. [8] captures the equilibrium CO concentration at later times. The steep rise of the CO profiles at the beginning in the case of the model by Zhang et al. [8] is responsible for the shorter pyrolysis time predictions.

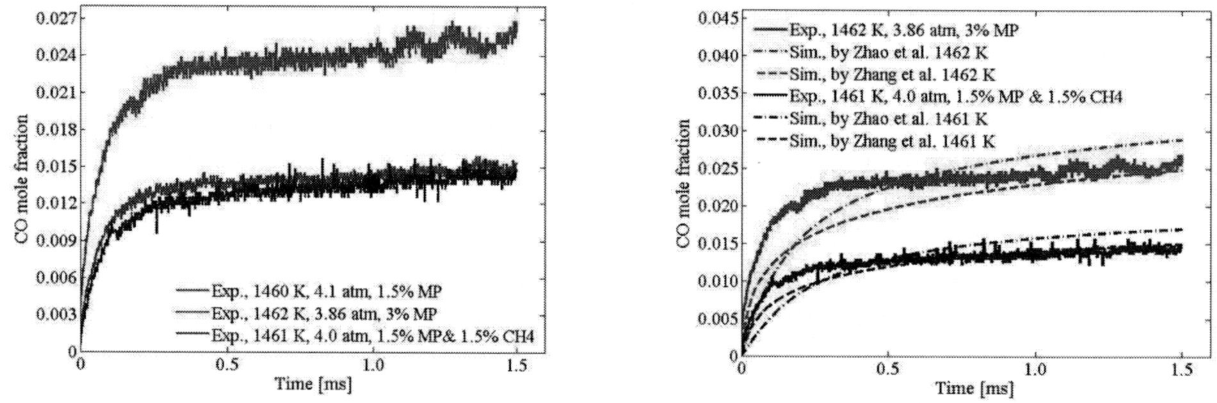

(a) (b)

Figure 3: a) CO time history during pyrolysis of MP and its equal blend with CH_4. b) Comparison of CO histories with simulations.

4. Conclusions

This work explored the ignition and pyrolysis of methyl propanoate (MP), methane and their mixtures. Shock tube ignition delay times, pyrolysis time scales and CO concentration time histories have obtained. The study brings out the enhancement of methane ignition through inclusion of methyl propanoate. With regards to MP pyrolysis, the presence of methane is observed not to significantly influence the pyrolysis time. The evolution of CO during pyrolysis of pure or blended MP shows a rapid initial rise and gradual approach to equilibrium.

Comparison of experimental results with model predictions show that current models predict ignition delay reasonably well with noted discrepancies in the case of pyrolysis and CO production.

5. Acknowledgements

Support from the Syracuse University's College of Engineering and Computer Science and Syracuse Center of Excellence for Environmental and Energy Systems is acknowledged.

6. References

[1] L. Coniglio, H. Bennadji, P. Glaude, O. Herbinet, F. Billaud, Combustion chemical kinetics of biodiesel and related compounds (methyl and ethylesters): Experiments and modeling - advanced and future refinements, Progress in Energy and Combustion Science 39 (2013) 340-382.

[2] L. Zhao, M. Xie, L. Ye, Z. Cheng, J. Cai, Y. Li, F. Qi, L. Zhang, An experimental and modeling study of methyl propanoate pyrolysis at low pressure, Combustion and Flame 160 (2013) 1958-1966.

[3] A. Farooq, D. Davidson, R. Hanson, C. Westbrook, A comparative study of the chemical kinetics of methyl and ethyl propanoate, Fuel 134 (2014)26-38.

[4] D. Felsmann, H. Zhao, Q. Wang, I. Fraf, T. Tan, X. Yang, E. Carter, Y. Ju, K. Kohse-Höinghaus, Contributions to improving small ester combustion chemistry: Theory, model and experiments, Proc. Combust. Inst. 36 (2017) 543-551.

[5] H. Ning, J. Wu, L. Ma, W. Ren, D. Davidson, R. Hanson, Chemical kinetic modeling and shock tube study of methyl propanoate decomposition, Combustion and Flame 184 (2017) 30-40.

[6] Y. Wang, D. Lee, C. Westbrook, F. Egolfopoulos, T. Tsotsis, Oxidation of small alkyl esters in flames, Combustion and Flame 161 (2014) 810-817.

[7] B. Akih-Kumgeh, J. Bergthorson, Structure-reactivity trends of c1-c4 alkanoic acid methyl esters, Combustion and Flame (2011) 1037-1048.

[8] Z. Zhang, E. Hu, L. Pan, Y. Chen, J. Gong, Z. Huang, Shock-tube measurements and kinetic modeling study of methyl proanoate ignition, Energy and Fuels 28 (2014) 7194-7202.

[9] K. Kumar, C. Sung, B. Weber, J. Bunnell, Autoignition of methyl propanoate and its comparison with methyl ethanoate and methyl butanoate, Combustion and Flame 188 (2018) 116-128.

[10] L. Wei, P. Geng, A review on natural gas/diesel dual fuel combustion, emissions and performance, Fuel Processing Technology 142 (2016) 264-278.

[11] J. Naber, D. Siebers, S. D. Julio, C. Westbrook, Effects of natural gas composition on ignition delay under diesel conditions, Combustion and Flame 99 (1994) 192-200.

[12] A. Kakaee, A. Paykani, M. Ghajar, The inuence of fuel composition on the combustion and emission characteristics of natural gas fueled engines, Renewable and Sustainable Energy Reviews 38 (2014) 64-78.

[13] W. Yang, H. An, J. Li, L. Duan, Impact of methane addition on the performance of biodiesel fueled diesel engine, Applied Energy 160 (2015) 784-792.

[14] S. Hiremath, S. Khandal, N. Banapurmath, V. Math, V. Gaitonde, Comparative analysis of performance of dual fuel and homogeneous charge compression ignition engines fuelled with honne oil methyl ester and compress natural gas, Fuel 196 (2017) 134-143.

[15] S. Jouzdani, A. Zhou, B. Akih-Kumgeh, Propanol isomers: Investigation of ignition and pyrolysis time scales, Combustion and Flame 176 (2017) 229-244.

Spring Technical Meeting
Eastern States Section of the Combustion Institute
March 4-7, 2018
State College, Pennsylvania

Shock tube and CO laser-absorption measurements during cyclopentanone oxidation

Erik Ninnemann[1], Samuel Barak[1], Owen Pryor[1], William H. Green[2], Kuiwen Zhang[3], William J. Pitz[3], Subith Vasu[1,]*

[1]Center for Advanced Turbomachinery and Energy Research (CATER), Mechanical and Aerospace Engineering, University of Central Florida, Orlando, FL USA

[2]Massachusetts Institute of Technology (MIT), Cambridge, MA 02139, USA

[3]Lawrence Livermore National Laboratory (LLNL), Livermore, CA, USA

***Corresponding Email:** subith@ucf.edu

Abstract: The chemical kinetics of cyclopentanone, an attractive fuel for spark ignition engines, was studied in a double-diaphragm, heated shock tube at pressures of 1, 6, and 8atm over the temperature range of 1105-1327K. A stoichiometric fuel-oxygen mixture with a fuel concentration of 0.1% was chosen to investigate the pressure, emission, and carbon monoxide time-histories during cyclopentanone oxidation. CO mole fractions were measured behind reflected shock waves with a continuous wave distributed feedback quantum cascade laser at 2183.22cm^{-1}, with no interference from other intermediates. Measured values were compared to two recent chemical kinetic mechanisms. Models slightly under predict the rate and over predict the total concentration of CO formed during the exothermic process, over the entire temperature ranged investigated. However, at higher pressures, the modeled CO time histories for both mechanisms predict very well the amount of CO formed, and the rate at which it is produced.

Keywords: laser absorption, shock tube, cyclopentanone, modeling, chemical kinetics

1. Introduction

Biofuels are an attractive alternative to fossil fuels because they have the potential to reduce greenhouse gas emissions. Cyclopentanone is an advanced biofuel candidate that is considered as part of the U.S. Department of Energy's Co-Optimization of Fuels and Engines [1] (Co-Optima). However before it can be used in engines, fundamental combustion measurements (e.g., ignition delay times, flame speeds, etc.) need to be taken at engine relevant conditions. Therefore in this work, carbon monoxide time-histories and ignition delay times were measured behind reflected shockwaves in a shock tube for a stoichiometric mixture of 0.1%C_5H_8O/O_2 balanced in argon in a temperature range of 1105-1327K and pressures of 1, 6 and 8 atm.

2. Experiments

Carbon monoxide time-histories were measured in a double-diaphragm, heated, electropolished high pressure shock tube facility at the University of Central Florida (UCF) with an internal diameter of 14.17cm (details in [2-5]). The driven section houses 8 equally spaced window plugs located 2.00cm from the end-wall for emission, pressure, and spectroscopic diagnostics. The test pressure in these experiments was monitored via a Kistler 603B1 piezoelectric transducer, and the emission signal captured with a Thorlabs PDA25K detector. For these experiments, time zero is defined as the minimum of the Schlieren spike due to the reflected shock wave. The speed of the incident shock wave is calculated via 5 equally spaced PCB 113B26 transducers connected to four time interval counters (Agilent 53220A) placed along the last 1.50m of the driven section. From the calculated velocity, the temperature (T_5) and pressure (P_5) are obtained through quasi 1-D normal shock relations.

A continuous wave distributed feedback quantum cascade laser (Thorlabs DFB QCL) was used to measure carbon monoxide time-histories at 2183.22cm^{-1} during the oxidation of cyclopentanone. According to estimates using kinetic simulations, the top intermediates in the exothermic reaction studied are H_2O, CO_2, and C_2H_4, none of which absorb at the chosen wavelength to measure CO absorption. Experimental CO mole fractions were obtained through the Beer-Lambert law $X_{CO} = \frac{\alpha RT}{\sigma LP}$ where $\alpha = -\ln(\frac{I}{I_o})$ is the absorbance, R is the universal gas constant (J/mol*K), T is the temperature of the species (K), σ is the cross section of the absorbing species (m^2/mol), L is the path length (m), and P is the total pressure of the mixture (Pa). Cross section data was obtained from the HITRAN 2016 database [6] at each tests given initial test temperature and pressure (T_5 and P_5, respectively).

Due to the longer ignition delay times experienced at lower temperatures, driver gas tailoring was implemented to extend test time to observe CO formation during combustion at these lower temperatures. Driver gas tailoring mitigates the interaction between the reflected shock wave and the contact surface, which is the region produced between the driver and driven gases when the diaphragm ruptures. To prepare each tailored mixture, two Teledyne-Hastings HFC-D-303A mass flow controllers were used to regulate the flow of helium and nitrogen into the driver section according to gas dynamic equations. For these experiments, the mole fraction of helium ranged from X_{He}=0.643 to X_{He}=1.000.

Cyclopentanone mixtures were prepared manometrically with a 100-Torr (MKS Instruments/Baratron E27D; accuracy of 0.12% of reading) and a 10,000 Torr (MKS Instruments/Baratron 628D; accuracy of 0.25% of reading) full-scale range capacitance type manometers. Mixtures were contained in a 0.033m^3 teflon-coated mixing tank with a magnetically driven stirrer, and were mixed for at least one hour prior to use. With a vapor pressure of 11.4 Torr at ambient conditions for cyclopentanone, the mixing facility was heated to allow for higher pressure mixtures to be created. The mixing tank was heated to 43°C where the vapor pressure of cyclopentanone is 30.23 Torr [7], partial pressures of cyclopentanone were kept at or below 70% of this value to ensure the fuel did not condense. In order to verify the initial concentration of fuel in each prepared mixture, the cross section of cyclopentanone was measured using FTIR spectroscopy (details in [8]). Cross section measurements were taken over the range of 2850-3050cm^{-1} with a resolution of 2cm^{-1}, a path length of 10.0cm, a pressure of 5.44 Torr, and temperature of 295K. Using the cross section data from these measurements, the initial mole fraction of cyclopentanone was verified through laser absorption at 2938.32cm^{-1} at the beginning of each shock tube test, with an uncertainty of 1.1%. Ignition delay times in this study are defined from time zero to the peak of the broadband emission profile. For the 6 and 10 atm data the emission profile was also captured at the 2.00cm location.

Simulations were conducted in Chemkin Pro [9] using a closed homogeneous batch reactor model with adiabatic and constant volume assumptions. Several of the experiments performed in the UCF shock tube showed evidence of a constant linear pressure rise, dP_5/dt, prior to ignition. For these individual experiments the linear pressure rise was quantified in terms of %/ms to obtain pressure profiles for input into Chemkin Pro. All simulations were performed under the initial test temperature (T_5) and the calculated pressure profile, with an origin at the initial test pressure (P_5). In order to compare the predicted ignition delay times to experimentally obtained ignition delay times, the peak of the modeled time dependent OH profile was used.

3. Results and Discussion

Figure 1 shows a typical pressure trace and carbon monoxide time-history during the oxidation of 0.1% cyclopentanone in oxygen at stoichiometric conditions near 1 atm. The experimental pressure trace is compared to the theoretical pressure profile obtained with the LLNL mechanism (private communication), modeled with constant internal energy and volume assumptions and facility dependent effects. The two traces match extremely well. The measured CO mole fraction is also compared against the LLNL mechanism and Thion mechanism [10].

Both simulated CO time-histories over predict the time at which the maximum CO concentration is occurs. Results from the current study show the peak CO mole fraction occurs at 2179μs, while the LLNL and Thion mechanisms predict the peak time to be at 4172μs and 4171μs, respectively. Both mechanisms also severely over predict the total carbon monoxide yield in the exothermic reaction. Carbon monoxide yield was measured to be 2339ppm out of the theoretical possible 5000ppm, assuming all carbon atoms went to CO formation. The LLNL and Thion mechanism predict the CO yield to be 3509ppm and 4297ppm, respectively.

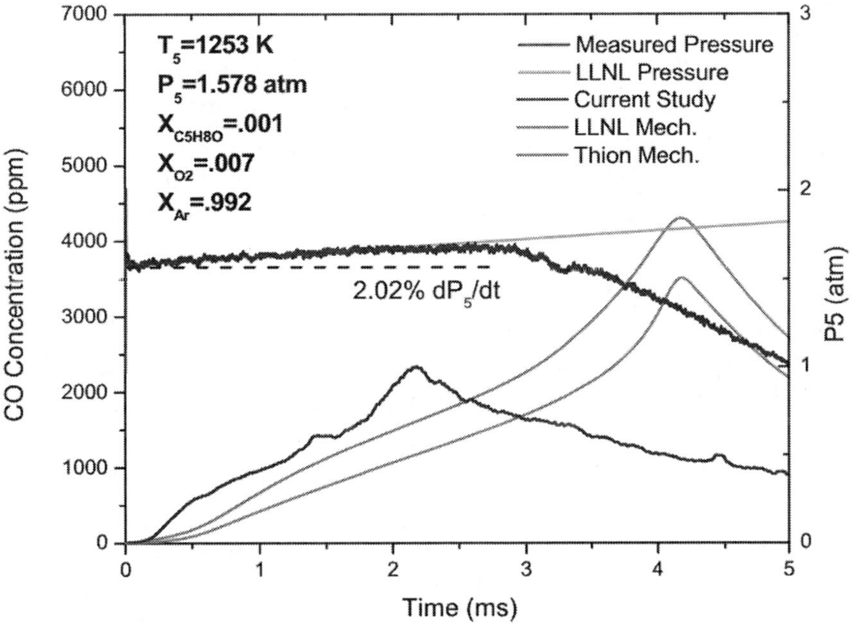

Figure (1). Measured pressure and CO mole fraction with comparisons to the LLNL and Thion mechanisms near 1 atm.

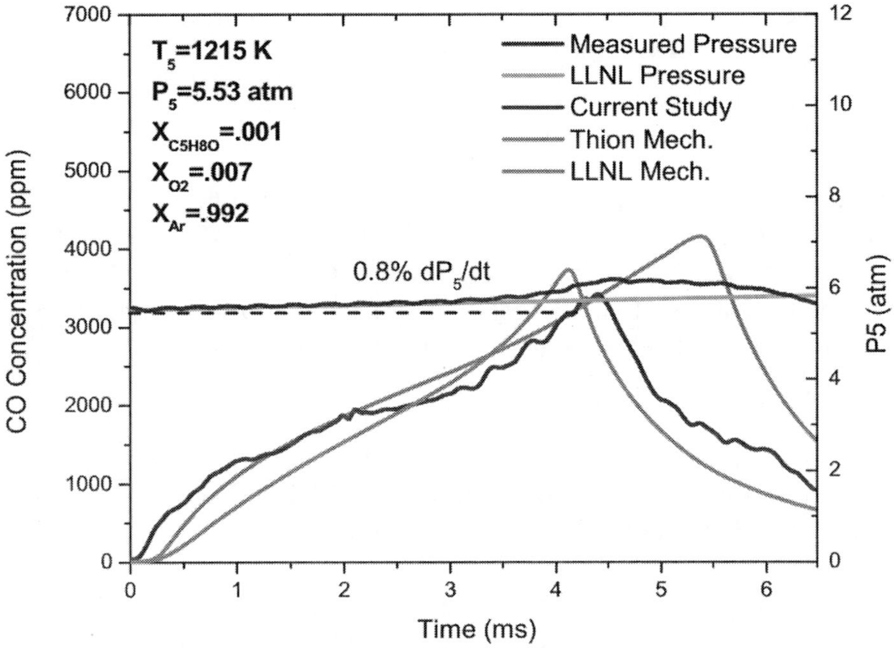

Figure (2). Measured pressure and CO mole fraction with comparisons to the LLNL and Thion mechanisms near 6 atm.

In contrast to the 1atm previously reported, simulations with the Thion and LLNL mechanisms at elevated pressures (6 and 8atm) matched experimental measurements very well. Figure 2 shows these simulations compared to the measured values for a test conducted near 6 atm. Maximum CO yield in this case was found to be 3420ppm which is excellently matched by the LLNL mechanism which predicts a yield of 3732ppm. However, the Thion mechanism again over predicts the yield with a concentration of 4155ppm. The measured peak CO concentration was observed at 4405μs in contrast to the LLNL and Thion mechanisms which predicted a time of 4123μs and 5372μs, respectively. At this elevated pressure, it is clear that the LLNL mechanism predictions are closer to current data. At even higher pressures of 8 atm (not shown here) both mechanisms do very well at predicting the overall CO yield and time of peak CO concentration. Measured CO yield was 3698ppm while the modeled predictions with the LLNL and Thion mechanisms predict 3738ppm and 4185ppm, respectively.

4. Conclusions

Carbon monoxide time-histories were measured behind reflected shock waves during cyclopentanone oxidation – a Co-Optima biofuel- and compared to predictions of two chemical kinetic mechanisms. At lower pressures, both mechanisms drastically over predicted the total yield and time at which the peak CO formation was observed. At high pressures however, the LLNL mechanism showed better agreement and predicted the total yield and peak CO time within just a few percent. To further the chemical kinetic knowledge and to improve upon the current existing mechanisms, additional experiments are needed over a wider temperature range and at even higher temperatures than the ones presented in this work. Work is currently underway to extend the experimental strategy to other Co-Optima biofuels as well.

Acknowledgements

The work by authors at LLNL was performed under the auspices of the U.S. Department of Energy (DOE), Contract DE-AC52-07NA27344. The work at LLNL, UCF, and MIT was conducted as part of the Co-Optimization of Fuels & Engines (Co-Optima) project sponsored by the U.S. Department of Energy (DOE) Office of Energy Efficiency and Renewable Energy (EERE) under the Bioenergy Technologies Office Award Number DE-EE0007982.

References

[1] J. Farrell, Co-Optimization of Fuels and Engines, ; NREL (National Renewable Energy Laboratory (NREL), Golden, CO (United States))2016.

[2] E. Ninnemann, B. Koroglu, O. Pryor, S. Barak, L. Nash, Z. Loparo, J. Sosa, K. Ahmed, S. Vasu, New insights into the shock tube ignition of H2/O2 at low to moderate temperatures using high-speed end-wall imaging, Combustion and Flame 187 (2018) 11-21.

[3] Z.E. Loparo, J.G. Lopez, S. Neupane, W.P. Partridge, K. Vodopyanov, S.S. Vasu, Fuel-rich n-heptane oxidation: A shock tube and laser absorption study, Combustion and Flame 185 (2017) 220-233.

[4] G. Barari, O. Pryor, B. Koroglu, J. Lopez, L. Nash, S.M. Sarathy, S.S. Vasu, High Temperature Shock Tube Experiments and Kinetic Modeling Study of Diisopropyl Ketone Ignition and Pyrolysis, Combustion and Flame 177 (2017) 207-218.

[5] B. Koroglu, S.S. Vasu, Measurements of Propanal Ignition Delay Times and Species Time Histories Using Shock Tube and Laser Absorption, Int. J. Chem. Kinet. 48 (2016) 679-690.

[6] I.E. Gordon, L.S. Rothman, C. Hill, R.V. Kochanov, Y. Tan, P.F. Bernath, M. Birk, V. Boudon, A. Campargue, K.V. Chance, B.J. Drouin, J.M. Flaud, R.R. Gamache, J.T. Hodges, D. Jacquemart, V.I. Perevalov, A. Perrin, K.P. Shine, M.A.H. Smith, J. Tennyson, G.C. Toon, H. Tran, V.G. Tyuterev, A. Barbe, A.G. Császár, V.M. Devi, T. Furtenbacher, J.J. Harrison, J.M. Hartmann, A. Jolly, T.J. Johnson, T. Karman, I. Kleiner, A.A. Kyuberis, J. Loos, O.M. Lyulin, S.T. Massie, S.N. Mikhailenko, N. Moazzen-Ahmadi, H.S.P. Müller, O.V. Naumenko, A.V. Nikitin, O.L. Polyansky, M. Rey, M. Rotger, S.W. Sharpe, K. Sung, E. Starikova, S.A. Tashkun, J.V. Auwera, G. Wagner, J. Wilzewski, P. Wcisło, S. Yu, E.J. Zak, The HITRAN2016 molecular spectroscopic database, Journal of Quantitative Spectroscopy and Radiative Transfer 203 (2017) 3-69.

[7] A. Barhala, D. Dragoescu, M. Teodorescu, I. Wichterle, Isothermal (vapour+ liquid) equilibria in the binary mixtures (1, 2-dichloroethane and 1, 1, 1-trichloroethane with cyclopentanone) within the temperature range (298.15 to 313.15) K, The Journal of Chemical Thermodynamics 38 (2006) 617-623.

[8] B. Koroglu, S. Neupane, O. Pryor, R.E. Peale, S.S. Vasu, High temperature infrared absorption cross sections of methane near 3.4 μm in Ar and CO2 mixtures, Journal of Quantitative Spectroscopy and Radiative Transfer 206 (2018) 36-45.

[9] ANSYS Chemkin-Pro, http://www.ansys.com/products/fluids/ansys-chemkin-pro, San Diego, CA., 2017.

[10] S. Thion, C. Togbé, G. Dayma, Z. Serinyel, P. Dagaut, Experimental and Detailed Kinetic Modeling Study of Cyclopentanone Oxidation in a Jet-Stirred Reactor at 1 and 10 atm, Energy & Fuels 31 (2017) 2144-2155.

Spring Technical Meeting
Eastern States Section of the Combustion Institute
March 4-7, 2018
State College, Pennsylvania

Premixed Flame Propagation in Mixtures of Cyclopentanone/Air

*Gihun Kim, Bader Almansour, Richard Blair, Kareem Ahmed, Jayanta Kapat, Subith S. Vasu**

Center for Advanced Turbomachinery and Energy Research (CATER), University of Central Florida, 4000 Central Florida Blvd, Orlando, FL 32816, USA
**Corresponding Author Email: subith@ucf.edu*

Abstract: There are several promising biofuel candidates for automotive applications that reduce greenhouse gases, improve air quality, and reduce the dependence on foreign oil. Currently, gasoline fuel properties such as octane number, pressure and temperature sensitivities limit the operating range of reciprocating engines. Because of those limitations, and the need to operate at a higher load to increase engine efficiencies, advanced fuels must be used. Cyclopentanone (C_5H_8O) is among one of the biofuel candidates considered for t U.S. Department of Energy's Co-Optimization of Fuels and Engines (Co-Optima). In this study, laminar burning velocity (LBV) of cyclopentanone was measured in a spherical combustion chamber at initial conditions of 408 K, and atmospheric pressure for various equivalent ratios (Φ) from 0.7 to 1.4. LBV is defined as the velocity at which unburned gases moves in the normal direction to the combustion wave surface. This is a crucial property in spark ignition engines because it affects the combustion duration. High-speed schlieren photography method was used to capture the flame propagation events. The flame stability was assessed based on Markstein length, and both linear and nonlinear extrapolation methods were utilized and compared.
Keywords: Biofuel, Cyclopentanone, Laminar Burning Velocity, Chemical Kinetics, Ignition

1. Introduction

The laminar burning velocity (LBV) is one of the most important parameters to characterize the performance of a fuel inside combustion engines. It is a crucial parameter for engine design, modeling turbulent combustion, and validation of chemical kinetic mechanisms [1]. Recently, there have been several investigations into promising biofuel candidates for applications that reduce greenhouse gases, improve air quality, and reduce the dependence on foreign oil. Cyclopentanone (C_5H_8O) is among one of the biofuel candidates considered for the U.S. Department of Energy's Co-Optimization of Fuels and Engines (Co-Optima) [2]. In the literature, several LBV studies exist for various biofuels [3, 4], but very few for cyclopentanone. Bao et al. [5] studied laminar flame characteristics of cyclopentanone and compared it with ethanol and gasoline.

2. Experimental Method

The experimental setup is mainly composed of a spherical chamber, mixing tank, optical components, ignition system, and data acquisition. See Fig. 1 for a schematic of the experimental setup, more details can be found in our earlier work [6, 7]. The setup has four optical access windows to visualize spark ignition, flame kernel development, and flame propagation using high-speed schlieren photography. A Light Emitting Diode (LED) was used to expand and collimate to a 2 inches diameter before passing it perpendicular to the ignition system electrodes. Bi-Convex lens and a pinhole were constructed to remove diffracted rays to produce a clear image based on the density gradient. A Phantom v12.1 high-speed camera was used to capture the flame propagation events with a resolution of 512 X 512 pixels at a rate of 20,000fps.

The Ignition circuit was controlled by LabVIEW program to provide spark ignition mechanism. A modified NGK-BR4HS spark plug was used to provide a spark at the center of the combustion chamber. The Kistler 603B1 dynamic pressure transducer was used to record combustion pressure at 100 kHz. The entire experimental setup was placed inside an oven achieve uniform initial temperature distribution. The uncertainty in laminar burning velocity was estimated to be 5.8%.

Due to concerns regarding condensation of cyclopentanone, current experiments were performed at 408K, which was higher than the boiling point of the fuel. Both combustion chamber and the mixture tank were evacuated to 0.1 torr using an oil rotary pump. Mixtures of fuel and artificial air were prepared by using the partial pressure method, which was measured by using two Baratron gauges to an accuracy of 0.01 torr. The data were recorded using the National Instruments DAQ system. A 5 minutes waiting time was implemented before igniting the mixture to ensure homogeneity and stagnant status The artificial air mixture was composed of O_2 and N_2 and the ratio was 1: 3.76. A minimum of 3 experiments were conducted for every equivalence ratio (Φ) and average results are presented in plots.

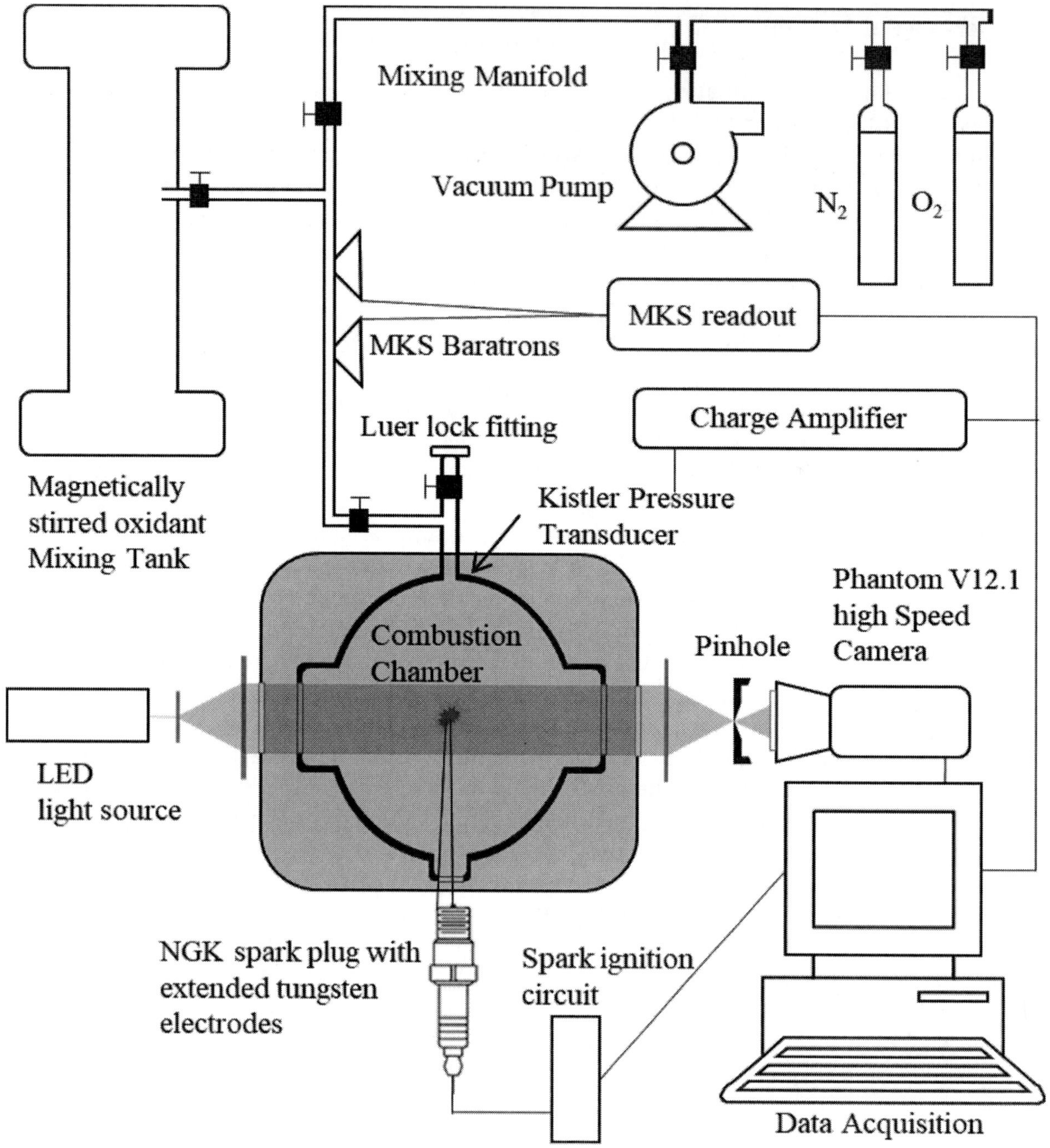

Figure 1 Schematic of the experimental setup

3. Results and Discussion

The present experiments were conducted at initial pressure of 1 atm and temperature of 408K. O'Donovan et al. [8] and Bradley et al. [9] derived flame radius growth equations by using assumptions, which are that the flame is smooth, thin, spherical, both the burned and unburned mixtures are ideal gases, and there is no buoyancy effect in during combustion process. Hill et al. [10] and Takizawa et al. [11] described LBV by the derivation of the flame radius (R_f) and the definition of LBV. The LBV can be calculated by Equation (1), which is found by using the flame radius (R_f) and the definition of the LBV. S_u, ρ_u, and R_o are laminar burning velocity, the density of unburned mixture, and vessel radius, respectively.

$$S_u = \frac{R_O^3}{3R_f^2}\left(\frac{P_i}{P}\right)^{\frac{1}{\gamma}}\frac{dX}{dt} \tag{1}$$

The results of cyclopentanone laminar burning velocity measurements are shown in Fig. 2 along with other fuels from the literature. The maximum laminar burning velocity of cyclopentanone is 83.3 cm/s at an equivalence ratio of 1.1. It is clear that the LBV of cyclopentanone is much faster than that of other fuels. It is possible that this could be due to the higher initial temperature in current experiments than what was used in other studies. More tests with various initial temperatures are needed in the future. Also, comparisons with predictions of cyclopentanone chemical kinetic mechanisms are currently underway.

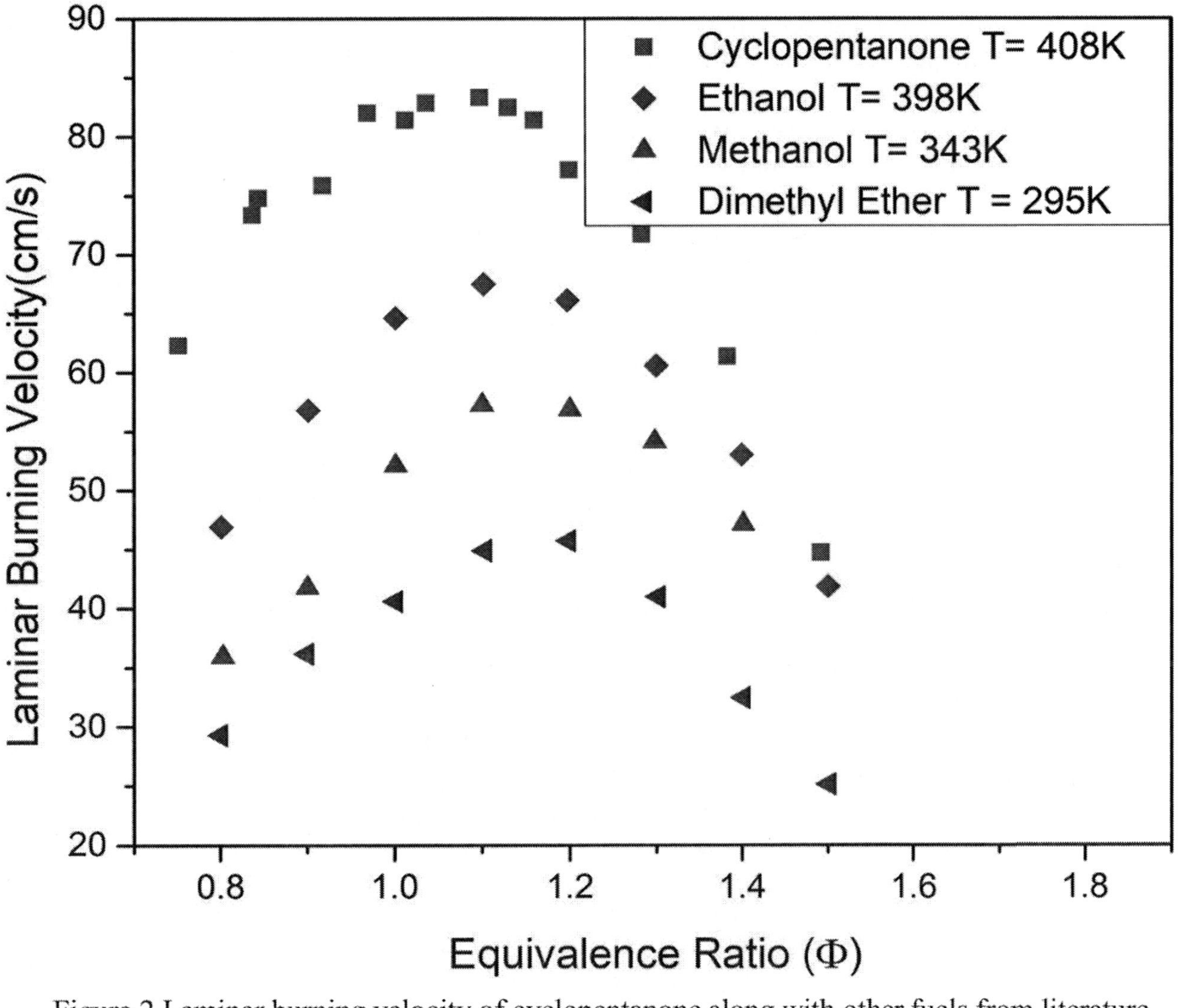

Figure 2 Laminar burning velocity of cyclopentanone along with other fuels from literature.

4. Conclusions

In this study, the laminar burning velocity (LBV) of cyclopentane was conducted in a spherical combustion chamber at an initial T=408 K, and P=1 atm for various equivalence ratios by using constant volume methods. Measurements showed the classical LBV vs equivalence ratio trend with the peak occurring around $\Phi = 1$. Current data were compared with other fuels form the literature. Current measurements are considered preliminary and further study is required in the future at higher pressure condition and various initial temperatures. Also, comparisons with predictions by chemical kinetic mechanisms will be evaluated.

5. Acknowledgements

This material is based upon work supported by the U.S. Department of Energy's Office of Energy Efficiency and Renewable Energy (EERE) under Award Number DE-EE0007984 (CoOptima).

Disclaimer: This report was prepared as an account of work sponsored by an agency of the United States Government. Neither the United States Government nor any agency thereof, nor any of their employees, makes any warranty, express or implied, or assumes any legal liability or responsibility for the accuracy, completeness, or usefulness of any information, apparatus, product, or process disclosed, or represents that its use would not infringe privately owned rights. Reference herein to any specific commercial product, process, or service by trade name, trademark, manufacturer, or otherwise does not necessarily constitute or imply its endorsement, recommendation, or favoring by the United States Government or any agency thereof. The views and opinions of authors expressed herein do not necessarily state or reflect those of the United States Government or any agency thereof.

6. References

[1] J. Buffam, K. Cox, H. Schiess, Measurement of Laminar Burning Velocity of Methane-Air Mixtures Using a Slot and Bunsen Burner, WORCESTER POIYTECHNIC INSTITUE, (2008).
[2] J. Farrell, Co-Optimization of Fuels and Engines, ; NREL (National Renewable Energy Laboratory (NREL), Golden, CO (United States))2016.
[3] D. Bradley, R. Hicks, M. Lawes, C. Sheppard, R. Woolley, The measurement of laminar burning velocities and Markstein numbers for iso-octane–air and iso-octane–n-heptane–air mixtures at elevated temperatures and pressures in an explosion bomb, Combustion and flame 115 (1998) 126-144.
[4] C.A. Daly, J.M. Simmie, J. Würmel, N. DjebaÏli, C. Paillard, Burning velocities of dimethyl ether and air, Combustion and flame 125 (2001) 1329-1340.
[5] X. Bao, Y. Jiang, H. Xu, C. Wang, T. Lattimore, L. Tang, Laminar flame characteristics of cyclopentanone at elevated temperatures, Applied Energy 195 (2017) 671-680.
[6] B. Almansour, S. Alawadhi, S. Vasu, Laminar Burning Velocity Measurements in DIPK-An Advanced Biofuel, SAE International Journal of Fuels and Lubricants 10 (2017).
[7] B. Almansour, L. Thompson, J. Lopez, G. Barari, S.S. Vasu, Laser Ignition and Flame Speed Measurements in Oxy-Methane Mixtures Diluted With CO2, Journal of Energy Resources Technology 138 (2016) 032201.

[8] K. O'Donovan, C.J. Rallis, A modified analysis for the determination of the burning velocity of a gas mixture in a spherical constant volume combustion vessel, Combustion and Flame 3 (1959) 201-214.
[9] D. Bradley, C. Morley, Chapter 7 Autoignition in spark-ignition engines, in: M.J. Pilling (Ed.), Comprehensive Chemical Kinetics, Elsevier1997, pp. 661-760.
[10] P. Hill, J. Hung, Laminar burning velocities of stoichiometric mixtures of methane with propane and ethane additives, Combustion science and technology 60 (1988) 7-30.
[11] K. Takizawa, A. Takahashi, K. Tokuhashi, S. Kondo, A. Sekiya, Burning velocity measurement of fluorinated compounds by the spherical-vessel method, Combustion and Flame 141 (2005) 298-307.

Spring Technical Meeting
Eastern States Section of the Combustion Institute
March 4-7, 2018
State College, Pennsylvania

Sooting tendencies of aromatic hydrocarbons with oxygen-containing side-chains

Brian P. Beekley[1], Charles S. McEnally[1,], Peter C. St. John[2], Seonah Kim[2], Abhishek Jain[3], Hyunguk Kwon[3], Yuan Xuan[3], Lisa D. Pfefferle[1]*

[1]Yale University, New Haven CT 06520 USA
[2]National Renewable Energy Laboratory, Golden CO 80401 USA
[3]Pennsylvania State University, University Park 16802 USA
[*]Corresponding Author Email: charles.mcenally@yale.edu

Abstract: Yield Sooting Index (YSI) was determined for 13 oxygen-containing aromatic compounds by measuring line-of-sight spectral radiance (LSSR) in a laminar non-premixed methane/air flame doped with 1000 ppm of each compound. The compounds included anisole (methoxybenzene), other aromatic ethers, and phenols. The presence of oxygen greatly reduced soot formation in some cases, but increased it in others. Thus oxygenated aromatics are potentially valuable as fuel components with intrinsically low particulate emissions, but the specific structure is important. Compounds containing an aromatic ring with an adjacent oxygen atom tended to have a lower sooting tendency than did compounds with oxygen atoms elsewhere in the side chain. This observation was quantitatively explored using Density Functional Theory (DFT) calculations. The bond dissociation energies for nine aromatic hydrocarbons and oxygenates were calculated using GGA B3LYP density functions with 6-31G(d) basis sets. When possible, unimolecular dissociation to phenoxy radical was favored and resulted in much lower YSIs than similar compounds where the oxygen atom was not adjacent to the aromatic ring. Structures that dissociated to benzyl radical had much higher sooting tendencies, sometimes by over a factor of two in comparison to phenoxy-forming compounds.
soot, IC engines, alternative fuels, biofuels

1. Introduction

Biomass-derived fuels offer many benefits compared to traditional petroleum-based fuels: when used properly they can reduce overall greenhouse gas emissions since their plant sources grow by removing CO_2 from the atmosphere [1], they can be produced domestically in any country and therefore promote energy security, and they are a perpetually-renewing resource with no `peak'. Of greatest interest to this work, they may also reduce emissions of carbonaceous soot particles. Soot emissions are the second largest source of climate change [2] and contribute to ambient particulate matter that causes over 3 million deaths worldwide each year [3]. Biofuels usually contain oxygenated hydrocarbons [4], and the partially oxidized carbon atoms in these molecules can hinder the formation of the multi-ring aromatic molecules that nucleate into soot particles [5 – 9].

A particularly interesting category of biofuels are oxygenated aromatics derived from lignin. Lignin is a non-edible portion of plants that confers them with structural rigidity. It constitutes roughly 20% of hardwoods and straws, 30% of softwoods, and 40% of agricultural residues such as cotton and hazelnut shells [10, 11]. Chemically, it is an amorphous polymer composed of the three monolignols, each of which contains a benzenoid ring, a hydroxyl group, a propenoxy group, and up to two methoxy groups. Given this structure, liquid fuels derived by thermal or biochemical processing of lignin tend to contain large quantities of oxygenated aromatics such as anisole, methylphenol, and benzaldehyde [10, 11].

The objective of the research reported here was to measure quantitative sooting tendencies for 15 oxygenated aromatics having a range of specific chemical structures. The number and variety of compounds was large enough to determine whether adding an oxygen atom to an aromatic causes a fixed changed in the sooting tendency, or whether it depends on the chemical details. Sooting tendencies have been reported earlier for many oxygenates such as alcohols and esters, but the only measurements for oxygenated aromatics are our measurements for anisole and phenetole.

2. Methods

The sooting tendencies of the oxygenated aromatics were measured using the yield-based approach we developed earlier [12]. We sequentially doped 1000 ppm of n-heptane, toluene, and each oxygenated aromatic into the fuel of a methane\air nonpremixed flame and then measured the soot level in each flame with line-of-sight spectral radiance (LSSR). The results were rescaled into a yield sooting index (YSI) by

$$YSI_{TF} = (YSI_T - YSI_H) * \frac{LSSR_{TF} - LSSR_H}{LSSR_T - LSSR_H} + YSI_H \tag{1}$$

where TF, T, and H are the test fuel, toluene, and heptane. This rescaling factors out uninteresting details such as the specific dopant concentration and the burner dimensions. More importantly, it allows the new results to be quantitatively compared with a database we have published that contains YSI for hundreds of other hydrocarbons [13, 14]. These earlier results span a wide range of sooting tendencies – from small oxygenates like methanol to four-ring polycyclic compounds like pyrene – so many of them had to be measured at dopant concentrations that differed from the 1000 ppm used in this study. The parameters YSI_T and YSI_H are constants that define the scale; their values – 170.9 and 36.0 – were taken from the database [13].

Quantum chemistry calculations were performed to identify the most likely initial reaction pathways for the oxygenated aromatics. Gaussian 09 was used to fully optimize all stationary points using the composite G4 method for Density Functional Theory (DFT) calculations [15]. We performed optimizations with the hybrid GGA B3LYP density functions with the 6-31G(d) basis sets. Harmonic vibrational frequencies were calculated for all optimized structures including transition states to verify energy minima, possessing zero imaginary frequency for optimized structures and one imaginary frequency for transition states. To verify corresponding reactants and products from transition state, the intrinsic reaction coordinate (IRC) calculations using B3LYP/6-31(G) were also performed.

The relative importance of unimolecular versus bimolecular consumption reactions were determined for toluene in a toluene-doped flame by numerical simulations with a detailed chemical kinetic model, using the FlameMaster code [16]. The numerical framework is based on modified flamelet equations [17]. Briefly, the flamelet equations are re-derived for species mass fractions along the centerline of the experimental flames. These equations take into account the effects of multidimensional diffusion and convection of species in mixture fraction space due to non-unity Lewis numbers. The modified flamelet equations take as input the temperature, convective velocity, and scalar dissipation rate profiles calculated from the direct simulation of the undoped flame. The numerical sooting tendencies are then calculated using the PAH dimer production rate generated from the flamelet solutions doped by the test species. A detailed chemical kinetic model containing 754 species and 7491 reactions (forward and backward counted separately) was used for the YSI simulations.

3. Results and Discussion

Table 1 lists the measured YSI. A detailed uncertainty analysis is in progress, but the uncertainties are estimated to be ±10 YSI units. Each measurement was repeated at least three times, and this uncertainty includes the deviation of the repeat measurements and systematic uncertainties due to uncertainty in the liquid-phase mass density of the dopants. The table also lists the calculated bond dissociation energies (BDE) for the weakest bond in some of the molecules, and the products formed by dissociation of the weakest bond. Results are shown for compounds with a single side-chain; calculations of the remaining BDE are in progress.

Table 1: Measured YSI and calculated BDE (kJ/mol) of the weakest bond for several oxygenated aromatics and related compounds.

ID	Name	CAS No.	Formula	YSI	BDE	Products
HEP	n-heptane	142-82-5	C_7H_{16}	36		
TOL	toluene	108-88-3	C_7H_8	171	373	benzyl + H
ANI	anisole	100-66-3	C_7H_8O	111	271	phenoxy + CH_3
BOH	benzyl alcohol	100-51-6	C_7H_8O	189	335	benzyl + OH
3MP	3-methylphenol	108-39-4	C_7H_8O	103		
BZA	benzaldehyde	100-52-7	C_7H_6O	119		
EBZ	ethylbenzene	100-41-4	C_8H_{10}	216	322	benzyl + CH_3
PHE	phenetole	103-73-1	$C_8H_{10}O$	112	280	phenoxy + C_2H_5
BME	benzyl methyl ether	538-86-3	$C_8H_{10}O$	202	298	benzyl + OCH_3
1PE	1-phenylethanol	98-85-1	$C_8H_{10}O$	142	308	α-hydroxybenzyl + CH_3
2PE	2-phenylethanol	60-12-8	$C_8H_{10}O$	207	311	benzyl + CH_2OH
2EP	2-ethylphenol	90-00-6	$C_8H_{10}O$	120		
3EP	3-ethylphenol	620-17-7	$C_8H_{10}O$	138		
2XY	2-xylene	95-47-6	C_8H_{10}	200		
4XY	4-xylene	106-42-3	C_8H_{10}	202		
2MA	2-methylanisole	578-58-5	$C_8H_{10}O$	122		
3MA	3-methylanisole	100-84-5	$C_8H_{10}O$	126		
4MA	4-methylanisole	104-93-8	$C_8H_{10}O$	123		
3MB	3-methylbenzyl alcohol	587-03-1	$C_8H_{10}O$	243		
24D	2,4-dimethylphenol	105-67-9	$C_8H_{10}O$	144		

Figure 1 shows the YSI for the C_7H_8O isomers (lighter brown columns) and compares them with benzene and toluene (darker blue columns). These are all compounds that can be formed by adding an oxygen atom to toluene. The results show that the oxygen atom can significantly reduce the sooting tendency, but the effects depend on its specific location. For example, putting the oxygen atom between the benzenoid ring and the methyl group, as in anisole (ANI), produces a sooting tendency much lower than toluene (TOL) and close to benzene. On the other hand, putting the oxygen atom between the carbon atom and one of the hydrogen atoms in the side-chain, as in benzyl alcohol (BOH), increases the sooting tendency beyond toluene. Finally, putting the oxygen atom between the benzenoid ring and one of its hydrogen atoms, as in 3-methylphenol (3MP), reduces the sooting tendency even closer to benzene than anisole. This latter observation shows that when the aromatic contains both oxygenated and unoxygenated side-chains, the oxygenated side-chain dominates the sooting tendency.

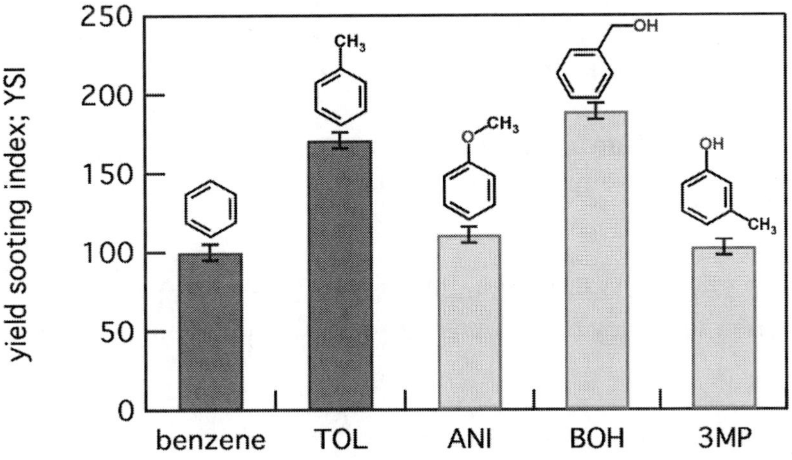

Figure 1: YSI measured for benzene, toluene, and oxygenated aromatics formed by adding an O-atom to toluene.

Similar observations are made for oxygenated aromatics produced by adding oxygen atoms to ethylbenzene or xylene (see Table 1). Collectively these results indicate that oxygenated aromatics can provide fuels with low sooting tendencies relative to regular aromatic fuels. However, the specific structural details are critical and the fuels have to be chosen carefully or they may have the opposite effect.

5. Acknowledgements

This material is based upon work supported by the U.S. Department of Energy's Office of Energy Efficiency and Renewable Energy (EERE) under the Bioenergy Technologies Office (BETO) and Vehicle Technologies Office (VTO) Program Award Number DE-EE0007983. This work is also supported by the National Science Foundation (NSF) under Grant No. CBET 1604983.

6. References

[1] Tilman, D.; Socolow, R.; Foley, J. A.; Hill, J.; Larson, E.; Lynd, L.; Pacala, S.; Reilly, J.; Searchinger, T.; Somerville, C.; Williams, R. Beneficial Biofuels — The Food, Energy, and Environment Trilemma. Science 2009, 325, 270–271.

[2] Bond, T. C. et al. Bounding the role of black carbon in the climate system: A scientific assessment. Journal of Geophysical Research-Atmospheres 2013, 118, 5380–5552.

[3] Lim, S. S. et al. A comparative risk assessment of burden of disease and injury at- tributable to 67 risk factors and risk factor clusters in 21 regions, 1990-2010: a system- atic analysis for the Global Burden of Disease Study 2010. Lancet 2012, 380, 2224–2260.

[4] Kohse-Hoeinghaus, K.; Osswald, P.; Cool, T. A.; Kasper, T.; Hansen, N.; Qi, F.; West- brook, C. K.; Westmoreland, P. R. Biofuel Combustion Chemistry: From Ethanol to Biodiesel. Angewandte Chemie-International Edition 2010, 49, 3572–3597.

[5] Westbrook, C.; Pitz, W.; Curran, H. Chemical kinetic modeling study of the effects of oxygenated hydrocarbons on soot emissions from diesel engines. Journal of Physical Chemistry A 2006, 110, 6912–6922.

[6] Pepiot-Desjardins, P.; Pitsch, H.; Malhotra, R.; Kirby, S. R.; Boehman, A. L. Structural group analysis for soot reduction tendency of oxygenated fuels. Combustion and Flame 2008, 154, 191–205.

[7] McEnally, C. S.; Pfefferle, L. D. Sooting Tendencies of Oxygenated Hydrocarbons in Laboratory-Scale Flames. Environmental Science & Technology 2011, 45, 2498–2503.

[8] Lemaire, R.; Lapalme, D.; Seers, P. Analysis of the sooting propensity of C-4 and C-5 oxygenates: Comparison of sooting indexes issued from laser-based experiments and group additivity approaches. Combustion and Flame 2015, 162, 3140–3155.

[9] Abboud, J.; Schobing, J.; Legros, G.; Bonnety, J.; Tschamber, V.; Brillard, A.; Leyssens, G.; Lauga, V.; Iojoiu, E. E.; Da Costa, P. Impacts of oxygenated com- pounds concentration on sooting propensities and soot oxidative reactivity: Application to Diesel and Biodiesel surrogates. Fuel 2017, 193, 241–253.

[10] Wang, S.; Dai, G.; Yang, H.; Luo, Z. Lignocellulosic biomass pyrolysis mechanism: A state-of-the-art review. Progress in Energy and Combustion Science 2017, 62, 33–86.

[11] Dhyani, V.; Bhaskar, T. A comprehensive review on the pyroly- sis of lignocellulosic biomass. Renewable Energy 2017, In Press, https://doi.org/10.1016/j.renene.2017.04.035.

[12] McEnally, C. S.; Pfefferle, L. D. Improved sooting tendency measurements for aromatic hydrocarbons and their implications for naphthalene formation pathways. Combustion and Flame 2007, 148, 210–222.

[13] McEnally, C. S.; Das, D. D.; Pfefferle, L. D. Yield sooting Index Database Volume 2: Sooting Tendencies of a Wide Range of Fuel Compounds on a Unified Scale. 2017; DOI:10.7910/DVN/7HGFT8.

[14] Das, D. D.; St. John, P. C.; McEnally, C. S.; Kim, S.; Pfefferle, L. D. Measur- ing and predicting sooting tendencies of oxygenates, alkanes, alkenes, cycloalka- nes, and aromatics on a unified scale. Combustion and Flame 2018, In Press, https://doi.org/10.1016/j.combustflame.2017.12.005.

[15] Frisch, M.; Trucks, G.; Schlegel, H.; Scuseria, G.; Robb, M.; Cheeseman, J. Gaussian 09 Revision D.01. Gaussian Inc. Wallingford CT, 2009.

[16] Pitsch, H.; Barths, H.; Peters, N. SAE Technical Paper 962057. SAE Technical Paper 1996.

[17] Xuan, Y.; Blanquart, G. Numerical modeling of sooting tendencies in a laminar co-flow diffusion flame. Combustion and Flame 2013, 160, 1657–1666.

2018 Spring Technical Meeting
Eastern States Section of the Combustion Institute
Hosted by The Pennsylvania State University
March 4-7, 2018

Numerical investigation of pressure dependence of yield sooting tendencies

Y. Xuan[1], *C.S. McEnally*[2], *A. Jain*[1], *H. Kwon*[1], *L.D. Pfefferle*[2]

[1]*Mechanical and Nuclear Engineering, The Pennsylvania State University, University Park, PA*
[2]*Chemical and Environmental Engineering, Yale University, New Haven, CT*

The sooting tendencies of various fuels have been historically quantified via the Threshold Sooting Index (TSI), which is based on spoke point measurement. More recently, a Yield-based Sooting Index (YSI) has been proposed with lower measurement uncertainties, high throughput, and applicability to low-volatility fuels. YSI is experimentally measured in an atmospheric-pressure, laminar, non-premixed co-flow methane/air flame generated with a Yale Co-flow Burner. As such, YSI quantifies soot propensity under ambient pressure, which is much lower than the pressure levels in practical devices. In this work, we numerically investigate the pressure dependence of YSI for a wide range of fuels. We first validate our numerical models in a series of high pressure laminar co-flow non-premixed flames with pressures ranging from 5atm to 20atm. We then use these numerical models to predict the YSI of a wide range of fuels under pressures from 1atm to 20atm. Finally, the predicted YSI values under these pressure levels are compared against the experimental measurements of YSI under 1atm to quantify the pressure dependence of YSI.

1 Introduction

Soot, an undesirable by-product of incomplete hydrocarbon combustion has proven to be detrimental to human health and the environment [1]. The implementation of strict legislation that places limits on both particle size and total mass has led to greater efforts in understanding and characterizing soot yield. This has also spurred an interest in developing alternate fuels derived from biological sources that will serve as replacements for traditional fuels. There is therefore a need to develop methods that can succesfully predict sooting tendencies of newer fuels based on their molecular structure as a strong dependence has been observed between soot emissions and the fuel molecular structure [2, 3].

The Yield-based Sooting Index (YSI) [4] has thus been introduced as a metric for measuring sooting tendencies of hydrocarbon species. The YSI are linear functions of the maximum soot volume fraction measured on the centerline of an axisymmetric co-flow diffusion flame with the fuel stream doped with a test hydrocarbon. The YSI overcomes the difficulties of the smoke point based traditional Threshold Sooting index (TSI) [5] approach by being solely a function the chemistry and independent of the burner device used. There have been numerous studies to quantify the sooting tendencies of various fuels and surrogates using the YSI [4, 6–8] technique and it has been proven to have high throughput, lower measurement uncertainties and has been found to be applicable to low-volatility fuels. Currently YSI is measured in an atmospheric-pressure, laminar, non-premixed co-flow methane/air flame generated with a Yale Co-flow burner. The use of this burner confines the quantification of fuel sooting propensities to ambient atmospheric pressure.

However, practical combustion devices like diesel engines and airplane gas turbines operate at elevated pressures to increase efficiency. It has been found in numerous studies [9–16] that soot emissions increase significantly as pressure increases. There is, therefore a need to incorporate pressure dependence into YSI calculations to make it useful for obtaining sooting tendencies of fuels used in practical combustion device operating conditions. As a result, the study is performed with the following objectives. First, to check the validity of the numerical models against a series of laminar flames with pressures ranging from 1-20 atm. Second, to use these validated numerical models to predict the YSI of a wide range of fuels under the same pressure range (1-20 atm). The third and final objective of this paper is to understand and quantify the pressure dependence on YSI.

2 Simulation Framework and Test cases

2.1 Numerical Methods

The flow solver used in this work is the NGA code [17]. The NGA solver uses staggered variables and allows for accurate, robust, and flexible simulations of both laminar and turbulent reactive flows in complex geometries and has been applied in a wide range of test problems [18, 19]. The species mass fractions and temperature are transported using the BQUICK scheme, which ensures that the physical bounds of appropriate quantities are numerically preserved throughout the simulation without adding significant artificial diffusion [20]. A recently-developed computationally-efficient, semi-implicit, iterative method is used for the time-integration of chemical source terms for the transport equations of gas-phase species. This method has been shown to be free of lagging errors, as efficient as an explicit time-integration per time step, and capable of using large, stable time-step sizes [21]. These numerical methods guarantee globally second-order accuracy in both space and time [21].

2.2 Chemical and soot model

A semi-detailed chemical kinetic model containing 47 species and 290 reaction (forward and backward reactions counted separately) is used. This mechanism was obtained from the originally-developed full mechanism [22, 23] with 185 species and 1903 reactions using a multi-step approach combining automatic and manual chemistry reduction [24]. This chemical model has been previously used to simulate laminar and turbulent sooting flames under various configurations [18, 24].

The soot model employed in the current study is taken from [19]. Soot particles are described as fractal aggregates using a bi-variate approach based on the total volume, V, and total surface area, S. The population of the soot particles is statistically described using a bi-variate soot Number Density Function (NDF), n (V; S), and is approximated by two delta functions, following previous work [19]. Key soot moments are evolved using the following transport equation,

$$\frac{\partial M_{x,y}}{\partial t} + \nabla.(\mathbf{u}^* M_{x,y}) = \dot{\omega}_{x,y}, \tag{1}$$

where

$$\mathbf{u}^* = \mathbf{u} - 0.556\frac{\nu}{T}\nabla T, \tag{2}$$

is the velocity vector accounting for thermophoresis effects [24] and ν is the kinematic viscosity. The source terms, $\dot{\omega}_{x,y}$, include the contributions from soot nucleation, coagulation, condensation, surface growth, and oxidation to the evolution of the soot population, which is determined using the Direct Quadrature Method of Moments (DQMOM)

2.3 Model validation

The models described above are validated against the experimental flames found in Ref. [14]. The flames are axisymmetric co-flow laminar CH_4/air diffusion flames established on the burner whose details can be found in [14]. The fuel mass flow rate is 0.55 mg/s while the air flow rate is maintained at 0.4 g/s. The inlet temperature for both fuel and air is 300 K. Numerical simulations are performed for the 1, 5, 10 and 20 atm flames.

From the experiments, it was observed that the visible flame height increases from 1 atm to 5 atm, and remains constant thereafter till 20 atm at $\simeq$ 10 mm [14]. The increase in pressure is also accompanied by a thinning of the flame. These trends are captured well by the numerical simulations. Flame heights of the numerically computed 5 atm, 10atm, and 20atm flames are 9.9, 9.8 and 14.9 mm respectively. Simulated flame heights are calculated by measuring the axial location along the centerline where the soot volume fraction drops to 0.01 ppm. The slight difference found in the 20 atm flame can be attributed to the fact that the flame has not reached numerical convergence and needs to be simulated further. The experimental measurements also showed a significant increase in the maximum soot volume fraction with increasing pressure [14]. The numerical predictions reproduce the trend of the increase in maximum soot volume fraction with increasing pressure. These are depicted in Figure 1. However there is a $\simeq 0.5$ difference between the magnitude of the maximum soot volume fraction, which may occur due to uncertainties embedded in the boundary conditions, kinetic and soot models.

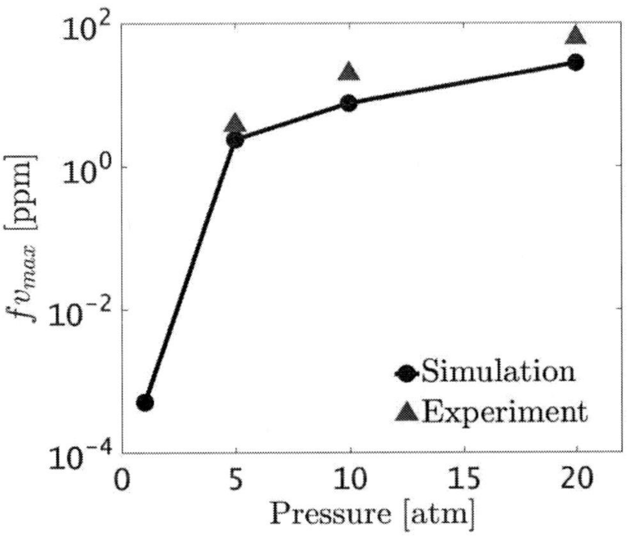

Figure 1: Variation of maximum soot volume fraction with pressure.

3 YSI simulations at high pressures

The concept of YSI was introduced to quantify the likelihood of forming soot from a given fuel, which can be seen as an inherent fuel property, independent of the specific combustion characteristics, such as equivalence ratio, temperature, or pressure. However, due to the large difference between YSI measurement pressure condition (1 atm) and the pressure levels in combustion devices (around 20 atm in HCCI engines), it is necessary to examine the pressure dependence of YSI to better connect YSI to fuel sooting properties under device-relevant conditions. Although experimentally assessing pressure effects on YSI would lead to significant cost and effort (e.g., installation of a pressure vessel), it can be easily done using numerical simulations.

We have numerically computed the baseline YSI flame [4] under a series of pressure levels from 1 atm (experimental condition) to 10 atm using multi-dimensional simulations with the semi-detailed chemical kinetic model and the soot model described in Section 2.2. Following the numerical procedure for YSI computation described in [8], we have successfully recomputed the YSI values for a range of chemical compounds under 2 atm. The results obtained are compared against their corresponding computed values at 1 atm and the experimental values measured at 1 atm. It is shown that pressure dependence is quite weak under this pressure condition. It seems that although the absolute level of soot formation increases significantly with pressure from 1 to 2 atm, this pressure dependence factors out when absolute soot concentrations are rescaled into YSIs. YSI simulations at even higher pressures (up to 20 atm) will be performed to further quantify pressure effects of YSIs for a wider range of species in future work.

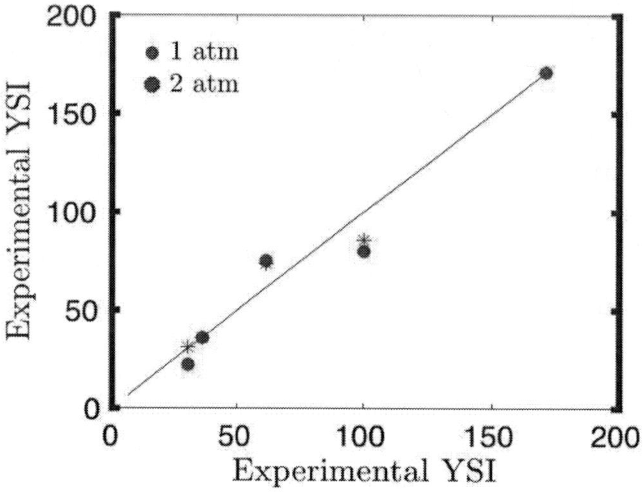

Figure 2: Predicted YSIs for a range of chemical compounds at 1 and 2 atm..

4 Conclusion

A recent, high throughput YSI method has been proposed to quantify the sooting tendencies of fuel species. However, this method is limited by experimental apparatus to ambient pressure conditions. The predicted 1 atm sooting behavior has been found to differ vastly from the reported soot yields found at higher pressures that are pertinent to practical combustion devices. Thus, in this study we first test the validity of the numerical methods mentioned in Sections 2.1 and 2.2 against a series of high pressure laminar coflow diffusion flames and find that these methods are successfully able to reproduce flame behavior and sooting trends for the range of pressures studied. These numerical methods have then been used to simulate the baseline YSI flame under a series of pressure levels from 1 to 10 atm. Furthermore, YSI values were computed for various species at 2 atm and compared to their 1 atm counterparts. A weak pressure dependence was observed at this pressure condition and even though the magnitude of soot yield is greater at 2 atm, the rescaled YSI values do not carry the pressure dependency.

Acknowledgments

This material is based upon work supported by the U.S. Department of Energy's Office of Energy Efficiency and Renewable Energy (EERE) under the Bioenergy Technologies Office (BETO) and Vehicle Technologies Office (VTO) Program Award Number DE-EE0007983.

References

[1] Brian S Haynes and H Wagner. *Progress in energy and combustion science*, 7 (1981) 229–273.

[2] Yefu Wang, Anandkumar Makwana, Suresh Iyer, Milton Linevsky, Robert J Santoro, Thomas A Litzinger, and Jacqueline OConnor. *Combustion and Flame*, (2017) .

[3] Anandkumar Makwana, Yefu Wang, Suresh Iyer, Milton Linevsky, Robert J Santoro, Thomas A Litzinger, and Jacqueline O'Connor. *Combustion and Flame*, (2017) .

[4] Charles S McEnally and Lisa D Pfefferle. *Combustion and flame*, 148 (2007) 210–222.

[5] I. Glassman. *Combustion.* (second ed.), Academic Press, Orlando, pp. 370373, 1987.

[6] Charles S McEnally and Lisa D Pfefferle. *Proceedings of the Combustion Institute*, 32 (2009) 673–679.

[7] Dhrubajyoti D Das, Charles S McEnally, Thomas A Kwan, Julie B Zimmerman, William J Cannella, Charles J Mueller, and Lisa D Pfefferle. *Fuel*, 197 (2017) 445–458.

[8] Yuan Xuan and Guillaume Blanquart. *Combustion and Flame*, 160 (2013) 1657–1666.

[9] Fengshan Liu, Kevin A Thomson, Hongsheng Guo, and Gregory J Smallwood. *Combustion and Flame*, 146 (2006) 456–471.

[10] Irvin M Miller and Howard G Maahs. (1977) .

[11] WL Flower and CT Bowman. *Symposium (International) on Combustion*, 21 (1988) 1115–1124.

[12] Wonnam Lee and Yong Dae Na. *JSME International Journal Series B Fluids and Thermal Engineering*, 43 (2000) 550–555.

[13] LL McCrain and WL Roberts. *Combustion and Flame*, 140 (2005) 60–69.

[14] Kevin A Thomson, Ömer L Gülder, Elizabeth J Weckman, Roydon A Fraser, Greg J Smallwood, and Dave R Snelling. *Combustion and Flame*, 140 (2005) 222–232.

[15] Hyun I Joo and Ömer L Gülder. *Proceedings of the Combustion Institute*, 32 (2009) 769–775.

[16] Ömer L Gülder, Gorngrit Intasopa, Hyun I Joo, Paul M Mandatori, Décio S Bento, and Marie E Vaillancourt. *Combustion and Flame*, 158 (2011) 2037–2044.

[17] Olivier Desjardins, Guillaume Blanquart, Guillaume Balarac, and Heinz Pitsch. *Journal of Computational Physics*, 227 (2008) 7125–7159.

[18] Y Xuan and G Blanquart. *Combustion and Flame*, 166 (2016) 113–124.

[19] Y Xuan and G Blanquart. *Proceedings of the Combustion Institute*, 35 (2015) 1911–1919.

[20] Marcus Herrmann, Guillaume Blanquart, and Venkat Raman. *AIAA journal*, 44 (2006) 2879.

[21] B Savard, Y Xuan, B Bobbitt, and Guillaume Blanquart. *Journal of Computational Physics*, 295 (2015) 740–769.

[22] G Blanquart, P Pepiot-Desjardins, and H Pitsch. *Combustion and Flame*, 156 (2009) 588–607.

[23] K Narayanaswamy, G Blanquart, and H Pitsch. *Combustion and Flame*, 157 (2010) 1879–1898.

[24] Fabrizio Bisetti, Guillaume Blanquart, Michael E Mueller, and Heinz Pitsch. *Combustion and Flame*, 159 (2012) 317–335.

2018 Spring Technical Meeting
Eastern States Section of the Combustion Institute
Hosted by The Pennsylvania State University
March 4-7, 2018

Sensitivity analysis and uncertainty propagation in numerical simulations of yield sooting tendencies

Y. Xuan[1], *K. Mohhan*[1], *A. Jain*[1], *H. Kwon*[1]

[1]*Mechanical and Nuclear Engineering, The Pennsylvania State University, University Park, PA*

The sooting tendencies of various fuels have been historically quantified via the Threshold Sooting Index (TSI), which is based on spoke point measurement. More recently, a Yield-based Sooting Index (YSI) has been proposed with lower measurement uncertainties, high throughput, and applicability to low-volatility fuels. YSI of a test species is experimentally measured in an atmospheric-pressure, laminar, non-premixed co-flow methane/air flame generated with a Yale Co-flow Burner by doping the baseline methane flame with a small concentration of the test species. The YSI concept interfaces well with computations since it is a perturbation-based approach with nearly identical temperature and velocity fields in the YSI flames compared to those in the undoped methane/air flame. In this work, we use a flamelet-based model to computationally simulate the YSI of fuels relevant to diesel and gasoline combustion. The numerically predicted YSI values are compared against experimental measurements. For fuels with large discrepancies in their simulated YSI, sensitivity analysis is performed to highlight deficiencies in the chemical reaction pathways. Numerical uncertainties are then introduced in the rate constants of these reactions and propagated through YSI simulations to quantify the impact of these reactions on the YSI predictions.

1 Introduction

Soot and greenhouse gas emissions from internal combustion engines are known to have an adverse effect on the environment and the health of human beings. This has led to the implementation of specific guidelines controlling the mass and distribution of particulate emissions, particularly soot. Alternative fuels consiting of bio-derived matter are being considered as replacements for current widely used fuels. Predicting the sooting tendencies of these alternative fuels has been a major concern to the automakers due to the need to adhere to enacted legislation.

Traditional approaches to quantify the sooting tendencies of various fuels have been based on the the Threshold Sooting Index (TSI) [1]. It characterizes a particular hydrocarbon's sooting tendency by experimentally measure the height of the hydrocarbon's jet flame, H, at the smoke point. This methods works well for smaller hydrocarbon's but has difficulties in measuring smoke heights for heavily sooting species like Polycyclic Aromatic Hydrocarbons (PAH). To overcome these challenges, the Yield-based Sooting Index (YSI) [2] has been introduced as a metric for measuring sooting tendencies of hydrocarbon species. The YSI are linear functions of the maximum soot volume fraction measured experimentally on the centerline of an atmospheric-pressure, laminar, non-premixed co-flow methane/air flame generated with a Yale Co-flow Burner with the methane stream doped with small concentrations of the test hydrocarbon. There have been numerous studies to quantify the sooting tendencies of various fuels and surrogates using the YSI [2–5] technique and it has been proven to have high throughput, lower measurement uncertainties and has been found to be applicable to low-volatility fuels.

The YSI method also proves to easily interface with numerical computations since it is a perturbation based approach and the temperature and velocity profiles of the doped methane/air flame compare nearly identically with their undoped counterpart. The most accurate way to mimic the experimental YSI technique would be to carry out high fidelity Direct Numerical Simulations (DNS) with finite-rate chemistry for each dopant separately and measure the obtained soot yield. However, doing this would prove to be prohibitively expensive computationally. The flamelet-based model, therefore serves as a feasible alternative that is computationally efficient and has proven to be popular in evaluating soot formation and evolution in laminar flames [6, 7]. Recently, modified flamelet equations have been developed that incorporate multidimensional diffusion and strong convection effects that are excluded in the steady-state flamelet model. These have been found to successfully capture the experimentally observed species mass fractions correctly along with their soot yield [5].

For this work, we use the modified flamelet equations to computationally simulate the YSI of fuels relevant to diesel and gasoline combustion. The YSI calculated from the numerical simulations are further compared against experimental measurements to identify fuels that diverge greatly from their experimental YSI. Sensitivity analysis is then performed on these identified fuels to highlight deficiencies in the chemical reaction pathways and numerical uncertainties are introduced into the rate constants of particular reactions and propagated through the YSI simulations to quantify the impact of these reactions on the YSI prediction.

2 Framework

2.1 Numerical Methods

The baseline flame is a CH_4/air coflow laminar diffusion flame established on the Yale Co-flow burner. The fuel stream consists of $55\%CH_4$ and $45\%N_2$. Further details on the burner configuration and inlet conditions can be found in the following references [2, 5].

The baseline flame simulations are done using the NGA solver [8]. The NGA code uses staggered variables and allows for accurate, robust, and flexible simulations of both laminar and turbulent reactive flows in complex geometries and has been applied to a wide range of test problems [9, 10]. The temperature, convective velocity, and scalar dissipation rate profiles calculated from the baseline (undoped) flame simulation are extracted and used as input for the modified flamelet equations.

The modified flamelet equations take into account the effects of multidimensional diffusion and convection of species in mixture fraction space due to non-unity Lewis numbers. The modified flamelet equations are solved using the FlameMaster code [11] following previous work [5]. The numerical sooting tendencies in the form of YSI measurements are then obtained using the PAH dimer production rate generated from the flamelet solutions doped by the test species. YSI measurements are performed for toluene, n-heptane, benzene, iso-octane, methanol and hexene in this study. Additionally, testing is done for ethanol for the chemical mechanisms containing these species. These species form the essential components for fuels widely used in gasoline and diesel engines. They are also important parts of any fuel surrogate formulation that aim to mimic the sooting tendencies of widely used fuels.

2.2 Chemical kinetic models

Multiple chemical kinetic mechanisms are used to test the dependencies of the the fuel YSI on the underlying chemical mechanism. The first mechanism considered is the CaltechMech chemical kinetic model. It consists of 185 species and 1903 reactions (forward and backward reactions counted separately), and takes into account all major pathways of polycyclic aromatic hydrocarbons (PAH) formation [12, 13]. This chemical model has been extensively validated for multiple fuels [14], including the one used in this study. This model has been used in multiple configurations, including laminar and turbulent flames [5].

The second and third chemical models are reduced kinetic models which were derived from the parent mechanism developed by LLNL for gasoline surrogates [15] which contains 1550 species and 11612 reactions. These chemical models are obatined from the reduction of the detailed parent mechanism using techniques described in [16]. The second kinetic model is a skeletal mechanism containing 323 species which was developed by the researchers at the University of California targeting the ignition-delay of specific fuel mixtures. [17]. The third model contains 679 species and 6854 reactions and is available at [18]. The reduced mechanisms have not been validated for soot formation, as very few aromatic growth reactions were included. Therefore, in this work, the reduced mechanisms were merged with a detailed aromatic growth sub-mechanism [13]. The designation LLNL-Model I and LLNL-Model II will be used to refer to the 323 and 679 species mechanisms hereafter in this study.

3 Results

3.1 YSI calculations

YSI calculations are performed for the 5 test species, namely toluene, n-heptane, benzene, iso-octane, methanol and 1-hexene using the three chemical kinetic models. A comparison is then made between experimental and numerical YSI measurements. Figure 1 depicts results using the Caltechmech chemical kinetic mechanism. The straight line drawn in the graph is the line that indicates correspondence with experimental values. It is important to note that the points indicating n-heptane and toluene are on the straight line as they are used as reference fuels. This means that the numerical value of YSI is equivalent to the experimental value of the YSI. Therefore, the mechanism cannot be judged for its efficiency by using the numerical YSI values of these 2 fuels. CaltechMech performs well for most of the test species as indicated by the closeness to the equivalence line. It however does a poor job in calculating the YSI value for methanol which may occur due to deficiencies in the CaltechMech mechanism as it has not been optimized for alcohols.

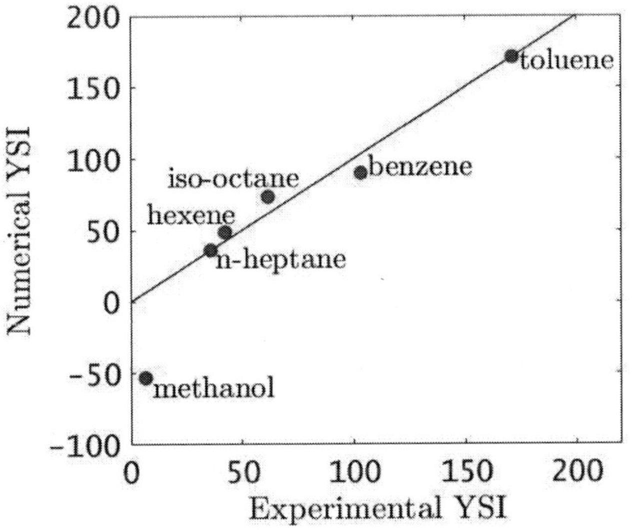

Figure 1: Numerical YSI and Experimental YSI comparison for CaltechMech.

Figure 2 shows a comparison of the simulated and experimental YSI values when simulaing using the LLNL-Model I (323 species mechanism). The LLNL-Model I shows significant departure from the experimental value for all the test species. This may occur as the reduction techniques used to obtain this mechanism were targeted towards surrogate fuels not containing the test species used here.

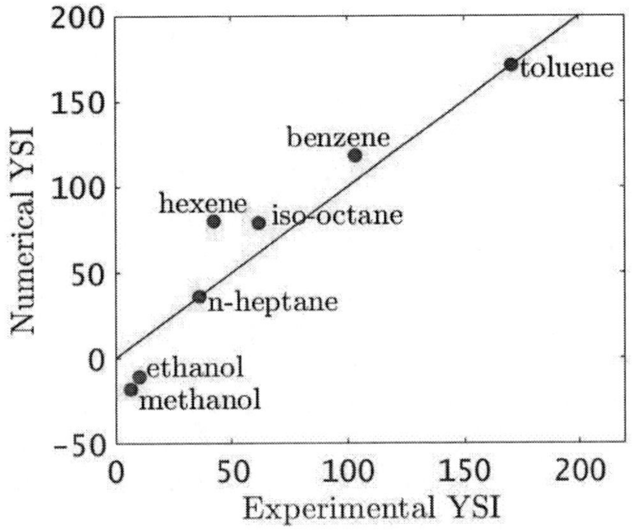

Figure 2: Numerical YSI and Experimental YSI comparison for LLNL-Model I.

The LLNL-Model II (679 species) was designed targeting iso-octane, n-heptane, toluene, 2-pentene and ethanol as pure components and their mixtures. It therefore performs well in the numerical YSI calculations and compares favorably with the experimental values for most test species. This is shown in Figure 3. However, it shows slight departure for ethanol and iso-octane.

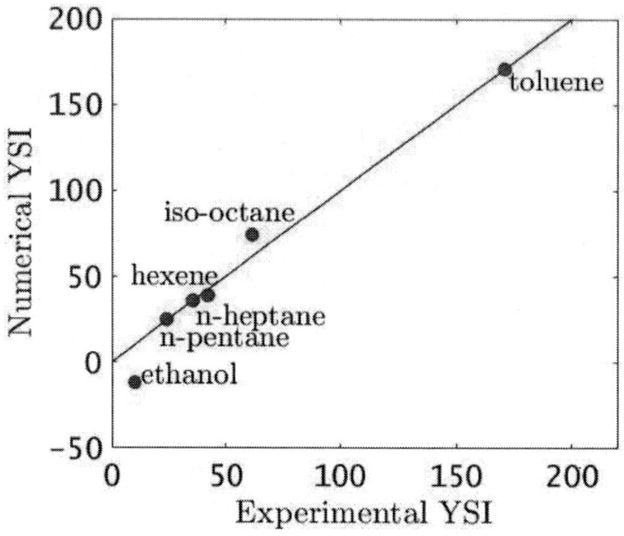

Figure 3: Numerical YSI and Experimental YSI comparison for LLNL-Model II.

3.2 Reaction sensitivity analysis and chemical kinetic uncertainty propagation

The relatively large difference between the experimentally measured and computationally predicted YSI values for ethanol and iso-octane observed in Fig. 3 is likely due to large chemical kinetic uncertainties embedded in the chemical model used (LLNL-Model II). To isolate the deficiencies in the chemical mode, reaction sensitivity analysis is performed in the flamelet-based YSI calculation when ethanol and iso-octane are used as dopant. From the sensitivity analysis of flamelet doped with ethanol, the following two reactions involving ethanol,

$C_2H_5OH- > C_2H_5 + OH$

$C_2H_5OH- > C_2H_4 + H_2O,$

are found to be the most sensitive reactions to the production of soot. From the sensitivity analysis of flamelet doped with is-ooctane, the following three reactions involving iso-octane or one of the direct byproduct of iso-octane decomposition,

$C_7H_{15}- > I - C_3H_7 + I - C_4H_8$

$C_7H_{15}- > C_7H_{14} + H$

$I - C_8H_{18}- > C_7H_{15} + CH_3$

are found to be the most sensitive reactions to the production of soot. The rate coefficients (pre-exponential factors) of the most sensitive reactions identified above are then perturbed within 50% range of uncertainty and propagated through the flamelet-based YSI calculations, to assess the impact of these reaction on the YSI predictions of iso-octane and ethanol. The resulting numerical uncertainty ranges of YSI for both ethanol and isooctane are plotted together with the previously computed YSI values in Fig. 4. It is found that the 50% uncertainty introduced in the identified most sensitive reactions leads to around 4% uncertainty in the computed YSIs for iso-octane and ethanol. Larger perturbations in the reaction rates of the sensitive reactions will be numerically introduced in future calculations.

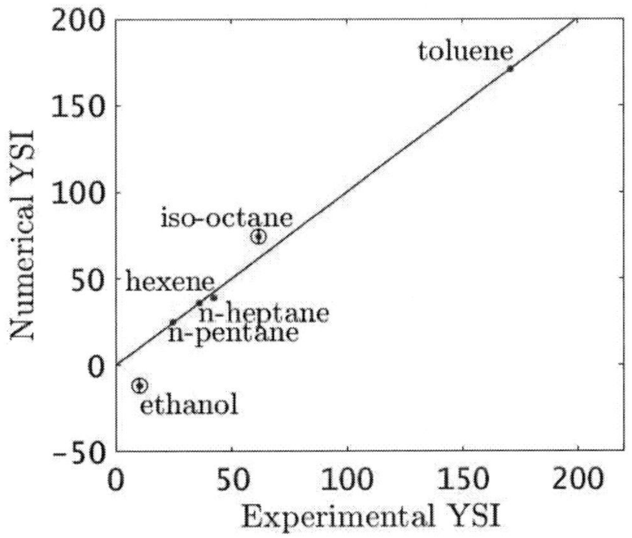

Figure 4: Comparison between computed YSI using the LLNL-Model II mechanism with numerical uncertainties for iso-octane and ethanol against Experimental YSI.

4 Conclusion

A recent Yield-based Sooting Index (YSI) has been developed to evaluate sooting tendencies of various hydrocarbon species. In this work, a flamelet-based model is used to simulate YSI's of species relevant to diesel and gasoline combustion. Three chemical kinetic mechanisms, namely CaltechMech, LLNL-Model I, and LLNL-Model II are used with the flamelet model for YSI calculations. The numerical values are then compared against experimental YSI measurements to evaluate the dependence of YSI's on the chemical kinetic model. Of the three models used, the LLNL-Model II proves to be the best in accurately predicting the numerical YSI of the fuel species. However, there are still differences in simulated and experimental YSI values for ethanol and iso-octane. Sensitivity analysis was subsequently conducted to identify key reactions that lead to 4% uncertainty in computed YSIs. These reactions are highlighted in Section 3.2.

Acknowledgments

This material is based upon work supported by the U.S. Department of Energy's Office of Energy Efficiency and Renewable Energy (EERE) under the Bioenergy Technologies Office (BETO) and Vehicle Technologies Office (VTO) Program Award Number DE-EE0007983

References

[1] I. Glassman. *Combustion.* (second ed.), Academic Press, Orlando, pp. 370373, 1987.

[2] Charles S McEnally and Lisa D Pfefferle. *Combustion and flame*, 148 (2007) 210–222.

[3] Charles S McEnally and Lisa D Pfefferle. *Proceedings of the Combustion Institute*, 32 (2009) 673–679.

[4] Dhrubajyoti D Das, Charles S McEnally, Thomas A Kwan, Julie B Zimmerman, William J Cannella, Charles J Mueller, and Lisa D Pfefferle. *Fuel*, 197 (2017) 445–458.

[5] Yuan Xuan and Guillaume Blanquart. *Combustion and Flame*, 160 (2013) 1657–1666.

[6] K Claramunt, R Consul, D Carbonell, and CD Pérez-Segarra. *Combustion and Flame*, 145 (2006) 845–862.

[7] BC Connelly, MB Long, MD Smooke, RJ Hall, and MB Colket. *Proceedings of the Combustion Institute*, 32 (2009) 777–784.

[8] Olivier Desjardins, Guillaume Blanquart, Guillaume Balarac, and Heinz Pitsch. *Journal of Computational Physics*, 227 (2008) 7125–7159.

[9] Y Xuan and G Blanquart. *Combustion and Flame*, 166 (2016) 113–124.

[10] Y Xuan and G Blanquart. *Proceedings of the Combustion Institute*, 35 (2015) 1911–1919.

[11] H Pitsch. *RWTH Aachen, Aachen, Germany*, (1998) .

[12] G Blanquart, P Pepiot-Desjardins, and H Pitsch. *Combustion and Flame*, 156 (2009) 588–607.

[13] K Narayanaswamy, G Blanquart, and H Pitsch. *Combustion and Flame*, 157 (2010) 1879–1898.

[14] Guillame Blanquart. Caltechmech. http://www.theforce.caltech.edu/CaltechMech/.

[15] Marco Mehl, William J Pitz, Charles K Westbrook, and Henry J Curran. *Proceedings of the Combustion Institute*, 33 (2011) 193–200.

[16] Tianfeng Lu and Chung K Law. *Progress in Energy and Combustion Science*, 35 (2009) 192–215.

[17] Marco Mehl, Jyh-Yuan Chen, William J Pitz, S Mani Sarathy, and Charles K Westbrook. *Energy & Fuels*, 25 (2011) 5215–5223.

[18] Lawrence Livermore National Labs. Gasoline surrogate. https://combustion.llnl.gov/mechanisms/surrogates/gasoline-surrogate.

Spring Technical Meeting
Eastern States Section of the Combustion Institute
March 4-7, 2018
State College, Pennsylvania

Analyzing the Robustness of the Yield Sooting Index as a Measure of Sooting Tendency

Matthew J. Montgomery1[1,], Dhrubajyoti D. Das[1], Charles S. McEnally[1], Lisa D. Pfefferle[1]*

[1]*Chemical Engineering, Yale University, New Haven CT 06520 USA*
**Corresponding Author Email: matthew.montgomery@yale.edu*

Abstract: Soot formation from combustion is linked with many issues, ranging from global warming to human health complications. To tackle these problems, measures of sooting tendency that are accurate and robust with respect to flame conditions are needed to facilitate the development of cleaner fuels and combustion technologies. The objective of this research is to analyze the accuracy and robustness of the Yield Sooting Index (YSI), a measure of sooting tendency, under different temperatures and air-to-fuel equivalence ratios (λ). In this work, the YSI of various cyclic hydrocarbons and jet fuels were measured in several methane/air co-flow flames doped with small quantities (0.85% by mass of total fuel) of each test compound using two soot diagnostic techniques, color-ratio pyrometry and line-of-sight spectral radiance. The flame temperature and λ-values of these doped co-flow flames were varied by premixing the base methane fuel with varying amounts of nitrogen and/or air. The sensitivity of the measured YSI for the test compounds to varying conditions was analyzed, and our results show that YSI values were invariant across all investigated flame conditions for all dopants, suggesting YSI is robust across a wide range of premixed air levels and temperatures relevant to soot formation.

Keywords: *Particulates; Premixed flame; Jet fuels; Sooting tendency*

1. Introduction [12pt]

Combustion processes account for most soot emissions around the world, and these contribute to health and environmental problems through many routes [1-4]. In order to curb soot emissions from technologies such as diesel and jet engines, the relationship between fuel chemical structure and soot emissions must be understood. Towards this end, many researchers have aimed to develop and improve sooting tendencies relevant to engines which can be used to quantify the propensity of a particular fuel to form soot. The earliest measure of sooting tendency to be widely-adopted is smoke point, which is the height of a wick burner flame at conditions where soot begins to be emitted from the tip [5, 6, 7]. Another measure of sooting tendency, called the threshold sooting index (TSI), normalizes the smoke point to values for reference compounds [8, 9, 10]. Subsequent studies have developed approaches to obtain more reliable TSI measurements through improved methods of determining and defining the smoke point, such as by employing fuel uptake rate measurement with threshold imaging methodology [11] or by utilizing normalized smoke points [12]. The oxygen extended sooting index (OESI) introduced by Barrientos et al. [13] better captures the chemical effects of oxygen in fuel structure. Lemaire et al. [14] have also developed the fuel equivalent sooting index (FESI), which uses measurements in turbulent spray flames to characterize sooting tendency.

The yield sooting index (YSI) is an alternative approach to quantifying sooting tendencies based on converting the maximum soot concentration in a fuel-doped methane/air co-flow laminar flame into an index describing the fuel's propensity to soot [15, 16, 17]. The small amount of dopant used in this approach minimizes effects due to fuel dilution and allows YSI to effectively isolate and measure the effect of the fuel chemistry on soot formation. Laser-induced incandescence has been traditionally used to evaluate the maximum soot concentration for YSI measurements [15, 16]; recent works have employed color-ratio pyrometry [18, 19] to evaluate the maximum soot concentration. The YSI methodology has been used to characterize the sooting tendency of over 400 different compounds, including small hydrocarbons, oxygenates, large aromatics, and real diesel and jet fuels [15-16,18-19].

One issue with most sooting tendency measurements is that they are performed under a very limited set of flame conditions, whereas in real devices parameters such as the air-to-fuel equivalence ratio (λ) and characteristic temperature vary significantly [20-21]. This calls into question the applicability of these sooting tendency measures in the development of combustion technologies, since sooting indices determined under one set of these parameters may not extend to another set of conditions. For measures utilizing the smoke point, it is difficult to assess the effect of partial premixing or temperature changes because the wick lamps traditionally used in these measurements [7] do not allow these parameters to be systematically varied.

In order for sooting indices to be used in applications such as developing surrogate fuels or aiding in the transition to cleaner engines and fuels, robust measures which can accurately describe the sooting behavior of fuels across a wide range of combustion conditions are necessary. The objective of this research is to assess whether YSI is consistent across a range of combustion conditions relevant to soot formation in a series of doped nonpremixed and partially premixed flames. The combustion conditions varied in this work include λ and the adiabatic flame temperature T_{ad}. The fuels tested include three cyclic hydrocarbons and three jet fuels. The cyclic hydrocarbons were methylcyclopentane, cyclohexane, and cyclohexene. They were chosen based on the hypothesis that their sooting behavior would be particularly sensitive to changes in λ since their chemical consumption mechanisms involve competition between ring-breaking unimolecular dissociation reactions and ring-preserving abstraction reactions with O_2 [22, 23]. The jet fuels were practical fuel samples obtained from Wright-Patterson Air Force Base and were used to examine the impact of changing flame conditions on the sooting behavior of real fuels.

2. Methods / Experimental

Atmospheric-pressure coflow laminar premixed flames were generated with a Yale Coflow Burner [24]: the fuel mixture – methane, nitrogen, primary air, and a dopant – flowed out of a 4.0 mm inner-diameter tube and reacted with secondary air that flowed from the annular region between this tube and a 76 mm inner-diameter tube. Figure 1 shows the burner setup. For a detailed burner description see [25]. The gaseous reactants flowed from 99.99+ % purity cylinders (methane, nitrogen) or a compressor (air). Electronic mass flow controllers (MKS 1179A and 1559A, Omega 5514) actively governed the flowrates of these reactants and were calibrated with soap bubble meters for the specific process gases. The dopants examined were either pure cyclic hydrocarbons (methyl cyclopentane, cyclohexane, and cyclohexene) or jet fuels (POSF 4658, POSF 10264, and POSF 10289) (Table I). Their YSI values have been previously characterized [15, 16]. The dopants were injected into the gaseous methane/nitrogen/air mixture by a syringe pump (KDS 100). The performance of this pump was characterized by measuring its linear rate-of-travel with calipers in place of a syringe; the pump was sufficiently accurate across the range of flowrates used in this study to contribute negligibly to uncertainties in the dopant mass fraction in the flame. The syringe needle entered the fuel line through a septum in a stainless steel tee. Resistive tapes heated the fuel line and the burner fuel tube to 175 °C, with the fuel tube temperature maintained to within ±1 °C under PID control; given this heating the dopants vaporized rapidly upon injection and then were swept to the burner by the other fuel components. The nominal flow rates of methane and secondary air were chosen such that the volumetric flow rate of methane Q_m = 282 mL/min and volumetric flow of secondary air Q_{air} = 50,000 mL/min. The flow rates of the dopants were chosen such that the mass fraction of dopant in the total dopant + methane base fuel mixture was Y_{dop}=0.85%. The volumetric flow rate of the primary nitrogen Q_{N2} and primary air Q_{pa} were independently varied in order to explore different flame temperatures and λ values, respectively. Adiabatic flame temperatures were calculated using the NASA CEA program [28]. Though actual flame temperatures are lower than the adiabatic flame temperature, this calculation captures the relative differences in flame temperatures induced by changes in the nitrogen dilution levels in the fuel.

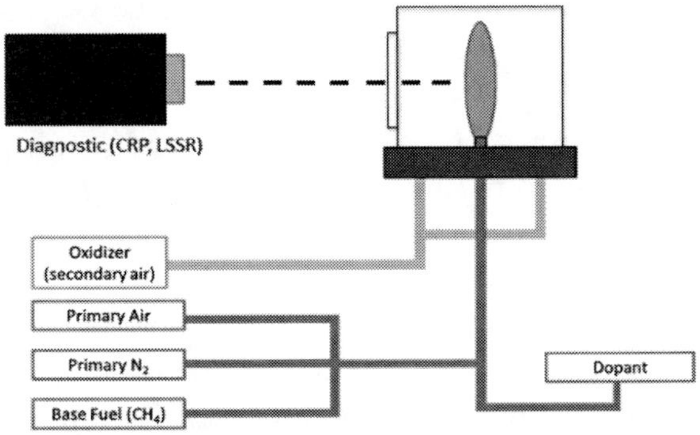

Figure 1- Burner setup used for the experiments described in this paper

We measured absolute maximum soot concentrations using color-ratio pyrometry. Detailed descriptions of this technique are available in the literature [19, 29]. Briefly, radiative emissions from soot in a doped flame are imaged by a digital camera, then spatial maps of the soot volume fraction f_v distribution in the flame are generated. The maximum soot concentration in these f_v distributions are used in determining the YSI of a dopant fuel. The setup consists of: (1) an anodized aluminum chimney with a transparency film window minimizes backscattering and allows light from the flame to be imaged; (2) a Schott BG-7 filter permits the transmission of light in the 400 to 700 nm range and provides a better balance between the red, green, and blue intensities from the flame [29]; (3) a Nikon D90 consumer camera captures light from the flame that is transmitted through the filter and is subsequently saved in the camera's .NEF 12-bit compressed raw format. For each dopant, 100 flame images were captured over a 600 second period, demosaiced using OMA [30], and then the reconstruction procedure and following temperature and soot calculations were performed using Python code.

In order to validate our results across different experimental methods for determining sooting tendency, line-of-sight spectral radiance (LSSR) at 660 nm was employed to measure relative soot concentrations. In this method the integrated radiant signal from soot particles along a path through the centerline of the flame is measured. We include "line-of-sight" in the name of this technique because the depth-of-field of our detection system (~8 mm) is comparable to the flame diameter; thus the measured radiance comes from all flame locations along the detection system's axis, not from a single point. The basic experimental setup consists of: (1) a UV silica window mounted in the chimney transmits light emitted by the flame to the rest of the measurement system; (2) a fused silica biconvex lens focusses some of this light onto a 1 mm diameter circular aperture with unity magnification; (3) an interference filter (Thorlabs FB660-10, 10 nm FWHM, λ_{center} = 660 ± 2 nm) and an infrared-blocking filter (Schott KG2) isolate the portion of this light at 660 nm; and (4) a red-enhanced photomultiplier tube (PMT; Oriel 77348) detects it. An A/D converter (LeCroy LT342, 1 mΩ input impedance) samples the PMT output at 5 Hz. Each sample is an average of 50,000 8-bit readings recorded 2 μsec apart. The noise in the measurements is negligible compared to other sources of experimental error.

3. Results and Discussion

In this study, we evaluated the maximum peak soot volume fraction for nonpremixed and partially premixed methane flames doped with three cyclic hydrocarbons and three jet fuels under nine different flame conditions which span a range of adiabatic flame temperatures and λ values. For a subset of conditions (T_{ad}, λ) = (2224 K, 0), (2224 K, 0.02), (2224 K, 0.04), (2185 K, 0.02), the maximum soot concentrations were arbitrarily evaluated via color-ratio pyrometry using the "peak-region" method. In this method, local soot volume fractions f_v in the regions of the flame with f_v above the 90th percentile are averaged; this average is used as the representative peak soot volume fraction $f_{v,max}$ in the flame. This has been shown to be a robust method of quantifying the peak soot levels in similar flames [31]. Since YSI is based upon relative differences in sooting tendency between different doped flames, the near-peak method yields a measure of sooting level relevant for determining YSI. For the remaining five flame conditions, the LSSR technique was used to measure the relative differences in sooting tendency of the different fuels. The measured LSSR signal depends on the maximum soot concentration, as well as other parameters such as soot temperature, air-to-fuel ratio, soot emissivity, and other scaling variables. Besides the soot concentration, these other parameters have a weak dependence on the type of dopant used at any given flame condition, and therefore cancel out when the LSSR signal is converted into YSI.

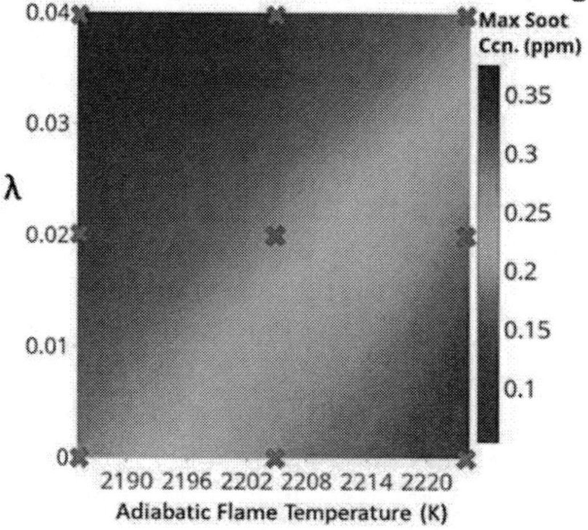

Figure 2-2D soot concentration map demonstrating the range of conditions examined in this study

Nine different flame conditions were considered in this work. Figure 2 shows a 2D soot concentration map of POSF 10264 across the range of air-to-fuel ratios and adiabatic flame temperatures encountered in these experiment, with X's denoting the particular flame conditions analyzed for measurements used in Fig. 3. POSF 10264 has a YSI of 49.6, which is almost exactly halfway between that of hexane (0) and benzene (100). It can be seen from Figure 2 that the range of flame conditions chosen were sufficient to span a wide range of sooting conditions for the POSF 10264-doped flames. The YSIs of all six hydrocarbons and fuels listed in Section 2.1 were then determined for each set of conditions and compared to a reference YSI. The results are summarized in Fig. 3.

In Fig. 3, the YSIs are compared as follows: a standard set of YSIs are measured under conditions (T_{ad}, λ) = (2224K, 0) for each test compounds and used as the reference YSI values on the horizontal axes of the nine scatter plots. The YSIs measured under a particular set of (T_{ad}, λ) conditions are represented on the vertical axes of the nine scatter plots. Thus, for conditions (T_{ad}, λ) = (2224K, 0), all data points lie on the line y=x (denoted in the plots by a dashed line) by definition. This mediates the problem of comparing the literature YSI of the test compounds, which were measured on either a mass or mole basis [16, 18], to the YSI measurements in these studies. As seen in Fig. 3, the YSIs of the various fuels measured across various λ values and adiabatic flame temperatures agree well with their corresponding reference YSI. The good agreement between the YSIs measured across the range of conditions examined in this study and their reference YSIs suggests that the sooting tendency relationship between fuel compositions and structures holds across a range of air-to-fuel ratios and flame temperatures.

The three cyclic compounds were chosen for this study based on the hypothesis that the sooting behavior of cyclic compounds could be strongly impacted by the air-to-fuel ratio. Under oxygen-rich environments, O_2 can potentially extract protons from cyclic compounds, leading to the formation of cyclic alkenes, dienes, and eventually benzene [22, 23]. This would manifest as a greater sooting propensity for cyclic compounds under increasing oxygen-rich conditions relative to the sooting propensity of hexane or benzene, thereby skewing the YSI value of these cyclic compounds at different air-to-fuel ratios. However, from Fig. 3, it is evident that the YSI of the cyclic compounds did not deviate significantly from their reference YSI with changes in air-to-fuel ratio. This suggests that any increases in soot concentration through the H-abstraction pathway by O_2 are insignificant under the range of conditions examined. As the air-to-fuel ratios encountered in these studies led to practically no soot formation at higher flame temperatures and λ values (Fig. 2), it follows that these pathways would not contribute significantly to soot formation pathways at higher air-to-fuel ratios, as they would be outcompeted with hydrocarbon oxidation pathways to CO_2 and H_2O.

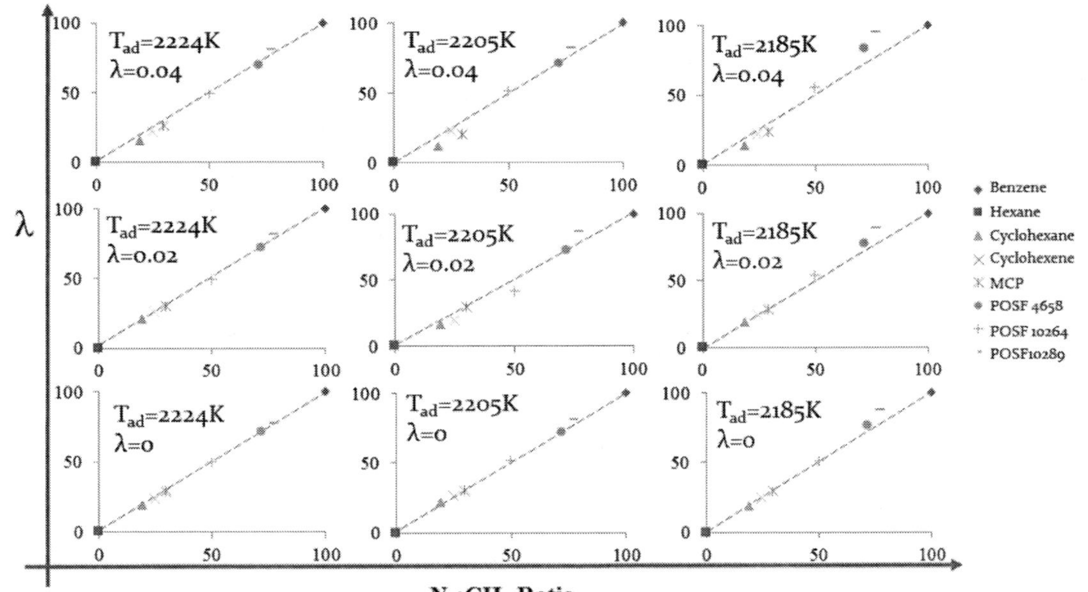

Figure 3- Reference YSI vs. Experimentally Measured YSI Across Various Air-to-fuel ratios (λ) and N_2:CH_4 ratios (Q_{N2}/Q_m)

4. Conclusions

In this work, YSI measurements of various fuels doped into a nonpremixed and partially premixed methane/air coflow flame under a range of flame temperatures and air-to-fuel ratios were reported and validated by two different techniques. Color-ratio pyrometry and LSSR were demonstrated to be feasible methods for determining the relative differences in maximum sooting tendency in doped methane flames. The relative differences in the maximum sooting tendencies of the various fuels were preserved across nine different combustion conditions, yielding YSI measurements which agreed well with their corresponding literature values. The hypothesis that an increase in oxygen content would lead to an increase in the relative sooting tendency for the cyclic compounds was not realized, and is likely not to affect YSI measurements across a moderate range of air-to-fuel ratios. This work is expected to be of relevance to surrogate formulation applications, as well as to validate computational models on sooting tendency.

5. Acknowledgements

This material is based upon work supported by the U.S. Department of Energy's Office of Energy Efficiency and Renewable Energy (EERE) under the Bioenergy Technologies Office (BETO) and Vehicle Technologies Office (VTO) Program Award Number DE-EE0007983. This work is also supported by the National Science Foundation (NSF) under Grant No. CBET 1604983. We thank Tim Edwards (Wright-Patterson AFB) for useful conversations and for sending us fuel samples.

6. References

1) M.P. Fraser, Z.W. Yue, R.J. Tropp, S.D. Kohl, J.C. Chow, Atmospheric Environment 36 (2002) 5751-5758.
2) M. Zheng, L.G. Salmon, J.J. Schauer, L. Zeng, C.S. Kiang, Y. Zhang, G.R. Cass, Atmospheric Environment 39 (2005) 3967-3976.
3) S.S. Lim, , et al. Lancet 380 (2012) 2224-2260.
4) T.C. Bond, S.J. Doherty, D.W. Fahey, J. Geophys. Res. 118 (2013) 5380-5552.
5) R.A. Hunt, Ind. Eng. Chem. 45 (1953) 602-606.
6) A. Gomez, G. Sidebotham, I. Glassman, Combust. Flame 58 (1984) 45-57.
7) ASTM International, D1322-15: Standard test method for smoke point of kerosene and aviation turbine fuel, West Conshohocken PA USA, 2015.
8) H.F. Calcote, D.M. Manos, Combust. Flame, 49 (1983) 289-304.
9) P. Pepiot-Desjardins, H. Pitsch, R. Malhotra, S.R. Kirby, A.L. Boehman, Combust. Flame 154 (2008) 191-205.
10) A. Mensch, R.J. Santoro, T.A. Litzinger, S.-Y. Lee, Combust. Flame 157 (2010) 1097-1105.
11) R.J. Watson, M.L. BoteroC.J. Ness., N.M. Morgan, M. Kraft, Fuel 111 (2013) 120-130.
12) L. Li, P.B. Sunderland, Combust, Sci. Technol. 184 (6) (2012) 829-841.
13) E.J. Barrientos, M. Lapuerta, A.L. Boehman, Combust. Flame 160 (2013) 1484-1498.
14) R. Lemaire, D. Lapalme, P. Seers, Combust. Flame 162 (2015) 3140-3155.
15) C.S. McEnally, L.D. Pfefferle, Combust. Flame, 148 (2007) 210-222.
16) C.S. McEnally, L.D. Pfefferle, Environ. Sci. Technol. 45 (2011) 2498-2503.
17) M. Kashif, P. Guibert, J. Bonnety, G. Legros, Combust. Flame 161 (2014) 1575-1586.
18) D.D. Das, C.S. McEnally, T.A. Kwan, J.B. Zimmerman, W.J. Cannella, C.J. Mueller, L.D. Pfefferle, Fuel 197 (2017) 445-458.
19) D.D. Das, W.J. Cannella, C.S. McEnally, C.J. Mueller, L.D. Pfefferle, Proc. Combust. Inst. 36 (2017) 871-879.
20) T. Kamimoto, M.-h. Bae, SAE Tech. Pap. 880423 (1988).
21) A.B. Dempsey, S.J. Curran, R.M. Wagner, Int. J. Engine Res. 17 (2016) 897-917.
22) H.R. Zhang, E.G. Eddings, A.F. Sarofim, C.K. Westbrook, Proc. Combust. Inst. 32 (2009) 377-385.
23) W.K. Stuckey, J. Heicklen, J. Chem. Phys., 46 (1967) 4843-4846
24) M.D. Smooke, C.S. McEnally, L.D. Pfefferle, R.J. Hall, M.B. Colket, Combust. Flame 117 (1999) 117-139.
25) J. Gau, D. Das, C. McEnally, D. Giassi, N. Kempema, M. Long, Yale Coflow Burner Information and CAD Drawings, 2017. (https://doi.org/10.6084/m9.figshare.5005007.v1)
26) B.-D. Gabi, O. Sadot, O. Igra, 30th International Symposium on Shock Waves 1: ISSW30-Volume 1, Springer International Publishing, New York, 2017, p. 311
27) S. Dooley, S.H. Won, M. Chaos, J. Heyne, Y.G. Ju, F.L. Dryer, K. Kumar, C.J. Sung, H.W. Wang, M.A. Oehlschlaeger, R.J. Santoro, T.A. Litzinger, Combust. Flame 157 (2010) 2333-2339.
28) B.J. McBride, S. Gordon, Computer Program for Calculation of Complex Chemical Equilibrium Compositions and Applications II. User's Manual and Program Description, Report No. E-8017-1, National Aeronautics and Space Administration, 1996
29) P.B. Kuhn, B. Ma, B.C. Connelly, M.D. Smooke, M.B. Long, Proc. Combust. Inst. 33 (2011) 743-750.
30) P.A.M. Kalt, M.B. Long, OMA – Image processing for Mac OS X, 2015, (http://www.oma-x.org)
31) D.D. Das, P.S. John, C.S. McEnally, S. Kim, L.D. Pfefferle, engrXiv, 2017, (DOI 10.17605/OSF.IO/9WQFN)

Spring Technical Meeting
Eastern States Section of the Combustion Institute
March 4–7, 2018
State College, Pennsylvania

A Multi-Moment Sectional Method to Predict the Soot Size Distribution

Suo Yang[1,*] *and Michael E. Mueller*[1]

[1]*Department of Mechanical and Aerospace Engineering, Princeton University, NJ 08544*
[*]*Corresponding author:* suoy@princeton.edu

Abstract: A new Multi-Moment Sectional Method (MMSM) is developed to predict the evolution of the soot size distribution. Unlike conventional sectional methods, in MMSM, multiple moments are solved within each section, and the soot number density function within each section is reconstructed from a polynomial profile or, for the last section, an exponential profile. The MMSM approach is combined with a detailed univariate soot model and applied to the prediction of the soot size distribution in an ethylene laminar premixed flame. Compared to a pure sectional method, MMSM is shown to achieve a higher rate of convergence for the predicted total number density and particle size distribution.
***Keywords:* Soot, Particle size distribution, Sectional method, Method of moments**

1. Introduction

One of the major challenges in modeling soot is the large number of particles. This requires a statistical description of the evolution of the soot population using a Number Density Function (NDF) rather than tracking individual particles. However, the resulting Population Balance Equation (PBE), which governs the evolution of the NDF, is a high-dimensional equation that depends not only on space and time but also the internal coordinates used to describe soot particles (volume, surface area, etc.). Therefore, a stastical model is required for tracking the evolution of the NDF. These statistical models can be grouped into three categories: Monte Carlo [1], sectional methods [2], and methods of moments [3, 4]. Monte Carlo methods can track the complete NDF very accurately but are expensive and limited to simple 0D and 1D problems. Sectional methods can provide accurate predictions of the NDF with lower computational cost than Monte Carlo but only if a single internal coordinate (e.g., volume or mass of soot particle) is utilized to describe soot particles. However, even with one single coordinate, $O(100)$ sections are required to adequately resolve the NDF, which is still very expensive for multi-dimensional calculations. Converesly, the computational cost of methods of moments is very low compared to both Monte Carlo and sectional methods since only moments of the NDF are tracked, which has made them popular for application in unsteady, multi-dimensional turbulent flow calculations. Some recent methods of moments have even attempted to provide some limited information about the NDF. However, no methods of moments can provide direct information about the NDF itself consistent with the underlying governing physics of the PBE.

In this study, a new approach, termed the Multi-Moment Sectional Method (MMSM), is developed and applied to the statistical modeling of soot. Within each section, a polynomial (for the sections of finite support) or exponential (for the last section of semi-infinite support) determined from the local moments within each section is used to reconstruct NDF within a section. The formulation of the new approach is first presented followed by application to a laminar premixed flame. The rate of convergence of the method compared to a pure sectional method is analyzed.

2. Multi-Moment Sectional Method

The governing equation for the local xth-order moments M_x^i in the ith section is

$$\frac{\partial M_x^i}{\partial t} + \frac{\partial u_j M_x^i}{\partial x_j} - \frac{\partial}{\partial x_j}\left(0.55\frac{\nu}{T}\frac{\partial T}{\partial x_j}M_x^i\right) = \dot{M}_x^i, \tag{1}$$

where the third term on the left hand side is the thermophoresis of soot particles [5] and molecular diffision is neglected [6]. The source terms on the right hand side include contributions from various physical and chemical processes affecting soot evolution. In the MMSM approach, these moment source terms are closed by reconstructing the local NDF.

For soot, due to the rapid coagulation process, a geometrically increasing section size is commonly adopted in methods with fixed sections, in which a spacing factor $f_s = \Delta V_{i+1}/\Delta V_i$ is fixed with $f_s > 1$ [2]. Except for the last section, each section has a finite size ΔV_i, so a polynomial approximation of the NDF is introduced in each section. In this work, only two moments per section will be considered, so the NDF is locally approximated as linear. For the last section (i.e., $i = N$), which is semi-infinite, the NDF is reconstructed with an exponential profile.

In the detailed soot model, soot nucleation occurs via the collision of two PAH dimers [7] at a fixed volume V_0 that corresponds to four napthalene molecules. The nucleation source term is only nonzero in the first section, and the volume of the first section is set to the dimer size.

Quadrature-based approximations to integrals for the local moment source terms are utilized, so the coagulation source term can be recast as a discrete sum over all of the quadrature points [8]:

$$\dot{M}_0^i = \sum_{\substack{k\le j\le i \\ (V_{j\pm}+V_{k\pm})\in\Omega_i}} \left(1-\frac{\delta_{j,k}}{2}\right)\beta_{j\pm,k\pm} w_{j\pm} N_{j\pm} w_{k\pm} N_{k\pm} - w_{i\pm} N_{i\pm} \textstyle\sum_{k=0}^{N} \beta_{i\pm,k\pm} N_{k\pm}, \tag{2}$$

$$\begin{aligned}\dot{M}_1^i = & \sum_{\substack{k\le j\le i \\ (V_{j\pm}+V_{k\pm})\in\Omega_i}} \left\{\left(1-\frac{\delta_{j,k}}{2}\right)\beta_{j\pm,k\pm}\left(V_{j\pm}+V_{k\pm}\right) w_{j\pm} N_{j\pm} w_{k\pm} N_{k\pm}\right\} \\ & - w_{i\pm} N_{i\pm} \textstyle\sum_{k=0}^{N} \beta_{i\pm,k\pm}\left(V_{i\pm}+V_{k\pm}\right) N_{k\pm}.\end{aligned} \tag{3}$$

In this expression, the $w_{i\pm}$ and $V_{i\pm}$ correspond to the weights and abscissae of the 2-point Gauss-Legendre or 2-point Gauss-Laguerre quadrature rules, which are determined from the linear and exponential profiles, respectively. Each of the sums above is abbreviated notation for four terms that considers all four combinations of the quarature points that is, $j+/k+$,$j+/k-$, $j-/k+$, and $j-/k-$ for the birth rates and $i+/k+$,$i+/k-$, $i-/k+$, and $i-/k-$ for the death rates. The collision frequency $\beta_{i,j}$ is modeled as the haromic mean of the free molecular and continuum regimes collision frequencies [9].

In this work, the treatment of the growth and oxidation terms follows the conventional three-point approach for fixed-sectional methods [2] but utilizies the quadrature points directly:

$$\dot{M}_0^i = \frac{A_{i-1}\left[w_{(i-1)\pm}I_{(i-1)\pm}N_{(i-1)\pm}\right]}{V_{i-1}} + \frac{B_i\left[w_{i\pm}I_{i\pm}N_{i\pm}\right]}{V_i} + \frac{C_{i+1}\left[w_{(i+1)\pm}I_{(i+1)\pm}N_{(i+1)\pm}\right]}{V_{i+1}}, \tag{4}$$

$$\dot{M}_1^i = \dot{M}_0^i V_i, \tag{5}$$

where the $\pm$ represents a summation over the two quadrature points of the corresponding section. The coefficients A, B, and C are given by Park and Rogak [2]. The growth rates $I_i = dV_i/dt$ correspond to growth and destruction processes: condensation, surface growth, and oxidation.

3. Results and Discussion

Both a basic Sectional Method (SM) and the MMSM approach are implmented within FlameMaster [10]. The two methods are used to compute the evolution of the soot NDF in a burner-stabilized laminar premixed flat flame investigated experimentally by Abid et al. (Flame No. C4 in [11]) at atmospheric pressure. The fuel is ethylene, and the oxidizer is a mixture of oxygen and argon. The unburned gases have an equivalence ratio of 2.07 (C_2H_4: 16.3%; O_2: 23.7%; Ar: 60%), and the cold gas velocity is 6.53 cm/s. Transmission Electron Microscopy (TEM) images indicate that aggregates are not formed in this flame, justifying the use of the univariate spherical soot particle model. In the simulations, the experimental temperature is directly enforced, and the gas-phase chemistry is moded using the chemical mechanism of Naraynaswamy *et al.* [12].

For the basic Sectional Method (SM), the simulations required at least 16 setions in order to obtain a solution; for MMSM, solutions could be obtained with as few as a single section, but no fewer than 4 sections, that is, 8 total variables, will be considered here. Up to 128 total variables were considered (128 sections with SM; 64 sections with MMSM) to test the relative rates of convergence of the two methods. As the number of sections was increased, the ratio V_N/V_1 was fixed, requiring a decrease in f_s with an increasing number of sections.

Comparisons in global soot quantities between the experimental measurements and the SM and MMSM approaches are shown in Fig. 1 for both the soot number density and the soot volume fraction. In the experimental measurements, particles smaller than 2.5 nm are not measured [11], so these are subtracted from the population for comparison with the experimental measurements. For both methods, the soot volume fraction compares favorably with the experimental measurements, with a sufficient number of variables; both methods tend to overpredict the total number density by approximately a factor of five compared to the nominal experimental measurements.

For SM compared to MMSM, once converged, both methods give essentially the same soot volume fraction and the same number density. For SM, all of the numbers of sections considered gave the same volume fraction; for MMSM, the volume fraction is overpredicted when the number of variables is extremely small, much smaller than anything possible with SM. For the number density, for both SM and MMSM, 64 and 128 variables are visually indistinguishable, except during the very early stages of soot inception. Except for the same early stages, with only 32 variables, MMSM is also nearly the same as the converged solution.

The Particle Size Distribution Function (PSDF) is determined from the reconstructed NDF in both SM and MMSM and compared to the experimental measurements in Fig. 2 at two locations above the burner where the predicted soot volume fraction agrees well with the measured value.

Closer to the burner $H = 3.5$ mm where the PSDF is unimodal, the PSDF from both SM and MMSM agree reasonably well with the experimental data except for the smallest number of sec-

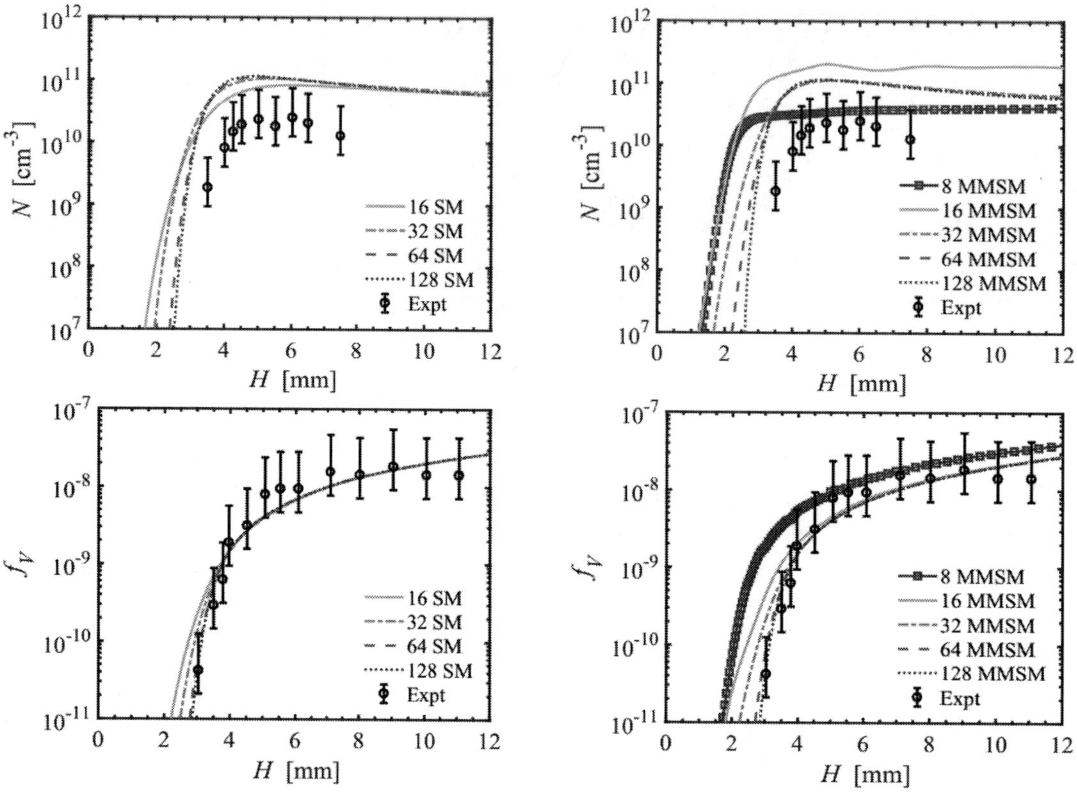

Figure 1: Global soot quantities through the flame with particles smaller than 2.5 nm removed: number density N (top) and volume fraction f_V (bottom). Lines are simulations (SM, left; MMSM, right), and symbols are experimental measurements with uncertainties. The total number of variables for each method are indicated.

tions for each method. At this location, the last section in MMSM, which has a presumed exponential decay profile, is at a larger size than the soot particles with non-zero number density. As a result, for a small number of sections, the last section with non-zero number density in Fig. 2 can actually have an increasing number density within the section, which is unexpected. However, this behavior diminishes as the number of sections is increased.

Further from the burner $H = 5.5$ mm, the experimental PSDF becomes distinctly bimodal, and this is not captured with either SM or MMSM, which still predict a unimodal distribution albeit with the apparent development of the second mode and the "trough." This discrepancy is an issue with the underlying chemical mechanism and soot model, not the statistical model. Nonetheless, the PSDF predicted by SM and MMSM are qualitatively similar, with two exceptions. First, the non-constant NDF within each section with MMSM allows for the resolution of some features of the PSDF at scales finer than the section spacing. For any number of sections, unlike closer to the burner, the last non-zero section exhibits a decrease in the PSDF with increasing diameter, consistent with the experimental data. Second, the behavior of MMSM in the vicinity of the developing "trough" results in a sawtooth-shape. More specifically, between 3 nm and 5 nm, the MMSM PSDF exhibits an increasing PSDF within each section, even though the number density there are essentially the same.

While both SM and MMSM provide convergent solutions, the results of the previous two sub-

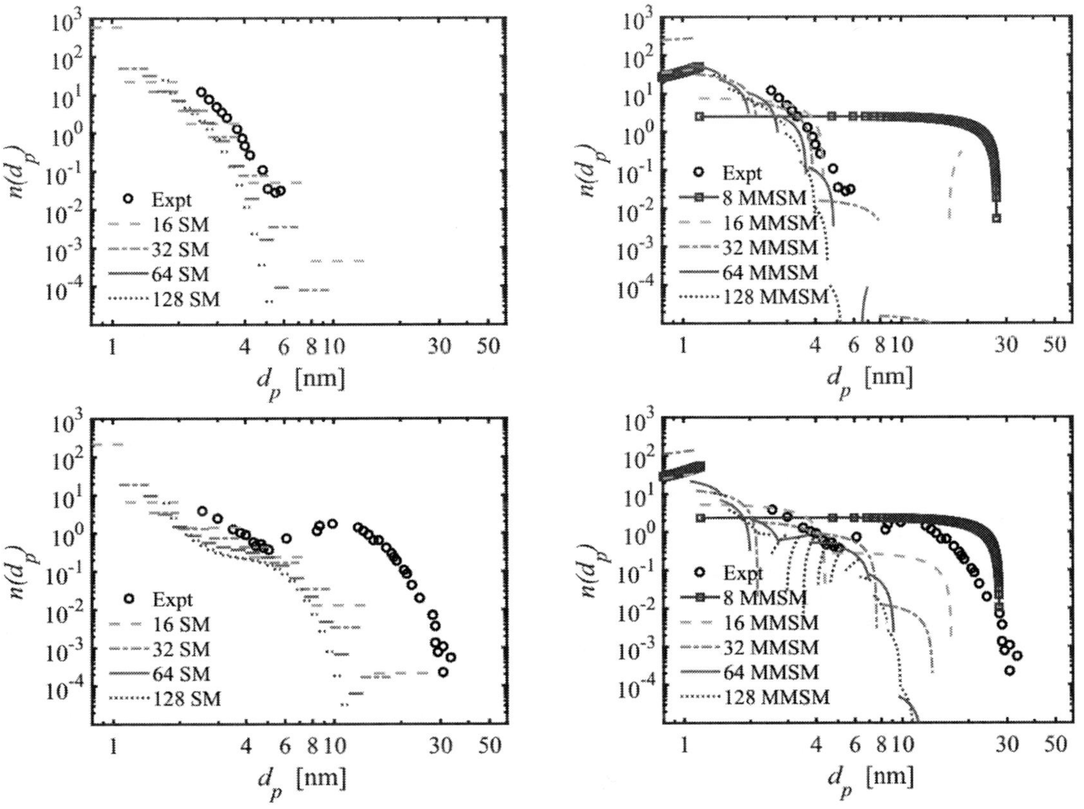

Figure 2: Particle Size Distribution Functions (PSDFs) at two locations in the flame: $H = 3.5\,\mathrm{mm}$ (top) and $H = 5.5\,\mathrm{mm}$ (bottom). Lines are simulations (SM, left; MMSM, right), and symbols are experimental measurements. The total number of variables for each method are indicated.

sections demonstrate that the rate of convergence is different. Formal rates of convergence for the L_2 domain integral of the total soot number density are order 1.3 and 2.7 for SM and MMSM, respectively. In other words, MMSM convergences with about twice the formal order as SM. For the convergence of the PSDF, L_2 differences were computed between PSDFs with a factor of two difference in the number of sections to define a formal rate of convergence; only the section centers common to both resolutions were considered in the L_2 differences. Using these L_2 differences, the formal rate of convergence for SM was order 1.5 and 1.4 at $H = 3.5\,\mathrm{mm}$ and $H = 5.5\,\mathrm{mm}$, respectively; the form rate of convergence for MMSM was order 1.9 and 2.4 at $H = 3.5\,\mathrm{mm}$ and $H = 5.5\,\mathrm{mm}$, respectively. Like the total number density, for the PSDF, MMSM converges with up to approximately twice the formal order as SM.

4. Conclusions

A computationally efficient Multi-Moment Sectional Method (MMSM) has been developed to predict the evolution of the soot Number Density Function (NDF). MMSM is the combination of a method of moments and a sectional method. Multiple moments are considered within each section, and these moments are utilized to reconstruct the local NDF within the section. In this work, only two moments per section were considered, which corresponds to a local linear reconstruction. The MMSM was combined with a detailed univariate soot model and applied to the simulation of a laminar premixed ethylene flame that forms only spherical particles. Compared to a basic sectional method, MMSM was shown to converge with approximately twice the formal order, both for global soot quantities such as the number density and for the size distribution. This increase in formal accuracy indicates that MMSM could provide a solution as good as a basic sectional method but at lower computational cost with fewer total variables.

5. Acknowledgements

This research was funded by the Princeton University School of Engineering and Applied Science through the Alfred Rheinstein Faculty Award.

References

[1] M. Balthasar and M. Kraft, A stochastic approach to calculate the particle size distribution function of soot particles in laminar premixed flames, Combust. Flame 133 (2003) 289–298.

[2] S. H. Park and S. N. Rogak, A novel fixed-sectional model for the formation and growth of aerosol agglomerates, J. Aerosol Sci. 35 (2004) 1385–1404.

[3] M. Frenklach, Method of moments with interpolative closure, Chem. Eng. Sci. 57 (2002) 2229–2239.

[4] D. L. Marchisio and R. O. Fox, Solution of population balance equations using the direct quadrature method of moments, J. Aerosol Sci. 36 (2005) 43–73.

[5] L. Waldmann and K. H. Schmitt, Thermophoresis and diffusiophoresis of aerosols, in: C. N. Davies (Ed.), Aerosol Science, Academic Press, London, 1966.

[6] F. Bisetti, G. Blanquart, M. E. Mueller, and H. Pitsch, On the formation and early evolution of soot in turbulent nonpremixed flames, Combust. Flame 159 (2012) 317–335.

[7] G. Blanquart and H. Pitsch, A joint volume-surface-hydrogen multi-variate model for soot formation, in: H. Bockhorn, A. D'Anna, A. F. Sarofim, and H. Wang (Eds.), Combustion Generated Fine Carbonaceous Particles, KIT Scientific Publishing, 2009, chap. 27, pp. 437–463.

[8] T. T. Nguyen, F. Laurent, R. O. Fox, and M. Massot, Solution of population balance equations in applications with fine particles: mathematical modeling and numerical schemes, J. Comput. Phys. 325 (2016) 129–156.

[9] S. E. Pratsinis, Simultaneous nucleation, condensation, and coagulation in aerosol reactors, J. Colloid Interface Sci. 124 (1988) 416–427.

[10] H. Pitsch, FlameMaster, a C++ computer program for 0D combustion and 1D laminar flame calculations, (1998).

[11] A. D. Abid, N. Heinz, E. D. Tolmachoff, D. J. Phares, C. S. Campbell, and H. Wang, On evolution of particle size distribution functions of incipient soot in premixed ethylene–oxygen–argon flames, Combust. Flame 154 (2008) 775–788.

[12] K. Narayanaswamy, G. Blanquart, and H. Pitsch, A consistent chemical mechanism for oxidation of substituted aromatic species, Combust. Flame 157 (2010) 1879–1898.

Sub Topic: Soot

Spring Technical Meeting
Eastern States Section of the Combustion Institute
March 4–7, 2018
State College, Pennsylvania

Subfilter Transport Modeling for Large Eddy Simulation of Turbulent Nonpremixed Sooting Flames

Suo Yang[1,*], *Jeffry K. Lew*[1], *and Michael E. Mueller*[1]

[1]*Department of Mechanical and Aerospace Engineering, Princeton University, NJ 08544*
[*]*Corresponding author:* suoy@princeton.edu

Abstract: Soot is formed by combustion of rich mixtures and is rapidly oxidized before being transported by turbulence into lean mixtures. Furthermore, different soot evolution mechanisms are dominant over distinct regions of mixture fraction (Z). For these reasons, a new subfilter PDF is proposed to account for this distribution of soot in Z space. In this model, the sooting mode of a bimodal soot subfilter PDF is locally activated only at rich mixture fractions where surface growth is locally faster than the oxidation. At the same time, Direct Numerical Simulation (DNS) studies of turbulent nonpremixed jet flames have revealed that Polycyclic Aromatic Hydrocarbons (PAH) are confined to spatially intermittent regions of low scalar dissipation rates due to their slow formation chemistry. The length scales of these regions are on the order of the Kolmogorov scale or smaller, where molecular diffusion dominates over turbulent mixing irrespective of the large-scale turbulent Reynolds number. A strain-sensitive transport model is developed to identify such species. Using a conventional nonpremixed "flamelet" approach, these species are then modeled with their molecular Lewis numbers, while the remaining species are modeled with an effective unity Lewis number. These two models are implemented within a Large Eddy Simulation (LES) framework, applied to a series of turbulent nonpremixed sooting jet flames, and validated via comparisons with experimental measurements of soot volume fraction. Compared to previous models, the soot volume fraction is significantly increased and in much better agreement with the experimental measurements.
***Keywords:* Soot, Large Eddy Simulation (LES), Subfilter PDF, Differential diffusion**

1. Introduction

In Large Eddy Simulation (LES), subfilter soot-turbulence-chemistry interactions can be statistically described by a subfilter density-weighted joint PDF $\tilde{P}$ of the thermochemical variables ξ_i and soot variables M_j. By Bayes' theorem, the joint PDF $\tilde{P}(\xi_i, M_j)$ can be decomposed into a thermochemical PDF $\tilde{P}(\xi_i)$ and a PDF $P(M_j|\xi_i)$ of the soot scalars conditioned on ξ_i:

$$\tilde{P}(\xi_i, M_j) = \tilde{P}(\xi_i) P(M_j|\xi_i), \tag{1}$$

where $\tilde{P}(\xi_i)$ is typically modeled as a beta distribution for the mixture fraction Z [1, 2].

In previous studies, Polycyclic Aromatic Hydrocarbons (PAH) were found to have much slower formation chemistry than the primary heat-releasing chemistry, which is described by ξ_i [3]. Since PAH are soot precursors, soot itself has even longer characteristic time scales. For this reason, Mueller and Pitsch [4] argued that the time scales of ξ_i and M_j are disparate enough such that the latter is only weakly dependent on the former. As a result, the conditional subfilter PDF $P(M_j|\xi_i)$ can be simplified to a marginal PDF of only the soot scalars $P(M_j)$.

Previous experiments and Direct Numerical Simulation (DNS) studies [3, 5] have shown that soot is confined to thin, spatially intermittent structures. To capture this property, Mueller and Pitsch [4] proposed a subfilter PDF consisting of two delta functions to represent the non-sooting and sooting modes within an LES filter width:

$$P(M_j) = \omega\delta(M_j) + (1-\omega)\delta(M_j - M_j^*), \tag{2}$$

where ω is the subfilter intermittency and M_j^* is formulated to recover the filtered values of the soot scalars $\overline{M}_j$ upon integration of $\tilde{P}(\xi_i, M_j)$. For the remainder of this paper, Eq. (2) is referred to as the Z-Uniform Soot Subfilter PDF (ZUSSP).

However, the above argument can be violated during surface growth near the flame and during soot oxidation, when the direct interactions between the major gas-phase chemistry and soot are stronger. For non-smoking flames, there should be zero soot in fuel-lean regions of the flame due to rapid oxidization [3], which is not qualitatively captured by ZUSSP. Therefore, the subfilter soot-turbulence interactions and combustion chemistry interactions should not be independent, and Eq. (2) should contain a connection to ξ_i. Specifically, the soot subfilter PDF needs to account for the local ratios of fuel and oxidizer as described by Z, since all soot-dependent processes should be restricted to fuel-rich locations where soot exists. For this reason, a Z-Activated Soot Subfilter PDF (ZASSP) is developed in this study.

In nonpremixed flamelet models, the typical assumption is that all species and temperatures have effectively unity Lewis numbers. This assumption is based on the reasoning that turbulence mixes indiscriminately at sufficiently large Reynolds numbers. Predictions of the flame temperature and the mass fractions of species with relatively fast kinetics from these flamelet solutions agree well with experiments and DNS. However, DNS studies [3] have suggested that this assumption may be inappropriate for PAH, which are very sensitive to the local scalar dissipation rate due to their slow formation chemistry. PAH are confined only to regions of low scalar dissipation rate that are on the order of the Kolmogorov length scale or smaller [6]. At such scales, transport is solely dictated by molecular diffusion. It is clear that there is a simultaneous presence of transport by molecular diffusion for species that are confined to very small length scales and transport by turbulent eddies for species that participate in fast chemistry. A model for gas-phase species transport that attempts to account for both unity and non-unity Lewis numbers is developed in this work and will be referred to as the Strain-Sensitive Transport Approach (SSTA).

In this study, ZASSP and SSTA are implemented within a LES framework, applied to a series of turbulent nonpremixed sooting jet flames, and validated via comparisons with experimental measurements of soot volume fraction.

2. Integrated Modeling Framework

In this work, turbulent combustion is modeled through the Radiative Flamelet/Progress-Variable method (RFPV) [2], which decouples the computation of the detailed flame structure from the computation of the flow field. The flame structure is evaluated *a priori*, parameterized with a reduced set of variables (mixture fraction Z, progress-variable C, and radiative heat loss parameter H), and stored in a database that is accessed during LES. Radiation is modeled with an optically thin gray gas approach. Since gas-phase PAH are unable to immediately adjust to the rapidly changing scalar dissipation field due to their relatively slow kinetics, their lumped mass fraction is

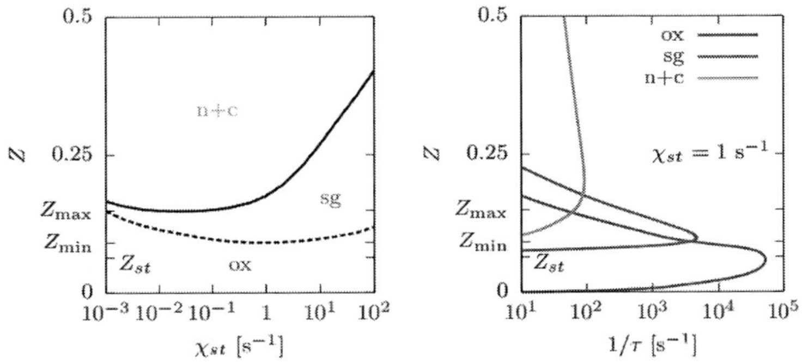

Figure 1: Regime diagram (left) and time scales (right) of soot evolution processes: nucleation (n), condensation (c), surface growth (sg), and oxidation (ox).

obtained from a transport equation instead of from the RFPV database [7]. The evolution of soot is modeled by the Hybrid Method of Moments (HMOM) [8]. Instead of directly solving for the Number Density Function (NDF) of the soot particles, HMOM solves for the avarage quantities, known as the moments M_j of the NDF.

2.1 Z-Activated Soot Subfilter PDF

In the Z-Activated Soot Subfilter PDF (ZASSP),

$$P(M_j|Z) = \omega\delta(M_j) + (1-\omega)\delta(M_j - M_j^*(Z)), \tag{3}$$

where the sooting mode is activated only for values of Z that are larger than a cut-off value Z_{cut} through a Heaviside step function $H(\cdot)$:

$$M_j^*(Z) = M_j^{**}H(Z - Z_{\text{cut}}), \tag{4}$$

where M_j^{**} is again selected such that $M_j^*(Z)$ recovers $\overline{M}_j$ upon integration against the PDF.

To determine the optimized Z_{cut}, Figure 1 shows a regime diagram (left) and a plot of time scales (right) for different soot evolution process. Based on which process has the fastest time scale τ, the mixture fraction (Z)-scalar dissipation rate (χ) space is divided into three regimes. Soot is created in the nucleation+condensation regime, transported to the surface growth regime by turbulence, and destroyed when it reaches the oxidation regime. According to Fig. 1, even at stoichiometric mixture fractions, the rate of soot oxidation is much faster than any other process, and the sooting mode should exist only in the regions where surface growth is faster than oxidation. However, for different local scalar dissipation rates χ, the Z values with equal surface growth rate $1/\tau_{\text{sg}}$ and oxidation rate $1/\tau_{\text{ox}}$ are different. Therefore, the optimized Z_{cut} should be a χ-dependent function $Z_{\text{adapt}}(\chi) = Z(\chi|\tau_{\text{sg}} = \tau_{\text{ox}})$.

2.2 Strain-Sensitive Transport Approach

The Strain-Sensitive Transport Approach (SSTA) applies molecular transport to species with relatively slow chemistry. Selection of these species is through a strain-sensitivity parameter

$$\zeta_k \equiv \rho\chi/\dot{m}_k^+, \tag{5}$$

where $\dot{m}_k^+$ is the chemical *production* rate of species k.

A new definition of the effective species Lewis number $\check{Le}_k$ that depends on ζ_k is required:

$$\check{Le}_k(\zeta_k) = \frac{Le_k}{Le_k + H(\min(\zeta_k) - 1) \cdot (1 - Le_k)}, \tag{6}$$

where Le_k is the molecular Lewis number of species k and $H(\cdot)$ is the Heaviside step function. Integration into the flamelet equations simply involves replacing Le_k with Eq. (6).

3. Results and Discussion

In this section, ZASSP and SSTA are validated against experimental data from two turbulent non-premixed jet flames, both targets of the International Sooting Flame (ISF) Workshop. The first flame is Adelaide Jet Flame 2 [9], a turbulent, nonpremixed, $C_2H_4/H_2/N_2$ [40/41/19 by volume] jet flame at atmospheric pressure. The flame is maintained at a jet exit Reynolds number of 15,000 and the inner jet diameter is 5.8 mm. The fuel average jet velocity is 42.4 m/s and the co-flowing air mean velocity is 1.1 m/s. The second flame is the Sandia Flame [10], a turbulent, nonpremixed, C_2H_4 piloted jet flame with an exit Reynolds number of 20,000 and an inner jet diameter of 3.2 mm. The fuel average jet velocity is 54.7 m/s and the co-flowing air mean velocity is 0.6 m/s.

Grid filtered LES is carried out with NGA [11], a finite difference code for low-Mach number turbulent reacting flows. The computational domains of both flames consist of a structured grid with $192 \times 96 \times 32$ points in the streamwise, radial, and circumferential directions, respectively. For the Adelaide Flame, this corresponds to a domain with lengths $200D$ in the streamwise direction and $60D$ in the radial direction, where D is the inner jet diameter. For the Sandia Flame, this corresponds to a domain with lengths $300D$ in the streamwise direction and $70D$ in the radial direction. The grid is stretched in both directions. The database of solutions to the steady flamelet equations with radiative losses are computed with FlameMaster [12]. The detailed chemical mechanism contains 158 species and 1804 reactions [13, 14].

Figure 2 shows the comparison between experimental data and LES predictions using ZASSP and ZUSSP, in terms of the time-averaged centerline soot volume fraction f_V. The same transport model (SSTA) is used in all of the LES. For the Adelaide Jet Flame 2, the LES with ZUSSP underpredicts f_V by roughly two orders of magnitude, while the LES with ZASSP using $Z_{\text{cut}} = Z_{\text{adapt}}$ only underpredicts f_V by a factor of five. For the Sandia Flame, the LES with ZUSSP underpredicts by roughly one order of magnitude, while the LES with ZASSP using $Z_{\text{cut}} = Z_{\text{adapt}}$ overpredicts f_V by a factor of two. In both flames, ZASSP with $Z_{\text{cut}} = Z_{\text{adapt}}$ predicts f_V roughly 20 times larger than ZUSSP and is in significantly better agreement with the experimental measurements.

Figure 3 presents the comparison between SSTA and $Le = 1$, also in terms of the time-averaged centerline f_V. For the reason explained above, ZASSP with $Z_{\text{cut}} = Z_{\text{adapt}}$ is applied in all cases. For the Adelaide Jet Flame 2, the location of the maximum value predicted by $Le = 1$ is too far downstream, while that predicted by SSTA is consistent with the experimental data. For the Sandia Flame, both SSTA and $Le = 1$ provide predictions within the same order of magnitude as the experimental data. However, SSTA overpredicts while $Le = 1$ underpredicts, and they differ from each other by a factor slightly larger than two. $Le = 1$ predicts a peak value closer to the experimental one than SSTA in this case, but predicts the location of the peak too far downstream, as was seen in Adelaide Jet Flame 2. Overall, SSTA provides a better prediction than $Le = 1$.

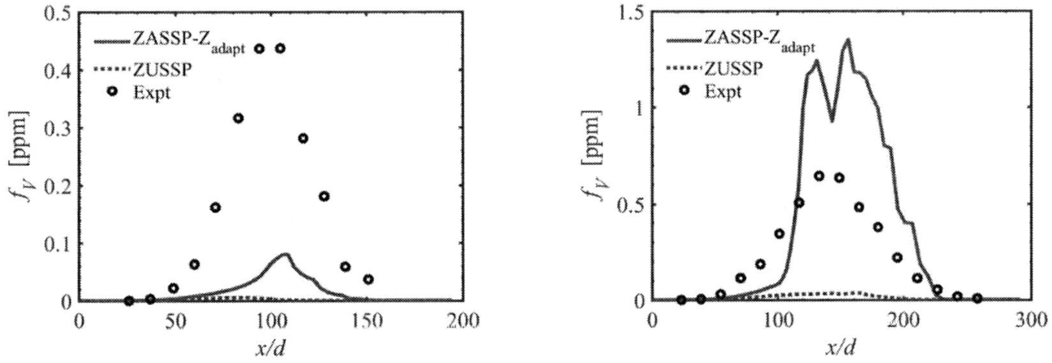

Figure 2: Comparison between LES with the Z-Activated and Z-Uniform Soot Subfilter PDFs, in terms of the soot volume fraction f_V in the Adelaide Jet Flame 2 (left) and in the Sandia Flame (right). The transport model (SSTA) is the same for all LES.

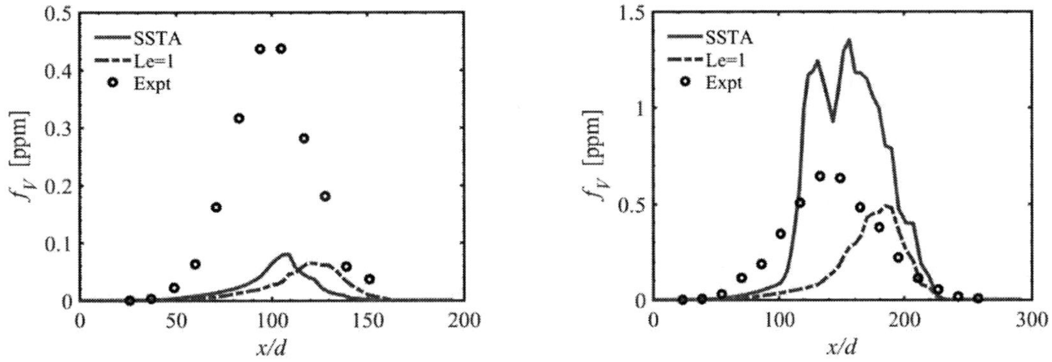

Figure 3: Comparison between LES with the Strain-Sensitive Transport Approach (SSTA), and effective unity Lewis number ($Le = 1$), in terms of the soot volume fraction f_V in the Adelaide Jet Flame 2 (left) and in the Sandia Flame (right). ZASSP with $Z_{cut} = Z_{adapt}$ is used in all LES.

4. Conclusions

In this study, a new subfilter PDF is proposed to account for the distribution of soot in mixture fraction space. In particular, the sooting mode of a bimodal subfilter PDF is locally activated only at rich mixture fractions where surface growth is locally faster than oxidation. In addition, a strain-sensitive transport model is developed to identify molecular diffusion dominated species, while applying effective unity Lewis number for other species. These two models are implemented within an integrated LES framework, which is applied to two turbulent nonpremixed sooting jet flames. Compared to previous models, the soot volume fraction is increased by roughly a factor of 20 and is in much better agreement with the experimental measurements.

5. Acknowledgements

The authors gratefully acknowledge valuable support in the form of computational time on the TIGRESS high performance computer center at Princeton University, which is jointly supported by the Princeton Institute for Computational Science and Engineering (PICSciE) and the Princeton University Office of Information Technology's Research Computing department.

References

[1] A. W. Cook and J. J. Riley, A subgrid model for equilibrium chemistry in turbulent flows, Phys. Fluids 6 (1994) 2868–2870.

[2] M. Ihme and H. Pitsch, Modeling of radiation and nitric oxide formation in turbulent non-premixed flames using a flamelet/progress variable formulation, Phys. Fluids 20 (2008) 055110.

[3] A. Attili, F. Bisetti, M. E. Mueller, and H. Pitsch, Formation, growth, and transport of soot in a three-dimensional turbulent non-premixed jet flame, Combust. Flame 161 (2014) 1849–1865.

[4] M. E. Mueller and H. Pitsch, Large eddy simulation subfilter modeling of soot-turbulence interactions, Phys. Fluids 23 (2011) 115104.

[5] A. Attili, F. Bisetti, M. E. Mueller, and H. Pitsch, Damköhler number effects on soot formation and growth in turbulent nonpremixed flames, Proc. Combust. Inst. (2015), pp. 1215–1223.

[6] P. Vaishnavi, A. Kronenburg, and C. Pantano, On the spatial length scales of scalar dissipation in turbulent jet flames, J. Fluid Mech. 596 (2008) 103–132.

[7] M. E. Mueller and H. Pitsch, LES model for sooting turbulent nonpremixed flames, Combust. Flame 159 (2012) 2166–2180.

[8] M. E. Mueller and H. Pitsch, Hybrid Method of Moments for modeling soot formation and growth, Combust. Flame 156 (2009) 1143–1155.

[9] S. M. Mahmoud, G. J. Nathan, Z. T. Alwahabi, Z. W. Sun, P. R. Medwell, and B. B. Dally, The effect of exit strain rate on soot volume fraction in turbulent non-premixed jet flames, Proc. Combust. Inst. (2017), pp. 889–897.

[10] J. Zhang, C. R. Shaddix, and R. W. Schefer, Design of "model-friendly" turbulent non-premixed jet burners for C2+ hydrocarbon fuels, Rev. Sci. Instrum. 82 (2011) 074101.

[11] O. Desjardins, G. Blanquart, G. Balarac, and H. Pitsch, High order conservative finite difference scheme for variable density low Mach number turbulent flows, J. Comput. Phys. 227 (2008) 7125–7159.

[12] H. Pitsch, FlameMaster, a C++ computer program for 0D combustion and 1D laminar flame calculations, (1998).

[13] G. Blanquart, P. Pepiot-Desjardins, and H. Pitsch, Chemical mechanism for high temperature combustion of engine relevant fuels with emphasis on soot precursors, Combust. Flame 156 (2009) 588–607.

[14] K. Narayanaswamy, G. Blanquart, and H. Pitsch, A consistent chemical mechanism for oxidation of substituted aromatic species, Combust. Flame 157 (2010) 1879–1898.

Spring Technical Meeting
Eastern States Section of the Combustion Institute
March 4-7, 2018
State College, Pennsylvania

Simulations of Soot Formation in High-Pressure Transient Spray Flames

*S. Ferreyro Fernandez, D.C. Haworth**

Department of Mechanical & Nuclear Engineering, The Pennsylvania State University, University Park, PA 16802, USA
Corresponding Author Email: dch12@psu.edu

Abstract: A transported probability density function (PDF) method is exercised to generate insight into soot processes in transient high-pressure turbulent n-dodecane spray flames, under diesel-engine-relevant conditions. The emphasis is on a split-injection case, where the fuel is injected in two pulses of 0.5 ms each, separated by a dwell time of 0.5 ms during which no fuel is injected. PDF model results are compared with experimental measurements and with results from a locally well-stirred reactor (WSR) model that neglects unresolved turbulent fluctuations in composition and temperature. Computed total soot mass is highly sensitive to the modeling of unresolved turbulent fluctuations. To achieve the best agreement between model and experiment and to capture the highly intermittent nature of soot in the turbulent flame, it is necessary to accurately represent mixing and the high effective Schmidt number (low diffusivity) of soot particles. This is accomplished in the PDF framework using a mixing model that enforces locality in the gas-phase composition space, and not mixing the transported soot variables in the soot model. The results suggest that mixing is at least as important as kinetics in controlling soot formation and evolution in high-pressure turbulent flames.

Keywords: *Turbulent combustion, Soot modeling, High-pressure spray combustion, PDF method*

1. Introduction

In a recent paper [1], unsteady Reynolds-averaged simulations of soot formation in high-pressure transient turbulent spray flames were compared with experimental measurements to provide insight into soot formation processes under diesel-engine-relevant conditions. A transported composition probability density function (PDF) method was used to account for unresolved turbulent fluctuations in composition and temperature in an unsteady Reynolds-averaged (URANS) formulation. Results for several PDF mixing model variants were compared to explore the role of mixing on soot formation, and PDF model results were compared with those from a locally well-stirred reactor (WSR) model to quantify the influence of turbulence-chemistry interactions. Key findings were: 1) It is essential to account for the influences of unresolved turbulent fluctuations, while accounting for the spatial and temporal coherence of the fluctuations may be of secondary importance, since the main effects with respect to overall soot levels and soot spatial distributions were captured in URANS/PDF. 2) It is equally important to account for the high effective Schmidt number (low diffusivity) of soot particles and the highly intermittent nature of the soot distribution in a turbulent flame. 3) Mixing is at least as important as kinetics in controlling soot formation and evolution in high-pressure turbulent flames. These conclusions were robust to the specific choice of gas-phase chemical mechanism and soot model, although the quantitative results varied depending on the specific models used.

In this paper, additional results are presented that further corroborate the findings in [1]. These include results for additional model variants, and results for a split-injection case that was not considered earlier.

2. Flame Configuration and CFD Model

The configuration is ECN Spray A [2,3]. A liquid spray of n-dodecane is injected into a high-temperature (900 K), high-pressure (gas density 22.8 kg/m^3) ambient gas mixture (15% O_2, 75.15% N_2, 6.22% CO_2, 3.62% H_2O) using a single-hole (90 μm diameter) common-rail fuel injector (injection pressure 150 MPa). The baseline Spray A condition is a relatively long duration single-injection case where fuel is injected continuously over a period of approximately 5.5 ms. Results for a split-injection case were presented in [4]. There the fuel injector is opened for 0.5 ms, then closed for 0.5 ms, then opened again for 0.5 ms, and finally closed for the duration of the combustion event. This gives a very different evolution of soot compared to the single-injection case, as shown in the following.

An unsteady Reynolds-averaged Navier–Stokes (URANS) formulation is adopted, using solvers based on the OpenFOAM v2.3.x toolbox [5]. A segregated pressure-based finite-volume method is used to solve the coupled mean momentum, pressure and enthalpy equations, with second-order spatial discretizations and first-order implicit time discretization. A two-equation k–ε turbulence model is used. The liquid fuel injection and spray evolution are modeled using a stochastic Lagrangian parcel method. For the gas phase, a chemically reacting ideal-gas mixture is considered, using a 133-species chemical mechanism [6]. Soot is modeled using either a semi-empirical two-equation model [7] (with acetylene-based inception) or a method of moments with interpolative closure (MOMIC) [8] (with PAH-based inception). A variant of MOMIC is also considered, where a separate treatment is used for the smallest particles, together with a different interpolation scheme for the negative moments of the soot-particle-size distribution function. Here the modified scheme is denoted as HMOM (a hybrid method of moments).

A Lagrangian-particle-based transported composition PDF method [9] is used to account for unresolved turbulent fluctuations in composition and temperature, with the Euclidean minimum spanning tree (EMST) mixing model [10], which enforces locality of mixing in composition space. The value of the mixing model coefficient is $C_\phi = 1.5$. To mimic the high effective Schmidt number of soot particles, soot particle quantities are not mixed. For comparison purposes, some results are presented for a different value of C_ϕ and for a WSR model. Further details can be found in [1] and in [11].

3. Results and Discussion

The influence of unresolved turbulent fluctuation on soot formation and evolution can be divided into two components: the contributions from fluctuations in temperature and gas-phase species, and the contribution from fluctuations in soot scalars. To isolate and quantify these effects, results for three model scenarios are compared with experiment in Fig. 1: WSR model (all fluctuations ignored), PDF model (fluctuations in temperature, gas-phase species, and soot model quantities considered), and a PDF model but with soot quantities computed using local cell-mean values of temperature and gas-phase species. Both of the PDF-based models capture the initial transient rise and subsequent falloff to a quasi-steady value that is observed in the experiments, although the levels are a factor of two-two-three lower when the soot calculation is based on the local cell-mean temperature and composition. The WSR model fails to capture the initial transient altogether, as seen earlier in [1]. Also, for the WSR model, there is relatively more computed soot downstream of the experimental field of view.

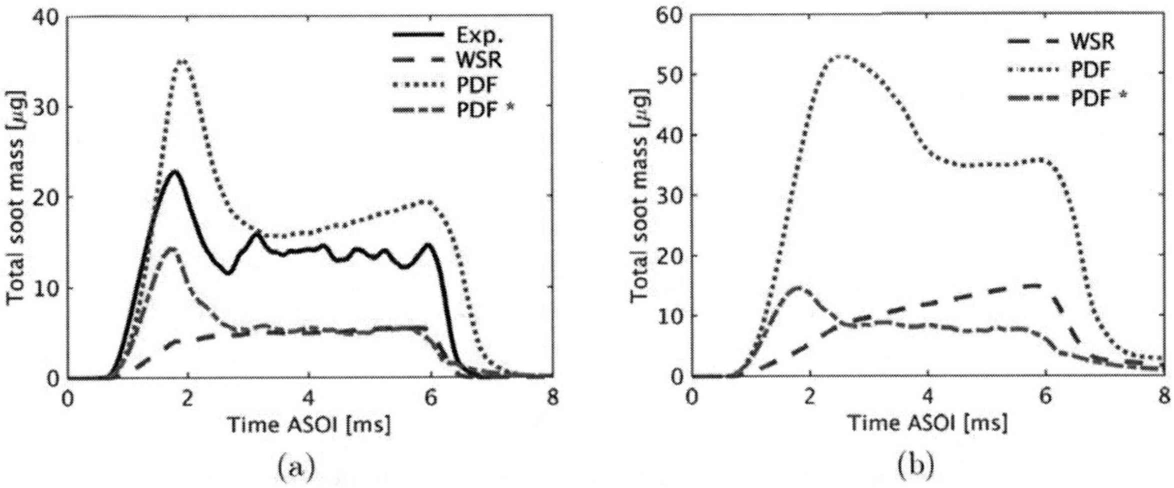

Figure 1: Measured and computed (using WSR and PDF models) total soot mass versus time after start of injection for the single-injection case. Results are for the MOMIC/PAH-based soot model. For the PDF case marked with an asterisk (*), soot quantities are calculated using cell-mean values of temperature and gas-phase species mass fractions. a) Total soot mass in the experimental field of view. b) Total soot mass in the full vessel.

Computed total soot mass in the experimental field of view as a function of time for the split-injection case is presented in Fig. 2a for the WSR model, and in Fig. 3a for the PDF model. The corresponding total numbers of computed soot particles in the full computational domain are shown in Figs. 2b and 3b, respectively. The WSR model results are similar to those shown in Fig. 1. The WSR model fails to capture the rapid initial transient rise, and underpredicts the total soot mass; it does show the correct phasing for the second peak, however. HMOM total soot mass is essentially the same as that from MOMIC. However, some differences can be seen in the numbers of soot particles between MOMIC and HMOM. These differences are attributed to the improvement in the calculation of the coagulation source terms in HMOM compared to MOMIC [11]. The two-equation soot model yields approximately half of the total soot mass compared to MOMIC or HMOM, and a very small number of soot particles, mainly due to the assumption of a uniform particle size. No particle-number experimental data are available for comparison.

On the other hand, the PDF-based models tend to over-predict the rapid initial transient rise and the total soot mass. MOMIC and HMOM yield asymmetric peaks in total soot mass, with the later peak being higher than the earlier one, but both peaks occur somewhat later than is observed experimentally. The total soot mass results (especially the earlier peak) are sensitive to the specification of C_ϕ, to an extent that is greater than in the single-injection case (not shown here – see [1]). Compared to the single-injection case, in the split-injection case the chemical mechanism has to operate at lower temperatures and at more fuel-lean conditions, which affect the PAH predictions, in particular. There are also differences in the turbulence and mixing that remain to be unraveled. In general, the sensitivities of computed soot levels to turbulence-chemistry interactions and mixing that were observed in the single-injection case in [1] appear to be amplified in the split-injection case.

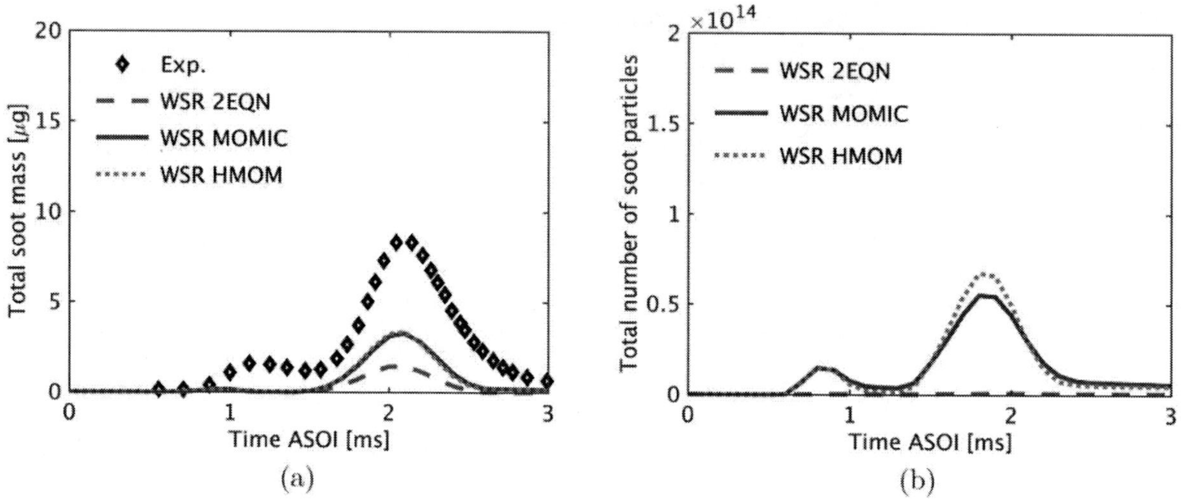

Figure 2: Measured and computed (using WSR model) soot quantities versus time after start of injection for the split-injection case. Computed results are shown for three different soot models. a) Total soot mass in the experimental field of view. b) Total number of soot particles in the full vessel.

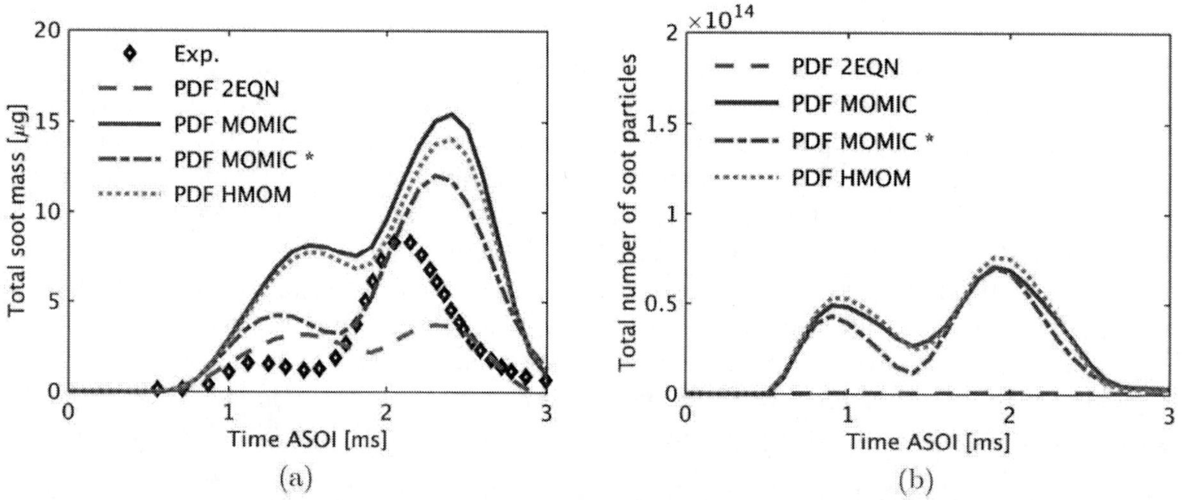

Figure 3: Measured and computed (using PDF model) soot quantities versus time after start of injection for the split-injection case. Computed results are shown for three different soot models. All PDF results are for $C_\phi = 1.5$, except for the case marked with an asterisk (*), where $C_\phi = 2.5$. a) Total soot mass in the experimental field of view. b) Total number of soot particles in the full vessel.

4. Conclusions

Given the large uncertainties in gas-phase chemistry and soot modeling, and the difficulties of making quantitatively accurate soot measurements in high-pressure turbulent combustion systems, the differences in results obtained using different model variants are deemed to be more significant than the absolute level of agreement with experiment.

The principal conclusions that were drawn in [1] for the single-injection case carry over to the split-injection case, and the sensitivities of computed soot levels to model variations are even stronger in the split-injection case. The modeling of turbulence-chemistry interactions is especially important for soot. It is essential to account for unresolved turbulent fluctuations in temperature and composition, for the high effective Schmidt number (low diffusivity) of soot particles, and for the highly intermittent nature of the soot distribution in the turbulent flame. Mixing is at least as important as chemical kinetics in soot formation and evolution in high-pressure turbulent combustion systems, where the chemistry is relatively fast.

5. Acknowledgements

This research has been funded by the U.S. National Science Foundation through grants CBET-1258613 and CBET-1604446. Funding was also provided by the Department of Energy, Office of Energy Efficiency and Renewable Energy (EERE) and the Department of Defense, Tank and Automotive Research, Development, and Engineering Center (TARDEC), under Award Number DE-EE0007278.

6. References

[1] S. Ferreyro Fernandez, C. Paul, A. Sircar, A. Imren, D.C. Haworth, S. Roy, M.F. Modest, Soot and spectral radiation modeling for high-pressure turbulent spray flames, Combust. Flame 190 (2018) 402-415.

[2] Engine Combustion Network, available at https://ecn.sandia.gov (2017).

[3] L.M. Pickett, C.L. Genzale, G. Bruneaux, L.-M. Malbec, L. Hermant, C. Christiansen, J. Schramm, Comparison of diesel spray combustion in different high-temperature, high-pressure facilities, SAE Int. J. Engines 3 (2010) 2156-2181 .

[4] S.A. Skeen, J. Manin, L.M. Pickett, E. Cenker, G. Bruneaux, K. Kondo, T. Aizawa, F. Westlye, K. Dalen, A. Ivarsson, T. Xuan, J.M. Garcia-Oliver, Y. Pei, S. Som, W. Hu, R.D. Reitz, T. Lucchini, G. D'Errico, D. Farrace, S.S. Pandurangi, Y.M. Wright, M.A. Chishty, M. Bolla, E.R. Hawkes, A progress review on soot experiments and modeling in the engine combustion network (ECN), SAE Int. J. Engines 9 (2016) 883-898 .

[5] OpenFOAM open source software for computational fluid dynamics, available at http://www.openfoam.org (2017).

[6] A. Frassoldati, G.D. Errico, T. Lucchini, A. Stagni, A. Cuoci, T. Faravelli, A. Onorati, E. Ranzi, Reduced kinetic mechanisms of diesel fuel surrogate for engine CFD simulations, Combust. Flame 162 (2015) 3991-4007.

[7] K.M. Leung, R.P. Lindstedt, W.P. Jones, A simplified reaction mechanism for soot formation in nonpremixed flames, Combust. Flame 87 (1991) 289-305.

[8] M. Frenklach, Method of moments with interpolative closure, Chem. Eng. Sci. 57 (2002) 2229-2239.

[9] D.C. Haworth, Progress in probability density function methods for turbulent reacting flows, Prog. Energy Combust. Sci. 3 6(2010) 168-259.

[10] S. Subramaniam, S.B. Pope, A mixing model for turbulent reactive flows based on Euclidean minimum spanning trees, Combust. Flame 115 (1998) 487–514.

[11] S. Ferreyro Fernandez, Advanced soot and radiation models for laminar and turbulent flames, Ph.D. thesis, The Pennsylvania State University, University Park, PA, USA (2018).

Sub Topic: Turbulent Flames

Spring Technical Meeting
Eastern States Section of the Combustion Institute
March 4-7, 2018
State College, Pennsylvania

Large Eddy Simulations of Staged Pressurized Oxy-Combustion

F.N. Karaismail[1], A. Gopan[2], R. L. Axelbaum[2], I. Celik[1], B.M. Kumfer[2]*

[1] *Department of Mechanical and Aerospace Engineering, West Virginia University, Morgantown, WV 26508, USA*

[2] *Department of Energy, Environmental & Chemical Engineering, Consortium for Clean Coal Utilization, Washington University in St. Louis, MO 63130, USA*

**Corresponding Author Email: fcayan@mix.wvu.edu*

Abstract: Oxy-fuel combustion is one of the popular methods for reducing carbon dioxide emissions from fossil fuel combustion systems. However, a large amount of flue gas recycle is required in oxy-fuel combustion to control the heat flux by lowering the flame temperature, but this poses a significant efficiency burden on the process. In non-premixed oxy-gas combustion systems, the high radiative heat flux results from the large soot volume fraction. One of the ways to control the radiative heat flux is to reduce the soot volume fraction, without compromising on the flame temperature. To do so, the inert can be strategically distributed between oxidizer and fuel to change the stoichiometric mixture fraction, Z_{st}, of the flame, which has been shown to reduce soot and radiative heat flux. In this study, Large Eddy Simulations (LES) is used to further characterize this suppression of soot inception. The experimental reactor at Washington University in St. Louis is modeled for methane combustion, and mixture fraction maps are generated to provide the flame structure, which has a strong influence on soot inception within the reactor. The mixture fraction maps are obtained by extracting the corresponding data from arbitrary lines across the turbulent flame for different Z_{st} values at the inlet. The important parameters affecting soot inception, such as the C/O ratio and the temperature limits for soot formation, are calculated by means of a simple model. Parametric studies have been performed and variances over space and time are calculated.

Keywords: Oxy-fuel combustion, Large Eddy Simulations, stoichiometric mixture fraction, soot formation

1. Introduction

Oxy-fuel combustion is one of the promising technologies with the benefits of high thermal efficiency, high concentration of CO_2 in flue gas readily available for carbon capture and improved flame stability [1]. The most prominent feature of this technology is that pure oxygen rather than air is used as oxidizer. However, the flame temperatures resulting from combustion in pure oxygen, when not controlled by recycled flue gas, can be quite high. In order to eliminate the detrimental effects of the heat flux resulting from high flame temperatures on the boiler tubes, large amount of recycled flues gas should be used. This poses a significant efficiency burden on the process [2].

In non-premixed oxy-fuel combustion systems, the limitation on high heat flux due to elevated temperatures could be eliminated by controlling the soot formation and growth without reducing the temperature [3]. One way is to strategically distribute an inert, N_2 gas, between oxidizer and fuel to change the stoichiometric mixture fraction, Z_{st}, of the flame, given by

$$Z_{st} = \left(1 + \frac{\nu_{O_2} W_{O_2} Y_{F,f}}{\nu_F W_F Y_{O_2,o}}\right)^{-1} \qquad (1)$$

where Y is the mass fraction, ν is the stoichiometric coefficient, W *is* the molecular weight and subscripts f and o denote fuel and oxidizer inlet streams. Combination of oxygen enrichment and fuel dilution at the inlet by changing mixture fraction, Z_{st}, which has been shown to reduce soot and radiative heat flux, provides a means of controlling soot formation without changing the flame temperature.

In previous studies, [3][4], empirical data derived from laboratory scale experiments were used to determine soot inception limits in terms of temperature and C/O ratio. For design analysis purposes, it is desirable to reduce the empirical input and rely more on simulations. With this in mind, Large Eddy Simulation (LES) is used in this study to model methane combustion for the experimental reactor at Washington University in St. Louis to further characterize soot inception regions. Mixture fraction maps are generated by extracting the corresponding LES data from arbitrarily selected lines across the turbulent flame obtained from simulations for different Z_{st} values at the inlet. Some important parameters affecting soot inception, such as the C/O ratio and the temperature are calculated using the model presented in [4] as the basis. Parametric studies have been performed and variances over space and time are calculated and the results are presented in the preceding sections.

2. Methods

Large eddy simulation (LES) of the experimental reactor at WUSTL was utilized to generate the mixture fraction maps to provide the flame structure for the analysis of soot formation within the reactor. The simulations were performed using the readily available models in ANSYS Fluent with nominally 2nd order numerical schemes and equilibrium-pdf combustion model combined with the Smagorinsky sub-grid-scale (sgs) turbulence model. No assessment of the LES quality was made other than visual inspection of the turbulent fluctuations. In this regard, the results presented here must be viewed as 'preliminary'.

Flame structure is examined by modifying the stoichiometric mixture fraction, Z_{st}, at the inlet, following the study of Kumfer *et al.* [4]. The mixture fraction maps are obtained by extracting the corresponding data from an arbitrary line across the flame and the important parameters affecting the soot inception such as C/O ratio, temperature, fuel and oxygen concentrations are calculated. The actual location of the line has proven to be not an important factor in obtaining the mixture fraction map (see Figure 1).

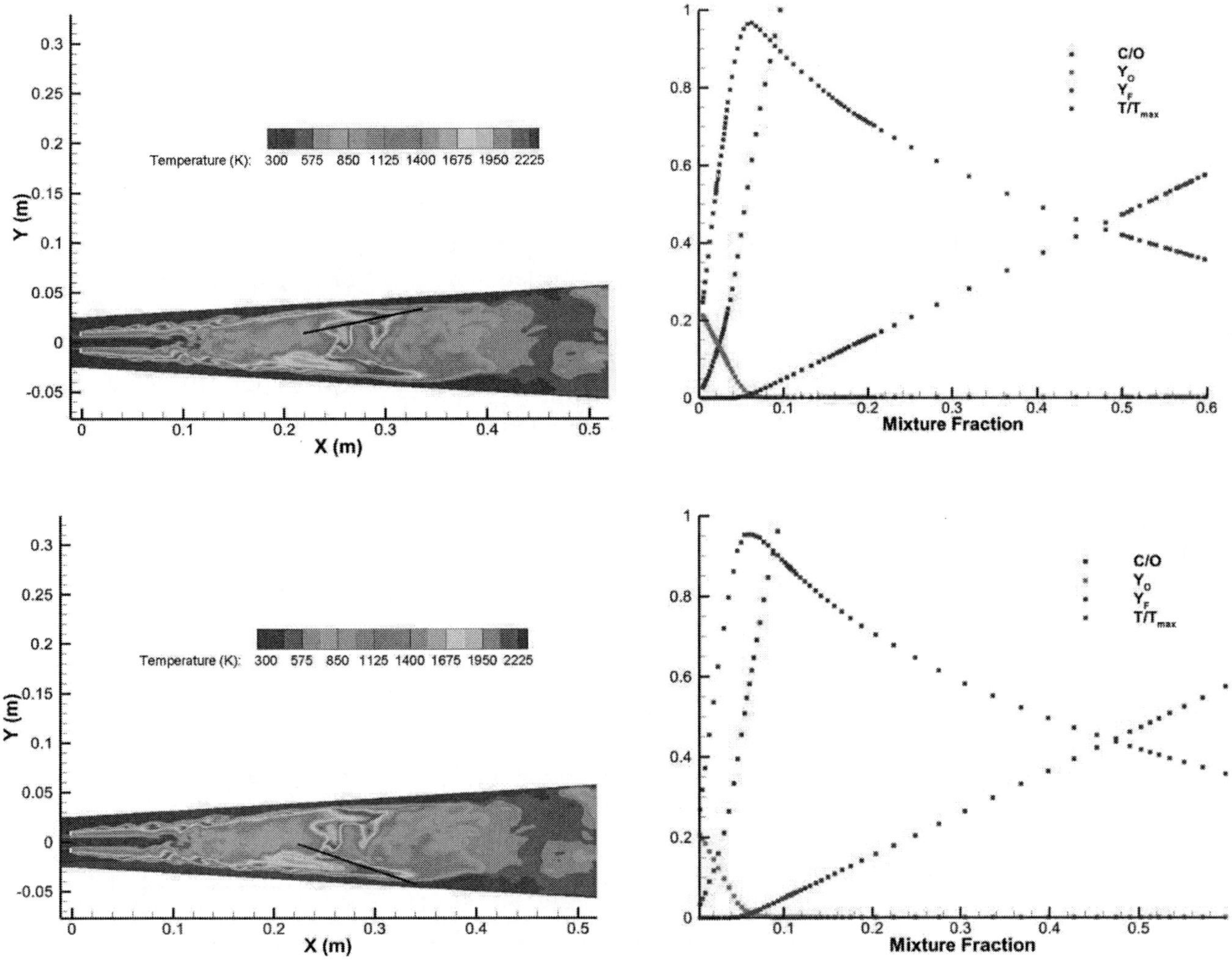

Figure 1: Temperature contour plots and mixture fraction maps obtained from two arbitrary lines across the flame for $Z_{st} = 0.055$

Following [4], two temperature limits, T_{kin} and $T_{C/O}$, are used to identify the boundaries of the soot inception regions. T_{kin}, the temperature above which the kinetics are favorable for soot formation is calculated from

$$T_f = T_{kin} + \frac{Z_{st}}{C_0.\tau} exp\left(\frac{E}{RT_{kin}}\right) \quad (2)$$

where, T_f is the local flame temperature (K), τ (s) is the time scale determined using local diffusivities, C_0 is an experimentally determined model constant, R is the universal gas constant and E is activation energy for soot inception taken as 31 kcal/mol. C_0 is assumed to be a constant with a value of 780 $K^{-1}s^{-1}$ currently; a more definite value can be determined via calibration using experimental data. It should be noted that the combined parameter $C_0\tau$ is to be considered as an empirical model parameter, which can only be determined using experimental validation since the calculation of the time scale τ is not trivial.

$T_{C/O}$ is the high temperature limit at which the local gas composition is no longer favorable for soot formation and is determined using the following model equation:

$$T_f = T_{C/O} + (T_{C/O} - T_0)Z_{st}\left[2\frac{m+n/4}{m}(C/O)_{cr} - 1\right] \quad (3)$$

where m and n are the numbers of C and H atoms in fuel, respectively, and T_0 (K) is the temperature at the inlet. $(C/O)_{cr}$ is the critical carbon-to-oxygen ratio for soot inception. For the model to be implemented $(C/O)_{cr}$ needs to be determined. The right hand side (RHS) of Eq. 3 is evaluated along the line selected across the flame using locally calculated variables for $T_{C/O}$, Z_{st} and C/O ratio. Then the intersection with the local flame temperature, i.e. the maximum temperature along the selected line, is determined. At the intersection, there exists a unique solution that is physically meaningful for $T_{C/O}$, Z_{cr} and $(C/O)_{cr}$ which satisfy Eq. 3.

3. Results and Discussion

The simulations are performed considering three different stoichiometric mixture fractions. The inlet mass fractions used are summarized in Table 1.

Table 1: Inlet mass fractions used for simulations

Z_{st}	$Y_{F,f}$	$Y_{O_2,o}$
0.055	1.0	0.233
0.27	0.185	0.272
0.56	0.126	0.648

Attention was focused on the near-burner zone and flame structure shortly after ignition. The methane and air streams were fed at 300 K with mass flow rates of 0.455 kg/h pure methane, 23.3 kg/h air at 300 K at an operating pressure of 1 bar. The reactor wall temperature was set to a fixed temperature of 400 K for all the simulations.

Figure 3 illustrates the proposed calculation procedure where the intersection of the RHS of Eq. 3 with the local flame temperature is for different Z_{st} values of 0.055, 0.27 and 0.56. Table 2 summarizes the values obtained using the proposed calculation procedure.

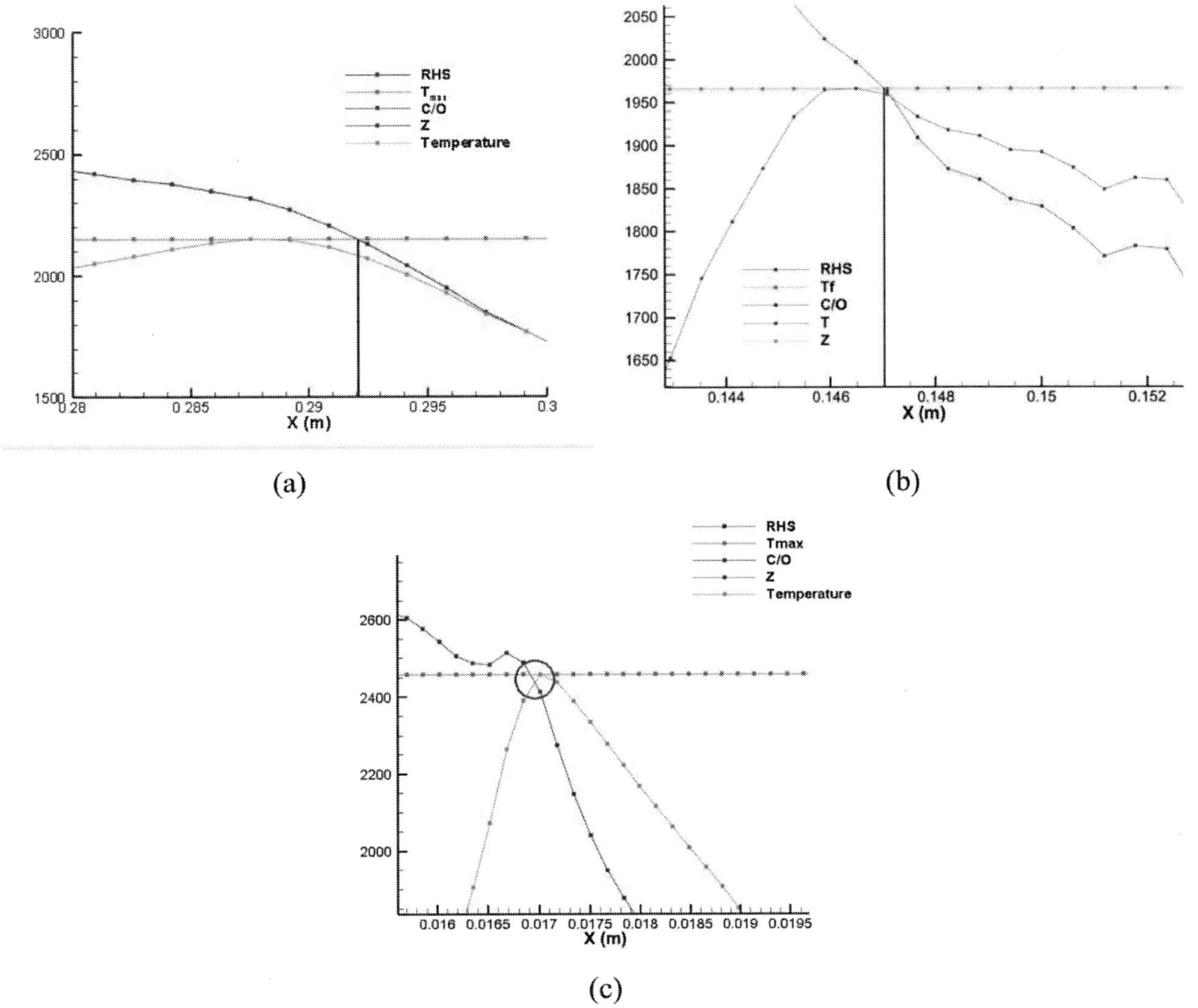

Figure 3: Calculation of parameters affecting the soot inception for (a) $Z_{st} = 0.055$, $T_f = 2148.8$ K (b) $Z_{st} = 0.27$, $T_f = 1985.5$ K, (c) $Z_{st} = 0.56$, $T_f = 2455.7$ K

Table 2: Critical values obtained using the proposed calculation the procedure

Z_{st}	$T_{C/O}$ (K)	$(C/O)_{cr}$	Z_{cr}
0.055	2138	0.33	0.051
0.27	1959.5	0.31	0.269
0.56	2054.6	0.35	0.58

As can be seen from Table 2, $(C/O)_{cr}$, which is considered essentially a fuel property, does not change much with Z_{st}. This is a good indication that the model is implemented correctly.

The mixture fraction maps with the soot inception limit are given in Figure 4. As can be seen from the figure, as Z_{st} is increased the region for soot inception narrows in its extent, as expected.

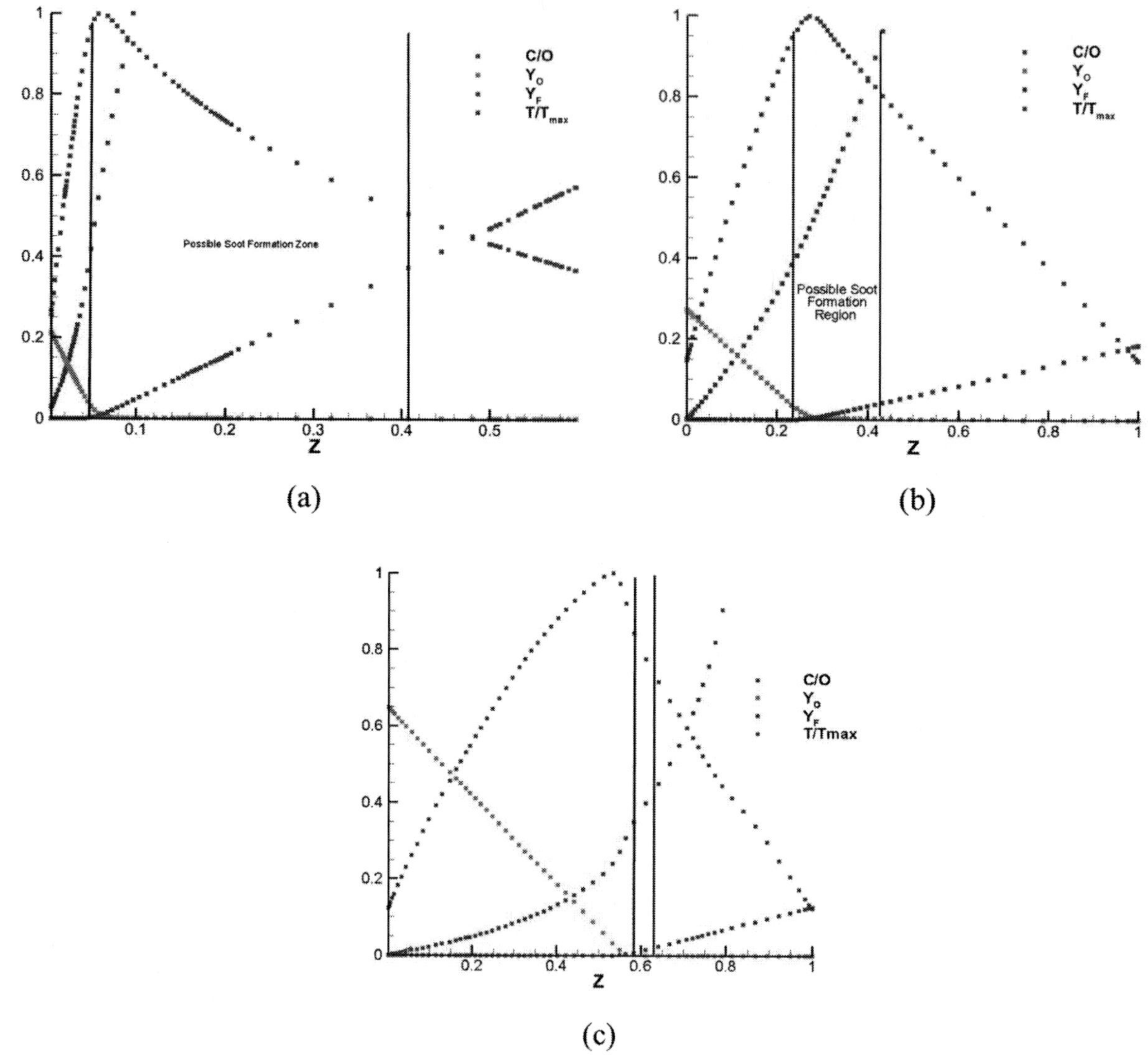

Figure 4: Mixture Fraction Maps with possible soot inception region for (a) $Z_{st} = 0.055$ (b), $Z_{st} = 0.27$, (c) $Z_{st} = 0.56$

Calculations are repeated for the different lines across the flame (see Figure 5 for $Z_{st} = 0.055$) and the variance over space is calculated and the results are presented in Table 3 for all cases.

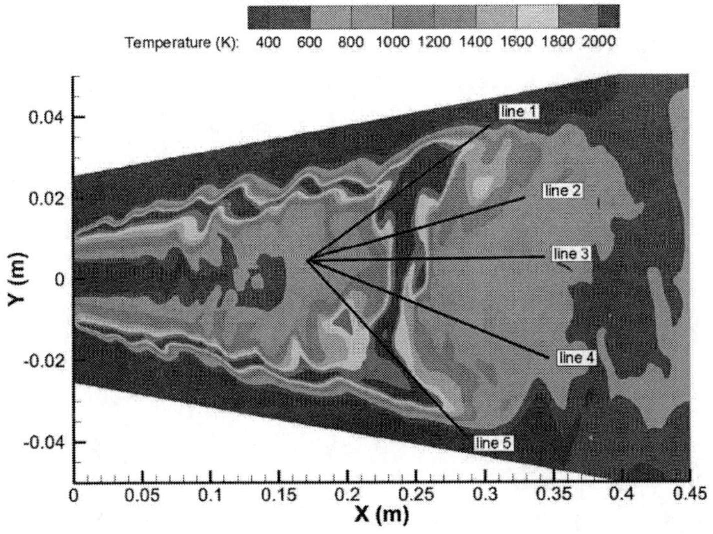

Figure 4: Arbitrary lines across the flame for $Z_{st} = 0.055$ to calculate variance over space

Table 3: Variance of calculated properties over space

Case	Z_{st}	T_f	$T_{C/O}$	$(C/O)_{cr}$	Z_{cr}	T_{kin}
1	0.055	2150 ± 2	2144 ± 2	0.34 ± 0.01	0.057 ± 0.01	1086 ± 2
2	0.27	2051 ± 3	2044 ± 3	0.30 ± 0.02	0.31 ± 0.02	1661 ± 3
3	0.56	2458 ± 10	2055 ± 10	0.35 ± 0.03	0.59 ± 0.03	1804 ± 10

As can be seen from Table 3, the variances for calculated boundary temperatures are within 2-10°C and between 0.01-0.03 for the critical values like C/O ratio and mixture fraction Z.

After the solutions attained a statistically stationary state the calculations are repeated at various times, namely at 1.8 s, 2.4 s, 3.0 s, 3.6 s and 4.2 s. The variance over time is calculated and the results are given in Table 4.

Table 4: Variance of the calculated properties over time.

Case	Z_{st}	T_f	$T_{C/O}$	$(C/O)_{cr}$	Z_{cr}	T_{kin}
1	0.055	2150 ± 4	2144 ± 4	0.34± 0.02	0.057± 0.02	1086 ±4
2	0.27	2051 ± 5	2044 ± 5	0.30±0.02	0.31±0.02	1661±5
3	0.56	2458±12	2055±12	0.35±0.03	0.59±0.03	1804±12

As can be seen from Table 4, the variances for calculated boundary temperatures are within 4-12°C and between 0.02-0.03 for the critical values like C/O ratio and mixture fraction Z.

4. Conclusions

The experimental reactor at Washington University in St. Louis is modeled for methane combustion using Large Eddy Simulations (LES) to characterize the conditions for soot inception. Mixture fraction maps are generated to provide the flame structure, which has a strong influence on soot inception within the reactor. The mixture fraction maps are obtained by extracting the corresponding data from arbitrary lines across the turbulent flame for different Z_{st} values at the inlet. The important parameters affecting soot inception, such as the C/O ratio and the temperature limits for soot formation, are calculated by means of a previously proposed model but without empirical input. Parametric studies performed to calculate variances over space and time indicated that the variances for calculated boundary temperatures are within 4-12°C and between 0.02-0.03 for the critical values affecting soot inception region, namely C/O ratio and mixture fraction Z. It is shown that some critical parameters associated with soot inception can be directly deduced using LES data. This is encouraging for future study of soot control strategies for pilot-scale demonstration units.

5. Acknowledgements

This research was conducted with support from the US-China Clean Energy Research Center – Advanced Coal Technology Consortium (CERC-ACTC) under DOE contract DE-PI0000017.

6. References

[1] H. Nikzat, H. Pak, T. Fuse, Y. Hu, K. Ogyu, N. Kobayashi, M. Hasatani, Characteristics of pulverized coal burner using a high-oxygen partial pressure, Trans IChemE, Part A, 82(A1) (2004) 99-104.
[2] A. Gopan, B. M. Kumfer, J. Phillips, D. Thimsen, R. Smith, R. L. Axelbaum, Process design and performance analysis of a staged, pressurized oxy-combustion (SPOC) power plant for carbon capture, Appl. Energy, 125 (2014) 179-188
[3] A. Gopan, Z. Yang, B.M. Kumfer, R.L. Axelbaum Effects of inert placement (Zst) on soot and radiative heat flux in turbulent diffusion flames, Energy Fuels, 31 (2017) 7617-7623.
[4] B.M. Kumfer, S.A. Skeen and R.L. Axelbaum, Soot inception limits in laminar diffusion flames with application to oxy-fuel combustion, Combust. Flame, 154 (2008) 546-556.

Spring Technical Meeting
Eastern States Section of the Combustion Institute
March 4-7, 2018
State College, Pennsylvania

Reconciliation of Carbon Oxidation Rates and Activation Energies Based on Changing Nanostructure

Singh, Madhu[1], Chethan K. Gaddam[1], Randy L. Vander Wal[1*], Xu Chen[2], Aleksey Yezerets[2], Krishna Kamasamudram[2]

[1] *John and Willie Leone Family Department of Energy and Mineral Engineering and the EMS Energy Institute, Penn State University, University Park, PA 16802, USA*
[2] *Cummins Inc., 1900 McKinley Avenue, Columbus, IN 47201, USA*
[*] *Corresponding Author Email: ruv12@psu.edu*

Abstract: HRTEM reveals the nanostructure of the model carbons and changes therein due to oxidation. The variation in oxidation rates during the oxidation process highlights the dependence of oxidation rates on carbon nanostructure and in particular, the carbon atom site types. Changes in nanostructure necessarily occur with change in the relative proportion of site types. In turn, the oxidation rate changes accordingly. An inside-out burning mode, onion peeling, graphitization and edge site creation are behaviors observed during the course of oxidation of the model carbons. The rate for arc-generated soot falls off exponentially while that for M1300, a carbon black exhibits a parabolic increase. R250, another carbon black maintains a constant rate, lower than that for graphitized multi-walled carbon nanotubes (MWCNTs) or graphitic carbon onions. The former exhibit exclusive interior oxidation while the latter graphitic materials exhibit only edge site oxidation. XPS confirmed the interpretation of graphitic lamella that was restricting further oxidation for the arc soot, by showing an increase in the sp^2/sp^3 content. Raman spectra are related to oxidation induced defects in the MWCNTs and onions.

Keywords: *Soot, Oxidation, TEM, Nanostructure*

1. Introduction

Nanostructure reflects the next physical scale following atomic-level bonding. Nanostructure in carbon refers to the details of the lamella sheets (aka graphene segments) that comprise the material. Parameters of nanostructure include the in-plane layer dimensions, L_a, L_b, the stacking dimension L_c, interplanar separation distance d, and tortuosity or curvature related to the waviness or curvature of the lamella segments [1,2]. Various diagnostic tools can quantify values of each metric [3–8]. Recently image analysis algorithms have been applied to quantify these measures as observed from high-resolution transmission electron microscopy (HRTEM) images [9,10].

Elsewhere it has been recognized that carbon blacks are often poor substitutes for diesel soot given the chemical variability and hetero-elements often present in the latter material [11]. Alternatively, carbon blacks are devoid of chemical heterogeneity, nanostructure and morphology, thereby avoiding convolving these factors. Our motivation here was to study synthetic carbons, as model "soots" to investigate the role of nanostructure upon burning mode and rate. The particular carbon forms were chosen for their uniform composition and morphology, and diversity of nanostructure to best trace oxidation dependence upon nanostructure. Comparative techniques routinely used to assess carbon structure and bonding, i.e. X-ray photoelectron spectroscopy (XPS) and Raman spectroscopy are compared to the nascent and partially oxidized materials for comparison to HRTEM and analyses for nanostructure.

2. Methods / Experimental

Five model carbons were comprehensively studied in this work including three commercial carbon blacks Monarch 1300 (Cabot®), Regal 250 (Cabot®), arc soot (MER Corp.) and two graphitic carbons: graphitized multi-walled carbon nanotubes (MWCNTs) and carbon nano-onions. The MWCNTs were produced via a CVD process, followed by high temperature heat treatment for purification and resulting graphitization [12]. Carbon nano-onions are synthesized by heat-treating R250 carbon black in a high temperature furnace under a He atmosphere at 3000 ^{0}C for a one hour duration [2]. These model carbons are nearly pure carbon with a C:H ratio >10:1 and negligible heteroelement content. Choice of these different model carbons is based on their distinctly different (initial) nanostructure.

2.1 *Oxidation reactor*

A simplified bench flow reaction system was used in this work to measure the activation energies (E_a) of the model carbons. More details on the reaction system, pre-treatment and the overall process can be found elsewhere [13].

2.2 *Thermogravimetric analysis*

A thermogravimetric analyzer (TA instruments, SDT Q600) with recording software was used to measure the oxidation rates of the model carbons. The TGA measured and recorded weight changes over a wide range of temperatures in controlled gas environments. Samples for HRTEM were generated by terminating the oxidation process after reaching a conversion (weight loss) of ~50% to explore the nanostructural changes of these carbons with oxidation, and by reaching a conversion of ~80% for Raman analysis.

2.3 *Transmission electron microscopy*

For HRTEM imaging, soot was deposited on a lacey C/Cu grid by depositing a drop of the soot suspension, created by sonication in ethanol. TEM images were taken using a Phillips CM200 having nominal resolution of 0.14 nm with Gatan Image Filter for digital imaging featuring live Fourier transforms having nominal resolution of 0.14 nm. The instrument was operated at 200 keV using a LaB6 filament. All images were digitally recorded with a slow scan CCD camera (image size 1024 x 1024 pixels), and image processing was carried out using Gatan image software v. 3.4. Two separate analysis algorithms conducted quantification of the HRTEM images. Fringe length is calculated by the first routine, while the second program calculated tortuosity. Further details of these programs have been provided elsewhere [2,10]. The parameters extracted from the skeleton images are denoted as fringe length, and fringe tortuosity. Lattice fringe length is a measure of the physical extent of the atomic carbon layer planes as seen in the HRTEM image. Larger lengths correspond to a higher level of organization and hence the material is considered to more closely resemble graphitic structure. Whereas fringe tortuosity is defined here as the ratio of the actual fringe length (L) to the shortest distance between the endpoints of the carbon segment, which provides a measure of the undulation of the layer planes [9].

2.4 *Raman spectroscopy*

A confocal Raman spectrometer (WITec CRM200) was used to determine the degree of graphitization of the model carbons. The excitation laser was an Argon ion laser (λ = 488 nm, source power 10 mW). The laser was focused onto the sample using a microscope objective lens (40x) that has a numerical aperture of 0.6. All spectra were collected at the same source power level via a software package (WITec project). Spectra were further post-processed using Matlab (Mathworks Inc.). Spectra were fitted after multipoint baseline correction by two Gaussian-shaped G and D1 bands at about 1580 and 1360 cm^{-1}. The integrated intensity ratio I_D/I_G was used for the determination of the crystallite size (L_a) [14,15].

3. Results and Discussion

With progressive oxidation, rate change trends differ between samples. These are proposed to arise from different nanostructures. As observed in Figure 1, the rate for R250 is relatively constant. There is little change in nanostructure as shown by HRTEM images and verified by comparative image analysis relative to R250, nascent, and at different oxidation stages (< 10%), despite the unusual burning mode [16]. The modest rate increase can be understood in terms of increasing porosity development of the perimeter shell, as it otherwise (likely) limits reactant and product transport to and from the interior.

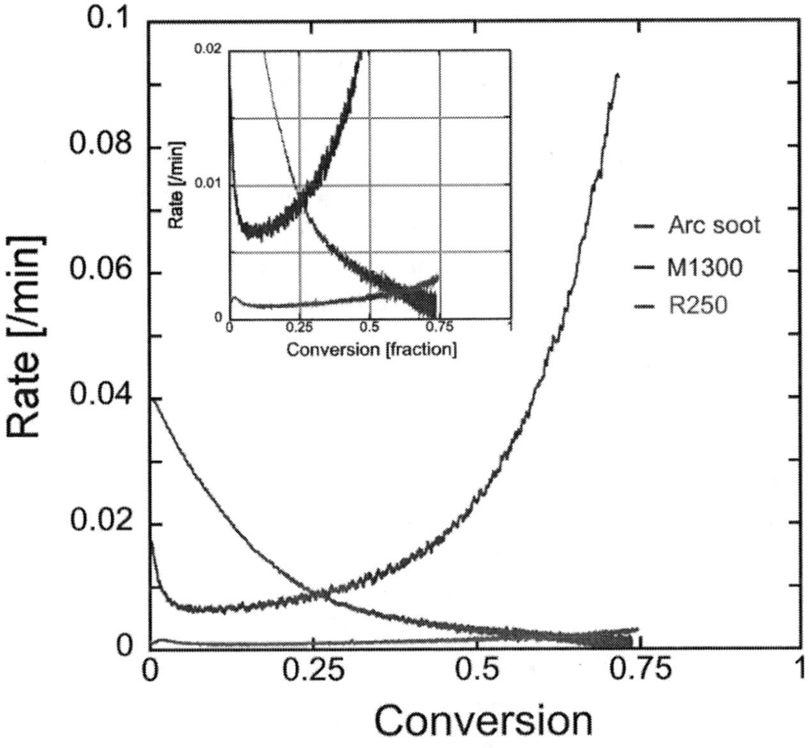

Figure 1. Change of oxidation rate with conversion for the three model carbons: Arc soot, M1300, and R250.

The dramatic decrease in the oxidation rate of arc soot is related to both a preferential burnout of amorphous carbon and concurrent graphitization of outer shells, as measured by analysis of HRTEM images. In fact the oxidation activates the formation of the extended outer lamella. Without removal of cross-links, adventitious carbon and alkyl substituents, graphene segments would not be "unlocked", able to reorient and expand. The rapid rise in the M1300 soot also reflects the initial nanostructure. The fullerenic nanostructure throughout the particle contains extended lamella dominated by basal carbon atoms. Once oxidation of a shell starts, the number of reactive edge sites rapidly multiplies while also exposing underlying shells to oxygen. Additionally these interior fullerenic shells with higher curvature will increase in reactivity, due to higher curvature imposing bond strain. The inset in Figure 1 shows a graph for these three model carbons with expanded scale showing the different trends in oxidation rates for these varied carbons. As for the MWCNTs and nano-onions, they exhibit similar trends. Composed of extended, contiguous lamella, these forms initially present only basal plane sites. With very low curvature, these positions are largely unreactive under the conditions studied.

The arc soot oxidation rate is consistent with initial burnout of amorphous material followed by oxidation of ever increasing graphitic content. The highly disordered matrix provides material for growth of lamella upon segment reorientation, realignment upon removal of cross-links, pendant side groups, etc. The fringe analysis shows that the decrease in oxidation rate is not due solely to preferential (initial) removal of amorphous material but also a concurrent shift in the distribution to longer lamella, due to graphitization, a form of oxidative induced densification, as has been noted for other carbons [17,18,19,20].

The rate variation for the M1300 is consistent with increasing exposure of strained layers and disordered carbon, as outer fullerenic shells are removed. Exposing only basal plane positions, oxidation proceeds slowly at first. Reflecting bond strain as a result of curvature, these sites are however more reactive than flatter lamella as in the R250 carbon. As an initial vacancy (hole) is created in a shell, edge sites are formed. With progressive oxidation an increasing number of perimeter edge sites are formed and oxidation accelerates geometrically with the hole radius. Once an opening is created in a basal plane, it not only creates edge sites for subsequent oxidation but also exposes an interior shell – with higher curvature and reactivity. The combination of these two processes in repeated succession accounts for the dramatic increase in rate over the course of burnout. Fringe analysis shows little difference between the nascent and 50% burnout forms – indicating absence of oxidation-induced graphitization. Such lack of change is consistent with the fullerenic nanostructure – a recognized form of non-graphitizing carbon. That the fringe distribution does not shift towards smaller mean values suggests that shells oxidize in a successive, rather than concurrent manner, thereby leaving inner (measureable) shells intact.

By comparison the R250 has a relatively constant oxidation rate, with preferential interior oxidation, progressing in an outward direction. Such a burnout mode suggests very different reactivity between interior and peripheral lamella. That no significant change in the "shell" nanostructure is observed suggests that these lamellae are not sufficiently freed to realign nor that there is available material for so-called vegetative growth [21]. This is consistent with it being not a flame-formed soot with high hydrogen content but rather a carbon black formed at very high temperatures. In fact by elemental analysis H-atom content is small (< 1%) [22]. Moreover, any growth would be countered by edge site oxidation.

The increases in rates for graphitized carbons such as the onions and MWCNTs likely reflect an increasing number of edge sites, but at a comparatively slower rate relative to the other model carbons, as HRTEM images did not show an appreciable change in size. Both materials are highly graphitic, exposing only edge sites of well-stacked, extended lamella. Oxidative attack occurs preferentially at these edge sites. These sites are not multiplied by geometry nor are they under high strain, except at bends, as for the onions. Due to their highly graphitic form, the lamellae are not turbostratically stacked but maintain a c-axis spacing near that of graphite. Therein steric considerations can further inhibit oxidation, as for graphite under similar conditions [23]. Only upon substantial burnout are additional edge sites made assessable, finally allowing for a rate increase, as observed.

4. Conclusions

A particularly intriguing aspect of this study is that the activation energy does not necessarily remain constant over the course of oxidation. Dependent upon initial nanostructure, and changes in nanostructure lead to varying activation energies. As illustrated by the present work, the carbons can become either more or less reactive towards oxidation, or remain roughly constant. Variations in each case are a reflection of a varying proportion of edge and basal carbon sites. If nanostructure changes during the course of oxidation the relative proportion of these sites will change with this then being manifested in activation energy values. Surprising examples include oxidation assisted lamella formation and growth. The relative spatial distribution of these carbon types also impacts oxidation rates. Extended lamella oriented with preferential exposure of basal sites can shield edge sites associated with internal, more disordered carbon. For some nanocarbons, these peripheral lamellae form during oxidation. In other nanocarbons initial oxidation can create edge sites along strained lamella junctions leading to an acceleration of oxidation rate by creating edge sites.

The interpretation of the nanostructural changes contributing to activation energy variation during oxidation and differences in activation energies between the different carbons are enabled by HRTEM. The physical structure provides the basis for interpretation of changes and differences in terms of nanostructure – itself being a fundamental manifestation of carbon atom site type.

5. Acknowledgements

The assistance of the EMS Energy Institute is gratefully acknowledged. This work was supported through NSF CBET- 1342920 with the Pennsylvania State University, University Park, PA 16802.

6. References

[1] Vander Wal, R. L., Tomasek, A. J. (2003). Soot oxidation: dependence upon initial nanostructure. *Combustion and Flame*, *134*(1), 1-9.

[2] Vander Wal, R. L., Tomasek, A. J., Street, K., Hull, D. R., Thompson, W. K. (2004). Carbon nanostructure examined by lattice fringe analysis of high-resolution transmission electron microscopy images. *Applied spectroscopy*, *58*(2), 230-237.

[3] Jawhari, T., Roid, A., Casado, J. (1995). Raman spectroscopic characterization of some commercially available carbon black materials. *Carbon*, *33*(11), 1561-1565.

[4] Sadezky, A., Muckenhuber, H., Grothe, H., Niessner, R., Pöschl, U. (2005). Raman microspectroscopy of soot and related carbonaceous materials: spectral analysis and structural information. *Carbon*, *43*(8), 1731-1742.

[5] Schmid, J., Grob, B., Niessner, R., Ivleva, N. P. (2011). Multiwavelength Raman microspectroscopy for rapid prediction of soot oxidation reactivity. *Analytical chemistry*, *83*(4), 1173-1179.

[6] Knauer, M., Schuster, M. E., Su, D., Schlögl, R., Niessner, R., Ivleva, N. P. (2009). Soot structure and reactivity analysis by Raman microspectroscopy, temperature-programmed

oxidation, and high-resolution transmission electron microscopy. *The Journal of Physical Chemistry A*, *113*(50), 13871-13880.
[7] Al-Qurashi, K., Boehman, A. L. (2008). Impact of exhaust gas recirculation (EGR) on the oxidative reactivity of diesel engine soot. *Combustion and Flame*, *155*(4), 675-695.
[8] Patel, M., Ricardo, C. L. A., Scardi, P., Aswath, P. B. (2012). Morphology, structure and chemistry of extracted diesel soot—Part I: Transmission electron microscopy, Raman spectroscopy, X-ray photoelectron spectroscopy and synchrotron X-ray diffraction study. *Tribology International*, *52*, 29-39.
[9] Vander Wal, R. L., Tomasek, A. J., Pamphlet, M. I., Taylor, C. D., Thompson, W. K. (2004). Analysis of HRTEM images for carbon nanostructure quantification. *Journal of Nanoparticle Research*, *6*(6), 555-568.
[10] Yehliu, K., Vander Wal, R. L., Boehman, A. L. (2011). Development of an HRTEM image analysis method to quantify carbon nanostructure. *Combustion and Flame*, *158*(9), 1837-1851.
[11] Watson, A. Y., Valberg, P. A. (2001). Carbon black and soot: two different substances. *AIHAJ-American Industrial Hygiene Association*, *62*(2), 218-228.
[12] Andrews, R., Jacques, D., Qian, D., Dickey, E. C. (2001). Purification and structural annealing of multiwalled carbon nanotubes at graphitization temperatures. *Carbon*, *39*(11), 1681-1687.
[13] Yezerets, A., Currier, N. W., Kim, D. H., Eadler, H. A., Epling, W. S., Peden, C. H. (2005). Differential kinetic analysis of diesel particulate matter (soot) oxidation by oxygen using a step–response technique. *Applied Catalysis B: Environmental*, *61*(1), 120-129.
[14] Tuinstra, F., Koenig, J. L. (1970). Raman spectrum of graphite. *The Journal of Chemical Physics*, *53*(3), 1126-1130.
[15] Knight, D. S., White, W. B. (1989). Characterization of diamond films by Raman spectroscopy. *Journal of Materials Research*, *4*(2), 385-393.
[16] Jaramillo, I. C., Gaddam, C. K., Vander Wal, R. L., Lighty, J. S. (2015). Effect of nanostructure, oxidative pressure and extent of oxidation on model carbon reactivity. *Combustion and Flame*, *162*(5), 1848-1856.
[17] Song, J., Alam, M., Boehman, A. L., Kim, U. (2006). Examination of the oxidation behavior of biodiesel soot. *Combustion and flame*, *146*(4), 589-604.
[18] Hurt, R. H., Dudek, D. R., Longwell, J. P., Sarofim, A. F. (1988). The phenomenon of gasification-induced carbon densification and its influence on pore structure evolution. *Carbon*, *26*(4), 433-449.
[19] Easler, T. E., Bradt, R. C., Walker, P. L. (1990). Gasification induced densification of form coke. *Fuel*, *69*(1), 124-125.
[20] Dukhan, A., Kantorovich, I. I., Barziv, E., Kandas, A., Sarofim, A. F. (1996, January). Structural changes of char particles during chemically controlled oxidation. In *Symposium (International) on Combustion* (Vol. 26, No. 2, pp. 3111-3118). Elsevier.
[21] Emmerich, F. G. (1995). Evolution with heat treatment of crystallinity in carbons. *Carbon*, *33*(12), 1709-1715.
[22] Kamegawa, K., Nishikubo, K., Yoshida, H. (1998). Oxidative degradation of carbon blacks with nitric acid (I)—changes in pore and crystallographic structures. *Carbon*, *36*(4), 433-441.
[23] Allendorf, H. D., Rosner, D. E. (1965). High-temperature kinetics of graphite oxidation by dissociated oxygen. *AIAA Journal*, *3*(8), 1522-1523.

Spring Technical Meeting
Eastern States Section of the Combustion Institute
March 4-7, 2018
State College, Pennsylvania

Analysis of East Asian Soot-based Inksticks through Raman Spectroscopy

Jennifer Giaccai[1,2], J. Houston Miller[1,]*

[1]Department of Chemistry, George Washington University, Washington D.C., U.S.A.
[2]Department of Conservation and Scientific Research, Freer Gallery of Art and Arthur M. Sackler Gallery, Smithsonian Institution, Washington D.C., U.S.A.
[*]Corresponding Author Email: houston@gwu.edu

Abstract: Carbon-based black pigments have been used extensively in art and cultural heritage objects. In East Asia particularly, soot has been the primary black pigment in the form of Chinese ink and inksticks. Valuable information about soot's chemical morphology can be obtained by Raman spectroscopy. Analysis of the D and G bands in Raman spectra of carbonaceous materials has long been used to identify the amount of disorder, but a more detailed application of Raman has yet to be applied to artists' pigments. In this study, a number of East Asian inksticks and their component soots (soot from pine wood or vegetable oils and, more recently, industrial carbon black) were examined. The D and G peaks of the Raman spectra were fit with both two-band and five-band peak fitting routines. The fitting results were used in principal component analysis (PCA) to differentiate between different Chinese ink sources.

Keywords:* *Raman Spectroscopy, Soot, Principal Component Analysis (PCA), Polyaromatic Hydrocarbons (PAH)

1. Introduction

Tuinstra and Koenig first used Raman to examine the crystallite size of various graphites,[1] a technique which was later applied to a wider range of carbon-based materials.[2] Raman has been used specifically for examining the size of polyaromatic hydrocarbon molecules (PAH) that compose the primary particles of soot for a number of years.[3–5] The Raman G band arises from the unbroken graphene-like structure in the center of the PAH molecules, while the asymmetric stretch which composes the D band only becomes spectroscopically allowed at edges of or defects within the graphene plane. Due to studies on the structures of graphenes using Raman spectroscopy, analysis of the relative intensity of the G band and D band of soot can be used to determine the average conjugation length of the PAH molecules composing a soot sample.[6]

In order to determine the I_D/I_G and conjugation length, the fit of the D and G band peaks must be as accurate as possible. Previous studies in our lab found that for pure soot a Lorentzian fit for the D band and Breit-Wigner-Fano fit for the G band provided the best overall fit and thus the most accurate I_D/I_G ratio. However, preliminary studies on a variety of artists' materials suggested that these results needed to be revisited when examining black pigments.[7] Other studies of combustion products have suggested that the minor D_2, D_3 and D_4 peaks can be fit under the D and G band peak envelope to provide more information and a better overall fit to the Raman spectra.[3,6]

Soot has been used in artists' pigments for as long as humans have been making art.[8] Understanding the materials that works of art are made of can help determine authenticity, trace trade routes, understand the working methods of the artist, or aid in conservation. Being able to distinguish when parts of a manuscript or work of art were created by more than one hand can help curators and conservators understand the methods of manufacture in a manuscript workshop or painting studio. Understanding if more than one hand is involved may also have repercussions on authorship of a manuscript or attribution of a painting.

In East Asia, soot is mixed with a proteinaceous animal skin glue as well as small amounts of additives to form and dry ink sticks.[9,10] The ink sticks can then be used as needed by grinding with water. Ink sticks have been used for thousands of years and are still used today by artists' and calligraphers. If the methods investigated here are useful in distinguishing between different soot sources, this methodology could be useful to environmental scientists in distinguishing between different sources of black carbon pollution.

2. Methods / Experimental

East Asian ink sticks are made from three main sources of soot-based black pigments: pine soot, lamp black from vegetable oils or, more recently, commercially produced carbon black. The reference collection of the Freer and Sackler Galleries contains a few ink sticks labeled with the type of soot used to make the ink sticks as well as the suppliers of the inksticks. A selection of these also have the pigment used to make the ink stick available separately. Table 1 shows the pigments and inksticks included in this study.

Table 1: Reference materials examined in this study. 'X'es indicate separate samples analyzed.

Soot Type	Sample Source	Pigment	Ink Stick	Ink	Paint Out
Carbon black	Columbia Raven 1250	x			
	Columbia Raven 410	x			
	ISAF 220	x			
	Chu Pong-In factory		x x		
Vegetable oil lamp black	Kobaien	x	x x	x	x
	Cheong Hak-Jin		x		
Pine soot	Shanghai		x		
	Kobaien		x	x	x
	Japanese	x			

Samples were analyzed with a Kaiser Optical Systems RXN-1 Raman spectrometer with an Invictus external-cavity-stabilized diode excitation laser at 785 nm and a Mark II microprobe with a 48 μm spot size. The laser power was varied from 8 - 50mW. The maximum laser power that could be used without causing any laser-induced incandescence in the sample over the course of the measurement was selected. Five-10s scans were coadded for each spectrum. Before peak fitting, a linear baseline was fit and subtracted.

A peak fitting program in Python was used where the D and G bands were fit with Lorentzian profiles and residuals minimized with a least-squares fit. When only the D and G peaks were fit, the G band could be fit with either a Lorentzian or a Breit-Wigner-Fano peak profile (BWF). After initial tests, the Q factor in the BWF peak profile was held constant at -10. The program could be modified to include the D_3 peak or the D_2, D_3 and D_4 peaks in addition to the major D and G bands, see Figure 1.

Initial conditions and bounds for all fit parameters needed to be determined before a peak-fitting routine could be automated. Initial conditions were determined by monitoring both the overall sum of squares of residuals and the proximity of the initial condition being varied to the final fit parameter. Ideally initial conditions were chosen that had the smallest residual while also maintaining a close proximity to the final parameter. Once initial conditions were chosen, they were applied to all Raman spectra acquired from all samples. Initial conditions for peak location and width were based on previous soot-based studies.[6] Initial conditions for D and G band peak intensities were determined by calculating the maximum intensity of the Raman spectrum in the region of the D and G peak maxima. This was then multiplied by an initial intensity multiplier (IIM) for each peak. The IIM for the minor D_2, D_3 and D_4 peaks was multiplied by the initial D (for D_3 and D_4) or G (for D_2) band intensity rather than directly related to the Raman intensity.

Due to consistent difficulties in finding ideal initial conditions for the five-peak fit, a second peak fit cycle was added in which the initial conditions used were those determined from first five-peak fit routine. For the two- and three-peak fitting routines the IIM used to determine the initial intensity for both D and G peaks was identical. The added contributions to the G peak from the D_3 and D_2 peaks required that the G IIM in the five-peak fit be set 25% lower in intensity than that of the D IIM.

The peak positions, widths, and intensities for all peaks were recorded for analysis with PCA. Eigenvector Solo software was used for PCA where the peak widths, positions, and intensity ratios are input as scaled and mean-centered independent variables.

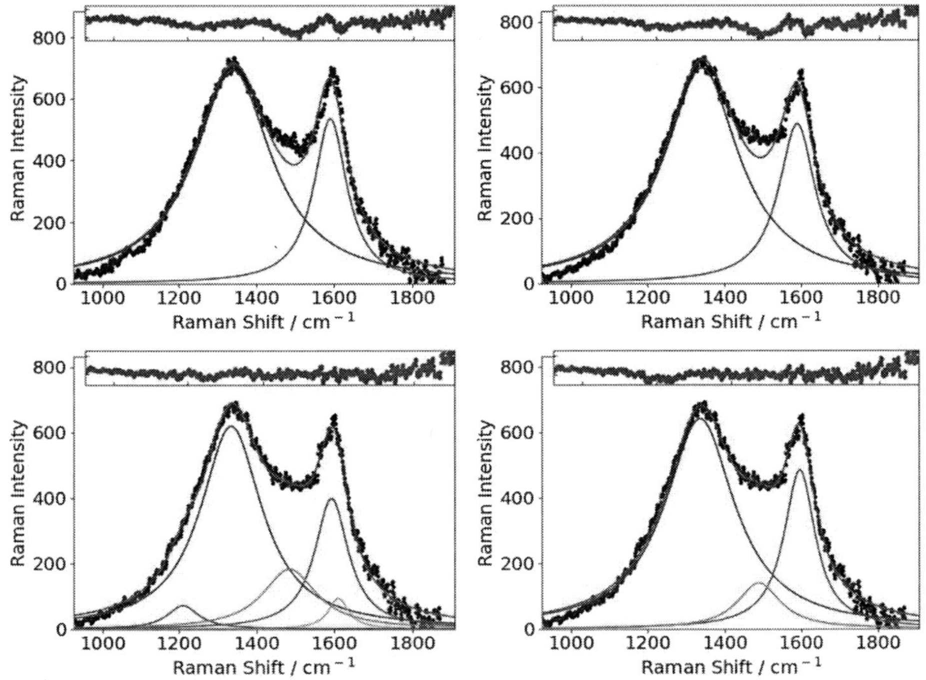

Figure 1: Peak fits from Ishida pine soot ink stick. Clockwise from top left: two Lorentzian peak fit, Lorentzian and BWF line fit, three Lorentzian peak fit, and five Lorentzian peak fit. Black: Raman spectrum, red: overall peak fit, blue: D peak fit, green: G peak fit, magenta: D_4 peak fit, cyan: D_3 peak fit, yellow: D_2 peak fit. Residuals are shown in top bar.

3. Results and Discussion

As the number of parameters in the peak fitting increased, the spectra became increasingly harder to fit (see Table 2). This is related to the number of parameters available, and therefore, the difficulty of determining appropriate initial settings for the parameters as the number of parameters increased. Effectively, not all of the parameters in the five Lorentzian peak fit can be treated as completely independent.

Table 2: Parameters and statistics of various peak fitting routines.

	Lor./BWF	2 Lor.	3 Lor.	5 Lor.
# of parameters to fit	6*	6	9	15
% of spectra fit	97	99	86	61
Average ID/IG and standard deviation for Ishida pine soot ink stick	1.34 (+/- 0.02)	1.34 (+/- 0.02)	1.33 (+/- 0.02)	1.43 (+/- 0.08)

* Q factor of BWF peak fit is held constant at -10.

Comparison of the sum of the squares of the residuals was used to evaluate success of fitting the overall Raman spectrum. Figure 2 demonstrates that as the number of fitting parameters increases, the residuals decrease. If only the two major peaks are fit, the Lorentzian/BWF peak fit gives better results than using two Lorentzian peaks. The tail of the BWF line shape is oriented to contribute to the valley between the D and G peaks, therefore lessening the need to add the third D_3 peak into the overall fitting routine, however addition of a third peak produces an overall better fit than either of the two-peak fitting routines. Similarly, addition of the D_2 and D_4 peaks into the five-peak fitting routine provides the best overall spectrum fit.

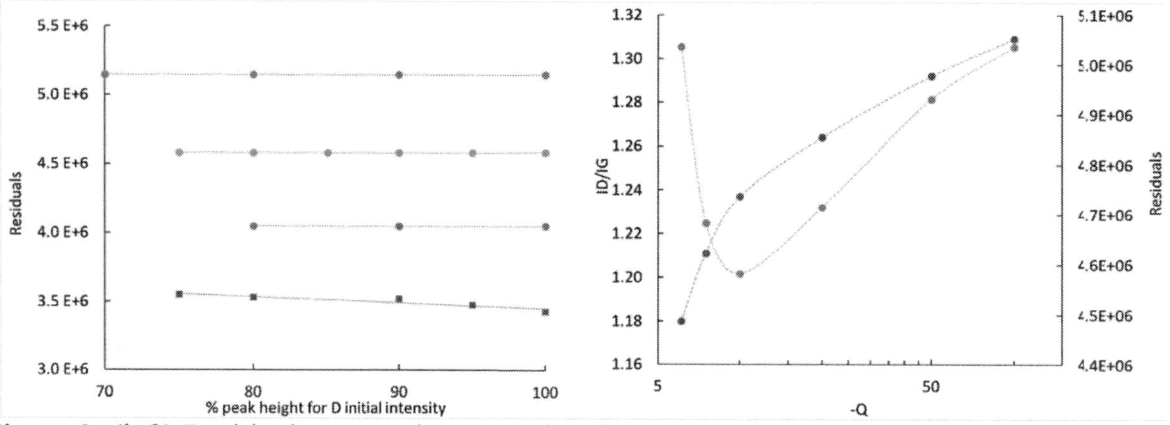

Figure 2: (left) Residuals versus the correction factor for D IIM for two Lorentzian (orange), Lorentzian/BWF (gray), three Lorentzian (green), and five Lorentzian (blue) peak fits. (right) Residuals (orange) and I_D/I_G (blue) with variation in Q in the Lorentzian/BWF peak fit.

Examination of the residuals in Figure 2 (right) demonstrates that an ideal value of -10 can be determined for Q with the Lorentzian/BWF line fit. It should also be noted the effect changing the Q factor has on the I_D/I_G.

The I_D/I_G can also be affected by the initial conditions for the individual peak parameters passed to the minimization routine. Figure 3 shows the differences that can be observed in individual peak fits, particularly the G band, as the initial conditions passed to the five-peak fitting minimization routine are varied. Although the overall spectrum fit is generally good, the changes in the G band with the variation of the initial conditions for the minor D_3 and D_4 peaks, are notable. For the spectrum shown below the variation in the I_D/I_G ratio is minimal. Over the six different initial conditions tested, the I_D/I_G ranged from 1.39 to 1.48. It is particularly interesting to observe that variations in the D_4 initial conditions show large variation in the D_2 and therefore G band intensity. This suggests that the D_2 and G bands, at least in this class of samples, are two broad to be effectively separated and fit.

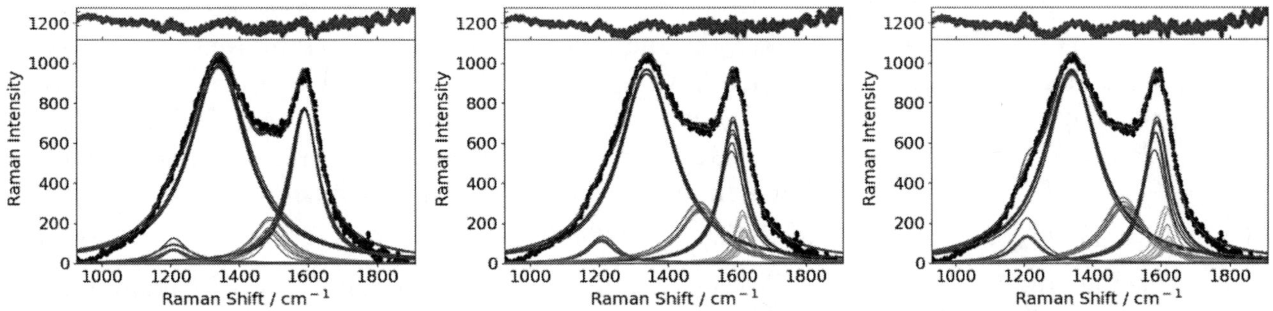

Figure 3: Variation in peak fit as initial conditions are varied for (left) D and G peak intensities, (center) D_3 intensity and (right) D_4 intensity. Black: Raman spectrum, red: overall peak fit, blue: D peak fit, green: G peak fit, magenta: D_4 peak fit, cyan: D_3 peak fit, yellow: D_2 peak fit. Residuals are shown in top bar.

PCA on ink sticks show that some separation is possible, and easier in PCA of the Lor./BWF results. Between the three principal components in the analysis, most inksticks could be separated, Figure 4. The broad groupings for some samples are generally attributed to lower quality inks, as independently determined by conservators and calligraphers at the Freer Gallery of Art.

4. Conclusions

The choice of peak fitting routine will have an impact on I_D/I_G ratio and therefore conjugation length of PAH in soots determined by Raman spectroscopy. Overall both two- and three- peak fitting routines were fairly robust in effectively fitting a wide variety of pigment samples with minimal human intervention. Five-peak fitting does provide the best fit to the Raman spectrum overall but is not robust and produces significant variation in fitting the G and D_2 peak parameters depending on initial conditions of the peak fit. In addition, PCA of the peak fitting results provided better separation of the various ink samples using two-peak fitting results.

PCA is effective in separating ink samples using parameters from Lor./BWF peak fitting. While ink sticks from lamp black overlay both carbon black and pine soot based ink sticks, the separation of individual ink sticks, especially those of higher quality, show promise for this method. Testing these procedures on museum collections is planned in the near future.

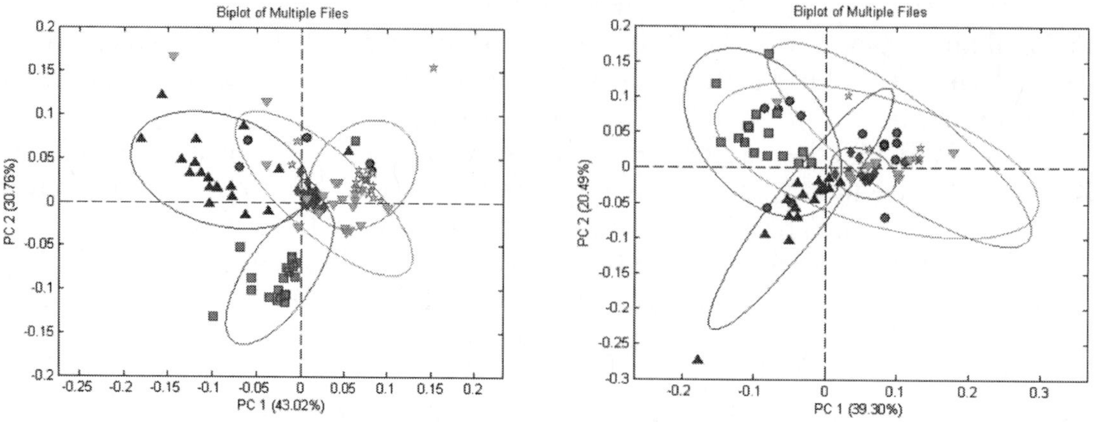

Figure 4: PCA of data from ink sticks and pigments showing separation into individual samples. Less overlap is seen in the PCA of the Lor./BWF results (left) than in the five-peak results (right). Further separation is apparent in PC 3 vs. PC 1, not shown. All data is from ink sticks: (red) Ishida pine soot, (green) Kobaien pine soot, (blue) Kobaien lamp black (cyan) Ulsan lamp black (pink) Ulsan carbon black.

5. Acknowledgements

This material is based upon work supported by the U.S. National Science Foundation under grant number CBET- 1706757 with Song-Charng Kong serving as technical monitor. Support for J.A.G. was provided by the David G. White Fellowship.

6. References

[1] F. Tuinstra, J.L. Koenig, Characterization of graphite fiber surfaces with Raman spectroscopy, J. Compos. Mater. 4 (1970) 492–499.

[2] A.C. Ferrari, J. Robertson, Interpretation of Raman spectra of disordered and amorphous carbon, Phys. Rev. B. 61 (2000) 14095–14107.

[3] A. Sadezky, H. Muckenhuber, H. Grothe, R. Niessner, U. Pöschl, Raman microspectroscopy of soot and related carbonaceous materials: Spectral analysis and structural information, Carbon. 43 (2005) 1731–1742.

[4] J.D. Herdman, B.C. Connelly, M.D. Smooke, M.B. Long, J.H. Miller, A comparison of Raman signatures and laser-induced incandescence with direct numerical simulation of soot growth in non-premixed ethylene/air flames, Carbon. 49 (2011) 5298–5311.

[5] M. Commodo, A. D'Anna, G. De Falco, R. Larciprete, P. Minutolo, Illuminating the earliest stages of the soot formation by photoemission and Raman spectroscopy, Combust. Flame. 181 (2017) 188–197.

[6] M. Saffaripour, L.-L. Tay, K.A. Thomson, G.J. Smallwood, B.T. Brem, L. Durdina, M. Johnson, Raman spectroscopy and TEM characterization of solid particulate matter emitted from soot generators and aircraft turbine engines, Aerosol Sci. Technol. 51 (2017) 518–531.

[7] J. Giaccai, J.H. Miller, Chemical Composition of Carbon Inksticks Revealed through Raman Spectroscopy, in: 10th US Natl. Combust. Meet., College Park, MD, 2017.

[8] J.W. Winter, E.W. Fitzhugh, Pigments based on carbon, in: B.H. Berrie (Ed.), Artists Pigments Handb. Their Hist. Charact., National Gallery of Art, Washington, 2007: pp. 1–37.

[9] J.W. Winter, Chinese Ink, Exped. Mag. 31 (1989) 52–53.

[10] J. Winter, East Asian Paintings: Materials, Structures and Deterioration Mechanisms, Archetype Publications Ltd., London, 2008.

Spring Technical Meeting
Eastern States Section of the Combustion Institute
March 4-7, 2018
State College, Pennsylvania

Informing TiRe-LII assumptions of soot nanostructure and optical properties for estimation of soot primary particle diameter

*Madhu Singh, Randy L. Vander Wal**

John and Willie Leone Family Department of Energy and Mineral Engineering and the EMS Energy Institute, The Pennsylvania State University, University Park, 16802, PA, USA
**Corresponding Author Email: ruv12@psu.edu*

Abstract: Time-resolved laser-induced incandescence for primary particle size determination is tested using three model carbon blacks. Optical properties change as does the nanostructure upon laser annealing whereas aggregate morphology and primary particle size remain equivalent to the original material, as shown by transmission electron microscopy (TEM). Primary particle diameters found from fitting experimentally measured time-resolved laser-induced incandescence (LII) signals with existing models do not match the particle diameters as directly visualized by TEM. The accommodation coefficient is shown to be a crucial parameter which can result in substantial variations in simulated conductive cooling profiles for particle sizing. Aggregate structure in the form of intra-aggregate connectivity and shielding is an additional underlying cause for erroneous particle sizing, not presently captured by LII models.
Keywords: Laser induced incandescence, Time-resolved LII, TEM, Primary particle size

1. Introduction

Time-resolved laser-induced incandescence (TiRe-LII) is a non-intrusive, in-situ optical diagnostic technique used to estimate soot primary particle size [1]. In TiRe-LII, soot is heated to incandescence temperatures (~3700 °C) with a pulsed laser with fluences that do not cause soot vaporization [2]. The decay of the incandescence signal is dependent on the conductive cooling rate of the particles, which in turn is a function of the particles' surface area to volume ratio [3,4]. Conductive cooling to the surroundings is considered the primary mode of heat dissipation after a period of ~ 100 ns following the laser pulse [5]. An inherent assumption in TiRe-LII theory is that the particles must lie within the Rayleigh size regime of the wavelength of light used to irradiate the soot ($\pi d_p/\lambda < 0.3$) [6]. For the 1064 nm pulsed Nd:YAG laser used in this work, particles of diameter < 102 nm satisfy the Rayleigh approximation, acting as volumetric absorbers, and hence emitters, of light. Particle sizing is thus dependent upon time-resolved detection of the incandescence signal and consequent inference of the particle's time-temperature-history (TTH) as it cools.

To better understand particle sizing by TiRe-LII, models have been developed to shed light on the experimental parameters driving the heating and physical processes governing the cooling of the (primary) particle [7]. In-spite of using similar model conditions and calculations, vast differences are observed in the calculated LII signal and predicted particle size when compared across ten different TiRe-LII models [7]. These models neglect nanostructure changes in soot when irradiated with a pulsed laser to high (~3500 °C) temperatures and the subsequent effect on the emissive properties of soot.[8] Just as different TiRe-LII models predict different particle sizes, the dependence of soot emissivity with wavelength varies with the correlation used [7,9]. The vast majority of this effort to determine wavelength dependent emissivity has focused on nascent soot. Here, "nascent" refers to fully formed soot or commercially obtained carbon black that is unmodified prior to laser heating. However, when irradiated with a high energy pulsed laser, soot has been observed to undergo significant physical transformations [8]. The degree of annealing and lamellae rearrangement has been observed to increase with increasing laser fluence [10]. Because heat-treatment changes soot structure at the nanometer length scale and this in turn influences its absorption/emission properties [10–12], it's inferred cooling rate by pyrommetry also changes.

Related works [13–19] have also emphasized the effect of aggregate size and soot morphology on modeled LII temporal profiles to determine primary particle size. As the number of primary particles per aggregate increase, the effect of shielding by outer particles becomes appreciable, decreasing the aggregate's accessible surface area below that of the sum of its primary particles [13]. The result is larger aggregates cooling slower despite being heated to the same peak temperature [16]. The thermal accommodation coefficient, a parameter that accounts for the "temperature jump" between the particle and the surrounding gas molecule when these are not in thermal equilibrium [20], is yet another parameter central to TiRe-LII modeling with inherent uncertainty and a spectrum of values, ranging from 0.07 – 1.0 [21–23]. Traditionally defined as a function of temperature [20], LII studies often use a temperature-independent or a constant accommodation coefficient to model heat dissipation via conduction, with little consensus on the value of the coefficient being used.

Thus, multiple factors plague results from TiRe-LII modeling and this work aims to provide insight into assumptions regarding these. Using carbon black as a soot surrogate, LII TTHs in relation to primary particle diameters are observed experimentally. These experimental time-temperature profiles are compared to profiles from a numerical model that simulates an expected LII signal using known experimental conditions. The effect of laser irradiation on the particle's nanostructure is shown using TEM and consequent changes in its absorption/emission properties are shown to effect its temperature decay with time.

2. Methods / Experimental

The carbon blacks used in this study are Black Pearls 800 and Regal 250 obtained from Cabot Corporation and Flammrus 101 obtained from Degussa AG (now Orion Engineered Carbons LLC) respectively.

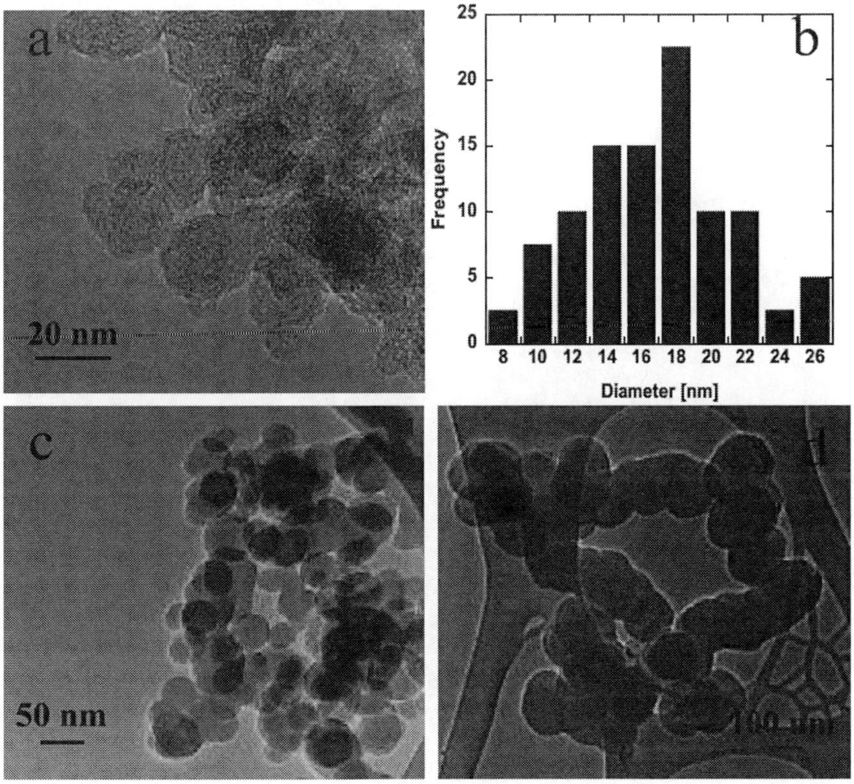

Figure 1: TEM micrographs of nascent carbon blacks illustrating their different primary particle diameters – (a) 18 nm and its (b) associated histogram, (c) 45 nm and (d) 112 nm (To note is the different magnification scales for each image).

The average primary particle diameters are 18, 45, and 112 nm, shown in Figure 1. This set of carbon blacks was selected to provide a large range of primary particle diameters and for possessing a narrow distribution of diameters within a given grade, as illustrated by the associated histogram for the 18 nm average diameter sample. A high-speed bladed mixer suspended the carbon blacks as an aerosol. A nitrogen sweep carried this aerosol to an optically transparent quartz tube with an additional nitrogen shroud to prevent carbon from building up on the quartz walls (Figure 2). A laser fluence of 200 mJ/cm^2 from a Continuum Surelite-III Q-switched Nd:YAG laser heated the carbon aerosol with a Princeton Instrument PI-MAX® intensified camera fitted to a spectrograph recording the incandescence signal.. The spectrograph-camera system was calibrated for intensity and wavelength with a black body tungsten lamp and a Hg-Ar lamp, respectively [24]. The system is capable of collecting spectra using a 2 ns gate width, enabling the collection of detailed TTHs by stepping out the gate delay. The spectra were then fit to black-body to determine temperatures at different instances in time.

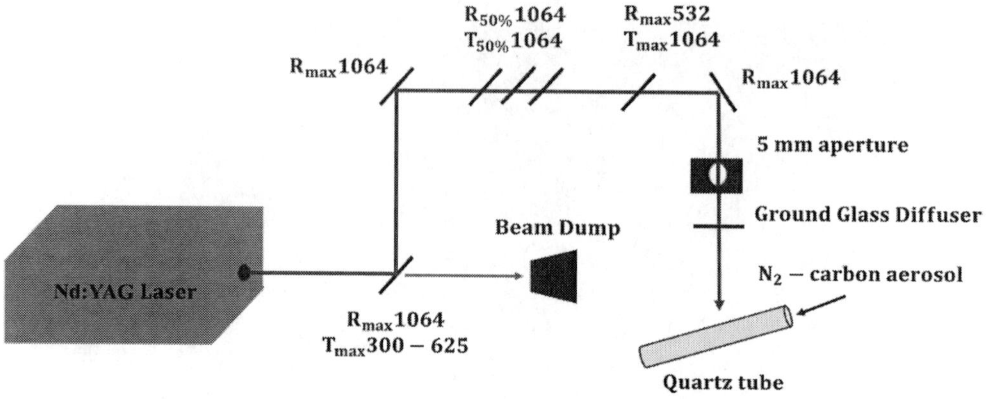

Figure 2: A schematic of the experimental setup.

3. Results and Discussion

Experimentally measured TTHs for the three model carbons are shown in Figure 3a post 100 ns, when conduction is the dominant mode of heat loss. As observed the large 112 nm particles have the slowest cooling rate, the 18 nm particles cooling fastest, with the 45 nm particle cooling rate in between. This ordering illustrates the conductive cooling rate having a direct correlation to their respective primary particle sizes. Temperatures have been calculated by fitting the incandescence signal to Planck's black-body approximation. This trend highlights the importance of volume to surface area ratio for conductive cooling, experimentally verifying the basis of TiRe-LII for particle sizing.

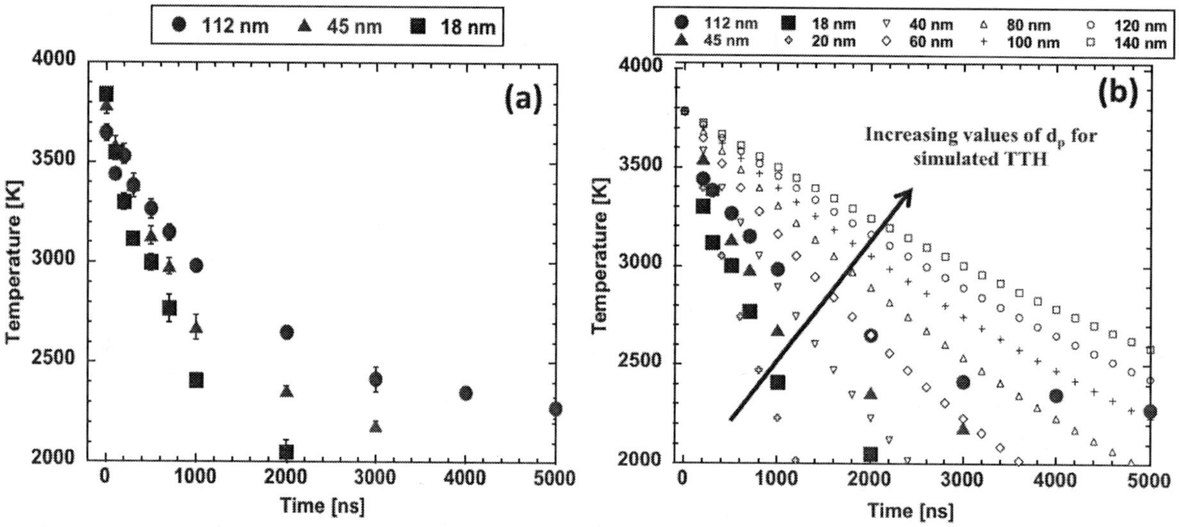

Figure 3: (a) Experimental Time-Temperature-Histories of laser heated model carbons with different primary particle diameters. (b) Experimental vs. LIISim [25] simulated time-temperature profiles for a series of primary particle sizes.

A comparison of experimental TTH with simulated temporal profiles using monodisperse particles in the conduction transition regime with LIISim [25] is shown in Figure 3b. LIISim provides a web-based interface for simulating LII signals with selectable experimental parameters and the gas-kinetic regime for heat conduction. These can be divided into the free molecular, continuum or the transition regime. The mismatch between known primary particle size and that predicted by the simulation increases as d_p increases, i.e. experimental and simulated conductive cooling profiles are in agreement for $d_p = 18$ nm, but this agreement worsens at larger values of d_p. These comparisons use a thermal accommodation coefficient value of 0.3 [26] for the simulated time-temperature profiles. This disagreement may be attributed to a number of factors and assumptions inherent to the various models used for particle sizing with TiRe-LII.

Studies on LII have interpreted changes in soot structure and particle size using absorption/emissivity values [7,9] to calculate the primary particle size of soot. Some prior studies have inferred a possible increase in primary particle size, ranging from 6 – 40% based on changes in the absorption/scattering ratio [27]. It has been speculated that aggregates fuse together under rapid heating [28] yet other works have mentioned no fragmentation of aggregates at sublimation temperatures [29]. A possible unifying explanation for these different conclusions is that the particle nanostructure has changed, i.e. annealed. TEM imaging shows that the aggregate morphology is preserved (Figure 4b) and that the particle diameter does not measurably change due to laser heating (Figure 4a vs. 4c). Figure 4b and 4c in particular show apparent mass loss from and/or material rearrangement in the particle interior without a change in primary particle diameter, indicating that the size distributions of nascent and laser heated material are equivalent for laser fluences below those that lead to sublimation.

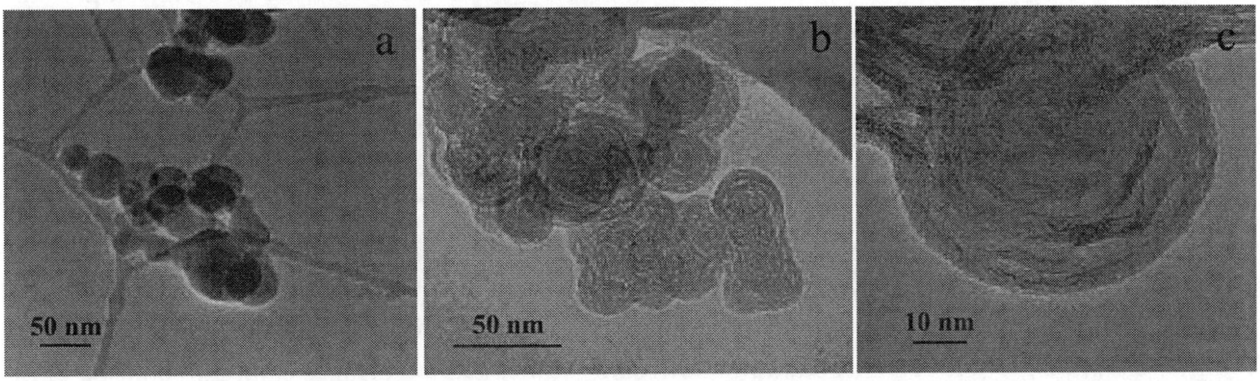

Figure 4: TEM micrographs of (a) nascent and (b,c) laser heated carbon black with an average particle diameter of 45 nm (geometric mean).

The partial amalgamation of particles shown in Figs. 1 and 4 can result in a significant difference in calculation of particle diameter numerically, as the surface area of the particle undergoing conductive cooling is less than that of a true isolated sphere. The effects of aggregation compared within and across each of the three regimes, i.e., free molecular, continuum and transition, for 20 and 200 primary particles per aggregate using LIISim[25] are shown in Figure 5a.

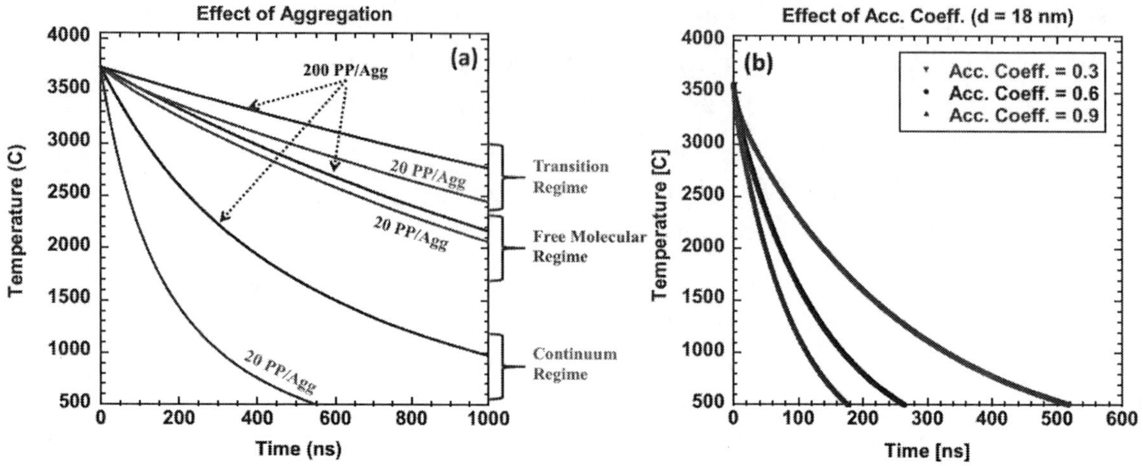

Figure 5: (a) Effect of particle aggregation and its variation across different heat conduction models and (b) effect of accommodation coefficient. Time-temperature profiles were generated using the LIISim web interface.[25]

Figure 5a illustrates that the effect of particle aggregation, with all else being constant, may (or may not) result in a significant change of the particle's conductive cooling profile, depending upon the heat conduction regime. For the three carbon blacks studied, there remain significant differences in particle connectivity within the aggregates. Particle connectivity remains unaccounted for under the assumption of point-contacting spheres, resulting in different primary particle diameters calculated using TiRe-LII models when compared to experimentally determined values.

Literature suggests an effective value of $\alpha = 0.25$ based on ease of model implementation [21] or, as in other studies, $\alpha = 0.3$ is likely assumed as an average of values ranging from 0.2 – 0.5 derived from measurements of N_2 at 1220 K or NO at 700 K scattered from graphite surfaces [26]. Most of these measurements are at temperatures significantly lower than those observed on laser annealing and assume α to be temperature-independent. As demonstrated in Figure 5b, all else being constant, different accommodation coefficient values can result in a dramatic change in the particle's TTH which in turn will influence particle size calculations.

4. Conclusions

Time-resolved LII has the capability to resolve primary particle size within carbon (soot) aggregates. Yet most TiRe-LII models are based upon an energy balance that has several assumptions regarding optical and material properties. TEM clearly reveals substantial particle annealing, not surprising given the temperature attained but perhaps unexpected given the short duration at which the particles remain above 2000 °C. Preservation of morphology and primary particle size as shown by TEM is consistent with the theory of particle sizing by TiRe-LII. However, most TiRe-LII models do not yield meaningful particle diameter values by the usual formalism, as shown by comparison to TEM measured values and images of materials that lack point-contacting sphere-like particles. Shielding effects due to the fractal-like nature of aggregates further add to uncertainty in size measurements. Moreover there is no single value of the thermal accommodation coefficient which applies to the different carbons and associated range of primary particle sizes. However, it is also important to understand that the various available models differ in their assumptions and some models do a better job at reproducing LII signals under a wide range of conditions. There still is room for improvement, and some understanding of the combustion system and the types of particles expected is needed prior to drawing conclusions of particle size using TiRe-LII.

5. Acknowledgements

The authors acknowledge support by the National Science Foundation (NSF), Chemical, Bioengineering, Environmental, and Transport Systems (CBET), under Grant Number 1236757. TEM was performed using the facilities of the Materials Research Institute at The Pennsylvania State University.

6. References

1. H. A. Michelsen. Probing soot formation, chemical and physical evolution, and oxidation: A review of in situ diagnostic techniques and needs. *Proc. Combust. Inst.* **36,** 717–735 (2017).
2. H. A. Michelsen, C. Schulz, G. J. Smallwood, and S. Will. Laser-induced incandescence: Particulate diagnostics for combustion, atmospheric, and industrial applications. *Prog. Energy Combust. Sci.* **51,** 2–48 (2015).
3. R. L. Vander Wal and M. Y. Choi. Pulsed laser heating of soot: morphological changes. *Carbon* **37,** 231–239 (1999).
4. R. L. Vander Wal, T. M. Ticich, and A. Brock Stephens. Can soot primary particle size be determined using laser-induced incandescence? *Combust. Flame* **116,** 291–296 (1999).
5. H. A. Michelsen et al. Modeling laser-induced incandescence of soot: enthalpy changes during sublimation, conduction, and oxidation. *Appl. Phys. B Lasers Opt.* **93,** 645–656 (2008).
6. T. L. Farias, Ü. Ö. Köylü, and M. G. Carvalho. Range of validity of the Rayleigh--Debye--Gans theory for optics of fractal aggregates. *Appl. Opt.* **35,** 6560–6567 (1996).
7. H. A. Michelsen et al. Modeling laser-induced incandescence of soot: a summary and comparison of LII models. *Appl. Phys. B* **87,** 503–521 (2007).
8. R. L. Vander Wal. Laser-induced incandescence: excitation and detection conditions, material transformations and calibration. *Appl. Phys. B* **96,** 601–611 (2009).
9. F. Goulay, P. E. Schrader, and H. A. Michelsen. Effect of the wavelength dependence of the emissivity on inferred soot temperatures measured by spectrally resolved laser-induced incandescence. *Appl. Phys. B* **100,** 655–663 (2010).
10. R. L. Vander Wal, T. M. Ticich, and A. B. Stephens. Optical and microscopy investigations of soot structure alterations by laser-induced incandescence. *Appl. Phys. B Lasers Opt.* **67,** 115 (1998).
11. C. Jäger, T. Henning, R. Schlögl, and O. Spillecke. Spectral properties of carbon black. *J. Non. Cryst. Solids* **258,** 161–179 (1999).
12. B. Apicella, M. Alfè, R. Barbella, A. Tregrossi, and A. Ciajolo. Aromatic structures of

carbonaceous materials and soot inferred by spectroscopic analysis. *Carbon* **42,** 1583–1589 (2004).

13. D. E. Rosner and P. Tandon. Prediction and correlation of accessible area of large multiparticle aggregates. *AIChE J.* **40,** 1167–1182 (1994).
14. A. V. Filippov, M. Zurita, and D. E. Rosner. Fractal-like Aggregates: Relation between Morphology and Physical Properties. *J. Colloid Interface Sci.* **229,** 261–273 (2000).
15. C. M. Sorensen. Light Scattering by Fractal Aggregates: A Review. *Aerosol Sci. Technol.* **35,** 648–687 (2001).
16. F. Liu, G. J. Smallwood, and D. R. Snelling. Effects of primary particle diameter and aggregate size distribution on the temperature of soot particles heated by pulsed lasers. *J. Quant. Spectrosc. Radiat. Transf.* **93,** 301–312 (2005).
17. F. Liu, B. J. Stagg, D. R. Snelling, and G. J. Smallwood. Effects of primary soot particle size distribution on the temperature of soot particles heated by a nanosecond pulsed laser in an atmospheric laminar diffusion flame. *Int. J. Heat Mass Transf.* **49,** 777–788 (2006).
18. R. P. Bambha and H. A. Michelsen. Effects of aggregate morphology and size on laser-induced incandescence and scattering from black carbon (mature soot). *J. Aerosol Sci.* **88,** 159–181 (2015).
19. H. Bladh et al. Probing the smallest soot particles in low-sooting premixed flames using laser-induced incandescence. *Proc. Combust. Inst.* **35,** 1843–1850 (2015).
20. F. O. Goodman. Thermal accommodation coefficients. *J. Phys. Chem.* **84,** 1431–1445 (1980).
21. S.-A. Kuhlmann, J. Reimann, and S. Will. On heat conduction between laser-heated nanoparticles and a surrounding gas. *J. Aerosol Sci.* **37,** 1696–1716 (2006).
22. H. A. Michelsen. Derivation of a temperature-dependent accommodation coefficient for use in modeling laser-induced incandescence of soot. *Appl. Phys. B* **94,** 103 (2008).
23. X. López-Yglesias, P. E. Schrader, and H. A. Michelsen. Soot maturity and absorption cross sections. *J. Aerosol Sci.* **75,** 43–64 (2014).
24. J. P. Abrahamson, M. Singh, J. P. Mathews, and R. L. Vander Wal. Pulsed laser annealing of carbon black. *Carbon* **124,** 380–390 (2017).
25. M. Hofmann, B. Kock, and C. Schulz. *A Web-Based Interface for Modeling Laser-Induced Incandescence (LIISim).* (2006).
26. H. A. Michelsen. Understanding and predicting the temporal response of laser-induced incandescence from carbonaceous particles. *J. Chem. Phys.* **118,** 7012–7045 (2003).
27. H. A. Michelsen, P. E. Schrader, and F. Goulay. Wavelength and temperature dependences of the absorption and scattering cross sections of soot. *Carbon* **48,** 2175–2191 (2010).
28. E. Cenker, G. Bruneaux, T. Dreier, and C. Schulz. *Determination of small soot particles in the presence of large ones from time-resolved laser-induced incandescence. Applied Physics B* **118,** (2014).
29. H. A. Michelsen et al. Particle formation from pulsed laser irradiation of soot aggregates studied with a scanning mobility particle sizer, a transmission electron microscope, and a scanning transmission x-ray microscope. *Appl. Opt.* **46,** 959 (2007).

Spring Technical Meeting
Eastern States Section of the Combustion Institute
March 4-7, 2018
State College, Pennsylvania

Soot derivatization for source identification

*Madhu Singh, Randy L. Vander Wal**

John and Willie Leone Family Department of Energy and Mineral Engineering and the EMS Energy Institute, The Pennsylvania State University, University Park, 16802, PA, USA
**Corresponding Author Email: ruv12@psu.edu*

Abstract: Soot source identification by laser derivatization (SSILD) is based on the premise that soots originating from different combustion sources differ in their nanostructure, owing to the variation in formation conditions. Combustion conditions – temperature, pressure, fuel, fuel-to-air-ratio, fuel-air mixing – are specific to each source and differ between sources. Thus, soot produced from different sources will reflect its formation conditions. HRTEM imaging in concert with image analysis algorithms is used to quantify nanostructure parameters. Contrasting nanostructure can identify the soot origin when sufficiently distinct. Where similar nanostructure or more often, poorly contrasted nanostructure is observed, subtle differences may be magnified by laser derivatization to enhance lamellae recognition and quantification.
Keywords: Soot, HRTEM, Nanostructure, LII

1. Introduction

Soot is a black, carbonaceous substance and an unwanted product of incomplete combustion of hydrocarbons. Incomplete combustion of fossil fuels and organic matter is the major source of soot that is released into the environment [1]. Gasoline and diesel powered engines, wood burning in home fireplaces, and open burning of waste, furnaces burning coal and coke production processes are principal sources of soot emissions. Soot is a common term for combustion derived PM 2.5 particle pollution[2]. Emission of carbonaceous soot is a health hazard and many studies have linked soot exposure to various cancers and respiratory problems[3]. Thus, source identification is important to pollution receptor models that require knowledge of source contributions at receptor sites.

Combustion produced soot shows varied nanostructure and chemistry that is dependent upon fuel and combustion conditions under which soot is formed[4]. Combustion conditions – temperature, pressure, fuel, fuel-to-air-ratio, fuel-air mixing – are specific to each source and differ between sources. Thus, soot produced from different sources will vary, reflecting its formation conditions. These differences in the structure and chemistry of soot can, in principle, be used to identify its origin (source) by characterizing the collected soot sample to differentiate them on the basis of their composition and nanostructure[4]. However, these structural aspects of soot lamellae in its "nascent" or "as is" formed state are subtle while soot sample quantities are often insufficient for characterization using a traditional thermo-physical analytical techniques. In order to enhance nanostructure differences in soots from different sources, a process of pulsed laser annealing is proposed accentuate nanostructure, referred to as laser derivatization.

Derivatization is a well-known process by which a compound is chemically changed, producing a new compound that has properties amenable to a particular analytical method. Laser derivatization is used here in a similar manner to enhance subtle differences in soot nanostructure upon laser heating. Due to the extremely rapid heating (~ 5 nanoseconds to reach 3700°C) and cooling (microsecond) time scales observed when soot is irradiated with a Q-switched laser pulse while at atmospheric pressure. Consequently soots anneal incompletely due to apparent kinetic limitations on these timescales[5]. Determining the direction and extent of these changes are small differences in initial nanostructure and composition that become magnified by the fast heating and preserved by the rapid quenching[4].

In this study, soot samples from known sources are imaged via transmission electron microscopy (TEM) before and after laser heat treatment. Image processing algorithms quantify nanostructure characteristics including fringe length, fringe tortuosity, fringe spacing, stacking and curvature for nascent and laser heated soot. These metrics collectively comprise the identification marker or "fingerprint" for soot. Unknown samples can then be compared against information available from existing samples to identify its source or have a close estimate of what the source may be, both qualitatively and quantitatively. Quantitative comparison for identification uses statistical analyses. By this approach, SSILD uses the changed soot nanostructure post laser heat treatment to identify the soot source.

2. Methods / Experimental

The primary harmonic, 1064 nm, of a 10 Watt Q-switched Continuum Surelite- III Nd:YAG laser is used to anneal soot samples directly on 3 mm diameter lacy carbon TEM grids. The pulse duration is fixed at 8 ns and the pulse energy is kept at 125 mJ/cm^2 for this study to avoid sample ablation at higher fluences. [6] Upon exiting the laser, the 1064 nm beam travels to lens with maximum reflection for 1064 nm to filter out any residual wavelengths by transmitting them to a beam dump. The beam is steered towards the sample with the help of mirrors or partial reflectors to further filter out residual 532 nm light prior to striking the sample. The original 9 mm diameter Gaussian beam is diffused using a ground glass diffuser to form a top-hat like beam and its wings are clipped using a 5 mm aperture to ensure uniform heating of the TEM grid placed in the center of the beam path. The sample, prepared on a 300-mesh copper lacey carbon TEM grid, is held in a steel sample holder with recessed holes which slides in and out of a steel chamber. Laser heating is performed under an inert gas, Argon, with the sample holder forming a vacuum tight seal with the sample chamber. A schematic of the experimental setup is shown in Figure 1. TEM images were obtained using an FEI Talos F200X S/TEM.

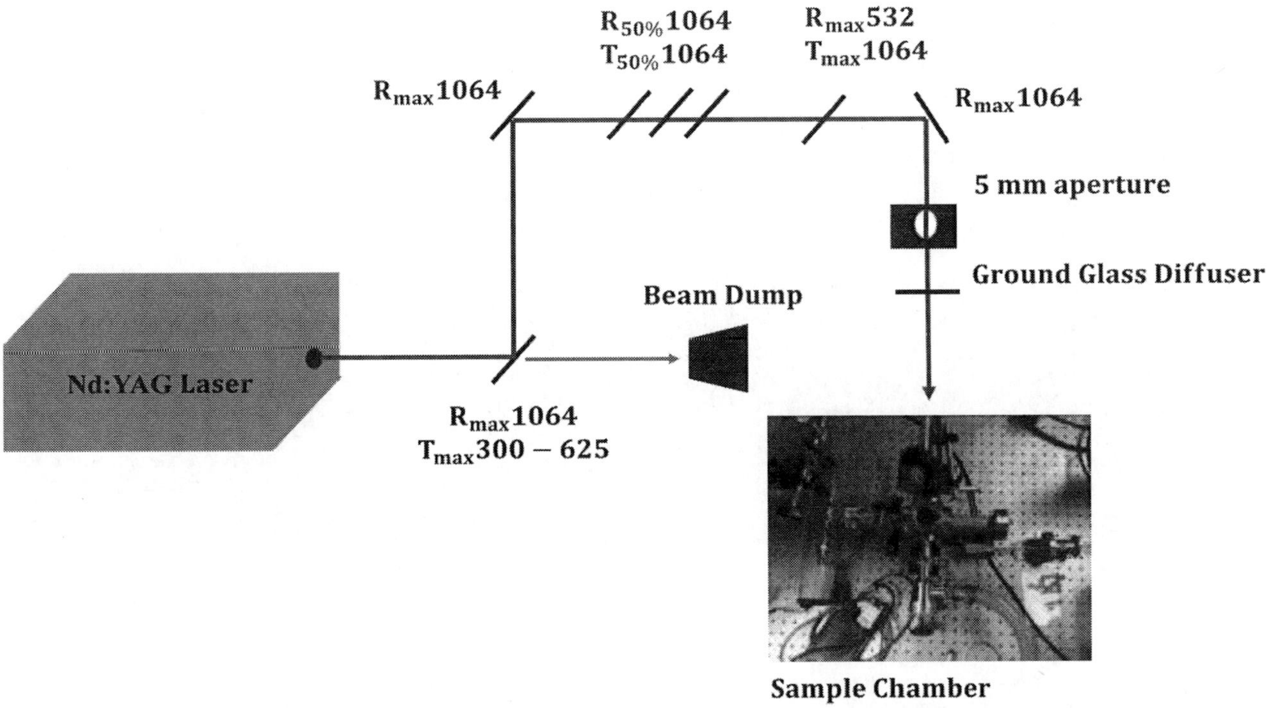

Figure 1: A schematic of the Nd:YAG laser setup

3. Results and Discussion

Soot samples from three sources – an ultra-low sulfur diesel soot from a 1999 1.7L Mercedes Benz direct injection common rail diesel engine, soot from a cookstove burning pine wood and soot collected from a wood fired boiler – were obtained, imaged and analyzed in their nascent (as is) form as well as after laser annealing, shown in figs. 2-4, respectively.

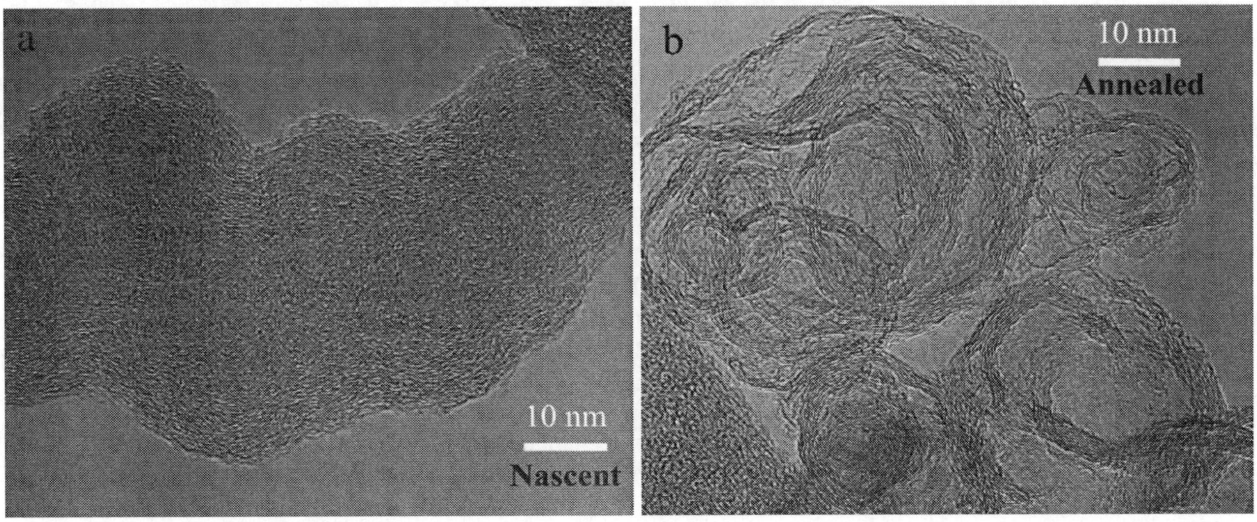

Figure 2: TEM micrographs showing the nanostructure of (a) nascent and (b) laser-annealed ultra-low sulfur diesel soot (ULSD) soot

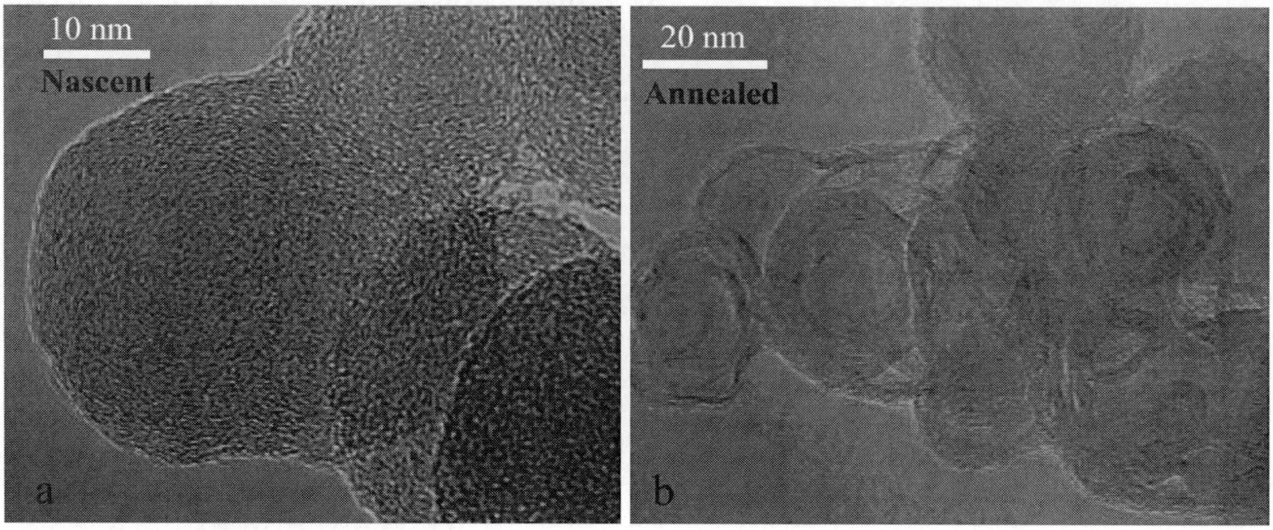

Figure 3: TEM micrographs showing nanostructure of (a) nascent and (b) laser-annealed cookstove soot formed by burning pine wood

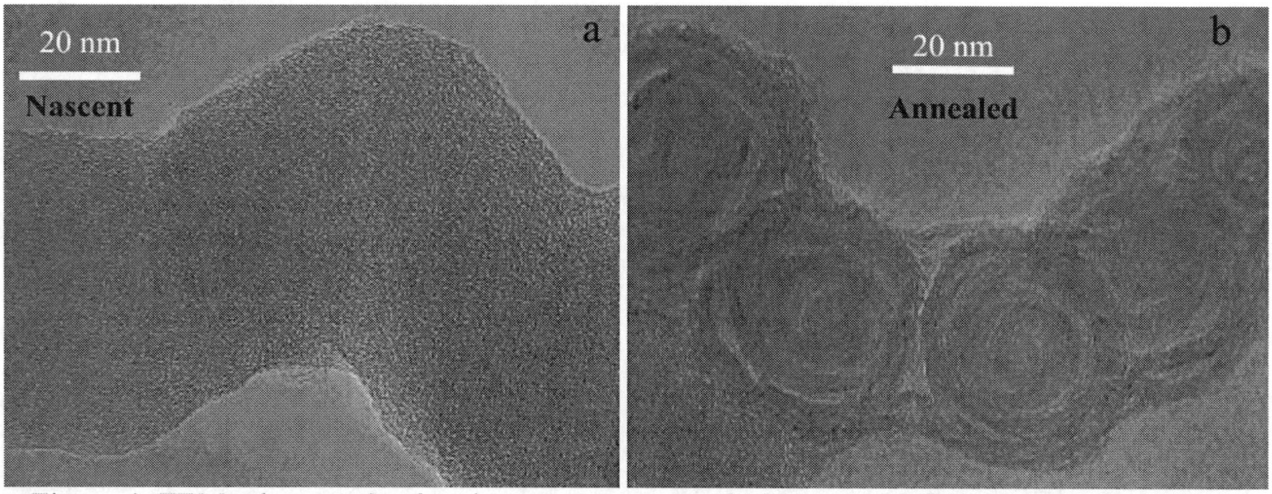

Figure 4: TEM micrographs showing nanostructure of (a) nascent and (b) laser-annealed soot from a wood fired boiler

To quantify nanostructure, TEM images were analyzed using an in-house image analysis algorithm. Previous studies [4,5,7,8] have shown the applicability of the algorithm for image analysis of soot and that it is statistically meaningful. Nanostructure metrics include length, curvature, number of fringes per stack and spacing between opposing planes. Curvature is measured by tortuosity or the radii of curvature. The algorithm analyzes many (i.e. 100-1000s) individual graphene segments from each primary particle. These metrics extracted from the soot samples of interest are presented as a comparison of fringe length (Figure *5*) and fringe tortuosity (Figure *6*).

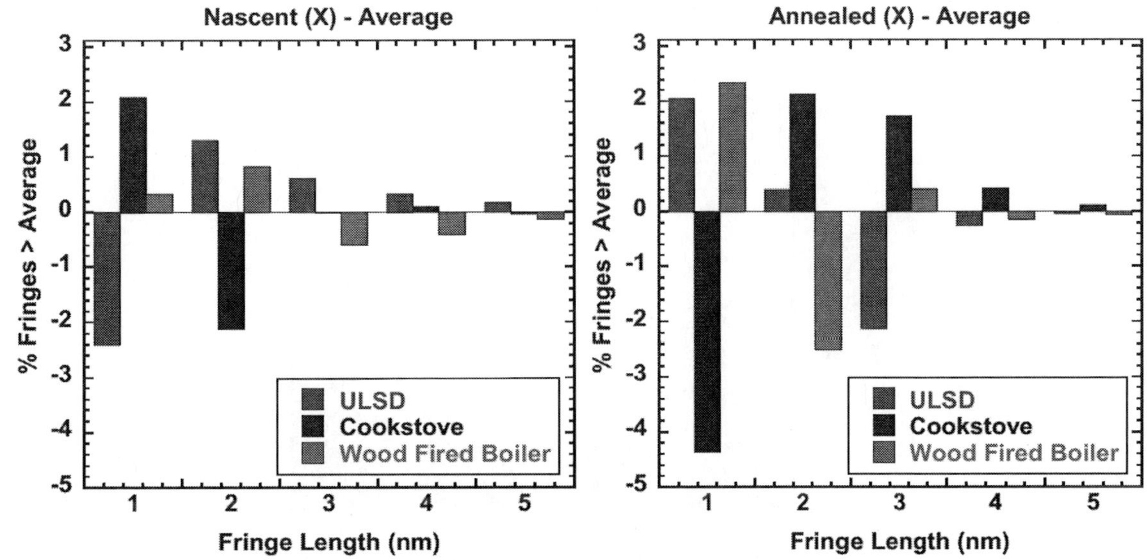

Figure 5: Fringe length statistics for three soot samples and their deviation from the average of the three sources for nascent and laser annealed soots (X = fringe statistics of nascent/annealed soot, respectively)

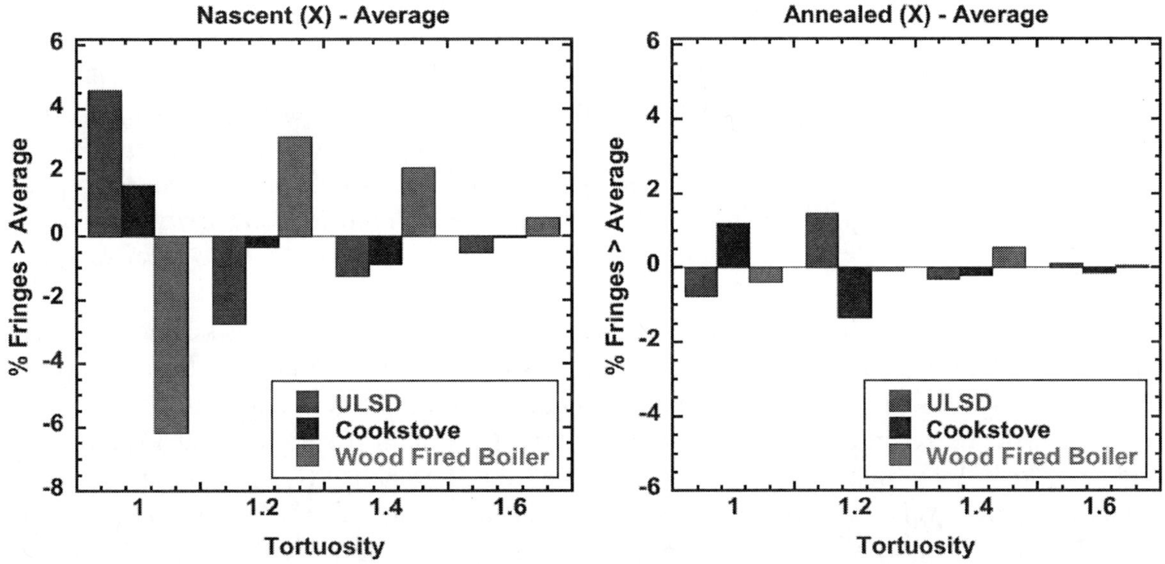

Figure 6: Fringe tortuosity statistics for three soot samples and their deviation from the average of the three sources for nascent and laser annealed soots (X = fringe statistics of nascent or annealed soot, respectively)

The algorithm used to extract statistics on fringe stacks [9] calculates the number of carbon lamellae per stack based on three user defined parameters – the maximum allowed spacing between two layers, the angle between the layers and the centroid to centroid distance of the layers. Here, values of 3.8 Å, 10° and 8 Å were selected as the controlling parameters.

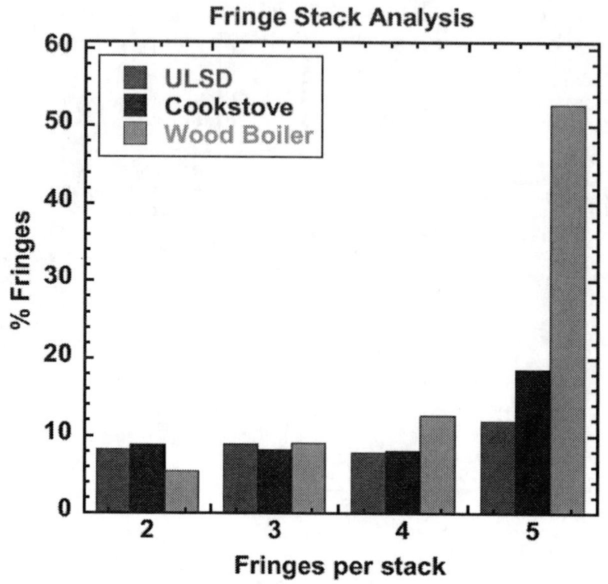

Figure 7: Statistical analysis of the number of fringes per stack across the three soot samples after laser annealing

Figs. 5-7 show clear changes in fringe statistics of length, tortuosity and stacking between the three samples tested. Wood fired boiler soot, for instance, shows an increased deviation from the average trend post laser annealing, while tortuosity is markedly different for the nascent samples when compared to their average. Fringe tortuosity decreases for all samples after annealing with fringes becoming straighter and longer, also supported by TEM micrographs in Figs. 2-4. Fringes per stack for annealed samples also quantitatively distinguish the soots for 5 or more fringes per stack indicative of ULSD having the least stack height when compared to cookstove and wood fired boiler soots.

Such statistics of nanostructural parameters characterizing these soots will add to a database of multiple soot samples from a variety of sources spanning a range of combustion conditions. As observed from Figs. 5-7, for each individual sample, the combination of length, tortuosity, stacking and other (yet to be calculated) metrics for nanostructure quantification form a unique pattern, serving as a "key" by which individual soots can be identified and differentiated from other soots and sources, respectively. This recognition key, by utilizing each soot's unique pattern, can then be applied to identify an unknown soot sample against existing soots in the database by which to identify its source and/or combustion conditions pertaining to its formation.

4. Conclusions

Soots from three different sources are shown to vary in their nanostructure as seen by TEM. Laser annealing accentuates differences in nanostructure while retaining initial morphology owing to time at annealing temperatures being only 10's of nanoseconds, thereby kinetically limiting material transformation. Quantification of nanostructural parameters by TEM image analysis further aids qualitative image differentiation by studying their variation across soot samples. Source identification by laser derivatization is thus a promising technique to identify soot contributions from sources to a receptor site, an analysis useful for pollution models based on source apportionment.

5. Acknowledgements

The authors acknowledge support by the National Science Foundation (NSF), Chemical, Bioengineering, Environmental, and Transport Systems (CBET), under Grant Number 1236757. TEM was performed using the facilities of the Materials Research Institute at The Pennsylvania State University.

6. References

1. G. Shrestha, S. J. Traina, and C. W. Swanston. Black Carbon's Properties and Role in the Environment: A Comprehensive Review. *Sustainability* **2,** 294–320 (2010).
2. C. I. Davidson, R. F. Phalen, and P. A. Solomon. Airborne Particulate Matter and Human Health: A Review. *Aerosol Sci. Technol.* **39,** 737–749 (2005).
3. C. A. Pope et al. Exposure to Fine Particulate Air Pollution Is Associated with Endothelial Injury and Systemic Inflammation. *Circ. Res.* (2016).
4. C. K. Gaddam. Electron Microscopic and Spectroscopic Characterization for Soot Source Differentiation by Laser Derivatization. (2015).
5. J. P. Abrahamson. Pulsed laser annealing of carbon. (The Pennsylvania State University, 2017).
6. J. P. Abrahamson, M. Singh, J. P. Mathews, and R. L. Vander Wal. Pulsed laser annealing of carbon black. *Carbon* **124,** 380–390 (2017).
7. R. L. Vander Wal, V. M. Bryg, and M. D. Hays. Fingerprinting soot (towards source identification): Physical structure and chemical composition. *J. Aerosol Sci.* **41,** 108–117 (2010).
8. K. Yehliu, R. L. Vander Wal, and A. L. Boehman. Development of an HRTEM image analysis method to quantify carbon nanostructure. *Combust. Flame* **158,** 1837–1851 (2011).
9. E. Luow. Structure and combustion reactivity of inertinite-rich and vitrinite rich South African coal chars: quantification of the structural factors contributing to reactivity differences. (The Pennsylvania State University, 2013).

Sub Topic: Soot

Spring Technical Meeting
Eastern States Section of the Combustion Institute
March 4-7, 2018
State College, Pennsylvania

The effects of oxygenated fuels on soot particle nanostructure

*Junqing Zhu**, *Charles S. McEnally, Lisa D. Pfefferle*

Yale University, New Haven CT 06520 USA
**Corresponding Author Email: junqing.zhu@yale.edu*

Abstract: The internal nanostructure of soot particles provides clues to their formation mechanisms, affects their oxidation rates in flames, and controls the regeneration time of filters that trap them. The objective of the experiments reported here was to determine whether the nanostructure of soot particles produced from burning oxygenated fuels was the same or different from that of particles produced by regular hydrocarbon fuels. A thermophoretic sampling device has been constructed that allows TEM grids to be inserted into a flame for a controlled duration. High-speed imaging verified that the exposure time is accurate to within 4 msec for typical exposure times of 100 msec, and that the transit time in and out of the flame is negligible compared to the exposure time. The imaging also shows that the probing process does not perturb the upstream portion of the flame. The specific systems under study are fuel-doped nonpremixed coflow methane/air flames. The dopants include regular hydrocarbons found in regular gasoline, such as heptane, isooctane, toluene, and cyclohexane, and oxygenated hydrocarbons that have been proposed as gasoline blendstocks, such as isobutanol, anisole, cyclopentanone, and 2-methylfuran.
soot, IC engines, alternative fuels, biofuels

1. Introduction

The long-term objective of this project is to determine the effects of fuel composition on soot nanostructure by doping methane flames with various fuel compounds, sampling particles from the flames by thermophoretic deposition, then analyzing their structure with high resolution transmission electron microscopy (HRTEM). Prior research has shown that variation in fuel structure can alter the reactivity of soot particles, and therefore their oxidation rates in flames and in particulate filters [1-5]. Figure 1 shows an example of an HRTEM image of a soot particle sampled from our undoped methane flame.

The short-term objective addressed in this paper is to validate the thermophoretic sampling (TS) procedure. TS is widely used for studying soot particles, but it is an intrusive approach that requires inserting a relatively large solid object into the flame. There are very few studies that have compared results of TS measurements and in situ diagnostics. In earlier work we demonstrated good agreement between optical measurements and soot volume fractions determined from TEM of particles sampled on the centerline of an ethylene/air nonpremixed flame[6]. However, this comparison was limited to only a few locations due to the tedious nature of the TEM analysis. In this work we have used visible light photography to determine the TS mass deposition rates with much greater spatial resolution, and have compared these to deposition rates calculated from in situ data measured with color-ratio pyrometry (CRP).

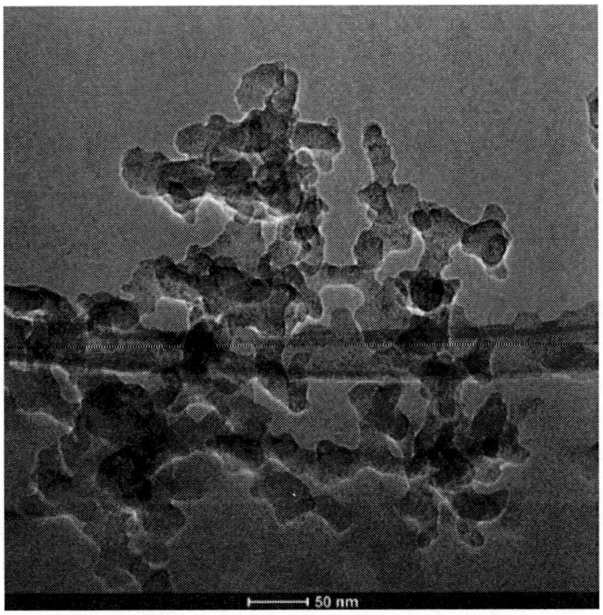

Figure 1: TEM image of a soot particle sampled from a nonpremixed coflow methane/air flame.

2. Methods / Experimental

Burner

A Yale Coflow burner was used here to establish atmospheric pressure, axisymmetric, nonpremixed, laminar methane/air coflow flames. The fuel mixture was methane and nitrogen, which flows out of a 4 mm inner diameter tube and reacts with air flowing from an annular region between the tube and a 7.4 cm inner diameter concentric aluminum tube. The normal flow rates of the different gases were 282 cm3/min, 112 cm3/min and 50,000 cm3/min for CH4, N2 and air, respectively, resulting in a visible flame of around 53 mm in height.

Thermophoretic Sampling (TS)

In order to collect soot particles from the flames, a pneumatic sampling apparatus was assembled and used. A directional control solenoid pilot valve (Model V61R517A-A213JB made by IMI NORGREN) controlled by a home-made C++ program and an electronic circuit. A stainless steel cylinder with a piston and a rod (Model SR-043-DPY-00MC) was connected with the solenoid pilot valve. A spatula with a white paper tape (around 45 mm × 4 mm) on it was attached to the rod through a hexagonal prism made with aluminum. The spatula is thin enough (0.055 mm) that the perturbation to the flame was minimized. Particles from different height above burner (HAB) was sampled on the white tape.

Through the C++ program on a computer, exposure time, the number of exposures and pause time before next exposure were tunable. A set of suitable values of these parameters were chosen so that the deposition of soot particles was enough for analysis and the paper tape would not carbonize because of high temperature.

After soot particles were deposited on the white tape, the hexagonal prism along with the tape on the spatula was detached from the device and immobilized to a certain position under the camera. Considering the fact that grey scale analysis is sensitive to light and sun light is always changing, a lamp was used in a fixed position as a constant light source during photo capture. For the sake of transforming pixels into real millimeter during grey scale analysis, a printed paper ruler was attached on the same height as the white tape.

Camera

An iPhone 7 plus was used here to take videos and photos. Since the tape is inserted into and out of the flame, length of transit time during insertion and retraction is a concern because deposition in anywhere other than the desired position would introduce error. In order to confirm the transit and sampling time, the slow mode (SLO-MO) record with a 720p video quality at 240 frame per second (fps) was used. According to the video, sampling took 64 frames (267 ms) while both insertion and retraction took 4 frames (17 ms), which is only around 6% of the exposure time. With such a short transit time, collected soot particles/aggregates during insertion and retraction was negligible. What's more, the video also showed that the upstream of the flame was nearly unperturbed.

As for photos, the best resolution that iPhone 7 plus provides under regular "PHOTO" mode is 4032×3024. Photos were taken after the tape along with detached hexagonal prism was fixed under the camera. All the photos were transferred to a computer for further analysis.

Grey Scale Analysis

Grey scale analysis was carried out by ImageJ (programed and developed by NIH). By definition, 255 represents pure white and 0 represents pure black. All photos are added into a stack and transformed into 32-bit grey scale images. Because of carbonization on the bottom of the tape, a 20-mm-long line which was 2 mm above the tape bottom was chosen for analysis. The grey scale distribution along the horizontal position was obtained. In order to get the data that represents soot deposition, the average background grey scale was subtracted from the deposition tape data.

Normalized Grey Scale Contour Plot

In order to eliminate the data shift derived from spatula vibration, burner centerlines for each HAB data need to be calculated and aligned. Since the flame is axisymmetric, the approximate centerline was obtained by averaging the position of 20% largest data points. After subtracting the centerline from the previous horizontal positions, all the positions were now radial position (r) relative to the burner centerline. All data points were normalized by the largest data and those from r = -6 to 6 were plotted into the normalized grey scale contour plot.

Color Ratio Pyrometry (CRP)

Two-dimensional soot volume fraction (f_v) and temperature maps were measured using an in-situ unperturbed method, color-ratio pyrometry, which has been used in lots of reports [7-10]. Details about CRP principles and setup are in our previous work [11, 12].

Normalized Mass Flux Contour plot

Based on previous reports [13], the thermophoretic mass flux of soot particles to the sampling spatula, j_w'', can be estimated as follows:

$$j_w'' \cong D_T \frac{Nu_x}{2x}\left[1-\left(\frac{T_w}{T_g}\right)^2\right]\rho_p f_v$$

where D_T represents thermophoretic diffusivity, and Nu_x stands for the local Nusselt number for heat transfer at x, the vertical position on the sampling spatula measured from the lower edge, and is given as $Nu_x = 0.332\,Pr^{\frac{1}{3}}\,Re_x^{\frac{1}{2}}$, where Pr ≈ 0.7 [14] and $Re_x = \frac{u_g x}{\nu_f}$ (see, e.g., Rosner [15]).

T_w and T_g are the spatula surface and gas temperatures, respectively. ρ_p is the intrinsic soot density, and f_v is the local soot volume fraction. ρ_p was taken as 2.2 g/cm^3, the density of graphite from CRC handbook. The value of x used here was fixed at 0.2 cm, and T_w was taken to be room temperature (298 K). Both f_v and T_g were obtained from the spatial temperature and soot volume fraction maps determined from CRP. The value of D_T was taken to be the Fickean diffusion coefficient of N_2 in itself at 298.15K [16]. All other parameters were derived from DIPPR database.

In order to compare two contour plots, the thermophoretic data was also normalized by the largest data and plotted in the same range for HAB and r.

3. Results and Discussion

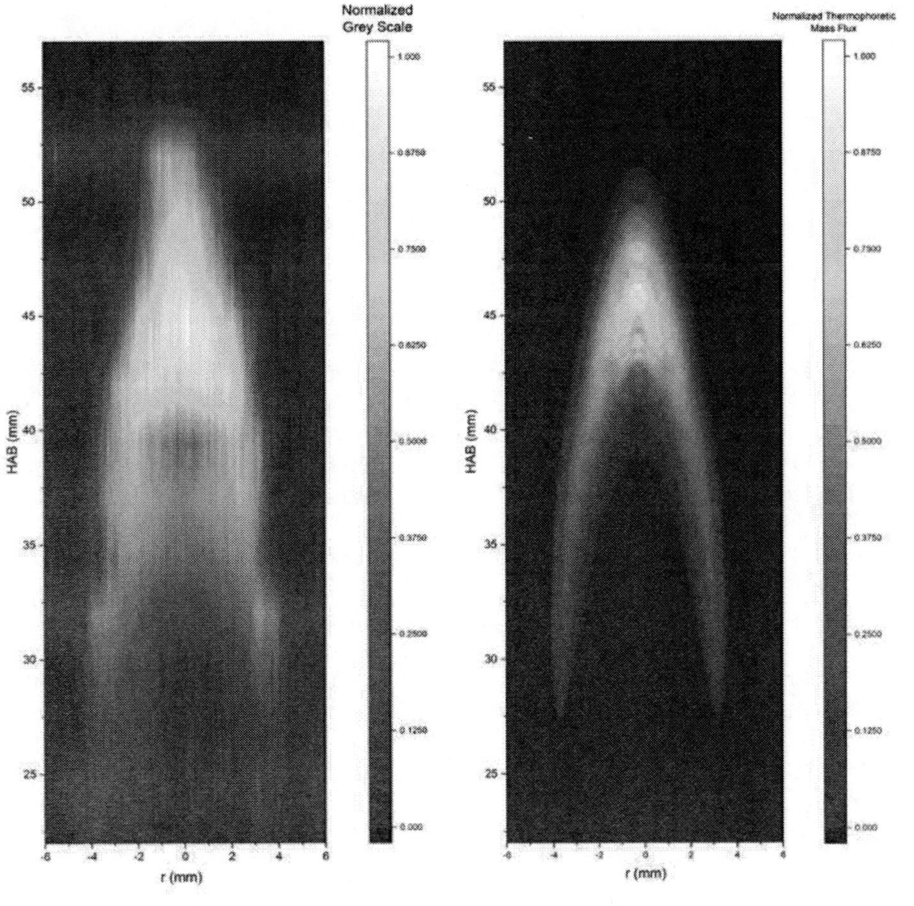

Figure 2: Normalized contour plots of grey scale (left) from thermophoretic sampling and thermophoretic mass flux (right) from CRP.

As showed in Figure 2, left is normalized grey scale plot derived from thermophoretic sampling, and right is the normalized thermophoretic mass flux derived from the in-situ technique, CRP. Two contour plots have similar distribution. Moreover, the maximum signals for both measurements are around HAB = 45 mm. As for the lowest height that have soot detected, the value from thermophoretic sampling is around 22.5 mm, as one can tell from the plot and the actual tape. On the other hand, the lowest HAB where CRP could detect soot is around 27.5 mm. This might be because some of the nascent soot particles are "liquid-like" [17, 18] and would not be behaving as true blackbodies. Since incandescence from soot is required for the CRP analysis, these liquid-like particles may be unable to be detected by the diagnostic. But on the whole, the data indicates that the thermophoretic sampling is reliable.

4. Conclusions

Two normalized contour plots were made from thermophoretic sampling and CRP. Data shows that the distribution was very similar. Both measurements show the largest signal around 45mm. However, the lowest value for thermophoretic sampling is around 22.5mm while the lowest HAB where CRP could detect sign is around 27.5 mm. Further investigation is needed to explain this difference.

5. Acknowledgements

This material is based upon work supported by the U.S. Department of Energy's Office of Energy Efficiency and Renewable Energy (EERE) under the Bioenergy Technologies Office (BETO) and Vehicle Technologies Office (VTO) Program Award Number DE-EE0007983. This work is also supported by the National Science Foundation (NSF) under Grant No. CBET 1604983.

6. References

[1] R.L. Vander Wal, A.J. Tomasek, Soot nanostructure: dependence upon synthesis conditions, Combust Flame 136 (2004) 129-140.

[2] C.-H. Huang, R.L. Vander Wal, Effect of Soot Structure Evolution from Commercial Jet Engine Burning Petroleum Based JP-8 and Synthetic HRJ and FT Fuels, Energ Fuel 27 (2013) 4946-4958.

[3] A. Strzelec, T.J. Toops, C.S. Daw, Oxygen reactivity of devolatilized diesel engine particulates from conventional and biodiesel fuels, Energ Fuel 27 (2013) 3944-3951.

[4] H. Ghiassi, P. Toth, J.S. Lighty, Sooting behaviors of n-butanol and n-dodecane blends, Combust Flame 161 (2014) 671-679.

[5] I.C. Jaramillo, C.K. Gaddam, R.L. Vander Wal, J.S. Lighty, Effect of nanostructure, oxidative pressure and extent of oxidation on model carbon reactivity, Combust Flame 162 (2015) 1848-1856.

[6] Ü.Ö. Köylü, C.S. McEnally, D.E. Rosner, L.D. Pfefferle, Simultaneous measurements of soot volume fraction and particle size/microstructure in flames using a thermophoretic sampling technique, Combust Flame 110 (1997) 494-507.

[7] P.B. Kuhn, B. Ma, B.C. Connelly, M.D. Smooke, M.B. Long, Soot and thin-filament pyrometry using a color digital camera, Proceedings of the Combustion Institute 33 (2011) 743-750.

[8] B. Ma, M.B. Long, Absolute light calibration using S-type thermocouples, Proceedings of the Combustion Institute 34 (2013) 3531-3539.

[9] H. Guo, J.A. Castillo, P.B. Sunderland, Digital camera measurements of soot temperature and soot volume fraction in axisymmetric flames, Appl Optics 52 (2013) 8040-8047.

[10] D.R. Snelling, K.A. Thomson, G.J. Smallwood, Oslash, L. G-uacute, lder, E.J. Weckman, R.A. Fraser, Spectrally Resolved Measurement of Flame Radiation to Determine Soot Temperature and Concentration, Aiaa J 40 (2002) 1789-1795.
[11] D.D. Das, W.J. Cannella, C.S. McEnally, C.J. Mueller, L.D. Pfefferle, Two-dimensional soot volume fraction measurements in flames doped with large hydrocarbons, Proceedings of the Combustion Institute 36 (2017) 871-879.
[12] D.D. Das, C.S. McEnally, T.A. Kwan, J.B. Zimmerman, W.J. Cannella, C.J. Mueller, L.D. Pfefferle, Sooting tendencies of diesel fuels, jet fuels, and their surrogates in diffusion flames, Fuel 197 (2017) 445-458.
[13] A.D. Eisner, D.E. Rosner, Experimental studies of soot particle thermophoresis in nonisothermal combustion gases using thermocouple response techniques, Combust Flame 61 (1985) 153-166.
[14] J.P. Holman, Heat transfer, McGraw-hill New York1986.
[15] D.E. Rosner, Transport processes in chemically reacting flow systems, Courier Corporation2012.
[16] H. Vugts, A. Boerboom, J. Los, Diffusion coefficients of isotopic mixtures of CO and N2, Physica 50 (1970) 593-605.
[17] A.D. Abid, N. Heinz, E.D. Tolmachoff, D.J. Phares, C.S. Campbell, H. Wang, On evolution of particle size distribution functions of incipient soot in premixed ethylene–oxygen–argon flames, Combust Flame 154 (2008) 775-788.
[18] A.D. Abid, E.D. Tolmachoff, D.J. Phares, H. Wang, Y. Liu, A. Laskin, Size distribution and morphology of nascent soot in premixed ethylene flames with and without benzene doping, Proceedings of the Combustion Institute 32 (2009) 681-688.

Sub Topic: Soot/Laminar Flames

Spring Technical Meeting
Eastern States Section of the Combustion Institute
March 4-7, 2018
State College, Pennsylvania

Effects of hydrogen addition on the structure of *n*-dodecane laminar, co-flow flame

A. Makwana[1*], *M. Linevsky*[1], *S. Iyer*[1], *R. Santoro*[1], *T. Litzinger*[1], *J. O'Connor*[1]

[1]*Mechanical and Nuclear Engineering, Pennsylvania State University, University Park, PA, USA*
[*]*Corresponding Author Email: axm571@psu.edu*

Abstract: The goal of this work is to understand the effects of hydrogen addition on *n*-dodecane flame structure in an axisymmetric, co-flow, laminar flame at atmospheric pressure. Hydrogen is added to the *n*-dodecane flame to increase the fuel H/C ratio from 2.16 to 4.0. The flames investigated are non-premixed and the total carbon flow rate is kept constant to facilitate comparison amongst the flames. Additionally, to isolate the chemical and dilution effects of hydrogen addition on flame structure and soot volume fraction, flames diluted with helium are also studied. Laser-induced incandescence and laser extinction are applied to obtain two-dimensional soot volume fraction. The experimental results indicate that visible flame length increases by 25% with the addition of hydrogen. However, the increase in visible flame length is marginal when the same volume fraction of helium is added. Additionally, the peak soot volume fraction of the *n*-dodecane flame doesn't change significantly with addition of either hydrogen or helium. In hydrogen diluted flames, the location where LII is first detected shifts to higher heights above the burner as compared to helium diluted flames.
Keywords: Hydrogen, n-dodecane, soot volume fraction, temperature

1. Introduction

The increase in fuel demand, instability in cost of the conventional fuels, and stricter environmental rules has led to an increased interest in use of alternative fuels. Hydrogen is one alternative fuel that is being considered because it is a clean fuel. A number of studies in the past have experimentally and numerically investigated effects of hydrogen addition on soot formation in methane [1, 2] or ethylene [3, 4] co-flow flames. Gülder *et al.* [3] investigated the effect of hydrogen and helium addition on soot formation in ethylene, propane, and butane co-flow diffusion flames. In the ethylene flame, the authors found that hydrogen was more effective in reducing soot formation as compared to helium. The authors concluded that hydrogen, in addition to dilution and thermal effects, chemically inhibits soot formation in ethylene flames. In propane and butane flames, hydrogen and helium reduced the soot formation by similar levels. The authors implied that in propane and butane fuels, hydrogen does not have any chemical effect on soot formation processes. In the methane flame [1], hydrogen was found to be less effective than helium in inhibiting soot formation. Therefore, it was postulated that in methane flames, hydrogen addition enhances soot formation due to chemical effects. However, the overall soot formation decreased as compared to the base methane flame due to the dilution effects. These studies show that the effect of hydrogen addition on soot formation processes is dependent on the fuel type.

Most of the work until now has focused on understanding the effects of hydrogen addition on soot formation in gaseous fuels. The aim of the current work is to aid in understanding the effects of hydrogen addition on the flame structure and soot formation in a co-flow, laminar, *n*-dodecane diffusion flame. In order to isolate the relative impact of chemical and dilution effects, flames with helium added to *n*-dodecane are also studied.

2. Methods / Experimental

2.1 Burner and Fuel matrix

The fuel matrix is tested in a co-flow, laminar flame configuration. The burner used in the current study is a modified version of the Santoro burner. This modified burner can be operated to obtain stable flames using pre-vaporized liquid fuels. The detailed design of this burner is discussed in Ref. [5]. Table 1 shows the experimental fuel matrix used in the current work. A non-premixed, *n*-dodecane flame is the base flame. A nitrogen flow rate of 100 sccm is used to carry the vaporized *n*-dodecane fuel from the vaporization chamber to the tube exit. To investigate the effects of hydrogen addition on the flame structure, the hydrogen fuel flow rate is systematically changed to increase the fuel H/C ratio from 2.16 to 4.0. The carbon flowrate of *n*-dodecane is kept constant at 0.144 slpm for all the flames. Additionally, to isolate the chemical and dilution effects of hydrogen addition on the flame structure and soot volume fraction, flames diluted with helium are also studied. At a particular level of hydrogen/helium addition, the helium flow rate is fixed such that the inlet mole fraction of *n*-dodecane:helium is identical to the *n*-dodecane:hydrogen mole fraction.

Table 1: Experimental fuel matrix.

Flame	Primary fuel	Liquid Fuel flow rate (ml/min)	Second component	Fuel mole fraction (second component)	H/C ratio (for H_2 added flame)
C_{12}	*n*-dodecane (C_{12})	0.144	N/A	0%	2.16
66%-He 66%-H2		0.144	H_2/He	66.9%	2.50
83%-He 83%-H2		0.144	H_2/He	83.2%	3.00
88%-He 88%-H2		0.144	H_2/He	88.9%	3.50
91%-He 91%-H2		0.144	H_2/He	91.5%	4.00

2.2 Experimental methods

Laser-induced incandescence (LII) and laser extinction (LE) are applied to obtain two-dimensional (2-D) soot volume fraction. Furthermore, temperature is obtained in the non-sooting region of the flame using custom-built R-type thermocouples, similar to the one used in Ref. [5]. The setup, data collection, and experimental uncertainty of LII, LE and temperature measurement is identical to that described in Ref. [5]. An Nd:YAG laser (Spectra Physics GCR 270-10, 10Hz, 7ns, 532 nm) is used to irradiate the soot particles. The laser beam is expanded into a laser sheet using a plano-convex cylindrical lens and focused on the center of the burner using a convex lens. The LII signal from soot is collected using a Princeton Instruments ICCD camera, using a 100 ns gate width after a gate delay of 20 ns from laser pulse. A filter centered at 430 nm with 10 nm FWHM is used to collect the LII signal [5]. The calibration process of LII using LE is described in Wang *et al.* [5]. A value of $\widetilde{m} = 1.57 - 0.56i$ is used as refractive index of soot ($\widetilde{m}$) [5] and the line-of-sight extinction profiles are deconvoluted using Fourier inversion technique [6]. For the current optical setup, 3350 LII counts is equivalent to 1 ppm soot. The spatial resolution of LII setup is 1 mm/camera pixel. The measurement uncertainty in soot volume fraction is estimated to be ±15%. The measurement uncertainty for temperature measurement and spatial location of thermocouple bead are ±100 K and ±5 mm, respectively.

3. Results and Discussion

3.1 Soot volume fraction

Figure 1 presents the comparison of the 2-D soot volume fraction profiles for *n*-dodecane and hydrogen added to *n*-dodecane flames. Figures 1 (a)-(e) show soot volume fraction distribution for flames with H/C ratio of 2.16, 2.5, 3.0, 3.5, and 4.0, respectively. An isocontour of 1 ppm of soot volume fraction is plotted on all the figures. Comparison of Figs. 1 (a) and (e) show that the flame height, as defined by the location where soot volume fraction reaches 1 ppm, increases by 25% due to hydrogen addition to the *n*-dodecane flame. The increase in flame height due to hydrogen addition has been reported in ethylene flames [4].

Figure 1 shows that the soot volume fraction distribution is uniform across the cross-section for all the five flames. However, due to hydrogen addition to the *n*-dodecane flame, the first location where LII signal is first detected shifts to higher height above the burner (HAB) in both the annular region and on the centerline. Quantitatively, within measurement uncertainty of ±15%, the peak soot volume fraction is similar, 4 ppm, in all the five flames.

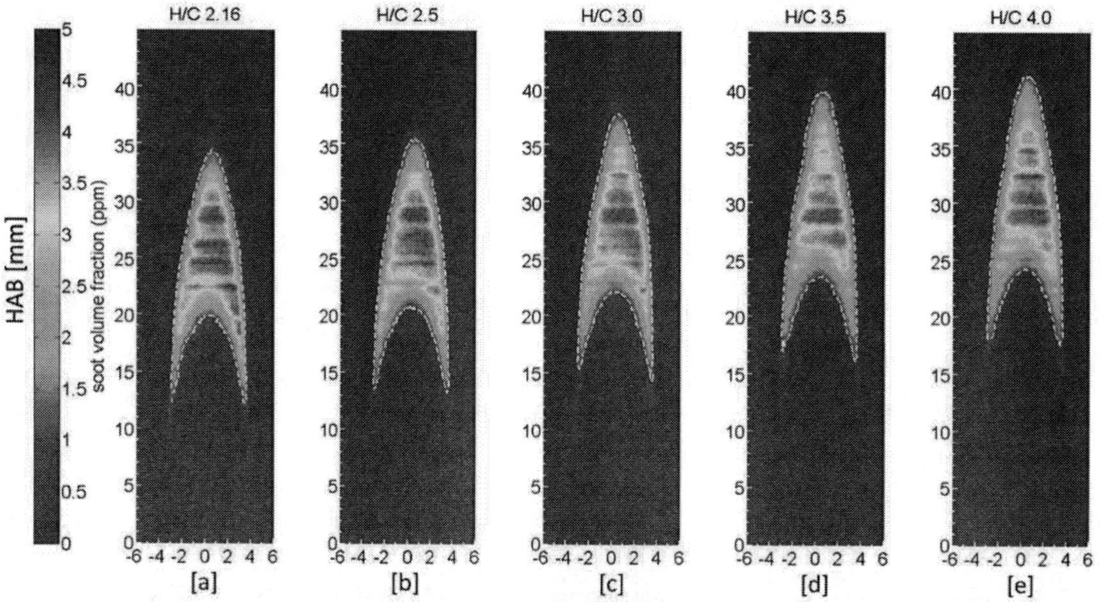

Figure 1: Comparison of 2-D soot volume fraction [ppm] obtained for: (a) *n*-dodecane flame, H/C: 2.16; (b) *n*-dodecane + H_2 flame, H/C: 2.5; (c) *n*-dodecane + H_2 flame, H/C: 3.0; (d) *n*-dodecane + H_2 flame, H/C: 3.5; and (e) *n*-dodecane + H_2 flame, H/C: 4.00. Note: The dotted line is an *iso*-contour of 1 ppm.

Figure 2 (a)-(e) shows the comparison of the 2-D soot volume fraction profiles for *n*-dodecane and helium-added *n*-dodecane flames. An iso-contour of 1 ppm is plotted on the figures to show the spatial extent of the LII signal. In contrast to the hydrogen-added flames discussed earlier, Figure 2 shows that adding helium increases the flame height marginally by 1.5 mm. Similar to the hydrogen-added *n*-dodecane flames, the soot volume fraction distribution is uniform across the cross-section in helium added *n*-dodecane flames. Additionally, Fig. 2 shows that the peak soot volume fraction is similar, within the measurement uncertainty, in the five flames. Furthermore, due to the addition of helium to *n*-dodecane, the location where LII signal is first detected shifts to higher HAB.

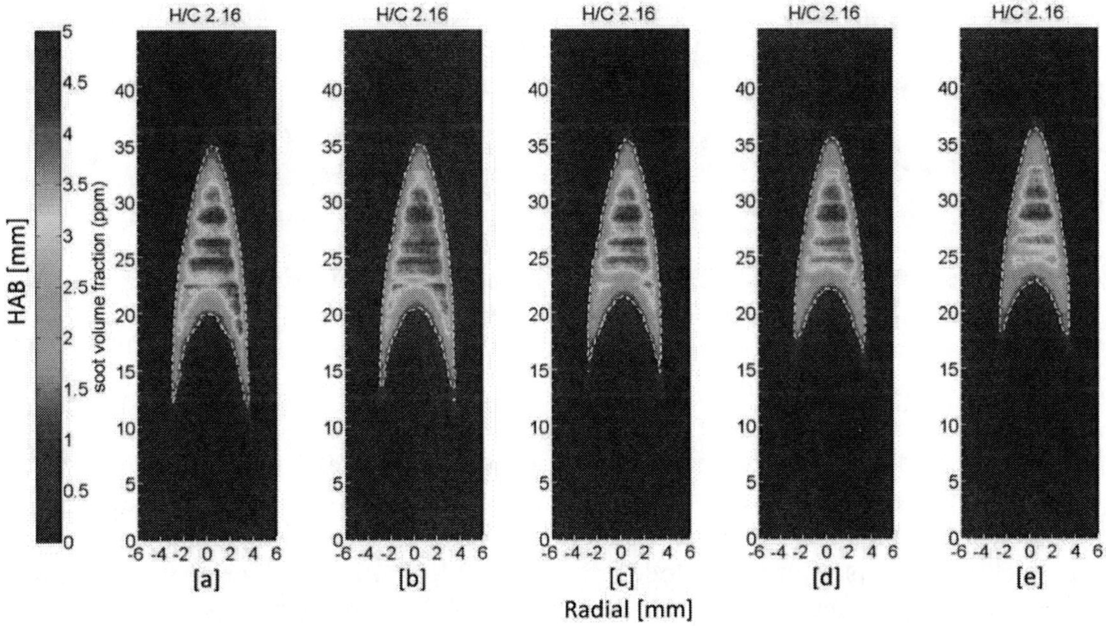

Figure 2: Comparison of 2-D soot volume fraction [ppm] obtained for: (a) *n*-dodecane flame, H/C: 2.16; (b) *n*-dodecane + He flame, H/C: 2.16, 66%-He; (c) *n*-dodecane + He flame, H/C: 2.16, 83%-He; (d) *n*-dodecane + He flame, H/C: 2.16, 88%-He; and (e) *n*-dodecane + He flame, H/C: 2.16, 91%-He. Note: The dotted line is an *iso*-contour of 1 ppm.

Figure 3 (a) and (b) shows the centerline soot volume fraction profiles for the hydrogen and helium added flames, respectively. In hydrogen-added flames, Fig. 3(a) shows that both soot formation and soot oxidation processes are delayed. However, in helium-added flames, Fig. 3 (b) shows that only soot formation processes are shifted at higher HAB. Moreover, due to helium addition, the height by which LII signal is shifted to higher HABs is smaller as compared to that observed in the hydrogen added flames.

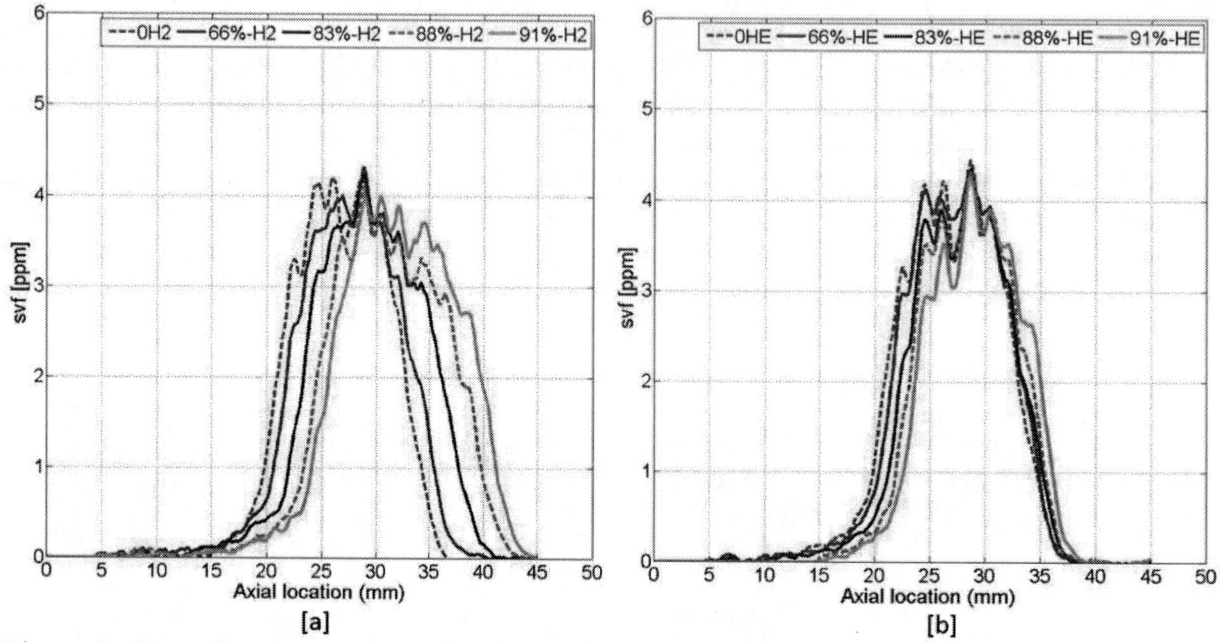

Figure 3: Centerline soot volume fraction as a function of axial location: (a) Hydrogen-added flames; (b) Helium-added flames. Note: The profile is averaged 1mm.

3.2 Flame temperature

Figure 4 (a) shows the radial temperature profiles measured at 5 mm HAB for five flames: *n*-dodecane base flame, and two flames with either H_2/He added to the *n*-dodecane. The qualitative trends in temperature measurement are similar to those reported in previous work with co-flow flames [4]. The comparison of the *n*-dodecane flame, Fig. 4 (a), the flames with hydrogen show that the peak temperature increases with the increase in hydrogen addition to the fuel. Furthermore, the location where the peak temperature occurs shift away from the flame centerline. In contrast, in the flames with helium, the location of peak temperature does not change as compared to the base *n*-dodecane flame. Additionally, the peak temperature in the helium added flame decreases. The temperature in H_2 added flame increases since hydrogen acts as a fuel while temperature in He added flame decreases due to the dilution effects of helium.

Figure 4 (b) shows the centerline temperature until 14 mm HAB in the *n*-dodecane flame and two levels of H_2/He added to the *n*-dodecane flame. The temperature in the flames with H_2 and He decreases as compared to the *n*-dodecane flame as a result of dilution. At the highest level of H_2 or He addition, the flame with H_2 has a higher temperature as compared to the flames with helium until 9 mm HAB; at further downstream locations, the temperature are similar in the flames with hydrogen or helium. The higher temperature on the flame centerline in the flames with H_2 as compared to the flames with He is possibly due to higher peak temperature in the flames with H_2, as discussed in Fig. 4 (a), which could result in higher radial heat transfer between the flame front and the centerline.

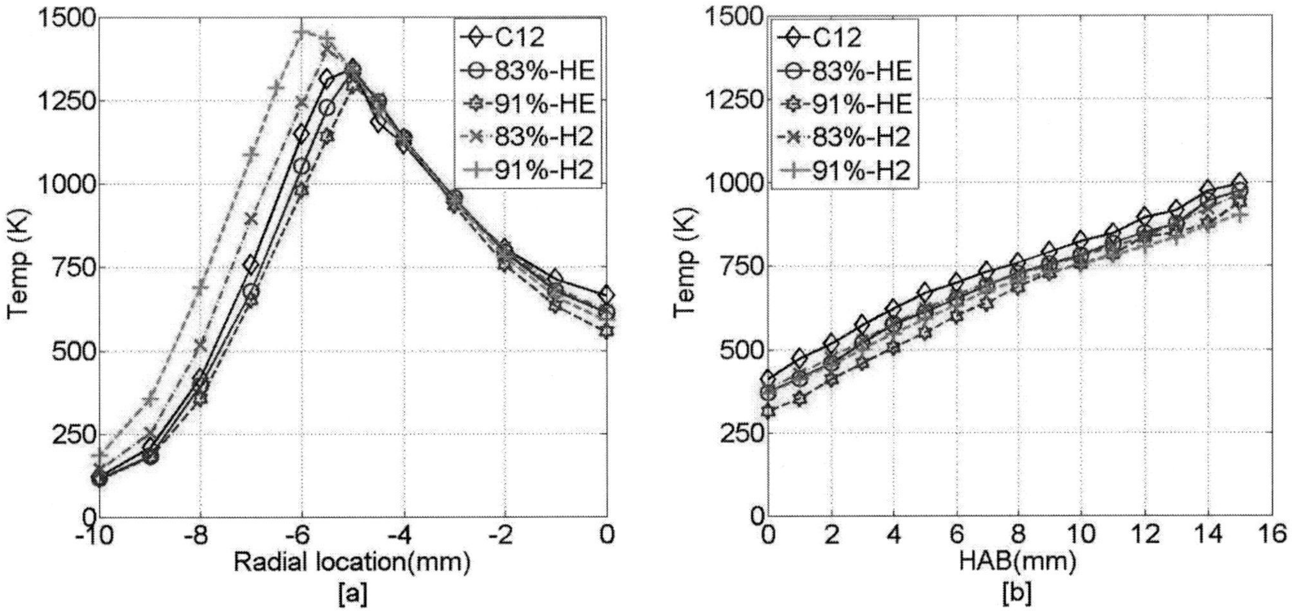

Figure 4: Comparison of temperature measurements: (a) Radial profiles at 5mm HAB (b) Axial profile until 14 mm HAB.

4. Conclusions

The effects of addition of hydrogen on the structure of *n*-dodecane flame is investigated in a laminar, co-flow flame configuration. The experimental results show that due to the addition of hydrogen, soot field shifts downstream and the flame height increases by 25%. Furthermore, the soot volume fraction distribution and peak soot volume fraction are similar in the *n*-dodecane flame, and flames with hydrogen added to *n*-dodecane. To isolate the dilution and chemical effects of hydrogen, flames with helium added to *n*-dodecane flame are also investigated. The results show that the flame height increases only marginally due to the addition of helium. Additionally, the soot volume fraction is uniform across the cross-section in flames with hydrogen or helium. At the highest level of hydrogen or helium addition, the temperature measurements on the flame centerline show that flames with hydrogen have higher temperature than flames with helium until 9 mm HAB. To understand the differences in the temperature and the soot field of helium and hydrogen added flames, efforts are underway to simulate these flames and to obtain aromatics-LIF. The LIF results and simulation will help to identify the differences in the spatial distribution of key species and soot precursors.

5. Acknowledgements

This work was supported by the Strategic Environmental Research and Development Program (WP-2145) under the direction of Dr. Robin Nissan and Mr. Bruce Sartwell.

6. References

[1] F. Migliorini, Study of combustion process of hydrogen-hydrocarbon mixtures, Ph. D thesis, Politecnico di Milano, Italy, 2008.
[2] F. Liu, Y. Ai, W. Kong, Effect of hydrogen and helium addition to fuel on soot formation in an axisymmetric coflow laminar methane/air diffusion flame, international journal of hydrogen energy 39 (2014) 3936-3946.
[3] Ö. Gülder, D. Snelling, R. Sawchuk. Influence of hydrogen addition to fuel on temperature field and soot formation in diffusion flames. In: editor^editors. Symposium (International) on Combustion; 1996: Elsevier. p. 2351-2358.
[4] Z. Sun, B. Dally, G. Nathan, Z. Alwahabi, Effects of hydrogen and nitrogen on soot volume fraction, primary particle diameter and temperature in laminar ethylene/air diffusion flames, Combustion and Flame 175 (2017) 270-282.
[5] Y. Wang, A. Makwana, S. Iyer, M. Linevsky, R.J. Santoro, T.A. Litzinger, J. O'Connor, Effect of fuel composition on soot and aromatic species distributions in laminar, co-flow flames. Part 1. Non-premixed fuel, Combustion and Flame, (2017).
[6] C.J. Dasch, One-dimensional tomography: a comparison of Abel, onion-peeling, and filtered backprojection methods, Applied Optics 31 (1992) 1146-1152.

Sub Topic: IC and GT Engines

Spring Technical Meeting
Eastern States Section of the Combustion Institute
March 4-7, 2018
State College, Pennsylvania

Soot Characteristics of Light Naphtha in a Constant Volume Combustion Chamber Using Two-Color Pyrometry

Zengyang Wu[1], Libing Wang[1], Jihad A. Badra[2], William L. Roberts[3], Tiegang Fang[1,]*

[1]Department of Mechanical and Aerospace Engineering
North Carolina State University
Raleigh, NC 27695
[2]Fuel Technology Division, R&DC
Saudi Aramco
Dhahran, Saudi Arabia
[3]Clean Combustion Research Center
King Abdullah University of Science and Technology
Thuwal, Saudi Arabia
[*]Corresponding Author Email: tfang2@ncsu.edu

Abstract: This study investigates the soot processes of light naphtha (LN) under diesel engine conditions with an outwardly opening piezo gasoline direct injection (GDI) fuel injector. Five ambient oxygen concentrations varying from 10% to 21% were selected to simulate both conventional diesel engine operation environment and low oxygen concentration combustion environment. A two-color pyrometry was applied to measure flame temperature and soot concentration (i.e., KL factor) distribution under different ambient conditions. KL factor is a measure of soot particulate concentration in which K is a measure of soot number density, and L is a measure of the path length along which the measurement is made. It is found that flame temperature ranges from 1500K to 2300K under selected conditions. A higher oxygen concentration could accelerate fuel combustion rate. High temperature areas are mainly located at the center or downstream of the flame. Flame upstream always has lower temperature than other regions. High KL factors are observed at flame periphery where flame temperature is lower than 1800K and has rapid temperature variation. Accumulated KL factor was calculated to evaluate overall soot amount. The lowest accumulated KL factors are observed at 1000 K ambient temperature condition when the ambient oxygen concentration is 21%.

Keywords: two-color pyrometry, auto-ignition, soot, naphtha

1. Introduction

Currently, gasoline (spark ignited engines) and diesel (compression ignited engines) are two major road transport fuels. The primary difference between these is the auto-ignition quality [1] where diesel is more reactive and easier to ignite than gasoline. Currently, more stringent emissions standards and the demand for higher fuel economy have driven the auto makers to search for alternative engine and fuel solutions.

Gasoline compression ignition (GCI) engine technology is one of the promising alternative solutions. Compared to diesel fuels, gasoline has longer ignition delay time which allows it to better mix with the in-cylinder air resulting in overall leaner charge. The combustion of this leaner gasoline/air mixture usually results in lower soot emissions [2-3]. In addition, the leaner mixture yields lower combustion temperatures (LTC) and hence lower NOx emissions [4]. Furthermore, the compression and lean combustion mode eliminates throttle losses yielding higher thermodynamic efficiencies compared to gasoline spark ignited engines [5, 8]. Overall, GCI engine technology seems to be a promising alternative that meets the future emissions and fuel consumption regulations without the need to make the after-treatment system more complicated and costly.

In general, the auto-ignition quality of a fuel at a given engine condition is described by an octane index that defined as OI = (1-K) RON + K MON [6], in which RON and MON are the Research and Motor Octane numbers, respectively and K is dictated by the engine operating conditions. Approaches like direct injection, higher compression ratios, downsizing and turbocharging reduce the unburnt gas temperatures for a given pressure and push the value of K downwards [7]. Meanwhile, the occurrence and development of auto-ignition are strongly affected by the residual gas and thermal distributions. It is shown that a heterogeneous combustion has large spatial and temporal variations even with a homogeneously premixed charge [9]. The thermal inhomogeneous distribution correlates well with that of the residual gas, and both a higher residual gas fraction level and a higher temperature inhomogeneity advances the auto-ignition occurrence [10 - 11].

While gasolines with higher octane numbers can enable more efficient future spark ignition (SI) engines, low octane gasoline-like fuels might be desirable in compression ignition (CI) engines. Compared to commercial gasoline and diesel fuels, blends of various refinery streams with low octane numbers (RON) in the 50–80 range have recently been considered attractive alternatives to provide suitable chemical characteristics (longer ignition delay than diesel) in GCI engines at lower production cost and well-to-tank CO2 emissions. In addition, the utilization of the valuable high-RON streams and octane boosters is not part of the refining process of producing low-octane fuels. Hao et al. [12] found that compared with the conventional pathway, the low-octane gasoline-GCI pathway leads to a 24.6% reduction in energy consumption and a 22.8% reduction in GHG emissions. The research group in Saudi Aramco extensively investigated the combustion of various fuels in GCI engines [13-15].

In present study, auto-ignition of light naphtha (LN) was investigated under different ambient conditions. Two-color pyrometry was employed to measure flame temperature and KL factor distributions. Average flame temperature, KL and accumulated KL factors were calculated to evaluate the effect of ambient conditions on flame characteristics.

2. Experimental setup

Experiments were conducted in a constant volume combustion chamber with six ports under simulated engine combustion environments. The optical access was enabled by a quartz window installed on one port. Other ports were sealed by solid stainless steel plugs with gas inlet, outlet, fuel injector, spark plug, pressure sensor, and other necessary components. The inner diameter of the quartz window is 100 mm, and the total chamber inner volume is 0.95 L. More details of the chamber can be found in the previous publications [16-17]. In order to generate an auto-ignition environment, a premixed combustion of acetylene/air was used prior to the fuel injection. Three different streams of acetylene, O_2/N_2 (50%/50%) and dry air were used to formulate appropriate mixtures to obtain the desired ambient temperatures and oxygen concentrations. Premixed reactants were ignited using a spark plug. For all the experiments, gas density in chamber was set to 15 kg/m^3. Chamber pressure was around 44 bar for 1000 K when fuel injection was triggered. Five ambient oxygen concentrations ranging from 10% to 21% were selected to simulate both the conventional diesel engine and low-oxygen concentration environments. The global air fuel ratios at different oxygen concentrations were 143, 171, 214, 257 and 300, respectively. The ambient temperature was 1000 K for all experiments. An outwardly opening piezoelectric fuel injector was utilized in this study. Fuel injection pressure difference was maintained at 10 MPa (100 bar) by a gasoline common rail fuel system through all experiments. Injection duration was kept as 1.4 ms by an external pulse generator with 50 mg fuel mass per injection. Temperature of chamber walls and fuel injector body were heated up to 100 °C. A Kistler 6041B transducer coupled with a Kistler 5004 charge amplifier was utilized to measure the transient pressure during premixed combustion, and the ambient temperature was calculated based on the pressure data using a corrected ideal gas law. Two-color pyromery was employed in this study to calculate flame temperature and KL factor distribution. The implementation of two-color pyrometry was described in detail in ref [17]. A high-speed digital video camera (Phantom V4.3) was used to capture the flame luminosity at the two bands.

3. Results and Discussion

Five oxygen concentrations (10% to 21%) were selected to simulate conventional diesel engine to low oxygen concentration environments at 1000 K. Transient flame temperature and KL factor distribution were calculated and analyzed based on two-color measurement. Average flame temperature and average KL factor at selected time steps were calculated to explore temperature and KL factor development during fuel combustion.

3.1 Soot and temperature distribution

Conventional diesel-like fuel compression ignition is considered at 1000 K ambient temperature, 21% ambient oxygen concentration and 15 kg/m^3 gas density. LN flame temperature and KL factor distribution results at different timings are presented in Figure 1 (a) and (b). Under this condition, flame peak temperature reaches more than 2300K, and flame temperature of most parts are higher than 2000K before 8.651 ms after start of injection trigger (ASOIT). Meanwhile, the inner side of the flame has higher temperature than flame periphery, and the lowest flame temperature is found at flame upstream. After 8.651 ms ASOIT, flame temperature decreases until the end of fuel combustion. Flame begins to shrink at 8.651 ms ASOIT, and extincts at 12.741 ms ASOIT. Comparing with other oxygen concentration conditions, flame at 21% oxygen concentration condition has a more confined hollow-cone shape of flame at the first three images. This is a result of the shorter ignition delay and faster fuel combustion rate at this ambient condition,ed and both of them shorten the flame spread areas. KL factor distributions at 1000K and 21% oxygen concentration are presented in Figure 1 (b). KL factor across the flame area could be negligible until 10.651 ms ASOIT. Only a few discrete areas are found with high KL factor by the end of combustion. In addition, flame temperatures at areas that have high KL factor are lower than 1800K with high temperature gradients. This may indicate that soot is more likely to exist at temperatures lower than 1800K.

3.3 Time resolved results

3.3.1 Averaged flame temperature

To evaluate the effect of ambient oxygen concentration on flame temperature development, the average flame temperature at each time step is calculated. The average flame temperatures are shown in Figure 2. Developments of average flame temperature are similar. For each condition, the average flame temperature first increases monotonically for the first half of combustion process, and then decreases until the end of combustion. Differences are observed in the average temperature value at each time step, the timing to reach temperature peak, and temperature drop rate. In the first half of combustion, a higher oxygen concentration results in a higher average flame temperature at the same time step. However, the average flame temperature drops faster for the flame at a higher ambient oxygen concentration in the second half of combustion. Moreover, flames at a higher ambient oxygen level reach the peak average temperature earlier.

3.3.2 Averaged and accumulated KL factor

Average KL factor at each ambient condition is calculated during the combustion process and results are shown in Figure 3(a). A higher average KL factor is observed at a lower ambient oxygen concentration condition at the beginning of fuel combustion. Then the average KL factor continues decreasing during the first half of combustion. In the second half of combustion, KL factor first increase rapidly and then drop to around zero by the end of combustion. The effect of ambient oxygen concentration on average KL factor is different in different time stages. For the first half of combustion, a higher average KL factor is observed at lower ambient oxygen concentrations. For the second half of combustion, a higher ambient oxygen concentration, on the contrary, leads to a higher average KL factor. In addition, flame under a higher ambient oxygen concentration reaches the peak value earlier than at lower ambient oxygen concentration. However, it should be noted that ambient oxygen concentration has little effect on the absolute values of the average KL factor peaks which are scattered around 0.08 a.u. under different ambient oxygen concentrations.

In order to evaluate the overall soot formation during the combustion process, accumulated KL factors are calculated for each ambient condition. Accumulated KL factor is defined as the sum of KL factors of the whole flame at each time step. In general, accumulated KL factor could be regarded as an indicator of amount of soot formation during the combustion. Results of accumulated KL factor development at different oxygen concentrations are presented in Figure 3(b). It is found that the development of the accumulated KL factor at different oxygen concentration is similar. For each oxygen concentration, the accumulated KL factor keeps increasing until the end of combustion. By the end of combustion, the accumulated KL factor drops rapidly. Furthermore, a lower oxygen concentration results in a higher accumulated KL factor for most part of the combustion process. Flame at 21% oxygen concentration always has the lowest accumulated KL factor during the whole combustion process.

4. Summary

In this study, two-color measurement was conducted for LN under different ambient conditions. Major findings are summarized as follows:

1. A higher oxygen concentration could accelerate fuel combustion rate. In general, LN flame temperature ranges from 1500K to 2300K. A higher flame temperature peak could be observed at a higher ambient oxygen concentration condition.

2. For each ambient condition, high flame temperature areas are found at flame center or flame downstream during most parts of combustion, and they gradually move downstream with time. Flame upstream always has lower temperature than other flame regions.

3. LN flames have low KL factor values for all the conditions. High KL factors are observed at flame periphery where the flame temperature is lower than 1800 K.

4. For each condition, the average flame temperature first increases then decreases with time. Flame at higher ambient oxygen levels has higher average flame temperature during the increasing stage. After the peak, the average flame temperature at a higher ambient oxygen concentration drops faster. A lower oxygen concentration leads to a higher accumulated KL factor. The lowest accumulated KL factor is found at 21% oxygen concentration condition for all time steps.

5. Acknowledgements

This research was supported in part by the Saudi Aramco Public R&D Center through the Clean Combustion Research Center of the King Abdullah University of Science and Technology under the FUELCOM program.

6. References

[1] Heywood, John B. Internal combustion engine fundamentals. New York: Mcgraw-hill, 1988.
[2] Kalghatgi, G., et al. "Low NOx and low smoke operation of a diesel engine using gasoline-like fuels", Proc. of the ASME IC Engine Division 2009 Technical Conf., ICES2009-76034, 2009.
[3] Hildingsson, L., et al., "Fuel Octane Effects in the Partially Premixed Combustion Regime in Compression Ignition Engines," SAE Technical Paper 2009-01-2648, 2009
[4] Sjöberg, M. and Dec, J., "Influence of Fuel Autoignition Reactivity on the High-Load Limits of HCCI Engines," SAE Int. J. Engines 1(1):39-58, 2009, doi:10.4271/2008-01-0054.
[5] Han, Xiaoye, Kelvin Xie, Ming Zheng, and William De Ojeda. Ignition control of gasoline-diesel dual fuel combustion. No. 2012-01-1972. SAE Technical Paper, 2012.
[6] Risberg, Per, Gautam Kalghatgi, and Hans-Erik Ångstrom. Auto-ignition quality of gasoline-like fuels in HCCI engines. No. 2003-01-3215. SAE Technical Paper, 2003.
[7] Kalghatgi, G.T. Auto-ignition quality of practical fuels and implications for fuel requirements of future SI and HCCI engines. No. 2005-01-0239. SAE Technical Paper, 2005.
[8] Kalghatgi, G.T.,et al., Advantages of fuels with high resistance to auto-ignition in late-injection, low-temperature, compression-ignition combustion. SAE Technical Paper, 2006-01-3385.
[9] Richter, M., et al., "The Influence of Charge Inhomogeneity on the HCCI Combustion Process," SAE Technical Paper 2000-01-2868, 2000.
[10] Li, N., et al., CFD study on effects of thermal and residual gas inhomogeneous distribution on auto-ignition of gasoline HCCI combustion. No. 2010-01-0160. SAE Technical Paper, 2010.
[11] Thirouard, B., Cherel, J., and Knop, V., "Investigation of Mixture Quality Effect on CAI Combustion," SAE Technical Paper 2005-01-0141, 2005.
[12] Hao, H., Liu, F., Liu, Z., and Zhao, F., "Compression ignition of low-octane gasoline: Life cycle energy consumption and greenhouse gas emissions," Appl. Energy, 181:391-398, 2016.
[13] Viollet, Y., et al., "Compression Ratio and Derived Cetane Number Effects on Gasoline CI Engine Running with Naphtha Fuels,"SAE Int. J. Fuels Lubr., 7(2):412-426, 2014.
[14] Badra, J.A., et al., "Numerical Simulations of Hollow-Cone Injection and GCI Combustion With Naphtha Fuels," J. Energy Resour. Technol., 138(5):052202-052202, 2016.
[15] Badra, J., Viollet, Y., Elwardany, A., Im, H.G., et al., "Physical and chemical effects of low octane gasoline fuels on compression ignition combustion," Appl. Energy, 183:1197-1208, 2016.
[16] Jing, W., Z. Wu, W.L. Roberts, and T. Fang. "Spray combustion of biomass-based renewable diesel fuel using multiple injection strategy in a CVCC." Fuel 181 (2016): 718-728.
[17] Jing, W., Z. Wu, W.L. Roberts, and T. Fang. "Effects of Fuel Quantity on Soot Formation Process for Biomass-Based Renewable Diesel Fuel Combustion." J. Eng. Gas Turbines Power 139, no. 10 (2017): 102803.

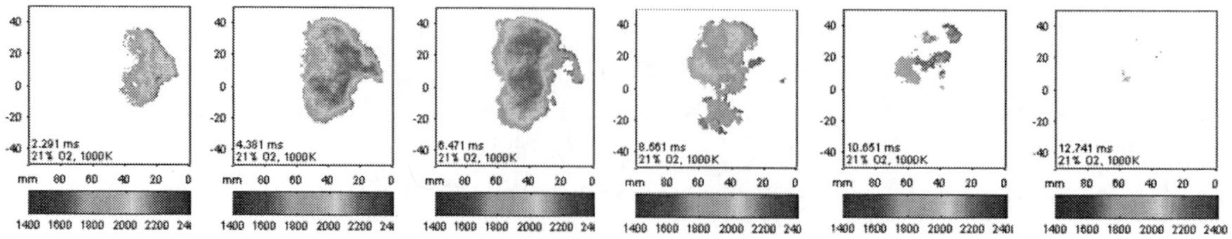

(a) Flame temperature distribution at 21% 1000K

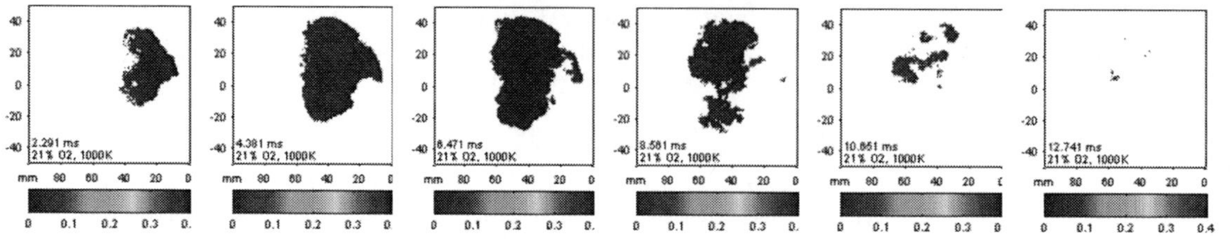

(b) KL factor distribution at 21% 1000K

Figure 1 Flame temperature and KL factor distribution at 21% and 1000K condition.

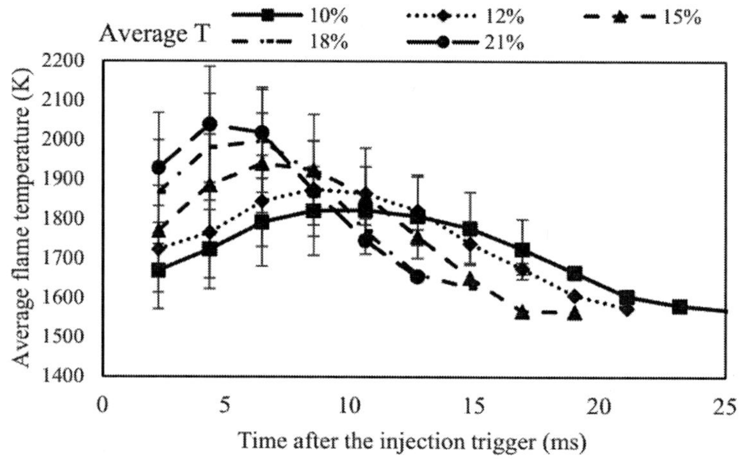

Figure 2 Flame average temperature at different oxygen concentrations.

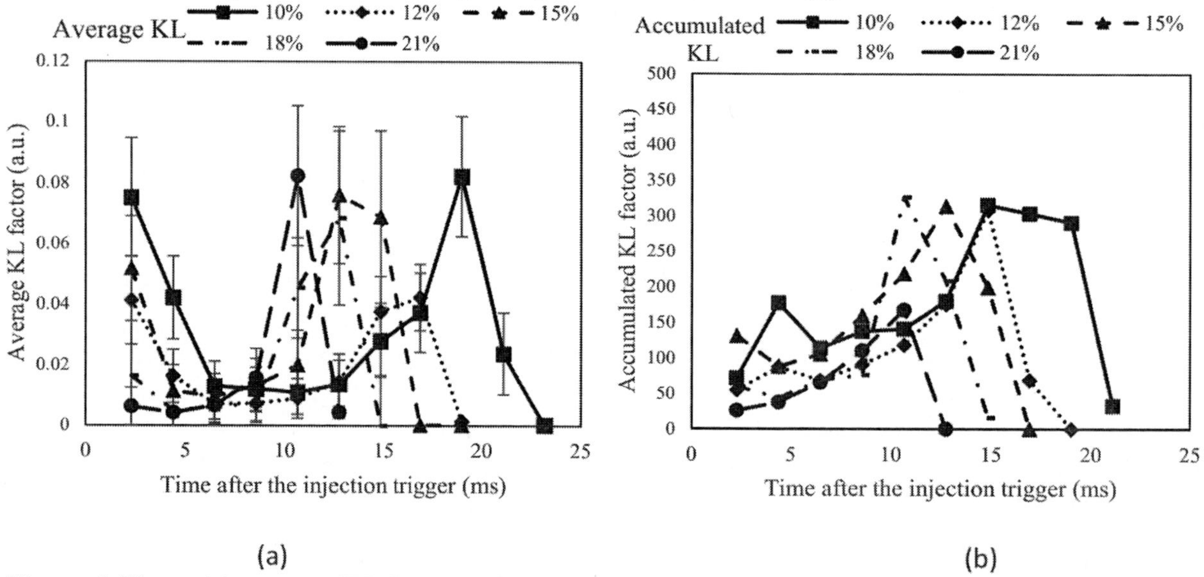

Figure 3 Flame (a) average KL factor and (b) accumulated KL factor at different oxygen concentrations.

Sub Topic: Soot

Spring Technical Meeting
Eastern States Section of the Combustion Institute
March 4-7, 2018
State College, Pennsylvania

Quantification of Nanostructure Changes by HRTEM and Fringe Analyses during NO_2-O_2 Oxidation

Mek Srilomsak[1, 2], Madhu Singh[2], Katsunori Hanamura[1], Randy L. Vander Wal[2,]*

[1] Graduate School of Engineering, Department of Mechanical Engineering, Tokyo Institute of Technology, 2-12-1 O-Okayama, Meguro, Tokyo, 152-8550, Japan
[2] John and Willie Leone Department of Energy and Mineral Engineering and the EMS Energy Institute, the Pennsylvania State University, University Park, PA, 16802, United States.
[*]Corresponding Author Email: ruv12@psu.edu

Abstract: Development of the regeneration process on diesel particulate filters (DPF) requires a better understanding of soot oxidation phenomena, especially its relation to soot nanostructure. Nitrogen dioxide (NO_2) is known to play an essential role in passive regeneration by partially oxidizing soot at low temperatures, in the presence of oxygen (O_2) in the exhaust. However, change in soot nanostructure due to oxidation by NO_2-O_2 mixtures has not received much attention. High-resolution transmission electron microscopy (HRTEM) reveals nanostructure of model carbons and changes during oxidation. This work focuses on passive regeneration of the DPF by oxidation of soot at normal exhaust gas temperatures (300-400°C). HRTEM imaging of partially oxidized model carbons (R250, M1300, arc-generated soot, graphitic carbon onion, and flame soot) under NO_2-O_2 mixtures is done to investigate physical changes in nanostructure to test for correlation with the oxidation behavior. Each model carbon presents a specific type of nanostructure – as identified in varied diesel soots. Fringe analysis of HRTEM images of carbon nanostructure is performed to reveal the difference in the structural metrics of fringe lengths, curvature, spacing and tortuosity of the resultant structures. The variation in oxidation rates of different samples highlights the dependence of the oxidation on the nanostructure. These results are compared with the oxidation and nanostructure of a diesel soot. .

Keywords: HRTEM, DPF, Carbon, Soot, Fringe

1. Introduction

Development of regeneration process of diesel particulate filter (DPF) is still a challenging issue because of complex nature of soot and its unpredictability due to varying fuel compositions and engine operating conditions. A better understanding of soot oxidation is required, especially in relation with its carbon nanostructure. Nanostructure refers to spatial metrics of the carbon lamellae that comprise the material, otherwise referred to as graphene segments. Parameters of the nanostructure include in-plane layer dimensions (L_a, L_b), stacking height (L_c), interplanar spacing (d), and tortuosity or curvature of the lamellae [1, 2]. The nascent nanostructure not only impacts but can govern oxidation behavior for various types of carbon (including diesel soot) under O_2 oxidation at moderate temperatures, such as those in a DPF. Varied studies have shown this impact of nanostructure on oxidation rate [3-5]. Given the complexity of real soots (e.g. heteroatoms, volatile content) many groups have studied reaction kinetics of varied carbon materials as surrogate soots to compare with the engine soot. [6-9]. As soot surrogates, a variety carbons including carbon blacks, arc-generated soot, and laboratory produced flame soot have been proposed to represent particular physical characteristics of the diesel soot nanostructure, e.g., particle size, surface area or aggregate size. Increasingly, nanostructure is being recognized as the better comparative [10-12]. In a previous study of our group, model carbons bracketing the nanostructure types for their impact on soot oxidation behavior. Gaddam et al. employed microscopic and spectroscopic techniques to establish correlations between nanostructure and reactivity of fullerenic, graphitic, mixed nanostructure carbon blacks and an amorphous, arc-generated "soot" [13]. The study discussed microscopic visualization analysis of the carbons at different stages of oxidation by oxygen gas (O_2), establishing relations between observed nanostructure changes and oxidation rate variations, reconciling activation-energy.

Of the studies performed to-date, more have focused on DPF active regeneration due to oxidation by O_2 [14-16] rather than the passive regeneration (soot oxidation by NO_2-O_2 mixtures at relatively lower temperature). Yet NO_2-O_2 mixtures are routinely produced in diesel exhaust steams under passive regeneration due to the high reactivity of soot to such mixtures at normal exhaust gas temperature (300-400°C) [17, 18]. NO_2 is known to play an essential role by partially oxidizing soot at low temperatures, in the presence of O_2. As discussed by A. Setiabudi et al. [19], soot oxidation is initiated by NO_2-soot reaction to create surface oxygen complexes (SOCs) as intermediates. The intermediates, to oxygen, subsequently produce carbon dioxide (CO_2) and/or carbon monoxide (CO). As correspondence, K. O. Lee and coworkers reported that in the presence of O_2, increasing NO_2 promotes soot oxidation at a relatively lower temperature (200-580°C) [20].

It is widely reported that soot oxidation can be accompanied by both physical and chemical transformation towards a more graphitic structure. Typically, soot oxidation by O_2 gas will preferentially remove amorphous carbon relative to graphitic carbon, as well as loosely stacked and curved fringe segments. However, such knowledge for oxidation by NO_2-O_2 gas mixtures is still limited, having received little attention until recently. To-date, few studies have used electron microscopy along with fringe analysis to examine the changes in soot nanostructure due to NO_2-O_2 oxidation in conjunction with using thermogravimetric analysis (TGA) [21]. High-resolution transmission electron microscopy (HRTEM) provides images from which nanostructure can be extracted. Such images facilitate a direct measure of the length, curvature and stacking of graphitic fringes. In a recent investigation, Vander Wal et al. reported [22], HRTEM images – revealing differences in soot nanostructure for different types of soot primary particles before and after partial oxidations. Fringe analysis algorithms were then employed to relate reactivity to the nanostructure, building upon prior works [23-25].

In this study, TGA and HRTEM are employed to investigate oxidation behavior of surrogate diesel soots (carbon blacks, arc-generated soot, and flame soot) and their changes in nanostructure during NO_2-O_2 oxidation, respectively. The results are compared with the characteristics of soot generated from a diesel engine. Fringe analysis of HRTEM images for nanostructure is then performed to reveal the differences in the structural details of fringe length, curvature, spacing, and tortuosity.

2. Experimental

The five model carbons used to represent nanostructure types as observed in diesel soots included Monarch 1300 (Cabot®), Regal 250 (Cabot®), carbon nano-onions (annealed R250), arc generated soot (MER Corp.), and a diesel burner soot (flame soot). The carbon nano-onions were synthesized by heat-treating R250 carbon black in a high-temperature furnace under a He atmosphere at 3000°C for 1-hour duration [2]. Diesel soot and flame soot samples were provided by Cummins Inc.

2.1 **Thermogravimetric analysis**

A thermogravimetric analyzer (TA instruments, SDT Q600) with recording software was used to observe the oxidation rates of all soot samples. The TGA recorded weight changes in controlled gas environments over isothermal tests. Experiments were conducted with approximately 3 mg of each sample. The samples were pretreated with nitrogen gas (N_2) until the desired oxidation temperature of 425°C was attained at a ramp rate of 20°C/min. Upon reaching this temperature, the samples were exposed to NO_2-O_2 gas mixture flow. 900 ppm (0.09%) NO_2, 10.5% O_2 by volume are introduced with N_2 making up the balance. The mass loss curve was obtained as oxidation proceeded isothermally approximately every half a second. Oxidation rates were normalized to the weight at the starting of oxidation (weight loss in the inert atmosphere was not considered). Additional samples were generated by terminating the oxidation process after reaching conversion (weight loss) at 2 burnouts, ~ 35-40%, ~ 75-80% samples to explore changes in the nanostructure of the samples due to NO_2-O_2 oxidation using transmission electron microscopy.

2.2 **Transmission electron microscopy**

For HRTEM visualization, each soot sample was deposited on a lacey C/Cu grid as a drop of the sample suspension, prepared by sonication in methanol. TEM images were taken using a FEI Talos F200X S/TEM with a nominal resolution of 0.14 nm; Gatan Image Filter was used for digital imaging featuring live Fourier transforms. The instrument was operated with a 200 keV FEG. Image processing was done using Gatan image software v3.4.

Two separate analysis algorithms were used for quantification of HRTEM images; the first program calculated fringe length, while the second analyzed the tortuosity. Further details of these programs can be found elsewhere [2, 24]. The parameters extracted from the HRTEM images are fringe length and fringe tortuosity. Lattice fringe length determines the physical extent of the atomic carbon layer planes. Longer lengths correspond to a higher degree of organization and hence the material is considered to be more graphitized. On the other hand, fringe tortuosity is defined here as the ratio of the actual fringe length to the shortest distance between the endpoints of a carbon segment, which provides a measure of the undulation of the layer planes [25]. Amorphous nature and stack height were also observable from the HRTEM images [7, 13].

3. **Results and Discussion**

HRTEM of nascent and partially oxidized samples

1. M1300

Nascent M1300 exhibits multiple closed shell structures (fullerene-like), with the void interior appeared especially of those on the perimeter as shown in Fig 1(a); such a character could be obtained as well in a EuroIV test diesel engine [26]. Moreover, the M1300 was investigated to have a specific surface area of the primary particle similar to diesel soot generated from a heavy-duty diesel engine [7]. A previous study suggested that the M1300 oxidized uniformly under O_2 gas oxidizer, though the loss of outer-shell structure, short fringes, and highly curved shells observed. Partially oxidized M1300 by NO_2-O_2 mixtures of this study is expected to exhibit a unique burning mode. Curved outer shells are more likely to be absent as compared to those observed in the previous work since NO_2 gas is known to play a significant role in being reactive at the outer surface [19]. Eventually, this will result in a significantly shrunken primary particle following a shrinking-sphere model, thus reducing the diameter with increasing level of burnout.

2. R250

Nascent R250 consists of short lamellae arranged concentrically, with some radial variation in its structure, as presented in Fig. 1(b). Lamellae are short and spatially disorganized within the particle interior. R250 exhibits an internal burning mode after partial oxidation by O_2 gas (shell-like structure), which is a regression of amorphous areas. Similar observations have been reported for soot collected from various operating conditions of diesel engines [23]. Upon NO_2-O_2 mixture oxidation, R250 is still expected to exhibit the internal burning mode. However, development of shell-like structure is likely more retarded as compared to results of oxidation by O_2.

3. Carbon nano-onions

The nano-onions (graphitized R250) exhibit a central void structure with graphitized outer shells as presented in the Fig. 1(c). The material, which shows only edge site oxidation during O_2 oxidation, is anticipated to exhibit the same character of structural changes during NO_2-O_2 oxidation, but, with a more significant degree of external burning since NO_2 is known to create mainly surface oxygen complexes (SOCs) around the outer shells [19].

4. Arc generated soot

Nascent structure of arc soot is mainly amorphous, though some lamellae are recognizable as seen in Fig 1(d). There is no visible stacking or spatial order evident from HRTEM images, given lamellae spatial scatteredness; as can be observed from soot generated from spark-ignited direct injection engines [27, 28]. O_2 oxidation of arc soot [13] exhibits preferential oxidation of amorphous material resulting in concurrent oxidation-induced graphitization of the remaining material. On the other hand, during NO_2-O_2 oxidation, the arc soot is anticipated to have a higher degree of external burning than the core area due to relatively lower oxidation temperature, and the outer surface reactivity nature of NO_2-O_2 mixtures.

5. Flame soot and diesel soot

Nascent flame soot exhibits amorphous in the core, while, outer shell constructed by curved highly lamellae as illustrated in Fig. 1(e). On the other hand, nascent structure of the diesel soot consists of relatively less curvatures, though some amorphous core is also recognizable as seen in Fig 1(d).

NO_2-O_2 oxidation rates of all types of sample are also investigated by TGA in order to establish correlations between nanostructure and reactivity.Comparisons of all aforementioned results with diesel soot is a future work of this study with a purpose to realize particular physical characteristics of the diesel soot nanostructure, i.e., particle size, surface area or aggregate size.

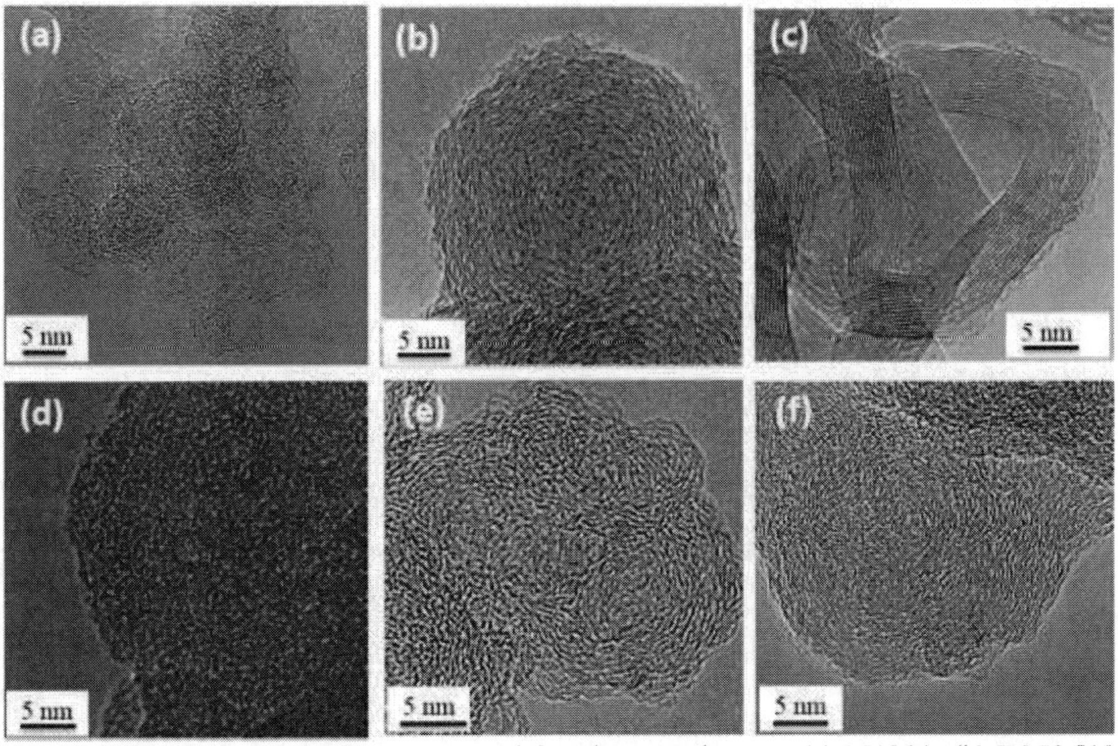

Figure 1. HRTEM images of nascent model carbons and soots: (a) M1300, (b) R250 [13], (c) carbon nano-onions [13], (d) arc soot [13], (e) flame soot, (f) diesel soot

4. Conclusions

HRTEM investigations of five surrogate soot samples, eventually compared with diesel soot, highlight the differences in nanostructure during NO_2-O_2 oxidation, dependent on particular physical characteristics of each surrogate. The variation in oxidation rates confirms the dependence of the NO_2-O_2 oxidation on the nanostructures. In consequence, roles of NO_2 on the changes in carbon nanostructure are discussed.

5. Acknowledgements

This work was supported by the EMS Energy Institute at the Penn State University, University Park, PA. Soot samples provided by Cummins Inc. are also appreciated.

6. References

[1]R.L. Vander Wal, A.J. Tomasek, Soot oxidation: dependence upon initial nanostructure, Combust. Flame 134 (2003) 1-9.
[2] R.L. Vander Wal, A.J. Tomasek, K. Street, D.R. Hull, W.K. Thompson, Carbon Nanostructure Examined by Lattice Fringe Analysis of High-Resolution Transmission Electron Microscopy Images, Appl. Spectrosc. 58 (2004) 230-237.
[3] J. Song, M. Alam, A.L. Boehman, U. Kim, Examination of the oxidation behavior of biodiesel soot, Combust. Flame 146 (2006) 589-604
[4] K. Yehliu, R.L. Vander Wal, O. Armas, A.L. Boehman, Impact of fuel formulation on the nanostructure and reactivity of diesel soot, Combust. Flame 159 (2012) 3597-3606.
[5] J.B. Donnet, Structure and reactivity of carbons: from carbon black to carbon composites, Carbon 20 (1982) 267-282.

[6] J.P.A. Neeft, T.X. Nijhuis, E. Smakman, M. Makkee, J.A. Moulijn, Kinetics of the oxidation of diesel soot, Fuel 76 (1997) 1129-1136.
[7] L. Pahalagedara, H. Sharma, C. Kuo, S. Dharmarathna, A. Joshi, S.L. Suib, A.B. Mhadeshwar, Structure and oxidation activity correlations for carbon blacks and diesel soot, Energy Fuels 26 (2012) 6757-6764.
[8] H. Seong, S. Choi, Oxidation-derived maturing process of soot, dependent on O2–NO2 mixtures and temperatures, Carbon 93 (2015) 1068-1076.
[9] I. Atribak, A. Bueno-López, A. García-García, Uncatalysed and catalysed soot combustion under NOx + O-2: Real diesel versus model soots, Combust. and Flame 157 (2010) 2086–2094.
[10] D. Zhang , Y. Ma, M. Zhu, Nanostructure and oxidative properties of soot from a compression ignition engine: The effect of a homogeneous combustion catalyst, Proc. Combust. Inst. 34 (2013) 1869–1876.
[11] Y. Zhang, R. Zhang, S. Kook, Nanostructure Analysis of In-flame Soot Particles under the Influence of Jet-Jet Interactions in a Light-Duty Diesel Engine, SAE Int. J. Engines 8(5), (2015) 2213-2226.
[12] P. Karin, Y. Songsaengchana, S. Laosuwana, C. Charoenphonphanicha, N. Chollacoopb, K. Hanamura, Nanostructure Investigation of Particle Emission by Using TEM Image Processing Method, E. Proc. 34 (2013) 757 – 766.
[13] C.K. Gaddam, R.L. Vander Wal, X. Chen, A. Yezerets, K. Kamasamudram, Reconciliation of carbon oxidation rates and activation energies based on changing nanostructure, Carbon 98 (2016) 545-556.
[14] Oki, H., Karin, P., and Hanamura, K., Visualization of Oxidation of Soot Nanoparticles Trapped on a Diesel Particulate Membrane Filter, SAE Int. J. Engines 4(1), (2011) 515-526.
[15] K. Hanamura, P. Karin, L. Cui, P. Rubio, T. Tsuruta, T. Tanaka, T. Suzuki, Micro- and macroscopic visualization of particulate matter trapping and regeneration processes in wall-flow diesel particulate filters, Int. J. Engine Research 10(5), (2009) 305 – 321.
[16] P. Karin, Microscopic visualization and characterization of particulate matter trapping and oxidation in diesel particulate filters and membrane filters, Doctoral thesis, Tokyo Institute of Technology, (2010).
[17] R. Matarrese, L. Castoldi, L. Lietti, Oxidation of model soot by NO_2 and O_2 in the presence of water vapor, Chemical Engineering Science 173 (2017) 560–569.
[18] A. Strzelec, T.J. Toops, C.S. Daw, R. Vander Wal, Investigation of burning mode for diesel particulate oxidation: contrasting O_2 & NO_2, Cambridge Particle Meeting 2013, May 13, Cambridge, Great Britain.
[19] A. Setiabudi, M. Makkee, J.A. Moulijn, The role of NO2 and O2 in the accelerated combustion of soot in diesel exhaust gases, Appl. Catal. B 50 (2004) 185–194.
[20] K.O. Lee, H. Seong, S.M. Choi, Detailed analysis of kinetic reactions in soot oxidation by simulated diesel exhaust emissions, Proc. Combust. Inst. 34 (2013) 3057–3065.
[21] Stratakis GA, Stamatelos AM. Thermogravimetric analysis of soot emitted by a modern diesel engine run on catalyst doped fuel. Combust. and Flame 132 (2003) 157–69.
[22] R.L. Vander Wal, A TEM Methodology for the Study of Soot Particle Structure, Combust. Sci. and Technol. 126 (1997) 333-351.
[23] R.L. Vander Wal, A. Yezerets, N.W. Currier, D.H. Kim, C.M. Wang, HRTEM Study of diesel soot collected from diesel particulate filters, Carbon 45 (2007) 70–77.
[24] K. Yehliu, R.L. Vander Wal, A.L. Boehman, Development of an HRTEM image analysis method to quantify carbon nanostructure, Combust. Flame 158 (2011) 1837-1851.
[25] R.L. Vander Wal, A.J. Tomasek, M.I. Pamphlet, C.D. Taylor, W.K. Thompson, Analysis of HRTEM images for carbon nanostructure quantification, J. Nanoparticle Res. 6 (2005) 555-568.
[26] D.S. Su, J.-O. Müllera, R.E. Jentofta, D. Rotheb, E. Jacobs, R.Schlögla, Fullerene-like Soot from EURO-IV Diesel Engine: Consequences for Catalytic Automotive Pollution Control, Topics in Catalysis, 30/31 (2004), 241-245.
[27] C.K. Gaddam, R.L. Vander Wal, Physical and chemical characterization of SIDI engine particulates, Combust. Flame 160 (2013) 2517-2528.

Sub Topic: Soot

[28] S. Choi, C. L. Myung, S. Park, Review on characterization of nano-particle emissions and PM morphology from internal combustion engines: Part 2, Int. J. Automotive Tech. 15 (2014) 219–227.

Spring Technical Meeting
Eastern States Section of the Combustion Institute
March 4-7, 2018
State College, Pennsylvania

Three Stage Quasi-steady Droplet Burning Behavior of n-Alkane Droplets at Elevated Pressure Conditions: Hot, Warm and Cool Flame Combustion

Tanvir Farouk[a], Daniel Dietrich[b] and Frederick L. Dryer[a,c]

[a]Department of Mechanical Engineering
University of South Carolina
Columbia, SC 29208, USA

[b]NASA Glenn Research Center
Cleveland, OH 44135, USA

[c]Department of Mechanical and Aerospace Engineering
Princeton University
Princeton, NJ 08544, USA

**Corresponding Author Email: tfarouk@sc.edu*

Abstract: Transient, isolated *n*-heptane droplet combustion is simulated at elevated pressure for helium-diluent substituted-air mixtures. We report the presence of unique quasi-steady, three-stage burning behavior of large sphero-symmetric *n*-alkane droplets at these elevated pressures and helium substituted ambient fractions. Upon initiation of reaction, hot-flame diffusive burning of large droplets is initiated that radiatively extinguishes to establish cool flame burning conditions in nitrogen/oxygen "air" at atmospheric and higher pressures. However, at elevated pressure and moderate helium substitution for nitrogen ($X_{He} > 20\%$), the initiated cool flame burning proceeds through two distinct, quasi-steady-state, cool flame burning conditions. The classical "*Hot flame*" (~1500K) radiatively extinguishes into a "*Warm flame*" burning mode at a moderate maximum reaction zone temperature (~ 970K), followed by a transition to a lower temperature (~765 K), quasi-steady "*Cool flame*" burning condition. The reaction zone ("flame") temperatures are associated with distinctly different yields in intermediate reaction products within the reaction zones and surrounding near-field, and the flame-standoff ratios characterizing each burning mode progressively decrease. The presence of all three stages first appears with helium substitution near 20%, and the duration of each stage is observed to be strongly dependent on helium substitutions level between 20 – 60%. For helium substitution greater than 60%, the hot flame extinction is followed by only the lower temperature cool flame burning mode. In addition to the strong coupling between the diffusive loss of both energy and species and the slowly evolving degenerate branching in the low and negative temperature coefficient (NTC) kinetic regimes, the competition between the low-temperature chain branching and intermediate-temperature chain termination reactions control the "*Warm*" and "*Cool*" flame quasi-steady conditions and transitioning dynamics. Experiments onboard the International Space Station with *n*-dodecane droplets confirm the existence of these combustion characteristics and predictions agree favorably with these observations.

Keywords: Cool Flames, Three-stage, Elevated Pressure, n-Alkane, NTC

1. Introduction

The chemical characteristics of high-temperature flames for both premixed and non-premixed systems have been thoroughly investigated by theoretical, experimental and computation methods. The comparative characteristic time scales for convective, diffusive, radiative, and chemical heat generation dictate the existence and burning intensity of these flames and their ignition, quasi-steady burning, and extinction dynamics.

Most liquid fuels of practical interest exhibit a wide range of chemical kinetic time scales generally classified into low, intermediate, and high temperature behaviors. The low and intermediate-temperature behaviors exhibit characteristics related to peroxy and peroxide chemistry of large hydrocarbon radicals that result in degenerate branching and negative temperature coefficient (NTC) behaviors. Hydrogen peroxide formed as an intermediate begins to rapidly decompose as reaction temperature increases, and combined with significant β scission of the larger radicals leads to auto-thermal acceleration of chemical rates and, eventually to high temperature kinetic behavior dominated by small radical chain branching chemistry. These chemistries are manifest in cool and hot flame dynamic behaviors. "*Cool flame*" phenomena have been observed for homogenous vapor phase conditions in static reactor [1], flow tube, and opposed flow configurations [2], and for isolated droplet configuration associated with auto-ignition research in drop towers [3].

In recent ground-based, opposed flow, pre-vaporized, laminar diffusion flame configurations, ozone was added into the oxidizer stream [4]. The rapid decomposition of ozone produced active atomic oxygen, substantially reducing the induction timescale for initiating low-temperature, exothermic fuel oxidation chemistry. The reduced time scales led to flame initiation and quasi-stable "*Cool flame*" diffusive burning at intermediate reaction temperatures characteristic of the NTC kinetic regime. Recent experiments [5] and numerical simulations [6] have shown that in the opposed flow configuration for dimethyl ether, there can exist a multistage premixed flame at near-limit conditions. The occurrence of this multistage flame structure is attributed to the result of two different sets of chain-branching reactions involving small radical chemistry and peroxy chemistry at intermediate and low temperatures respectively. However, neither experimental or computational endeavors have revealed whether similar stable multistage flame behaviors could exist for multiphase (e.g. isolated microgravity droplet) configurations.

Below, we report results from numerical simulations of *n*-heptane and *n*-dodecane droplet combustion at elevated pressure where the ambient nitrogen is progressively replaced with helium to augment diffusive losses. Modeling results show that at moderate helium substitution, the initiated burn proceeds through three distinct stages. For large initial drop size, the classical hot flame radiatively extinguishes into a "*Warm flame*" burning moderate/intermediate reaction zone temperature condition, followed by a transition to a lower temperature quasi-steady "*Cool flame*" burning condition. The simulations further demarcate the conditions under which quasi-steady, three-stage burning behaviors exist, and the dynamic transitions that progress to yield these different burning behaviors.

2. Experimental setup and procedure

Experiments were performed in the Multi-User Droplet Combustion Apparatus (MDCA) insert in the ISS Combustion Integrated Rack (CIR), a full-sized rack facility in the ISS-Destiny module. The hardware and software are nearly identical to that detailed in a previous publication [7]. In the present experiments, the field-of-view (FOV) of the backlit images of the droplet increased to 5 cm x 5 cm, resulting in slightly less spatial resolution but increased ability to record longer duration burning histories within the FOV. A new intensified camera (1392 x 1040 pixel array) equipped with a rotating filter wheel and an increased FOV (90 mm x 68 mm) recorded the narrow band radiative emissions associated with "*Hot*" and "*Cool flame*" conditions. The rotating filter wheel contained a 430 nm filter to visualize the CH* chemiluminescence during "*Hot flame*" burning, and a wideband filter to detect formaldehyde chemiluminescence (390 – 490 nm) during "*Cool flame*" burning. A separate, wide angle radiometer was used to record radiant emission history. The radiometer signal was used to determine the initiation of "*Hot flame*", automatically power off and retract the igniter from the vicinity of the droplet. When the radiometer signal showed "*Hot flame*" extinction, the intensified camera intesifier gain increases and "*Cool flame*" filter rotated into position to record the formaldehyde chemiluminescence during cool flame burning. The observed "*Cool flame*" emission could not be observed on the color video camera on the MDCA and was of poor quality even on the intensified camera. The "*Cool flame*" images were improved using median filtering and image-averaging over 500 ms [8], and an automated image processing procedure was then applied to determine flame diameter from line profile across the center of the flame image. Here we report two relevant measures of the flame, the peak-to-peak flame diameter representing the diameter of peak chemiluminescence and an FWHM diameter that represents the outer diameter of the flame where the grayscale value of the flame chemiluminescence decreased to one-half the peak value.

3. Numerical modeling

The mathematical model employed to describe the transient vaporization, ignition, and combustion histories of isolated pure and multi-component fuel droplets both in the high and low-temperature regime under microgravity conditions was presented and validated in earlier publications [9, 10]. The model assumes sphero-symmetry because of microgravity conditions. The conservation equations for mass, species, and energy are solved for the liquid and gas phases. Both Fickian and Soret diffusion are taken into account for evaluating multicomponent species diffusion velocities. The radiative effects are predicted via a spectrally resolved statistical narrow band (SNB) radiation model. Liquid phase property data are taken from the Daubert and Danner database [11]. Comprehensive kinetic mechanisms consisting of 652 species, 2827 reactions [12] and 583 species and 2609 reactions [13] were employed to describe the gas phase chemical reactions of *n*-heptane and *n*-dodecane respectively..

4. Results and discussion

First, we discuss the simulation results for an *n*-heptane droplet of $d_o = 4.00$ mm burning at 3 atm pressure. The ambient composition consists of $X_{O2} = 21\%$ and balance of nitrogen that is progressively subsituted with helium. Three representative helium substitution cases (i.e. $X_{He} =$ 20, 40 and 60%) are presented. These cases are chosen to be within substitution range where inception and cessation of three-stage burning behavior are observed. For complete substitution ($X_{He} = 79\%$), the presence of a three-stage burning behavior is not predicted. Figure 1 shows the predicted droplet diameter, FSR, burning rate (K_b) and peak gas temperature (T_{max}) evolution histories. For all three cases, the FSR initially grows, representative of an increasing "*Hot flame*" diameter. The flame steadily grows in diameter with a decrease in the flame temperature to maximum diameter (i.e. maximum FSR) and then undergoes radiative extinction. In all the three cases, following the first-stage "*Hot flame*", the system transitions to a second-stage, but unlike previous observations [9, 10] this second-stage is associated with an intermediate-temperature "*Warm flame*" burn occurring at ~ 970 K (Fig. 1b). The system undergoes a second transition to a third-stage burn characterized by a lower temperature (~ 765 K) "*Cool Flame*" burn. It is apparent from FSR evolution history that each of the burning transitions is associated with a contraction of the flame (reduction in FSR). Despite, the large variation in the flame structure, there are no distinctive changes to be noted in the predicted droplet diameter regression history.

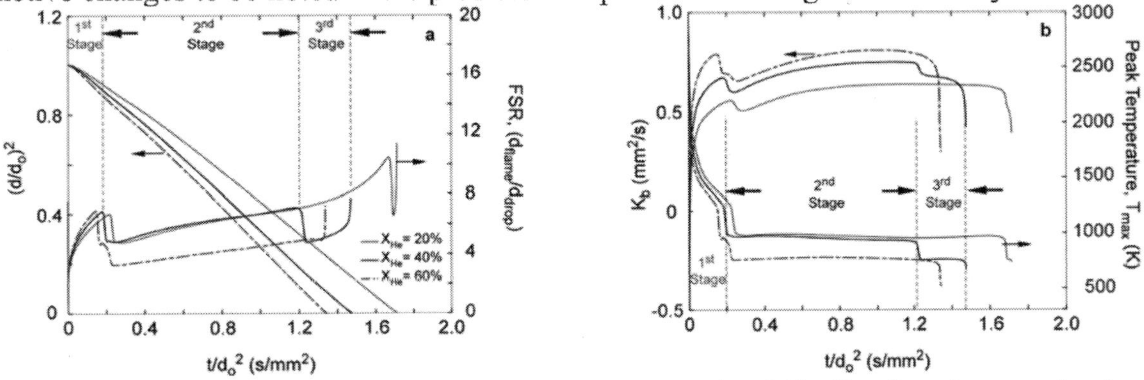

Figure 1. Evolution of a) droplet diameter, FSR and b) predicted K_b , T_{max} for *n*-heptane droplet combustion (d_o = 4.00 mm, P = 3 atm, X_{O2} = 21%, denoted X_{He} and balance X_{N2}). The three-stage burning behavior is demarcated with the vertical dashed lines only for the X_{He} = 40% case.

The helium substitution in the ambient strongly dictates the three-stage burning characteristics. For X_{He} = 20%, the intermediate-temperature second-stage burn continues for ~ 23 s and is followed by a very brief ~ 0.5 s low-temperature third-stage. As the X_{He} fraction is increased, the duration of the second-stage is decreased and the third-stage increases. As the X_{He} approaches 60%, the second-stage is suppressed to ~ 1 s, followed by a third-stage "*Cool flame*" burn for ~ 18 s. The shorter and longer burn durations are also evidenced in the K_b evolution (Fig. 1b). The higher helium substitution augments the diffusive losses which dictates second and third-stage burn. The first-stage remains relatively insensitive to helium substitution with the transition to the second-stage at ~ 3 s. However, helium substitution increases K_b and hence results in shorter total burn times (Fig. 1b). The higher thermal diffusivity of helium leads to improved heat transfer to the droplet surface which contributes to the increasing K_b.

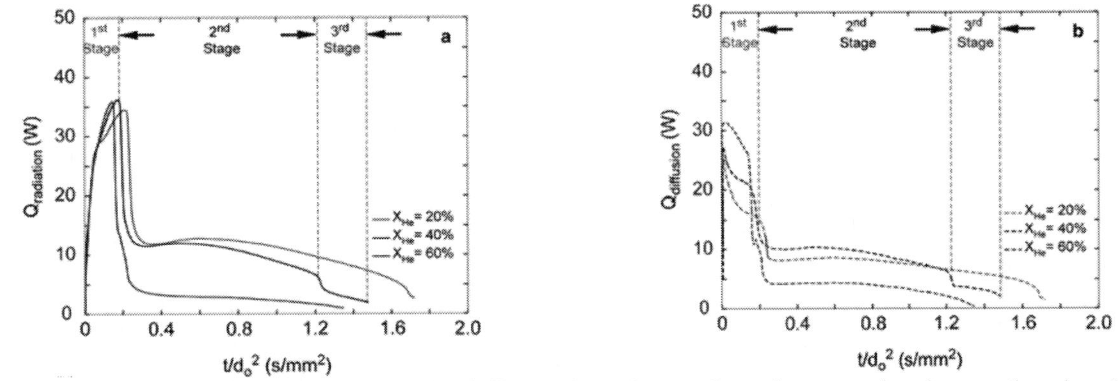

Figure 2. Evolution of a) radiative and b) diffusive heat losses for *n*-heptane droplet combustion (d_o = 4.00 mm, P = 3 atm, X_{O2} = 21%, denoted X_{He} and balance X_{N2}). The three-stage burning behavior is demarcated with the vertical dashed lines only for the X_{He} = 40% case.

Figure 2 presents the predicted temporal evolution of the integrated net radiant and diffusive heat losses for the 3 atm cases. The net radiative and diffusive heat loss evolution is obtained by integrating the predicted radiative fluxes at all wavelength and the diffusive heat fluxes from the flame location to the far-field. For all the cases, the radiative heat loss dominates the first-stage burn and the transition to the second-stage takes place when the radiant heat reaches ~ 35 W. The growth of the flame radius in the first-stage is such that for all the cases the radiant heat attains its maximum at ~ 3 s. As the system transitions to the second-stage "*Warm flame*" burn the radiative heat loss decreases but still remains significant and only diminishes as the burn evolves to the third-stage. This is in contrast to atmospheric pressure, *n*-heptane/air "*Cool flame*" droplet burning where radiative losses were considerably less than diffusive losses. In comparison, the diffusive heat loss is lower in the first-stage but is comparable to the radiative loss in the second-stage. It is only in the low-temperature "*Cool flame*" burn that the diffusive heat loss to the far-field surpasses the radiative contribution. It can be seen that with an increase in helium concentration the diffusive heat loss increases which is due to the higher thermal diffusivity of helium. The sharp rise in the early transient of the diffusive heat loss is a result of the prescribed thermal ignition profile.

Data for an elevated pressure, helium substituted droplet experiment with *n*-dodecane as fuel (**FL**ame **EX**tinguishment experiment 1268) are compared with simulations in Fig. 3. Similar to the *n*-heptane results discussed above, a three-stage burning behavior is observed. While the predicted droplet diameter regression and the transition time to the second-stage agree very well with measurements, the predicted FSR and radiant emission results considerably differ from the measurements. Sooting observed in the experimental hot ignition and flame transition phase are likely responsible for the disparities during the first-stage "*Hot flame*" burn. Furthermore, the predictions show a longer second-stage burn that transitions to a short third-stage "*Cool flame*".

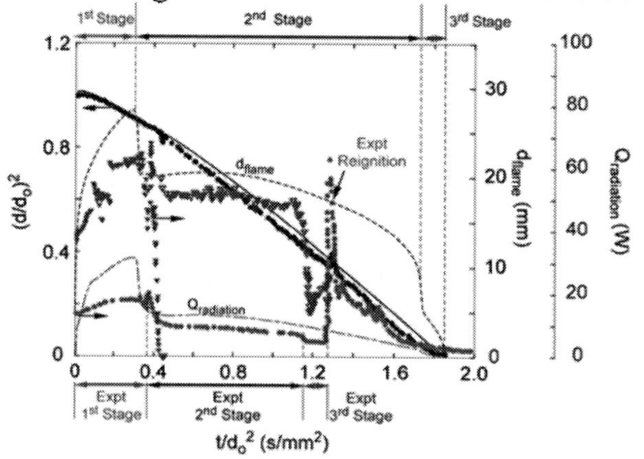

Figure 3. Measured and predicted evolution of droplet diameter, flame diameter and radiant heat for an *n*-dodecane droplet combustion (d_o = 4.35 mm, 2.75 atm, X_{O2} = 21%, X_{He} = 29% and balance N_2, FLEX 1268). The duration of the first, second and third-stage burn of the experiment and model is demarcated at the top and bottom of the figure.

Experimentally the duration of the second-stage "*Warm flame*" is a factor of ~ 2 shorter than predictions, and a re-initiation and radiant extinction of hot flame burning is also observed. Closer consideration of the kinetic model for *n*-dodecane oxidation show that the low-temperature and NTC regime properties are identical to those for *n*-heptane, likely because these kinetics for the *n*-dodecane model are based upon similarity to and data for the *n*-heptane low and NTC kinetics. However, the overall rates of isomerization for radical-O_2 adducts yield degenerate branching should be faster for *n*-dodecane than for n-decane, based upon the statistical number of five and six membered ring formations possible with increased *n*-alkane carbon number [14]. It is likely that the above disparities of second and third-stage behaviors for *n*-dodecane experiment and predictions may result from uncertainties in low and intermediate-temperature *n*-dodecane kinetics.

5. Summary

Experimental and simulation data for large-diameter *n*-alkane droplet combustion at elevated pressure and helium diluent substitution are reported. A unique three-stage burning behavior is observed. Initiation of "*Hot flame*" burning transitions through radiation extinction to nearly quasi-steady "*Warm*" and then "*Cool flame*" burning conditions. The simulations elucidate the strong dependence of observations on helium diluent substitution and ambient pressure that promote different burning conditions where net heat generation rate from low or NTC kinetic behaviors and net total heat loss rate from radiation and diffusive transport are nearly balanced. The second-stage "*Warm flame*" condition results from a balance of heat release at kinetic conditions within the NTC kinetic regime against heat loss, which is dominated by radiant loss. The third-stage condition results from an inability to achieve higher heat release rates to balance net heat loss rates as flame temperatures decrease toward the turnover temperature condition. The "*Warm flame*" condition can no longer be stabilized and dynamically transitions to a nearly quasi-steady balance of heat generation and loss at flame temperatures lower than the turnover temperature and within the low-temperature kinetic regime. In this third-stage "*Cool flame*" condition, diffusive losses are found to be greater than radiative contributions to the total heat loss rate. The time and spatial histories of species for *n*-alkane droplet combustion indicative of incomplete hydrocarbon consumption (reduced oxidation of CO to CO_2) and the shift in intermediate oxygenated alkyl fragments demarcate the kinetic characteristics for all three burning modes and support that the "*Warm*" and "*Cool flame*" burning modes lie in the NTC, and low-temperature kinetic regimes, respectively.

6. References

[1] R. Fairlie, J.F. Griffiths, K.J. Hughes, H. Pearlman, Cool flames in space: experimental and numerical studies of propane combustion, Proc. Combust. Inst. 30 (2005) 1057-1064.

[2] R. Foresti, Stabilization and temperature measurements of flat cool flames, Symposium (International) on Combustion 5 (1955) 582 - 589.

[3] M. Tanabe, M. Kono, J. Sato, J. Koenig, C. Eigenbrod, H. Rath, Effects of natural convection on two stage ignition of an n-dodecane droplet, Symposium (International) on Combustion 25 (1994) 455 - 461.

[4] C.B. Reuter, S.H. Won, Y. Ju, Experimental study of the dynamics and structure of self-sustaining premixed cool flames using a counterflow burner, Combust Flame 166 (2016) 125-132.

[5] C.B. Reuter, S.H. Won, Y. Ju, Flame structure and ignition limit of partially premixed cool flames in a counterflow burner, Proc. Combust. Inst. 36 (2017) 1513-1522.

[6] Y. Ju, C.B. Reuter, S.H. Won, Numerical simulations of premixed cool flames of dimethyl ether/oxygen mixtures, Combust Flame 162 (2015) 3580-3588.

[7] D.L. Dietrich, V. Nayagam, M.C. Hicks, P.V. Ferkul, F.L. Dryer, T. Farouk, B.D. Shaw, H.K. Suh, M.Y. Choi, Y.C. Liu, C.T. Avedisian, F.A. Williams, Droplet Combustion Experiments Aboard the International Space Station, Microgravity Science and Technology 26 (2014) 65-76.

[8] V. Nayagam, D.L. Dietrich, F.A. Williams, Partial-Burning Regime for Quasi-Steady Droplet Combustion Supported by Cool Flames, AIAA Journal 54 (2016) 1235-1239.

[9] T.I. Farouk, F.L. Dryer, Isolated n-heptane droplet combustion in microgravity: "Cool Flames" – Two-stage combustion, Combust Flame 161 (2014) 565-581.

[10] T.I. Farouk, D. Dietrich, F.E. Alam, F.L. Dryer, Isolated n -decane droplet combustion – Dual stage and single stage transition to "Cool Flame" droplet burning, Proc. Combust. Inst. 36 (2017) 2523-2530.

[11] T.E. Daubert, R.P. Danner, Physical and Thermodynamic Properties of Pure Chemicals: Data Compilation, Hemisphere Publishing, New York, 1989.

[12] M. Mehl, W. Pitz, C. Westbrook, H. Curran, Kinetic modeling of gasoline surrogate components and mixtures under engine conditions, Proc. Combust. Inst. 33 (2011) 193 - 200.

[13] S. Sarathy, C. Westbrook, M. Mehl, W. Pitz, C. Togbe, P. Dagaut, H. Wang, M. Oehlschlaeger, U. Niemann, K. Seshadri, P. Veloo, C. Ji, F. Egolfopoulos, T. Lu, Comprehensive chemical kinetic modeling of the oxidation of 2-methylalkanes from C7 to C20, Combust Flame 158 (2011) 2338 - 2357.

[14] F.L. Dryer, Chemical kinetic and combustion characteristics of transportation fuels, Proc. Combust. Inst. 35 (2015) 117 - 144.

Spring Technical Meeting
Eastern States Section of the Combustion Institute
March 4-7, 2018
State College, Pennsylvania

Computational Study of Oscillatory Cool Flame Dynamics for Submillimeter Sized n-Heptane Droplet

Fahd E. Alam[a], Ali Charchi Aghdam[a], Frederick L. Dryer[a,b] and Tanvir I. Farouk[a,]*

[a] *Department of Mechanical Engineering, University of South Carolina, Columbia, SC 29208, USA*

[b] *Department of Mechanical and Aerospace Engineering, Princeton University, Princeton, NJ 08544, USA*

**Corresponding Author Email: tfarouk@sc.edu*

Abstract: Computational analysis of atmospheric pressure oscillatory cool flame of isolated n-heptane (n-C_7H_{16}) droplet ($D_o = 0.5$ mm) under selective ozone-seeded nitrogen-oxygen environment is reported in this paper. We employed an evolutionary one-dimensional droplet code capable of incorporating relevant physics and detailed chemistry for this study. It is found that a combination of low oxygen index and selective ozone seeding initially attempts to establish a stable cool flame which then dynamically transitions to an oscillating cool flame. The inception of oscillation is triggered by the depletion of fuel-vapor oxidizer layer near the droplet surface. The computations indicate that the competition between the reaction classes- (a) degenerate chain branching and (b) chain termination/propagation - along with continuous fuel and oxygen leakage through the flame location contributes to an oscillatory droplet burning with ever-increasing amplitude. Analysis based on single full oscillatory cycle reveals that the reaction progression matrices (evolution of heat and species) for QOOH→chain propagation/termination reactions directly scales with the gas phase temperature. However, interestingly QOOH→degenerate branching reactions experiences three distinct stages within the same oscillatory cycle. We concluded that the kinetic processes dynamically cross through the conditions characterizing the negative temperature coefficient (NTC) turnover temperature separating low temperature and NTC kinetic regimes.

Keywords: *Cool flame, droplet combustion, ozone kinetics, n-heptane, flame oscillation.*

1. Introduction

Computational studies of isolated large diameter n-alkane droplet cool flame experiments [1, 2] on board the international space station (ISS) [3] offer great potential in understanding the fundamental insights of cool flame dynamics and low-temperature chemistry. These endeavors are different than cool flame studies in gas phase premixed/non-premixed flame research [4-6] and involve distinct heat-release/loss rate driven phenomena applicable to isolated sphero-symmetric droplet combustion [7-9].

In general, the ISS cool flame experiments [3] require large initial diameter droplets ($D_o > \sim 3.0$ mm) that undergo an initial radiative flame extinction following a prolonged cool flame burning. In rare instances, cool flame burning was able to be established directly, without incurring an initial hot flame mode and subsequent radiative extinction [9]. Subsequent analyses showed that the overall ignition energy and the heat release from vapor/oxidizer layer oxidation near the drop surface have to be sufficiently low to initiate direct cool flame for large diameter droplets.

Applying the innovative ozone-augmented (O_3) ambient concept described in Won et al. [5], Fahd and coworkers [10] recently demonstrated the possibility of achieving direct and sustained cool flame droplet burning for even submillimeter sized (D_o = 0.5 mm) cases. Here, we extend these earlier predictions to further elucidate the oscillatory droplet burning behaviors of smaller droplets. It should be noted that this study is different then it's large diameter counterpart studies [8, 9, 11, 12].

Here we numerically investigate the oscillatory cool flame behavior of O_3 assisted 0.5 mm n-C_7H_{16} droplet at 1 atm and 298 K ambient condition for two reduced oxygen indices (X_{O2} = 10% and 15%). First, the establishment of directly induced, near quasi-steady (QS), cool flame burning is briefly introduced; and then the conditions responsible for the oscillating cool flame behavior are discussed. The detailed flame structures are analyzed, and the role of underlying thermo-kinetic/transport properties on oscillatory behaviors is elucidated.

2. Numerical modeling

An evolutionary sphero-symmetric, multi-component droplet combustion code has been employed for this study, the details of which can be found elsewhere [13, 14] and references therein. The results presented here (for n-C_7H_{16} initial droplet diameter 0.5 mm) are obtained by employing a numerically reduced kinetic model (130 species, 565 elementary reactions) [15] which has been further refined for droplet combustion simulations. Reaction sets of O_3 kinetics from Reuter et al. [6] were appended to the n-C_7H_{16} models. Lastly, n-C_7H_{16} liquid phase properties were evaluated based on the correlations reported in Daubert and Danner [16]. The results presented are for a spatial resolution of 200 grid points (50 in the liquid and 150 in the gas phase) that confirmed a grid independent solution.

3. Results and discussions

Figure 1 reports the temporal evolution of major droplet combustion parameters including droplet diameter regression, burning rates (K_o) and peak gas temperature (T_{max}) for two different oxygen indices with same ozone loading (5%). For the X_{O2} = 21%, the model predicts an almost instantaneous establishment of a near quasi-steady cool flame burn. For low flash point fuel like n-C_7H_{16} (T_{flash} = 269.3 K), a significant vapor/oxidizer mixture is initially present near the droplet surface. The initiation energy leads to the decomposition of ozone, liberating atomic O which then reacts with the fuel vapor (RH + O ➔ R˙+ OH), followed by RH + OH ➔R˙+ H_2O and associated heat release. Chemically induced heat release as well as energy supplied by the initiation source provides the subsequent heating of the liquid surface [10] which then transits the partially premixed reaction near the surface to diffusive cool flame burning mode. The predicted average quasi-steady state (QSS) cool flame temperature is ~760 K. The simulation predicts a continually increasing flame-location trend (i.e., FRS) due to the thermal buffering of the far field.

For X_{O2} = 10% case, a QS cool flame is predicted to occur after an induction time of ~0.12 s, with a lower flame temperature of ~710 K. The reduction in oxygen level increases the FSR and as a consequence, T_{max} and K_o decrease significantly from its base counterpart case (i.e., X_{O2} = 21%). Initial O_3 decomposition stimulates the gas phase oxidation of the fuel/air mixture and the local heat release attains gas temperatures near the drop surface exceeding 1100 K. The dynamic balance of diffusive-thermo-kinetic terms, as discussed in [11], stabilizes the flame temperature at around ~710 K for a brief period (till t~0.18 s) and then the droplet enters into the oscillatory/pulsing regime. Notably, the FSR, T_{max}, and K_o progressively grow in magnitude with each successive pulsation (c.f. subplot (a-c)). The impression of this transient flame behavior is also reflected in the Stefan flux at the droplet surface. A distinct feature of the oscillation is that at every successive cycle, the cyclic amplitudes of FSR and T_{max} increase but maintain a mutual out of phase trend. During each cycle, a maximum of FSR corresponds to a minimum in peak gas phase flame temperature and vice versa over the burning progression. The differences between the minimum and maximum FSRs and local maximum and the minimum flame temperatures grow over subsequent cycles with a notable decrease in cycle frequency. For example, at ~0.42 s, the peak gas temperature increases to ~738 K from a plateau of ~696 K, dropping again to ~692 K in the next cycle after reaching a value of ~ 742 K. During this time period, the FSR oscillates from ~9.6 to ~6.5 and then ~10.5 to ~ 6.7. A similar prediction trend is also observed for higher oxygen index (i.e. X_{O2} = 15%) exhibiting lower amplitude and higher frequency.

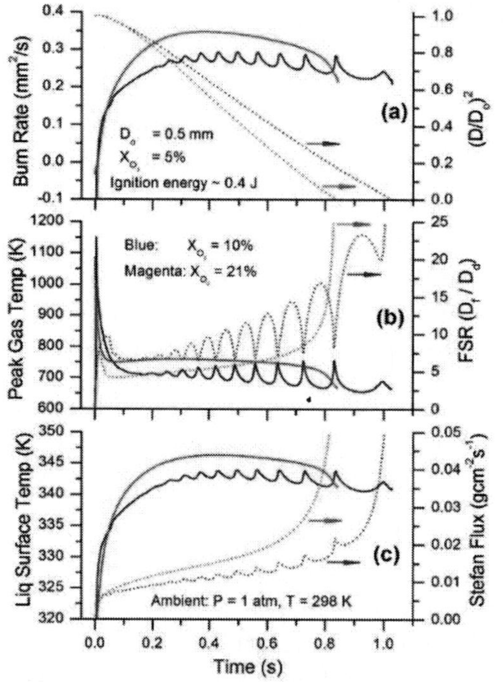

Fig. 1. Temporal evolution of (a) droplet diameter, burning rate, (b) peak gas temperature, flame stand-off ratio and (c) droplet surface temperature and Stefan flux at the droplet surface for an n-heptane droplet exposed to X_{O2} = 10% and 21% condition.

Figure 2 illustrates the spatiotemporal evolution of the gas phase temperature and selective species to delineate further insights into the oscillatory flame dynamics. As a consequence of the rapid chemical energy heat release upon ignition, the temperature rises ~1150 K and the flame outwardly expands through the flammable mixture (subplot b). The resultant is a gradual decay in flame temperature. To facilitate thermal input back to the droplet surface to sustain fuel vaporization, the flame eventually contracts back and repositions itself closer to the surface with an FSR of ~ 6.5. During this period a cool flame burn (T_{max} ~710 K) is stabilized for a brief period of time before transitioning to a continuous the oscillatory flame mode. It is obvious from the X_{O2} distribution (subplot e, h) that ozone dissociation (primarily via O_3+N_2 → O_2+O+N_2) augments the local effective oxygen index from ~0.10 to ~0.16, which is then maintained at that level as the system attempts to attain stable burning. As the oscillatory burning proceeds, the peak X_{O2} in the surrounding is maintained at ~ 0.14.

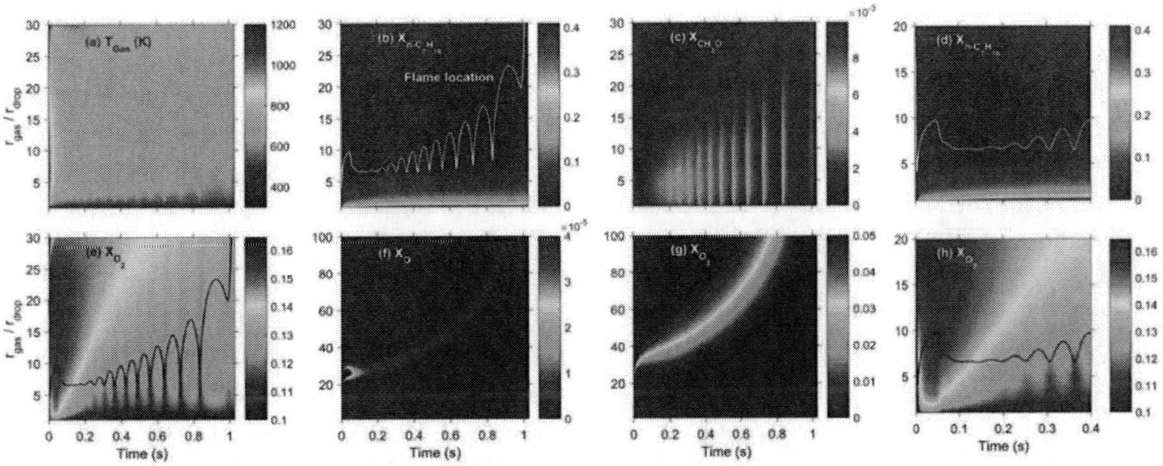

Fig. 2. Spatiotemporal evolution of (a) T_{gas}, (b) $X_{n\text{-}C7H16}$ with FSR, (c) X_{CH2O}, (e) X_{O2} with FSR, (f) X_{O}, and (g) X_{O3} for n-C_7H_{16} droplet combustion. Subplot (d) and (h) shows $X_{n\text{-}C7H16}$ and X_{O2} distribution during the early transients (d_o = 0.50 mm, X_{O2} = 10%, X_{O3} = 5%, balance X_{N2}, 1 atm).

The pulsation behavior is initiated as the reaction zone X_{O2} approaches ~0.12. The local unavailability of the oxidizer drives the flame to move outward during which unreacted X_{O2} diffuses inward to the droplet surface. As the X_{O2} increases in the reaction zone, the flame temperature increases and it contracts, depleting the local X_{O2}, and the cyclic process repeats. As the pulsation progresses, flame expansion and contraction increases, enabling more X_{O2} to diffuse through the reaction zone resulting in a progressive increase of T_{max} at each pulse.

The dynamics of the oscillatory flame is further analyzed through the temperature and selective low-temperature species evolution for one complete cycle with a time window ~0.558-0.638 s, presented in Fig. 3. The figure illustrates T_{max} and maximum mole fraction of $\Sigma C_7H_{14}O_3$ and ΣC_7H_{14} profiles. These two species (with cumulative isomers) are considered to represent the extent of influence of hydroperoxyheptyl radical ($C_7H_{14}OOH$) consumption from the competition between (a) chain propagation/termination and (b) degenerate chain branching reactions. The reaction class (a) converts $C_7H_{14}OOH$ primarily to n-C_7H_{14} (chain termination via HO_2 elimination) and $C_7H_{14}O$+OH (chain propagation). To the contrary, the degenerate chain branching reactions convert $C_7H_{14}OOH$ to $C_7H_{14}O_3$ through molecular O_2 addition and internal hydrogen abstraction. Here T_{max} decreases from a higher value of ~746 K to a saddle value of ~682 K and then increases to a higher value of ~750 K before entering the next cycle. Mole fraction evolution of n-C_7H_{14} can be directly correlated with the temperature field while the evolution of $C_7H_{14}O_3$ passes through three distinct stages, color labeled as 1–3 in Fig. 3.

In general, under atmospheric pressure condition, the favorable maximum temperature for $O_2C_7H_{14}OOH$ production hovers around 700 K. As can be seen in the figure, the cycle starts from a higher temperature with a diminishing production rate of $O_2C_7H_{14}OOH$ (precursor for $C_7H_{14}O_3$) and the subsequent result is reflected on the mole fraction of $C_7H_{14}O_3$. Subsequently, a reduction in heat generation *via* the exothermic O_2QOOH = $C_7H_{14}O_3$+OH reaction class is discernible. Moreover, the initially high temperature starts to decompose $C_7H_{14}O_3$ through the endothermic reaction class $C_7H_{14}O_3$ = R'CHO+R'(CO)R''+OH (stage 1). This results in a sharp decrease in temperature as the reaction zone cannot sustain the same field temperature from the contribution of exothermic chain branching (QOOH = OH+QO) alone. Therefore, the peak n-C_7H_{14} mole fraction also decreases with time. With the continuation of temperature decreasing the slope of the $C_7H_{14}O_3$ profile changes and at ~0.57 s reaches a saddle point '**A**' (time-coordinate) is reached, beyond which the $C_7H_{14}O_3$ starts to increase. The gas phase temperature falls below a 'threshold value' that supports the formation of $C_7H_{14}O_3$. This reactivates the exothermic O_2QOOH = $C_7H_{14}O_3$+OH reaction-class and is reflected in the temperature slope change after ~0.58 s. Thus, the simultaneous contribution of heat release from both the degenerate chain branching and chain propagation routes enables the peak temperature to recover. Such dual heat contribution continues to progress up to point '**B**' ~0.63 s where the temperature increases sufficiently to force to shut down the $C_7H_{14}O_3$ production. In essence, in the pulsing regime, the kinetics processes switch back and forth crossing the NTC turnover temperature with every pulsation. This dynamic competition between these two reaction classes over the entire burn time can also be understood in terms of the spatiotemporal evolution of ketohydroperoxide ($C_7H_{14}O_3$) and cyclic ether ($C_7H_{14}O$) presented in Fig. 4, along with an inlay of the rescaled peak gas temperature evolution.

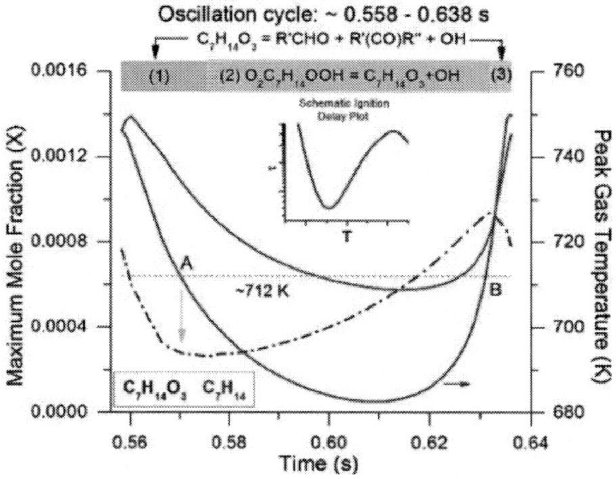

Fig. 3. Temporal evolution of peak $X_{C7H14O3}$ and $X_{n\text{-}C7H14}$ and T_{max} for an oscillation cycle at ~0.558-0.638 s for the base case. Solid-dash line demarcates the temperature responsible for driving the $C_7H_{14}O_3$ kinetics: threshold temperature for $C_7H_{14}O_3$ kinetics.

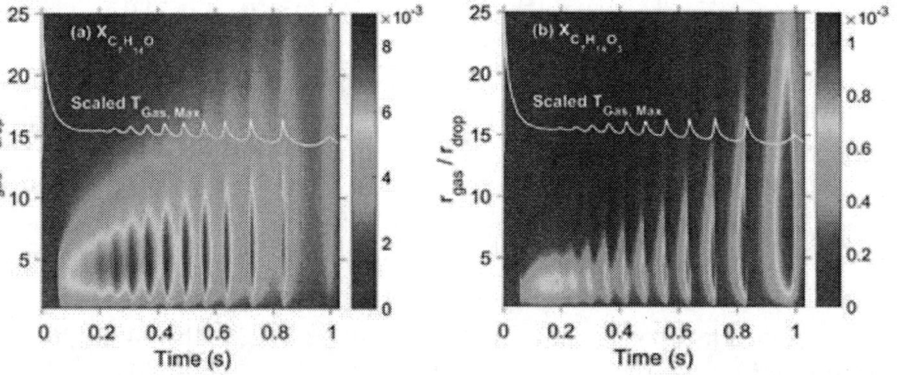

Fig. 4. Spatiotemporal evolution of (a) X_{C7H14O} and (b) $X_{C7H14O3}$ for n-heptane droplet combustion (d_o = 0.50 mm, X_{O2} = 10%, X_{O3} = 5%, balance X_{N2}, 1 atm). r_{gas}/r_{drop}=1 denotes the droplet surface.

A scaled peak gas temperature fluctuation is also overlaid for visual correlation of the species and temperature.

4. Summary

The dynamics of the oscillatory cool flame of isolated submillimeter sized n-heptane droplets was numerically investigated under reduced oxygen index condition at atmospheric pressure. At low oxygen indices, a quasi-steady cool flame burn was unattainable and the system transitioned to a continuous pulsing, cool flame droplet combustion mode. Unlike its stable counterpart, the oscillatory cool flame initially attempts to maintain a stable cool flame for a brief period of time prior to the switch to oscillatory behavior. Such transitions occur due to the depletion of the O_2 near the droplet surface followed by an expansion of the flame to the outer zone. The simulation shows that the oscillation magnitude progressively grows both in time and space until the fuel is completely depleted. Analyses further indicate that two factors contribute this anomalous cool flame behavior- (i) dynamic interaction of degenerate chain branching and chain termination/propagation reaction classes of QOOH associated with the low temperature and NTC kinetic regimes, and (ii) continuous fuel leakage across the flame location. It is found that over a single oscillation, the progression of QOOH➔chain propagation/termination reactions directly correlate with the gas phase temperature while QOOH➔degenerate branching reactions undergo three distinct stages within the same oscillation.

5. Acknowledgements

The financial support from the National Aeronautics and Space Administration (NASA) through the grant NNX17AF97A is acknowledged.

6. References

[1] V. Nayagam, D.L. Dietrich, P.V. Ferkul, M.C. Hicks, F.A. Williams, Combust. Flame 159 (2012) 3583-3588.
[2] V. Nayagam, D.L. Dietrich, M.C. Hicks, F.A. Williams, Combust. Flame 162 (2015) 2140-2147.
[3] D. Dietrich, V. Nayagam, M. Hicks, P. Ferkul, F. Dryer, T. Farouk, B. Shaw, H. Suh, M. Choi, Y. Liu, C.T. Avedisian, F. Williams, Microgravity Sci. Technol. 26 (2014) 65-76.
[4] R. Fairlie, J.F. Griffiths, K.J. Hughes, H. Pearlman, Proc. Combust. Inst. 30 (2005) 1057-1064.
[5] S.H. Won, B. Jiang, P. Diévart, C.H. Sohn, Y. Ju, Proc. Combust. Inst. 35 (2015) 881-888.
[6] C.B. Reuter, S.H. Won, Y. Ju, Combust. Flame 166 (2016) 125-132.
[7] A. Cuoci, M. Mehl, G. Buzzi-Ferraris, T. Faravelli, D. Manca, E. Ranzi, Combust. Flame 143 (2005) 211-226.
[8] T.I. Farouk, F.L. Dryer, Combust. Flame 161 (2014) 565-581.
[9] T.I. Farouk, D. Dietrich, F.E. Alam, F.L. Dryer, Proc. Combust. Inst. 36 (2017) 2523-2530.
[10] F.E. Alam, S.H. Won, F.L. Dryer, T.I. Farouk, Combust. Flame (2017, Recently accepted).
[11] T.I. Farouk, M.C. Hicks, F.L. Dryer, Proc. Combust. Inst. 35 (2015) 1701-1708.
[12] A. Cuoci, A. Frassoldati, T. Faravelli, E. Ranzi, Proc. Combust. Inst. 35 (2015) 1621-1627.
[13] A.J. Marchese, F.L. Dryer, V. Nayagam, Proc. Combust. Inst. 116 (1999) 432-459.
[14] T. Farouk, F.L. Dryer, Combust. Theory Modell. 15 (2011) 487-515.
[15] W. Sun, S.H. Won, T. Ombrello, C. Carter, Y. Ju, Proc. Combust. Inst. 34 (2013) 847-855.
[16] T.E. Daubert, R.P. Danner, Physical and Thermodynamic Properties of Pure Chemicals, Hemisphere Publishing Corp. , New York, USA, 1989.

Sub Topic: Sprays and Droplets

Spring Technical Meeting
Eastern States Section of the Combustion Institute
March 4–7, 2018
State College, Pennsylvania

Motion of a Merged Drop Upon Impacting a Liquid Pool

Abhishek Saha[1], *Yanju Wei*[1,2], *Xiaoyu Tang*[1], *and Chung K. Law*[1,*]

[1]*Department of Mechanical and Aerospace Engineering, Princeton University, New Jersey, United States of America*
[2]*School of Energy and Power Engineering, Xi'an Jiaotong University, Xi'an, China*
[*]*Corresponding author:* `cklaw@princeton.edu`

Abstract: Drop impact on liquid films is relevant to many natural and industrial processes. Examples include the impact of sprayed fuel droplets on the fuel films formed from previously deposited droplets on the walls and piston head of internal combustion engines, and the impact of rain drops on ocean and wetted soil surfaces. Since the impacted drop can contain liquids with different temperature and chemical species than the impacted film, it is the motion of the merged drop and the associated mixing that initiates many post-merging thermal and chemical processes. In this paper we report an experimental study on the morphological evolution of an impacting drop inside the liquid pool after coalescence, using high speed laser induced fluorescence technique. We shall show that the drop motion can be divided into three stages, where inertia, capillary and viscous effects alternatingly play dominant roles to characteristically modulate the drop penetration. Furthermore, we shall also evaluate the relevant time and length scales of these three stages, and propose a unified description of the drop motion in the impacted liquid. Finally, we will use the experimental data from a wide range of drop diameter and impact speed to validate the proposed scaling.
Keywords: ***Drop Impact, Merging***

1. Introduction

When an impacting drop merges with a liquid pool, it inherently generates vortex rings due to the strong shear between the moving mass from the drop and the surrounding stagnant liquid. In general, formation of the vortex enhances the mixing between the two liquid masses, which is the precursor to subsequent processes critical in natural and industrial applications. For example, the impaction and subsequent mixing of rain drops on river or ocean promotes the transport of atmospheric chemicals to the water bodies on the terrestrial surfaces. The burning of the hypergolic fuels in rocket engines rely largely on the mixing efficiency and liquid phase reaction, which is controlled by the contact area of the fuel and the oxidizer, as well as the mixedness at their interfaces. In inkjet printing, both in 2D and 3D, the immediate mixing of injected ink material with the pre-deposited layer is critical to promoting the polymeric reactions and thus bonding, leading to the uniformity of printed patterns. Thus, an understanding of the motion and mixing of the drop after subsequent to merging is of interest in many practical situations.

The generation of vortices during the drop impact on a liquid pool has been studied for decades [1–3]. Some of the studies focused on the initial stage of the vortex ring generation [4], which largely affects the splashing behavior [5]. For the long-time penetration process, Rodriguez and Mesler [6] studied the effect of drop shape on the penetration of the vortex ring, while Durst [7]

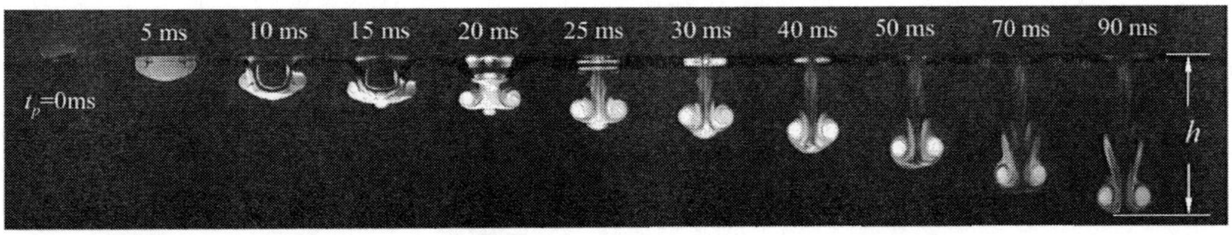

Figure 1: A typical penetration sequence of a water drop after merging with the liquid pool. $D = 2.0$ *mm*, $We = 16.8$.

studied the evolution of the diameter of the vortex ring and the maximum penetration depth as functions of the impact speed. However, a quantitative description and scaling of the vortex penetration process is lacking. Recognizing such a need, in this work, we study herein the propagation front of the vortex ring for an extended time and define three stages of penetration, each of which is controlled by inertial, capillary or viscous effects. We will analyze the controlling mechanisms in each stage and provide scalings which predict the penetration behavior reasonably well.

2. Experimental Method

To observe the vortex behavior inside the liquid pool, Laser Induced Fluorescence of dyed drop is performed to distinguish the liquid mass of the drop from that of the pool. Particularly, a trace amount of Rhodamine-6G dye was used to dope the drop, which then lands on the liquid pool where a vertical thin green laser sheet intersects. A synchronized high-speed camera with band-pass filter captured the fluorescence images from the drop. For this study, water was chosen to be the test liquid instead of oil, since the dye does not dissolve in oil. A large pool thickness (10 times larger than drop radius) was selected to study the vortex penetration for longer time without any effects from the bottom substrate. Three drop sizes ($D = 2, 3, 3.5$ *mm*) and impact speeds with We in the range of 5 to 40 were used.

3. Results and Discussion

3.1 General description of the penetration process

When a drop impacts a liquid surface, it squeezes the interfacial gas layer, which renders merging difficult from the resistance against the drop motion. If the impact inertia is large enough, eventually the gas layer is squeezed out and the interfaces come close enough for the van der Waals force to induce merging. Since we focus on the motion of the drop after merging, we only consider the experiments with sufficiently high impact inertia to induce instantaneous merging of the interfaces upon the impact.

The motion of a merged drop during a typical impact process can be illustrated through a series of fluorescence images, shown in Fig. 1, which demonstrate several notable characteristics. Specifically at very early stage of the impact, when the drop starts penetrating into the stagnant liquid pool, a vortex ring is generated around the outer periphery of the drop due to the large shear rate (Fig. 1, $t = 5$ *ms*) and becomes larger as the drop penetrates into the pool. When the rear end of the drop falls below the liquid surface, a crater is developed on the top surface of the liquid pool, which pushes the drop downwards until the liquid surface starts relaxing (Fig. 1, $t = 20$ *ms*) to

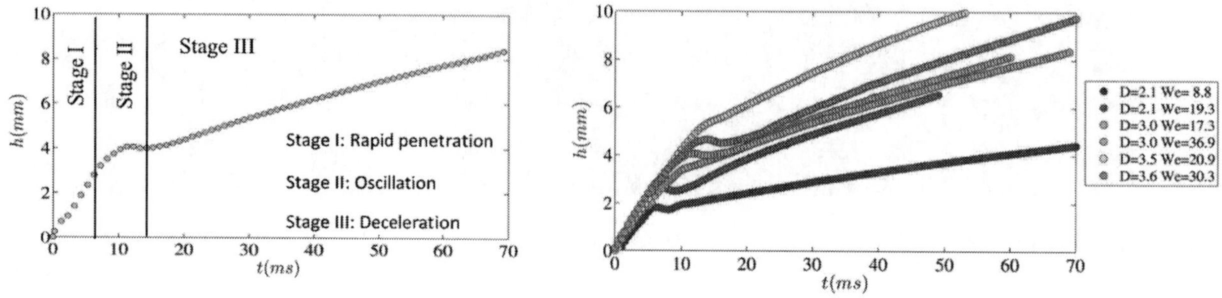

(a) (b)

Figure 2: Penetration depth evolution for (a) one experiment, (b) multiple experiments ($D = 2 - 3.5\ mm$, $We = 8 - 50$).

retain its original unperturbed state. Subsequently, the vortex ring becomes smaller as the vortices approach each other and continue the penetration.

Although several dynamic and characteristic analysis can be performed on this process, here we are interested in the overall downward motion or penetration process of the drop. Specifically, the primary quantity we are interested in is the penetration depth, $h(t)$, which is the distance between the bottom-most point of the drop and the unperturbed liquid surface at a given time, as defined in Fig. 1. The temporal evolution of $h(t)$ is plotted for an experiment in Fig. 2a, where we can clearly identify three distinct stages of the penetration process. That is, immediately after the impact, a steady increase of $h(t)$ with almost a constant slope is observed. The drop motion in this stage, termed as *rapid penetration*, is primarily controlled by the impact inertia and as such the slope is almost constant. Next, we observe a non-monotonic penetration process with $h(t)$ increasing and then slight decreasing with time, resembling an oscillatory behavior. We shall show later that this stage of *oscillation* is due to the wave created by the impact on the liquid surface. Finally, we observe a monotonic penetration process with steady decay in the slope at the final stage or *deceleration* due to viscous dissipation. Evolution of the penetration depth is compared for different drop diameters and We in Fig. 2b, where we observe no clear similarity, which suggests the three stages are characterized by different length and time scales.

3.2 Inertial effect

At the rapid penetration stage, since the liquid surface is not readily deformed and viscous dissipation is not effective in a short time, impact inertia is the controlling mechanism. We can simply use the inertial time scale, $t_{iner} = 2R/U$ as the characteristic time and the drop radius (R) as the characteristic length to normalize the penetration data. In Fig. 3a, the penetration history is plotted in the normalized form for a wide range of drop diameters and We, which clearly shows collapse of the data until $t/t_{iner} \approx 1$, at which the data from different experiments diverge. It is also noted that the slope of the normalized data is close to unity, suggesting that penetration velocity of half of the impact velocity, $U_{p,0} = U/2$. Mechanistically, after the impact, the drop displaces the liquid in the pool, and in the process part of the impact kinetic energy is converted to the kinetic energy of the pool. So the penetration process in this stage scales linearly with time and can be expressed as

$$h(t) = Ut/2 \ \ for\, t \leq t_{iner} \qquad (1)$$

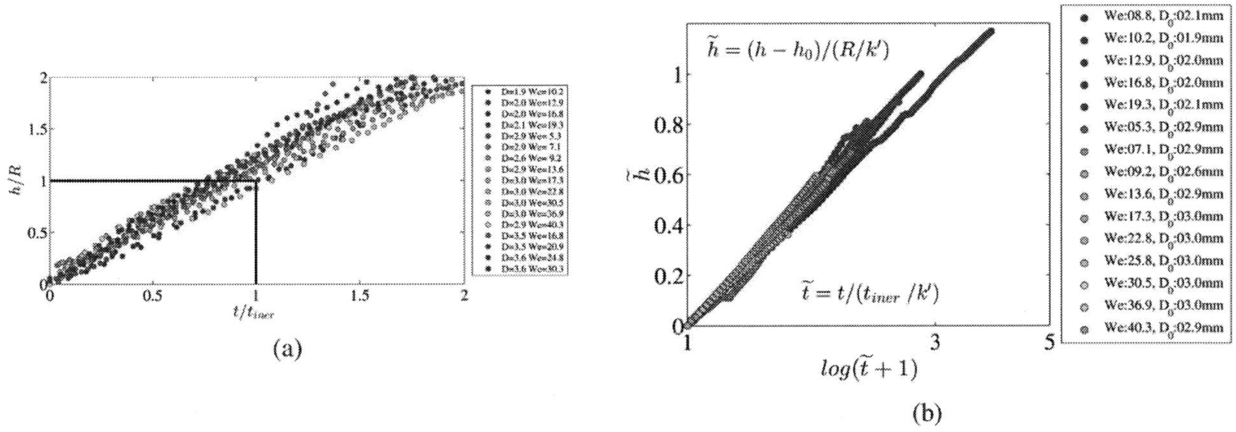

Figure 3: (a) Normalized penetration depth h/R vs normalized time t/t_{iner}. $t_{iner} = 2R/U$ (b) Penetration depth evolution considering the viscous drag effect.

3.3 Capillary Oscillation

Due to the initial impact inertia, as the drop penetrates into the liquid pool, it also creates a crater on the surface of the liquid pool. The crater eventually reaches a maximum depth due to capillary effect, and then the highly distorted liquid surface retracts to attain its original shape. This retraction decelerates the drop motion and in some cases the drop can even momentarily move upwards as seen in Fig. 2a. The capillary oscillation of the pool surface can be characterized by two parameters, the maximum crater depth $H_{md}/R \approx We^{1/2}$ and the time taken to form the maximum crater depth $t_{md} \approx t_{cap}$.

To evaluate the effect of crater formation on the drop motion, we first formulate the crater formation process, which can be expressed as a capillary wave moving through the surface of the pool. Since the crater (a cavity in the pool surface) formation begins when the top of the merged drop moving with velocity, U, reaches the unperturbed liquid surface, one can express the characteristic velocity of the crater front as, $U_c = U\cos[\omega(t - t_{iner})]$, where $\omega = 2\pi/t_{cap}$ is the capillary frequency. One can integrate the crater front velocity to arrive at the equation for the crater depth, $h_c = \frac{U}{\omega}\sin[\omega(t_p - t_{iner})]$. The effect of the capillary wave as manifested by the formation and relaxation of the crater modulates the drop penetration, as long as the drop is in close vicinity of the crater. Conservatively, one can assume that the effect will last for half of the capillary-period when the pool surface reaches its initial state during the retraction process. Thus, the drop penetration process under the capillary effect can be expressed as,

$$h(t) = R + \frac{Ut_{cap}}{2\pi}\sin(\frac{2\pi}{t_{cap}}(t - t_{iner})) \;\; for\, t_{iner} \le t \le t_{iner} + t_{cap}/2 \qquad (2)$$

3.4 Viscous drag

When the drop moves away from the influence of the capillary wave on the pool surface, it almost resembles the motion of a sphere in a stagnant liquid. The dye is observed to be localized in a zone encompassing the vortex ring, which we assume a spherical shape to describe the merged drop. Since the mixing of the dye with the surrounding liquid via the vortex is still not significant,

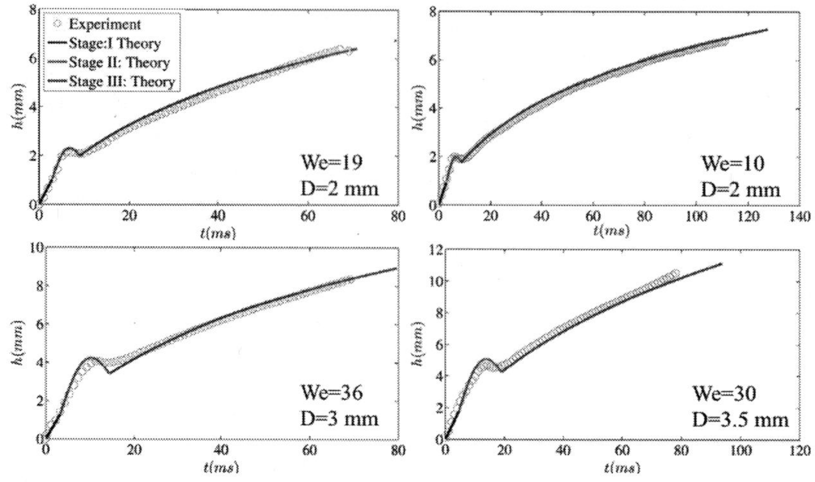

Figure 4: Comparison of the penetration depth evolution between experimental data and theoretical prediction. (a) $D = 2\ mm$, $We = 19$. (b) $D = 2\ mm$, $We = 10$. (c) $D = 3\ mm$, $We = 36$. (d) $D = 3.5\ mm$, $We = 30$.

the entertainment of the dye-free water from the pool is neglected here. Thus the motion of the drop can be modeled as a sphere moving through a stagnant liquid, and as such we can write $\rho \cdot \frac{4}{3}\pi R_S^3 \frac{dU_p}{dt} = -\pi R_S^2 \cdot \frac{1}{2}\rho U_p^2 \cdot C_D$, where C_D is the drag coefficient. Integrating this expression we get,

$$h(t) = h_0 + \frac{1}{k} ln(1 + U_{p,0}kt) = h_0 + \frac{R}{k'} ln(1 + k'\frac{t}{t_{iner}}) \;\; for\, t \geq t_{iner} \qquad (3)$$

where, $k' = 3C_D/(8R_S/R)$. From experiments we observe that the size of the sphere, R_s/R is rather constant and is close to unity. For large Re, C_D is also constant. We find that for $C_D = 0.3$, the penetration history in the third stage of *deceleration* for all experimental conditions can be collapsed reasonably well (Fig. 3b).

3.4.1 Unified description of the Penetration process

Since we have identified three mechanisms that control the drop motion through Eqns. 1–3, we can now simply combine them to arrive at a unified description of the penetration process. It is interesting to note that there is no single time scale that can be used for all three stages of motion. For example, while t_{iner} is the relevant time scale for both stages I and III, in stage II both t_{iner} and t_{cap} are present. Since the changes in the drop diameter alter both the t_{iner} and t_{cap}. while changes in the impact speed alters t_{iner}, each experimental condition contains a unique set of t_{cap} and t_{iner}. Thus, the penetration history at all the stages cannot be collapsed by using a single non-dimensional time. Instead, in Fig. 4, we compare the dimensional experimental $h(t)$ vs t plot with the theoretical prediction. The reasonably good agreement between the experimental results and theoretical prediction demonstrates that the theory captures the physics well.

4. Conclusions

In this paper, we have experimentally visualized and analyzed the post-merging motion of a drop impacting on a liquid pool. We have used laser induced fluorescence of the dyed drop to differentiate it from the liquid of the pool. Vortex rings were observed to form and penetrate in the liquid pool. Measurement of the penetration history of the drop depicts three stages, namely, *rapid penetration*, *oscillation* and *deceleration*. Furthermore, three mechanisms were identified which controls the motion of the merged drop (or dyed mass) in these stages. Initially the penetration is controlled by the drop impact inertia, which is predominant in the *rapid penetration* stage. The impact creates a crater at the pool surface, leading to a capillary wave which subsequently affects the drop motion in the *oscillation* stage. Finally, the merged drop is slowed down by the viscous drag due to the relative motion between the drop and the liquid pool. This drag is effective in both *oscillation* and *deceleration* stages. Equations of motion based on these individual effects and subsequently for three stages have been derived, which show good agreement with the experimental data. Furthermore, theoretical analysis shows the importance of both capillary and inertial time scales at different stages and, thus, a single timescale representing the whole penetration process cannot be defined.

5. Acknowledgements

The work at Princeton University was supported by the Army Research Office under Grant (W911NF-16-0449) and by the Xerox Corporation under a UAC grant.

References

[1] D. S. Chapman and P. R. Critchlow, Formation of vortex rings from falling drops, J. Fluid Mech. 29 (1967) 177–185.

[2] P. N. Shankar and M. Kumar, Vortex rings generated by drops just coalescing with a pool, Phys. Fluids 7 (1995) 737–746.

[3] J. R. Saylor and N. K. Grizzard, The optimal drop shape for vortices generated by drop impacts: the effect of surfactants on the drop surface, Exp. Fluids 36 (2004) 783–790.

[4] B. S. Dooley, A. E. Warncke, M. Gharib, and G. Tryggvason, Vortex ring generation due to the coalescence of a water drop at a free surface, Exp. Fluids 22 (1997) 369–374.

[5] J. San Lee, S. J. Park, J. H. Lee, B. M. Weon, K. Fezzaa, and J. H. Je, Origin and dynamics of vortex rings in drop splashing, Nat. Commun. 6 (2015).

[6] F. Rodriguez and R. Mesler, The penetration of drop-formed vortex rings into pools of liquid, J. Colloid Interface Sci. 121 (1988) 121–129.

[7] F. Durst, Penetration length and diameter development of vortex rings generated by impacting water drops, Exp. Fluids 21 (1996) 110–117.

Spring Technical Meeting
Eastern States Section of the Combustion Institute
March 4-7, 2018
State College, Pennsylvania

Droplet Size and Velocity Measurements in a Heptane/Propane Spray Flame

Brian T. Fisher *, Steven G. Tuttle, Alfredo D. Tuesta*[1]

Chemistry Division, Naval Research Laboratory, Washington, DC, USA
**Corresponding Author Email: brian.fisher@nrl.navy.mil*

Abstract: Two-phase fuel spray diffusion flames are relevant to many applications (e.g., oil well fires), but are not well understood or characterized. This paper describes experiments in a laboratory-scale, piloted spray burner that uses an open-pipe atomizer (OPA) to deliver a two-phase fuel flow. With propane and heptane, respectively, as the vapor and liquid fuel constituents, a phase Doppler anemometry (PDA) instrument was used to measure droplet sizes and velocities for a range of fueling conditions. From these data, axial and radial profiles of size and velocity were constructed and relevant droplet statistics were analyzed. While most of the data showed expected behavior, at higher propane flow rates there was an unexpected emergence of bimodal axial velocity distributions in the shear layer at a height equivalent to approximately 15 fuel tube diameters. The exact cause of this behavior is not yet known and remains a topic of ongoing study.
Keywords: Spray Flame, Open Pipe Atomizer, Phase Doppler Anemometry, Droplet

1. Introduction

Turbulent spray flames are commonly found in a variety of combustion applications, including diesel engines, gas-turbine engines, and oil and gas fires. There are additional features that make oil and gas fires unique, however. For one, two fuel phases (liquid and gas) are simultaneously present in the spray, with the liquid representing a large fraction of the fuel. Researchers have studied two-phase fuel spray flames to gain insight on their behavior, though these studies have been somewhat sparse. Purdue University researchers have measured flame heights and examined radiation and emission behavior of spray flames using toluene and crude oils as liquid fuels and primarily methane as the gaseous fuel [1-4]. These studies were all conducted using an effervescent atomizer, which injects the gaseous fuel into the liquid fuel inside a tube at a relatively short distance from an exit orifice. In a real-world oil/gas fire, however, both fuel phases travel together a long distance through an open pipe, leading to an annular dispersed flow at the pipe exit [5].

At the Naval Research Laboratory, a laboratory-scale, piloted spray burner with an open-pipe atomizer has been set up to study two-phase, turbulent spray flames that simulate real-world oil/gas fires and enable detailed optical measurements of flame behaviors and characteristics. This paper reports phase Doppler anemometry measurements of droplet sizes and velocities in a spray flame using propane and heptane, respectively, as the vapor and liquid fuel constituents.

---

[1] National Research Council (NRC) Postdoctoral Associate

2. Experimental Setup

The centerpiece of this study is the spray burner, shown in Figure 1. The outer section of the burner, which includes a series of perforated metal plates and a honeycomb disk (25 mm thick with 0.76-mm cells), delivers an approximately uniform outlet flow of premixed hydrogen (H_2) and air. The central tube, which is a simple open-pipe atomizer (OPA), delivers the two-phase fuel flow of interest for the spray flame. The H_2/air mixture issuing from the outer section is used to support a flat flame just above the exit of the honeycomb. The flat flame serves as a pilot that helps ignite and stabilize the central fuel jet. The diameter of the overall burner exit is 99 mm (3.9 inches) and the central fuel tube has an outside diameter of 6.4 mm (0.25 inches) and an inside diameter of 0.84 mm (0.033 in). A three-dimensional traverse system is used to translate the burner to obtain spatially-resolved measurements and construct profiles.

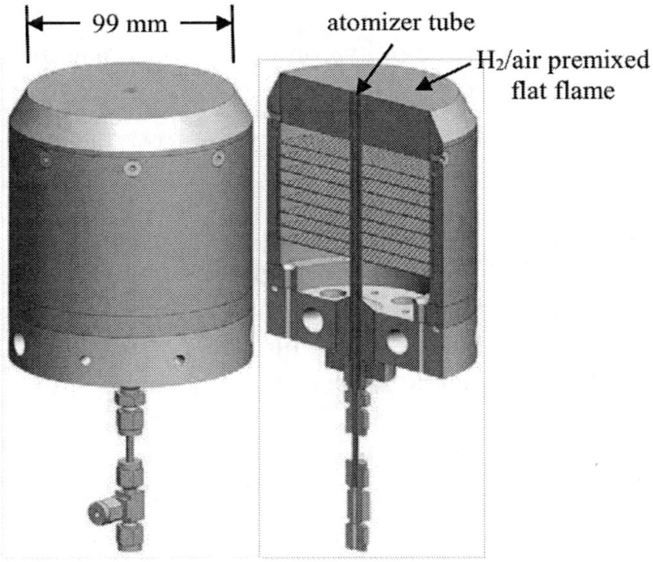

(a)

(b)

Figure 1: Schematic of spray burner (a) and photograph of heptane/propane spray flame (b).

In this study, the two-phase fuel consisted of propane (gas) and heptane (liquid). Propane flow is controlled by an Alicat Scientific mass flow controller and heptane flow is controlled by an Ismatec piston pump. Though not shown in detail in this paper, a custom mixing tee was designed to combine the gas and liquid flows and create the desired annular flow in the central fuel tube. In addition, a pulsation dampener was used to smooth out flow oscillations and deliver an approximately steady liquid flow. As shown in Table 1, three experimental cases were studied, in which propane flow rate was varied and heptane flow rate remained constant.

Table 1: Burner conditions for PDA droplet measurements.

Propane flow rate [g/s]	*Heptane flow rate [g/s]*	*Liq/Gas Mass Ratio*
0.1	0.4	4.0
0.2	0.4	2.0
0.3	0.4	1.3

A two-dimensional phase Doppler anemometry (PDA) system by Dantec Dynamics was used to measure droplet diameters and velocities in the spray flame. Measurement principles of PDA are well-documented in the literature and thus are not discussed here. The details of the PDA system used in this study also can be found in the literature [6]. The PDA software was configured to collect 10,000 samples at each spatial location, but with a 10-second time limit. Typically, a precipitous decrease in data collection rate is observed when the measurement reaches and goes beyond the outer boundary of the spray cone. The diameter statistic of particular interest is the Sauter mean diameter (D_{32}), calculated according to Eq. (1):

$$SMD = D_{32} = \frac{\Sigma N_i D_i^3}{\Sigma N_i D_i} \tag{1}$$

In this expression, N is the number of droplets in a particular statistical bin of the probability distribution and D is the diameter of that bin. Sauter mean diameter (SMD) statistically represents the volume-to-surface area ratio of the droplet distribution, and is an important spray parameter. The two velocity components measured by the PDA system are the vertical (axial) component and one of the horizontal components. The horizontal velocity component is considered to be a radial velocity, since the burner is nominally axisymmetric.

3. Results and Discussion

Profiles of SMD are shown in Figure 2, including axial profiles at the burner nominal centerline ($r = 0$ mm) and radial profiles at a height of 30 mm above the burner ($z = 30$ mm). For each flow rate, SMD decreases with increasing axial distance due to mass and heat transfer until it levels off by ~25 mm. The fact that SMD levels off rather than continuing to decrease may be due to the competing effects of heat/mass transfer and droplet coalescence. At 30-mm height, which is in the SMD plateau region, SMD is minimum at the centerline, increases as radial distance increases, and levels off by ~4 mm away from center (noting that the actual flame center is at approximately −1 mm rather than at the nominal center of 0 mm). In addition, SMD converges such that there is little difference between the 0.2- and the 0.3-g/s cases. Though it remains to be proven, it is possible that there is a transition from subsonic flow at 0.1 g/s to supersonic flow at the higher flow rates, and that this leads to shock structures that rapidly rip apart the droplets.

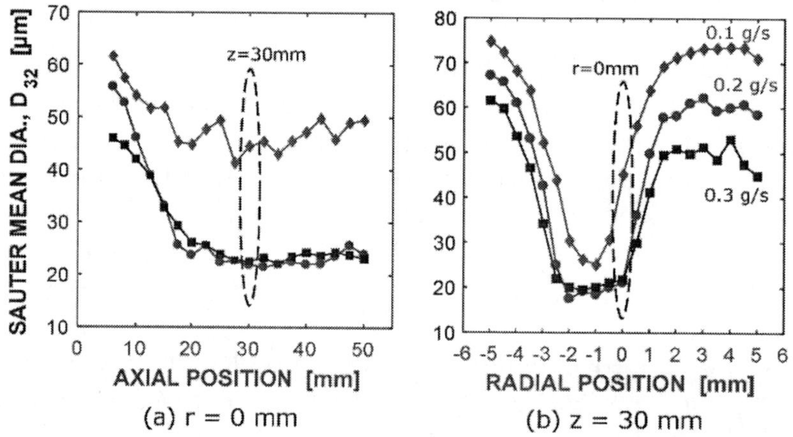

Figure 2: Sauter mean diameter profiles. Axial profiles (a) correspond to the burner centerline (r = 0 mm). Radial profiles (b) correspond to height of z = 30 mm above the burner.

Profiles of the axial (U) and radial (V) velocity components are shown in Figure 2, where each value is averaged over all (typically 10,000) data samples at each location. Radial profiles show that U is highest at the centerline and decreases with increasing radial distance. As axial distance increases, the droplets initially accelerate (i.e., U increases) rapidly, level off, then steadily decelerate. Compared to U, V is extremely low, indicating that the droplets are moving almost entirely in the axial direction. As expected, both U and V increase as propane flow rate increases.

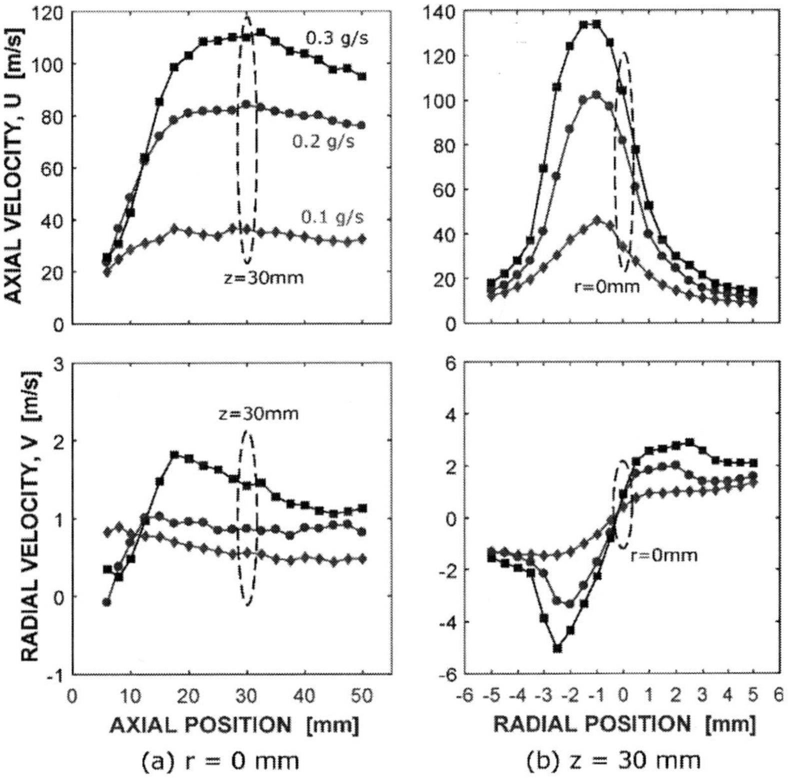

Figure 3: Velocity profiles. Axial profiles (a) correspond to the burner centerline ($r = 0$ mm). Radial profiles (b) correspond to height of $z = 30$ mm above the burner.

Droplet size distributions (probability density functions, or PDFs) are shown in Figure 4 for selected locations. The PDFs have the expected overall shape, and it is clear that they become increasingly broad as radial distance increases. Consistent with the SMD values shown in Figure 2, the size distributions are quite similar at the two different heights (20 mm and 30 mm), while they broaden, and thus lead to larger SMDs, with increasing flow rate.

Axial velocity distributions (PDFs) are shown in Figure 5 for selected locations. For the lowest propane flow rate (0.1 g/s), the PDFs have the expected shape and the shifts from higher to lower velocities with increasing radial distance are consistent with the average U values shown in the top panel of Figure 3b. For the higher propane flow rates, however, unexpected behavior emerges. At the true centerline of the spray flame ($r = -1$ mm), the PDFs have the familiar single-mode shape. With increasing radial distance, however, the PDFs become bimodal. The effect is more pronounced for the highest flow rate, and is increasingly pronounced as radial distance increases. The bimodal behavior was not observed below $z = 12$ mm at any radial location, thus it is not due to oscillating or pulsating liquid flow from the piston pump. It is speculated that the higher propane flow rates lead to supersonic velocities, and that the bifurcation in droplet velocity is related to some kind of unsteady shock effect in the shear layer. Further work and in-depth analysis is needed to better understand the bimodal velocity and determine its cause.

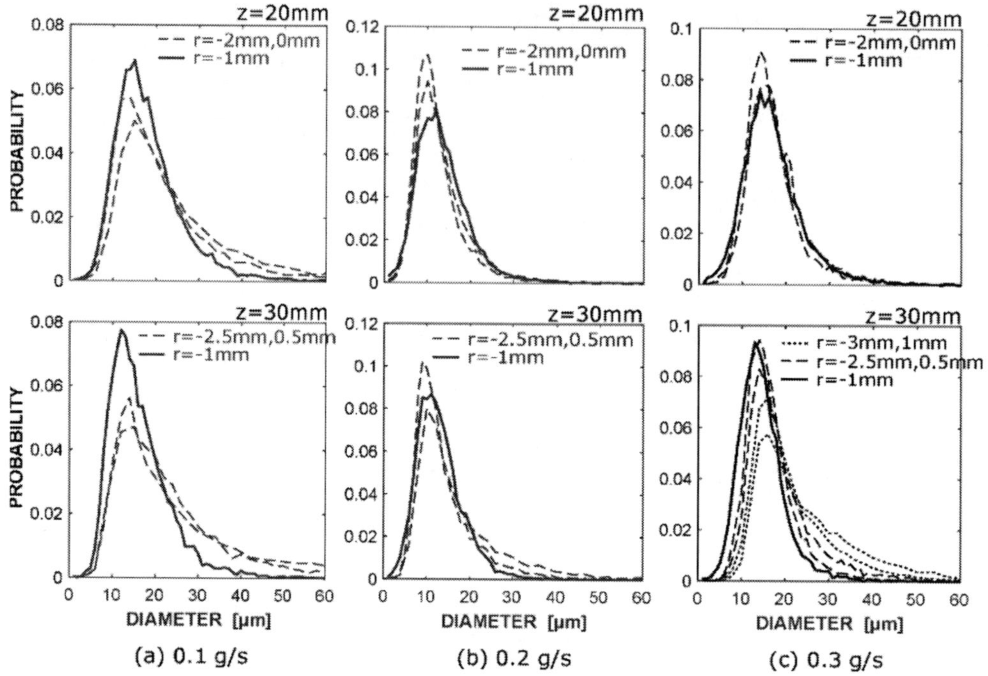

Figure 4: Diameter probability density functions. The true centerline of the spray is $r = -1$ mm.

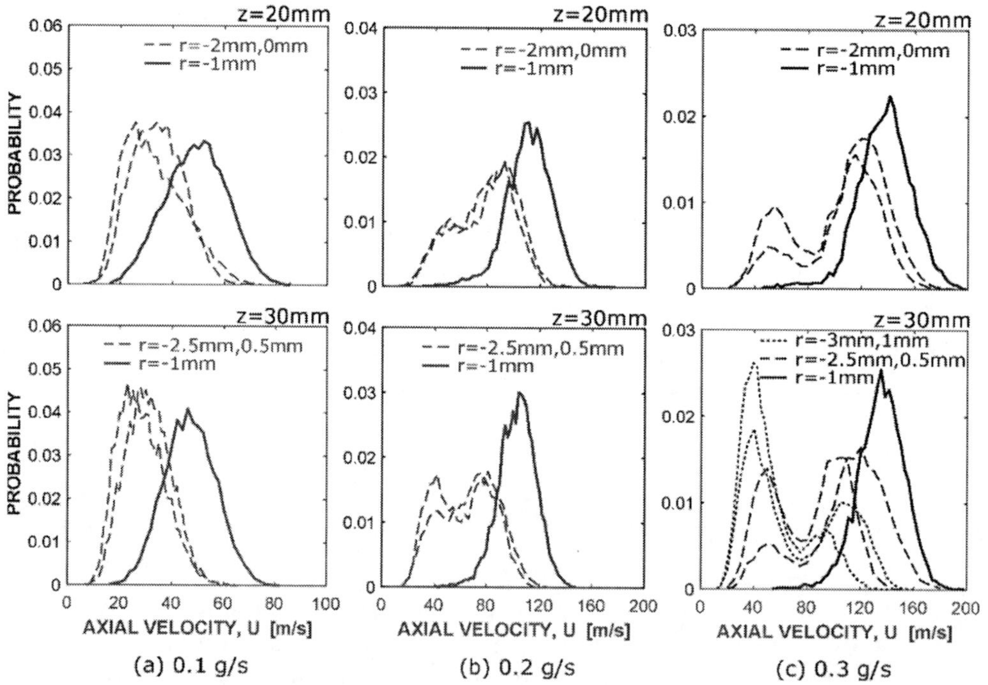

Figure 5: Velocity probability density functions. The true centerline of the spray is $r = -1$ mm.

4. Summary and Conclusions

A phase Doppler anemometry instrument was used to measure droplet sizes and velocities in a heptane/propane turbulent spray flame, with a premixed H_2/air flat flame used as a pilot flame to ignite and stabilize the spray flame. The spray flame architecture used in this study simulates a real-world gas/oil fire, but in a laboratory-scale burner that enables detailed optical measurements of flame behavior. Axial and radial profiles of droplet Sauter mean diameter and axial and radial velocity followed the generally expected trends for a spray flame, and droplet size distributions showed the expected single-mode shape. At higher propane flow rates, however, axial velocity distributions displayed unique and unexpected bimodal behavior. Bimodal velocity distributions emerged in the shear layer, and not at the centerline, at heights above 12 mm (equivalent to ~15 fuel-tube diameters). The bimodal behavior was apparently not due to liquid flow oscillations, since it was not observed at all locations. It is difficult to draw any definitive conclusion from this limited data set, but since the bimodal behavior was observed only at higher flow rates, it is speculated that there was a transition to supersonic flow at these flow rates, and that the bimodal behavior resulted from some kind of unsteady shock effect in the shear layer. This remains a topic of ongoing and future work, and more data and in-depth analysis will be needed to understand the bimodal velocity and determine its cause. To the authors' knowledge, this behavior is unique and has not been observed before in spray studies with conventional atomizers.

5. Acknowledgments

This study was funded by the U.S. Department of the Interior, Bureau of Safety and Environmental Enforcement through Interagency Agreement El4PG00060 with the Naval Research Laboratory. The authors thank the National Research Council for administering the postdoctoral fellowship program.

6. References

[1] P. Dutta, Y.R. Sivathanu, J.P. Gore, An investigation of oil and gas well fires and flares, Report No. NIST-GCR-94-653, National Institute of Standards and Technology, Gaithersburg, MD, USA, 1994.

[2] P. Dutta, J.P. Gore, Y.R. Sivathanu, P.E. Sojka, Global properties of high liquid loading turbulent crude-oil + methane/air spray flames, Combust. Flame 97 (1994) 251-260.

[3] R. Wade, Y.R. Sivathanu, J.P. Gore, A study of two phase high liquid loading et fires, Report No. NIS-GCR-95-678, National Institute of Standards and Technology, Gaithersburg, MD, USA, 1995.

[4] J.P. Gore, S.M. Skinner, D.W. Stroup, D. Madrzykowski, D.D. Evans, Structure and radiation properties of large two phase flames, ASME Heat Transfer in Combustion Systems Winter Annual Meeting, San Francisco, CA (1989).

[5] C. Berna, A. Escriva, J.L. Munoz-Cobo, L.E. Herranz, Review of droplet entrainment in annular flow: Characterization of the entrained droplets, Prog. Nucl. Energ. 79 (2015) 64-86.

[6] B.T. Fisher, M.R. Weismiller, S.G. Tuttle, K.M. Hinnant, Effects of fluid properties on spray characteristics of a flow-blurring atomizer, J. Eng. Gas Turb. Power 140 (2018) 0415111-0415118.

Spring Technical Meeting
Eastern State Section of the Combustion Institute
March 4-7, 2018
State College, Pennsylvania

Droplet Flamelet-Generated Manifold Modeling for Two-Phase Turbulent Combustion

Brian T. Bojko *Paul E. DesJardin*

Department of Mechanical and Aerospace Engineering,
University at Buffalo, the State University of New York (SUNY Buffalo)
Buffalo, NY 14260-4400

A Droplet Flamelet-Generated Manifold (DFGM) is developed to account for the effects of finite-rate chemistry of individual droplet combustion during simulations of a turbulent combustion environment. A spherically symmetric droplet model is developed for methanol using the hydrocarbon and nitrogen kinetic mechanisms developed at UC San Diego to account for chemical reaction rates. The inclusion of finite-rate chemistry allows for the capturing of the transition from diffusion to kinetically controlled combustion as the droplet diameter decreases. The droplet model is used to create a DFGM by successively solving the 1D flame equations at varying drop sizes, where the source terms for energy, mixture fraction (Z), and progress variable are cataloged as a function of Z and droplet diameter. A unique coupling of the spherical and planar FGMs is developed and is used to account for individual and group combustion processes simultaneously. Three combustion models are considered when modeling a methanol spray flame; i) an evaporative model coupled with an unsteady planar Flamelet-Generated Manifold (UFGM), ii) using only the DFGM method, and iii) the DFGM coupled with the planar UFGM. The models are compared against one another as well as experimental data for a methanol spray flame with an annular air jet. The DFGM model is shown to agree well with burn rates and normalized flame radii from individual droplet burning experiments. Good overall agreement is observed between the experimental data and the DFGM and coupled UFGM models for temperature, particle size distribution, and OH concentration.

1 Introduction

The flamelet-generated manifold (FGM) method has been used to describe premixed flames [1] and non-premixed flames [2] with extensions to multi-dimensional turbulent combustion environments. The FGM approach is applied in this study due to its ability to reduce complex chemistry in a tabular form through the use of a progress-variable to describe the extent of combustion as the flame transitions between states caused by local flame strain. Previous multiphase FGM studies include n-heptane [3], α-methylnaphthalene/n-decane [4], n-dodecane [5], acetone [6], and Jet-A [7] spray flames. In these studies the droplets are assumed to only evaporate, therefore burning occurs in an external surface sheath surrounding the spray core, or a group combustion mode of burning. The focus of this study is to account for both modes of burning so the relative effects of group versus droplet burning can be better understood. To explore the use of the DFGM, three combustion submodels are investigated. First, an evaporating methanol droplet model is coupled with a planar unsteady FGM (UFGM) [8] to model combustion for interfaces with mass blowing. A Lagrangian droplet description is used to evaluate the mass, momentum, and energy transfer between the evaporating methanol droplet and the gas phase, similar to the multiphase spray formulation of Refs. [3–7, 9] and consistent with previous approaches discussed above. This model is referred to as, the "evapFGM" model and corresponds to high values of group combustion (G) [11]. The second model considers the DFGM on its own to account for droplet combustion only (low G), denoted as "DFGM". Finally, the third approach couples the DFGM with the UFGM from the first case and is referred to as, "coupFGM" and is able to account for locally varying G. All three cases (evapFGM, DFGM, coupFGM) are assessed and compared to an experimental dataset for a methanol spray flame with an annular air jet that exhibits individual droplet combustion investigated by Friedman et al. [10].

2 Model Formulation

2.1 Two-Phase LES

The multiphase system is formulated by phase-averaging of the instantaneous multiphase equations over a representative volume that is compactly defined by the filter function, $G(\boldsymbol{x}-\boldsymbol{x}')$, with the normalization property, $\int_{V_\infty} G dV = 1$. Application of the filtering operator in the context of a reaction progress-variable description for combustion results in the following system of equations,

$$\partial(\phi\bar{\rho})/\partial t + \nabla\cdot(\phi\bar{\rho}\tilde{\boldsymbol{u}}) = c^{\ddagger} \tag{1a}$$

$$\partial(\phi\bar{\rho}\tilde{\boldsymbol{u}})/\partial t + \nabla\cdot(\phi\bar{\rho}\tilde{\boldsymbol{u}}\tilde{\boldsymbol{u}}) = \nabla\cdot[\phi(-\bar{p}\underset{\sim}{\boldsymbol{I}} + \underset{\sim}{\bar{\tau}} + \underset{\sim}{\boldsymbol{T}}_{\boldsymbol{uu}})] + \phi\bar{\rho}\tilde{\boldsymbol{g}} + \boldsymbol{m}^{\ddagger} \tag{1b}$$

$$\partial(\phi\bar{\rho}\widetilde{Z})/\partial t + \nabla\cdot(\phi\bar{\rho}\widetilde{Z}\tilde{\boldsymbol{u}}) = \nabla\cdot[\phi(\underset{\sim}{\boldsymbol{T}}_{\boldsymbol{u}Z} + \bar{\boldsymbol{q}}_Z)] + c_Z^{\ddagger} \tag{1c}$$

$$\partial(\phi\bar{\rho}v_z)/\partial t + \nabla\cdot(\phi\bar{\rho}v_z\tilde{\boldsymbol{u}}) = 2\bar{\rho}D_T(\nabla\widetilde{Z})^2 - \bar{\rho}\widetilde{\chi} \tag{1d}$$

$$\partial(\phi\bar{\rho}\widetilde{C})/\partial t + \nabla\cdot(\phi\bar{\rho}\widetilde{C}\tilde{\boldsymbol{u}}) = \nabla\cdot[\phi(\underset{\sim}{\boldsymbol{T}}_{\boldsymbol{u}C} + \bar{\boldsymbol{q}}_C)] + < \dot{S}_{C,d} > + \phi\,\overline{\dot{S}_{C,g}} \tag{1e}$$

$$\partial(\phi\bar{\rho}\tilde{e}_t)/\partial t + \nabla\cdot(\phi\bar{\rho}\tilde{\boldsymbol{u}}\tilde{h}_t) = \nabla\cdot[\phi(\underset{\sim}{\boldsymbol{T}}_{\boldsymbol{u}h_t} + \tilde{\boldsymbol{u}}\cdot\underset{\sim}{\bar{\tau}} - \bar{\boldsymbol{q}})] - \phi\sum_{i=1} h^o_{f_i}\overline{\dot{\omega}_i'''} + \phi\bar{\rho}\tilde{\boldsymbol{u}}\cdot\tilde{\boldsymbol{g}} + e^{\ddagger} \tag{1f}$$

where, ϕ is the gas volume fraction, $\widetilde{Z}$ is mixture fraction, v_Z is mixture fraction variance and $\widetilde{C}$ is a reaction progress variable. The filtered viscous stress tensor ($\underset{\sim}{\bar{\tau}}$), species diffusion ($\bar{\boldsymbol{q}}_i$), and heat fluxes ($\bar{\boldsymbol{q}}$) are modeled using Newton's, Fick's, and Fourier's laws respectively and are expressed in terms of resolved quantities, thereby neglecting any subgrid scale (SGS) fluctuations. The second-order correlation quantities, $\underset{\sim}{\boldsymbol{T}}_{\alpha\beta}$, in Eqs. (1b) - (1f) represent unknown SGS correlations for variables α and β, and are closed using dynamic Smagorinsky and gradient diffusion models. The terms $c^{\ddagger}$, $c_Z^{\ddagger}$, $c_C^{\ddagger}$, $\boldsymbol{m}^{\ddagger}$, and $e^{\ddagger}$ represent phase interaction terms for mass, mixture fraction, progress variable, momentum and energy phase-exchange processes between the droplets and gas phases. In (1f), $\overline{\dot{\omega}_i'''}$ is the species source/sink term from homogeneous reactions originating from both individual droplet and group combustion modes of burning. The heat released from burning droplets appears in the gas-phase equations by decomposing the homogeneous reaction term of Eq. (1f) into flames from individual burning droplets and from secondary burning from the evaporating fuel that mixes with the air *via* group combustion [11, 12], *i.e.*, $-\phi\sum_{i=1}^{N} h^o_{f_i}\overline{\dot{\omega}_i'''} =< \dot{S}_{E,d} > +\phi\,\overline{\dot{S}_{E,g}'''}$ where, $< \dot{S}_{E,d} >$ is the average (filtered) energy production rate from individual droplets and $\overline{\dot{S}_{E,g}'''}$ is the net heat released (per unit volume gas) from group combustion. Closure of $< \dot{S}_{C,d} >$, $\overline{\dot{S}_{C,g}'''}$, $< \dot{S}_{E,d} >$ and $\overline{\dot{S}_{E,g}'''}$ are determined using FGM discussed in section 2.2. The droplet field is decomposed into groups of droplets, or parcels, with each parcel having identical droplet properties. The number of droplets per parcel is selected to best balance computational efficiency while maintaining the appropriate mass loading of the fuel spray. Each parcel is governed by a system of ordinary differential equations, describing conservation of mass, species, momentum, and energy of an isolated lumped droplet. Ranz-Marshall correlations and standard drag laws for flow over a sphere are used to extend quasi-steady evaporation and burning to flows with finite slip velocities.

2.2 FGM Formulations

The UFGM and the DFGM are created solving the compressible 1D conservation equations for a reacting gas in planar and spherical coordinates, respectively. For these 1D simulations, the UCSD hydrocarbon and nitrogen mechanism consisting of 70 species and 322 reactions is employed [13]. For the UFGM, an expanding diffusion flame initially at a high scalar dissipation rate is considered and allowed to diffuse outward to the low strain limit. A progress-variable defined as $C = \sum_{i=1}^{N} \alpha_i (Y_i)/\sum_{i=1}^{N} \alpha_i (Y_{eq,i})_{st}$, is constructed where the weighting factors, α_i, are determined using a cost function approach [14] and a downhill simplex method [8]. A single flame index, Λ, is used to catalog each flame, where $\Lambda = C_{st} = C(Z = Z_{st})$. This ensures statistical independence between the flame index and Z for determining statistical distributions necessary for LES closure terms. The transient flames are cataloged at discrete time steps then are mapped to Z and C.

The UFGM solutions are used to construct the filtered source/sink terms, *via*, $\overline{\dot{S}'''_{i,g}} = \int_\Lambda \int_Z \dot{S}'''_{i,g}(Z,\Lambda) P(Z,\Lambda) dZ d\Lambda$ where $P(Z,\Lambda) = P(\Lambda|Z)P(Z)$ is the joint PDF of Λ and Z. A beta PDF is used to described the statistical distribution of Z and a Dirac delta is used to described Λ, so that $P(\Lambda|Z) = \delta(\Lambda - \widetilde{C})$. The filtered values are obtained for all dependent variables, including the progress variable, $\widetilde{C}$, before source terms are mapped in terms of $\widetilde{Z}$, $\widetilde{v}_z$, and $\widetilde{C}$ only. For the DFGM, a series of steady flame solutions are conducted for varying values of D_d ranging from $200\mu m$ down to $\simeq 40\mu m$ where upon flame extinguishment occurs. The flames are tabulated as a function of $Z(r)$ and droplet diameter (D_d). For each flame, the radial distribution rates of energy ($\dot{S}'''_E$) and C ($\dot{S}'''_C$), along with T_s and $\dot{m}_d$ are known. To obtain a point source representation for the droplets, the progress variable or energy source term required is determined by numerically integrating $\dot{S}'''_i$ ($i = \{E, C\}$) over the droplet solution to obtain $\dot{S}_{i,d} = 4\pi \int_{r_s}^{r_\infty} r^2 \dot{S}'''_i dr$, where r_∞ is selected as the value of r for when Z from the DFGM equals Z_∞, therefore $\dot{S}_{i,d}(Z_\infty, D_d)$. In the limit of $Z_\infty \to 0$, $\dot{S}_{i,d}$ corresponds to that of an isolated droplet burning in air. As the far-field value of mixture fraction approaches the value of Z at the surface, $Z_\infty \to Z_s$, then $\dot{S}_{i,d} \to 0$ corresponding to the fuel rich flame limit due to the high concentration of products in the far-field. The integrated values of $\dot{S}_{i,d}(Z_\infty, D_d)$ are stored in the flamelet table. The required droplet source terms are determined from the joint PDF description of Z_∞ and D_d, with D_d acting as a flame index, using the relation, $< \dot{S}_{i,d} >= \int_{D_d} \int_{Z_\infty} \dot{S}_{i,d}(Z_\infty, D_d) P(Z_\infty, D_d) dZ_\infty dD_d$ where $P(Z_\infty, D_d)$ is the joint PDF of Z_∞ and D_d. Assuming statistical independence between Z_∞ and D_d, along with $P(Z) \approx \delta(Z_\infty - \widetilde{Z_\infty})$ and $P(D_d) \approx \sum_d G\delta(D_d - D_d^+)$ then $< \dot{S}_{i,d} >= \sum_d G\dot{S}_{i,d}(\tilde{Z}_\infty, D_d^+)$, where D_d^+ is the local Lagrangian representation of the computational parcel.

3 Results and Discussion

A methanol spray flame surrounded by an annular air jet was investigated by Friedman et al. [10] using phase-doppler interferometry (PDI) to determine droplet characteristics, planar laser-induced fluorescence (PLIF) for imaging OH concentration, and $75\mu m$ type S thermocouples to determine temperature profiles. A standard pressure-swirler nozzle is used to inject methanol droplets in a hollow-cone configuration. The nozzle is surrounded by a brass fitting with a diameter of $50.4\,mm$

that is surrounded by another tube with a diameter of $63.6\,mm$ to create an annular air inlet to surround the spray. A single mass flow rate of liquid methanol fuel is set at $0.42\,g/s$. Two different volumetric flow rates of the surrounding air jet are chosen at $Q_g = 4.77$ ($Q1$), and $9.52\ l/s$ ($Q2$) corresponding to Reynolds numbers of $13,000$ and $26,000$, respectively. Measurements of gas temperature, OH concentration, along with liquid Sauter mean diameter (D_{32}), and volume flux are collected at $25\,mm$, $60\,mm$, and $100\,mm$ downstream of the injector.

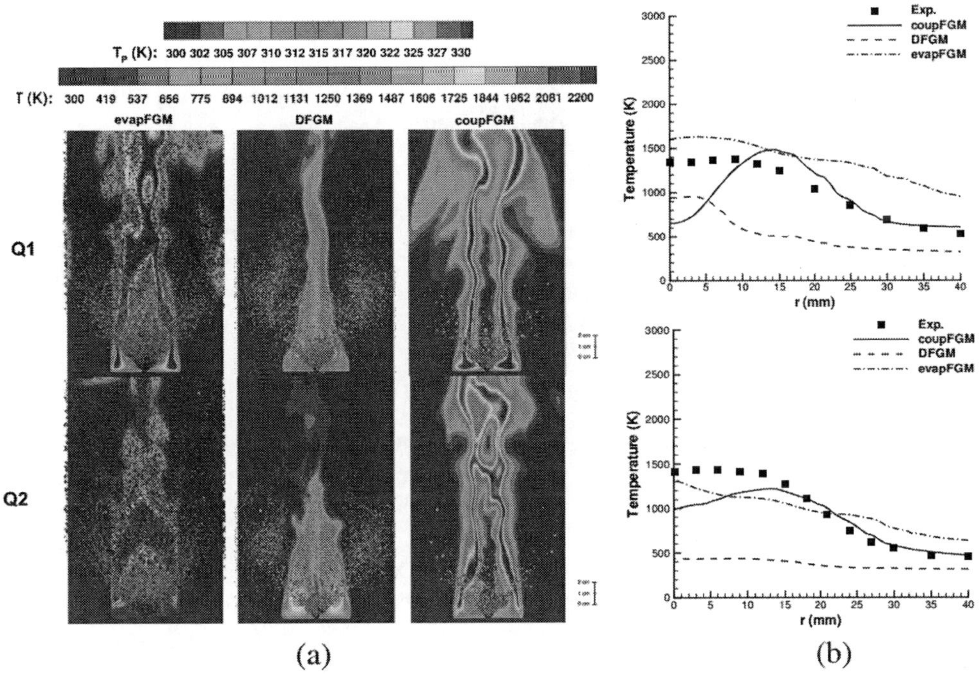

Figure 1: Temperature comparisons showing (a) instantaneous centerline slices of centerline gas and droplet temperatures and (b) time-averaged comparisons to data at $100\,mm$ downstream. $Q_1 = 4.77\ l/s$ and $Q_2 = 9.52\ l/s$.

Figure 1(a) shows instantaneous snapshots of centerline gas and droplet temperatures. Each column represents a different model with the left column being the evapFGM, the center column is the DFGM, and the right column is the coupFGM. The rows represent the different flow rates, $Q1$, and $Q2$. The evapFGM approach displays a higher concentration of droplets than the other two models due to only modeling droplet evaporation, which have lower mass loss rates than a burning droplet. The DFGM model exhibits lower temperatures due to droplets quickly reaching the extinguishment limit, $\sim 40\,\mu m$, and not including any group combustion. When coupled with the gas phase chemistry, higher temperatures are possible and are seen between the center of the jet and the annular flow, but tend to be lower in the center due to higher mixture fractions and lower group combustion. Good qualitative agreement is observed between the spray angles and combustion regions when compared to the experimental images. Figure 1(b) shows time averaged radial temperature comparisons at $z = 100\,mm$ downstream. Experimental averages for temperature and droplet sizes are taken using a ~ 2050 sample data set taken at a $250\,Hz$ sampling rate [15]. Simulation averaging is started after approximately five flow-through (FT) times defined as, $FT = 185\,mm/U_o$, where $185\,mm$ is the domain length from the injection point to the outlet and U_o is the annular air-jet velocity. The averaging is done using 100 snapshots over a one-second period, which corresponds to approximately 22 and 45 FT times for the $Q1$ and $Q2$ cases respectively. Overall the coupFGM model best matches the temperature data, with the DFGM being better in the centerline for the lowest measuring planes.

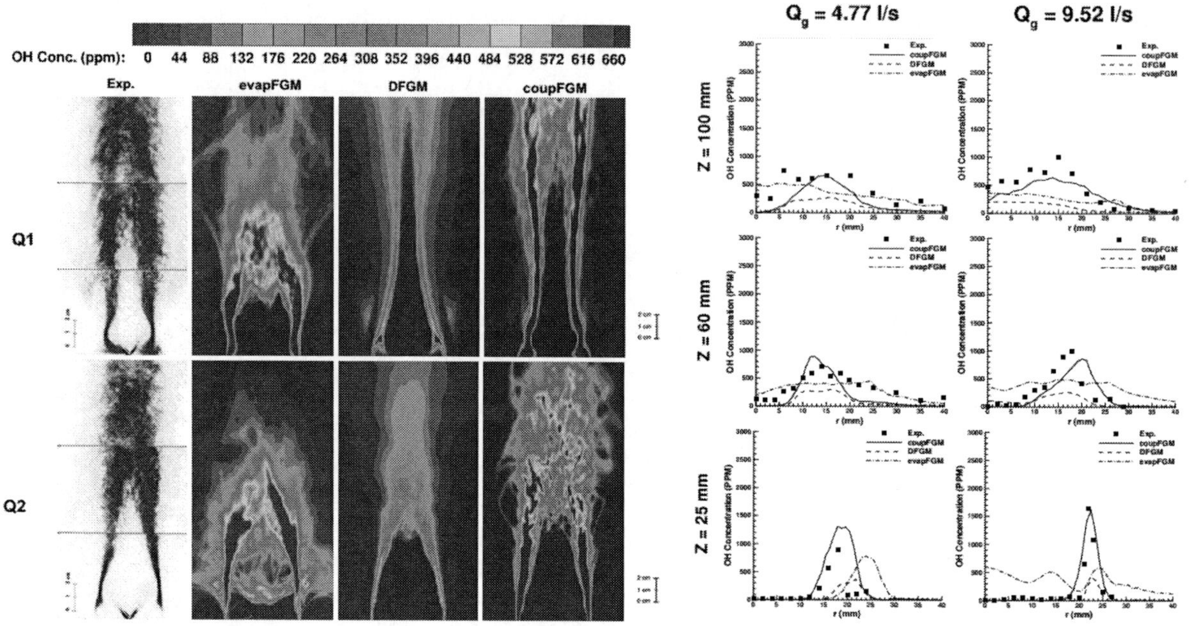

(a) (b)

Figure 2: OH comparisons showing averaged (a) centerline and (b) radial profiles for $Q_1 = 4.77\ l/s$ and $Q_2 = 9.52\ l/s$.

Figure 2(a) shows slices of the OH signature averaged over 100 instantaneous slices. Experimental images were taken once every $3-5$ seconds due to limitations in saving the files to the computer disk [15] and 100 simulation snapshots were taken during one second as described previously. The annular flow rate exhibits an effect on the flame shape such that, increasing Q_g results in OH converging to a point above the spray. This converging point decreases in height and the base of the PLIF widens with increasing flow rates. The DFGM and coupFGM slices for $Q1$, appear to be "tulip"/bulbous shaped with some OH attachment to the bluff body, which agree well with the experimental image. As the flow rate increases to $Q2$, the DFGM and coupFGM again appear to be similar with a more triangular base, a shorter point of convergence, and flame attachment to the outer edge of the bluff body. The OH contours show good overall qualitative agreement and the level of agreement is comparable to those of Kim et al. [15]. Figure 2(b) shows the comparison between the measured OH concentrations and the modeled values for increasing flow rate from left to right at the three different measurement planes, increasing in distance from bottom to top. It is seen from this figure, that the general shape and magnitudes of the OH concentrations match up relatively well, even though the experimental values can be as much as 50% in error due to the level of assumptions necessary to obtain OH concentrations from PLIF [10]. The $z = 25\,mm$ plane shows the best agreement when comparing the broadness and peak values of the PLIF, where the coupFGM model tends to capture the shape and magnitude well. As the distance from the injection point is increased to $z = 60\,mm$ the coupFGM model agrees well with the magnitudes, but does not appear to broaden as well as the evapFGM model. In the final plane, the coupFGM model matches the magnitudes and shape reasonably well. The trends from the OH and the combustion temperatures for the LES agree relatively well, showing similar trends between the energy release associated with the presence of OH and the evaluated temperatures. Conversely, the experimentally measured temperatures tend to be higher in the core where OH measurements are relatively low, emphasizing the effects of not accounting for the radiation on the thermocouple. Overall, the best agreement is observed between the coupFGM model and the experimental data, especially for $Q2$ case where the shape and magnitudes match well. This implies the combination of individual droplet and group combustion is important.

4 Conclusion

A new droplet flamelet-generated manifold (DFGM) methodology is developed which expands on previously developed FGM modeling for spray combustion by including the effects of individual droplet burning. Three modeling approaches are explored of i) an evaporating droplet coupled with a UFGM (evapFGM), ii) a DFGM model, and iii) the DFGM coupled with the UFGM to account for group combustion (coupFGM). The evapFGM is used to describe spray combustion in the dilute limit when mostly droplet evaporation and group combustion should be present. The DFGM model is considered on its own to account for the finite-rate effects of individual droplet combustion. Lastly, the coupFGM accounts for droplet and group combustion modes of burning by considering the limits of the droplet source terms for energy and progress variable (C) when immersed in increased levels of mixture fraction (Z). Using the same numerical setup and only interchanging the three combustion models, LES is compared to the experimental methanol spray with an annular air jet explored by Friedman et al. [10]. Overall good qualitative agreement with data of gas temperature and OH are observed.

5 Acknowledgments

Support for this work has been provided by NSF under grant number 1704447 and NAVAIR through the STTR phase II program under contract N68335-10-C-0418

References

[1] J. A. van Oijen and L. P. H. de Goey. *Combust. Theor. Model.*, 6 (2002) 463–478.

[2] L. M. Verhoeven, W. J. S. Ramaekers, J. A. Van Oijen, and L. P. H. deGoey. *Combust. Flame*, 159 (2012) 230–241.

[3] C. Bekdemir, L. M. T. Somers, and L. P. H. de Goey. *Proc. Combust. Inst.*, 33 (2011) 2887–2894.

[4] J. Tillou, J.-B. Michel, C. Angelberger, and D. Veynante. *Combust. Flame*, 161 (2014) 525–540.

[5] A. Wehrfritz, O. Kaario, V. Vuorinen, and B. Somers. *Combust. Flame*, 167 (2016) 113–131.

[6] M. Chrigui, J. Gounder, A. Sadiki, A. R. Masri, and J. Janicka. *Combust. Flame*, 159 (2012) 2718 – 2741. Special Issue on Turbulent Combustion.

[7] E. Knudsen, Shashank, and H. Pitsch. *Combust. Flame*, 162 (2015) 159 – 180.

[8] B. T. Bojko and P. E. DesJardin. *Combust. Flame*, 173 (2016) 296–306.

[9] R. W. Bilger. *Combust. Flame*, 158 (2011) 191–202.

[10] J. A. Friedman and M. Renksizbulut. *Combust. Flame*, 117 (1999) 661–684.

[11] H. H. Chiu and T. M. Liu. *Combust. Sci. Technol.*, 17 (1977) 127–142.

[12] Q. Zhou and S. C. Yao. *Int. J. Heat Mass Tran.*, 35 (1992) 121–129.

[13] UCSD. Chemical-kinetic mechanisms for combustion applications. San Diego Mechanism web page, http://web.eng.ucsd.edu/mae/groups/combustion/mechanism.html, 2016.

[14] M. Ihme, L. Shuun, and J. Zhang. *J. Comput. Phys.*, 231 (2012) 7715–7721.

[15] W. T. Kim, K. Y. Huh, J. A. Friedman, and M. Renksizbulut. *Combust. Sci. Technol.*, 171 (2001) 119–139.

Spring Technical Meeting
Eastern States Section of the Combustion Institute
March 4–7, 2018
State College, Pennsylvania

Reynolds-Averaged Navier-Stokes Simulations of a Piloted Heptane/Propane Spray Flame

David A. Kessler[1,*], *Brian T. Fisher*[2], *Alfredo D. Tuesta*[2], *and Steven G. Tuttle*[2]

[1]*Laboratories for Computational Physics and Fluid Dynamics, Naval Research Laboratory, Washington, DC, USA*
[2]*Combustion and Reacting Transport Section, Naval Research Laboratory, Washington, DC, USA*
[*]*Corresponding author:* david.kessler@nrl.navy.mil

Abstract: Experiments conducted at the Naval Research Laboratory have shown interesting flame and droplet dynamics in a small-scale piloted spray burner. Liquid heptane droplets are generated using an open-pipe atomizer in which pressurized propane gas is used to drive a liquid film through a straight, vertical feed pipe. The liquid filaments that emerge from the open-end of the pipe are readily atomized by the strong shear layer formed between the high-speed propane jet and surrounding environment. After ignition, the resulting heptane/propane spray flame is stabilized using a lean, premixed hydrogen/air flame that surrounds the jet. The unique combination of physics controlling the spray flame generated by this burner poses a number of modeling challenges. In this paper, we discuss some of these challenges and present preliminary Reynolds-averaged Navier-Stokes simulations designed to provide a representative model of the laboratory burner.
Keywords: ***turbulent jet flame, spray combustion, two-phase combustion, partially-premixed RANS model***

1. Introduction

A variety of industrial and energy production processes can result in the accidental discharge of highly pressurized two-phase jets consisting of a mixture of liquid and gaseous fuels. Ignition of such two-phase jets poses a serious fire and pollution hazard. In particular, the efficiency with which the liquid phase component burns is unknown. Several previous studies have made progress toward characterizing these types of large scale fires [1], but still very little is known about the behavior of such flames. From a theoretical standpoint, the sheer size, extreme physical conditions, and complex physical and chemical processes that occur in the discharge jet and turbulent flame brush make simulations impossible to do from first principles. Instead, an engineering model is needed that accounts for the coupling between the fluid dynamics, chemistry, and thermodynamics that can provide reliable predictions for large-scale flame behavior and burning characteristics.

In this work, we describe an ongoing effort to develop and validate a computationally-feasible two-phase, multiple-fuel, turbulent jet flame model. At the Naval Research Laboratory, a bench-scale, piloted spray burner with an open-pipe atomizer has been set up to enable detailed optical measurements of flame behaviors and characteristics of turbulent spray flames designed to simulate real-world oil/gas fires. We describe preliminary simulation results for a baseline piloted propane jet flame and a liquid n-heptane/propane jet flame that mimic the laboratory experiments that are described in a companion paper [2].

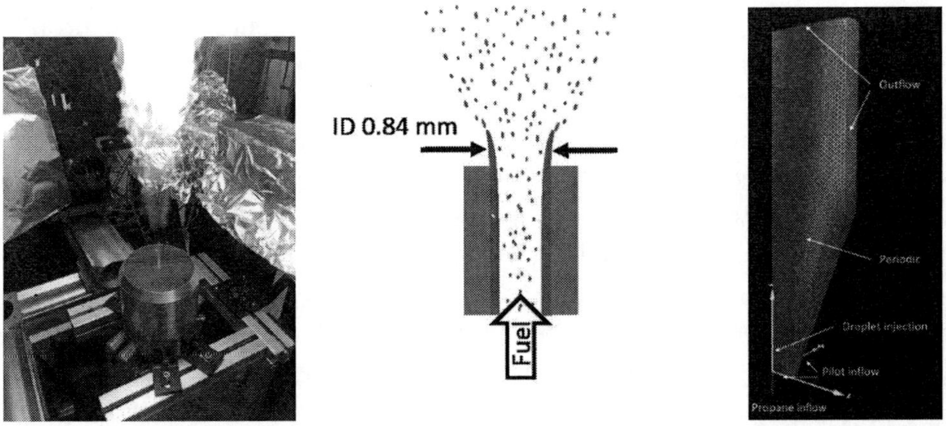

(a) (b) (c)

Figure 1: (a) Image of the laboratory two-phase jet burner, (b) notional representation of the discharging jet, and (c) model computational domain.

2. Methods

An image of the open-pipe-atomizer (OPA) spray burner is shown in Figure 1. A complete description of the burner is given in Fisher et al. [2], and here we only list the features most relevant to developing a predictive computational model of its operation. Both gaseous and liquid fuels are discharged through a common central feed tube with inner diameter of 0.84 mm. For combined oil/gas flames, liquid n-heptane is injected into a stream of gaseous propane far upstream of the open end of the atomizer, and the two components travel together through the feed tube. The single- or two-phase jet discharges into the open atmosphere and is ignited. The resulting flame is stabilized using a lean premixed hydrogen-air pilot flame that completely surrounds (and is concentric with) the central feed tube.

In order to simplify the modeling of the spray-flame burner, we assume a one-way coupling between the premixed hydrogen-air pilot flame and the central jet flame. We note that the pilot flame anchors very close to the burner surface, which suggests that a one-dimensional, laminar flame analysis is sufficient to set the inflow conditions for a fixed co-flow of the pilot flame combustion products. The pilot flame inflow stream thus contains the product species oxygen, nitrogen, and water vapor. We assume a nominal pilot-flame equivalence ratio of $\phi = 0.398$, which corresponds to an adiabatic flame temperature of 1400 K and mole fractions of 0.1167, 0.1543, and 0.729 for the three species, respectively. The injection velocity is taken to be 0.5 m/s, which is consistent with hydrogen-air flame speeds.

The inflow boundaries for the hot pilot-flame products were offset from the exit of the feed tube by 2 cm to prevent any numerical artefacts created in the jet mixing layer from interacting with the boundary conditions. An illustration of the model system and computational domain used in this work is shown in Figure 1. We assume that the burner and resulting mean flow field are axisymmetric with zero swirl and simulate a 15 degree sector of the full three-dimensional system. Rotationally-periodic conditions are applied on the lateral faces of the domain. Propane is injected into the central tube at the bottom of the domain with a stagnation temperature of 300 K and gauge stagnation pressure of 30 kPa. The length of the propane feed tube (inner diameter of 0.082 cm) is 7 cm, which is sufficient to generate a fully-developed turbulent pipe flow profile. The system discharges through an outflow boundary at the top of the domain (0.5 m above the pilot-flame) at which we apply characteristic boundary conditions and a mean pressure equal to atmospheric.

We discretize the computational domain using an unstructured tetrahedral mesh that has been locally refined near the feed-tube exit and inside the mixing layer between the propane jet and the pilot flame oxidizer inlet stream. The minimum grid spacing was 0.02 mm in the most refined regions of the mesh and grows to approximately 2 mm near the right-most edge of the domain. This level of resolution was found to be sufficient to provide converged numerical solutions. Liquid n-heptane droplets are injected from a surface located 0.1 mm downstream of the central feed tube. We assume the liquid phase discharges through the open pipe as a thin film along the pipe walls. Immediately upon discharge, the sheet breaks up into 20 micron diameter droplets whose injection locations are randomly chosen from an annular region with inner diameter of 0.7 mm and outer diameter of 0.8 mm. A uniform droplet mass flow rate of 0.4 g/s (relative to the full burner cross section). The droplets are injected with an initial velocity of 30 m/s with injection times randomly chosen within an interval of 1e-3 s. Liquid droplets are further atomized according the KHRT model, which accounts for breakup due to both Kelvin-Helmholz and Rayleigh-Taylor instabilities.

The multiphase reacting flow is simulated using an Eulerian-Lagrangian approach in which the gas-phase is modeled using the Favre-averaged reacting Reynolds-Averaged Navier-Stokes equations for the mixture density ρ, momentum $\rho\mathbf{u}$, and total enthalpy $\rho\overline{H}$, using the realizable k-ε model for the Reynolds stresses, and the liquid-phase is modeled as a collection of discrete Lagrangian droplets that are convected by and exchange mass, momentum, and energy with the continuous gas-phase. We assume that the only forces acting on the droplets are gravity and drag. The drag force is computed assuming the droplets remain spherical and using a correlation for the drag coefficient, $C_D = a_1 + a_2/Re + a_3/Re^2$, where a_1, a_2, and a_3 are given by [3].

The droplets are assumed to be composed of a single-component liquid that vaporizes according to a diffusion-controlled model, for which the flux of vapor into the gas phase is proportional to the difference in vapor concentration at the droplet surface ($C_{i,s}$) and the bulk gas ($C_{i,\infty}$), $N_i = k_c(C_{i,s} - C_{i,\infty})$. We assume thermodynamic equilibrium at the droplet surface and compute $C_{i,s}$ using the saturation pressure at the current droplet temperature. The mass transfer coefficient is calculated using a correlation for the particle Sherwood number, $k_c d_p / D_{i,m} = Sh = 2.0 + 0.6 * Re_d^{1/2} Sc^{1/3}$, where d_p is the droplet diameter, $D_{i,m}$ is the diffusion coefficient of the vapor in the bulk gas mixture, Sc is the liquid substance Schmidt number, and Re_d is the particle Reynolds number. The rate of change of particle mass is governed by

$$\frac{\mathrm{d}m_p}{\mathrm{dt}} = -N_i A_p M_{w,i}, \tag{1}$$

where A_p is the surface area of the droplet and $M_{w,i}$ is the molecular weight of the liquid fuel. The particle temperature is governed by

$$m_p c_p \frac{\mathrm{d}T_p}{\mathrm{dt}} = h A_p (T_{gas} - T_p) - \frac{\mathrm{d}m_p}{\mathrm{dt}} h_L, \tag{2}$$

where c_p is the liquid-phase specific heat coefficient, h is a convective heat transfer coefficient, and h_L is the latent heat of the liquid. This model neglects the effects of droplet heating due to radiation, which are likely to be non-negligible in this system. This effect, along with more sophisticated drag and vaporization models, will be included in future refinements of this modeling effort.

The turbulent flame brush is modeled as a partially-premixed flame that assumes chemical equilibrium throughout both premixed and non-premixed reaction zones. Here, we parameterize the premixed flame model using a single reaction progress variable, $c = \frac{\sum_k \alpha_k (Y_k - Y_k^u)}{\sum_k \alpha_k (Y_k^{eq} - Y_k^u)}$, for which we utilize CO and CO_2 as tracer species, such that α_{CO} and α_{CO_2} are taken to be unity and all other species zero. The non-premixed combustion is characterized by two mixture fractions $f_1 = \frac{Z_i - Z_{i,ox}}{Z_{i,C_3H_8} - Z_{i,ox}}$ and $f_2 = \frac{Z_i - Z_{i,ox}}{Z_{i,C_7H_{16}} - Z_{i,ox}}$, for which the Z_i are the conserved atomic mass fractions of the elements present in the propane, n-heptane, and oxidizer (subscript "ox") streams, and the system enthalpy. We construct a joint probability density function (pdf) for these four variables $p(f_1, f_2, c, \overline{H})$ as a product of the three independent pdf's for f_1, f_2, c, and $\overline{H}$. We model $p(\overline{H})$ as a delta function and $p(f_i)$ and $p(c)$ as assumed-shape beta functions for which we solve transport equations for the mean and variance of f_i and the mean of the progress variable c,

$$\frac{\partial}{\partial t}(\rho \overline{f_i}) + \nabla \cdot (\rho \mathbf{u} \overline{f_i}) = \nabla \cdot \left(\left(\frac{k}{C_p} + \frac{\mu_t}{\sigma_l} \right) \nabla \overline{f_i} \right) + S_{D_{f,i}}, \quad (3)$$

$$\frac{\partial}{\partial t}\left(\rho \overline{f_i'^2}\right) + \nabla \cdot \left(\rho \mathbf{u} \overline{f_i'^2}\right) = \nabla \cdot \left(\left(\frac{k}{C_p} + \frac{\mu_t}{\sigma_l} \right) \nabla \overline{f_i'^2} \right) + C_g \mu_t \left|\nabla \overline{f_i}\right|^2 - C_d \frac{\varepsilon}{k} \overline{f_i'^2}, \quad (4)$$

$$\frac{\partial}{\partial t}(\rho c) + \nabla \cdot (\rho \mathbf{u} c) = \nabla \cdot \left(\left(\frac{k}{C_p} + \frac{\mu_t}{\mathrm{Sc}_t} \right) \nabla \right) + \overline{S}_c, \quad (5)$$

and the mean enthalpy,

$$\frac{\partial}{\partial t}(\rho \overline{H}) + \nabla \cdot (\rho \mathbf{u} \overline{H}) + S_{D_H}. \quad (6)$$

The source term in equation 5 is defined by $\overline{S}_{Y_c} = \rho_u U_t |\nabla c|$, where ρ_u is the density of the unburned gas mixture and U_t is the turbulent flame speed computed using the Zimont turbulent flame speed closure[4], $U_t = A u'^{3/4} U_l^{1/2} \alpha^{-1/4} l_t^{1/4}$, α is the thermal diffusivity of the unburned mixture, U_l is the laminar flame speed, u' is root mean square of the turbulent velocity, $l_t = C_D (u')^3/\varepsilon$ is a turbulence length scale, ε is the turbulence dissipation rate, and A and C_D are model constants chosen to be 0.52 and 0.37, respectively. The source terms $S_{D_{f_i}}$ and S_{D_H} are the contributions to the mean mixture fractions and enthalpy from the discrete-phase droplets. Here, S_{D_f} is equal to zero for f_1, since the droplets only adds mass to the secondary f_2 fuel stream. The source terms are computed as the average contribution over individual grid cells.

The compressible RANS equations along with transport equations for ε and the turbulence kinetic energy k, and Eqns. [3–6] are solved using Fluent V18.2's coupled pressure-based solver. We utilize a second-order upwind spatial discretization along with a multi-dimensional gradient limiter and a second-order interpolation scheme for pressure. The steady versions of the governing equations are solved using an iterative procedure with under-relaxation. The discrete-phase equations Eqns. [1–2] are integrated in time using a trapezoidal integration method. Coupling between phases is enforced every 0.001 s of the particle trajectories.

3. Results

We begin by considering the gas-phase only propane jet to establish a baseline shape, temperature, and anchoring position for the gas-phase flame. The mass flow rates considered in the laboratory

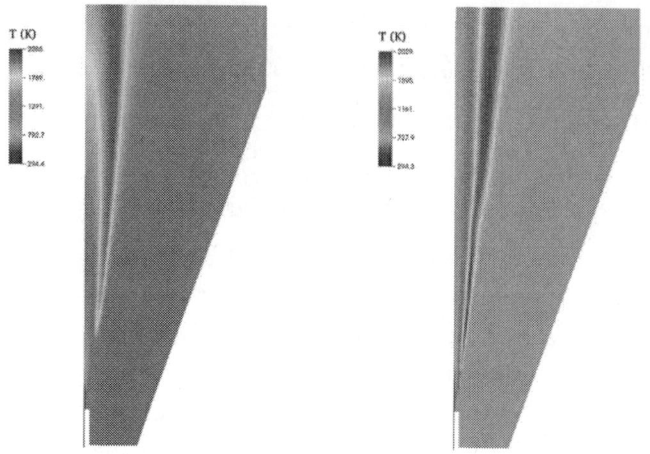

(a) (b)

Figure 2: Temperature profiles on a planar cross-section of the a) single-phase propane jet flame and b) two-phase propane-n-heptane jet flame.

burner shown (cf. Figure 1) range from 0.1–0.3 g/s, corresponding to flow conditions that vary from weakly compressible to underexpanded. In this study, we consider the lowest mass flow rate of 0.1 g/s. The maximum jet velocity reaches 136 m/s as it discharges, corresponding to a Mach number of approximately 0.5. The jet momentum is sufficient to cause the flame to lift off from the burner lip. The resulting temperature profile is shown in Figure 2(a). The lift-off height for this case is approximately 40 mm above the burner, which agrees well with the laboratory measurements.

We next consider a two-phase jet flame for which the propane is also injected with a mass flow rate of 0.1 g/s. Choosing the laboratory conditions used by Fisher et al. [2], we inject liquid n-heptane with a mass flow rate of 0.4 g/s. As discussed earlier, we use a simplistic droplet injection model which assumes the annular liquid sheet that discharges from the central feed tube rapidly breaks up into 20 micron diameter droplets evenly distributed over an annular surface just downstream of the discharge point. The droplets begin vaporizing immediately resulting in a transfer of n-heptane to the gas-phase. Figure 3 shows the distribution of the secondary mixture fraction f_2, which is computed from the n-heptane mass fraction, and the primary mixture fraction f_1, based on the propane mass fraction. As expected, the gaseous n-heptane gathers predominantly toward the center of the jet, and toward the periphery mixes with the gaseous propane where it reacts with the heated oxidizer stream. The resulting temperature field is shown in Figure 2.

4. Conclusions

Preliminary calculations show the addition of liquid n-heptane droplets into a piloted gas-phase propane jet flame significantly alters the flame structure. The resulting jet flame is cooler in the core region but reacts more rapidly in the propane-air mixing layer, causing the flame to anchor approximately 20 mm closer to the burner than the single-phase propane jet flame. We also observe an overall decrease in maximum flame temperature and a broader flame structure. We qualify these observations by pointing out the limitations of the n-heptane injection model used in these calculations results in a much narrower spray cone than that observed in the companion experi-

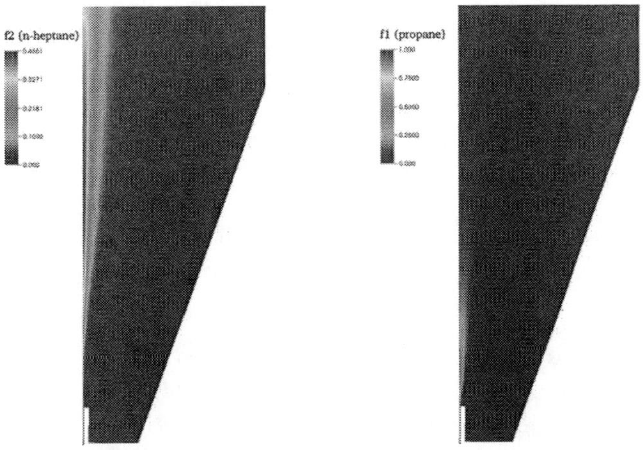

(a) (b)

Figure 3: Mixture fraction profiles on a planar cross-section of the two-phase propane-n-heptane jet flame for the a) secondary fuel stream f_1 (n-heptane) and b) primary fuel stream f_2 (propane).

ments. A higher-fidelity physical model of the liquid sheet breakup is under development and will be discussed in the oral presentation and in future publications.

5. Acknowledgements

This study was funded by the US Department of the Interior, Bureau of Safety and Environmental Enforcement through Interagency Agreement E14PG00060 with US Naval Research Laboratory. The authors thank the DoD High Performance Computing Modernization Program for providing the computational resources.

References

[1] L. B. J., H. G., M. R. Acton, and G. Chamberlain, An Overview of the Nature of Hydrocarbon Jet Fire Hazards in the Oil and Gas Industry and a Simplified Approach to Assessing the Hazards, Process Safety Environ. Protect. 85 (2007) 207–220.

[2] B. T. Fisher, S. G. Tuttle, and A. D. Tuesta, Droplet Size and Velocity Measurements in a Heptane/Propane Spray Flame, Eastern States Section of the Combustion Institute Meeting (2018).

[3] S. A. Morsi and A. J. Alexander, An Investigation of Particle Trajectories in Two-Phase Flow Systems, J. Fluid. Mech. 55 (1972) 193–208.

[4] V. Zimont, W. Polifke, M. Bettelini, and W. Weisenstein, An Efficient Computational Model for Premixed Turbulent Combustion at High Reynolds Numbers Based on a Turbulent Flame Speed Closure, J. Gas Turbines Power 130 (1998) 526–532.

Spring Technical Meeting
Eastern States Section of the Combustion Institute
March 4-7, 2018
State College, Pennsylvania

A Numerical Investigation of the Chemical Kinetics in the Low Temperature Oxidation Process of n-Heptane Spray (I)

Yu Li[1,], Haiwen Ge[2], Hailin Li[1]*
1 West Virginia University, Morgantown, WV 26506, USA
2 Department of Mechanical Engineering, Texas Tech University, Lubbock, TX, USA
**Corresponding Author Email: yuli1@mix.wvu.edu*

Abstract: This paper investigates the instantaneous chemical kinetics process of n-heptane spray at low temperature oxidation stage. The air entrainment is found to play an important role on the instantaneous chemical kinetics process of n-heptane spray. There exists a region where $C_7H_{15}O_2$ radical dissociated to C_7H_{15}-2, which has not been reported by the past literature. The combination of flow field caused by high pressure injection and air entrainment leads a significantly high reactivity region inside the spray. The OH radical is found to play an important role during the low temperature combustion stage. The instantaneously heat release rate analysis shows that endothermic reaction is mainly happened at the dissociation of large radicals, and the exothermic reaction is dominated by reactions featuring with oxygen addition and H abstraction by OH radical.
Keywords: CFD, chemical kinetics, n-heptane spray, instantaneous ROP.

Introduction

N-heptane has cetane number and ignition characteristic comparable to diesel fuels and has been widely recognized as a convenient surrogate for diesel fuel in both experimental and numerical study [1]. The zero-dimensional (0D) model has been recognized as a convenient tool for understanding the auto-ignition process of diesel fuel. It provides important information such as rate of production for key species, heat release rate of key reactions, etc.. However, the 0D model can not reproduce the combustion combustion process for real situation especially in spray combustion where fuel/air mixing plays an important role. The computational fluid dynamics (CFD) model coupled with chemical kinetics mechanism has been well known as a powerful tool to investigate the combustion process. It helps to understand the spatial distribution of key parameters such as temperature, equivalence ratio. However, there is limited research report the instantaneously chemical kinetics process for CFD model.

This paper presents a numerical study regarding the instantaneously chemical kinetics process inside n-heptane spray during low temperature combustion (LTC) stage. The computational model is developed by commercial software CONVERGE and validated against the experimental data provide by engine combustion network (ECN). The temperature, pressure, and species concentration are exported and further post-processed using a recent developed post-processing tool [2]. The result showed that the air entrainment caused by high pressure injection has a strong effect on instantaneous chemical kinetics process inside the n-heptane spray.

Numerical Simulation Model

The initial condition of numerical model is shown in Table. 1, the detail information of the experimental data can be found in [3]. The Reynolds-averaged Navier–Stokes (RANS)-based turbulence model is applied in this study. The model used an adaptive mesh refinement (AMR) method to achieve high-resolution mesh, a maximum cells number of 2,000,000 is applied. Fig. 1 shows the comparison of simulated liquid penetration and vapor penetration against the experimental data. Both the simulated liquid and vapor penetration distance is well predicted; thus, this model will be applied to investigate the n-heptane combustion in this study.

Table 1 experimental condition for penetration distance validation.

Temperature (K)	1000
O2 volume fraction	21
Injection pressure (MPa)	150
Density (kg/m^3)	14.8
Injection duration (ms)	6.8
Injection mass (mg)	17.8
Nozzle diameter (mm)	0.1
Discharge coefficient (Cd)	0.8

Fig. 1 comparison of simulated liquid and vapor penetration against experimental data

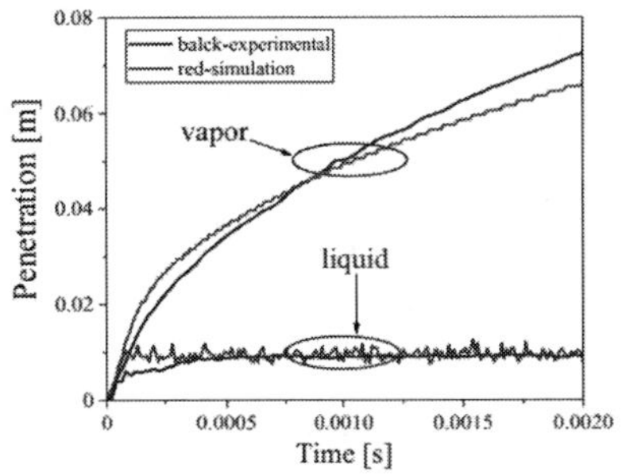

The temperature, pressure, and species concentrations at each computational cell calculated using the CFD model are exported to a 'column format' file by CONVERGE. A MATLAB code was developed to read the temperature, pressure and species concentrations in 'column format' file and transfer them to CHEMKIN [4] for ROP data calculation. The calculated ROP data were analyzed and visualized by in-house MATLAB code. The detailed description of the post-processing method and ROP evaluation method can be found in [2].

Instantaneous Species Creation/Destruction Pathway

In this research, the instantaneous species destruction pathway (ISDP) was derived by examining the instantaneous contribution of each reaction to the destruction of each species using equation (2):

$$\beta_{d,i,j} = \frac{\omega_{d,i,j}}{\omega_{d,i}} \quad (1)$$

Where $\omega_{d,i,j}$ represents the destruction rate of the ith species by the jth reaction, $\omega_{d,i}$ is the summation of destruction rate of reactions that destruct i^{th} species. Accordingly, $\beta_{d,i,j}$ is the normalized contribution of the jth reaction to i^{th} species destruction.

Representative creation/destruction reaction

Among all reactions involved in the creation of the i^{th}species in each computational cell, there exists a reaction that has the largest creation rate. Such a reaction has the largest contribution to the creation of the ith species and is defined as the representative creation reaction (RCR) of the ith species. Similarly, the reaction that has the highest destruction rate makes the largest contribution to the destruction of the ith species, and is defined as the representative destruction reaction (RDR) of the ith species. This research examines the RCR and RDR dominating the formation and consumption of pentyl-dioxidanyl radical ($C_5H_{11}O_2$) noted as RCR_{PE} and RDR_{PE}, respectively. The analysis of the RCR_{PE} and RDR_{PE} derives the $C_5H_{11}O_2$ formation and consumption pathway in n-heptane spray during LTC.

Table. 2 Reactions discussed in this research

R1	$C_7H_{16}+OH=C_7H_{15}\text{-}2+H_2O$	R12	$2HO_2=H_2O_2+O_2$
R2	$C_7H_{16}+HO_2= C_7H_{15}\text{-}2+H_2O_2$	R13	$H_2O_2(+M)=2OH(+M)$
R3	$C_7H_{15}\text{-}2+O_2=C_7H_{15}O_2$	R14	$CH_2O+OH=H_2O+HCO$
R4	$C_7H_{15}O_2=C_7H_{14}O_2H$	R15	$CH_2O+HO_2=H_2O_2+HCO$
R5	$C_7H_{14}O_2H+O_2=C_7H_{14}O_2HO_2$	R16	$HCO+O_2=CO+HO_2$
R6	$C_7H_{14}O_2HO_2=C_7KET_{21}+OH$	R17	$C_2H_4+OH=CH_2O+CH_3$
R7	$C_7KET_{21}=C_5H_{11}CO+CH_2O+OH$	R18	$C_2H_5+O_2=C_2H_4+HO_2$
R8	$C_7H_{15}\text{-}2=C_6H_{12}+CH_3$	R19	$CH_3+HO_2=CH_3O+OH$
R9	$C_7H_{15}\text{-}2=C_3H_6+C_4H_9$	R20	$C_5H_{11}CO=C_5H_{11}+CO$
R10	$C_4H_9=C_2H_4+C_2H_5$	R21	$C_5H_{11}=C_2H_5+C_3H_6$
R11	$H_2+OH=H+H_2O$		

Results and discussion

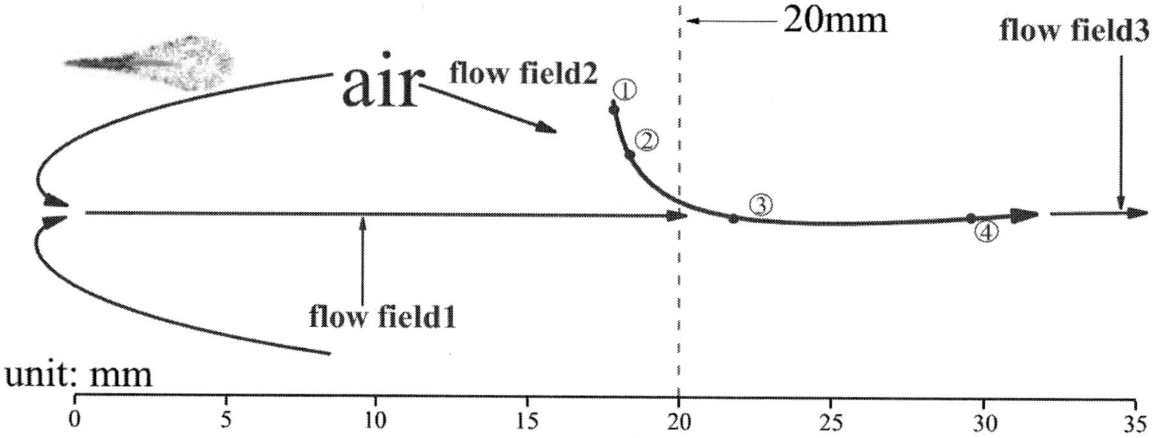

Fig. 2 the flow field inside n-heptane spray at 0.4ms.

Fig. 2 shows the flow field inside the n-heptane spray at 0.4ms. Fig. 3 shows the distribution of temperature, n-heptane and OH concentration, n-heptane destruction rate, $C_7H_{15}O_2$ RCR/RDR, exothermic HRR. In order to understand the effect of air entrainment on instantaneous chemical kinetics process inside the n-heptane spray, four representative points along flow field2 shown in Fig. 2 are selected, the feature of these points are listed below: point1: spray periphery; point2: RDR_{PE}-R3 region; point3: highest OH concentration; point4: highest HRR after 25mm.

Fig. 4 shows the temperature, mixture fraction, n-heptane destruction rate, heat release rate, and mole concentration of $C_7H_{15}\text{-}2$, O_2, $C_7H_{15}O_2$, OH and H_2O_2 radical at points 1-4 marked in Fig. 2. Fig. 5 (a)-(d) shows the ISDP at points 1-4. The discussion of these points is shown below.

Temperature [K]	n-heptane mole concentration	nC_7H_{16} destruction rate (log)

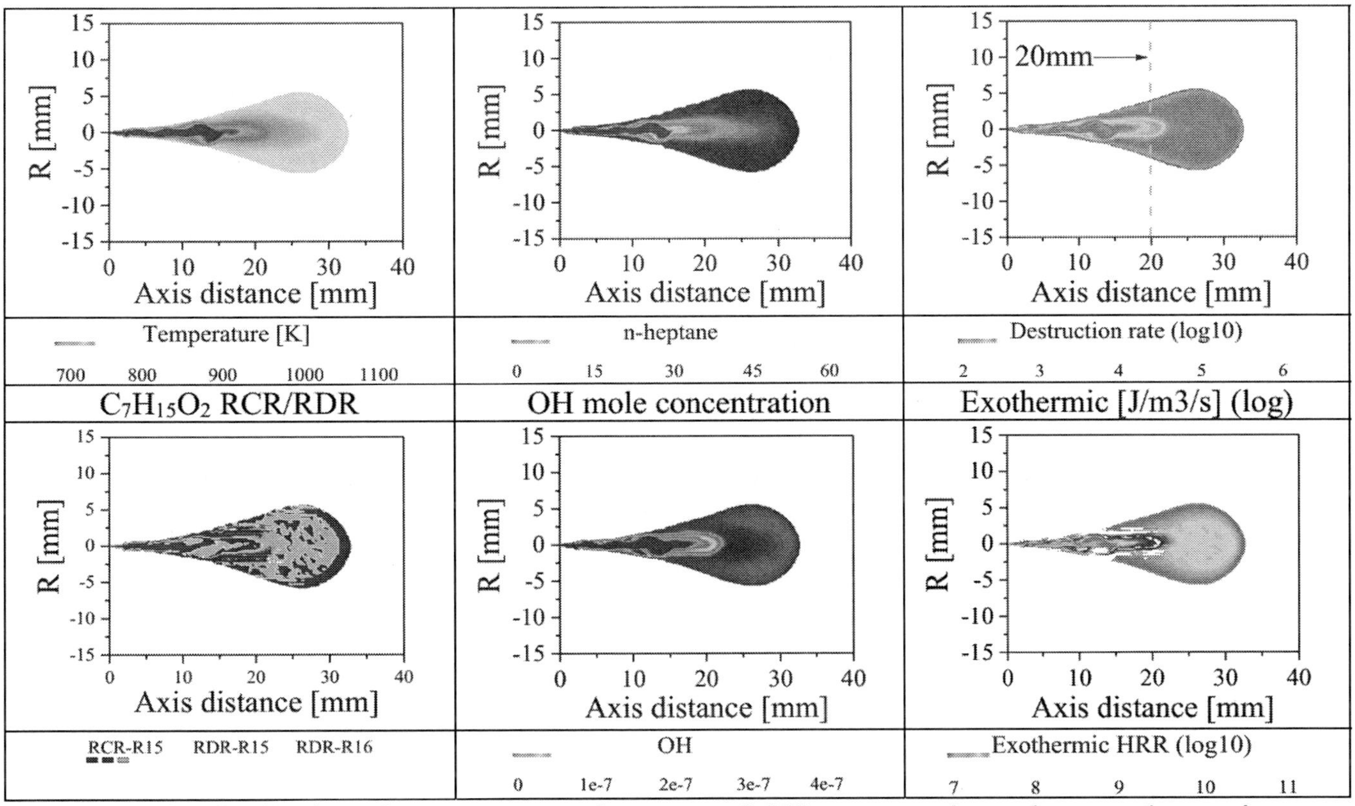

Fig. 3 the distribution of temperature, n-heptane and OH concentration, n-heptane destruction rate, $C_7H_{15}O_2$ RCR/RDR, exothermic HRR.

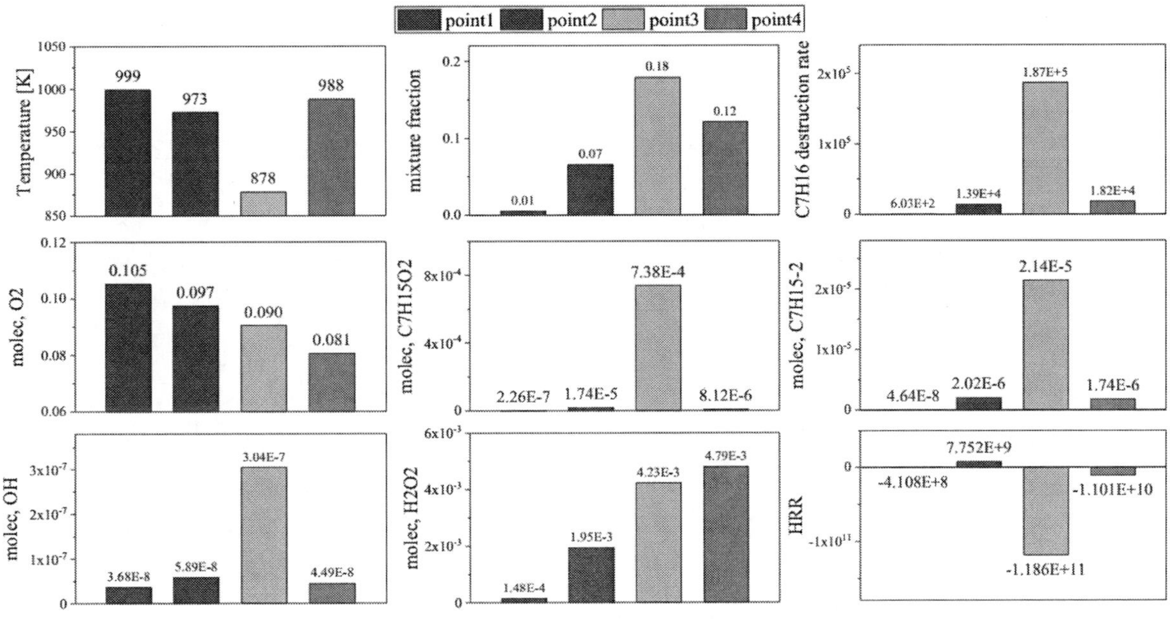

Fig. 4 Temperature, mixture fraction, n-heptane destruction rate (unit: kmol/m^3/s), heat release rate (HRR) (unit: J/m^3/s), mole concentration (unit: kmol/m^3) of C_7H_{15}-2, O_2, $C_7H_{15}O_2$, OH and H_2O_2 at points 1, 2, 3, and 4.

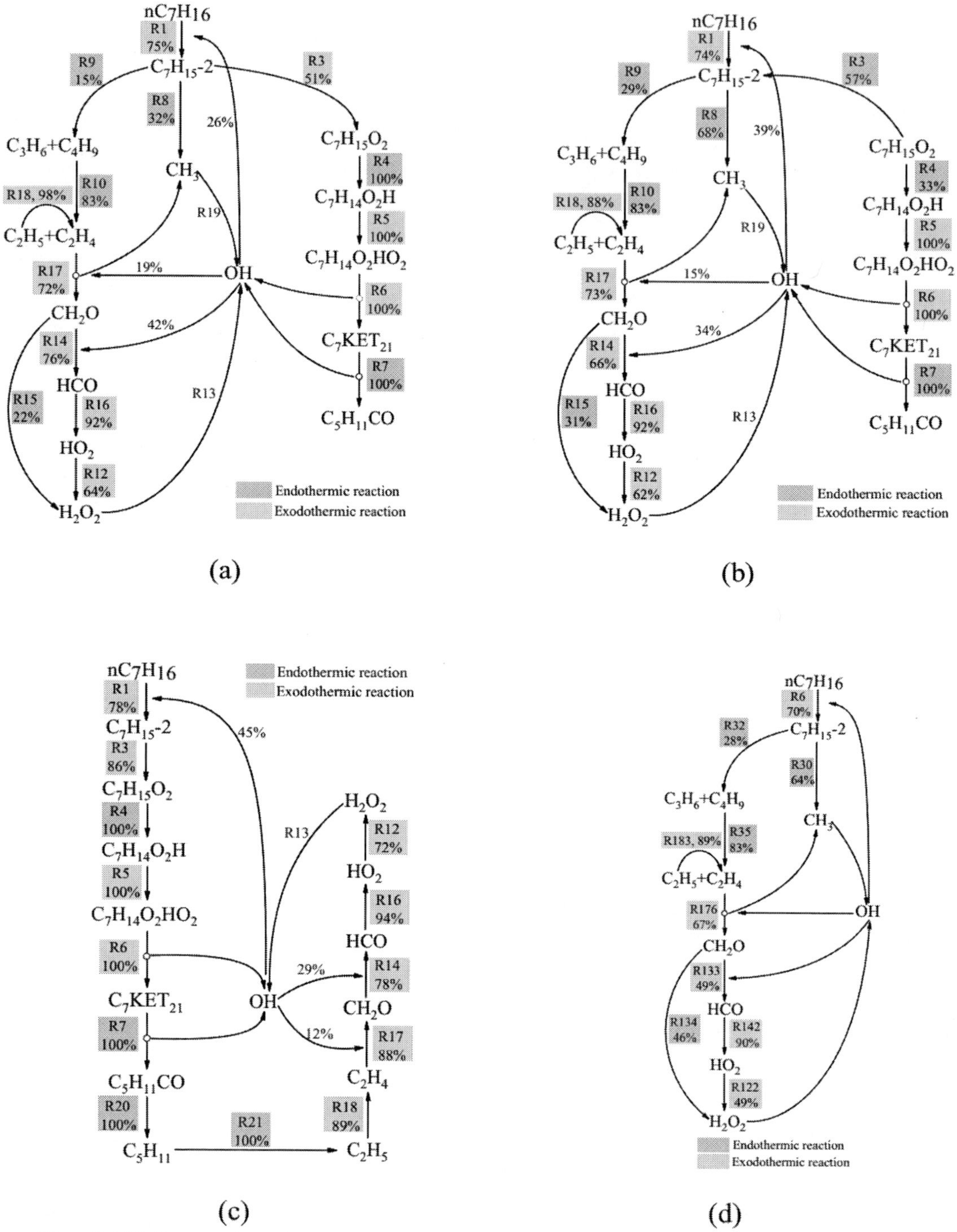

Fig. 5 ISDP at points 1, 2, 3, 4. Reactions marked with blue indicate endothermic, reactions marked with red indicate exothermic.

Point1: the destruction of n-heptane is dominated by H abstraction of OH radical through R1. The destruction of heptyl radical is dominated by R3 featuring with oxygen addition (51%). The reaction R8 (32%) and R9 (15%) also make notable contribution to heptyl radical destruction, which indicates that there exists a competition between R3, R8 and R9 for heptyl radical destruction at point1. There are three major chain branching pathways (CBP) observed in Fig. 8

which are referred as CBP1, CBP2, and CBP3 in this research. The CBP1, CBP2, and CBP3 are concluded below:

CBP1: nC_7H_{16} -> C_7H_{15}-2 -> $C_7H_{15}O_2$ -> $C_7H_{14}O_2H$ -> $C_7H_{14}O_2HO_2$ -> C_7KET_{12} -> $C_5H_{11}CO$

CBP2: nC_7H_{16} -> C_7H_{15}-2 -> C_4H_9 -> C_2H_4 -> CH_2O -> HCO -> HO_2 -> H_2O_2

CBP3: nC_7H_{16} -> C_7H_{15}-2 -> CH_3

The CBPs 1, 2, 3 produce OH radical by R6/R7, R13, and R19, respectively. The OH radical produced are in turn take part in the H abstraction of n-heptane by R1, and reactions R14 and R17 which belong to CBP2. The chemical reactivity is enhanced by this way which has been widely recognized as the LTC chain branching in numerous literature in past decades.

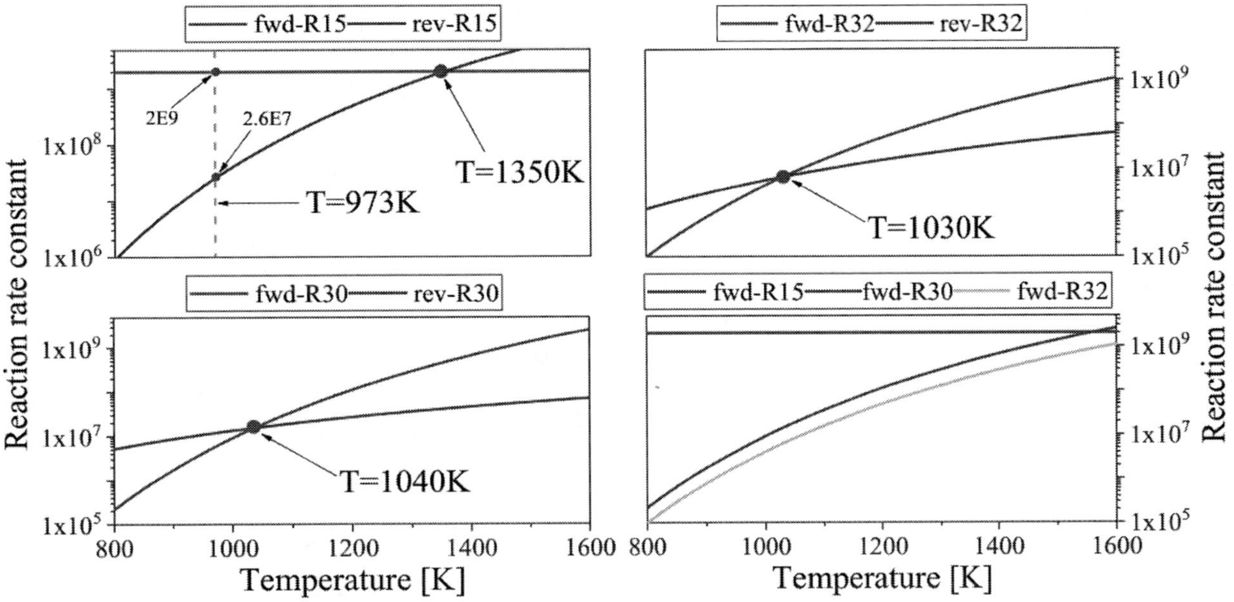

Fig. 6 comparison of forward and reverse RRC of R3, R8 and R9, and the comparison of forward RRC of R3, R8 and R9.

Point2: Fig. 8 compares the forward and reverse RRC of R3, R8 and R9, the RRC between R3 and R8/R9 is found with different characteristic. As shown in Fig. 4, the mole concentration of $C_7H_{15}O_2$ radical is significantly higher than that of C_7H_{15}-2 radical at points 1, 2, 3, 4. The progress variable of R3 is negative, and the $C_7H_{15}O_2$ radical is dissociating to C_7H_{15}-2 and O_2 through R3. In order to proof such phenomenon that has not been reported in the past literature, the progress variable for R3 is examined by equation (2):

$$q_{15} = k_{f_3}[C_7H_{15} - 2][O_2] - k_{r_3}[C_7H_{15}O_2] \qquad (2)$$

Where $k_{f_{15}}$ and $k_{r_{15}}$ is the forward and reverse RRC of R3, respectively. The temperature at point2 is 973K (Fig. 4). As shown in Fig. 8, the forward RRC of R3 (2E9) is 77 times higher than the reverse RRC of R3 (2.6E7) at point2. However, as indicated by Fig. 4, the concentration of $C_7H_{15}O_2$ radical is about 9 times larger than the C_7H_{15}-2 radical, which leads the term

$[C_7H_{15}-2][O_2]$ (2E-7) about 89 times smaller than term $[C_7H_{15}O_2]$ (1.74E-5) in equation (2). As a result, the calculated progress variable of R3 at point2 is negative and leads the reverse reaction of R3. Thus, the formation of RDR_{PE}-R15 region can be explained by three factors which have strong relationship with the air entrainment. (1) The temperature in RDR_{PE}-R3 region increases due to the mixing with ambient air of 1000K. This may lead the increased reverse RRC of R3 and inhibit the net production and consumption rate of $C_7H_{15}O_2$ and C_7H_{15}-2 radical, respectively. (2) The concentration of O_2 is lower in RDR_{PE}-R3 region than that in the periphery of n-heptane spray, which may also weaken the forward reaction of R3. (3) The mixing between LTC products produced at the spray periphery with RDR_{PE}-R3 region enhances the chain branching, which leads more chemical heat release and a higher temperature and promotes the factor (1).

The formation of RDR_{PE}-R3 region represented by point2 has a strong impact on the ISDP. As shown in Fig. 5 (b), the consumption of $C_7H_{15}O_2$ radical is dominated by R3 and R4 with contribution of 57% and 33%, respectively. Accordingly, the reaction R3 becomes endothermic instead of exothermic in point1 (Fig. 4). Meanwhile, the destruction of heptyl radical is dominated by R8 and R9 with contribution of 68% and 29%, respectively. This indicates that the chain branching at point2 is mainly attribute to CBP2 and CBP3. The reactions R6 and R7 belong to CBP1, reaction R13 belongs to CBP2, and reaction R19 belongs to CBP3 are observed producing OH radical. As shown in Fig. 4, the mixture fraction, destruction of n-heptane, and the mole concentration of OH radical at point2 are higher than that in point1, which indicates that the chemical reactivity at point2 is stronger than that at point1.

Point3: As shown in Fig. 2, the vf2 merges into vfl at point3 due to the entrainment effect caused by high pressure injection. Before the n-heptane fuel at vfl reaches point3, the mixture fraction and n-heptane mole concentration are very high. On the contrary, the temperature, concentration of low temperature combustion products including OH, H_2O_2, and n-heptane destruction rate is significantly low. This indicates that the chemical reactivity at point 2 is significantly weakened as it is merging into vfl, leading to an obviously different ISDP compared with that at point2, as shown in Fig 5. The chain branching at point3 is dominated by CBP1. The destruction of heptyl radical is dominated by oxygen addition via R3 (78%) despite a slightly lower O_2 concentration than that at point2. The change of ISDP at point3 is also caused by the scalar mixing between vfl and vf2. With the merging of mixtures belongs to vfl with that belongs to vf2, the mixture at point2 is cooled (Fig. 6, decrease to 878K) by that at the spray centerline since the temperature of injected n-heptane fuel is as low as 373K. This may lead the enhanced forward reaction of R3 by decreasing the value of term k_{r_3}/k_{f_3}, and weakened forward reaction of R8 and R9 by increasing the values of terms $k_{r_{30}}/k_{f_{30}}$, $k_{r_{32}}/k_{f_{32}}$, respectively, as shown in Fig. 6. The mixing between LTC products at point2 with the rich n-heptane/air mixture at the spray centerline also has a strong effect on the local chemical reactivity by following steps: (1) the mixing with OH radical produced at point2 increases the local OH concentration at the high concentration n-heptane/air mixture at the spray centerline, thus increases the progress variable of R1, R14 and R17; (2) the enhanced reaction rate of R1, R14 and R17 release more chemical heat, leading to the increase of local temperature; (3) the higher temperature increases the rate of reactions belongs to CBP1, leading the improvement of chain branching at point3; (4) the increases reaction rates of R6, R7 and R13 leads to a higher production rate of OH radical and go on promote the 1st step.

Point4: As the injection continues, the n-heptane/air mixture as well as the LTC products at point3 is pushed toward the injection tip. As the movement of point3 toward point4, the chemical heat continuously releases, leading to the gradually increased temperature. This is supported by a higher temperature at point4 (988K) than that at point3 (878K), as shown in Fig. 4. The increased temperature strongly affects the ISDP at point4 compared with that at point3. As shown in Fig. 5 (d), the heptyl radical destruction at point4 is dominated by R8 (64%) and R9 (28%). In comparison, the contribution of R3 to heptyl radical destruction is negligible. This can also be explained by examining the expression of progress variable for these three reactions. As shown in Fig. 8, the increased temperature may lead the inhibited forward reaction of R3 by increasing the value of term k_{r_3}/k_{f_3}, and weakened forward reaction of R8 and R9 by decreasing the values of terms $k_{r_{30}}/k_{f_{30}}$, $k_{r_{32}}/k_{f_{32}}$, respectively. Meanwhile, the decreased O_2 concentration at point4 also inhibits the forward reaction rate of R3. Therefore, the CBP1 at point4 becomes negligible, leading to less OH radical production by R6 and R7. However, the destruction of n-heptane fuel, as well as CBP2 continues to consume OH radical through R1, R14 and R17, leading to the formation of H_2O_2 radical pool through reaction pathway: C_2H_4 -> CH_2O -> HCO -> HO_2 -> H_2O_2 before the initiate of auto-ignition. This is supported by a significantly high H_2O_2 concentration and low OH concentration at point4, as shown in Fig. 4.

Conclusion

This paper investigates the instantaneous chemical kinetics process of n-heptane spray at low temperature oxidation stage. The air entrainment is found to play an important role on the instantaneous chemical kinetics process of n-heptane spray. There exists a region where $C_7H_{15}O_2$ radical dissociated to C_7H_{15}-2, which has not been reported by the past literature. The combination of flow field caused by high pressure injection and air entrainment leads a significantly high reactivity region inside the spray.

Acknowledge

The author acknowledges the support of Convergent Science to this research by providing the Converge code. The authors acknowledge the High-Performance Computing Center (HPCC) at Texas Tech University at Lubbock for providing HPC resources that have contributed to the research results reported within this paper.

Reference

[1] Curran, H. J., et al. "A comprehensive modeling study of n-heptane oxidation." Combustion and flame 114.1 (1998): 149-177.

[2] Li Y, Li H, Guo H, et al. A numerical study on the chemical kinetics process during auto-ignition of n-heptane in a direct injection compression ignition engine[J]. Applied Energy, 2018, 212: 909-918.

[3] Engine Combustion Network (ECN). Diesel sprays data search utility. <http://www.sandia.gov/ecn/cvdata/dsearch/frameset.php> [accessed Feb. 2015].

[4] Design, R. CHEMKIN Theory Manual [J]. San Diego, CA, 2007.

Spring Technical Meeting
Eastern States Section of the Combustion Institute
March 4-7, 2018
State College, Pennsylvania

A "Scale-up" Swiss-roll Combustor and Its Application in Waste Gas Incineration

Joel Crawmer[1], Chien-Hua Chen[1], Bradley Richard[1], Howard Pearlman[1], Paul Ronney[2], Thomas Edwards[3]*

[1]Advanced Cooling Technologies, Inc., Lancaster, PA
[2]University of Southern California, Los Angeles, CA
[3]Temple University, Philadelphia, PA
[*]Corresponding Author Email: joel.crawmer@1-act.com

Abstract: Waste Volatile Organic Compounds (VOCs) from a variety of processes can be incinerated, often with supplemental fuel, to convert the VOCs to less harmful combustion products before being vented to the atmosphere, to meet emission regulations. In this work, a novel combustion device was developed to incinerate waste VOC streams with reduced supplemental fuel consumption. The device, termed a "Swiss-roll", embeds a combustion chamber inside of a spiral heat exchanger to recover heat from the combustion exhaust stream to the premixed inlet reactant stream. With efficient heat recuperation, a combustion reaction is stabilized at ultra-lean, super-adiabatic conditions, resulting in reduced supplemental fuel consumption and harmful emissions. While Swiss-roll combustors are typically applied in small-scale combustion, this work emphasizes advantages even at larger scales, due to exceptional thermal efficiency and the associated fuel savings. In this work, a relatively large-scale prototype incinerator, roughly 1ft. tall by 1ft. in diameter, was designed, fabricated, and tested. Experimental results include promising scalability, a significant extension in the fuel's lean flammability limit beyond conventional limits, high destruction of toluene VOCs, and low carbon monoxide and nitrogen oxide emissions. The results of this work and potential commercial applications in waste VOC incineration are discussed in detail.

Keywords: Swiss-roll, Heat Recuperation, Volatile Organic Compound, Incineration

1. Introduction

In a variety of industrial processes, there are waste volatile organic compounds (VOCs) that can not be directly vented to the atmosphere due to their ability to react with NOx to produce ozone, a known component of smog. Some recent research has focused on innovative vapor recovery units (VRUs) to recover waste VOCs for future reuse [1]. In applications where vapor recovery is not feasible or economical, the VOC-laden streams are simply incinerated using a flare or thermal oxidizer, with strict regulations on the resulting combustion emissions. However, these combustion processes require supplemental fuel if the chemical enthalpy of the waste VOC stream is below the lean flammability limit. Furthermore, non-premixed incineration systems operating at high temperature are prone to thermal NOx formation, which is a harmful environmental pollutant. The fuel expense and environmental impact can be reduced if incineration systems are designed to address these specific challenges.

A unique combustor, termed a "Swiss-roll", embeds a combustion chamber inside of a spiral heat exchanger to extend the flammability limits of the input fuel using efficient heat recuperation. The Swiss-roll combustor was first proposed by Dr. Felix Weinberg in 1974, and many studies since have demonstrated its unique performance capabilities [2-8]. Specifically, the Swiss-roll transfers thermal energy from the combustion exhaust to the inlet reactants, raising the total enthalpy of the inlet stream (Figure 1A). With this excess enthalpy, the inlet reactants can self-sustain combustion in the center of the Swiss-roll at super-adiabatic conditions for a given equivalence ratio, as shown in Figure 1B.

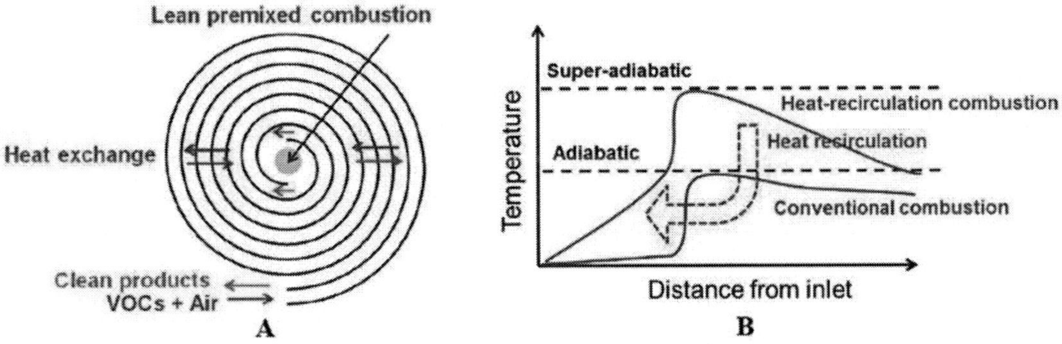

Figure 1: (A) A schematic of Swiss-roll incinerator. (B) Temperature profiles of the combustor with and without heat recirculation.

Since the reactants gain excess enthalpy from the combustion products, they have sufficient thermal energy to build a stable combustion zone with smaller chemical enthalpy (fuel) inputs. In other words, the excess enthalpy reaction extends the flammability limits of the fuel and enables ultra-lean, self-sustained combustion, minimizing the amount of supplemental fuel needed to incinerate fuel-lean waste VOC streams. Also, the combustion temperatures inside the Swiss-roll are lower than typical combustion temperatures without preheated reactants. At reduced temperatures, the thermal NOx formation time scale is increased and the exhaust does not have sufficient residence time in the hot combustion zone to form NOx. The Swiss-roll design is extremely thermally efficient since the reaction zone, which is most prone to heat loss, is surrounded by the heat exchanger. The compact, thermally efficient design, with no moving parts, makes the Swiss-roll potentially feasible for waste gas incineration applications.

Past Swiss-roll research has been predominately focused on small-scale combustion applications, where heat losses can have a dramatic effect on reaction stability and performance. However, most waste gas incineration applications would require a much larger Swiss-roll combustor than previously studied. This work attempts to scale-up the Swiss-roll combustor and characterize performance of larger devices.

2. Methods / Experimental

As a proof of concept exercise for the Swiss-roll incineration technology, a prototype combustor was designed, fabricated, and experimentally tested at the Advanced Cooling Technologies, Inc. (ACT) facility. The prototype constructed in this work includes a unique center combustion zone in an attempt to improve upon conventional Swiss-roll designs. A standard Swiss-roll's combustion zone is defined by the open region in the center of the device, where the inlet channel ends and the exhaust channel begins. In this design, both the inlet and exhaust channel are open to the combustion zone along the entire height of the channel, leaving a large recirculation zone in the center of the device due to the swirling flow mechanics during operation. While the standard design is easier to construct, ideally the flow would utilize the entire combustion zone volume, without a recirculation zone, to reduce flow velocity and resist the blow-off extinction limit. With this in mind, ACT redesigned the center combustion zone to operate with a more ideal, plug flow reactor type, flow pattern. The innovative design forces preheated reactants from the inlet channel to enter the combustion zone at one end, turn 90 degrees to flow axially along the reactor height, and the leave center combustion zone at the opposite end as shown in Figure 2A. When this flow pattern is used in conjunction with a flow straightener, the reacting flow evenly fills the entire volume of the combustion zone.

ACT incorporated the new center design into a working prototype Swiss-roll incinerator for experimental testing. The entire prototype, Figure 2C, stands roughly 15.5 inches tall and is 14 inches in diameter. The device is comprised of a cylindrical center combustion zone, about 4 inches in dimeter by 11.75 inches tall, surrounded by a 3.5 turn, countercurrent spiral heat exchanger with 3/8 inch wide by 11.75 inches tall channels, Figure 2B. The top and bottom faces of the Swiss-roll body were sealed and insulated using successive layers of wire-reinforced ceramic blanket, ceramic board, stainless-steel plate, stainless-steel honeycomb, and an aluminum plate. Threaded rods were inserted through alignment holes on the periphery of the aluminum plates to clamp the layers together and provide an airtight seal between the channels.

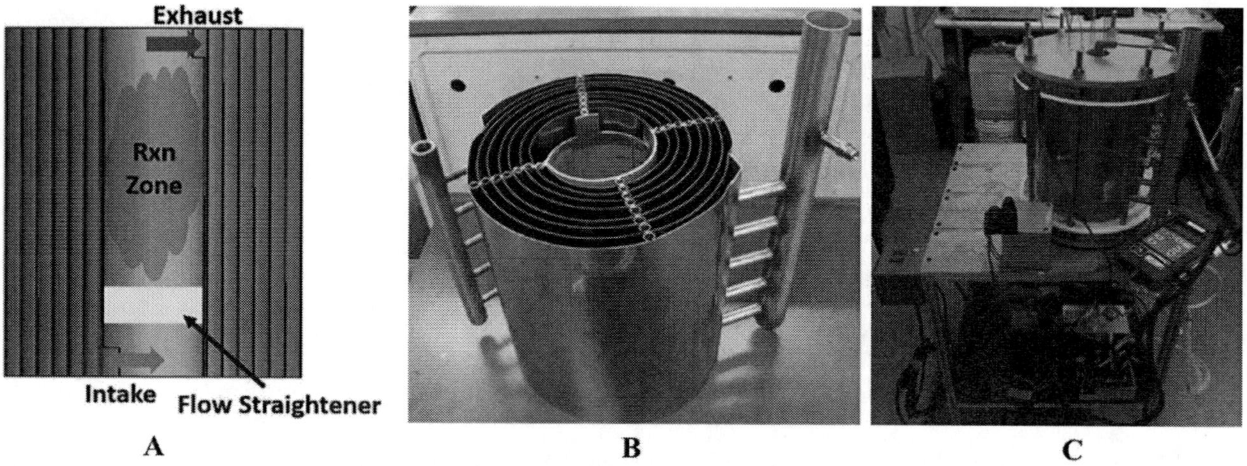

Figure 2: (A) A schematic of the end-to-end center design used to eliminate recirculation zones in the center combustion zone. (B) The Swiss-roll prototype body, without sealing/insulating layers, which was built with the end-to-end combustion zone flow path. (C) The entire Swiss-roll prototype assembly integrated with the experimental test setup.

Experimental testing of the prototype focused on completely characterizing the device's operational range to set a baseline for sizing the Swiss-roll incinerator. Specifically, the lean operating limit, destruction efficiency, and emissions were evaluated up to 800 SLPM (28 SCFM) total process flow rates. A schematic of the experimental test setup is shown in Figure 3.

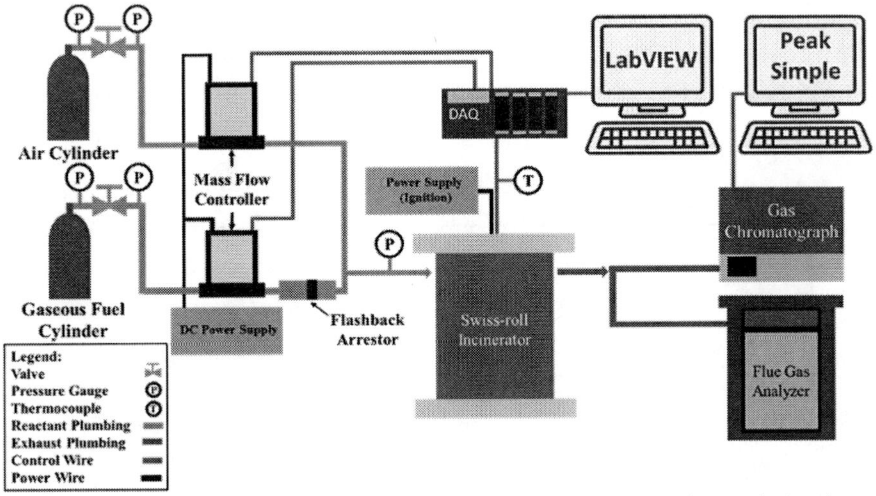

Figure 3: A schematic of the experimental setup used to test the Swiss-roll incinerator prototype.

For a given constant total flow rate (air and fuel) the lean limit was found by slowly reducing the fuel percentage until the reaction extinguished or the carbon monoxide level began to rise significantly. The lowest fuel percentage that could self-sustain a combustion reaction in the device, marked the lean limit operating condition for a given flow rate. Repeating this test procedure over a range of flow rates, created a lean limit operating curve for the device. During testing, the carbon monoxide and nitrogen oxide emissions were measured using a KANE flue gas analyzer to observe their respective trends across the lean operating limit of the device.

ACT measured the VOC destruction efficiency by adding toluene vapor to the inlet reactant stream using a temperature-controlled storage canister. Liquid toluene, held at 10°C by a recirculating chiller, released vapor into the reactant stream, which passed through the vapor head space in the storage canister. The amount of toluene vapor injected was estimated using the vapor pressure of toluene at a given temperature. Gaseous samples were taken from the inlet and exhaust manifolds of the Swiss-roll during operation and the gas composition was evaluated by a gas chromatograph using a flame ionization detector (FID). By making the rough assumption that the total number of moles in the system is conserved, the destruction efficiency was calculated using the ratio of toluene mole percentages between the exhaust and reactants.

3. Results and Discussion

Experimental testing proved the scaled-up Swiss-roll incinerator prototype extended the lean flammability limit of the input fuel for a wide range of flow conditions while maintaining high destruction efficiency and low emissions. The experimentally generated lean extinction limit plot, using propane fuel, is shown in Figure 4. Like previous studies, the Swiss-roll lean extinction limit curve in this work takes a parabolic shape, since more chemical energy is needed to combat heat losses in the low Reynolds number regime and the blow-off limit in the high Reynolds number regime.

While the performance of the scaled-up device is similar to previous experimental and computational findings at smaller combustor scales, the results in this work show an extended blow-off limit capability enabled by the end-to-end center combustion zone design. Since the flow velocity in the center combustion zone is reduced by eliminating the traditional recirculation zone, the device can operate at larger inlet Reynolds number flow conditions while avoiding the blow-off extinction limit. This result is important for potential large-scale applications, such as waste gas incineration, where high (>1000 SCFM) process flow rates are commonplace.

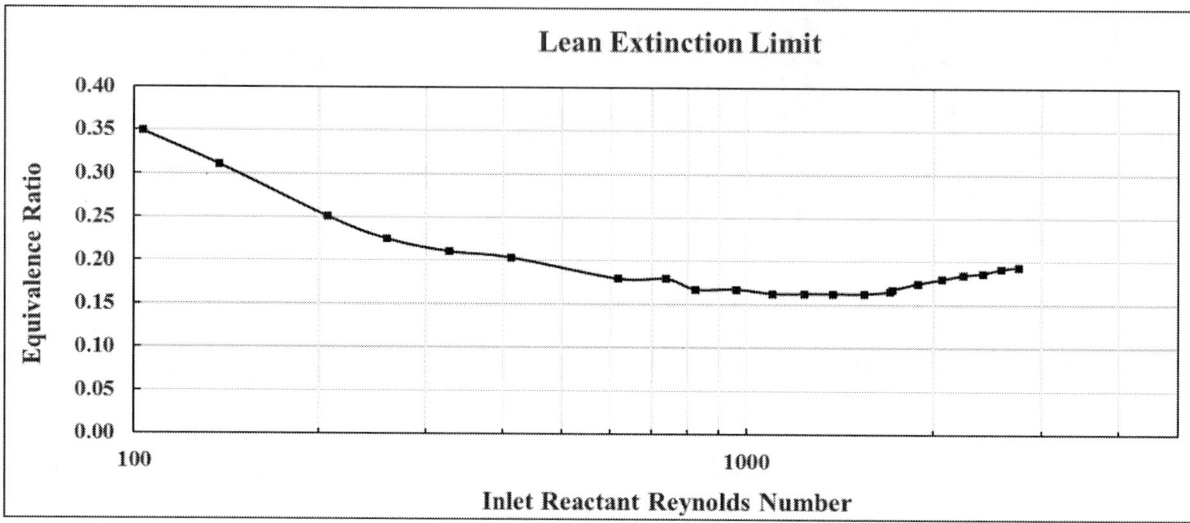

Figure 4: The lean extinction limit curve for the Swiss-roll incinerator prototype.

When considering the Swiss-roll combustor for waste gas incineration applications, the destruction efficiency and resulting emissions from the device are the most important performance metrics. The gas chromatograph results in Figure 5 show the Swiss-roll incinerator's high destruction efficiency during testing with toluene vapor and propane supplemental fuel. Based on the GC measurements at the inlet and exhaust of the combustor, the prototype demonstrated a destruction efficiency of just less than 99%. In addition, the measured carbon monoxide and nitrogen oxide emissions at the lean extinction limits across the entire tested flow range were less than 75 ppm and 5 ppm respectively.

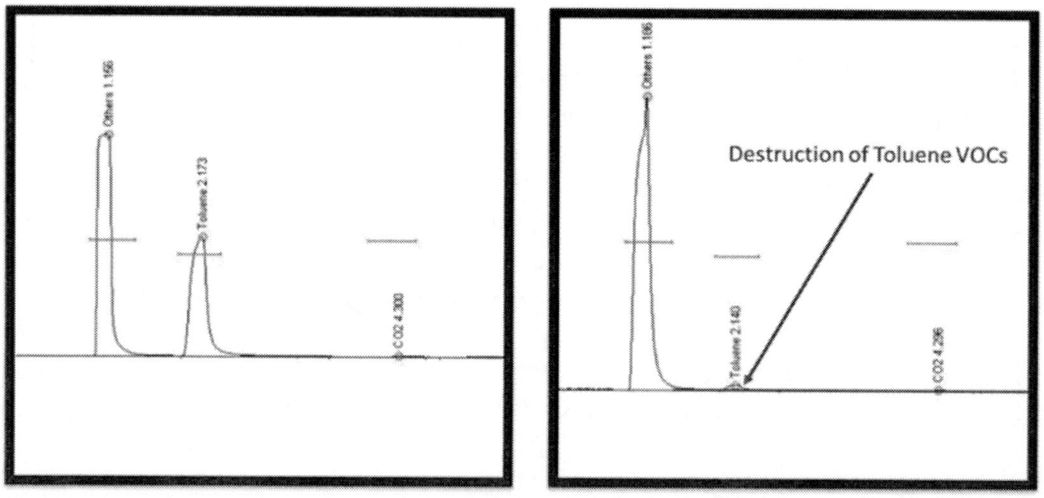

Figure 5: LEFT – The GC species composition measurement from the Swiss-roll inlet reactant sample showing toluene. RIGHT – The Swiss-roll exhaust sample GC measurement showing a significant reduction in toluene after thermal incineration.

4. Conclusions

The scaled-up Swiss-roll prototype fabricated and experimentally tested in this work demonstrated extended lean flammability limits, high destruction efficiency, and low harmful emissions. The combustion performance is similar to traditional smaller Swiss-roll devices, proving the scalability of the technology for large-scale applications. In addition, the end-to-end center design developed in this work effectively extended the blow-off extinction limit in Swiss-roll combustors, allowing the device to operate a higher inlet Reynolds number conditions. These developments are important for potential waste gas incineration applications, where large flow rates are often used to treat large amounts of waste VOCs with the least amount of supplemental fuel possible. A large-scale Swiss-roll can potentially excel in this application with an extremely thermally efficient combustion system.

5. Acknowledgements

This research was funded by the Department of Energy through the Small Business Innovation Research Fund (SBIR) under contract number DE-SC-0011884.

6. References

[1] K. Chue, Y. Park, J. Jeon, Development of adsorption buffer and pressure swing adsorption (PSA) unit for gasoline vapor recovery, Korean Journal of Chemical Engineering 21(3) (2007) 676-679.

[2] S. Lloyd, F. Weinberg, A burner for mixtures of very low heat content, Nature 251 (1974) 47-49.

[3] A. Jones, S. Lloyd, F. Weinberg, Combustion in heat exchangers, Proceedings of the Royal Society of London Section A 360 (1978) 97-115.

[4] P. Ronney, Analysis of non-adiabatic heat-recirculating combustors, Combustion and Flame 135 (2003) 421-439.

[5] C. Chen, P. Ronney, Three-dimensional effects in counterflow heat-recirculating combustors, Proceedings of the Combustion Institute 33 (2011) 3285-3291.

[6] C. Chen, P. Ronney, Scale and geometry effects on heat-recirculating combustors, Combustion Theory and Modelling 17 (2013), 888-905.

[7] C. Chen, P. Ronney, Geometrical effects on swiss-roll heat-recirculating combustors, PhD Thesis (2011).

[8] Joel Crawmer, et al., 10th U. S. National Combustion Meeting, College Park, MD, April 23-26, 2017, https://www.1-act.com/wp-content/uploads/2017/05/Crawmer-An-Innovative-Volatile-Organic-Compound-Incinerator.pdf

Spring Technical Meeting
Eastern States Section of the Combustion Institute
March 4-7, 2018
State College, Pennsylvania

Swiss-roll JP-8 Fuel Reformer with Direct Center Fuel Injection and Mixing Chamber Design

JoelCrawmer[1], Chien-Hua Chen[1], Bradley Richard[1], Howard Pearlman[1], Paul Ronney[2]*

1Advanced Cooling Technologies, Inc., Lancaster, PA
[2]The University of Southern California, Los Angeles, CA
[*]Corresponding Author Email: joel.crawmer@1-act.com

Abstract: Reforming higher hydrocarbons, such as the military logistic fuel JP-8, into synthesis gas has been heavily researched to enable the use of fuel cells in military operations while conforming to the single fuel policy. Reforming JP-8 is specifically challenging due to its notoriously high sulfur content, which poses serious poisoning challenges for catalyst-based systems. A non-catalytic fuel reforming system can be advantageous when reforming fuels with non-ideal compositions. The proposed system studied in this work uses a "Swiss-roll" combustion reactor to recuperate heat losses using a spiral heat exchanger, which enables self-sustained thermal partial oxidation at super-adiabatic temperatures near 1200°C (gas temperature). While the effective heat recirculation achieves high reforming efficiency without using catalysts, one challenge is the auto-ignition of the preheated reactants before they enter into the desired reaction zone. In this work, an advanced Swiss-roll fuel reformer was designed, fabricated, and experimentally tested to address this challenge. In particular, a new approach to Swiss-roll reactant injection, by directly injecting the fuel to the center reaction zone, was integrated with a mixing chamber design to ensure the reactants are completely mixed before the partial oxidation reaction without auto-igniting. The experimental results and potential applications are discussed in detail.

Keywords: Swiss-roll, Thermal Partial Oxidation (TPOX), Non-Catalytic Fuel Reforming

1. Introduction

Fuel cells are a promising technology for efficient electrical power generation from hydrogen, or hydrocarbon fuels, with no moving parts, harmful emissions, or noise. Promising commercial applications have already been realized in Fuel Cell Electric Vehicles (FCEV) and stationary power generation plants. However, enabling fuel cells for widespread use, especially military applications, presents challenges due to the availability of the fuel source. Most commercial fuel cell systems use a reliable source of high purity hydrogen, or methane, as the input chemical energy, but for military applications these fuel infrastructures are not available. In fact, the U.S. Army has instituted a Single Fuel Policy built around JP-8 fuel, meaning all systems requiring a fuel source must operate on JP-8 to simplify fuel logistics. Therefore, new technologies for military applications must be proven to adhere to this policy.

Since fuel cells operate using hydrogen (or syngas) a fuel reformer must be used in order to enable operation with JP-8 fuel. Reforming JP-8 fuel using a precious metal catalyst has been studied in the past, but there are still challenges to overcome due to catalyst deactivation from coking or sulfur poisoning. Thermal partial oxidation is an ideal reforming technique for JP-8 fuel due to its high tolerance to undesired species, but its low exothermicity at fuel-rich reforming conditions makes it difficult to the achieve the high-temperature reaction conditions required.

A Swiss-roll reactor, first discovered in 1974 by Dr. Felix Weinberg, uses heat recuperation to raise the temperature of a given reaction to super-adiabatic conditions [1-2]. The device surrounds a reaction chamber with a spiral heat exchanger to effectively reduce the propensity for thermal losses. In this way, a Swiss-roll fuel reformer enables the desirable TPOX reforming mechanism for JP-8 fuel. Past research has shown the Swiss-roll's capability in producing a robust, super-adiabatic reaction zone with no external energy input at steady state [3-7]. The reactor is inherently sulfur tolerant, self-sustaining, compact, and has no moving parts.

2. Methods / Experimental

A Swiss-roll JP-8 fuel reformer was designed, fabricated, and tested in this work to address the aforementioned challenges with Swiss-roll TPOX fuel reforming. In particular, the design included direct fuel injection to the Swiss-roll reaction zone with an integrated mixing chamber to eliminate fuel auto-ignition in the spiral heat exchanger while maintaining a premixed gas-phase reaction to reduce soot formation. Based on this design, a stainless-steel Swiss-roll reactor was fabricated via additive manufacturing using a technique called Direct Metal Laser Sintering (DMLS). A model of the design and the Swiss-roll part integrated with the experimental test setup are shown in Figure 1.

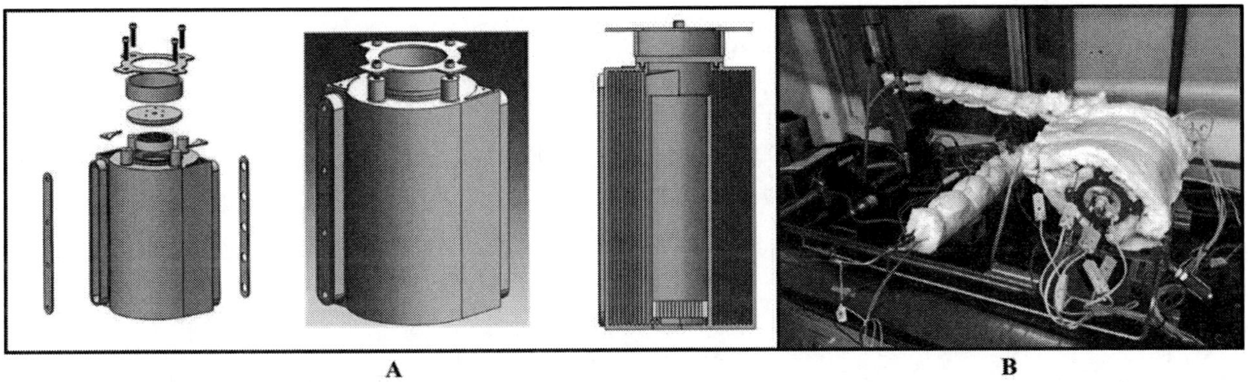

Figure 1: (A) A 3-D model of the Swiss-roll fuel reformer prototype design. The vertical section view shows the mixing chamber integrated at the bottom of the center combustion zone. (B) The Swiss-roll fuel reformer prototype integrated with the experimental test setup.

The Swiss-roll prototype was tested at a variety of fuel-rich reaction conditions, using JP-8 fuel, to characterize the TPOX reforming performance of the device. Compressed air was supplied to a mass flow controller to precisely control the amount of air entering the reactor. Similarly, a second mass flow controller metered propane into the reactor to preheat the device at startup, before injecting JP-8. A peristaltic JP-8 pump forced a prescribed amount of liquid fuel through a fuel heater to vaporize the fuel before injection into the Swiss-roll center. Strategically placed thermocouples measured the temperature throughout the reactor system to create a thermal profile for the device, while a single pressure transducer upstream of the Swiss-roll reactor measured the pressure drop across the system. A LabVIEW control system allowed the user to prescribe the input flow conditions, total flow rate and fuel percentage, and displayed and recorded the real-time temperature and pressure measurements. The reformate species composition was measured for a range of input reaction conditions, using a SRI Gas Chromatograph (GC), to characterize fuel reforming performance.

3. Results and Discussion

The results from JP-8 fuel reforming tests show successful conversion of the target fuel to syngas using high-temperature, gas-phase TPOX reforming with no significant soot formation. The thermal and flow profiles in Figure 2, show the temperature of the reaction at the prescribed input fuel and air settings for each experimental condition. The Swiss-roll reactor was able to produce a stable, high-temperature reaction zone (often between 900-1000°C), at a variety of flow conditions.

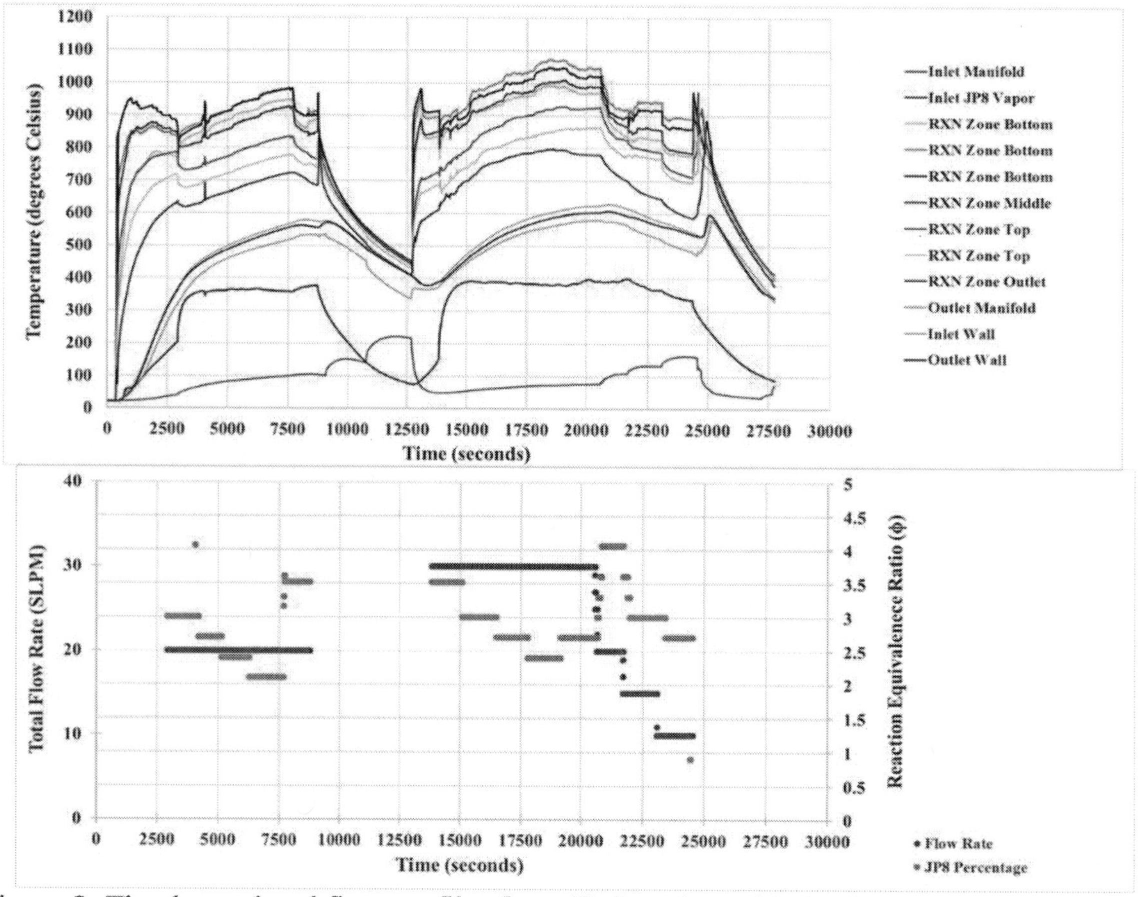

Figure 2: The thermal and flow profiles from JP-8 testing with the Swiss-roll fuel reformer prototype.

For each set of flow parameters, a reformate sample was extracted and analyzed by the GC. The species yield for each of the two main constituents of syngas (hydrogen and carbon monoxide) is shown in Figures 3 and 4 respectively. For a rough comparison, theoretical thermodynamic equilibrium concentrations were calculated using an isooctane/oxygen/nitrogen reaction using GASEQ software.

In general, lower equivalence ratio tests for a given flow rate produced the closest yields to estimated thermodynamic. This was due in part to the higher reaction temperatures achieved and the lower maximum theoretical equilibrium values. However, at a total flow rate of 20 SLPM, equivalence ratios 3.5 and 4 produced higher yields and were closer to theoretical equilibrium than values measured at lower equivalence ratios for the same flow rate. The reaction condition which was closest to thermodynamic equilibrium for both hydrogen and carbon monoxide concentrations was 30 SLPM total flow rate and ϕ=2.4. At this set point, the hydrogen yield was nearly 75% of equilibrium while the carbon monoxide yield reached about 98% of equilibrium. Non-coincidently, this condition also produced the highest temperature during testing at nearly 1100°C. As predicted, the reformate yield, and therefore the reactor performance, is highly dependent on reaction conditions and the temperature in the reaction zone.

Despite not reaching estimated thermodynamic equilibrium, the chemical enthalpy in the fuel reforming system was mostly conserved when considering all major flammable species measured. Figure 5 shows the reaction chemical efficiency defined as the chemical enthalpy in the reformate over the chemical enthalpy in the initial JP-8 feedstock. For 12 of the 13 total reformate samples analyzed, the chemical efficiency was over 60%. The highest chemical efficiencies (>80%) were measured at a total flow rate of 30 SLPM with ϕ ranging from 2.4 to 3.

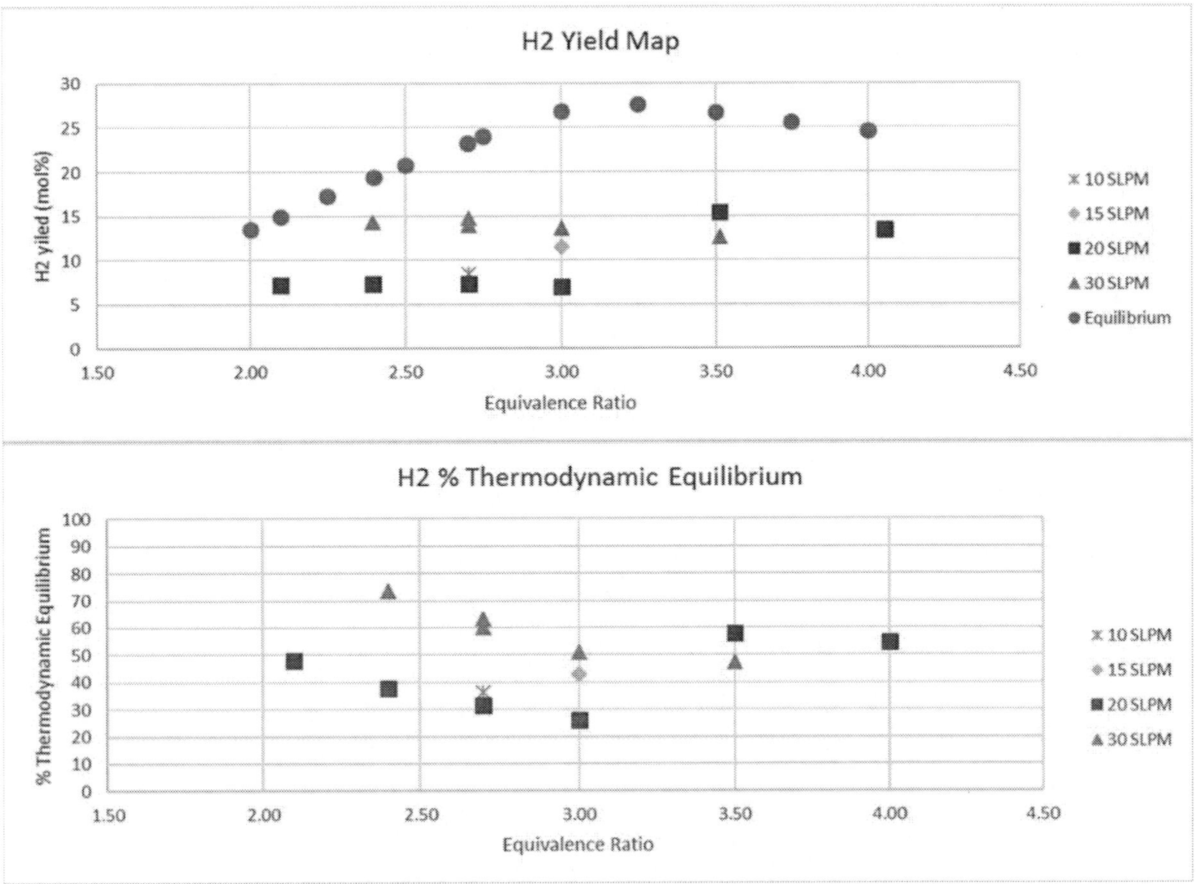

Figure 3: The hydrogen yield in the Swiss-roll reformate as a function of reaction equivalence ratio at different total flow rates.

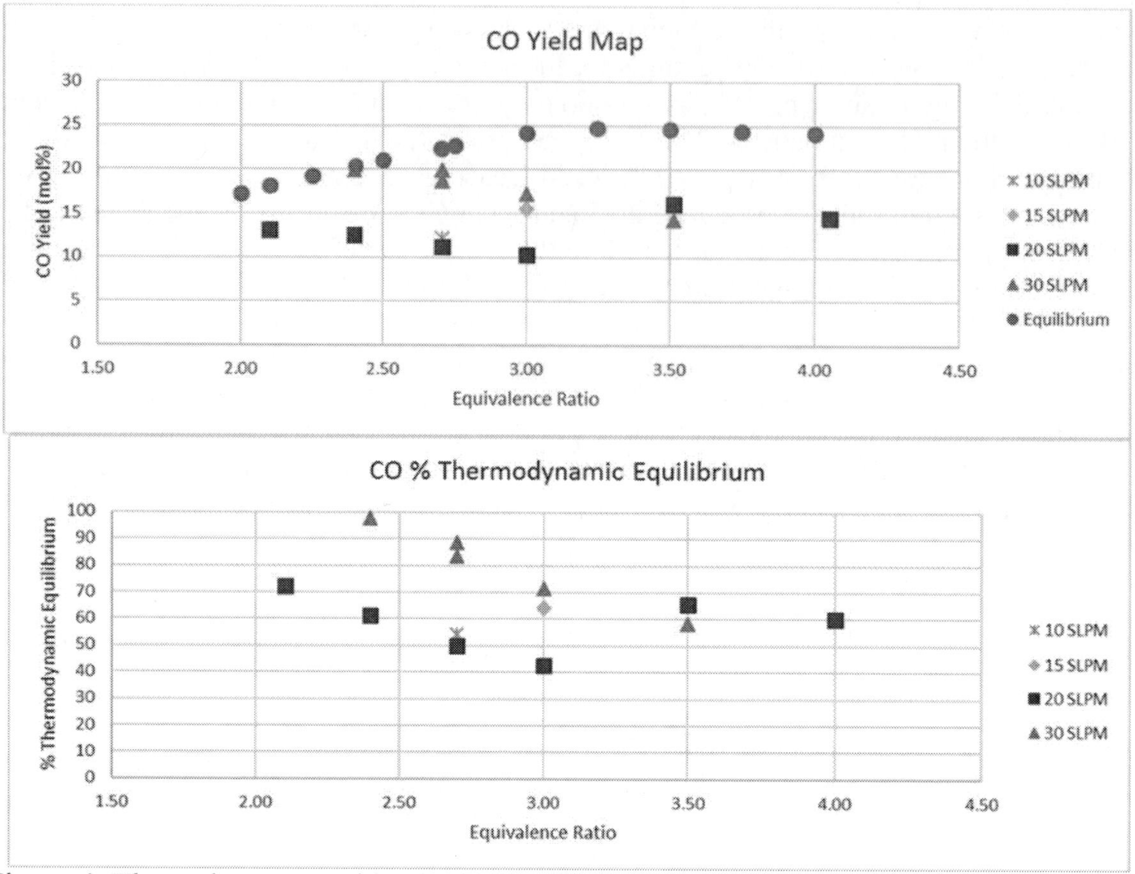

Figure 4: The carbon monoxide yield in the Swiss-roll reformate as a function of reaction equivalence ratio at different total flow rates.

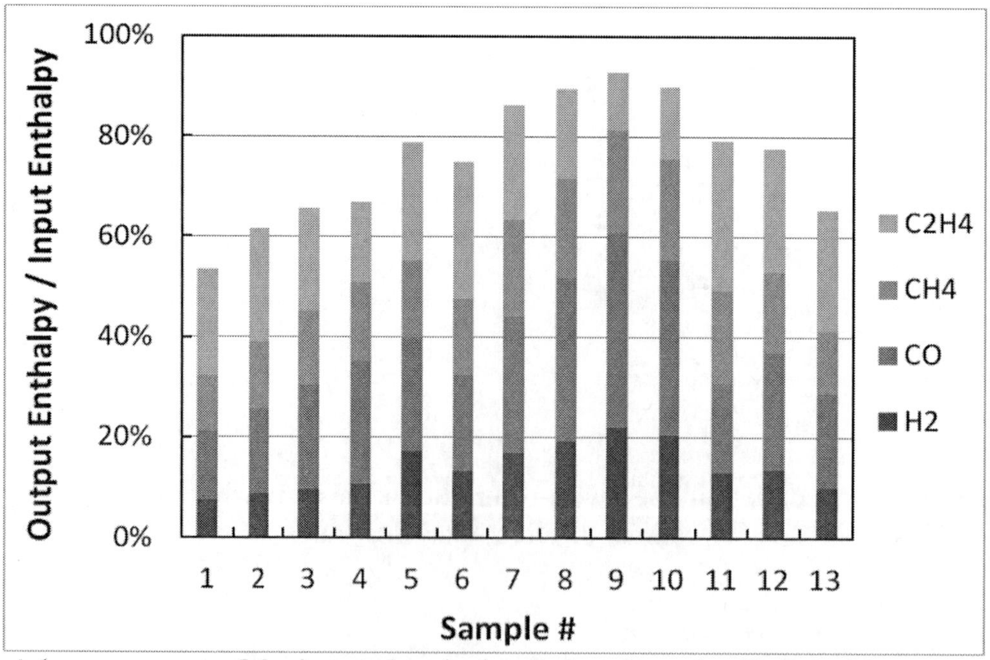

Figure 5: A large amount of the input chemical enthalpy from the JP-8 fuel is conserved in the Swiss-roll reformate after TPOX fuel reforming.

4. Conclusions

The Swiss-roll reactor tested in this work was able to successfully reform JP-8 fuel using TPOX with heat recuperation. Hydrogen and carbon monoxide yields at about 75% and 98% of thermodynamic equilibrium were demonstrated respectively, while retaining over 80% of the input chemical enthalpy. In addition, no significant soot accumulation was visibly observed during testing. Future work is needed to further characterize the device over a wider range of reaction conditions, with an emphasis on building and sustaining a higher temperature reaction zone to push the reformate toward thermodynamic equilibrium.

5. Acknowledgements

This research was funded by the Department of Defense U.S. Army through the Small Business Innovation Research Fund (SBIR) under contract number W56HZV-16-C-0016.

6. References

[1] S. Lloyd, F. Weinberg, A burner for mixtures of very low heat content, Nature 251 (1974) 47-49.

[2] A. Jones, S. Lloyd, F. Weinberg, Combustion in heat exchangers, Proceedings of the Royal Society of London Section A 360 (1978) 97-115.

[3] P. Ronney, Analysis of non-adiabatic heat-recirculating combustors, Combustion and Flame 135 (2003) 421-439.

[4] C. Chen, P. Ronney, Three-dimensional effects in counterflow heat-recirculating combustors, Proceedings of the Combustion Institute 33 (2011) 3285-3291.

[5] C. Chen, P. Ronney, Scale and geometry effects on heat-recirculating combustors, Combustion Theory and Modelling 17 (2013), 888-905.

[6] C. Chen, P. Ronney, Geometrical effects on swiss-roll heat-recirculating combustors, PhD Thesis (2011).

[7] Ryan Zelinsky et al., 10th U. S. National Combustion Meeting, College Park, MD, April 23-26, 2017, https://www.1-act.com/wp-content/uploads/2017/05/Zelinsky-Swiss-Roll-Style-Combustion-Reactor-for-Non-Catalytic-Reformi....pdf

Spring Technical Meeting
Eastern States Section of the Combustion Institute
March 4-7, 2018
State College, Pennsylvania

Microwave Plasma Processing of Methane: Optical Diagnostics and Carbon Characterizations

Randy Vander Wal[1], Arupananda Sengupta[1], Evan Musselman[2], Kurt Zeller[2], George Skoptsov[2]*

[1]*John and Willie Leone Family Dept. of Energy and Mineral Engineering and the EMS Energy Institute, Penn State University, University Park PA 16802 USA*
[2]*H Quest Vanguard, Inc., U-PARC Bldg. B11 William Pitt Way, Pittsburgh, PA 15238 USA*
[*]*Corresponding Author Email: ruv12@psu.edu*

Abstract: This paper presents optical spectroscopic diagnostics in H Quest Vanguard's microwave-plasma technology applied to coal and material characterization results from its application to H_2 generation from natural gas. Plasma diagnostics for process control will be illustrated. Chemical analyses are also presented of coal-to-liquid (CTL) products.

Keywords: Coal, Microwave, Plasma, Fuels, Chemicals

1. Introduction

Microwave-driven plasma-mediated reaction engineering represents process intensification for energy efficient and economical hydrocarbon conversion processes. Examples include coal-to-liquids and natural gas conversion to hydrogen and premium carbon products. Optical emission spectroscopy and analysis provides species identification and determination of temperatures in different energy modes in the non-equilibrium plasma.

2. Methods / Experimental

Microwave energy, variable between 1- 4 kW is coupled into a flow reactor. Reactive gas feeds using hydrogen or methane depend on desired hydrogenation and/or methylation reactions for product upgrading in the Wave Liquefaction™ (WL™) process applied to coal. Pulverized coal is fed by a screw coal feeder and entrained within the incoming gas stream. Gas ratios are adjusted for the particular hydrocarbon feedstock conversion. A complimentary application of the microwave-enhanced pyrolysis technology is methane decomposition to H_2 and premium carbon products. Reactor residence times are sub-second and the process proceeds at ambient pressure. Filters, traps and cyclone stages downstream of the reactor perform gas/liquid/solid separations.

2.1 Optical Emission Spectroscopy

Spectroscopic data collection was performed using a fiber-optically coupled Ocean Optics HR2000 spectrometer fitted with a UV-NIR grating for a spectral range of 900 nm. All optical emission spectra are spectrally calibrated using a Hg/Ar lamp and corrected for instrument response function using a NIST-traceable tungsten lamp as spectral calibration standard. C2* temperatures are determined using custom spectral band-fitting algorithms, taking into account the electronic baseline and underlying blackbody radiation. Spectral fits are performed at 1 nm spectral resolution. Carbon particle temperatures are determined by fitting Planck's radiation law with black body conditions. Ar* emission lines are identified by reference to the NIST spectroscopic database.

2.2 Gas/Liquid/Carbon Analyses

A suite of characterization tools have been applied to the light gases, liquid and solid carbon products. Gas analyses are performed online with a 2-channel gas chromatograph, augmented with a 200-amu mass spectrometer. Analyses of the captured liquids were performed using a gas chromatography-mass spectrometer. For the solid carbon products scanning electron microscopy (SEM) has been applied to coal chars (solid by-product of coal pyrolysis) for cenosphere analysis. For the carbon products from methane decomposition, scanning electron microscopy (SEM) was applied for assessment of morphology, transmission electron microscopy (TEM) for (nano)structural assessment and identification of different carbon forms (i.e. sp^2 phases), X-ray diffraction (XRD) for evaluation of graphitic structure, thermo-gravimetric analysis (TGA) for bulk determination of oxidative reactivity of the carbons as a means by which to assess graphitic content and gas (N2) adsorption analyses for texture – i.e. surface area and porosity. Elemental analysis was also performed for HCNS and O by difference.

2. Results and Discussion

Figure 1 shows an optical emission spectra of C_2* ($d^3\pi_g$ – $a^3\pi_u$) from the H Quest MW pilot scale reactor (with addition of CH_4 and H_2 to the process gas feed) along with simulated spectra for the indicated temperatures, where the fitting resolution is 1 nm. Optical diagnostics are central to reaction characterization in MW-enhanced pyrolysis and hold particular value for species identification and temperature determination [2]. Observed intensity ratios or spectra band shapes can yield temperature by Boltzman analysis using spectral constants. Moreover,

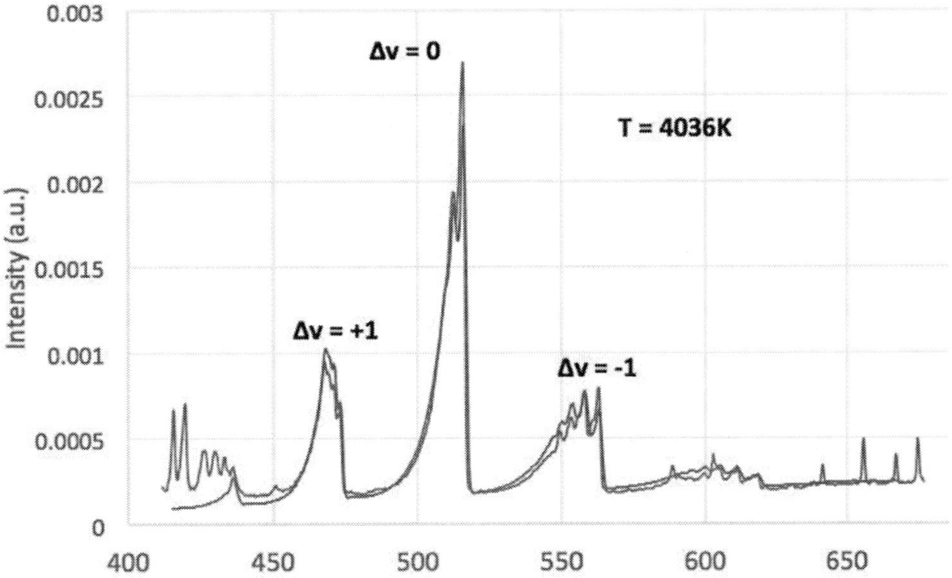

Figure 1. Swan band emission from C_2 (blue) and simulated spectra fit (red) for T = 4036 K.

optical emission serves to identify reactive species and intermediates, albeit indirectly inferred by the observation of their electronically excited counterparts, e.g. CH* and C2* radicals. For example, the presence of key atomic or diatomic radicals can support postulates of electron impact dissociation, and radical mediated bond insertion or radical capping reactions and provide mechanistic insights from the temperatures associated with the different degrees of freedom – electronic, vibrational, such as from the C_2^* ($d^3\pi_g - a^3\pi_u$) Swan band emission.

Microwave (MW)-activated natural gas decomposition for H_2 generation produces no CO2 and requires zero process water, in contrast to the presently dominant industrial process – steam reforming. As the microwave decomposition process is continuous, has short residence times, and has instant start-up / shutdown capabilities, it opens a new path for renewable energy storage by coupling electrical energy into chemical bond energy – with H_2 being particularly versatile as fuel or chemical precursor. Moreover, MW-activated natural gas decomposition does not require a catalyst (carbon or otherwise) but rather produces carbon in what is hypothesized to be a self-catalytic process. Therein MW-driven decomposition of methane achieves green hydrogen production and production of value-added carbon materials. Initial assessment of these carbon materials requires characterization by microscopic and spectroscopic techniques. Illustrative examples of carbon products by high-resolution transmission electron microscopy (HRTEM), Raman and X-ray diffraction will be shown – Fig. 2 shows a representative Raman spectrum of the carbon product. Complementary thermo-gravimetric analyses assess the oxidative reactivity of the carbon products. Optical diagnostics such as multi-wavelength pyrometry relate these material characteristics to reactor conditions and can be applied for process control.

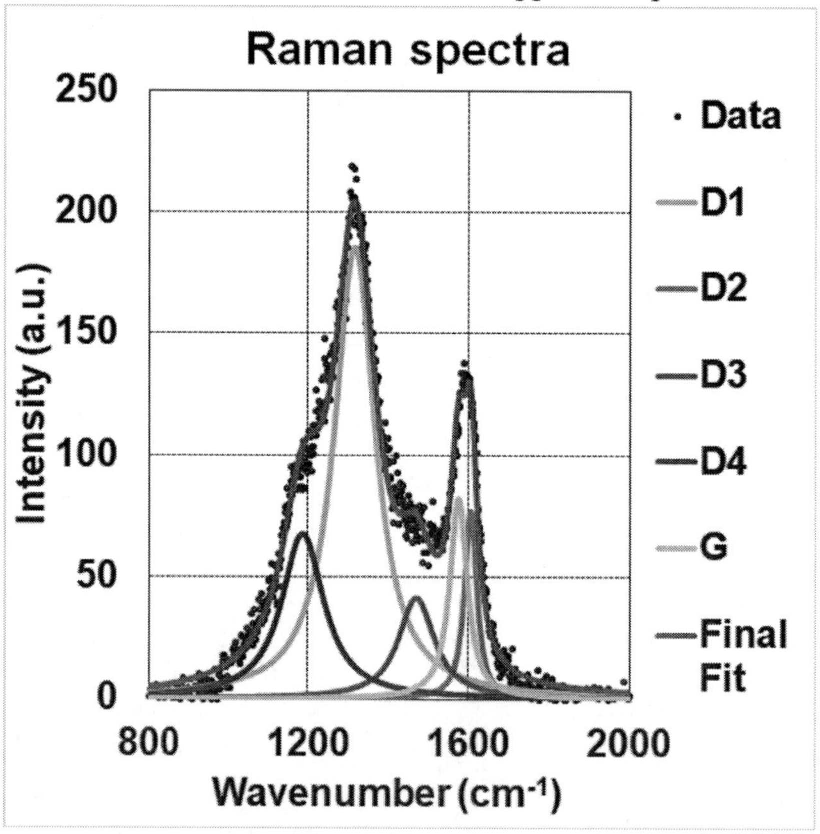

Figure 2. Raman spectrum and 5-fold peak deconvolution of generated carbon product.

3. Conclusions

H Quest Vanguard is developing a family of microwave plasma processes targeting conversion of a broad spectrum of hydrocarbon feedstocks such as coal and natural gas to value-added materials, chemicals and fuels [1,2]. The prototype coal process was originally proposed in response to the DARPA initiative for GHG-emission-free and cost-effective production of JP-8 jet fuel for the US Air Force from the domestic coal resources. Natural gas can be used in single-stage reactor as a hydrogen source, eliminating the need for external hydrogen production units and the associated CO_2 production, water consumption, and capital costs, and providing excess hydrogen sufficient for downstream hydrotreating. The MW process achieves localized high heating rates (measured at 10,000K/sec) resulting in flash devolatilization and pyrolysis followed by fast quenching, preserving the primary pyrolysis products' and volatiles' molecular structure.

5. Acknowledgements

H Quest Vanguard, Inc. is a privately held technology company, based in Pittsburgh, Pennsylvania, focused on the development and commercialization of novel hydrocarbon conversion technologies. This material is based on work supported by the Department of Energy, Office of Science through subaward agreement no. 186949 with H Quest Vanguard, Inc. under the Prime Award DE-SC0015895 Phase I SBIR.

6. References

[1] Skoptsov, George L., and Alan A. Johnson. Method for processing hydrocarbon fuels using microwave energy. U.S. Patent 9,682,359, issued June 20, 2017.

[2] Strohm, J.J., Linehan, J.C., Roberts, B.Q., McMakin, D.L., Sheen, D.M., Griffin, J.W. and Franz, J.A., Battelle Memorial Institute, 2012. Heavy Fossil Hydrocarbon Conversion And Upgrading Using Radio-Frequency or Microwave Energy. U.S. Patent Application 13/401,216.

Spring Technical Meeting
Eastern States Section of the Combustion Institute
March 4-7, 2018
State College, Pennsylvania

Development of a Constant Volume Combustion Chamber for Material Synthesis

Mohammadrasool Morovatiyan[1,], Martia Shahsavan[1], J. Hunter Mack[1]*

[1]Department of Mechanical Engineering, University of Massachusetts Lowell
[*]Corresponding Author Email: Mohammadrasool_Morovatiyan@uml.edu

Abstract: A constant volume combustion chamber (CVCC) was constructed to enable material synthesis procedures that are sensitive to temperature, pressure, and ambient species concentrations. Material synthesis processes require specific operating conditions in order to carry out the desired chemical reactions and property transformations, including the creation of paper-templated metals and nanoparticles. The 1.13 liter combustion chamber includes a test stand for conducting the material synthesis experiments. A premixed fuel-air mixture is ignited at a desired equivalence ratio in order to produce the required synthesis conditions. In comparison to furnaces and ovens, this approach provides greater flexibility for materials synthesis procedures. Computational modeling using adaptive mesh refinement, alongside preliminary experimental testing results, confirms that the CVCC can provide the appropriate conditions to synthesize paper-templated metals. The approach demonstrates that the CVCC can be a viable alternative to a furnace for use in materials synthesis applications.

Keywords: Constant Volume Combustion Chamber, Premixed Combustion, Material Synthesis, Paper-templated Structure, Furnace

1. Introduction

Constant volume combustion chambers (CVCC) are typically used in order to investigate fundamental aspects of combustion phenomena such as premixed ignition [1-2], diffusion flames, laminar flame speed [3], turbulent flame speed, autoignition [4-5], injection strategies [6-8], and emissions formation [9]. In this study, a small test stand is placed at the bottom of the cylindrical CVCC for conducting materials synthesis experiments, initially focused on the creation of paper-templated metals. In order to reach the desired conditions, e.g. temperature, and pressure, for fabricating the fibrous metallic structures, a premixed fuel-air mixture is combusted at a desired equivalence ratio. In initial experiments, methane is used as a fuel with air as the oxidizer. Combustion of the premixed charge in the CVCC yields high temperatures, high pressures, and species that are important in the material synthesis reactions.

Previous CVCC research by multiple groups has yielded a variety of design configurations that vary depending on their application, especially in regards to optical access. In this study, the combustion chamber is cylindrical and a test stand is placed at the bottom to enable the materials synthesis experiments. The material synthesis procedures are assisted by combustion in the CVCC. During the combustion, the temperature is raised up to 1400 K at the flame sheet and the pressure up to 10 bar. Applying a wide range of equivalence ratios aids to provide the required conditions for the material synthesis procedures. In the procedures, the reduction of the metal ions to elemental metal are favorable. In order to reduce the elemental ions, some specific gas species participate in the reduction process which are produced through the combustion. This combustion-assisted approach allows the material synthesis procedures to be encountered with new initial conditions in order to fabricate complex metallic structures.

Initial tests are based on a method called paper-templating [10], which uses a paper or fiber template that is loaded with an aqueous solution of metallic ions. The template is then heated in a furnace under specific conditions, resulting in metallic structures that have the morphology of the template material. These fibrous metallic structures have high surface areas per unit of weight/volume and are permeable to gas and liquids, which makes them attractive materials for several applications including catalysis processes. In contrast to a furnace, a constant volume combustion chamber (CVCC) can potentially provide greater operational flexibility, thus enabling synthesis operations unobtainable in a furnace or oven.

A CVCC, in comparison to a furnace, provides (i) higher maximum temperatures, (ii) higher pressures, and (iii) varied ambient species concentrations. Potentially, these conditions can facilitate processes that use different metals and more intricate template structures.

2. Methods / Experimental

2.1 Design / Experimental setup

The 1.13 liter cylindrical combustion chamber, shown in Figure 1, has an internal height of 152.4 mm and an internal diameter of 97.18 mm. Two flanges are welded to the cylinder that then connect to end-caps, using up to 8 bolts. A test stand is assembled on interior side of the bottom end-cap using 4 spacers and bolts. The whole assembly is mounted on a frame to make the bottom flange easily accessible. The top end-cap contains a thermocouple (OMEGA KMTXL-062G-6) a pressure transducer (KISTLER 601CAA), a gas inlet/outlet, and the spark plug.

(a)

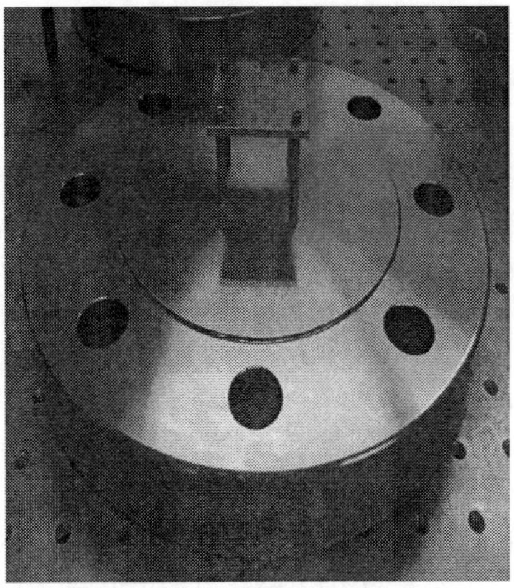
(b)

Figure 1: (a) The CVCC chamber, (b) The test stand on the bottom end-cap

Using a gas delivery system, the CVCC is filled with a mixture of methane-air at a desired equivalence ratio. The spark plug is triggered by an ignition coil (40,000 V) using a custom National Instruments LabVIEW program, which also handles data acquisition. The experimental set-up is shown in Figure 2.

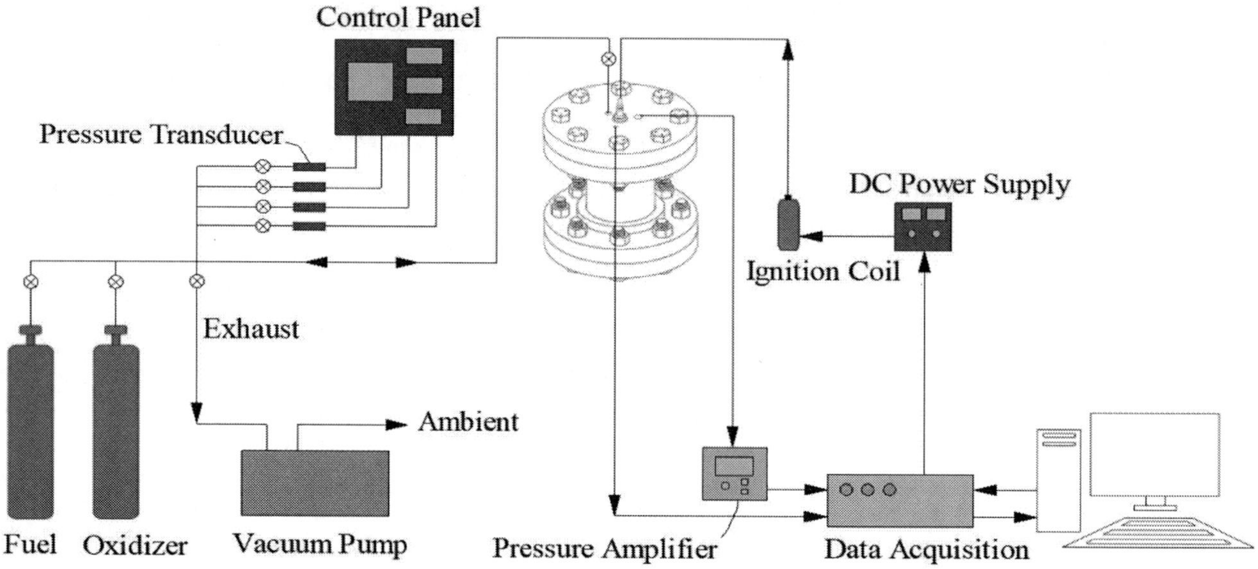

Figure 2: Experimental set up of the CVCC system

2.2 Numerical setup

A three-dimensional transient simulation of the premixed methane-air charge in a CVCC is implemented using CONVERGE computational code [11]. A 21-species, 84 reaction CHEMKIN mechanism for methane combustion (GRI MECH 1.2) [12] is applied. The combustion model uses a G-equation energy source (Equation 1) as the triggering energy for spark in a premixed flame-front tracking model. In the combustion model, Gulder model and Peters model are utilized for laminar flame speed model and turbulent flame speed model, respectively. For turbulence, a Reynolds Averaged Navier-Stokes (RANS) with Re-normalization group (RNG) k-ϵ model is applied. A variable time-step method is taken into account to observe the combustion process and heat transfer behavior properly. To capture all the events, three methods of generating cells are applied to the geometry: base grid, adaptive mesh refinement (AMR), and fixed embedding. The base grid is set to 5 mm. The AMR method is used for velocity and temperature profiles from the onset of combustion, with a same maximum embedding level of 2 for both profiles and sub-grid criterions of 1 and 5, respectively. Fixed embedding is temporarily implemented for the spark plug and its electrode through the end of the spark event, plus a permanent embedding next to the test stand with scales of 3 and 2, respectively. The use of AMR generates smaller meshes in higher gradient regions and greatly enhances computational efficiency. In this study, three initial wall conditions for the CVCC are examined: (i) no insulation, (ii) insulation, and (iii) adiabatic. The initial temperature and pressure of the mixture are 300 K and 1 bar, respectively. The simulation is ran for a sufficient time period in order to observe both combustion and heat transfer processes.

The G-equation is derived by Peters [13] as an unsteady transport equation following the expansion of a propagating flame interface. The following equation is typically used for tracking the laminar flame position.

$$\rho\frac{\partial G}{\partial t} + \rho\vec{u}.\nabla G = \rho U_L|\nabla G| - \rho D\kappa|\nabla G| \tag{1}$$

Where ρ is fluid density term, $\vec{u}$ is the fluid velocity vector, D is the diffusivity term, U_L is the laminar flame speed, and κ is the flame stretching term.

For turbulent flames, the equation (2) is averaged using the Favre- and Reynolds-averaged method:

$$\tilde{\rho}\frac{\partial \tilde{G}}{\partial t} + \tilde{\rho}\tilde{\vec{u}}.\widetilde{\nabla G} = \widetilde{(\rho U_T)}|\nabla\tilde{G}| - \tilde{\rho}D_t'\tilde{\kappa}|\nabla\tilde{G}| \tag{2}$$

Where $\tilde{\rho}$ is the average fluid density, $\tilde{\vec{u}}$ is the average fluid velocity vector, $\tilde{\kappa}$ is the flame stretching term defined by $\tilde{\kappa} = \nabla.\vec{n}$ in which $\vec{n} = -\frac{\nabla G}{|\nabla G|}$, D_t' is a diffusivity term defined by $D_t' = \sqrt{\frac{C_\mu C_S}{2Sc_t}}l_t u'$ in which C_μ and C_S are modeling constants, Sc_t is turbulent Schmidt number, l_t is the turbulent length scale, and u' is the turbulent velocity scale.

3. Results and Discussion

3.1 Temperature

After the spark plug is triggered, the flame propagates downward through the chamber towards the test stand. The average temperature and pressure increase as the flame front progresses. The simulation results are plotted for 120 ms, from onset of ignition. Figure 3 illustrates the maximum, mean, and test stand temperatures for three cases (no insulation, insulation, and adiabatic). With regards to the test stand temperature, as the flame propagates downward and reaches the test stand, the temperature rapidly increases to the flame temperature. After the flame front passes, the temperature continues to change, dependent on the initial boundary conditions. With no insulation, the temperature slowly decreases to room temperature. As insulation is added, the temperature drops at a slower rate due to less heat transfer. The decrease, and subsequent rise in test stand temperature is due to the flame front passing the top portion, proceeding around the edges, and subsequently heating the back side of the test stand.

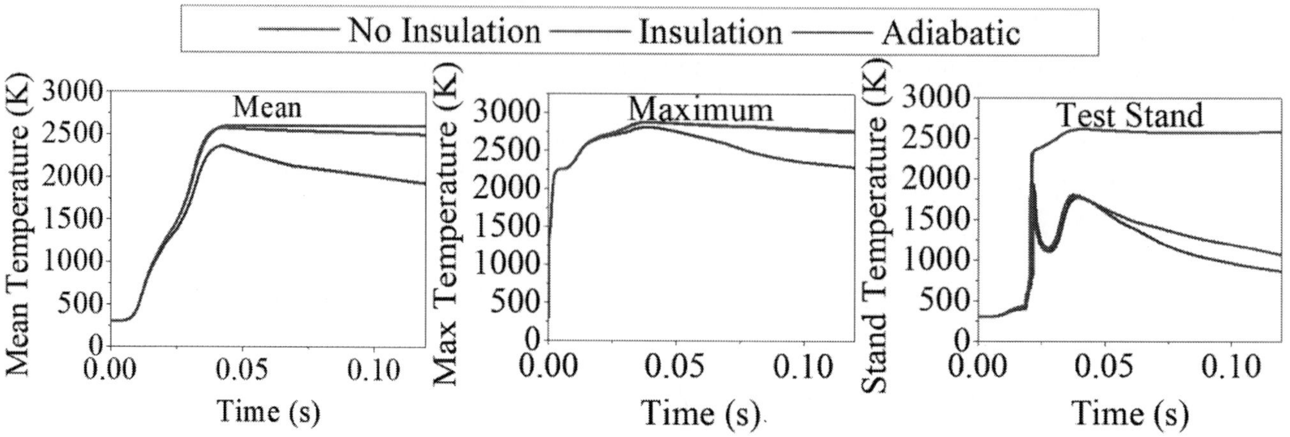

Figure 3: (a) mean, (b) maximum, and (c) test stand temperature for a period of 120 ms after spark triggering

3.2 Pressure

The pressure history for the combustion of the methane and air mixture in the CVCC is plotted in Figure 4(a). The comparison shows that applying an adiabatic condition to the chamber wall keeps the pressure higher than in the other two cases. In the case with a constant wall temperature condition, a very rapid drop in pressure is observed due to the rapid decrease in temperature.

3.3 Species

Figure 4(b) illustrates the time history of selected species for the no insulation and adiabatic cases. The CO and H_2 species concentration are explicitly significant for property transport in material synthesis procedures since both compounds are compelling reducing agents and are commonly used for the reduction of specific compounds [10].

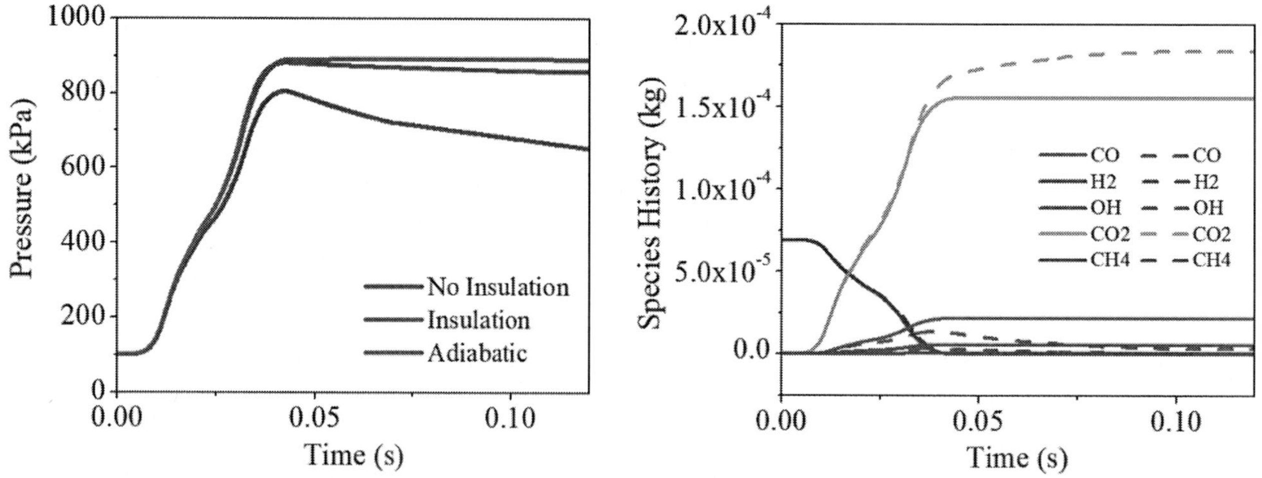

Figure 4: (a) maximum pressure, and (b) species history versus time: solid lines show the adiabatic condition, the dashed lines represent the no insulation case.

4. Conclusions

The use of a constant volume combustion chamber (CVCC) provides a viable alternative to a furnace for some materials synthesis operations. The CVCC approach yields higher maximum temperatures and pressures, while allowing for varied ambient species concentrations, dependent on the equivalence ratio of the combustion process. An increase in insulation allows for longer durations of the high-temperature, high-pressure-conditions, as expected. Simulations indicate that the combustion process produces elevated levels of H2 and CO, two species important to several materials synthesis reactions. By allowing for a wider range of operating conditions, the CVCC can enable materials synthesis operations inaccessible to a furnace or oven.

5. Acknowledgements

This authors would like to recognize the University of Massachusetts Lowell for supporting this research.

6. References

[1] O. Askari, Z. Wang, K. Vien, M. Sirio, and H. Metghalchi, "On the flame stability and laminar burning speeds of syngas/O 2/He premixed flame," *Fuel,* vol. 190, pp. 90-103, 2017.

[2] M. E. Feyz, S. I. Pishbin, M. Ghazikhani, and S. M. Razavi, "Parametric assessment of a low-swirl burner using the exergy analysis," *Energy,* vol. 79, pp. 117-126, 2015.

[3] E. Rokni, A. Moghaddas, O. Askari, and H. Metghalchi, "Measurement of Laminar Burning Speeds and Investigation of Flame Stability of Acetylene (C2H2)/Air Mixtures," *Journal of Energy Resources Technology,* vol. 137, no. 1, p. 012204, 2015.

[4] D. Kang, V. Kalaskar, D. Kim, J. Martz, A. Violi, and A. Boehman, "Experimental study of autoignition characteristics of Jet-A surrogates and their validation in a motored engine and a constant-volume combustion chamber," *Fuel,* vol. 184, pp. 565-580, 2016/11/15/ 2016.

[5] H. Kuszewski, A. Jaworski, A. Ustrzycki, K. Lejda, K. Balawender, and P. Woś, "Use of the constant volume combustion chamber to examine the properties of autoignition and derived cetane number of mixtures of diesel fuel and ethanol," *Fuel,* vol. 200, pp. 564-575, 2017/07/15/ 2017.

[6] M. Shahsavan and J. H. Mack, "Mixedness Measurement in Gaseous Jet Injection," *engrXiv.,* no. engrxiv.org/w5ebh, 18 May 2017.

[7] M. Shahsavan and J. H. Mack, "The Effect of Heavy Working Fluids on Hydrogen Combustion," *engrXiv.,* no. engrxiv.org/94s8d, 18 May 2017.

[8] M. Rahimi Boldaji, A. Sofianopoulos, S. Mamalis, and B. Lawler, "CFD Simulations of the Effect of Water Injection Characteristics on TSCI: A New, Load-Flexible, Advanced Combustion Concept," no. 58318, p. V001T03A019, 2017.

[9] D. H. Qi, B. Chen, D. Zhang, and C. F. Lee, "Optical study on the combustion characteristics and soot emissions of diesel–soybean biodiesel–butanol blends in a constant volume chamber," *Journal of the Energy Institute,* vol. 89, no. 4, pp. 807-820, 2016/11/01/ 2016.

[10] D. C. Christodouleas *et al.*, "Fabrication of Paper-Templated Structures of Noble Metals," *Advanced Materials Technologies,* vol. 2, no. 2, pp. 1600229-n/a, 2017, Art. no. 1600229.

[11] K. Richards, P. Senecal, and E. Pomraning, "CONVERGE (Version 2.4.), Convergent Science," *Inc., Madison, WI,* 2017.

[12] M. Frenklach *et al.*, "GRI-Mech---An Optimized Detailed Chemical Reaction Mechanism for Methane Combustion," November 1 1995, Art. no. GRI-95/0058.

[13] N. Peters, Turbulent combustion, Cambridge University Press, Cambridge, United Kingdom, 2000.

Spring Technical Meeting
Eastern States Section of the Combustion Institute
March 4–7, 2018
State College, Pennsylvania

Nitrogen Oxide Evolution in Oxy-Coal Combustion

Sai Krishna Sirumalla[1], *Aidin Panahi*[2], *Abhir Purohit*[1], *Andrew Baugher*[1], *Yiannis A. Levendis*[2], *and Richard H. West*[1,*]

[1]*Department of Chemical Engineering, Northeastern University, Boston, MA, USA*
[2]*Department of Mechanical & Industrial Engineering, Northeastern University, Boston, MA, USA*
[*]*Corresponding author:* r.west@northeastern.edu

Abstract: This paper investigates emissions of NO_x from pulverized coal burning in O_2/CO_2 environments. Such environments are pertinent to oxy-coal combustion, a promising "clean-coal" technology. The replacement of the inert nitrogen gas in air with carbon dioxide, which has different physical properties, alters the combustion conditions in the furnace. Hence, the purpose of this work is to theoretically examine the effects of (a) the oxygen concentration in the O_2/CO_2 gases, and (b) the resulting combustion temperatures, on the evolution of NO_x. To achieve these goals a previously published kinetic model was used, which assumes that fuel-bound nitrogen is released along with the tars during coal devolatilization and converts mostly to hydrogen cyanide. A sizable fraction of hydrogen cyanide is then converted to NO. Flame simulations were performed using Cantera to investigate the relative impacts of temperature and oxygen mole fraction, and to understand the causes of the observed trends.
Keywords: ***Coal, NOx, Fuel-Nitrogen, Hydrogen Cyanide, Oxy-Combustion.***

1. Introduction

Combustion-generated pollutants, such as acid gases, particulates, and air toxics, as well as greenhouse gases such as carbon dioxide, are a serious concern. Technologies to capture the pollutants, with various degrees of efficiency, have been developed and implemented [1]. One promising approach is carbon capture and sequestration (CCS) or carbon capture and utilization (eg. for underground oil recovery)[2]. Oxy-coal combustion, where coal combustion occurs in neat oxygen provided by an air separation plant, can greatly facilitate CCS as, upon drying, flue gases consist mainly of carbon dioxide. To reduce temperatures, flue gases are recycled, so the oxidizer stream is essentially a mixture of O_2 and CO_2.

In addition to facilitating carbon dioxide capture, oxy-coal combustion can drastically reduce the generation of nitrogen oxides, NO_x, which consist of mostly NO and little NO_2[3]. Since combustion takes place in the absence of atmospheric N_2 no thermal NO_x is generated; the detected amount of NO_x originates from fuel nitrogen. It has been shown that increasing the concentration of O_2 increases the NO_x generation during coal oxy-combustion, but it remains unclear whether this is driven by the increased temperature, increased oxygen concentration, or both. This study aims to separate the effects of temperature and oxygen partial pressure. This separation was investigated by first assuming that fuel-bound nitrogen is released along with the tars during coal devolatilization and converts mostly to hydrogen cyanide, HCN. Then to assess the conversion of HCN to NO a published kinetic model for HCN oxidation was used. This model was developed by Dagaut and Glarborg [4] and enhanced with reactions pertinent to CO_2 by Glarborg and co-workers; see Alzueta et al.[5] and Hashemi et al. [6].

Hashemi et al.[6] have worked on developing the nitrogen chemistry of coal oxy-combustion along with a model for different reaction mechanisms. According to the model, differences in the NO yield between the oxy-fuel combustion and the conventional combustion of pulverized coal can mostly be attributed to the recycling of NO (re-burning effect) and to changes in the mixing patterns between fuel and oxygen. Alzueta et al.[5] formulated the model for calculating NO_x concentration by combustion of HCN as intermediate. It was also found that the higher availability of oxygen increases the HCN conversion, due to the increased radical formation. The kinetic and stoichiometric data for modeling is collected from various research papers and theses on nitrogen chemistry and mechanisms in coal oxy-combustion [7–11].

2. Methods

2.1 Kinetic Model

Hashemi et al. [6] published a detailed kinetic model in CHEMKIN format but the transport data of species which is required to perform flame simulations was not made available in their published model. Transport data was found by comparing the species names using RMG-models [12] database and the transport data for pseudo species in Hashemi model were found in Ergut et al [13].

2.2 Thermogravimetric analysis of coal

The volatilization temperature of an HVA bituminous black coal (Pittsburgh No.8 from Pennsylvania (PSOC-1451)) was determined by thermogravimetric analysis(TGA), using TA instruments' Q500 Series thermogravimetric analyzer. A sample of bituminous coal was loaded on a platinum sample pan which was further loaded on a sensitive thermobalance in the analyzer. The volatilization temperature was determined by ramping the temperature to 900 °C at the rate of 20 °C/min while recording the weight change. The weight change data with increasing temperature was analyzed using TA Universal Analysis software.

2.3 Counter current diffusion flame simulations

Upon injection of a bituminous coal particle in a furnace, it heats up and devolatilizes. Upon ignition of the volatiles, an envelope diffusion flame appears surrounding the particle, as depicted in Fig. 1. This flame was simulated herein as a one-dimensional counter-current diffusion flame. Simulations were carried out using the open-source software Cantera[14] via the Python interface. The solutions are converged on approximately 120 grid points with convergence criteria `set_refine_criteria( ratio=3.0,slope=0.1,curve=0.1,prune=0.04)`, without radiation, and a domain width of 0.02 m. The simulations are carried out for bituminous coal by varying oxidizer inlet temperature and oxidizer composition. The inlet fuel composition is assumed to be composed of the coal volatile matter composition determined by Hashemi et al. Table 1 shows the molar composition of bituminous coal.

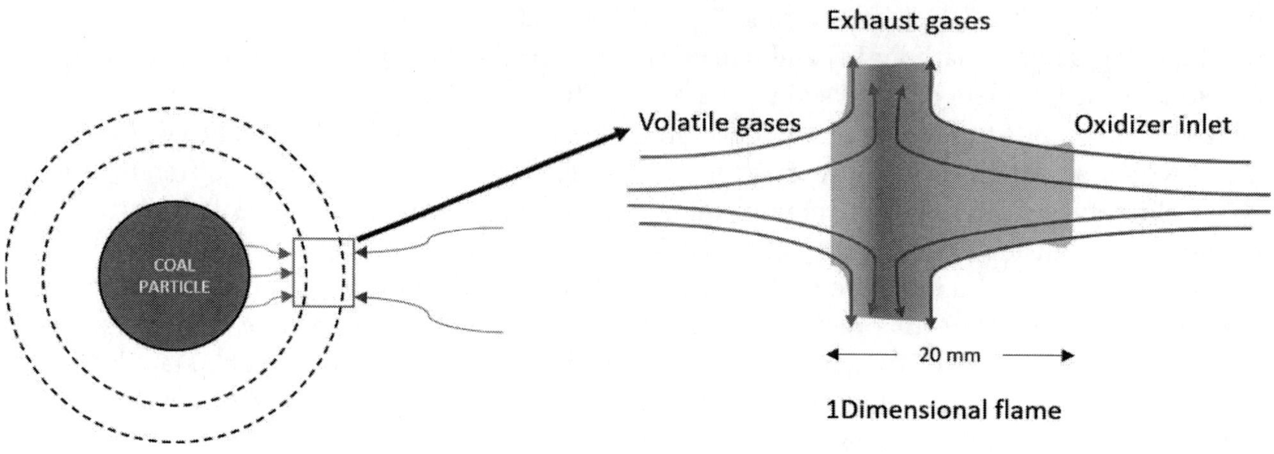

Figure 1: Schematic of simulated counter-current diffusion flame

Species	SOOT	H_2	CO_2	CO	CH_4	C_2H_4	H_2O	HCN
Molar composition	28.36	38.52	3.26	8.53	3.76	1.76	14.56	1.25

Table 1: Estimated molar composition of volatiles for bituminous coal

3. Results

3.1 Thermogravimetric analysis

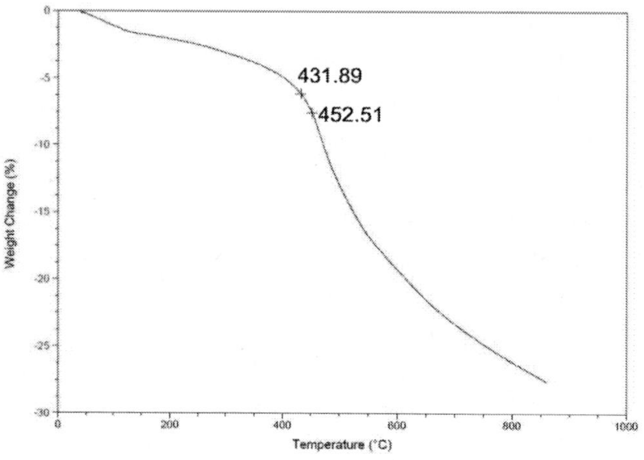

Figure 2: Weight change profiles for bituminous coal

As shown in Figure 2, during the thermogravimetric analysis, the weight of coal decreases with temperature. Initially the weight drops due to evaporation of water in both coals. The weight of bituminous coal sample starts to decrease sharply between 430 °C and 450 °C. Such behavior can be attributed to volatilization of gases from this coal. Hence we assumed the devolatilization temperature of the bituminous coal to be 442 °C. This determines the fuel feed temperature in the flame simulations.

3.2 Effect of inlet oxidizer temperature on NO concentration in flame

Counter-current diffusion flame simulations were carried out by changing the oxidizer inlet temperature, which is the furnace gas temperature, but keeping the oxidizer composition constant. This was done to quantify the effect of furnace gas temperature on NO generation and particle flame temperature. The oxidizer composition was kept constant at O_2:0.20,CO_2:0.80 while the inlet temperatures of the oxidizer gas were set at [1300, 1400, 1500] in K. Figure 3.a, shows the effect of oxidizer inlet temperature on the NO concentration in the diffusion flame. Figure 3.b shows the effect of the oxidizer inlet temperature on the flame temperature across the width of flame and the peak temperatures in the flame are reported.

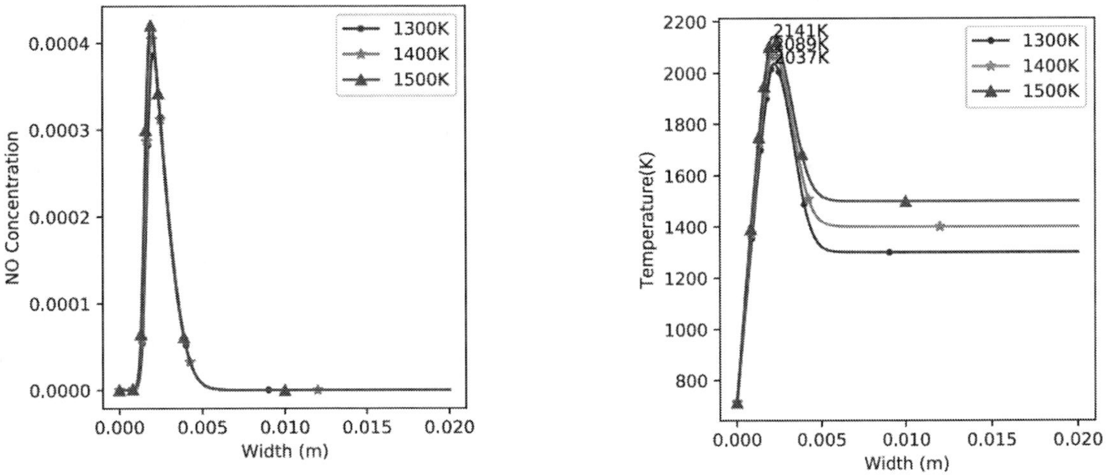

(a) NO concentration profile

(b) Flame temperature profile

Figure 3: NO species and flame temperature profiles by varying oxidizer inlet temperature

3.3 Effect of oxidizer inlet composition on NO concentration in flame

Counter-current diffusion flame simulations were carried out by changing the oxidizer inlet composition but keeping the oxidizer inlet temperature, which is the furnace gas temperature, constant to quantify the effect of oxidizer composition on NO generation and flame temperature. The oxidizer inlet temperature is kept constant at 1400 K while the O_2 and CO_2 concentrations at inlet were set at [0.20,0.30,0.40], [0.80,0.70,0.60] respectively. Figure 4.a shows the effect of the oxidizer inlet temperature on the NO concentration in the flame. Figures 4.b shows the effect of the oxidizer inlet composition along the width of the flame and the peak temperatures in the flame are reported. The difference between the peak temperatures in the flames is considerable when the oxidizer composition is changed as shown in 4.b .

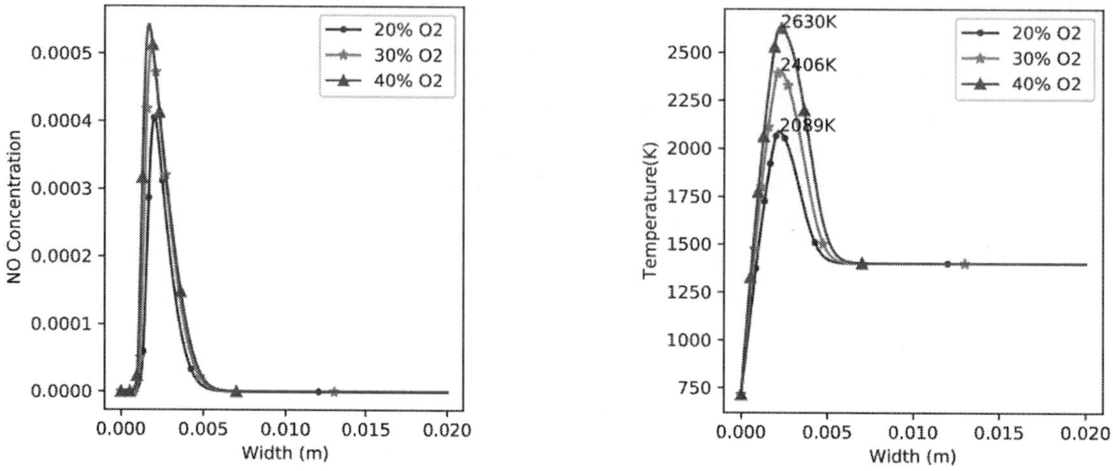

(a) NO concentration profile (b) Flame temperature profile

Figure 4: NO species and flame temperature profiles by varying oxidizer inlet composition

A comparison of the results shown in Figures 3 and 4 shows that the change in oxygen mole fraction in the range of [0.20-0.40] is more influential on both the flame temperature and on the NO formation than the gas temperature variation in the range of 1300-1500 K.

4. Conclusions

From the theoretical investigation of the NOx concentrations in a diffusion flame and of the temperature profile of this flame, it can be concluded that the oxidizer inlet temperature has a significant effect on the peak NO concentration during oxy-combustion of coal at a fixed oxidizer inlet composition as shown in figure 3, whereas the oxidizer inlet composition has a more significant effect on flame temperatures as shown in 4. These conclusions are valid for the ranges of the two parameters (gas temperature and oxygen concentration) input to this model. Further theoretical and experimental work is needed to determine what type of reactions and at what temperatures are producing more NOx and how to minimize the production on NO while also retaining combustion efficiency.

Acknowledgments

Support is gratefully acknowledged from the Departments of Chemical Engineering and of Mechanical and Industrial Engineering at Northeastern University.

References

[1] E. Rokni, H.-H. Chi, and Y. Levendis, In-Furnace Sulfur Capture by Co-Firing Coal with Alkali-Based Sorbents, 139 (2017).

[2] B. Abraham, J. Asbury, E. Lynch, and A. Teotia, Coal-oxygen process provides for enhanced recovery, Oil Gas J. 80 (1982).

[3] F. L. Horn and M. Steinberg, Control of carbon dioxide emissions from a power plant (and use in enhanced oil recovery), Fuel 61 (1982) 415–422.

[4] P. Dagaut, P. Glarborg, and M. U. Alzueta, The oxidation of hydrogen cyanide and related chemistry, Progress in Energy and Combustion Science 34 (2008) 1–46. DOI: 10.1016/j.pecs.2007.02.004.

[5] J. Giménez-López, A. Millera, R. Bilbao, and M. U. Alzueta, HCN oxidation in an O_2/CO_2 atmosphere: An experimental and kinetic modeling study, Combustion and Flame 157 (2010) 267–276. DOI: 10.1016/j.combustflame.2009.07.016.

[6] H. Hashemi, S. Hansen, M. B. Toftegaard, K. H. Pedersen, A. D. Jensen, K. Dam-Johansen, and P. Glarborg, A model for nitrogen chemistry in oxy-fuel combustion of pulverized coal, Energy and Fuels 25 (2011) 4280–4289. DOI: 10.1021/ef200853t.

[7] G. Scheffknecht, L. Al-Makhadmeh, U. Schnell, and J. Maier, Oxy-fuel coal combustion-A review of the current state-of-the-art, International Journal of Greenhouse Gas Control 5 (2011) 16–35. DOI: 10.1016/j.ijggc.2011.05.020.

[8] B. Buhre, L. Elliott, C. Sheng, R. Gupta, and T. Wall, Oxy-fuel combustion technology for coal-fired power generation, Progress in Energy and Combustion Science 31 (2005) 283–307. DOI: 10.1016/j.pecs.2005.07.001.

[9] F. Normann, K. Andersson, B. Leckner, and F. Johnsson, High-temperature reduction of nitrogen oxides in oxy-fuel combustion, Fuel 87 (2008) 3579–3585. DOI: 10.1016/j.fuel.2008.06.013.

[10] A. J. Mackrory, a Mechanistic Investigation of Nitrogen Evolution in Pulverized Coal Oxy-Fuel Combustion, (2008) 1–368.

[11] E. Vainio, A. Brink, M. Hupa, H. Vesala, and T. Kajolinna, Fate of fuel nitrogen in the furnace of an industrial bubbling fluidized bed boiler during combustion of biomass fuel mixtures, Energy and Fuels 26 (2012) 94–101. DOI: 10.1021/ef201145j.

[12] V. R. Lambert and R. H. West, Identification, Correction, and Comparison of Detailed Kinetic Models, 9th US National Combustion Meeting Cincinnati, OH, (2015).

[13] A. Ergut, S. Granata, J. Jordan, J. Carlson, J. B. Howard, H. Richter, and Y. A. Levendis, PAH formation in one-dimensional premixed fuel-rich atmospheric pressure ethylbenzene and ethyl alcohol flames, Combustion and Flame 144 (2006) 757–772.

[14] D. G. Goodwin, H. K. Moffat, and R. L. Speth, *Cantera: An Object-oriented Software Toolkit for Chemical Kinetics, Thermodynamics, and Transport Processes*, http://www.cantera.org, Version 2.3.0, 2017, DOI: 10.5281/zenodo.170284.

Spring Technical Meeting
Eastern States Section of the Combustion Institute
March 4-7, 2018
State College, Pennsylvania

Effect of variability in biomass properties on biomass devolatilization

Himanshu Goyal[1], *Perrine Pepiot*[2]

[1]*Robert F. Smith School of Chemical and Biomolecular Engineering, Cornell University, Ithaca, New York 14853, USA*
[2]*Sibley School of Mechanical and Aerospace Engineering, Cornell University, Ithaca, New York 14853, USA*

Efficient conversion of biomass into biofuels is still a challenge, and predictive modeling tools have the potential to accelerate the development and deployment of new conversion technologies, such as biomass thermochemical conversion. The high variability in the physical properties of biomass and their influence on biomass devolatilization hinder the deployment of detailed particle models for biomass devolatilization. In this work, we focus on the modeling of biomass devolatilization and evaluate the impact of variability in the physical properties of biomass on the validation of devolatilization models against experiments. For this purpose, we develop a one-dimensional spherically symmetric intra-particle heat and mass transfer model that describes the devolatilization of a porous solid biomass particle coupled with a lumped chemical kinetic model. Thermal conductivities of biomass and char, and specific heat capacity of biomass are varied and the predictions of the devolatilization model are compared with the experiments of maple wood sphere devolatilization.
Keywords: biomass devolatilization, intra-particle processes, variability in biomass properties

1 Introduction

Advanced biofuel technologies, based on sustainable, non-food sources of feedstocks, are required to meet aggressive volume goals for biofuel deployment [1–3]. Thermochemical conversion provides significant advantages, such as the ability to robustly handle a wide range of feedstocks, and to produce various types of transportation fuels. Currently, design and scale-up of thermochemical conversion reactors are mostly empirically-based, relying heavily on expensive and lengthy pilot-scale reactor studies. Yet, measurements in those reactors are unlikely to be detailed enough to provide the knowledge, especially at the particle scale, necessary to efficiently optimize biomass conversion [4]. Mathematical modeling and simulation tools provide a complementary, flexible and affordable framework to investigate the controlling chemical and physical processes, with the potential to play a determining role in the development and deployment of the technology.

While considerable progress has been achieved for both gasification and pyrolysis routes for biofuels production [5], several technical challenges still need to be addressed to enable mass industrialization of these processes, especially in the description of chemical processes, before simulation tools can be utilized to their full potential. A strong need for modeling efforts in biomass devolatilization has been recognized in the literature [6]. But the chemistry of this conversion process is extremely complicated to model due to the high variability of the feedstock and the complex structure of biomass particles [7]. These difficulties have hindered the development of detailed kinetic models for biomass devolatilization chemistry.

There is a large deviation in the reported values of several physical properties of biomass, such as thermal conductivity and specific heat capacity of wood and char [8]. These properties are known to have a significant influence on biomass devolatilization. Blasi [9] showed that physical properties of biomass greatly influence the conversion time and the activity of secondary reactions but have negligible effect on the primary products distribution. Considering the large deviation in the physical properties of biomass and their influence on biomass devolatilization, it becomes imperative to include this variability while validating any detailed biomass devolatilization model with experiments and/or optimizing the chemical kinetic parameters. In this work, we develop a one-dimensional spherically symmetric biomass devolatilization model using the lumped chemical kinetic model developed by Corbetta et al. [10]. The predictions of the resulting model are compared with the experiments of Park et al. [11] under varying physical properties of biomass.

In the following, the intra-particle model for biomass devolatilization is presented in Section 2, considering the detailed heat and mass transport inside the biomass. Comparisons of the particle model are made with the experiments in Section 3, providing an assessment on the influence of biomass physical properties on the validation of the devolatilization model.

2 Methods

In this section, a short description of the governing equations of intra-particle model is presented. Biomass devolatilization chemistry is represented by the chemical kinetic mechanism developed by Corbetta et al. [10]. A detailed description of the intra-particle can be found in reference [12].

2.1 Description of the intra-particle model

A one-dimensional spherically symmetric model of a devolatilizing biomass is numerically solved to resolve the intra-particle processes of biomass. A short description of the governing equations and the boundary conditions is provided here.

The temporal mass balance of solid components of biomass is given by

$$\dot{\rho}_{s,i} = \dot{\omega}_{s,i} \tag{1}$$

where $\dot{\rho}_{s,i}$ is the rate of change of density of solid component $i = 1 \ldots n_s$ and $\dot{\omega}_{s,i}$ is the reaction source term for solid component i obtained from the chemical kinetic scheme. Mass conservation for the gas phase is given by

$$\frac{\partial (\varepsilon \rho)}{\partial t} + \frac{1}{r^2}\frac{\partial (r^2 \rho U)}{\partial r} = \Sigma R_j \,, \tag{2}$$

where ε is the particle porosity, ρ is the gas density, U is superficial velocity of the gas, and R_j is the source term for gas phase species $j = 1 \ldots n_g$. Mass conservation for the individual gas species is given by

$$\frac{\partial(\varepsilon \rho y_j)}{\partial t} + \frac{1}{r^2}\frac{\partial(r^2 U \rho y_j)}{\partial r} = \frac{1}{r^2}\frac{\partial}{\partial r}\left(r^2 D \frac{\partial \rho y_j}{\partial r}\right) + R_j\,, \tag{3}$$

where y_j is the mass fraction of gas species j and D is effective diffusivity taken as a constant. Assuming a local thermal equilibrium between gas and solid phase is assumed, the energy conservation equation can be written as

$$(\rho C_p)_p \frac{\partial T}{\partial t} + (\rho C_p)_g U \frac{\partial T}{\partial r} = \frac{1}{r^2}\frac{\partial}{\partial r}\left(r^2 \lambda \frac{\partial T}{\partial r}\right) + Q\,, \tag{4}$$

where T is the particle temperature, $(\rho C_p)_p$ is the total heat capacity of the particle, $(\rho C_p)_g$ is the heat capacity of the gas phase, Q is the heat source term, and λ is the effective thermal conductivity [13].

At the particle surface, species flux is imposed by convective mass transfer and thermal energy flux is imposed by convective and radiative heat transfer.

$$D \left.\frac{\partial \rho y_j}{\partial r}\right|_s = \alpha(\rho y_{j,s} - \rho y_{j,\infty}) \tag{5}$$

$$\lambda \left.\frac{\partial T}{\partial r}\right|_s = h(T_s - T_\infty) + e\sigma(T_f^4 - T_s^4) \tag{6}$$

where α is the convective heat transfer coefficient, h is the convective mass transfer coefficient, e is surface emissivity, $y_{j,s}$ is the mass fraction of gas species j at the particle surface, $y_{j,\infty}$ is the ambient mass fraction of gas species j, T_∞ is the ambient temperature, T_f is the furnace temperature, and T_s is the particle surface temperature.

At the particle center, a zero gradient condition is imposed due to symmetry:

$$\left.\frac{\partial T}{\partial r}\right|_{r=0} = \left.\frac{\partial \rho y_j}{\partial r}\right|_{r=0} = 0 \tag{7}$$

A low Mach number assumption is made to decouple the effect of pressure on density. Gas phase is assumed to be ideal. Therefore, ρ can be calculated from the equation of state, as $\rho = \frac{PM}{RT}$. Here, P is the thermodynamic pressure assumed to be constant, M is the molecular weight of the gas phase, R is the ideal gas constant. The resulting set of coupled non-linear governing equations is numerically solved to get the instantaneous state of the biomass particle. All the properties of biomass are used from Corbetta et al. [10].

3 Results and Discussion

The intra-particle model described in the previous section is used to simulate the experiments of Park et al. [11], in which devolatilization of a single maple wood sphere of diameter 2.54 cm is studied for a range of reactor temperatures. We simulate the devolatilization experiments for

T = 688 K and 879 K under three varying biomass properties: wood thermal conductivity, λ_w, char thermal conductivity, λ_c, and wood specific heat capacity, C_w, which are known to affect the devolatilization process and have high variability in their reported experimental values. The value of each property for the base case is taken from Corbetta et al. [10]. The minimum and maximum values for most of the properties are extracted from the literature. The corresponding references are provided along with the values in Table 1. The maximum value of λ_c is approximated to be 50% higher than the base case, whereas for the minimum value of C_w, a linear approximation in temperature similar to the base case is assumed with the constant part used from Gupta et al. [14].

Table 1: Range of biomass properties considered for the biomass devolatilization simulations.

Parameter	Units	Values		
λ_w	W/m.K	0.1 [11]	0.19 [10]	0.7 [15, 16]
λ_c	W/m.K	0.05 [8]	0.14 [10]	0.2
C_w	J/kg.K	$1000 + T$	$1500 + T$ [10]	$1113 + 4.9T$ [17]

Comparisons are made between experimental and simulated values of particle surface and center temperature, solid mass fraction, and lumped product yields shown in Fig. 1. In column 1 and 2 of Fig. 1, dashed lines represent the lower and upper limit for particle surface and center temperature, respectively, and in column 3 error bars represent the possible deviation in the solid mass fraction evolution. Figure 1 shows that there is a significant variation in the predictions of particle center temperature and solid mass fraction with respect to λ_w and C_w, whereas little variation is observed with respect to λ_c. Particle surface temperature does not vary much for all the three physical properties considered. Figure 2 shows the comparison between the experimental and simulated product yields. It is interesting to note that no noticeable difference is observed in the product yields for all the three physical properties considered. Increasing λ_w and λ_c increases the heating rate and devolatilization rate of biomass particle, while increasing C_w slows down the heating rate and devolatilization rate.

4 Conclusions

In this work, a one-dimensional spherically symmetric intra-particle model for biomass devolatilization is used to simulate the devolatilization experiments of a spherical maple wood particle under varying biomass properties. Wood thermal conductivity, λ_w, char thermal conductivity, λ_c, and biomass specific heat capacity, C_w are varied in the range of their reported values. It is found that λ_w and C_w have significant effect on the comparison of the particle model predictions with the experiments, especially for the particle center temperature and solid mass fraction. λ_c is found to have very little impact on the particle model predictions for the range of variability considered. Moreover, none of these parameters have any noticeable effect on the product yields. As single particle devolatilization experiments are widely used for chemical kinetic model development, this work reinforces the importance of considering the possible deviations of the model predictions based on the variability in the biomass properties. This can have a significant consequence on the selection of Arrehnius parameters for a devolatilization chemistry model.

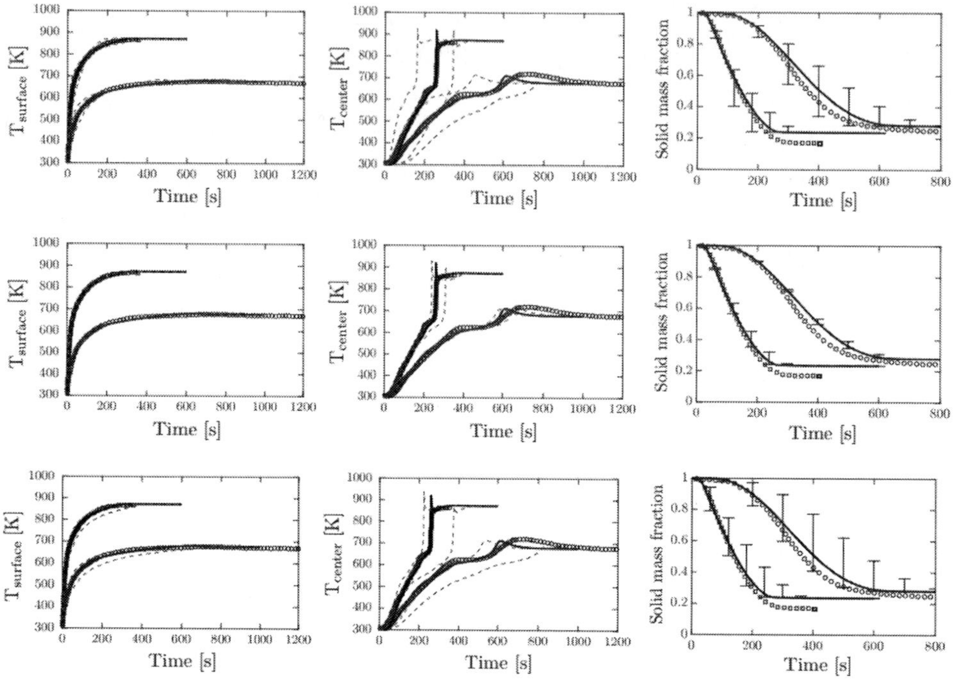

Figure 1: Particle center and surface temperature, and solid mass fraction for devolatilization of a spherical maple wood particle: Comparison between the simulation results (lines) and experiments (Park et al. [11], symbols). First, second, and third rows show the simulation results with variations of λ_w, λ_c, and C_w, respectively. Solid lines represent the base case simulations and dashed lines represent the simulation results for lower and upper limits of biomass properties. Different symbols indicate different temperatures (square: 688 K, circle: 879 K)

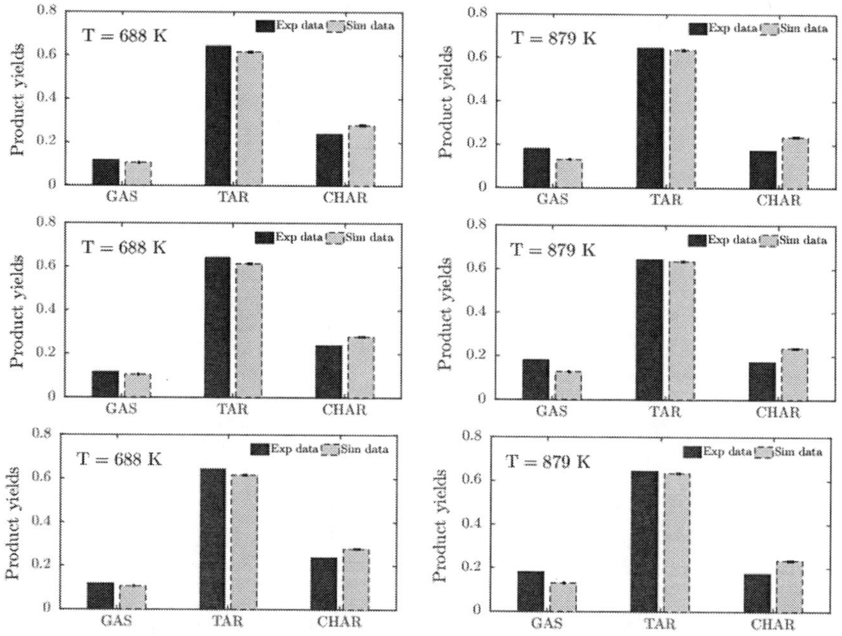

Figure 2: Comparison between the lumped product yields obtained from the base case simulations and experiments (Park et al. [11]). First, second, and third rows show the simulation results with variations of λ_w, λ_c, and C_w, respectively.

5 Acknowledgements

This material is based in part upon work supported by the National Science Foundation under Grant CBET 1638837.

References

[1] A. E. Farrell, R. J. Plevin, B. T. Turner, A. D. Jones, M. O'hare, and D. M. Kammen. *Science*, 311 (2006) 506–508.

[2] G. W. Huber, S. Iborra, and A. Corma. *Chemical reviews*, 106 (2006) 4044–4098.

[3] T. D. Foust, K. N. Ibsen, D. C. Dayton, J. R. Hess, and K. E. Kenney. *Biomass recalcitrance: deconstructing the plant cell wall for bioenergy*, (2008) 7–37.

[4] M. A. Van der Hoef, M. van Sint Annaland, N. G. Deen, and J. A. M. Kuipers. *Annu. Rev. Fluid Mech.*, 40 (2008) 47–70.

[5] A. Demirbas. *Progress in energy and combustion science*, 33 (2007) 1–18.

[6] Alberto Gómez-Barea and Bo Leckner. *Progress in Energy and Combustion Science*, 36 (2010) 444–509.

[7] Paulo Eduardo Amaral Debiagi, Giancarlo Gentile, Matteo Pelucchi, Alessio Frassoldati, Alberto Cuoci, Tiziano Faravelli, and Eliseo Ranzi. *Biomass and Bioenergy*, 93 (2016) 60–71.

[8] KW Ragland, DJ Aerts, and AJ Baker. *Bioresource technology*, 37 (1991) 161–168.

[9] Colomba Di Blasi. *Fuel*, 76 (1997) 957–964.

[10] M. Corbetta, H. Frassoldati, A.and Bennadji, M. J. Smith, K.and Serapiglia, G. Gauthier, T. Melkior, E. Ranzi, and E. M. Fisher. *Energy & Fuels*, 28 (2014) 3884–3898.

[11] W. C. Park, A. Atreya, and H. R. Baum. *Combustion and Flame*, 157 (2010) 481–494.

[12] Himanshu Goyal and Perrine Pepiot. Integrating intra-particle processes in large scale simulation of biomass thermochemical conversion. In *9th US National Combustion Meeting, Cincinnati, Ohio*, 2015.

[13] C. Di Blasi. *Chemical engineering science*, 51 (1996) 1121–1132.

[14] Murlidhar Gupta, Jin Yang, and Christian Roy. *Fuel*, 82 (2003) 919–927.

[15] Takamaro Maku. *Bulletin Wood Res. Inst*, 13 (1954) 1–80.

[16] Giancarlo Gentile, Paulo Eduardo Amaral Debiagi, Alberto Cuoci, Alessio Frassoldati, Eliseo Ranzi, and Tiziano Faravelli. *Chemical Engineering Journal*, 321 (2017) 458–473.

[17] Frederick Dunlap. *The specific heat of wood*. Number 110. US Dept. of Agriculture, Forest Service, 1912.

Spring Technical Meeting
Eastern States Section of the Combustion Institute
March 4-7, 2018
State College, Pennsylvania

A study on pulverized coal ignition using a two-stage flat-flame burner with a transition from a reducing to oxidizing environment

Adewale Adeosun[1], Dishant Khatri[1], Akshay Gopan[1], Zhiwei Yang[1], Tianxiang Li[1], Richard L. Axelbaum[1,]*

[1]Department of Energy, Environmental & Chemical Engineering, Washington University in St. Louis, St. Louis, Missouri-63130, USA
[*]Corresponding Author Email: axelbaum@wustl.edu

Abstract: Coal particles experience a transition from a reducing to oxidizing environment in the near-burner region of pulverized coal (pc) burners. For the first time, we report a fundamental study of ignition of a coal-particle stream experiencing a flame environment that transitions from a reducing to an oxidizing environment (termed reducing-to-oxidizing transition). High-speed videography is used to observe the particles in situ, and scanning electron microscopy is used to characterize the sampled particles. The effects of particle size on ignition are presented for four size bins for PRB subbituminous coal at nominal gas temperatures of 1800 K. The results show that the ignition behavior of particles in reducing-to-oxidizing conditions is different from that in purely oxidizing conditions. The reducing-to-oxidizing transition increases ignition delay time, on average, by more than 100% over those of oxidizing conditions. However, for particles below 150 μm diameter, ignition delay time is relatively independent of particle size in either oxidizing or reducing-to-oxidizing conditions due to high heating rate. Hetero-homogeneous ignition mode is dominant in both environment, except for the 125-149 μm particle size range in oxidizing environment, where homogeneous-to-heterogeneous ignition occurs.
Keywords: reducing-to-oxidizing transition, hetero-homogeneous ignition, coal, ignition

1. Introduction

Most fundamental theoretical and experimental studies of coal ignition have concentrated on ignition of isolated coal particles [1-3]. However, experimental evaluation of particle-stream ignition is more relevant to pulverized coal (pc) burners because particle-stream experiments consider the effects of particle-particle interactions on early-stage coal combustion processes. Also, particle stream ignition affects other processes, such as char burnout and pollutant formation [4, 5]. Thus, a better understanding of ignition for a coal particle stream will provide valuable information for designing more efficient burners, and also help to address specific challenges such as flame instability and loss-on-ignition in pulverized coal boilers [6].

Ruiz et al. [7] conducted an experimental study in a laminar flow reactor and evaluated ignition delay time for a continuous particle stream of high-volatility bituminous coal injected into a 9 vol.% oxygen environment for gas temperatures between 1023 and 1150 K. Their work, which is based on photographs of luminous emission from the burning particles, showed that ignition delay time first decreases with increasing particle density, and then increases with further particle loading. Liu et al. [8] conducted experiments on the ignition properties for a variety of coals at different particle loadings, using a laminar entrained flow reactor at gas temperatures between 1230 K and 1320 K. A particle size of 75-105 μm was used for their experiments and the oxygen volume percent in nitrogen was varied from 12% to 20%. They found that ignition delay time is very sensitive to coal particle size, coal feed rate, and the temperature of the coflow stream into which the coal stream is introduced. They also found that there is a minimum in ignition delay time as coal feed rate increases.

While these experimental studies provide important data for model validation, they do not simulate the combustion environment that is typical of conventional pc burners, where the coal experiences a characteristic transition from a reducing to oxidizing environment in the near-burner region. Typical of a pc burner, the locally reducing environment in the near-burner region is created by a high density of particles undergoing devolatilization and the fast depletion of oxygen in this region [6, 9]. Beyond this region, the particles move into an oxidizing environment. Such a transition is herein referred to as a reducing-to-oxidizing transition (R-O).

To this end, and different from any existing platforms, a new burner with two flat-flame regions is designed; hence the name two-stage Hencken flat-flame burner. A unique feature of this design is that this burner has an inner flat flame, which can be operated in a fuel-rich mode, and an outer flat flame, which is operated in a fuel-lean mode. Thus, the combination creates a reducing-to-oxidizing transition for the particle stream because the particles first experience the reducing environment of the inner flat flame and only later experience the oxidizing environment of the outer flat flame. The burner can also be operated as a purely oxidizing environment (O-O), by ensuring the inner flat flame is the same stoichiometry as the lean outer flat flame. Using this new two-stage flat-flame burner and with the aim of improving upon the limited data on ignition of coal particle streams, this paper presents new experimental findings isolating the effects of particle size, and reducing-to-oxidizing transition (R-O) on ignition of a coal stream. Results are presented for gas temperatures of 1800 K.

2. Experimental

Figure 1 shows a cross section of the two-stage Hencken burner. The flat-flame burner is designed with a honeycomb structure to ensure flow uniformity and with a circular cross section to ensure symmetry in temperature and gas composition. The up-fired burner, with an 80-mm diameter, is composed of 570 hypodermic tubes creating an equivalent number of non-premixed flamelets. Fuel and oxidizer gases are introduced in separate chambers and are isolated by gasket material. A central tube with a 2 mm ID passes through the burner for coal feeding. Like the work of others [10, 11], a quartz tube around the perimeter of the burner isolates the process and provides optical access. A fine stainless-steel wire mesh on top of the quartz tube minimizes disturbances from ambient air and ensures steady convective flow of the post-flame combustion gases. Further details on the design and characterization can be found in [12].

High-speed, back-lit high-resolution videography is employed to image the particles under oxidizing and R-O conditions for four different particle sizes. Images of thermal emission from burning char particles or volatiles are obtained using a NAC MEMRECAM HX-7 high-speed, high-resolution camera. To capture particle motion, the camera is set to 8,000 frames per second (fps) and exposure time of 50 μs (1/20,000 s^{-1}) for all experiments. The camera is also fitted with a SIGMA APOMACRO 180 mm lens to obtain image magnification of 1:4. To view non-emitting coal particles, a blue LED light source with a light diffuser is used as background light. The resulting shadow allows for observation of particles. The radiation emission intensity from the red sensors of the camera's RGB channels can be used to get luminous emission during both volatile and char combustion. A new Matlab image-processing code is written to extract profiles of particle emission intensity along the reactor from the exit of the coal feeder up to a height of 9 cm from the burner. The code allows for a quantitative evaluation of the coal stream ignition. The ignition characteristics are determined for each combustion condition by averaging radially-binned vertical profiles of the particle emission intensity along the reactor using 6000 images.

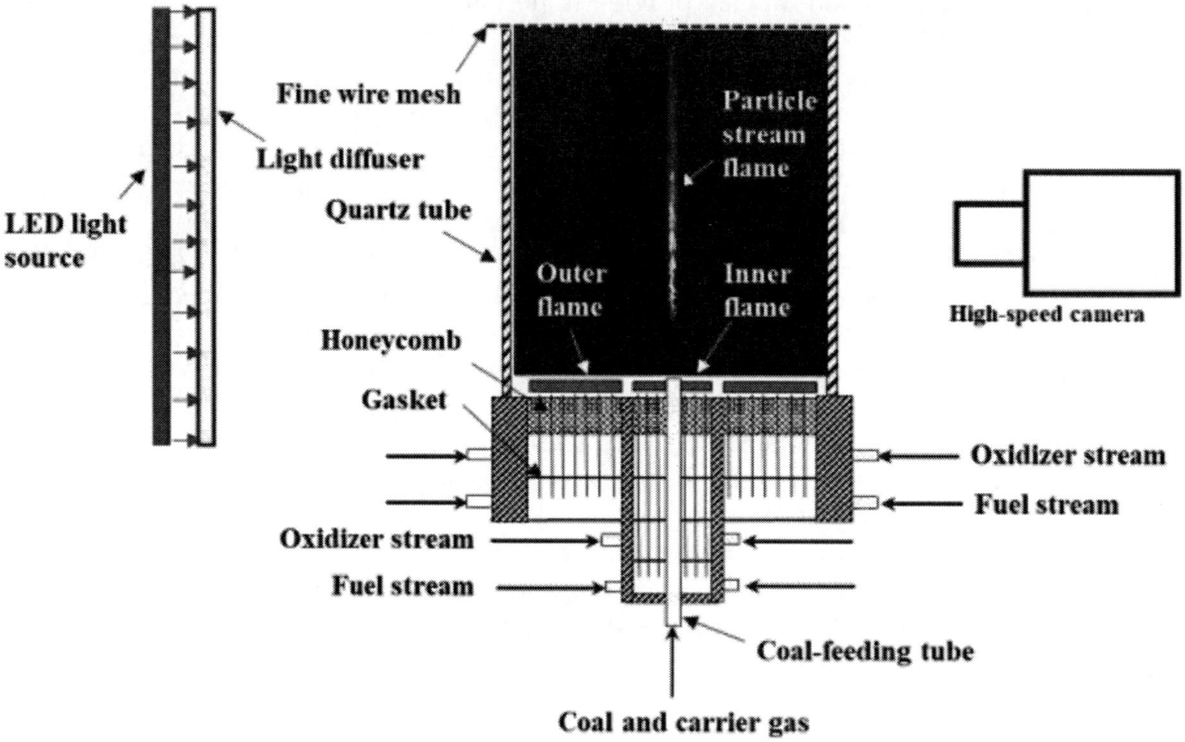

Fig. 1: Cross section of a two-stage Hencken flat-flame burner, showing the inner and outer flat flames with an image of a coal particle stream flame superimposed.

3. Results and Discussion

3.1 Effect of particle size coal stream ignition in an oxidizing environment

The video frames in Fig. 2 show the combustion of coal particle streams for the four size classifications in an environment having a 20 vol.% O_2 composition (O-O) and a nominal gas temperature of 1800 K. The particles are fed into the reactor through a central coal feeding tube, and the background in the figure is from the blue LED light source. The coal particles leaving the tube can be seen as black particles in the backlit image. The released volatiles ignite, creating a bright volatile flame. In fact, the length of the volatile flame can be seen directly in Fig. 2 (as labeled) and by the broad region of high emission intensity in Fig. 3. The length increases with particle size due to the increase in volatile mass with particle size. Given that the diameters of luminosity of the particles emerging from the volatile flames are about the same as the particle diameters, the emerging particles are assumed to be char particles and, as expected, the char burnout is fastest for the smallest particles.

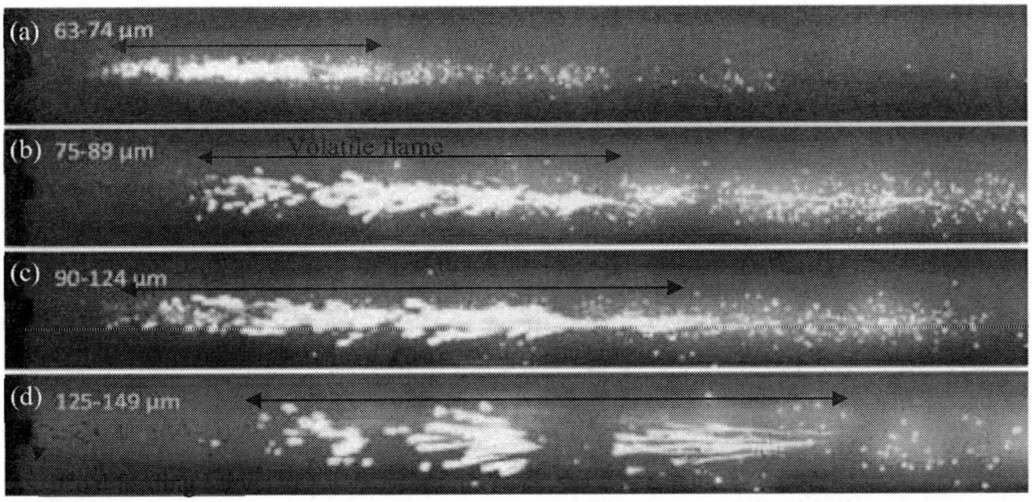

Fig. 2: Coal stream ignition for different particle sizes in 1800 K, O-O environments.

Figure 3 presents a plot of normalized intensity with height, obtained from averaging signals from 6000 frames. For the 63-74 μm particles in Fig. 2a, no observable time separation between thermal emissions from volatiles oxidation and char oxidation in the emission intensities is observed, suggesting hetero-homogeneous ignition [3, 13]. Figures 2b&c show similar qualitative observations with respect to the mode of ignition for 75-89 μm and 90-124 μm particle sizes, respectively. However, the 125-149 μm particle particles initially ignite homogeneously. Subsequently, after the first flame extinguishes due to depletion of volatiles, a second ignition involving char oxidation occurs, leading to the two-peak normalized emission intensity profile in Fig. 3. Prior to the extinction of the volatile flame in Fig. 2(d), long tar tail combustion, oriented along the gas flow streamline, is seen for the 125-149 size μm particles. The tail flame is tar combustion, as indicated in previous studies [14, 15].

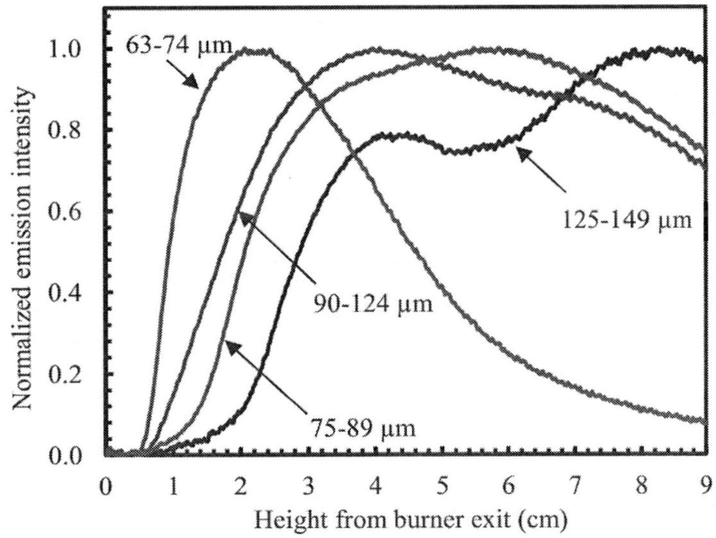

Fig. 3: Normalized emission intensities for different particle sizes in an 1800 K, O-O environment

3.2 Effect of particle size coal stream ignition in R-O environment

As noted above, there are marked differences in the emission intensity profiles under oxidizing conditions (Fig. 3)., The volatiles ignite locally as they are released, under oxidizing conditions, leading to distinct ignition curves as a function of size. This is obvious and expected since the rate of heating, and thus volatile release, is dependent on particle size. However, ignition occurs at about the same height for all particle sizes in the R-O environment. Figure 4 shows clearly this observation with the plots of emission intensities in the R-O condition. Volatiles begin to release within the reducing zone due to the high heating rate, but the volatiles cannot ignite until after the reducing zone, when the oxygen concentration has increased sufficiently for combustion. As the particles stream through the high-temperature flame sheet and mix with oxygen, the volatiles ignite, leading to the observed similarity in the emission intensity profiles for the different particle sizes. Similar standoff heights suggest that ignition in the R-O environment is strongly influenced by the location at which volatiles and char mix with sufficient oxygen, which is just at the flame sheet location (i.e., 2 cm height).

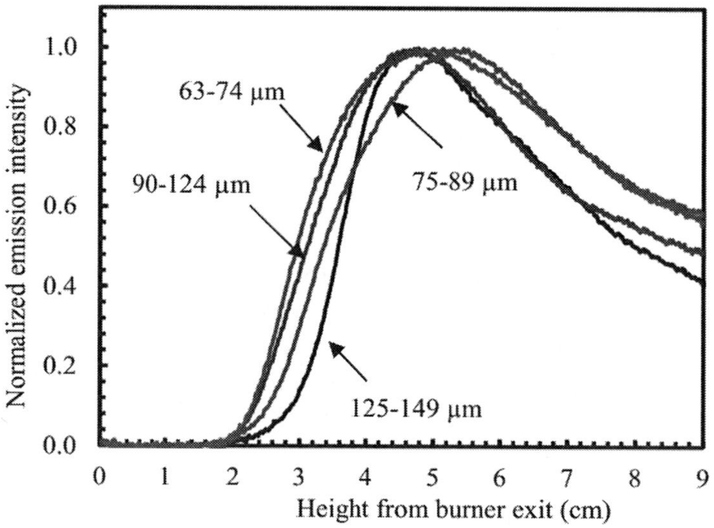

Fig. 4: Normalized emission intensities for different particle sizes in an 1800 K, R-O environment

3.3 Effects of particle sizes on ignition delay times in O-O and R-O environments

Figure 5 compares ignition delay times for various particle size classifications in both the oxidizing environment and the R-O environment. Particle tracking is used to obtain particle residence time. In oxidizing environment, the ignition delay time increases with size from 63-74 μm to 75-89 μm. The ignition delay time for 90-124 μm particles is even lower than that of 75-89 μm particles due to its characteristically lower minimum ignition temperature. This is consistent with other results in Fig. 2 and Fig. 3. The trend in the ignition times in the O-O is the same in the R-O, though less noticeably. Generally, by considering the differences in ignition times for different particle sizes, ignition delay time is relatively independent of particle size either in an oxidizing condition or in a reducing-to-oxidizing condition. However, the reducing-to-oxidizing transition increases ignition delay times, on average, by more than 100% over those of oxidizing conditions.

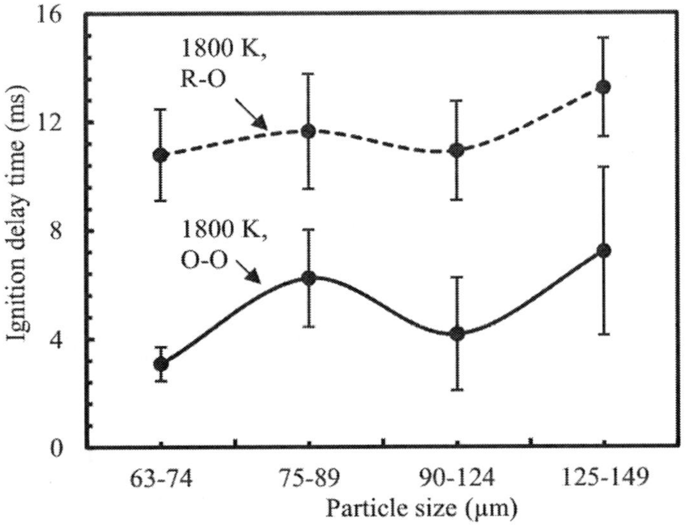

Fig. 5: A comparison of ignition delay times with particle sizes in O-O and R-O environments

3.4 Micrographs of particles sampled from O-O and R-O environments

Scanning electron microscopy (SEM) was used to characterize the 63-74 μm particles to gain insights into the controlling modes of ignition. Corresponding to a time that is post ignition [16], samples are obtained at the same residence time of 50 ms. The scanning electron micrographs presented are representative of many particles surveyed. can be gained from these micrographs. Figure 6a shows that at the higher gas temperature, the surface is smooth, suggesting fusion of mineral matter after complete combustion of char [17]. However, Fig. 6b shows that combustion is not complete at 50 ms for the R-O environment, with the particle displaying a significantly rough surface. Thus, the incomplete combustion is an indication that particles in the R-O environment have longer homogeneous ignition delay times and longer char burnout times than those in the oxidizing environment, a consequence of the delayed exposure to oxygen.

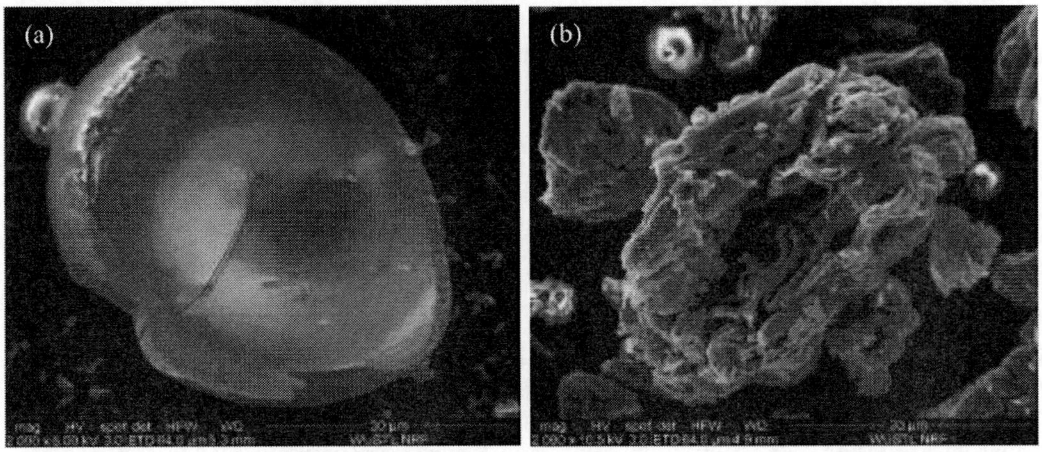

Fig. 6: SEM images of 63-74 μm particles in (a) an oxidizing condition, and (b) a reducing-to-oxidizing condition at a 1800 K gas temperature.

4. Conclusion

In an effort to replicate the characteristics of the near-burner region of a conventional pulverized coal furnace, we employ a two-stage Hencken flat-flame burner, and study the effects of the reducing-to-oxidizing environment on ignition modes and ignition delays for coal particle streams as a function of particle size. The results showed that ignition modes in flames undergoing a reducing-to-oxidizing transition can be quite different from those in oxidizing conditions, especially for small particles. Ignition is influenced significantly by the amount of volatiles released, depending on particle size and the local oxygen concentration. In oxidizing conditions, the hetero-homogeneous ignition mode is dominant at high gas temperature, except for the 125-149 μm particles where the homogeneous-to-heterogeneous ignition mode is seen.

5. Acknowledgements

The Department of Energy under Award Number DE-FE0009702 and the Consortium for Clean Coal Utilization at Washington University in St. Louis funded this research.

6. References

[1] J. Riaza, R. Khatami, Y.A. Levendis, L. Álvarez, M.V. Gil, C. Pevida, F. Rubiera, J.J. Pis, Single particle ignition and combustion of anthracite, semi-anthracite and bituminous coals in air and simulated oxy-fuel conditions, Combustion and Flame 161 (2014) 1096-1108.

[2] C.R. Shaddix, A. Molina, Particle imaging of ignition and devolatilization of pulverized coal during oxy-fuel combustion, Proceedings of the Combustion Institute 32 (2009) 2091-2098.

[3] R.H. Essenhigh, M.K. Misra, D.W. Shaw, Ignition of coal particles: A review, Combustion and Flame 77 (1989) 3-30.

[4] C. Sheng, Y. Li, X. Liu, H. Yao, M. Xu, Ash particle formation during O2/CO2 combustion of pulverized coals, Fuel Processing Technology 88 (2007) 1021-1028.

[5] X. Wang, H. Guo, F. Liu, R. Hu, M. Wang, Effects of CO2 on sulfur removal and its release behavior during coal pyrolysis, Fuel 165 (2016) 484-489.

[6] L.D. Smoot, P.J. Smith, Coal combustion and gasification, Springer Science & Business Media2013.

[7] M. Ruiz, K. Annamalai, T. Dahdah, An experimental study on group ignition of coal particle streams, Heat and Mass Transfer in Fires and Combustion Systems 148 (1990) 19-26.

[8] Y. Liu, M. Geier, A. Molina, C.R. Shaddix, Pulverized coal stream ignition delay under conventional and oxy-fuel combustion conditions, International Journal of Greenhouse Gas Control 5 (2011) S36-S46.

[9] Z. Huque, M.R. Ali, R. Kommalapati, Application of pulse detonation technology for boiler slag removal, Fuel Processing Technology 90 (2009) 558-569.

[10] R.-G. Kim, D. Li, C.-H. Jeon, Experimental investigation of ignition behavior for coal rank using a flat flame burner at a high heating rate, Experimental Thermal and Fluid Science 54 (2014) 212-218.

[11] Y. Yuan, S. Li, G. Li, N. Wu, Q. Yao, The transition of heterogeneous–homogeneous ignitions of dispersed coal particle streams, Combustion and Flame 161 (2014) 2458-2468.

[12] A. Adeosun, Q. Huang, T. Li, A. Gopan, X. Wang, S. Li, R.L. Axelbaum, Characterization of a new Hencken burner with a transition from a reducing-to-oxidizing environment for fundamental coal studies, Review of Scientific Instruments, (2017) Under review.

[13] H. Jüntgen, K.H. Van Heek, An update of german non-isothermal coal pyrolysis work, Fuel Processing Technology 2 (1979) 261-293.

[14] W.J. McLean, D.R. Hardesty, J.H. Pohl, Direct observations of devolatilizing pulverized coal particles in a combustion environment, Symposium (International) on Combustion 18 (1981) 1239-1248.

[15] W.R. Seeker, G.S. Samuelsen, M.P. Heap, J.D. Trolinger, The thermal decomposition of pulverized coal particles, Symposium (International) on Combustion 18 (1981) 1213-1226.

[16] Y. Yuan, S. Li, F. Zhao, Q. Yao, M.B. Long, Characterization on hetero-homogeneous ignition of pulverized coal particle streams using CH∗ chemiluminescence and 3 color pyrometry, Fuel, doi:http://dx.doi.org/10.1016/j.fuel.2015.11.032(2015).

[17] M. Neville, R.J. Quann, B.S. Haynes, A.F. Sarofim, Eighteenth Symposium (International) on CombustionVaporization and condensation of mineral matter during pulverized coal combustion, Symposium (International) on Combustion 18 (1981) 1267-1274.

Spring Technical Meeting
Eastern States Section of the Combustion Institute
March 4-7, 2018
State College, Pennsylvania

Explosive Dust Characteristics Evaluation of Pulverized Pittsburgh Coal Using ASTM E1226-12a

Jacob Miller[1], Phillip Mulligan, PhD.[1], Catherine Johnson, PhD.[1]

[1]Missouri University of Science and Technology, Rolla, MO, USA
**Corresponding Author Email: jlmill@mst.edu*

Abstract: The National Fire Protection Association standard 652 (NFPA) defines combustible dust as a finely divided combustible particulate solid that presents a flash fire or explosion hazard when suspended in air over a range of concentrations and requires the evaluation of suspect dust. American Society for Testing Materials International (ASTM) developed test method E1226 (ASTM E1226) to provide a standard test method characterizing the explosibility of particulate solids of combustible materials suspended in air. The Siwek 20 liter apparatus was calibrated with niacin as part of CaRo 15 held by Adolf Kuhner AG. A maximum explosion pressure (P_{max}) of 8.4 bar was recorded during the calibration which is within the provided reference values of 7.3 to 9.0 bar. Pulverized Pittsburgh Coal is listed as a calibration material within the ASTM E1226 standard. When evaluated within the Siwek 20-L vessel the Pulverized Pittsburgh Coal a P_{max} value of 7.7 bar was recorded. The reference value provided for Pulverized Pittsburgh Coal is 7.0 bar. The discrepancy between the recorded and reference value has raised the question how different vessels, energy requirements and testing procedures influence recorded test values for Pulverized Pittsburgh Coal. The research presented herein investigates the explosive characteristic of Pulverized Pittsburgh Coal dust using the latest ASTM E1226-12a test standard and compares the results to the Unites States Bureau of Mines' PRL 20L vessel. Pulverized Pittsburgh Coal dust concentrations, ranging from 30 to 1,500 g/m3, were tested in a Siwek 20 liter apparatus, an ASTM E1226 compliant vessel. The Pulverized Pittsburgh Coal dust is characterized by maximum explosion pressure (P_{max}), the maximum rate of pressure rise (dP/dt) max, and explosibility index (K_{st-max}). The data presented in this paper vary from the USBM data with a higher P_{max} and K_{st}. The data variances are indicative of the impact of the different testing apparatus and procedures.
Keywords: ASTM E1226, explosive dust, coal dust, Siwek 20 liter

1. Introduction

Dust explosions are a hazard that effects many industrial processes and facilities. With the recent passage of NFPA 652: Standard on the Fundamentals of Combustible Dust,(NFPA 652) in 2015 new scrutiny is being placed on mitigating the risks and hazards associated with the handling of combustible dusts. NFPA 652 defines combustible dusts and sets the testing requirements to determine the combustibility of the dusts created or modified during processing. The definition of a combustible dust is defined as "a finely divided combustible particulate solid that presents a flash-fire or explosion hazard when suspended in air or the process-specific oxidizing medium over a range of concentrations [1]."

NFPA 652 stipulates facilities that have suspect combustible dust are required to complete a Dust Hazard Analysis (DHA). The DHA should identify combustible and explosive dust hazards and document how those hazards are to be controlled or mitigated [1]. In order to properly assess the hazards associated with combustible dusts the physical characteristics and properties of the dust must be known.

NFPA 652 specifically designates ASTM E1226: Standard Test Method for Explosibility of Dust Clouds (ASTM E1226) as a test procedure meeting the specifications to examine the explosive characteristic of an airborne dust. ASTM E1226 lists Pittsburgh seam bituminous coal dust as a calibration material for 20 liter chambers [2].

It is known to the authors that the United States Bureau of Mines (USBM) conducted coal dust explosion research and published reports related to dust explosions. The published research was reviewed as a potential source of the coal dust explosion characteristics reference data within ASTM E1226. It was noted that the testing parameters used by the USBM were different than current testing standards. An evaluation of Pittsburgh coal dust was conducted to investigate if the differences had an impact on the resulting explosion characteristics. The dust explosion characteristics maximum mean value of peak pressure, rate of pressure change and explosive index are reported as P_{max}, $(dP/dt)_{max}$; and $K_{st\text{-}max}$.

2. Background

Physical properties such as; dust particle size, dust concentration, turbulence, moisture, initiation pressure, and initiation energy are known to effect the dust explosion characteristics P_{max}, $(dP/dt)_{max}$, and $K_{st\text{-}max}$. An increase in turbulence, initiation pressure and initiation energy are known to increase dust explosion characteristics. While an increase in particle size and moisture decrease the characteristics. Dust concentration has a varied effect. Initially as dust concentrations increase dust explosion characteristics increase until an optimum mixture of dust to oxygen is reached. After an optimum mixture is reached any further increase in dust concentration reduces explosion characteristics. [3]

The USBM conducted a series of coal dust explosion tests in the 1980's and 1990's using their internally designed PRL 20L chamber, see Figure 1. The USBM used pulverized coal dust from a Pittsburgh coal seam. All tests were conducted at an ignition pressure of 1-atm using a pyrotechnic igniter of 2.5 kilojoule (kJ) to initiate the coal dust explosions.

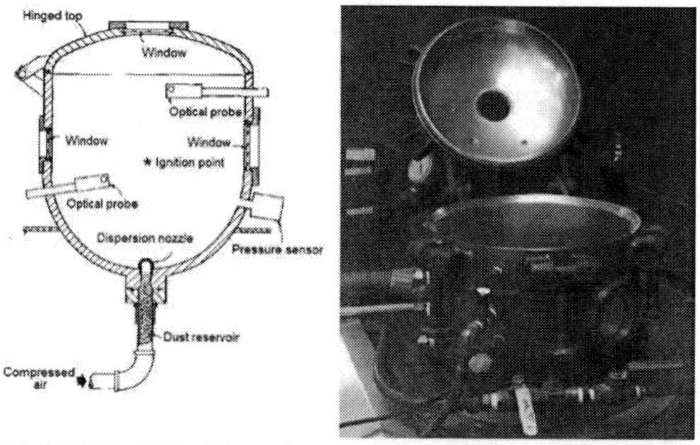

Figure 1. USBM PRL 20L Chamber, cross-sectional [4] and external view [5]

The standard pulverized Pittsburgh coal (PPC) dust used by the USBM was described as being 80% minus 200 mesh (<75 μm), 13% minus 20 μm and a median particle diameter of 48 μm [6].

The peak dust characteristic values reported, Figure 2 circled in red, are approximately 6.9 bar (P_{max}) and 36 bar*m/s ($K_{st\text{-}max}$) [7]. The K_{st} values increased with increased dust concentrations until approximately 300 to 500 g/m^3, Figure 2a. As the dust concentration increases the variability of K_{st} appears to increase and a trend is hard to discern. Figure 2b is a plot of the peak pressure measured at different dust concentrations. Between dust concentrations of 0-300 g/m^3 pressure rises as dust concentrations increase. Around 300 g/m^3 the trend changes and the pressure remains relatively constant at approximately 6.5 bar.

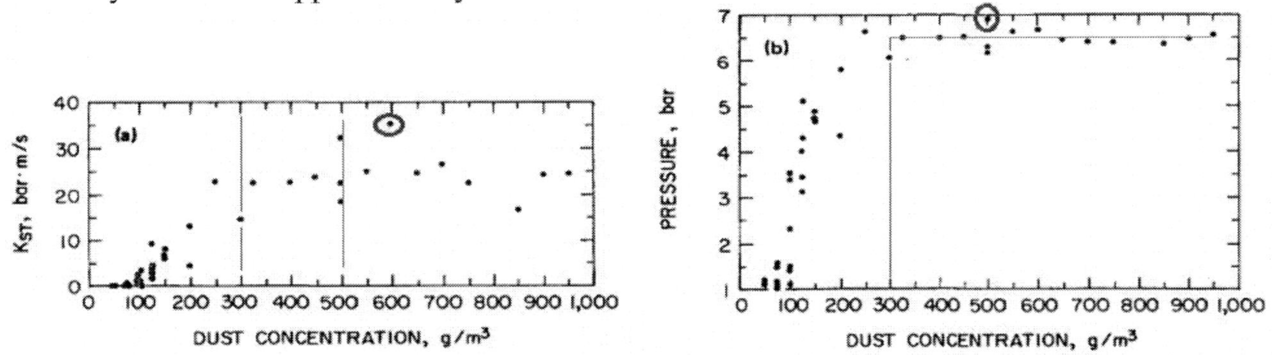

Figure 2. Explosibility Data for Pittsburgh Seam Coal Dust, modified from [7]

The reference values for Pittsburgh coal supplied within ASTM E1226 are different. P_{max} is stated to be 7.0 bar, $(dP/dt)_{max}$ is 430 bar/s, and K_{st-max} is 117 bar*m/s [2]. When compared to the USBM data the P_{max} is about the same. The difference in K_{st-max} is remarkable with the USBM having a value 81 bar*m/s less than the supplied ASTM E1226 reference value. The difference between the two K_{st-max} values impelled the authors to conduct an explosive dust characteristics evaluation of PPC following ASTM E1226-12a.

3. Method

A Siwek 20 liter apparatus was used to measure the dust explosion parameters of PPC. This vessel meets the requirements specified in ASTM E1226 and can be seen in Figure 3. The explosion chamber is spherical with an internal volume of 20 liters. The explosion chamber is evacuated with a vacuum to -0.6 (gauge) before testing. Inside the explosion chamber a rebound nozzle, located at the bottom of the vessel disperses the dust in the chamber. The rebound nozzle is connected to an isolation valve which isolates the explosion chamber from the dust reservoir. The dust reservoir contains one sample of the dust to be evaluated. The dust reservoir is 0.6 liter in volume and is pressurized to 20 bar (gauge) for testing.

The dust is accelerated and dispersed when the isolation valve is opened. The pressure differential between the explosion vessel and dust reservoir creates a flow of air that transports the dust through the rebound nozzle into the explosion chamber. The isolation valve closes to prevent the explosion from back flowing into the storage reservoir. The dust cloud created by the air flow and interaction with the rebound nozzle is ignited with two 5kJ Sobbe chemical igniters placed at the center of the spherical explosion chamber.

Dust samples were not sifted, sorted, or processed to change particle size or shape. To minimize particle segregation that may have occurred during shipping the PPC was blended using a shaker-mixer technique. All test samples were pulled from the same lot of PPC to minimize particle size variation. Dust moisture levels were below 5% and kept consistent by storing all samples and material in sealed containers.

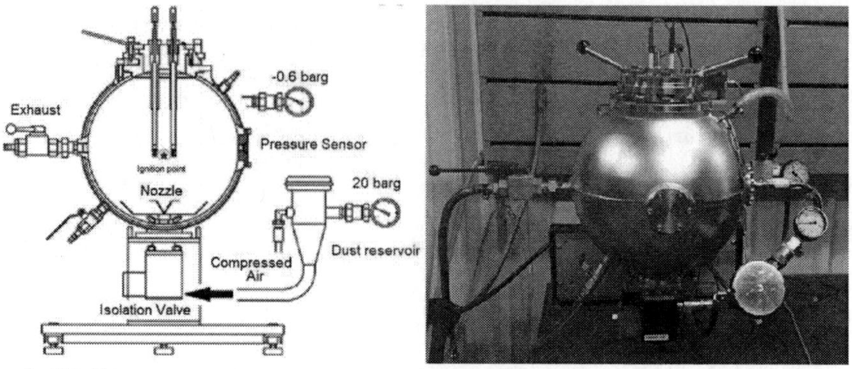

Figure 3. Siwek 20 liter apparatus, cross-sectional diagram,modified [8] and external view

Two Kistler pressure transducers are used to observe the pressure evolution of the explosion. The 20 liter apparatus control system automatically runs the evaluation by controlling the pressurization of the loading chamber, opening/closing of the isolation valve, time delay to ignitor ignition, ignitor ignition, and data recording. The control system automatically makes explosive pressure corrections as specified in X1.8 and X1.9 of ASTM E1226.

The test procedure outlined within the Siwek 20 liter apparatus manual [8] recommends an initial test series with dust masses and concentrations. The first test series starts at a low dust concentration of 60 g/m^3 and increases in steps to a high dust concentration of 1500 g/m^3 (dust concentrations 60, 125, 250, 500, 750, 1,000, 1,250, 1,500 g/m^3).

The dust explosion characteristics for each test shot are peak preassure, rate of pressure increase and explosive index are reported as P_m, $(dP/dt)_{ex}$; and K_{st}. The maximum value for the explosion pressure (P_m), and the rate of pressure increase, $(dP/dt)_{ex,}$ can clearly be determined over the wide range of dust concentrations evaluated. Each concentration was tested three times. The maximum mean value of peak pressure, rate of pressure change and explosive index from the complete evaluation are reported as P_{max}, $(dP/dt)_{max}$; and K_{st-max}.

4. Results and Discussion

The Siwek 20 liter apparatus was calibrated with niacin following the procedures within the Calibration Round Robin of 2015 sponsored by Adolf Kuhner AG [9]. Niacin's explosive characteristics recorded during the calibration are 8.4 bar (P_{max}) and 250 bar*m/s (K_{st-max}). Adolf Kuhner reported that reference values of 8.2 +/- 10% (P_{max}) and 245 +/- 10% bar*m/s (K_{st-max}) [10]. The University's evaluation of the supplied niacin is within the published reference values. The Siwek 20 liter apparatus is therefore presumed to be calibrated and test results are acceptable.

The data collected during the University PPC evaluation can be seen in with published data from the USBM [7]. In Figure 4a peak pressures recorded in the Siwek 20 liter are consistently higher until the concentration level of 900 g/m^3 is reached. Concentrations above 900 g/m^3 resulted in lower peak pressure when compared to USBM data. The maximum corrected pressure reached in this research was 7.7 bars compared to approximately 6.6 bar reported by the USBM [7]. The trend lines vary as well. The research conducted by the authors shows a more prominent pressure peak developing between concentrations of 250 and 500 g/m^3.

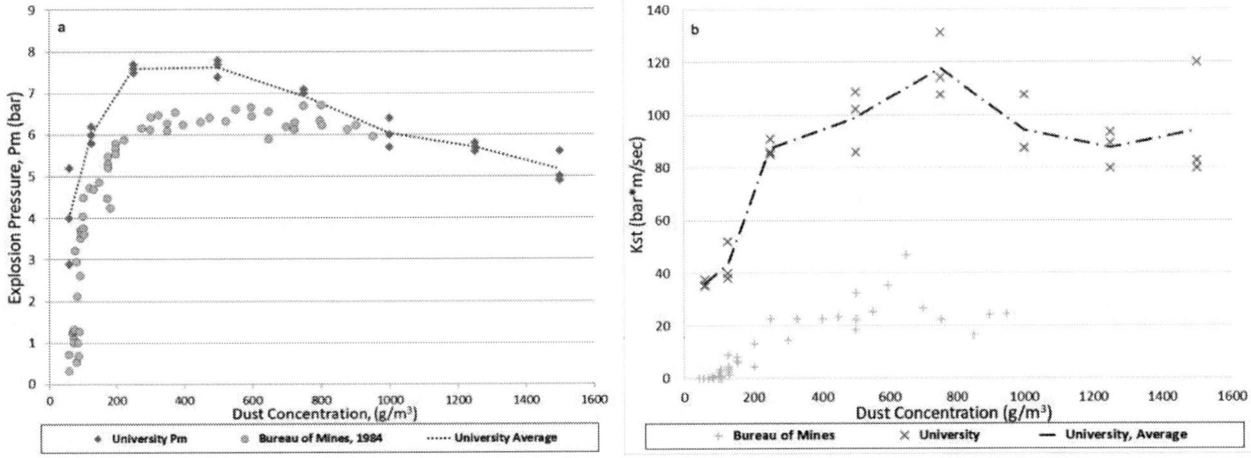

Figure 4. University Data with Referenced Data [7]

The K_{st}, explosive index, also varies between the USBM and the University data, shown in Figure 4b. The K_{st-max} in the USBM data of around 30 bar*m/s [7], while the University's K_{st-max} is 118 bar*m/s. The trends are similar but the University data has a magnitude of 3-4 times the values seen in the USBM data. Dp/dt data is not shown but $(dP/dt)_{max}$ for the University is 435 bar/s.

The data gathered by the University is very similar to the reference data specified in ASTM E1226. ASTM E1226 supplies a P_{max} value of 7.0 bar, a (dP/dt)max equal to 430 bar/s, and a K_{st-max} of117 bar*m/s [2]. The University's test found a P_{max} of 7.7 bar, a $(dP/dt)_{max}$ equal to 435 bar/s, and a K_{st-max} equal to 118 bar*m/s. These values pass the reproducibility requirement In section 15 of ASTM E1226, which allows for a 10% difference in P_{max} and a 20% difference in K_{st-max} when comparing test data from different laboratories. The USBM's P_{max} value 6.6 bar is the only explosive characteristic that meets ASTM E1226 reproducibility requirement.

The difference between the USBM and University K_{st} data is explained by Cashdollar in 2000 when he wrote, "Note that the turbulence level was lower in the PRL 20L chamber for these tests than that recommended in ASTM E1226. At the higher turbulence level recommended in ASTM Standard E1226, the maximum $(dP/dt)V^{1/3}$, K_{st} data for this Pittsburgh coal would be roughly three times higher."

Other variances in data trends shown in Figure 4 may be due to a couple factors. The 10kJ chemical igniters that were used during the University's ASTM standard study could account for the higher P_{max}, and K_{st-max} when compared to the USBM data that used 2.5kJ ignitors. The University's clearly defined pressure peak and could also be a result of the Siwek 20 liter apparatus creating a more turbulent flow than the USBM PRL 20 liter vessel.

The differences between the USBM and ASTM E1226 testing parameters have a clear influence on test data. These differences could have a direct impact on the design dust explosion mitigation systems. Historic data related to USBM and their PRL 20L should be questioned and reviewed as companies create Dust Hazard Analysis to fulfill NFPA 652's requirements

5. Conclusions

Dust explosions are a major safety concern industry. The explosion and fires following dust explosions can result in significant property damage and loss of life. The information presented in this paper can be directly related to industry standards and applied to explosion mitigation engineering practices.

A P_{max} of 7.7 bar was calculated during the University explosibility testing of PPC dust. The maximum rate of pressure change, $(dP/dt)_{max}$, was 435 bars/s with a resulting K_{st-max} maximum of 118 m bar/s. The values are very similar to the reference data supplied by ASTM E1226.

The University and ASTM E1226 explosive dust characteristic are both greater than the data published by the USBM. have P_{max} value of approximately 6.6 bar published by the USBM. The trends of K_{st} values are similar between the USBM data and the ASTM data. The USBM has a peak K_{st} of around 30 bar*m/s while the ASTM study values are magnitude of 3-4 times the values seen in the USBM data.

Future research will focus on testing how increased dust concentrations, up to 5000 g/m3, affect the peak pressure and KST data. Future work will also explore how the dust concentrations tested in the 20-liter vessels correspond to dust concentrations observed in coal mines and processing facilities. Other historical USBM dust testing of combustible dust and dust clouds could be evaluated following current ASTM standards.

6. References

[1] NFPA 654, "Standard for the Prevention of Fire and Dust Explosions from the Manufacturing, Processing, and Handling of Combustible Particulate Solids," National Fire Protection Association, 2017.

[2] ASTM E1226-12a, West Conshohoken, PA: ASTM International, 2012.

[3] R. Echoff, Dust Explosions in the Process Industries (3rd Edition), Gulf Professional Publishing, 2003.

[4] K. Cashdollar, "Overview of dust explosibilty characteristics," vol. 13, pp. 183-199, 2000.

[5] C. K. Man and M. L. Harris, "Participation of large particles in coal dust explosions," *Journal of Loss Prevention in the Process Industries 27,* pp. 49-54, 2014.

[6] Cashdollar, "Coal dust explosibility," *Journal Loss Prevention in the Process Industries,* vol. 9, pp. 183-199, 1996.

[7] Cashdollar, Kenneth; Hertzberg, Martin, "20-l explosibility test chamber for dusts and gases," *Review of Scientific Instruments,* pp. Vol. 56, No. 596, 1985.

[8] KuhnerAG, "20L Apparatus Manual, KSEP 7.0," Kuhner AG, Birsfelden, Switzerland, 2011.

[9] Adolf Kuhner AG, "Manual CaRo 15," Birsfelden, Switzerland, 2015.

[10] Adolf Kuhner AG, "Final Report Calibration-Round-Robin CaRo15," Birsfelden, Switzerland, 2016.

[11] K. Cashdollar, M. Sapko, M. Hertzberg and E. Weiss, "Laboratory and Mine Dust Explosion Research at the Bureau of Mines," in *Industrial Dust Explosions*, Wst Conshohocken, PA, ASTM International, 1987, pp. 107-123.

Spring Technical Meeting
Eastern States Section of the Combustion Institute
March 4-7, 2018
State College, Pennsylvania

Coal particle ignition in a combustion environment with a reducing-to-oxidizing transition

*Dishant Khatri, Adewale Adeosun, Akshay Gopan, Zhiwei Wang, Richard L. Axelbaum**

Department of Energy, Environmental & Chemical Engineering, Consortium for Clean Coal Utilization, Washington University in St. Louis, St. Louis, Missouri-63130, USA
**Corresponding Author Email: axelbaum@wustl.edu*

Abstract: A transition from a reducing-to-oxidizing (R-O) environment is characteristic of what a coal particle experiences in the near-burner region of pulverized coal (pc) furnaces. The R-O transition can influence early-stage coal combustion processes such as ignition, aerosol formation, and char burnout. However, fundamental studies have focused on either oxidizing conditions (mimicking the post-flame region) or reducing conditions (mimicking the devolatilization region). The effect of this transition on early-stage coal combustion has, until now, not been considered. Here, the role of this transition on single-particle ignition is evaluated. PRB sub-bituminous coal was used, with a particle size of 125-149 μm and two nominal gas temperatures of 1300 K and 1800 K. The experimental findings for purely-oxidizing conditions with 20 vol. % oxygen are compared with those of the R-O environment. Single particles were tracked using high-speed, high-resolution videography. Emission intensities of the particles were used to evaluate the prevailing ignition modes in both oxidizing and R-O environments. Homogeneous-to-heterogeneous mode of ignition is prevalent for purely oxidizing conditions for both nominal gas temperatures of 1300 K and 1800 K. However, hetero-homogeneous ignition is favored in R-O environment at 1800 K and heterogeneous ignition at 1300 K gas flame temperature.
***Keywords:* reducing-to-oxidizing transition, ignition, homogeneous-to-heterogeneous ignition, hetero-homogeneous ignition**

1. Introduction

The average residence time of a coal particle in a conventional pulverized coal (pc) boiler is about two seconds [1]. However, the early-stage processes of coal combustion (i.e., processes occurring within the first tens of milliseconds) influence the ignition processes in the boiler [2,3]. During the initial stages of coal combustion, volatiles are released and char is formed [4]. The oxidations of the volatiles and char are directly affected by these early-stage processes [5]. Since the oxidation reactions are important for burner stabilization and efficient coal combustion [4], fundamental studies of early-stage processes, such as ignition, can provide valuable information for more efficient boiler design and optimization.

In a review of various fundamental coal ignition studies, Essenhigh et al. [6] noted that most studies have focused on understanding the mechanisms that control the ignition of single coal particles. Homogeneous ignition refers to the process whereby the released volatiles ignite in the gas phase with a flame surrounding the particle, while heterogeneous ignition refers to direct particle surface combustion. Howard and Essenhigh [7] studied single-particle ignition and showed that coal particles can experience both homogeneous and heterogeneous ignition, depending on the particle size and local combustion conditions. In addition though, two other mechanisms for ignition are possible: first one is termed hetero-homogeneous ignition in which a particle experiences both homogeneous and heterogeneous ignition because the particle heating time is of the same order of magnitude as the pyrolysis time. The other is where the coal particle initially ignites homogeneously, and subsequently, after a time delay, the particle ignites heterogeneously, termed homogeneous-to-heterogeneous ignition.

While there are many studies on coal particle ignition [8-10], fundamental studies of hetero-homogeneous ignition and homogeneous-to-heterogeneous ignition are lacking, especially under conditions that simulate the reducing-to-oxidizing transition (R-O) environments found in practical pc burners. The locally reducing environment in the near-burner region is created by coal devolatization occurring in a region where the particle number density is high [4]. Beyond this region, the particles move into an oxidizing environment. Wall et al. [9] have noted that, depending on the injected coal mass and coal carrier gas, particles transitioning from a reducing to oxidizing environment can strongly affect the combustion characteristics and the controlling mechanisms for ignition. In this work, a new two-stage Hencken burner was designed to study single particle ignition of coal under conditions that simulate the reducing-to-oxidizing transition that can occur during pc combustion [11], and in a temperature range relevant to the early-stage processes of coal combustion [12]. Herein, we employ the burner, along with high speed videography, to study the role of the reducing-to-oxidizing transition and the ambient gas temperature.

2. Experimental

2.1 Hencken flat-flame burner

Figure 1 shows a schematic representation of the two-stage Hencken flat-flame burner. This burner has an inner methane flat-flame, operated in a fuel-rich mode, and an outer methane flat-flame, operated in a fuel-lean mode, creating a reducing-to-oxidizing transition that mimics the near-burner environment of a typical pc furnace. The up-fired burner was designed with a circular cross section to ensure symmetry in temperature and gas composition. The burner, with an 80-mm diameter, is composed of 570 hypodermic tubes inside of a honeycomb core, creating an equivalent number of non-premixed flamelets. Fuel and oxidizer gases are introduced in separate chambers and are isolated by gasket materials. A central tube with a 2 mm ID passes through the burner for coal feeding. A quartz tube is set above the burner to isolate the process and provide optical access. With the quartz tube and a Hencken burner about twice the size of typically reported burners, the centerline profile is less affected by heat loss, giving a better profile along the burner height. A fine stainless-steel wire mesh on top of the quartz tube minimizes disturbances from ambient air and ensures steady convective flow of the post-flame gas.

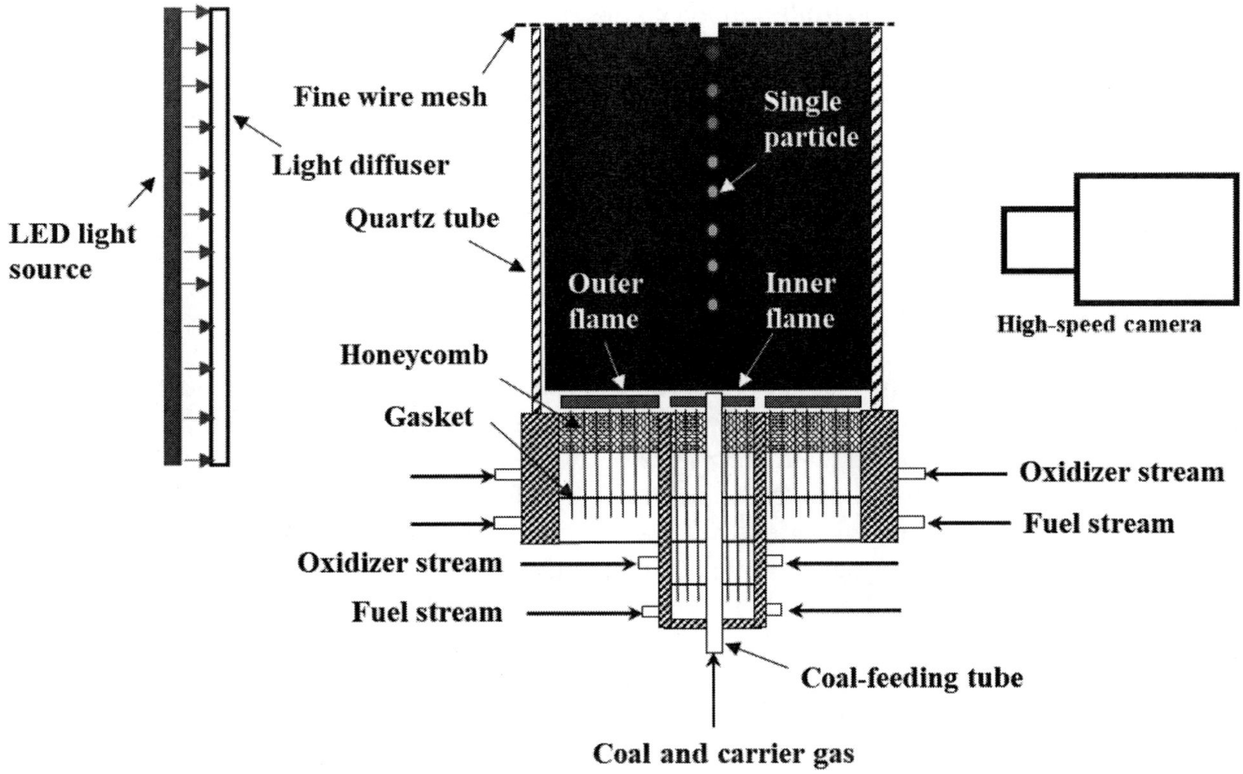

Fig. 1: Cross section of a two-stage Hencken flat-flame burner, showing the inner and outer flames

2.2 High speed photography

Images of visible light emission were obtained using a NAC MEMRECAM HX-7 high-speed video camera. The camera was fitted with a SIGMA APOMACRO 180 mm lens, used to obtain image magnification of 1:4. To view non-emitting coal particles, a blue LED light with a light diffuser was used as a background light source. The resulting shadow allowed for observation of both non-emitting coal or char. Utilizing the background light allowed the blue sensors of the camera's RGB channels to be used to continuously observe the particle because the amount of emission in the blue wavelength is small except at high temperatures. The radiation emission intensity from the red sensors of the camera's RGB channels was used to observe luminous emission during both volatile and char combustion, from which qualitative and quantitative information on single particle ignition can be obtained. A Matlab image-processing code was written to allow a single particle to be tracked from the burner surface up to a height of 6 cm from the burner.

2.3 Coal

Table 1 shows the proximate and ultimate PRB subbituminous coal particles used for this study. All the coal particles have been screened to 124-149 μm.

Table 1: Properties of PRB sub-bituminous coal

Proximate analysis (%wt)				Ultimate analysis (%daf)					HHV
M	VM	FC	Ash	C	H	O	N	S	(MJ/kg)
10	39	42.87	8.14	74.66	5.45	18.24	1.08	0.57	27.45

3. Results and Discussion

The effects of temperature on single particle ignition were studied in an oxidizing condition with a nominal oxygen composition of 20 vol.% (hereafter referred to as the oxidizing condition). Emission intensities were extracted for many particles and Fig. 2 shows a representation of a typical particle ignition trend along the central axis. The signal magnitude may differ among individual particles, but the characteristic times and trends are repeatable. For ease of comparison, the emission profiles are normalized by their respective peak intensities. There are two peaks in the emission intensity profiles for both temperatures. The first peak corresponds to volatile oxidation after homogeneous ignition. The second wider peak is due to heterogeneous ignition and char oxidation after an induction time. Induction time is the characteristic time for combustion products to diffuse away from the particle surface and for oxygen and heat to diffuse to the particle surface. The light intensity shows some modulation during the induction time while the particle surface is heated to the required temperature for heterogeneous ignition. The modulation is from particle rotation and/or subsequent ejection of material as the char surface-temperature rises. After heterogeneous ignition in the oxidizing condition at 1800 K, the emission intensity remains fairly constant with time, after which the intensity declines. As char oxidation progresses, the particle size decreases, intensity declines because of the spreading of the ash coverage of the exterior surface and thus the lower emissivity. Also, char oxidation rate will decrease as carbon content decreases with time.

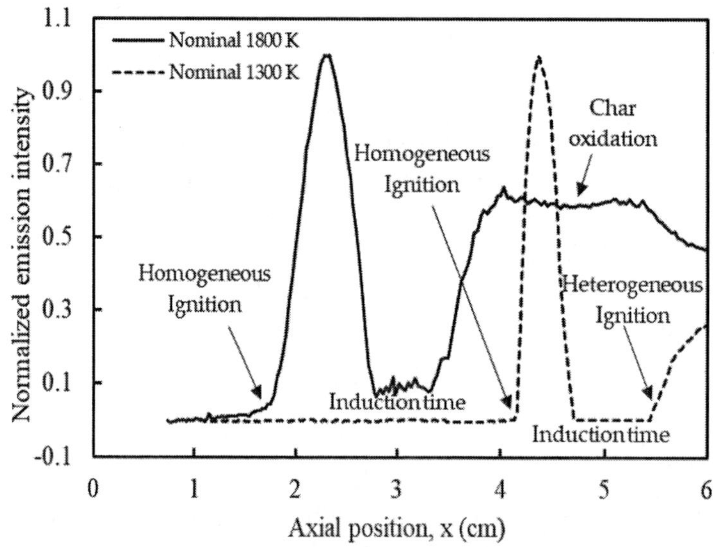

Fig. 2: Normalized light emission intensity vs. height from the burner for 20 vol.% oxidizing condition at 1800 K and 1300 K

Under reducing-to-oxidizing conditions, particle devolatilization is fast since the particle experiences a very high heating rate at 1800 K. But, the volatiles diffuse away from the surface of the unburned coal particle since there is no oxygen in the reducing zone. Since the lower flammability limit is very low in this high-gas-temperature environment, volatiles released from the particles in the oxidizing environment are burned on the surface of the particle. Hence, under 1800 K nominal gas environment, volatile oxidation occurs on the surface of the particle, along with char oxidation. Thus, unlike the homogeneous-to-heterogeneous ignition events observed for the two temperatures under oxidizing conditions, a single peak corresponding to hetero-homogeneous ignition is observed for R-O condition as shown in Fig. 3 for nominal 1800 K gas condition.

In Fig. 3, under 1300 K gas temperature, volatile reaction can be observed but it is quenched almost immediately because the volatiles that are released after the reducing zone experience a gas stream that initially has little oxygen. Particle surface reaction begins, i.e., heterogeneous ignition, once the particle interacts with enough oxygen.

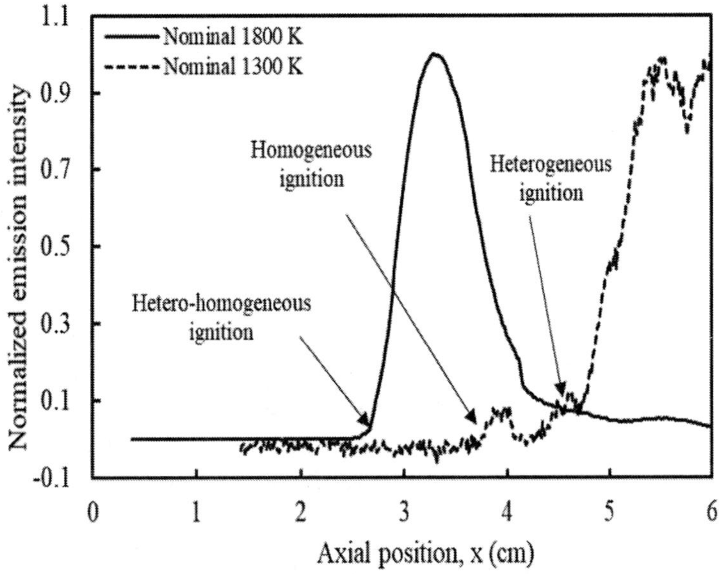

Fig. 3: Normalized light emission intensity vs. height from the burner for reducing-to-oxidizing condition at 1800 K and 1300 K

Figure 4 shows the ignition and induction times for the conditions studied in this work. The R-O environments have longer ignition delay times in comparison to oxidizing environments. For the nominal 1300 K temperature, the characteristic ignition delay times for both oxidizing and reducing-to-oxidizing conditions are approximately 30 ms and 36 ms respectively, an average of 20% increase in R-O environment. For 1800 K nominal gas temperature, the ignition delay time for the R-O environment increases by an average of 4 ms over those of oxidizing environment, a more than 40% increase. Interestingly though, while the R-O environments cause longer ignition delay times than in oxidizing environments at the same nominal gas temperatures, the induction times in R-O environments are shorter than in oxidizing environments. Under the conditions of this work including temperatures and combustion environments, these experimental findings indicate that the characteristics of ignition in R-O environments can be quite different from those in oxidizing environments for PRB coal particles.

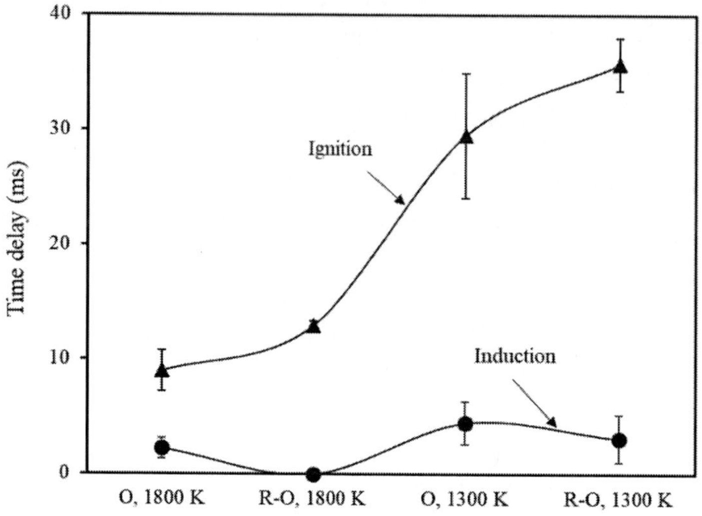

Fig. 4: Effect of reducing-to-oxidizing transition and temperature on ignition and induction times

4. Conclusion

A better understanding of particle ignition, as an early-stage process of coal combustion, can improve boiler design and efficiency when studied in conditions relevant to practical pc burners. Such conditions include a reducing-to-oxidizing transition. The ignition mode of PRB sub-bituminous coal has been investigated at nominal gas temperatures of 1300 K and 1800 K in both reducing-to-oxidizing transition and oxidizing conditions. Single particle ignition was studied experimentally using high speed videography. The work presented here shows that the reducing-to-oxidizing transition can affect ignition behavior of single coal particle (both the mechanism and delay time). Under our experimental conditions, it was found that reducing-to-oxidizing transition conditions promote hetero-homogeneous ignition in a high temperature environment. The traditionally studied oxidizing conditions with 20 vol. % oxygen lead to homogeneous-to-heterogeneous ignition mode.

5. Acknowledgement

This research was funded by the Department of Energy (Award Number DE- FE0009702), the U.S./China CERC-ACTC, and the Consortium for Clean Coal Utilization at Washington University in St. Louis.

6. References

[1] K. Annamalai, W. Ryan, Interactive processes in gasification and combustion—II. Isolated carbon, coal and porous char particles, Progress in Energy and Combustion Science 19 (1993) 383-446.

[2] Q. Gao, S. Li, Y. Yuan, Y. Zhang, Q. Yao, Ultrafine particulate matter formation in the early stage of pulverized coal combustion of high-sodium lignite, Fuel 158 (2015) 224-231.
[3] Q. Gao, S. Li, Y. Yuan, Y. Zhao, Q. Yao, Role of Minerals in the Evolution of Fine Particulate Matter during Pulverized Coal Combustion, Energy & Fuels 30 (2016) 1815-1821.
[4] L.D. Smoot, P.J. Smith, Coal combustion and gasification, Springer Science & Business Media2013.
[5] L. Chen, S.Z. Yong, A.F. Ghoniem, Oxy-fuel combustion of pulverized coal: Characterization, fundamentals, stabilization and CFD modeling, Progress in Energy and Combustion Science 38 (2012) 156-214.
[6] R.H. Essenhigh, M.K. Misra, D.W. Shaw, Ignition of coal particles: A review, Combustion and Flame 77 (1989) 3-30.
[7] J.B. Howard, R.H. Essenhigh, Mechanism of solid-partical combustion with simultaneous gas-phase volatiles combustion, Symposium (International) on Combustion 11 (1967) 399-408.
[8] X. Du, K. Annamalai, The transient ignition of isolated coal particle, Combustion and Flame 97 (1994) 339-354.
[9] T. Wall, R. Gupta, V. Gururajan, D.-k. Zhang, The ignition of coal particles, Fuel 70 (1991) 1011-1016.
[10] Y. Yuan, S. Li, G. Li, N. Wu, Q. Yao, The transition of heterogeneous–homogeneous ignitions of dispersed coal particle streams, Combustion and Flame 161 (2014) 2458-2468.
[11] A. Adeosun, Q. Huang, T. Li, S. Li, R.L. Axelbaum, Design and characterization of a two-stage Hencken burner for combustion of solid fuels 10th U. S. National Combustion Meeting, Organized by the Eastern States Section of the Combustion Institute, College Park, Maryland, 2017.
[12] Z. Huque, M.R. Ali, R. Kommalapati, Application of pulse detonation technology for boiler slag removal, Fuel Processing Technology 90 (2009) 558-569.
[13] T. Maffei, R. Khatami, S. Pierucci, T. Faravelli, E. Ranzi, Y.A. Levendis, Experimental and modeling study of single coal particle combustion in O_2/N_2 and Oxy-fuel (O_2/CO_2) atmospheres, Combustion and Flame 160 (2013) 2559-2572.
[14] Y.A. Levendis, K. Joshi, R. Khatami, A.F. Sarofim, Combustion behavior in air of single particles from three different coal ranks and from sugarcane bagasse, Combustion and Flame 158 (2011) 452-465.
[15] H. Jüntgen, K.H. Van Heek, An update of german non-isothermal coal pyrolysis work, Fuel Processing Technology 2 (1979) 261.

Spring Technical Meeting
Eastern States Section of the Combustion Institute
March 4-7, 2018
State College, Pennsylvania

Predicting particle deposition for flow over a circular cylinder in combustion environments

Akshay Gopan[1], Zhiwei Yang[1], Richard L. Axelbaum[1,]*

[1]Department of Energy, Environmental & Chemical Eng., Washington University in St. Louis, 1 Brookings Drive, St. Louis, MO 63130, USA
[*]Corresponding Author Email: axelbaum@wustl.edu

Abstract: Particle deposition on heat exchanger tubes is a serious concern in solid fuel combustion and gasification systems, such as power plants and syngas coolers. In order to predict deposition rates, which affect heat transfer efficiencies, several detailed computational fluid dynamic (CFD) models have been developed. However, these models are computationally expensive and cannot be used for quick determination of deposition and/or slagging tendencies. On the other hand, particle impaction efficiency correlations have been developed. While not as accurate as the detailed models, they are easier to use and indicate the impaction of particles on the heat exchanger tubes. However, since deposition and slagging are not just functions of particle impaction rates, but also their sticking propensity, which is related to the particle temperature at impact, the impaction efficiency correlations fail to provide sufficient information. In this work, similar correlations for particle temperature at impact have been developed, based on a non-dimensional parameter that captures the flow and boundary conditions, as well as particle properties. When used alongside the impaction efficiency correlations, the new correlations developed here can provide a reasonable estimate of the deposition and slagging tendencies, at negligible computational expense.
Keywords: ash, deposition, slagging, coal combustion

1. Introduction

Particulate deposition on heat exchanger tubes is a serious concern during solid fuel combustion and gasification [1,2]. Accumulation of particulate matter (ash) on the heat exchanger surfaces increases the heat transfer resistance, which reduces steam production [2,3], increases corrosion and erosion of the tubes, and is currently one of the major causes of unscheduled boiler shutdowns. Hence, developing a clear understanding of particle deposition is critical to improving the design and operation of boilers.

In this work, we are mainly concerned with deposition that occurs during flow over a circular cylinder. The particle impaction rate on a circular cylinder is a classical aerosol problem encountered in applications ranging from filters to boiler tubes. Weber et al. [4] showed that particle impaction on cylinders can be predicted fairly accurately using RANS simulations. However, a high level of grid refinement is required, leading to significant computational expense.

Simpler, empirical models exist that only require free-stream information, and hence can be used for predicting impaction rate, with negligible computational expense [5], fairly accurately. Since particle deposition is a combination of both impaction and sticking, the sticking probability needs to be predicted as well. A number of models exist to predict the sticking probability [6–8]. One critical parameter required as input to all the sticking models is the particle temperature at impact. This key parameter is extremely difficult to determine experimentally, and as will be shown in the presentation, accurate numerical simulations are extremely expensive.

To this end, the objective of the current work is to develop a simple, yet robust model to predict the particle temperature at impaction. The approach is to develop the appropriate non-dimensional parameters that can collapse particle temperature data obtained from detailed computational fluid dynamic (CFD) simulations under a wide range of conditions.

2. Geometry and CFD Methods

The computational domain, shown in Fig. 1, is a fundamental "test geometry" that can be easily extrapolated to practical systems. The domain and the discretization scheme are identical to those used by Weber et al. [4], and the grid resolution matches or exceeds the minimum requirement for accurate prediction of the impaction rate [4].

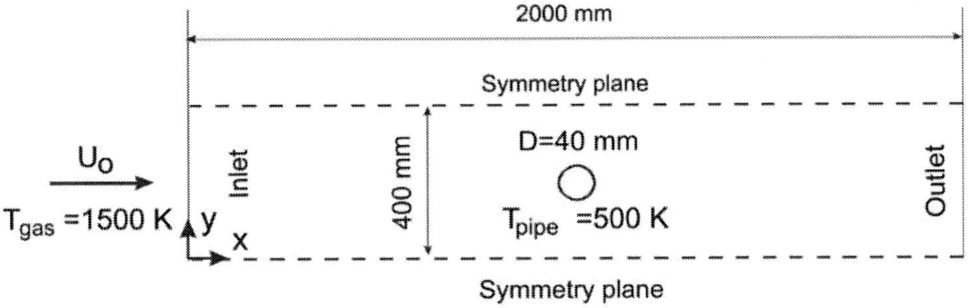

Figure 1. Computational domain. Reproduced from Weber et al. [4].

The inlet flow is a gaseous stream at 1500 K. For the base case, the gas has a density of 0.245 kg/m^3, viscosity of 46x10^{-6} Pa-s, thermal conductivity of 0.079 W/m-K, and heat capacity (c_{pg}) of 1200 J/kg-K. The gas entering the computational domain is at a uniform free stream velocity (U_0) of 5 m/s in the base case. The gas flows for all the cases considered here have a laminar boundary layer on the tube [9]. Particles enter the domain with the velocity and temperature same as the gas. The particle density and specific heat capacity are 2450 kg/m^3 and 1680 J/kg-K respectively.

For the base case, the mesh was created by dividing the tube circumference into 1256 parts, and in the normal direction the cell size was uniform (0.1 mm) up to one tube radius from the tube surface. Further away, the size of the cells was increased at a small rate. The high near-wall resolution was to ensure fine resolution (12 cells) of the viscous sub-layer at the stagnation point.

In this study, we use a RANS-based steady solver with the continuous phase (gas) solved with an Eulerian method and the discrete phase (particles) solved using a Lagrangian approach. Here, the *k-ε* model with enhanced wall treatment is used to model the eddy viscosity. Since the impaction is driven by inertia, the eddy viscosity model has negligible impact [4].

3. Model Development and Discussion

Particle deposition in an inertially impacting flow is usually studied using a non-dimensional time scale comparison, leading to the development of the Stokes number (*Stk*), defined in Eq. 1.

$$Stk = \frac{\rho_p d_p^2 C_c u_0}{18\mu_g D} \qquad (1)$$

where ρ_p is the particle density, d_p is the particle diameter, C_c is the Cunningham slip correction factor, μ_g is the gas dynamic viscosity, u_0 is the free-stream velocity, and D is the tube diameter.

For *Stk* >> 1, the particles tend to not flow with the fluid, and thus they will impact the surface. For small particles, *Stk* << 1, the particle relaxation time is short, and hence the particles remain entrained in the flow. However, it was found that this method does not fully characterize the process—the curves separate at different Reynolds numbers [10]. Aw generalized Stokes number was developed by accounting for the non-linear drag and incorporated as a correction (Eq. 2) [11]:

$$Stk_{eff} = Stk\ \psi_f(Re_0) \tag{2}$$

where Re_0 is calculated on the basis of the free-stream velocity. This correction is especially important when considering the polydispersity of particles in a typical industrial combustor.

Due to the transport analogy between flow (or motion) and heat transfer, a similar approach as above can be utilized to analyze the particle cooling in the vicinity of the "cold" surface of the tube. A characteristic particle cooling time ($\tau_{T,p}$) can be defined as

$$\tau_{T,p} = \frac{m_p c_{p,s}}{hA_p + \frac{\varepsilon_p \sigma A_p(\theta_R^4 - T_p^4)}{T_g - T_p}} \approx \frac{m_p c_{p,s}}{hA_p} = \frac{\rho_p d_p^2 c_{p,s}}{6Nu_p k_g} \tag{3}$$

where T_p , m_p, $c_{p,s}$, ε_p, and A_p are, respectively, the temperature, mass, specific heat capacity, and surface area of the particle, h is the convective heat transfer coefficient, T_g and θ_R are the gas and radiation temperatures, respectively. And, Nu_p and k_g are, respectively, the particle Nusselt number and gas thermal conductivity. The radiative part in the denominator of the equation is neglected in the final form, based on an order of magnitude analysis for the typical conditions encountered in the applications of interest here.

Non-dimensionalizing the characteristic time for particle cooling with the time for fluid cooling provides us with a heat transfer analog for the Stokes number, which we call here the thermal Stokes number (Stk_{th}), as shown in Eq. 4. A correction for the thermal Stokes number is introduced here, and an effective thermal Stokes number defined as in Eq. 5.

$$Stk_{th} = \left. \frac{\rho_p d_p^2 c_{p,s}}{6Nu_p k_g} u_0 \middle/ \frac{2D}{Nu_D} \right. \tag{4}$$

$$Stk_{th}^{eff} = Stk_{th}\ \psi_{th} \tag{5}$$

Finally, since the quantity of interest is the particle temperature at impaction, a non-dimensional temperature is defined as:

$$T^+ = \frac{T_p - T_w}{T_g - T_w} \tag{6}$$

where T_w is the wall (cylinder) temperature. Here, T^+ ranges from 0 to 1.

Mesh sensitivity studies indicate that a minimum of 12 cells in the viscous boundary layer are required to reasonably accurately predict particle temperature at impaction. Even at this condition, an understanding of the physics is required for the correct interpretation of the results. These details will be discussed in the presentation.

3.1. Effect of gas and particle properties

Since a range of particle sizes and gas flow conditions can be expected for various practical systems, it is important to study the effects of the variations in the particle and gas properties, and flow conditions. A number of different variations in the flow conditions, and the particle and gas properties are studied here, and the description of the cases simulated is mentioned in Table 1. As an example, in Fig. 3, the effect of changing the non-dimensional particle specific heat capacity, $c^*_{p,s}\left(= c_{p,s}/(k_g/\mu_g)\right)$ is plotted on a T^+ vs Stk^{eff}_{th} graph. Fig. 3a shows the average temperature of the impacting particles. However, since the particles are injected along the projected area plane of the tube, the different particles traverse different path lengths to reach the tube, and the average temperature is not representative of the worst-case scenario. To also clearly show the worst-case scenario, in Fig 3b, the particle temperature at impact is shown only for those particles that were released at the centerline of the tube, about 5 tube diameters away. These particles spend the least amount of time in the boundary layer and hence represent particles with the maximum temperature.

Table 1. The flow cases considered. Case A is the base case in this study, and in case of variation of a parameter from the base case, the exact parameters varied are mentioned.

Case	Re_{tube}	Pr_g	$c^*_{p,s}\left(=\frac{c_{p,s}}{k_g/\mu_g}\right)$
A	1065	0.7	0.98
B	1065	0.7	0.31 $(c_{p,s} = 531\,J/kg\,/K)$
C	1065	0.7	1.65 $(c_{p,s} = 2829\,J/kg\,/K)$
D	213 $(u_0 = 1\ m/s;\ \rho_g = 0.245\ kg/m^3)$	0.7	0.98
E	213 $(u_0 = 5\ m/s;\ \rho_g = 0.049\ kg/m^3)$	0.7	0.98
F[(a)]	4261 $(u_0 = 20\ m/s;\ \rho_g = 0.245\ kg/m^3)$	0.7	0.98
G	1065	1.4 $(c_{p,g} = 503\,J/kg\,/K)$	0.98

[(a)]Note: To compensate for the reduced boundary layer thickness, the near-tube wall-normal cell resolution was refined to 0.05 mm, along with a circumferential resolution of 2512 nodes, resulting in a total of 2.97M cells for the entire computational domain.

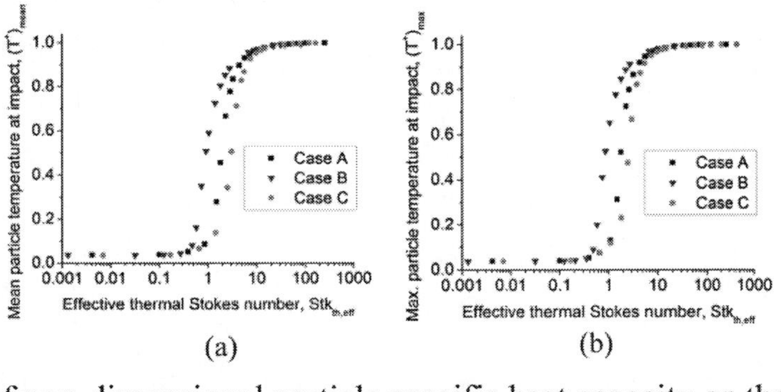

Figure 3. Effect of non-dimensional particle specific heat capacity on the (a) mean particle temperature, and the (b) maximum particle temperature at impact, when plotted against the effective thermal Stokes number.

As can be seen, the curves are all of the same functional form, however, they are shifted. The minimum for all the curves is zero as $Stk^{eff}_{\text{th}} \to 0$, while the maximum is one as $Stk^{eff}_{th} \to +\infty$. At the extremes, the curves all collapse into the same minimum and maximum value because when the Stk_{th}^{eff} is very small, the particle thermal relaxation time is much smaller than that of the gas, hence it reaches the same temperature as that of the tube surface, and for a very large Stk_{th}^{eff}, the particle cools much slower than the gas and remains essentially at the free-stream temperature. However, in the middle, these limit conditions do not hold, and the curves separate because the non-dimensional particle cooling time does not capture the entire physics of the system. Specifically, it does not take into account the residence time of the particles in the "cooling region" before impaction. The characteristic residence time is related to the particle Stokes number. So here we define another parameter:

$$\phi = \frac{Stk_{th}^{eff}}{Stk^{eff}} = \frac{3\mu_g c_{p,s} Nu_D \psi_{th}}{Nu_p k_g C_c \psi_f} = \frac{3c_{p,s}^* Nu_D \psi_{th}}{Nu_p C_c \psi_f} \quad (7)$$

which is the ratio of the thermal to the particle movement relaxation times. We also propose a new non-dimensional parameter, instead of the thermal Stokes number defined earlier, that characterizes this characteristic residence time:

$$G = Stk_{th}\phi^{\alpha} \quad (8)$$

The exponent, α, needs to be determined using experimental data, or, as in this case, accurate simulation data, so that all the curves collapse into one. Using the data from all the cases listed in Table 1, this exponent is obtained as $\alpha \approx -0.6$ for a circular cylinder. The same exponent holds for both the average temperature at impact and the maximum temperature at impact.

To check the robustness of this method, several parameters (Table 1) are varied—the non-dimensional particle specific heat (how fast a particle can cool for a given temperature profile), the gas Prandtl number (temperature profile in the boundary layer), and the flow Reynolds number (viscous boundary layer, and the temperature profile), and the impact temperature results are used to generate empirical correlations (Eq. 9). Unlike Stk_{th}^{eff}, the new parameter, G, captures the dominant physics, leading to a data collapse into a single curve (Fig. 4).

Since the final form in Fig. 4 is an S-curve, a logistic equation can be used to represent it:

$$T_{p,max}^{+} = 1 - \frac{1}{1 + 10^{2.945(\log_{10}(G)+0.1535)}} \quad (16a)$$

$$T_{p,mean}^{+} = 1 - \frac{1}{1 + 10^{2.3871(\log_{10}(G)+0.0661)}} \quad (16b)$$

where the coefficients were obtained from the simulations listed in Table 1. A comparison of the modeled curve (Eq. 9) and the individual data points is shown in Fig. 4. Both, the mean and the maximum temperature agree very well with their modeled curves.

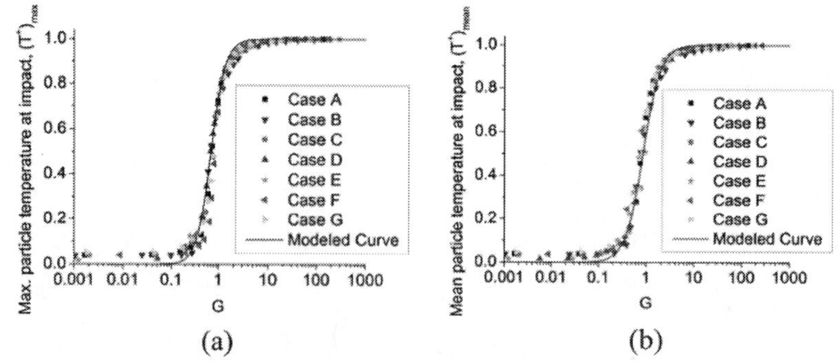

Figure 5. Comparison of the modeled curve (correlation developed in this work) and the individual data points. (a) Maximum, and (b) mean particle temperature at impact.

4. Conclusions

This is the first study to clearly characterize the temperature of particles impacting on a surface from non-isothermal flows. A simple algebraic equation for temperature at impaction has been developed as a function of newly defined non-dimensional parameters that characterize the particle and flow properties and conditions. The equation requires only the free-stream information, and can work accurately without expensive fine-meshing to resolve the viscous sub-layer in the heat exchanger tube vicinity—clearly a non-trivial task.

Using the model presented in this work, along with the previously developed engineering models for impaction, particle deposition rates and fouling/slagging tendencies can be characterized with extreme simplicity (and without computational expense). It not only provides means to reduce computational expense when running CFD simulations for boilers or syngas coolers, but also helps to characterize the deposition tendencies in the design conceptualization phase, or for simple back-of-the-envelope type calculations, which was hitherto not possible. Finally, this work can easily be used as a template to develop similar models for other practical configurations to reduce complexity and computational cost.

6. References

[1] Z. Zhan, A.R. Fry, J.O.L. Wendt, Proc. Combust. Inst. 36 (2017) 2091–2191.

[2] U. Kleinhans, C. Wieland, S. Babat, G. Scheffknecht, H. Spliethoff, Proc. Combust. Inst. 36 (2017) 2341–2350.

[3] S. Balakrishnan, R. Nagarajan, K. Karthick, Energy 81 (2015) 462–470.

[4] R. Weber, N. Schaffel-Mancini, M. Mancini, T. Kupka, Fuel 108 (2013) 586–596.

[5] R.A. Wessel, J. Righi, Aerosol Sci. Technol. 9 (1988) 29–60.

[6] S. Srinivasachar, C.L. Senior, J.J. Helble, J.W. Moore, Twenty-Fourth Symp. Combust. Combust. Inst. (1992) 1179–1187.

[7] C. Wieland, B. Kreutzkam, G. Balan, H. Spliethoff, Appl. Energy 93 (2012) 184–192.

[8] T. Mao, D.C.S. Kuhn, H. Tran, AIChE J. 43 (1997) 2169–2179.

[9] M.M. Zdravkovich, P.W. Bearman, Flow Around Circular Cylinders—Volume 1: Fundamentals, Oxford Science Publications, 1998.

[10] R.J. Brun, W. Lewis, P.J. Perkins, J.S. Serafini, (1955).

[11] R. Israel, D.E. Rosner, Aerosol Sci. Technol. 2 (1982) 45–51.

Spring Technical Meeting
Eastern States Section of the Combustion Institute
March 4-7, 2018
State College, Pennsylvania

Spectroscopic Emission Measurements and System Level Modeling of a Two-Stage Wood-Fired Hydronic Heater: Effects of Non-Homogeneous Fuel Decomposition

Joshua M. Weisberger, Joseph P. Richter, Joseph C. Mollendorf, Paul E. DesJardin

Department of Mechanical and Aerospace Engineering, University at Buffalo, the State University of New York, Buffalo, New York

This study investigates the time-dependent non-homogeneous decomposition of solid biomass fuels in a wood-fired heating appliance. Throughout a run (heating cycle), flue exhaust gases are monitored using a Bosch oxygen sensor and a tunable diode laser absorption spectroscopy system. The data is used to derive an experimental pyrolysis model to calculate the instantaneous fuel composition (carbon-normalized H, O, and N). The fuel H/O atomic ratio remains nearly constant until all the H and O have been released. The results from the experimental pyrolysis model are used as input to a simplified three-zone (chamber) system level model to calculate time-accurate bulk emissions and temperatures in a domestic two-stage wood-fired hydronic heater. Pyrolysis gas and carbon char fuel streams are separately defined in the three chambers. Wood pyrolysis products and char are determined via a multi-phase chemical equilibrium solver at a constant pressure and specified pyrolysis temperature. Comparison of experimentally measured emissions to the system level model predictions show good agreement, indicating that coupling a non-homogeneous description of the decomposing fuel with a simplified model can provide accurate emissions predictions from a boiler run in batch-mode.

1 Introduction

In the northeastern US, emissions from wood-fired heaters is one of the leading sources of reduced air quality. To provide stricter guidelines on the certification of these systems, the Environmental Protection Agency (EPA) has recently defined specific new regulations for wood burning appliances including wood stoves, masonry heaters, pellet-burning stoves and hydronic heaters [1]. Reduction in harmful pollutants resulting from the combustion of cellulosic based biomass materials is the primary goal of many government agencies and research developments. Therefore, with the aim to better understand and reduce pollution, many studies focus on the characterization of emissions from specific biomass appliances, boilers, and large scale (>1 MW) power plants [2, 3].

A recent review of biomass for boilers may be found by Saidur *et al.* [4] where biomass appliances are grouped into pellet [5], wood chips [6, 7], cord-wood, wood briquettes [8], and coal [9] based systems. For pellet and wood chip feed systems, the fuel flow rate is predictable and controlled, allowing for quasi-steady-state operation. Consequently, efficiencies and emissions are better understood and are more easily optimized. Less is known about cord-wood systems since they are often run as an unsteady batch process. For batch-type systems, few studies developing system level models of domestic wood boilers have been published. Significant benefits could be gained with a system level model approach, providing a computationally low-cost method to accurately model lumped predictions of temperatures and emissions. Analyzing batch-run systems is further complicated by the non-homogeneous nature of the biomass fuel decomposition.

Emission measurements can be used to understand the fuel decomposition process and validate the system level models. In this study, laser absorption measurements are used to develop an experimental pyrolysis model that computes the instantaneous atomic composition of the fuel as it decomposes. The results from the experimental pyrolysis model are used in a system level model, where temperature and complete emissions information are computed and compared to experimentally measured values.

The rest of the study is organized as follows: Section 2 describes the experimental pyrolysis model, Section 3 describes the system level modeling, Section 4 presents results with discussion, and Section 5 concludes.

2 Experimental Pyrolysis Model

A two-stage wood-fired hydronic heater (boiler) is used as the test-bed for the measurements in the current study [10]. A key finding from the earlier study is the importance of the non-homogeneous decomposition of the fuel. The global reaction mechanism used in this study can be seen in Eq. (1), where the unknowns being solved for are the carbon-normalized atomic hydrogen, oxygen, and nitrogen of the fuel (x, y, and z) at each time step during a run. O_2 concentration is obtained using a Bosch LSU 4.9 ZrO_2 wide-band oxygen/lambda sensor. The CO_2 and H_2O concentrations are measured using a combined direct absorption spectroscopy (DAS) and wavelength modulation spectroscopy (WMS) measurement setup. The solution of these atomic values at every time step during a typical boiler run constitutes an experimental pyrolysis model, as it determines the non-homogeneous decomposition of the fuel.

$$C_1H_xO_yN_z + a\,(O_2 + 3.76\,N_2 + \gamma\,H_2O) \rightarrow c\,H_2O + e\,CO_2 + 3.76\,a\,N_2 + g\,O_2 \qquad (1)$$

Equation (1) has eight unknowns. The unknown fuel composition includes the fuel moisture content, and provides three equations in the form of atom balances for H, O, and N. The moisture coefficient γ is calculated from the ambient humidity and temperature. Two equations come from DAS/WMS measurements. One equation comes from Bosch O_2 measurements. Closure of the system comes from a summation of the mole fractions to unity.

Laser Absorption Measurements

A schematic of the flue and laser absorption measurement setup can be seen in Fig. 1(a). The setup operates as a pitch-catch system, with each side housed in nitrogen-purged boxes to avoid any absorption of laboratory gases outside the flue gases of interest. Two wedged sapphire windows provide optical access to the flue gases, while also minimizing unwanted etaloning effects. On the pitch side, a distributed feedback Nanoplus laser is controlled using an ILX Lightwave LDC-3742C laser diode driver and thermoelectric cooler along with a Tektronix AFG3101 arbitrary waveform generator. The laser passes through the flue gases and is acquired as channel 1 (CH1) on a detector (termed the *absorbing data*). Back on the pitch side, the beam is split twice, with one portion of the beam acquired on a detector to provide the *non-absorbing* laser data (channel 2/CH2). The other portion of the beam passes through a solid Germanium etalon with a free spectral range of $0.75\ GHz$, and into a third detector (channel 3/CH3). This data is termed the *etalon* data, and provides the wavenumber calibration needed for the analysis. A Measurement Computing PCI-DAS 4020/12 data acquisition card is used to acquire data from the detectors.

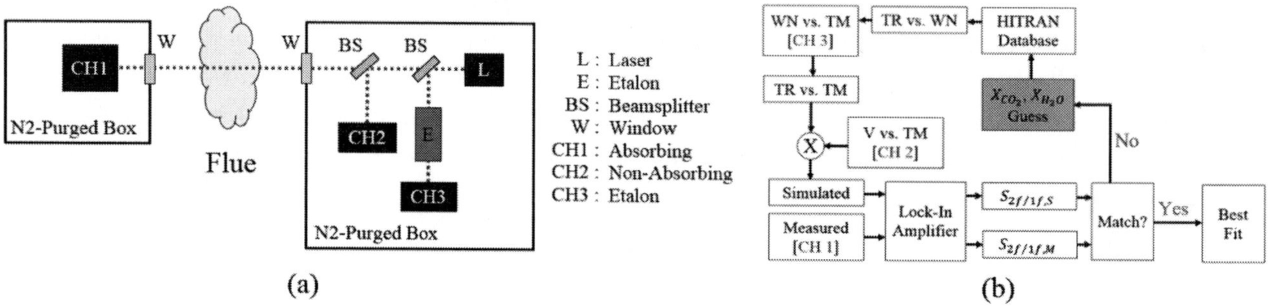

Figure 1: (a) Schematic of WMS setup for boiler flue measurements and (b) WMS data analysis flowchart.

For scanned-WMS measurements, the laser modulation is composed of two different frequencies, one slow and one fast. The slow-scan is a sawtooth function that ramps the wavenumber across an absorption feature at $20\ Hz$. A typical wavenumber scan range for these experiments is approximately $3.5\ cm^{-1}$. Superimposed on the slow-scan ramp is a high frequency sinusoidal modulation at $10\ kHz$. This modulation rapidly scans the wavenumber up and down as the slow scan ramps through the wider wavenumber range. By modulating at high frequencies, the information about the absorption due to flue gases is shifted to higher frequencies, minimizing the noise associated with low frequency operation and increasing sensitivity of the measurements.

Laser Data Analysis

The three channels of data acquired during the boiler run are analyzed using the steps outlined in Fig. 1(b) in order to determine the concentrations of CO_2 and H_2O in the flue gas. This method is based on the analysis in Sun [11], where a $2f/1f$ calibration-free method is used. The data analysis process starts at the gray box by guessing a value for the both X_{CO_2} and X_{H_2O}. A simulation of the transmission versus wavenumber is obtained using data from the HITRAN spectroscopic database [12]. The etalon data from CH3 gives the wavenumber as a function of time. Combining both the HITRAN simulation and the CH3 data gives the transmission from the simulation as a function of time. To simulate the WMS absorbing data, the voltage versus time data from CH3 is multiplied by the transmission versus time data just obtained.

The simulated data (from CH2 and CH3), along with the measured absorbing voltage versus time (CH1), are passed through a digital lock-in amplifier (LIA) which consists of a mixer and a low pass filter. Identical parameters are used in the LIA for both the simulated and measured signals. The resulting LIA signals ($S_{2f/1f,\mathrm{S}}$ and $S_{2f/1f,\mathrm{M}}$) are compared using least squares errors. If they do not match within some error tolerance, the guesses for the flue gas concentrations are updated and the process continues again. If they do match within some tolerance, then the best-fit gas concentrations have been found. This process is repeated for every data point during a boiler run.

3 System Level Modeling

A system level model has been developed based on quasi-steady flow assumptions and well-stirred reactor theory. This study builds on a previously developed single-chamber model by incorporating energy conservation equations derived individually for three separate "zones" in the hydronic heater, each with steel and water (b), solid fuel (f), and gas phase (g) systems [13]. Conservation equation for the primary chamber including the solid fuel charge proceed as follows:

$$m_{g1} C_{v,g1} \frac{dT_{g1}}{dt} = \sum_{fuel} \dot{m}_{i,g1} h_{i,pyr} - \dot{Q}_{gb1} + \dot{Q}_{air1} - \dot{Q}_{prod1} + \dot{m}_f L_p \quad (2)$$

$$m_{b1} C_{b1} \frac{dT_{b1}}{dt} = \dot{Q}_{gb1} - \dot{Q}_{del1} \quad (3)$$

where $\dot{Q}_{del} = C_w \dot{m}_w (T_{w,out} - T_{w,in})$ is the delivered heat to the thermal load (heat exchanger) which is experimentally determined from measurements of the water temperature and flow rate going in ($T_{w,in}$) and out ($T_{w,out}$) of the boiler, and segmented for each respective chamber based initially on chamber surface area ratios, later refined to best fit experimental measurements. The quantities $\dot{Q}_{fg}$, $\dot{Q}_{gb}$ and $\dot{Q}_{prod}$ represent the energy transfer rates from the fuel to the gas, from the gas to the boiler, and that flowing out of each zone, respectively. The heat transfer from the gas to the boiler is modeled using a mixed convective and radiative condition assuming black surfaces: $\dot{Q}_{gb} = A_{gb}[h_{gb}(T_g - T_b) + \sigma(T_g^4 - T_b^4)]$ where h_{gb} is the heat transfer coefficient across the gas-boiler interface with area A_{gb}, and $\sigma = 5.567 \times 10^{-8} W/m^2 - K^4$ is the Stefan-Boltzmann constant. Since the thermal inertia of the solid fuel is small relative to that of the boiler, the fuel is assumed to be at its pyrolysis temperature where L_p is the latent heat of pyrolysis and $\dot{m}_f$ is the total fuel mass loss rate that is experimentally measured. Ash in oak is typically less than 0.5% by mass of the dry fuel [14], and by ignoring this small fraction the mass flow rate of the fuel into the gas phase can be written as; $-\dot{m}_f = \sum_{fuel} \dot{m}_{g,i}$ and substituting this relation into the second term in Eq. (2) results in: $\sum_{fuel} \dot{m}_{g,i} h_i = -\dot{m}_f \sum_{fuel} Y_i h_i(T_p)$ where Y_i are the mass fractions of the pyrolysis gases and h_i is the enthalpy of the ith species at the wood pyrolysis temperature, T_{pyr}. The energy flow of gases in and out of the heater are defined as: $\dot{Q}_{air} = \dot{m}_{air} h_{air}$ and $\dot{Q}_{prod} = \dot{m}_{flue} \sum_{flue} Y_i h_i(T_g)$, respectively, where $\dot{m}_{prod}$ is the total exhaust gas flow rate for that zone defined as: $\dot{m}_{prod1} = \dot{m}_{air} + \dot{m}_f$, $\dot{m}_{prod2} = \dot{m}_{air} + \dot{m}_{prod1}$, and $\dot{m}_{flue} = \dot{m}_{prod2}$. Conservation equations for chamber 2 and chamber 3 follow similarly. The model is driven by two experimental inputs, $\dot{m}_f$ and $\dot{Q}_{del}$ separated into their respective fuel stream (pyrolysis and char) and their respective zones. These system of equations are integrated in time using a semi-implicit time marching scheme. In this approach, temperature time derivatives are discretized using first-order forward differences, *i.e.*, $dT/dt \simeq (T^{n+1} - T^n)/\Delta t$ and all source term quantities on the right hand side are evaluated at the $n+1$ time level. The system of non-linear, implicit equations are solved using an algorithm that employs inverse quadratic interpolation with a combination of the bisection and secant method.

4 Results and Discussion

From the experimental pyrolysis model, the carbon-normalized atomic ratios of hydrogen, oxygen, and nitrogen are computed. A figure of the H/C and O/C for three wood types is plotted in

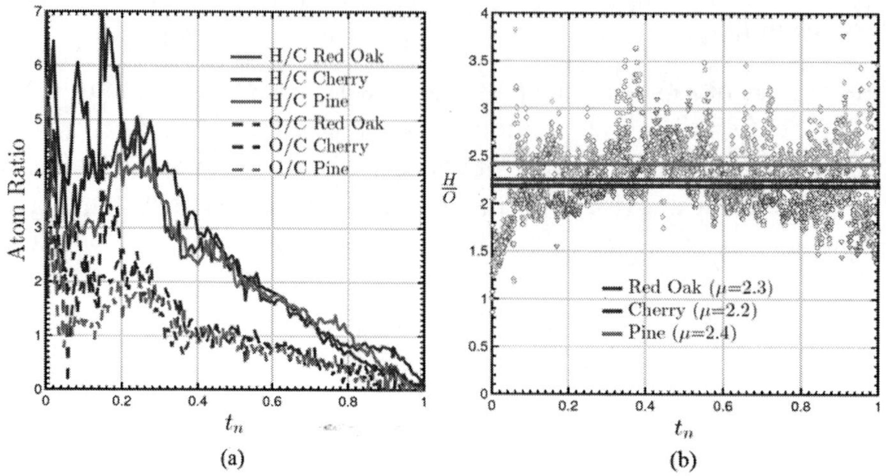

Figure 2: (a) Carbon-normalized atom ratios for H, O, and N, and (b) H/O ratios for three types of fuels.

Fig. 2(a) versus a normalized time, t_n, defined as the time at which all the H and O have been driven from the fuel (mole fraction of H_2O is equal to that of the ambient air introduced by the blower). Red oak, cherry, and pine are used to provide different initial fuel compositions and loading geometry. The first part of the run corresponds to boiler start-up, and can be a chaotic period as temperatures increase and operation stabilizes. At a normalized time of approximately 0.3, the atomic ratios of all three fuels stabilize in relation to each other, and continue to steadily decrease until all the H and O have been driven from the fuel. Figure 2(b) shows that release of H and O is nearly constant during a run. After this point and until the end of the run (when no fuel remains), the fuel is composed mainly of carbon.

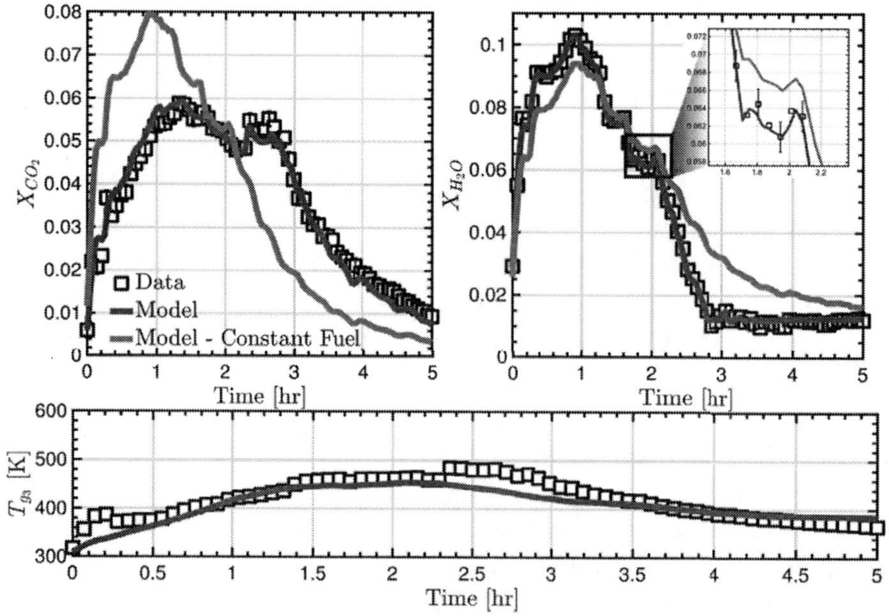

Figure 3: Comparisons of CO_2 and H_2O data from a Testo gas analyzer with predicted values from the system level model.

In Fig. 3, comparisons between data taken by the Testo gas analyzer during a typical run (using red oak) and the gas concentrations in the flue from the system level model can be seen along with the corresponding flue temperature below. The gray line corresponds to the concentrations derived using a constant fuel assumption, where the carbon-normalized composition is assumed to be $C_1H_{1.7}O_{0.72}N_{0.0001}$ throughout the entire run. At approximately 3 hours into the run the hydrogen in the fuel has been consumed, and the resulting effect is a steady level of H_2O derived from air humidity. While significant differences between the data and the constant fuel model are evident, there is excellent agreement between the data and non-homogeneous fuel model, indicating that a simplified model can be used to accurately infer flue gas emissions for a non-homogeneously decomposing fuel in a batch-run boiler.

5 Conclusions

A pyrolysis model derived experimentally from emission measurements in the flue of a two-stage wood-fired hydronic boiler provides time-accurate, non-homogeneous fuel decomposition information. The instantaneous fuel composition during a run is used in a three-zone system level model, which predicts temperatures and gas concentrations in the three chambers of the boiler. The results from this simplified model are compared to measurements from a Testo gas analyzer unit. The data from both the system level model and gas analyzer agree well, and indicate that a simplified model can be used to predict emissions from a batch-run boiler.

Acknowledgments

Support for this work has been provided by NSF under grant #1704447 and by the New York State Energy Research and Development Authority (NYSERDA) under contract number #32966.

References

[1] T. Butcher and N. Russell. Review of epa method 28 outdoor wood hydronic heat test results. Technical Report 11-17, New York State Energy Research and Development Authority (NYSERDA), 2011.

[2] Jacobo Porteiro, Joaquin Collazo, David Patino, Enrique Granada, Jorge Carlos Moran Gonzalez, and José Luís Míguez. *Energy & Fuels*, 23 (2009) 1067–1075.

[3] M Rabaçal, U Fernandes, and M Costa. *Renewable Energy*, 51 (2013) 220–226.

[4] R. Saidur, E.A. Abdelaziz, A. Demirbas, M.S. Hosszin, and S. Mekhilef. *Renewable and Sustainable Energy Reviews*, 15 (2011) 2262–2289.

[5] J. Chaney, H. Liu, and J. Li. *Energy Conversion and Management*, 63 (2012) 149–156.

[6] H Sefidari, N. Razmjoo, and M. Strand. *Fuel*, 135 (2014) 120–129.

[7] M. Tamura, S. Watanabe, N. Kotake, and M. Hasegawa. *Fuel*, 134 (2014) 544–553.

[8] M. M. Roy and K. W. Corscadden. *Applied Energy*, 99 (2012) 206–212.

[9] A. Ryfa, R. Buczynski, M. Chabinski, A. Szlek, and R. A. Bialecki. *Applied Thermal Engineering*, 73 (2014) 794–804.

[10] J. Richter, B. Bojko, J. Mollendorf, and P. DesJardin. *Renewable Energy*, 96 (2016) 400–409.

[11] K Sun, X Chao, R Sur, C S Goldenstein, J B Jeffries, and R K Hanson. *Measurement Science and Technology*, 24 (2013) 125203.

[12] L. S. Rothman. *JQSRT*, 110 (2009) 533–572.

[13] J. Richter, B. Bojko, J. Mollendorf, and P. DesJardin. *Renewable Energy*, 96 (2016) 400–409.

[14] Ayhan Demirbas. *Progress in energy and combustion science*, 30 (2004) 219–230.

Spring Technical Meeting
Eastern States Section of the Combustion Institute
March 4-7, 2018
State College, Pennsylvania

Experimental Investigation of the Stabilization and Structure of Turbulent Cool Diffusion Flames

Christopher B. Reuter[1,], Omar R. Yehia[1], Sang Hee Won[2], Matthew K. Fu[1], Katherine Kokmanian[1], Marcus Hultmark[1], and Yiguang Ju[1]*

[1]Department of Mechanical and Aerospace Engineering, Princeton University, Princeton, NJ 08540
[2]Department of Mechanical Engineering, University of South Carolina, Columbia, SC 29208
[*]Corresponding Author Email: cbreuter@princeton.edu

Although recent studies of laminar cool flames have provided important advances in understanding the low-temperature chemistry of both hydrocarbons and oxygenates, there has been limited experimental insight into how interactions between turbulence and chemistry occur in cool flames. To address this, a new Co-flow Axisymmetric Reactor-Assisted Turbulent (CARAT) burner has been developed and characterized in this investigation for the purpose of directly studying turbulent cool flames. A methodology for establishing stable turbulent cool diffusion flames under well-defined conditions is proposed. The structure of dimethyl ether flames is examined using both formaldehyde planar laser-induced fluorescence and Rayleigh scattering. It is found that weak turbulence produces wrinkled turbulent cool flames in which fluctuations occur mainly on the fuel side of the flame. However, at increased levels of turbulence, large pockets of unburned reactants appear in the vicinity of the cool flame, and structural fluctuations extend to both sides of the flame. This study offers a well-defined experimental platform for the study of turbulence-chemistry interactions at low temperatures.
Keywords: Turbulent diffusion flame, turbulent cool flame, cool diffusion flame, dimethyl ether

1. Introduction

Over the past several years, experimental investigations of cool flame phenomena have proliferated. Cool flames have been studied under well-defined conditions in microgravity droplets [1, 2], counterflow burners [3-5], micro flow reactors [6], and Hencken burners [7]. The results of these studies have shown that the low-temperature chain-branching sequence (peroxy chemistry) is capable of producing enough radicals and heat release to sustain stable cool flames across a variety of conditions. These experiments have also provided important validation targets for the development of chemical kinetic models at low temperatures [8].

However, despite the ubiquity of turbulent combustion in real engines, turbulent cool flame experiments have been extremely limited in number. Gokalp [9] and Koliatis [10] studied premixed cool flames within reactors subject to various degrees of turbulence. Some of the authors' previous experiments [11, 12] involved turbulent premixed flames that were strongly affected by low-temperature ignition, but no cool flame fronts were actually visible in those experiments. Recent numerical simulations have shown that cool flames can play an important role in the ignition [13-15], stabilization [16], and propagation [17] of turbulent flames. However, numerical prediction of cool flame behavior is extremely sensitive to the chemical kinetic model employed. Therefore, the experimental measurement of turbulent cool flames can provide insights into the modeling of turbulence-chemistry interactions at low temperatures.

In the present study, a new Co-flow Axisymmetric Reactor-Assisted Turbulent (CARAT) burner is employed for the investigation of turbulent cool diffusion flames. Dimethyl ether (DME) is chosen as the fuel due to its relatively strong low-temperature chemistry and known Rayleigh scattering cross-section [18]. The two-dimensional turbulent cool diffusion flame structure is examined through the formaldehyde planar laser-induced fluorescence technique, while one-dimensional radial temperature profiles are produced using Rayleigh scattering. The effects of increases in turbulence on the cool diffusion flame structure are also discussed.

2. Experiment

The Co-flow Axisymmetric Reactor-Assisted Turbulent (CARAT) burner is a generational update to the Reactor-Assisted Turbulent Slot (RATS) burner [11, 12]. Like the RATS burner, the CARAT burner is a Bunsen-type burner with an extended reactor section capable of controlling the pre-flame residence time and temperature. Two grids located in the reactor section generate turbulence within the main flow. The burner main exit is 15 mm in diameter (D). A small annulus (2-mm gap width) surrounding the main exit enables the use of a pilot flow. The CARAT burner also includes a large exterior channel that enables a heated or vitiated co-flow, similar to the well-known Cabra burner [19]. The co-flow passes through a ceramic honeycomb mesh, which allows for flow laminarization and well-defined boundary conditions.

The turbulence in the main flow has been characterized by a nanoscale thermal anemometry probe (NSTAP) [20] in combination with a Dantec Dynamics Streamware Pro Constant Temperature Anemometry system. Due to its small size (60 μm long, 2 μm wide, 100 nm thick), the NSTAP is capable of extremely high spatial and temporal resolution. The NSTAP was positioned at the centerline of the main exit and operated at a sampling frequency of 250 kHz. Figure 1 shows the range of turbulence intensities (u'/U) and integral length scales (l) measured by the probe as a function of the bulk Reynolds number ($Re_D = UD/\nu$). The integral time scales (and therefore the integral length scales) were determined by extrapolating the power spectrum to zero frequency. Since the turbulent flow field characterization was performed at room

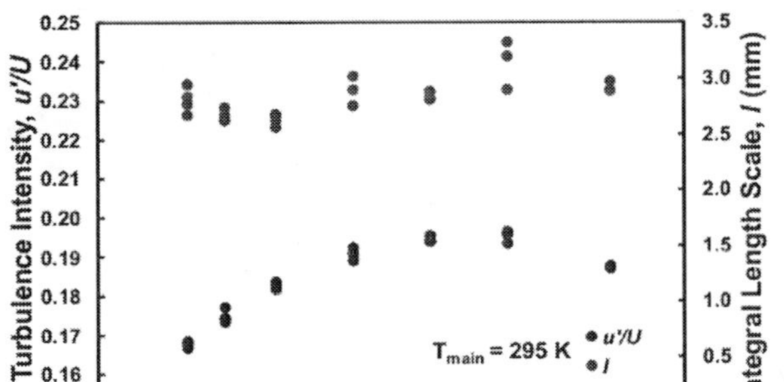

Fig. 1. CARAT burner turbulent flow field characterization in terms of the turbulence intensity (u'/U) and integral length scale (l).

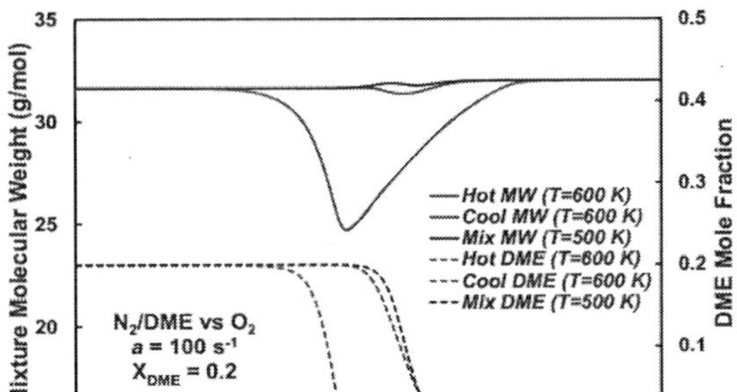

Fig. 2. Computed mixture properties for hot flames, cool flames, and unreacting mixtures in a laminar counterflow configuration.

temperature (295 K), the effects of elevated temperatures on u'/U and l are accounted for by modifying the mixture viscosity appropriately to determine $Re_D(T)$.

To visualize the turbulent cool flame structure, formaldehyde (CH_2O) planar laser-induced fluorescence (PLIF) is used. The third harmonic (355 nm) of an Nd:YAG laser (Quantel, Q-smart 850) excites the CH_2O at an energy of approximately 200 mJ/pulse. The laser operates at a frequency of 5 Hz. Two band filters, passing frequencies between 400 and 450 nm, are positioned in front of an ICCD camera (Princeton Instruments, PI-MAX 4) to isolate the CH_2O fluorescence. The laser beam is expanded into an approximately 200-μm-thick vertical sheet and positioned at the centerline of the burner.

Rayleigh scattering is performed in order to quantify the turbulent cool flame temperature. The second harmonic (532 nm) of a 10-Hz Nd:YAG laser (Spectra-Physics, lab-170-10) is directed through the flame at an energy of approximately 400 mJ/pulse. Accounting for changes in the Rayleigh scattering cross-section within the flame is often the most challenging aspect of the technique [18]. However, the cool flame's tendency to exhibit massive fuel leakage eases this difficulty. Using the example of a 1-D opposed-flow flame [21, 22], it can be seen in Fig. 2 that an unburning mixture with boundary temperatures of $T = 500$ K and a cool flame with boundary temperatures of $T = 600$ K have nearly the same mixture properties. The presence of a hot flame, on the other hand, causes significant changes which must be accounted for through extensive corrections of the Rayleigh signal. Therefore, in this study, the cool flame temperature is determined by directly comparing its Rayleigh scattering signal to an unburning flow field with identical mixture properties at a known temperature.

3. Stabilization of Turbulent Cool Diffusion Flames

In this study, turbulent cool diffusion flames are established using dimethyl ether (DME) as the fuel. Under the conditions given in Fig. 3, cool flames can be stabilized in the recirculation zone between the main flow (DME/N_2) and the pilot flow (O_2). The range of conditions for which a cool flame can be stabilized is highly temperature dependent. Specifically, DME turbulent cool diffusion flames were seen to form readily for $T = 600$ K boundary temperatures but not for $T = 550$ K or $T = 500$ K. The use of pure oxygen in the pilot flow extends the range of conditions for which turbulent cool diffusion flames can be stabilized but is not strictly necessary, as cool flames were also able to be formed with air in the pilot flow. Notably, increasing the main flow velocity (U) or decreasing the DME mole fraction (X_{DME}) does not result in a sharp transition between stable turbulent cool diffusion flames and blowoff. Rather, it was seen that a range of conditions exist for which lifted turbulent cool diffusion flames result, some of which transiently reach blowout over the course of several minutes. For all measurements in this study, the pilot velocity is 1.0 m/s, and the co-flow velocity is approximately 0.55 m/s.

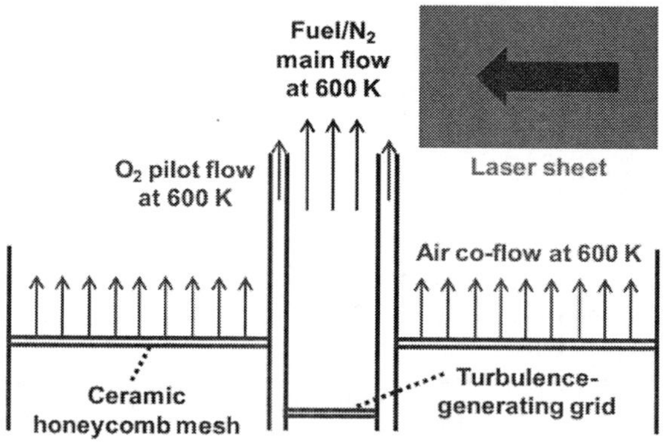

Fig. 3. Experimental schematic for producing turbulent cool diffusion flames in the CARAT burner.

4. Flame Structure Measurements through CH_2O PLIF

Due to its unique chemistry, DME is an extremely attractive fuel for the CH_2O PLIF technique. Multiple reaction pathways are available for DME to decompose into CH_2O, and as a result the signal-to-noise ratio for CH_2O PLIF of DME flames is typically quite high. Figure 4 shows the DME turbulent cool diffusion flame structure for a main flow velocity of U = 1.50 m/s and a DME mole fraction of X_{DME} = 0.2. All three streams (main, pilot, co-flow) exit the burner at a temperature of T = 600 K (as elucidated in Fig. 3). The turbulent cool flame is clearly wrinkled but unbroken (Fig. 4a), with the majority of the wrinkling occurring on the fuel side of the flame. The higher likelihood of fuel-side fluctuations is confirmed quantitatively by the values of the standard deviation of the CH_2O signal in Fig. 4c.

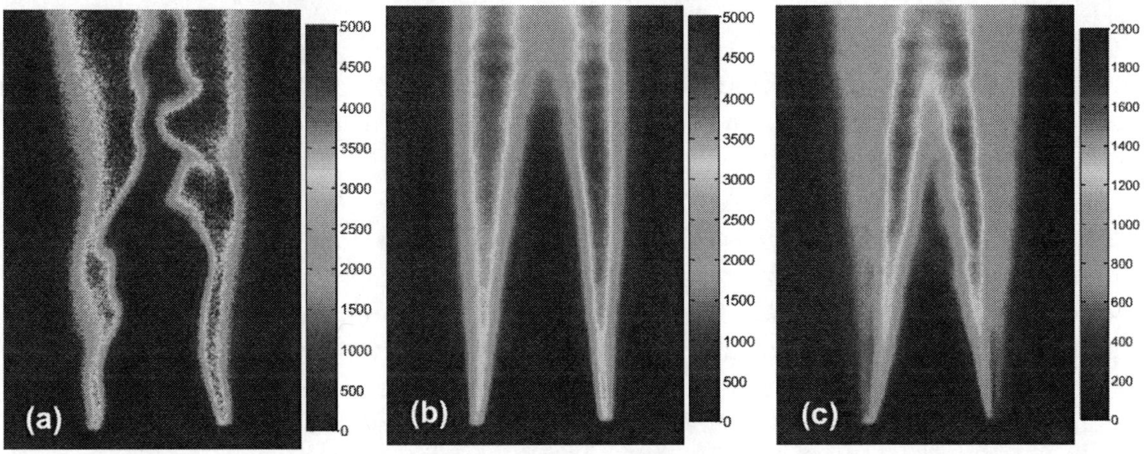

Fig. 4. CH_2O PLIF measurements for DME turbulent cool flames at U = 1.50 m/s. Instantaneous values are shown in (a), mean values in (b), and standard deviations in (c).

When the main flow velocity is increased to U = 2.75 m/s, pockets of unburned reactants and flame islands become much more common (Fig. 5a). The strength of the average CH_2O signal (Fig. 5b) decreases noticeably compared to the U = 1.50 m/s case, especially in the regions closest to the burner exit. Fig. 5c also reveals that the standard deviation of the CH_2O signal generally decreases on the fuel side of the flame. However, fluctuations are seen to substantially increase on the oxidizer side, particularly for the regions furthest from the burner exit. Thus, at increased levels of turbulence, two distinct maxima in the fluctuating values occur, which resembles measurements of DME turbulent hot diffusion flames [23].

5. Flame Structure Measurements though Rayleigh Scattering

Figure 6 shows the DME turbulent cool diffusion flame temperature distribution at x/R = 2.2 for a main flow velocity of U = 1.50 m/s and a DME mole fraction of X_{DME} = 0.2 (identical conditions to Fig. 4). As alluded to previously, the cool flame temperatures are determined by comparing the raw Rayleigh scattering signal at T = 600 K (cool flame) to the raw signals from a T = 500 K unburning case and a T = 550 K unburning case. Besides the exit temperatures, all three cases have identical boundary conditions (X_{DME}, U_{main}, U_{pilot}, etc.). The temperature of the central N_2/DME stream at an axial location of x/R = 2.2 is determined to be $T \approx$ 595 K, which is quite close to the value of T = 600 K measured by thermocouple at x/R = 0. At x/R = 2.2, the maximum mean temperatures occur near $r/R \approx \pm 0.9$. The maximum temperature is slightly asymmetric, peaking at $T \approx$ 725 K on left side and $T \approx$ 700 K on right side.

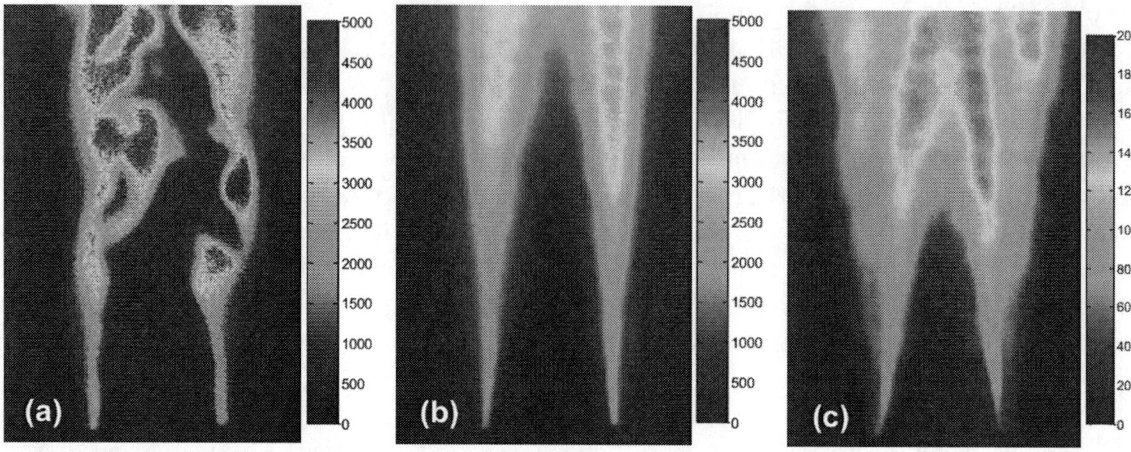

Fig. 5. CH_2O PLIF measurements for DME turbulent cool flames at U = 2.75 m/s. Instantaneous values are shown in (a), mean values in (b), and standard deviations in (c).

Like the CH_2O PLIF measurements at the same condition (Fig. 4), the maximum standard deviation in the temperature ($T' \approx 90$ K) occurs closer to fuel side ($r/R \approx \pm 0.8$) than the maximum mean temperature. Interestingly, the maximum temperature fluctuations in Fig. 6 are very close to the values obtained from thermocouple measurements in an *n*-heptane turbulent cool premixed flame [9]. In this study, the minimum temperature fluctuations ($T' \approx 20$ K) occur in the unburning regions away from the flame. Together, these measurements give $T'/T \approx 0.03$ in the unburning regions and up to $T'/T \approx 0.13$ in the cool flame itself.

6. Conclusion

A new Co-flow Axisymmetric Reactor-Assisted Turbulent (CARAT) burner has been utilized to study turbulent cool diffusion flames. Measurements with a nanoscale thermal anemometry probe (NSTAP) reveal that the burner is capable of producing turbulence intensities from $0.16 < u'/U < 0.20$ over the range of conditions studied. By heating the co-flow (air), pilot (oxygen), and main (nitrogen/dimethyl ether) streams to $T = 600$ K, turbulent cool diffusion flames are able to be established over a range of conditions. Formaldehyde planar laser-induced fluorescence measurements indicate that weak turbulence causes wrinkling to occur primarily on the fuel side of the cool flame. However, increased turbulence is seen to produce significant fluctuations on both sides of the flame. Measurements of the cool flame temperature, obtained through Rayleigh scattering, confirm these structural observations for weak turbulence. The experimental methodology described in this study provides a well-defined platform for the study of turbulence-chemistry interactions at low temperatures.

7. Acknowledgements

The authors recognize the support of the NETL UTSR Program under DOE Award No. DE-FE0011822. CBR and MKF are supported by the NDSEG fellowship program of the U.S. Department of Defense. ORY is

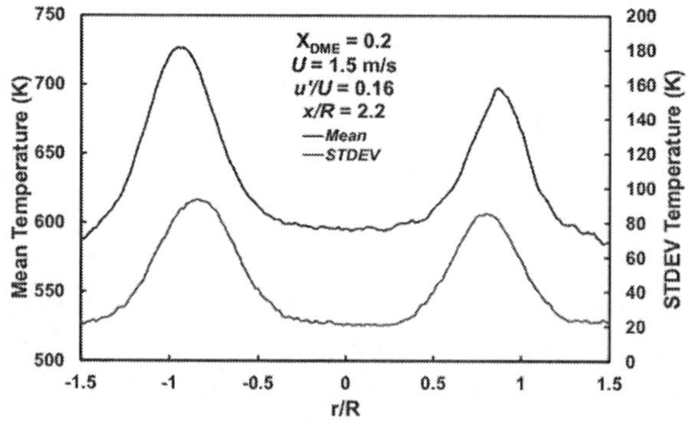

Fig. 6. Measured mean and standard deviation of temperature for a U = 1.50 m/s DME turbulent cool flame at x/R = 2.2.

grateful for support from the Daniel and Florence Guggenheim Foundation Fellowship at Princeton University.

8. References

[1] V. Nayagam, D.L. Dietrich, P.V. Ferkul, M.C. Hicks, F.A. Williams, Can cool flames support quasi-steady alkane droplet burning?, Combust. Flame 159 (2012) 3583-3588.
[2] T.I. Farouk, M.C. Hicks, F.L. Dryer, Multistage oscillatory "Cool Flame" behavior for isolated alkane droplet combustion in elevated pressure microgravity condition, Proc. Combust. Inst. 35 (2015) 1701-1708.
[3] S.H. Won, B. Jiang, P. Diévart, C.H. Sohn, Y. Ju, Self-sustaining n-heptane cool diffusion flames activated by ozone, Proc. Combust. Inst. 35 (2015) 881-888.
[4] C.B. Reuter, S.H. Won, Y. Ju, Experimental study of the dynamics and structure of self-sustaining premixed cool flames using a counterflow burner, Combust. Flame 166 (2016) 125-132.
[5] S. Deng, D. Han, C.K. Law, Ignition and extinction of strained nonpremixed cool flames at elevated pressures, Combust. Flame 176 (2017) 143-150.
[6] S. Kikui, T. Kamada, H. Nakamura, T. Tezuka, S. Hasegawa, K. Maruta, Characteristics of n-butane weak flames at elevated pressures in a micro flow reactor with a controlled temperature profile, Proc. Combust. Inst. 35 (2015) 3405-3412.
[7] M. Hajilou, T. Ombrello, S.H. Won, E. Belmont, Experimental and numerical characterization of freely propagating ozone-activated dimethyl ether cool flames, Combust. Flame 176 (2017) 326-333.
[8] C.B. Reuter, M. Lee, S.H. Won, Y. Ju, Study of the low-temperature reactivity of large n-alkanes through cool diffusion flame extinction, Combust. Flame 179 (2017) 23-32.
[9] I. Gökalp, G.M.L. Dumas, R.I. Ben-Aïm, Temperature field measurements in a premixed turbulent cool flame, Symp. (Int.) Combust. 18 (1981) 969-976.
[10] D.I. Kolaitis, M.A. Founti, A tabulated chemistry approach for numerical modeling of diesel spray evaporation in a "stabilized cool flame" environment, Combust. Flame 145 (2006) 259-271.
[11] S.H. Won, B. Windom, B. Jiang, Y. Ju, The role of low temperature fuel chemistry on turbulent flame propagation, Combust. Flame 161 (2014) 475-483.
[12] B. Windom, S.H. Won, C.B. Reuter, B. Jiang, Y. Ju, S. Hammack, T. Ombrello, C. Carter, Study of ignition chemistry on turbulent premixed flames of n-heptane/air by using a reactor assisted turbulent slot burner, Combust. Flame 169 (2016) 19-29.
[13] A. Krisman, E.R. Hawkes, M. Talei, A. Bhagatwala, J.H. Chen, A direct numerical simulation of cool-flame affected autoignition in diesel engine-relevant conditions, Proc. Combust. Inst. 36 (2017) 3567-3575.
[14] T. Echekki, S.F. Ahmed, Turbulence effects on the autoignition of DME in a turbulent co-flowing jet, Combust. Flame 178 (2017) 70-81.
[15] R.N. Dahms, G.A. Paczko, S.A. Skeen, L.M. Pickett, Understanding the ignition mechanism of high-pressure spray flames, Proc. Combust. Inst. 36 (2017) 2615-2623.
[16] A. Krisman, E.R. Hawkes, M. Talei, A. Bhagatwala, J.H. Chen, Characterisation of two-stage ignition in diesel engine-relevant thermochemical conditions using direct numerical simulation, Combust. Flame 172 (2016) 326-341.
[17] Y. Minamoto, J.H. Chen, DNS of a turbulent lifted DME jet flame, Combustion and Flame 169 (2016) 38-50.
[18] F. Fuest, R.S. Barlow, J.-Y. Chen, A. Dreizler, Raman/Rayleigh scattering and CO-LIF measurements in laminar and turbulent jet flames of dimethyl ether, Combust. Flame 159 (2012) 2533-2562.
[19] R. Cabra, T. Myhrvold, J.Y. Chen, R.W. Dibble, A.N. Karpetis, R.S. Barlow, Simultaneous laser raman-rayleigh-lif measurements and numerical modeling results of a lifted turbulent H2/N2 jet flame in a vitiated coflow, Proc. Combust. Inst. 29 (2002) 1881-1888.
[20] S.C.C. Bailey, G.J. Kunkel, M. Hultmark, M. Vallikivi, J.P. Hill, K.A. Meyer, C. Tsay, C.B. Arnold, A.J. Smits, Turbulence measurements using a nanoscale thermal anemometry probe, J. Fluid Mech. 663 (2010) 160-179.
[21] A.E. Lutz, R.J. Kee, J.F. Grcar, F.M. Rupley, OPPDIF: A Fortran program for computing opposed-flow diffusion flames, Sandia National Laboratories Report SAND96-8243, 1997.
[22] X. Yang, X. Shen, J. Santner, H. Zhao, Y. Ju, Princeton HP-Mech. http://engine.princeton.edu/mechanism/HP-Mech.html, 2017.
[23] K.N. Gabet, H. Shen, R.A. Patton, F. Fuest, J.A. Sutton, A comparison of turbulent dimethyl ether and methane non-premixed flame structure, Proc. Combust. Inst. 34 (2013) 1447-1454.

Spring Technical Meeting
Eastern States Section of the Combustion Institute
March 4-7, 2018
State College, Pennsylvania

A Study of JP-10 Pyrolysis by Molecular Beam Mass Spectrometry with Comparison to Literature Model

*Gregory P. Simms and Harsha K. Chelliah**

Department of Mechanical and Aerospace Engineering, University of Virginia, Charlottesville, Virginia 22904, USA
Corresponding Author Email: Harsha@virginia.edu

Abstract: A 4 mm inner diameter quartz laminar flow reactor was used to investigate the pyrolytic cracking of exo-tetrahydrodicyclopentadiene ($C_{10}H_{16}$, JP-10) over an intermediate temperature range of 950-1200 K, at a residence time of approximately 120 ms. An argon carrier gas was preheated to 500 K and JP-10 was introduced by liquid syringe pump and ultrasonic atomizing nozzle, and the diluted fuel was subsequently heated to the target temperature in the microflow tube reactor. The gaseous pyrolysis products were sampled directly from the reactor exit via a supersonic molecular beam. The collisionless beam was then ionized by electron impact, and separated by mass in a quadrupole mass spectrometer. The methods of calibration and data reduction in electron impact molecular beam mass spectrometry are discussed. The major decomposition products were found to be C_2-C_4 alkenes, molecular hydrogen, methane, aromatics, and C_5 ringed species. Experimental results are compared to with numerical simulations using a recent JP-10 model created by reaction mechanism generation at MIT.
Keywords: Molecular Beam Mass Spectrometry, JP-10, Pyrolysis, Microflow Tube Reactor

1. Introduction

An attractive candidate for hypersonic aviation fuel is JP-10, which is almost entirely composed of *exo*-tetrahydrodicyclopentadiene ($C_{10}H_{16}$) [1]. As a polycyclic hydrocarbon, JP-10 has desirable properties when compared to *n*-alkanes, which are the principal component of kerosene-based fuels. Among those advantages are higher energy density, larger heat sink capacity, and lower viscosity [2]. The endothermic potential of fuels has long been understood to be a key to sustained hypersonic flight [3]. In advanced thermal management systems, the fuels recover energy lost from combustion through endothermic cracking. The small pyrolysis products (predominantly hydrogen and alkenes) are known to help additionally in the eventual combustion of the fuel by reducing ignition delay, as described by Colket et al. [4]. A fuel's viscosity is relevant to its ability to be atomized during injection into a combustor, and while JP-10 is superior to kerosene-based fuels in this regard, the products of thermal cracking of JP-10 may further enhance the atomization of the injected fuel [5]. An additional particular interest of JP-10 pyrolysis is its tendency to form five-membered carbon rings, which have been identified as key contributors to polycyclic aromatic hydrocarbon growth and subsequent coke formation in cooling channels [6]. Finally, from an experimental and modeling perspective, JP-10 is a fortunate choice because it is essentially a single-component fuel.

Due to the intense interest in the combustion community on this particular fuel, there exist several experimental and modeling studies on JP-10. A very recent work by Zhao et al. [7] provides a comprehensive review of the relevant works. Gao et al. [8] have employed reaction mechanism generation (RMG) to develop the only detailed JP-10 chemical kinetic model available in the literature. The model contains over 600 species and 15,000 reactions and was validated using shock tube studies from the combustion group at Ghent University [8]. The shock tube employed for these experiments carries very significant uncertainties due to the small tube diameter and boundary layer effects, so further tuning the model requires accurate data with well-characterized uncertainties.

In this study, analysis of the products of JP-10 pyrolysis was performed by Molecular Beam Mass Spectrometry (MBMS). The molecular beam is formed through a sampling cone whose tip is placed directly at the reactor exit. An orifice in the cone separates the reactor from a first stage chamber which is pumped to a vacuum of approximately 0.1 torr. A second stage cone and a third stage orifice expand the gas over two stages to approximately 10^{-4} and 10^{-7} torr, respectively. The successive expansions accelerate the gas to supersonic speeds and the orifices collimate the gas into what is well-approximated as a collionless molecular beam. The beam is then ionized by electron impact and separated by mass in a quadrupole mass spectrometer. The resultant mass spectrum is then reduced into mole fraction data as detailed in section 2.2, and the uncertainties, while not explored in detail here, are generally around 10%.

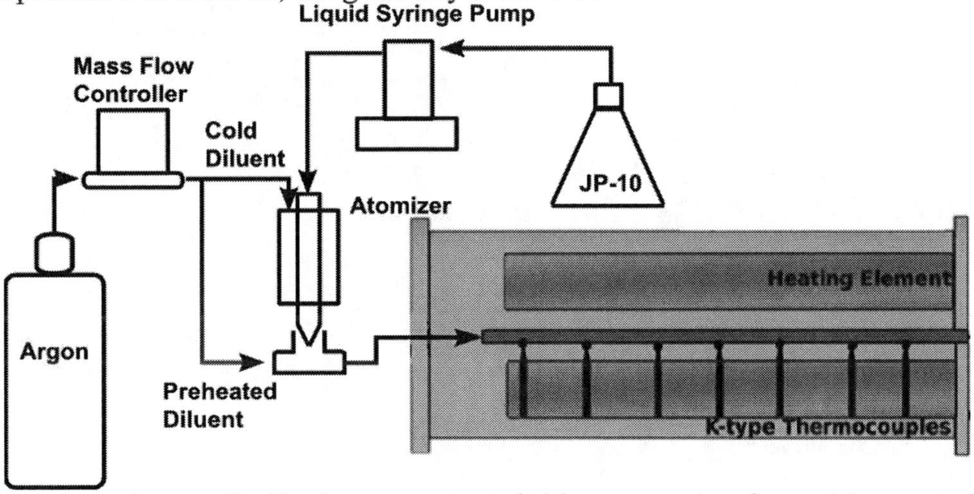

Figure 1 – The experimental setup: a liquid syringe pump extracts fuel from a reservoir and pumps it into an atomizer. The main argon flow is preheated above 500 K to ensure complete vaporization of the fuel, and a smaller argon flow serves to cool the atomizer. The fuel/diluent mixture is then carried to the reactor and heated to a controlled temperature.

2. Methods / Experimental

2.1 Tube Reactor and Experimental Condictions: Flow reactor experiments performed at UVA on ethane, *n*-butane, and *n*-dodecane are described in detail in Shrestha et al. [9]. For the JP-10 experiments detailed here, a simplified version of the UVA microflow tube reactor was used. The reactor used is a circular quartz tube with a 4 mm inner diameter. Five high-temperature modular ceramic heaters were used in conjunction with a PID system to regulate temperature along the 45.7 cm major axis of the reactor. A carrier gas of Argon was supplied by mass flow controllers (Sierra 100 series, 1 % full scale uncertainty) and preheated to 500 K. JP-10 was introduced via liquid syringe pump and an ultrasonic atomizer to ensure complete vaporization of the liquid in the carrier. The fuel/Argon mixture was heated in the reactor to 950 – 1200 K for analysis. The mass flow was kept constant for all experiments at 1.0 SLPM (at 300K and 1 atm) Argon and 0.015 mL/min JP-10, which corresponds to 0.25% molar fraction JP-10. A diagram of the experimental setup is shown in Fig. 1. Given the mass flow rates, the approximate residence time in the high temperature region for each experiment is 120 ms.

2.2 Molecular Beam Mass Spectrometer: The makeup of the gasses exiting the flow reactor was analyzed by MBMS (Extrel VeraSpec MBS, with MAX quadrupole mass spectrometer, integrated quadrupole control, and Merlin command system). A schematic of the sampling system is shown in Fig. 2a, which displays the three differentially pumped stages. The first stage is pumped by two Pfeiffer Penta Series pumps, and the second stage features a Pfeiffer turbopump, backed up by the second Pfeiffer Penta. The third stage, which maintains pressures below 10^{-7} torr, has a turbo pump backed by a Duo Series Pfieffer pump. A cross-sectional view of the first two stages of sampling cones is displayed in Fig. 2b, where neutral molecules travel left to right towards the third stage, where they are ionized and then separated by mass. A key difficulty in reducing the mass spectrum to mole fraction data is the fact that the ionizing

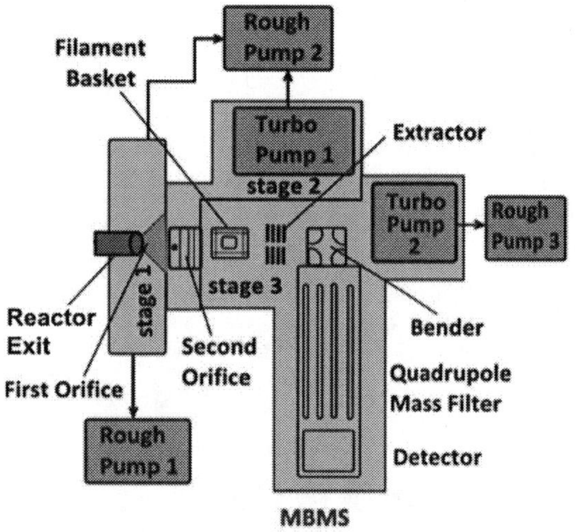

Figure 2a – MBMS schematic shows the three differentially pumped stages.

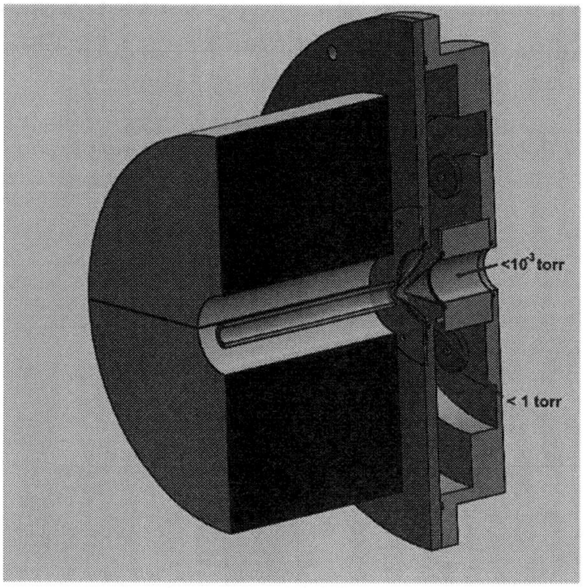

Figure 1b - Tube Reactor/MBMS Interface with detail of two consecutive skimmer cones.

electron stream destabilizes as well as ionizes the molecules in the neutral molecular beam. For example, ethylene ionized by electron impact does not simply create a peak at mass 28. The C-H bonds can be broken in the electron impact, and then ethylene fragments may become ionized, creating mass peaks at 27, 26, 25, and 14. This fragmentation pattern is a function of the molecule's ionization cross section, which depends on the energy of the bombarding electrons. The fragmentation pattern of ethylene is displayed in Fig. 3. In order to analyze the pyrolytic products, these fragmentation patterns needed to be determined for each quantified species. Another difficulty of electron impact ionization is that the electrons are generated thermionically by heating a tungsten filament, and thus have a Boltzmann-type distribution of energies. Therefore, it is difficult to separate species by ionization energy, as is done with photoionization

[10], and isomers are therefore difficult to separate. For the stable species quantified in this report, the fragmentation pattern was determined by direct calibration. Hydrogen, methane,

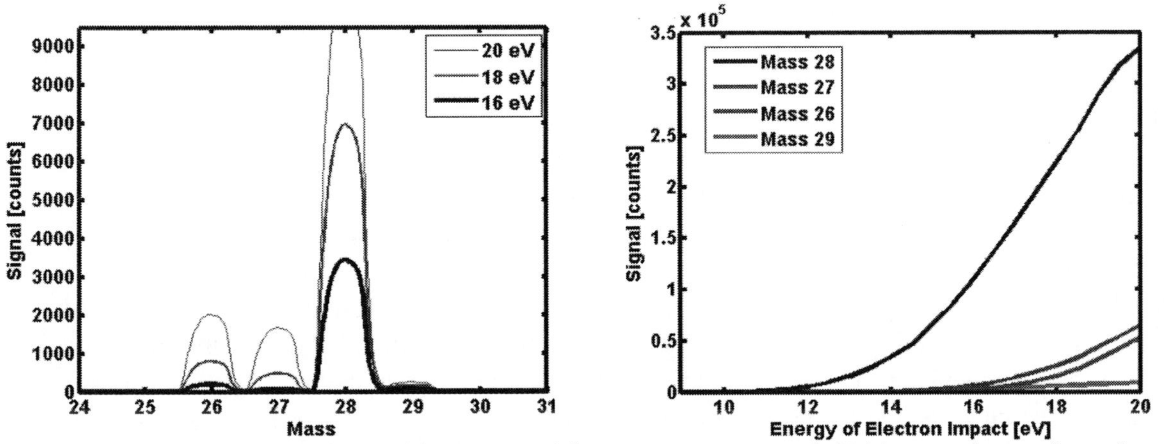

Figure 2 - Signal strength of ethylene and fragments at different ionizing energies (left). Mass signal strength as a function of electron energy (right).

acetylene, ethylene, and ethane were calibrated directly in-house using Sierra mass flow controllers. Allene/propyne (these isomers are indistinguishable in these experiments), propylene, 1,3-butadiene, butene, benzene and toluene were calibrated using commercially mixed gas standards from AirGas. Finally, some species were only available in liquid form and were calibrated by dissolving in methanol before being introduced through the atomizer via syringe pump. These include cyclopentene/pentadiene (these isomers are indistinguishable), cyclopentadiene, styrene and indene. Finally, the quadrupole mass filter can separate mass in the range of 1-120, but JP-10 has a mass of 136. However, JP-10 fragments easily and has been identified by its primary peak at mass 95. As with the other calibrated species, this fragmentation pattern was determined by introducing different JP-10/Argon mixtures into the reactor. To determine the composition of the molecular beam, a top-down approach was utilized. The calibration curve of indene, the largest molecule identified, was matched to the curve of the experimental result at each temperature. Then, the fragments which were due to indene were subtracted from the data file. Next, a similar procedure was carried out with Styrene on the remaining masses and its fragments were removed. This process was repeated for each product down to hydrogen. A demonstration of this fragment removal process is displayed in Fig. 4

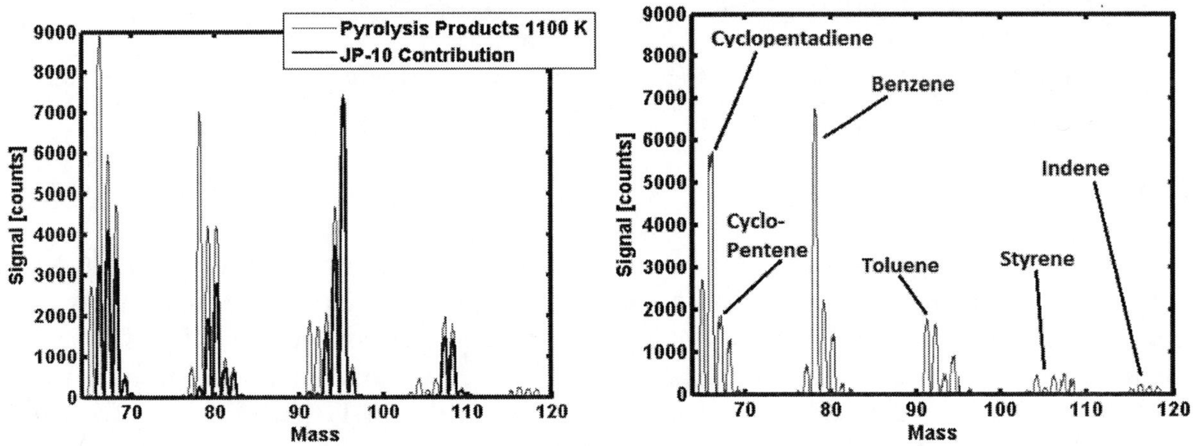

Figure 3 - The contribution of JP-10 and its fragments (dark line) is fit to the mass 95 peak of the (light line) overall pyrolysis products (left). The remaining spectrum after removing JP-10 and its fragments (right).

The method of determining the mole fractions of unstable/uncalibrated species is well described by Oβwald et al. [10]. But it should be noted that all of the species quantified in this report were directly calibrated using the process described above.

3. Results and Discussion

Selected speciation data from JP-10 pyrolysis are presented in Fig. 5 and are compared directly to the RMG model developed by the Green group at MIT [9]. The chemical-kinetic model was implemented along with a time-temperature profile using the open-source software package, Cantera [11]. The first significant (>1 ppm) products appear at 950 K, at which point the measured JP-10 mole fraction is within the measurement uncertainty of the original mixture. This result is in reasonable agreement with the model and also with the data produced in the ALS/NSL experiments [7]. It has been theorized that C_5-ringed species are key intermediates to soot formation [6], and the experiments here support that conclusion, as benzene growth begins at higher temperatures than both cyclopentene and cyclopentadiene, but at lower temperatures than acetylene, which is also considered a key soot/PAH precursor.

The preliminary results show that there is reasonable agreement between the model and the experimental data, and, in particular, there is excellent qualitative agreement on the trends of the intermediate species like butene, cyclopentene, cyclopentadiene, propene and butadiene. Possible identification of key radical species with MBMS remains a challenge.

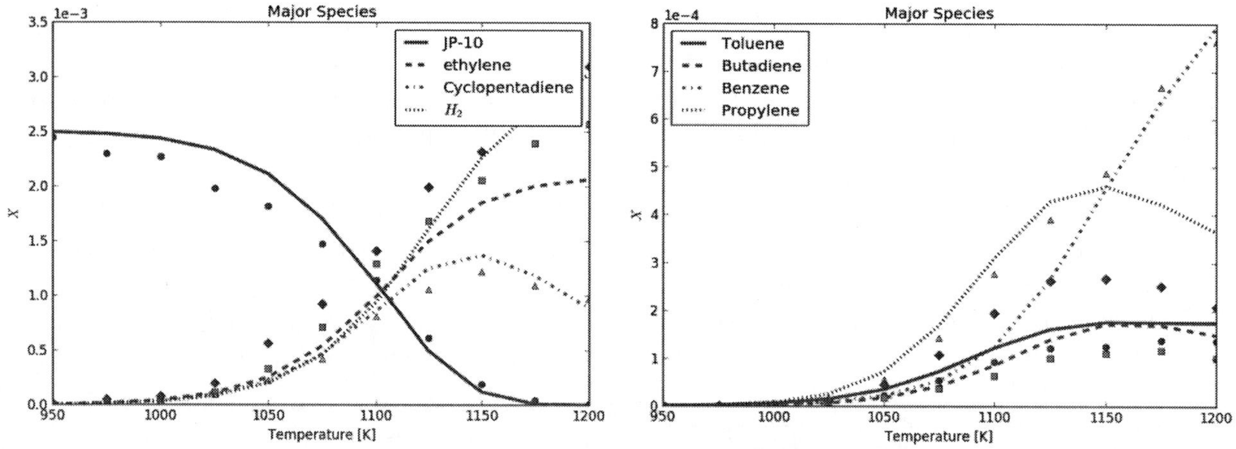

Figure 4 - Left: JP-10 (circles), ethylene(squares), hydrogen (diamonds) and cyclopentadiene (triangles) compared to model predictions. Right: Benzene (triangles), propylene (diamonds), toluene (circles) and butadiene (squares) compared to model predictions.

4. Summary

Liquid JP-10 was injected into a preheated Argon carrier stream and heated to 950-1200 K for approximately 120 ms. The product gasses were sampled at the reactor exit by a 3D-printed Inconel cone which led to successive differentially pumped stages and electron-impact MBMS. The data were reduced by a direct-calibration method into mole fractions and compared directly to the predictions of a model developed by the reaction kinetics group at MIT [9]. The model and data are in reasonable agreement, though the data do suggest there is room for improvement in some of the reaction rate constants.

A sensible next step for this data set would be to use the relative ionization cross section method, as described by Oβwald et al., to determine the contributions to the signal of other uncalibrated species [10]. Also, a direct comparison will be made to the results of the NSL/ALS experiments.

5. Acknowledgements

The work at UVa is supported by grants from Rolls-Royce and AFOSR, with Dr. MS Anand and Dr. Chiping Li as the Technical Monitors, respectively. Greg Simms gratefully acknowledges the support of the Rolls-Royce fellowship.

6. References

[1] T.J. Bruno, M.L. Huber, A. Laesecke, E.W. Lemmon, R. A. Perkins, Thermochemical and thermophysical properties of JP-10. Report No. NISTIR 6640:325, National Institute of Standards and Technology, Boulder, CO, USA, 2006.

[2] H.S. Chung, C.S.H. Chen, R.A. Kremer, J.R. Boulton, G.W. Burdette, Recent developments in high-energy density liquid hydrocarbon fuels, Energy & Fuels 12 (1999) 641-649.

[3] H. Lander, A.C. Nixon, Endothermic fuels for hypersonic vehicles, Journal of Aircraft 8 (1971) 200-207.

[4] M.B. Colket, L.J. Spadaccini, Scramjet fuels autoignition study, Journal of Propulsion and Power 17 (2001) 315-323.

[5] R. Jiang, G. Liu, X. Zhang, Thermal cracking of hydrocarbon aviation fuels in regenerative cooling microchannels, Energy & Fuels 27 (2013) 2563-2577.

[6] D.H. Kim, J.A. Mulholland, D. Wang, A. Violi, Pyrolytic hydrocarbon growth from cyclopentadiene, The Journal of Physical Chemistry A 114 (2010) 12311-12416.

[7] L. Zhao, T. Yang, R. Kaiser, T. Troy, B. Xu, M. Ahmed, J. Alarcon, D. Belisario-Lara, A.M. Mebel, Y. Zhang, et al., A vacuum ultraviolet photoionization study on high temperature decomposition of JP-10, Physical Chemistry Chemical Physics (2017), doi: 10.1039/C7CP01571B

[8] C.W. Gao, A.G. Vandeputte, N.W. Yee, W.H. Green, R.E. Bonomi, G.R. Magoon, H. Wong, O.O. Oluwole, D.K. Lewis, N.M. Vandewiele, et al., JP-10 combustion studied with shock tube experiments and modeled with automatic reaction mechanism generation, Combust. Flame 162 (2015) 3115-3129.

[9] U. Shrestha, M.J. Rahimi, G.P. Simms, H.K. Chelliah, Fuel pyrolysis in a microflow tube reactor–Measurement and modeling uncertainties of ethane, n-butane, and n-dodecane pyrolysis, Combust. Flame 177 (2017) 10-23.

[10] P. Oβwald, M. Kohler, An atmospheric pressure high-temperature laminar flow reactor for investigation of combustion and related gas phase reaction systems, Rev. Sci. Instr. 86 (2015) 10509.

[11] D.G. Goodwin, H.K. Moffat, R.L. Speth, Cantera: An object-oriented software toolkit for chemical kinetics, thermodynamics, and transport processes, http://www.cantera.org, 2017. Version 2.3.0. doi:10.5281/zenodo.170284

Spring Technical Meeting
Eastern States Section of the Combustion Institute
March 4-7, 2018
State College, Pennsylvania

Investigation of Ethylene Ozonolysis Reaction's Temporal Behavior and Products Using Plug Flow Reactor

Bin Wu[1], Xiang Gao[1], Wenting Sun[1]

[1]Aerospace Engineering, Georgia Institute of Technology, 270 Ferst Dr., Atlanta, GA 30332
**Corresponding Author Email: binwu@gatech.edu*

Abstract: The ozonolysis reaction of ethylene ($C_2H_4 + O_3$=products) has been studied, using a plug flow reactor (PFR) and a gas chromatography (GC). Some recognizably stable products are investigated quantitatively under 3 different stoichiometry by the GC, which are excessive C_2H_4 (χ_{C2H4}: $\chi_{O3} \approx$ 10:1), excessive O3 (χ_{C2H4}: $\chi_{O3} \approx$ 1:10) and stoichiometric (χ_{C2H4}: $\chi_{O3} \approx$ 1:1) conditions. As a result, concentrations of recognizable species are given individually for all 3 cases, which are concentrations of C_2H_4, O_2, HCHO, H_2, CO, CO_2, CH_3CHO, CH_3OH, HCOOH, and C_2H_4O. Comparisons have also been made and some differences been discovered and discussed. A specific system is designed in order to quench the O_3 when sampling, so that one can acquire data at different positions along the centerline of PFR. Nitric oxide (NO) is chosen as the O_3 scavenger because of its fast reaction with O_3.
Keywords: Ethylene, Ozonolysis, Gas Chromatography

1. Introduction

It has been shown by many that the addition of non-equilibrium plasma could help enhance and control combustion process. Unfortunately, high pressure could cause plasma thermal instability, non-uniform structure [1], and collisional thermalization of the discharge [2]. However, as one of the products from non-equilibrium plasma, ozone (O_3) is a long-lived species even at relatively high pressure. O_3 addition has also shown great potential in enhancing and controlling combustion process [3-12] by producing activate O atoms via the O_3 decomposition reaction. Recently, ozonolysis reactions (the rapid direct exothermic reactions between O_3 and un-saturated hydrocarbons) attracts attention as an alternative way to enhance and control combustion. Xiang et al. [13-15] observed great enhancement in flame propagation speed using a non-premixed lifted C_2H_4 jet flame with O_3 addition, which creates autoignitive conditions via ozonolysis reaction of C_2H_4. It is promising to apply ozonolysis to control the combustion at harsh conditions, for instance ultra-lean combustion in energy industry and extremely short residence time in scram jet.

However, the detailed chemical pathways of ozonolysis reactions are still not clear. As the simplest alkene in nature, the reaction between C_2H_4 and O_3 has been widely studied both theoretically and experimentally for a long time, since alkenes can be a sink of O_3 in the atmosphere and a significant source of hydroxyl radical (OH) along with other energetic free radicals, which could finally lead to air pollution in urban area [16-18]. The first step of ozonolysis of C_2H_4 would be the cleavage of carbon double bond and formation of 1, 2, 3-trioxolanes which also called primary ozonide (POZ), by the cycloaddition of O_3 onto alkenes double bond. In the case of C_2H_4 and O_3, the unstable POZ will rapidly decompose into a carbonyl compound which is formaldehyde (HCHO) and a carbonyl oxide which is called Criegee Intermediate (CI), this CI CH_2OO from C_2H_4 ozonolysis is the simplest CI that ever exists, and has been experimentally observed by Taatjes et al. [19]. In liquid phase C_2H_4 ozonolysis reaction, most of HCHO and CH_2OO will then recombine into a secondary ozonide (SOZ) 1, 2, 4-trioxolane as proposed by

Ciregee, however, almost none SOZ has been observed in gas phase reactions [20, 21], which is believed due to the low concentrations of HCHO and CH_2OO in gas phase compared with in liquid phase.

The high energy CH_2OO which is suggested to be a zwitterion by infrared spectrum [22] will then have two different further pathways. One pathway is unimolecular decomposition generating different pairs of products, which are CO_2+H_2, CO_2+2H, $CO+H_2O$, HCO+OH and HCOOH established on experiments [23-26]. The other pathway is stabilization by colliding with other bodies and producing CH_2OO radical. Those stabilized CH_2OO radicals can then further react with many other species or itself bimolecularlly, such as alkenes, organic acids, ketones, aldehydes, SO_X, NO_X and etc. Such bimolecular reactions mentioned above could produce many air pollutants like H_2SO_4 efficiently and cause severe air pollution. For this reason, the CH_2OO related bimolecular reactions have drawn great public attentions. However, experimental studies cannot really happen until recently in 2012, Welz et al. developed a reliable method to produce enough amount of CH_2OO by the reaction between CH_2I and O_2 [27]. The completely proposed mechanism of C_2H_4 ozonolysis is as followed in Fig. 1,

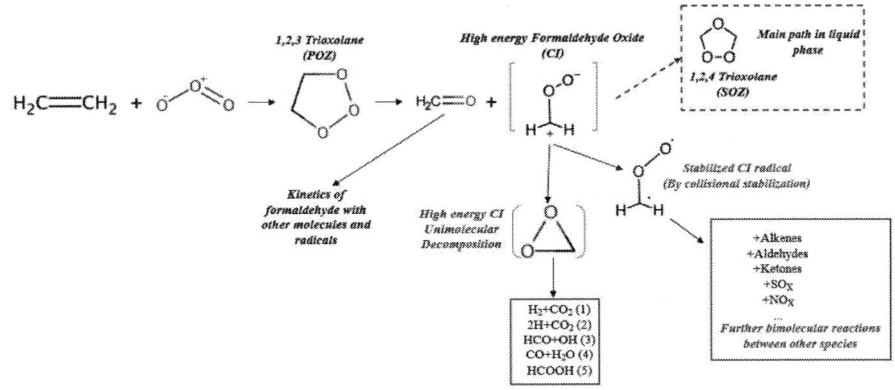

Figure 1. Detailed mechanism of C_2H_4 Ozonolysis Reaction

However, former experimental and theoretical studies have restricted themselves to relatively low temperature and pressure to simulate Earth's atmosphere, even though few are at room temperature and pressure, but still not align with combustion conditions. Furthermore, those studies set concentrations of C_2H_4 and O_3 in an extremely low level (only ppb no more than 100 ppm), nevertheless, concentration of O_3 can be easily above 1000ppm in ozone assisted combustion, not to mention the concentration of fuels. Therefore, side reactions will definitely play an important role in the complete mechanism. Those side reactions give uncertainty to the final products of C_2H_4 ozonolysis reaction, and a re-check of the branching ratios of reactions (1)-(5) is necessary, so that reliable parameters can be obtained and examined for accurate numerical simulations in the future.

2. Methods

In this work, a plug flow reactor (PFR) is used to study the ozonolysis of C_2H_4 at 300K and 1atm. Similar experimental set-up has been used as mentioned in [28]. A pre-cleaned 1.59mm OD stainless steel sampling probe is used to do sampling in situ from the test section of PFR. INFICON Micro GC Fusion® Gas Analyzer (precision up to 5 PPM) does the chromatographic analysis. For most gaseous species, GC is calibrated using commercial calibration standards; and liquid species (e.g. HCOOH, CH_3OH, etc.) calibrations are done by high purity liquid standards from Sigma-Aldrich and Fisher Scientific.

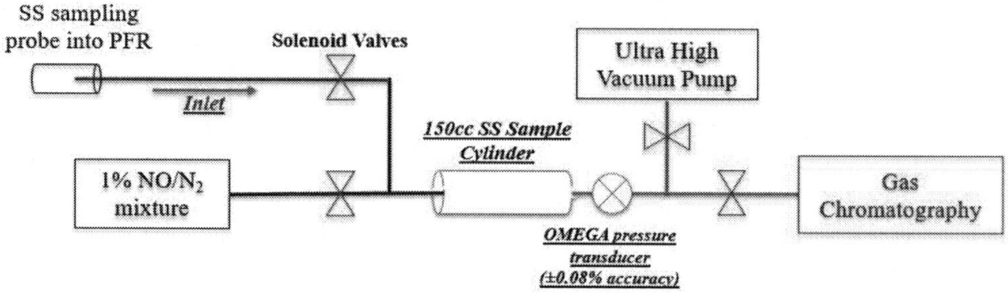

Figure 2. Schematic of NO Sampling System

Due to nature of alkene ozonolysis reactions, the reaction between C_2H_4 and O_3 would be extremely fast especially when concentrations are high, this would give great difficulty in experimental measurements. To further study the C_2H_4 ozonolysis reaction in combustion applications, it is essential to acquire data along the PFR at different axial positions, so that temporal behavior of the reaction can be studied. Nevertheless, due to the relatively long sampling time of GC, which is around 20s, no clear temporal data can be obtained for now since the reaction would almost be done within 20s. To do so, O_3 needs to be quenched rapidly as sampling along the test section in PFR. Nitric oxide (NO) can be used as a promising O_3 scavenger because of its fast reaction with O_3 [29] producing NO_2+O_2, even 100 times faster than C_2H_4 ozonolysis reaction. Besides, NO has low reactivity with all other stable products from ozonolysis reaction. A sampling system using NO as the scavenger has been designed and the schematic has been shown in Fig. 3 as above.

For operation, first using ultra high vacuum pump to draw vacuum in the whole system, with solenoid valve which connects to PFR closed. Next put a small amount of 1% NO/N_2 mixture into system, the partial pressure of this mixture can be obtained by high accuracy OMEGA pressure transducer. Since the total pressure in sampling system is still below 1 atm, then open the solenoid valve that connects to PFR, as a result ozonolysis products would be drew into sample cylinder driven by pressure difference. After the pressure equalization, again measure the total pressure by pressure transducer. Ideally, as long as the ozonolysis products mixed with NO/N_2 mixture, the unreacted O_3 will be scavenged by NO forming NO_2 and O_2, and then the mixture in the whole system would be approximately chemically frozen. By sampling at different axial positions in PFR with this sampling system, it's theoretically possible to acquire temporal data of C_2H_4 ozonolysis reaction.

In this paper, three different groups of concentrations are designated for $C_2H_4/O_2/O_3/N_2$, to investigate the behavior of products and figure out their possible production mechanism under different stoichiometry, which are (the excessive is in terms of comparison between C_2H_4 and O_3),

a) ***Excessive C_2H_4, with $C_2H_4/O_2/O_3/N_2$ = 22023/93590/1910/balance (ppm)***
b) ***Excessive O_3, with $C_2H_4/O_2/O_3/N_2$ = 873/139093/8658/balance (ppm)***
c) ***$C_2H_4:O_3 \approx 1:1$, with $C_2H_4/O_2/O_3/N_2$ = 1982/93590/1910/balance (ppm)***

3. Results and Discussions

Concentrations of all recognized species both reactants and products are given in Table 1, in units of ppm. The consumption of C_2H_4 for excessive O_3, excessive C_2H_4 and 1:1 cases are 786.7ppm, 1648.7ppm and 1127.3ppm, and as products, the concentrations of HCHO are 512.4ppm, 1883.3ppm and 1282.7ppm. Although some errors could exsist, still major inconsistancy can be seen for those 3 cases. As the carbonyl compound directly from ozonolysis reaction, the comsumption of C_2H_4 and production of HCHO should be almost equalized, if only the ozonolysis reaction is considered. For the lattar 2 cases this statement can stand, and the difference is +14.23%, +13.79% relative to C_2H_4 consumption for excessive C_2H_4 and 1:1 cases. But for excessive O_3 case, the difference is -34.87%. Direct reaction between HCHO and O_3 is indeed possible, but based on the study from [30], this reaction rate would be way too slow at 300K. However, with abundant O_3, many radicals and species with high reactivity such as HO_2 and H_2O_2 can be produced repidly, from the mechanism of O_3+H=OH+O_2 (dominant path compare with the lattar one) [31], O_3+H=HO_2+O [32, 33]and O_3+OH= HO_2+O_2 [29], then HO_2 can further react with O_3 by the mechanism of HO_2+O_3=OH+2O_2 [29]. Through this route the abundant O_3 can be a source of reactive radicals, and by some chances, they can be a sink of HCHO which lead to H_2O and CO_2 as observed by GC. It has also been observed by Neeb et al. [21] that the decrease in HCHO correspoding to increase in CO. And they proposed the mechanism as followed, HCHO+OH=HCO+H_2O, HCO+O_2=CO+HO_2 and HO_2+O_3=OH+2O_2, the excessive O_3 will drive this loop forward and scavenge HCHO. In the GC results, it is indeed that the excessive O_3 case has most CO production. Moreover from column C, we can find that for all 3 cases, HCHO and HCOOH are positively related, higher concentration HCHO corresponds more HCOOH, this gives no contradiction to the proposed CH_2OO+HCHO=HCOOH+HCHO mechanism by [21].

	Species/Elements	Excessive O_3	Excessive C_2H_4	1:1
Before ozonolysis	$(C_2H_4)_0$	786.7	20842.0	1812.3
	$(O_2)_0$	141575.9	96118.1	94954.3
	$(O_3)_0$ (from monitor)	9107.2	1981.8	1957.8
Stable products from C_2H_4 ozonolysis	C_2H_4	0.0	19193.3	685.0
	O_2	144792.0	96355.0	97532.0
	H_2	104.0	239.5	132.0
	CO	546.0	387.5	350.0
	CO_2	425.0	321.7	220.7
	H_2O	1711.0	563.0	817.0
	CH_3OH	0.8	15.1	1.4
	CH_3CHO	70.3	89.4	20.0
	HCHO	512.4	1883.3	1282.7
	HCOOH	15.5	165.0	78.3
	C_2H_4O	7.5	7.5	7.5

Table 1. Concentrations of recognized species from chromatographs

The production of H_2 for all 3 cases are 104ppm, 239.5ppm and 132ppm, in the order as mentioned above (excessive O_3, excessive C_2H_4 and 1:1). The ratios between comsumed C_2H_4 and produced H_2 ($H_{2,\,produced}/C_2H_{4,\,comsumed}$) are 13.22%, 14.53% and 11.71% which are similar. In the reacting system we studied, C_2H_4 ozonolysis dominates and side reactions regards C_2H_4 would be too weak at 300K and 1atm, besides no other source for H element except C_2H_4. Therefore, almost all H_2 produced is is from ozonolysis reaction, and none noticable H_2 comsumption by other redicals or species. Recombination of H radicals into H_2 is not likely since abundant O_3 and O_2 exsist in the chemical pool, especially the reaction O_3+H is even faster then ozonolysis reaction, and the pressure is too low for fast three-body recombination to happen.

Unlike H_2 and HCHO, both CO_2 and H_2O would have even more sources other than ozonolysis. It's possible that side reactions contribute some CO_2 and H_2O comparable to the main path, i.e.ozonolysis, since high-concentration reactants are deployed. However, with a relatively detailed kinetic mechanism and numerical algorithm, such as genetic algorithm, it is definitely promising to find out the branching ratios of CH_2OO decomposition.

4. Conclusion

Stable major products from ozonolysis of C_2H_4 are measured and evaluated with 3 different stoichiometric cases. And based on former work of [28], much more unknown peaks in chromatographs are recognized. Until now, detected stabe products are HCHO, H_2, CO, CO_2, CH_3CHO, CH_3OH, HCOOH and C_2H_4O. The differences in chromatographs have been noticed for 3 cases, most likely due to the side reactions regards abundant O_3 or C_2H_4. However, more GC measurements need to be repeated so that error analysis can be done, until then the dataset is not reliable enough to be used on parameter calculations, as our ultimate goal. Until now the NO sampling system has been built but not fully tested, so more experiments need to do to confirm the system is functional. Furthermore, as mentioned before, the reactions of stabilized "cold" CH_2OO between other species and radicals are still unclear and would beyond the ability of GC, so the stabilized CH_2OO also needs to be scavanged, in order that solely CH_2OO decomposition would happen in reacting system as we desired.

5. Acknolwdgement

This work is supported by the Air Force Office of Scientific Research (FA9550-16-1-0441), with Dr. Chiping Li as technical monitor.

6. References

[1] Sun, W., and Ju, Y., "Nonequilibrium plasma-assisted combustion: a review of recent progress," *J. Plasma Fusion Res* Vol. 89, No. 4, 2013, pp. 208-219.
[2] Sun, W., "Non-equilibrium plasma-assisted combustion." Princeton University, 2013.
[3] Nomaguchi, T., and Koda, S., "Spark ignition of methane and methanol in ozonized air," *Symposium (International) on Combustion*. Vol. 22, 1989, pp. 1677-1682.
[4] Tachibana, T., Hirata, K., Nishida, H., and Osada, H., "Effect of ozone on combustion of compression ignition engines," *Combustion and Flame* Vol. 85, No. 3, 1991, pp. 515-519.
[5] Nishida, H., and Tachibana, T., "Homogeneous charge compression ignition of natural gas/air mixture with ozone addition," *Journal of propulsion and power* Vol. 22, No. 1, 2006, pp. 151-157.
[6] Foucher, F., Higelin, P., Mounaïm-Rousselle, C., and Dagaut, P., "Influence of ozone on the combustion of n-heptane in a HCCI engine," *Proceedings of the Combustion Institute* Vol. 34, No. 2, 2013, pp. 3005-3012.
[7] Ombrello, T., Won, S. H., Ju, Y., and Williams, S., "Flame propagation enhancement by plasma excitation of oxygen. Part I: Effects of O3," *Combustion and Flame* Vol. 157, No. 10, 2010, pp. 1906-1915.
[8] Halter, F., Higelin, P., and Dagaut, P., "Experimental and detailed kinetic modeling study of the effect of ozone on the combustion of methane," *Energy & Fuels* Vol. 25, No. 7, 2011, pp. 2909-2916.
[9] Wang, Z. H., Yang, L., Li, B., Li, Z. S., Sun, Z. W., Aldén, M., Cen, K. F., and Konnov, A. A., "Investigation of combustion enhancement by ozone additive in CH4/air flames using direct laminar burning velocity measurements and kinetic simulations," *Combustion and Flame* Vol. 159, No. 1, 2012, pp. 120-129.
[10] Gao, X., Zhang, Y., Adusumilli, S., Seitzman, J., Sun, W., Ombrello, T., and Carter, C., "The effect of ozone addition on laminar flame speed," *Combustion and Flame* Vol. 162, No. 10, 2015, pp. 3914-3924.
[11] Vu, T. M., Won, S. H., Ombrello, T., and Cha, M. S., "Stability enhancement of ozone-assisted laminar premixed Bunsen flames in nitrogen co-flow," *Combustion and Flame* Vol. 161, No. 4, 2014, pp. 917–926.
[12] Zhang, Y., Zhu, M., Zhang, Z., Shang, R., and Zhang, D., "Ozone effect on the flammability limit and near-limit combustion of syngas/air flames with N 2, CO 2, and H 2 O dilutions," *Fuel* Vol. 186, 2016, pp. 414-421.
[13] Gao, X., Zhai, J., Sun, W., Ombrello, T., and Carter, C., "The Effect of Ozone Addition on Autoignition and Flame Stabilization," *54th AIAA Aerospace Sciences Meeting*. 2016, p. 0960.
[14] Gao, X., Sun, W., Ombrello, T., and Carter, C., "The Effect of Ozonolysis Activated Autoignition on Jet Flame Dynamics," *55th AIAA Aerospace Sciences Meeting*. 2017, p. 0960.
[15] Gao, X., Yang, S., Wu, B., and Sun, W., "The effects of ozonolysis activated autoignition on non-premixed jet flame dynamics: a numerical and experimental study," *53rd AIAA/SAE/ASEE Joint Propulsion Conference*. 2017, p. 4774.

[16] Docherty, K. S., and Ziemann, P. J., "Effects of stabilized Criegee intermediate and OH radical scavengers on aerosol formation from reactions of β-pinene with O 3," *Aerosol Science & Technology* Vol. 37, No. 11, 2003, pp. 877-891.
[17] Finlayson-Pitts, B. J., and Pitts, J. N., "Tropospheric air pollution: ozone, airborne toxics, polycyclic aromatic hydrocarbons, and particles," *Science* Vol. 276, No. 5315, 1997, pp. 1045-1051.
[18] Grosjean, E., and Grosjean, D., "Formation of ozone in urban air by photochemical oxidation of hydrocarbons: Captive air experiments in Porto Alegre, RS," *Journal of the Brazilian Chemical Society* Vol. 9, No. 2, 1998, pp. 131-143.
[19] Taatjes, C. A., Meloni, G., Selby, T. M., Trevitt, A. J., Osborn, D. L., Percival, C. J., and Shallcross, D. E., "Direct observation of the gas-phase Criegee intermediate (CH2OO)," *Journal of the American Chemical Society* Vol. 130, No. 36, 2008, pp. 11883-11885.
[20] Martinez, R. I., Huie, R. E., and Herron, J. T., "Mass spectrometric detection of dioxirane, H2COO, and its decomposition products, H2 and CO, from the reaction of ozone with ethylene," *Chemical Physics Letters* Vol. 51, No. 3, 1977, pp. 457-459.
[21] Neeb, P., Horie, O., and Moortgat, G. K., "The ethene– ozone reaction in the gas phase," *The Journal of Physical Chemistry A* Vol. 102, No. 34, 1998, pp. 6778-6785.
[22] Su, Y.-T., Huang, Y.-H., Witek, H. A., and Lee, Y.-P., "Infrared absorption spectrum of the simplest Criegee intermediate CH2OO," *Science* Vol. 340, No. 6129, 2013, pp. 174-176.
[23] Herron, J. T., and Huie, R. E., "Stopped-flow studies of the mechanisms of ozone-alkene reactions in the gas phase. Ethylene," *Journal of the American Chemical Society* Vol. 99, No. 16, 1977, pp. 5430-5435.
[24] Horie, O., and Moortgat, G., "Decomposition pathways of the excited Criegee intermediates in the ozonolysis of simple alkenes," *Atmospheric Environment. Part A. General Topics* Vol. 25, No. 9, 1991, pp. 1881-1896.
[25] Su, F., Calvert, J. G., and Shaw, J. H., "A FT IR spectroscopic study of the ozone-ethene reaction mechanism in oxygen-rich mixtures," *The Journal of Physical Chemistry* Vol. 84, No. 3, 1980, pp. 239-246.
[26] Thomas, W., Zabel, F., Becker, K., and Fink, E., "A mechanistic study on the ozonolysis of ethene," *Proceedings of the 1st European Symposium on Physico-Chemical Behaviour of Atmospheric Pollutants*. 1993, pp. 207-212.
[27] Welz, O., Savee, J. D., Osborn, D. L., Vasu, S. S., Percival, C. J., Shallcross, D. E., and Taatjes, C. A., "Direct kinetic measurements of Criegee intermediate (CH2OO) formed by reaction of CH2I with O2," *Science* Vol. 335, No. 6065, 2012, pp. 204-207.
[28] Wu, B., Gao, X., and Sun, W., "The Investigation of Ozonolysis Reactions of Ethylene at Combustion Environment Using a Flow Reactor," *53rd AIAA/SAE/ASEE Joint Propulsion Conference*. 2017, p. 4853.
[29] Baulch, D., Cox, R., Hampson Jr, R., Kerr, J. A., Troe, J., and Watson, R., "Evaluated kinetic and photochemical data for atmospheric chemistry," *Journal of Physical and Chemical Reference Data* Vol. 9, No. 2, 1980, pp. 295-472.
[30] Wang, F., Sun, H., Sun, J., Jia, X., Zhang, Y., Tang, Y., Pan, X., Su, Z., Hao, L., and Wang, R., "Mechanistic and kinetic study of CH2O+ O3 reaction," *The Journal of Physical Chemistry A* Vol. 114, No. 10, 2010, pp. 3516-3522.
[31] DeMore, W. B., Sander, S. P., Golden, D., Hampson, R. F., Kurylo, M. J., Howard, C. J., Ravishankara, A., Kolb, C., and Molina, M., "Chemical Kinetics and Photochemical Data for Use in Stratospheric Modeling. Evaluation No. 12," 1997.
[32] Howard, C. J., and Finlayson-Pitts, B. J., "Yields of HO2 in the reaction of hydrogen atoms with ozone," *The Journal of Chemical Physics* Vol. 72, No. 6, 1980, pp. 3842-3843.
[33] Finlayson-Pitts, B., and Kleindienst, T., "The reaction of hydrogen atoms with ozone: Evidence for a second reaction path producing HO2 and O (3 P)," *The Journal of Chemical Physics* Vol. 70, No. 10, 1979, pp. 4804-4806.

Spring Technical Meeting
Eastern States Section of the Combustion Institute
March 4-7, 2018
State College, Pennsylvania

Catalytic ignition and pressure dependence of methane/air combustion over palladium oxide

Ran Sui[1,2], Wenkai Liang[1], John Mantzaras[2,], Chung K. Law[1,*],*

[1]Department of Mechanical and Aerospace Engineering, Princeton University, Princeton NJ 08544, USA

[2]Laboratory of Thermal Processes and Combustion, Paul Scherrer Institute, Villigen PSI 5232, Switzerland

[*]Corresponding Authors Emails: ioannis.mantzaras@psi.ch, cklaw@princeton.edu

Abstract: The catalytic ignition and kinetics of CH_4/air oxidation over PdO were studied using wire microcalorimetry. Ignition temperatures and heat release rates were measured and global kinetic parameters were extracted. Key reactions rates were amended and a detailed surface reaction scheme for atmospheric pressure was proposed. In a subsequent step, the pressure dependence of CH_4/air catalytic oxidation over PdO was investigated in the range 1-12 bar in a channel-flow catalytic reactor. In situ Raman measurements of major gas phase species concentrations assessed the heterogeneous reactivity, while planar laser induced fluorescence (LIF) of the OH radical monitored homogeneous combustion. Simulations were performed using a 2-D code with detailed hetero-/homogeneous reaction mechanisms. Comparisons between Raman measured and numerically predicted profiles of major species allowed for the construction of a global catalytic step valid for 1-12 bar. The CH_4 catalytic reaction rate had an overall pressure dependence $\sim p^{1-n}$ where the exponent n was itself a monotonically increasing function of pressure. This resulted in non-monotonic pressure dependence, a behavior in stark contrast to other noble metals such at Pt and Rh.

***Keywords:* Catalytic ignition, Pressure dependence of catalytic reactivity, Methane total oxidation over PdO**

1. Introduction

The total oxidation of methane over palladium-based catalysts is of key importance for pollutant abatement in natural-gas-driven vehicles [1], for gas turbines employing hybrid hetero-/homogeneous combustion concepts [2] and for gas-to-liquid (GTL) off-shore compact conversion systems [3]. Palladium is the most active noble metal for the total oxidation of methane at temperatures up to 900 K [4]. The present work first reviews a study of catalytic ignition and kinetics of fuel-lean CH_4/air oxidation over PdO using a wire microcalorimetry setup. Subsequently the pressure dependence of CH_4 catalytic reactivity over PdO at 1-12 bar is investigated using in situ Raman measurements of major gas-phase species concentrations and planar laser induced fluorescence (LIF) of the OH radical.

2. Wire microcalorimetry

The catalytic ignition and kinetics of CH_4/air oxidation over PdO at atmospheric pressure were studied using wire microcalorimetry experiments [5] and numerical simulations [6] over the temperature range ~600-800 K. Ignition temperatures and heat release rates of different fuel-lean CH_4/air mixtures were measured and global kinetic parameters were extracted.

A schematic of the wire microcalorimetry is illustrated in Fig. 1. A Pd/PdO wire is positioned in a steady, heated, low-speed gas cross-flow. By passing an electric current through the wire, the wire is heated and the temperature is determined via a balance between the Joule heat generation and the heat losses. Thus, by passing a given electric current I and by measuring the wire electric resistance R, the power required to maintain a certain temperature T is calculated as $P = I^2R$, where T is determined from the wire resistance. A nonreactive mixture is firstly used to determine the power P_1 necessary to maintain the wire at a given temperature. Afterwards, a reactive mixture is created by adding small amounts of fuel into the stream and another power P_2 is determined. As the mixture compositions are very similar, the extent of heat loss is considered to be constant for both cases at the same temperature. Once catalytic combustion is ignited, less current is needed to maintain the same wire temperature T, leading to a reduced power input $P_2 < P_1$. The first observed temperature where $P_2 \neq P_1$ determines the catalytic ignition temperature (T_{ig}) and the difference in the power input $\Delta P = P_1 - P_2$ is the catalytic heat release at temperature T. By continuously increasing T, a relation $\Delta P(T)$ is deduced, from which further information on the global catalytic reaction rate can be extracted.

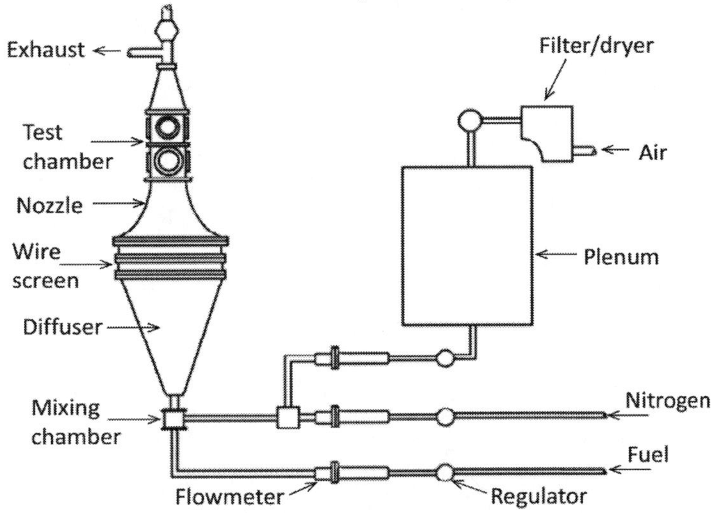

Figure 1: Schematic of the microcalorimetry flow system

Advanced surface chemical analyses characterized a 1-2 μm porous PdO layer on the Pd/PdO wire under the experimental conditions and effective catalyst surface areas were further determined [6, 7]. It was found that the overall catalytic oxidation rate was dominated by the oxygen adsorption and desorption, which determined the palladium surface site density and further the CH_4 adsorption. Reaction rates were amended and a detailed surface reaction scheme for atmospheric pressure was proposed [7].

3. High-pressure channel-flow reactor

The pressure dependence of CH_4/air catalytic oxidation over PdO was subsequently investigated at pressures 1-12 bar in the channel-flow catalytic reactor of Paul Scherrer Institute. The test-rig (see Fig. 2) includes a high-pressure, optically accessible catalytic channel-flow reactor, a laser setup for 1-D Raman measurements and a setup for planar LIF of OH.

The channel enclosure comprised two horizontal Si[SiC] ceramic plates and two 3-mm-thick vertical quartz windows, as shown in Fig. 2(a). The plates had a length of 300 mm, a width of 104 mm, a thickness of 9 mm, and were placed 7 mm apart. The plate inner surfaces were coated with a 1.5-µm-thick non-porous Al_2O_3 layer support and then with a 2.2-µm-thick Pd layer using Plasma Vapor Deposition (PVD). The reactor assembly was mounted inside a high-pressure cylindrical steel tank equipped with two 350×50×35 mm^3 quartz side windows. Two additional quartz windows at the rear tank flange and the reactor exhaust provided optical accesses for the LIF laser beam. The outer plate surfaces were heated by two adjustable-power resistive coils to control the catalyst surface temperatures. S-type thermocouples positioned along the *x-y* symmetry plane and embedded 0.9 mm beneath the catalyst monitored the surface temperatures. High purity CH_4 was injected counterflow to the O_2/N_2 mixture inside a special steel conical section (see Fig. 2(b)). Afterwards the flow mixture passed through a mesh and two grids, ensuring mixture homogeneity and flow uniformity.

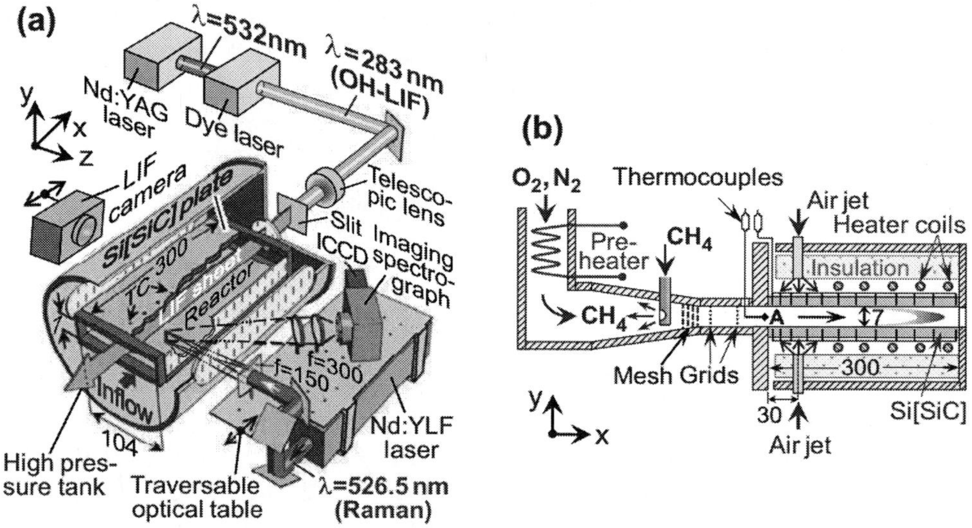

Figure 2: (a) High-pressure rig and Raman/LIF optics, (b) reactor details (dimensions in mm).

The laser setups have been detailed in [8] and are only briefly described here (see Fig. 2(a)). For Raman, a frequency-doubled pulsed Nd:YLF laser provided a radiation at 526.5 nm, with a pulse duration and energy of 120 ns and 40 mJ, respectively. The laser beam was focused into a ~0.3 mm thick vertical line that spanned the 7 mm channel height. The Raman scattered light was collected by a spectrograph equipped with an intensified CCD camera. The laser, the optics and the spectrograph were traversed axially, allowing measurements over $8 \leq x \leq 120$ mm. For OH-LIF, the 532 nm second harmonic of a pulsed Nd:YAG laser was frequency-doubled to generate the 283 nm excitation beam, which was converted into a vertical light sheet that propagated counterflow along the reactor central plane. An intensified CCD camera recorded the fluorescence at 90° through the side quartz windows.

Numerical simulations were performed with a Navier-Stokes 2-D steady CFD code (see [9]). A staggered grid with 480×68 points (in -*x* and -*y*, respectively) produced grid-independent solutions. Temperature profiles at the channel walls were constructed by fitting curves through the thermocouple measurements. The temperature, axial velocity and species mass fractions were uniform at the inlet. Detailed surface and gas-phase reaction mechanisms were taken from [10] and [11], respectively. Surface and gas-phase chemical reaction rates were evaluated using Chemkin [12, 13]. Mixture-average diffusion was used together with the Chemkin transport database [14].

Pressures 1-12 bar, equivalence ratios $\varphi = 0.27$-0.44 and wall temperature 700-950 K were investigated (see Table 1). The flows were laminar with inlet Reynolds numbers (based on the channel height) Re_{IN} = 650-1150. Before each test, the catalyst was oxidized in a 350 K air-flow for 30 min.

Table 1: Experimental conditions.

Case	p	φ	U_{IN}	T_{IN}	CH_4	O_2
1	1	0.43	3.73	439	5.4	25.0
2	1	0.44	4.26	439	5.4	24.8
3	3	0.35	1.61	438	4.9	27.8
4	3	0.40	1.21	430	5.0	25.2
5	5	0.27	0.92	440	5.5	40.7
6	5	0.40	0.72	427	6.0	30.0
7	8	0.29	0.51	444.	5.8	39.6
8	8	0.40	0.52	445	6.0	30.0
9	10	0.30	0.41	426	5.9	39.9
10	10	0.40	0.41	441	6.0	30.0
11	12	0.30	0.24	426	5.9	39.9
12	12	0.39	0.33	374	5.2	26.4

4. Pressure dependence of CH4/air oxidation over PdO

Comparisons between Raman-measured and predicted transverse profiles of CH_4 and H_2O mole fractions at 1 and 3 bar and at five axial positions are presented in Fig. 3. As attested by both OH-LIF measurements and simulations, homogeneous ignition was not achieved for all investigated cases. Hence, the CH_4 and H_2O transverse profiles were solely established by catalytic chemistry. The computed transport-limited (TL) transverse profiles for the last axial position are also included, providing the physically attainable lowest (highest) bounds for the CH_4 (H_2O) mole fractions. Kinetically controlled CH_4 conversion is evident in Fig. 3, which is a key requirement for assessing the catalytic reactivity. The wall temperatures for Cases 1 and 3 over the x-distances of Fig. 3 were 772-886 K and 778-896 K, respectively.

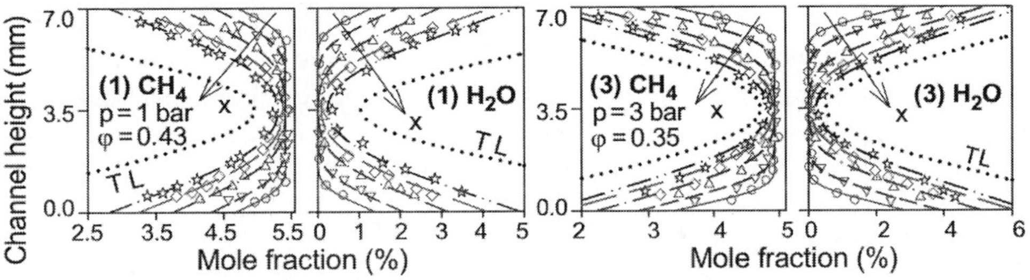

Figure 3: Raman-measured (symbols) and predicted (lines) transverse profiles of CH_4 and H_2O mole fractions for selected cases at 1 and 3 bar; x = 8 mm, 21 mm, 51 mm, 81 mm and 111 mm. Dotted lines marked TL are transport-limited simulations at x = 111 mm.

Simulations at 1 bar in Fig. 3 agreed very well with the measurements at all axial positions, reproducing the measured CH_4 and H_2O mole fractions and the near-wall bending of these species. At 3 bar, however, deviations between simulations and measurements became evident. The overprediction of the catalytic reactivity progressively aggravated with increasing pressure from 5 to 12 bar. This is evidenced in Fig. 4 where the predicted CH_4 (H_2O) mole fractions were substantially lower (higher) than the measurements. Characteristically, at p = 5, 8,

10 and 12 bar the predicted methane mole fractions in Fig. 4 were 27%, 36%, 50% and 56% lower than the corresponding measurements at the last x-position and topmost y-location.

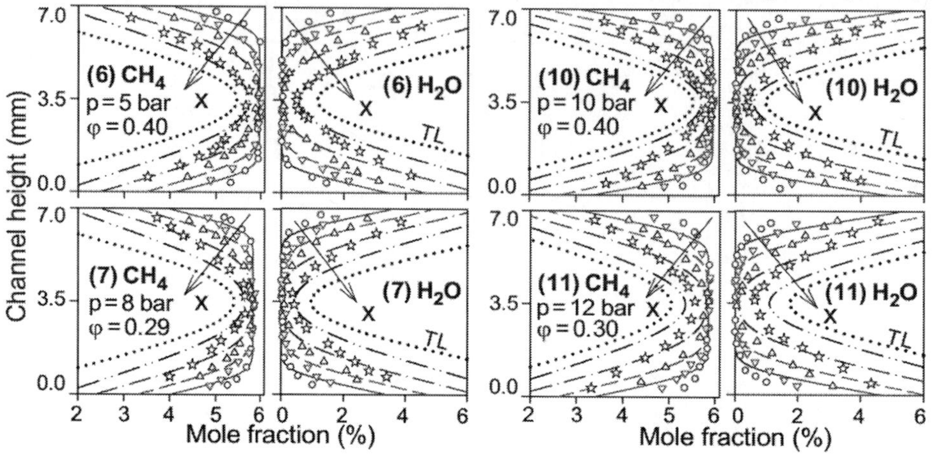

Figure 4: Raman-measured (symbols) and predicted (lines) transverse profiles of CH_4 and H_2O mole fractions for selected cases at 5-12 bar; x = 8 mm, 21 mm, 51 mm and 111 mm.

A one-step reaction for methane total oxidation, first-order with respect to CH_4 was shown in the past valid for platinum [9] and rhodium [15]:

$$\dot{s}_{CH4} = A\,(p/p_o)^{-n}\exp(-E_a/RT_w)[CH_4]_w, \tag{1}$$

For Pt $n = 0.53$ (valid over 1-16 bar, p_o= 1 bar) while for Rh $n = 0.70$ (2-12 bar, p_o= 2 bar).

The positive pressure exponent n decelerated the increase of the reaction rate with rising pressure, since $\dot{s}_{CH4} \sim p^{1-n}$ (given that $[CH_4]_w \sim p$ in Eq. (1)). The global step in Eq. (1) with fitted parameters E_a = 58 kJ/mol, $A = 1.7\times10^4$ cm/s and p_o = 1 bar reproduced excellently the predictions of the detailed mechanism [10] at 1 bar and also the corresponding Raman measurements. At elevated pressures the accelerated drop of the reaction rate with rising pressure necessitated a pressure-dependent exponent n. Simulations were in quite good agreement with the Raman measurements for the deficient methane reactant at all pressures, when $n = 0.58$ (3 bar), 0.78 (5 bar), 0.92 (8 bar), 0.98 (10 bar) and 1.02 (12 bar).

The reactivity drop with rising pressure was in qualitative agreement with two earlier studies on supported-Pd at $p \geq 5$ bar [16, 17]. The reduction in CH_4 conversion with rising pressure along with the constant mass flow rates in [16, 17] indicated a pressure-induced kinetic inhibition. This pressure inhibiting effect was likely linked to transformations in the catalyst structure as the partial pressure of O_2 changed and to interactions between the catalyst and the methane surface reactions.

5. Conclusions

The catalytic ignition and kinetics of CH_4/air over PdO were studied using wire microcalorimetry experiments and numerical simulations. Ignition temperatures and heat release rates were measured and global kinetic parameters were extracted, key reactions rates were amended, and finally a detailed surface reaction scheme for atmospheric pressure was proposed. Subsequently, the pressure dependence of CH_4/air catalytic oxidation over PdO was investigated at pressures 1-12 bar. A global catalytic step was constructed. The catalytic reactivity of CH_4 over PdO exhibited a pressure dependence ~p^{1-n} with a pressure-dependent exponent n. This resulted in a monotonic drop of the catalytic reaction rate with rising pressure for 3-12 bar, a behavior opposite to that observed in other noble metals [9, 15].

6. Acknowledgements

This research was funded by Swiss National Science Foundation EPM fellowship and EU project HRC-Power.

7. References

[1] M. Hussain, F.A. Deorsola, N. Russo, D. Fino, R. Pirone, Abatement of CH_4 emitted by CNG vehicles using Pd-SBA-15 and Pd-KIT-6 catalysts, Fuel 149 (2015) 2-7.
[2] R. Carroni, T. Griffin, J. Mantzaras, M. Reinke, High-pressure experiments and modeling of methane/air catalytic combustion for power generation applications, Catal. Today 83 (2003) 157-170.
[3] G. Arzamendi, P.M. Dieguez, M. Montes, J.A. Odriozola, E.F. Sousa-Aguiar, L.M. Gandia, Methane steam reforming in a microchannel reactor for GTL intensification: A computational fluid dynamics simulation study, Chem. Eng. J. 154 (2009) 168-173.
[4] M. Lyubovsky, L.L. Smith, M. Castaldi, H. Karim, B. Nentwick, S. Etemad, R. LaPierre, W.C. Pfefferle, Catalytic combustion over platinum group catalysts: fuel-lean versus fuel-rich operation, Catal. Today 83 (2003) 71-84.
[5] T.C. Zhang, D.L. Zhu, N. Yao, F. Qi, C.K. Law, A wire microcalorimetric study of catalytic ignition of methane-air mixtures over palladium oxide, Proc. Combust. Inst. 33 (2011) 1819-1825.
[6] T.C. Zhang, Y.X. Xin, Z.Y. Ren, F. Qi, C.K. Law, Catalytic oxidation of methane over PdO in wire microcalorimetry, Combust. Flame 160 (2013) 149-154.
[7] Y.X. Xin, S. Lieb, H. Wang, C.K. Law, Kinetics of catalytic oxidation of methane over palladium oxide by wire microcalorimetry, J. Phys. Chem. C 117 (2013) 19499-19507.
[8] R. Sui, E. Es-sebbar, J. Mantzaras, R. Bombach, Homogeneous ignition during fuel-rich $H_2/O_2/N_2$ combustion in platinum-coated channels at elevated pressures, Combust. Flame 180 (2017) 184-195.
[9] M. Reinke, J. Mantzaras, R. Schaeren, R. Bombach, A. Inauen, S. Schenker, High-pressure catalytic combustion of methane over platinum: in situ experiments and detailed numerical predictions, Combust. Flame 136 (2004) 217-240.
[10] T. Shimizu, H. Wang, Temperature-dependent gas-surface chemical kinetic model for methane ignition catalyzed by in situ generated palladium nanoparticles, Proc. Combust. Inst. 33 (2011) 1859-1866.
[11] J. Warnatz, R.W. Dibble, U. Maas, Combustion, Physical and Chemical Fundamentals, Modeling and Simulation, Springer-Verlag, New York, USA, 1996.
[12] M.E. Coltrin, R.J. Kee, F.M. Rupley, Surface Chemkin: A Fortran package for analyzing heterogeneous chemical kinetics at the solid surface-gas phase interface, Report No. SAND90-8003C, Sandia National Laboratories, USA, 1996.
[13] R.J. Kee, F.M. Rupley, J.A. Miller, Chemkin II: A Fortran chemical kinetics package for the analysis of gas-phase chemical kinetics, Report No. SAND89-8009B, Sandia National Laboratories, USA, 1996.
[14] R.J. Kee, G. Dixon-Lewis, J. Warnatz, M.E. Coltrin, J.A. Miller, A Fortran computer code package for the evaluation of gas-phase multicomponent transport properties, Report No. SAND86-8246, Sandia National Laboratories, USA, 1996.
[15] R. Sui, J. Mantzaras, R. Bombach, A. Denisov, Hetero-/homogeneous combustion of fuel-lean methane/oxygen/nitrogen mixtures over rhodium at pressures up to 12 bar, Proc. Combust. Inst. 36 (2017) 4321-4328.
[16] K. Persson, A. Ersson, A.M. Carrera, J. Jayasuriya, R. Fakhrai, T. Fransson, S. Jaras, Supported palladium-platinum catalyst for methane combustion at high pressure, Catal. Today 100 (2005) 479-483.
[17] J. Requies, M.C. Alvarez-Galvan, V.L. Barrio, P.L. Arias, J.F. Cambra, M.B. Guemez, A.M. Carrera, V.A.D. la Peha O'Shea, J.L.G. Fierro, Palladium-manganese catalysts supported on monolith systems for methane combustion, Appl. Catal., B 79 (2008) 122-131.

Sub Topic: IC and GT Engines

Spring Technical Meeting
Eastern States Section of the Combustion Institute
March 4-7, 2018
State College, Pennsylvania

Impact of Cyclo-Alkanes on Ignition Propensity Measured as Derived Cetane Number in Multi-Component Mixtures

Dalton Carpenter, Stuart Nates, Seung Jae Lim, Frederick L. Dryer, Sang Hee Won,
Department of Mechanical Engineering, University of South Carolina, Columbia SC, USA
Corresponding Author Email: doc@email.sc.edu,

Abstract: Cyclo-alkane is one of hydrocarbon molecular classes pertinent to real jet fuels. Compared to n- and iso-alkanes, the ring structure present in cyclo-alkanes exhibits distinct chemical kinetic behaviors. At high temperatures, thermal decomposition reactions dominate, producing n-alkyl radicals similar to the oxidation reactions of n-alkanes thus promoting the reactivity. Whereas in low temperatures, the presence of the ring structure essentially suppresses the formation of alkylhydroperoxy radicals (QOOH) from alkylperoxy radicals (RO_2), thus exhibiting similar reactivity to iso-alkanes. In case of petroleum-derived jet fuels, the fraction of cyclo-alkanes are typically less than 30 %, and their chemical functionalities have been successfully represented by an appropriate mixture of n- and iso-alkanes in surrogate formulation. However, in case of alternative jet fuels and high-performance fuels (e.g. JP-10), there is a desire to increase cyclo-alkane fraction further due to its unique advantages, such as the increase of heat contents and better endothermic performances. In this regard, it is of importance to evaluate the impacts of cyclo-alkanes on fuel chemical properties particularly on the multi-component mixtures. To do this, the ignition characteristics of cyclo-alkanes and their mixtures with other molecular classes are investigated by measuring DCN values from an Ignition Quality Tester (IQT). To quantify the role(s) of cyclo-alkane functionality, chemical functional group approach is used by separately defining the CH_2 groups in cylco- and n-alkanes and a quantitative structure-property relationship (QSPR) regression model is constructed based on chemical functional group descriptor against the DCN measurement database. A feature sensitivity analyses are performed to identify relative significance of cyclo-CH_2 functional group in autoignition propensity of multi-component fuels by varying the cyclo-alkane fractions.
Keywords: Cyclo-alkane, Derived Cetane Number, Ignition Quality Tester, Ignition Delay Time

1. Introduction

Petroleum-derived jet fuels contain hundreds of individual hydrocarbon molecules, thus exhibiting significant high-order complexity due to not only the number of molecules involved but also the variation in molecular classes (n-alkanes, iso-alkanes, cyclo-alkanes, alkenes, and aromatics) [1-17]. As a typical approach to convert this high-order complexity to reduced-order model, the surrogate approach has been extensively utilized. A surrogate mixture can be defined *either* by matching the distribution of molecular classes *or* by matching the key fuel properties through the combustion property reference indicators [1-8].

Extensive studies [1-17] have demonstrated the applicability of the surrogate approach. Our previous studies [1-8] have shown that the surrogate mixtures can be successfully formulated by matching four combustion property targets (CPTs); hydrogen to carbon molar ratio (H/C ratio), derived cetane number (DCN), average molecular weight (MW), and threshold sooting index (TSI). In this surrogate formulation, surrogate components were chosen to represent the typical molecular classes (n-alkane, iso-alkane, and alkylated aromatics), and their composition was determined by matching CPTs. The fidelity of surrogate mixture, to emulate the global combustion behaviors of a target petroleum-derived jet fuel, has been extensively evaluated through the comprehensive experimental campaign, including homogenous ignition delays in both high and low temperature conditions, laminar flame speed, extinction limits of both premixed and non-premixed flames, and speciation profiles from both single pulse shock tube and flow reactors. The applicability of the surrogate approach has been also verified with alternative jet fuel [4].

Fundamental aspects for the experimentally observed success in surrogate approach have been recently elucidated by employing the concept of chemical functional groups. The chemical functional group approach can be regarded as a low-dimensional descriptor, which can define the fuel reaction kinetic characteristics as they pertain to distinct combustion behaviors. A "chemical functional group" is the concept of regarding a molecule as group of atoms (molecular fragments) that yield distinctively different chemical (kinetic) behaviors [6]. Simplified chemical functional group descriptions of complex multi-component mixtures permits the construction of quantitative structure-property relationship (QSPR) regression models, relating the functional group composition to combustion behaviors [6, 18]. Recently, a QSPR regression model has been proposed for Derived Cetane Number (DCN) [6] with chemical functional group descriptor and utilized to investigate how the CPT-based surrogate approach can constrain the chemical kinetic characteristics in multi-component surrogate formulation. The chemical functional group descriptor used in the analysis are CH_2, $(CH_2)_n$, CH_3, CH, C, and benzyl-type functional group, representing the n-alkyl, iso-alkyl, and aromatic functionalities.

Considering that cyclo-alkane is one of major molecular classes (up to 30 % in liquid volume) in petroleum-derived jet fuels, it is noteworthy that the global combustion behaviors of petroleum-derived jet fuel can be well emulated with a surrogate mixture without including cyclo-alkane. Our previous study has attempted to elucidate the role(s) of cyclo-alkane functionality [1] and suggested that the specific contribution of cyclo-alkane functionality can be only found during low- to high-temperature transition regime in flow reactor experiments, whereas other global combustion behaviors (ignition delays and flame extinction) are not affected by inclusion or exclusion of cyclo-alkane in surrogate mixture. However, with the recent focus on high performance fuels, there is a desire to increase the cyclo-alkane fraction to capitalize on its unique characteristics such as the better endothermic and/or cooling performance. Therefore, it is of importance to extend the current surrogate approach and chemical functional group descriptor to appropriately evaluate the role(s) of cyclo-alkanes regarding global combustion behaviors of both petroleum-derived and alternative jet fuels. Unfortunately, the role of cyclo-alkanes in multi-component fuel mixtures has not been well understood and there is a lack of experimental database (e.g. DCN) to be used for QSPR regression analysis.

In this regard, the objectives of the present study are 1) to establish comprehensive DCN measurement database of cyclo-alkanes and their mixtures with other molecular classes and 2) to perform the QSPR regression analysis to evaluate the role(s) of cyclo-alkane functionality for fuel ignition propensity measured as DCN. To do this, the chemical functional group descriptor is extended by adding $(CH_2)_{cyclo}$ to reflect the cyclo-alkyl functionality.

2. Methods / Experimental

The experiment was conducted using Ignition Quality Tester (IQT) developed by Advanced Engine Technology Ltd. The IQT measures the derived cetane number (DCN) as per the ASTM D 6890 test methods [19]. It can test a wide range of diesel fuels and provide the ignition delay time by monitoring the pressure time-history after spray injection into a heated and pressurized constant-volume combustion chamber. The measured ignition delay times are later converted to the DCN values through the correlation. The IQT evaluates the DCN value of fuel sample through averaging 32 measurements after performing 8 pre-injections to achieve steady-state operation. The chamber temperature was set 576.4 degrees Celsius in order to calibrate the IQT to be with in the acceptable range for n-heptane's DCN of 53.8. The IQT used an ultra-high purity mixture of 79% nitrogen and 21% oxygen from Praxair with 99.993% purity. All of the cyclo-alkanes and n-alkanes were produced from Sigma-Aldrich with a purity rating over 99.9%. The Decalin used in this work is a mixture between cis- and trans-.

The IQT was then used to test mixtures of cyclo-alkanes and their mixtures with n-heptane and iso-octane to investigate the potential interactions. Mole fractions of around 0.2, 0.4, 0.6, and 0.8 were used to accomplish this. The mole fractions were determined by mass to the 4th decimal place. Once these mixtures had been created, they were run in the IQT. Each mixture had 8 pre-injections that were not used in the determination of the DCN, but were used as a way to ensure all temperature and pressures with the system were within allowed tolerance with the standards. After these pre-injections, the mixtures were run for another 32 injections that were recorded and used to determine the DCN of the mixtures. Both the pre-injections and injections are considered to be one single run. Each experiment was conducted in three separated runs.

Each of the cyclo-alkane and n-alkane mixtures were conducted on the same test day. These test day started off with two runs of n-heptane and/or a diesel check fuel to ensure the IQT was still within calibration. The same diesel check fuel was run in the middle of multiple runs to ensure calibration was constant and another n-heptane and/or diesel check fuel was run after the experiment to again ensure the IQT was within acceptable calibration range. Each run had its ignition delay, DCN, and standard deviation for each recorded. The DCN of the mixtures is what this paper's analysis was based from.

3. Results and Discussion

Table 1: DCN of selected cyclo-alkanes using ASTM D 6890

Fuel	DCN	Uncertainty	Note
Cyclohexane (C_6H_{12})	21.4	+/- 2.592	Present work, 20.0 in [20]
Cyclooctane (C_8H_{16})	23.8	+/- 1.61	Present work, 22.3 in [20]
Methylcyclohexane (C_7H_{14})	24.9	+/- .64	Present work, 24.4 in [20]
Decalin ($C_{10}H_{18}$)	N/A	-	From [20]
n-Butylcyclohexane ($C_{10}H_{20}$)	47.6	-	From [20]

Table 1 summarizes the DCN values of neat cyclo-alkanes. DCN values of cyclohexane and cyclooctane are measured in the present work, and those of other cyclo-alkanes are from [20]. The DCN values measured in this study for cyclohexane and cyclooctane are very close to the values reported in [20] within the measurement uncertainty, confirming the fidelity of the current measurements. Direct comparison between cyclohexane and cyclooctane indicates that the increase of carbon number in the cyclo-alkane corresponds to an increase in the DCN of the pure compound. The DCN value of n-butylcyclohexane is the highest among the cyclo-alkanes, indicating that the ignition propensity is considerably increased due to the n-alkyl chain, which can be represented by $(CH_2)_n$ group [6]. Whereas, the DCN value of methylcyclohexane is similar to those of cyclohexane and cyclooctane, indicating insignificant role of methyl (CH_3) group on ignition propensity.

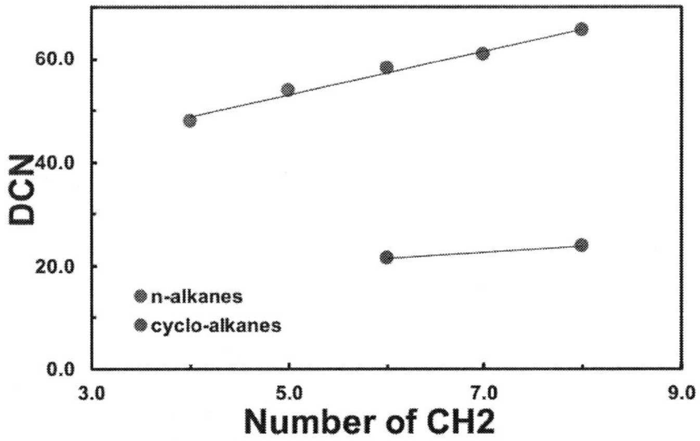

Figure 1: Comparison of DCN between $(CH_2)_n$ and $(CH_2)_{cyclo}$

The role of CH_2 groups in cyclic structures can be further evaluated by comparing the role(s) of CH_2 groups in n-alkanes. Figure 1 compares the changes of DCN value in a function of the number of CH_2 groups between n- and cyclo-alkanes. The DCN values of n-alkanes increase considerably as increasing the number of CH_2 group, whereas those of cyclo-alkanes exhibit very minor dependency on the number of cyclic CH_2 group. This different behavior can be explained by the distinct role(s) of CH_2 groups between n- and cyclo-alkanes during the low-temperature oxidation processes relevant to the conditions for DCN measurements. In case of n-alkanes, the possibility to undergo low-temperature chain branching reactions is directly coupled by the number of available CH_2 group, which governs the reactions of $R+O_2 = RO_2$ and consequent isomerization reactions to form QOOH. However, in case of cyclo-alkanes, the cyclic structure suppresses the formation of QOOH, thus inhibiting low-temperature chain branching reactions. Furthermore, the high activation energy for cyclic structure decomposition (ring opening) reactions also suppresses the ignition process.

To evaluate the role(s) of cyclic CH_2 group further in terms of its potential interaction with other chemical functional groups, mixtures of cyclohexane, cyclooctane, methylcyclohexane, decalin, and n-butylcyclohexane with n-heptane are taken using the IQT. An additional mixture of methylcyclohexane and iso-octane (2,2,4-trimethylpentane) are also taken to determine the interactions between cyclic CH_2 and CH_3 groups. Figure 2(a) shows the measured DCN values of the mixtures in a function of n-heptane mole fraction in the mixtures. Since the DCN values of all cyclo-alkanes are lower than that of n-heptane (53.8 by definition), the measured DCN values of the mixtures exhibit monotonic increase of DCN values as increasing the n-heptane fraction. To clearly evaluate the interactions between n- and iso-alkanes, the DCN values are normalized as below and the results are depicted in Fig. 2(b).

$$DCN_{normalized} = \frac{DCN_{mix} - DCN_{mix,min}}{DCN_{mix,max} - DCN_{mix,min}} \quad (1)$$

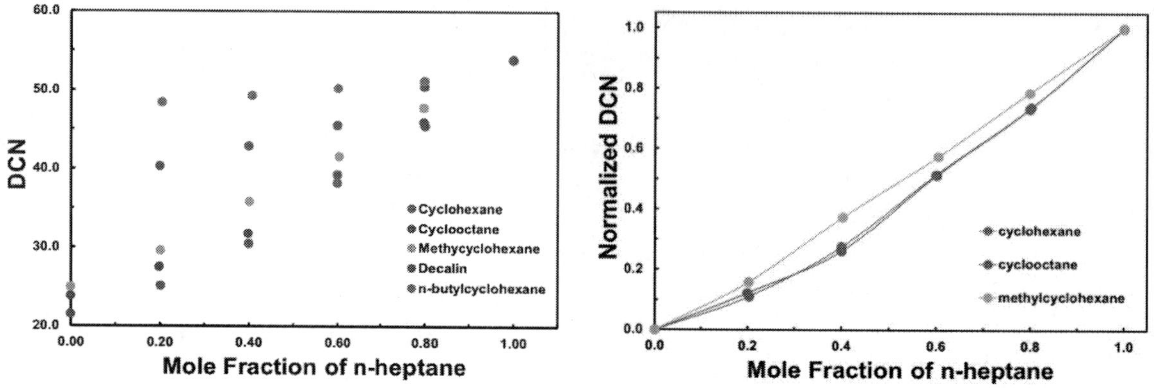

Figure 2. (a) Comparison of DCN values for cyclo-alkane and n-heptane mixtures, (b) Comparison of normalized DCN to mole fraction n-heptane

Here, we only compare the mixtures having the DCN value of neat cyclo-alkanes directly measured in the present work to avoid any potential bias due to the measurement uncertainty. The normalized DCN values of cyclohexane and cyclooctane exhibit indiscernible difference, indicating the identical contributions of cyclic CH_2 group in the mixtures. In case of methylcyclohexane and n-heptane mixtures, the normalized DCN values shows slightly higher values compared to those of cyclohexane and cyclooctane, indicating that the CH_2 groups in n-alkanes promote the ignition propensity as described above.

Previously, the role(s) of chemical functional groups have been investigated by constructing a quantitative structure-property relationship (QSPR) regression analysis [6]. The chemical functional groups used were CH_2, $(CH_2)_n$, CH_3, CH, C and benzyl-type groups, and the regression analysis pointed out the significance of $(CH_2)_n$ groups in DCN characteristics. Here, to account for the distinct behavior of cyclic CH_2 group, we assign it as $(CH_2)_{cyclo}$ separately. Using the results above, the QSPR regression analysis is performed to clearly identify the role(s) of $(CH_2)_{cyclo}$ groups on DCN characteristics.

Table 2: QSPR Regression coefficients based on functional groups.

Regression coefficients		
	ai	**bi**
CH2	7.455	-0.050
(CH2)n	10.406	0.771
CH3	15.411	0.025
CH	1.965	0.091
C	-4.326	-0.433
Benzyl	-10.181	1.570
(CH2)c	0.075	2.808

The equation below is used as a transfer function for QSPR regression analysis and Table 2 summarizes the regression coefficients to predict the DCN for all of the function groups.

$$DCN_{regression} = \sum_{1}^{i} a_i n_i^{x_i} \tag{2}$$

As shown in Fig. 3, in general, the QSPR regression model is able to capture most of the impacts that cyclo-alkanes have in mixtures with n-alkanes. As it has been seen with lower amounts of the cyclo-alkanes in mixtures shown in Fig. 2, the QSPR regression model is capable of predicting DCN reasonably well. However, it fails to predict the DCN values of neat cyclohexane and cyclooctane. These molecular arrangements still need to be investigated as to why the model is unable to predict them, and this will be conducted in further research. Nevertheless, the regression results show reasonable prediction capability of all fuel mixtures considered here through the chemical functional group descriptor.

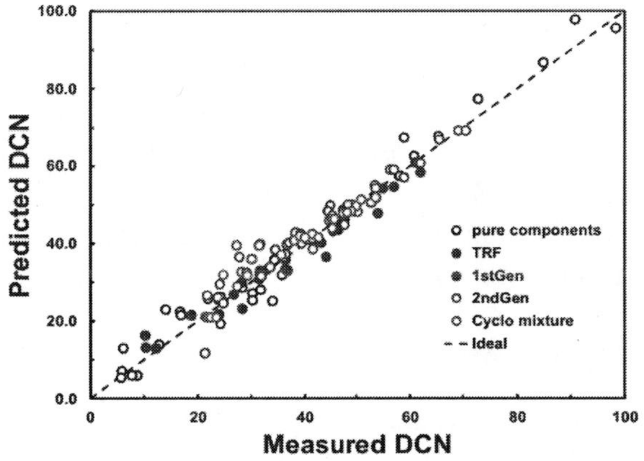

Figure 3. QPSR regression results

The significance of cyclo-alkanes in the surrogate formulation for petroleum-derived jet fuel was experimentally evaluated previously by comparing the two surrogate mixtures against target jet fuel, Jet-A POSF 4658 [1]. The inclusion of 22.5 % mole fraction of methylcyclohexane in surrogate provide a minor improvement, compared to the surrogate without including methylcyclohexane. The experimental results have shown that the cyclo-alkane functionality can be property represented by the mixtures of n- and iso-alkanes, as long as the cyclo-alkane fraction is limited to the fraction typical to petroleum-derived jet fuels (~ 40 % max.). However, there is a desire to increase the fraction of cyclo-alkanes due to its unique chemical and physical properties. Particularly, the higher activation energy for thermal decomposition reactions makes the high cyclo-alkane containing fuel attractive as a high-performance fuel to be used for coolant. In this regard, it is of importance to understand the significance of cyclic CH_2 group quantitatively, when the cyclo-alkane fraction is increased. To investigate this, the feature sensitivity analysis is performed by increasing the methylcyclohexane fraction in the surrogate mixture (Surrogate 1) formulated for Jet-A POSF 10325 in [6]. The surrogate 1 is composed of n-dodecane (0.490 in mole fraction), iso-octane (0.210), and 1,3,5-trimethylbenzene (0.300).

Figure 4 compares the normalized sensitivity coefficients of chemical functional groups for the mixtures varying the mole fraction of methylcyclohexane, using the QSPR regression model constructed for DCN above. When methycyclohexane is mixed with multi-component surrogate mixture, the sensitivity analysis indicates that the ignition propensity measured by DCN is dominated by the $(CH_2)_n$ functional group. Obviously, the role(s) of CH_2 (isolated CH_2 group), CH_3, and CH groups are found to be insignificant. Compared to the previous sensitivity analysis without considering $(CH_2)_{cyclo}$ functional groups [6], the present analysis exhibits strong sensitivity to the C group. This originates from the fine molecular structure of cyclo-alkanes, requiring to use

C group for full description of cyclo-alkane structure along with $(CH_2)_{cyclo}$. As a result, the normalized sensitivity coefficients of both C and $(CH_2)_{cyclo}$ groups increase as increasing the methylcyclohexane fraction. Nevertheless, it is noteworthy that the ignition propensity is always dominated by $(CH_2)_n$ over $(CH_2)_{cyclo}$ until the mole fraction of methylcyclohexane becomes larger than 0.6. In summary, the results presented here indicates that one should pay attention on the role(s) of cyclo-alkanes in fuel ignition characteristics once the cyclo-alkane fraction becomes larger than 0.6 in mole fraction in the formulation of high performance fuel.

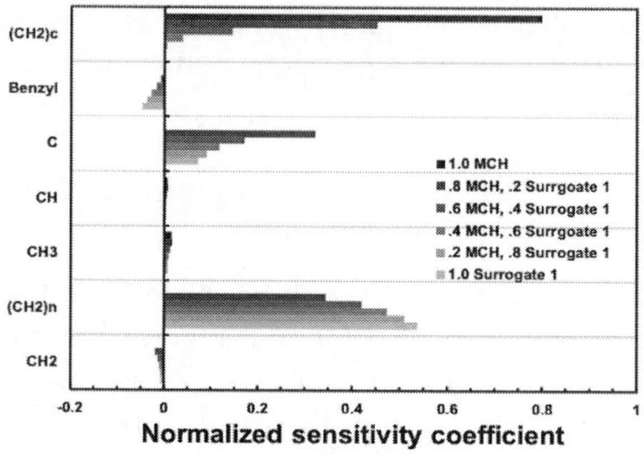

Figure 4. Comparing sensitivity of functional groups of mixture methylcyclohexane and Surrogate 1

4. Conclusions

The impact of the cyclo-alkane function group on the DCN in multicomponent mixtures has been investigated in this research. Other research has shown that this functional group has minor improvements on predicting DCN when including cyclo-alkanes in a surrogate mixture. The findings in this work has shown that for cyclo-alkane mixtures of less than 0.6 mole fraction the $(CH_2)_{cyclo}$ group has little impact on the DCN compared to the influence of the$(CH_2)_n$ group. As the mole fraction of the cyclo-alkanes increases to more than 0.6, the impact of $(CH_2)_n$ is reduced and $(CH_2)_{cyclo}$ increases, but the role of $(CH_2)_n$ is still significant in the reaction and in determining the DCN of the mixture. As more fuel high performance fuel is incorporated with cyclo-alkanes the $(CH_2)_{cyclo}$ functional group will play a larger role in the combustion process.

5. Acknowledgements

This work was supported by the Air Force Research Laboratory (AFRL) and Universal Technology Corporation (UTC) (Grant Number FA8650-14-D-2411)

6. References

[1] S. Dooley, J. Heyne, S.H. Won, P. Dievart, Y. Ju, F.L. Dryer, Importance of a Cycloalkane Functionality in the Oxidation of a Real Fuel, Energy Fuels 28 (2014) 7649-7661.
[2] S. Dooley, S.H. Won, M. Chaos, J. Heyne, Y. Ju, F.L. Dryer, K. Kumar, C.-J. Sung, H. Wang, M.A. Oehlschlaeger, R.J. Santoro, T.A. Litzinger, A jet fuel surrogate formulated by real fuel properties, Combust. Flame 157 (2010) 2333-2339.
[3] S. Dooley, S.H. Won, J. Heyne, T.I. Farouk, Y. Ju, F.L. Dryer, K. Kumar, X. Hui, C.-J. Sung, H. Wang, M.A. Oehlschlaeger, V. Iyer, S. Iyer, T.A. Litzinger, R.J. Santoro, T. Malewicki, K. Brezinsky, The experimental evaluation of a methodology for surrogate fuel formulation to emulate gas phase combustion kinetic phenomena, Combust. Flame 159 (2012) 1444-1466.
[4] S. Dooley, S.H. Won, S. Jahangirian, Y. Ju, F.L. Dryer, H. Wang, M.A. Oehlschlaeger, The combustion kinetics of a synthetic paraffinic jet aviation fuel and a fundamentally formulated, experimentally validated surrogate fuel, Combust. Flame 159 (2012) 3014-3020.

[5] F.L. Dryer, S. Jahangirian, S. Dooley, S.H. Won, J. Heyne, V.R. Iyer, T.A. Litzinger, R.J. Santoro, Emulating the Combustion Behavior of Real Jet Aviation Fuels by Surrogate Mixtures of Hydrocarbon Fluid Blends: Implications for Science and Engineering, Energy Fuels 28 (2014) 3474-3485.
[6] S.H. Won, F.M. Haas, S. Dooley, T. Edwards, F.L. Dryer, Reconstruction of chemical structure of real fuel by surrogate formulation based upon combustion property targets, Combust. Flame 183 (2017) 39-49.
[7] S.H. Won, P.S. Veloo, S. Dooley, J. Santner, F.M. Haas, Y. Ju, F.L. Dryer, Predicting the global combustion behaviors of petroleum-derived and alternative jet fuels by simple fuel property measurements, Fuel 168 (2016) 34-46.
[8] F.L. Dryer, Chemical kinetic and combustion characteristics of transportation fuels, Proceedings of the Combustion Institute 35 (2015) 117-144.
[9] C. Allen, D. Valco, E. Toulson, T. Edwards, T. Lee, Ignition behavior and surrogate modeling of JP-8 and of camelina and tallow hydrotreated renewable jet fuels at low temperatures, Combust. Flame 160 (2013) 232-239.
[10] P. Dagaut, A. El Bakali, A. Ristori, The combustion of kerosene: Experimental results and kinetic modelling using 1- to 3-component surrogate model fuels, Fuel 85 (2006) 944-956.
[11] E.G. Eddings, S. Yan, W. Ciro, A.F. Sarofim, Formulation of a Surrogate for the Simulation of Jet Fuel Pool Fires, Combust. Sci. Technol. 177 (2005) 715-739.
[12] T. Edwards, L.Q. Maurice, Surrogate Mixtures to Represent Complex Aviation and Rocket Fuels, J. Propul. Power 17 (2001) 461-466.
[13] S. Honnet, K. Seshadri, U. Niemann, N. Peters, A surrogate fuel for kerosene, Proceedings of the Combustion Institute 32 (2009) 485-492.
[14] S. Humer, A. Frassoldati, S. Granata, T. Faravelli, E. Ranzi, R. Seiser, K. Seshadri, Experimental and kinetic modeling study of combustion of JP-8, its surrogates and reference components in laminar nonpremixed flows, Proceedings of the Combustion Institute 31 (2007) 393-400.
[15] D. Kim, J. Martz, A. Violi, A surrogate for emulating the physical and chemical properties of conventional jet fuel, Combust. Flame 161 (2014) 1489-1498.
[16] C.J. Mueller, W.J. Cannella, T.J. Bruno, B. Bunting, H.D. Dettman, J.A. Franz, M.L. Huber, M. Natarajan, W.J. Pitz, M.A. Ratcliff, K. Wright, Methodology for Formulating Diesel Surrogate Fuels with Accurate Compositional, Ignition-Quality, and Volatility Characteristics, Energy Fuels 26 (2012) 3284-3303.
[17] C. Pera, V. Knop, Methodology to define gasoline surrogates dedicated to auto-ignition in engines, Fuel 96 (2012) 59-69.
[18] M. Dahmen, W. Marquardt, A Novel Group Contribution Method for the Prediction of the Derived Cetane Number of Oxygenated Hydrocarbons, Energy Fuels 29 (2015) 5781-5801.
[19] ASTM Standard D6890, 2007a, Standard Test Method for Determination of Ignition Delay and Derived Cetane Number (DCN) of Diesel Fuel Oils by Combustion in a Constant Volume Chamber, ASTM International, West Conshohocken, PA, 2007.
[20] J. Yanowitz, M.A. Ratcliff, R.L. McCormick, J.D. Taylor, M.J. Murphy, Compendium of experimental cetane numbers, NREL (National Renewable Energy Laboratory (NREL), Golden, CO (United States)), 2017.

Spring Technical Meeting
Eastern States Section of the Combustion Institute
March 4-7, 2018
State College, Pennsylvania

Surrogate Formation Based on Chemical Functional Group Analysis

Stuart Nates, Dalton Carpenter, Seung Jae Lim, Frederick Dryer, Sang Hee Won*

[1]*Department of Mechanical Engineering, University of South Carolina, Columbia SC 29208*
Corresponding Author Email: natessd@email.sc.edu

Abstract: Chemical kinetic characteristics of real fuels exhibit high-dimensional complexity due to the excessive number of molecules and molecular classes. As a method of projecting this high-dimensional complexity, which is pertinent to real fuels, to the low-dimensional description, either a surrogate approach or detailed experiments have been utilized, can guide the construction of chemical kinetic models for real fuel. Although the validity of the surrogate approach has been extensively demonstrated by a wide range of canonical experiments for whole fuel/air mixtures, the use of empirical fuel property indicators still worries whether or not the surrogate mixture truly captures the more complex chemical behaviors coupled with fuel physical properties (e.g. distillation). Particularly, a recent experiment has shown that the near-limit combustion behaviors (e.g. lean blow off in gas turbine combustor) are strongly governed by the chemical characteristics of the front (light) end in fuel boiling characteristics. Thus, it is of importance to develop an alternative approach, which can fundamentally characterize fuel chemical properties along the fuel distillation curve. Using Nuclear Magnetic Resonance (NMR) spectra, it is possible to quantify the specific chemical functional groups present in a sample. To demonstrate the applicability of chemical functional group approach in conjunction with NMR spectra interpretation, a surrogate formulation approach based on NMR spectra is demonstrated by using a 12-component model fuel and comparing the synthetic NMR spectra between target model fuel and surrogate mixture. To account for the variation of chemical properties along the fuel distillation curve, the distillation curve is introduced in a way that would split the real fuel (Jet-A POSF 10325) into five different bins of 20% liquid volume of the original fuel. Using these five bins and the NMR approach, it correctly matches the chemical functional groups and their variation along the distillation curve.
Keywords: Surrogate, Jet Fuel, Chemical Functional Group, NMR

1. Introduction

Surrogate approach has been a widely used scheme to convert the high-dimensional complexity typical in real liquid fuels to a low-dimensional representative mixture, which can emulate the combustion behaviors of a target liquid fuel [1-18]. Surrogate approach is particularly important, when one tries to develop either a detailed or reduced-order chemical kinetic model for both engine development and characterization through computational fluid dynamic simulation. In this regard, extensive studies have been performed to develop an effective surrogate formulation methodology *either* based on molecular classes found in liquid fuel [11-15, 18] *or* through the target fuel characterization with combustion reference indicators [1-8]. Our previous works have shown that a surrogate mixture can be successfully formulated by matching four combustion property targets (CPTs); hydrogen to carbon molar ratio (H/C ratio), derived cetane number (DCN), average molecular weight (MW), and threshold sooting index (TSI) [1-8]. Extensive and comprehensive experimental verifications have been conducted, demonstrating the effectiveness of the surrogate approach to emulate the global combustion behaviors of pre-vaporized fuel/air mixtures.

In spite of the success of surrogate approach, the use of empirical reference indicators (e.g. DCN and TSI) exhibits the technical limitations, primarily originating from the ambiguity on relative significance of each CPT regarding the target combustion behaviors. A typical approach has been also relying on the user's empirical judgement, resulting in weighting specific CPTs over the others [1-8]. In this regard, the recent study [6] has introduced a concept of chemical functional group and elucidated how the CPTs imposed to formulate a surrogate can constrain the chemical kinetic characteristics through the statistical analysis. Chemical function group approach considers the complex molecular diversity with a few functional groups (either the molecular fragment or atom) composing individual fuel molecule in multi-component fuel mixtures. The chemical functional groups utilized in [6] are CH_2, $(CH_2)_n$, CH_3, CH, C, and benzyl-type fragments. The ability to define the complex molecular diversity pertinent in multi-component fuel mixture with a few chemical functional groups allows constructing a quantitative structure-property relationship (QSPR) regression models for various combustion behaviors (e.g. DCN [6, 19] and homogenous ignition delay times [20]). The QSPR regression model can predict target combustion behaviors effectively if the chemical functional groups utilized can appropriately reflect the role(s) of chemical functionalities governing oxidation/pyrolysis reactions during combustion processes. The feature sensitivity analysis with a QSPR regression model for DCN confirms that n-alkyl, iso-alkyl, and aromatic functionalities can be represented by the abundance of $(CH_2)_n$, CH_3, and benzyl-type functional groups.

Considering that the chemical functional group distribution defines the global combustion behaviors explicitly, a question arises whether a surrogate can be formulated based on chemical functional groups. In fact, the chemical functional groups discussed above can be directly measured by interpreting nuclear magnetic resonance (NMR) spectra [21]. Accurate quantification of key chemical functional group might require further development. Nevertheless, the ^1H NMR spectrum fruitfully represents the abundances of the key chemical functional groups mentioned above. In this regard, the objective of this study is to seek the possibility to utilize the ^1H NMR spectrum as a target for formulating a surrogate mixture. To demonstrate this possibility, a synthetic NMR spectrum of the model fuel utilized in [6] is calculated and used to address the fidelity of the approach.

2. Methods and Verification

The ^1H NMR spectrum of pure component can be obtained from a website by the Institute of Chemical Sciences and Engineering, which allows for the creation of synthetic NMR spectra [22] (https://www.nmrdb.org/). In case of multi-component mixture, the NMR spectrum can be simply calculated by summating each component NMR spectrum multiplied by its mole fraction. Here, the 12-component model fuel utilized in [6] is used as a target fuel for the approach verification.

The flow chart of the optimization process is shown in **Fig. 1**. First, the optimization process to calculate the composition of multi-component mixtures is composed with 3 steps. The step 1 populates 10000 mixtures randomly by varying the mixture fraction of each component from 0 to

1, while ensuring their summation to be unity. Then, the synthetic NMR spectrum of each mixture is generated as mentioned above. The NMR spectrum is integrated in entire chemical shift range, so that the cumulated curve in a function of chemical shift coordinate. Each cumulated curve is normalized by the maximum value for the comparison to the target spectrum. Optimization process compares the cumulated curve of NMR spectrum from randomly generated mixtures to that of target NMR spectrum. In this comparison, the error between two spectra is defined as the summation of differences between two spectra. The mixture having the minimal error is chosen, if its error value is less than the specified value (typically 15 %), otherwise the optimization process produces another 10000 random mixtures until it finds the solution with the error smaller than specified value.

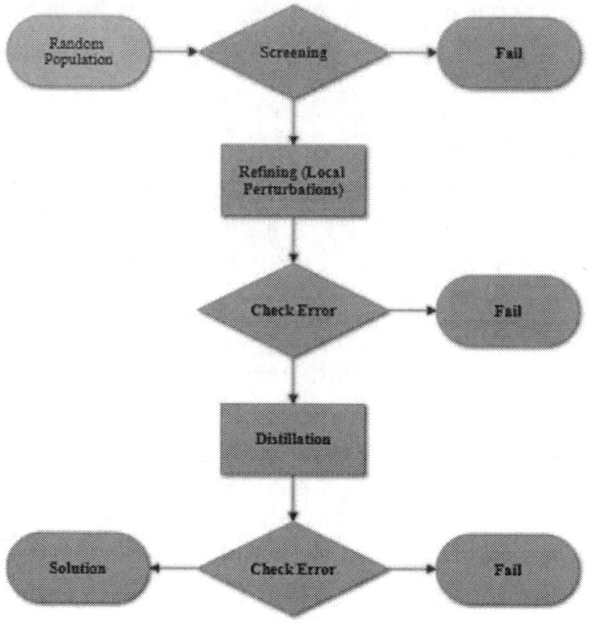

Figure 1. Schematic of NMR Based Surrogate Formulation

In step 2, the mole fraction of each component of the mixture found in step 1 is locally perturbed with a specified range to accurately reproduce the cumulated curve of the target NMR spectrum through identical approach in Step 1. **Figure 2** shows the comparison between target cumulated NMR spectrum and the generated one, showing that the gross feature of NMR spectrum can be well matched with Step 2.

Step 3 calculates the distillation curve and compares three distillation temperatures (T_{10}, T_{50}, and T_{90}) between the solution from step 2 and the target mixture, as shown in **Fig. 3**. Another local perturbation is performed to match three distillation temperatures. Here, the component mole fractions are perturbed within the categorized three molecular classes (n-alkanes, iso-alkanes, and aromatics) to ensure this process holds the relative fractions of molecular classes. After the third step, the cumulative NMR spectrum is again compared to that of the target; otherwise the optimization process repeats the step 2. In each step, if the optimization process

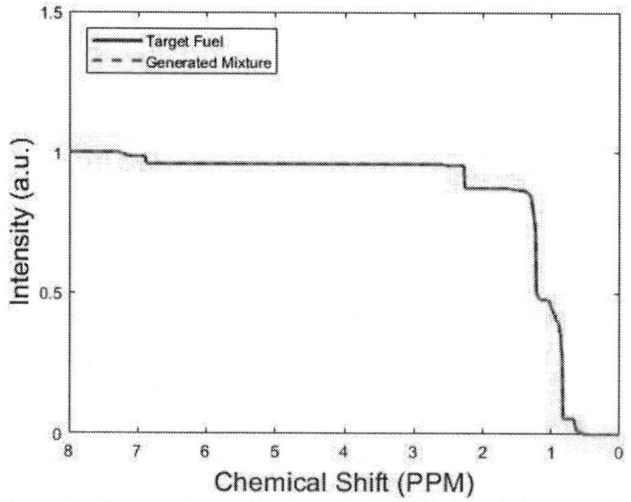

Figure 2. Comparison of Integrated NMR Spectra After Step 2

fails to reproduce the target cumulated NMR spectrum and the three distillation temperatures, it starts the optimization process again from the generation of random 10000 mixtures at the beginning, as indicated in **Fig. 1**.

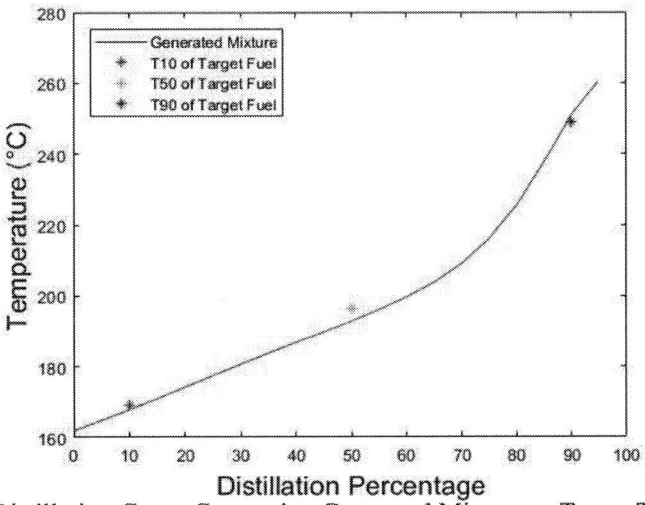

Figure 3. Distillation Curve Comparing Generated Mixture to Target Temperatures

3. **Results and Discussion**

As mentioned above, a known 12-component mixture (Model fuel used in [6]) was used as the target fuel. **Figure 4** shows the result of the optimization method of creating a synthetic NMR spectrum of a mixture, which is identical to the target spectrum. Encouraged by a successful reproduction of NMR spectra through the optimization of composition mentioned above, the approach is further tested with a real petroleum-derived jet fuel, Jet-A POSF 10325 and the results are discussed below.

Applicability to real fuel sample

After the 12-component target mixture was properly matched and the distillation values were within a reasonable limit, the next step to checking the validity of this method is to test it using real fuels, Jet-A POSF 10325.

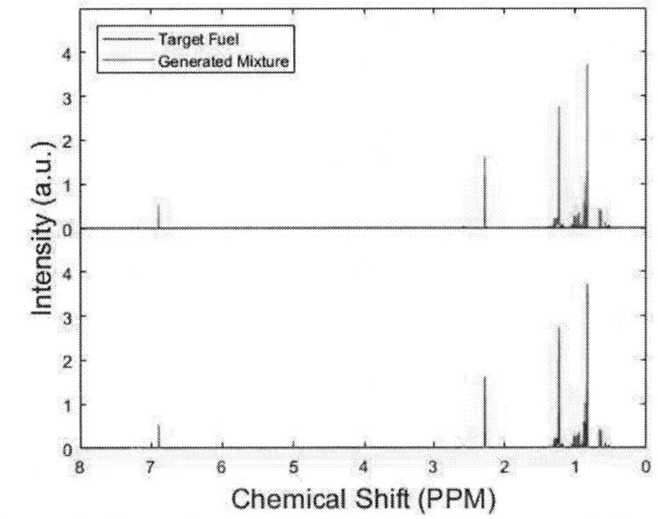

Figure 4. Comparison of NMR Spectra of Generated Surrogate Mixture and Target Fuel

Compared to the Model fuel used for approach verification, the real petroleum-derived jet fuel contains considerable fraction of cyclo-alkane, which are not considered as surrogate components currently. On the other hand, a different method of determining error in optimization process is needed to compare the measured and synthetic spectra, because the measured spectrum is of lower resolution than its synthetic counterpart. Consequently, the cumulated spectrum of target real fuel cannot be exactly matched as shown in **Fig. 5**.

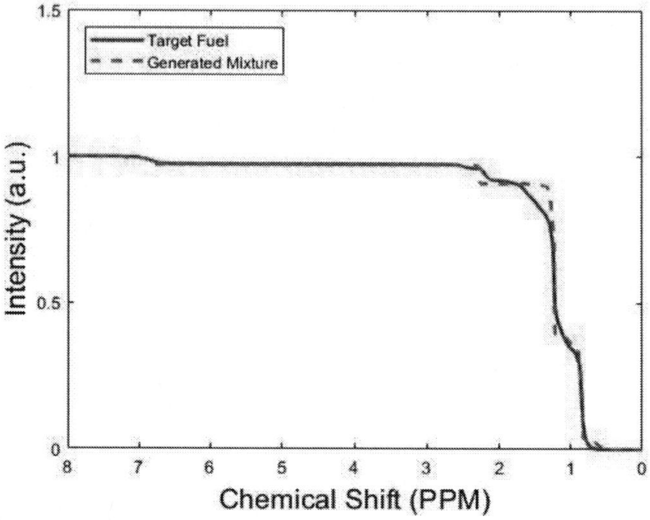

Figure 5. Comparison of Integrated NMR Spectra of Real Jet Fuel and Generated Mixture

Figure 6 compares the NMR spectra between the measured target Jet-A POSF 10325 and the optimized mixture to be considered as a surrogate mixture. Although not shown in a figure, the three distillation temperatures are closed matched with a surrogate mixture, exhibiting less than 8°C difference. The optimization process shows that the chemical kinetic characteristics represented by the NMR spectrum and the three distillation temperatures can be reasonably reproduced by using 9 surrogate components. Validity of the optimized surrogate mixture can be evaluated by comparing the abundances of key chemical functional groups, CH_2, CH_3, and benzyl-type groups. Following the specific chemical shift according to each functional group [20, 21], the peak areas of key chemical functional groups are compared between target real fuel and the surrogate mixture. To further evaluate how well this method can generate a surrogate mixture that recreates a target fuels NMR spectrum, the error of each major peak was also calculated. It was found that the surrogate generated for POSF 10325 exhibited a 4.6% difference in the peak one region which was defined as 0.5 to 1. Peak 2 exhibited a 4.7% difference which had a domain between 1 and 2. The third region that was defined was between 2 and 2.9 and consisted of several small peaks. The error that was found for the third region was 16% however this region is not of major significance in the overall error of the surrogate. The fourth series of peaks which contains the aromatic components was found to have an error of 8.6%. Altogether the total percent difference of the surrogate mixture was 0.81% with a

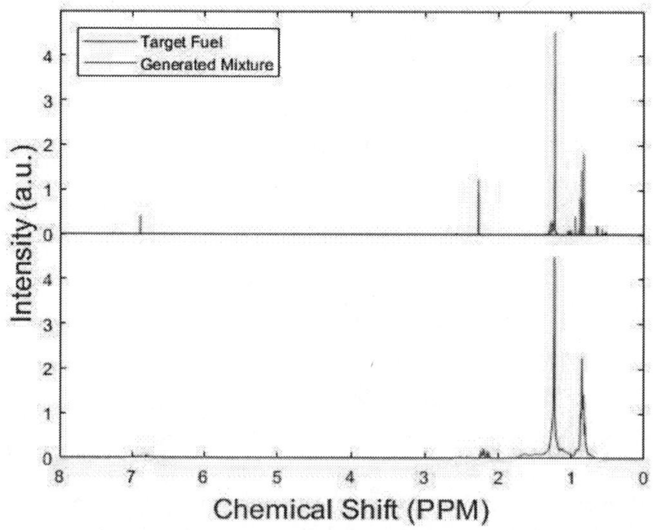

Figure 6. Comparison of NMR Spectra of Real Jet Fuel and Generated Mixture

maximum deviation from the distillation curve at T_{50} and was found to be 7.5 °C different. Recent rig-scale experimentations have shown that the near-limit combustion behaviors are controlled by the complex interactions between fuel chemical and physical properties. Particularly, the preferential vaporization characteristics have been found to be significant. The preferential vaporization impacts originate from the dynamic changes of the fuel chemical properties along the distillation curve. In this regard, the approach shown here can be further extended to consider the preferential vaporization characteristics, which cannot be considered by the previous surrogate approach, which only considers the gloss characteristics of fuel chemical properties by combustion property indicators. In this regard, the approach presented here has the potential to consider the preferential vaporization, once the dynamic changes of chemical property are directly measured by interpreting the NMR spectra along the distillation curve.

4. Conclusions

Using NMR, it is possible to reproduce the overall chemical functional group distribution of a 12-component target fuel. It was found that the distillation calculation needed to be checked against the target temperatures as to correctly determine a mixture that will represent the target fuel along the distillation curve. Applying this technique leads to a mixture that approximately matches the functional group distribution of the target fuel. This mixture's distillation curve also satisfactorily matches the target fuel's temperatures. When applying this technique to real fuels it is important to match the resolution of the target spectrum to that of the synthetic NMR spectra as to not skew the mixture. The process of matching the NMR spectrum and distillation curve of the target fuel potentially allows for preferential vaporization to be accounted for. This method also allows for the determination of chemical composition at different locations of the distillation curve.

5. Acknowledgements

This work was supported by the Air Force Research Laboratory (AFRL) and Universal Technology Corporation (UTC) (Grant Number FA8650-14-D-2411).

6. References

[1] S. Dooley, J. Heyne, S.H. Won, P. Dievart, Y. Ju, F.L. Dryer, Importance of a Cycloalkane Functionality in the Oxidation of a Real Fuel, Energy Fuels 28 (2014) 7649-7661.

[2] S. Dooley, S.H. Won, M. Chaos, J. Heyne, Y. Ju, F.L. Dryer, K. Kumar, C.-J. Sung, H. Wang, M.A. Oehlschlaeger, R.J. Santoro, T.A. Litzinger, A jet fuel surrogate formulated by real fuel properties, Combust. Flame 157 (2010) 2333-2339.

[3] S. Dooley, S.H. Won, J. Heyne, T.I. Farouk, Y. Ju, F.L. Dryer, K. Kumar, X. Hui, C.-J. Sung, H. Wang, M.A. Oehlschlaeger, V. Iyer, S. Iyer, T.A. Litzinger, R.J. Santoro, T. Malewicki, K. Brezinsky, The experimental evaluation of a methodology for surrogate fuel formulation to emulate gas phase combustion kinetic phenomena, Combust. Flame 159 (2012) 1444-1466.

[4] S. Dooley, S.H. Won, S. Jahangirian, Y. Ju, F.L. Dryer, H. Wang, M.A. Oehlschlaeger, The combustion kinetics of a synthetic paraffinic jet aviation fuel and a fundamentally formulated, experimentally validated surrogate fuel, Combust. Flame 159 (2012) 3014-3020.

[5] F.L. Dryer, S. Jahangirian, S. Dooley, S.H. Won, J. Heyne, V.R. Iyer, T.A. Litzinger, R.J. Santoro, Emulating the Combustion Behavior of Real Jet Aviation Fuels by Surrogate Mixtures of Hydrocarbon Fluid Blends: Implications for Science and Engineering, Energy Fuels 28 (2014) 3474-3485.

[6] S.H. Won, F.M. Haas, S. Dooley, T. Edwards, F.L. Dryer, Reconstruction of chemical structure of real fuel by surrogate formulation based upon combustion property targets, Combust. Flame 183 (2017) 39-49.

[7] S.H. Won, P.S. Veloo, S. Dooley, J. Santner, F.M. Haas, Y. Ju, F.L. Dryer, Predicting the global combustion behaviors of petroleum-derived and alternative jet fuels by simple fuel property measurements, Fuel 168 (2016) 34-46.
[8] F.L. Dryer, Chemical kinetic and combustion characteristics of transportation fuels, Proceedings of the Combustion Institute 35 (2015) 117-144.
[9] C. Allen, D. Valco, E. Toulson, T. Edwards, T. Lee, Ignition behavior and surrogate modeling of JP-8 and of camelina and tallow hydrotreated renewable jet fuels at low temperatures, Combust. Flame 160 (2013) 232-239.
[10] P. Dagaut, A. El Bakali, A. Ristori, The combustion of kerosene: Experimental results and kinetic modelling using 1- to 3-component surrogate model fuels, Fuel 85 (2006) 944-956.
[11] E.G. Eddings, S. Yan, W. Ciro, A.F. Sarofim, Formulation of a Surrogate for the Simulation of Jet Fuel Pool Fires, Combust. Sci. Technol. 177 (2005) 715-739.
[12] T. Edwards, L.Q. Maurice, Surrogate Mixtures to Represent Complex Aviation and Rocket Fuels, J. Propul. Power 17 (2001) 461-466.
[13] S. Honnet, K. Seshadri, U. Niemann, N. Peters, A surrogate fuel for kerosene, Proceedings of the Combustion Institute 32 (2009) 485-492.
[14] S. Humer, A. Frassoldati, S. Granata, T. Faravelli, E. Ranzi, R. Seiser, K. Seshadri, Experimental and kinetic modeling study of combustion of JP-8, its surrogates and reference components in laminar nonpremixed flows, Proceedings of the Combustion Institute 31 (2007) 393-400.
[15] D. Kim, J. Martz, A. Violi, A surrogate for emulating the physical and chemical properties of conventional jet fuel, Combust. Flame 161 (2014) 1489-1498.
[16] C.J. Mueller, W.J. Cannella, T.J. Bruno, B. Bunting, H.D. Dettman, J.A. Franz, M.L. Huber, M. Natarajan, W.J. Pitz, M.A. Ratcliff, K. Wright, Methodology for Formulating Diesel Surrogate Fuels with Accurate Compositional, Ignition-Quality, and Volatility Characteristics, Energy Fuels 26 (2012) 3284-3303.
[17] C. Pera, V. Knop, Methodology to define gasoline surrogates dedicated to auto-ignition in engines, Fuel 96 (2012) 59-69.
[18] M.L. Huber, E.W. Lemmon, T.J. Bruno, Surrogate Mixture Models for the Thermophysical Properties of Aviation Fuel Jet-A, Energy Fuels 24 (2010) 3565-3571.
[19] M. Dahmen, W. Marquardt, A Novel Group Contribution Method for the Prediction of the Derived Cetane Number of Oxygenated Hydrocarbons, Energy Fuels 29 (2015) 5781-5801.
[20] K. Dussan, S. Dooley, S.H. Won, F.L. Dryer, A Nuclear Magnetic Resonance Orientated Combustion Property Regression 9th U.S. National Combustion Meeting, Hilton Cincinnati Netherland Plaza, Cincinnati, OH, May 17-20, (2015).
[21] G.F. Pauli, B.U. Jaki, D.C. Lankin, Quantitative 1H NMR: Development and Potential of a Method for Natural Products Analysis, J. Nat. Prod. 68 (2005) 133-149.
[22] A.M. Castillo, L. Patiny, J. Wist, Fast and accurate algorithm for the simulation of NMR spectra of large spin systems, J. Magn. Reson. 209 (2011) 123-130.

Sub Topic: Reaction Kinetics

Spring Technical Meeting
Eastern States Section of the Combustion Institute
March 4-7, 2018
State College, Pennsylvania

Pyrolysis of Fuel Mixtures at Supercritical Conditions: A ReaxFF Molecular Dynamics Study

Sharmin Shabnam, Chowdhury Ashraf, Abhishek Jain, Yuan Xuan and Adri C.T. van Duin *

Department of Mechanical and Nuclear Engineering, The Pennsylvania State University, University Park, PA 16802, USA

**Corresponding Author Email: acv13@psu.edu*

Abstract: Combustion devices such as rocket engines, gas turbines and HCCI engines frequently operate at a pressure higher than the critical pressure of the fuel or the oxidizer. These high-pressure conditions can significantly alter their chemical properties and application of existing chemical kinetics models will be significantly limited as they are developed only considering temperature dependence on the reaction rates via simple Arrhenius-type rate laws while neglecting pressure dependence on combustion pathways. This significantly prohibits transferring the models directly from low pressure to high pressure regime. In this presentation, we will demonstrate how the ReaxFF reactive force field method can be used as a valuable tool to study combustion kinetics of fuels and fuel mixtures at supercritical condition. The objective of this study is to investigate how the presence of a highly reactive fuel can alter the properties of a much less reactive fuel during pyrolysis. We will compare our results with continuum simulation results and elaborate why the continuum results fail to capture the phenomenon predicted by the ReaxFF simulations. This study enables us to identify the pressure/temperature regime and the mixing conditions where the simple first order kinetics and Arrhenius type relations do not prevail and the probable reasons behind this characteristic. Our results indicate that, ReaxFF force field based molecular dynamics simulations can provide important atomistic insights on the combustion properties of fuel mixtures at supercritical conditions where the experiments are difficult to perform.
Keywords: Pyrolysis, Arrhenius parameters, Fuel mixture, Toluene, n-Dodecane

1. Introduction

Majority of the chemical kinetic models for fuel pyrolysis are developed for single component fuels while the 'real fuels' are mostly a mixture of various hydrocarbons. Again, most of these models are developed for low pressure/temperature condition[1], whereas the combustion devices frequently work at extremely high pressures[2,3]. Since experiments are difficult to conduct at supercritical pressure/temperature regime, and the quality of CFD simulations are highly dependent on the chemical kinetic model one choses to implement, we need a computationally feasible method which can simulate complex combustion reactions at these conditions. In this study, we applied ReaxFF-MD to investigate the effect of blending a highly reactive fuel (n-dodecane) with a less reactive one (toluene) on fuel pyrolysis at high pressure/density condition. We compared our results with continuum model simulations and discuss why the continuum results differ from the ReaxFF ones. This paper is organized as follows, in Section 2, ReaxFF method is described while simulation setups for both ReaxFF and continuum methods are described in section 3. Results are discussed in section 4 and lastly, conclusions are summarized in Section 5.

2. ReaxFF Reactive Force Field Method

ReaxFF is a bond order[4] based empirical potential, which is capable of bond formation and breaking during the simulation. A more detail description of ReaxFF energy terms will be found in previous literature[5,6]. This method was previously used to study different combustion related problems including pyrolysis and combustion of n-dodecane[7], coal[8] and JP-10 fuel[9].

3. Simulation Methodology

3.1 ReaxFF method

To validate the quality of the force field[10], we first calculated the Arrhenius parameters for each of the fuel component separately at two different densities. For the single component systems, we placed 40 molecules of each type in a cubic box with a dimension required to produce the desired densities. For fuel mixtures, we kept the number of toluene molecules fixed at 40 and added different number of n-dodecane molecules with a mixing ratio (α) defined as the ratio of the mole fraction of n-dodecane and mole fraction of toluene in the mixture.

Table 1: Simulation conditions for fuel mixtures

Mixing ratio, α	Density, ρ (kg/dm^3)	Temperature, T (K)
0.0, 0.025, 0.125, 0.25, 0.50 and 1.00	0.2 and 0.4	2000K to 2600K, interval 100K

After energy minimization, and low temperature equilibration for 10ps, 10 different initial configurations were selected to perform a series of NVT-MD simulations for 200ps at different temperatures with time-step of 0.1 fs. Table 1 shows the simulation conditions and Figure 1 shows a snapshot of a representative n-dodecane/toluene mixture with mixing ratio α = 0.125. Further simulation details can be found in the literature[6,10].

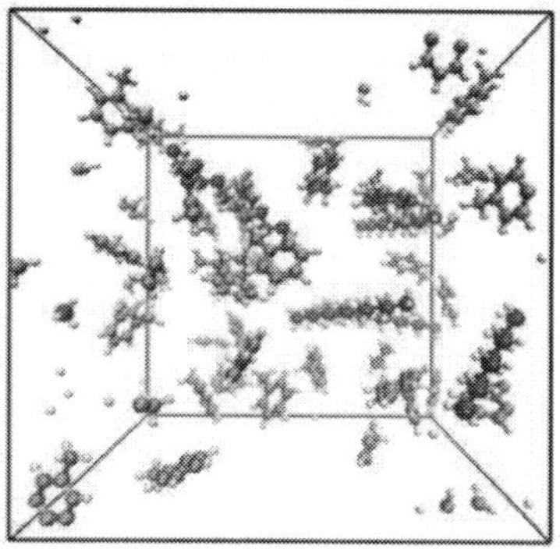

Figure 1: Snapshot of n-dodecane/toluene system with a mixing ratio of 0.125 and density of 0.2 kg/dm^3. The red spheres indicate the carbon atoms of the n-dodecane, while the cyan spheres indicate the carbon atoms of toluene. Hydrogen atoms are indicated by white spheres.

3.2 Continuum Method

Zero-dimensional (time-evolving), continuum-scale pyrolysis simulations of the same series of test cases (see Table 1) are also performed in this study. A cubic equation of state is used to account for real gas effects, and a widely-used chemical kinetic mechanism[11] with Arrhenius-type rate coefficients is used. This chemical mechanism contains 179 species and 1895 chemical reactions, and it has been extensively validated for jet fuel surrogate components under lower pressure and temperature ranges[12].

4. Results and Discussion

4.1. Validation of force field against single component systems

We considered first-order kinetics to calculate the Arrhenius parameters. The consumption rate of the fuel molecules is found to be a function of temperature and we determined the rate constant at each temperature with the following integrated first order rate law equation:

$$\ln(N_0) - \ln(N_t) = kt \quad (1)$$

where N_0 is the number of molecules initially in the system and N_t is the number of molecules at any time t. Next, we used these rate constants from different temperatures to determine the activation energy (E_a) and the pre-exponential factor (A) with the help of a linear fit of the Arrhenius plot from the equation described as:

$$k = A \exp(-E_a/RT) \quad (2)$$

where, R is the universal gas constant.

Table 2. Arrhenius parameters for single component systems.

Molecule	ReaxFF			Continuum		Experiment[13,14]	
	Density (kg/dm^3)	E_a (kcal/mol)	A (1/s)	E_a (kcal/mol)	A (1/s)	E_a (kcal/mol)	A (1/s)
Toluene	0.2	95.71	28.3×10^{16}	95.17	6.46×10^{16}	88.9-97.0	$0.28\text{-}1 \times 10^{16}$
	0.4	89.90	11.59×10^{16}				
n-Dodecane	0.2	60.94	0.95×10^{16}	62.15	0.86×10^{16}	61.32	0.12×10^{16}
	0.4	63.11	1.53×10^{16}				

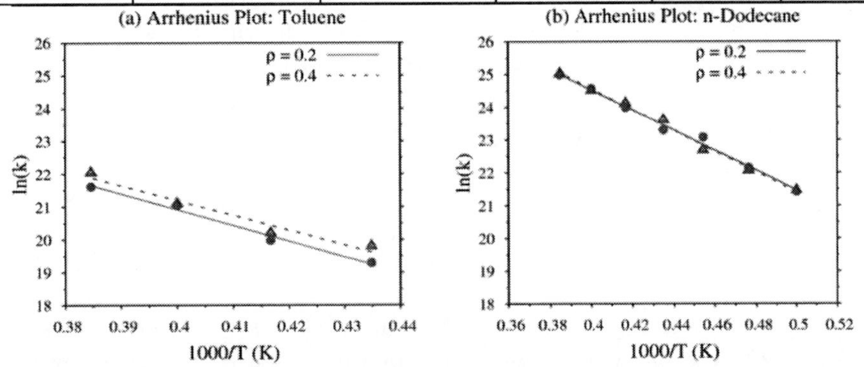

Figure 2: Arrhenius plots for calculated and fitted rate constants of (a) toluene and (b) n-dodecane as single component systems with $\rho = 0.2$ kg/dm^3 and 0.4 kg/dm^3. The points represent simulation results while the lines are fitted to estimate Arrhenius parameters.

The results obtained from ReaxFF-MD simulations for single component systems listed in Table 2 show good agreement with those obtained from the experiments thus validating the quality of the force field for such pyrolysis problems. Since we used first order kinetics to calculate the Arrhenius parameters, the difference in pressure/density between experimental and ReaxFF studies should not affect the activation energies. Table 2 also lists the Arrhenius parameters calculated from continuum simulations, which shows reasonable agreement with both ReaxFF and experimental activation energies.

4.2. Effect of the presence of a highly reactive fuel molecule in pyrolysis of a less reactive fuel

Rate constants of toluene decomposition during ReaxFF simulations at various temperatures and mixing ratios are calculated to generate the Arrhenius plot as shown in figure 3(a). With increasing mixing ratio, the fitted straight lines shift upward in figure 3(a) indicating that the rate constants of toluene decomposition at any given temperature increases when reactive molecules are introduced in the system. According to this, at high temperature (2500K, 2600K) the rate constant is less affected by the inclusion n-dodecane in the mixture, whereas at low temperature and intermediate temperatures (2000K-2400K), the rate constants are mostly affected. Hence, with the increase of n-dodecane molecules in the system, lower temperatures become accessible to calculate reliable rate constants as more toluene molecule decomposes. Similar trend was observed for ρ= 0.4 kg/dm^3 case.

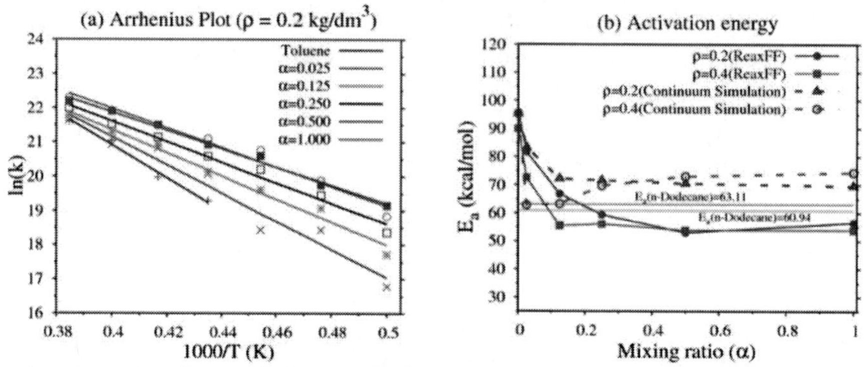

Figure 3: (a) Arrhenius plot for n-dodecane/toluene systems with different mixing ratios with ρ = 0.2 kg/dm^3. (b) Comparison of activation energies between ReaxFF and Continuum simulations at varying mixing ratios of n-dodecane/toluene mixtures for ρ = 0.2 and 0.4 kg/dm^3.

Figure 3(b) compares the activation energies of toluene decomposition between ReaxFF and continuum simulations for various mixing ratios. For both the densities, the presence of a small amount of n-dodecane significantly affects the activation energy of toluene decomposition. Further increase in mixing ratio predicts an activation energy of toluene which is even below of the activation energy of n-dodecane pyrolysis itself. However, the activation energy of toluene from continuum simulation in all of the mixtures remains higher than n-dodecane only system (62.15 kcal/mol) and appears to reach a plateau at higher mixing ratios. At higher density, the activation energy of toluene falls below the activation energy of n-dodecane even at low mixing ratio (α = 0.125) and further increasing the mixing ratio shows no significant effect. The continuum simulations predict that the activation energy of toluene decomposition for n-dodecane/toluene mixture always to be higher than that of n-dodecane decomposition. The differences of the ReaxFF and continuum simulation can be attributed to the fact that the continuum model uses a reaction mechanism which is developed and validated for low-pressure/temperature systems, while our simulation conditions are mostly supercritical.

4.3. Analysis of initiation reaction mechanism and product distribution

In order to investigate why the fuel mixtures behave differently at high mixing ratios and high density, we move our focus to analyze the initial reaction mechanism and product distribution of the two single component systems. Figure 4 shows the initiation reaction mechanisms and their relative percentages observed in ReaxFF-MD simulations for the pyrolysis simulation of toluene and n-dodecane. In accordance with experimental observation[14], toluene pyrolysis proceed through two major pathways. One the other hand, pyrolysis of n-dodecane (figure 4(b)) is mostly dominated by C-C bond fission to produce two alkyl radicals, which ultimately leads to high radical pool build up consistent with experiment[13].

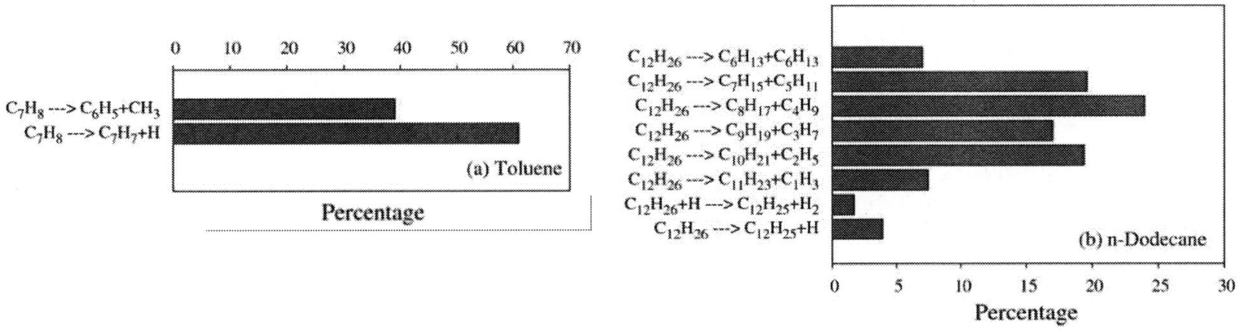

Figure 4: Initiation reaction mechanism and their relative percentages occurred in (a) toluene and (b) n-dodecane pyrolysis in a single component system obtained from ReaxFF–MD simulations at 2600 K (density = 0.2 kg/dm^3).

Figure 5 shows the evolution of radical species of the single component systems at 2600K. The total number of radicals generated during toluene pyrolysis is very small but n-dodecane pyrolysis generated a huge pool of alkyl radicals. Since, C-C bond scission requires relatively low energy, n-dodecane can generate radicals even at lower temperatures. This helps the toluene molecules to decompose at low temperatures, which significantly effects toluene activation energy. That is why, a small amount of n-dodecane can significantly reduce toluene activation energy when simple first order kinetics and pressure-independent Arrhenius equation is used.

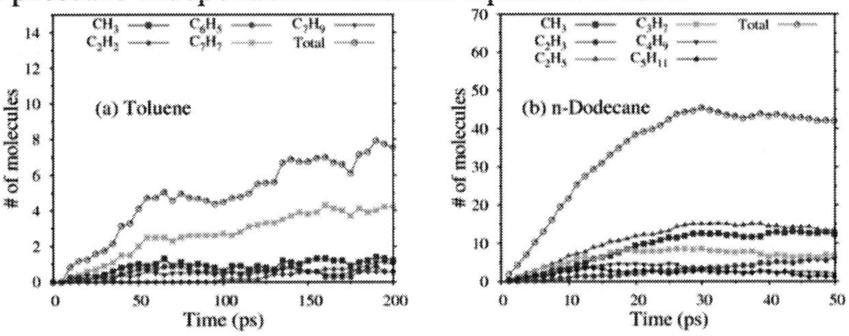

Figure 5: Time evolution of major radical species during the pyrolysis of (a) toluene, (b) n-dodecane using ReaxFF-MD. Number of radicals is averaged over 10 independent simulations performed at T = 2600K with a density of 0.2 kg/dm^3.

However, at a relatively lower density and lower number of n-dodecane in the system, active alkyl radicals have more space to move around and longer time for them to find a toluene molecule to collide. Thus, simple first order kinetics and Arrhenius type relations seems to work under such conditions. But when density is higher, the radicals can easily find a toluene molecule to react as they are diffusion limited at that density. At this point, our analysis does not hold anymore as it is based on first order kinetics assuming that a toluene molecule decompose by itself while the actual decomposition route is through the secondary reaction with generated alkyl radicals.

5. Conclusions

In this work, we implemented the ReaxFF-MD simulations and studied the pyrolysis of toluene and n-dodecane mixture systems. The Arrhenius parameters calculated from ReaxFF-MD simulations for the individual systems compare well with experimental literature as well as with the continuum simulations. However, for the fuel mixtures, the activation energy of toluene at higher mixing ratios falls below the activation energy of n-dodecane. With the help of initiation reaction mechanisms and product distribution, the reason can be attributed to the generation of highly active radical pool by the more reactive fuel molecules. Overall, the ReaxFF-MD simulations can provide useful atomistic-level information which could be useful to develop simpler kinetic models for fuel combustion.

5. Acknowledgements

This research was funded by the Air Force Office of Scientific Research (AFOSR) under grant no. FA9550-14-1-0355.

6. References

[1] J.M. Simmie, Detailed chemical kinetic models for the combustion of hydrocarbon fuels, Prog. Energy Combust. Sci. 29 (2003) 599–634. doi:10.1016/S0360-1285(03)00060-1.

[2] G.P. Sutton, History of Liquid Propellant Rocket Engines, American Institute of Aeronautics and Astronautics, 2006. doi:10.2514/4.868870.

[3] K.R. Cox, W.G. Chapman, The Properties of Gases and Liquids, 5th Edition By Bruce E. Poling (University of Toledo), John M. Prausnitz (University of California at Berkeley), and John P. O'Connell (University of Virginia). McGraw-Hill: New York. 2001. 768 pp. $115.00. ISBN 0-07-011682-2., J. Am. Chem. Soc. 123 (2001) 6745–6745. doi:10.1021/ja0048634.

[4] D.W. Brenner, Empirical potential for hydrocarbons for use in simulating the chemical vapor deposition of diamond films, Phys. Rev. B. 42 (1990) 9458–9471.

[5] A.C.T. van Duin, S. Dasgupta, F. Lorant, W.A. Goddard, ReaxFF: A Reactive Force Field for Hydrocarbons, J. Phys. Chem. A. 105 (2001) 9396–9409. doi:10.1021/jp004368u.

[6] K. Chenoweth, A.C.T. van Duin, W.A. Goddard, ReaxFF Reactive Force Field for Molecular Dynamics Simulations of Hydrocarbon Oxidation, J. Phys. Chem. A. 112 (2008) 1040–1053. doi:10.1021/jp709896w.

[7] Q.-D. Wang, J.-B. Wang, J.-Q. Li, N.-X. Tan, X.-Y. Li, Reactive molecular dynamics simulation and chemical kinetic modeling of pyrolysis and combustion of n-dodecane, Combust. Flame. 158 (2011) 217–226. doi:10.1016/j.combustflame.2010.08.010.

[8] F. Castro-Marcano, A.M. Kamat, M.F. Russo, A.C.T. van Duin, J.P. Mathews, Combustion of an Illinois No. 6 coal char simulated using an atomistic char representation and the ReaxFF reactive force field, Combust. Flame. 159 (2012) 1272–1285.

[9] K. Chenoweth, A.C.T. van Duin, S. Dasgupta, W.A. Goddard III, Initiation Mechanisms and Kinetics of Pyrolysis and Combustion of JP-10 Hydrocarbon Jet Fuel, J. Phys. Chem. A. 113 (2009) 1740–1746. doi:10.1021/jp8081479.

[10] C. Ashraf, A.C.T. van Duin, Extension of the ReaxFF Combustion Force Field toward Syngas Combustion and Initial Oxidation Kinetics, J. Phys. Chem. A. 121 (2017) 1051–1068.

[11] C. Ashraf, A. Jain, Y. Xuan, A.C.T. van Duin, ReaxFF based molecular dynamics simulations of ignition front propagation in hydrocarbon/oxygen mixtures under high temperature and pressure conditions, Phys. Chem. Chem. Phys. 19 (2017) 5004–5017.

[12] The FORCE: CaltechMech, (n.d.). http://www.theforce.caltech.edu/CaltechMech/ (accessed June 9, 2017).

[13] G. Liu, Y. Han, L. Wang, X. Zhang, Z. Mi, Supercritical Thermal Cracking of N-Dodecane in Presence of Several Initiative Additives: Products Distribution and Kinetics, Energy Fuels. 22 (2008) 3960–3969. doi:10.1021/ef800323d.

[14] M.B. Colket, D.J. Seery, Reaction mechanisms for toluene pyrolysis, Symp. Int. Combust. 25 (1994) 883–891. doi:10.1016/S0082-0784(06)80723-X.

Sub Topic: Sprays & Droplets

Spring Technical Meeting
Eastern States Section of the Combustion Institute
March 4-7, 2018
State College, Pennsylvania

Fuel Distillation Affects Spray Flame Blowout Thresholds

Jay A. Lefkowitz and Francis M. Haas*

Mechanical Engineering Department, Rowan University, Glassboro, NJ 08028
**Corresponding Author Email: lefkow81@students.rowan.edu*

Abstract: Emulation of prevaporized real fuel combustion behaviors through use of surrogates has been widely demonstrated in the literature. However, many combustion applications utilize atomized fuel sprays, and for these configurations, the prevaporized combustion assumption of fuel property homogeneity becomes tenuous. Our previous work uses a simplified ASTM D86-relevant distilling droplet model to demonstrate the real potential for vaporization-coupled deviations from single-valued surrogate fuel combustion properties that are characteristic of prevaporized combustion. Further, simulations reveal surrogate-to-surrogate differences in one or more of the combustion properties (such as ignitibility, H/C/O atomic ratios, average molecular weight, and sooting propensity) among sets of effectively equivalent prevaporized surrogates as each similar fuel distills. The present experimental work supports these model-based observations with flame blowout measurements from a custom-built annular spray burner rig. Initial measurements demonstrate the sensitivity of the new experimental setup by comparing chemistry-driven blowout behavior of 1) fully linear and fully branched alkanes of nearly equal chain length (nC7, iC8) and with otherwise similar physical properties, and 2) spray-coupled blowout in the chemically similar homologous family of n-alkanes (nC7, nC10, nC12). Subsequently, sets of essentially equivalent prevaporized jet fuel and gasoline surrogates suggested in the literature are tested to examine blowout threshold variations driven by distillation behavior. The noted variations in blowout limits (and also allowing for other limiting combustion behaviors not studied here) highlights the need to consider the coupling between distillation and prevaporized combustion properties when formulating and evaluating surrogate fuels.

Keywords: Preferential Vaporization, Blowout, Spray Combustion, Surrogate Fuel

1. Introduction

Petroleum-derived real fuels used in many propulsion applications are generally complex mixtures that contain large numbers of distinct chemical species. Combustion of these fuels involves fuel oxidation via an even more complex pool of combustion reaction intermediates, which hinders the high fidelity emulation of many combusting flow properties [1].

To mediate the complexity of real petroleum fuel combustion, many studies (e.g., [2-4]) have created fuel blends composed of few distinct chemical species in specific proportions that act as "surrogates" for the real fuel. Often, these surrogate blends are formulated using several targeted prevaporized combustion properties (CPs) with similar values shared between the particular target real fuel and its surrogate(s) [2, 4-7]. However, a variety of real combustion applications employ fuel sprays and involve a multiphasic combustion environment. Fuel distillation effects expected during multiphase combustion may challenge the prevaporized assumption supporting many of these surrogate formulation techniques and re-open the question of whether or not effectively equivalent prevaporized surrogates achieve their intended real fuel emulation.

Previous work [8] has resolved the distillation behavior of these surrogates using a low-order distillation model analogous to distillation by the ASTM D86 method. The model simulates ideal, equilibrium-limited, multicomponent batch distillation of a representative spherical fuel droplet in effective isolation, as indicated schematically in Figure 1. Here, a predicted arbitrary combustion property with value represented by color evolves for both the droplet and its equilibrium vapor cloud (separately) as the droplet distills. If the droplet is moving, as in a spray, the indicated property variation is equivalent to spatial stratification of fuel components and their convolved combustion behaviors. This framework suggests that as a surrogate fuel distills, preferential vaporization stratifies chemical composition and associated combustion properties, indicating the potential incongruence between effectively equivalent surrogate fuels formulated with species of dissimilar volatility.

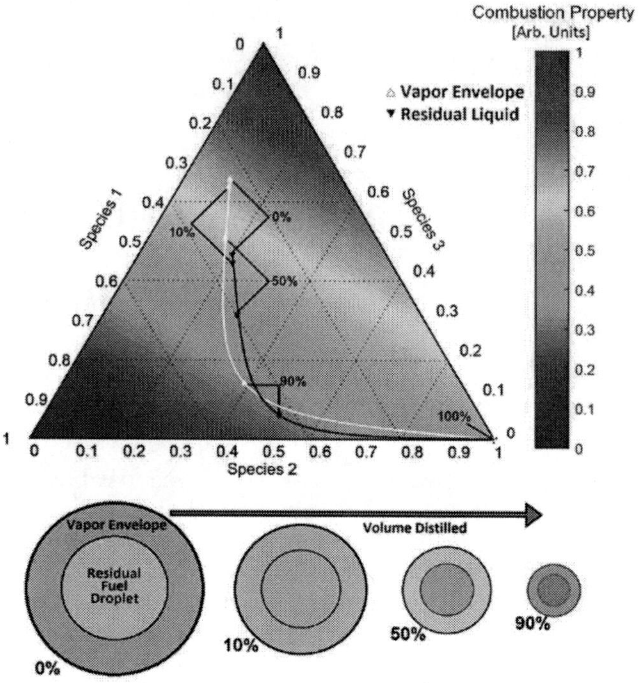

Figure 1: Distillation and combustion property evolution scheme from [8]. (Above) Ternary plot representing dissimilar combustion property evolution for liquid and vapor phases of a surrogate fuel. (Below) The associated droplet representation progressing through dissimilar liquid and vapor combustion properties as it undergoes equilibrium-limited vaporization.

To test the preferential vaporization hypothesis presented in [8], the current work experimentally compares a form of lean blowout (LBO) threshold among blends of "effectively equivalent" fuel surrogate mixtures developed for prevaporized combustion applications. The blowout metric serves to stress the fuels to limit conditions and exacerbate effects of property stratification. Accordingly, this metric may be a useful tool for developing and comparing fuel surrogates.

2. Experimental Methods

The atmospheric pressure spray burner apparatus developed for present blowout experiments is pictured in Figure 2. An annular co-flow of ambient air is supplied to the system through a variable speed air blower (1). Co-flow velocity is incrementally varied using a Tenma 72-7270 decade resistance box which precisely varies the power delivered to the blower. The air intake charge temperature is measured using a thermocouple (2), which permits flow rate correction for ambient temperature variation. Blower flow rate has been measured with a hand-held anemometer (Holdpeak 866B) and correlated to applied resistance over a broad range of ambient conditions. The blowout threshold (velocity) measurements subsequently presented result from this correlation. Co-flow passes through a series of removable and reconfigurable flow conditioners (3) and into the quartz combustion chamber (5) where a standard, interchangeable home heating oil spray nozzle (4) feeds a steady flow of atomized fuel to the system. Fuel is stored in a pressure resistant stainless steel vessel (6). A fueling valve (7) enables rapid refueling between tests, while a purge valve (9) permits draining and cleaning of the fuel tank to avoid chemical contamination between tests of dissimilar fuels. Steady fuel flow is delivered to the spray nozzle by supplying a constant, 100 psig regulated air pressure (8) above the liquid in the tank. This pressure is the spray nozzle design pressure for all nozzles used in present experiments. Experiments are monitored and recorded using separate CCTV cameras for liftoff (10) and blowout (11). Camera outputs can be fed to post-processing software for analysis. Blackout panels (12) surrounding the experiment provide the requisite versatility in light exposure to enable both liftoff and blowout tests to be performed with limited modification to the apparatus.

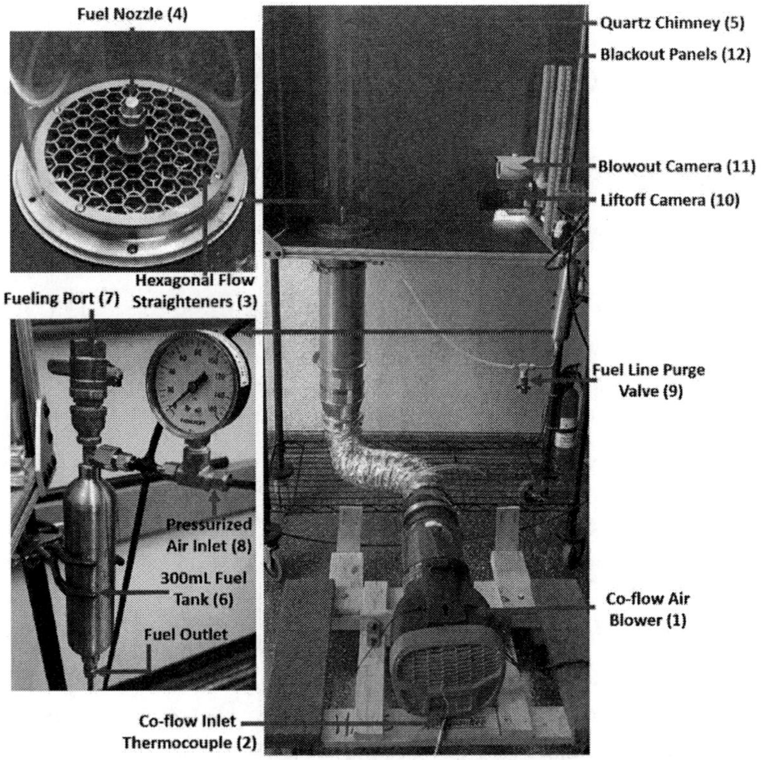

Figure 2: Spray burner rig configuration

The present work considers determination of blowout thresholds for fuels/mixtures. The blowout threshold – here, reported as the average cold-flow velocity across the quartz chimney – is determined by using the resistance decade box to incrementally increase the power delivered to the co-flow blower until the flame destabilizes and blows off. The resulting increased air flow rate tends to destabilize the pre-ignited flame by both increasing the local velocity against which the existing flame must propagate to achieve flame holding and by globally leaning the air-fuel mixture. Individual blowout events are unambiguous, though the blowout threshold appears to vary about a distribution assumed here to be normal. In practice, present determination of blowout threshold differs from conventional LBO tests in which the co-flow is held essentially constant while fuel supply rate is gradually reduced [9]. However, the present method is more suitable for this apparatus and the facilities that house it, and the measurements it produces nevertheless appear to support interpretation with which to explore the hypothesis advanced in the Introduction.

3. Results and Discussion

The combustion response of the fuels tested in the spray burner apparatus is necessarily dependent on uncontrollable variations in ambient conditions as well as particular experimental configurations (e.g., nozzle selection, flow straightener arrangement). To enforce consistency among data presented here, each (separate) set of measurements below in Figures 3-5 was taken as a series of determinations made with the same experimental configuration and in quick succession under similar atmospheric conditions. Accordingly, insights from dataset to dataset reasonably extrapolate, but direct quantitative comparison is unadvisable.

Key to understanding potential consequences of preferential vaporization on the combustion behaviors of multicomponent fuels is examination of chemical and physical fuel properties on relevant combustion performance metrics. Figure 3 uses pure component fuels to examine chemical and physical property dependencies of the blowout threshold, while also demonstrating the sensitivity of the present experiment to these parameters. The n-alkane series (nC7, nC10, nC12) represents species with highly similar prevaporized combustion chemistry [13, 14] and varying physical characteristics (e.g., normal boiling point). Blowout velocity varies modestly as a function of physical property effects (c.f. Table 1) on the complex spray combustion environment. On the other hand, nC7 and iC8 are species with similar physical properties and considerably different prevaporized combustion behavior. Figure 3 shows blowout thresholds that appear to reflect the relatively higher reactivity accorded to n-alkanes compared to their permethylated isomers over a broad range of combustion conditions. Taken as a whole, it is evident from Figure 3 that both chemical and physical properties contribute to blowout behavior. However, assuming fuels share similar volatility characteristics, the role of prevaporized chemistry seems to be significantly more influential than the modest variations in other physical properties. Though based only on pure component behavior, this experimental observation supports, if incompletely, the premise that preferential vaporization of more or less reactive species may influence limiting combustion behaviors such as LBO.

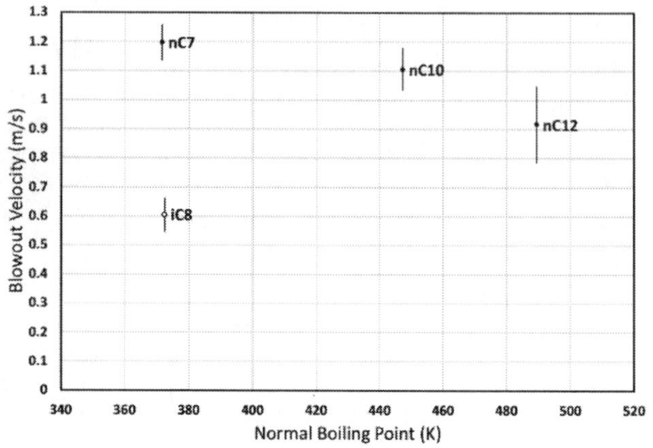

Figure 3: Blowout velocities of n-alkanes and iso-octane (iC8) measured using one consistent experimental configuration. Error bars are 1σ.

Having demonstrated that the blowout experiment is sufficiently sensitive to discriminate among fuel chemical and physical property variations, additional experiments were pursued to see if differences could be experimentally detected between essentially equivalent prevaporized surrogates for Jet-A [15], and separately, 95 RON gasoline [4]. Distillation-induced combustion property stratification for these surrogates was previously examined in [8], which suggested that the surrogates within respective fuel sets would behave differently – similarly to nC7 and iC8 in Figure 3 – based on distillation coupling to overall combustion reactivity (as measured by an *effective* (D)CN or RON "reactivity" index).

Table 1: Selected properties of pure component fuels used in this study [10-12]

Fuel	MW (g/mol)	H/C	(D)CN	RON	NBP (K)	ρ (kg/m^3)	μ × 10^4 (Pa.s)	σ × 10^2 (N/m)
n-heptane (nC7)	100.2	2.29	53.8	0	372	679	4.14	2.03
iso-octane (iC8)	114.2	2.25	18	100	372	688	4.79	1.84
n-decane (nC10)	142.3	2.20	65.5	-30	447	726	8.51	2.34
n-dodecane (nC12)	170.3	2.17	78	-40	489	745	13.54	2.48
n-hexadecane (nC16)	226.4	2.13	100		560	770	31.00	2.72
toluene (Tol)	92.1	1.14	0.2	120	384	862	5.58	2.79
1,3,5-trimethylbenzene (135TMB)	120.2	1.34	8		438	861	6.00	2.82
Ethanol (EtOH)	46.1	3.00	2.2	108	351	786	10.90	2.19

density, viscosity, and surface tension for liquid at 298 K

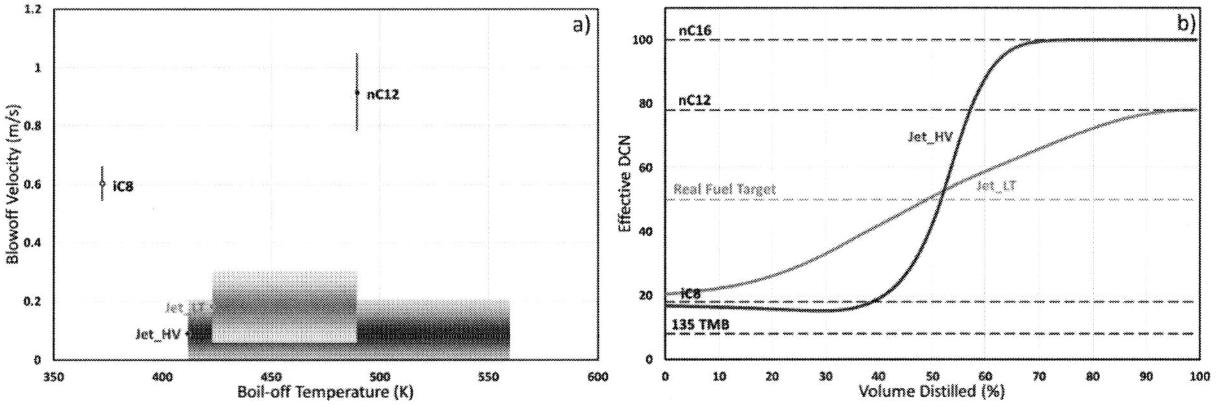

Figure 4: (a) Blowout velocities for Jet-A surrogates and some of their constituent pure components. Width of the surrogate fuel streaks indicate experimental error. (b) Distillation-resolved reactivity progression as indicated by an effective DCN value.

Figure 4a shows blowout thresholds for light (Jet_LT, Surrogate 1 in [15]) and heavy (Jet_HV, (Surrogate 2) distillation endpoint Jet-A surrogates developed in [15], and compares them to their common iC8 and nC12 components measured using the same experimental configuration (c.f. Figure 3). The multicomponent surrogates are characterized by a range of distillation temperatures, and cross-comparison of surrogates and pure components along this coordinate indicates that blowout behavior is clearly not governed by the endpoint component (e.g., Jet_LT vs. nC12). Both surrogates actually blowout far easier than the pure component alkanes indicated in the figure. This is likely due to the low reactivity and somewhat lower (than iC8) volatility of the relatively similar quantities of 135TMB present in each surrogate. In principle, the blowout threshold for this pure component fuel could be measured and added to this dataset to further clarify interpretation; however, facility limitations did not permit pure component aromatic testing due to heavy soot production.

A discrepancy in blowout threshold – albeit within overlapping 1σ error bars – is observed even though these surrogates are considered to be essentially equivalent when undergoing prevaporized combustion. The difference in blowout limits follows the preferential vaporization phenomenology discussed in [8]. The relative compositional difference between Jet_LT and Jet_HV results in Jet_HV swapping high-reactivity nC12 for heavier, similar reactivity nC16 while also having more light, low-reactivity iC8. Following the arguments for Figure 3, the *a priori* prediction is that that Jet_HV will be easier to blowout than Jet_LT, which is what is observed in Figure 4a. Figure 4b reiterates the predicted [8] change in reactivity (represented by an effective (D)CN) as the surrogates distill. Compared to Jet_LT, less reactive, more volatile components distill off first for Jet_HV and depress reactivity of the region (ostensibly) near the base of the spray flame. A reduced blowout threshold results from this reduced reactivity failing to maintain kinematic balance between downward flame propagation and upward bulk flow. Heat is not sufficiently transferred to vaporize heavier components of the fuel, leading to further reduced reactivity and an instability that results in blowout.

Blowout limits for a pair of gasoline surrogates (Sur95t, Sur95o) described in [4] and several of their constituent pure components were also measured. Figure 5 shows that the difference in fuel nozzle employed has a significant impact on blowout behavior via combustion-coupled spray dynamics and fuel volumetric flow. This makes a one-to-one comparison between Figures 5a and 5b difficult even with nC7 measurements acting as a reference; nevertheless, blowout thresholds

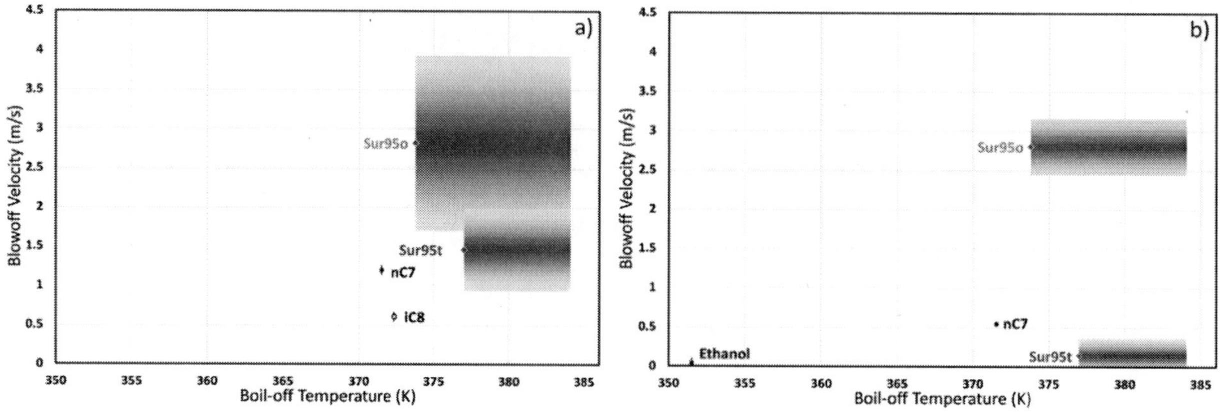

Figure 5: (a) Blowout velocities for 95 RON gasoline surrogates and some of their constituent pure components. Same experimental configuration as in Figs. 3 & 4. (b) Alternative experimental configuration.

for the surrogate fuels trend similarly in both configurations. In both cases, the ethanol-containing Sur95o resisted blowout at velocities greater than either Sur95t or the high-reactivity nC7. At present, these data appear anomalous but for the consistency, and further investigation in warranted. Nonetheless, the data again show differences in multiphase combustion behavior for essentially equivalent prevaporized surrogate fuels.

4. Conclusions

Blowout data obtained in the present spray burner experiments indicates blowout threshold dissimilarities between pairs of essentially equivalent surrogates developed to emulate prevaporized combustion of jet fuel and gasoline. These results suggest that the preferential vaporization inherent in multicomponent surrogate fuel sprays leads to a stratified combustion environment that may exhibit behaviors unlike those for other surrogates sharing the same prevaporized combustion characteristics, which is an observation that is consistent with computational predictions from previous work [8]. While much of this experimental work is still preliminary, the observation that fuel distillation affects spray flame blowout thresholds is supported for differing fuel design targets (Jet-A and 95 RON gasoline) and combustion environments (0.4 GPH solid cone versus 0.5 GPH hollow cone fuel spray nozzles). Further experiments are planned to more authoritatively demonstrate distillation behavior as a discriminant for effectively equivalent surrogates developed for prevaporized combustion.

5. Acknowledgements

The authors thank Robert Timmerman of Delavan Spray Technologies for providing several spray nozzles of varying flow rate and spray geometry rating.

6. References

[1] W.J. Pitz, N.P. Cernansky, F.L. Dryer, F.N. Egolfopoulos, et al., Development of an experimental database and chemical kinetic models for surrogate gasoline fuels, Report No. 0148-7191, SAE Technical Paper, 2007.

[2] S. Dooley, S.H. Won, J. Heyne, T.I. Farouk, et al., The experimental evaluation of a methodology for surrogate fuel formulation to emulate gas phase combustion kinetic phenomena, Combustion and Flame 159 (2012) 1444-1466.

[3] D. Kim, J. Martz, A. Violi, A surrogate for emulating the physical and chemical properties of conventional jet fuel, Combustion and Flame 161 (2014) 1489-1498.

[4] C. Pera, V. Knop, Methodology to define gasoline surrogates dedicated to auto-ignition in engines, Fuel 96 (2012) 59-69.

[5] S. Dooley, S.H. Won, M. Chaos, J. Heyne, et al., A jet fuel surrogate formulated by real fuel properties, Combustion and Flame 157 (2010) 2333-2339.

[6] F.L. Dryer, S. Jahangirian, S. Dooley, S.H. Won, et al., Emulating the Combustion Behavior of Real Jet Aviation Fuels by Surrogate Mixtures of Hydrocarbon Fluid Blends: Implications for Science and Engineering, Energy & Fuels 28 (2014) 3474-3485.
[7] M. Mehl, J.Y. Chen, W.J. Pitz, S.M. Sarathy, et al., An Approach for Formulating Surrogates for Gasoline with Application toward a Reduced Surrogate Mechanism for CFD Engine Modeling, Energy & Fuels 25 (2011) 5215-5223.
[8] J.A. Lefkowitz, F.M. Haas, Distillation-resolved evolution of key combustion properties, 10th US National Combustion Meeting, College Park, MD, 2017, pp. paper 1C18.
[9] A.H. Lefebvre, Gas turbine combustion: alternative fuels and emissions, 3rd ed., CRC Press, Hoboken, NJ, 2010.
[10] J. Yanowitz, M.A. Ratcliff, R.L. McCormick, J.D. Taylor, et al., Compendium of Experimental Cetane Numbers, Report No. NREL/TP-5400-61693, NREL, 2014.
[11] M. Frenkel, R.D. Chirico, V. Diky, X. Yan, et al., ThermoData Engine (TDE): software implementation of the dynamic data evaluation concept, Journal of chemical information and modeling 45 (2005) 816-838.
[12] CloudFlame Octane Number Calculator. cloudflame.kaust.edu.sa/fuel/octane_calc (accessed Jan. 21 2018).
[13] D.F. Davidson, S.C. Ranganath, K.Y. Lam, M. Liaw, et al., Ignition delay time measurements of normal alkanes and simple oxygenates, Journal of Propulsion and Power 26 (2010) 280-287.
[14] C. Ji, E. Dames, Y.L. Wang, H. Wang, et al., Propagation and extinction of premixed C5–C12 n-alkane flames, Combustion and Flame 157 (2010) 277-287.
[15] S.H. Won, F.M. Haas, S. Dooley, T. Edwards, et al., Reconstruction of chemical structure of real fuel by surrogate formulation based upon combustion property targets, Combustion and Flame 183 (2017) 39-49.

Spring Technical Meeting
Eastern States Section of the Combustion Institute
March 4–7, 2018
State College, Pennsylvania

Addressing discrepancies in hydrogen abstraction by OOH radical via automatic transition state theory calculations

Nathan Harms[1] *and Richard H. West*[1,*]

[1]*Department of Chemical Engineering, Northeastern University, Boston, MA, USA*
[*]*Corresponding author:* r.west@northeastern.edu

Abstract: Kinetic modeling of combustion chemistry has made substantial progress over recent years and has driven the development of new models. However, with the production of new models, disparities in reaction kinetics arise. We propose a workflow that identifies hydrogen abstraction reactions by hydroperoxyl radical (·OOH) in different models, performs DFT calculations on these reactions, and finds discrepancies in kinetics automatically. Abstraction by the ·OOH radical was chosen because this class of reaction forms hydrogen peroxide, which readily dissociates to produce highly unstable hydroxyl radicals. From 72 combustion models, we have identified 645 hydrogen reactions by OOH radical, and performed automated transition state calculations using AutoTST [1], a module included in the Reaction Mechanism Generator (RMG) [2]. AutoTST uses a training set of transition state calculations and a functional group decision tree to create estimates of the transition state geometries. It then creates estimates of the transition state geometry, performs a series of partial optimizations to converge on a finalized geometry, and implements intrinsic reaction coordinate calculations to verify the transition state. CanTherm [3], a canonical transition state calculator included with RMG, is then used to obtain kinetic parameters. Rates calculated by AutoTST were swapped into existing mechanisms for the combustion of butanol and heptane to identify reactions with large effects on ignition delay. These reactions are studied in depth to identify erroneous kinetic models.

1. Introduction

1.1 Generation of new models

Dozens of recent models, which describe tens of thousands of reactions between thousands of intermediate species, are capable of explaining complicated combustion phenomena, allowing increasingly accurate engine simulations. The construction of these models involves determining many thousands of thermodynamic and kinetic parameters using experimental measurements, theoretical calculations, estimations, or educated guesses. Most required parameters are not known, and are estimated or guessed. Given that many methods exist for determining reaction rates of identical reactions, discrepancies arise and need to be addressed.

1.2 Automated transition state theory calculations

To address discrepancies in the literature, we used our Automated Transition State Theory method (AutoTST) to perform automated transition state theory (TST) calculations. The geometries and thermochemical properties of the reactants and products can been found using the automated

"Quantum-Chemical Calculation of Thermochemical Properties" (QMTP) module [2] included in RMG [4]. AutoTST then provides estimates of transition state (TS) geometries using a group contribution method [5, 6] and embedding tools from the open-source cheminformatics program RDKit [7]. These TS geometries are created by editing portions of the bounds matrix corresponding to the reaction core. The reaction core (for bimolecular hydrogen abstraction reactions) represents the hydrogen being abstracted, the radical atom performing the abstraction, and the atom initially bound to the hydrogen being abstracted. AutoTST uses a hierarchal tree to provide guesses for the distances between each of the three atoms of the reaction core and edits the bounds matrix accordingly. Next, the TS geometry undergoes a series of partial optimizations using the Gaussian 09 quantum package [8] to arrive at a final transition state. Finally, the TS is verified using intrinsic reaction coordinate calculations [9], and kinetic parameters are determined using CanTherm, a canonical transition state calculator [3].

1.3 Ignition delay calculations

Hydrogen abstraction by ·OOH was selected for this study because these reactions are the first reaction in all combustion models and this class of reaction forms hydrogen peroxide, which readily dissociates to produce highly unstable hydroxyl radicals. Given this importance, these reactions often have a significant effect on ignition delay.

2. Methods

72 kinetic models from the literature were were imported into a common database and then searched for hydrogen abstraction reactions by ·OOH radical. Identified reactions were then passed through the AutoTST framework. AutoTST created estimates of TSs which were optimized in Gaussian 09 [8] using the M06-2X DFT functional [10, 11] with the MG3S basis set (equivalent to 6-311+G(2df,2p) for systems containing C, H, and O) [12, 13].

Successfully calculated reaction kinetic parameters were swapped into existing models from Lawrence Livermore National Laboratory (LLNL) for the combustion of butanol [14] and heptane [15] to generate alternative models allowing us to determine the impact of using our kinetics on ignition delay predictions. Ignition delays were calculated between 500 and 2500 K (at 500 K increments) at 10,000 Pa. Alternative models for n-butanol were also compared to experimental data using PyTeCK [16].

3. Results and discussion

645 reactions were identified as bimolecular hydrogen abstraction reactions by ·OOH radical. Of these, 595 have completed calculations, and kinetic parameters for 425 reactions were calculated and verified. Of the successful AutoTST calculated reactions, 56 were present in the butanol model and 118 were in the heptane model. The AutoTST calculated kinetic parameters were swapped into these models to generate alternative models to be tested.

To assess the impact of swapped models against parent models, we implemented the following metric:

$$I_i = \frac{1}{N_j} \sum_{j=1}^{N_j} \left| (\log \tau_{0,j} - \log \tau_{i,j}) \right| \tag{1}$$

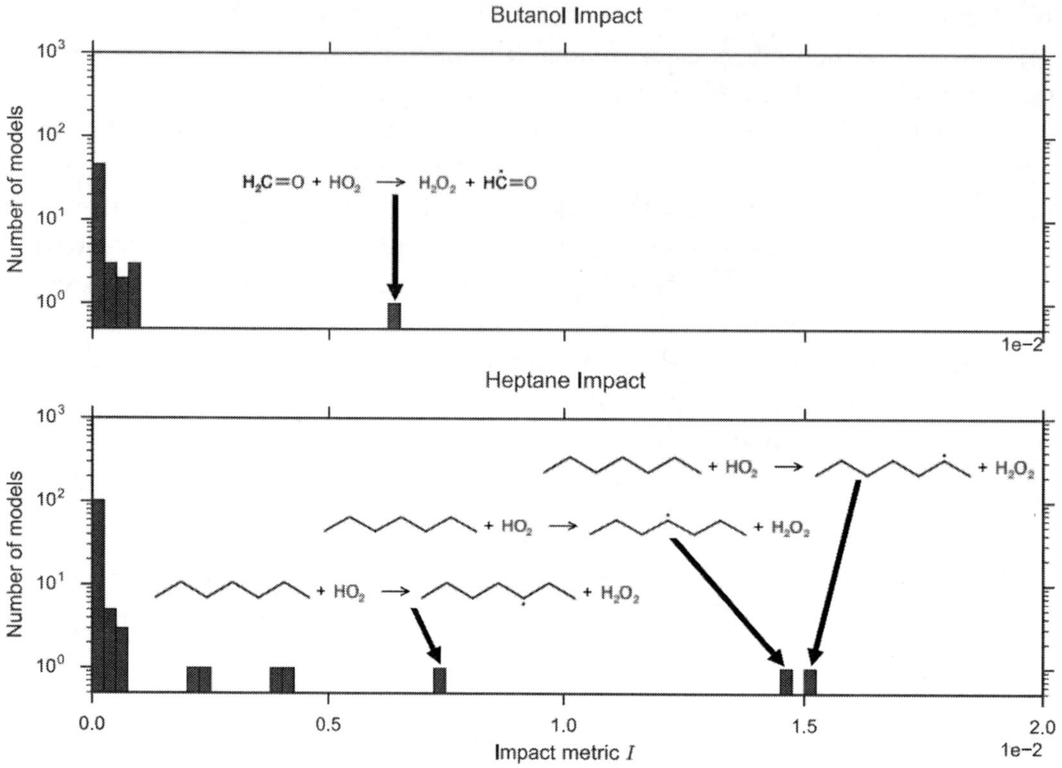

Figure 1: Histograms of the impact metric I from 56 and 118 swapped models for the combustion of butanol and heptane, respectively. Most swaps resulted an impact of less than 0.005 while 1 and 3 reactions were above this value for their respective models. These reactions are shown on their respective plots.

Where I_i is the impact of the ith swapped model on the ignition delay compared to the parent model, N_j is the number of temperatures that ignition delay was calculated at, $\tau_{0,j}$ is the ignition delay of the parent model at the jth temperature and $\tau_{i,j}$ is the ignition delay of the ith swapped model at the jth temperature. From our analysis, we identified 1 reaction that had a relatively high impact on the ignition delay of butanol and 3 reactions for heptane. The impact of all swapped models are on figure 1.

Ignition delay calculations of swapped models were compared to 4 experimental data sets [17–20] using PyTeCK [16], a Python package to compare chemical kinetic models against experimental data. PyTeCK compares experimental data against chemical kinetic models to return an error metric as described in [21]

Overall, a majority of our swapped models had an error metric that differed by less than 0.01 compared to the parent model for butanol combustion. The reaction change that had the biggest effect on overall error was $CH_3CH_2CH_2CH_2OH + \cdot OOH \rightleftharpoons CH_3CH_2CH_2\dot{C}HOH + H_2O_2$. The error metric for all swapped models compared to experimental data is displayed in figure 2.

3.1 Case Study: hydroperoxyl radical abstracting from diatomic hydrogen

In order to test the use of AutoTST as a work flow to identify discrepancies and resolve errors in more literature models, we chose to perform a brief investigation of the following reaction: H_2 +

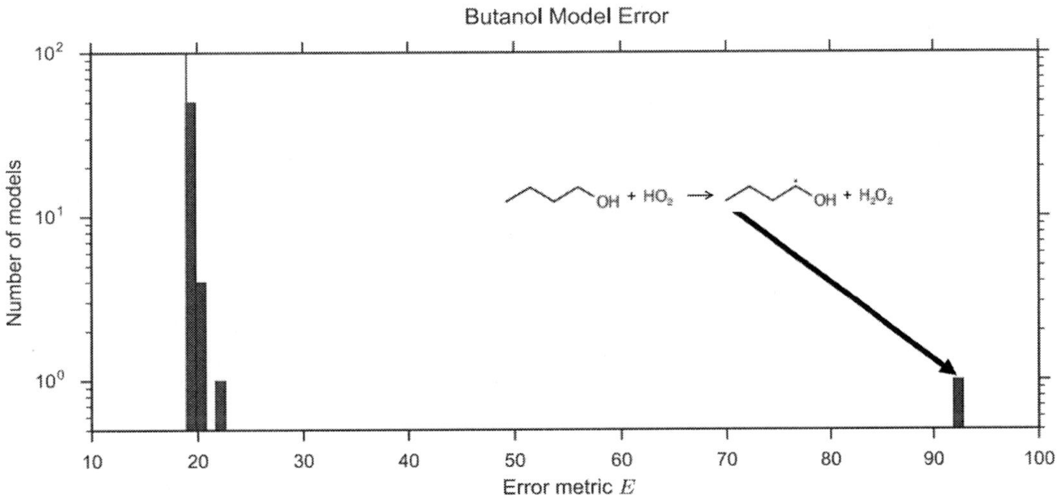

Figure 2: Histogram of error metric E from 56 models for butanol combustion, values ranged from 18.92 to 92.82. Most alternative models changed E by less than 0.01 from the original value of 18.98 (shown by the red line).

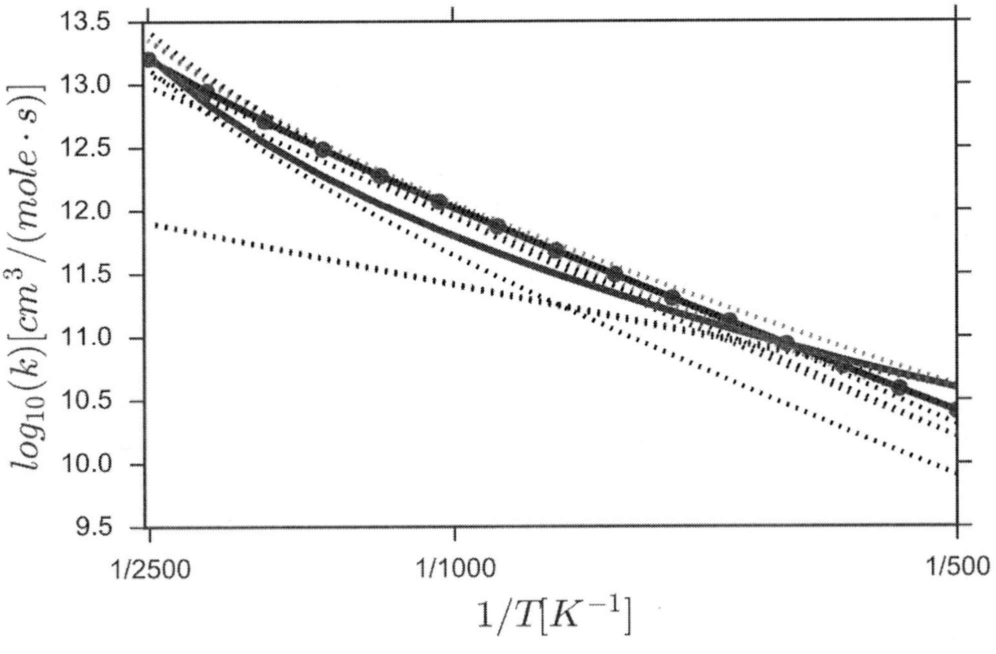

Figure 3: Arrhenius plot of the reaction $H_2 + \cdot OOH \rightleftharpoons \cdot H + H_2O_2$. 68 models contained kinetic parameters for this reaction and their reaction rates are plotted as dashed lines. AutoTST calculated kinetics are given by the red line, the original butanol combustion model is given the blue line, and the original heptane combustion model is given by the green dots.

$\cdot OOH \rightleftharpoons \cdot H + H_2O_2$. This reaction presents itself in 68 out of the models that we studied and we had close agreement with the kinetics in both models for butanol and heptane combustion. To identify erroneous models, an Arrhenius plot was generated with the 68 models containing this reaction shown in figure 3.

The kinetics determined through the AutoTST workflow align well with the general trend of most models; however, there are some discrepancies with some models at extreme low and extreme high temperatures. At the time of publication, we have not fully investigated these discrepancies, but this will be presented at a later date.

4. Conclusions

Through our work flow, we have identified kinetic parameters for 599 hydrogen abstraction reactions by hydroperoxyl radical with a success rate of 65%. Of the reaction kinetics that we calculated, 56 and 118 were present in LLNL combustion models for butanol and heptane, respectively. We then swapped our kinetic parameters into the parent models and quantified the impact of our parameters on the ignition delay. We saw that a majority of our kinetics have minimal impact while 4 reactions in total had a relatively large impact (1 of the 56 calculated reactions from the butanol model and 3 of the 118 reactions from the heptane model). We also compared our swapped models of butanol combustion to experimental data using PyTeCK, and most substitutions caused minimal change in the overall experimental error metric. Only 1 of 56 calculated reactions from the butanol model introduced a large error; this will be investigated further.

For the two models studied here, the workflow further validates the models: most of our automated TST calculations have minimal impact on ignition delays, and a few that do, seem to make the agreement with experimental ignition delay times (for butanol) worse. However, when applied to models containing more mistakes, discrepancies, or inaccuracies, we expect the workflow will help identify these deficiencies and become a tool to improve, rather than just validate, the models.

In addition, the study on the case reaction ($H_2 + \cdot OOH \rightleftharpoons \cdot H + H_2O_2$) highlights a unique use of this tool: verification and analysis of models to identify errors. Being able to automatically perform DFT calculation to spot-check important reactions allows us to better understand the chemistry happening in these models and will bring us closer to the truth. A more thorough analysis of these reactions allows us to sift through the literature and identify and identify the sources of these numbers. Moving forward, we plan to perform this analysis on the 5 reactions highlighted in our impact and error assessment; expand the scope of this project to include hydrogen abstraction reactions by other abstractors to highlight the versatility of our tool; and, improve additional models through our the impact and error analysis.

Acknowledgements

This material is based upon work supported by the National Science Foundation under Grant No. 1605568, and the Department of Chemical Engineering at Northeastern University.

References

[1] P. L. Bhoorasingh, B. L. Slakman, F. Seyedzadeh Khanshan, J. Y. Cain, and R. H. West, Automated Transition State Theory Calculations for High-Throughput Kinetics, The Journal of Physical Chemistry A 121 (2017) 6896–6904. DOI: 10.1021/acs.jpca.7b07361.

[2] G. R. Magoon and W. H. Green, Design and implementation of a next-generation software interface for on-the-fly quantum and force field calculations in automated reaction mechanism generation, Comput. Chem. Eng. 52 (2012) 35–45.

[3] J. W. Allen and W. H. Green, *CanTherm: Open-source software for thermodynamics and kinetics*, Included in: Reaction Mechanism Generator, v2.0.0. 2016 , URL: http://reactionmechanismgenerator.github.io.

[4] C. W. Gao, J. W. Allen, W. H. Green, and R. H. West, Reaction Mechanism Generator: Automatic construction of chemical kinetic mechanisms, Comput. Phys. Commun. 203 (2016) 212–225. DOI: 10.1016/j.cpc.2016.02.013.

[5] P. L. Bhoorasingh and R. H. West, Transition state geometry prediction using molecular group contributions, Phys. Chem. Chem. Phys. 17 (2015) 32173–32182. DOI: 10.1039/C5CP04706D.

[6] P. Bhoorasingh, B. Slakman, F. S. Khanshan, J. Cain, and R. West, Kinetic data for manuscript describing the AutoTST algorithm for automated Transition State Theory calculations of chemical reaction rates. figshare.com (2016) 10.6084/m9.figshare.4234160.

[7] G. Landrum, *RDKit: Open-source cheminformatics*, URL: http://rdkit.org.

[8] M. J. Frisch et al., *Gaussian 09*, Gaussian, Inc., Wallingford, CT, 2009.

[9] K. Fukui, The path of chemical reactions-the IRC approach, Acc. Chem. Res. 14 (1981) 363–368.

[10] Y. Zhao and D. G. Truhlar, The M06 suite of density functionals for main group thermochemistry, thermochemical kinetics, noncovalent interactions, excited states, and transition elements: two new functionals and systematic testing of four M06-class functionals and 12 other functionals, Theor. Chem. Acc. 120 (2008) 215–241.

[11] Y. Zhao and D. G. Truhlar, A new local density functional for main-group thermochemistry, transition metal bonding, thermochemical kinetics, and noncovalent interactions, J. Chem. Phys. 125 (2006) 194101.

[12] M. J. Frisch, J. A. Pople, and J. S. Binkley, Self-consistent molecular orbital methods 25. Supplementary functions for Gaussian basis sets, J. Chem. Phys. 80 (1984) 3265–3269.

[13] T. Clark, J. Chandrasekhar, G. n. W. Spitznagel, and P. V. R. Schleyer, Efficient diffuse function-augmented basis sets for anion calculations. III. The 3-21+G basis set for first-row elements, Li-F, J. Comput. Chem. 4 (1983) 294–301.

[14] S. M. Sarathy, S. Vranckx, K. Yasunaga, M. Mehl, P. Oßwald, W. K. Metcalfe, C. K. Westbrook, W. J. Pitz, K. Kohse-Höinghaus, R. X. Fernandes, and H. J. Curran, A comprehensive chemical kinetic combustion model for the four butanol isomers, Combust. Flame 159 (2012) 2028–2055. DOI: 10.1016/j.combustflame.2011.12.017.

[15] M. Mehl, W. J. Pitz, C. K. Westbrook, and H. J. Curran, Kinetic modeling of gasoline surrogate components and mixtures under engine conditions, Proc. Combust. Inst. 33 (2011) 193–200. DOI: 10.1016/j.proci.2010.05.027.

[16] K. E. Niemeyer, *PyTeCK v0.2.1*, 2017.

[17] J. T. Moss, A. M. Berkowitz, M. A. Oehlschlaeger, J. Biet, V. Warth, P.-A. Glaude, and F. Battin-Leclerc, An Experimental and Kinetic Modeling Study of the Oxidation of the Four Isomers of Butanol, J. Phys. Chem. A 112 (2008) 10843–10855. DOI: 10.1021/jp806464p.

[18] I. Stranic, D. P. Chase, J. T. Harmon, S. Yang, D. F. Davidson, and R. K. Hanson, Shock tube measurements of ignition delay times for the butanol isomers, Combust. Flame 159 (2012) 516–527. DOI: 10.1016/j.combustflame.2011.08.014.

[19] I. L. R. Bec, Y. Zhu, D. F. Davidson, and R. K. Hanson, Shock Tube Measurements of Ignition Delay Times for the Butanol Isomers Using the Constrained-Reaction-Volume Strategy, International Journal of Chemical Kinetics 46 (2014) 433–442. DOI: 10.1002/kin.20859.

[20] Y. Zhu, D. F. Davidson, and R. K. Hanson, 1-Butanol ignition delay times at low temperatures: An application of the constrained-reaction-volume strategy, Combust. Flame 161 (2014) 634–643. DOI: 10.1016/j.combustflame.2013.06.028.

[21] S. K. Sirumalla, M. A. Mayer, K. E. Niemeyer, and R. H. West, Assessing impacts of discrepancies in model parameters on autoignition model performance: A case study using butanol, Combustion and Flame 190 (2018) 284–292. DOI: 10.1016/j.combustflame.2017.11.018.

Spring Technical Meeting
Eastern States Section of the Combustion Institute
March 4–7, 2018
State College, Pennsylvania

Multiscale Informatics of Reactions involved in H_2O_2 Decomposition in the Presence of Dopants

Carly E. LaGrotta [1], *Mark C. Barbet* [1], *Lei Lei*[1], *and Michael P. Burke* [2,*]

[1]*Department of Mechanical Engineering, Columbia University, New York,New York, 10027, USA*
[2]*Department of Mechanical Engineering,Department of Chemical Engineering, and Data Science Institute, Columbia University, New York,New York, 10027, USA*
[*]*Corresponding author:* mpburke@columbia.edu

Abstract: Substantial recent improvements in the accuracy of ab initio theoretical kinetic calculations suggest the need for more advanced means of comparing theoretical and experimental data, and quantifying the information content from both, in a manner that considers each on an equal footing with their respective uncertainties. Building on uncertainty quantification methodologies developed previously, we have recently developed an uncertainty quantification approach that directly incorporates theoretical kinetics calculations into the model structure. In so doing, this MultiScale Informatics (MSI) approach can quantitatively integrate data from ab initio calculations and macroscopic experimental observables within a multiscale physics-based model structure. The resulting kinetics model from the MSI approach consists of optimized molecular parameters (with constrained uncertainties), which can be related through theoretical kinetics calculations to rate constants (with constrained uncertainties), which in turn can be related through physical models to macroscopic observables (with constrained uncertainties). Here we extend our previous MSI analysis of the H_2O_2 decomposition system (consisting of five controlling reactions) to include the additional controlling reactions for H_2O_2 decomposition under doping with other species such as CH_4 and H_2. This analysis produces optimized values and constrained uncertainties imposed by various data sources on reactions of HO_2 with other radicals highlighted as important reactions for combustion predictions in recent modeling studies. The results demonstrate the utility of theoretical data for quantifying uncertainties under extrapolation as well as the importance of incorporating raw experimental data for accurate quantification of their information content.
Keywords: Optimization, Uncertainty Quantification, Multi-scale modeling, Dopants

1. Introduction

In the past decade, there have been remarkable improvements in computational chemistry methodologies, which have facilitated improvements in the accuracy of *ab initio* thermochemical and kinetic calculations. Routine theoretical kinetics calculations have now progressed to a point where they are capable of accuracies of about a factor of two at combustion temperatures, while even more advanced kinetics calculations are capable of agreement with experimental data within 20%. To put these uncertainties in perspective, the best possible experimental predictions, under the most favorable of circumstances, still have an uncertainty of 10% [1]. It therefore appears useful to incorporate both theoretical and experimental data, along with their associated uncertainties, in the development of high-accuracy kinetic models.

Leveraging the uncertainty quantification methodologies [2–8], we have developed an approach that directly incorporates theoretical and kinetics calculations, as well as their respective uncertainties into a model structure and uncertainty quantification framework [8–10]. This MultiScale Informatics (MSI) approach allows the integration of the constraints imposed by *ab initio* calculations, rate constant determinations, and macroscopic experimental observations within a multi-physics-based model structure. This approach results in a kinetics model that consists of optimized molecular parameters, which can be related through theoretical kinetics calculations to rate constants, which in turn can be related through physical models to macroscopic observables, each with their respective constrained uncertainties. The MSI approach has been shown to yield high accuracy rate constants, with values and error bars that are consistent with later independent studies, making it a powerful tool for deriving high accuracy kinetic data from complex reaction systems.

Recent studies have indicated that uncertainties in H_2O_2/CH_3 and H_2O_2/CH_4 reactions have important ramifications for modeling of a variety of combustion systems. As natural gas becomes an increasingly viable energy source, understanding methane and hydrogen combustion systems and the uncertainties associated with them, becomes increasingly important [11, 12]. For example, the uncertainties in R6 and R7 (Table 1) were found to be among the largest contributors to the uncertainties in NO_x production in a natural gas turbine relevant flame [13] such that improved quantification, and reduction of, uncertainties in these reactions are of interest to computational engine design.

Here we present a MultiScale Informatics investigation of reactions involved in H_2O_2 decomposition in the presence of H_2 and CH_4 dopants in order to assess the extent to which our previous MSI analysis for H_2O_2 decomposition can be leveraged to obtain high-accuracy rate constants for reactions of CH_4 and H_2 with H_2O_2 from experimental measurements for these systems. In this study, we present results from uncertainty-weighted sensitivity analysis in order to discern the parameters influencing predictions and, therefore, the parameters that can be constrained by experimental measurements at those conditions. The results presented in this study constitute a first step towards potentially using these experiments to provide constraints on rate constants for these important reactions.

2. Methods

The objective of the MSI approach is to find optimized values and quantified uncertainties for a set of parameters that consist of molecular parameters within theoretical kinetics calculations, rate parameters, and physical model parameters within simulations of experimental observables. The mathematical implementation takes advantage of a linear "surrogate model" of the system in the neighborhood of the predicted active parameters, as described in [14]. The optimized values of the active parameters and the covariance matrix are then calculated using a standard weighted least-squares procedure, as used by Ruscic in the Active Thermochemical Tables, [7]. The optimization is performed in an iterative manner starting from the nominal parameters, and then solving iteratively to converge on the optimized values for the parameters. A more detailed description of this procedure can be found in [8].

The active parameters in this study include theoretical kinetics parameters, rate constant parameters, and physical model parameters. The target values include molecular properties from *ab initio* calculations, rate constant determinations, macroscopic experimental observables, and reported values for the experimental conditions. The inclusion of these different target classes as well as a more detailed description can be seen in [8].

2.1 Implementation

In this study, we extend our previous MSI implementation for the H_2O_2 system for reactions to include parameters and data relevant to H_2O_2 decomposition in the presence of H_2 and CH_4. The initial approach to understand this system included data from [8, 15–20], results for which are presented in [21]. This study additionally draws on experimental results from [22] when considering the CH_4 system and [23] when considering the H_2 system.

Here we will attempt to discern the parameters whose uncertainties influence predictions of experimental observables, which will allow us to determine the extent to which current experimental data can be used to constrain model parameters for reactions of CH_4 and H_2 with H_2O_2. The discussion below mainly focuses on results from uncertainty-weighted sensitivity analysis. These uncertainty weighted sensitivity coefficients indicate the parameters whose uncertainties influence uncertainties in the predictions and likewise, which parameters can be constrained by inclusion of data for those observables.

3. Results and Discussion

Table 1: Relevant Reactions

1	$H_2O_2(+M) <=> OH + OH(+M)$	12	$H + HO_2 <=> H_2O + O$
2	$H_2O_2 + OH <=> HO_2 + H_2O$	13	$CH_3 + H (+M) <=> CH_4 (+M)$
3	$HO_2 + HO_2 <=> H_2O_2 + O_2$	14	$CH_3 + O2 <=> CH_2O + OH$
4	$HO_2 + OH <=> H_2O + O_2$	15	$CH_2O + O_2 <=> HCO + HO_2$
5	$OH + OH <=> O + HO_2$	16	$CH_4 + O <=> CH_3 + OH$
6	$CH_3 + HO_2 <=> CH_4 + O_2$	17	$CH_3 + O_2 <=> CH_3O + O$
7	$CH_3 + HO_2 <=> CH_3O + OH$	18	$CH_3 + OH <=> CH_2(S) + H_2O$
8	$CH_4 + OH <=> CH_3 + H_2O$	19	$2 CH_3 (+M) <=> C_2H_6 (+M)$
9	$H + HO_2 <=> 2OH$	20	$H_2 + OH <=> H + H_2O$
10	$H + H_2O_2 <=> H_2O + OH$	21	$H + H_2O_2 <=> H_2 + HO_2$
11	$H + HO_2 <=> H_2 + O_2$	22	$H + H_2O_2 <=> H_2O + OH$

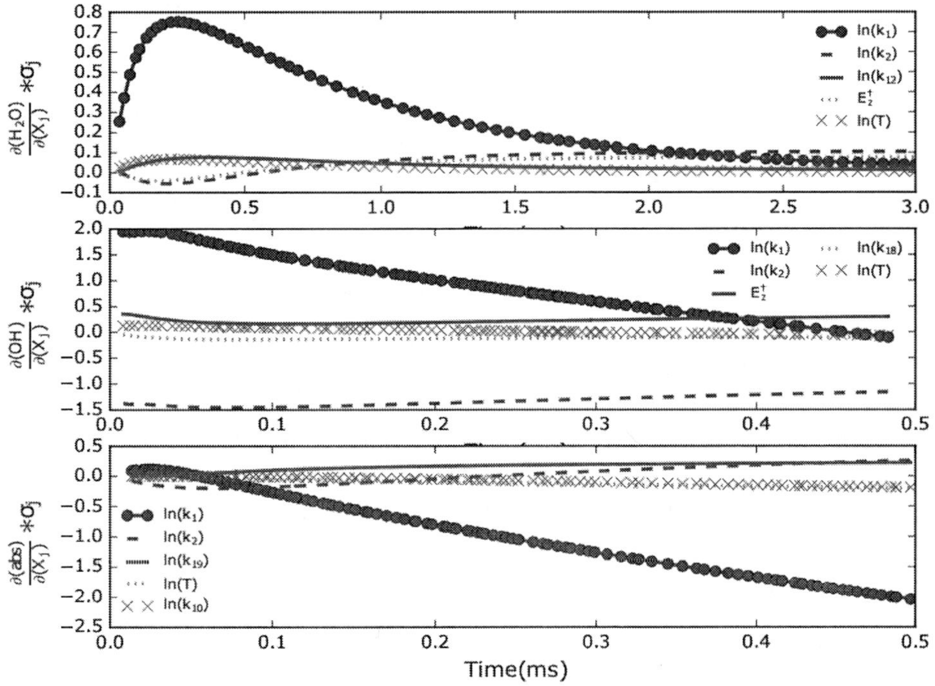

Figure 1: Uncertainty weighted sensitivity analysis of H_2O, OH, and absorbance at 227nm [15]

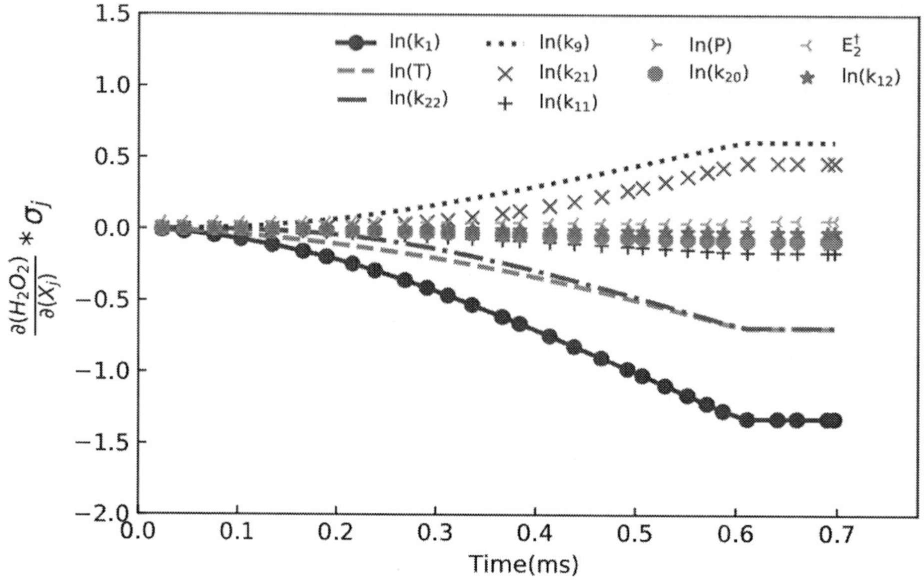

Figure 2: Uncertainty weighted sensitivity analysis of the ten most sensitive parameters to H_2O_2 time history for H2 doping [23]

Uncertainty-weighted sensitivity analysis of the time profiles for H_2O, OH and absorbance at 227 nm from shock-heated $H_2O_2/CH_4/Ar$ experiments are shown in Fig. 1. This analysis reveals substantial contributions from multiple sources to the uncertainty in the predictions of the experimental observables. In particular, in addition to being sensitive to parameters for reactions R1-R8, predictions of H_2O, OH, and absorption at 227 nm are sensitive to parameters for R12, R18, and R19 as well as physical model parameters – such that consideration of their uncertainties would also be required for interpreting the experimental data. An uncertainty weighted sensitivity analysis was also performed for shock-tube H_2O_2/H_2 experiments presented in Fig.2. As shown, the rate constants of R9 and R11, as well as R20-R22 appear in the top ten most sensitive parameters for the system doped with H_2, which is consistent with the analysis presented in [23] – such that experimental data under these conditions may constrain uncertainties for these reactions.

4. Conclusions

The results generated here present the first step toward extending previous MSI analysis for the H_2O_2 system to also include H_2O_2 doped with CH_4 and H_2, in order to optimize and quantify uncertainties for reactions of CH_3 and H with HO_2. Uncertainty weighted sensitivity analysis was performed to evaluate the extent to which available experimental data for these systems constrain uncertainties for these reactions. The analysis reveals that experimental observables in the H_2O_2/CH_4 system predominantly constrain rate constants for reactions within the H_2O_2 system (R1-R5) rather than $CH_3 + HO_2$ (R6-R7); however, experimental observables in the H_2O_2/H_2 system may provide meaningful constraints on rate constants for $H + HO_2/H_2O_2$ reactions.

5. Acknowledgements

The authors gratefully acknowledge the National Science Foundation for support of this research through a grant from the Combustion and Fire Systems program (#CBET-1733968). Acknowledgment is also made to the Donors of the American Chemical Society Petroleum Research Fund for support of this research through a Doctoral New Investigator (DNI) award. The authors would also like to thank Drs. Ahren Jasper and Stephen Klippenstein for providing the input files for the CH_3+HO_2 theory calculations.

References

[1] S. Klippenstein, Proc. Combust. Inst. 36 (2017) 77–111.

[2] M. Frenklach, A. Packard, P. Seiler, and R. Feeley, Collaborative data processing in developing predictive models of complex reaction systems, Int. J. Chem. Kinet. 36 (2004) 57–66.

[3] M. Frenklach, Transforming data into knowledge-Process Informatics for combustion chemistry, Proc. Combust. Inst. 31 (2007) 125–140.

[4] H. Wang and D. Sheen, Combustion kinetic model uncertainty quantification, propagation and minimization, Prog. En. Combust. Sci. 47 (2015) 1–31, URL: http://dx.doi.org/10.1016/j.pecs.2014.10.002.

[5] T. Turányi, T. Nagy, I. Zsély, and et al., Int. J. Chem. Kinet. 44 (2012) 284–302.

[6] C. Olm, T. Varga, É. Valkó, H. Curran, and T. Turányi, Uncertainty quantification of a newly optimized methanol and formaldehyde combustion mechanism, Combust. Flame 186 (2017) 45–64, URL: http://dx.doi.org/10.1016/j.combustflame.2017.07.029.

[7] B. Ruscic, R. Pinzon, M. Morton, G. Von Laszevski, S. Bittner, S. Nijsure, K. Amin, M. Minkoff, and A. Wagner, Introduction to active thermochemical tables: Several "Key" enthalpies of formation revisited, J. Phys. Chem. A 108 (2004) 9979–9997.

[8] M. Burke, S. Klippenstein, and L. Harding, A quantitative explanation for the apparent anomalous temperature dependence of OH+HO2=H2O+O2 through multi-scale modeling, Proc. Combust. Inst. 34 (2013) 547–555, URL: http://linkinghub.elsevier.com/retrieve/pii/S1540748912000429.

[9] M. Burke, C. Goldsmith, S. Klippenstein, O. Welz, H. Huang, I. Antonov, J. Savee, D. Osborn, J. Zádor, C. Taatjes, and L. Sheps, Multiscale Informatics for Low-Temperature Propane Oxidation: Further Complexities in Studies of Complex Reactions, J. Phys. Chem. A 119 (2015) 7095–7115.

[10] M. Burke, Harnessing the Combined Power of Theoretical and Experimental Data through Multiscale Informatics, Int. J. Chem. Kinet. 48 (2016) 212–235.

[11] M. Sangwan and L. Krasnoperov, Kinetics of the gas phase reaction CH3 + HO2, J. Phys. Chem. A 117 (2013) 2916–2923.

[12] S. Orhan Akansu, Z. Dulger, N. Kahraman, and T. Veziroglu Nejat, Internal combustion engines fueled by natural gas—hydrogen mixtures, J. of Hydrogen Energy 29 (2004) 1527–1539.

[13] A. C. A. Lipardi, J. M. Bergthorson, and G. Bourque, J. Eng. Gas Turb. Power 138 (2015) 051506.

[14] M. Frenklach, Systematic Optimization of A Detailed Kinetic Model Using a Methane Ignition Example, Combust. Flame (1984) 69–72.

[15] Z. Hong, D. Davidson, K.-Y. Lam, and R. Hanson, A shock tube study of the rate constants of HO2 and CH3 reactions, Combust. Flame 159 (2012) 3007–3013, URL: http://linkinghub.elsevier.com/retrieve/pii/S0010218012001484.

[16] A. Jasper, S. Klippenstein, and L. Harding, Theoretical rate coefficients for the reaction of methyl radical with hydroperoxyl radical and for methylhydroperoxide decomposition, Proc. Combust. Inst. 32 I (2009) 279–286, URL: http://dx.doi.org/10.1016/j.proci.2008.05.036.

[17] C. Kappel, K. Luther, and J. Troe, Shock wave study of the unimolecular dissociation of H2O2 in its falloff range and of its secondary reactions, Phys. Chem. Chem. Phys. 4 (2002) 4392–4398, URL: http://xlink.rsc.org/?DOI=b204364e.

[18] Z. Hong, A. Farooq, E. Barbour, D. Davidson, and R. Hanson, Hydrogen peroxide decomposition rate: a shock tube study using tunable laser absorption of H(2)O near 2.5 microm. J. Phys. Chem. A 113 (2009) 12919–12925.

[19] Z. Hong, K.-Y. Lam, R. Sur, S. Wang, D. Davidson, and R. Hanson, On the rate constants of OH + HO2 and HO2+HO 2: A comprehensive study of H2O2 thermal decomposition using multi-species laser absorption, Proc. Combust. Inst. 34 (2013) 565–571, URL: http://dx.doi.org/10.1016/j.proci.2012.06.108.

[20] Z. Hong, S. Vasu, D. Davidson, and R. Hanson, Experimental study of the rate of OH + HO2 H2O + O2 at high temperatures using the reverse reaction, J. Phys. Chem. A 114 (2010) 5520–5525.

[21] C. LaGrotta, M. Barbet, L. Lei, and M. Burke, Pending Publication ().

[22] N. Srinivasan, J. Michael, L. Harding, and S. Klippenstein, Experimental and theoretical rate constants for CH4 + O2 CH3 + HO2, Combust. Flame 149 (2007) 104–111.

[23] H. Kiljewski and J. Troe, Study of the Pyrolysis of H2O2 in the Presence of H2 and CO by Use of UV Absorption of HO2, J. Chem. Kinetics 3 (1971) 223–235.

Spring Technical Meeting
Eastern States Section of the Combustion Institute
March 4-7, 2018
State College, Pennsylvania

Studies of Pentane Oxidation and Pyrolysis in Nanosecond-pulsed Plasma Discharges using In-situ Laser Diagnostics

Aric Rousso[1], Xingqian Mao[1,2], Qi Chen[2], Yiguang Ju[1]*

[1]*Department of Mechanical and Aerospace Engineering*
Princeton University, New Jersey 08544, USA
[2]*School of Mechanical, Electronic and Control Engineering,*
Beijing Jiaotong University, Beijing 100044, China
[*]*Corresponding Author Email:* arousso@princeton.edu

This study explores plasma assisted pentane oxidation and pyrolysis in a nanosecond pulsed discharge using laser absorption spectroscopy and gas chromatography. An updated pentane model is developed with C0-C2 substitution by HP-Mech, a plasma chemistry sub-mechanism, and new electron impact dissociation reactions added. Using temperature as a model input, the electron impact cross sections of pentane are adjusted so the model accurately predicts the pyrolysis steady state and time-dependent species profiles, and then tested on the oxidation case. Results show favorable prediction of pentane consumption and H_2O and CH_2O formation. CH_4 is slightly under-predicted by the new model, while C_2H_2 is overpredicted, likely due to under-consumption, perhaps by radicals such as OH. The addition of electron impact dissociation reactions slows the overall oxidative pathway by a reduction in OH concentration and prevents over-prediction by adjusting the branching ratio towards smaller hydrocarbons instead of oxidative species.
Keywords: Plasma Assisted Combustion, Laser Diagnostics, Non-equilibrium plasma, low temperature oxidation

1. Introduction

Plasma assisted combustion (PAC) is an active area of research within the combustion community. Plasma has demonstrated the ability to enhance ignition behavior and accelerate slow low temperature chemistry, vital for ignition control in new, reduced emission engine designs that rely on two-stage combustion phenomenon. A nanosecond repetitively-pulsed discharge is advantageous for these conditions, as it generates high energy electrons and subsequent radical species without the large thermal heating that comes with conventional sparks, as well as providing a way to decouple the thermal and kinetic mechanisms for laboratory studies [1]. Previous experiments of smaller fuels, such as hydrogen [2,3], methane [4], and ethylene [3,5] have been carried out to identify kinetic enhancement pathways important to PAC as well as validate modeling efforts. Larger hydrocarbons, often liquid fuels requiring vaporization, require further study as they more closely match fuels currently in use, and often demonstrate complex low temperature chemistry that might interact with plasma pathways.

Our previous experimental work on n-heptane PAC [6] demonstrated significant fuel consumption and product species formation at low temperatures, but significant mismatch between the model and experiment even for main species such as H_2O and CH_2O. This suggests a lack of understanding of plasma reaction kinetics, missing pathways and a need for more experimental data of relevant parameters, such as electron impact cross sections. It is also critical to provide validation targets for plasma combustion kinetic models.

This study, therefore, will focus on the low temperature plasma assisted dissociation of pentane, both oxidation and pyrolysis, in a nano-second pulsed discharge using time dependent, *in-situ* laser absorption spectroscopy and steady state GC sampling measurements. An updated kinetic model with new electron impact dissociation reactions is developed and compared to the old model and experimental results to demonstrate improvement and highlight areas of further development.

2. Experimental and Numerical Methods

The experimental apparatus, diagrammed in Figure 1, has been reported in previous works [4–6], with slight modifications for the current experiment. Helium is used as a diluent for oxygen/pentane mixtures to prevent streamer and non-uniformity formation within the plasma. Fuel loading is kept at a constant of 2.8% with two conditions considered: a stoichiometric case with 75% helium dilution (0.028 C_5H_{12}, 0.22 O_2, 0.75 He mole fraction), and a pyrolysis case (0.028 C_5H_{12} 0.972 He mole fraction). Total flow rates are held at a constant of .2 m/s, resulting in a residence time of 0.22 seconds in the discharge region, using a Harvard Apparatus PHD 2000 Programmable syringe pump fed through a heated vaporization chamber for pentane. The reactor is kept at initial room temperature, 60 Torr pressure for all conditions. A nanosecond-pulsed power supply (FID GmbH FPG 30-50MC4) is used for plasma generation, which is operated in both continuous and burst modes via an external digital pulse generator (SRS DG535). The burst mode is utilized for all time dependent measurements with a total burst of 1500 pulses at 30 kHz, and a total plasma time of 0.05 seconds. Data sets are averaged over 25 runs, and demonstrated very high repeatability with little fluctuation. All time dependent data has been reported in number density (molecule/cm^3), as the exact overall number density due to rapidly changing temperature and reactive flow is unknown, increasing uncertainty in accurate conversion to mole fraction. All steady state GC measurements are reported in mole fraction as the sampling process removes the previous restraint.

A 24 pass Herriott cell is built into the walls of the reactor for mid-infrared laser absorption spectroscopy (TDLAS). Pentane is quantified using the PPNL gas database [7]. Due to its broadband absorptivity, the laser is parked at 1379 cm^{-1} and calibration curves are taken for both concentration and the temperature effect on the absorption cross section. Other species are quantified using the Beer Lambert law and the HITRAN database [8]. Quantification of water is at both 1338.55 cm^{-1} and 1339.15 cm^{-1} to utilize the two line absorption technique for temperature quantification [9], CH_4 at 1341.62 cm^{-1}, and C_2H_2 at 1342.35 cm^{-1}, all utilizing a Continuous Wave External Cavity Mode Hop Free (CW-EC-MHF) QCL from Daylight Solutions. CH_2O is quantified at 1726.79 cm^{-1} using a DFB-QCL from Alpes Lasers (sbcw3176). The addition of a DFB-QCL from Nanoplus (DFB-260290) allows for rapid quantification of OH radical at 3565.28 cm^{-1}. The scan rate of all lasers is monitored using a 50.8 mm Germanium etalon (FSR ≈ 0.74 GHz at 1345 cm^{-1}) and data is recorded on a Tektronix DPO 7104C oscilloscope. Steady state measurements using a gas chromatograph and thermal conductivity detector (TCD; Inficon 3000) are carried out for the pyrolysis case to provide validation targets for the model. Samples are extracted using a quartz sampling probe inserted just downstream of the plasma itself, and have a measurement uncertainty of 5%.

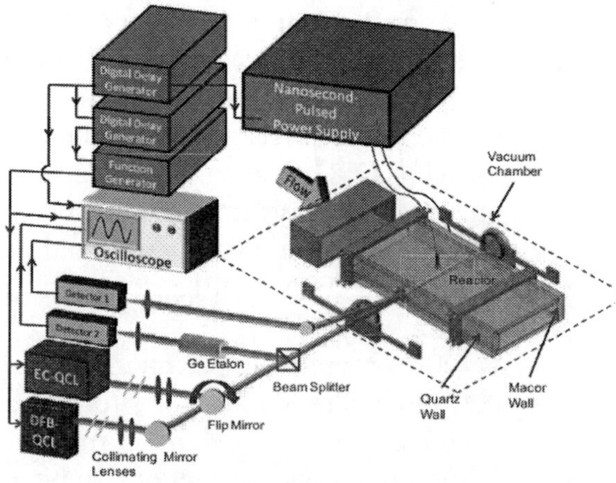

Figure 1: Experimental Setup

A zero-dimensional model was utilized to simulate the plasma assisted combustion process. The code couples the ZDPlasKin [10] and the CHEMKIN [11] packages to follow the time evolution of the species density and gas temperature in plasma assisted combustion. The detailed numerical method is described in Ref [4]. The pentane oxidation mechanism was reduced from J. Bugler's pentane mechanism [12] by Princeton CHEM-RC [13,14]. To check the validity of the reduced model, ignition delay times were calculated between the detailed and reduced mechanism, with errors less than 10% at both atmospheric pressure and 60 Torr. The plasma species included in the model are $O_2(a^1\Delta_g)$; $O_2(b^1\Sigma_g^+)$; $O_2(c^1\Sigma_u^-)$, $O_2(C^1\Delta_u)$, $O_2(A^3\Sigma_u^+)$ (all grouped as O_2(4.5eV)); $O(^1D)$; O_3; He^*; $C_5H_{12}^+$; $C_5H_{11}^+$; O_2^+; He^+; e. Electron impact excitation, dissociation, and ionization reactions are solved by BOLSIG+ [15] incorporated into ZDPlasKin, which updates at every time step to provide electron parameters and rate coefficients. The cross section data of O_2 and He is obtained from the Phelps [16] and Biagi databases [17] on Lxcat respectively. Since the pentane electron impact cross section data is not well known, the data of C_2H_6 [18] is used as a substitute. This cross section, plus the analogous electron impact reactions for pentane to those in the literature, are used for an initial modeling effort (original mechanism). To improve the prediction capabilities, the C0-C2 chemistry of the original mechanism was replaced by HP-Mech [19] and new direct electron impact reactions with pentane are added to the model (new mechanism), with updated electron impact cross sections, generated by fitting to the pyrolysis data. The method for calculating the excitation and ionization energies can be found in Ref. [20], which was used to extrapolate the threshold energy of pentane electron impact dissociation and ionization from C1-C4 values. Quenching, charge exchange, and recombination reactions are also considered in the model, with a final model size of 368 species and 2589 reactions. In this simulation, the experimental time-dependent temperature and pressure were used as the initial parameters to calculate the PAC at every time step. The reduced electric field is assumed to be a square wave and set to 180 Td, based on previous experiments [4,5].

3. Results and Discussion

Temperature in the reactor was measured throughout the burst using the two line absorption technique with H_2O [20] for the two cases considered, as well as a case without pentane for the heat addition due to joule heating by the plasma. These results can be seen in Figure 2, with curve profiles used as a fit for each set of data. These profiles were used as a model input parameter as the energy coupling from the plasma in these conditions is difficult to model accurately. The heat release due to combustion results in ~80K increase in temperature. As expected, the pyrolysis case is endothermic, resulting in a lower temperature throughout the burst than even the joule heating case. Interestingly, a maximum temperature is reached before the end of the discharge, unlike previous experiments conducted, possibly due to the high thermal diffusivity of helium in raised temperatures increasing heat loss to the walls.

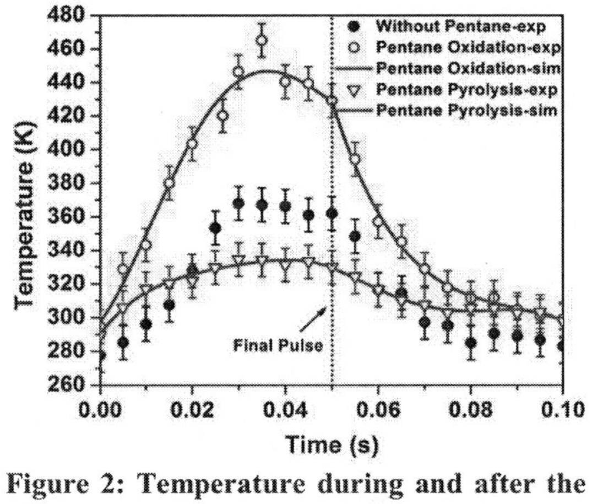

Figure 2: Temperature during and after the 1500 pulse, 30 kHz plasma burst for the oxidative, pyrolysis, and without pentane conditions. The fitting used by the model is shown by the solid lines.

Figure 3 shows the time dependent number density of pentane during the discharge for the pyrolysis and oxidative conditions respectively. The experimental results are plotted with the old and new simulations, as well as the change in number density solely due to the temperature change with no reaction (thin black line), to demonstrate the actual consumption quantity due to chemistry. The oxidative condition has significantly increased consumption, likely due to the oxidative plasma reactions creating species such as O and $O(^1D)$.

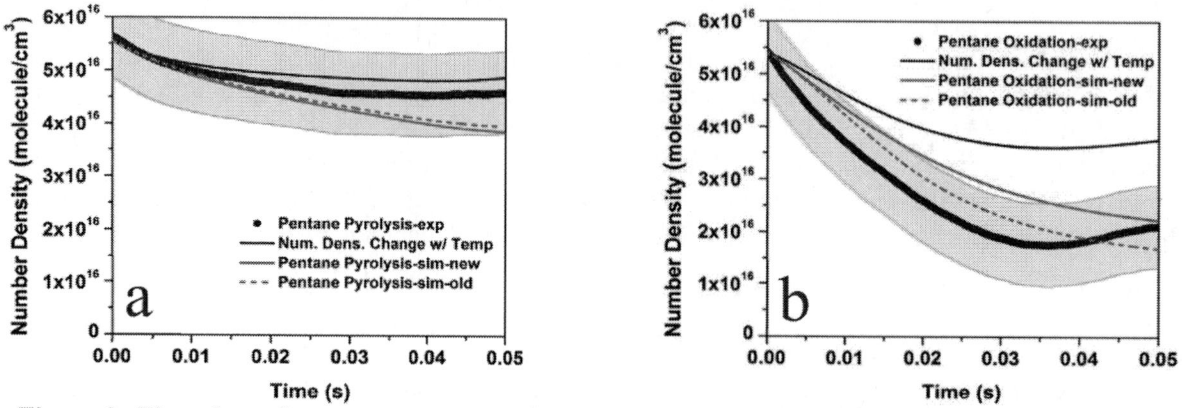

Figure 3: Time-dependent measurement and model prediction of pentane number density for the a) pyrolysis (0.028/0.972 mixture of C_5H_{12}/He) and b) oxidative (0.028/0.22/0.75 mixture of $C_5H_{12}/O_2/He$) cases respectively.

The flattening out of the data near the end of the burst is likely due to absorption by other larger fuel fragments, which would cause an apparent "decrease" in consumption. The model does a good job within the experimental error of predicting pentane fuel dissociation.

Figure 4 shows the time dependent number density of a) CH_4 and b) C_2H_2 for the oxidation and pyrolysis cases respectively. The old and new model can be seen by the dashed and solid lines. For the pyrolysis case, this data, along with the steady state GC data (select species shown in Figure 5) of H_2, CH_4, C_2H_2, C_2H_4, C_2H_6, C_3H_6, and C_5H_{12} was used to generate the new electron impact reactions and update the electron impact cross sections. The new model slightly reduces CH_4 production in the pyrolysis case by providing new electron impact dissociation pathways, which reduces the branching ratio to methane. This also causes the drastic increase in prediction in the C_2H_2 case, where before the experimental number density was underpredicted by over 4 orders of magnitude. The same is true for the steady state species measured in Fig. 5. Most of the direct electron impact reactions in the old model cause H-abstraction from C_5H_{12}, which recombine to form H_2. The lack of direct dissociation to species such as C_2H_4 and C_3H_6 result in orders of magnitude under-prediction. By adding direct electron impact pathways to these species, it both reduces H_2 concentration and increases formation of these smaller fuel species. Even in the pyrolysis case, significant concentrations of other fuels are formed from pentane, showing applications for fuel reforming, and now with greater model predictive capabilities the plasma can be adjusted to maximize outputs for future experiments.

For the oxidative case, the same model was applied with identical plasma input energy. As discussed, the additional consumption in the oxidative case in Fig. 3 comes from excited oxidative radicals (O, OH, $O(^1D)$) interactions with pentane. The actual consumption of pentane, once the change in number density is accounted for, is an order of magnitude larger than the pyrolysis case. Looking at

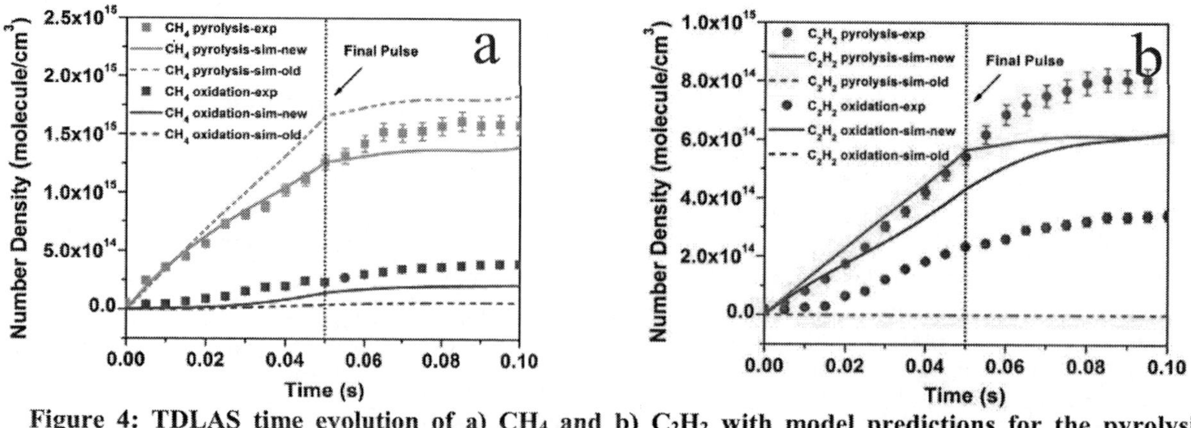

Figure 4: TDLAS time evolution of a) CH_4 and b) C_2H_2 with model predictions for the pyrolysis (0.028/0.972 mixture of C_5H_{12}/He) and oxidation case (0.028/0.22/0.75 mixture of $C_5H_{12}/O_2/He$).

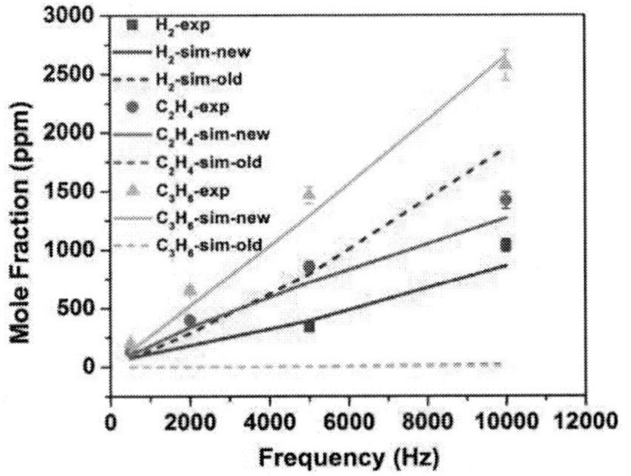

Figure 5: Steady state GC pyrolysis measurements of H_2, C_2H_4 and C_3H_6 for 0-10kHz continuous plasma pulse frequencies for 0.028/0.972 mixture of C_5H_{12}/He.

Fig. 4, the predictions for both CH_4 and C_2H_2 are greatly improved. The CH_4, which comes from CH_3 radical recombination, is still slightly underpredicted. Considering the electron impact dissociation matches well with the pyrolysis data, it is likely that one of the CH_3 recombination reactions is underpredicted. The C_2H_2 number density in the oxidative case is overpredicted. Interestingly, there is a drastic trend difference in the CH_4 trends for the oxidative and pyrolysis case, while the C_2H_2 number density ends up almost identically. This suggests that the oxidative case is missing consumption reactions for C_2H_2, as the production reactions are mainly from electron impact dissociation, which should not change between the two cases. Path flux shows consumption is currently mainly due to O and H radicals: given the high concentration and significance in pentane consumption, C_2H_2 interaction with OH might be under-predicted.

Finally, Figure 6 shows the time dependent measurements of the oxidative species H_2O and CH_2O. The old model significantly overpredicted both species. Path flux shows this is mainly due to a significant reduction in the overall OH concentration in the model. This reduction, plus the addition of new electron impact dissociation reactions, led to less contribution of H abstraction reactions of C_5H_{12} by OH, which results in a lowering of H_2O number density. A similar result can be seen with CH_2O, with less OH changing the dominant pathways of both formation and consumption. The increase in electron impact dissociation reactions led to increasing concentrations of C_3H_7 and C_4H_9, which oxidize and break down to form CH_2O, instead of a larger percentage from fuel radical RO. Overall, the addition of electron impact dissociation reactions drastically changed the oxidative chemistry as well and brought predictions by the model more in line with experimental data.

4. Conclusion

Measurements of both oxidation and pyrolysis have been carried out for nanosecond-pulsed plasma discharge dissociation of pentane, and compared to a 0-D model. Using temperature as a model input, as well as the C_2H_6 literature cross sections, new electron impact dissociation reactions were added to a pentane model and C0-C2 reactions updated with HP-Mech. The cross sections of these electron impact reactions were adjusted to fit the experimental pyrolysis data, and then utilized to predict the oxidative condition. The results showed favorable comparison for pentane consumption for both the oxidative and pyrolysis conditions. A slight under-prediction is found for CH_4 and a slight over-prediction for C_2H_2. This is likely the result of under-prediction in one of the CH_3 recombination reactions, and of the C_2H_2 reaction with OH respectively. The new model prediction of H_2O and CH_2O agree excellently with experimental data, due to the reduction in OH production from the added importance of the electron impact dissociation reactions. This data shows the importance of direct electron impact dissociation and provides an updated model for predictions of major species and intermediates for these nano-second plasma discharges with pentane.

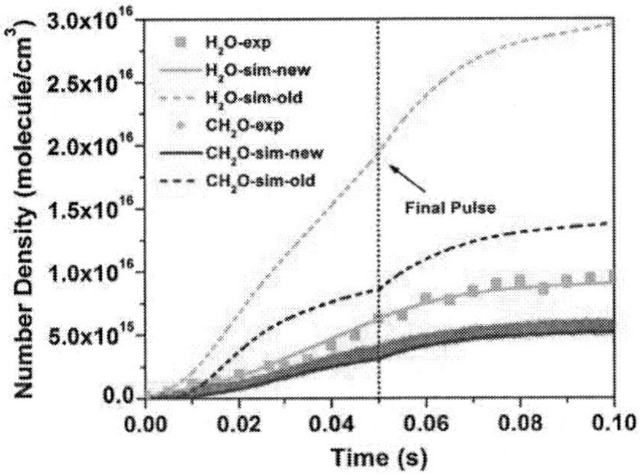

Figure 6: TDLAS time evolution of H_2O and CH_2O with model predictions for the oxidation case (0.028/0.22/0.75 mixture of C_5H_{12}/O_2/He).

5. Acknowledgments

The authors would like to thank the grant support from Exxon Mobile Inc. ACR is supported through the National Defense Science and Engineering Graduate (NDSEG) Fellowship.

6. References

[1] S. Yang, S. Nagaraja, W. Sun, V. Yang. Multiscale modeling and general theory of non-equilibrium plasma-assisted ignition and combustion. J. Phys. Appl. Phys. 50 (2017) 433001. doi:10.1088/1361-6463/aa87ee.

[2] S. Nagaraja, V. Yang, Z. Yin, I. Adamovich. Ignition of hydrogen–air mixtures using pulsed nanosecond dielectric barrier plasma discharges in plane-to-plane geometry. Combust. Flame 161 (4) 1026–37. doi:10.1016/j.combustflame.2013.10.007.

[3] Z. Yin, I. Adamovich. Ignition Delay and Time-Resolved Temperature Measurements in Nanosecond Pulse Hydrogen-Air and Ethylene-Air Plasmas at Elevated Initial Temperatures (2011). doi:10.2514/6.2011-1212.

[4] J.K. Lefkowitz, P. Guo, A.C. Rousso, Y. Ju. Species and temperature measurements of methane oxidation in a nanosecond repetitively pulsed discharge. Phil Trans R Soc A 373 (2015).

[5] J.K. Lefkowitz, M. Uddi, B.C. Windom, G. Lou, Y. Ju. In situ species diagnostics and kinetic study of plasma activated ethylene dissociation and oxidation in a low temperature flow reactor. Proc. Combust. Inst. 35 (2015) 3505–12. doi:10.1016/j.proci.2014.08.001.

[6] A. Rousso, S. Yang, J. Lefkowitz, W. Sun, Y. Ju. Low temperature oxidation and pyrolysis of n-heptane in nanosecond-pulsed plasma discharges. Proc. Combust. Inst. 36 (2017) 4105–12. doi:10.1016/j.proci.2016.08.084.

[7] S.W. Sharpe, T.J. Johnson, R.L. Sams, P.M. Chu, G.C. Rhoderick, P.A. Johnson. Gas-Phase Databases for Quantitative Infrared Spectroscopy. Appl. Spectrosc. 58 (2004) 1452–61.

[8] I.E. Gordon, L.S. Rothman, C. Hill, R.V. Kochanov, Y. Tan, P.F. Bernath, et al. The HITRAN2016 molecular spectroscopic database. J. Quant. Spectrosc. Radiat. Transf. 203 (2017) 3–69. doi:10.1016/j.jqsrt.2017.06.038.

[9] A. Farooq, J.B. Jeffries, R.K. Hanson. In situcombustion measurements of H2O and temperature near 2.5 μm using tunable diode laser absorption. Meas. Sci. Technol. 19 (2008) 075604. doi:10.1088/0957-0233/19/7/075604.

[10] S. Pancheshnyi, B. Eismann, G.J.M. Hagelaar, L.C. Pitchford. ZDPlasKin. University of Toulouse, LAPLACE, CNRS-UPS-INP, Toulouse, France: 2008.

[11] A.E. Lutz, R.J. Kee, J.A. Miller. SENKIN: A FORTRAN program for predicting homogeneous gas phase chemical kinetics with sensitivity analysis. Sandia Natl. Lab. Rep. SAND87-8248 (1988).

[12] J. Bugler, B. Marks, O. Mathieu, R. Archuleta, A. Camou, C. Grégoire, et al. An ignition delay time and chemical kinetic modeling study of the pentane isomers. Combust. Flame 163 (1) 138–56. doi:10.1016/j.combustflame.2015.09.014.

[13] W. Sun, Z. Chen, X. Gou, Y. Ju. A path flux analysis method for the reduction of detailed chemical kinetic mechanisms. Combust. Flame 157 (2010) 1298–307. doi:10.1016/j.combustflame.2010.03.006.

[14] X. Gou, W. Sun, Z. Chen, Y. Ju. A dynamic multi-timescale method for combustion modeling with detailed and reduced chemical kinetic mechanisms. Combust. Flame 157 (2010) 1111–21. doi:10.1016/j.combustflame.2010.02.020.

[15] G.J.M. Hagelaar, L.C. Pitchford. Solving the Boltzmann equation to obtain electron transport coefficients and rate coefficients for fluid models. Plasma Sources Sci. Technol. 14 (2005) 722–33. doi:10.1088/0963-0252/14/4/011.

[16] A. Phelps. Phelps database (n.d.). www.lxcat.net.

[17] S.F. Biagi. Biagi-v7.1 database (n.d.). www.lxcat.net.

[18] M. Hayashi. Electron Collision Cross-Sections for Molecules Determined from Beam and Swarm Data. Swarm Stud. Inelast. Electron-Mol. Collis., Springer, New York, NY; 1987, p. 167–87. doi:10.1007/978-1-4612-4662-6_33.

[19] H. Zhao, J. Fu, F.M. Haas, Y. Ju. Effect of prompt dissociation of formyl radical on 1,3,5-trioxane and CH2O laminar flame speeds with CO2 dilution at elevated pressure. Combust. Flame 183 (2017) 253–60. doi:10.1016/j.combustflame.2017.05.005.

[20] N.B. Anikin, S.M. Starikovskaia, A.Y. Starikovskii. Oxidation of saturated hydrocarbons under the effect of nanosecond pulsed space discharge. J. Phys. Appl. Phys. 39 (2006) 3244–52. doi:10.1088/0022-3727/39/15/006.

Spring Technical Meeting
Eastern States Section of the Combustion Institute
March 4–7, 2018
State College, Pennsylvania

Low-temperature Oxidation of n-Dodecane in a Microflow Tube Reactor: Temperature and Residence Times Effects

Khanh Dang[1] *and Harsha K. Chelliah*[1,*]

[1]*Department of Mechanical and Aerospace Engineering, University of Virginia, Charlottesville, VA 22904-4746, USA*

[*]*Corresponding author:* harsha@virginia.edu

Abstract: The influence of the class of reactions involving isomerization and decomposition of peroxy-alkyl-hydroperoxy radical (•OOQOOH) on low temperature oxidation of n-dodecane is investigated here using a novel micro-flow tube reactor (MFTR) and chemical kinetic pathway analyses. At a reactor temperature of 625 K and a residence time of 60 ms, sensitivity analyses show that the decomposition of carbonyl-hydroperoxides (CHP) formed from •OOQOOH generates a second hydroxyl radical, which in turn can explain the low temperature reactivity of n-dodecane observed in the flow reactor. As the temperature is increased to about 775-800 K, the HO_2 termination reactions leading to formation of stable H_2O_2, signifies the end of cool reaction regime. By considering several recent chemical kinetic models for n-dodecane oxidation, the consequences of extending high-temperature chemical kinetic models to low and NTC temperature regime conditions are clearly highlighted. In addition to determining the temperature at which the low temperature branching vs termination reactions dominate in n-dodecane oxidation, the present MFTR with excellent control of temperature and residence time can yield valuable speciation data for a broad range of fuel oxidation investigations.

***Keywords:* Microflow tube reactor, n-Dodecane, Oxidation, Low-temperature, Residence time**

1. Introduction

Understanding low-temperature oxidation chemistry of hydrocarbons is important for the designing of combustion devices, especially modern internal combustion engines with homogeneous charge compression ignition [1] and gas turbine engines with low temperature flow regions, e.g. preheat region of turbulent flames [2]. The optimization of such combustion devices to improve fuel consumption efficiency and minimize of emissions requires well-validated chemical kinetic models for a range of compositions, temperatures, pressures, and residence times. Extensive literature exists on low-temperature oxidation kinetics of primary reference fuel (PRF) components of gasoline fuels, namely n-heptane and iso-octane [3–6]. However, only limited experimental data and well-validated chemical kinetic models exist for the low- to NTC-temperature regime of n-dodecane oxidation, a key surrogate of diesel and jet fuels [7]. This paper focuses on addressing this deficiency by considering a recently developed micro-flow tube reactor [8] with excellent low-temperature and long residence time operability conditions.

In the low- to NTC-temperature regime (650-750 K), the competition between chain branching and chain termination by •QOOH reaction pathways, namely (i) keto-hydroperoxide formation/decomposition with two OH radicals and (ii) HO_2 elimination with olefinic species formation, on the overall oxidation of PRF of gasoline is well documented [6]. However, with regard to n-dodecane oxidation, only limited investigations have focused on the low-temperature •QOOH chemistry. This is mostly due to the primary importance of n-dodecane in surrogate jet fuels with relatively high-temperature and short residence time gas-turbine engine applications. Furthermore, with somewhat poorly validated •QOOH reaction pathways of n-dodecane, current chemical kinetic models of n-dodecane [9–12] clearly show significant deviation when predicting the ignition behavior in the NTC region. Thus, the objective of this study is to obtain extensive species time histories during the oxidation of n-dodecane and understand the current deficiencies of n-dodecane kinetic models. For this purpose, a temperature range of 650-750 K was explored with residence times of up to 120 ms, under atmospheric pressure and diluted lean ($\phi = 0.5$) mixture condition.

2. Experimental Approach

A micro-flow tube reactor (MFTR) setup that was recently developed for fuel pyrolysis investigations [8], was used here for n-dodecane oxidation study. As illustrated in Fig. 1, this setup consists of 4 mm ID, 0.37 m long hot section, heated by seven modular heaters with PID controllers with a maximum temperature limit of 1300 °C . The oxygen and main nitrogen diluent flow were preheated to the target reactor temperature in a 1.25 m long helical section with the same ID of 4 mm (not shown). A partially preheated mixture of n-dodecane and secondary nitrogen was introduced via two opposed 1 mm ID quartz tubing just downstream of the helical section as shown in Fig. 1. A porous quartz frit placed immediately downstream from the junction with side tubes, facilitating rapid mixing of fuel and the oxidizer streams by creating a small mixing volume of 0.2 cm^3.

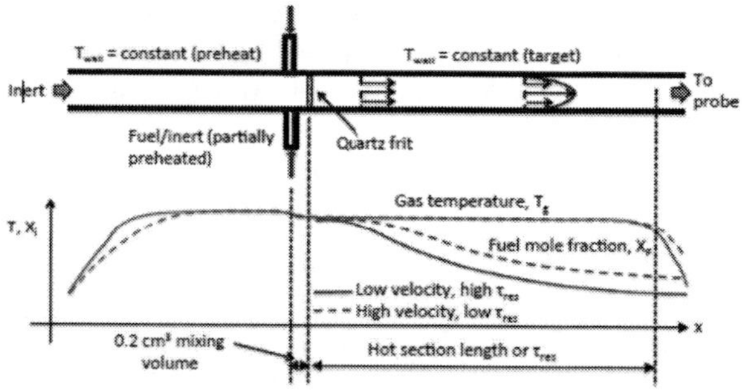

Figure 1: Illustration of the microflow tube reactor setup with preheating of nitrogen/oxygen mixture before fuel addition from side tubes, followed by the constant temperature hot section, and probing at the exit plane (note: the illustration is not to scale).

Given the small scale of the MFTR and low vapor pressure of the fuel, a key challenge was delivering only a small amount of pre-vaporized n-dodecane to the reactor, e.g. 0.01-0.05 mL/min. A liquid syringe pump (Teledyne ISCO Model 500D) was employed, with an ultrasonic atomizer generating a fine mist of 10-15 μm droplets mixed with preheated nitrogen carrier gas (up to 500 K). The gas flow was controlled by a series of mass flow controllers (Sierra 100 with 1% accuracy at full scale). The combustion products exiting the reactor were continuously extracted using a quartz micro-probe and quantified using a gas chromatograph (GC) system (Shimadzu Model 2014). Species O_2, H_2, N_2, CO, CO_2, and C_1-C_4 light hydrocarbons were analyzed in two different lines, with each line consisting of its own sample loop, heated multiport valves, and three detectors

- two thermal conductivity detectors (TCDs) and one flame ionization detector (FID) (see [8] for further details). For each data point presented, experiments were repeated at least five times to obtain a meaningful repeatability uncertainty. The total propagated experimental uncertainty (2σ) values are reported in the figures.

For the present low temperature experiments with low fuel conversion, the reactant mixture was diluted in 90% nitrogen in order to (i) minimize the temperature increase due to the exothermicity of oxidation reactions, and (ii) ensure accurate quantification of species mole fractions. The temperature of the mixture inside the hot section was measured for each target temperature case, with the longest residence time considered (i.e., 120 ms) to ensure that temperature increase was less than 5-10 K.

3. Chemical Kinetic Models

Several n-dodecane chemical kinetic models from the literature were considered in the present analysis. The first model considered was developed by Westbrook et al. [3] (identified as LLNL model) to describe the pyrolysis and oxidation of C_8 to C_{16} n-alkanes at low and high temperature conditions. The second model considered was by Mze-Ahmed et al. [11], consisting of 1377 species in 5864 reactions and validated against jet-stirred reactor speciation data. The third model considered was CRECK-POLIMI model by Ranzi et al. [10], which was validated over a broad range of temperatures and fuels. Other models considered include an updated JetSurF 1.0 model by Banerjee [12] and simplified LLNL model by Narayanaswamy et al. [13] The open-source Cantera package was used to integrate the coupled reacting system of equations [14].

4. Results and Discussion

Experiments investigating low-temperature oxidation of n-dodecane were conducted in a diluted O_2/N_2 mixture. A temperature range of 650-750 K was explored with residence times of 30 to 120 ms, under atmospheric pressure and lean condition ($\phi = 0.5$). Figure 2 shows a comparison of several C_2-C_4 species as well as CO and CO_2, with predicted results as a function of the flow reactor temperature for a flow residence time of 120 ms, a pressure of 1 atm, and a mixture composition of $X_{C_{12}H_{26}} = 0.003$, $X_{O_2} = 0.097$, and $X_{N_2} = 0.9$. While the agreement of present data with the literature kinetic models is not perfect, the data is seen to capture the low temperature branching vs termination pathway effects reasonably well, as indicated by the lower and upper bounds of the temperature. At this residence time, the model by Mze-Ahmed et al. [11] shows reactivity at a slightly higher temperature at which the •QOOH branching reaction pathway contributes. Such observed temperature sensitivity highlights the importance of speciation data gathered by the MFTR, especially for optimizing kinetic rate parameters.

Figure 3 compares the same speciation data and predictions with three models, but as a function of time and at a temperature of 650 K and a pressure of 1 atm. At this low temperature, only the comprehensive model by Westbrook et al. [3] and the CRECK-POLIMI model by Ranzi et al. [10] accurately predict the resident time at which the low temperature •QOOH chain branching reactions start to dominate. The peak species mole fractions measured are clearly not in good agreement, indicating a deficiency in the chain-propagation rate parameters.

Since the LLNL model by Westbrook et al. [3] was seen to closely agree with present experimental data and to capture the branching vs. termination effects at the low temperature regime, this

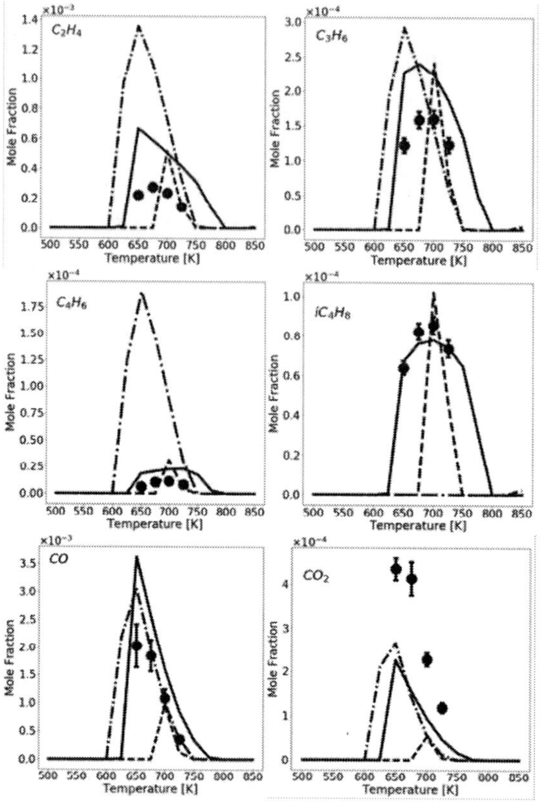

Figure 2: Species yield from n-dodecane oxidation as a function of temperature, for reactor residence time of 120 ms and lean mixture with 90% nitrogen dilution. Solid line - LLNL model [3]; dash-dot line - POLIMI model [10]; dash-line - Mze-Ahmed et al. [11].

model is used here for further analysis of rate-controlling chemical reaction pathways. The open-source Cantera code [14] was used to extract species sensitivities (using a brute force method) and reaction flux analyses. Specifically, sensitivity and flux analyses were performed at the wings of the temperature limits where the competition between chain branching and termination reactions are seen to dominate, i.e., at temperatures of 625 K and 800 K. At early stages of low temperature oxidation, namely at residence time of 5 ms and temperature of 625 K and pressure of 1 atm, the most sensitive reaction is the initiation step, $nC_{12}H_{26} + O_2 \rightarrow nC_{12}H_{25} + HO_2$. As the reaction proceeds to 60 ms, the importance of ketohydroperoxide decomposition to aldehydes, ketones, and the second hydroxyl radical is evident from reactions $OC_{12}H_{24}OOH \rightarrow C_2H_5COCH_2 + nC_7H_{15}CHO + OH$ and $OC_{12}H_{24}OOH \rightarrow CH_3COCH_2 + nC_8H_{17}CHO + OH$. Detailed reaction flux analysis also reveals these key reaction pathways. As the temperature of the MFTR is increased to about 775-800 K, i.e., end of the cool temperature regime, chain termination reactions of HO_2 dominate and shut off the reactivity. The sensitivity analyses performed at 800 K and at 60 ms clearly reveal the importance of $HO_2 + C_{12}H_{26} \rightarrow C_{12}H_{25} + H_2O_2$.

5. Conclusions

A micro-flow tube reactor with excellent control of temperature and residence time was utilized to analyze the low temperature chain branching vs termination reaction pathways of n-dodecane

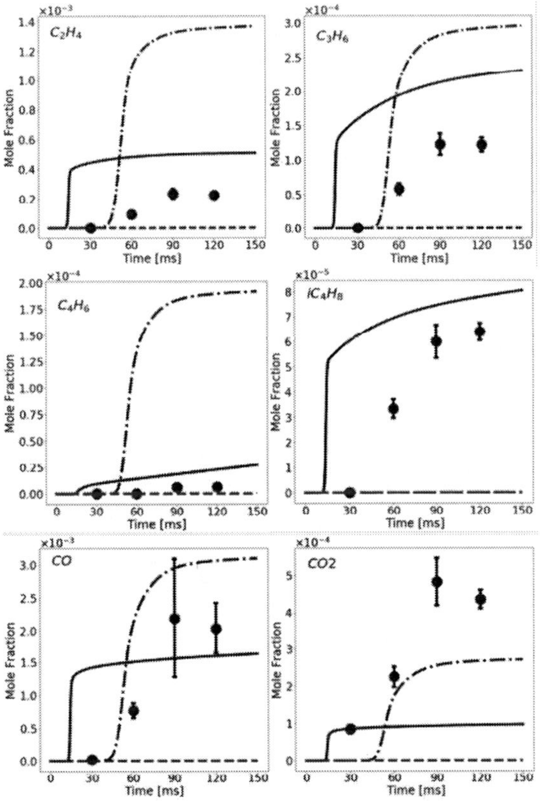

Figure 3: Species yield from n-dodecane oxidation as a function of time, for a reactor temperature of 650 K, pressure of 1 atm, and lean mixture with 90% nitrogen dilution. Solid line - LLNL model [3]; dash-dot line - POLIMI model [10]; dash-line - Mze-Ahmed et al. [11].

oxidation. Analogous to primary reference fuels of gasoline, for the temperature range of 625 to 650 K, the importance of reactions involving the decomposition of $OOC_{12}H_{24}OOH$ into aldehydes, ketones, and second hydroxyl radicals was identified. Kinetic models that included this class of reactions (e.g. LLNL and CRECK) were able to accurately predict the experimentally observed temperature bounds at which n-dodecane reactivity was detected by the micro-flow tube reactor. Furthermore, C_2-C_4, CO_2, and CO species measured by the flow reactor in the temperature range of 650 to 750 K and residence times of 30 to 120 ms were in reasonable agreement with model predictions.

References

[1] J. E. Dec, Advanced compression-ignition engines—understanding the in-cylinder processes, Proceedings of the Combustion Institute 32 (2009) 2727–2742.

[2] L. Esclapez, P. C. Ma, E. Mayhew, R. Xu, S. Stouffer, T. Lee, H. Wang, and M. Ihme, Fuel effects on lean blow-out in a realistic gas turbine combustor, Combustion and Flame 181 (2017) 82–99.

[3] C. K. Westbrook, W. J. Pitz, O. Herbinet, H. J. Curran, and E. J. Silke, A comprehensive detailed chemical kinetic reaction mechanism for combustion of n-alkane hydrocarbons from n-octane to n-hexadecane, Combustion and Flame 156 (2009) 181–199.

[4] O. Herbinet, B. Husson, Z. Serinyel, M. Cord, V. Warth, R. Fournet, P.-A. Glaude, B. Sirjean, F. Battin-Leclerc, Z. Wang, M. Xie, Z. Cheng, and F. Qi, Experimental and modeling investigation of the low-temperature oxidation of n-heptane, Combustion and Flame 159 (2012) 3455–3471.

[5] K. Zhang, C. Banyon, J. Bugler, H. J. Curran, A. Rodriguez, O. Herbinet, F. Battin-Leclerc, C. B'Chir, and K. A. Heufer, An updated experimental and kinetic modeling study of n-heptane oxidation, Combustion and Flame 172 (2016) 116–135.

[6] N. Atef, G. Kukkadapu, S. Y. Mohamed, M. A. Rashidi, C. Banyon, M. Mehl, K. A. Heufer, E. F. Nasir, A. Alfazazi, A. K. Das, C. K. Westbrook, W. J. Pitz, T. Lu, A. Farooq, C.-J. Sung, H. J. Curran, and S. M. Sarathy, A comprehensive iso-octane combustion model with improved thermochemistry and chemical kinetics, Combustion and Flame 178 (2017) 111–134.

[7] S. Dooley, S. H. Won, M. Chaos, J. Heyne, Y. Ju, F. L. Dryer, K. Kumar, C.-J. Sung, H. Wang, M. A. Oehlschlaeger, R. J. Santoro, and T. A. Litzinger, A jet fuel surrogate formulated by real fuel properties, Combustion and Flame 157 (2010) 2333–2339.

[8] U. Shrestha, M. Rahimi, G. Simms, and H. Chelliah, Fuel pyrolysis in a microflow tube reactor–Measurement and modeling uncertainties of ethane, n-butane, and n-dodecane pyrolysis, Combustion and Flame 177 (2017) 10–23.

[9] H. Wang, E. Dames, B. Sirjean, D. A. Sheen, R. Tango, A. Violi, J. Y. W. Lai, F. N. Egolfopoulos, D. F. Davison, R. K. Hanson, C. T. Bowman, C. K. Law, W. Tsang, N. P. Cernansky, D. L. Miller, and R. O. Lindstedt, *A high-temperature chemical kinetic model of n-alkane (up to n-dodecane), cyclohenxane, methyl-, ethyl-, n-propyl, and n-butyl-cyclohexane oxidation at high temperatures*, Version 2.0, 2010.

[10] E. Ranzi, A. Frassoldati, R. Grana, A. Cuoci, T. Faravelli, A. Kelley, and C. Law, Hierarchical and comparative kinetic modeling of laminar flame speeds of hydrocarbon and oxygenated fuels, Progress in Energy and Combustion Science 38 (2012) 468–501.

[11] A. Mzé-Ahmed, K. Hadj-Ali, P. Dagaut, and G. Dayma, Experimental and Modeling Study of the Oxidation Kinetics of n-Undecane and n-Dodecane in a Jet-Stirred Reactor, Energy & Fuels 26 (2012) 4253–4268.

[12] S. Banerjee, R. Tangko, D. A. Sheen, H. Wang, and C. T. Bowman, An experimental and kinetic modeling study of n-dodecane pyrolysis and oxidation, Combustion and Flame 163 (2016) 12–30.

[13] K. Narayanaswamy, P. Pepiot, and H. Pitsch, A chemical mechanism for low to high temperature oxidation of n-dodecane as a component of transportation fuel surrogates, Combustion and Flame 161 (2014) 866–884.

[14] D. G. Goodwin, H. K. Moffat, and R. L. Speth, *Cantera: An Object-oriented Software Toolkit for Chemical Kinetics, Thermodynamics, and Transport Processes*, `http://www.cantera.org`, Version 2.3.0, 2017, DOI: {10.5281/zenodo.170284}.

Spring Technical Meeting
Eastern States Section of the Combustion Institute
March 4-7, 2018
State College, Pennsylvania

Effects of sp^2 Carbon on Low-Temperature Oxidation of Cyclic Hydrocarbons

Jacob C. Davis[†1], *Alanna L. Koritzke*[†2], *Rebecca L. Caravan*[3], *Matthew G. Christianson*[2], *David L. Osborn*[3], *Craig A. Taatjes*[3], *and Brandon Rotavera*[1, 2, *]

[1] *College of Engineering*
[2]*Department of Chemistry*
University of Georgia
Athens, GA, 30602, USA
[3]*Combustion Research Facility, MS 9055*
Sandia National Laboratories
Livermore, CA, 94551, USA
**Corresponding Author Email:* rotavera@uga.edu
[†]*Authors contributed equally*

Abstract
Cycloalkanes comprise a significant portion of petroleum-derived transportation fuel, and cyclohexane is a representative molecule from the hydrocarbon class. Chemical kinetics models developed for cyclohexane combustion consistently predict appreciable formation of cyclohexene as a primary intermediate from oxidation reactions at temperatures below 1000 K. However, despite sub-mechanisms for primary intermediates playing central roles in chemical kinetics models, detailed chemistry is often simplified (e.g. neglecting subsequent reactions of $\dot{R} + O_2$).

Experimental identification of the species involved in the reaction of primary intermediates enhances the robustness of sub-mechanisms within the larger framework of chemical kinetics models. Accordingly, multiplexed photoionization mass spectrometry (MPIMS) coupled with tunable VUV ionization is used to examine the reaction mechanisms of $\dot{Q}OOH$ radicals derived from cyclohexene oxidation in order to assess how the reactions relevant to chain-branching are influenced by a C=C bond in a cyclic hydrocarbon. Products from $\dot{R} + O_2$ reactions were measured at 500, 600 and 700 K using Cl-initiated oxidation under pseudo-first-order conditions with $[O_2] \sim 10^{16}$ cm^{-3}.

Chain-branching in cyclohexene is evident despite the relatively lower concentrations herein, as indicated by temporally-resolved ion signal at the exact mass-to-charge ratio of ketohydroperoxide (KHP), *m/z* 128. The corresponding photoionization spectra are nearly identical to analogous species produced by *n*-butane and cyclohexane oxidation, where O_2 concentrations were approximately 10^2 higher than the present experiments. Due to the presence of KHP at such lower $[O_2]$, it is postulated that the presence of a C=C bond promotes $RO\dot{O} \rightarrow \dot{Q}OOH$ isomerization while also inhibiting unimolecular decomposition compared to oxidation of cyclohexane under similar conditions; the net effect being facilitation of $\dot{Q}OOH + O_2$ reactions and subsequent ketohydroperoxide formation. In addition to potentially facilitating KHP formation, the presence of a single C=C bond in a cyclic hydrocarbon allowed for various $\dot{Q}OOH$ ring-opening pathways. Evidence of ring-opening products were observed including ethene, formaldehyde, 1,3-butadiene, acrolein, and 3-butenal.

Keywords: Cyclohexane, Primary Intermediates, Resonance Stabilization

1. Introduction

Cycloalkanes (naphthenes) comprise a significant portion of transportation fuel [1, 2]. Cyclohexane is one component of the naphthene class, and is a base molecular structure of advanced biofuels, including cyclohexanol [3, 4], and cyclohexanone [3, 5]. The development of low-temperature combustion strategies [6-8] and the use of advanced biofuels in transportation systems demands robust chemical kinetics models with predictive capabilities for reactions of peroxy radicals (ROȮ), which dominate chain chemistry at temperatures below 1000 K. The role of peroxy radicals in combustion centers principally on the formation of hydroperoxyalkyl (Q̇OOH) radicals from isomerization of ROȮ and subsequent reaction pathways including unimolecular decomposition to primary intermediates, i.e. cyclic ethers and conjugate alkenes, and second-O_2-addition (Q̇OOH + O_2) leading to ketohydroperoxide formation.

Cyclohexene is a primary intermediate in low-temperature oxidation reactions of cyclohexane, formed from cyclohexyl + O_2 (**Figure 1a**), in what is a chain-inhibiting step because of the relatively unreactive co-product HOȮ. Cyclohexenyl radicals formed from abstraction of allylic hydrogen by ȮH or other radicals, are resonance-stabilized by electron delocalization (**Figure 1b**), and upon subsequent reaction with O_2 can form Q̇OOH radicals with or without resonance-stabilization (**Figure 2**). Besides resonance stabilization, sp^2 carbon atoms also allow radical-addition reactions to become important [9], primarily ȮH-addition, which in the case of cyclohexene yields 1-cyclohexanol-2-yl that can form an ȮH-substituted peroxy radical after reaction with O_2. The compounding effects of resonance-stabilization in cyclohexenyl and related Q̇OOH radicals and the consideration of radical addition reactions adds complexity to the low-temperature oxidation mechanisms of cyclohexene, which also holds relevance for kinetic modeling of cyclohexane oxidation.

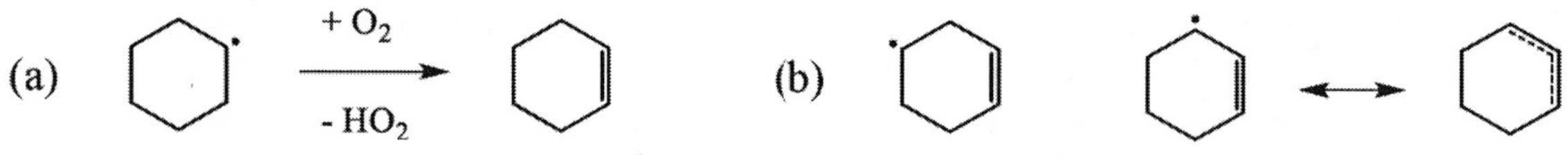

Figure 1. (a) Formation of cyclohexene from cyclohexyl + O_2. (b) Molecular structure of cyclohex-1-en-4-yl and cyclohex-1-en-3-yl radicals formed via H-abstraction. The presence of sp^2 carbon enables resonance-stabilization in cyclohex-1-en-3-yl. Vinylic cyclohex-1-en-1-yl is neglected.

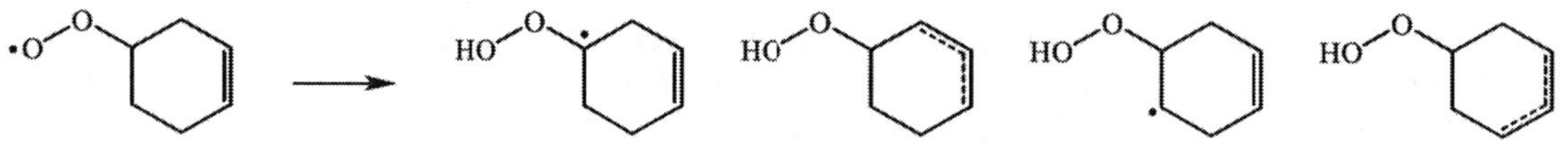

β-ROȮ *α*-Q̇OOH *β'*-Q̇OOH *β*-Q̇OOH *γ*-Q̇OOH

Figure 2. Allylic and alkylic Q̇OOH isomers derived from isomerization of cyclohex-1-en-4-peroxy.

Several models that include ROȮ and related Q̇OOH chemistry for cyclohexane [10-13] report cyclohexene as a major intermediate – formed in higher concentrations than cyclic ethers [10]. Rotavera et al. [14] measured a branching fraction of 20% for cyclohexyl + $O_2 \rightarrow$ cyclohexene + HOȮ at 700 K and 2 atm. MPIMS experiments on cyclohexene oxidation identified product formation from Q̇OOH-mediated reactions [15]. However, the specific products and related mechanisms responsible for the observed mass peaks were not identifiable due to the use of a H_2-discharge lamp as the ionization source, which produces a spectrum from approximately 8 – 10 eV. However, the results confirmed time-resolved cyclic ether formation from cyclohexenyl + O_2, indicating that $\dot{Q}OOH$ formation from cyclohexene is non-negligible. In addition to [15], combustion-relevant studies on cyclohexene oxidation include shock-tube [16] and RCM [17] ignition delay time measurements, chemical kinetics model development [16, 18], and time-dependent product formation from $\dot{O}H$ + cyclohexene in absence of oxygen [19]. Despite such rich $\dot{Q}OOH$ chemistry, because of a lack of experimental benchmarks most comprehensive chemical kinetics models neglect the complete description of primary intermediates such as cyclohexene, and rather consider a restricted number of reaction pathways involving lumped radicals.

Accordingly, the present measurements of product formation from reactions of cyclohexenylperoxy radicals aid in the development of chemical kinetics models for low-temperature combustion of cyclohexane and the abovementioned advanced biofuels. MPIMS experiments were conducted on Cl-initiated cyclohexene oxidation from 500 – 700 K to identify species formed from $\dot{R}$ + O_2 reactions, concentrating primarily on $\dot{Q}OOH$-mediated reaction pathways. Several products were identified, such as conjugate alkenes, species formed via $\dot{Q}OOH$ ring-opening channels, and ketohydroperoxide.

2. Experimental Approach

MPIMS experiments were conducted at the Chemical Dynamics Beamline of the Advanced Light Source [20] in a slow-flow quartz reactor [21] using pulsed-photolytic chlorine atom-initiated oxidation with constant initial reactant number densities, highly dilute in He (**Table 1**), at 10 Torr and three temperatures (500 K, 600 K, and 700 K). The procedure is similar to that in [16]. Photolysis of Cl_2, using unfocused 351-nm light from an excimer laser, generated $\dot{C}l$ atoms instantaneously and homogeneously along both the radial and longitudinal axes of the reactor. The $\dot{C}l$ atoms react via RH + $\dot{C}l$ → $\dot{R}$ + HCl to form the initial $\dot{R}$ radicals, which subsequently undergo pseudo-first-order reaction with O_2. Only allylic or alkylic hydrogen abstraction was considered under the present low-temperature conditions. To minimize side chemistry unrelated to $\dot{R}$ + O_2, pseudo-first-order conditions were employed for RH + $\dot{C}l$ such that $\dot{R}$ + $\dot{C}l$ reactions remain minimal, and the > 10^3 excess of $[O_2]_0$ relative to $[\dot{C}l]_0$ forces $\dot{R}$ + O_2 to be the dominant reaction. Initial depletion of $[RH]_0$ by $\dot{C}l$ is near 2%. The cyclohex-1-en-3-yl : cyclohex-1-en-4-yl ratio is approximated as 1:1 because $\dot{C}l$ is not selective in abstraction reactions involving hydrocarbons, and no site-specific H-abstraction rate coefficients for cyclohexene + $\dot{C}l$ appear in the literature.

Table 1. Initial reactant concentrations (cm^{-3}) utilized for Cl-atom-initiated oxidation experiments. [He] balance is not listed. Pseudo-first-order conditions were used: $[RH]_0$:$[\dot{C}l]_0$ = 60 and $[O_2]_0$:$[\dot{R}]_0$ = 4300.

pressure	$[RH]_0$	$[O_2]_0$	$[\dot{C}l]_0$
10 Torr	$4.1 \cdot 10^{14}$	$3.0 \cdot 10^{16}$	$7.0 \cdot 10^{12}$

In order to quantify species concentrations and assign isomeric contributions to the peaks in the mass spectra, absolute photoionization cross-sections $\sigma(E)$ were measured at 600 K for several species in separate experiments: 1,3-cyclohexadiene, 1,4-cyclohexadiene, 2-cyclohexen-1-one, 3,4-epoxycyclohex-1-ene, 4,5-epoxycyclohex-1-ene, 1,5-hexadien-3-ol, cyclohex-2-en-1-ol, and cyclohex-3-en-1-ol. The $\sigma(E)$ are defined relative to propene in accord with the procedure described in [22].

Relative yields were calculated as the concentration of a species (e.g. formaldehyde) relative to the sum of the concentration of the two conjugate alkenes (1,3- and 1,4-cyclohexadiene) using least-squares fitting of the corresponding absolute photoionization spectra time-integrated 20 ms post-photolysis. In both cases, $[\dot{R}]_0 \cong [\dot{C}l]_0$ because $\dot{C}l$ atoms are consumed exclusively by RH to produce $\dot{R}$. However, radical recycling (e.g. $\dot{O}H$ from chain-propagation and chain-branching) appears to contribute to product formation given an observed secondary depletion of cyclohexene.

3. Results and Discussion

The molecular structures of species detected in the photoionization experiments are identified using a combination of exact mass determination and least-squares fitting of photoionization spectra. Once identified, plausible reaction mechanisms were determined to explain the formation pathway(s) using bond-breaking schemes.

3.1 Photoionization mass spectra

The difference mass spectra in **Figure 3a**, measured from Cl-initiated oxidation experiments at 600 K and 10 Torr, show the mass peaks considered herein. The negative signals reflect depleted cyclohexene $[C_6H_{10}]^+$ and related fragment ions. The *m/z* 80 signal corresponds to conjugate alkene formation. Chlorinated products $R^{35}Cl$ and $R^{37}Cl$ from Cl-addition to cyclohexenyl radicals were observed at *m/z* 116 and *m/z* 118, which precluded normalizing product formation to the initial radical concentration. Calibration of the mass spectra was performed using species of known molecular formula and exact mass, spanning *m/z* 28 (ethene) to *m/z* 118 (RCl), where the former was confirmed via photoionization spectral analysis and the latter by the 116:118 peak ratio being 3:1 – representing the ^{35}Cl:^{37}Cl isotopic abundance. Exact mass determinations for other species, where the molecular formula was unknown, used calibrated coefficients that convert time-of-flight to mass. The mass peaks depicted in **Figure 3b** provide an example of such determinations, which, as an example, distinguish nominal *m/z* 56 as either C_4H_8 (*m/z* 56.063) or C_3H_4O (*m/z* 56.026).

In addition to the oxidation experiments, separate measurements were conducted in the absence of oxygen to assess dimer formation and whether any formation of products resulted from ring-opening of initial cyclohexenyl radicals. Ethene forms directly from cyclohex-1-en-3-yl decomposition and but-1-en-4-yl forms directly from cyclohex-1-en-4-yl, yet neither was observed in absence of O_2. However, dimer formation was observed at *m/z* 162, which in addition to the formation of RCl isomers prevented quantifying branching fractions from $\dot{R} + O_2$ (since some unknown portion of $\dot{R}$ reacts with other $\dot{R}$ radicals or with $\dot{C}l$). Therefore, product formation is quantified as a relative yield (relative to the sum of conjugate alkene isomers at 500 K).

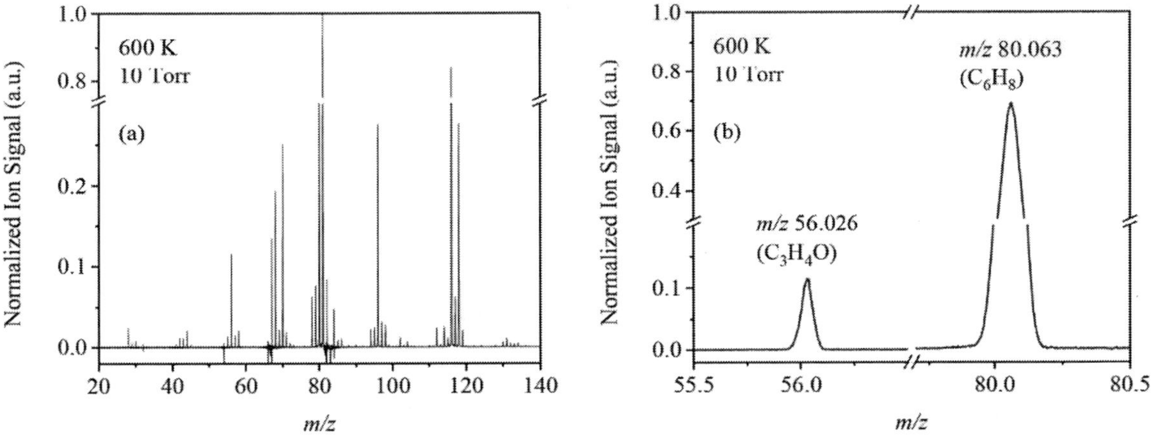

Figure 3. Difference mass spectra from Cl-initiated oxidation of cyclohexene at 600 K and 10 Torr, integrated 20-ms post-photolysis over the photon energy range 8.0 – 11.0 eV: (a) complete mass spectrum, (b) exact mass determinations of nominal *m/z* 56 and *m/z* 80.

3.2 Conjugate alkene isomer ratios and relative yields

Two conjugate alkene isomers, of nominal mass 80 amu, form from cyclohexene: 1,3-cylohexadiene and 1,4-cyclohexadiene. The branching ratio is determined at a given temperature from least-squares fitting of the photoionization spectrum to absolute reference spectra of the two isomers. In **Figure 4**, the solution from the fitting routine is represented as a combination of the two spectra, and at the listed conditions (600 K, 10 Torr) gives an isomeric ratio of approximately 3:1 for 1,3-cyclohexadiene:1,4-cyclohexadiene. Similar to the results in **Figure 4**, isomer ratios of the conjugate alkenes were determined for all temperatures (**Table 2**). 1,4-cyclohexadiene dominates at 500 K, where the isomer ratio is approximately 1:7, whereas at 700 K, only 1,3-cyclohexadiene is formed.

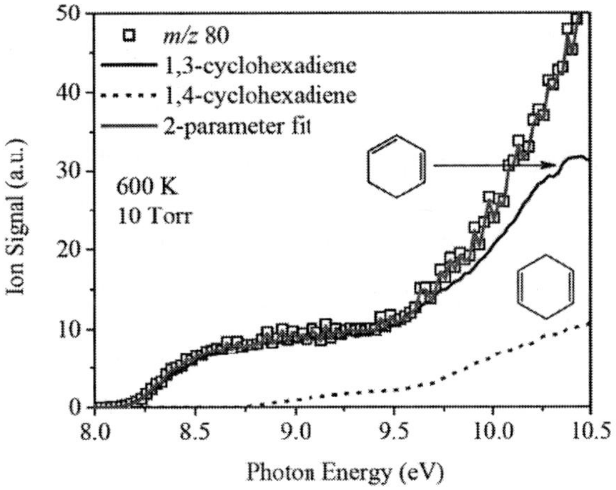

Figure 4. Photoionization spectra of *m/z* 80 ion signal from Cl-initiated oxidation of cyclohexene at 600 K and 10 Torr compared with photoionization spectra of the two conjugate alkenes 1,3-cyclohexadiene and 1,4-cyclohexadiene. Least-squares fitting of the *m/z* 80 ion signal yields an isomeric ratio (1,3-cyclohexadiene:1,4-cyclohexadiene) of approximately 3:1.

Table 2. Isomeric ratios of 1,3-cylohexadiene and 1,4-cylohexadiene, measured 20-ms post-photolysis, indicate larger selectivity of the former species with increasing temperature. Listed uncertainties are 1σ statistical errors in the fit of the photoionization spectra.

temperature	 1,3-cyclohexadiene	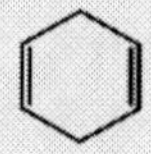 1,4-cyclohexadiene
500 K	0.13 ± 0.02	0.87 ± 0.03
600 K	0.74 ± 0.02	0.26 ± 0.03
700 K	1.00 ± 0.07	–

Relative yields were determined for ethene (*m/z* 28), formaldehyde (*m/z* 30), 1,3-butadiene (*m/z* 54), acrolein (*m/z* 56), and cyclohexadiene isomers (*m/z* 80), and quantified relative to the sum of 1,3-cyclohexadiene and 1,4-cyclohexadiene concentration at 500 K (**Table 3**). Therefore, the yields reflect the concentration relative to the concentration of cyclohexadiene isomers at 500 K, which is chosen because the temperature dependence is expected to be monotonically positive. Temperature dependence of the formation of other species, e.g. acrolein, is inferred relative to that expectation. For all cases, contribution to ion signals from the relevant ^{13}C isotope peaks were subtracted prior to fitting in order to ensure accuracy in the spectra being numerically fit.

Relative yields were quantified for six species in **Table 3**. Species that could not be quantified were identified either by photoionization spectral analysis, exact mass determination, or both methods. For all temperatures, secondary depletion of cyclohexene was observed throughout the kinetic window 20-ms post-photolysis, which may be partly ascribed to consumption of cyclohexene by $\dot{O}H$ radicals formed from chain-branching reactions involving ketohydroperoxides.

Table 3. Exact mass determinations, molecular formulae, and yield of species relative to the sum of the conjugate alkenes at 500 K. Listed uncertainties are 1σ statistical errors in the fit of the photoionization spectra.

m/z	formula	species	relative yield
28.031	C_2H_4	ethene	0.31 ± 0.01 (500 K) 1.18 ± 0.05 (600 K) 1.05 ± 0.05 (700 K)
30.011	CH_2O	formaldehyde	0.18 ± 0.01 0.54 ± 0.02 0.53 ± 0.02
54.046	C_4H_6	1,3-butadiene	0.02 ± 0.00 0.05 ± 0.00 0.08 ± 0.00

56.026	C_3H_4O	acrolein	0.85 ± 0.04 1.59 ± 0.07 0.82 ± 0.04
70.040	C_4H_6O	3-butenal	–
80.063	C_6H_8	1,3-cyclohexadiene	0.13 ± 0.01 1.01 ± 0.05 2.68 ± 0.12
		1,4-cyclohexadiene	0.87 ± 0.04 0.35 ± 0.02 0.00
128.047	$C_6H_8O_3$	ketohydroperoxide	*section 3.4*

3.3 – Q̇OOH ring-opening reactions

Multiple species in **Table 3** are linked to ring-opening of hydroperoxycyclohexenyl (Q̇OOH) radicals, ethene, formaldehyde, 1,3-butadiene, and 3-butenal, which form on the same kinetic timescale as other species from Ṙ + O_2 reactions. No ion signal at the masses of the species above were detected in absence of O_2, which indicates that formation results from oxidation reactions. By comparison with mass spectra from cyclohexane oxidation [14] measured under the same conditions as the present experiments no ring-opening reactions were evident. For example, ethene, formaldehyde, and acrolein were not detected in [14]. The difference between the results from cyclohexane oxidation and cyclohexene oxidation indicates that sp^2 carbon facilitates ring-opening reactions in Q̇OOH.

3.4 – Ketohydroperoxide (m/z 128)

Ion signal indicating the production of ketohydroperoxide was detected and the exact mass (128.047 amu) corresponds to the molecular formula $C_6H_8O_3$. Several Q̇OOH species may facilitate second-O_2-addition reactions, particularly where resonance-stabilization is present. Reverse isomerization and ring-opening are also feasible. As noted in Savee et al. [23], reaction rates of Q̇OOH with O_2 may differ substantially to rates of Ṙ + O_2, from which kinetics are inferred by analogy for use in chemical kinetics models.

The photoionization spectra are similar for all three temperatures; **Figure 5** shows the spectrum at 500 K and comparison to ketohydroperoxide spectra from *n*-butane oxidation [24, 25] and cyclohexane oxidation [26]. Arbitrary scaling is used in **Figure 5**, which reveals that the four spectra are identical. Eskola et al. [24] confirmed that ketohydroperoxide in *n*-butane is largely from γ-Q̇OOH. Given the spectral similarity, the cyclohexene-derived ketohydroperoxides may also stem from γ-Q̇OOH. Of particular note is the detection of ketohydroperoxide at such low O_2 concentrations. In prior measurements [24, 25], O_2 concentrations were $\sim 10^{18}$ cm^{-3}, $\sim 10^2$ higher than in the present experiments. The formation of ketohydroperoxide in cyclohexene oxidation, which is not seen in cyclohexane oxidation at the same oxygen concentration [26], points to the sp^2 carbon playing an important role.

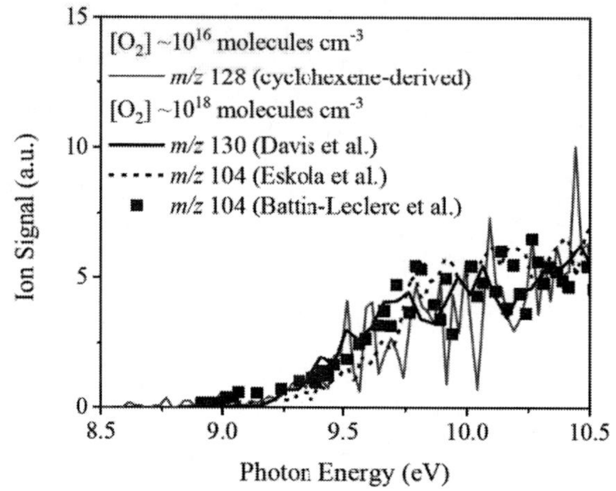

Figure 5. Photoionization spectra of ketohydroperoxides derived from oxidation of cyclohexene (*m/z* 128, present experiments, $[O_2] = 3 \cdot 10^{16}$ cm^{-3}), cyclohexane (*m/z* 130 [26], $[O_2] = \sim 10^{18}$ cm^{-3}), and *n*-butane (Eskola et al. [24] and Battin-Leclerc et al. [25], $[O_2] = \sim 10^{18}$ cm^{-3}).

4. Conclusion

Reaction mechanisms of cyclohexene, a primary intermediate of cyclohexane, were studied using Cl-initiated oxidation in MPIMS experiments at 10 Torr and from 500 – 700 K. Product formation from $\dot{Q}$OOH-mediated reactions included conjugate alkenes and species formed via ring-opening reactions. The latter included simple $\dot{Q}$OOH ring-opening, where only one C–C β-scission step occurs (in addition to O–O bond-scission releasing $\dot{O}$H), a pathway that leads to bimolecular products, and where multiple C–C β-scission steps occur resulting in several smaller species being formed (e.g. ethene, formaldehyde, and acrolein). In addition, ketohydroperoxide formation was detected despite relatively low oxygen concentration (ca. 10^{16} cm^{-3}). The ketohydroperoxide photoionization spectrum measured in cyclohexene oxidation is essentially identical to the analogous molecules observed in *n*-butane oxidation at higher O_2 concentrations (~10^{18} cm^{-3}).

More broadly, the results herein confirm the relevance of $\dot{Q}$OOH-mediated reactions in cyclohexene oxidation, including chain-propagation, chain-inhibition, and second O_2-addition reactions. Expanding the boundaries of sub-mechanisms for primary intermediates may increase the fidelity and mechanistic rigor of chemical kinetic models. In reactive environments where concentrations of such intermediates approach parity with the depleting closed-shell molecule, such as cyclohexane, the details of the oxidation chemistry may not be a minor perturbation on larger reaction mechanisms unfolding. In such cases, understanding and accounting for initiation and other consumption reactions of primary oxidation intermediates is likely to become important.

Acknowledgements

CAT, DLO, RLC and the development and maintenance of the MPIMS kinetics machine are supported by Office of Chemical Sciences, Biosciences, and Geosciences, Office of Basic Energy Sciences, of the U.S. Department of Energy (BES/USDOE). The Advanced Light Source is supported by the Director, Office of Science, BES/USDOE under Contract DE- AC02-05CH11231 at Lawrence Berkeley National Laboratory. Sandia National Laboratories is a multimission laboratory managed and operated by National Technology and Engineering Solutions of Sandia, LLC., a wholly owned subsidiary of Honeywell International, Inc., for the U.S. Department of Energy's National Nuclear Security Administration under contract DE-NA0003525

References

[1] C. Pera, V. Knop, *Fuel*, 96 (2012) 59-69.

[2] W.J. Pitz, C.J. Mueller, *Prog. Energy Comb. Sci.*, 37 (2011) 330-350.

[3] D.D. Laskar, B. Yang, H. Wang, J. Lee, *Biofuels, Bioproducts and Biorefining*, 7 (2013) 602-626.

[4] P. Varanasi, P. Singh, M. Auer, P.D. Adams, B.A. Simmons, S. Singh, *Biotechnology for Biofuels*, 6 (2013) 1-9.

[5] H.-Y. Zheng, Y.-L. Zhu, Z.-Q. Bai, L. Huang, H.-W. Xiang, Y.-W. Li, *Green Chemistry*, 8 (2006) 107-109.

[6] J.E. Dec, *Proc. Combust. Inst.*, 32 (2009) 2727-2742.

[7] B. Lawler, D. Splitter, J. Szybist, B. Kaul, *Applied Energy*, 189 (2017) 122-132.

[8] M.P.B. Musculus, P.C. Miles, L.M. Pickett, *Prog. Energy Comb. Sci.*, 39 (2013) 246-283.

[9] I.O. Antonov, J. Kwok, J. Zádor, L. Sheps, *The Journal of Physical Chemistry A*, 119 (2015) 7742-7752.

[10] Z. Serinyel, O. Herbinet, O. Frottier, P. Dirrenberger, V. Warth, P.A. Glaude, F. Battin-Leclerc, *Combust. Flame*, 160 (2013) 2319-2332.

[11] E.J. Silke, W.J. Pitz, C.K. Westbrook, M. Ribaucour, *The Journal of Physical Chemistry A*, 111 (2007) 3761-3775.

[12] C. Cavallotti, R. Rota, T. Faravelli, E. Ranzi, *Proc. Combust. Inst.*, 31 (2007) 201-209.

[13] F. Buda, B. Heyberger, R. Fournet, P.-A. Glaude, V. Warth, F. Battin-Leclerc, *Energy & Fuels*, 20 (2006) 1450-1459.

[14] B. Rotavera, J.D. Savee, I.O. Antonov, R.L. Caravan, L. Sheps, D.L. Osborn, J. Zádor, C.A. Taatjes, *Proc. Combust. Inst.*, 36 (2017) 597-606.

[15] B. Rotavera, R.L. Caravan, C.A. Taatjes, Sandia National Laboratories, Subsequent R + O_2 Chemistry of Intermediates Formed in Low-Temperature R + O_2 Reactions: Potential Importance in Modeling Autoignition Behavior, SAND2017-9099, 2017.

[16] G. Dayma, P.A. Glaude, R. Fournet, F. Battin-Leclerc, *International Journal of Chemical Kinetics*, 35 (2003) 273-285.

[17] O. Lemaire, M. Ribaucour, M. Carlier, R. Minetti, *Combust. Flame*, 127 (2001) 1971-1980.

[18] M. Ribaucour, O. Lemaire, R. Minetti, *Proc. Combust. Inst.*, 29 (2002) 1303-1310.

[19] G. Meloni, T.M. Selby, D.L. Osborn, C.A. Taatjes, *The Journal of Physical Chemistry A*, 112 (2008) 13444-13451.

[20] S.R. Leone, M. Ahmed, K.R. Wilson, *Phys Chem Chem Phys*, 12 (2010) 6564-6578.

[21] D.L. Osborn, P. Zou, H. Johnsen, C.C. Hayden, C.A. Taatjes, V.D. Knyazev, S.W. North, D.S. Peterka, M. Ahmed, S.R. Leone, *Rev. Sci. Inst.*, 79 (2008) 104103-104110.

[22] O. Welz, J. Zador, J.D. Savee, M.Y. Ng, G. Meloni, R.X. Fernandes, L. Sheps, B.A. Simmons, T.S. Lee, D.L. Osborn, C.A. Taatjes, *Phys Chem Chem Phys*, 14 (2012) 3112-3127.

[23] J.D. Savee, E. Papajak, B. Rotavera, H. Huang, A.J. Eskola, O. Welz, L. Sheps, C.A. Taatjes, J. Zádor, D.L. Osborn, *Science*, 347 (2015) 643-646.

[24] A.J. Eskola, O. Welz, J. Zádor, I.O. Antonov, L. Sheps, J.D. Savee, D.L. Osborn, C.A. Taatjes, *Proc. Combust. Inst.*, 35 (2015) 291-298.

[25] F. Battin-Leclerc, O. Herbinet, P.-A. Glaude, R. Fournet, Z. Zhou, L. Deng, H. Guo, M. Xie, F. Qi, *Angewandte Chemie International Edition*, 49 (2010) 3169-3172.

[26] J.C. Davis, A.L. Koritzke, R.L. Caravan, I.O. Antonov, A. Jasper, D.L. Osborn, J. Zádor, L. Sheps, C.A. Taatjes, B. Rotavera, *in preparation.*

The Kinetic Study of Excited Singlet Oxygen Atom O(^{1}D) Reactions with Acetylene

Chao Yan[*1], *Cheyenne Teng*[2], *Timothy Chen*[1], *Aric Rousso*[1], *Gerard Wysocki*[2], *Yiguang Ju*[1]

[1]*Department of Mechanical and Aerospace Engineering*
Princeton University, Princeton, NJ 08544, USA
[2]*Department of Electrical Engineering*
Princeton University, Princeton, NJ 08544, USA
[*]*Corresponding Author Email: chaoy@princeton.edu*

Abstract: The present work studies the branching ratio and rate constants of important reactions for the O(^{1}D) + C_2H_2 kinetic pathways in a Herriott cell photolysis flow reactor. Mid-infrared tunable diode laser absorption spectroscopy (TDLAS) and Faraday rotation spectroscopy (FRS) are utilized for quantitative *in situ* measurements of OH, HO_2, and O_3. The branching ratios of reactions O(^{1}D) + C_2H_2 → C_2H_2O (1a), O(^{1}D) + C_2H_2 → HCCO + H (1b) and O(^{1}D) + C_2H_2 → CH_2 + CO (1c) are fitted to the experimental data. Time dependent OH radical concentration suggests the branching ratio of channel 1a, 1b and 1c are 56%, 22% and 22%, respectively. The results show that low temperature C_2H_2 oxidation can be greatly improved by adding two key elementary reactions: C_2H_2 + OH → C_2H_2OH and C_2H_2OH + O_2 → C_2H_2O + HO_2. The rate constants for the reaction C_2H_2OH + O_2 → C_2H_2O + HO_2 is found to be $k_{C2H2OH+O2} = 1\times10^{12}$ cm^3 mole^{-1} s^{-1} (at 296 K and 60 Torr). The present experimental results demonstrate the potential of current setup for more accurate chemical kinetics study in both low-temperature combustion and plasma-assisted combustion.
Keywords: Singlet oxygen atom; Reaction kinetics; Photolysis flow reactor; Faraday rotation spectroscopy.

1. Introduction

Over the past years, numerous studies [1-3] have shown that O(^{1}D) atoms contribute significantly to non-equilibrium plasma-assisted combustion processes by enhancing ignition, reducing emissions, and reforming fuels. In particular, O(^{1}D) plays an important role in driving low temperature chain branching, leading to chemical kinetics involving several key intermediate species such as OH, HO_2, RO_2. As such, understanding of the reaction kinetics of O(^{1}D) with fuels is of great importance.

For unsaturated hydrocarbons such as acetylene (C_2H_2), the mechanism of its reaction with O(^{1}D) might different with saturated hydrocarbons because the difference in chemical structure. Unfortunately, current literature studies of O(^{1}D) reaction with unsaturated hydrocarbons are rare. If one assumes this reaction mechanism is similar to that of ground state atomic oxygen (O) with C_2H_2, there are four potential reaction channels as following:

$$O(^1D) + C_2H_2 \rightarrow C_2H_2O \qquad R_{1a} \qquad (3)$$
$$O(^1D) + C_2H_2 \rightarrow HCCO + H \qquad R_{1b} \qquad (4)$$
$$O(^1D) + C_2H_2 \rightarrow CH_2 + CO \qquad R_{1c} \qquad (5)$$
$$O(^1D) + C_2H_2 \rightarrow C_2H + OH \qquad R_{1d} \qquad (6)$$

In the reaction of O(^{1}D) with C_2H_2, channel 1b and 1c are major channel and channel 1d is minor [4]. However, the reaction pathways branching ratios for O(^{1}D) + C_2H_2 remain unknown.

While TDLAS has been widely employed in kinetic studies to measure concentrations of active intermediate radicals via absorption in the fingerprint spectral region (mid-Infrared), accurate measurements of radicals (e.g. RO_2, HO_2, OH) is difficult due to spectral contamination from H_2O_2, H_2O, and other hydrocarbons in mid-infrared (IR) and ultraviolet (UV) regions. Therefore, we employ Faraday Rotation Spectroscopy (FRS) to selectively detect paramagnetic molecules, such as HO_2, OH, HCO and NO, while dramatically eliminating spectral interferences from non-paramagnetic species such as the hydrocarbons. Brumfield et al. demonstrated the application of mid-IR FRS for quantitative measurements of HO_2 and OH of dimethyl ether oxidation in an atmospheric flow reactor [5, 6].

In this study, time dependent measurements of the reaction rates between $O(^1D)$ and C_2H_2 were carried out in a UV photolysis flow reactor using multi-pass infrared laser diagnostics on $C_2H_2/O_2/O_3/He$ mixtures. By comparing kinetic model simulations with time-dependent experimental measurements of O_3 and OH by UV and infrared TDLAS, the important chain branching ratios for the various channels of reaction R1 were determined. Moreover, utilizing FRS measurements of HO_2, the rate constant of the dominant elementary reactions for its formation in the $O(^1D) + C_2H_2$ system was quantified, providing more insight into the important pathways of $O(^1D)$ interactions with hydrocarbon species.

2. Experimental Section

2.1 Experimental setup

The experimental apparatus couples Nd:YAG laser photolysis and UV-IR spectroscopy with a Herriott cell multi-pass flow reactor (Fig. 1).

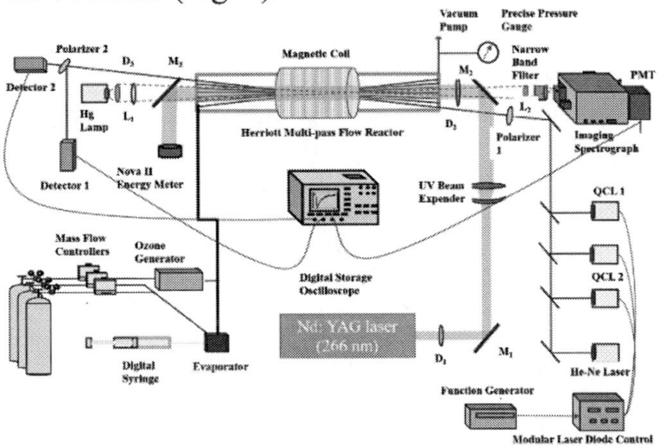

Fig. 1 Experimental setup schematics

The reactor is made of quartz tubing with an inner diameter of 56 mm. A pair of spherical mirrors (250 mm focal length) separated by 913 mm were installed at the both ends of the reactor to form the multi-pass cell (MPC). Two dichroic mirrors mounted at 45° are placed at the entrance and exit of the flow reactor. The photolysis beam was cropped by a 25-mm diaphragm before entering the reactor. The measured beam profile indicated that the uniformity across the reactor cross-section was ± 8.4 % from the mean value. Since residual light from photolysis laser and the fluorescence from any surface may influence O_3 measurement, a 254 nm mercury line filter (Semrock) was installed before the monochromator (Acton 2500i) to filter the 266 nm light. The monochromator was accurately calibrated using a mercury line, and a PMT (Hamamatsu R7154) with amplifier (Hamamatsu C6271) performs the measurement after monochromator.

Currently the MPC is set up in a 21-pass configuration, resulting in an effective pathlength of ~7.6 m. The time resolved absorption signals of OH radicals and HO_2 radicals are measured using DFB-QCLs centered around the spectral regions of 3568.52 cm^{-1} and 1396.91 cm^{-1}, respectively.

2.2 Faraday Rotation Spectroscopy (FRS)
FRS selectively detects paramagnetic species, which offers improved immunity to interference gasses and therefore is suitable for in-situ measurements of radical species for kinetic studies. When a linearly polarized laser light propagates co-linearly to an applied magnetic field, magnetically induced circular birefringence (MCB) causes rotation of the polarization plane of the transmitted light (the Faraday Effect).

As Fig. 1 shows, a solenoid is wrapped around the reactor and a DC current is applied to the solenoid to generate nearly uniform magnetic field at the center of the coil. A pair of wire grid polarizers (labeled Polarizer #1 and #2 in Fig. 1) are located at the front and end of the reactor. Since common mode noise (including shared absorption features from non-paramagnetic species such as C_2H_2, CH_3OH, CH_4, and H_2O) can be suppressed using balanced detection [1], this setup is especially suitable for *in situ* HO_2 and OH detection.

2.3 In situ actinometry
In this work, the absolute concentration of $O(^1D)$ was quantified by measuring the photon flux inside the reactor [7, 8] using the absorption cross-section and the photolysis reaction pathway of O_3. The major source of uncertainty in determining absolute concentration of $O(^1D)$ is quantifying the photolysis laser intensity inside the reactor. Therefore, we used *in situ* laser light actinometry based on ozone depletion, which is relatively stable at these conditions, with detailed characterized cross-section at the monitoring wavelength of 253.65 nm (Mercury line). The photon fluence F (photons cm^{-2}) can be calculated based on the absorbance (Abs):

$$F = \frac{Abs}{\sigma O3{,}253.65 * L * [O3] * \sigma O3{,}266} \qquad (7)$$

Where the ozone absorption cross-section at 253.65 nm and 266 nm are $\sigma_{O3,253.65}$ and $\sigma_{O3,266}$, L = 91.3 cm is the length of the reactor, and $[O_3]$ is the concentration. (Fig. 2)

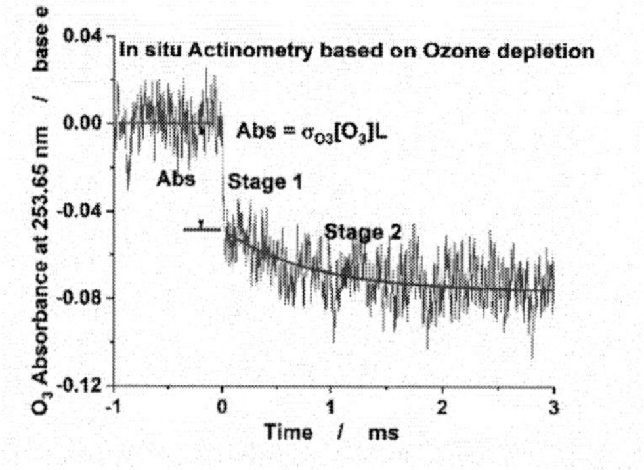

Fig. 2 In situ actinometry based on ozone depletion in $O_2/O_3/He$ gas mixtures.

3. Results and Discussion

3.1 The temporal profiles of OH radicals
Since we observed QCL laser frequency fluctuations up to 0.003 nm, which was 4 times smaller than the full width at half maximum (FWHM) of OH absorption profiles at 60 Torr and 296 K according to HITRAN simulations. Without active laser frequency stabilization, this caused considerable uncertainties in determining the precise laser frequency, therefore the following procedure was developed to determine the apparent absorption cross-section of OH radicals at target wavelength using well studied chemical reactions. The absolute concentration of O_3 and $O(^1D)$ were determined by *in situ* actinometry. With excess amount of H_2O molecules, the reaction

$O(^1D) + H_2O \rightarrow 2\ OH$ ends within 1 μs. Therefore, we can calibrate for the apparent absorption cross-section of OH using the predicted concentration as shown in Fig. 3.

OH radical absorption measured at 3568.52 cm^{-1} under 60 Torr and ambient temperature is shown in Fig. 4. The mechanism was constructed based on the HP-Mech and Ozone mechanism[9, 10] with additional reactions involving $O(^1D)$. This mechanism was used to model and simulate the experimental profiles. The key reactions of this mechanism are listed in Table 1.

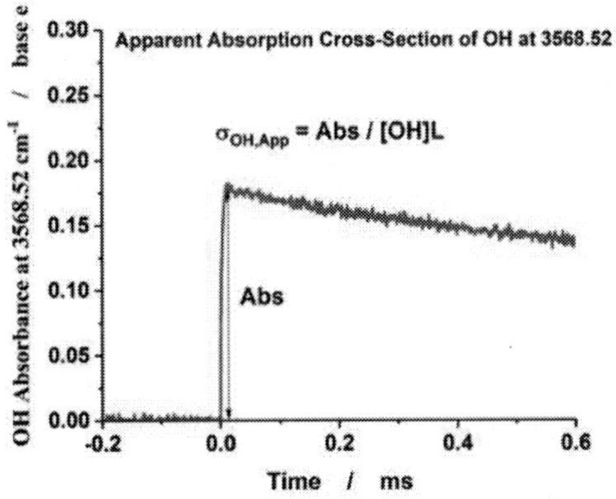

Fig. 3: The apparent absorption cross-section of OH radicals at 3568.52 cm^{-1} in a gas mixture of $O_3/O_2/H_2O/He$ under 60 Torr and 296 K.

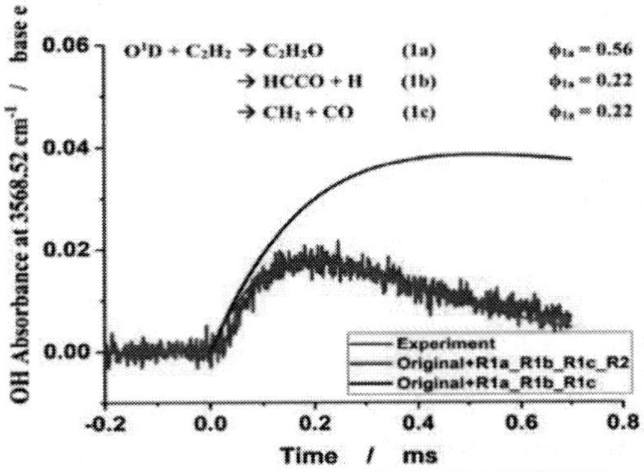

Fig. 4: Temporal profile of OH radicals in a gas mixture of $C_2H_2/O_2/O_3/He$. $[C_2H_2] = 1.32 \times 10^{16}$, $[O_2] = 4.94 \times 10^{15}$ and $[O_3] = 2.95 \times 10^{14}$ molecules cm^{-3}. Measurement performed at 60 Torr and 296 K.

Table 1: Rate constants of key elementary reactions used in the simulation. Units are in moles, cubic centimeters, seconds and cal/mole.

Number	*Reaction*	*A*	*n*	*Ea*	*Reference*	*Comments*
1a	$O(^1D)+C_2H_2 \rightarrow C_2H_2OH$	1.0e14	0	0	This study	56% of k_{1total}
1b	$O(^1D)+C_2H_2 \rightarrow HCCO+H$	4.0e13	0	0	This study	22% of k_{1total}
1c	$O(^1D)+C_2H_2 \rightarrow CH_2+CO$	4.0e13	0	0	This study	22% of k_{1total}
2	$C_2H_2+OH+M \rightarrow C_2H_2OH+M$	2.28e19	0	1071	Ref.[11]	
3	$O_3+H \rightarrow OH+O_2$	8.43e13	0	934	HP-Mech	
4	$CH_2+O_2 \rightarrow CO+OH+H$	5.67e12	0	1779	HP-Mech	
5	$HCCO+O_2 \rightarrow CO_2+CO+H$	4.78e12	-0.14	1150	HP-Mech	
6	$HO_2+OH \rightarrow H_2O+O_2$	2.89e13	0	-497	HP-Mech	
7	$C_2H_2OH+O_2 \rightarrow HO_2+C_2H_2O$	1.0e12	0	0	This study	
8	$HO_2+HO_2 \rightarrow H_2O_2+O_2$	1.30e11		-1629	HP-Mech	

In Fig. 4, a large discrepancy has been found when comparing the experimental data with the original HP-Mech mechanism. Reaction 2 in Table 1 is the dominant channel of the reaction C_2H_2 + OH in low temperature. After adding reaction 2 into the mechanism, the agreement between simulation and experimental results has been obviously improved. As shown in Fig. 4, no OH radials were observed right after the laser photolysis, which indicated that the channel $O(^1D) + C_2H_2 \rightarrow C_2H + OH$ (1d) can be neglected. Since there is no rate constant value of reaction 1, a conservative overall rate constant $k_1 = 1.8 \times 10^{14}$ cm^3 $mole^{-1}$ s^{-1} was estimated. According to the path flux analysis based on the updated mechanism, channels 1b and 1c have similar sensitivities of OH production within several chain branching reactions, while channel 1a also displays significant sensitivity but in the opposite direction to consume OH. As a result, combining a rough adjustment of the ratio of $k_{1a}/(k_{1b}+k_{1c})$ and a fine adjustment of the ratio of k_{1b}/k_{1c}, the agreement between simulation and experimental data was greatly improved. The derived branching ratios of channels 1a, 1b and 1c are 56%, 22% and 22%, respectively.

3.2 The temporal profiles of HO_2 radicals

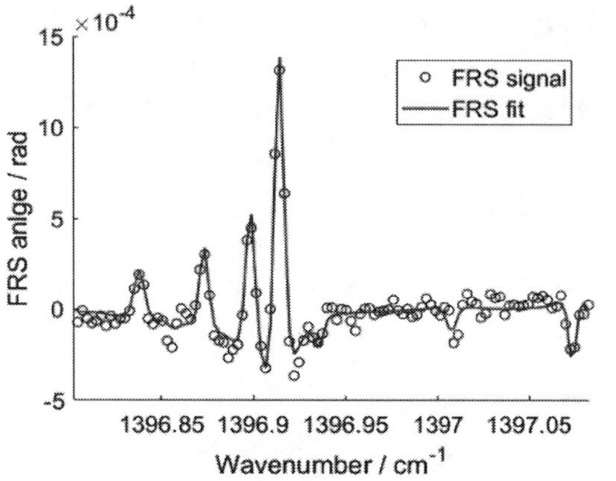

Fig. 5. FRS spectrum averaged over 100 photolysis events and the fitted model using HITRAN parameters.

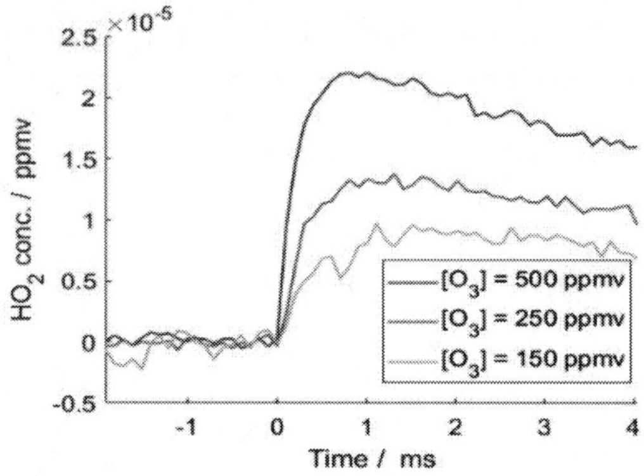

Fig. 6. HO_2 concentration retrievals using DC-FRS with balanced detection.

Concentration retrieval of HO_2 radicals during photolysis was measured using least-means-squares fitting of the obtained FRS spectrum. Fig. 5 shows close agreement between measured signal spectrum and the simulated FRS spectrum using HITRAN parameters. Measurements of HO_2 formation from reaction of $O(^1D)$ with C_2H_2 were repeated for varying levels of ozone concentrations to demonstrate feasibility of FRS measurements for kinetic studies. Fig. 6 shows that as ozone level increases, HO_2 production increases accordingly as expected.

The path flux analysis indicated that the original mechanism constructed using HP-Mech and Ozone with additional reactions involving $O(^1D)$ has little sensitivity to the production of HO_2, suggesting the absence of some important reactions as shown in Fig. 7. According AramcoMech 2.0 [12], the reaction between singlet 2-hydroxyvinyl radicals, SC_2H_2OH, and O_2, produces HO_2 and C_2H_2O or HCCOH. However, we suggest that an alternative reaction (reaction 7 listed in Table 1) should be included in our model. In fact, Fig. 7 shows dramatically improved agreement with experimental observations after including reaction 7, and the corresponding rate constant was determined using the HO_2 temporal profile:

$$k_{C2H2OH+O2} = 1\times10^{12}\ cm^3\ mole^{-1}\ s^{-1}\ (296\ K\ and\ 60\ Torr) \qquad (8)$$

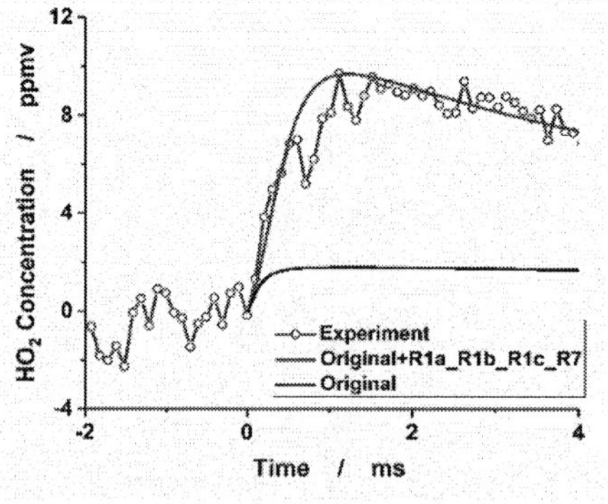

Fig. 7: The temporal profile of HO_2 concentration with a gas mixture of $C_2H_2/O_2/O_3$/He. $[C_2H_2] = 1.32 \times 10^{16}$, $[O_2] = 4.94 \times 10^{15}$ and $[O_3] = 2.95 \times 10^{14}$ molecules cm^{-3}.

4. Summary

UV photolysis of ozone in a flow reactor, coupled with mid-infrared tunable laser absorption spectroscopy (TDLAS) and Faraday rotation spectroscopy (FRS), was utilized for quantification of the important pathways of $O(^1D)$ reactions with C_2H_2. Time resolved measurements of O_3 as well as the active intermediate species OH and HO_2 were compared to a kinetic model of the $C_2H_2/O_2/O_3$/He mixture. Using the experimental data, the branching ratios of the reaction $O(^1D)$ with C_2H_2 was derived. An updated kinetic model was developed and validated using the new branching ratios of $O(^1D)$ with C_2H_2 and adding several missing reactions involving C_2H_2 with OH. The updated model showed significant improvement in reproducing experimental data. One important reaction of C_2H_2OH with O_2 was adopted in the updated kinetic model and more future investigation about this reaction is required.

5. Acknowledgement: This project is supported by the NSF CBET-1507358 and Princeton SEAS innovation grant.

6. Reference

[1] Y. Ju, W. Sun, Plasma assisted combustion: Dynamics and chemistry, Progress in Energy and Combustion Science 48 (2015) 21-83.
[2] Y. Suo, N. Sharath, S. Wenting, Y. Vigor, Multiscale modeling and general theory of non-equilibrium plasma-assisted ignition and combustion, Journal of Physics D: Applied Physics 50 (2017) 433001.
[3] S. Yang, X. Gao, V. Yang, W. Sun, S. Nagaraja, J.K. Lefkowitz, Y. Ju, Nanosecond Pulsed Plasma Activated C2H4/O2/Ar Mixtures in a Flow Reactor, Journal of Propulsion and Power 32 (2016) 1240-1252.
[4] D.L. Baulch, C.J. Cobos, R.A. Cox, C. Esser, P. Frank, T. Just, J.A. Kerr, M.J. Pilling, J. Troe, R.W. Walker, J. Warnatz, Evaluated Kinetic Data for Combustion Modelling, Journal of Physical and Chemical Reference Data 21 (1992) 411-734.
[5] B. Brumfield, W. Sun, Y. Ju, G. Wysocki, Direct In Situ Quantification of HO2 from a Flow Reactor, The Journal of Physical Chemistry Letters 4 (2013) 872-876.
[6] B. Brumfield, W. Sun, Y. Wang, Y. Ju, G. Wysocki, Dual modulation Faraday rotation spectroscopy of HO2 in a flow reactor, Opt. Lett. 39 (2014) 1783-1786.
[7] M. Sangwan, C. Yan, E.N. Chesnokov, L.N. Krasnoperov, Reaction CH3 + CH3 → C2H6 Studied over the 292–714 K Temperature and 1–100 bar Pressure Ranges, The Journal of Physical Chemistry A 119 (2015) 7847-7857.
[8] E.J. Dunlea, A.R. Ravishankara, Kinetic studies of the reactions of O(1D) with several atmospheric molecules, Physical Chemistry Chemical Physics 6 (2004) 2152-2161.
[9] X. Shen, X. Yang, J. Santner, J. Sun, Y. Ju, Experimental and kinetic studies of acetylene flames at elevated pressures, Proceedings of the Combustion Institute 35 (2015) 721-728.
[10] H. Zhao, J. Fu, F.M. Haas, Y. Ju, Effect of prompt dissociation of formyl radical on 1,3,5-trioxane and CH2O laminar flame speeds with CO2 dilution at elevated pressure, Combustion and Flame 183 (2017) 253-260.
[11] K.W. McKee, M.A. Blitz, P.A. Cleary, D.R. Glowacki, M.J. Pilling, P.W. Seakins, L. Wang, Experimental and Master Equation Study of the Kinetics of OH + C2H2: Temperature Dependence of the Limiting High Pressure and Pressure Dependent Rate Coefficients, The Journal of Physical Chemistry A 111 (2007) 4043-4055.
[12] M.U. Alzueta, M. Borruey, A. Callejas, A. Millera, R. Bilbao, An experimental and modeling study of the oxidation of acetylene in a flow reactor, Combustion and Flame 152 (2008) 377-386.

Spring Technical Meeting
Eastern States Section of the Combustion Institute
March 4–7, 2018
State College, Pennsylvania

On the Relative Importance of HONO versus HNO_2 in Low-temperature Combustion

Mark E. Fuller[1] *and C. Franklin Goldsmith*[1,*]

[1]*School of Engineering, Brown University, Providence, RI 02912, USA*
[*]*Corresponding author:* franklin_goldsmith@brown.edu

Abstract: In this work, NO_2-doped ignition experiments are examined to determine whether both HONO and HNO_2 must be treated as unique species or if an appropriate lumping strategy exists without sacrificing model fidelity. Kinetic mechanisms are taken from literature and updated with revised rate constants for several reactions involving HONO and HNO_2. Model predictions for full mechanisms with both HONO and HNO_2 are compared against variants in which either HNO_2 is removed from the mechanism or HNO_2 is replaced with HONO. Ignition delay time is modestly affected by the omission of HNO_2; replacement of HNO_2 with HONO results in a significant increase in reactivity and decrease in the predicted ignition delay. The recommendation is that for larger fuels, both HONO and HNO_2 should be included in the mechanism.
***Keywords:* HONO, Nitrogen Chemistry, NOx, Low-temperature Combustion**

1. Introduction

Hydrogen transfer reactions involving NO_2, both hydrogen abstraction from a fuel molecule and disproportionation from a fuel-derived radical, result in the formation of both HONO and HNO_2. In these reactions, the typical branching fraction of HONO is between 60% and 90% of the total flux and is dependent on the source of the hydrogen atom[1].

Both HONO and HNO_2 ultimately decompose predominantly to OH + NO. Unimolecular decomposition is the primary channel for high temperatures, but under lower temperatures, both HONO and HNO_2 can build up to sufficient concentrations to enable bimolecular reactions[2].

Recent studies strongly suggest that current models do not accurately capture low-temperature nitrogen chemistry and fuel-NO_x interactions: Giménez-López et al. [3] conducted flow reactor studies of $C_2H_4/O_2/NO$ mixtures under high pressure (60 bar) and temperatures of 600 K to 900 K. The experiments found significant removal of NO_x; this was not predicted by the kinetic mechanism. Experimental studies of low-temperature compression ignition (LTCI) engines with 2-ethylhexyl nitrate (EHN) as the cetane enhancer have also found discrepancies between model and experiment with roughly one-third of the fuel-bound nitrogen found in the exhaust as NO_x[4–6].

In addition to developing revised rate constants, in order to facilitate efficient mechanism development, we seek to address the question of whether it is necessary to include HNO_2 in mechanisms for combustion kinetics. The published mechanisms of Mathieu et al. [7] and Glarborg et al.[8] are examined herein as two independent and validated mechanisms. Each mechanism was used as a basis for four cases: First, the published mechanism ("Original"). Second, replacement of the published HONO/HNO_2 reactions with a new, theory-derived HONO/HNO_2 submechanism and thermodynamic properties for the species in the new HONO submechanism are taken from the Active Thermochemical Tables, version v1.122b[8, 9] ("Modified"). Third, elimination of all reactions containing HNO_2 from the second case mechanism ("No HNO_2"). Fourth, replacement of all HNO_2 reactions with HONO, treating these reactions as duplicates ("Duplicate").

The third case is expected to decrease reactivity by underpredicting the flux through RH + NO_2 $\rightleftharpoons$ R + HONO. The fourth case is expected to run counter to the effect of the third case. The four cases for each of the two mechanisms are examined for both H_2 and CH_4 ignition delays.

2. Computational Methods

Revised rate constants were calculated following the recommended electronic structure methods of Chai and Goldsmith [1]. One revision from Chai and Goldsmith is that this current work does not treat cis- and trans-HONO separately, but instead as a single isomer with a hindered internal rotation for cis- trans- conversion. Ongoing work by the authors suggests that cis-HONO and trans-HONO are no longer distinct species under engine-relevant conditions, and that they should be treated as a single HONO species in combustion mechanisms.

All kinetic simulations were performed using CANTERA[10].

The ignition delay time was defined as the time at which the simulated concentration of OH was maximum.

3. Results

3.1 Computational Kinetics

The potential energy surfaces for additional/isomerization/elimination pathways of H and CH_3 to HONO and HNO_2 were developed. The results of these surfaces were combined with the abstraction reactions for H_2 + NO_2 and CH_4 + NO_2 from Chai and Goldsmith [1]. Rates for HONO isomerization were also recalculated. These computational results were combined to form an updated set of parameters for insertion into literature mechanisms.

3.2 Ignition Delay Simulations

Mixture compositions and initial conditions for the ignition delay studies were adapted from published studies.

For the H_2 ignition delays, the initial composition was 1.0% H_2, 1.0% O_2, 0.16% NO_2, and the remainder Ar at 1.56 atm, as described in Zhang et al. [11]; these results are presented in Figure 1. Modification of the published mechanism (solid black) with the updated rates calculated for this study (dash blue) shows close agreement in both cases. Elimination of HNO_2 from the mechanism (dash-dot orange) has a small effect, slightly reducing the mixture reactivity. Replacement of 'HNO2' with 'HONO' ("Duplicate", dotted red), however, significantly increases reactivity and lowers the ignition delay. This effect is discussed below in Section 4.

For the CH_4 ignition delays, the initial composition was 9.49% CH_4, 19.0% O_2, 56.7% N_2, 14.6% Ar, and 0.15% NO_2 at 9 atm, as described in Mathieu et al. [12]; these results are presented in Figure 2. The effect of the mechanism modifications for methane is more apparent than for

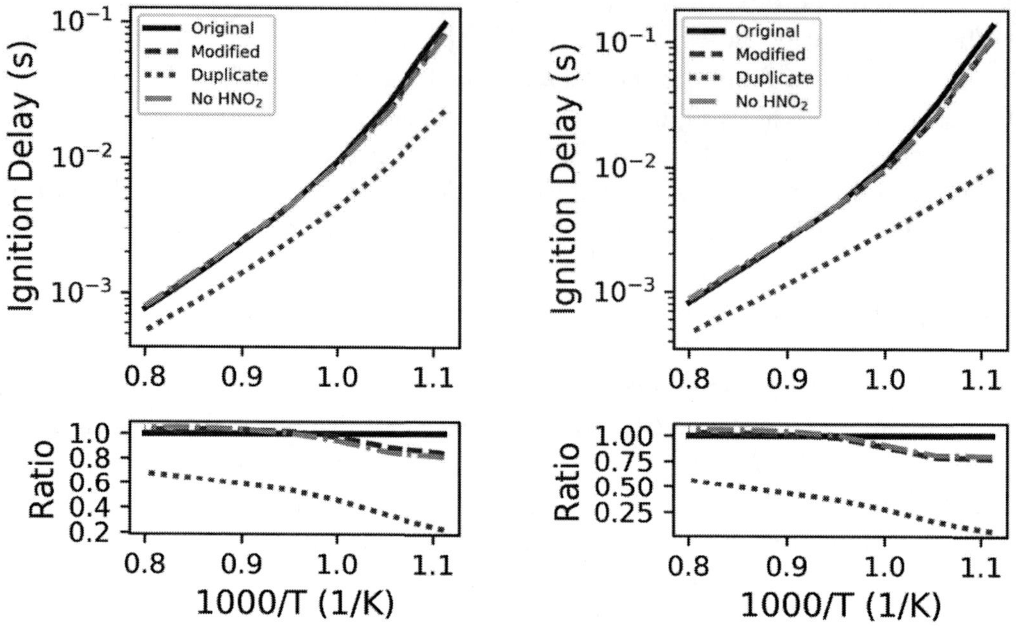

Figure 1: Ignition delay for H_2, $\phi = 0.5$ doped with 1600 ppm NO_2 at 1.56 atm with the Glarborg et al. mechanism, left, and Mathieu et al. mechanism, right

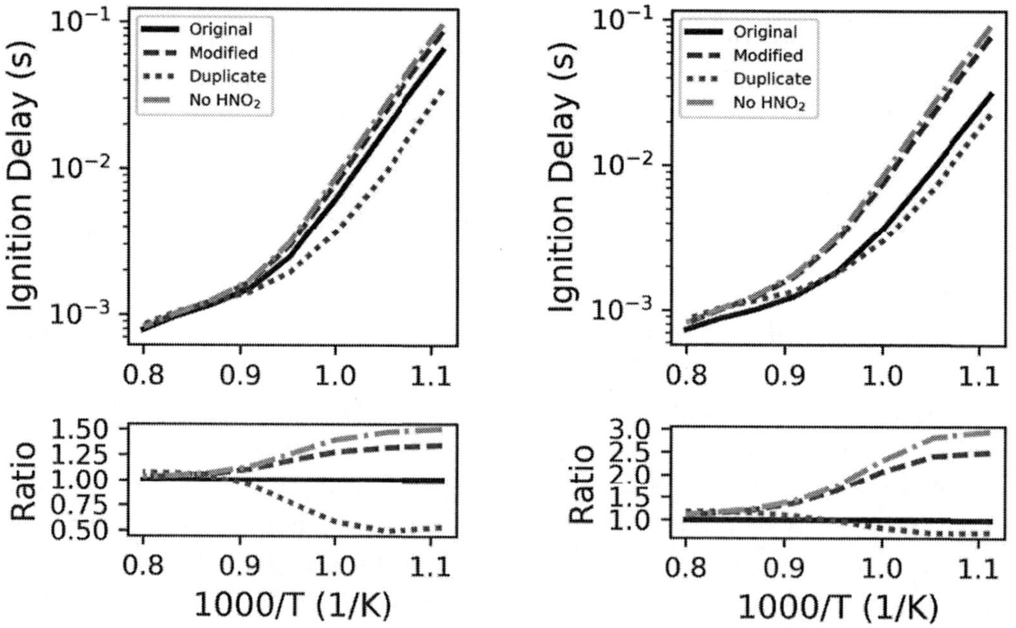

Figure 2: Ignition delay for CH_4, $\phi = 1.0$ doped with 0.15% NO_2 at 9 atm with the Glarborg et al. mechanism, left, and Mathieu et al. mechanism, right

hydrogen. The changes to the Glarborg mechanism, Figure 2(a), have a similar trend to Figure 1(a); the combining of HNO_2 with HONO ("Duplicate") decreases the ignition delay by nearly a factor of two. In contrast, the Mathieu mechanism, Figure 2(b), shows much less of an effect under the "Duplicate" strategy. The Mathieu mechanism also shows significantly more sensitivity to the revised rate constants, with the ignition delay time increasing by up to 80% with the updated rates. Regardless of mechanism, the ignition delay time is sensitive to the HONO/HNO_2 submechanism for $T < 1400$ K. Beyond this temperature, the contribution of NO_2 decreases, and the conventional high-temperature oxidation chemistry begins to dominate.

4. Discussion

In the case of HNO_2 elimination from the mechanism (dash-dot orange), there is a slight, but essentially negligible, increase in the predicted ignition delay time. Elimination of HNO_2 results in a reduction in the net flux of H_2 + NO_2 and from there to a corresponding reduction in the net rate of chain branching via decomposition of HNO_2. Replacement of HNO_2 with HONO has a much greater impact on the model prediction than the preceding would suggest. This effect is owed to the reactions proceeding in the endothermic direction. Proceeding in the exothermic direction, the rate constant for H + HNO_2 is approximately an order of magnitude greater than H + HONO; in the endothermic direction, H + HNO_2 is nearly an order of magnitude less than H + HONO. In the "duplicate" simulations, the forward (exothermic) rates are combined with the larger value of H + HNO_2 dominating. The endothermic reaction rate is then determined from thermodynamic properties, leading to the overprediction of the net rate for $H_2 + NO_2 \rightarrow H + HONO$. As the decomposition of HONO is chain-branching, this over-prediction enhances the reactivity of the mixture. The same effect observed when H is replaced with CH_3.

A second effect which is observed for CH_4 is the different response of the Glarborg and Mathieu models to the updated rates developed in this work. Both mechanisms use the rate constant of Dean and Bozzelli [2] for $CH_3 + HONO \rightleftharpoons CH_4 + NO_2$, but different values for $CH_3 + HNO_2 \rightleftharpoons CH_4 + NO_2$. In the case of Glarborg et al., that rate is also drawn from [2], but Mathieu et al. use the value of Yamaguchi et al.[13]. As compared the present value, the rate constant of Yamaguchi et al. is roughly double. As compared with Dean and Bozzelli, the value of Yamaguchi et al. is an order of magnitude greater. This difference in the rates employed in the published mechanisms accounts for the differing responses to the updated values from this work.

Overall, "lumping" together HNO_2 and HONO is not a successful strategy leading as it does to an over-estimate of chain branching from the net flux of RH + NO_2. The approach of eliminating HNO_2 has a much smaller effect, but this is believed to only be true for H_2 and CH_4 and not for larger fuels: Examining the results of Chai and Goldsmith [1], the branching fraction for $RH + NO_2 \rightleftharpoons R + HNO_2$ increases with the size of RH, from ~10% for H_2 and CH_4, to ~40% for C_4H_8 and C_4H_{10}. Consequently, inclusion of HNO_2 is preferred for larger mechanisms and should consistently include rates for both HONO and HNO_2.

5. Conclusions

The kinetic implications of HONO versus HNO_2 are examined. New and revised rate constants for reactions of interest involving HONO and HNO_2 in conjunction with H_2 and CH_4 as fuels are developed and implemented in published mechanisms. Examination of ignition delay modeling results confirms that while HONO and HNO_2 could lead to different products, this is not observed to have a significant effect. Efforts to eliminate HNO_2 as a unique specie were examined. The deletion of HNO_2 from the mechanism does not have a large impact on ignition delay for the fuels considered, but is expected to more severely retard ignition as the size of the hydrocarbon fuel molecule increases due to an increasing branching fraction of HNO_2 formed as a result of hydrogen abstraction by NO_2. The alternate approach of replacing HNO_2 with HONO results in a significant increase in the mixture reactivity due to an overestimate of the net contribution of RH + NO_2.

6. Acknowledgments

This work is supported by the U.S. National Science Foundation through Award Number CBET-1553366, with Dr. Song-Charng Kong as the program manager. MEF gratefully acknowledges help from Aaron Danilack and Xi Chen with the electronic structure theory calculations.

References

[1] J. Chai and C. F. Goldsmith, Rate coefficients for fuel + NO_2 : Predictive kinetics for HONO and HNO_2 formation, Proc. Comb. Inst. 36 (2017) 617–626. DOI: 10.1016/j.proci.2016.06.133.

[2] A. M. Dean and J. W. Bozzelli, "Combustion Chemistry of Nitrogen", in: Gas-Phase Combustion Chemistry, Springer New York, New York, NY, 2000, chap. 2, pp. 125–341, DOI: 10.1007/978-1-4612-1310-9_2.

[3] J. Giménez-López, M. Alzueta, C. Rasmussen, P. Marshall, and P. Glarborg, High pressure oxidation of C_2H_4/NO mixtures, Proc. Comb. Inst. 33 (2011) 449–457. DOI: 10.1016/j.proci.2010.05.098.

[4] A. B. Dempsey, N. R. Walker, and R. D. Reitz, Effect of Cetane Improvers on Gasoline, Ethanol, and Methanol Reactivity and the Implications for RCCI Combustion, SAE International Journal of Fuels and Lubricants 6 (2013) 170–187. DOI: 10.4271/2013-01-1678.

[5] D. A. Splitter and R. D. Reitz, Fuel reactivity effects on the efficiency and operational window of dual-fuel compression ignition engines, Fuel 118 (2014) 163–175. DOI: 10.1016/j.fuel.2013.10.045.

[6] A. M. Ickes, S. V. Bohac, and D. N. Assanis, Effect of 2-Ethylhexyl Nitrate Cetane Improver on NO_x Emissions from Premixed Low-Temperature Diesel Combustion, Energy & Fuels 23 (2009) 4943–4948. DOI: 10.1021/ef900408e.

[7] O. Mathieu, B. Giri, A. Agard, T. Adams, J. Mertens, and E. Petersen, Nitromethane ignition behind reflected shock waves: Experimental and numerical study, Fuel 182 (2016) 597–612. DOI: 10.1016/j.fuel.2016.05.060.

[8] P. Glarborg, J. A. Miller, B. Ruscic, and S. J. Klippenstein, Modeling Nitrogen Chemistry in Combustion, Prog. Energy Combust. Sci. (2017) (in press).

[9] B. Ruscic and D. H. Bross, *Active Thermochemical Tables (ATcT) values based on ver. 1.122 of the Thermochemical Network.* `https://atct.anl.gov/`.

[10] D. G. Goodwin, H. K. Moffat, and R. L. Speth, *Cantera: An Object-oriented Software Toolkit for Chemical Kinetics, Thermodynamics, and Transport Processes*, `http://www.cantera.org`, Version 2.2.1, 2016.

[11] Y. Zhang, O. Mathieu, E. L. Petersen, G. Bourque, and H. J. Curran, Assessing the predictions of a NO_x kinetic mechanism on recent hydrogen and syngas experimental data, Combust. Flame 182 (2017) 122–141. DOI: `10.1016/j.combustflame.2017.03.019`.

[12] O. Mathieu, J. M. Pemelton, G. Bourque, and E. L. Petersen, Shock-induced ignition of methane sensitized by NO_2 and N_2O, Combust. Flame 162 (2015) 3053–3070. DOI: `10.1016/j.combustflame.2015.03.024`.

[13] Y. Yamaguchi, Y. Teng, S. Shimomura, K. Tabata, and K. Suzuki, Ab Initio Study for Selective Oxidation of Methane with NO_x (x= 1, 2), J. Phys. Chem. A 103 (1999) 8272–8278.

Spring Technical Meeting
Eastern States Section of the Combustion Institute
March 4–7, 2018
State College, Pennsylvania

A Computational Investigation into the Kinetics of NO + CH_2CCH and its Effect on NO Reduction

Aaron D. Danilack[1] *and C. Franklin Goldsmith*[1,*]

[1]*School of Engineering, Brown University, Providence, RI, 02912, USA*
[*]*Corresponding author:* franklin_goldsmith@brown.edu

Abstract: NO preferentially reacts with radical species over closed-shell species during combustion. Unsaturated radicals, such as vinylic radicals, can combine with and subsequently reduce NO, often to HCN. As an initial investigation into the kinetics of NO + unsaturated radicals, the reaction of NO and the propargyl radical (CH_2CCH) is presented. The stationary points on the $C_3H_3N_1O_1$ potential energy surface are analyzed using the compound method ANL0, with key regions of the potential energy surface computed using multi-reference methods. The temperature- and pressure-dependent rate constants are computed using the RRKM/Master Equation. The results suggest that the dominant bimolecular products are CH_2CO + HCN and CO + CH_3CN. The new rate constants are implemented in a detailed chemical kinetic mechanism for the oxidation of C_2H_4 by O_2 + NO. Analysis of a constant temperature, constant pressure batch reaction suggests that NO + CH_2CCH could be an important pathway for both NO reduction and CH_2CCH oxidation in reburn chemistry, especially at lower temperatures and under fuel-rich conditions.
Keywords:* *Combustion, NO_x, Propargyl, Chemical Kinetics

Introduction

Reduced emissions of nitrogen oxides (NO_x= NO + NO_2) is a critical goal in the development of advanced internal combustion engines [1]. State-of-the-art chemical kinetic models can play a pivotal role in the design process by quantifying the rates of reactions that involve NO and NO_2 [2]. Due to the high stability of NO, reactions between this radical and closed-shell species are comparatively slow, and most NO-consuming pathways involve reaction with another radical. Notably, these reactions all lead to the reduction of NO, typically to HCN.

One potential sink for NO is the reaction with the propargyl radical, CH_2CCH. The recombination of propargyl is critical to the formation of the first aromatic ring and thence early soot precursors [3]. Furthermore, propargyl is difficult to oxidize with molecular oxygen [4], and thus the oxidation of CH_2CCH by NO has important implications for kinetic models of nascent soot chemistry. The aim of the present work is to predict the temperature- and pressure-dependent rate coefficients for the reaction of NO + CH_2CCH. Owing to the two resonance structures of propargyl, the initial addition of NO could occur at either the tail (–CH) of the radical, or at the head (–CH_2):

$$CH_2CCH + NO \rightarrow CH_2CCHNO \quad (R1)$$
$$\rightarrow CHCCH_2NO \quad (R2)$$

Each of these initial adducts could undergo a sequence of isomerization and decomposition steps leading to other $C_3H_3N_1O_1$ unimolecular intermediates or to various bimolecular products:

$$CH_2CCH + NO \rightarrow c\text{-}ONCHCHCH \quad (R3)$$
$$\rightarrow CO + CH_3CN \quad (R4)$$
$$\rightarrow CH_2CO + HCN \quad (R5)$$
$$\rightarrow \text{other products?}$$

Methods

Stationary points on the $C_3H_3N_1O_1$ potential energy surface (PES) were explored initially using the density functional ωB97XD [5]. Geometry optimization, normal mode analysis, and single-point calculations on the stationary points were refined with various electronic structure methods. This detailed PES resulted in 12 unimolecular intermediates and 15 transition states leading to 5 bimolecular products. Due to the complexity of this reaction system, the full PES was divided into two separate PES (one for each entry channel) to facilitate visualization. Figure 1 depicts the reaction pathways when NO adds to the -CH end of propargyl, and Figure 2 shows the reaction pathways when NO adds to the $-CH_2$ end.

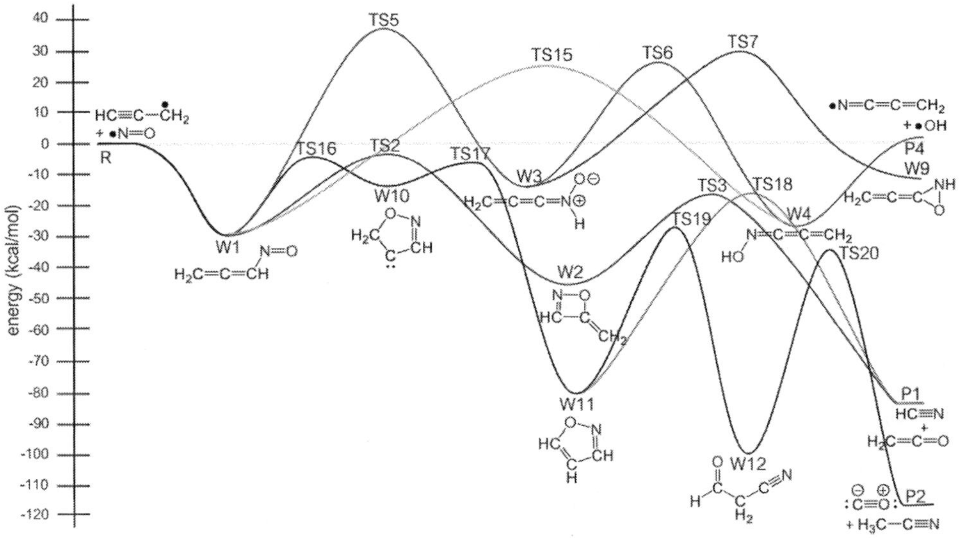

Figure 1: Entry channel 1: NO addition to the -CH end of propargyl.

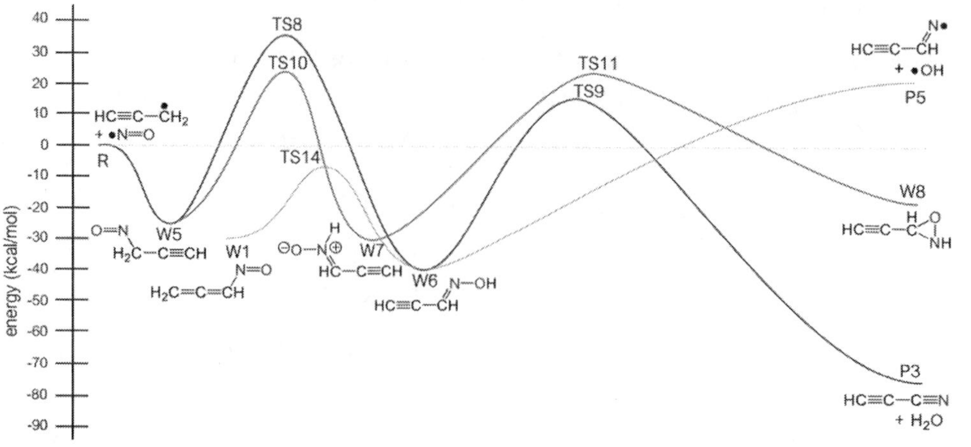

Figure 2: Entry channel 2: NO addition to the $-CH_2$ end of propargyl.

For the $C_3H_3N_1O_1$ intermediates and transition states that are below the reactants in potential energy, the geometries, frequencies, and final energies were redone using the compound method ANL0 [6], which should be accurate to $2\sigma \leq 0.3$ kcal/mol. Rate constants were computed from transition state theory (TST) calculations using the RRKM/ME code MESS [7, 8], which is part of the computational kinetics package PAPR developed by Argonne National Laboratory [9].

Results and Discussion

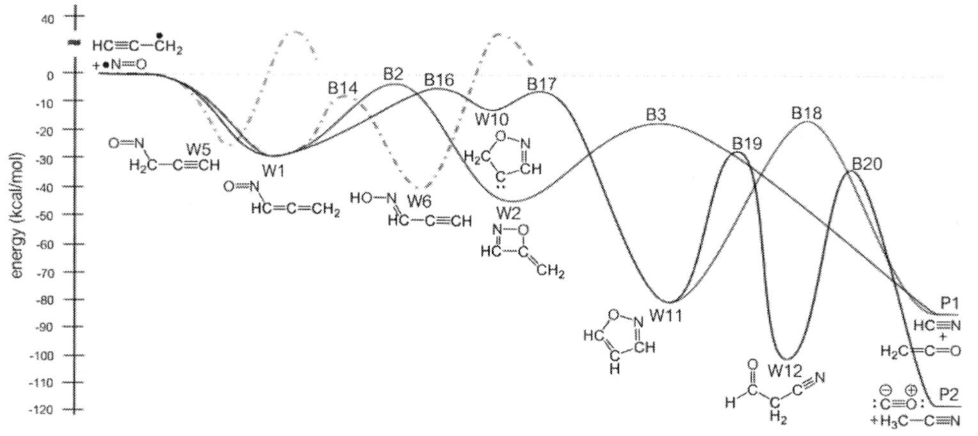

Figure 3: Submerged PES for NO + CH_2CCH.

A PES of the ANL0 energies for the kinetically significant pathways of NO + CH_2CCH is shown in Figure 3. It can be seen from Figure 3 that all submerged reactions of interest proceed through NO addition to the -CH end of propargyl.

Figure 4 presents the branching fractions as a function of temperature at three different pressures for the primary species of interest in Figure 3. The two dotted lines correspond to the unimolecular intermediates that do not lead to subsequent reduction of NO, W5 (cyan) and W6

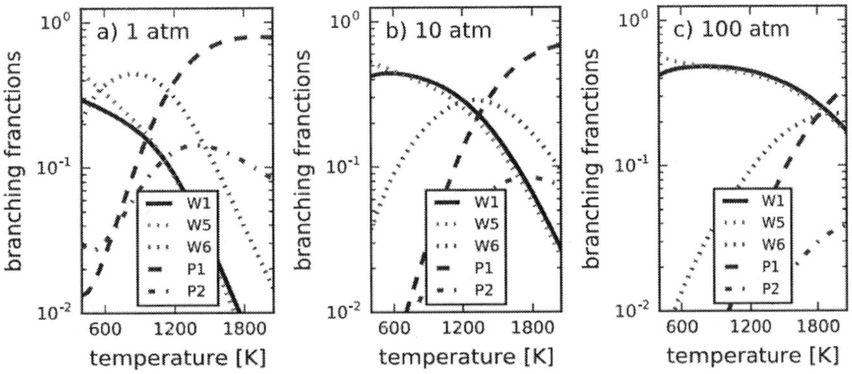

Figure 4: Branching fractions for NO + CH_2CCH at 1, 10, and 100 atm.

(magenta). At low temperatures, the majority of the flux goes to the two entrance channels, W1 (black) and W5, with the flux to W5 slightly higher.

Of particular interest is the temperature at which the two bimolecular product channels, P1 (dashed blue) and P2 (dash-dot red), become the dominant products. At 0.1 atm (not shown), this crossover temperature happens at 900 K. By 100 atm, the formally direct pathway to bimolecular products does not dominate until 1800 K. This crossover temperature is significant, since it indicates the conditions under which this reaction is expected to lead to a net reduction of NO (or oxidation of CH_2CCH).

Kinetic Simulations

To assess the impact of the title reaction on NO_x reduction, kinetic simulations were performed using CANTERA [10]. The detailed chemical kinetic model for $C_2H_4/O_2/NO$ developed by Giménez-López et al. was used as a starting mechanism [2]. Then, a submechanism for the formation of propargyl from smaller molecular weight species was included, as no propargyl chemistry is provided in Ref. [2]. This CH_2CCH submechanism consists of 26 elementary reactions that connect CH_2CCH with the smaller molecular weight species already present in Ref. [2]. Rate coefficients for these reactions were taken from a recently developed kinetic mechanism compiled by Jim Miller and coworkers [11]. Next, a submechanism consisting of the reactions in Figure 3 was included, the rate constants of which were computed with TST.

The kinetic simulations consisted of a constant-temperature, constant-pressure batch reactor. The oxidizer composition was 1 mole O_2, 3.76 mole Ar, and 0.2 mole NO. For 1 mole of C_2H_4, equivalence ratios of $\phi = 1.0$, 2.0, and 3.0 were considered at pressures of $P = 1$, 10, and 100 atm between $700 < T < 1600$ K. For each unique combination of T, P, and ϕ, the simulation was run until 99.9% of the C_2H_4 was consumed.

To quantify the effect of the title reaction on NO reduction, the fraction of the total NO flux that proceeds through the four reactions R1, R2, R4, and R5 is computed. At each timestep t in the simulation, CANTERA returns the `net_rates_of_progress` for each reaction, $r_i(t)$. We selected the subset of this array for those reactions that involve NO. The characteristic rate of progress for all NO chemistry at the timestep t is defined as the L2-norm of the net rates of progress for all the reactions involving NO:

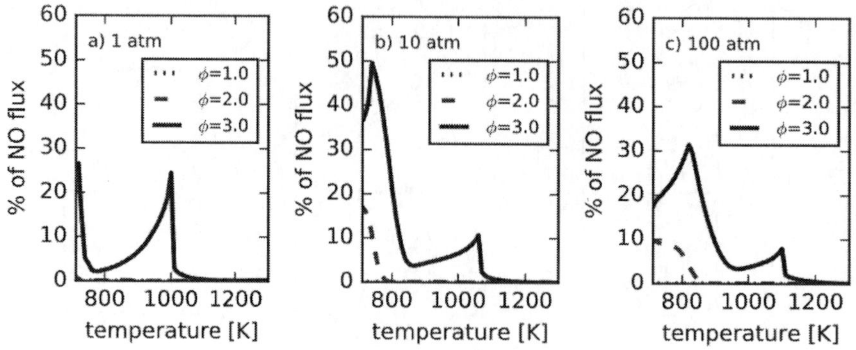

Figure 5: Percentage of total NO flux that goes through the NO + CH_2CCH system, as described in Equation (2), at three different pressures and three different equivalence ratios.

$$r_{\text{all NO}}(t) = \sqrt{\sum_{i}^{N_{\text{rxns with NO}}} |r_i(t)|^2} \tag{1}$$

A second characteristic rate of progress was computed for only the four reactions between NO and CH_2CCH, $r_{\text{NO+CH}_2\text{CCH}}(t)$. The ratio of these two characteristic rates was then integrated with respect to time:

$$r_{\text{NO}}^{\text{cumulative}} = \frac{1}{t_{end}} \int_0^{t_{end}} \frac{r_{\text{NO + CH}_2\text{CCH}}(t)}{r_{\text{all NO}}(t)} dt \tag{2}$$

where t_{end} is the time at which the conversion of C_2H_4 is 99.9%.

Figure 5(a-c) presents this fraction as a percentage of the cumulative flux at three different pressures. For $\phi = 1$ (dotted blue lines), the title reaction is negligible under these conditions. So long as the mixture is stoichiometric, the net rate of formation for propargyl from C_2H_4 is quite low, and therefore CH_2CCH will not be much of a sink for NO. For $\phi = 3$ (solid black lines), in contrast, the amount of propargyl generated in the reactor is substantial, and the reduction of NO by CH_2CCH is now a major fraction of the total NO flux. These results suggest that under fuel-rich conditions, the reaction of NO with CH_2CCH could account for a significant fraction of the total NO chemistry. We speculate that this reaction could be important in reburn chemistry for larger fuels, where NO and CH_2CCH are likely to be present at the same time.

Conclusions

A computational investigation into the kinetics of NO with CH_2CCH is presented. High-accuracy electronic structure calculations reveal multiple pathways that enable the reduction of NO (and oxidation of CH_2CCH) to CH_2CO + HCN and CO + CH_3CN. The new rate constants for NO + $CH_2CCH \rightleftharpoons$ products are implemented into a detailed combustion mechanism. Simulations of C_2H_4 oxidation in a constant-(T,P) batch reactor reveal that CH_2CCH is a major sink for NO under lower temperature, fuel-rich conditions. This study will be continued by examining the effects of other radicals, such as vinylic derivatives, on NO reduction.

Acknowledgements

This work is supported by the U.S. National Science Foundation through Award Number CBET-1553366, with Dr. Song-Charng Kong as the program manager. ADD is supported by the U.S. Office of Naval Research through ONR Grant Number N00014-16-1-2769, with Dr. Maria Medieros as the program manager.

References

[1] A. K. Agarwal, A. P. Singh, and R. K. Maurya, Evolution, challenges and path forward for low temperature combustion engines, Progress in Energy and Combustion Science 61 (2017) 1–56.

[2] J. Giménez-López, M. U. Alzueta, C. L. Rasmussen, P. Marshall, and P. Glarborg, High pressure oxidation of C_2H_4/NO mixtures, Proceedings of the Combustion Institute 33 (2011) 449–457.

[3] N. Hansen, J. A. Miller, S. J. Klippenstein, P. R. Westmoreland, and K. Kohse-Höinghaus, Exploring formation pathways of aromatic compounds in laboratory-based model flames of aliphatic fuels, Combustion, Explosion, and Shock Waves 48 (2012) 508–515.

[4] D. K. Hahn, S. J. Klippenstein, and J. A. Miller, A theoretical analysis of the reaction between propargyl and molecular oxygen, Faraday Discussions 119 (2001) 79–100.

[5] J.-D. Chai and M. Head-Gordon, Long-range corrected hybrid density functionals with damped atom–atom dispersion corrections, Physical Chemistry Chemical Physics 10 (2008) 6615.

[6] S. J. Klippenstein, L. B. Harding, and B. Ruscic, Ab initio Computations and Active Thermochemical Tables Hand in Hand: Heats of Formation of Core Combustion Species, The Journal of Physical Chemistry A 121 (2017) 6580–6602.

[7] Y. Georgievskii, J. A. Miller, M. P. Burke, and S. J. Klippenstein, Reformulation and solution of the master equation for multiple-well chemical reactions, The Journal of Physical Chemistry A 117 (2013) 12146–12154.

[8] Y. Georgievskii and S. J. Klippenstein, MESS*: Master Equation System Solver 2016.3.23*, `http://tcg.cse.anl.gov/papr/codes/mess.html/`.

[9] Y. Georgievskii, J. A. Miller, M. P. Burke, and S. J. Klippenstein, PAPR*: Predictive Automated Phenomenological Rates v1*, `http://tcg.cse.anl.gov/papr/`.

[10] D. G. Goodwin, H. K. Moffat, and R. L. Speth, CANTERA*: An Object-oriented Software Toolkit for Chemical Kinetics, Thermodynamics, and Transport Processes*, Version 2.3.0, 2017.

[11] J. A. Miller, N. Hansen, J. Zador, A. W. Jasper, M. P. Burke, C. F. Goldsmith, R. Sivaramakrishnan, P. Glarborg, and S. J. Klippenstein, Combustion Chemistry in the Twenty-First Century: Developing a Theory-Informed Chemical Kinetic Model for the Small-Hydrocarbon Fuels. Progress in Energy and Combustion Science, Progress in Energy and Combustion Science (invited; in progress) ().

Spring Technical Meeting
Eastern States Section of the Combustion Institute
March 4-7, 2018
State College, Pennsylvania

Multidimensional Numerical Investigation of NO_x Formation in a Burner Coupled Flow Tube Configuration

Sheikh F. Ahmed[a], Adhiraj Dasgupta[b], Frederick L. Dryer[a,c] and Tanvir I. Farouk[a]

[a]Department of Mechanical Engineering
University of South Carolina
Columbia, SC 29208, USA
[b]Combustion Science and Engineering Inc.
Columbia, MD 21045, USA
[c]Department of Mechanical and Aerospace Engineering
Princeton University
Princeton, NJ 08544, USA
**Corresponding Author Email: tfarouk@sc.edu*

Abstract: Multidimensional simulations have been conducted to simulate atmospheric pressure flat flame/McKenna-burner-driven-flow tube experiments targeted to obtain NO_x speciation data for syngas combustion applications. In prior work, we demonstrated that multidimensional transport exists in the post flame region, resulting in significant discrepancies with uni-dimensional predictions. Here we utilize a multidimensional laminar reacting flow solver to simulate the fully coupled flame and post flame regions to further elucidate the significance of earlier modeling assumptions on interpreting the post flame experimental data on observed NO_x evolution. The model is used to simulate a lean, premixed syngas/air flame and post flame region under atmospheric condition for which trace amounts of NO_x are added to the initial flow. The spatial evolution of NO_x species (NO and NO_2) in the flame and in the post-combustion zone suggest two distinct regions of behavior: the region encompassing the flame structure itself and the post flame region in which reaction temperature decays from the maximum flame temperature. The predicted NO_2 profiles show radial inhomogeneities that vary downstream of the burner surface having a maximum concentration close to but slightly offset from its centerline. The location of the peak NO_2 is dictated by the radial gradient of the temperature resulting from the wall cooling effects. Post flame uni-dimensional interpretations with and without transport fail to reasonably describe the noted spatial variations of NO_x kinetics in the different regions of the NO_x evolution.
Keywords: Multidimensional, Burner stabilized flame, NO_x recycling, NO-NO_2 distribution, Inhomogeneity

1. Introduction

A Flat flames stabilized on porous plug burners have gained attention to the scientific community due to the simple burner geometry, easily accessible flame structure using various measurement techniques and the spatial and temporal uniformity of the exhaust gas region [1]. These unique characteristics of flat flame burners aid in their widespread acceptance not only in fundamental laboratory experiments on flames, post combustion analysis or kinetic study but also in understanding and modeling of complex flows, such as turbulent flames [2]. In most of the flow reactor/tube modeling one dimensional analyses are conducted and the simulation of the post-flame/reaction zone is performed by initializing it with a burner-stabilized flame solution, or in separate flame and post-flame regimes of solution. In an effort to numerically simulate high pressure flow reactor driven by the effluent of a McKenna burner, Hunderup et al. [3] divided the flow regime into a flame zone and a post-flame zone. The significant species at the end of the flame zone were fed as reactants into the post-flame zone to simulate the post-combustion NO-NO_2 conversion. A similar strategy is employed to simulate the Stanford variable pressure flow reactor [4] where a flat-flame-burner-generated hot vitiated gas is passed through a constant diameter reaction zone in which the reactivity or speciation data is recorded. Both these modeling approaches assumes a centerline species distribution profile that is representative of a plug flow condition. However, such assumptions are valid for systems where multi-dimensional transport is insignificant (e.g. plug flow velocity profile having very small viscous boundary layer effect). For configurations operating in the laminar region the assumption that the variation in the system is one-dimensional, starts to fail when radial and axial diffusive and convective coupling effects start to become prominent. All the above findings accentuate the fact that a multidimensional simulation of a burner-coupled flow tube configuration with an appropriate molecular transport model is necessary to investigate the laminar reactive flow close to the flame as well in the post-flame regions.

The multidimensional laminar reactive flow modeling with a detailed chemical kinetic mechanism is a challenging problem due to the inherent stiffness of the large number of highly non-linear governing equations that are involved and required for capturing the physicochemical processes accurately. Additional challenges arise in optimizing the fineness of the spatial discretization to accurately capture the flame front without prohibitive requirements of memory and CPU time. Several efforts have been made in the last two decades to simulate multidimensional reacting flows. Ern et al. [5] proposed a velocity-vorticity based numerical formulation of an axisymmetric laminar diffusion flame. Ombrello et al. [6] investigated the deviation between one-dimensional numerical simulation of temperature and species profiles in PREMIX and two-dimensional flame simulation using UNsteady Ignition and COmbustion with ReactioNs (UNICORN) code. Cuoci et al. [2] developed and validated a computational framework, laminarSMOKE for the modeling of multidimensional laminar flames. The complexity that radial and axial diffusive coupling may impose in interpreting DME oxidation in a laminar flow tube was investigated by Dryer et al. [7] employing a detailed CFD code. Most recently, Asgari et al. [8] proposed a multidimensional model to simulate the post-combustion species and temperature distribution for syngas/NO_x combustion, highlighting the significance of radial and axial diffusive and convective coupling to chemical kinetics in laminar flow.

In spite of the efforts reported so far to accurately model reactive flows, a burner-coupled flow tube configuration has not yet been modeled with the capability to investigate the physicochemical aspects close to the flame and in the post-flame region. In addition, in most of the literature, the burner was reported solely as a source of post-flame gases and emission e.g. NO-NO_2 conversion only as a post-flame event. In the present work, a pressure-based finite volume code is implemented to model a flow tube configuration that is driven by a flat flame McKenna burner for generating a reacting flow stream [8] to investigate chemical kinetic processes. The model is particularly employed for a detailed kinetic analysis of NO_x formation and recycling at different temperature regions (flame and post-flame regimes) inside the burner-coupled flow tube. A mixture-averaged molecular transport model is also implemented for calculating the transport coefficients. The OpenFOAM [9] platform is used for conducting the simulations. The general-purpose numerical framework, proposed in this work was utilized to obtain insight into the combustion phenomena in a multi-dimensional laminar flat flame both qualitatively and quantitatively.

2. Modeling Approach

The computational domain illustrated in Fig. 1 starts from the burner outlet and covers a 200 mm long post-combustion zone having a constant diameter of 82.6 mm. A two-dimensional axi-symmetric geometry specified by a wedge of small angle (5 degree) has been generated and meshed using Gmsh [11], a finite element mesh generation software. A structured non-uniform mesh, having higher spatial resolution near the flame region divides the domain into 1200 elements with only one mesh cell in the azimuthal direction.

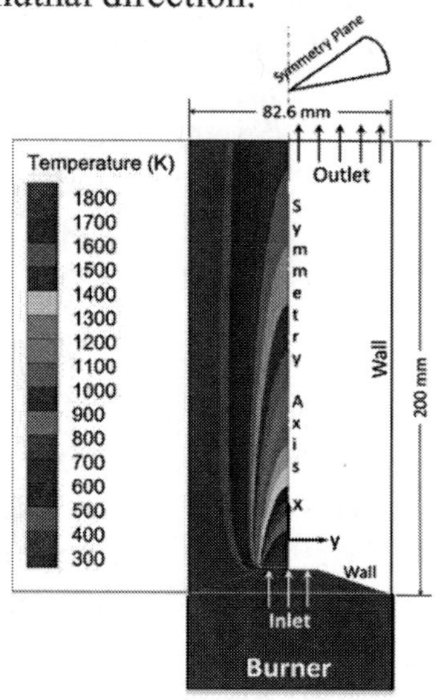

Figure 1: Schematic illustration of the computational domain of the burner-coupled model (right half) and the temperature field of the base case (NO_{in} = 100 ppm) at atmospheric pressure (left half). The flame structure is also visible from the high temperature zone near the inlet.

The solver employed in the present investigation for laminar flat flame simulation is based on an object-oriented open-source CFD code OpenFOAM [9]. This is a modified version of reactingFOAM - a turbulent diffusion flame solver. The reactive laminar flow under investigation is mathematically described by the transient forms of the conservation equations of total mass, mixture momentum, mass fraction of individual species and mixture energy. A computationally less expensive mixture-averaged molecular transport model [12] is incorporated. The gas temperature during combustion reaches high enough (~1900 K) for radiation heat transfer to take place. Therefore, an optically thin radiation model [13, 14] is included in the code as a baseline radiation model.

$$q_r = -4\sigma K_p(T^4 - T_\infty^4) - Cf_vT^5 \quad (1)$$

Here, σ is the Stefan-Boltzmann constant (5.6705×10^{-8} $Wm^{-2}K^{-4}$) and K_p is the Planck-mean absorption coefficient for the gas mixture. The second term in the right-hand side of Eqn (1) represents the radiation from the soot, which is negligible in the current investigation. The ambient temperature T_∞ is considered as 305K in this study.

The Planck-mean absorption coefficient for the gas mixture is expressed as [14]-

$$K_p = \sum_i^M P_i a_{P,i} \quad (2)$$

Here, P_i is the partial pressure of the gas mixture (in atm) and $a_{p,i}$ is the individual gas species absorption coefficient (in m^{-1} atm^{-1}).
The current code can run both in serial and parallel mode. For parallel operation in distributed-memory machines, the geometry and the associated fields are split into a number of sub-domains using the 'Simple' type domain decomposition method in which the domain along axial direction is split into the number of pieces similar to the number of processors used in the parallel calculation. The Message Passing Interface (MPI) is employed as the parallel communication protocol for the solver to communicate between the number of processors used in the parallel computing. A lean ($\Phi = 0.5$) premixed H_2/CO/Air mixture with two different initial NO_x doping- 100 ppm of NO and 100 ppm of NO_2 are introduced at the domain inlet with a flow rate of 4.21 L/m to generate the flat flame at atmospheric pressure. An initial high temperature region, 14.2 mm long and 12.7 mm wide having a maximum temperature of 2500 K is prescribed as an initial condition to ensure ignition of the fuel-air mixture. This limiting temperature is considered based on the curve fitting temperature ranges of the Planck-mean absorption coefficient for gas species CO in the radiation model [13]. A recent comprehensive and validated NO_x mechanism for syngas combustion proposed by Ahmed et al. [15] is employed as the kinetic reaction mechanism which includes 78 species and 442 reactions, culminated from extensive analysis [16] of the available NO_x formation pathways.

3. Results and Discussion

The multidimensional simulation is performed to obtain an insight into the spatial structure of NO_x profile as well as the different pathways contributing to this structure. Figure 2 shows the computed temperature, OH and normalized NO_x centerline profiles where the experimental measurements [8] of the axial variation of temperature and normalized NO_x concentrations for 100 ppm NO doping are compared with the numerical predictions. The normalization of the NO_x concentration is based on the amount of initial NO or NO_2 feedings in the mixture. The simulation reasonably predicts the trend in the gradual decrease in temperature downstream of the flow tube indicating the inherent two-dimensionality of the post-combustion gases. A similar trend of temperature and OH concentration profiles are observed for the 100 ppm NO_2 doping.

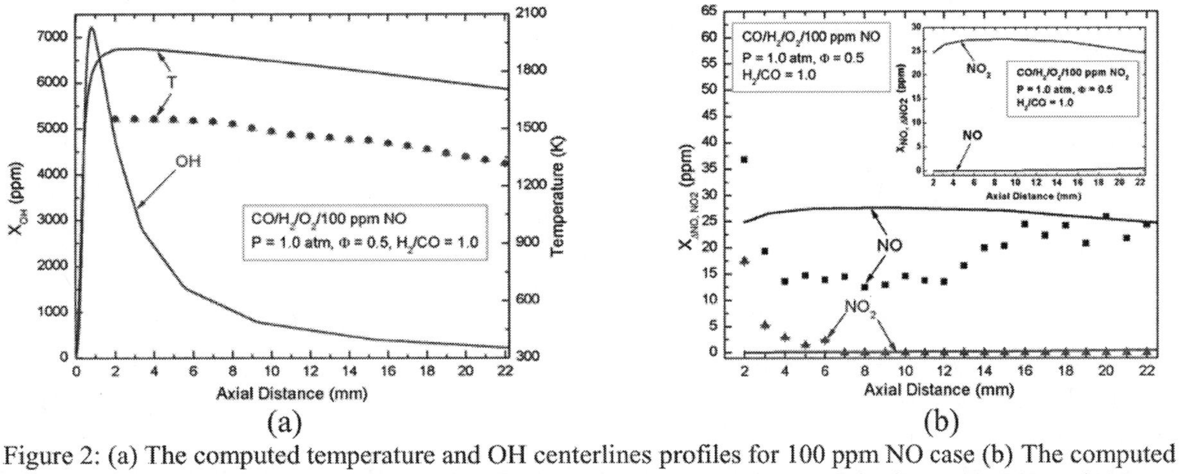

Figure 2: (a) The computed temperature and OH centerlines profiles for 100 ppm NO case (b) The computed normalized NO_x centerline profiles for 100 ppm NO and 100 ppm NO_2 (as inset) feedings. The NO and NO_2 mole fractions for the 100 ppm NO and NO_2 cases respectively are normalized based on the initial NO_x feedings. For both the cases, black and blue lines represent NO and NO_2 profiles respectively. The experimental measurements are from the post-flame investigations of Asgari et al. [13].

The flat flame is stabilized and anchored at the nozzle exit because of heat loss to the burner surface. A qualitative idea of the flame zone can be obtained through the predicted temperature contours presented in Fig. 1 where the highest temperature zone near the inlet demonstrates the flame. This study shows the dominance of the Zeldovich NO_x mechanism [17] in the flame zone due to high local temperature and air as the oxidizer in the combustion system. It consists of the high temperature decomposition of atmospheric molecular nitrogen by the reactions- $N_2 + O = NO + N$ and $N + O_2 = NO + O$. The two-dimensional flux analysis in the present study also highlights the prominence of the rates of the above decomposition reactions at the high temperature zones near the flame, which is illustrated in Fig. 3 below.

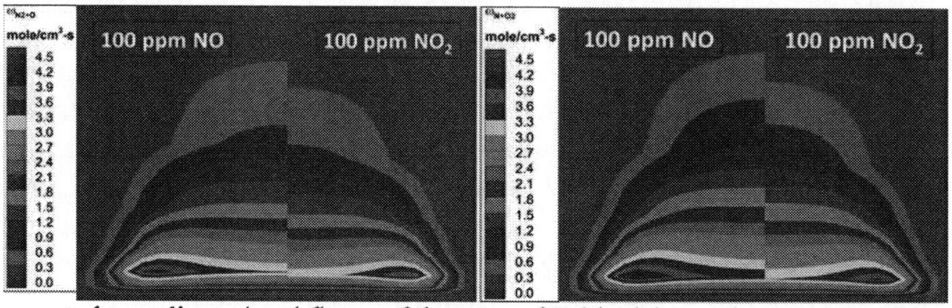

Figure 3: The computed two-dimensional fluxes of the rates of Zeldovich reaction rates for 100 ppm NO and 100 ppm NO_2 feeding.

The NO production closely follows the spatial temperature distribution, i.e. higher temperature leads to higher NO concentration, however significant amount of NO_2 is observed to accumulate close to but slightly offset from the burner centerline and vary downstream of the burner surface. To analyze this observance, the radial profile of the temperature and concentrations of NO_x, CO, CO_2 and dominant radicals such as O, H, OH were extracted along radial slices at different axial locations. It is evident from the temperature field in Fig. 1 that both the axial and the radial variation in peak NO_2 is dictated by the highest temperature gradient. The post-flame gases, further downstream of the tube gradually cool down due to heat loss dictated by the inherent two-dimensionality of the system. The total nitrogen-closure in that zone is ensured not only by NO-NO_2 conversion reactions- $NO + O(+M) = NO_2(+M)$, $NO_2 + OH = NO + HO_2$, $NO_2 + O = NO + O_2$, but also by NO-HNO conversions- $NO + H(+M) = HNO(+M)$, $HNO + OH = NO + H_2O$, $HNO + O = NO + OH$, $HNO + H = NO + H_2$.

4. Conclusions

The underlying assumption of flow homogeneity to simulate a flow reactor/tube configuration becomes invalid at low Reynolds number laminar flow. These low speed flows inherit a radial and axial diffusive and convective coupling to the reaction kinetics which results in a multi-dimensional variation along the test configuration. The current work employs a laminar reactive flow solver in a OpenFOAM framework for a multidimensional numerical simulation of a laminar flow through a tube. A unique feature of such a configuration is the coupling of a flat flame McKenna burner with the tube which essentially delineates the kinetics involved at different domain locations.

The simulations for a syngas/NO_x/Air stream indicates two distinct zones- flame and post-flame locations where the temperature and the NO_x kinetics involved are different. The high temperature zone near the flame is dominated by thermal NO formation mechanism, the temperature decaying post combustion zone substantiates the N-atom closure through $NO-NO_2$ as well as NO-HNO conversion paths. The current study finds that the NO profile closely follows the axial temperature distribution whereas the NO_2 concentration in the domain shows radially inhomogeneous variation that also changes along the axis. The peak NO_2 concentration is found to be dictated by the highest temperature gradients in the domain. Such spatial variation of NO_x kinetics in the domain cannot be captured by uni-dimensional simulations with or without considering transport.

5. Acknowledgements

This work was funded by the Department of Energy under award number DE-FE0012005. The authors highly acknowledge the financial support.

6. References

[1] S. Prucker, W. Meier, W. Stricker, A flat flame burner as calibration source for combustion research: Temperatures and species concentrations of premixed H2/air flames, Rev. Sci. Instrum. 65 (1994) 2908-2911.
[2] A. Cuoci, A. Frassoldati, T. Faravelli, E. Ranzi, A computational tool for the detailed kinetic modeling of laminar flames: application to C 2 H 4/CH 4 coflow flames, Combust. Flame 160 (2013) 870-886.
[3] J.W. Hunderup, R.J. Roby, An Experimental Investigation of the Conversion of NO to NO2 at High Pressure, doi:10.1115/95-GT-306(1995) V003T006A050.
[4] K.M. Walters, C.T. Bowman, Vitiated ethane oxidation in a high-pressure flow reactor, Combust. Flame 156 (2009) 1886-1897.
[5] A. Ern, C.C. Douglas, M.D. Smooke, Detailed Chemistry Modeling of Laminar Diffusion Flames On Parallel Computers, The International Journal of Supercomputer Applications and High Performance Computing 9 (1995) 167-186.
[6] T. Ombrello, C. Carter, V. Katta, Burner platform for sub-atmospheric pressure flame studies, Combust. Flame 159 (2012) 2363-2373.
[7] F.L. Dryer, F.M. Haas, J. Santner, T.I. Farouk, M. Chaos, Interpreting chemical kinetics from complex reaction–advection–diffusion systems: Modeling of flow reactors and related experiments, Prog. Energ. Combust. Sci. 44 (2014) 19-39.
[8] N. Asgari, S.F. Ahmed, T.I. Farouk, B. Padak, NOx formation in post-flame gases from syngas/air combustion at atmospheric pressure, International Journal of Hydrogen Energy 42 (2017) 24569-24579.
[9] OpenFOAM (2010), URL http://www.openfoam.org.
[10] A. Dasgupta, Numerical Simulation of AxiSymmetric Laminar Diffusion Flames with Soot, The Pennsylvania State University, 2015.
[11] C. Geuzaine, J.F. Remacle, Gmsh: A 3-D finite element mesh generator with built-in pre-and post-processing facilities, International Journal for Numerical Methods in Engineering 79 (2009) 1309-1331.
[12] R.J. Kee, G. Dixon-Lewis, J. Warnatz, M.E. Coltrin, J.A. Miller, A Fortran computer code package for the evaluation of gas-phase multicomponent transport properties, Sandia National Laboratories Report SAND86-8246 13 (1986) 80401-81887.
[13] H. Guo, F. Liu, G.J. Smallwood, Soot and NO formation in counterflow ethylene/oxygen/nitrogen diffusion flames, Combustion Theory and Modelling 8 (2004) 475-489.
[14] R. Barlow, A. Karpetis, J. Frank, J.-Y. Chen, Scalar profiles and NO formation in laminar opposed-flow partially premixed methane/air flames, Combust. Flame 127 (2001) 2102-2118.
[15] S.F. Ahmed, J. Santner, F.L. Dryer, B. Padak, T.I. Farouk, Computational Study of NO x Formation at Conditions Relevant to Gas Turbine Operation, Part 2: NO x in High Hydrogen Content Fuel Combustion at Elevated Pressure, Energy Fuels 30 (2016) 7691-7703.
[16] J. Santner, S.F. Ahmed, T. Farouk, F.L. Dryer, Computational Study of NO x Formation at Conditions Relevant to Gas Turbine Operation: Part 1, Energy Fuels 30 (2016) 6745-6755.
[17] Y.B. Zeldovich, Acta Physicochim 21 (1946).

Sub Topic: IC and GT Engines

Spring Technical Meeting
Eastern States Section of the Combustion Institute
March 4-7, 2018
State College, Pennsylvania

Kinetic Study of NO_x Formation for Synthetic Natural Gas Combustion under Gas Turbine Relevant Conditions

Fahd E. Alam[a], Sheikh F. Ahmed[a], Frederick L. Dryer[a,b] and Tanvir I. Farouk[a]

[a]Department of Mechanical Engineering, University of South Carolina, Columbia, SC 29208, USA.
[b]Department of Mechanical and Aerospace Engineering, Princeton University, Princeton, NJ 08544, USA
**Corresponding Author Email: tfarouk@sc.edu*

Abstract: The faster ignition of C_1-C_5 hydrocarbons in presence of small amounts of NO and the rapid NO-NO_2 interconversion in the presence of hydrocarbons (HC), known as mutual sensitization, have been an active area of research. A concrete understanding of the mechanistic coupling among the NO_x species and fuel fragments, especially during the combustion of C_1-C_2 hydrocarbons and their blends (e.g. natural gas) is required to address the stricter NO_x regulation-particularly for the stationary gas turbine industry. The mutual interaction of HC and NO_x have been investigated for a wide range of fuels under diverse operating and experimental conditions. However, the majority of these experimental studies were performed for the homogeneous systems and subsequent kinetic model formulations have been evolved and have been validated against these targets. A rarity of kinetic models validated against both the homogeneous and transport dependent experimental targets has motivated this study to present a newly proposed and validated natural gas/NO_x kinetic model. To achieve this goal, a recently developed synthetic gas model (*Ahmed et al., Energy & Fuels, 2016*) has been extended further to accommodate C_1-C_2+NO_x chemistry with comprehensive validation against shock tubes, stirred reactors, plug flow reactors, laminar flame speed and opposed diffusion flame speciation targets encompassing wide range of equivalence ratio (0.5-2.0) and pressure (1-60 atm). Rate constants of several important reactions involving HONO and HNO_2 species are updated based on the recent ab-initio calculations. Furthermore, the role of transport dependent validation is critically assessed. It is believed by the authors that a high-fidelity detailed chemical kinetic model to predict NO production and HC-NO_x mutual sensitization is necessary to culminate in a reduced model that is amenable to multidimensional CFD simulation.
Keywords:* *Mutual sensitization, Transport dependent validations, NO_x recycling, multidimensional simulation

1. Introduction

The mutual sensitization of hydrocarbon and NO_x has been an active area of research. The high temperature shock tube study by Slack and Grillo [1] was one of the earlier works to investigate the sensitization effects of NO_x on CH_4. The jet stirred reactor (JSR) experiments performed by Dagaut and co-workers [2-5] account for a wide range of fuels, including methane, ethane, ethylene and blends of natural gas. The detailed chemical kinetic models, assembled in those studies agreed reasonably well with the corresponding measurements. Sivaramakrishnan et al. [6] in their high pressure shock tube and JSR experiments on CH_4-C_2H_6 blends demonstrated the significant differences in reactivity between methane and natural gas blends in the presence of NO. Those findings were consistent as well with their assembled chemical kinetic models describing the observed changes in reactivity. Considering the potential relevance of high-pressure oxidation of natural gas in most practical combustion devices (engines, gas turbines etc.) the interaction of methane and oxides of nitrogen (NO_x) were investigated by Rasmussen et al. [7] under well-defined conditions in a high-pressure laminar flow reactor at pressures from 20 to 100 bars. The interpretation of the above mentioned high-pressure measurements in terms of a detailed chemical kinetic model agreed reasonably well. The interaction between C_2H_4 and NO under high-pressure and intermediate temperature flow reactor conditions were investigated both in terms of experimental measurements and a kinetic modeling study by Giménez-López et al. [8]. Recently, Deng et al. investigated the effects of NO_x on the oxidation of methane/NO_2 [9] and ethane/NO_2 [10] mixtures based on high-pressure shock tube ignition delay measurements. The mechanistic behaviors of those fuel mixtures were elucidated in terms of the proposed chemical kinetic models.

Despite the significance of NO predictions in high-pressure counter-flow partially premixed and non-premixed flames as well as premixed freely propagating flames based on the minimization of pollutant emissions from practical devices, such as gas-turbine engines, those transport-dependent venues are yet to be used widely as model validation targets alongside the homogeneous systems such as shock tubes, plug flow reactors, jet stirred reactors etc. The current investigation emphasizes on assembling a comprehensive detailed chemical kinetic model to describe NO_x kinetics in C_1-C_2 oxidation. Updated rate constants of several reactions involving HONO, HNO_2 species are updated based on the recent ab-initio calculations. Predictions from the model are compared against multiple experimental data sets over a wide range of experimental venues including both homogeneous and transport-controlled systems that cover pressures from 1 to 60 bar and equivalence ratios from 0.5 to 2.0. The authors believe that a detailed synthetic natural gas/NO_x chemical kinetic model which can reasonably predict the NO production and the HC-NO_x mutual sensitization is useful to develop a reduced model that can eventually contribute to the multidimensional CFD simulations of reacting flows.

2. Model Formulation Approach

The proposed natural gas/NO_x model consists of several submechanisms: C_0-C_2, NO_x, HC-NO_x interaction reactions and H/N/O submechanism. The hydrocarbon part of the proposed model is adopted from the small hydrocarbon and oxygenated fuel combustion model of Aramco Mech [11]. The reaction mechanism reported by Ahmed et al. [12] serves as the base set for NO_x and H/N/O kinetics. The detailed reaction mechanism for small hydrocarbon combustion with NO_x kinetics, proposed by Konnov [13] serves as the base set for the HC-NO_x interaction reactions with additional parameter revisions and elementary reaction inclusions. Thermochemical parameters in this model are adopted from the Burcat database [14].

3. Model Performance

To assess the assembled model, a wide range of experimental data from homogeneous systems such as shock tube, plug flow reactor, and stirred reactor configurations as well as transport-dependent systems such as counter-flow partially premixed and non-premixed flames and premixed freely propagating flames are considered.

Ignition Delay Times

The performance of the model is first compared against one of the global combustion targets, ignition delay times over a wide range of pressures. Shock tube ignition delay experiments of Deng et al. [9, 10] are simulated. In these shock tube experiments to promote/augment the effects of NO_2 on methane ignition, Deng et al. [9] added increasing amounts of NO_2 (NO_2:CH_4 of 30:70, 50:50 and 70:30) to a stoichiometric $CH_4/O_2/Ar$ mixture. They observed significant promotion of reactivity of methane upon the addition of NO_2 at all pressures, with the most significant effects at the highest pressure. The comparison between model predictions and measurements are presented for an exemplar case of 30:70 blending of NO_2:CH_4 for three different initial pressures (1.2, 4.0 and 10.0 atm), which appears in Fig. 1. In a more recent investigation of Deng et al. [10], the sensitization of NO_2 on ignition of ethane is measured based on shock tube ignition delay of $NO_2/C_2H_6/O_2/Ar$ mixtures over a wide range of pressures and equivalence ratios. A similar pressure is observed for ethane ignition, as that of $CH_4/O_2/Ar$ mixture, which is shown in Fig. 2. The kinetic model predicts reasonably well the experimental measurements for both the fuels.

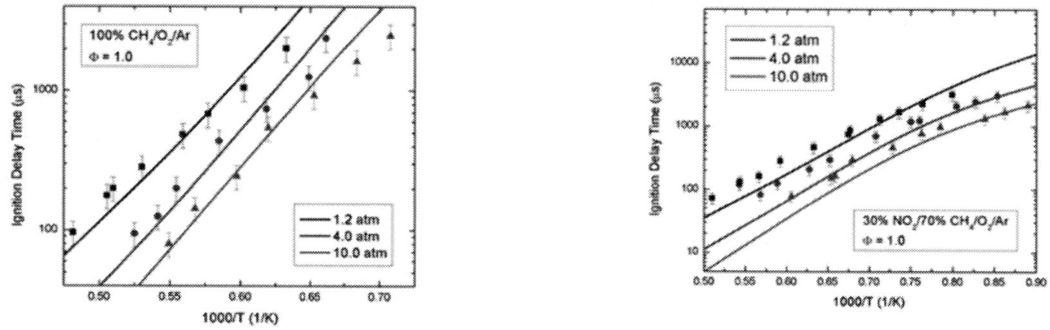

(a) (b)

Figure 1: Effects of pressure on τ_{ign} for stoichiometric (a) 100% $CH_4/O_2/Ar$ and (b) 30% NO_2/70% $CH_4/O_2/Ar$ mixtures. Lines represent numerical simulations and symbols represent measurements behind reflected shock waves [9].

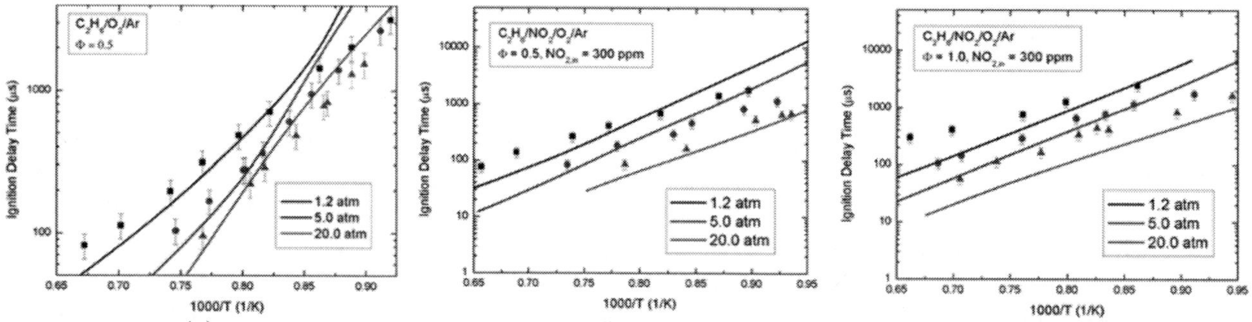

(a) (b) (c)

Figure 2: Effects of pressure on τ_{ign} for (a) $C_2H_6/O_2/Ar$ mixture with $\Phi = 0.5$, (b) C_2H_6/O_2/300 ppm NO_2/Ar mixture with $\Phi = 0.5$, and (c) C_2H_6/O_2/300 ppm NO_2/Ar mixture with $\Phi = 1.0$. Lines represent numerical simulations and symbols represent measurements behind reflected shock waves [10].

Plug Flow Reactor Experiments under Dilute Conditions

The performance of the model is further compared against detailed plug flow reactor (PFR) species evolution with time and temperature for a variety of PFR configurations based on initial mixing methods. The model is employed to simulate the experimental measurements of Giménez-López et al. [8], where species concentration for a $C_2H_4/O_2/NO$ system is measured in a flow tube at high pressure (60 bar) and various initial temperatures under a prescribed temperature distribution with isothermal conditions being maintained in the specified test section. Under reducing condition with excess air ratio of 0.2, Fig. 3(a), the kinetic model reasonably captures the consumption of C_2H_4 and the product formation of CO and CO_2. The major deviation is evident for the prediction of NO_x recycling at temperatures above 700 K where NO production is overpredicted.

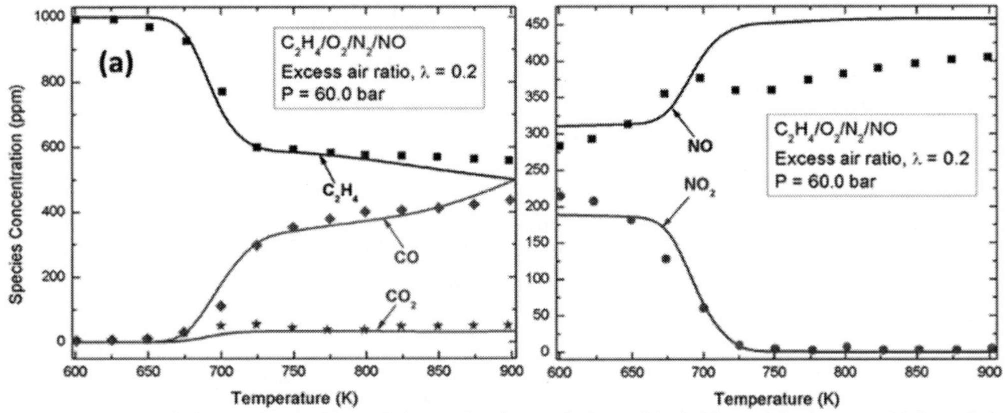

Figure 3: Experimental data [8] and model prediction of $C_2H_4/O_2/NO$ oxidation at 60 bar for reducing (excess air ratio, $\lambda = 0.2$) conditions.

In order to analyze the major high-pressure NO_x recycling pathways for three different fuel loadings, flux analyses were performed at 850 K. Under reducing condition, Fig 4(a), the recycling can take place (i) directly by the reactions $NO + C_2H_5O_2 = NO_2 + C_2H_5O$ (R1), $NO + CH_3O_2 = NO_2 + CH_3O$ (R2), $NO_2 + CH_3 = NO + CH_3O$ (R3), $NO_2 + C_2H_5 = NO + C_2H_5O$ (R4), or (ii) by the intermediate formation of HONO: $HNO + NO_2 = HONO + NO$ (R5), $HONO(+M) = NO + OH(+M)$ (R6). Besides, an NO-HNO interconversion takes place directly by the reactions: $HCO + NO = HNO + CO$ (R7), $NO + H(+M) = HNO(+M)$ (R8), $HNO + HNO = H_2NO + NO$ (R9), or by the intermediate formation of HONO: $HNO + NO_2 = HONO + NO$ (R10). Under stoichiometric condition, Fig 4(b), the direct recycling reactions changes to $NO + HO_2 = NO_2 + OH$ (R11), $NO_2 + H = NO + OH$ (R12). At the same time, two additional HONO production pathways appear: (a) $NO_2 + CH_2CHO = HONO + CH_2CO$ (R13) and, (b) $NO_2 + CH_2O = HONO + HCO$ (R14). The oxidizing condition favors two NO_2-NO conversion paths: $NO_2 + NO_2 = NO + NO + O_2$ (R15) and $NO_2 + CO = CO_2 + NO$ (R16), shown in Fig 4(c). The intermediate formation and isomerization of HNO_2 by the reactions: $NO_2 + HO_2 = HNO_2 + O_2$ (R17), $HNO_2(+M) = HONO(+M)$ (R17) adds another NO_2-NO conversion path under excess oxygen.

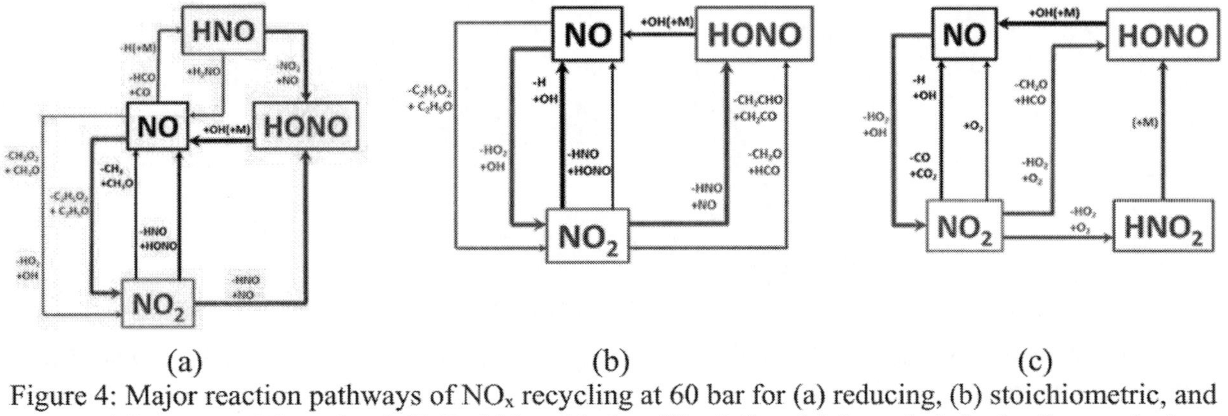

Figure 4: Major reaction pathways of NO_x recycling at 60 bar for (a) reducing, (b) stoichiometric, and (c) oxidizing conditions for $C_2H_4/O_2/NO$ oxidation. The "+" and "-" symbols in the flux analysis represent formation and consumption of the species associated with the symbol.

Transport Dependent Validations

In this section, we present model performances against experimental NO mole fraction measurements of Naik and Laurendeau (opposed-diffusion flame) [15] and Konnov (burner stabilized premix flame) [16]. These studies explore NO evolution for CH_4 and C_2H_4 flames. The detailed kinetic model of Ranzi et al. [17] and Sivaramakrishnan et al. [6] are also considered as their transport data set are readily available. The kinetic code OPPDIF [18] and PREMIX [19] are employed for the computations. Radiative heat loss from the flame is simulated assuming the gas phase mixture as an 'optically thin' medium.

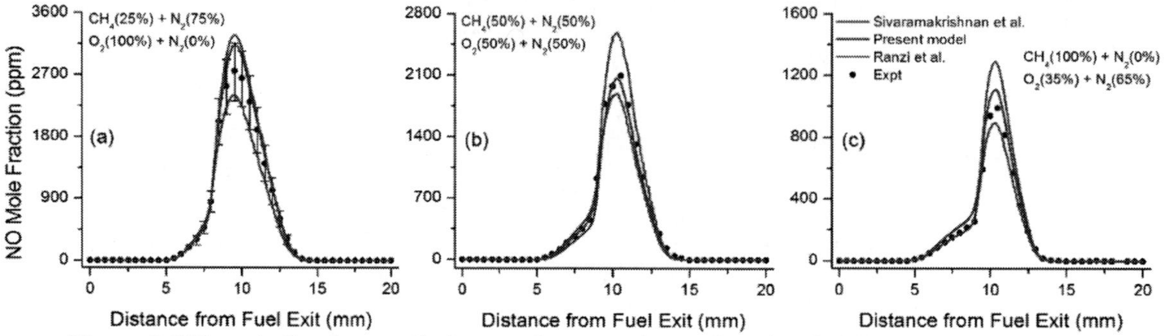

Figure 5: NO mole fraction predictions against experimental study of Naik et al. [15] at 1 atm.

Figure 5 consolidates the model predictions for NO mole fraction from Naik et al. for three different fuel/oxidizer loading. As can be seen, the current model predicted the NO evolution with satisfactory agreement for all three loading. Interestingly, the kinetic model of Sivaramakrishnan et al. under predicted the centerline-midplane NO for both the high and low fuel loading. The model prediction for the same location NO mole fraction is higher for Ranzi et al. The validation exercise is then extended towards laminar premixed flat flame burner experiment data of Konnov [16] for ethelyne flame. Figure 6 reports the laminar burning velocity (cm/s) experimental data against equivalence ratio and associated numerical simulation. Both the present study and Ranzi et al. model capture the experimental trend within data the scattering whereas the model of Sivarakrishnan et al. over-estimates the burning velocity, especially in the fuel rich zone.

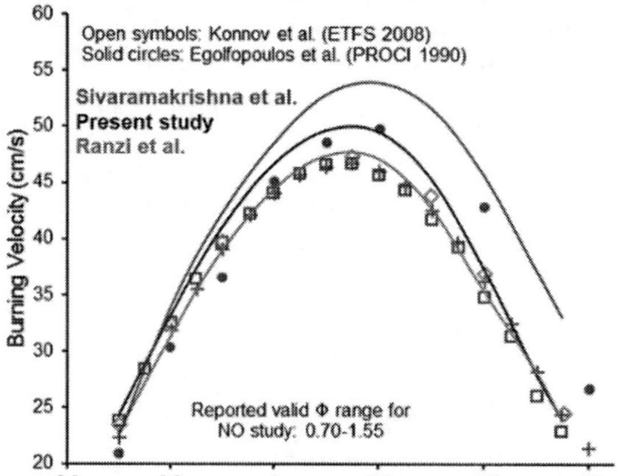

Figure 6: Ethylene burning velocity predictions against experimental data reported in [16] at 1 atm.

4. Summary

The stricter NO_x regulation for gas turbine operations requires a concrete understanding of the mechanistic coupling among NO_x species and fuel fragments, especially during the combustion of

C_1-C_2 hydrocarbons and their blends. The mutual interaction of HC and NO_x have been investigated for a wide range of fuels under diverse operating and experimental conditions. However, the majority of these experimental studies were performed for the homogeneous systems and subsequent kinetic model formulations have been evolved and validated against these targets. In order to meet the strict NO_x regulation imposed, a kinetic model that predicts both homogeneous and transport dependent venues is required. In light of that, the present investigation proposes a natural gas/NO_x combustion model application for a wide range of pressure, temperature and equivalence ratio. Rate constants of several important reactions involving HONO and HNO_2 species are updated based on the recent ab-initio calculations. Furthermore, the role of transport dependent validation is critically assessed. A detailed synthetic natural gas/NO_x chemical kinetic model reasonably predicting both homogeneous and transport dependent experimental venues is important to culminate in a reduced model compatible with multidimensional CFD modeling.

5. Acknowledgements

This work was funded by the Department of Energy under award number DE-FE0012005. The authors highly acknowledge the financial support.

6. References

[1] M.W. Slack, A.R. Grillo, Shock tube investigation of methane-oxygen ignition sensitized by NO2, Combust. Flame 40 (1981) 155-172.
[2] P. Dagaut, O. Mathieu, A. Nicolle, G. Dayma, Experimental Study and Detailed Kinetic Modeling of the Mutual Sensitization of the Oxidation of Nitric Oxide, Ethylene, and Ethane, Combust. Sci. Technol. 177 (2005) 1767-1791.
[3] P. Dagaut, A. Nicolle, Experimental study and detailed kinetic modeling of the effect of exhaust gas on fuel combustion: mutual sensitization of the oxidation of nitric oxide and methane over extended temperature and pressure ranges, Combust. Flame 140 (2005) 161-171.
[4] P. Dagaut, G. Dayma, The high-pressure reduction of nitric oxide by a natural gas blend, Combust. Flame 143 (2005) 135-137.
[5] P. Dagaut, G. Dayma, Mutual Sensitization of the Oxidation of Nitric Oxide and a Natural Gas Blend in a JSR at Elevated Pressure: Experimental and Detailed Kinetic Modeling Study, The Journal of Physical Chemistry A 110 (2006) 6608-6616.
[6] R. Sivaramakrishnan, K. Brezinsky, G. Dayma, P. Dagaut, High pressure effects on the mutual sensitization of the oxidation of NO and CH4-C2H6 blends, Physical Chemistry Chemical Physics 9 (2007) 4230-4244.
[7] C.L. Rasmussen, A.E. Rasmussen, P. Glarborg, Sensitizing effects of NOx on CH4 oxidation at high pressure, Combust. Flame 154 (2008) 529-545.
[8] J. Giménez-López, M.U. Alzueta, C.T. Rasmussen, P. Marshall, P. Glarborg, High pressure oxidation of C2H4/NO mixtures, Proc. Combust. Inst. 33 (2011) 449-457.
[9] F. Deng, F. Yang, P. Zhang, Y. Pan, J. Bugler, H.J. Curran, Y. Zhang, Z. Huang, Towards a kinetic understanding of the NOx promoting-effect on ignition of coalbed methane: A case study of methane/nitrogen dioxide mixtures, Fuel 181 (2016) 188-198.
[10] F. Deng, Y. Pan, W. Sun, F. Yang, Y. Zhang, Z. Huang, An ignition delay time and chemical kinetic study of ethane sensitized by nitrogen dioxide, Fuel 207 (2017) 389-401.
[11] W.K. Metcalfe, S.M. Burke, S.S. Ahmed, H.J. Curran, A Hierarchical and Comparative Kinetic Modeling Study of C1 − C2 Hydrocarbon and Oxygenated Fuels, Int. J. Chem. Kinet. 45 (2013) 638-675.
[12] S.F. Ahmed, J. Santner, F.L. Dryer, B. Padak, T.I. Farouk, Computational Study of NO x Formation at Conditions Relevant to Gas Turbine Operation, Part 2: NO x in High Hydrogen Content Fuel Combustion at Elevated Pressure, Energy Fuels 30 (2016) 7691-7703.

[13] A. Konnov, Implementation of the NCN pathway of prompt-NO formation in the detailed reaction mechanism, Combust. Flame 156 (2009) 2093-2105.
[14] E. Goos, A. Burcat, B. Ruscic, Extended Third Millennium Thermodynamic Database for Combustion and Air-Pollution, 2013.
[15] S.V. Naik, N.M. Laurendeau, Laser-saturated and linear laser-induced fluorescence measurements of nitric oxide in counterflow diffusion flames under non-sooting oxygen-enriched conditions, Combust. Sci. Technol. 174 (2002) 1-21.
[16] A.A. Konnov, I.V. Dyakov, J. De Ruyck, The effects of composition on the burning velocity and NO formation in premixed flames of C2H4+O2+N2, Experimental Thermal and Fluid Science 32 (2008) 1412-1420.
[17] E. Ranzi, A. Frassoldati, R. Grana, A. Cuoci, T. Faravelli, A.P. Kelley, C.K. Law, Hierarchical and comparative kinetic modeling of laminar flame speeds of hydrocarbon and oxygenated fuels, Progress in Energy and Combustion Science 38 (2012) 468-501.
[18] A.E. Lutz, R.J. Kee, J.F. Grcar, F.M. Rupley, OPPDIF: A Fortran program for computing opposed-flow diffusion flames, United States, 1997-05-01, 1997.
[19] R. J Kee, J. F Grcar, M. D Smooke, J. A Miller, E. Meeks, PREMIX: a fortran program for modeling steady laminar one-dimensional premixed flames, 1985.

Sub Topic: Fire

Spring Technical Meeting
Eastern States Section of the Combustion Institute
March 4-7, 2018
State College, Pennsylvania

CMOS Based High-Speed Camera Pyrometry Measurements for Validation of Upward Flame Spread Modeling

*Siddhant S. Aphale, Paul E. DesJardin**

Mechanical and Aerospace Engineering, University at Buffalo, Buffalo, NY, USA
Corresponding Author Email: ped3@buffalo.edu

Abstract: This study presents the adaptation of two-color pyrometry technique to a CMOS high speed sensor. An Edgetronic SC1 high speed camera is characterized and calibrated to provide a correlation between digital pixel intensities and the physical irradiance. A radiometric self-calibration methodology relying on computing the dynamic range of the camera is applied to obtain the camera response function. Using the Planck's radiation law and calibration source data, the relative camera response is further transformed into physical function. The robustness of the calibration is verified by comparing measured temperatures to blackbody emissions and the calibration data of a tungsten halogen lamp. The measurement method is then applied to upward flame spread along Poly(methyl methacrylate) (PMMA). Measured temperatures are used to compute soot volume fraction and radiative heat flux and are shown to be consistent with previous studies.
Keywords:* *Two-color pyrometry, Upward flame spread, Optical diagnostics

1. Introduction

Accurate field temperature measurement plays a vital role in the study of combustion. Flame temperatures have direct influence on fuel vaporization, combustion rate, heat transfer and pollutant formation. The main challenges measuring dynamic flames are the hostile high temperature environment and required short measurement times. Intrusive temperature measurement techniques are cost effective and simple to use but its accuracy is challenged by fluctuating temperatures and unsteady behavior of flames as well as gradual degradation due to continuous exposure to higher temperatures. Ratio pyrometry is a well-established non-intrusive methodology to measure the flame temperatures. The idea behind this methodology is to measure the ratio of two wavelengths from same source and application of Planck's law and Wien's law to obtain temperature. Two-color ratio pyrometry involves calculating the signal ratio of two different colors using Planck's law within two narrow spectral bands [1].
CCD or CMOS sensor based cameras capture images and videos with very high speed and provide high resolution images. The raw image signal can be obtained on three channels red, green and blue, enabling the study of three channels separately. Camera settings can be customized for different aperture and exposure allowing to carefully focus on the subject and obtain as much data as possible. Before measuring the temperatures, cameras need to be characterized and calibrated to correlate digital pixel intensity and physical irradiance. In this study, a CMOS sensor based high speed camera is calibrated to perform two-color pyrometry on an upward flame spread experiment along PMMA and measured temperatures are compared to upward flame spread modeling.

2. Camera Radiometric-Self Calibration

The spectral irradiance for a radiating surface can be given by Planck's law as:

$$E_\lambda(\lambda,\ T) = \epsilon_\lambda \frac{2\pi hc^2}{\lambda^5\left(e^{\frac{hc}{\lambda k_b T}} - 1\right)} \quad (1)$$

where k_b is the Boltzmann's constant, c is the speed of light, λ is the wavelength, ϵ_λ is the emissivity of the material, T is the temperature of the surface and h is the Planck's constant. $E_\lambda(\lambda,\ T)$ is the spectral irradiance and have units of $W{\cdot}m^{-2}{\cdot}\mu m^{-1}$ [2]. To determine temperature from Planck's law, two main approaches can be followed – one-color pyrometry and ratio pyrometry. In one-color pyrometry, the spectral emittance must be known or assumed. In color-ratio pyrometry, knowledge of spectral emittance is not required, rather, only the assumption that the spectral emittance is constant over a narrow spectral range. The color-ratio approach relies on the use of a digital camera as the detector and the measured signal S_F can be given by [1]:

$$S_F = (2\pi hc^2)\Delta t \int_{\lambda_1}^{\lambda_2} \frac{\epsilon_\lambda \eta}{\lambda^5(e^{hc/\lambda k_b T} - 1)} d\lambda = E_\lambda(\lambda, T)\Delta t\eta \quad (2)$$

where η is the efficiency of detector, lens and filter transmittance. Taking the ratio of two signals at two spectral bands and rearranging Eq. (2), temperature can be determined as [3]:

$$T = \frac{C_2\left(\frac{1}{\lambda_R} - \frac{1}{\lambda_G}\right)}{ln\frac{E_\lambda(\lambda_G, T)}{E_\lambda(\lambda_R, T)} + ln\left(\frac{\lambda_G}{\lambda_R}\right)^5 + ln\left(\frac{\epsilon_\lambda(\lambda_G)}{\epsilon_\lambda(\lambda_R)}\right) + ln\left(\frac{c_{\lambda,R}}{c_{\lambda,G}}\right)} \quad (3)$$

where λ_R and λ_G are the central wavelengths corresponding to red and green signal of camera, and $E_\lambda(\lambda_R, T)$ and $E_\lambda(\lambda_G, T)$ are the irradiances of red and green signals, $\epsilon_\lambda(\lambda_G)$ and $\epsilon_\lambda(\lambda_R)$ are the spectral emissivities of red and green narrow band filter wavelengths, $c_{\lambda,R}$ and $c_{\lambda,G}$ are constant factors associated with the camera optics.

A typical CCD or CMOS based camera provides images or videos in the form of digital pixel intensities distributed on three different channels – red, green and blue. Pixel intensities represent digital signal of the physical photon energy, however the relationship between them is often proprietary. Pixel intensities are a nonlinear function of the exposure which is defined as the product of scene irradiance and the camera exposure time, *i.e.* $f(E_i\Delta t_j)$. Radiometric calibration technique developed by Debevec and Malik is utilized to compute this nonlinear function [4]. The calibration algorithm was developed to find out the dynamic range of the camera and relies on the reciprocity of the camera. Using the definition of reciprocity law and of exposure, pixel intensity can be written as:

$$P_{ij} = f(E_i \Delta t_j) \quad (4)$$

where P_{ij} is the pixel intensity at i^{th} location and j^{th} image, Δt_j are the exposure times and E_i is the scene irradiance at the sensor site. To obtain the nonlinear function f it is assumed that it increases monotonically and its inverse f^{-1} is well defined. Taking logarithm of Eq. (4) and writing $ln(f^{-1}) = g^*$ a linear least squares problem can be formulated as [3,4]:

$$\sigma = \sum_{i=1}^{N}\sum_{j=1}^{M}\{w(P_{ij})[g(P_{ij}) - \ln(E_i) - \ln(\Delta t_j)]\}^2 + \lambda \sum_{P=P_{min}}^{P_{max}} [w(P)g''(P)]^2 \qquad (5)$$

where $w(P_{ij})$ is a weighting function, λ is the smoothness factor, N is the total number of pixel locations and M total number of images. This is a linear least squares problem and can be solved robustly using singular value decomposition [3,4]. The function obtained solving Eq. (5) gives the relative camera function at the sensor and needs to be shifted to obtain the physical response function at the source. A calibrated tungsten halogen lamp at 2800K and a blackbody source are utilized for this purpose.

The Edgetronic camera is a CMOS based high speed camera capable of 1/250,000s fastest shutter speed equipped with a 50 mm Nikon lens. Eleven different videos are captured at eleven different shutter speeds and extracted frames are then used to solve Eq. (5). Figure 1 (a) shows the blackbody source and tungsten halogen lamp used to characterize the camera. The camera response functions shown in Fig. 1(b) are consistent with the initial assumption of monotonic response function and the physical irradiance corresponding to each pixel intensity can be computed for a known camera exposure time. The calibration is verified by capturing the images of blackbody source at 1073K and tungsten lamp and determining temperatures from the captured images and the calibration function at different exposure time settings. Digital pixel intensity from images is converted to scene irradiance using the camera response function. For two-color pyrometry, red and green narrow band filters are utilized having central wavelengths of 650nm and 532nm respectively. This verification exercise also proved that the calibration is independent of camera exposure time settings.

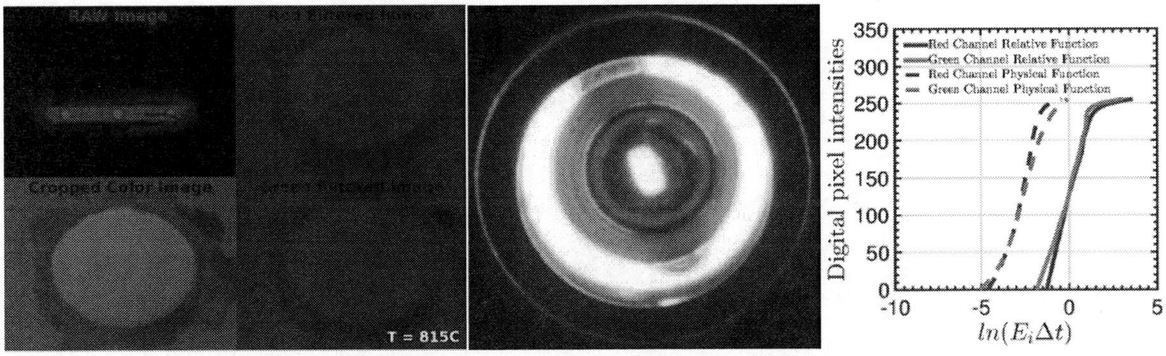

Figure 1: Source images captured using DSLR camera (a) blackbody source, (b) calibrated tungsten lamp, (c) Edgetronic high speed camera relative and physical response function.

3. Upward Flame Spread Experimental Setup & Temperature Measurement

The developed two-color pyrometry technique is applied to upward flame spread along Poly (methyl methacrylate) (PMMA). A PMMA sample 6 mm thick with a width of 90 mm and 145 mm long is suspended vertically. The camera is focused to obtain the side view of the burning sample. The camera shutter speed is set to 1/1000s and 800 frames per second are captured for 25 seconds of total exposure. The PMMA sample is ignited and simultaneously the camera is triggered to capture the burning sample flame spread. The captured video is further processed to extract all the frames and the temperature is measured for each frame using the developed two-color pyrometry algorithm. The camera is calibrated every time to obtain conversion from pixels to physical scale which in this case is 94 pixels/inch.

4. Results and Discussion

Figure 2 (a) shows the instantaneous captured frames at t = 1s, 8s, 25s and (b) show the corresponding temperature profiles using two-color pyrometry, respectively. The mean temperature measured over the three time instances is just below 1300K and the maximum temperature measured is 1670K at t = 25s - consistent with the previous modeling studies [5].

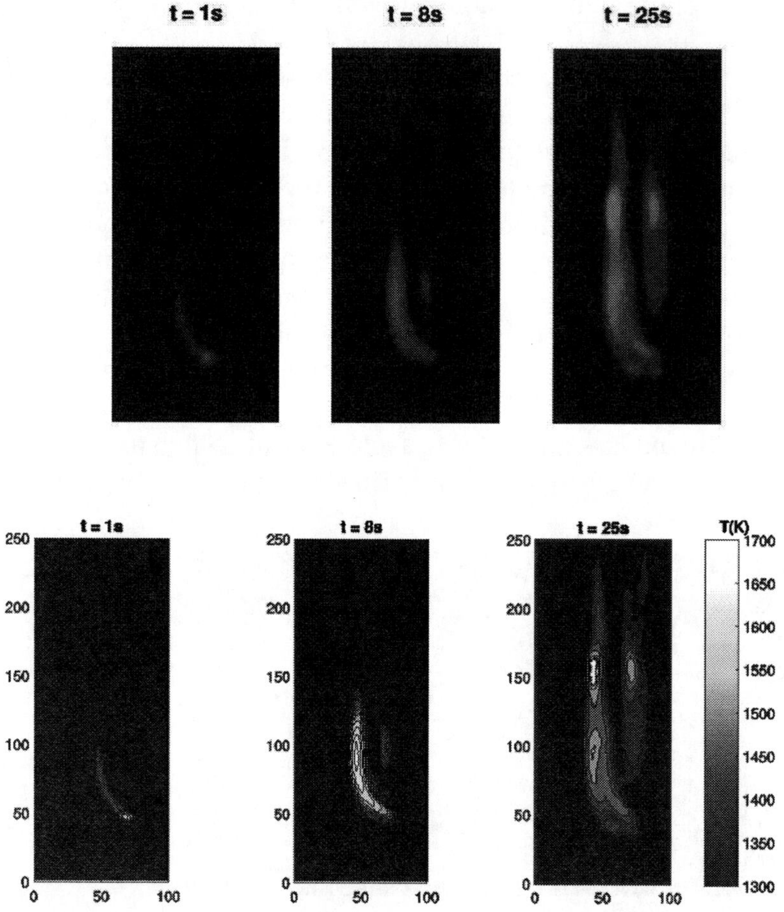

Figure 2: Upward flame spread frames extracted from high speed video and temperatures determined using two-color pyrometry - (a) flame images, (b) temperature contours.

Soot volume fraction (f_v) can be determined from temperature by assuming a homogeneous in-depth medium. Soot volume fraction requires absolute intensity calibration and tungsten lamp source is used for this purpose. Emissivity of tungsten bulb is computed based on the formulation provided in [6]. The soot volume fraction is given by [1,7]:

$$f_v = -\frac{\lambda_s}{CL} ln\left\{1 - \frac{E_{s,\lambda}}{E_{L,\lambda}} \epsilon_L(T_L, \lambda) exp\left[\frac{hc}{\lambda k_b}\left(\frac{1}{T_s} - \frac{1}{T_L}\right)\right]\right\} \tag{6}$$

where λ_s is the filter wavelength (650nm for this study), C is an empirical constant as given in [8], L is the absorption length along PMMA width, $\epsilon_L(T_L, \lambda)$ is the tungsten emissivity, E are the irradiance values, T are the temperature values and subscript S and L are for soot and calibration lamp respectively. The empirical constant *C* was computed by using complex index of refraction as given in [7,8]. The computed soot volume fraction contours are shown in Fig. 3. The overall shape is wispy and appear in thin streaks. The mean soot volume fraction is 12.4 ppm and the maximum is 12.5 ppm which are reasonable and are consistent to the previous studies [9,10].

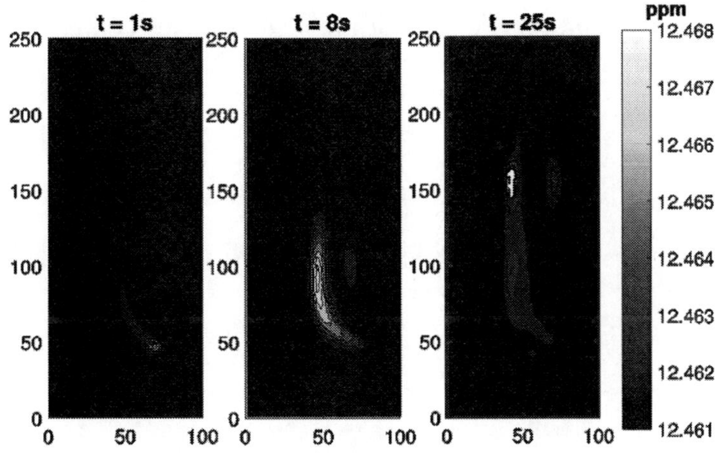

Figure 3: Soot volume fraction contours at t = 1s, 8s & 25s.

Radiative heat flux from the flame to PMMA wall is computed from the measured temperatures and soot volume fraction. Boundaries of the flame are traced using image processing and are homogeneously extruded in the third dimension assuming the profile seen in image is uniform in the in-depth direction. PMMA sample is divided into 30 x 33 grid points along the length and width, respectively to obtain grid convergent results. Inner flame boundary have ~90 points and outer boundary have ~150 points on it along which all possible rays are traced originating from PMMA grid points. Flame thickness is computed along these rays in between flame inner and outer boundary which is further used to compute the soot emissivity by using formulation given in [7,8]. Based on the calculated view factors and soot emissivity, radiative heat flux is computed and the instantaneous scatter plots at the left edge, center and right edge of the PMMA are shown in Fig. 4. It can be observed that the heat flux at the edges of the PMMA are lower than that at the center. The peak radiative heat flux at the wall is 10kW/m^2 at t = 25s - consistent with the prior studies [11].

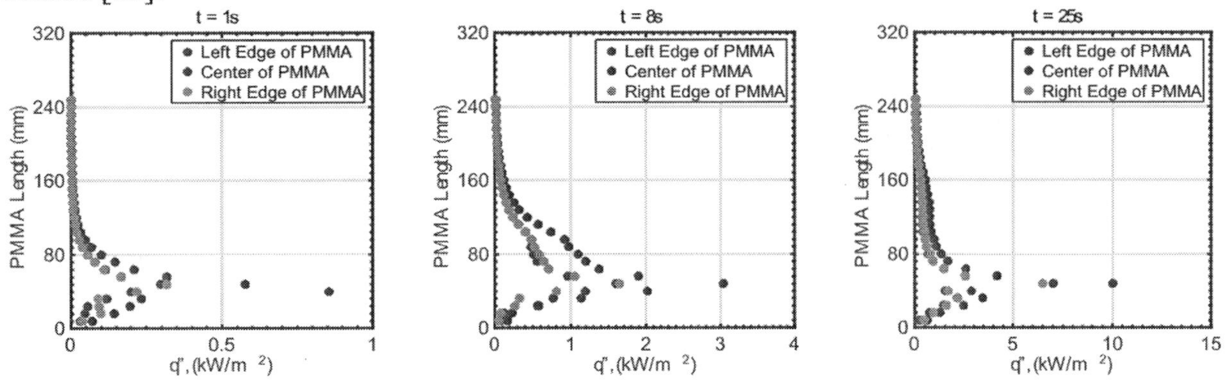

Figure 4: Vertical distribution of radiative heat flux along PMMA surface at t = 1s, 8s & 25s.

5. Conclusions

An Edgetronic high speed camera is characterized and calibrated using radiometric calibration technique to perform two-color pyrometry. The calibration technique is verified by measuring temperatures of calibration source. Two-color pyrometry is applied to upward flame spread along PMMA wall. Measured temperatures are qualitatively in good agreement with previous modeling studies. Soot volume fraction is computed and used in a radiation heat transfer model based on ray tracing and shows agreement with prior studies.

6. Acknowledgements

Support for this work has been provided by NSF under grant number 1704447.

7. References

[1] P. Kuhn, B. Ma, B.C. Connelly, M.D. Smooke, M.B. Long, Soot and thin-filament pyrometry using a color digital camera, Proc. Combust. Inst. 33 (2011) 743-750.
[2] T. Bergman, A. Lavine, F. Incropera, and D. Dewitt, Introduction to Heat Transfer, John Wiley and Sons, 2011.
[3] Z. Zhou, D. Tian, Z. Wu, Z. Bian & W. Wu, 3-D reconstruction of flame temperature distribution using tomographic and two-color pyrometric techniques, IEEE Trans. Instrum. Meas., 64(11) (2015) 3075-3084.
[4] P.E. Debevec and Jitendra Malik, Recovering high dynamic range radiance maps from photographs, ACM SIGGRAPH (2008) 31-40.
[5] W. Xie & P.E. DesJardin, An embedded upward flame spread model using 2D direct numerical simulations, Combust. Flame 156.2 (2009) 522-530.
[6] R. Larrabee, Spectral emissivity of tungsten, J. Opt. Soc. Am. 49.6 (1959)619-625.
[7] F. Cignoli, S. De Iuliis, V. Manta, G. Zizak, Two-dimensional two-wavelength emission technique for soot diagnostics, Appl. Opt., 40(30) (2001) 5370-5378.
[8] M. Modest, Radiative Heat Transfer, Academic Press, 2003.
[9] J. Consalvi, B. Porterie, M. Coutin, L. Audouin, C. Casselman, A. Rangwala, S. Buckley and J. Torero, Diffusion flames upwardly propagating over PMMA: theory, experiment and numerical modeling, Fire Safety Science, 8 (2005) 397-408.
[10] M. Lewis, P. Rubini, J. Moss, Field modeling of non-charring flame spread, Fire Safety Science, 6 (2000) 683-694.
[11] D. Drysdale, An Introduction to Fire Dynamics, John Wiley & Sons, 2011.
[12] J.Consalvi, Y. Pizzo, B. Porterie, J. Torero, On the flame height definition for upward flame spread, Fire Saf. J, 42 (2007) 384-392.

Spring Technical Meeting
Eastern States Section of the Combustion Institute
March 4-7, 2018
State College, Pennsylvania

Designing a gas sampling system for a 3 MW fire calorimeter using an experimental approach

**Javad Hashempour, Raymond Ranellone, Albert Simeoni, Nicholas Dembsey*

Department of fire protection engineering, Worcester Polytechnic institute, Worcester, MA 01609, USA
**Corresponding Author Email: jhashempour@wpi.edu*

Abstract: designing instrumentation is an imperative component in the construction process of full scale fire calorimeters. Selection of types and particular instrumentation positioning in exhaust duct of calorimeters, has a tremendous effect on accuracy and reliability of fire heat releases measured by the facility. In this study, a sampling system for collecting gas species for a non-fully mixed flow in the exhaust duct of a 3 MW calorimeter located at Worcester Polytechnic Institute, was designed using a simple experimental approach. Gas concentration inside the duct was initially assessed by conducting a series of experiments that enabled mapping the concentration level inside the duct. Later, different integration methods were exploited to average concentrations and calculate heat release rates. The results confirmed an I-shaped sampling system would capture more than 90% of gas concentrations, however more experiments and averaging methods of gas concentration are needed to reach a conclusion on design of sampling system.
Keywords: Instrumentation, Oxygen consumption, fire calorimetry, gas concentration

1. Introduction

Fire calorimetry is an essential method to assess material flammability and fire behavior, mainly by measuring fire heat release rate. Different methodologies were developed and attempted over time. However, measurements with oxygen consumption calorimetry received overwhelming attention due to its reliable and accurate measurements.

Oxygen Consumption (OC) calorimetry is based on capturing buyout gases of a fire combustion process and sample the gas mixture for detail species concentration analysis [1]. Many apparatuses, small to full scale sizes, were designed applying OC methods such as cone calorimeter [2], Fire Propagation Apparatus (FPA)[3], Furniture calorimeters and Full scale calorimeters that enables measurement of fires up to several Megawatts. In most OC calorimeters, a hood collector and an exhaust duct are the main features of design where the former assist with capturing combustion gases and the latter assist with measurement by instrumentation inside the duct. Therefore, the accuracy of such facilities is highly akin to proper design of instrumentation and an appropriate sampling method that correctly represent species concentration inside the exhaust duct.

For proper sampling of gas species, sample collections must be in an appropriate distance from the inlet of exhaust duct where the mixture of air and gas are fully mixed. Design studies on full scale facilities presented in the past list such as extensive woks by researchers in NIST [4], designing instrumentation in Carlton, Canada [5], and a technical report on Swedish Industrial calorimeter [6] where length of effective mixture cited as 8 to 10 diameter of exhaust duct. This also agrees with many fluid mechanic textbooks where length of fully developed turbulent mixture is ten times of a duct diameter [7]. Considering diameter of exhaust ducts in such facilities, the required length

for a fully mixed flow may reach to tens of meters which is not always achievable for many laboratory spaces.
This paper provides a method to design a sampling system for a non-fully mixed flow sampled at 2.5 of duct diameter of a 3 MW calorimeter which is located at Worcester Polytechnic Institute (WPI). The methodology consists of mapping gas concentration inside the duct by simple instrumentation that is manufactured in house. Different integration methods applied to average gas concentrations include O2, CO and CO_2 to calculate heat release rates and define a proper sampling system design which details are given in the following sections.

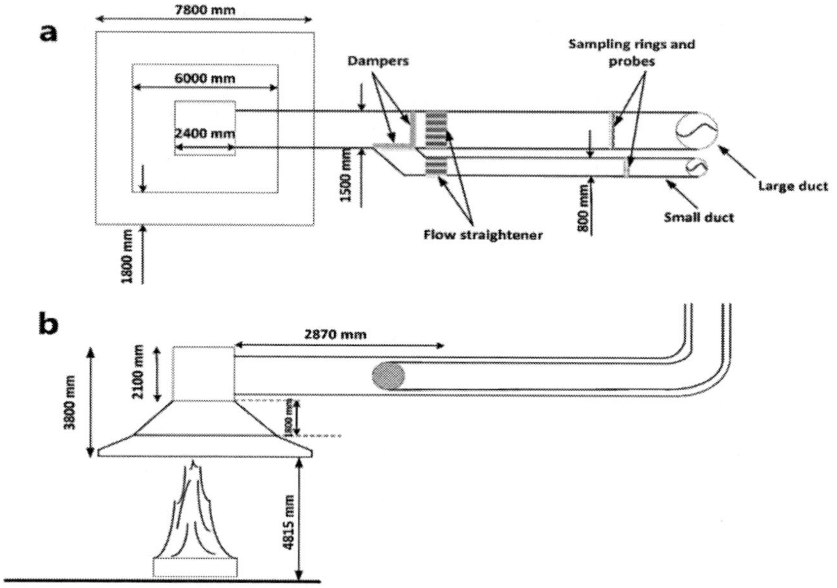

Figure 1 Scheme of WPI 3 MW fire calorimeter; a) top view; b) side view.

2. Methods

2.1. WPI 3 MW Fire calorimeter

The facility is located at Worcester Polytechnic Institute where it has been designed with maximum heat release capacity of 3 MW. Figure 1 provides top and side views of the facility including dimensions. The exhaust hood mounted with square size of 6 x 6 meter and covers specimen with the area size of 36 m2 (See figure 1). The hood has a movable extension attached which increases the size of hood to an area of 60.84 m^2. A round exhaust duct is connected from a side to the plenum. The suction flow, provided by several blowers positioned on the roof after passing through several carbon emission filters to inhibit the discharge of soot and carbon particles to the atmosphere.
The exhaust hood is connected to the ducts which then split into a small and a large duct with diameters of 0.8 m and 1.5m, respectively. The ducts are separated at their inlet using dampers at 2870 mm distance away from exhaust inlet in the plenum. A flow straightener consists of circular tubes with diameters of 75 mm (small duct) and 110 mm (large duct). The flow straighteners are mounted right after the dampers to smooth the flow of gas mixture flow and also enhance their mixture quality (See figure 1a). The dampers make operation of exhaust duct optional based on the size of fires and required exhaust flow rate.

According to the design, the calorimeter can be adjusted in three different modes of operations. Low range calorimetry only engages small duct and operates at 30 % of the calorimeter full capacity. Medium range calorimetry, with only large duct operational covers 70% of the calorimeter capacity and finally full range calorimetry, which is when both ducts are operational and the system works at its maximum capacity (100%). This study only considers design of sampling systems inside the large duct of calorimeter with medium range calorimetry setup (70 % of full capacity).

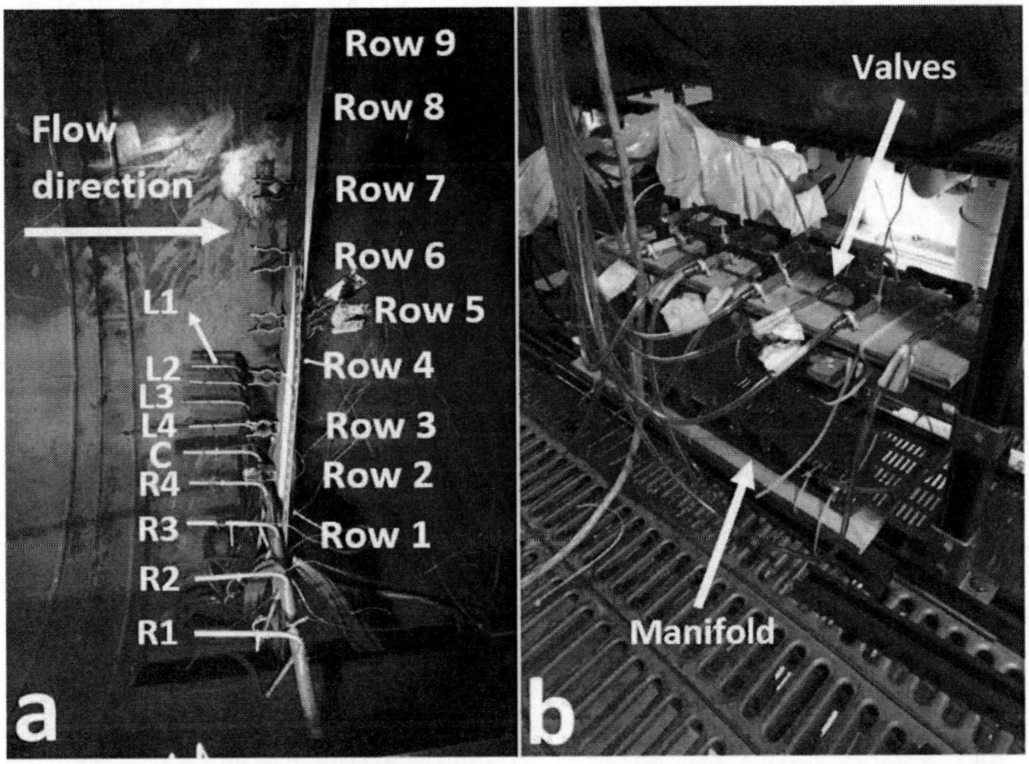

Figure 2 a) nozzles mounted inside the duct; b) valves and manifold outside the duct.

2.2.Measurement inside the duct

Species concentration was measured inside the duct for two different fires of 450 kW and 1000 kW. Special instrumentation was built in house to map gas concentration inside the duct. Figure 2 shows a picture of instrumentation and auxiliary valves. An array of nozzles mounted on a cable conduit in a way to collect samples across the duct within traverse interval of 10 cm which in overall covers. The nozzles made of 5 mm diameter stainless steel pipe which bended 90 degrees and then passed through the conduit. Each nozzle was connected to separate valves outside the duct (figure 3b) by nylon tubes. All valves were coupled to a manifold which directed sampled gases to a pump and then a gas analyzer in the control room. Array of nozzles were mounted on a wooden stand erected in the center of the duct and it was fixed on the stand by conduit brackets.
Ten different heights with a distance interval of 10 cm were attempted along the height of duct. The first row (Row 1) positioned with distance of 15 cm from duct floor and the last row (Row 10) had a distance of 115 cm from the floor. Sampling conducted for 60 seconds at each measurement point by opening an individual valve and keep others closed. The array moved up

manually when all sampling completed at one height and shutting the burner off. This procedure continued until all samples were collected.

Table 1: list of oxygen concentration measured at different locations inside the duct with fires of 450 kW and 1000 kW.

450 kW										
Position	L1	L2	L3	L4	C	R4	R3	R2	R1	R2
Row 1	20.46	20.45	20.45	20.45	20.46	20.47	20.48	20.50	20.49	20.50
Row 2	20.73	20.71	20.71	20.71	20.71	20.73	20.72	20.73	20.72	20.73
Row 3	20.73	20.70	20.69	20.69	20.69	20.69	20.70	20.71	20.72	20.71
Row 4	20.75	20.74	20.72	20.72	20.70	20.71	20.72	20.73	20.76	20.73
Row 5	20.73	20.71	20.70	20.70	20.70	20.70	20.72	20.74	20.74	20.74
Row 6	20.72	20.71	20.70	20.70	20.69	20.69	20.70	20.71	20.73	20.71
Row 7	20.66	20.64	20.64	20.64	20.63	20.63	20.65	20.66	20.65	20.66
Row 8	20.59	20.57	20.57	20.56	20.58	20.57	20.58	20.59	20.60	20.59
Row 9	20.54	20.53	20.51	20.51	20.51	20.51	20.52	20.54	20.59	20.54
Row 10	20.78	20.79	20.78	20.77	20.77	20.77	20.78	20.78	20.79	20.78
1000 kW										
Row 1	20.24	20.22	20.22	20.22	20.22	20.21	20.22	20.25	20.34	20.25
Row 2	20.49	20.45	20.43	20.44	20.43	20.42	20.42	20.46	20.51	20.46
Row 3	20.55	20.44	20.43	20.44	20.42	20.43	20.44	20.47	20.49	20.47
Row 4	20.52	20.49	20.48	20.49	20.48	20.48	20.48	20.52	20.55	20.52
Row 5	20.50	20.46	20.46	20.47	20.45	20.45	20.46	20.49	20.55	20.49
Row 6	20.50	20.48	20.46	20.47	20.46	20.46	20.47	20.49	20.52	20.49
Row 7	20.41	20.39	20.38	20.39	20.37	20.38	20.38	20.40	20.43	20.40
Row 8	20.33	20.32	20.30	20.31	20.31	20.31	20.33	20.35	20.36	20.35
Row 9	20.31	20.28	20.29	20.29	20.27	20.27	20.29	20.30	20.36	20.30
Row 10	20.57	20.56	20.54	20.55	20.54	20.54	20.54	20.56	20.57	20.56

3. Results and Discussion

Table 1 lists oxygen concentration measured by experiments with fire size of 450 kW and 1000 kW. Measurement with O_2 analyzer had uncertainty of ± 0.01, calculating based on five minutes monitoring of the analyzer. The values clearly indicate a stratification of oxygen concentration in both cases from lowest to highest part of the duct. Gas mixture in the lower half section of the duct shows higher oxygen level which implies more air accumulates in this region. However, it considerably drops in upper parts that is in agreement with tendency of low density smokes to move higher than air. The trend of concentrations for both cases show that oxygen levels across the duct is independent of fire sizes since minimum concentration level is obtained at lowest part of the duct (Row 1) and similar trend of changes is evident as sampling moves to highest part (Row 10). In contrast to lowest section, maximum concentration is recorded at highest part (Row 10) of the duct, near the wall. Also, the concentration level is nearly steady in traverse as values from right to left hand side recorded negligible change, as given in table 1. Based on the results, it can

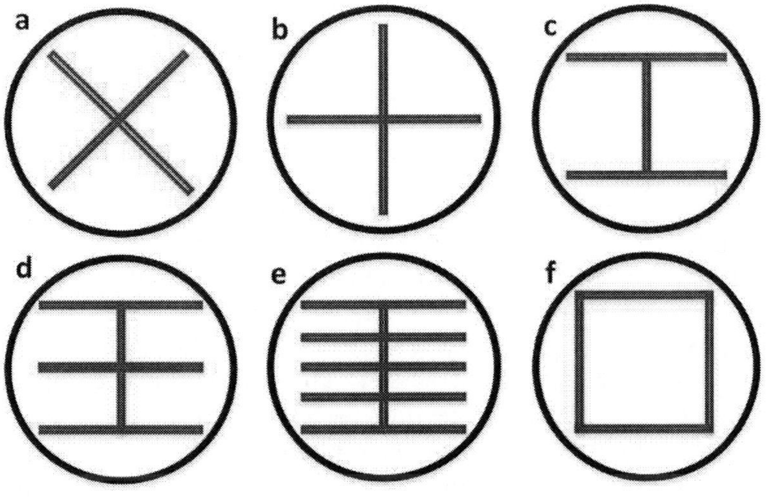

Figure 3 integration methods; a) Cross (C); b) Perpendicular (P) ; c) I-shaped (I); d) Ladder-3; e) Ladder-5; f) Square

be explicitly concluded that one design of gas sampling system would be sufficient for all ranges of fires.

Further, oxygen concentrations are integrated using different methods to find out an appropriate design for sampling system that provides accurate reading of heat release rate. Figure 3 illustrates six different integration methods applied in this study. Heat release rate calculation was based on methodology formulated by Parker [1] when O_2, CO and CO_2 concentration are measured and it is not replicated here. The gas concentrations were obtained from direct measurements through experiments and integrated according to figure 3. For example, concentrations for central measurements (C) of all rows and Row 5 were averaged in case of perpendicular shape or Row 1, Row 10 and central measurements (C) in case of I-shaped integration. In other cases, similar averaging was carried out using locations of patterns given in figure 3. Mass flow rate was calculated based on nominal volume flow rate for medium range calorimeter and average temperatures measured inside the duct.

Table 2 : heat release calculated for two different fires of 450 and 1000 Kw using different integration methods.

Integration Method	Cross	Perpendicular	I-shaped	Ladder-3	Ladder-5	Square
450 kW	379	338	426	386	394	386
1000 kW	849	834	910	868	883	841

Table 2 provides calculated heat release rates applying different integration methods for two attempted fires. Overall uncertainty obtained for heat release calculation is ±25 kW, based on ±0.01 in oxygen concentration and ± 1K in temperature reading using least square method. The worst scenario is calculated for gas concentrations integrated with perpendicular shape where relative variance for fire of 450 kW recorded as 24.8 % and 16.6% for fire of 1000 kW. Using cross integration improved heat release calculations and recorded a relative variance of about 15% for both cases. Best estimation is given with I-shaped integration where the variances decreased to low as 9% and 5.3% with fire size of 1000 kW and 450 kW, respectively. In other three cases, heat release calculation had a variance of greater than heat release rate calculated by I-shaped integration. One may notice that most of the integrations are affected by excessive air and low

concentration that led to poor estimation of heat release rate, as given in table 1, while I-shaped integration provides an equal averaging of minimum and maximum concentration levels.
According to table 2, it appears that sampling of gas concentrations in the form of I-shaped integration provides heat release rate with a reasonable accuracy, however, the design cannot be conclusive since it only limited to six integration methods. The technique presented here provided a reliable methodology for design process of sampling systems. In the future, other means of integration methods such as circular shapes or in combinations of above methods with more experiments can be considered to finalize a design for sampling system.

4. Conclusions

Proper sampling of gas mixture in oxygen consumption calorimeters greatly contributes to accuracy of fire heat release rate measurements. In current available designs, sampling occurs at distance where fully mixed flow inside the duct is accessible and noted as 8-10 times of duct diameter. In this paper, designing a sampling system for a non-fully mixed flow at a distance of 2.5 times of duct diameter was considered by a simple experimental technique. Gas concentration was mapped by mounting an array of nozzles inside the duct at different heights and gas concentrations sampled in the cross section of ducts under two fires conditions of 450 kW and 1000 kW. Then, heat release rate calculated using average gas concentration obtained by different integration methods. The concentration characteristic inside the duct could confirm a stratification due to low quality mixture of gas and air. This study found an I-shaped sampling system as best performance, however, the results was not conclusive due to small number of experiments and averaging methods. The methodology presented in this study can be exploited in the future to conduct further design studies with various integration methods and more experiments.

5. References

[1] W. J. Parker, "Calculations of the heat release rate by oxygen consumption for various applications," *J. Fire Sci.*, vol. 2, no. 5, pp. 380–395, 1984.

[2] ASTM 1354, "Standard Test Method for Heat and Visible Smoke Release Rates for Materials and Products Using an Oxygen Consumption Calorimeter." American Society for Testing and Materials (ASTM), West Conshohocken, PA, 2016.

[3] ASTM E2058, "ASTM E2058-13a Standard Test Methods for Measurement of Material Flammability Using a Fire Propagation Apparatus (FPA)," *Am. Soc. Test. Mater.*, pp. 1–30, 2013.

[4] R. A. Bryant *et al.*, "The NIST 3 megawatt quantitative heat release rate facility," 2003.

[5] Y. J. Ko, R. Michels, and G. V. Hadjisophocleous, "Instrumentation Design for HRR Measurements in a Large-Scale Fire Facility," *Fire Technol.*, vol. 47, no. 4, pp. 1047–1061, 2011.

[6] M. Dahlberg, "The SP industry calorimeter for rate of heat release measurments up to 10 MW," 1992.

[7] R. W. Fox, A. T. McDonald, and P. J. Pritchard, *Introduction to Fluid Mechanic*, 6th ed. John Wiley & Sons.Inc, 2004.

Spring Technical Meeting
Eastern States Section of the Combustion Institute
March 4-7, 2018
State College, Pennsylvania

Flight Test Demonstration of LED-based Fire Sensors for Space Propulsion Vehicles

Anthony Carmine Terracciano[1,2,*], *Kyle Thurmond*[1,2], *Michael Villar*[1,2], *Justin Urso*[1,2], *Erik Ninnemann*[1,2], *Akshita Parupalli*[1,2], *Zachary Loparo*[1,2], *Nick Demidovich*[3], *Jayanta S. Kapat*[1,2], *Subith S. Vasu*[1,2]

[1] *Mechanical and Aerospace Engineering, University of Central Florida, 4000 Central Florida Blvd., Orlando, FL 32816-2450, USA*

[2] *Center for Advanced Turbomachinery and Energy Research (CATER), University of Central Florida, 4000 Central Florida Blvd., Orlando, FL 32816-2450, USA*

[3]*FAA Office of Commercial Space Transportation, 800 Independence Ave SW, Washington DC 20591*

[*]*Corresponding Author Email:subith@ucf.edu*

Abstract: Gases which are indicative of combustion and pyrolysis have absorption features in the mid-infrared range. Therefore, it is possible to implement absorption spectroscopy as a detection method to provide real time monitoring of gases in spacecraft. A sensor developed at the University of Central Florida for this purpose was flown on a proof of concept test flight through partnerships with NASA's Columbia Scientific Balloon Facility (CSBF) and the Louisiana State University (LSU) High Altitude Student Payload (HASP). The test flight enabled autonomous operation of the sensor at an altitude of 105,000ft (32km). The sensor utilizes a 4.2 μm light-emitting diode (LED) source, amplitude modulation to characterize CO_2 concentration, and frequency modulation to characterize ambient temperature.

Keywords: Primary Life Support, Fire Detection, Crew Health Monitoring, Space Systems

1. Introduction

As space travel becomes more prevalent, the pressing concern of ensuring the safety of humans and equipment onboard remains. Hazardous gases within space vehicles pose dangers in the forms of toxic environments, and certain species also indicate the presence of fire. Carbon Dioxide (CO_2), for example, is often a byproduct of combustion and is also known to reduce cognitive performance at high enough concentrations, especially in closed environments [1]. Because of the critical role they possess, hazardous gas sensors are expected to perform reliably and accurately within these harsh environments. Current sensors used within space vehicles mainly employ laser-based absorption measurement systems because they are capable of collecting accurate time-resolved measurements of properties of a variety of gaseous species [2]. However, this method is expensive and much more delicate than preferred for the rough conditions in space. Light emitting diodes (LEDs) are an appealing alternative because they are less expensive, are robust enough to withstand extreme environments, and provide ultra stable output while maintaining a lower power consumption. LEDs also require significantly less complex electrical and thermal driver components. An LED-based sensor that operates on the principles of absorption spectroscopy has been designed and undergone several iterations of development. The latest iteration was tested for its abilities to measure CO_2 while aboard a weather balloon through the High Altitude Student Payload (HASP) group.

The LED sensor detected CO_2 through absorption spectroscopy, which uses the attenuation of light at certain wavelengths related to the oscillations of the gases being identified [3]. *Transmittance,* $\boldsymbol{T}$ *(unitless),* is the ratio of the *observed spectral irradiance,* $\boldsymbol{I_{\lambda,x}}$ *(W·cm⁻²)*, and *reference spectral irradiance* , $\boldsymbol{I_{\lambda,0,x}}$ *($W \cdot cm^{-2}$)* (1); these components are represented along with the source's native radiant flux, $\boldsymbol{I_{\lambda,0,x_0}}$ *($W \cdot cm^{-2}$)*. The total transmittance, $\boldsymbol{T}$, may be expressed in a more explicit form as shown in eq. (2). This form utilizes the *spectral line strength,* $\boldsymbol{S_\lambda}$ *($atm^{-1} \cdot cm^{-2}$), line shape function,* $\boldsymbol{\phi_\lambda}$ *(cm)*, and *mole fraction of the ith species,* $\boldsymbol{\chi_i}$ *(unitless)*, integrated over a *pathlength,* $\boldsymbol{x}$ *(cm)*, and *wavelength range,* $\boldsymbol{\lambda}$ *(nm)*; the total transmittance is the summation for each species exhibiting absorption within the wavelength range. These fundamental parameters may be lumped together for convenience and define a *wavelength-specific absorption coefficient,* $\boldsymbol{\alpha_{\lambda,i}}$ *(cm^{-1})*, as shown in eq. (3). For many molecular species of interest, the HITRAN database provides both $\boldsymbol{\phi_\lambda}$ and $\boldsymbol{S_\lambda}$ [4]. For species that cannot be found in the HITRAN database, it is necessary to measure the temperature-dependent absorption characteristics or to utilize other databases.

$$T_\lambda = \left(\frac{I_\lambda}{I_{\lambda,0}}\right) = \exp(-\alpha_{\lambda,i}\mathrm{L}) \tag{1}$$

$$T = \sum_i \int_{\lambda_0}^{\lambda_1} \int_{x_j}^{x_{j+1}} exp(-S_\lambda \cdot \phi_\lambda \cdot \chi_i)\, dx\, d\lambda \tag{2}$$

$$\alpha_{\lambda,i} = S_\lambda \cdot \phi_\lambda \cdot \chi_i \tag{3}$$

Within this text is a brief overview of the sensor's design and the balloon test procedure for the sensor's ability to detect CO_2 by establishing a noise floor and expected signal drift. The results obtained from the HASP flight are intended to be used to further improve the sensor so that it may advance in TRL and be utilized in primary life support.

2. Methods / Experimental

A previous iteration of this sensor had already been tested aboard a balloon flight, with the main goal of the design being to show that multiple gases could be analyzed simultaneously. The output of each LED was modulated at a different frequency and combined into a single beam that reached the detector. The setup for this is shown in Figure 1 below. While the sensor was able to function properly and collect data, the beam splitter design was determined to be ineffective because it further reduced the power output of the LEDs; this was not desirable because one of the main weaknesses of using LEDs instead of laser diodes is the lower power output. The design used, while sturdy, was also larger than preferred because of the required electronics. Therefore, the focus was placed on testing a single LED with a stronger signal while consuming less power overall and refining the electronics components.

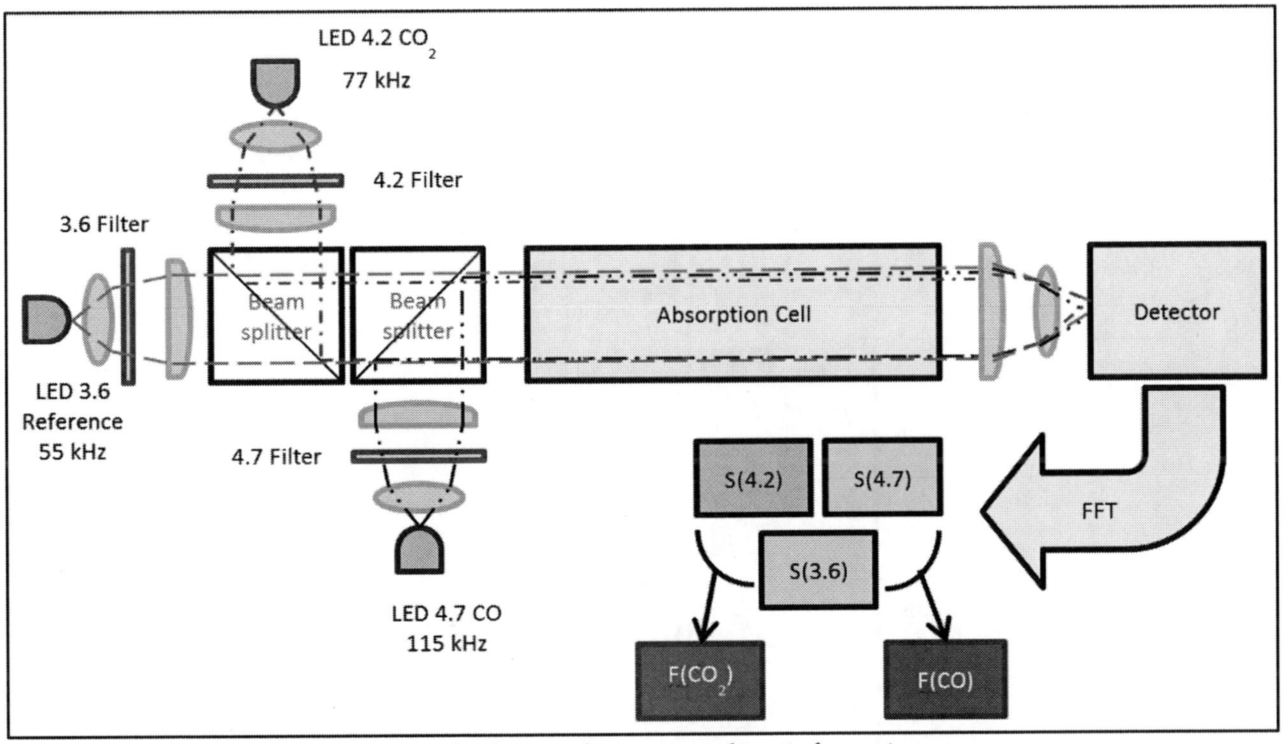

Figure 1: Sensor hardware and processing schematic for previous sensor

Figure 2 below shows the optical path from the LED to the photodiode detector for the current sensor design. The LED, centered around 4.2 μm, served as an emitter. The light first passes through a bi-convex collimating lens before reaching the 4.2 μm filter; the filtered light then passes through a plano-convex lens and a bi-convex lens in order to be collimated to 25mm in diameter before reaching a series of four condenser lenses. The condenser lenses serve to focus the beam onto the photodiode detector.

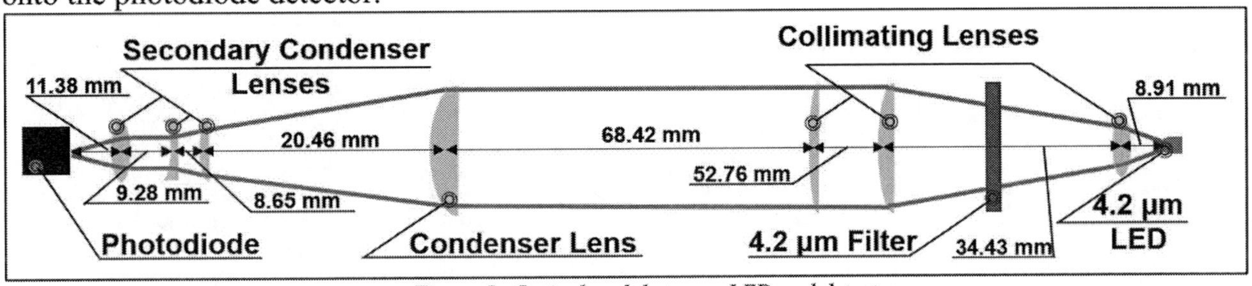

Figure 2: Optical path between LED and detector

A schematic depicting the power usage of the electrical systems is shown in Figure 3 below, while Table 1 shows the power breakdown. The sensor is designed to operate on 30VDC, 0.5A. Therefore, 5 separate DC/DC converters were required to ensure all electronics operated at optimum voltage levels. The National Instruments cRIO-9031 DAQ was used for data acquisition as well as for controlling the sensor and its components.

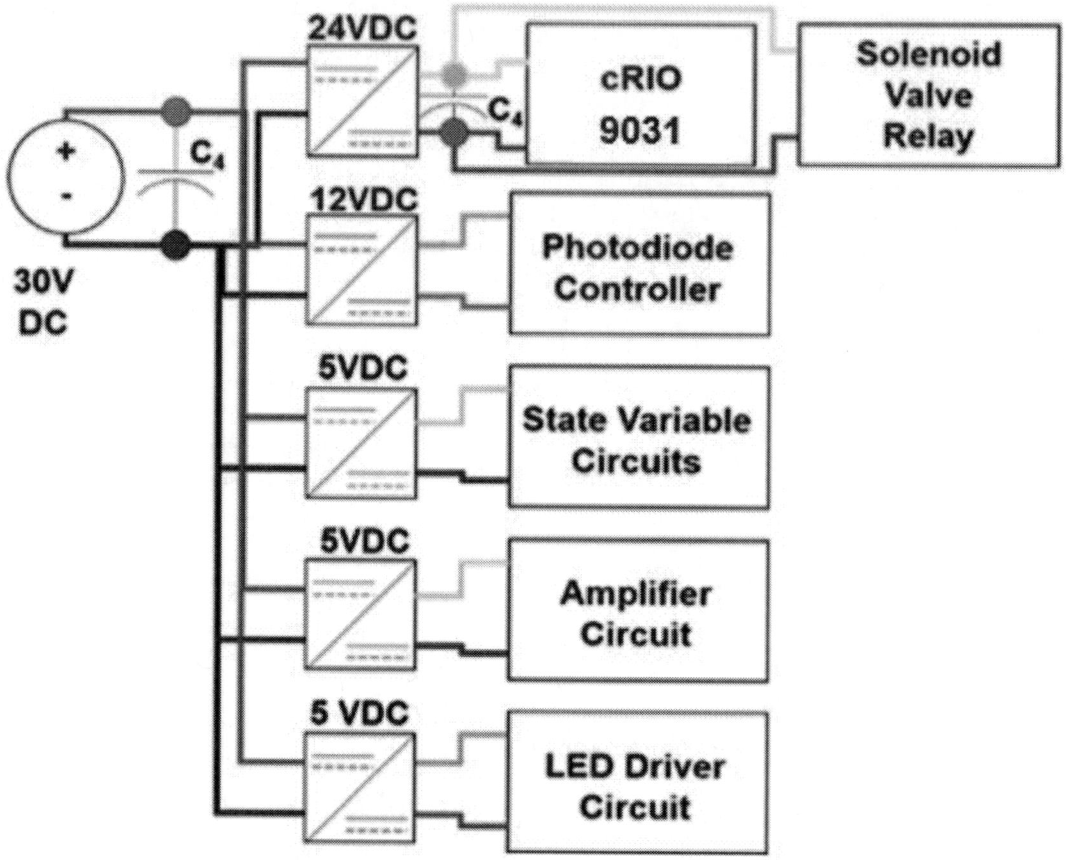

Figure 3: Power breakdown schematic

Table 1: Power Breakdown

Component	Quantity	Voltage	Max Current	Power	Notes
Photodiode Controller	1	12V	1A (2 stage TEC)	12W	Max power, including TEC operation
LED	1	5V	80 mA	0.40W	no TEC operation
cRIO-9031	1	12V	.8A	10W	Max power, continuous operation
Gas Cylinder Solenoid	1	24V	.4A	9.6W	Max power, sporadic operation. Supplemented with capacitors

Figure 4 below shows how the sensor was securely mounted to the HASP weather balloon's platform. Figure 4A shows the outer metal cover, which serves to both secure the sensor to the plate and to assist in radiating heat outward. All electronic and optical components were placed within the acrylic enclosure seen in Figure 4B. The enclosure has been tested to ensure that it is capable of performing in low pressure conditions without allowing air flow. Pressure and temperature within the acrylic enclosure were recorded via the cRIO. Temperatures were recorded using NTC thermistors incorporated into a Wheatstone bridge with a ceramic capacitor across the measurement terminals to mitigate burst noise.

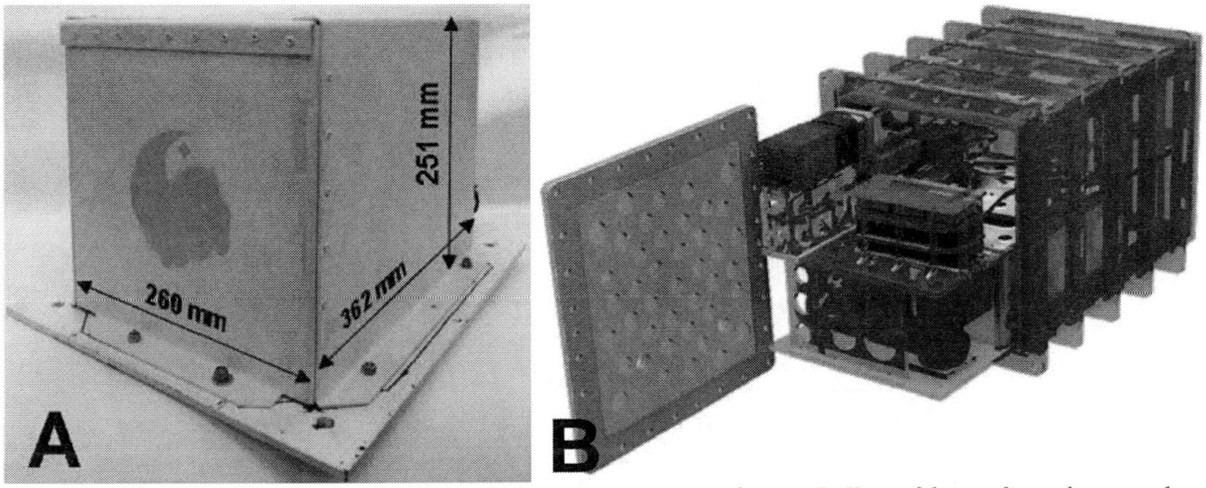

Figure 4: A: Sensor mounted to the HASP platform via an outer metal cover; B: View of the acrylic enclosure and components

3. Conclusions

A CO_2 measurement sensor was tested on a NASA-LSU sponsored high-altitude balloon test flight within an enclosure containing a fixed gas composition as a proof of concept demonstration for use of such a sensor to detect hazardous gases on spacecraft. The testing of the sensor was autonomous where amplitude modulation indicated CO_2 concentration, while frequency modulation was used to reveal temperature within the enclosure. Using ambient air within the sensor as a test gas, a linear curve fit for temperature and CO_2 signal intensity was found with respect to temperature. As this sensor validated its proof-of-concept while reducing power usage, the next iteration will focus on using a rotating grating and multiple LEDs centered around wavelengths of interest in order to test for detection of a larger selection of gases.

5. Acknowledgements

Research at UCF was supported by financial assistance from Federal Aviation Administration Center of Excellence for Commercial Space Transportation (FAA COE-CST) with Ken Davidian as program manager, Florida Space Institute (FSI), NASA Florida Space Grant Consortium (FSGC), the UCF Mechanical and Aerospace Department, and the UCF Office of Research and Commercialization. Z.L. thanks NSF GRFP for partially supporting this effort. The authors would like to thank Dr. Robert Peale (UCF) for assistance with environmental chamber tests.

6. References

1. Satish, U., et al., *Is CO2 an indoor pollutant? Direct effects of low-to-moderate CO2 concentrations on human decision-making performance.* Environmental health perspectives, 2012. **120**(12): p. 1671.
2. Teichert, H., T. Fernholz, and V. Ebert, *Simultaneous in situ measurement of CO, H 2 O, and gas temperatures in a full-sized coal-fired power plant by near-infrared diode lasers.* Applied Optics, 2003. **42**(12): p. 2043-2051.
3. Piatt, U. and J. Stutz, *Differential Optical Absorption Spectroscopy, Principles and Applications.* 2008, Springer, Heidelberg.
4. Rothman, L., et al., *The HITRAN database: 2012 edition.* J Quant Spectrosc Radiat Transfer, 2013.

Spring Technical Meeting
Eastern States Section of the Combustion Institute
March 4-7, 2018
State College, Pennsylvania

In Situ Burn Ignition Testing Methods and Results for California Crude Oils

Steven G. Tuttle, Brian T. Fisher, Christopher J. Pfützner, Thomas N. Loegel, and Katherine M. Hinnant
Chemistry Division, Naval Research Laboratory, Washington, DC
**Corresponding Author Email: steven.tuttle@nrl.navy.mil*

Abstract: Two experimental platforms have been developed and a range of chemical analytics have been applied to examine the ignition, flammability, heat release rates, and their dependence on oil constituents for a selection of California crude oils. The first platform is an actively-cooled, 1 m^2, square pan to examine ignition behavior at near-realistic geometric scales and water temperature. The second platform is an actively-cooled, 10 cm-diameter, 10-mm deep pan for a cone calorimeter to measure ignition behavior, burn efficiency, and heat release rate and their dependence on weathering and emulsification. The pan is surrounded by a water jacket to provide active cooling to mimic heat transfer to ocean water. The cone calorimeter architecture provides a controlled environment to determine how much heat transfers to an oil slick for a crude oil emulsion to adequately break emulsions and begin vaporizing flammable constituents before ignition is possible. Two suites of chemical analysis are performed in association with the flammability testing. The combination of the ignition tests and chemical analysis has revealed that crude oils that were previously considered inflammable can be ignited and burned. For brevity, only the cone calorimeter experiment and results will be discussed.
Keywords:* *Cone Calorimetry, Crude Oil, In Situ Burning

1. Introduction

In-situ burning (ISB) has become a standard oil spill response method to dispose of crude oil before oil slicks drift into sensitive coastal ecosystems and fisheries [1]. There are a number of incumbent difficulties with executing ISB. The first challenge is to mechanically move or chemically herd the slick into pools with sufficient thickness that the heat transfer to the ocean will not slow or stop hydrocarbon evaporation to extinguish the flames. Another is the formation of oil-seawater emulsions from benthic or surface mixing that cannot be ignited. Finally, at the end of a burn, residues frequently remain that are light enough to float but too heavy to ignite easily.

Crude oil undergoes a number of changes once it has spilled and interacts with the marine environment. As it spreads across the ocean surface, the increased surface area accelerates the evaporation of lightweight hydrocarbons by evaporative weathering [2], exposure to sunlight and air drives photochemical reactions [3, 4], and surface wave turbulence emulsifies the crude oil with saltwater [5]. All of these processes change the oil's physical, thermodynamic, and chemical properties. The emulsified crude and heavier fractions require additional thermal energy to break and evaporate, respectively, for ignition.

S.L. Ross Environmental Research, Ltd. conducted a number of tests on the suitability of ISB for California and other crude oils, which were then reported by McCourt *et al.* [6-9] in a number of venues. McCourt *et al.* [6-9] determined a weathering rate correlation and the emulsion stability by using standard methods [10]. Burn tests were conducted in a wave tank with water temperatures maintained at 20 °C. 2.5 L of oil or emulsion was rested on the water and was contained by a 40-cm diameter steel ring, suspended to the tank floor to form a 2-cm thick slick. Gelled gasoline (40 to 50 g) accelerated the ignition process. From their tests, they concluded that heavy crude oils, such as Santa Ynez, were inflammable if emulsified, while others were inflammable if weathered and emulsified.

From the description by McCourt *et al.* [6-9], there are a number of aspects of the testing that merited deeper examination. There was no metric used to indicate emulsion stability, since crude oil emulsions, in general, will separate given enough time. We also had questions about the size of the slick and the quantity of accelerant. In practice, it is assumed that the minimum acceptable oil slick thickness for ignition is 10 mm [11]. Furthermore, the current standard is to use 4 L of gelled diesel fuel to ignite oil slicks, which is approximately 4 kg of accelerant, or two orders of magnitude greater. If we assume a thickness of 10 mm, that is a pool area of 0.40 m^2 with a nominal diameter of 71 cm, which is much larger than the 40 cm slick used by McCourt *et al.* [6-9]. The work by Garo *et al.* [12] showed that the heat flux to the surface of a pool fire is nominally proportional to the square root of the diameter. At these scales, an accelerant pool two orders of magnitude smaller will produce a heat flux 32% of that produced by the nominal, practical accelerant pool. Since the heat flux back to the slick is what drives ignition and continued combustion, such a large gap suggests that the pool fire tests should be repeated with both larger slick and greater volumes of accelerant to establish whether an oil can be ignited or not.

The objective of this work is to establish the ignitability and burn behavior of neat, weathered, and emulsified California crude oils from the Santa Barbara channel using both a practical pool fire experiment and cone calorimetry. The pool fire experiments determine the ignitability in practical conditions while cone calorimetry provides comparative ignition, heat release rates, and burn efficiencies for a range of crude oils blends. The pool fire experiments, which will not be discussed in this paper, have shown that oils deemed inflammable can, in practice, be ignited and burned. For brevity, only the cone calorimetry experiment and results are reported in this paper.

2. Experimental Description

All of the five crude oils examined in this study are extracted from offshore wells along the Santa Barbara Channel off the coast of California, as shown in Table 1. They range from lightweight crude, such as Carpinteria, to heavy crude, such as Santa Ynez. The final test matrix calls for each oil to be tested neat, emulsified with 40% substitute ocean water [13], and then weathered for 48 hours and tested with 20%, 40%, and 60% ocean water emulsions. At present, all of the neat and 40% emulsion tests are complete while the weathering tests are underway.

We designed the experiment to mimic ocean conditions along the Santa Barbara channel during the month of August, when ocean temperatures, air temperatures, and hours of cloudless sunlight would be at the maximum. Such data can be compiled and averaged (Table 2) from NOAA buoy data that are posted online [14]. From these data, we determined the underlying water temperature and conditions for weathering and burn testing. For the data reported here, we only discuss the neat and 40% emulsified crude oil cone calorimetry burn tests.

Table 1: Crude oils tested in this investigation and their properties [6]. All are from the Santa Barbara Channel along the coast of California.

Crude Oil, Field	API (°)	Viscosity (cP) @ 15 °C	Asphaltene (%mass)	Emulsion Viscosity (cP)	Emulsion Water (% mass)	Platforms
Santa Ynez, Hondo Monterey	17.0	735	12	110,000	81	Heritage, Harmony, Hondo
Sockeye, Comingled	19.8	550	13	39,000	74	Gail
Santa Clara	20.0	304	13	2,700	61	Grace
Point Arguello - Comingled	21.0	533	16	180,000	82	Hidalgo, Harvest, Hermosa
Carpinteria	24.0	164	9	N/A	9	Hogan, Houchin, Henry

Table 2: Averages for the month of August as compiled from buoy 46053 in the Santa Barbara Channel.

Wind Speed (m/s)	Wind Direction (°)	Water Temp. (°C)	Salinity (g/Kg)	Current Speed (m/s)	Current Direction (°)
4.4	225.94	19.7	33	0.156 mph	146.69 °

Crude oil was emulsified by placing the neat crude oil, substitute ocean water, and a commercial surfactant (Span ® 85) in a vessel and mixing it with a high-shear mixer. The surfactant fraction is 2% by volume. The components were mixed until the emulsion remained stable for two hours. This follows a similar process used in previous work [15, 16].

The cone calorimetry experiments used a Fire Testing Technology instrument to measure ignition, flame spread, heat release rate, smoke release rate, products, and oxygen consumption calorimetry [17, 18]. Typically, these tests are conducted by placing a coupon of a solid material, such as those that are found in consumer products and buildings, on a load cell under a coiled heater element that radiates the sample at a fixed heat flux until the sample ignites and burns. To aid ignition, two electrodes suspended over the sample produce a steady arc that ignites flammable vapor and air. Once there is steady combustion, the two electrodes are removed. Such instruments are generally not used with liquids since they tend to boil and create an uncontrolled, uncontained burn. In the case of a crude oil slick on water, such a test creates a boilover phenomenon that is equally uncontrollable and uncontained when the underlying water boils [19-21].

To prevent boilover and maintain the water temperature at 19.7 °C, a sample burn pan was adapted from the work of van Gelderen *et al.* [22], as shown in Figure 1. Cooling water flows into and out of a jacket surrounding the liquid crude oil sample, through a chiller, and back to the sample pan.

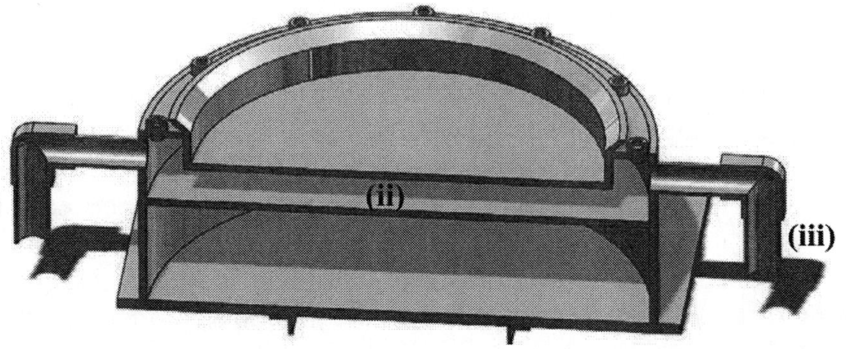

Figure 1: Cross-section of the cone calorimeter burn pan. Cooling water flows in (i) from the chiller, under the burn sample pan (ii), and returns out (iii) to the chiller.

We calculated the imposed calorimeter heat transfer to approximate the heat transfer from a 4 L pool of accelerant spread over a slick 10 mm thick. Using a 3% transfer fraction and the relation by Garo *et al.* [12], the radiative heat flux back to the slick is 16.3 kW/m^2. For simplicity and conservatism, we assumed that 15 kW/m^2 was the activating heat flux to the slick from the accelerant and used this value as the imposed cone heat flux.

3. Results

All of the neat and 40% emulsified cone calorimeter tests complete as of the writing of this report. Each test was completed three times to quantify scatter and uncertainty. In particular, we are interested in the heat release rate (Q) over the duration of the test and the burn efficiency (η_B), which is the ratio of the mass burned to the initial mass. The ignition energy, or the heat transferred to the slick, indicates how much heat is necessary to sufficiently break the emulsion and evaporate enough crude oil to create an ignitable concentration of hydrocarbon vapor. This is calculated by the product of the incident heat flux and the time between the start of the test and ignition.

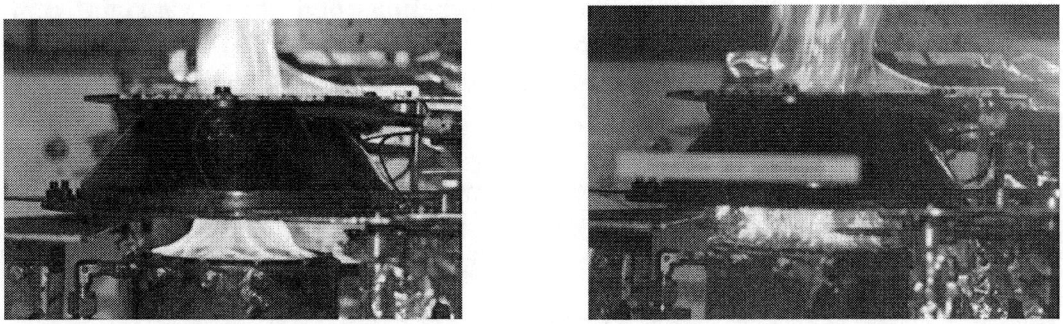

Figure 2: (L) Neat and (R) 40% emulsified Santa Ynez burning under the cone calorimeter with 15 kW/m^2. Notice the burning splatter from the emulsified slick as the water boiled and scattered oil droplets.

Figure 2 shows photographs of burning Santa Ynez crude oil under the cone calorimeter just after ignition. The neat crude exhibits the expected behavior of a small hydrocarbon pool fire while the emulsified crude splatters to form droplets that burn more rapidly. The emulsified crude oil tests are conducted with a window shield to contain the splatter (as indicated by the yellow strip on the shield, seen in the right photograph of Figure 2).

The ignition and burn behavior between the neat and emulsified crude is distinct. The neat crude oil ignites almost instantaneously once the calorimeter shutters are opened to allow radiative heat from the cone heater coils to reach the oil surface and start evaporation. In contrast, the emulsified crude oil requires time for the surface temperature to rise. During this warm up period, surface bubbles form and pop as some of the water boils and evaporates. Intermittently, flames form, propagate, and then extinguish over the oil surface as the vapor concentration increases. Eventually, the evaporation rate has sufficiently increased to produce a continuously flammable concentration. Once a steady flame forms, the additional radiative heat transfer increases the evaporation rate to rapidly boil the water and create a boilover effect that dramatically increases the heat release rate.

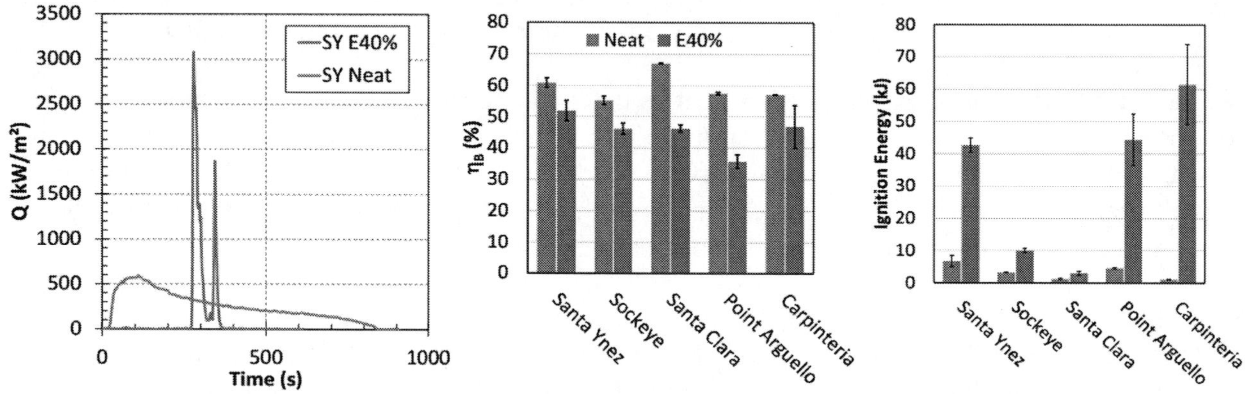

Figure 3: (L) Heat release rates, Q, during a test for Santa Ynez crude oil. (M) Burn efficiency, η_B, and (R) ignition energy comparison for all of the oils tested at this date. Oils in order of decreasing °API

The left plot in Figure 3 shows the heat release rate over time, with zero marking when the sample is exposed to the calorimeter. Comparing the neat and emulsified, there is a significant delay as the emulsion is heated, separates, and water boils before there is sufficient flammable vapor above the crude oil surface. Once ignited, boilover increases heat flux back to the surface, driving more rapid evaporation and greater heat release.

The burn efficiency (η_B) plot (middle, Figure 3) shows a clear dependence on emulsion, but there is not clear dependence on °API or any other common property. There must be additional oil properties that are driving the burn efficiency other than oil density.

The ignition energy plot (right, Figure 3) shows a clear dependence on the nature of the emulsion. All but the Sockeye and the Santa Clara require much larger ignition energies to break the emulsion. Though we have no explanation for the Sockeye crude, we do know that the Santa Clara contains an emulsion breaker that enhances oil and water separation.

4. Conclusions & Further Work

The preliminary results of crude oil cone calorimetry have revealed dramatically different behavior between neat and emulsified crude oil. The emulsified crude oil produced rapid, short, and large heat release rates as the water boiled off and atomized the crude oil, but it required more energy to break the emulsion, boil off water, and produce steadily flammable evaporation. Though significant energy must be transferred to the emulsion for ignition, these tests show ignition and burning is possible, even for heavy oils such as Santa Ynez.

Acknowledgements

This study was funded by the U.S. Department of the Interior, Bureau of Safety and Environmental Enforcement through Interagency Agreement E16PG00038 with the Naval Research Laboratory.

References

[1] I.A. Buist, J. McCourt, S. Potter, S. Ross, K. Trudel, In Situ Burning, Pure and Applied Chemistry 71 (1999) 43–65.

[2] M.F. Fingas, Evaporation Modeling, Oil Spill Science and Technology, Elsevier Science2010, pp. 202-242.

[3] M.F. Fingas, Photo-oxidation, Oil Spill Science and Technology, Elsevier Science2010, pp. 192.

[4] R.M. Garrett, I.J. Pickering, C.E. Haith, R.C. Prince, Photooxidation of Crude Oils, Environmental Science & Technology 32 (1998) 3719-3723.

[5] M.F. Fingas, Models for Water-in-Oil Emulsion Formation, Oil Spill Science and Technology, Elsevier Science2010, pp. 243-273.

[6] J. McCourt, I. Buist, Laboratory Testing to Determine Operational Parameters for In Situ Burning of Six U.S. Outer Continental Shelf Crude Oils, United States Department of the Interior, Herndon, Virginia, March 1998.

[7] J. McCourt, I. Buist, S. Buffington. 1999. Results of Laboratory Test on the Potential for using In Situ Burning on Seventeen Crude Oils. 23rd Arctic and Marine Oilspill Program (AMOP) Technical Seminar; 1999 1999; Ottawa, Ontario: Environment Canada. p. 917-922.

[8] J. McCourt, I. Buist, S. Buffington. 2001. Results of Laboratory Tests on the Potential for Using In Situ Burning on 17 Crude Oils. International Oil Spill Conference Proceedings; 2001. p. 529-532.

[9] J. McCourt, I. Buist, J.V. Mullin. 1999. Operational Parameters for In Situ Burning of Six U.S. Outer Continental Shelf Crude Oils. International Oil Spill Conference Proceedings; 1999 1999. p. 1261-1263.

[10] W. Zagorski, D. Mackay. 1982. Water in Oil Emulsions: a Stability Hypothesis. 5th Arctic and Marine Oil Spill Program Technical Seminar; 1982 1982; Ottawa, Ontario: Environment Canada. p. 61-74.

[11] I.A. Buist, S.G. Potter, B.K. Trudel, S.R. Shelnutt, A.H. Walker, D.K. Scholz, P.J. Brandvik, J. Fritt-Rasmussen, A.A. Allen, P. Smith, In Situ Burning in Ice-Affected Waters: State of Knowledge Report, Report No. Final Report 7.1.1, Arctic Response Technology,

[12] J.P. Garo, J.P. Vantelon, S. Gandhi, J.L. Torero, Determination of the Thermal Efficiency of Pre-boilover Burning of a Slick of Oil on Water, Spill Science & Technology Bulletin 5 (1999) 141-151.

[13] ASTM, Standard Practice for the Preparation of Substitute Ocean Water, ASTM International, West Conshohocken, PA, 2013.

[14] NOAA, National Data Buoy Center. http://www.ndbc.noaa.gov/.

[15] S.G. Tuttle, T.N. Loegel, K.M. Hinnant, A.D. Tuesta, B.T. Fisher. 2017. Fundamental Measurements of Crude Oil Spray Behavior. 40th Arctic and Marine Oil Spill Program (AMOP) Technical Seminar; 6-8 June 2017 2017; Alberta, Canada: Environmental Canada. p.

[16] S.G. Tuttle, K.M. Hinnant, T.N. Loegel, B.T. Fisher, A.D. Tuesta, M.R. Weismiller, Development of a Low-Emission Spray Combustor for Emulsified Crude Oil, Report No. NRL/MR/6185-17-9720, Naval Research Laboratory, Washington, DC, March 2017.

[17] ASTM, Test Method for Determining Material Ignition and Flame Spread Properties, ASTM International, West Conshohocken, PA, 2013.

[18] ASTM, Test Method for Heat and Visible Smoke Release Rates for Materials and Products Using an Oxygen Consumption Calorimeter, ASTM International, West Conshohocken, PA, 2017.

[19] M. Arai, K. Saito, R.A. Altenkirch, A Study of Boilover in Liquid Pool Fires Supported on Water Part I: Effects of a Water Sublayer on Pool Fires, Combustion Science and Technology 71 (1990) 25-40.

[20] H. Koseki, M.A. Kokkala, G.W. Mulholland, Experimental Study Of Boilover In Crude Oil Fires, Fire Safety Science 3 (1991) 865-874.

[21] P. Ping, J. Zhang, D. Kong, Z. Xu, H. Yang, Experimental study of the flame geometrical characteristics of the crude oil boilover fire under cross air flow, Journal of Loss Prevention in the Process Industries, doi:https://doi.org/10.1016/j.jlp.2017.12.005(2017).

[22] L. Van Gelderen, N.L. Brogaard, M.X. Sørensen, J. Fritt-Rasmussen, A.S. Rangwala, G. Jomaas, Importance of the Slick Thickness for Effective In-Situ Burning of Crude Oil, Fire Safety Journal 78 (2015) 1-9.

Sub Topic: Fire

Spring Technical Meeting
Eastern States Section of the Combustion Institute
March 4-7, 2018
State College, Pennsylvania

Evaluating Foam Degradation and Fuel Transport Rates through Novel Surfactant Firefighting Foams for the Purpose of AFFF Perfluorocarbon Replacement

Katherine Hinnant, Arthur Snow, Spencer Giles, and Ramagopal Ananth*

Chemistry Division, U.S. Naval Research Laboratory, Washington, D.C.

**Corresponding Author Email: Katherine.hinnant@nrl.navy.mil*

Abstract: Through a research program aimed at replacing perfluorocarbon surfactants in AFFF, we evaluated foam degradation and fuel transport through foams at elevated temperatures for various classes of fluorine-free surfactants. Foam solutions were made well above the critical micelle concentration for individual surfactants and surfactants mixed with a hydrocarbon surfactant. Foam degradation was measured by placing 4 cm of foam above a heated n-heptane pool and monitoring the change in foam height over time. Fuel transport through the foam was measured by placing 4 cm of foam above a heated fuel pool in a specially designed fuel flux apparatus that monitored the concentration of fuel vapors above the foam over time using a nitrogen sparger and an FTIR. Surfactants analyzed included ionic and non-ionic hydrocarbon surfactants and silicone surfactants. From the evaluated surfactants, none have matched the foam degradation or fuel transport performance of a simplified analytical AFFF or commercial AFFF. However, certain surfactants performed better than others indicating potential directions for future AFFF surfactant replacement.
Keywords: Fire suppression, surfactant, replacement, environment

1. Introduction

Aqueous film forming foams (AFFF) are used to suppress liquid pool fires that are unable to be suppressed by water alone. Foam is produced from an aqueous solution containing fluorocarbon and hydrocarbon surfactants that make stable foams that effectively cover the fuel surface and block the transport of fuel vapors to the fire above. Fluorocarbon surfactants are considered necessary for fast fire suppression, but are harmful to the environment and have been found in ground-water sources [1]. To eliminate the environmental impact of these suppression agents, we must alter the surfactants used, but maintain rapid fire suppression.

To assess the potential firefighting ability of novel surfactants, we evaluated the foam degradation and fuel transport rates through foams with a range of commercially available surfactants. Although extinction experiments provide instant performance metrics, little information is gained as to why the foam did or did not suppress the fire. We were interested in obtaining metrics that could be related to surfactant structure. In an extinction event, many mechanisms act simultaneously (foam degradation, fuel transport, surface cooling, etc. [2-4]) to suppress the flame. By studying individual mechanisms, such as foam degradation and fuel transport, we can better understand how surfactant structures effect mechanisms related to fire suppression.

2. Methods / Experimental

Surfactants

We evaluated five commercial surfactants and compared their individual performance and their performance as a fluorine-free formulation with a hydrocarbon surfactant and solvent to a reference and a commercial AFFF (Buckeye 3%).

Surfactant Solutions

Surfactant solutions were made of individual surfactants and a mixture of the surfactant and a hydrocarbon surfactant and solvent. A simplified mixture (a reference AFFF) with a fluorocarbon surfactant, hydrocarbon surfactant, and solvent was able to suppress a fire as effectively as a commercial AFFF at large-scale [5]. We compare the performance of the surfactants in a mixture to this reference AFFF and a commercial AFFF.

Surfactant solution concentrations were determined by measuring the critical micelle concentration (CMC) using a DuNouy ring tensiometer at 19°C. A 2 vol% surfactant solution was made and serial diluted by half to a concentration of 0.001 vol% and the surface tension was measured at each 12 concentrations. The critical micelle concentration was taken as the concentration before which the surface tension drastically increased. All surfactant solutions were made well above the CMC. This process was repeated for every fluorine-free candidate surfactant mixture. All formulations had CMC values between 0.008 and 0.07 vol%. Foam was then generated using the foam solution in a container with a nitrogen sparger, detailed elsewhere [3]. Nitrogen was supplied to produce a foam flow rate of 900 mL/min through a porous sparger, pore size 25-50 μm.

Foam Degradation

A 100 mL beaker, 5 cm in diameter, was placed in a water bath heated externally with heating tape to 60°C. 70 mL of n-heptane was poured into the beaker and allowed to heat to 60°C, monitored using a thermocouple in the fuel layer. Once the fuel was heated, foam was generated with nitrogen at a foam flow rate of 900 mL/min and then poured directly from the generator into the head space of the beaker, 4 cm in height. The foam height was leveled using a metal spatula ensuring a level surface at the top of the foam column. A ruler was placed next to the beaker and a video camera was used to record changes in foam height over time. A free online imaging software (ImageJ) was then used to measure the foam height at specified times. Each trial was repeated 3 times for precision.

Fuel Transport through Foam

A diagram of the fuel transport apparatus is shown below in Fig. 1.

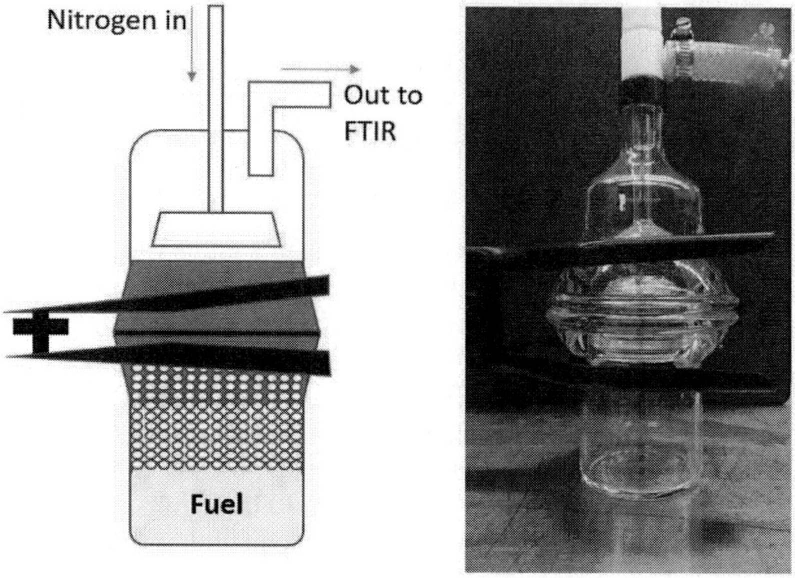

Fig. 1. Diagram of fuel transport apparatus on the left, image of actual apparatus on the right.

Fig. 1 shows that the apparatus is a two part cylindrical system with a bottom and top piece. The bottom piece holds the fuel and the foam while the top piece contains a nitrogen sparger that carries gas out of the apparatus and to an FTIR. The two part system allows for a level foam layer to be produced and eliminates artificial fuel concentrations that may exist in an apparatus before foam is applied. Although useful for measuring fuel transport, the closed system removes exposure to air and evaporation from the top of the foam layer. Therefore foam lifetimes are longer in the fuel flux apparatus compared to the open beaker as seen in the results and discussion section.

The bottom piece (5 cm in diameter, 8 cm tall) is kept in a water bath, heated externally with heating tape to 60°C. 65 mL of n-heptane is poured into the bottom piece and allowed to heat to 60°C, monitored using a thermocouple. Once the fuel is heated, foam is generated using nitrogen at a foam flow rate of 900 mL/min directly onto the fuel surface. As soon as foam touches the fuel surface, a stopwatch is started. The foam fills the headspace of the bottom piece (4 cm) and is leveled to form a flat surface using a spatula. A rubber O-ring is then placed on the bottom piece, and the top piece is lowered into place. A large screw clamp is then secured to ensure an air-tight seal. 500 mL/min of nitrogen is then swept into the container through a porous sparger in the top piece of the apparatus. An exit port connected to an FTIR is then opened allowing the carrier gas to exit the apparatus and flow into the FTIR. At this time, the stopwatch is stopped. The FTIR has a nitrogen bypass flowing at 100 mL/min to further dilute the carrier gas stream. n-heptane, carbon dioxide, and water concentrations are measured using the FTIR for the duration of the test, until the foam degrades, exposing the n-heptane pool below. The time recorded on the stopwatch between when foam is applied and when the FTIR collects data is added to the initial time of the measurement, it is on average 30 seconds.

The raw concentration data obtained by the FTIR is then converted into a flux by converting the ppm by volume output by the FTIR to mol/cm^{2}s using the total flow of nitrogen into the FTIR (600 mL/min) and the cross-sectional area of the flux apparatus (19 cm^2)

3. Results and Discussion

Fig. 2 and 3 below show the percent change in foam height overtime for individual surfactants and surfactants in a fluorine-free formulation respectively.

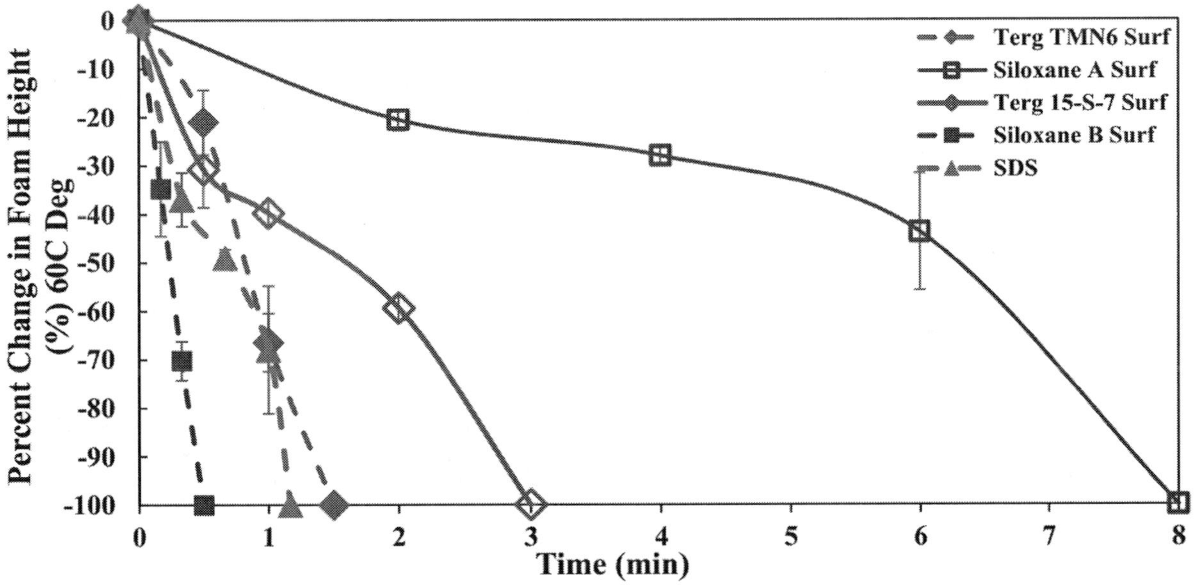

Fig. 2. Percent change in foam height for 4 cm of foam using individual surfactants over 60°C n-heptane.

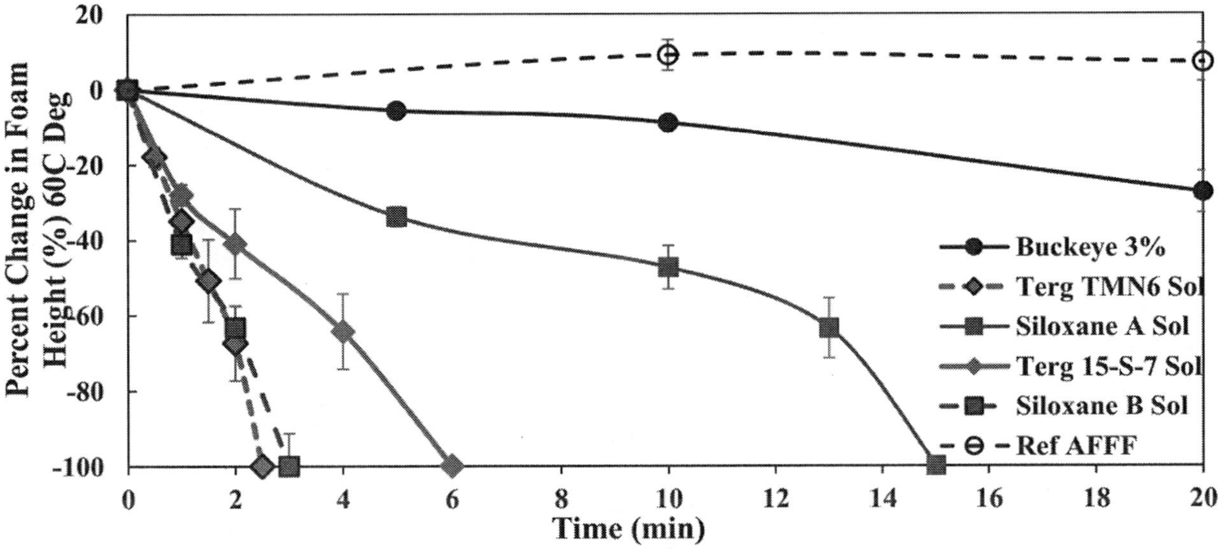

Fig. 3. Percent change in foam height for 4 cm of foam using surfactants in a fluorine-free formulation over 60°C n-heptane.

For all of the surfactants evaluated, foam lifetime was improved when the surfactant was in a fluorine-free formulation rather than when the foam only contained the individual surfactant. Tergitol TMN6 improved from 1.5 to 2.5 min, Tergitol 15-S-7 improved from 3.5 to 6 min, Siloxane B improved from 30 seconds to 3.5 min, and Siloxane A improved from 8 min to 15 min. Although an improvement in foam lifetime is seen, none of the surfactant solutions compare to the foam lifetime of the reference or commercial AFFF. The reference AFFF had a foam lifetime of 60 minutes while the commercial AFFF had a foam lifetime of 40 minutes. Siloxane

A has the longest foam lifetime of the surfactants evaluated suggesting the potential superiority of silicone surfactants. However, Siloxane B has the shortest foam lifetimes individually or in a formulation. Data on surfactant structure is available for Tergitol TMN6 and 15-S-7 with 15-S-7 having a more branched head group that TMN6 potentially stabilizing the surfactant, increasing the foam lifetime.

Fig. 4 and 5 show the n-heptane flux versus time for foams with individual surfactants and foams using a fluorine-free formulation respectively. SDS and the Siloxane B foam degraded so quickly that we were unable to obtain flux data for these foams.

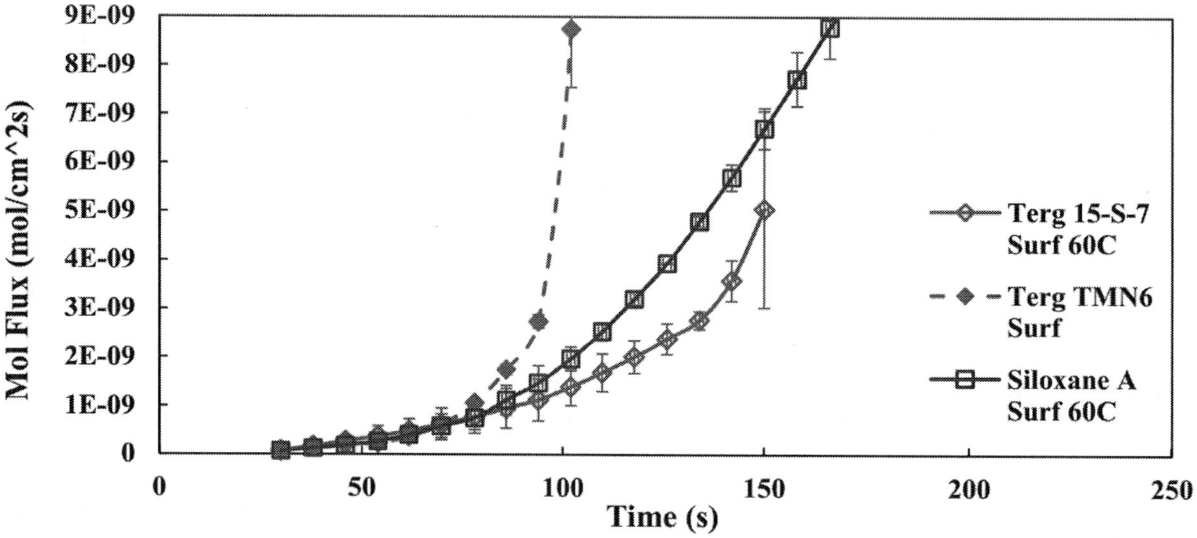

Fig. 4. Fuel transport through 4 cm of foam from individual surfactants over 60°C n-heptane.

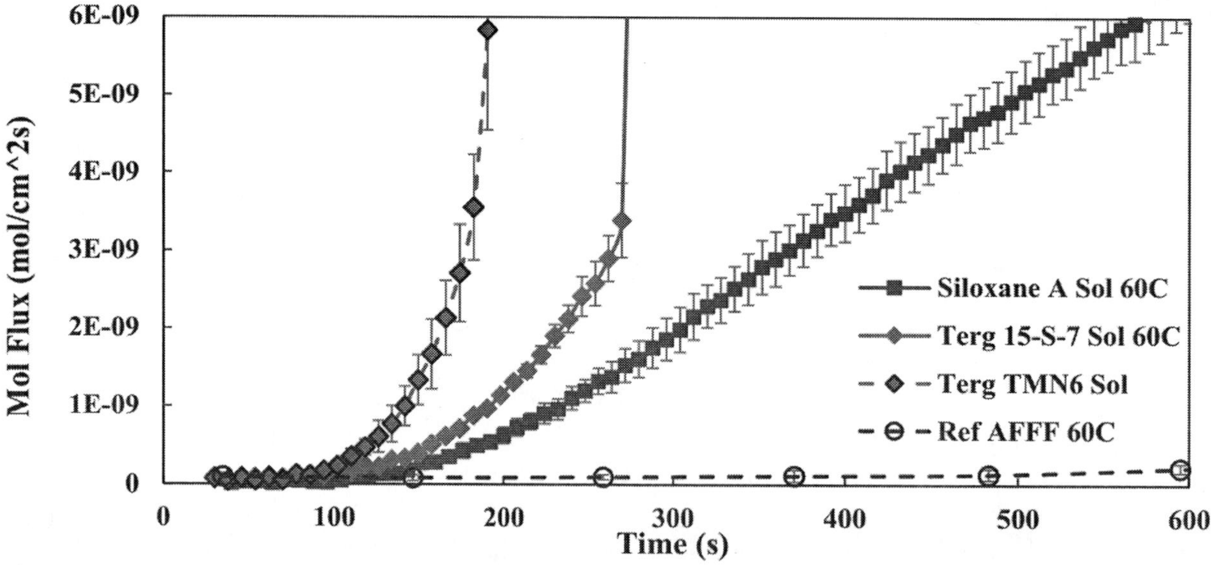

Fig. 5. Fuel transport through 4 cm of foam from surfactants in a fluorine-free formulation over 60°C n-heptane.

Fig. 4 and 5 show steady curves transitioning into a sharp incline in the time evaluated. This sharp incline physically refers to a point at which the foam completely degraded or a void in the foam layer formed exposing the fuel pool below, saturating the FTIR. Similar to the results of the foam degradation study, none of the surfactants evaluated have low fuel transport similar to the reference AFFF. At 200 seconds, fuel has yet to break through the 4 cm reference AFFF surface were as 30 ppm and 55 ppm of n-heptane have traveled through the 4 cm of the Siloxane A and Tergitol 15-S-7 formulations respectively. By 200 seconds, the Tergitol TMN6 solution has already degraded exposing the fuel pool. Fuel transport performance is improved for foams generated from a fluorine-free formulation.

4. Conclusions

Of the five evaluated surfactants, none compare to the foam degradation and fuel transport performance of a fluorocarbon surfactant. Fluorocarbon surfactants produce foams with longer foam lifetimes and less fuel transport. Siloxane A in a fluorine-free formulation appears to perform better in both foam degradation and fuel transport through the foam compared to the other surfactants examined; however, its proprietary formulation makes it difficult to optimize or make specific conclusions related to surfactant structure. Additional tests will be conducted with commercial surfactants to produce a larger database of foam degradation and fuel transport measurements that can be used to make conclusive arguments relating fire suppression mechanisms to surfactant structure.

5. Acknowledgements

This research was funded by the Office of Naval Research (ONR) through a 6.2 research program and the Strategic Environmental Research and Development Program (SERDP) under grant WP2739. We also acknowledge Dr. James Fleming for his advice during this research.

6. References

[1] E. F. Houtz, C. P. Higgins, J. A. Field, D. L. Sedlak, Persistence of perfluoroalkyl precursors in AFFF-Impacted groundwater and soil, Environ. Sci. Technol. 47(15) (2013) 8187-8195.

[2] K. M. Hinnant, M. W. Conroy, R. Ananth, Influence of fuel on foam degradation for fluorinated and fluorine-free foams, Colloids and Surfaces: A, 522 (2017) 1-17.

[3] B. Y. Lattimer, C. P. Hanauska, J. L. Scheffey, F. W. Williams, The use of small-scale test data to characterize some aspects of fire fighting foam for suppression modeling, Fire Safety Journal, 38(2) (2003) 117-146.

[4] M. W. Conroy, R. Ananth, Fuel surface cooling by aqueous foam: a pool fire suppression mechanism, Fire Technology, 51 (2015) 667-689.

[5] K. M. Hinnant, S. L. Giles A. W. Snow, J. Farley, J. W. Fleming, R. Ananth, An analytically defined fire suppressing foam formulation for evaluation of fluorosurfactant replacement, J. Surfactants and Detergents, Submitted Nov 2017.

Sub Topic: Fire

Spring Technical Meeting
Eastern States Section of the Combustion Institute
March 4-7, 2018
State College, Pennsylvania

Effect of Low Temperature Wall on the Extinction and Fuel Layer Distribution of Pool Fires

*Cong Li[1, *], Hamed Farmahini Farahani[1], Rui Yang[2], Ali S. Rangwala[1]*

[1]Department of Fire Protection Engineering, Worcester Polytechnic Institute, Worcester, MA 01609, USA
[2]Department of Engineering Physics, Tsinghua University, Beijing, 100084, China
[*]Corresponding Author Email: cli11@wpi.edu

Abstract: A series of experiments were carried out to study the effect of the low temperature boundary on the extinction of the pool fire flames. m-Xylene with initial thickness of 1mm and 2mm (on a water sublayer) was burned in a circular tray of 15.5 cm diameter while the wall of the tray was exposed to low temperatures (0 °C). Based on the experimental observations during burning of these liquids, the fuel layer thickness was observed to vary with respect to the radial distance from the center of the pool. The fuel was thinner at the center of the pool and increased its thickness near the tray wall. This was because of the low temperature boundary of the liquid and high temperature of the flames that gave rise to Marangoni stress on the free surface of the fuel. The Marangoni stress pushed the fuels towards the rim which led to the appearance of a ring-shaped fuel layer toward the end of the burning. The flames of the fuel ring were merged initially by the strength of air entrainment. Then the flames separated as the flame height decreased and formed a complete annular fire that was burning on the circumference of the tray. After a while the ring fire became intermittent until natural extinction. The findings of our study show that the extinction of pool fires with cold boundaries are delayed by appearance of the ring fire. This may be an influential factor in estimating the removal efficiency of *in-situ* burning in cold regions.
Keywords: pool fire, m-Xylene, Marangoni, ring fire

1. Introduction

In-situ burning (ISB) is a common method of cleaning oil spills on water. In open waters, the oil is contained using a boom and then burned. The method is efficient in removing relatively large amounts of oil quickly and inexpensively. In remote areas with presence of ice, such as those found in the Arctic, ISB may be the only method of choice. However, combustion of oil near ice boddies has a lot of special characteristics that are significantly different from pool fires that have been usually the subject of research for more than 50 years. In the presence of ice, ullage, ice melting, surface area increase, and movement of water sub-layer below the fuel alter the burning dynamics [1-3]. It is therefore important to investigate the controlling parameters of the fire dynamics to provide guidelines for ISB in the Arctic conditions.

Farahani et al. carried out the pool fires in a cylindrically shaped ice cavities of 5.7 cm diameter [4]. It was found that the geometry change of the ice (i.e. melting) was because of the heat transfer from the fuel layer. In particular, Marangoni stress was found to be the driving mechanism for transferring the heat toward the ice near the surface of the oil slick. Shi et al. performed large-scale crude oil fires in ice cavities [5]. The melting of ice and the boilover were proved to increase the burning rate of oil. van Gelderen et al. [6] investigated the burning efficiency and vaporization order of the components in crude oil. Bellino et al. analyzed the influence factors of mass loss rate and the combustion efficiency in an ice channel. The mass loss rate was proved to be primarily limited by ullage and fuel layer depth and less affected by channel width [7]. However, previous studies usually focus on the occurrence and development of pool fire, and few studies have focused on the evolution of the flame and the fuel during the extinction period.

In this study, m-Xylene pool fires with various wall temperatures were carried out in order to investigate the effect of temperature on fuel layer thickness and mass loss rate (MLR) of the fuel. The evolution of flame shape was observed and the fuel distribution was deduced accordingly. From the experimental phenomenon, there are various evolutions of both the flame and the fuel distribution before the extinction that could potentially improve removal efficiency of *in-situ* burning.

2. Experimental procedures

A steel tray with diameter of 15.5 cm and wall thickness of 0.1 cm was placed in the center of an external container. The space between the two trays was filled with the ice–water mixture and room temperature water to keep the wall temperature at 0 °C and 20 °C, respectively. For the experiments with the ice bath, m-Xylene with density of 0.864 g/mL with initial thickness of 1 mm and 2 mm was used as the liquid fuel. The ullage was set at 2 mm to eliminate the ullage impact [8]. An electric balance with accuracy of 0.01 g and sampling frequency of 5 Hz was used to record the mass loss. A DSLR camera was used to record the flame shape.

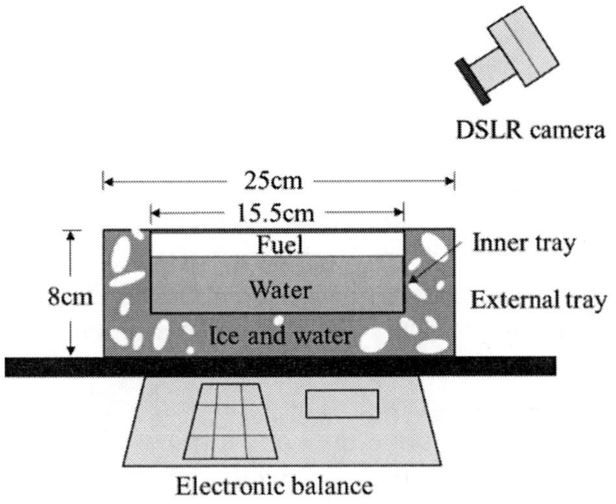

Figure 1: Schematic illustration of the experimental setup.

3. Results and Discussion

3.1. Mass loss rate (MLR)

Figure 2 (a) and (b) show the mass loss rate (MLR) for two fuel thicknesses of 1 and 2 mm during the duration of the experiments. According to the MLR trends, the whole combustion process can be divided into the following three stages:

(i) Pre-steady stage. The MLR increased rapidly after ignition and peaked at 30 s.

(ii) Quasi-steady stage. The pool fire grew slowly and the MLR showed a slight rise. The MLR with wall temperature of 0 °C was lower than 20 °C of because of heat losses from the side wall. In the experiments with initial fuel thickness of 1 mm, the duration of quasi-steady stage was very short.

(iii) Decay stage. After peaking at the quasi-steady stage, the MLR decreased until the extinction. The extinction time with wall temperature of 0 °C was later than 20 °C with the same initial fuel thickness.

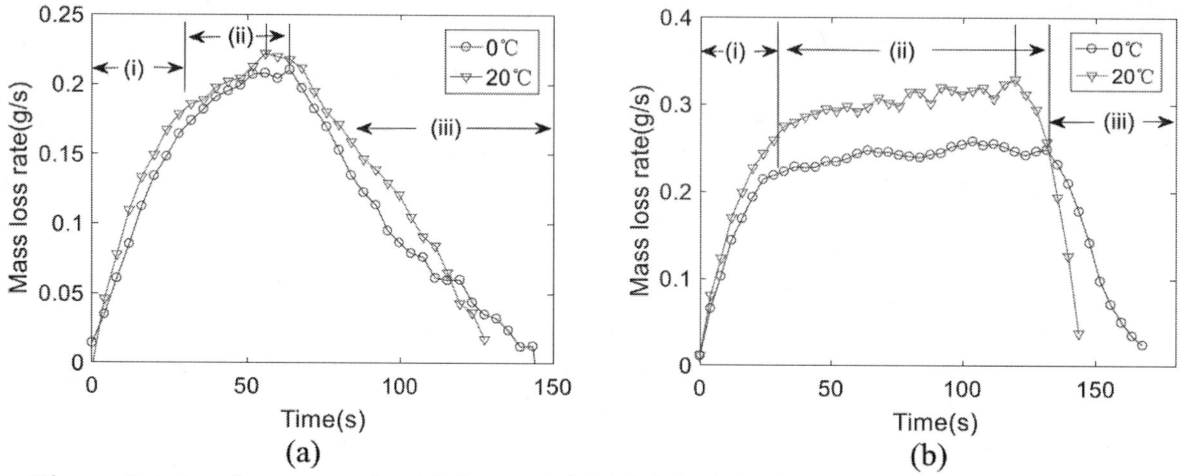

Figure 2: Mass loss rate of m-Xylene with initial fuel thickness of (a) 1 mm and (b) 2 mm.

As can be seen from Figure 2, the MLR for 1 mm experiments showed similar trend while for 2 mm tests the 0 °C experiment showed lower MLR than 20 °C. The similar trends for 1 mm tests means that the energy balance was not dependent on the boundary temperature (sides and water sublayer). The higher MLR for 2 mm test could be explained through the insulation effect of the thickness of the fuel layer.

3.2. Evolution of the flames

According to the images recorded by the DSLR camera, the evolution of the flames can be summed up as the following stages shown in Figure 3. (i) At the beginning of the combustion, the fire burned uniformly on the entire pool surface. Also, the flames were merged and puffing was observed above the pool center because of the air entrainment [9]. (ii) Later, it was observed that there was no burning in the center of the pool that created a dark area in the flames. The dark area is emphasized by white circles in Fig. 3. The flame existed only on the circumference of the tray and were long enough to merge above the pool surface. The phenomenon indicated that the fuel in the tray center migrated toward the rims while the remaining fuel was gathered near the tray wall. Eventually the flame length shortened and the fuel only burned at the circumference in the shape of a ring-fire. For m-Xylene (and most liquid fuels), the surface tension decreases with the increase of temperature, which results in the higher surface tension around the wall than the tray center. The fuel therefore moves toward the wall under this effect, that is also known as Marangoni stress [10]. The dimensionless Marangoni number that is used to describe the effect [11] is expressed by the following

$$Ma = \sigma_T R \Delta T_h / \mu\alpha \ , \tag{1}$$

where σ_T is the change of surface tension to temperature, R is the radius of pool, μ and α are the viscosity and thermal diffusivity of m-Xylene respectively, and ΔT_h represents the temperature difference between the tray center and wall. In the experiment with wall temperature of 0 °C, the Marangoni number is much larger than 20 °C tests because of the larger temperature difference. As a result, the ring fire lasted for a longer time as shown in Fig. 3. The duration of the ring fire with 1 mm thickness and 0 °C wall temperature was about 84 s, which was longer than 20 °C tests with 60 s duration. While in the experiments with 2 mm thickness, the duration at 0 °C was 40 s, which was also longer than tests at 20 °C with 30 s duration. (iii) With the expansion of the dark area at the tray center, both the width of the fuel ring and the MLR decreased. Correspondingly, the flame height was reduced and separated to form the complete annular fire as shown in the fourth image of Figure 3. (iv) And finally, the complete annular fire broke into some fractions until it went out.

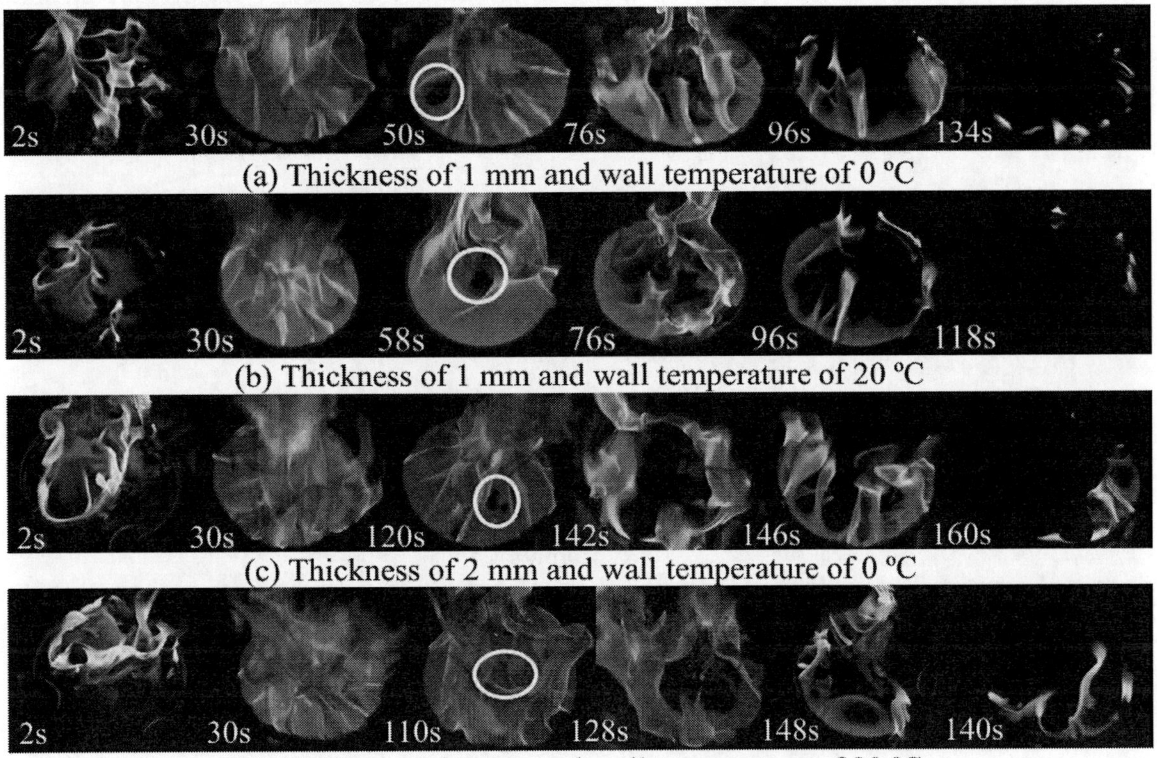

(a) Thickness of 1 mm and wall temperature of 0 °C

(b) Thickness of 1 mm and wall temperature of 20 °C

(c) Thickness of 2 mm and wall temperature of 0 °C

(d) Thickness of 2 mm and wall temperature of 20 °C.

Figure 3: Flame evolution of m-Xylene with various initial thickness and boundary temperature

3.3. Fuel thickness

The time instant of flame disappearance in the tray center was defined as the ring-fire formation time. In the experiment with 1 mm fuel layer thickness with wall temperature 0 °C and 25 °C, the formation time was at 50 s and 58 s from the ignition, respectively. While the time with 2 mm fuel layer thickness and 0 °C and 25 °C wall temperature was 120 s and 110 s, respectively. It should be noted that the ring-fire formation time is close to the peak time of MLR, that is to say, the flame evolutions (II), (III) and (IV) are all in the decay stage of MLR. According to the observation above, before the appearance of the fuel ring, the fuel was considered to be distributed over the entire surface. Then, at the instant of the ring fire, the fuel layer was thin enough for the Marangoni convection to drive the fuel preferentially to the circumference and the fuel width became narrower. The fuel layer thickness can be calculated based on the residual mass. In order to simplify the analysis, the width is assumed to shrink linearly after the ring-fire formation time.

The results of fuel thickness are shown in Fig. 4. The first dotted line section of every curve represents the uniform coverage of fuel where the fire is burning over the entire fuel surface. The solid line with various symbols means the fuel has formed an annular shape starting from the flame disappearance in the tray center visually. The fuel thickness steadily decreased with the consumption of the fuel mass. The thickness with 0 °C wall temperature was always larger than 20 °C, which indicates that the lower wall temperature leads to larger Marangoni stress that drove the fuel to the wall more forcefully. Further, when the flame goes out, the thickness of the residual fuel layer is slightly larger but not equal to zero. It indicates that the thickness will not decrease to zero until the extinction of the fire. This is because of the formation and shrinkage of the fuel ring during the ring-fire. Of course, further experiments with better observation and more accurate measurement need to be carried out in the future.

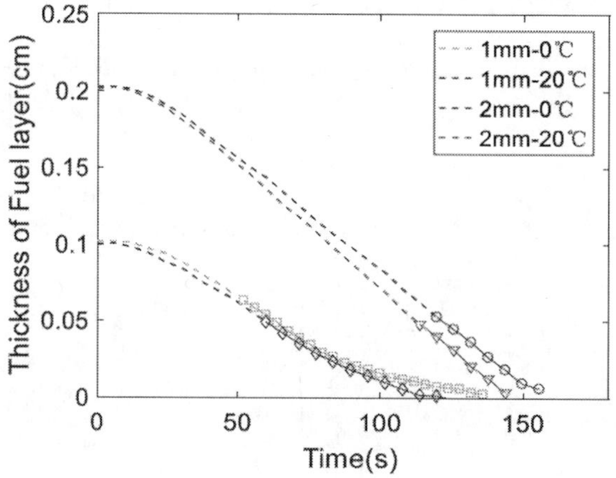

Figure 4: Fuel thickness based on the remaining mass.

4. Conclusions

In this paper, a series of pool fire experiments were carried out to study the effect of the low wall temperature of the tray on the fuel distribution and evolution of the flames in m-Xylene pool fires. m-Xylene with initial thickness of 1 mm and 2 mm were burned in circular trays of 15.5 cm diameter while the wall temperature was fixed at 0 °C and 25 °C.

It was observed that, the burning rate can be divided into three stages according to the MLR data, which includes the pre-steady stage, quasi-steady stage and decay stage. The flame evolution during the burning period showed the following shapes in turn: the uniform burning over the entire surface, burning around the wall with flames merged in the center, complete ring-shaped fire with flames being separated, and weak ring fire before extinction. The ring fire occurred because of radial fuel movement towards the circumference of the tray which at a critical thickness caused the pool fire to be converted to a ring-shaped fire. The ring-fire continued for a longer duration when the temperature of the wall was lower at 0 °C compared to that of 20 °C tests.

5. Acknowledgements

This study was funded by Tsinghua University doctoral student short-term study abroad fund.

6. References

[1] Farmahini Farahani, H., Shi, X., Simeoni, A., Rangwala, A.S., A study on burning of crude oil in ice cavities. Proc. Combust. Inst., 2015. 35(3): p. 2699-2706.
[2] Farahani, H.F., Alva, W.U.R., Rangwala, A.S., Jomaas, G., Convection-driven melting in an n-octane pool fire bounded by an ice wall. Combust. Flame, 2017. 179: p. 219-227.
[3] Ranellone, R.T., Tukaew, P., Shi, X., Rangwala, A.S., Ignitability of crude oil and its oil-in-water products at arctic temperature. Mar. Pollut. Bull., 2017. 115(1-2): p. 261-265.
[4] Farmahini Farahani, H., Jomaas, G., Rangwala, A.S., Effects of convective motion in n-octane pool fires in an ice cavity. Combust. Flame, 2015. 162(12): p. 4643-4648.
[5] Shi, X., Bellino, P.W., Simeoni, A., Rangwala, A.S., Experimental study of burning behavior of large-scale crude oil fires in ice cavities. Fire Safety J., 2016. 79: p. 91-99.
[6] Van Gelderen, L., Malmquist, L.M.V., Jomaas, G., Vaporization order and burning efficiency of crude oils during *in-situ* burning on water. Fuel, 2017. 191: p. 528-537.
[7] Bellino, P.W., Rangwala, A.S., Flynn, M.R., A study of *in-situ* burning of crude oil in an ice channel. Proc. Combust. Inst., 2013. 34(2): p. 2539-2546.
[8] Shi, X., Sahu, A.K., Nair, S., Raghavan, V., Rangwala, A.S., Effect of ullage on burning behavior of small-scale pool fires in a cavity. Proc. Combust. Inst., 2017. 36(2): p. 3113-3120.
[9] Weng, W.G., Kamikawa, D., Fukuda, Y., Hasemi, Y., Kagiya, K., Study on flame height of merged flame from multiple fire sources. Combust. Sci. Technol., 2004. 176(12): p. 2105-2123.
[10] Ely, J.F., Hanley, H.J.M., Prediction of Transport Properties. 2. Thermal Conductivity of Pure Fluids and Mixtures. Industrial & Engineering Chemistry Fundamentals, 1983. 1(22): p. 90-97.
[11] Yumoto, T., Takahashi, A., Handa, T., Combustion Behavior of Liquid Fuel in a Small Vessel Effect of Convective Motion in the Liquid on Burning Rate of Hexane in the Early Stage of Combustion. Combust. Flame, 1977. 30: p. 33-43.

Spring Technical Meeting
Eastern States Section of the Combustion Institute
March 4-7, 2018
State College, Pennsylvania

Liquid-Pool Fire Extinction Characteristics of Aqueous Foams Generated from Fluorine-free Surfactants

Ramagopal Ananth, Spencer Giles, Katherine Hinnant, Xiaohong Zhuang[1], Arthur Snow, James Fleming[2], John Farley*

Chemistry Division, U.S. Naval Research Laboratory, Washington, DC, USA
**Corresponding Author Email: Ramagopal.ananth@nrl.navy.mil*

Abstract: The fluorocarbon surfactants used for generating aqueous film forming foam (AFFF) are extremely effective for suppressing liquid pool fires. However, the fluorosurfactants must be replaced with environmentally friendly surfactants without sacrificing the fire suppression effectiveness. We prepared well-defined, fluorine-free, custom, formulations containing siloxane or hydrocarbon surfactant components. Using the custom formulations, we isolated the effect of the surfactant and related the differences in fuel transport and foam degradation with fire extinction. The benchtop fire extinction measurements show: (1) a direct correlation between fuel-transport rate through a 4-cm foam layer laid on hot n-heptane pool and fire extinction effectiveness, (2) variations in head group lead to dramatic variation in the fuel-blocking and fire extinction time, (3) one formulation based on a siloxane surfactant is better than the extinction times of an ICAO qualified fluorine-free foam, RF6. Molecular dynamics simulations and surfactant synthesis are being performed for well-defined surfactant structures to establish a quantitative relationship with extinction.

Keywords: fire suppression, surfactants, aqueous foams, fluorine-free

1. Introduction

Aqueous Film Forming Foams (AFFF) are used worldwide to fight pool fires since their invention by the U.S. Naval Research Laboratory (NRL) in the 1960s [1]. The perfluoro-octanyl based surfactants contained in AFFF provide rapid fire extinction performance that is critical to prevent weapons "cook-off" [2]. Recently, the perfluoro-octanyl based surfactants have been shown to be environmentally persistent and toxic [3]. Replacing the fluorocarbon surfactant with a fluorine-free surfactant will significantly reduce the environmental impact. However, existing alternatives to AFFF are unable to meet the stringent fire extinction time requirements [4] listed in U.S. DOD's Mil-F-24385F [5]. A perfluorocarbon surfactant tail is both oleophobic and hydrophobic unlike an equivalent hydrocarbon surfactant tail [6], which is only hydrophobic. Siloxane can be used to induce both the oleophobic and hydrophobic character to a fluorine-free surfactant's head and tail. In this paper, we chose siloxane and hydrocarbon surfactants with varying head and tail sizes and measured the surfactant's effect on heptane-pool fire extinction at various foam application rates. We relate the surfactant's effect on extinction with fuel vapor transport through a foam layer covering a heptane pool at 60 C that were measured in separate experiments [7].

[1]ASEE Post Doctoral Associate
[2]NOVA Scientific Inc., Alexandria, VA

2. Methods / Experimental

Commercial AFFF solutions contain numerous proprietary components preventing us from isolating the effect of a surfactant on fire extinction from other components in the solution. Previously, we developed a custom made reference AFFF formulation (RefAFFF) consisting of 0.3 % Capstone 1157 (Chemours Inc.), 0.2 % Glucopon 215UP (BASF Co.), and 0.5 % Diethylene glycol Butyl ether (DGBE, Sigma Aldrich Co.) in distilled water as a tool to evaluate new fluorine-free surfactants [8]. We defined the three components in RefAFFF analytically and showed that the RefAFFF has fire extinction times comparable to the commercial AFFF both in bench-scale and in 28 ft^2 Milspec testing [8].

In this paper, we prepared aqueous solutions of siloxane and five hydrocarbon surfactants at concentrations well above the critical micelle concentrations (CMC). We used the candidate surfactant solutions and the formulations to generate aqueous foams using a sparging method (with air passed through a porous disk, 25-50 μm pore size, immersed in solution at 3-cm depth) described elsewhere [9,10]. We measured the initial expansion ratio (volume of foam divided by volume of liquid in the foam), foam degradation and fuel transport rate through a 4-cm foam layer laid on top of hot (60 C) heptane pool, and fire extinction time for a heptane pool. The fuel temperature was chosen based on our previous measurements of foam-fuel interface temperature [9]. The fire extinction apparatus is shown in Fig. 1 and the other apparatus is described in [7].

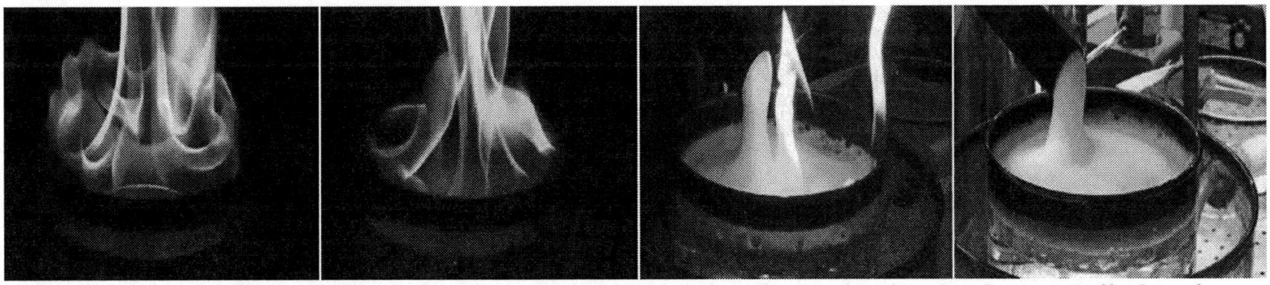

Figure 1. A pool (12-cm dia., n-heptane liquid, 2-cm lip) fire extinction by foam applied at the center at a constant flow rate; from 30-sec pre-burn to extinction.

Fire extinction was conducted on 12-cm n-heptane pool with 2-cm head space to accommodate the foam layer. Foam was applied at a constant flow rate near the center of the pool after a 30 s pre-burn as shown in Fig. 1. Extinction experiments were also performed on a 19-cm n-heptane pool with 1-cm head space and 60 s pre-burn time for high performing surfactant formulations. Foam flow rate was varied by varying the air flow through the sparging disc (25-50 μm pores) from 100 to 4000 mL/min and fire extinction time was measured. No extinction was recorded at extinction time greater than 90 s.

3. Results and Discussion

The surfactants were chosen for their structural differences in tail and head groups and they were not necessarily designed for the purpose of firefighting. We performed fire extinction on a larger (19-cm diameter) pool for Siloxane A formulation. Siloxane A is not viscous unlike RF6 which contains glucoside surfactants and xanthan gum additive. RF6 was qualified by the International Civil Aviation Organization (ICAO) and is sold in Europe, but it does not meet US Milspec because of 2x longer extinction time and a large viscosity of the solution [4]. Both RF6 and Siloxane A formulations underperformed relative to RefAFFF as shown in Fig. 2.

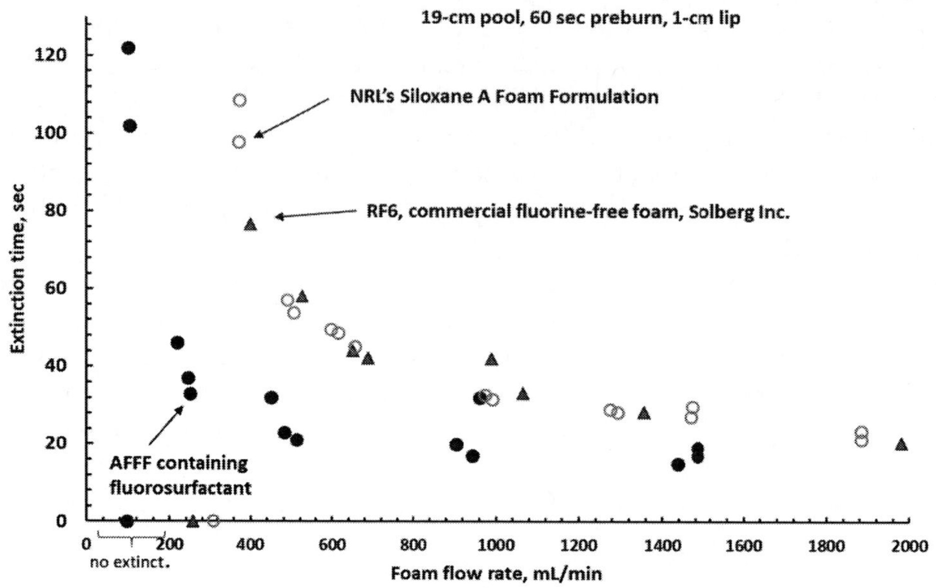

Figure 2. Comparison of fire extinction of NRl's siloxane formulation with RF6, a commercial fluorine-free foam, and AFFF with foam applied at a constant rate at the center of the pool.

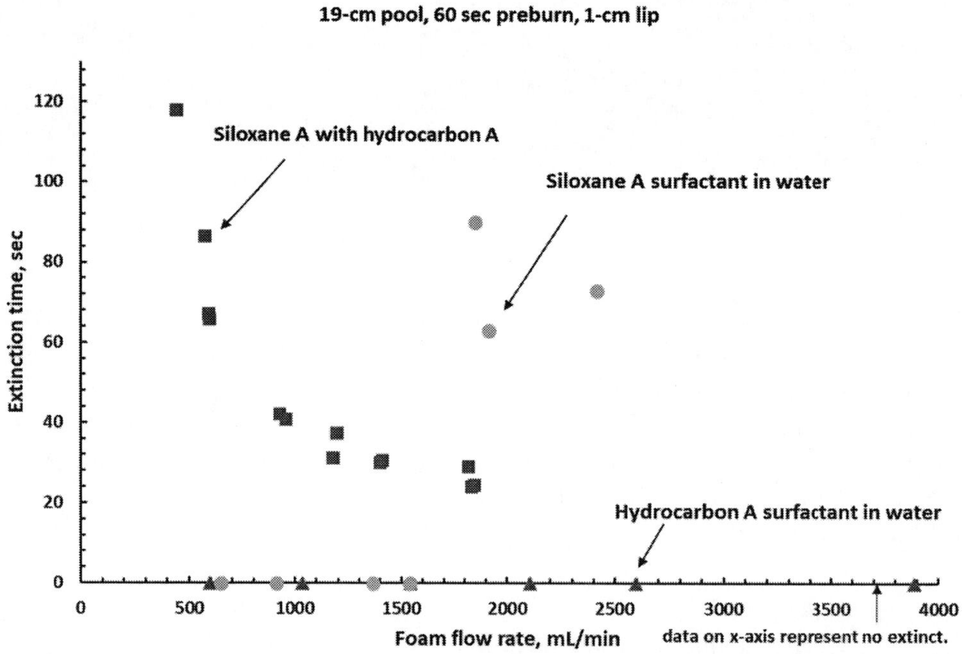

Figure 3. Synergistic effects of siloxane surfactant with a hydrocarbon surfactant on pool fire extinction.

We have discovered that there is a synergism between a hydrocarbon surfactant and a siloxane surfactant in terms of fire extinction. Fig. 3 shows that the fire extinction performance of NRL's siloxane formulation is better than that of the individual components in the formulation.

This is a key finding that has the potential to improve siloxane surfactant performance further. The hydrocarbon surfactant solution alone did not extinguish the fire on a 19-cm pool even at 3900 mL/min in 60 s. The superior performance is due to increased transport and degradation times.,

We measured the fuel transport rate through a 4-cm layer foam placed on top of hot (60 C) heptane pool with time [10]. We determined the time taken for the heptane vapor concentration at the foam surface to reach the lower flammability limit of 1 volume % for each of surfactants shown in Figure 4. So, larger the transport time, slower the fuel flux through the foam. Fig. 4 shows the foam flow rate needed for fire extinction time of 30 s relative to the flow rate for AFFF versus the transport time to achieve the lower flammability limit relative to that for AFFF. The smaller the foam flow rate needed for a 30 s extinction time, the more efficient is extinction. Fig. 4 shows a power law correlation between the relative foam flow at 30 s extinction and the relative transport rate. The scatter among the data points is partially attributed to significant and uneven degradation of the foam layer by the heptane fuel during the fuel flux measurements; some of the foams degraded very quickly before fuel flux could be measured. The correlation shows that the NRL's siloxane foam degradation has to be reduced by a factor of 4 to match the extinction performance of AFFF.

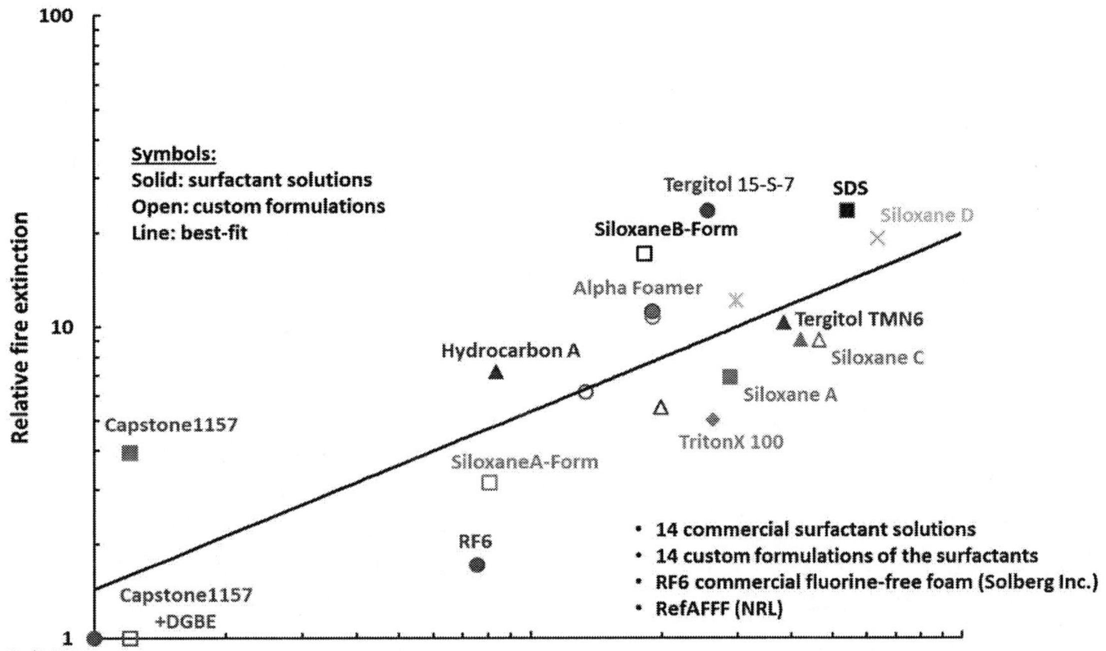

Figure 4. Fire extinction versus fuel transport through a 4-cm foam layer covering hot (60 C) heptane pool; extinction is expressed as minimum foam flow rate needed to achieve fire extinction time of 30 s relative to that of AFFF and fuel transport is expressed as time needed to achieve lower flammability limit (1 vol %) of n-heptane vapor at the foam surface relative to that for AFFF.

4. Conclusions

We have shown a potential relationship between surfactant structure and fire extinction time for several siloxane and hydrocarbon surfactant candidates as alternatives to AFFF; many were not necessarily designed for firefighting. We found that a siloxane surfactant exhibited synergistic extinction when combined with a hydrocarbon surfactant that matches the extinction times of a commercially available RF6 fluorine-free, firefighting, foam at different foam application rates. Fire extinction time, foam degradation, and fuel transport of NRL's siloxane formulation were much smaller than for the individual components. We also show a correlation between extinction performance and transport of heptane through a foam layer placed on a hot pool. We are performing surfactant synthesis and molecular dynamics simulations of the air-water and heptane-water interfaces to calculate the effect of surfactant's structure on the interfacial properties relevant to fire extinction.

5. Acknowledgements

This work is supported by the Office of Naval Research (ONR) through the Naval Research Laboratory and by Strategic Environmental Research and Development Program (SERDP) under grant WP2739.

6. References

[1] R.L. Turve and E.J. Jablonski, "Compositions and Methods for Fire Extinguishment and Prevention of Flammable Vapor Release", U.S. Patent 3, 258, 423, June 28, 1966

[2] H.W. Carhart, J.T. Leonard, R.L. Darwin, R.E. Burns, J.T. Hughes, E.J. Jablonski, "Aircraft Carrier Flight Deck Firefighting Tactics and Equipment Evaluation Tests", NRL Memorandum Rpt., 5952, February, 1987

[3] C.A. Moody and J.A. Field, "Perfluorinated surfactants and the environmental implications of their use in fire-fighting foams". Environmental Science and Technology, **34**, (2000) 3864-3869

[4] B.A. Williams, T. Murray, C. Butterworth, Z. Burger, R. Sheinson, J.W. Fleming, C. Whitehurst, and J. Farley, "Extinguishment and Burnback Tests of Fluorinated and Fluorine-free Foams with and without Film Formation", Suppression, Detection and Signaling Research and Applications (SUPDET 2011), Orlando, FL, March 22-25

[5] Military Specification, "Fire Extinguishing Agent, Aqueous Film-forming Foam (AFFF) Liquid Concentrate, for Fresh and Sea Water", MIL-F-24385F, January 7, 1992

[6] M.P. Kraft and J.G. Reiss, "Chemistry, physical chemistry, and uses of molecular fluorocarbon-hydrocarbon diblocks, triblocks, and related compounds for self-assembled colloid and interface engineering", Chemical Reviews, 109, (2009) 1714-1792

[7] K.M. Hinnant, A.W. Snow, S. Giles, R. Ananth, "Evaluating foam degradation and fuel transport rates through novel surfactant firefighting foams for the purpose of AFFF perflurocarbon replacement", Eastern States Section of the Combustion Institute Spring Technical Meeting, (2018), State College, PA, March 4-7, Paper# 36FI-0047

[8] K.M Hinnant, A.W. Snow, J. Farley, S. Giles, R. Ananth, "Development of an analytical AFFF formulation for the evaluation of alternative surfactants", 10th U.S. National Combust. Meeting, College Park, MD, April 23-26 (2017), 1C09

[9] M. W. Conroy, J.W. Fleming, R. Ananth, "Surface cooling of a pool fire by aqueous foams", Comb. Sci. Tech., 189, 806-840 (2017)

[10] R. Ananth, S. Mott, M. Waheed, M. Epstein, J.M. Smith, M.W. Conroy, J.W. Fleming, "Bench-scale apparatus for studying pool fire extingushment by class B foams", Eastern States Section of the Combustion Institute Fall Technical Meeting, (2013), Clemson, SC, October, Paper# 085FR-0045

Spring Technical Meeting
Eastern States Section of the Combustion Institute
March 4-7, 2018
State College, Pennsylvania

Mechanism Development of Aqueous Hydroxylammonium Nitrate under Thermal Decomposition Conditions

Kaiqiang Zhang[1] *and Stefan T. Thynell*[1,*]

Department of Mechanical and Nuclear Engineering
The Pennsylvania State University
University Park, Pennsylvania, 16802, U.S.A.
[*]*Corresponding Author Email: thynell@psu.edu*

Abstract A detailed chemical mechanism was recently developed for aqueous hydroxylammonium nitrate (HAN), a potential green propellant material, based on theoretical quantum mechanical (QM) calculations. The ωB97X-D density functional theory (DFT) with the SMD solvation model was implemented to optimize the molecular geometries, locate transition states, and compute the solution-phase free energies. The mechanism includes the nitration reactions between hydroxylamine and nitric acid, the subsequent nitrosation reactions between hydroxylamine and HONO, and the autocatalytic steps of H-abstraction by NO_2. To examine the mechanism, the kinetic modeling was performed with rate constants predicted by the conventional transition-state theory (CTST) and with consideration to effects from diffusion. The kinetic modeling predicted that the activation energy for 0.1 m HAN is 109 kJ/mol, compared to the experimentally reported 103±21 kJ/mol. NO_2, the major H-abstraction agent, only evolves substantially after over 99.9% of the hydroxylamine is consumed. The kinetic simulation also supports that the decomposition can be greatly accelerated with excess nitric acid, but largely suppressed with insufficient acid or excess hydroxylamine.
Key words: ***hydroxylammonium nitrate, DFT, kinetic modeling***

1. Introduction

Hydroxylammonium nitrate (HAN) has been of interest for a wide variety of propulsion applications. HAN was first considered as the oxidizer in several potential liquid-gun propellants [1, 2]. In rocket propulsion, HAN has recently been considered as a potential alternative to hydrazine in rocket thrusters [3, 4]. The use of HAN-based propellants would reduce the toxicity and the polluting emissions while maintaining the high energy density and specific impulse, and thus improve the overall safety and performance.

HAN is highly soluble in water, and most of its applications use HAN-water solutions. Many studies have attempted to characterize the thermal decomposition, ignition and combustion behavior of aqueous HAN using various techniques, including among others, strand burner [5, 6], flow reactor [7, 8], and fast-thermolysis apparatus [9]. The most commonly-cited global reaction for HAN decomposition is given by:

$$4[NH_3OH]NO_3 = 3N_2O + 2HNO_3 + 7H_2O \qquad (R1)$$

Other gaseous products were also reported under certain experimental conditions, including NO, NO_2, N_2 and possibly O_2, and the composition of the products depends on concentration and temperature conditions. The kinetic parameters also depend on the initial concentration.

According to Schoppelrei and Brill [7], the global activation energy (E_a) is 129±29 kJ/mol for 0.87-1.52 m solutions, whereas at higher concentration (1.58-1.74 m), E_a drops to 66±8 kJ/mol but with a larger pre-exponential factor (A). The decomposition rate was reported to decrease again in highly concentrated solutions [9].
The objective of this work is to develop a detailed mechanism for the decomposition of HAN solutions and to use the mechanism to explain experimental observations. The methods include quantum mechanical (QM) calculations, estimation of rate constants using transition state theory (TST), and kinetic modeling. A brief description of the elementary reactions will be presented, and the kinetic modeling results of dilute solutions under thermal decomposition conditions will be discussed.

2. Computational Methods

The proposed mechanism, which is composed of over 300 elementary reactions, was developed based on the QM molecular calculations using the Gaussian09 (g09) software package [10]. The geometries of all the involved stable species and transition states (TSs) were optimized using the ωB97X-D density functional theory (DFT) [11] with the implementation of the SMD implicit solvation model [12]. The 6-311++G** basis set was used. The solution-phase Gibbs free energies were obtained using the same method. All the reactions were considered reversible, and the rate constants were estimated using the thermodynamic formulation of the conventional transition-state theory (CTST):

$$k = \sigma\Gamma(T)\cdot k_bT/h\cdot \exp(-\Delta G^{\ddagger}/R_uT) \qquad (1)$$

The diffusion effect was considered for bimolecular reactions by applying the quasi steady-state assumption to the encounter complexes and estimating the diffusion rates with the Stokes-Einstein relation. The evolution of the sample's composition was predicted by solving the kinetic system composed of ODEs for all the stable species:

$$\mathrm{d}c_i/\mathrm{d}t = \dot{w}_i \qquad (2)$$

The sample temperature in the flow reactor experiment was estimated based on the extent of reaction of R1 and assuming a one-dimensional laminar flow:

$$\rho c_p\cdot dT/dt = Q\cdot \dot{w}_{N2O}/3 - 4h(T-T_w)/D \qquad (3)$$

where the reaction heat (Q) of R1 predicted by SMD-ωB97X-D is 193.0 kcal/mol and the specific heat (c_p) was estimated by the value of water (4.5 kJ/(kg·K)).

3. Results and Discussion

3.1 Mechanism Description

The mechanism considers proton transfer from NH_3OH^+ to NO_3^-. In dilute solutions, the process is most likely realized via fast acid-base equilibria with the participation of water:

$$NH_3OH^+ + H_2O = NH_2OH + H_3O^+ \qquad (R2)$$
$$HNO_3 + H_2O = NO_3^- + H_3O^+ \qquad (R3)$$

The acid dissociation constants (pK_as) were reported to be -1.5 and 5.94 [13] for NH_3OH^+ to HNO_3, respectively, which implies that the free energy change for the initial proton transfer reaction is about 10.1 kcal/mol, compared to the SMD-ωB97X-D predicted 10.2 kcal/mol.
The nitration reactions, that is, addition of a nitro (-NO_2) group to hydroxylamine, lead to the first formation of HONO. Potential nitration agents in HAN solution include HNO_3, NO_2^+, N_2O_5

($O_2NO\text{-}NO_2$), and after HONO formation, N_2O_4 ($ONO\text{-}NO_2$ and $O_2N\text{-}NO_2$) and N_2O_3 (ON-ONO and $O_2N\text{-}NO_2$). The following reactions are considered potentially important in determining the kinetics of the initial stage of decomposition:

$$HNO_3 + HNO_3 = N_2O_5 + H_2O \quad (\Delta G_f^{\ddagger} = 23.0\ \&\ \Delta G_r = 15.9\ \text{kcal/mol}) \quad (R4)$$
$$2HNO_3 + H_2O = N_2O_5 + 2H_2O \quad (27.2;\ 15.9) \quad (R5)$$
$$HNO_3 + H_3O^+ = NO_2^+ + 2H_2O \quad (16.2;\ 7.6) \quad (R6)$$
$$NO_2^+ + NO_3^- = N_2O_5 \quad (\Delta G_r = -0.5\ \text{kcal/mol}) \quad (R7)$$
$$NH_3OH^+ + N_2O_5 = NH_3ONO_2^+ + HNO_3 \quad (8.3;\ -6.1) \quad (R8)$$
$$NH_3OH^+ + NO_2^+ + H_2O = NH_3ONO_2^+ + H_3O^+ \quad (12.8;\ 2.2) \quad (R9)$$
$$NH_2OH + NO_2^+ = NH_2O(H)NO_2^+ \quad (\Delta G_r = -20.6\ \text{kcal/mol}) \quad (R10)$$
$$NH_2OH + NO_2^+ = HONH_2NO_2^+ \quad (\Delta G_r = 3.1\ \text{kcal/mol}) \quad (R11)$$

The nitration products are calculated to have exceptionally negative pK_{a}s (<-10). Accordingly, they would first undergo deprotonation before decomposition to HONO and HNO.

$$HONH_2NO_2^+/\ NH_2O(H)NO_2^+/\ NH_3ONO_2^+ \rightarrow HONO + HNO + H^+ \quad (R12)$$

The subsequent nitrosation reactions of hydroxylamine by HONO and its derivatives (NO_2^+, $ONO\text{-}NO_2$, ON-ONO, ON-NO, etc.) compose the major scavenging pathways, according to several previously proposed simplified mechanisms. In our detailed mechanism, the following reactions are considered to best describe the scavenging process:

$$HNO_3 + HONO = ONONO_2 + H_2O \quad (8.7;\ -0.2) \quad (R13)$$
$$HONO + H_3O^+ = NO^+ + 2H_2O \quad (3.4;\ -4.2) \quad (R14)$$
$$NO^+ + NO_3^- = ONONO_2 \quad (\Delta G_r = -4.9\ \text{kcal/mol}) \quad (R15)$$
$$NH_3OH^+ + ONONO_2 = NH_3ONO^+ + HNO_3 \quad (6.3;\ -2.0) \quad (R16)$$
$$NH_3OH^+ + NO^+ + H_2O = NH_3ONO^+ + H_3O^+ \quad (19.1;\ -4.9) \quad (R17)$$
$$NH_2OH + NO^+ = NH_2O(H)NO^+ \quad (\Delta G_r = -0.6\ \text{kcal/mol}) \quad (R18)$$
$$NH_2OH + NO^+ = HONH_2NO^+ \quad (\Delta G_r = -11.0\ \text{kcal/mol}) \quad (R19)$$
$$HONH_2NO^+/\ NH_2O(H)NO^+/\ NH_3ONO^+ \rightarrow HONNOH + H^+ \rightarrow N_2O + H_3O^+ \quad (R20)$$

HNO can either dimerize and then decompose into N_2O and H_2O, or react with NO_2^+ or NO^+ under strongly acidic conditions:

$$HNO + HNO \rightarrow HONNOH \rightarrow N_2O + H_2O \quad (R21)$$
$$HNO + NO_2^+ + H_2O = ONNO_2 + H_3O^+ \quad (\Delta G_r = -40.0\ \text{kcal/mol}) \quad (R22)$$
$$HNO + NO^+ + H_2O = ONNO + H_3O^+ \quad (\Delta G_r = -9.8\ \text{kcal/mol}) \quad (R23)$$

The homolytic dissociation of N_2O_x (x = 2 to 4) is expected to be the main source of NO and NO_2.

$$N_2O_4\ /\ N_2O_3\ /\ N_2O_2 \rightarrow NO,\ NO_2 \quad (R24)$$

Several previously proposed mechanisms attributed the autocatalytic behavior of HAN to the nitration effect of $ONO\text{-}NO_2$ [8, 14]:

$$NH_3OH^+ + ONONO_2 = NH_3ONO_2^+ + HONO \quad (29.2;\ 13.0) \quad (R25)$$

Our theoretical calculation predicts that R25 has a much higher free energy barrier compared to the corresponding nitrosation reactions (R16), and thus it is not kinetically plausible. The other potential pathway was considered, which is composed of a series of H-abstraction reactions by NO_2:

$$NH_2OH + NO_2 = NH_2O + HONO \qquad (13.1; -10.9)\ (R26)$$
$$NH_2O + NO_2 = HNO + HONO \qquad (22.2; -6.7)\ (R27)$$
$$HNO + NO_2 = NO + HONO \qquad (12.1; -27.2)\ (R28)$$

The overall mechanism is schematically summarized in Fig. 1.

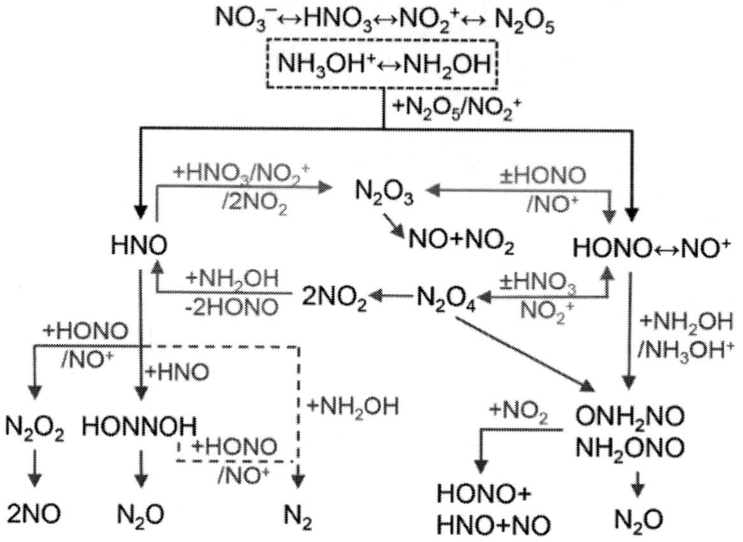

Figure 1: Schematic illustration of the mechanism developed for HAN decomposition.

3.2 Kinetic Simulation Results

The results for 0.1 m solution are presented in Fig. 2, along with the experimental data from Ref. [8]. Figure 2(a) shows the concentration evolution of hydroxylamine and N_2O at 493 K. The final composition of the products is very close to the stoichiometry in R1, with the concentration of the main minor species (NO) lower than 10^{-3} m. The maximum temperature rise is predicted to be smaller than 0.2 K. In the calculated profiles, a slight acceleration can be observed in the middle stage of decomposition. It is most likely caused by the rise of acidity, which leads to an increase in the rates of both the nitration and nitrosation steps. The autocatalytic effect by NO_2 is expected to be very low, considering the extremely low NO_2 concentration ($< 10^{-10}$ m) in most of the decomposition period. NO_2 only evolves notably ($> 10^{-6}$ m) at the very end of decomposition when over 99.99% of the initial hydroxylamine has been consumed. Figure 2(b) shows the first-order plots of the consumption of HAN up to 60% at various temperature conditions, from which the kinetic parameters in Arrhenius form can be derived. Our predicted activation energy is about 109 kJ/mol, compared to experimentally derived 103±21 kJ/mol [8].

In practical application, aqueous HAN is often not mixed equi-molar amounts of hydrylamine and acid in order to improve long-term stability. Figure 3 shows the calculated decomposition profiles for cases with doubled or half HNO_3 concentration, compared with the equi-molar HAN solution. Doubling HNO_3 concentration would reduce the half-life of hydroxylamine by 94%, which implies that the first-order rate increases by roughly 16 times, compared to the experimentally reported 10±3 times. This acceleration can be attributed to the multi-order characteristic of the initial nitration steps (R4-R6) on HNO_3, as well as a greatly increased acidity. For the half-HNO_3 solution, the decomposition is predicted to be largely suppressed by the high basicity, with the half-life of hydroxylamine prolonged by over 10^2 times, even though R1 suggests that essentially HAN has 50% excess HNO_3. By contrast, the calculated half-life of hydroxylamine in the solution of 0.1:0.05 NH_3OH^+: NO_3^- only increases by 130%.

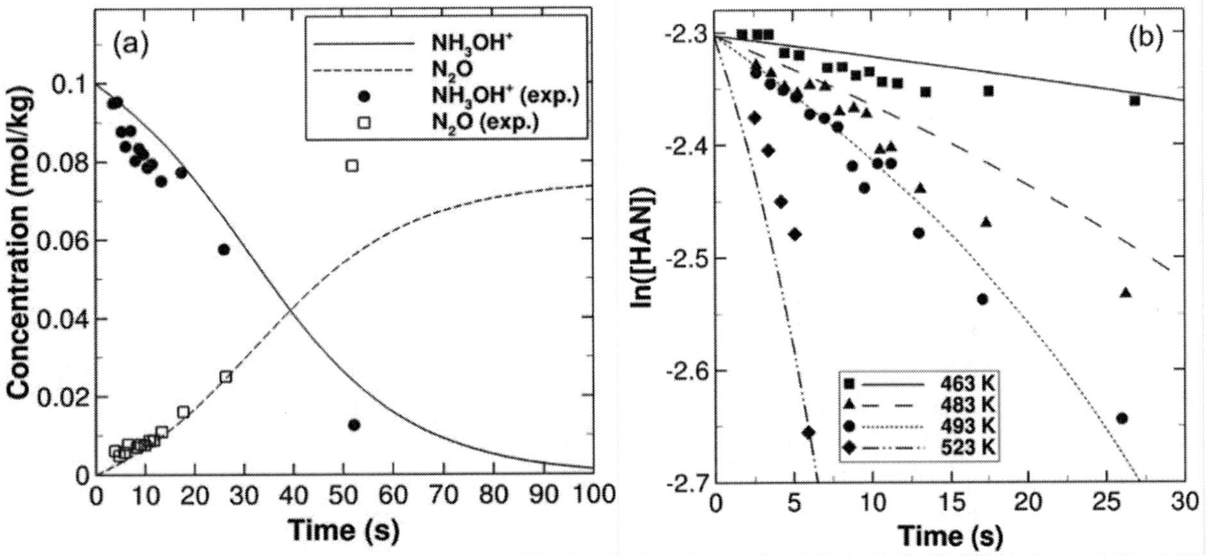

Figure 2: Results of kinetic modeling on the decomposition of 0.1 m HAN: (a) Evolution of hydroxylamine and N_2O at 493 K. (b) The consumption of HAN vs. time at various temperature conditions. Experimental data (dots) are from Ref. [9].

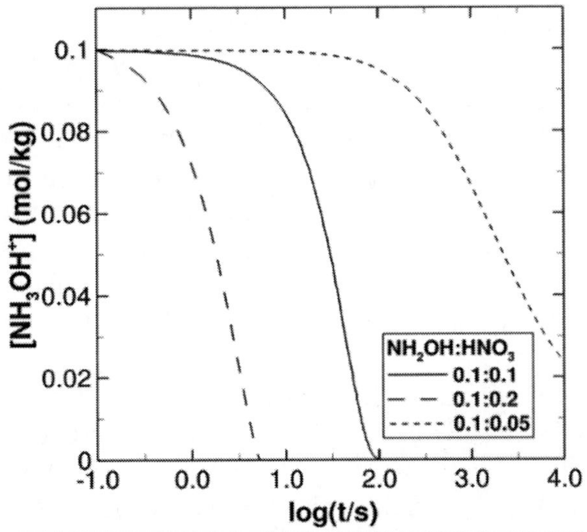

Figure 2: Results of kinetic modeling on the decomposition of 0.1 m HAN, 0.1 m HAN plus 0.1 m HNO_3, and 0.05 m HAN plus 0.05 m NH_2OH. Temperature is 493 K.

4. Conclusions

The newly developed mechanism reveals that the decomposition of aqueous HAN starts from proton transfer between two dominant ions (NH_3OH^+ and NO_3^-). HONO evolves from the nitration steps, and the majority of hydroxylamine is consumed via the nitrosation reactions by HONO derivatives, leading to the formation of N_2O. The kinetic modeling predicts that the decomposition of dilute HAN solution roughly follows the stoichiometry of R1. The predicted first-order activation energy is 109 kcal/mol. NO_2 only evolves notably near the complete consumption of hydroxylamine, and thus the detection of NO_2 can be regarded as a sign of the end of the decomposition. The kinetic modeling also predicts that the solution acidity has a strong impact on the decomposition rate, which can be accelerated by excess HNO_3 but suppressed by excess NH_2OH.

5. Acknowledgements

This research was funded by Air Force Office of Scientific Research under grant number FA9550-13-1-0004.

6. References

[1] N. Klein, Liquid propellants for use in guns - a review, Report No. BRL-TR-2641, US Army Ballistic Research Laboratory, Aberdeen Proving Ground, MD, USA, 1985.
[2] M.M. Decker, N. Klein, E. Freedman, C.S. Leveritt, J.Q. Wojciechowski, HAN-based liquid gun propellants: physical properties, Report No. BRL.-TR-2864, US Army Ballistic Research Laboratory, Aberdeen Proving Ground, MD, USA, 1987.
[3] R. Jankovsky, HAN-based monopropellant assessment for spacecraft, 32nd AIAA/ASME/SAE/ASEE Joint Propulsion Conference and Exhibit (1996), AIAA-paper 96-2893.
[4] T. Katsumi, H. Kodama, T. Matsuo, H. Ogawa, N. Tsuboi, K. Hori, Combustion characteristics of a hydroxylammonium nitrate based liquid propellant. Combustion mechanism and application to thrusters, Combust. Explos. Shock Waves 45 (2009) 442.
[5] S.R. Vosen, Concentration and pressure effects on the decomposition rate of aqueous hydroxylammonium nitrate solutionsa, Combust. Sci. Technol. 68 (1989) 85-99.
[6] S.R. Vosen, Hydroxylammonium nitrate-based liquid propellant combustion-interpretation of strand burner data and the laminar burning velocity, Combust. Flame 82 (1990) 376-388.
[7] J.W. Schoppelrei, M.L. Kieke, T.B. Brill, Spectroscopy of hydrothermal reactions. 2. Reactions and kinetic parameters of $[NH_3OH]NO_3$ and equilibria of $(NH_4)_2CO_3$ determined with a flow cell and FT Raman spectroscopy, J. Phys. Chem. 100 (1996) 7463-7470.
[8] J.W. Schoppelrei, T.B. Brill, Spectroscopy of hydrothermal reactions. 7. Kinetics of aqueous $[NH_3OH]NO_3$ at 463−523 K and 27.5 MPa by infrared spectroscopy, J. Phys. Chem. A 101 (1997) 8593-8596.
[9] H. Lee, S. Thynell, Confined rapid thermolysis/FTIR spectroscopy of hydroxylammonium nitrate, 33rd AIAA/ASME/SAE/ASEE Joint Propulsion Conference and Exhibit (1997), AIAA-paper 97-3232.
[10] M.J. Frisch, G.W. Trucks, H.B. Schlegel, G.E. Scuseria, M.A. Robb, J.R. Cheeseman, G. Scalmani, V. Barone, B. Mennucci, G.A. Petersson, et al., Gaussian 09, Revision D.03, Gaussian, Inc., Wallingford, CT, USA, 2009.
[11] J.-D. Chai, M. Head-Gordon, Long-range corrected hybrid density functionals with damped atom-atom dispersion corrections, Phys. Chem. Chem. Phys. 10 (2008) 6615-6620.
[12] A.V. Marenich, C.J. Cramer, D.G. Truhlar, Universal solvation model based on solute electron density and on a continuum model of the solvent defined by the bulk dielectric constant and atomic surface tensions, J. Phys. Chem. B 113 (2009) 6378-6396.
[13] W.M. Haynes, CRC Handbook of Chemistry and Physics, CRC Press, Boca Raton, 2009.
[14] J.C. Oxley, K.R. Brower, Thermal decomposition of hydroxylamine nitrate, Proc. SPIE Int. Soc. Opt. Eng. 0872 (1988) 63-70.

Development of gas-phase reaction mechanism for ammonium perchlorate using quantum mechanics calculations

*Tanusree Chatterjee and Stefan T. Thynell**

Department of Mechanical and Nuclear Engineering
The Pennsylvania State University
University Park, Pennsylvania, 16802, U.S.A.
**Corresponding Author Email: thynell@psu.edu*

Abstract - The chemical kinetics involved during the initiation and subsequent reactions of ammonium perchlorate (AP) in gas phase is studied using quantum mechanics (QM) based calculations. In this study, a gas-phase reaction mechanism is formulated where-in several new initiation steps are proposed between NH_3 and $HOCLO_3$. The developed reaction mechanism is further compared with existing gas-phase mechanisms developed by Beckstead and his group at Brigham Young University, and additional reactions proposed by Lin and his group at Emory University. Here-in, quantum mechanics calculations were performed using Gaussian 09 program package to identify reaction pathways via transition state theory (TST). Searches for transition states were performed using density functional theory (DFT) and low-level basis-sets such as B3LYP/6-31G(d). Once a transition state with a single negative frequency was identified, intrinsic reaction coordinate (IRC) calculations were then performed to link the reactants to the products. Higher level calculations were then performed using the compound method G4(MP2) to compute the thermodynamic properties of species at equilibrium or at transition states. This work elucidates how QM can contribute to further our understanding of the complex chemical kinetics involved during the decomposition of AP in the gas phase.

Key Words: Ammonium perchlorate, decomposition, gas phase, reaction mechanism

1. Introduction

Ammonium perchlorate (AP) is one of the most common oxidizers used in composite propellants for propulsion of rockets. AP contains ~30% excess of oxygen, which allows combustion products of monopropellant AP to react readily with the fuel. Furthermore, it has been observed that the burn rate of propellants can be controlled by varying the particle size distribution of AP [1]. This unique ability of AP provides greater control of propellant burn rate and makes it even more desirable among the various oxidizers used today to propel rockets. To understand the complex chemical kinetics involved during the combustion of AP, numerous experimental and numerical studies have been undertaken. However, due to the extremely small scale of the AP flame structure, it has been a challenge to measure the species profiles in the premixed gas-phase flame. Hence, minimal amount of experimental data pertaining to species concentrations and temperature are available [2]. Therefore, numerical simulations of the flame can be utilized in order to understand the AP combustion characteristics. During the past decades, several numerical models were developed to study AP combustion. Among which the Beckstead-Derr-Price (BDP) model [3] developed in 1970 provides the most widely accepted theoretical picture of AP combustion. However, in this model Beckstead et al. [2] used 1 global reaction to model the kinetics involved. Gradually over the years in order to develop better theoretical understanding of AP combustion, models shifted to using detailed gas-phase mechanism. In 1982, based on experimental data Ermolin et al. [4] developed the first detailed AP gas-phase mechanism consisting of 80 reactions and 24 species. The Ermolin mechanism served as the basis for the development of more refined gas phase mechanism developed by Jing et al. in 1998 [5]. Jing et al. [5] were able to accurately predict burn rate, temperature and species concentrations using their gas-phase mechanism. However, in recent years, the accuracy of several kinetic parameters of Ermolin gas-phase mechanism has been questioned. Furthermore, when the gas-phase mechanism by Jing et al. was incorporated into BYU combustion model, the predicted burn rates and final temperature did not match well with the experimental data [6]. Hence, it is evident that more rigorous theoretical based calculations are needed to improve the accuracy of kinetic parameters in order to predict the burning rates and final temperature better. Lin et al. [7] has performed numerous ab-initio studies pertaining to chlorine containing species. These studies were previously incorporated by Gross et al. [6] in their gas phase mechanism to study the combustion characteristics of AP and have shown significant improvement in prediction of burning rate and temperature. In this study, using quantum mechanics based calculations we have identified several new reactions involved during the initial phase of AP decomposition.

2. Computational Details

Gaussian 09 quantum chemistry software package [8] was used for carrying out all calculations. Searches for transition states were typically performed using density functional theory (DFT) and low-level basis sets, such as B3LYP/6-31G(d). Once a transition state was identified with its single negative frequency, intrinsic reaction coordinate calculations were performed to link the reactants to the products [9, 10]. Higher-level calculations were then performed using the G4(MP2) compound method for both the equilibrium species with only real vibrational frequencies, or transition state structures with a single negative frequency [11,12]. Here, the changes in the thermodynamic properties, such as enthalpy and free energy, were directly computed by the G4(MP2) compound method. In addition, these calculations were performed for a temperature of 298.15 K The thermodynamic properties of all the species are calculated using both G4(MP2) and B3LYP/6-311+G(3df,2p). Additionally, spin-forbidden reactions arise in the reaction mechanism development due to the Born-Oppenheimer assumption, and they require special attention. Such attention will be considered in the future.

3. Results and Discussion

Zhu and Lin [13] performed first-principles calculations to study the sublimation/decomposition of AP crystals to form NH_3 and $HOClO_3$ in the gas phase. Their calculations have shown that the enthalpy barrier involved in the process is ~47 kcal/mol. NH_3 and $HOClO_3$ then undergo a series of reactions to form final products (O_2, N_2, HCl and H_2O). In the mechanism used by Gross et al. (ref) to predict the combustion characteristics of AP, the only initiation reaction involving NH_3 and $HOClO_3$ is an unimolecular bond-breaking reaction of $HOClO_3$ dissociating into OH and ClO_3 having forward enthalpy barrier of 39.1 kcal/mol. No additional reactions involving NH_3 and $HOClO_3$ are considered. However, the enthalpic barrier is about 15 kcal/mol higher for the unimolecular bond-breaking process, suggesting that additional initiation reactions must be identified. In this study, based on QM calculations we propose 6 new initiation steps involving NH_3 and $HOClO_3$. The details of these reactions are listed in Table 1 and the transition state structures are shown in Fig. 1. It should be noted that Chen and McQauid have also identified additional initiation reactions between $HOClO_3$ and NH_3, but products are not listed.[14]

Table 1. Frequency of transition state (ν in cm^{-1}), enthalpy (ΔH_f) and Gibbs free energy (ΔG_f) barrier in the forward direction, enthalpy (ΔH_r) and Gibbs free energy of reaction (ΔG_r) for newly found initiation reactions. (Enthalpy and Gibbs free energies are in kcal/mol)

No.	Reaction	ν	ΔH_f	ΔG_f	ΔH_r	ΔG_r
R1	$NH_3+HOClO_3 \leftrightarrow NH_3O+HOClO_2$	386	46.6	56.4	23.1	21.7
R2	$NH_3+HOClO_3 \leftrightarrow H_2O+NH_2OClO_2$	656	45.5	55.5	-17.8	-18.9
R3	$NH_3+HOClO_3 \leftrightarrow$ trans-$NH_2OH+HOClO_2$	348	52.7	62.4	-1.6	-3.3
R4	$HOClO_3+HOClO_3 \leftrightarrow O_3ClOClO_3+H_2O$	137	49.5	60.7	7.7	8.4
R5	$HOClO_3+HOClO_3+H_2O \leftrightarrow O_3ClOClO_3+2H_2O$	151	38.5	60.2	7.7	8.4
R6	$HOClO_3+HOClO_3+NH_3 \leftrightarrow O_3ClOClO_3+H_2O+NH_3$	155	35.3	56.8	7.7	8.4

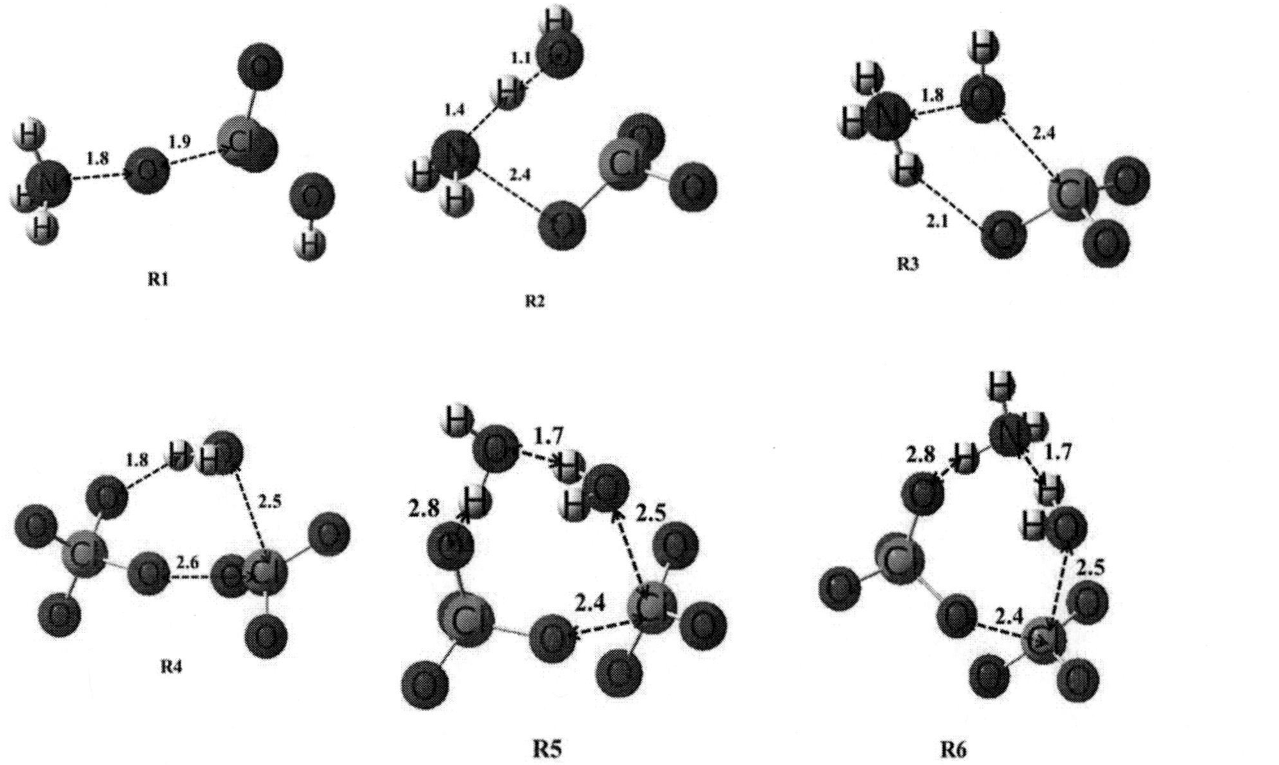

Fig.1. Molecular structure of transition states for 6 initiation steps (**R1-R6**) listed in Table 1. (Distances between atoms are in Angstrom)

From Table 1, it can be observed that QM calculations suggest that NH_3 and $HOClO_3$ react in the gaseous phase via 3 different ways. Among these 3 reactions, while the forward enthalpy barrier of reaction **R3** is similar to that of $HOClO_3$ bond-breaking reaction, reactions **R1** and **R2** have forward enthalpy barrier in the range of ~45-46 kcal/mol, which is ~7-8 kcal/mol less. Hence, it is evident that the newly found initiation reactions using QM calculations play important role during the initial phase of AP decomposition. Apart from **R1**, **R2** and **R3**, $HOClO_3$ can also undergo self-reaction to form H_2O and higher molecular weight ClO containing intermediate species, $O_3ClOClO_3$ via reaction **R4**. It is to be noted that in reaction **R5** and **R6**, H_2O and NH_3 act as a proton transfer agent in the formation of H_2O from $HOClO_3$.The forward enthalpy barrier of **R5** and **R6** is ~35 kcal/mol, which is significantly lower (~18 kcal/mol less) than that of unimolecular $HOClO_3$ bond-breaking reaction. Furthermore, since during the initial phase of AP decomposition, NH_3 concentration is significantly high, **R6** plays a major role. During the initiation steps (**R1-R6**), several important intermediates ($HOClO_2$, $O_3ClOClO_3$, NH_3O) are formed which react further with NH_3 or $HOClO_3$ to produce final products such as HCl, O_2 and H_2O. The details of these reactions are listed in Table 2 below. It can be observed from Table 2

Table 2. Details of newly found reactions producing final product, H_2O, O_2 and HCl

No.	Reaction	ν	ΔH_f	ΔG_f	ΔH_r	ΔG_r
R7	NH_3+$HOClO_2$↔NH_3O+HOClO	520	50.5	60.2	30.9	30.5
R8	NH_3+HOClO↔NH_3O+HOCl	428	37.4	47.4	1.1	1.3
R9	NH_3+HOCl↔NH_3O+HCl	525	37.8	48.0	21.0	22.5
R10	NH_3O+NH_3↔trans-NH_2OH+NH_3	1428	0.8	12.2	-24.7	-25.1
R11	trans-NH_2OH+$HOClO_3$↔$HONH_2O$+$HOClO_2$	379	42.2	52.5	0.4	-0.9
R12	trans-NH_2OH+$HOClO_2$↔$HONH_2O$+HOClO	454	30.4	42.2	8.2	7.8
R13	trans-NH_2OH+HOClO↔$HONH_2O$+HOCl	397	32.7	43.5	-20.7	-20.5
R14	trans-NH_2OH+HOCl↔$HONH_2O$+HCl	548	33.2	44.0	- 1.8	-0.3
R15	$HONH_2O$↔H_2O+HNO	955	15.2	15.0	-24.1	-34.9
R16	$HOClO_3$+$HOClO_2$↔$O_3ClOClO_2$+H_2O	156	48.0	59.5	0.6	1.2
R17	$HOClO_3$+HOClO↔$O_3ClOClO$+H_2O	175	14.8	26.8	0.4	0.4
R18	$HOClO_3$+HOCl↔O_3ClOCl+H_2O	387	46.2	58.0	-1.2	-0.5
R19	$HOClO_3$+HOCl↔$O_2ClOOCl$+H_2O	234	41.7	49.4	9.0	8.1
R20	$HOClO_2$+$HOClO_2$↔$O_2ClOClO_2$+H_2O	111	18.0	29.6	0.8	1.2
R21	$HOClO_2$+HOClO↔$O_2ClOClO$+H_2O	1087	15.3	27.3	-13.9	-13.1
R22	$HOClO_2$+HOCl↔O_2ClOCl+H_2O	344	44.7	56.0	-4.6	-4.4
R23	HOClO+HOClO↔O_2ClOCl+H_2O	979	16.8	28.0	-33.6	-32.7
R24	HOClO+HOCl↔ClOClO+H_2O	800	23.6	34.7	-4.6	-4.4
R25	HOCl+HOCl↔ClOCl+H_2O	327	12.0	22.3	-3.2	-2.9
R26	$O_3ClOClO_3$↔$O_2ClOClO_2$+O_2	421	46.7	46.9	17.5	4.5
R27	$O_2ClOOCl$↔ClOClO+O_2	214	29.3	28.2	18.6	8.0
R28	ClOClO↔ClOOCl	171	12.9	13.3	-7.8	-7.5
R29	ClOOCl↔Cl_2+O_2	284	11.0	11.0	-32.0	-41.0
R30	$O_2ClOClO_2$↔O_2ClOCl+O_2	169	19.6	20.8	5.5	-12.9
R31	O_2ClOCl↔ClOCl+O_2	238	24.8	25.2	12.4	2.4
R32	$O_3ClOClO_3$+NH_3↔NH_2OClO_2 (ion-pair)+$HOClO_3$	361	31.5	42.3	-25.5	-27.4
R33	$O_3ClOClO_2$+NH_3↔NH_2OClO_2(ion-pair)+$HOClO_2$	427	43.4	54.3	-18.4	-20.2
R34	O_3ClOCl+NH_3↔NH_2OCl+$HOClO_2$	272	37.1	46.6	-6.0	-8.2
R35	$O_2ClOClO_2$+NH_3↔NH_2OClO+$HOClO_2$	43	16.1	26.9	-25.4	-26.0
R36	$O_2ClOClO$+NH_3↔NH_2OClO+HOClO	757	34.4	45.1	-10.6	-11.6
R37	O_2ClOCl+NH_3↔NH_2OCl+HOClO	383	26.9	37.1	5.3	4.5
R38	ClOClO+NH_3↔NH_2OCl+HOCl	192	14.1	24.7	-23.7	-23.8
R39	ClOCl+NH_3↔NH_2OCl+HCl	831	5.6	11.6	-6.2	-5.1
R40	NH_2OClO_2 (ion-pair) ↔HNO+HOClO	1030	12.9	13.0	-1.2	-12.9
R41	NH_2OClO↔HNO+HOCl	700	7.3	7.2	-15.5	-26.7
R42	NH_2OCl↔HNO+HCl	568	23.8	23.5	-12.6	-22.2

that similar to reaction **R1,** $HOClO_2$ thus formed can react with NH_3 to further produce NH_3O and another new intermediate species, HOClO (**R7**). HOClO thus formed reacts with NH_3 in a similar way **(R8)** to produce HOCl, the primary intermediate required to form HCl during the initial phase of AP decomposition. Simultaneously, NH_3O in presence of NH_3 quickly transforms to trans-NH_2OH (**R10**) which reacts further with $HOClO_3$, $HOClO_2$, HOClO and HOCl via several reactions **(R11-R14)** to form $HONH_2O$ and HCl. In the process, intermediate species, $HOClO_x$ (x=0-2) are also regenerated. $HONH_2O$ thus formed via **R11-R14** then quickly decomposes to H_2O and HNO via **R15**. Once intermediate species such as $HOClO_2$, HOClO and HOCl are formed, similar to reaction **R4** they can react with each other via bimolecular reactions (**R16-R25**) to produce H_2O and higher molecular weight ClO containing species. Similar to **R5** and **R6**, NH_3 and H_2O act as catalyst in some of these reactions as well. As it can be observed from reactions **R26-R31** listed in Table 2, the ClO containing species thus formed through self-reactions between $HOClO_x$ (x=0-3) releases O_2 and produces lower molecular weight species which decomposes further to release O_2. Simultaneously, ClO containing species can also react with NH_3 to produce ion-pairs such as NH_2OClO_2, NH_2OClO and NH_2OCl. In the process, one of the final products HCl is formed **(R39)** and critical species such as $HOClO_3$, $HOClO_2$, HOClO and HOCl are also regenerated via reactions **R32-R38.** The ion-pairs thus formed decompose further to produce HCl, HNO and key intermediate species such as HOClO and HOCl (**R40-R42**).

4. Conclusion

QM calculations are performed to identify new initiation reactions involving NH_3 and $HOClO_3$. Along with initiation steps, several new reactions producing critical intermediates and final products such as O_2, H_2O, HCl are also studied. The results from this study suggest that the initiation of AB decomposition not only involves the unimolecular bond-breaking of $HOClO_3$, rather initiation can also possibly proceed via self-reaction of $HOClO_3$ and bimolecular reactions involving both NH_3 and $HOClO_3$. Furthermore, QM calculations show that NH_3 and H_2O play a major role as catalyst during the initial phase of AP decomposition. This study based on QM methods identify several new chlorine containing intermediate species such as $HOClO_x$ (x=0-3) and higher molecular weight ClO containing species which are likely to play major role during the process of AP decomposition. It is also evident from this study that self-reactions between $HOClO_x$ (x=0-3) species formed during the initial phase are crucial for regeneration of critical intermediates as well as for producing final products. This work though not complete takes us one step forward in deciphering the complex chemical kinetics involved during the initial phase of AP combustion. Nevertheless, there are many more possibilities of reactions which need to be further investigated.

5. Acknowledgement

This research was funded by Air Force Office of Scientific Research under grant number FA9550-13-1-0004.

6. References

[1] M. L. Gross, M. W. Beckstead, Steady-state combustion mechanisms of ammonium perchlorate composite propellants, J. Propuls. Power. 27 (2011) 1064-1078.

[2] M. L. Gross, M. W. Beckstead, Diffusion flame calculations for composite propellants using a vorticity-velocity formulation, J. Propuls. Power. 25 (2009) 74–82.

[3] M.W. Beckstead, R.L. Derr, C.F. Price, A model of composite solid-propellant combustion based on multiple flames, AIAA J. 8 (1970) 2200-2207.

[4] N.E. Ermolin, O.P. Korbeinichev, A.G. Tereshchenko, V.M. Fomin, Kinetic calculations and mechanism definition for reactions in an Ammonium Perchlorate flame, Combust. Explos. Shock Waves 18 (1982) 61-70.

[5] Q. Jing, M. W. Beckstead, M. Jeppson, Influence of AP solid-phase decomposition on temperature profile and sensitivity, 36th Aerospace Sciences Meeting and Exhibit (1998), AIAA Paper 98-0448.

[6] M. L. Gross, Two-Dimensional modeling of AP/HTPB utilizing a vorticity formulation and one dimensional modeling of AP and ADN, Brigham Young University, 2007. https://scholarsarchive.byu.edu/etd/1167/.

[7] M.C. Lin, M.C. Lin's Chemical Kinetics Research Group, http://www.chemistry.emory.edu /faculty/lin/.

[8] M. J. Frisch, G. W. Trucks, H. B. Schlegel, G. E. Scuseria, M. A. Robb, J. R. Cheeseman, G. Scalmani, V. Barone, B. Mennucci, G. A. Petersson, H. Nakatsuji, M. Caricato, X. Li, H. P. Hratchian, A. F. Izmaylov, J. Bloino, G. Zheng, J. L. Sonnenberg, M. Hada, M. Ehara, K. Toyota, R. Fukuda, J. Hasegawa, M. Ishida, T. Nakajima, Y. Honda, O. Kitao, H. Nakai, T. Vreven, J. A. Montgomery, Jr., J. E. Peralta, F. Ogliaro, M. Bearpark, J. J. Heyd, E. Brothers, K. N. Kudin, V. N. Staroverov, T. Keith, R. Kobayashi, J. Normand, K. Raghavachari, A. Rendell, J. C. Burant, S. S. Iyengar, J. Tomasi, M. Cossi, N. Rega, J. M. Millam, M. Klene, J. E. Knox, J. B. Cross, V. Bakken, C. Adamo, J. Jaramillo, R. Gomperts, R. E. Stratmann, O. Yazyev, A. J. Austin, R. Cammi, C. Pomelli, J. W. Ochterski, R. L. Martin, K. Morokuma, V. G. Zakrzewski, G. A. Voth, P. Salvador, J. J. Dannenberg, S. Dapprich, A. D. Daniels, O. Farkas, J. B. Foresman, J. V. Ortiz, J. Cioslowski, and D. J. Fox, Gaussian, Inc., Wallingford CT, 2013, Gaussian 09, Revision D.01.

[9] K. Fukui, The path of chemical reactions - the IRC approach, Acc. Chem. Res. 14 (1981) 363–368.

[10] H. P. Hratchian, H. B. Schlegel, Accurate reaction paths using a Hessian based predictor–corrector integrator J. Chem. Phys. 120 (2004) 9918-9924.

[11] L.A. Curtiss, P.C. Redfern, K.J. Raghavachari, Gaussian-4 theory using reduced order theory, Chem. Phys. 127 (2007) 124105.

[12] P. Pechukas, F. McLafferty, Quantum transition state theory, Chem. Phys. Lett. 27 (1974) 511-514.

[13] R. Zhu, M. C. Lin, Kinetics and mechanism of AP combustion initiation: a complete quantum chemical prediction for reactions in three phases, Trans. JSASS Aerospace Tech. Japan 10 (2012) 77- 84.

[14] C.-C. Chen, and M. J. McQuaid, "Modeling the deflagration of ammonium perchlorate at pressures from 300 to 30,000 psia. Part I: Gas-phase, finite rate, chemical kinetics mechanism development," Presented at the 46th JANNAF Combustion Subcommittee Meeting; 2014 Dec 8–11; Albuquerque, NM.

Sub Topic: Heterogenous Combustion

Spring Technical Meeting
Eastern States Section of The Combustion Institute
March 4-7, 2018
State College, Pennsylvania

Investigation into the Structure-Function Relation of 3D Printed Energetic Films: Single Layer Formulation Study

Miles C. Rehwoldt[1], H. Wang[1], Noah Eckman[1], Dylan Kline[1], Michael R. Zachariah [1,]*

[1] Department of Chemistry and Biochemistry and Department of Chemical and Biomolecular Engineering, University of Maryland, College Park, Maryland 20742, United States
[*]Corresponding Author Email: mrz@umd.edu

Abstract: Metallized energetic composite films incorporating high mass loading of aluminum, titanium, and tantalum nanoparticle fuels and relevant oxidizers were constructed via layer-by-layer additive manufacturing (3-D Printing) in order to study their structure-function relationship. Utilizing polyvinylidene fluoride (PVDF) as a potential energetic binder, homogenous layers of polymer encapsulated nanoparticles are extruded with various thicknesses and compositions into desired architectures. Thermal properties and combustion performances are measured using high speed videography, X-ray crystallography, and thermogravimetric analysis (TGA) in order to gain intuition into fundamental reaction mechanisms with respect to characteristic diffusion length scales and composition. Scanning Electron Microscopy (SEM) and high-speed analysis give insight into their distinct physical properties and possible methods for mass and energy transport.

Keywords: Films, Heterogenous Combustion, 3D Printing, Nanoparticles

1. Introduction

The fabrication of solid propellants employing metallic nanoparticles as an energy dense fuel has become a promising frontier in the search for new materials with the potential for transient, high yield energy release. In recent years, the availability and auspicious cost of nanoscale reactive metals has been the catalyst for a plethora of research striving for materials which offer a greater chemical energy density than traditional CHNO chemistry[1]. Although traditional metalized systems generally possess greater energy densities, their reaction kinetics are relative slow, limited by the diffusion length scales between fuel and oxidizers. Contemporary formulations utilizing nanoscale fuels and oxidizers look to reduce characteristic diffusion lengths between these components and increase their reactivity in order to approach that of molecular monopropellants such as 2,4,6-trinitrotoluene (TNT).

The majority of studies of this type, done by our group and several other groups, have utilized aluminum nanoparticles as the feature fuel for thermite based composites because of its light weight, superior combustion performance, and cost effectiveness compared to alternative metal fuel sources. However, recent mechanistic studies involving titanium and tantalum nanoparticles have demonstrated that both tantalum and titanium possess favorable fuel centric reaction mechanics for ignition which maintain a relative high combustion enthalpy and have the potential for increased reactivity stemming a high melting point which preserves nanostructure prior to ignition[2, 3]. These fuels offer ignition characteristics which could possibly enhance combustion performance and ignition control when added with aluminum and appropriate oxidizers.

In this study and in past studies undertaken by our group, applicable energetic composites of this type are generally formulated by incorporating high mass loadings of metallic nanoparticles and other relevant additives within a binder[4]. The formulation exists in a liquid state where a suitable solvent is utilized in order to first dissolve the polymer binder and allow for thorough mixing. Past studies subject the precursor solution to the aerosol deposition technique of electrospray which uses electric fields in order to direct a plume of charged, monodispersed precursor droplets to a rotating conductive substrate[5]. The deposited droplets wet the surface of the substrate and dry as a result of solvent evaporation, leaving behind a thin homogenous laminate of energetic materials. Although electrospray has been an effective technique at producing microstructures of various types by changing precursor formulations or electrospray parameters, its geometric reproducibility and scale up potential are limited[5]. The thickness of an electrospray film is highly dependent on the quality and stability of the electrospray Taylor Cone which itself is dependent on a number of physical and electrical parameters[5]. All of these factors result in a low degree of reproducibility with relatively large variations in the thickness profiles from run to run. The thickness of the resulting film is also limited by how large the electric field between the conductive substrate and the tip of an extrusion syringe can get while printing. Since the routinely established electric field is roughly 3 kV/cm, a decrease in the distance between the syringe and substrate as the film gets thicker increases the electric field strength, leading to a possible hazardous situation of an unintended dielectric break down.

In order to alleviate the short comings of electrospray deposition, the energetic precursor was utilized as a 3D printable ink in an additive manufacturing extrusion approach. Although this method only allows for the fabrication of laminates, film thickness is no longer limited by an electrical field deposition method. Previous restrictions on precursor formulations in terms of viscosity and employed solvent are now relaxed with a pressure driven printing technique providing a relatively large quantity of experimentally acceptable materials in a matter of minutes, instead of hours. The controlled translation and extrusion dynamics inherent to 3D printing also allows for increased degrees of freedom for energetic materials to be printed in a variety of patterns, shapes, and structures. With this approach, we hope to gain phenomenological intuition into possible formulations for enhanced combustion performance throttled by material architectures utilizing layers of various compositions and diffusion lengths scales. Ultimately, these materials will be tailored macroscopic objects built upon microscopic features incorporating particular fuels, oxidizers, binders, and other additives such as biocidal iodine or burn speed enhancing mesoporous silica[6, 7].

2. Methods/Experimental

Materials: The metallic nanoparticles used in this preliminary study include nano-aluminum (nAl) (~50nm) purchased from Novacentrix, nano-titanium (nTi) (30-80nm) purchased from US Research Nanomaterials, and nano-tantalum (nTa) (<50 nm) purchased from Global Advanced Metals. It was determined from thermogravimetric analysis (TGA) that these nanoparticles with their respective oxide layers result in particles which are 81%, 70%-75%, and

70% active by mass, respectively. Poly(vinylidene fluoride) powder (PVDF) (MW=534000) and dimethylformamide (DMF) (99.8 wt. %) solvent was purchased from Sigma Aldrich.

Film Synthesis: Synthesis of energetic composites was achieved using a System 30M 3D printer purchased from Hyrel 3D. Each precursor 3D printable ink considered was formulated by dissolving 500 mg of PVDF into 5 ml of DMF and adding the appropriate mass content of metal fuel, oxidizer, and additives. The fluoropolymer PVDF is primarily utilized for its solubility in DMF and other solvents, and because it is also an oxidizing source[4]. From previous studies, it was shown that PVDF decomposes to form HF, a strong oxidizing agent which readily reacts with nAl. The work probing the Al/PVDF interaction was conducted by DeLisio et al. and proposes a mechanism for the fluorination of aluminum which is essential to the ignition and propagation of the energetic composite in anaerobic environments[4, 8].

Fluorination of Aluminum: $2Al + 6HF_{(g)} \rightarrow 2AlF_3 + 3H_{2(g)}$ (1)

Using equation 1 and a guess at analogous reactions involving the possible fluorination of titanium (TiF_4) and tantalum (TaF_5), stoichiometric mixtures of PVDF and their respective metal fuels were formulated (Table 1). Precursor suspensions are then sonicated for 30-45 minutes and left to stir for 24hrs using a magnetic stir plate/bar. In preparation for printing, the precursor solution is poured into the back of a 22-gauge blunt needle attached 30cc disposable syringe, and mounted to the extrusion pump of the 3D printer. Before printing, the ink is primed and the glass printing bed heated to as high as 80°C. Since the ink is in liquid form, the printer must have a programed routine which allows sufficient time for the previous pass to dry fast enough for the layer to hold its shape by the time the next pass occurs. The drying process can be controlled by the temperature of the printing bed, the speed and extrusion rate of the print, and/or the volitivity of the chosen solvent. In order to study the 1D combustion performance of our materials, we chose to print single pass strands with a width which is roughly equivalent to the outer diameter of the extrusion needle, and a thickness which represents 5 total passes over the same area. In order to achieve a sufficient amount of material for characterization, multiple 3 cm long strands were harvested from a larger 25cm perimeter rectangular structure which was printed in a single run and carefully pealed off of the glass printing bed.

Table 1: Energetic Precursor Formulations

Fuel: (wt. %)	PVDF (wt. %)	Added Oxidizer: (wt. %)	Additive: (wt. %)
Al: 25	75	-----	-----
Ti: 33	67	-----	-----
Ta: 61.8	38.2	-----	-----

Characterization: Films in this study were characterized based on their thickness, reproducibility, morphology and composition, combustion performance in different atmospheric conditions, and the content of their condensed phase combustion products.

Cross-sections of each film are analyzed using Scanning Electronic Microscopy (SEM, Hitachi, SU-70 FEG-SEM) and compared to one another. For each film formulation, combustion was conducted in both air and argon. In air, film ends were simply anchored to a flat surface with tape and ignited via a butane lighter. In order to establish an argon atmosphere, a small cylindrical chamber with a gas valve inlet and outlet was purged with argon for 5-10 minutes. The chamber was developed with a mounting rod which would be used to hold the film and a resistively heated nichrome wire in place. Once the chamber has been purged and sealed, the nichrome wire acts as the initiator for primary ignition of the film at one of its ends. Combustion performance was assessed using a Phantom v12.0 high-speed digital camera running Phantom 692 software and operating between 2000-3000 fps with an exposure of 200μs. Although characteristic combustion speeds can be estimated by the knowing the length of the film and the elapsed time, videos can be further processed in Matlab and ImageJ to obtain time resolved position as a function of time. Argon atmosphere combustion products and products from TGA samples slowly heated to 1000°C at a rate of 10°C/min under an argon flow of 100 ml/min are then collected for X-ray diffraction analysis (XRD) in order to speculate possible reaction pathways for self-propagation.

3. Results and Discussion

Metal/PVDF Films: Immediately following film fabrication, films are prepared for SEM for cross-section analysis with the results shown in figure 1.

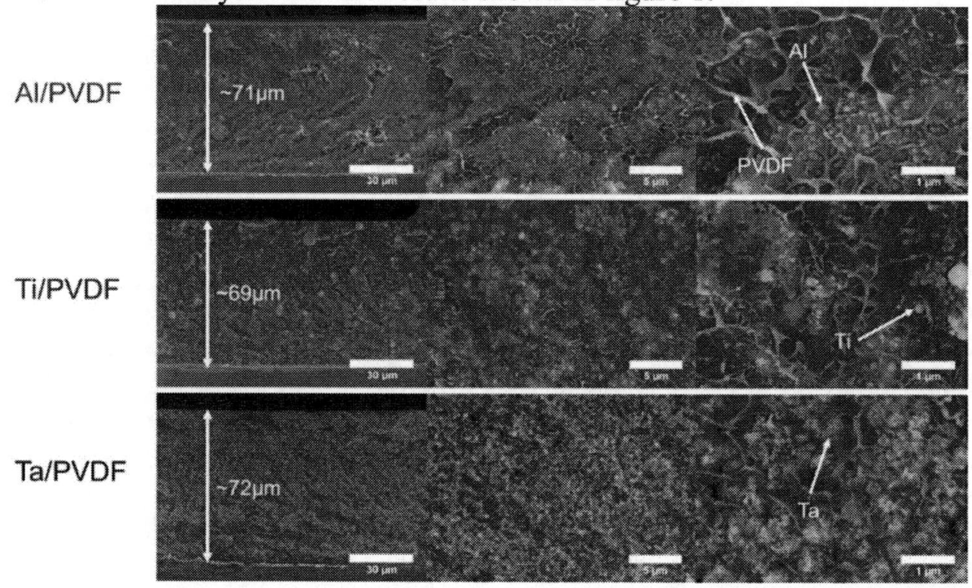

Figure 1: Cross-section of 3D printed Al, Ti, and Ta nanoparticles incorporated within PVDF

As figure 1 illustrations, the three main formulations incorporating the different metal nanoparticles look nearly identical in terms of surface morphology and geometry with well-defined uniform edges and homogenously dispersed nanoparticles within the polymer matrix. Ta/PVDF appears the most distinguishable with what appears to be a larger particle density compared to the other two films. This could be a consequence of stoichiometry in the amount of nTa used or a difference in the inherent physics of nTa agglomeration. Despite this difference, each formulation was printed with 5 layers and resulted in films with nearly identical thicknesses, roughly 70μm (~14μm/layer).

Although these films may appear the same physically, their combustion performance appears to be quite different as Figure 2 depicts.

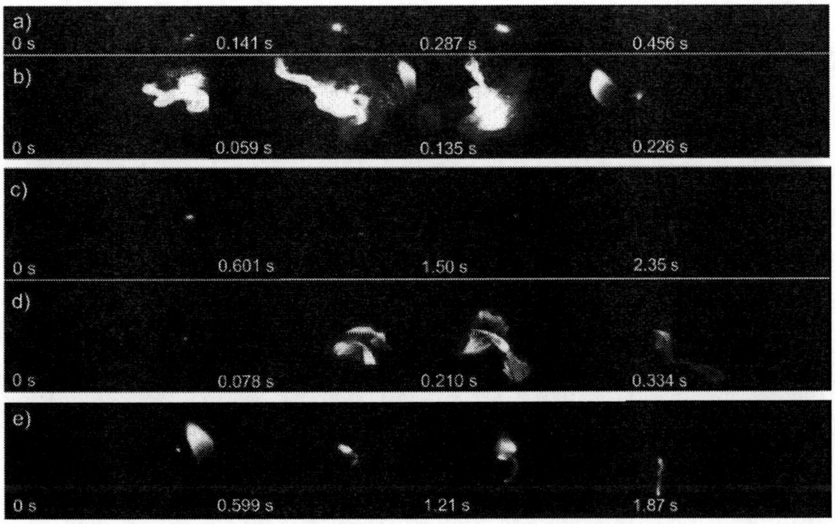

Figure 2: Combustion of a) Al/PVDF in argon, b) Al/PVDF in air, c) Ta/PVDF in argon, d) Ta/PVDF in air, and e) Ti/PVDF in air

Snap shots of the combustion of each film type uncover a clear hierarchy in terms of combustion performance. As expected, aluminum out performs each of the other metal fuels in both environments owing to its superior combustion enthalpy per gram and its ability to be strongly oxidized by HF as a result of PVDF decomposition. What appears most interesting is the performance of Ta/PVDF over Ti/PVDF and the lack of self-propagation of Ti/PVDF in the argon environment. Although titanium has a much higher theoretically normalized combustion enthalpy per gram compared to tantalum due to tantalums large molar mass, Ta/PVDF appears to be much more reactive in air and has the ability to self-propagate. Since the argon environment is theoretically devoid of oxygen contain oxidizers, there likely exists a reaction of nTa with PVDF which allows for self-propagation of the flame front. Ti must not have a significantly exothermic interaction with HF or any interaction with PVDF at all. Propagation in air is thus due to diffusion of oxygen from outside of the film with the PVDF binder acting as a non-interacting heat reservoir, absorbing heat from the main reaction. XRD of the combustion products shown in Figure 3 confirms aluminum fluorination, but also explicitly reveals tantalum fluorination in two forms which are indistinguishable from the perspective of XRD (TaF_3 and $TaFO_2$). Further studies must be performed to probe this interaction, but these preliminary results suggest that PVDF decomposition products oxidize both the tantalum core and oxide layer.

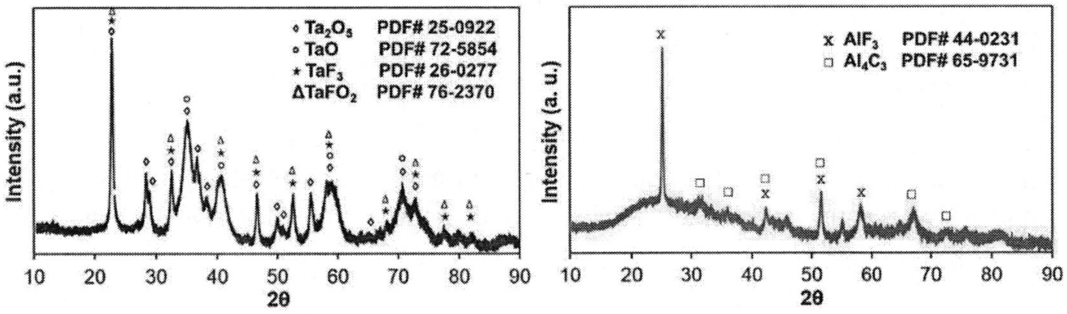

Figure 3: XRD of Ta/PVDF (left) and Al/PVDF (right) argon combustion products

4. Conclusions

This study preliminarily investigated the formulation and combustion performance of 3D printed energetic films incorporating high mass loadings of aluminum, titanium, and tantalum nanoparticles within the energetic polymer binder PVDF. It was found that each of the resulting two component films was homogenously mixed with a high degree of geometric reproducibility even between films with differing metal content. Analysis of combustion in air and argon environments not only confirmed the fluorination of aluminum, but also revealed the fluorination of tantalum and, unlike Ti/PVDF films, the ability of Ta/PVDF films to propagate in an oxygen free environment. Further probing of this fundamental interaction will need to conducted in order to effectively take advantage of these findings.

5. Acknowledgements

The authors gratefully acknowledge support from AFOSR.

6. References

[1] J. DeLisio, UNDERSTANDING THE RELATIONSHIPS BETWEEN ARCHITECTURE, CHEMISTRY, AND ENERGY RELEASE OF ENERGETIC NANOCOMPOSITES, Department of Chemical and Biomolecular Engineering and Department of Chemistry and Biochemistry, University of Maryland, College Park, College Park, Maryland, 2017, pp. 194.

[2] M. Rehwoldt, Y. Yang, H. Wang, S. Holdren, Z. Michael R., Ignition of Nano-scale Titanium/Potassium Perchlorate Pyrotechnic Powder: Reaction Mechanism Study, Physical Chemistry Chemical Physics: Works in preparation (2018).

[3] J.B. DeLisio, X.Z. Wang, T. Wu, G.C. Egan, R.J. Jacob, M.R. Zachariah, Investigating the oxidation mechanism of tantalum nanoparticles at high heating rates, Journal of Applied Physics 122 (2017).

[4] N.S. Cohen, R.W. Fleming, R.L. Derr, ROLE OF BINDERS IN SOLID-PROPELLANT COMBUSTION, Aiaa Journal 12 (1974) 212-218.

[5] T. Wu, X. Li, X. Hu, J. Delisio, W. Zhou, M.R. Zachariah, Direct-Deposition to Create High Particle Loading Propellants with Controlled Architecture: Combustion and Mechanical Properties, 54th AIAA Aerospace Sciences Meeting, AIAA SciTech, San Diego, California, USA, 2016.

[6] H. Wang, J. DeLisio, T. Wu, Y. Yang, J. Hu, M.R. Zachariah, Mesoporous Silica Spheres Incorporated Aluminum/Poly (Vinylidene Fluoride) for Enhanced Burning Propellants, Advanced Engineering Materials 1527-2648 (2017).

[7] H.Y. Wang, J.B. DeLisio, G.Q. Jian, W.B. Zhou, M.R. Zachariah, Electrospray formation and combustion characteristics of iodine-containing Al/CuO nanothermite microparticles, Combustion and Flame 162 (2015) 2823-2829.

[8] J.B. DeLisio, X.L. Hu, T. Wu, G.C. Egan, G. Young, M.R. Zachariah, Probing the Reaction Mechanism of Aluminum/Poly(vinylidene fluoride) Composites, Journal of Physical Chemistry B 120 (2016) 5534-5542.

Spring Technical Meeting
Eastern States Section of the Combustion Institute
March 4-7, 2018
State College, Pennsylvania

Liquid-Phase Decomposition of RDX: Formation of Oxy-s-triazine and 1,3,4-oxadiazole

Lalit Patidar[1], Mayank Khichar[1], Stefan T. Thynell[1,]*

[1]Department of Mechanical and Nuclear Engineering
The Pennsylvania State University, University Park, PA 16802
[*] Corresponding Author Email: thynell@psu.edu

Abstract: Cyclotrimethylene trinitramine (RDX) is widely used as an ingredient in solid propellants and explosives. As a result, RDX has been investigated extensively in many experimental and theoretical studies to elucidate its liquid-phase and gas-phase decomposition. The decomposition of liquid-phase RDX was studied using Fourier transform infrared (FTIR) spectroscopy of evolved gas-phase species from rapid thermolysis and of the condensate and residue formed from the decomposition, as well as time-of-flight mass spectrometry (ToFMS) of the evolved gas-phase species. Sub-milligram sample of RDX was heated at rates of about 2000K/s to a set temperature where decomposition occurred under isothermal conditions. Since RDX melts at around 205°C, decomposition studies were conducted at temperatures ranging from 265 to 305°C. Oxy-s-triazine ($C_3N_3H_3O$, m/z=97) and (1,3,4-oxadiazole ($C_2N_2H_2O$, m/z=70) were two of the compounds detected in the thermolysis tests. Chemical reaction mechanisms explaining the formation of oxy-s-triazine (OST) and 1,3,4-oxadiazole were formulated using quantum mechanics calculations guided by the acquired experimental data and thermochemical analysis. The quantum mechanics investigation is based on using density functional theory (DFT) at the B3LYP/6-311++G(d,p) level. The liquid-phase solvation effect is accounted via the Conductor-like Polarizable Continuum Model (CPCM) with water as solvent within the Gaussian program package.
Keywords: *Quantum mechanics, Energetic materials, Liquid-phase decomposition*

1. Introduction

RDX is a widely used solid propellant ingredient due to its low toxicity and capability of producing high specific impulse with less smoke generation and corrosion. The study of physical and chemical processes occurring during thermal decomposition under various operating conditions is of great importance for understanding ignition and combustion characteristics, as well as developing burn-rate modifiers for nitramine-based propellants. As a result, it has been extensively studied [1,2]. In spite of large number of studies, reaction mechanisms in the condensed phase of RDX still pose an important challenge. The main issue involves deciphering the chemical reactions that occur in the liquid phase and how these reactions affect the ignition and combustion of the propellant.

Numerous attempts have been made to decipher the condensed-phase combustion chemistry of RDX. Information on the chemical composition and kinetics of the "melt" phase of RDX and HMX was obtained by Brill and co-workers [3] using Fourier transform infrared spectroscopy (FTIR). Behrens and co-workers [4,5] employed simultaneous thermogravimetry modulated beam mass spectrometry (STMBMS) measurements, time-of-flight (TOF) velocity-spectra analysis to identify four primary reaction pathways in the liquid-phase decomposition of RDX.

Different sets of intermediates and products produced by these pathways were identified by various investigators by performing mass spectrometry work on RDX. In many of these works, which deal with decomposition, species having *m/z* value of 97 and 70 have been observed in the mass spectra. The results from experiments by Behrens [5] show that the formula of the *m/z* = 97 ion is $C_3H_3N_3O$ which is labeled as oxy-s-triazine or OST. The formula of the *m/z* = 70 ion is believed to be $C_2H_2N_2O$ with a structure suggested to be 1,2,5-oxadiazole [5]. Lee et al. [6] analyzed RDX samples using a triple quadrupole mass spectrometer and established that species at masses 97 and 70 were true decomposition products of RDX. In their gas chromatography-mass spectrometry tests, Schroeder et al. [7] have called the compound with molecular weight 70 as "unknown A" and have stated that it is probably 1,2,4-oxadiazole. In all these mass spectrometry tests, it has been confirmed that a species having molecular weight 97 and 70 are part of the RDX decomposition mechanism but the structures of these species have not been confirmed. Hence, no steps could be taken to update the chemical kinetics mechanism of RDX by investigating the chemical reactions that can lead to the formation of these two particular species.

Here we have performed quantum mechanics calculations for subsequent reactions of the intermediates formed by various initial steps and ring-opening reactions investigated in our earlier study [8]. Based on these calculations, mechanistic details for the formation of species with molecular weight 97 (oxy-s-triazine) and molecular weight 70 (an oxadiazole) are presented. Theoretical calculations are guided and supported by experimental results obtained using FTIR spectroscopy and ToFMS of fast thermolysis of RDX.

2. Computational Details

Gaussian 09 quantum chemistry software package is used for carrying out all calculations. Density functional theory (DFT) with hybrid functional B3LYP and 6311++G(d,p) basis set is used to optimize the geometries of reactants, products, intermediates and transition states. For liquid-phase calculations, implicit conductor-like solvation model CPCM is used with water as solvent. Intrinsic reaction coordinate (IRC) calculations are performed to verify that the identified transition state indeed connects the reactants to expected products.

3. Results and Discussions

The three isomers of OST and the three oxadiazole compounds, as shown in Fig. 1, were optimized and the converged structures showed only positive harmonic frequencies. To improve the accuracy, anharmonic frequency calculations were performed on the optimized structures of these six compounds using two functionals (B3LYP and B2PLYP) and three basis sets (6-311++G(d,p), N07D and aug-cc-PVTZ), and the results are compared against the experimentally assigned vibrational bands for 1,3,4-oxadiazole [9] as shown in Table 1. Barone and co-workers [10] have shown that B2PLYP/N07D model provides good quality harmonic frequencies and correctly describes anharmonic contributions. However, we have obtained results of similar accuracy with B3LYP/N07D model, which is significantly less expensive in terms of the computational cost. Hence, vibrational calculations for all the remaining structures were performed at B3LYP/N07D level.

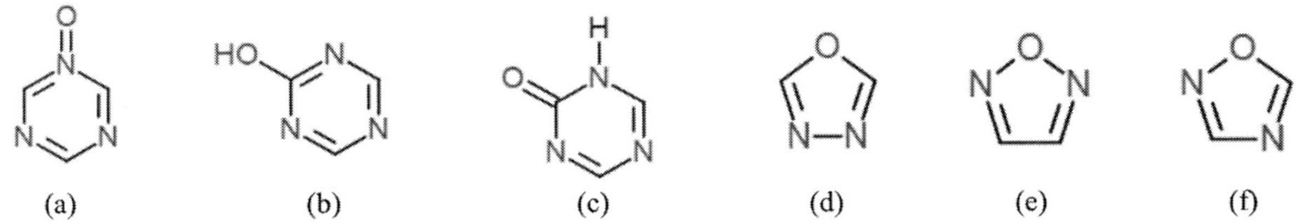

Figure 1: Structures of three isomers of OST (a) N-oxide: OST_N (b) 2-hydroxide: OST_C (c) keto tautomer of 2-hydroxide: OST_K and structures of oxadiazole compounds (d) 1,3,4-oxadiazole (e) 1,2,5-oxadiazole (f) 1,2,4-oxadiazole

Table 1: Predicted and measured wavenumbers (cm^{-1}) from infrared spectra of 1,3,4-oxadiazole

Measured (Anthonsen et al. [9]) (cm^{-1})	B3LYP			B2PLYP		
	6-311++G(d,p)	N07D	aug-cc-pVTZ	6-311++G(d,p)	N07D	aug-cc-pVTZ
3169	3155	3167	3159	3182	3198	3185
3167	3150	3161	3153	3177	3193	3181
1541	1508	1508	1505	1498	1503	1499
1534	1491	1488	1490	1476	1481	1476
1272	1246	1243	1256	1246	1242	1256
1215	1170	1162	1184	1179	1167	1195
1092	1063	1064	1068	1070	1070	1075
1078	1049	1061	1054	1061	1066	1064
951	958	966	951	957	963	945
925	930	935	931	928	934	928
920	915	919	911	917	919	913
852	875	847	890	879	851	893
840	865	821	866	885	826	869
657	661	659	666	660	660	663
625	627	618	634	627	624	634
Mean Unsigned error (MUE)	**20.4**	**18.2**	**19.8**	**20.8**	**20.0**	**19.9**

During liquid-phase decomposition of RDX, some intermediate species, which are formed very early, may not evolve into the gas phase in detectable quantities due to their high molecular weight. Instead, they may accumulate in the residue left from the liquid-phase decomposition or may condense out quickly. Confined Rapid Thermolysis (CRT)/FTIR studies along with ToFMS were performed to detect such species as described in detail by Kumbhakarna [11]. The spectrum of the condensed-phase residue is shown in Fig. 2(a), along with the spectrum of pure RDX. An additional weak band at 1704 cm^{-1} was observed when compared to the pure RDX spectrum.

On comparison with a number of reference spectra from NIST database, the band at 1704 cm^{-1} is most likely attributed to the ν(C=O) of a carbonyl group, for example, in various derivatives of pyrimidine (cytosine, thymine, uracil etc.). Behrens and Bulusu [4] strongly claim the presence of OST which has C-N six-membered heterocyclic ring along with C=O as shown in Fig. 1(c).

Hence, it is believed that this OST molecule exists in the liquid during the early reaction events and appears in the residue spectrum.

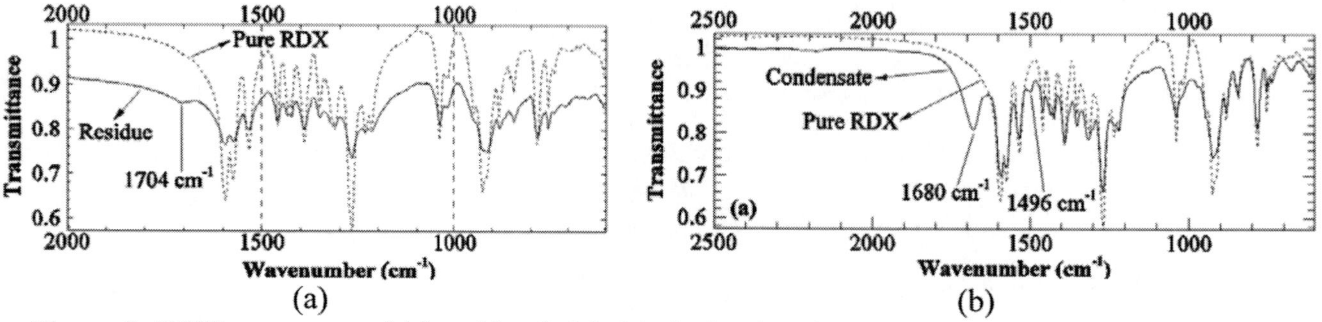

Figure 2: FTIR spectrum of (a) residue left behind after heating RDX at 220°C for 10 minutes in the CRT chamber (b) Condensate obtained after decomposition of 0.616 mg RDX at 285°C and pressed in KBr pellet [11].

In the absence of experimental data on the vibrational assignment of the three isomers of oxy-s-triazine, we have compared the calculated anharmonic frequencies to the FTIR spectrum of residue in Fig. 2(a). The OST_K isomer of oxy-s-triazine molecule which has m/z=97 and a C=O bond on its ring, has a ν(C=O) stretching at 1704 cm^{-1}. Since it is also the most stable of all three isomers, it is more likely to be the structure of OST. The calculated spectrum of OST_C also shows various bands, which are common with that of the residue as shown in Fig. 2(a). Although many of the bands from the calculated spectrum of OST_N are also present in the spectrum of the residue, they are also common with either OST_K or OST_C. Hence, we believe that the species with m/z=97 is most likely OST_K and it is formed via OST_C through keto-enol tautomerization.

Furthermore, the spectrum in Fig. 2(b) shows that RDX is present in the condensate since all the bands shown by pure RDX are visible in the condensate spectrum. Figure 2(b) also shows additional bands at 1680 and 1496 cm^{-1}. Comparison of this spectrum with the IR spectrum of pure liquid 1,2,4-oxadiazole and 1,2,5-oxadiazole reveals that these two isomers are unlikely to be present in the condensate. However, the IR spectrum of 1,3,4-oxadiazole shows a striking similarity with the condensate spectrum. Hence, out of the three plausible isomers of oxadiazole, we believe that the species with m/z=70 corresponds to 1,3,4-oxadiazole.

3.1 Pathways to the formation of OST

Of the three plausible structures of OST as shown in Fig. 1(a)-(c), the keto-tautomer of 2-hydroxide (OST_K) is the most stable. The 2-hydroxide (OST_C) is 5.7 kcal/mol (H_{298}) less stable and the *N*-oxide (OST_N) is 58.3 kcal/mol (H_{298}) less stable as compared to OST_K. Since experimental evidence suggests the presence of OST_K, we have investigated the routes to OST_K and its enol-tautomer OST_C as shown in Table 2.

INT175, INT128 and triazine (TAZ) are formed by successive HONO elimination pathway from RDX (reactions (1)-(3)) and they subsequently undergo hydrogen abstraction via NO_2 leading to the formation of INT174, INT127 and TAZ_{-H} respectively (reactions (4)-(6)). The carbon radical center in these intermediates undergo OH addition via *trans*-HONO and NO is released via highly exothermic reactions (7)-(9). Subsequently, OST_C is formed via reactions (9)-(11) and it is converted to OST_K via reaction (12).

Table 2: Pathways to the formation of OST (units are kcal/mol and cm^{-1}, T=298 K)

S. No.	Reaction	$\Delta G\ddagger_f$	$\Delta G\ddagger_b$	ΔH_r	ΔG_r	Freq. of TS
(1)	RDX = INT175 + *trans*-HONO	38.3	65.5	-13.3	-27.2	1400
(2)	INT175 = INT128 + *trans*-HONO	39.0	63.8	-11.3	-24.8	1380
(3)	INT128 = TAZ + *trans*-HONO	32.1	74.9	-30.8	-42.9	778
(4)	INT175 + NO_2 = INT174 + *cis*-HONO	28.6	28.4	0.6	0.2	1670
(5)	INT128 + NO_2 = INT127 + *cis*-HONO	26.5	32.1	-6.0	-5.6	1540
(6)	TAZ + NO_2 = TAZ_{-H} + *cis*-HONO	38.6	9.4	29.7	29.2	1250
(7)	INT174 + *trans*-HONO = INT191 + NO	28.9	49.4	-21.6	-20.5	621
(8)	INT127 + *trans*-HONO = INT144 + NO	28.4	46.2	-19.3	-17.8	623
(9)	TAZ_{-H} + *trans*-HONO = OST_C + NO	20.0	83.6	-65.6	-63.6	672
(10)	INT191 = INT144 + *trans*-HONO	37.7	65.6	-15.6	-27.8	1390
(11)	INT144 = OST_C + *trans*-HONO	37.8	91.7	-41.5	-53.9	743
(12)	OST_C = OST_K	35.4	40.9	-5.7	-5.8	1930

3.2 Pathways to the formation of 1,3,4-oxadiazole

Structure of 1,3,4-oxadiazole (see Fig. 1(d)) is less likely to be formed by ring opening and subsequent closure. Since $C_2N_2H_2O$ is observed in the condensate spectrum, we believe $C_2N_2H_2O$ is formed at a later stage from recombination reactions and it condenses out because of high molecular weight. Hence, pathways to the formation of this compound were investigated based on reactions of intermediates such as CH_2O, CH_2N, HONO etc., formed from initiation reactions. The proposed mechanism for the formation of $C_2N_2H_2O$ is shown in Table 3.

Radical recombination reaction (13) produces CH_2NNCH_2, which is attacked by NO_2 via reactions (14) and $ONOCH_2NNCH_2$ is formed. $ONOCH_2NNCH_2$ dissociates via highly exothermic reaction (15) and the resulting CH_2O and CH_2NN recombines to form the oxadiazole ring via reaction (16). Subsequent hydrogen abstractions via reactions (17) and (18) leads to the formation of 1,3,4-oxadiazole.

Table 3: Pathways to the formation of 1,3,4-oxadiazole (units are kcal/mol and cm^{-1}, T=298 K)

S. No.	Reaction	$\Delta G\ddagger_f$	$\Delta G\ddagger_b$	ΔH_r	ΔG_r	Freq. of TS
(13)	$CH_2N + CH_2N = CH_2NNCH_2$	0.0	28.4	-41.0	-28.4	NA
(14)	$CH_2NNCH_2 + NO_2 = ONOCH_2NNCH_2$	24.0	8.4	3.8	15.6	486
(15)	$ONOCH_2NNCH_2 = CH_2NN + CH_2O + NO$	14.8	53.8	-16.6	-39.0	1026
(16)	$CH_2NN + CH_2O$ = INT72	33.1	33.5	-13.5	-0.4	495
(17)	INT72 + NO_2 = INT71 + *cis*-HONO	26.7	31.9	-5.5	-5.2	1591
(18)	INT71 + *trans*-HONO = Oxadiazole134 + H_2O + NO	29.1	77.8	-38.9	-48.7	1736

4. Conclusions

Species corresponding to m/z=97 and m/z=70 are produced when RDX undergoes thermal decomposition. The bands observed in infrared spectra of the condensate at 1680 cm^{-1} and 1496 cm^{-1} indicate the presence of 1,3,4-oxadizole. Similarly, the band observed in infrared spectra of the residue point towards the possibility of the presence of heterocyclic compound with a carbonyl (C=O) group attached to it which is believed to be oxy-s-triazine. Based on the ab initio quantum chemical calculations, reaction schemes have been proposed that shows how OST and 1,3,4-oxadiazole results from RDX. Reactions for the consumption of these two species are being investigated.

Acknowledgements

This material is based upon work supported by, or in part by, the U. S. Army Research Laboratory and the U. S. Army Research Office under grant number W911NF-15-1-0202.

References

[1] T.L. Boggs, The thermal behavior of cyclotrimethylenetrinitramine (RDX) and cyclotetramethylenetetranitramine (HMX), Fundam. Solid-Propellant Combust. 90 (1984) 121–175.

[2] C.F. Melius, Thermomechanical modeling. II. Application to ignition and combustion of energetic materials, NATO ASI Ser., Ser. C. 309 (1990) 51–78.

[3] R.J. Karpowicz, T.B. Brill, In situ characterization of the “melt” phase of RDX and HMX by rapid-scan FTIR spectroscopy, Combust. Flame. 56 (1984) 317–325.

[4] R. Behrens, S. Bulusu, Recent advances in the thermal decomposition of cyclic nitramines, Mater. Res. Soc. Symp. Proc. 296 (1993) 13–24.

[5] R. Behrens, S. Bulusu, Thermal decomposition of energetic materials. 3. Temporal behaviors of the rates of formation of the gaseous pyrolysis products from condensed-phase decomposition of 1,3,5-trinitrohexahydro-s-triazine (RDX), J. Phys. Chem. 96 (1992) 8877–8891.

[6] Y. Lee, C.-J. Tang, T.A. Litzinger, A study of the chemical and physical processes governing CO2 laser-induced pyrolysis and combustion of RDX, Combust. Flame. 117 (1999) 600–628.

[7] M.A. Schroeder, R.A. Fifer, P.J. Kaste, S.A. Liebman, Thermal Decomposition of RDX in the Presence of Added K12B12H12, J. Propuls. Power. 17 (2001) 441–448.

[8] L. Patidar, S.T. Thynell, Quantum mechanics investigation of initial reaction pathways and early ring-opening reactions in thermal decomposition of liquid-phase RDX, Combust. Flame. 178 (2017) 7–20. doi:10.1016/j.combustflame.2016.12.024.

[9] J.W. Anthonsen, D.H. Christensen, J.T. Nielsen, O.F. Nielsen, Vibrational spectra and fundamental frequencies of 1, 3, 4-oxadiazole and its deuterated species., Spectrochim. Acta Part A Mol. Spectrosc. 32 (1976) 971–982.

[10] M. Biczysko, P. Panek, G. Scalmani, J. Bloino, V. Barone, Harmonic and anharmonic vibrational frequency calculations with the double-hybrid B2PLYP method: Analytic second derivatives and benchmark studies, J. Chem. Theory Comput. 6 (2010) 2115–2125.

[11] N.R. Kumbhakarna, Combustion analysis of RDX propellant and novel high-nitrogen propellant ingredients, The Pennsylvania State University, 2014. https://etda.libraries.psu.edu/catalog/21140.

Spring Technical Meeting
Eastern States Section of the Combustion Institute
March 4-7, 2018
State College, Pennsylvania

Computational study of condensed-phase kinetics during combustion of pure RDX

*Mayank Khichar, Lalit Patidar, Stefan T. Thynell**

Department of Mechanical and Nuclear Engineering
The Pennsylvania State University, University Park, PA 16802, USA
**Corresponding Author Email: thynell@psu.edu*

Abstract: A computational model for constant pressure combustion of RDX monopropellant has been developed to analyze the significance of condensed phase decomposition process. The model is based on conservation equations and takes into account detailed chemical kinetics in the condensed phase. Experimentally measured values of the melt layer thickness, the burn rate, and the propellant burning surface temperature are provided as input to the model. The extent of RDX decomposition in the melt layer is calculated and compared against values obtained in previous studies with global reaction mechanism for the condensed phase. Systematic sensitivity analysis of various model parameters, which affect RDX decomposition in condensed phase, is also performed. The results show that RDX mole fraction at the propellant burning surface is 90% at atmospheric pressure. This value is very sensitive to the kinetic rate parameters of the HONO-elimination step for RDX. The effect of experimental uncertainty in measurement of temperature and burn rate is comparatively small. The effect of higher pressures is also examined. It is found that at higher pressure the extent of RDX decomposition is significantly higher. This suggests that the significance of condensed phase decomposition process increases with pressure.
Keywords: RDX, constant pressure combustion, melt layer, reaction mechanism.

1. Introduction

Cyclotrimethylenetrinitramine (RDX) is a monopropellant that is widely used as an ingredient in rocket propellants and explosives. The understanding of physiochemical processes occurring during its decomposition is limited. Many studies have been conducted during the past five decades to improve the understanding of its ignition and combustion behavior. A mechanism describing the decomposition and secondary reactions for RDX in the gas phase was initially provided by Melius [1]. The mechanism was further improved to its current state by Yetter et al. [2] and Chakraborty et al. [3]. Based on experimental studies, Behrens and Bulusu proposed four pathways for condensed-phase decomposition of RDX. Patidar and Thynell [4] reviewed condensed-phase decomposition process using quantum mechanics methods, and identified many condensed-phase initiation pathways and secondary reactions. A comprehensive condensed-phase reaction mechanism consisting of 321 species and 500 elementary reactions with optimized rate parameters has been compiled and will be presented in a future publication.

Various experimental and computational efforts have been made to develop a better understanding of the combustion-wave structure of RDX. The entire combustion-wave structure can be segmented into three distinguishable regions: 1) solid phase, 2) melt layer, and 3) gas phase. Korobeinichev et al. [5] and Litzinger et al. [6] provided experimentally measured species profiles in the gas-phase region for pressures ranging from 0.1 to 3 atm. Zenin [7] used micro-thermocouple technique to measure different combustion-wave parameters for RDX. The melt layer thickness, the burn rate, and the propellant burning surface temperature along with various other parameters are provided for the pressure range of 1−90 atm. Several theoretical models for the combustion-wave were also developed [8–10]. These models differ in the consideration of physical phenomena occurring in the melt layer (also referred as foam layer). Liau and Yang [8] assumed that the decomposition products were collected in small bubbles in the melt layer, Prasad et al. [9] considered decomposition with products dissolved in the melt layer, whereas Anderson and Conner [10] considered evaporation only at the liquid-gas interface. The major drawback of these models is the uncertainty in the condensed-phase decomposition process. Due to the unavailability of detailed condensed-phase mechanism, Liau and Yang [8] and Prasad et al. [9] considered global reaction mechanism in the condensed phase.

The purpose of the present work is to analyze the significance of the condensed-phase decomposition process during constant pressure combustion of RDX. A computational model based on conservation equations has been developed for the melt layer. In contrast to previous combustion models, we have incorporated the detailed chemical kinetics in the condensed phase. Experimentally measured values of the melt-layer thickness, the burn rate, and the propellant burning surface temperature are provided as input to the computational model. The model is capable of calculating the evolution profiles of various species in the condensed phase. The extent of RDX decomposition in the melt layer is thus determined. Its comparison with the values obtained in previous studies with global reaction mechanism is provided in the results. Systematic sensitivity analysis results of various model parameters are examined in detail. The variation in significance of condensed-phase decomposition process with system pressure is also discussed.

2. Theoretical formulation

Among the three different regions in combustion-wave structure, we consider here the melt layer to analyze the significance of condensed-phase decomposition. The analysis of the melt layer is based on the species conservation equations. The equations take much simpler form when written in the frame of reference of liquid-gas interface (propellant burning surface). The propellant burning surface regress at burn rate (r_b). At steady state, the melt layer thickness (l_m), the burn rate, and the propellant burning surface temperature (T_s) remains constant and can be determined experimentally. The solid-liquid interface is defined by thermodynamic phase transition process and hence its temperature is equal to the melting point temperature of RDX. Due to the small melt-layer thickness, the temperature variation inside the melt layer can be approximated as linear. Temperature plots obtained by Liau and Yang [8] confirms the validity of this assumption.

Let $Y_{c,i}$ be the mole fraction of species i in the melt layer. The set of one-dimensional differential conservation equations for all the N_c species in the melt layer (condensed phase) can be written as,

$$\frac{\partial}{\partial t}\left(\rho_c Y_{c,i}\right) + \frac{\partial}{\partial x}\left(\rho_c r_b Y_{c,i}\right) = \dot{\omega}_{rxn,c,i}\, W_i, \qquad i = 1,2,\ldots,N_c, \tag{1}$$

where ρ_c is the condensed-phase density, W_i is the molecular weight of species i, and $\dot{\omega}_{rxn,c,i}$ is the net molar production rate of species i in the condensed phase per unit volume. Condensed-phase density does not change much in the temperature range considered here. Hence, for simplicity, we assume constant density. The net molar production term $\dot{\omega}_{rxn,c,i}$ is evaluated by considering rate law and species stoichiometry for each of the elementary reaction of species i in the condensed phase. The effective rate constants are calculated for all the elementary reactions by considering their both kinetic and diffusion rate constants. The kinetic rate constants are evaluated using the thermodynamic formulation of conventional transition state theory. Diffusion rate constants are estimated by extending the theory by Strehlow [11] for bimolecular reactions using quasi-static assumption.

The transient term in governing equations (1) is eliminated for steady state calculations. The resulting system of equations is treated as an initial value problem with marching along x −coordinate instead of time marching. The center of the coordinate axis (x=0) is aligned with the solid-liquid interface. Chemical reactions are assumed to be completely absent in the solid phase and hence mass fraction of RDX is taken to be 1 at x=0. The system is solved using DVODE routines developed by Brown et al. [12].

3. Results and Discussion

The model described in previous section was first simulated for atmospheric pressure using the experimental data provided by Zenin [7] for the melt-layer thickness, the burn rate, and the propellant burning surface temperature. The effect of uncertainty in various model parameters on extent of RDX decomposition in condensed phase at 1 atm was obtained. The uncertainty values in measured parameters are estimated using data provided by Zenin [7,13]. The activation energies for reactions in detailed condensed-phase reaction mechanism were evaluated using density functional theory with the hybrid functional B3LYP and 6-311++G(d,p) basis set and hence have uncertainty of ~3 kcal/mol. Lastly, the effect of pressure on condensed-phase decomposition was studied.

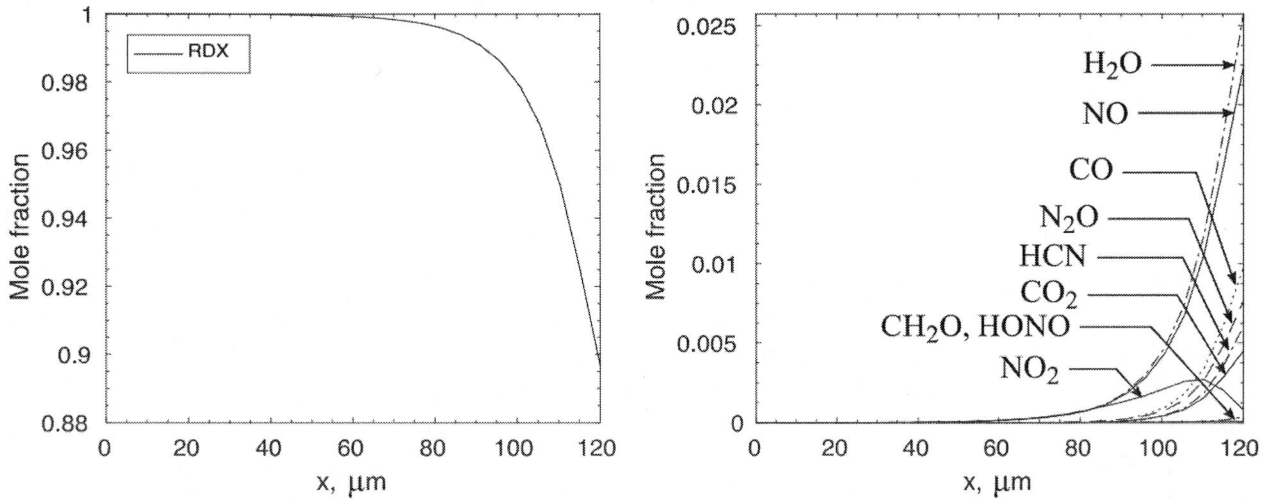

Figure 1. Species mole fraction variations inside melt layer during constant pressure combustion of pure RDX at 1 atm. x=0 and x=120 correspond to the solid-liquid interface and liquid-gas interface, respectively.

Figure 1 shows the obtained evolution profiles of various species in the melt layer during constant pressure combustion of RDX at 1 atm. Near solid-liquid interface (x=0) the temperature is low leading to slow decomposition rate of RDX and hence its mole fraction remains almost unity. Towards the liquid-gas interface (x=l_m) the temperature is higher and hence RDX decomposes quickly in this region. At propellant burning surface, RDX mole fraction reduces to 90% which corresponds to a mass fraction of 97%. This shows that a small amount of RDX is decomposed in condensed phase at 1 atm. The result is significantly different form earlier studies based on global condensed-phase mechanism. Liau and Yang [8] and Prasad et al. [9] evaluated the RDX mole fraction at the propellant burning surface and obtained 50% and 60%, respectively. Figure 1 also shows the variation of smaller species mole fractions in the melt layer. First stage oxidation products such as NO and H_2O are observed in relatively higher concentration. Decrease in NO_2 concentration occurs because of first-stage oxidation is also observed. CH_2O and HONO are quickly consumed as soon as they are formed.

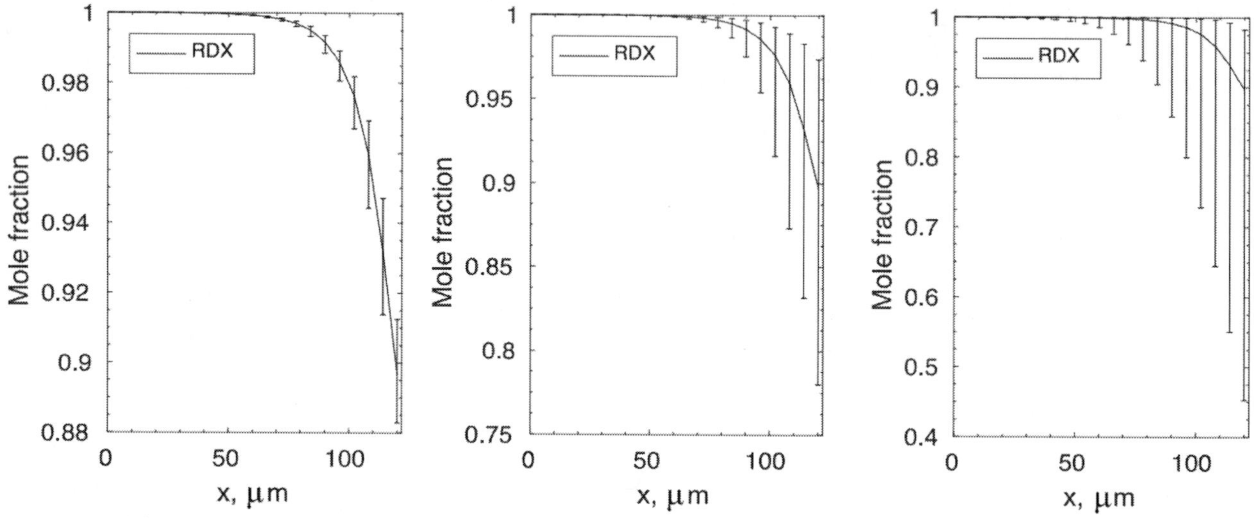

(a) $r_b = \pm$ 0.01 cm/s (b) $T_s = \pm$ 20 °C (c) $E_a = \pm$ 3 kcal/mol

Figure 2. Uncertainty in condensed-phase decomposition of RDX during constant pressure combustion at 1 atm.

Effect of uncertainty in various model parameters is studied. Figure 2 shows the range of uncertainty in condensed-phase decomposition of RDX introduced because of uncertainty in the burn rate (r_b), the propellant burning surface temperature (T_s), and activation energies (E_a) of chemical reactions. The calculations are performed for 1 atm pressure. Since there are 500 elementary reactions, only the result corresponding to most sensitive reaction is provided in Figure 2. The HONO-elimination step is found to be most sensitive for the condensed-phase region. The $N-NO_2$ homolytic cleavage step is not sensitive because of the barrier-less reverse reaction. Figure 2 shows that effect of uncertainty in r_b and T_s is limited to ~13% change in RDX mole fraction. But a change of 3 kcal/mol in the activation energy of most sensitive reaction can correspond to up to 45% change in RDX mole fraction. This explains the low mole fraction values of RDX at propellant burning surface obtained by Liau and Yang [8] and Prasad et al. [9]. The global condensed-phase reaction mechanism used by them has an activation energy of 36 kcal/mol compared to 38.3 kcal/mol for our condensed-phase HONO-elimination step.

Effect of external pressure on extent of RDX decomposition is studied for 1−90 atm range. Experimental data for r_b, T_s, and l_m are taken from Zenin [7]. With increase in pressure the thickness of melt layer decreases but its temperature increases. The burn rate also increases with pressure. The combined effect of these factors can be seen in Figure 3. The extent of RDX decomposition in the condensed-phase increases with increasing pressure. Using the global decomposition mechanism for the condensed phase, Liau and Yang [8] obtained the same RDX mole fraction at the propellant burning surface for both 1 and 90 atm case. That is, the importance of incorporating condensed-phase mechanism increases for high pressure cases. Figure 3 shows that in 1−90 atm range there is almost linear relationship between pressure and RDX mole fraction at the propellant burning surface, but not enough data points are present to support this observation.

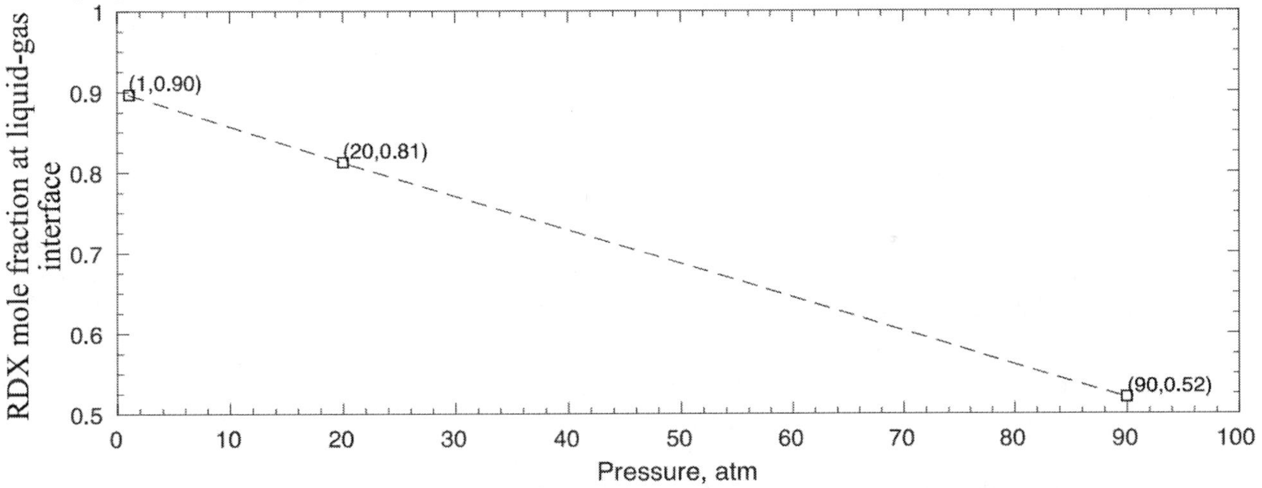

Figure 3. Effect of pressure on extent of RDX decomposition during constant pressure combustion.

4. Conclusions

A computational model for the melt layer formed during constant pressure combustion of RDX has been developed. The model is simulated for various pressure conditions to determine the extent of RDX decomposition in the condensed phase. The results show that the RDX mole fraction at the propellant surface is reduced to 90% at 1 atm. The mole fraction decreases to 52% as the pressure reaches 90 atm. This suggests that the condensed-phase decomposition mechanism becomes very important as the pressure increases. Hence, a detailed condensed-phase mechanism should be used instead of a global mechanism during simulation of constant pressure combustion of RDX. In the condensed phase, the extent of RDX decomposition is found to be most sensitive to the HONO-elimination step, which is the initiation step of the RDX decomposition. This indicates that HONO-elimination step is more important than $N-NO_2$ homolytic cleavage step in the condensed phase.

Our study also shows that the extent of RDX decomposition is very sensitive to activation energies of condensed-phase reactions. This suggests that there is a need to improve the existing condensed-phase quantum mechanics models. Bubble formation is experimentally observed in the melt layer and should be included in the melt layer model to improve its prediction capabilities. Furthermore, detailed gas-phase reaction mechanism should be used inside bubbles instead of using global reactions.

5. Acknowledgement

This research was funded by the U. S. Army Research Laboratory and the U. S. Army Research Office under grant number W911NF-15-1-0202.

6. References

[1] C.F. Melius, Thermochemical modeling: II. Application to ignition and combustion of energetic materials, in: Chem. Phys. Energ. Mater., Springer, 1990: pp. 51–78.

[2] R.A. Yetter, F.L. Dryer, M.T. Allen, J.L. Gatto, Development of gas-phase reaction mechanisms for nitramine combustion, J. Propuls. Power. 11 (1995) 683–697.

[3] D. Chakraborty, R.P. Muller, S. Dasgupta, W.A. Goddard, A detailed model for the decomposition of nitramines: RDX and HMX, J. Comput. Mater. Des. 8 (2001) 203–212. doi:10.1023/A:1020074113000.

[4] L. Patidar, S.T. Thynell, Quantum mechanics investigation of initial reaction pathways and early ring-opening reactions in thermal decomposition of liquid-phase RDX, Combust. Flame. 178 (2017) 7–20. doi:10.1016/j.combustflame.2016.12.024.

[5] O.P. Korobeinichev, L. V Kuibida, V.N. Orlov, A.G. Tereshchenko, K.P. Kutsenogii, R. V Mavliev, N.E. Ermolin, V.M. Fomin, I.D. Emel'yanov, Mass spectrometric probe study of the flame structure and kinetics of chemical reactions in flames, Mass Spectrosc. Chem. Kinet. (1985) 73–93.

[6] T.A. Litzinger, B.L. Fetherolf, Y. Lee, C.-J. Tang, Study of the gas-phase chemistry of RDX: experiments and modeling, J. Propuls. Power. 11 (1995) 698–703. doi:10.2514/3.23895.

[7] A. Zenin, HMX and RDX: Combustion mechanism and influence on modern double-base propellant combustion, J. Propul. Power. 11 (1995) 752–758. doi:10.2514/3.23900.

[8] Y.-C. Liau, V. Yang, Analysis of RDX monopropellant combustion with two-phase subsurface reactions, J. Propuls. Power. 11 (1995) 729–739. doi:10.2514/3.23898.

[9] K. Prasad, R.A. Yetter, M.D. Smooke, An eigenvalue method for computing the burning rates of RDX propellants, Combust. Sci. Technol. 124 (1997) 35–82. doi:10.1080/00102209708935640.

[10] W.R. Anderson, C.B. Conner, Comparison of gas-phase mechanisms applied to RDX combustion model, Proc. Combust. Inst. 32 (2009) 2123–2130.

[11] H. Strehlow, Rapid reactions in solution, VCH Weinheim etc., 1992.

[12] P.N. Brown, G.D. Byrne, A.C. Hindmarsh, VODE: A variable-coefficient ODE solver, SIAM J. Sci. Stat. Comput. 10 (1989) 1038–1051.

[13] A.A. Zenin, S. V. Finjakov, Response functions of HMX and RDX burning rates with allowance for melting, Combust. Explos. Shock Waves. 43 (2007) 309–319. doi:10.1007/s10573-007-0044-z.

Spring Technical Meeting
Eastern States Section of the Combustion Institute
March 4-7, 2018
State College, Pennsylvania

Impact of Chemically Termolecular Reactions on the Kinetics of Energetic Materials

Rodger E. Cornell[1,], Carly E. LaGrotta[1], Mark C. Barbet[1], Michael P. Burke[2]*

[1]*Department of Mechanical Engineering, Columbia University, New York, New York*
[2]*Department of Mechanical Engineering, Department of Chemical Engineering, and Data Science Institute, Columbia University, New York, New York*
[*]*Corresponding Author Email: rc3141@columbia.edu*

Abstract: Recent studies have shown that chemically termolecular reactions, mediated through ephemeral collision complexes, can significantly influence the global reactivity of various combustion systems. Specifically, it has been demonstrated that the reactions between H + O_2 collision complexes with other system radicals can significantly impact flame speed predictions in common combustion scenarios. The discovery of these major kinetic pathways has since prompted the question of how these chemically termolecular reactions will influence other reaction environments.

This study extends the investigation to RDX (1,3,5-trinitrohexahydro-s-triazine), for which a wealth of empirical and computational data exists, as a case study for energetic materials. Using rate constants for chemically termolecular reactions from recent master equation simulations, kinetic-transport simulations are employed to reveal the extent to which chemically termolecular reactions influence the reaction zone kinetics of RDX deflagrations and/or detonations. Results from preliminary studies, discussed herein, suggest that chemically termolecular reactions may have a meaningful impact on the kinetics of gas-phase RDX combustion, and further investigation is underway to better quantify and understand their effects.

Keywords: RDX, termolecular, kinetics, mechanism

1. Introduction

1,3,5-trinitrohexahydro-s-triazine, more commonly known as RDX, is a high-performance, military grade secondary explosive. Due to its high stability, relatively low production costs, and high energy content, RDX has been integrated into numerous propellant and explosive formulations since its inception is the 1930's. These favorable characteristics have led to its global adoption, which in turn has produced an extensive amount of empirical and computational data at various conditions.

Due to this widespread use, RDX has become one of the most studied and well characterized explosive materials to date. Over the past 40 years, there have been many efforts focused on the development of detailed kinetic models for the combustion of RDX in the gas phase [1-6] and models for condensed phase RDX decomposition [7-14].

Previous kinetic models to date for combustion, including those mentioned above for RDX, have been generally comprised of three types of phenomenological reactions in the gas-phase: unimolecular decomposition, bimolecular, and termolecular association reactions. Recent work of Burke and Klippenstein [15] has indicated that a new class of phenomenological reactions, chemically termolecular reactions, can also significantly influence global kinetics and should therefore be considered for potential inclusion to detailed kinetic models.

While the work Burke and Klippenstein [15] demonstrated that the reactions between H + O_2 collision complexes with other system radicals can significantly impact flame speed predictions in common combustion scenarios, it has been theorized that other combustion environments are likely affected in a similar manner by other chemically termolecular reactions. Here, we seek to quantify the potential role of this previously unconsidered class of reactions on RDX combustion as a case study for energetic materials.

2. Methods / Experimental

In this preliminary study, as a first step toward understanding the role of chemically termolecular reactions on energetic materials, calculations are performed under simplified conditions where the reaction occurs entirely in the gas phase, employing models based on the gas-phase detailed kinetic model from Chakraborty et al. [1]. Simulations using Cantera 2.3.0 were performed for mixtures consisting of gas-phase RDX at 10% mole fraction balanced with both argon and synthetic air (consisting of nitrogen and oxygen). The initial temperature and pressure of each mixture was set to 800 K and 1 atm, respectively, and the simulations were defined to be of constant volume, adiabatic, and zero-dimensional.

Simulations were performed for the original model of Chakraborty et al. [1] and modified versions of this model where chemically termolecular reactions have been appended using estimated rate constants. Comparing simulations for models with and without chemically termolecular reactions allows one to evaluate the role of chemically termolecular reactions in affecting predictions of key observables. Mole fractions of, NO, HCN, H_2O, N_2O, CO, and CO_2, known to be among the major species involved in RDX combustion [4,16], are shown for the original model [1] for RDX/air in Figure 1.

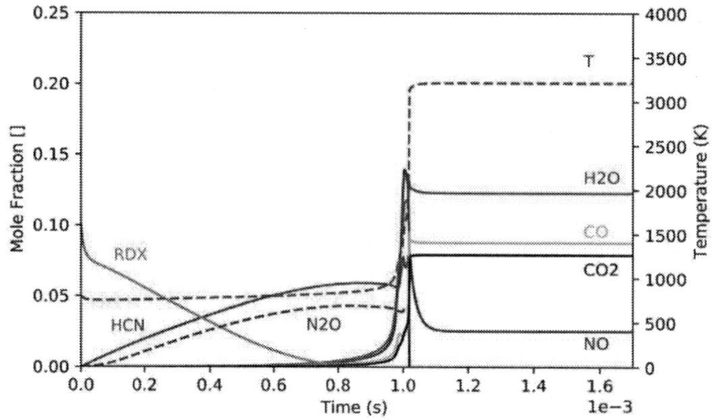

Figure 1: Mole fraction profiles from a constant volume simulation using the Chakraborty et al. RDX Mechanism. Initial mole fractions: 0.1 RDX, 0.7 N_2, 0.2 O_2.

A recently developed algorithm that autonomously discovers and estimates rate constants for potentially influential chemically termolecular reactions [17] was utilized here to create a list of potential chemically termolecular reactions to be appended to the model of Chakraborty et al. to evaluate their impact. These potential reactions are identified using the available kinetics information for all reactions currently in the mechanism. This analysis revealed 951 potential chemically termolecular reactions, all of which were added to the Chakraborty mechanism and the simulations were rerun. To determine which of these added reactions were most important to the system kinetics, a sensitivity analysis was performed to identify the most influential reactions for the previously specified constant volume conditions. Simulations were then rerun again with four of the most influential reactions appended individually to assess their individual impacts on predictions.

3. Results and Discussion

The results shown below were performed for simulations that only include the four reactions highlighted among the most sensitive reactions from the sensitivity analysis (listed in Table 1).

Table 1: Four chemically termolecular reactions identified to be of potential importance.

Number	Reaction	Designation
1	***H + HCN + HCN (+M) <=> CN + H2CNH (+M)***	***RDX+1***
2	***H + NO + NO2 (+M) <=> HONO + NO (+M)***	***RDX+2***
3	***H + NO + OH (+M) <=> H2O + NO (+M)***	***RDX+3***
4	***H + NO + O (+M) <=> NO + OH (+M)***	***RDX+4***

Simulations using this model, with the four selected chemically termolecular reactions included (shown in Figure 2), suggested significantly different concentration transients compared to the original model (shown in Figure 1).

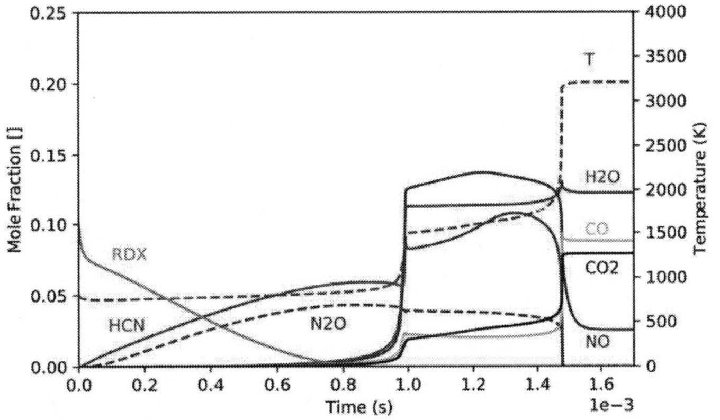

Figure 2: Mole fraction profiles from a constant volume simulation using the modified RDX Mechanism. Initial mole fractions: 0.1[RDX]/ 0.7[N_2]/ 0.2[O_2]

Comparison of Figure 1 with Figure 2 reveals that the chosen chemically termolecular reactions delay the onset of chemical equilibrium by approximately 0.5 ms. In an attempt to better understand why this occurs, further simulations were then performed with four separate models, each of which contains only one of the four chemically termolecular reactions. The observed delays in chemical equilibrium were then traced to the inclusion of certain reactions. Figure 3 shows the temporal variation in NO and CO concentrations between the various models (using model designations indicated in Table 1).

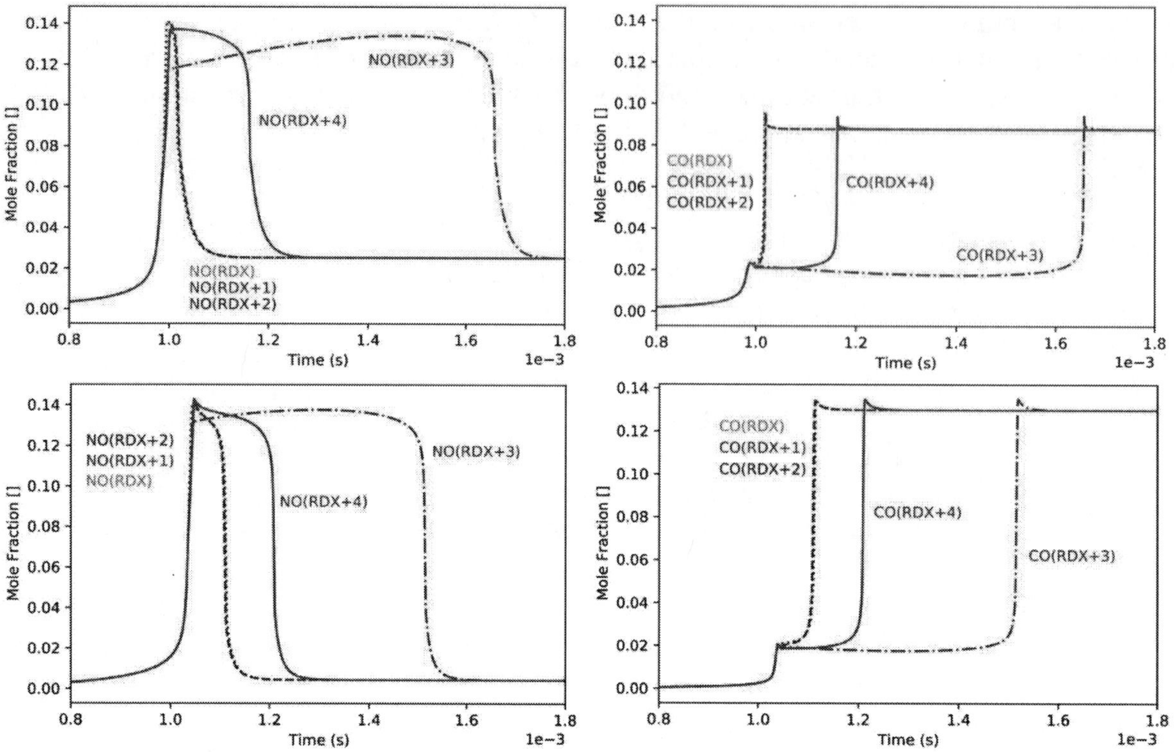

Figure 3: NO (left) and CO (right) mole fraction profiles from constant volume simulations using the original kinetic model of Chakraborty et al. and modified versions containing the four different chemically termolecular reactions, individually (see Table for model designation labels). Initial mole fractions: 0.1[RDX]/ 0.7[N_2]/ 0.2[O_2] (top), 0.1[RDX]/ 0.9[Ar] (bottom).

Figure 3 indicates that reactions 'RDX+1' and 'RDX+2' have little effect on major species mole fractions for the conditions here, while reactions 'RDX+3' and 'RDX+4' substantially delay the final stage of reaction. The results also suggest that the general trends observed for the impact of each chemically termolecular reaction are the same for both Ar and synthetic air – indicating that these reactions are of potential importance under both decomposition and combustion situations.

4. Conclusions

The results suggest that the four identified reactions listed in Table 1 may have a meaningful impact on the evolution of key species during RDX decomposition and combustion. Further investigation is necessary to arrive at more definite conclusions regarding the importance and quantitative impact of chemically termolecular reactions on decomposition and combustion of RDX.

5. Acknowledgements

The authors would like to acknowledge support from Columbia University and the DoD SMART Scholarship Program.

6. References

[1] D. Chakraborty, R.P. Muller, S. Dasgupta, W.A. Goddard, A detailed model for the decomposition of nitramines: RDX and HMX, Journal of Computer-Aided Materials Design 8 (2001) 203-212.

[2] C.F. Melius, Thermochemical modeling: I. application to decomposition of energetic materials, Chemistry and Physics of Energetic Materials (1990) 21-49.

[3] C.F. Melius, Thermochemical modeling: II. application to ignition and combustion of energetic materials, Chemistry and Physics of Energetic Materials (1990) 51-78.

[4] R.A. Yetter, F.L. Dryer, M.T. Allen, J.L. Gatto, Development of gas-phase reaction mechanisms for nitramine combustion, Journal of Propulsion and Power 11 (1995) 683-697.

[5] W.R. Anderson, C.B. Conner, Comparison of gas-phase mechanisms applied to RDX combustion model, Proc. Combust. Inst. 32 (2009) 2123-2130.

[6] T.A. Litzinger, B.L. Fetherolf, Y. Lee, C. Tang, Study of the gas-phase chemistry of RDX: experiments and modeling, Journal of Propulsion and Power 11 (1995) 698-703.

[7] K.V. Meredith, M.L. Gross, M.W. Beckstead, Laser-induced ignition modeling of HMX, Combust. Flame 162 (2015) 506-515.

[8] M.W. Beckstead, K. Puduppakkam, P. Thakre, V. Yang, Modeling of combustion and ignition of solid-propellant ingredients, Progress in Energy and Combustion Science 33 (2007) 497-551.

[9] A. Strachan, E.M. Kober, A.C.T. van Duin, J. Oxgaard, W.A. Goddard, Thermal decomposition of RDX from reactive molecular pathways, Journal of Chemical Physics 122 (2005) 054502.

[10] Y.-C. Liau, E.S. Kim, V. Yang, A comprehensive analysis of laser-induced ignition of RDX monopropellant, Combust. Flame 126 (2001) 1680-1698.

[11] L. Patidar. S.T. Thynell, Quantum mechanics investigation of initial reaction pathways and early ring-opening reactions in thermal decomposition of liquid-phase RDX, Combust. Flame 178 (2017) 7-20.

[12] T.P. Parr, D.M. Hanson-Parr, Solid propellant diffusion flame structure, Symp. (Int.) Combust. 26 (1996) 1981-1987.

[13] J.E. Davidson, M.W. Beckstead, A three-phase model of HMX combustion, Symp. (Int.) Combust. 26 (1996) 1989-1996.

[14] K.K. Irikura, Aminoxyl (Nitroxyl) radicals in the early decomposition of the nitramine RDX, Journal of Physical Chemistry A 117 (2013) 2233-2241.

[15] M.P. Burke, S.J. Klippenstein, Ephemeral collision complexes mediate chemically termolecular transformations that affect system chemistry, Nature Chemistry 9 (2017) 1078-1082.

[16] J.D. Cosgrove, A.J. Owen, The thermal decomposition of 1,3,5 Trinitro Hexahydro 1,3,5 Triazine (RDX) – Part I: The products and physical parameters, Combust. Flame 22 (1974) 13-18.

[17] M.C. Barbet, K. McCullough, M. P. Burke, A framework for automatic discovery of chemically termolecular reactions, submitted for publication.

Sub Topic: Fire

Spring Technical Meeting
Eastern States Section of the Combustion Institute
March 4-7, 2018
State College, Pennsylvania

An improved Test Method for Refrigerant Flammability Limits in a 12 L Vessel

*D.K. Kim, A.E. Klieger, P.Q. Lomax, C.G. McCoy, J.Y. Reymann, P.B. Sunderland**

Dept. of Fire Protection Engineering, University of Maryland, College Park, MD, 20742, USA
**Corresponding author email: pbs@umd.edu*

Abstract: An improved test method for refrigerant flammability limit measurements is presented. Such measurements are essential for determining the lower flammability limits of refrigerants, and thus their safety classifications. Predicated on expert interviews and experiments, several changes to ASTM E681 and related standards are recommended, as follows. The 12 L glass vessel should be replaced with transparent polycarbonate (or other transparent plastic). The orientation of the electrode supports and the temperature probe should be changed from vertical to horizontal to prevent flame quenching. All penetrations should be removed from the rubber stopper, it should be weighted for a total mass of 2.5 kg, and the initial pressure should be 90 kPa absolute. The flame angle should be plotted versus refrigerant concentration, whereby a least-squares line determines the flammability limit at a flame angle of 90°. These changes are relatively easy to implement and they improve the test precision and reproducibility without significantly changing previously established flammability limits.

Keywords: Combustion, fire, flame, 2L refrigerant

1. Introduction

An international drive toward improved sustainability of refrigeration systems [1–2] is motivating the adoption of refrigerants with low global warming potential (GWP) and low ozone depleting potential (ODP). Most of these refrigerants are mildly flammable, which is the main impediment to their adoption. As engineers balance refrigerant performance against sustainability and flammability, safety is always an important factor [3–5].

ANSI/ASHRAE 34 [6] establishes refrigerant flammability classifications based in part on the ASTM E681 [7] standard test method. These standards use visual observations of flame propagation in a 12 L spherical glass vessel to measure the lower flammability limits (LFLs) of refrigerants. Flammable conditions are defined as those for which a flame propagates outside a 90° cone angle, measured from the ignition point. This angle was chosen because it corresponded with refrigerant flammability limits in a 200 L cylindrical vessel [8]. The LFL measurements of ASTM E681 are essential in determining whether refrigerants or their blends are Class 1 (no flame propagation), Class 2 or 2L (LFL > 0.1 kg/m3), or Class 3 (LFL < 0.1 kg/m3). The ISO 817 [9] standard replaces the 0.1 kg/m^3 threshold with a refrigerant concentration of 3.5 vol. %.

Unfortunately, ASTM E681 suffers from limited precision and reproducibility. For example, it has led to published LFLs of R-32 (difluoromethane) in air of 13.48 vol. % [10], 14.4 vol. % [6, 11], 14.73 vol. % [12], and 14.8 vol. % [13]. Eight experts with extensive ASTM E681 refrigerant flammability experience were interviewed. They expressed several key concerns about this standard, as follows.

- Etching of the 12 L glass vessel requires frequent replacement and can obscure the flame behavior. Most A2L refrigerants contain hydrofluorocarbons (HFCs) and/or hydrofluoroolefins (HFOs), for which gaseous hydrogen fluoride (HF) can be the dominant product following combustion or heating [14–16].
- Electrode supports and the temperature probe in the flame path disturb the flames. The quenching distances of 2L refrigerants are typically far larger than those of hydrocarbons [12], e.g., 7.55 and 24.8 mm for R-32 and R-1234yf [17].
- Existing E681 apparatus have different venting behavior. Many laboratories consider flame observations after the onset of venting and/or after the flame completes its upward propagation to determine the LFL.
- The sample size is small in the pass/fail flammability criterion.

Motivated by these and other concerns, the objective of this study is to improve the precision and reproducibility of ASTM E681 for refrigerant flammability limit testing. New hardware and methods are developed and changes are recommended. Despite these changes the key strengths of the standard are maintained: the test apparatus is relatively inexpensive and easy to fabricate and the flames are observed visually.

2. Methods / Experimental

A standard ASTM E681 apparatus was built (see McCoy [13]; Lomax [18]). R-32 concentrations were established with partial pressures using an Ashcroft DG25 pressure gauge with a range of 0 – 102 kPa_a and a stated accuracy of ±0.5 kPa. Because the calibration was confirmed frequently at vacuum and at atmospheric pressure, the resulting R-32 concentrations have an estimated uncertainty of ±0.1 vol. %.

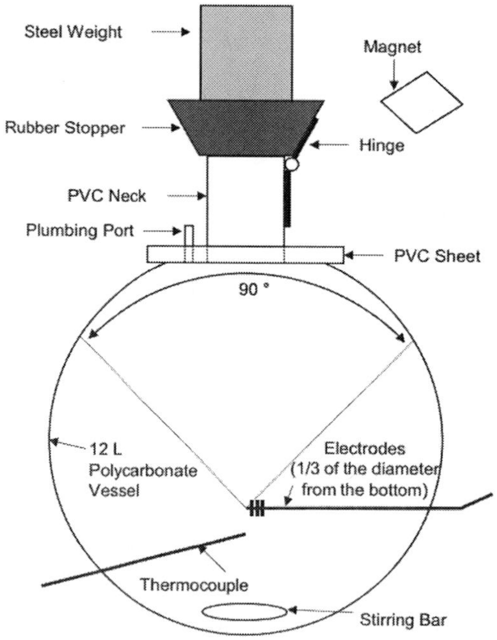

Figure 1: Schematic of the polycarbonate vessel design. The vessel as tested did not have a thermocouple, and used a rope and pulley instead of a magnet and hinge.

This apparatus was then modified to replace its glass vessel with polycarbonate. The spherical part of this vessel was a clear polycarbonate lighting globe (Edith Aiken Company, US$45). Its capacity and dimensions are similar to those of the standard 12 L glass vessel.

Figure 1 shows a schematic of the polycarbonate vessel design. A 13 mm thick polyvinylchloride (PVC) sheet with a 60 mm hole was attached to the top of the lighting globe with room-temperature-vulcanizing silicone adhesive. A PVC tube with an inner diameter of 53 mm was then attached to the PVC sheet (see Reymann [19]).

Two resealable holes were drilled in a rubber stopper, and two others in the wall of the polycarbonate vessel, to introduce the electrodes vertically or horizontally. The location, spacing, and orientation of the tungsten electrode tips were the same in both cases. Glass sheaths surrounding the stainless steel electrode supports are specified in E681, but these complicate the sealing and were not used here. These were not necessary because no electrical current was observed except between the electrode tips. Tests were performed with R-32, whose LFL is not sensitive to humidity [10], in dry air at 21 – 23 °C. The spark parameters were 15 kV, 30 mA, and 0.2 s [10, 20–21], where the duration was reduced from 0.4 s in accordance with ASHRAE 34 [6].

The tests were performed with a thin-film pressure transducer connected to a pressure tap in the vessel wall. The transducer was a PCB 1501B02EZ100psig with a response time of 1 ms, a stated range of 0 – 690 kPa_g, and an accuracy of ±1.7 kPa. It was found to maintain this accuracy for pressures as low as –34 kPa_g.

The polycarbonate vessel facilitates moving all penetrations from the rubber stopper to the vessel walls. However, the 270 g mass of the size 14 rubber stopper without these attachments is too low to prevent venting before the flames reach the vessel wall. It was found that a weighted stopper with a total mass of 2.5 kg was ideal. For typical refrigerant test conditions, this resulted in venting that occurred soon after the flame reached the vessel wall. The 2.5 kg mass also prevented leakage during hold times near atmospheric pressure. For tests at high elevation, a higher mass would be necessary.

Just prior to the end of flame propagation, an image of each flame was used to measure its flame angle – the angle subtended by the flame with respect to the electrode gap. These angles were measured using ImageJ software.

Figure 2: Color images of the vessels used. Shown here are: (a) a new glass vessel; (b) a glass vessel that has been etched by 10 flame tests followed by immediate flushing; and (c) a polycarbonate vessel following 68 similar tests. Behind each vessel is a 30 cm ruler.

3. Results and Discussion

A new 12 L glass flask is shown in Fig. 2a. Unfortunately, glass is prone to HF etching and is difficult to drill. Glass etching is clearly visible in Fig. 2b following 10 tests of R-32 in air near its

LFL. The polycarbonate vessel (Fig. 2c) is nearly as transparent as a new glass vessel, but showed no signs of etching, bubbling, or discoloration after 68 similar tests. It was readily drilled for gas-tight penetrations for the electrodes and plumbing.

Figure 3 illustrates the effects of electrode support orientation on two representative flames just before they reached the vessel wall. All other conditions were matched. As seen in Fig. 3, vertical electrode supports cause a large hole in top of the flame, a dimmer and less symmetric flame, and a reduced flame angle (defined below). Such disturbances can change the LFL determination and impair the test precision and reproducibility.

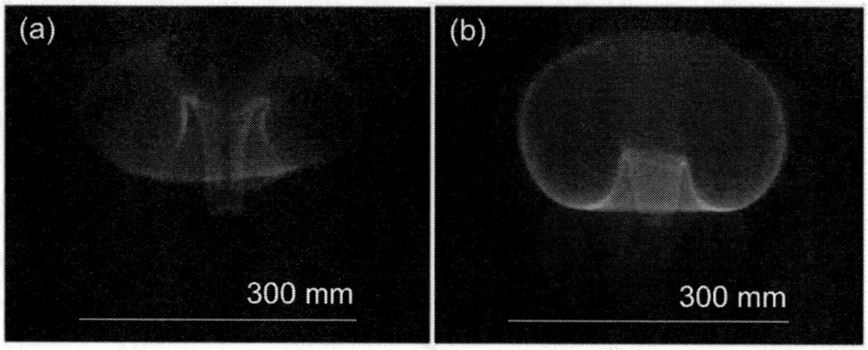

Figure 3: Effects of electrode support orientation, namely (a) vertical and (b) horizontal. The initial composition was 14.8 vol. % R-32 in air at 101 kPa_a. The images were recorded 0.35 s after ignition.

The pressures were measured with the time datum corresponding to the first video frame for which a spark was visible. With horizontal electrode supports the pressure increased faster, which confirms that vertical electrode supports weakened the flame.

With same time datum, it was determined when flame propagation stopped near the vessel wall and when venting started. Flame propagation was observed in the video record, and venting was identified using the pressure transducer and a microphone near the stopper. The flames reached the wall after approximately 0.36 s regardless of R-32 concentration, but venting started earlier with increasing R-32 concentration. In all cases flame propagation stopped before venting started. Similar behavior was observed at initial pressures of 81 and 101 kPa_a [22].

ASTM E681 [7] stipulates an initial pressure of 101 kPa_a. However, this has three drawbacks: it can result in venting before the completion of flame propagation; the mean pressure during a test is above atmospheric; and laboratories at high elevations cannot easily follow the standard. Therefore, testing was conducted at initial pressures of 81, 91, and 101 kPa_a. To be close to atmospheric pressure without a risk of venting during flame propagation, an initial pressure of 90 kPa_a is recommended.

Figure 4 shows a plot of flame angle versus R-32 concentration. A linear best fit line is shown and this has a R2 coefficient of determination of 0.92. This line's intersection with 90° is used here to determine the LFL [13, 19, 23]. The 95% confidence interval curves are also shown. This yielded an LFL of 14.8 vol. % with a 95% confidence interval of ±0.1 vol. %. This is slightly higher than the generally accepted R-32 LFL of 14.4 vol. % [6, 11]. The method of Fig. 6 incorporates several flame angle measurements into the determination of the LFL, compared to only two out of three tests above 90°, according to ASTM E681. Additionally, it yields a 95% confidence interval on the LFL.

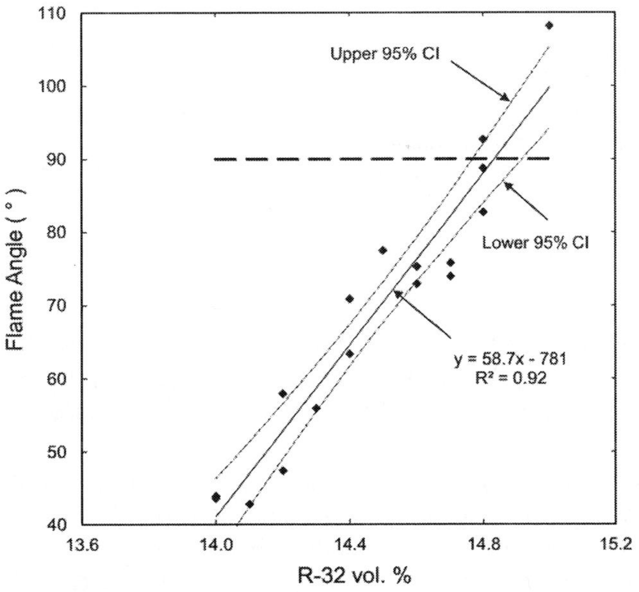

Figure 4: Flame angle plotted with respect to R-32 concentration. The initial pressure was 91 kPa_a.

The modifications proposed above result in only a small change in the measured LFL of R-32. The vertical electrodes decrease the LFL whereas preventing venting before the flame reaches the vessel walls increases it. At least for R-32, this maintains the agreement between these 12 L tests and the large-scale tests of Richard [8].

Conclusions

Predicated on expert interviews and experiments, this study recommends several changes to the ASTM E681 standard test method for refrigerants, as follows. The vessel material should be changed from glass to polycarbonate (or other transparent plastic) to facilitate vessel penetrations and to eliminate etching and frequent vessel replacement. The electrode supports and temperature probe should be oriented horizontally rather than vertically to minimize flame quenching. Venting should not occur before the flame stops propagating near the vessel wall. This can be accomplished with an initial pressure of 90 kPa_a and by having a stopper with no penetrations and a mass of 2.5 kg. The flame angle should be plotted versus refrigerant concentration for at least 6 concentrations near and on both sides of the LFL. The measurements should be fit with a least-squares line. Where this line intersects a flame angle of 90° is the LFL.

Acknowledgements

This research was funded by ASHRAE grant RP-1717, with Debra Kennoy serving as technical contact. Laboratory assistance was provided by V. Lecoustre, A. Hermann M. Kokot, and. N. Thomas.

References

[1] J.S. Brown, Fourth ASHRAE/NIST refrigerants conference: "moving towards sustainability", HVAC&R Research 19 (2013a) 101-102.

[2] S. Kujak, Flammability and new refrigerant options, ASHRAE J. 59 (2017) 16-24.

[3] W.-T. Tsai, An overview of environmental hazards and exposure risk of hydrofluorocarbons (HFCs), Chemosphere 61 (2005) 1539–1547.
[4] J.S. Brown, Introduction to hydrofluoro-olefin alternatives for high global warming potential hydrofluorocarbon refrigerants, HVAC&R Research 19 (2013b) 693-704.
[5] S. Kujak, K. Schultz, Insights into the next generation of HVAC&R refrigerant future, Science and Technology for the Built Environment 22 (2016) 1226-1237.
[6] ASHRAE 34, ANSI/ASHRAE Standard 34-2016, Designation and Safety Classification of Refrigerants. Atlanta, GA: American Society of Heating, Refrigerating and Air-Conditioning Engineers, Inc, 2016.
[7] ASTM 2015. ASTM E681-09, Standard Test Method for Concentration Limits of Flammability of Chemicals. West Conshohocken, PA: American Society of Testing and Materials, 2015.
[8] R.G. Richard, Refrigerant flammability testing in large volume vessels. ARTI Report 665-52400. Arlington, VA: Air-Conditioning, Heating, and Refrigeration Institute, 1998.
[9] ISO 817, Refrigerants — Designation and Safety Classification. Geneva: International Organization for Standardization, 2014.
[10] S. Kondo, K. Takizawa, K. Tokuhashi, Effects of temperature and humidity on the flammability limits of several 2L refrigerants, Journal of Fluorine Chemistry 144 (2012) 130-136.
[11] D. Wilson, R. Richard, Determination of refrigerant lower flammability limits in compliance with proposed addendum p to standard 34. ASHRAE Transactions 108 (2002) 739-756.
[12] I. Kul, D.L. Gnann, A.L. Beyerlein, D.D. DesMarteau, Lower flammability limit of difluoromethane and percolation theory, International Journal of Thermophysics 25 (2004) 1085-1095.
[13] C. McCoy, Improved Venting for Flammability Limit Testing Using ASTM E681 Apparatus, M.S. Thesis, Univ. of Maryland, Dept. of Fire Protection Engineering, 2016.
[14] C. Womeldorf, W.L. Grosshandler, Flame extinction limits in CH2F2/air mixtures, Combust. Flame 118 (1999) 25-36.
[15] A. Moghaddas, C. Bennett, E. Rokni, and H. Metghalchi, Laminar burning speeds and flame structures of mixtures of difluoromethane (HFC-32) and 1,1-difluoroethane (HFC-152a) with air at elevated temperatures and pressures, HVAC&R Research 20 (2014) 42-50.
[16] B. Feng, Z. Yang, R. Zhai, Experimental research on the concentration characteristics of R32 and R161 combustion product HF, Energy 125 (2017) 671-680.
[17] K. Takizawa, N. Igarashi, S. Takagi, K. Tokuhashi, and S. Kondo, Quenching distance measurement of highly to mildly flammable compounds, Fire Safety Journal 71 (2015) 58-68.
[18] P.Q. Lomax, Development of the ASTM E681 Standard. M.S. Thesis, Univ. of Maryland, Dept. of Fire Protection Engineering, 2016.
[19] J. Reymann, Improvements to Refrigerant Flammability Testing Through use of a Polycarbonate Apparatus, M.S. Thesis, Univ. of Maryland, Dept. of Fire Protection Engineering, 2017.
[20] S. Kondo, Y. Urano, A. Takahashi, K. Tokuhashi, Reinvestigation of flammability limits measurement of methane by the conventional vessel method with AC discharge ignition, Combustion Science and Technology 145 (1999) 1-15.
[21] D. Clodic, T. Jabbour, Method of test for burning velocity measurement of flammable gases and results, HVAC&R Research 17 (2011) 51-75.
[22] A.E. Klieger, Pressure Measurements for Flammability Limit Testing Using ASTM E681 Apparatus, MS. Thesis, Univ. of Maryland, Dept. of Fire Protection Engineering, 2017.
[23] K. Takizawa, K. Tokuhashi, S. Kondo, Flammability assessment of CH2=CFCF3: Comparison with fluoroalkenes and fluoroalkanes, Journal of Hazardous Materials 172 (2009) 1329–1338.

Sub Topic: Fire

Spring Technical Meeting
Eastern States Section of the Combustion Institute
March 4-7, 2018
State College, Pennsylvania

Deflagrations of Mildly Flammable Refrigerant-Air Mixtures in Closed Volumes

K. C. Gottiparthi[1], P. Papas[1], V. Sankaran[1], P. Verma[1], R. Lord[2] and L. Burns[2]*

[1]*United Technologies Research Center, East Hartford, CT 06108*
[2] *Carrier Corporation, Indianapolis IN 46231*
**Corresponding Author Email: papasp@utrc.utc.com*

Abstract: With current concerns around global climate change, new hydrofluorocarbons with low Global Warming Potential (GWP) are being evaluated as alternatives. These alternative refrigerants, however, are mildly flammable and pose safety concerns for the heating, ventilation, air conditioning, and refrigeration industry. Consequently, careful assessment of different flammability characteristics and risks for these refrigerants is essential for their safe use in different applications. In this study, we performed tests and numerical simulations for two refrigerants, difluoromethane (R-32) and 2,3,3,3-tetrafluoropropene (R-1234yf), in a closed 3m container to identify important characteristics related to deflagration propagation in closed volumes. The thermodynamic property data as well as chemical kinetic and transport descriptions for these reaction systems were calibrated based on available experimental flame speed data as well as our detailed chemical mechanism available in the literature. Large Eddy Simulations (LES) were performed to analyze global flame structure and propagation differences in uniformly mixed, stoichiometric R32-air and stoichiometric R-1234yf/air mixtures. For the R-1234yf case, results demonstrate important buoyancy-induced effects whereby initial deflagration "bubbles" form upon ignition from a hot surface point and propagate vertically upwards with relatively larger velocities than in the lateral direction. Unlike the R-32/air case, which produces a fireball propagating almost uniformly in all directions, the R-1234yf/air case initially develops jet-like flame features that propagate vertically upwards, impinges upon the enclosure ceiling and transitions to rapidly consume the entire remaining flammable mixture within the empty enclosure. The stages of flame propagation along with the comparisons with the experimental data are discussed.

Keywords: flammable refrigerant, buoyancy, premixed flame, confinement

1. Introduction

Studies on the combustion characteristics of hydrofluorocarbon (HFC) mixtures with air have been motivated by numerous practical interests, including use as fire suppressants and refrigerants. More recent research on new low Global Warming Potential (GWP) refrigerants that are also mildly flammable, as defined by the A2L safety group classification [1], are motivated by concerns from the Heating, Ventilation, Air Conditioning, and Refrigeration (HVACR) industry not commonly accustomed to dealing with flammability safety issues. Due to environmental regulations, these fluids such as 2,3,3,3-tetrafluoropropene (R-1234yf) are just now entering the marketplace despite the fact that their flammability characteristics are still not well understood. Although there have been some investigations of fundamental combustion properties such as the laminar flame speed and flammability limits [2-5], practical studies of HFC-air flame measurements in large or congested spaces, where turbulent effects become important, have not been undertaken. Given the current interest in employing mildly flammable refrigerants in both residential and commercial HVAC&R systems, understanding the severity of HFC-air events in practical geometries, the focus of the current study, has become imperative, in particular in large or confined spaces common to most applications.

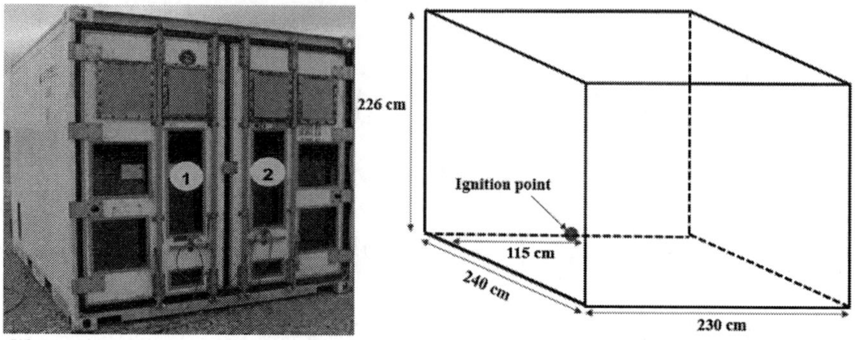

Figure 1: (Left) Closed testing container showing front doors and pressure transducer locations for PT1 and PT2. (Right) Schematic simulation domain and back wall ignition point location

2. Experimental Description

To characterize risks for using these mildly flammable HFCs, flammability testing in different HVAC&R configurations has been undertaken to quantify severity outcomes upon ignition in practical application accidents [6]. These data quantify the overpressures and/or turbulent deflagration speeds generated in practical application geometries; turbulent deflagration speeds for well-mixed, refrigerant-air mixtures in practical geometries have been shown to be orders of magnitude larger than their corresponding laminar flame speed values that are used in classifying A2L flammable refrigerants. Flammability testing [6] conducted in an empty container, reported here without congestion, for leaked refrigerants (R-32 and R-1234yf) has quantified the increased severity of a combustion event, as measured by the resulting overpressure, with increasing congestion or confinement as a consequence of increased induced turbulence. These data will be used for verification of numerical simulations to be discussed in the next section. The closed volume geometry is shown in Fig. 1, and was nearly cubical with dimensions of about 2.26 x 2.30 x 2.40 m. For the tests discussed, the ignition source was a glow plug located near the floor in front of the rear wall, generating a surface temperature exceeding 1000°C. Overpressure measurements were provided by two calibrated pressure transducers (Kulite Model XT-190-10D) at locations 1 and 2 shown in Fig. 1.

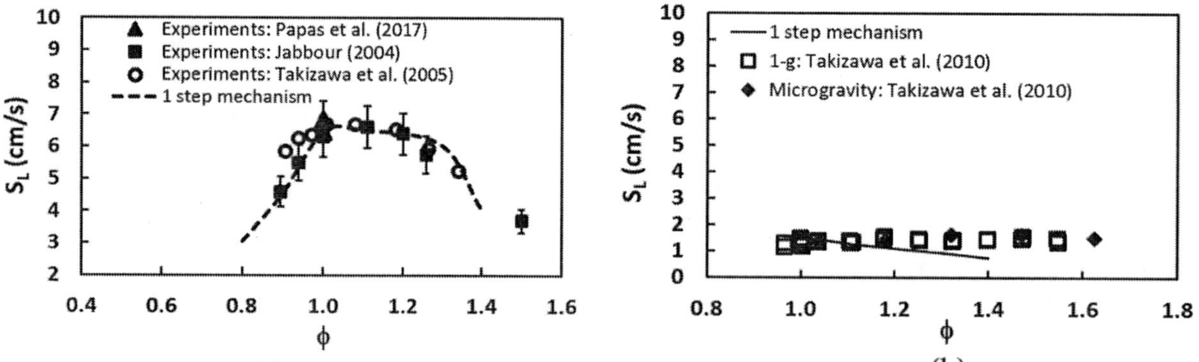

Figure 2: Variation of laminar flame speed with equivalence ratio for (a) R-32/air and (b) R-1234yf/air mixtures

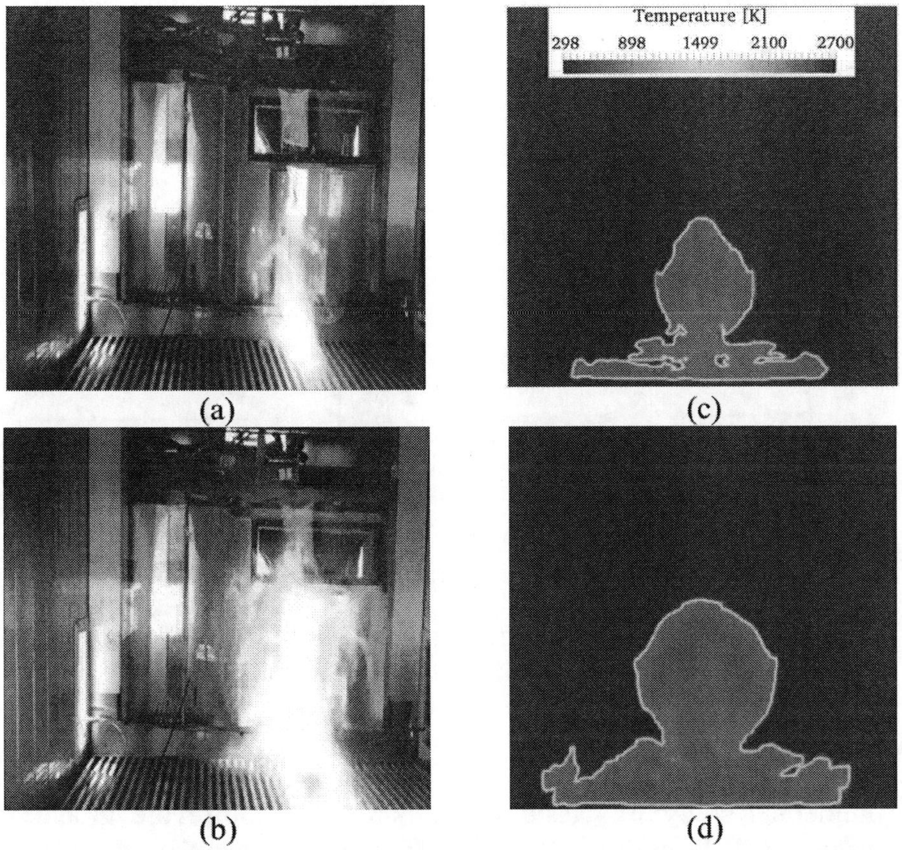

Figure 3: Evolution of a deflagration in a stoichiometric R32-air mixture in a closed volume. (a) and (b): Test images [Test#1], (c) and (d): Temperature isocontours from current numerical simulations

Different tests were conducted with uniform mixing of the refrigerant-air mixture with internal blowers prior to ignition. For the tests shown, the room was empty, quiescent, and without any obstructions; hence, these data served to verify the simulation methodology. Tests were performed for two different mildly flammable refrigerants, R-32 and R-1234yf, with air at near stoichiometric conditions. Although testing demonstrated the importance of turbulent effects for large-scale applications, the reported maximum laminar flame speeds of R-32 and R-1234yf are about 6.7 [2-3] and 1.2 cm/s [2, 4-5], respectively, and shown in Fig. 2. Since the maximum laminar burning velocity is less than 10 cm/s, both refrigerants fall into the A2L safety group classification designated by ASHRAE Standard 34 [1].

Comparison of testing results for R-32 (Test#1) and R-1234yf (Test#2) deflagrations demonstrated important differences. Figure 3(a) and (b) shows testing images of a stoichiometric R-32 deflagration (Test#1) near onset. The corresponding measurements for the temporal evolution of overpressure are shown in Fig. 4. The R-32 deflagration, which generated a peak overpressure of 3.3 psig at about 1 s after ignition, evolved in a manner typical for most hydrocarbon fuel-air systems whereby the initial propagation rates in the lateral and vertical directions were comparable. For the R-1234yf case (Test#2), test results demonstrated important buoyancy-induced effects whereby initial deflagration "bubbles" form upon ignition from the hot surface point and propagate vertically upwards with relatively larger velocities than in the lateral direction.

Unlike the R-32/air mixture case, which produced a fireball propagating almost uniformly in all directions, the R-1234yf/air mixture initially developed jet-like flame features that propagated vertically upwards (shown in Fig. 5(a) and (b)), impinged upon the enclosure ceiling and transitioned to rapidly consume the entire remaining flammable mixture within the empty enclosure.

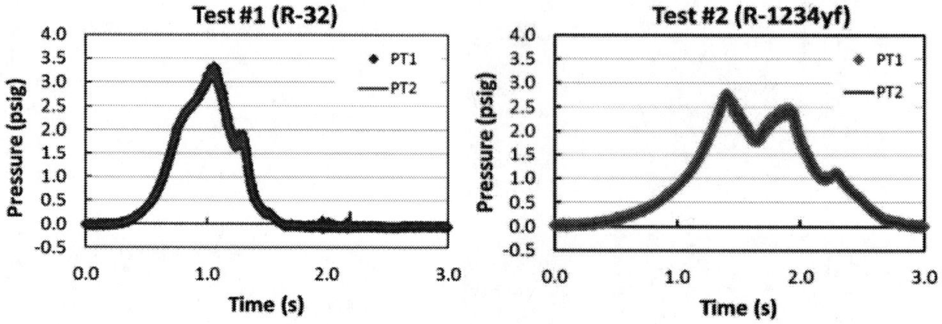

Figure 4: (Left) Overpressure evolution for stoichiometric R-32/air mixture. (Right) Overpressure evolution for stoichiometric R-1234yf/air mixture

These two distinct R-1234yf propagation stages resulted in a double peaked overpressure evolution shown in Fig. 4, where the maximum overpressure was about 2.8 psig.

3. Numerical Simulations

A preliminary numerical study was used to qualitatively capture the global flame structure differences between the R-32 and R-1234yf mildly flammable refrigerant deflagrations. The unsteady, compressible, filtered Navier-Stokes equations for reacting gas flows were solved [7-8]. The complete set of governing equations along with the turbulence closure used to perform Large Eddy Simulations (LES) in the current study can be found elsewhere [7]. An approximate Riemann solver with Monotone Upstream-centered Schemes for Conservative Laws (MUSCL) reconstruction and monotonized-central limiter was used to solve the governing equations. The Riemann solver used was a hybrid solver, which employed the Harten-Lax-van Leer (HLL) flux solver with contact wave (HLLC) everywhere except in the shock-capturing region, where in the direction transverse to the shock, the HLLE (HLL by Einfeldt [9]) Riemann solver was used. The scheme was second order accurate in both space and time. Detailed validation of this approach to simulate shock-turbulence interaction [8] and detonations/ explosions in multi-phase flows [7] has already been reported. Single-step Arrhenius kinetic descriptions were employed and developed based on past laminar flame speed measurements as well as a detailed chemical HFC mechanism [2]. The product species considered were CO_2 and HF for the 1-step R-32/air reaction, and for the R-1234yf/air reaction were CO_2, HF and CF_2O. Thermodynamic and transport data were taken from Papas et al. [2]. The expression for the overall reaction rate ($\dot{\omega}$) for R-1234yf is given as:

$$\dot{\omega} = 1.5 \times 10^{28} \chi_{R1234yf}^{1.0} \chi_{O_2}^{2.5} \exp(-43100/T),$$

where χ is the concentration of the reactants, and T is the temperature. For R-32 the overall reaction rate was determined as:

$$\dot{\omega} = 7.0 \times 10^{18} \chi_{R_{32}}^{1.0} \chi_{O_2}^{1.0} \exp(-38500/\mathrm{T}).$$

Although all container tests were well-mixed at near stoichiometric conditions, the laminar flame speeds were computed at different equivalence ratios (ϕ) using the single step expressions given and compared with the values obtained from full detailed kinetics [2] and experimental data shown in Fig. 2. Laminar flame speed values for both R-1234yf and R-32 show good agreement with literature results for $\phi = 1.0$. The domain shown in Fig. 1 was discretized using a structured grid, and configured to be fine in the regions with flame and flow structures, and coarsened (Δx_{max} = 5.0 cm) elsewhere in regions away from the flame structures. The minimum computational cell width, Δx_{min} = 1.0 cm, was chosen to resolve the laminar flame thickness in each case.

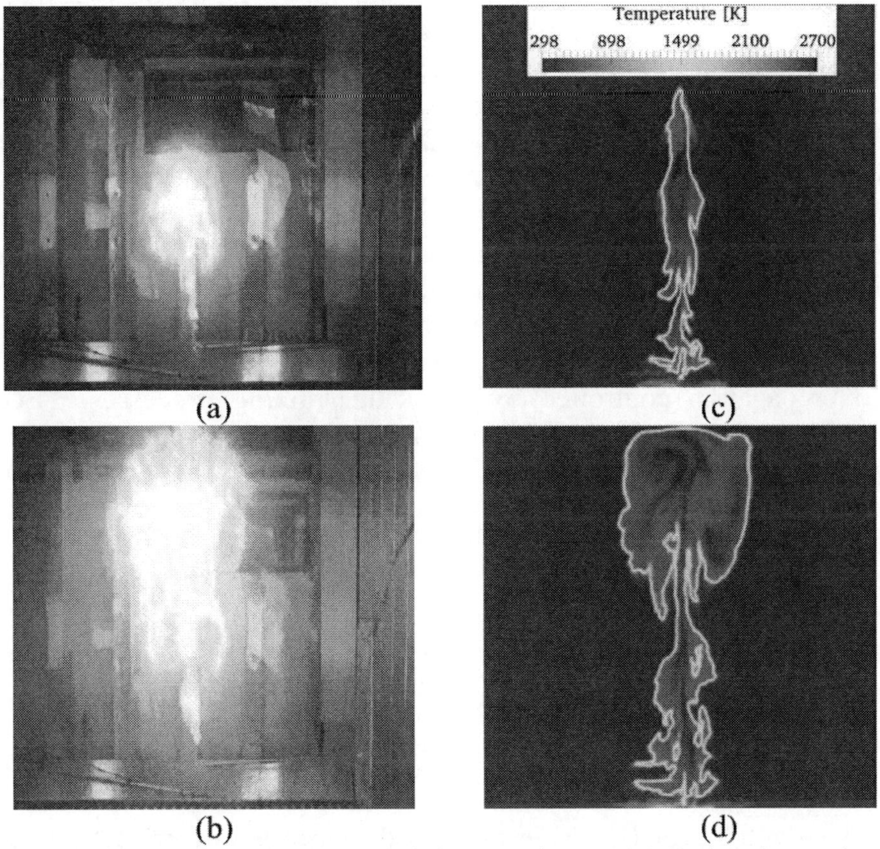

Figure 5: Evolution of a deflagration in a stoichiometric R1234yf-air mixture in a closed volume. (a) and (b): Test images [Test#2], (c) and (d): Temperature isocontours from current numerical simulations

As in tests, the initial gas pressure and temperature was 1 atm and 300 K, respectively. The adequacy of the chosen parameters and physical combustion models will be verified in future work. Although three-dimensional simulations are being considered, the first simulations reported here in Figs. 3 and 5 were two-dimensional using a 2.30 x 2.26 m domain for the closed volume. The flame was ignited at the center of the domain as indicated in Fig. 1, and the ignition source was simulated by means of a 4000 K hot spot. All enclosure walls incorporated adiabatic, no-slip reflecting surfaces.

4. Discussion and Conclusions

The preliminary two-dimensional LES results are shown in Figs. 3 and 5 in comparison with test images at comparable temporal evolution stages. Similar to testing results, the numerical simulation in Fig. 3 shows that the R-32 deflagration evolves in a manner whereby the initial lateral and vertical propagation rates are comparable. On the other hand, both the numerical and testing results for initial R-1234yf deflagration propagation reveal very unique features. Since the laminar flame speed for R-1234yf is extremely low, almost six times lower than the R-32 value, buoyancy-induced effects are important for the initial deflagration onset. Due to these effects, the vertical propagation rate is initially large in comparison to the lateral rate; consequently, flame "bubbles" or structures formed that preferentially propagated to the room ceiling relatively quickly in comparison to the container side walls. After impingement of these vertical jet-like features upon the container ceiling, the deflagration rapidly propagated laterally along the ceiling and throughout the room. These distinct propagation stages give rise to a double-peaked overpressure evolution with time, as shown in Fig. 4. Despite the almost six-fold difference in laminar flame speeds, the maximum peak pressures from the R-32 and R-1234yf deflagrations in our empty enclosure tests differ by only 18%. Future work is underway to perform three-dimensional LES to facilitate quantitative comparisons with experimental overpressure and flame propagation measurements, including testing with congestion not reported here. This work illustrates the importance for severity evaluations of actual geometries of concern such as in large or congested spaces. Under these conditions, the propagation of deflagrations in confined spaces, for example, is known to be accelerated through the effects of turbulence which can be much more significant in comparison to laminar flame propagation controlled by molecular mixing processes. Even for mildly flammable refrigerants characterized by laminar flame speeds < 2 cm/s, relatively fast deflagrations, i.e. large overpressures, can be generated for very congested geometries where downstream turbulence is generated as the flame front passes over obstacles in these situations.

5. References

[1] ANSI/ASHRAE Standard 34-2013. Designation and Safety Classification of Refrigerants. Atlanta: ANSI/ASHRAE. 1791 Tullie Circle, NE, Atlanta, GA 30329-2305, ISSN 1041-2336.

[2] P. Papas, S. Zhang, W. Kim, S. P. Zeppieri, M. B. Colket, P. Verma, Laminar Flame Speeds of 2,3,3,3-tetrafluoropropene Mixtures, Proc. Combust. Inst. 36 (2017) 1145-1154.

[3] T. Jabbour, Flammable Refrigerant Classification Based on the Burning Velocity, Ph.D. Thesis, L'Ecole des Mines de Paris, 2004.

[4] K. Takizawa, A. Takahashi, K. Tokuhashi, S. Kondo, A. Sekiya, Burning velocity measurement of fluorinated compounds by the spherical-vessel method, Combust. Flame 141 (2005) 298–307.

[5] K. Takizawa, K. Tokuhashi, S. Kondo, M. Mamiya, H. Nagai, Int. Symp. Next-gen. Air Cond. Refrig. Technol., Tokyo, Japan, 2010.

[6] P. Papas, P. Verma, L. Burns, W. Clough, R. Lord, "Turbulent Hydrofluorocarbon Deflagrations in Confined Spaces," to be submitted to the Journal of Fire Safety. Also, "Flammability Testing in HVAC Configurations using Low GWP, Refrigerants," NIST Workshop on the Exothermic Reaction of Halogenated Hydrocarbons, October 27-28, 2014.

[7] K. C. Gottiparthi, S. Menon, A Study of Interaction of Clouds of Inert Particles with Detonation in Gases, Combust. Sci. Technol. 184 (2012) 406-433.

[8] F. Genin, S. Menon, Studies of Shock/Turbulent Shear Layer Interaction using Large-Eddy Simulation, Comput. Fluids 39 (2010) 800-819.

[9] B. Einfeldt, SIAM J. Numer. Anal. 25 (1988) 294-318.

Sub Topic: Fire Research

Spring Technical Meeting
Eastern States Section of the Combustion Institute
March 4-7, 2018
State College, Pennsylvania

Predicted Burning Velocities of C1 and C2 Hydrofluorocarbon Refrigerant Flames with Air

Gregory Linteris[1], Valeri Babushok[1]
[1]Engineering Laboratory, National Institute of Standards and Technology, Gaithersburg, MD, 20899

Abstract: Due to their high global warming potentials, many existing working fluids for heating, cooling and refrigeration equipment will be phased out. Their replacements will be flammable or slightly flammable, and the burning velocity of refrigerant-air mixtures is being used as a metric to rank their flammability. To allow industry to estimate the flammability of new blends of agents, predictive tools for the burning velocity of refrigerants are being developed, and a kinetic mechanism is required. The NIST hydrofluorocarbon HFC mechanism was developed 20 years ago to describe hydrocarbon-air flames with added trace amounts of hydrofluorocarbon fire retardants (primarily CH_2F_2, CF_3H, CF_4, $C_2H_2F_4$, C_2HF_5, and C_2F_6). The mechanism has been modified to include new HFC compounds, more recent rate data, and rate data for new species. The modified mechanism is used to predict steady, planar, 1D, unstretched burning velocities for mixtures of each refrigerant with air, over a range of fuel-air equivalence ratio ϕ, for comparison with the available experimental data. Agreement is good, and major kinetic pathways and the radical populations are explored to uncover the general reaction properties of these new flames.

Keywords: Refrigerant flammability, Burning velocity, Low GWP refrigerants, C1-C2 HFCs

1. Introduction

Existing refrigerant working fluids in vapor-compression cooling equipment have high global warming potential (GWP), and are being phased out through international treaties (i.e., the Kigali Agreement, which is an extension of the Montreal Protocol.). Low-GWP Replacements have been developed, primarily by adding double bonds or hydrogens to the molecules, which makes them break down in the troposphere. Unfortunately, these properties also make them more flammable. Flammable refrigerants are a new challenge for the Heating, Ventilation, Air-conditioning, and Cooling (HVAC) industry, and new building standards are required for the safe use of the new compounds. Burning velocity has been adopted as part of the standard to characterize the new refrigerants.

To meet the challenges of high efficiency, good volumetric capacity, low toxicity, low ozone depletion potential, low GWP, and low flammability, industry will use blends of agents. Analytical methods exist for optimizing the blends for all these properties except for flammability. To allow industry to estimate the flammability of new blends of agents, predictive tools for the burning velocity of refrigerants are being developed, and a kinetic mechanism is required. The NIST hydrofluorocarbon HFC mechanism was developed 20 years ago to describe the addition of HFC fire suppressants to hydrocarbon-air flames. It is currently being updated and extended to apply to new refrigerants added at high concentrations in air, and will likely require additional reactions and species. As a first step in this process, the existing NIST HFC mechanism was applied to predict burning velocities of some pure C1 and C2 HFC compounds in air. Agreement was initially good for some compounds and poor for others and consequently, some initial improvements were made, including addition of new HFC intermediates and their reaction rates, more recent rate data, and updated thermodynamic data. The modified mechanism is then used to predict steady, planar, 1D, unstretched burning velocities for mixtures of each refrigerant with air, over a range of fuel-air equivalence ratio ϕ, for comparison with existing experimental data in the literature. The compounds of interest are the saturated C1 and C2 HFC compounds with varying fluorine to hydrogen ratio, which is characterized by F/(F+H), or the (moles F)/(mole F + moles H) in the reactant stream.

2. Kinetic Modeling

The starting kinetic model is from the NIST C_1-C_2 hydrofluorocarbon model [1, 2], which originally was combined with GRI2.0. The mechanism has been updated and expanded to include larger HFC's, $C_3H_2F_3Br$, and $C_6F_{12}O$, as described in Ref. [3]. Since the original work, a rather large amount of new kinetic data on the reactions of fluorine containing species has been published. The kinetic model has now been modified to include new species and recent reaction rate data. The thermodynamic data for fluorine-containing species has been also updated using the data of Burcat et al. [4]. The Chemkin and Cantera packages were used for combustion equilibrium calculations and for modeling of laminar flame propagation in mixtures of the present C1-C2-HFCs in dry air. It should be noted that the present kinetic model should be considered as a starting point for further development and refinement. Numerous changes to both the rates and the reactions may be made once a variety of experimental data and theoretical results are available for testing the mechanism.

3. Results and Discussion

Figure 1 and Figure 2 (lines) show, for a range of stoichiometries, the adiabatic flame T_{aft} temperature (upper curves), and the steady, adiabatic, 1D, planar, laminar, unstretched burning velocities S_u^0 calculated with the Sandia EQUIL and PREMIX codes (lower curves). The figures also show (points) the experimental data of Takizawa from outwardly propagating spherical flames in constant volume and constant pressure experiments [5-7]. Table 1 summaries for each compound the peak T_{aft} and S_u^0 from the experiments and simulations. Also shown are the stoichiometric volume fraction of each compound and the ratio of fluorine to hydrogen atoms, expressed as F/(F+H).

The stoichiometric concentration of the agents is generally higher than that for hydrocarbons, and increases as the fluorine loading in the molecule increases. The temperatures are similar, however for the flammable HFCs (R161, R41, R152, R152a, R143, R143a) the peak flame temperatures are in the range 2100 K $\le T_{aft} \le$ 2300 K, which is comparable, sometimes higher, than that of hydrocarbons, while for the non-flammable refrigerants (R23, R134a, R134, and R125), the peak T_{aft} is somewhat lower (< 1960 K). As illustrated, the agreement in the measured and predicted burning velocity is reasonable. It should be noted that agreement between predictions and measurements for fluoromethane (CH_3F) and fluoroethane (C_2H_5F) using the original NIST model was poor. Modification of enthalpy of formation for these compounds and kinetic data for several reactions was required to improve the predictions. Nonetheless, it should be noted that the stretch and radiation have not been included in the simulations or in the reduction of the experimental

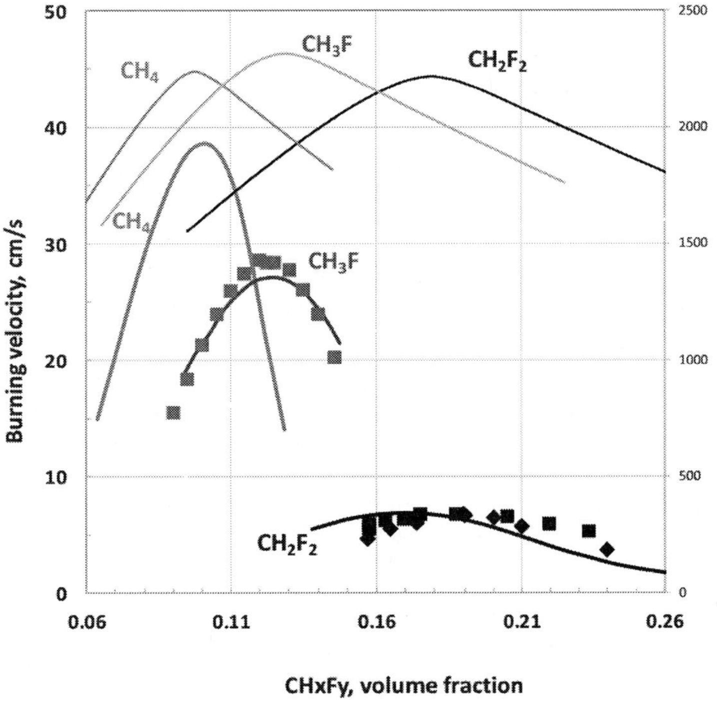

Figure 1: Burning velocity (left scale) and adiabatic flame temperature (right scale) for C1 hydrofluorocarbons in air.

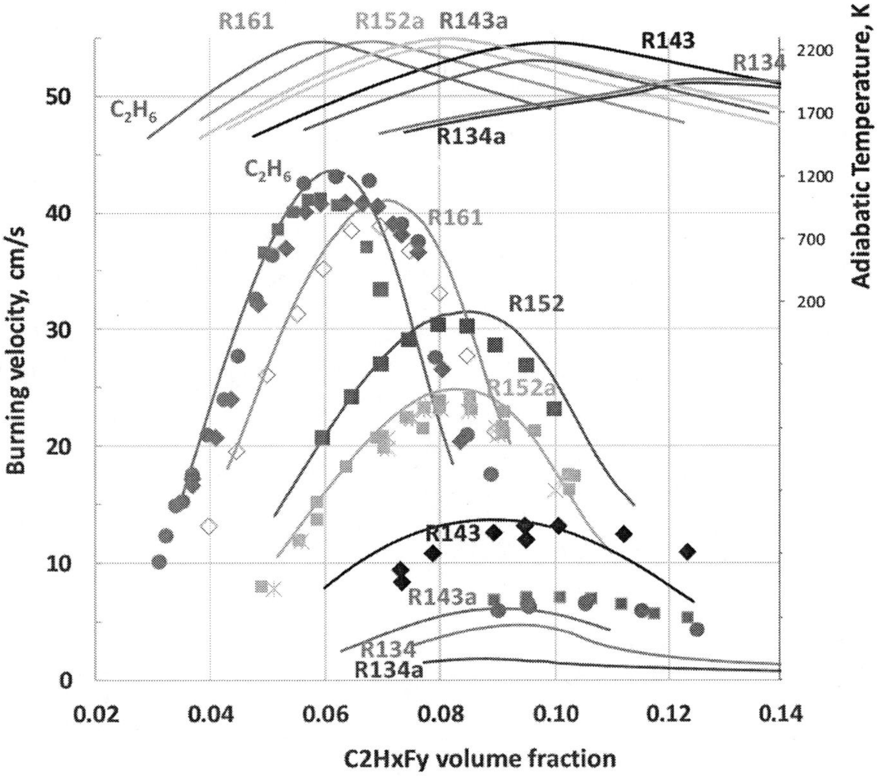

Figure 2: Burning velocity (left scale) and adiabatic flame temperature (right scale) for C2 hydrofluorocarbons in air.

data, and these might affect the results [8].

In Table 1, the data are listed from highest value of S_u^0 to lowest, and both T_{aft} and S_u^0 decrease with increasing fluorine loading in the refrigerant. This is shown in Figure 3 which gives T_{aft} and S_u^0 as a function of the fluorine loading the system, expressed as F/(F+H). Also, the asymmetrical isomers tend to be less flammable than the symmetrical ones, with both lower T_{aft} and S_u^0.

Table 1. Burning velocities and adiabatic combustion temperatures for stoichiometric refrigerant-air mixtures (initial temperature 298 K, 1 bar).

Refrigerant	Formula	T_{aft} K	S_u^0 max (Expt.) cm/s	S_u^0 max (Calc.) cm/s	$X_{stoic.}$ %	F/(F+H)
Flammable:						
R-170	C_2H_6	2265	40.9	43.1	5.66	0.00
R161	C_2H_5F	2265	38.3	41	6.54	0.17
R-50	$CH4$	2230	36.5	38.6	9.5	0.00
R-152	CH_2F-CH_2F	2278	30.1	32.1	7.75	0.33
R-41	CH_3F	2273	28.3	27.2	12.3	0.25
R-152a	CH_3-CHF_2	2220	23.6	24.9	7.75	0.33
R-143	CH_2F-CHF_2	2248	13.1	13.7	9.5	0.50
R-32	CH_2F_2	2207	6.7	7.3	17.4	0.50
R-143a	CH_3-CF_3	2115	7.1	6.1	9.5	0.50
Non-Flammable:						
R-134	CHF_2-CHF_2	1958		4.6	12.3	0.67
R134a	CH_2F-CF_3	1927		1.8	12.3	0.67
R-125	CHF_2-CF_3	1793		1.56 (at 400K)	17.4	0.83
R-23	CHF_3	1713		0.57 (at 400K)	29.6	0.75
R-116	C2F6	1389		0	29.58	1.00

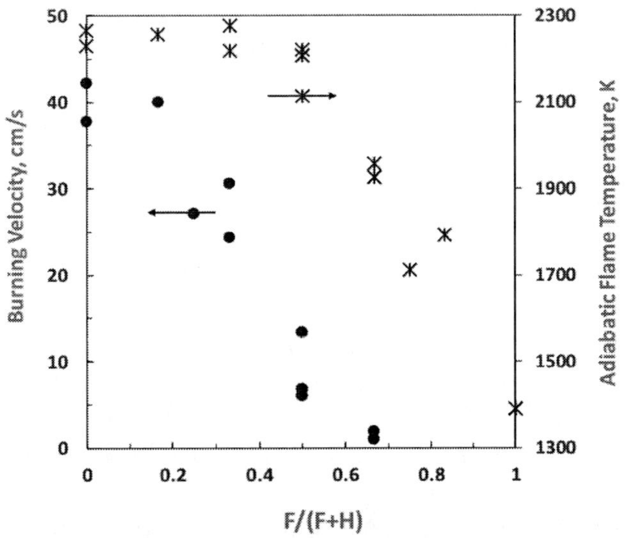

Figure 3: Burning velocity (left scale) and adiabatic flame temperature (right scale) as a function of the ratio of F atoms to H atoms in the initial refrigerant-air mixture.

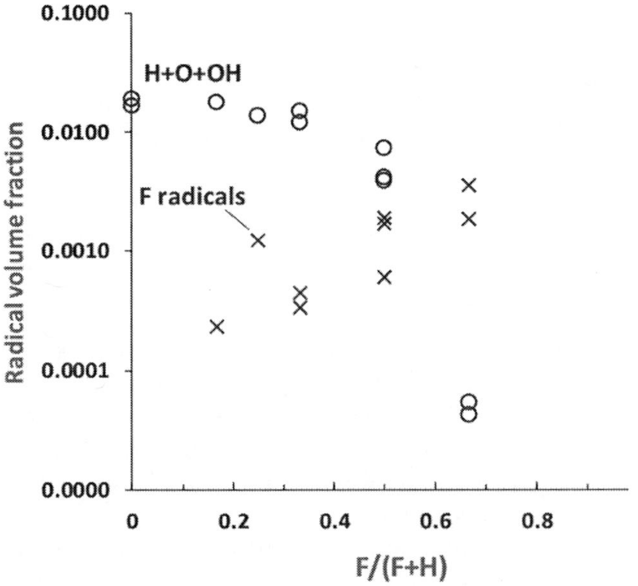

Figure 4: Volume fraction of total chain-branching radicals (H+O+OH) and F atoms as a function of the ratio of F atoms to H atoms.

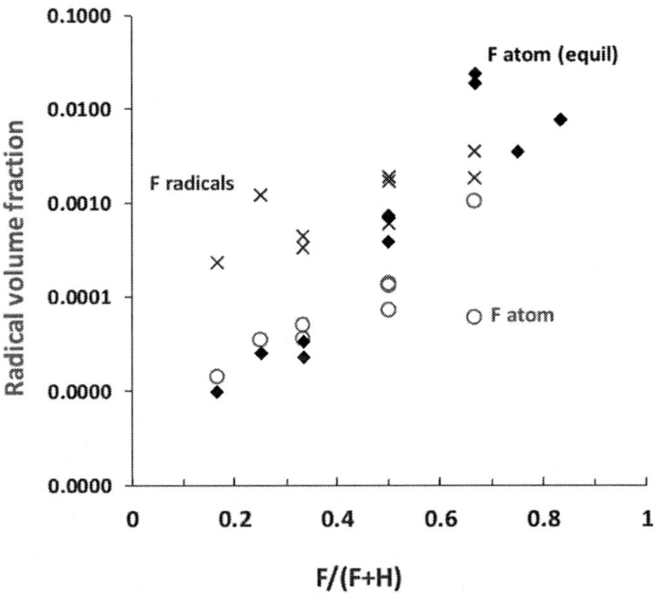

Figure 5: Volume fraction for all fluorine radicals (F-radicals) and F atoms in the flame reaction zone, and for F atoms at equilibrium as a function of the ratio of F to H atom.

Using the calculated flame structures, the variation in radical concentrations with fluorine loading is determined. **Error! Reference source not found.**Figure 4 shows the sum of the peak volume fraction of chain-carrying radicals (H, O, and OH) as well as that for F-containing radicals, as a function of the fluorine loading. As illustrated, the former drops off rapidly as the number of H atoms in the system becomes close to that of F atoms F/(F+H)=0.5, while the F radicals steadily increase. Hence, at higher fluorine loading, the chemistry becomes dominated by fluorine-containing radicals. Figure 5 shows that with increased fluorine loading, the volume fraction of both F-atoms (at the point of maximum OH volume fraction) and sum of the peak for F-containing radicals increases steadily, as does the equilibrium F atom volume fraction, which becomes higher than the value in the flame zone. For these flames equilibrium F atom volume fraction can be on the order of 1 %.

4. Conclusions

The laminar, planar, 1D, adiabatic burning velocity of saturated C1 and C2 HFC refrigerants (R41, R32, R161, R152, R152a, R143, R143a, R134, R134a were calculated using the NIST HFC mechanism available in the literature and compared to existing experimental data. The predictions were in significant disagreement for CH_3F and C_2H_5F, and mild disagreement for other compounds. Consequently, the NIST HFC mechanism was modified with additional reactions, using more recent rate data in the literature, with updated thermodynamic properties. After the changes, the agreement for these refrigerants, over a range of equivalence ratio, was good. The mechanism was then used to examine the properties of the refrigerant-air flames. Adiabatic temperatures of the refrigerant-air flames were comparable, sometimes higher, than similar hydrocarbons, and were lower for the non-flammable refrigerants. Burning velocity and flame temperature decreased as the F/H ratio in the reactants increased. The symmetrical isomers of the fluoroethanes (R152, R143, R134) have higher T_{aft} and S_u^0 than the asymmetrical isomers (R152a, R143a, R134a), consistent with differences in enthalpies of formation. Analysis of the flame structures revealed that with increasing F/H ratio, the chain-branching radical concentrations in the flame decrease, and fluorine-containing radical, particularly F atom, increase. At high enough F/H ratio, the F atom equilibrium values are even higher than those in the flame zone.

5. Acknowledgements

The research was supported by the US Dept. of Energy Building Technologies Office and by NIST.

6. References

[1] D. R. Burgess, M. R. Zachariah, W. Tsang, P. R. Westmoreland, Thermochemical and chemical kinetic data for fluorinated hydrocarbons, Prog. Energy Combust. Sci. 21 (1995) 453-529.

[2] D. Burgess, M. R. Zachariah, W. Tsang, P. R. Westmoreland, Thermochemical and Chemical Kinetic Data for Fluorinated Hydrocarbons, Report No. NIST Technical Note 1412, Gaithersburg, MD, 1995.

[3] V. I. Babushok, G. T. Linteris, D. R. Burgess, P. T. Baker, Hydrocarbon flame inhibition by C3H2F3Br (2-BTP), Combust. Flame 162 (2015) 1104-1112.

[4] E. Goos, A. Burcat, B. Ruscic, Extended third millennium thermodynamic database for combustion and air-pollution use with updates from active thermochemical tables. <Ftp://ftp.technion.ac.il/pub/supported/aetdd/thermodynamics/BURCAT.THR > (August 2012).

[5] K. Takizawa, A. Takahashi, K. Tokuhashi, S. Kondo, A. Sekiya, Burning velocity measurement of fluorinated compounds by the spherical-vessel method, Combust. Flame 141 (2005) 298-307.

[6] K. Takizawa, A. Takahashi, K. Tokuhashi, S. Kondo, A. Sekiya, Burning velocity measurement of HFC-41, HFC-152, and HFC-161 by the spherical-vessel method, J Fluorine Chem 127 (2006) 1547-1553.

[7] K. Takizawa, S. Takagi, K. Tokuhashi, S. Kondo, M. Mamiya, H. Nagai, Assessment of Burning Velocity Test Methods for Mildly Flammable Refrigerants, Part 1: Closed-Vessel Method, ASHRAE Trans. 119 (2013) 243-254.

[8] R. Burrell, J. L. Pagliaro, G. T. Linteris, Effects of stretch and thermal radiation on difluoromethane-air burning velocity measurements in constant volume spherically expanding flames, Proc. Combust. Inst. 37 (2018) submitted.

Sub Topic: Reaction Kinetics

Spring Technical Meeting
Eastern States Section of the Combustion Institute
March 4-7, 2018
State College, Pennsylvania

Development and Validation of a Mechanism for Flame Propagation in R-32/Air Mixtures

Donald R. Burgess, Jr.[1*], *Jeffrey A. Manion*[1], *Robert R. Burrell*[2], *Valeri I. Babushok*[2], *Michael J. Hegetschweiler*[2], *Gregory T. Linteris*[2]

[1]*Chemical Sciences Division, NIST, Gaithersburg, MD 20899*
[2]*Energy & Environment Division, NIST, Gaithersburg, MD 20899*
[*]*Corresponding Author Email: dburgess@nist.gov*

Abstract: A mechanism for the combustion of the refrigerant R-32 (CH_2F_2) in air mixtures was developed and validated through comparisons with measured flame speeds for a range of equivalence ratios (0.9 to 1.4) and pressures (1 to 3 bar) using a constant-volume spherical flame method. Premixed flame calculations were performed and analyzed to identify primary species and reactions contributing to flame speeds and combustion. We found that there were only three HFC reactions that contributed significantly to flame speeds. Their rate constants were optimized within uncertainty limits and the model showed excellent agreement (<3 %) with measured flames speeds.
Keywords: Chemical kinetics, Refrigerant flammability, Burning velocity, Hydrofluorocarbons

1. Introduction

The overall purpose of this work is to characterize the flammability of a set of fluoromethanes, fluoroethanes, and fluoropropenes, and their mixtures for use as refrigerant working fluids. Although it is possible to make measurements of the flammability of a single refrigerant under a limited set of conditions, it is not realistic to measure flammabilities of all possible formulations under a wide range of refrigerant-to-air ratios for different diluents, ambient temperatures, and humidity levels. The driving force for this work is the development of new refrigerant blends that simultaneously minimizes their global warming potentials (GWP) and flammabilities, while maximizing their refrigerant performance which is a function of thermodynamic and physical properties (e.g., critical temperature, vaporization enthalpy, thermal conductivity, saturation vapor pressure). The ability to have robust and accurate predictive tools that are benchmarked to high quality measurements of the flammability of these refrigerants for a set of specific mixtures under a range of conditions will allow industry to screen, optimize, and rank different blends to enable rapid development of new refrigerant formulations.

In the work reported here, the flammability of the refrigerant R-32 (CH_2F_2, difluoromethane) is studied using burning velocities (flame speeds). R-32 is a widely used refrigerant that is non-ozone depleting and has a moderate GWP. It, however, is mildly flammable, and is used with less flammable refrigerants in blends. This current work on R-32, a standard refrigerant, will provide a benchmark for extending flammability models based on elementary reaction kinetics to other refrigerants (both pure and blends) such as the HFC's R-125 (pentafluoroethane), R-134a (1,1,1,2-tetrafluoroethane), R-152a (1,1,-difluoroethane), and the hydrofluoroolefins HFO-1234yf (2,3,3,3-tetrafluoropropene) and HFO-1234ze (1,3,3,3-tetrafluoropropene).

2. Methods

In this work, we developed and validated a chemical kinetic mechanism/model to predict the flammability of the refrigerant R-32 (CH_2F_2). This requires understanding the chemistry on a microscopic level from the analysis of reaction pathways and comparison of the model predictions to measured flame speeds.

The full details of the measurements in this work are reported elsewhere [1] and are only summarized here. The flame speeds of R-32/air mixtures were measured using a constant-volume spherical-flame method. A spherical chamber about 15 cm diameter (~1.8 liters) was filled with R-32/air mixtures with equivalence ratios ranging from 0.9 to 1.4, and using different initial pressures on the order of (0.87 to 1.13) bar. The mixtures were then ignited by a spark at the center of the chamber, and the pressure rise (final pressures of about 7 to 11 bar) as a function of time was monitored. The final pressures were about 7 to 11 bar, but instabilities, interaction with the chamber wall, and other effects limited the reliable data to final pressures of about 1 to 3 bar. The pressure traces were then used to calculate flame radius, and thus flame speeds $S_u(T,P)$, using a thermodynamic spherical flame propagation model. In the data reduction procedures, corrections were made to account for thermal radiation of the burned gas [3] and flame stretch [4]. Buoyancy is unimportant for the present conditions [5] and was not considered.

In this work, the development and validation of a chemical kinetic mechanism (model) to predict the flammability of R-32 (CH_2F_2) was iterative. An initial chemical kinetic mechanism was developed based on compilations from the literature, updates to reflect more recent work, and an evaluation of reactions and their rate constants. Simulations were performed using the Sandia Premix code [6] and Cantera [7] and the results examined using reaction path analysis employing the graphics post-processor XSenkplot [8] to identify important species and reactions. The refined mechanism was used to simulate the flame speeds. Rough agreement between the computed and measured flames speeds was initially found, and the rate constants were then adjusted/optimized within their uncertainty limits (factors of about 1.3 to 2.0) to achieve best agreement.

3. Rate Constant Evaluation

We provide here a short discussion of the rate constants evaluated and utilized in this work. Key rate expressions are provided in Table 1. The rate expressions for hydrogen-oxygen chemistry and hydrocarbon/oxidized hydrocarbon chemistry were taken from GRI-Mech 3.0 [9], while those for H/O/F chemistry were based on our fits to rate constants reported in the literature [10-12].

The rate expression for $CH_2F_2 \rightarrow CHF + HF$ was derived from the shock tube measurements for the analogous reaction $CHF_3 \rightarrow CF_2 + HF$ by Schug and Wagner [13] and employing the very rough relative decomposition rates for CH_2F_2 and CHF_3 from the pyrolysis study of Politanskii and Shevchuk [14]. This scaled rate expression was then refined during our optimizations. It was found that the flame speeds could not be modeled with a pressure independent rate constant for this unimolecular reaction. It was necessary to use a pressure-dependent rate constant of the form $Rate = k_0[M] + k_1$, where the pressure dependent term $k_0[M]$ dominated – suggesting that the reaction was approaching the low pressure limit under (P, T) conditions in the flame.

The rate expressions for H abstractions from CH_2F_2 by the flame radicals H, O, OH, and F were evaluated in this work. In our evaluation, we utilized rate constants from the literature [15-21] and then fit them to extended Arrhenius rate expressions to provide rate constants over a wide range of temperatures. Depending upon the reaction, available data, and fitting procedures, we estimate uncertainty factors on the order of about (1.3 to 1.6) for these reactions.

The rate constants for $CHF_2 + O_2 \rightarrow CF_2O + OH$ and $CHF + O_2 \rightarrow CHFO + O$ were estimated by analogy to those for $CH_3 + O_2 \rightarrow CH_2O + OH$ and $CF_2 + O_2 \rightarrow CF_2O + O$, respectively [22-23]. Both of the former reactions are significantly more exothermic (~100 kJ mol^{-1}) than the reference reactions, and consequently should have much smaller barriers. For the $CHF_2 + O_2$ and $CHF + O_2$ reactions, we initially estimated activation energies on the order of (15 to 20) kJ mol^{-1} and (20 to 25) kJ mol^{-1}, but our optimizations compared to our measured flame speeds suggest somewhat smaller activation energies – on the order of (5 to 15) kJ mol^{-1} and (15 to 25) kJ mol^{-1}, respectively.

The rate expression for $CHF_2 + CHF_2 \rightarrow CHF{=}CF_2 + H$ used in this work is from the earlier ab initio transition state and RRKM calculations by Burgess et al [24]. Although this is a primary pathway, this reaction has a small impact on flames speeds (< 3 %) and combustion – it simply provides a pathway to products. We utilized the rate expression for $CHF + H_2O \rightarrow CH_2O + HF$ from the ab initio transition state and RRKM calculations of Zachariah et al [25]. This reaction is a major destruction pathway for CHF and the primary pathway leading to the formation of HO_2 ($CH_2O + OH \rightarrow CHO + H_2O$, followed by $CHO + O_2 \rightarrow CO + HO_2$). This reaction, however, contributes a negligible amount (<0.1 %) to changes in flame speeds – it is largely a conduit for establishing steady state concentrations during the pre-ignition process.

For the reactions R• (CHF_2, CF_2, CHF) + X (H, O, OH, F), we utilized rate constants from the literature. These reactions are generally very fast and contribute negligible amounts to flame speeds and little to the combustion chemistry other than to provide pathways to complete combustion. Similarly, reactions involving CHFO, CF_2O, and CFO, contribute negligible amounts to changes in flame speeds, and provide pathways to combustion products.

Table 1: Rate parameters for elementary reactions employed in the model. The rate constants have the Arrhenius form $k = Ae^{-E/RT}$ where A, E, and R are the pre-exponential, activation energy, and the gas constant, respectively. Units are mol, cm^3, s, kJ. Note: these are only rate expressions developed as part of this work. Rate expressions for other reactions used in the model were taken from the literature (see text).

Reaction	A	E
$CH_2F_2 + M \rightarrow CHF + HF + M$	5.9E+17	295.0
$CH_2F_2 \rightarrow CHF + HF$	4.9E+11	272.0
$CH_2F_2 + H \rightarrow CHF_2 + H_2$	2.5E+14	62.5
$CH_2F_2 + O \rightarrow CHF_2 + OH$	1.9E+14	56.3
$CH_2F_2 + OH \rightarrow CHF_2 + H_2O$	1.7E+13	30.0
$CH_2F_2 + F \rightarrow CHF_2 + HF$	3.1E+14	19.4
$CHF_2 + O_2 \rightarrow CF_2O + OH$	4.9E+10	14.6
$CHF + O_2 \rightarrow CHFO + O$	2.2E+13	20.9

4. Results and Discussion

Figure 1 shows the experimentally-derived flames speeds (points) and the modeled flame speeds (curves). The solid symbols show the flame speeds corrected using an optically thin radiation model, while the open symbols show the uncorrected flames for 1 bar (2 and 3 bar data not shown for clarity). The deviations between the experimental flame speeds and the model predictions are on the order of (1 to 3) %. This is excellent agreement given that the uncertainties in the flames speeds are on the order of (10 to 20) % due to uncertainties in the rate constants which are multiplicative factors of about 1.5 to 2.0. As seen in Figure 1, the flame speeds are roughly linear with pressure suggesting the rate constant $Rate = k_0[\mathrm{M}] + k_1$ is approaching the low pressure limit.

The optically thin (limit) radiation model correction increases flames speeds by about (19±2) % relative to those computed using adiabatic flame temperatures (shown for 1 bar) and is independent (<1 %) of equivalence ratio. When estimated radiation absorption corrections are used for each major species, flame speeds decrease by about (8±3) % relative to those using the optically thin radiation model (not shown). Given the insensitivity of the corrections to equivalence ratio, the rate constants for the primary reactions contributing to flame speeds can be easily adjusted to correct for absorption of radiation. The correction using the optically thin radiation model translates into changes in the rate constants of about 40% (this is shown for 1 bar). This is lower than the uncertainties in the rate constants which have uncertainty factors of about 1.4 to 2.0.

We note that under lean conditions (where flame stretch is most important) the measured flame speeds S_u(T,P) appear to be slightly higher (3 to 6) % than the modeled flame speeds. Recently, we have reprocessed the data better fitting the burning velocity / temperature-pressure curves S_u(T,P) and find lower burning velocities under lean conditions at 1 atm in good agreement (<3 %) with the model (these data are not shown, because we have not re-optimized model yet). In addition, we are currently in the process of optimizing the temperature and pressure dependencies of the rate constants by considering measurements that we have made using argon/O_2 mixtures (instead of air mixtures). The lower heat capacity of argon increases flame temperatures and thus increases flame speeds (by about a factor of 2). The refined experimental data and re-optimized rate constants are likely to provide better agreement between measured and model flame speeds.

Figure 2 shows the dependence of flame speeds for the three most important reactions as a function of equivalence ratio. The dependencies are on the order of (5 to 20) %, while all other reactions are less than 3 % and relatively independent of equivalence ratios. The quantity shown here is the percent change in flame speed relative to a change in rate constant. That is, a value of 20 % would mean increasing a rate constant by 10 % would result in an increase of 2 % in flame speed. The two most important reactions are the unimolecular decomposition of the refrigerant $CH_2F_2 \rightarrow CHF + HF$ and the subsequent reaction of its decomposition product CHF with oxygen in the mixture $CHF + O_2 \rightarrow CHFO + O$. Both of these reactions contribute to increases in flame speeds. In contrast, H abstraction from the refrigerant $CH_2F_2 + H \rightarrow CHF_2 + H_2$ causes a decrease in flame speeds – it inhibits the flame by consuming the radical H atom during pre-ignition that otherwise would be involved in radical chain branching leading to ignition.

Reaction pathway analysis and flame speed dependencies of rate constants show that burning of R-32 can be described by two quasi-separate reaction pathways. The first stage is pre-ignition driven by unimolecular decomposition of CH_2F_2 forming CHF. The subsequent second stage is a combustion pathway driven by H abstraction by flame radicals H, O, OH, and F forming CHF_2. Both CHF and CHF_2 then react quickly with the reactant O_2 to form the fluorocarbonyls CHFO and CF_2O, respectively. The pre-ignition pathway is primarily responsible for generation of initial flame radicals and hence drives flame propagation, while the combustion pathway occurs after ignition, mediated by flame radicals, and drives the process to products. Interestingly, reactions in the combustion pathway actually inhibit ignition by consuming radicals needed for chain branching – tying up the chemistry in relatively stable intermediates that are slow to decompose.

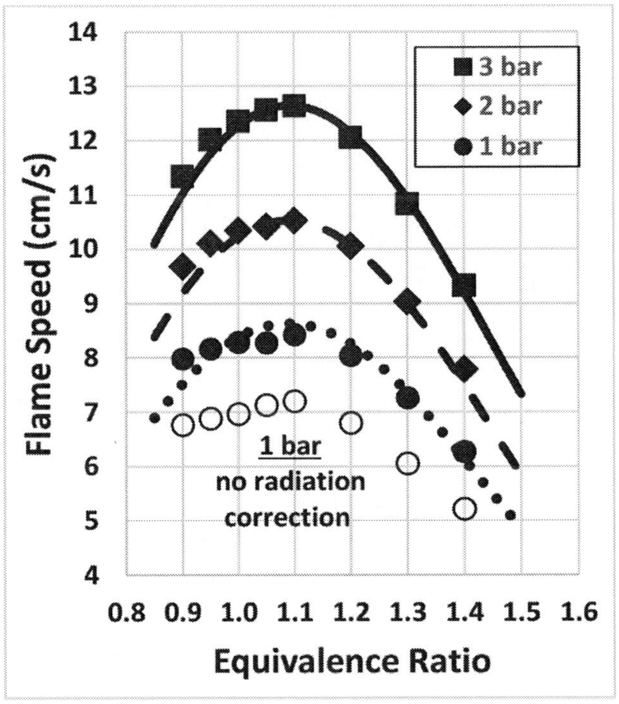

Figure 1: Comparison between measured and model flame speeds for CH_2F_2/air mixtures.

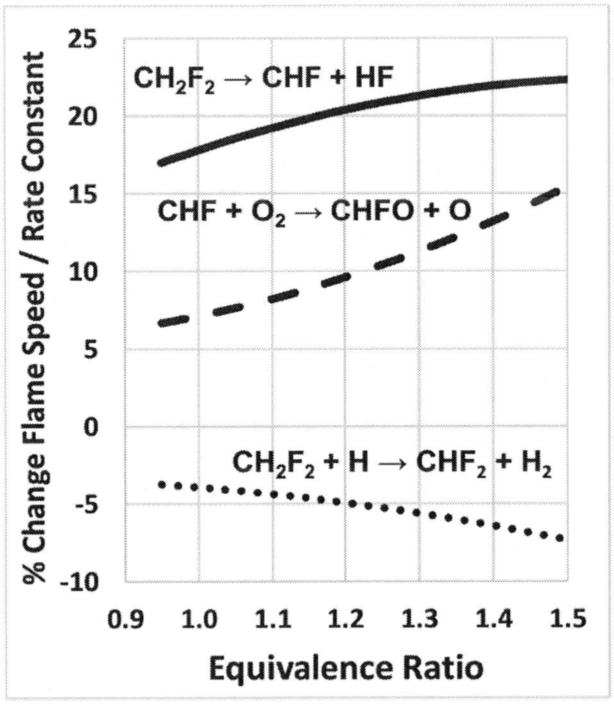

Figure 2: Dependence of flame speeds on the three most important rate constants.

5. Conclusions

In this work, a mechanism for the combustion of the refrigerant R-32 (CH_2F_2) in air mixtures was developed and validated through comparisons with measured flame speeds for a range of equivalence ratios (0.9 to 1.5) and pressures (1 to 3 bar) using a constant-volume spherical flame method. Premixed flame calculations were performed and analyzed to identify primary species and reactions contributing to flame speeds and combustion. We found that there were just three HFC reactions that contributed significantly to flame speeds. Their rate constants were optimized (within their uncertainty limits) and the model showed excellent agreement (<3 %) with measured flame speeds. Ongoing experiments and modeling will refine this model using other conditions such using Ar/O_2 mixtures (instead of air) to change flame temperatures and the addition of H_2O and H_2 to change the concentration of flame radicals. The radiation parameters will also be improved. This work will be also extended to other refrigerants: the fluoroethanes R-125, R-134a, and R-152a and the hydrofluoroolefins HFO-1234yf (2,3,3,3-tetrafluoropropene) and HFO-1234ze (1,3,3,3-tetrafluoropropene).

6. Acknowledgements

The research was supported by the US Dept. of Energy Building Technologies Office and by NIST.

7. References

[1] R. Burrell, J. L. Pagliaro, G. T. Linteris, Effects of stretch and thermal radiation on difluoromethane-air burning velocity measurements in constant volume spherically expanding flames, Proc. Combust. Inst. 37 (2018) submitted.

[2] C. Xiouris, T.L. Ye, J. Jayachandran, F. N. Egolfopoulos, Laminar flame speeds under engine-relevant conditions: Uncertainty quantification and minimization in spherically expanding flame experiments, Combust. Flame 163 (2016) 270-283.

[3] Z. Chen, Effects of radiation absorption on spherical flame propagation and radiation-induced uncertainty in laminar flame speed measurement, Proc. Combust. Inst. 36 (2017) 1129-1136.

[4] F.J. Wu, W.K. Liang, Z. Chen, Y.G. Ju, C.K. Law, Uncertainty in Stretch Extrapolation of Laminar Flame Speed from Expanding Spherical Flames, Proc. Combustion Inst. 35 (2015) 663-670.

[5] K. Takizawa, S. Takagi, K. Tokuhashi, S. Kondo, M. Mamiya, H. Nagai, Assessment of. Burning Velocity Test Methods for Mildly Flammable Refrigerants, Part 1: Closed-Vessel Method, ASHRAE Trans. 119 (2013) 243-254.

[6] R. J. Kee, J. F. Grcar, M. D. Smooke, J. A. Miller, PREMIX: A Fortran Program for Modeling Steady Laminar One-Dimensional Premixed Flames (Version 2.5b), Report No. SAND85-8240, Sandia National Laboratories, Livermore, California, 1992.

[7] D. G. Goodwin, H. K. Moffat, R. L. Speth, Cantera, Ver. 2.3.0, 2017. http://www.cantera.org

[8] D. Burgess, J. Racek, XSenkplot: An Interactive, Graphics Postprocesser for Numerical Simulations of Chemical Kinetics, National Institute of Standards and Technology, Gaithersburg, MD, 1996.

[9] G. P. Smith, D. M. Golden, M. Frenklach, N. W. Moriarty, B. Eiteneer, M. Goldenberg, C. T. Bowman, R. K. Hanson, S. Song, W. C. Gardiner, Jr., V. V. Lissianski, Z. Qin, GRI-Mech 3.0, 1999.

[10] A. Persky, H. Kornweitz, The kinetics of the reaction $F+H_2 \rightarrow HF+H$, IJCK 29 (1997) 67-71.

[11] PS Stevens, WH Brune, JG Anderson, Kinetic and mechanistic investigations of F atom + water and F atom + hydrogen over the temperature range 240-373 K, J. Phys. Chem. 93 (1989) 4068.

[12] C. D. Walther, H. G. Wagner, Uber die Reaktionen von F-atomen mit H_2O, H_2O_2 und NH_3, Ber. Bunsenges. Phys. Chem. 87 (1983) 403.

[13] Schug, K. P.; Wagner, H. Gg., Zum thermischen zerfall von CH_3F, Z. Phys. Chem. 86 (1973) 59.

[14] S. F. Politanski, U. V. Shevchuk, Thermal conversions of fluoromethanes. II. Pyrolysis of difluoromethane and trifluoromethane, Kinet. Catal. 9 (1968) 411.

[15] G. O. Pritchard, M. J. Perona, Some hydrogen atom abstraction reactions of CF_2H and CFH_2 radicals, and the C-H bond dissociation energy in CF_2H_2, Inter. J. Chem. Kinet. 1 (1969) 509-525.

[16] A. A. Westenberg, N. DeHaas, Rates of $H + CH_3X$ reactions, J. Chem. Phys. 62 (1975) 3321.

[17] A. Miyoshi, K. Ohmori, K. Tsuchiya, H. Matsui, Reaction rates of atomic oxygen with straight chain alkanes and fluoromethanes at high temperatures, Chem. Phys. Lett. 204 (1993) 241-247.

[18] A. Matsugi, H. Shiina, Kinetics of hydrogen abstraction reactions from fluoromethanes and fluoroethanes, Bull. Chem. Soc. Jpn. 87 (2014) 890-901.

[19] A. Persky, The temperature dependence of the rate constants for the reactions $F+CH_3F$ and $F+CH_2F_2$, Chem. Phys. Lett. 376 (2003) 181-187.

[20] L. Wang, J. Y. Liu, Z. S. Li, Ab initio direct dynamics studies on the hydrogen abstraction reactions of CF_2H_2 and CF_3H with F atom, Chem. Phys. 351 (2008) 154-158.

[21] J. B. Burkholder, J. Sander, J. R. Abbat, R. E. Huie, C. E. Kolb, M. J. Kurylo, V. L. Orkin, D. M. Wilmouth, P. H. Wine, Chemical Kinetics and Photochemical Data for Use in Atmospheric Studies, JPL Publication 15-10, Jet Propulsion Laboratory, Pasadena, 2015. http://jpldataeval.jpl.nasa.gov

[22] N. K. Srinivasan, M. C. Su, J. W. Sutherland, J. V. Michael, Reflected shock tube studies of high-temperature rate constants for CH_3+O_2, H_2CO+O_2, $OH+O_2$. J. Phys. Chem. A 109 (2005) 7902-7914.

[23] EL Keating, RA Matula, The high temperature oxidation of tetrafluoroethylene, JCP 66 (1977) 1237.

[24] D. Burgess Jr., M.R. Zachariah, W. Tsang and P.R. Westmoreland, Thermochemical and Chemical Kinetic Data for Fluorinated Hydrocarbons, Prog. Energ. Comb. Sci., 21 (1996) 453-529.

[25] M. R. Zachariah, P. R. Westmoreland, D. Burgess, Jr., Theoretical Prediction of Thermochemistry and Kinetics of Halocarbons, in ACS Volume 611 on Halon Replacements, 1995, pp. 358-373.

Spring Technical Meeting
Eastern States Section of the Combustion Institute
March 4-7, 2018
State College, Pennsylvania

The Effect of Non-Axisymmetric Fuel Staging on Flame Structure in a Multiple-Nozzle Model Turbine Combustor

Olivia Sekulich[1], Wyatt Culler[1], Jacqueline O'Connor[1]*

Department of Mechanical and Nuclear Engineering, The Pennsylvania State University, University Park, PA, United States
[*]Corresponding Author Email: jxo22@engr.psu.edu

Abstract: Fuel staging is often used to suppress combustion instabilities in a gas turbine engine. This strategy is accomplished by applying differing flow rates of fuel to different nozzles of a multi-nozzle combustor. As such, the mixture composition and critical flame properties like flame speed and heat release may vary from nozzle to nozzle. In this experiment, five identical nozzles are arranged with a central nozzle surrounded symmetrically by four others; this is commonly referred to as a "four-around-one" configuration. The fuel is staged by adding more fuel through the middle and one outer nozzle, where staging through the middle nozzle will be referred to as axisymmetric staging, and staging through an outer nozzle is non-axisymmetric staging. This staging is analyzed through the use of tomographic image reconstruction. Images are taken in 5 degree increments over a 180 degree span around the flame for each staging condition. We compare three cases: one without staging, one with axisymmetric staging, and another with non-axisymmetric staging. Fuel staging can cause deviations in the flame structure where neighboring flames meet each other. Non-axisymmetric versus axisymmetric staging are found to result in different flame shapes and stabilization methods due to the difference in the number of flames that are interacting with the staged nozzle. ***Keywords: Turbulent Flames, Fuel Staging, Flame Interaction***

1. Introduction

Most modern gas turbine engines use lean-premixed combustion in order to meet emissions requirements. While lean-premixed combustion successfully reduces harmful pollutants such as NO_x, it makes flames more susceptible to combustion instability. Combustion instability is caused by a coupling between heat release rate fluctuations and combustor acoustics in a feedback loop [1]. This coupling is undesirable as it reduces engine operability, increases emissions, and in rare cases, causes catastrophic hardware failure [1, 2]. Combustion instability may be suppressed using active or passive techniques, though passive techniques are preferred in industry due to their reliability. Fuel staging, which is when fuel is spatially nonuniformly distributed in the combustor, is a commonly used instability suppression mechanism [1-6].

The primary effect of fuel staging is to redistribute the heat release, and a recent computational study by Li *et al.* [7] illustrates how a non-axisymmetric distribution of heat release rates in a Rijike tube has a stabilizing influence. They additionally found that the higher the nonuniformity the higher the stabilizing influence. Samarasinghe *et al.* [8] recently proposed a mechanism as to how axisymmetric fuel staging suppresses instability in a model gas turbine combustor. They utilized a four-around-one multi-nozzle combustor (the same configuration used in this work) to study the effects of axisymmetric fuel staging on self-excited instabilities. They found that fuel staging altered the phase relationships between the staged, center flame and the un-staged, outer flames that resulted in destructive interference, thereby suppressing the instability.

Fuel staging may be conducted in an axisymmetric [3, 9] or non-axisymmetric [5, 6] manner, and it remains an open question as to which configuration is more effective in suppressing instability. It is likely that flame interaction plays a strong role in the effectiveness of different fuel staging strategies. To investigate the effect of flame interaction, this work utilizes tomographic reconstructions of axisymmetrically and non-axisymmetrically staged flames to study how the location of fuel staging changes flame shape.

2. Experimental Setup and Methods

The experimental facility is as pictured in Figure 1a and is described in more detail in Refs. [10, 11]. The combustor burns a preheated (200 °C), premixed mixture of natural gas and air with a bulk velocity of 26 m/s. A 260 mm diameter, 300 mm long quartz tube serves as the combustor liner. Each fuel nozzle consists of an annulus, a swirler of swirl number 0.7, and a recessed centerbody. The process of fuel staging is described in more detail in previous publications [8, 12], and is accomplished by injecting a small amount of fuel through fuel staging annulus where it flows to the swirler and enters the main flow path as in Figure 1b. Previous measurements by Orawannukul [13] indicate this extra fuel is well mixed by the nozzle exit. The operating conditions are summarized in Table 1.

Table 1: Staging equivalence ratios

Staging Configuration	Non-staged ϕ/Staged ϕ
No Staging	0.65/0.65
Axisymmetric Staging	0.68/0.85
Non-Axisymmetric Staging	0.68/0.85
Non-Axisymmetric Staging	0.68/0.90

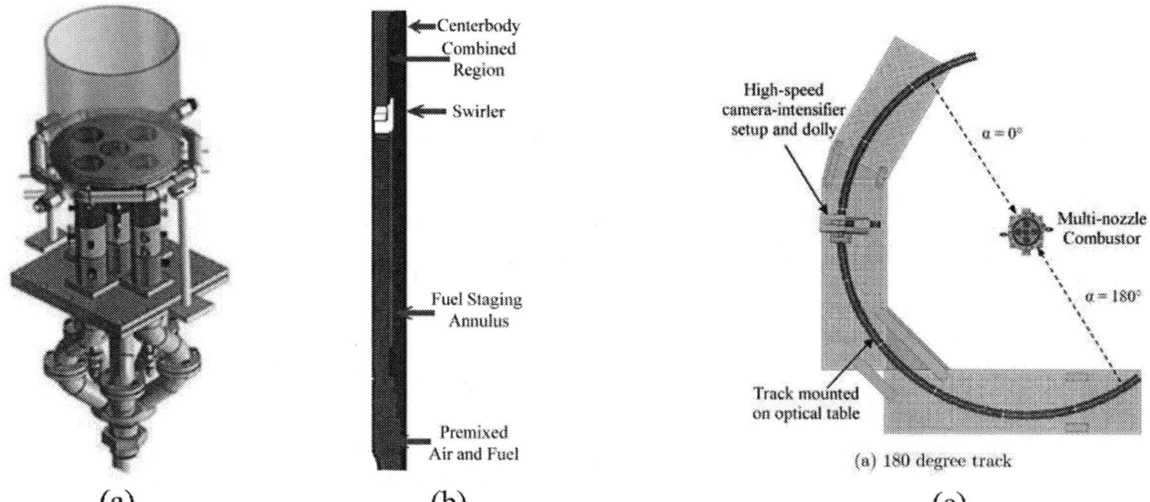

(a) (b) (c)

Figure 1: (a) Multinozzle Combustor (b) Cutaway of staging fuel path (c) track setup for tomographic reconstruction (reproduced from [11])

High speed images CH* are obtained using a Photron SA4 high speed camera coupled to an Invisible Vision Invisible Vision UVi 1850-10 intensifier, a Nikon AF Micro-Nikkor 60mm f/2.8

lens, and a 432 ±5 nm bandpass filter. A recess mounted PCB pressure transducer on the dump plane is used to measure the combustor pressure. K-type thermocouples are used to monitor dump plane and centerbody temperatures.

The setup for the tomographic reconstruction is shown in **Figure 1**c. Three dimensional reconstructions of the flame are obtained by calculating the inverse Radon Transform using a filtered back projection algorithm as described in Ref. [11]. High-speed CH* chemiluminescence images are recorded for one second at a framerate of 4 kHz in five-degree increments along the 180-degree section of track. This procedure results in a total of 36 angles for each reconstruction. The images are background subtracted, flat field corrected, cropped, and time-averaged at each angle before being processed using the inverse Radon transform.

3. Results and Discussion

The interior structures of the multi-nozzle flames are inspected in two-dimensional slices of the reconstructed figures. For each staging case (non-staged, axisymmetric, or non-axisymmetric) we consider six horizontal slices of the flame at increasing axial distances from the dump plane in increments of 0.5 in. Additionally, we consider two vertical slices displaying the XZ plane and YZ plane through the center nozzle. **Figure 2** shows the nozzle configuration where "F" is the front nozzle, "L" is the left nozzle, "B" is the back nozzle, and "R" is the right nozzle. The XZ plane is shown the by black line which slices through the front, center, and back nozzles while the XZ plane is shown by the green line and slices through the right, center, and left nozzles.

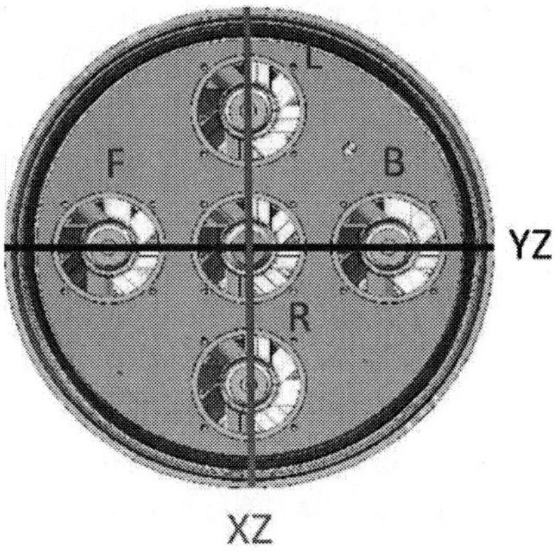

Figure 2: Nozzle Configuration

No Fuel Staging

Figure 3 (a – c) shows the flame structure of the non-staged case. Interaction between the flames is first observed at 1 in. from the dump plane, shown in Figure 3a, image 2. This area of flame interaction grows further downstream. At z = 1.5 in., the outer flames contact the combustor liner. The center flame develops a distinct, non-axisymmetric square shape at the same location, whereas the outer nozzle flames remain relatively axisymmetric. Further downstream, the flame shapes begin to twist counter-clockwise due to bulk swirl effects. This is especially evident at 3 in. downstream. This swirl causes regions of high chemiluminescence intensity, and therefore heat release, in a "star" shape around the center nozzle and in a non-axisymmetric "lima bean" shape around the outer nozzle flames. The greatest chemiluminescence intensity is found in the flame interaction regions. Figures 3b and 3c, which show cuts in the XZ and YZ planes respectively, highlight the structure of flame stabilization. The outer flames impinge on the wall of the combustor, resulting in high levels of heat release, and can be seen to stabilize in both the inner and outer shear layers in the regions near the wall. In the flame interaction regions, the flame shape is less certain. Pockets of high chemiluminescence can be seen between the branches of the flames stabilized in the inner shear layers, but the structure is quite different from that of the "M-flame" shape seen near the walls, suggesting that the flames in the interaction zones may not be stabilizing in the outer shear layer. The pocket structure may be the result of strong recirculation between flames.

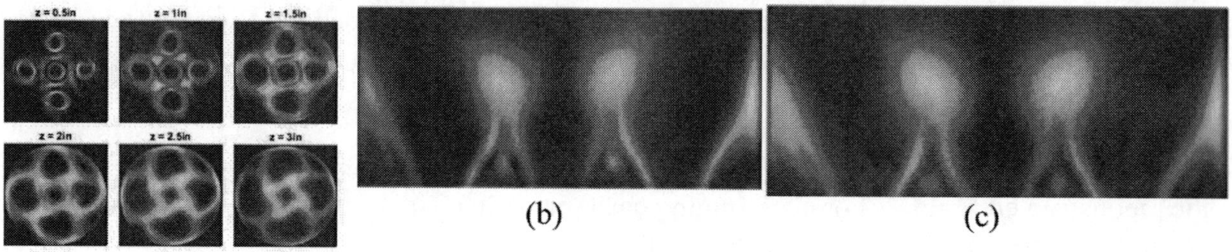

(b) (c)

(a)

Figure 3: Non-staged at $\phi = 0.65$ a) Z-direction b) XZ-plane c) YZ-plane

Axisymmetric Staging

The axisymmetric staging case is displayed in Figure 4 (a – c); the staging is evidenced by the higher chemiluminescence intensity of the center flame, as the equivalence ratio of that flame is higher than that of the outer flames. Despite the additional fuel, there is no clear indication of flame interaction until 1.5 in. from the dump plane; this is shown by the low chemiluminescence intensity in (Fig 4a). This staged flame structure differs from the non-staged case, which exhibits interaction as early as 1.0 in., and this difference is likely due to the change in flame shape from the variation in flame speed at higher equivalence ratios. Further downstream, the center flame shape follows a similar progression in shape seen in the non-staged case, where the high heat-release region forms a square and then a star farther downstream due to the swirl direction of the nozzles. Figures 4b and 4c show a significant change in the stabilization of the center flame with staging. In particular, the flame seems to stabilize in both the inner and outer shear layers, resulting in an "M-shape" flame in the interaction zone, as compared to the "V-shaped" flame seen in the non-staged case.

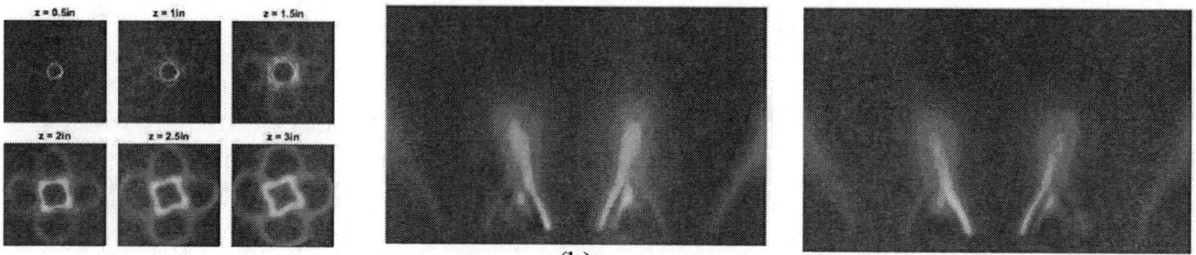

(b) (c)

(a)

Figure 4: Axisymmetric staging for $\phi_{Center} = 0.85, \phi_{Outer} = 0.68$
a) Z-direction b) XZ-plane c) YZ-plane

Non-axisymmetric Staging

Non-axisymmetric staging of the multinozzle combustor is displayed in Figures 5 (a – c) for a staging equivalence ratio of 0.85 and Figure 6 (a – c) for a staging equivalence ratio of 0.9. The axial development of the flame shapes is similar to the development in the axisymmetric staging case. The "lima bean" and "star" shapes are present in the outer and inner flames, respective, in both Figures 5a and 6a. The lima bean structure is especially pronounced at the staged nozzle because of the additional fuel that ultimately causes higher chemiluminescence signal in that flame. At an axial distance of 3 in., the star shape emerges around the center nozzle as the chemiluminescence intensity around the staged nozzle diminishes. Figures 5b and 6b show the XZ plane, where the flame stabilization is more evident; here the pocket of chemilumescence, likely from hot-gas recirculation, is present but little outer shear layer flame stabilization is seen. This result differs from the axisymmetric staging, where staging the center flame resulted in a change in the flame shape, whereas the shape does not change significantly in the outer flames when they are staged. For both non-axisymmetric cases, the XZ plane images (Fig 4b, 5b) show streaking and nonuniformity. This could be a result of error in the reconstruction though it is suggestive that these features are visible in both reconstructions.

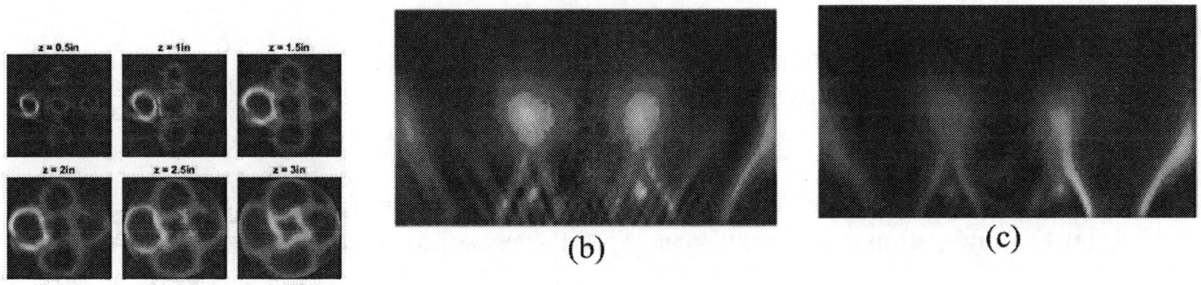

(a) (b) (c)

Figure 5: Non-axisymmetric Staging for $\phi_{Back} = 0.85, \phi_{Other} = 0.68$
a) Z-direction b) XZ-plane c) YZ-plane

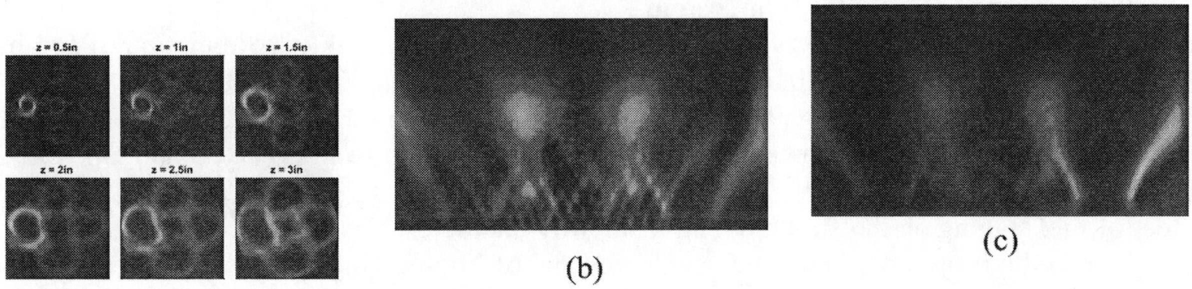

(a) (b) (c)

Figure 6: Non-axisymmetric Staging for $\phi_{Back} = 0.90, \phi_{Other} = 0.68$
a) Z-direction b) XZ-plane c) YZ-plane

4. Conclusions

This work examined the flame structure of non-staged, axisymmetrically staged, and non-axisymmetrically staged conditions. There were several similarities and differences found between the flame structures for different staging configurations. All cases exhibit flame shape variation with downstream distance. Non-staged flames tend to have significant flame interaction and partial V-shape flames. Axisymmetric staged flames have less noticeable flame interaction than the non-staged case, but more distinct M-shape flames are formed. Non-axisymmetric staged flames have a structure more similar to the non-staged case.

This work suggests that the location of fuel staging impacts flame shape. Different staging configurations can drastically change flame structure as well as flame-flame interaction, flame-wall interaction, and heat release rates. Future work will use OH laser-induced fluorescence measurements to better understand the connection between flame shape and instability suppression. Steps will be taken to fine tune the tomographic reconstruction techniques in order to create the most accurate figures. More data for each case will allow for precise comparisons of the various staging configurations as well as a better assessment of the non-axisymmetric discontinuities.

5. Acknowledgements

The authors would like to thank Dr. Janith Samarasinghe and Dr. Steve Peluso for their suggestions and insights on this work. The authors would also like to thank the Shuman Mechanical Engineering Scholars Program and the U.S. Department of Energy under award number DE-FE0025495 and contract monitor Mark Freeman for funding this work.

6. References

[1] Lieuwen, T. C., and Yang, V., 2005, Combustion Instabilites in Gas Turbine Engines, American Institute of Aeronautics and Astronautics.

[2] Lieuwen, T. C., and Yang, V., 2013, Gas Turbine Emissions, Cambridge University Press, New York.

[3] Bulat, G., Skipper, D., McMillan, R., and Syed, K., "Active Control of Fuel Splits in Gas Turbine DLE Combustion Systems," Proc. ASME Turbo Expo 2007: Power for Land, Sea, and Air.

[4] Davis, L. B., and Black, S. H., "Dry Low NOx Combustion Systems for GE Heavy-Duty Gas Turbines," GE Power Systems, Schenectady, NY.

[5] Hermann, J., Orthmann, A., Hoffmann, S., and Berenbrink, P., "Combination Of Active Instability Control and Passive Measures to Prevent Combustion Instabilities in a 260MW Heavy Duty Gas Turbine," RTO AVT Symposium.

[6] Cohen, J., Hagen, G., Banaszuk, A., Becz, S., and Mehta, P., 2011, "Attenuation of combustor pressure oscillations using symmetry breaking," AIAA Paper No. GT2011-0060.

[7] Li, L., Sun, X., Lioi, C., and Yang, V., 2016, "Effect of Azimuthally Nonuniform Heat Release on Longitudinal Combustion Instabilities," Journal of Propulsion and Power, 33(1), pp. 193-203.

[8] Samarasinghe, J., Culler, W., Quay, B. D., Santavicca, D. A., and O'Connor, J., 2017, "The effect of fuel staging on the structure and instability characteristics of swirl-stabilized flames in a lean premixed multi-nozzle can combustor," Journal of Engineering for Gas Turbines and Power.

[9] García-Armingol, T., Sobrino, Á., Luciano, E., and Ballester, J., 2016, "Impact of fuel staging on stability and pollutant emissions of premixed syngas flames," Fuel, 185, pp. 122-132.

[10] Samarasinghe, J., Peluso, S., Szedlmayer, M., De Rosa, A., Quay, B., and Santavicca, D., 2013, "Three-Dimensional Chemiluminescence Imaging of Unforced and Forced Swirl-Stabilized Flames in a Lean Premixed Multi-Nozzle Can Combustor," Journal of Engineering for Gas Turbines and Power, 135(10), p. 101503.

[11] Samarasinghe, J., 2016, "The Structure and Velocity-Forced Response of Swirl-Stabilized Flames in a Lean-Premixed Multi-Nozzle Can Combustor," Ph.D. Thesis, The Pennsylvania State University.

[12] Culler, W., Samarasinghe, J., Quay, B. D., Santavicca, D. A., and O'Connor, J., "The Effect of Transient Fuel Staging on Self-Excited Instabilities in a Multi-Nozzle Model Gas Turbine Combustor," Proc. ASME Turbo Expo 2017.
[13] Orawannukul, P., 2014, "An Experimental Study of Forced Flame Response in Technically Premixed Flames in a Lean Premixed Gas Turbine Combustor," Ph.D. Thesis, The Pennsylvania State University.

Spring Technical Meeting
Eastern States Section of the Combustion Institute
March 4-7, 2018
State College, Pennsylvania

The Influence of Mixedness on Ignition for Hydrogen Direct Injection in a Constant Volume Combustion Chamber

Martia Shahsavan[1,], Mohammadrasool Morovatiyan[1], J. Hunter Mack[1]*

[1] Department of Mechanical Engineering, University of Massachusetts Lowell, 1 University Ave, Lowell, MA, USA
[*]Corresponding Author Email: martia_shahsavan@student.uml.edu

Abstract: The ignition behavior of the fuel in non-premixed turbulent combustion applications such as diesel engines and gas turbines is dependent on the mixing rate of the injected fuel and the working fluid. In this study, three-dimensional modeling of hydrogen injection into a constant volume combustion chamber (CVCC) is used to investigate the correlation between the mixing rate and important parameters of non-premixed combustion, such as ignition delay. Mixedness is quantified using mean spatial variation, which reflects the homogeneity of the mixture, and mean scalar dissipation, which represents the local gradients of the scalar. The case studies include nitrogen and argon as working fluids; injection velocities and nozzle diameters are varied for comparison. For consistency, the injected mass is kept constant and the injection duration is adjusted accordingly. The results indicate that a strong correlation exists between ignition delay and the defined mixedness parameters. The cases with higher mixedness values lead to a shorter ignition delay and a higher maximum flame temperature. Changing the working fluid and injection parameters can effectively modify the mixedness, and consequently affect the ignition onset and flame properties.

Keywords: Mixedness, Ignition Delay, Hydrogen Injection, CVCC

1. Introduction

Recently, hydrogen as a potential alternative fuel for use in internal combustion engines has gained interest among researchers due to its wide flammability range and the elimination of major pollutants [1]. Direct injection can be an optimal utilization of hydrogen since it removes the low hydrogen density restriction and enables a higher volumetric efficiency [2]. Moreover, it prevents backfire, which is a major problem of premixed hydrogen combustion [3]. There are a number of studies on the use of hydrogen in direct injection applications. Controlling autoignition of hydrogen in an optical engine has been investigated by attempting various injection strategies [4]. Numerical modeling has indicated that dual fuel hydrogen-diesel internal combustion engines can operate more efficiently due to the power density enhancement [5]. Adding hydrogen to natural gas-diesel engines has been observed to provide more complete combustion at low loads and a reduction of unburned hydrocarbons and CO emissions [6]. Additionally, it has been shown that hydrogen as an additive to CNG increases the maximum engine speed [7].

The fundamentals of the gaseous jet should be well understood in order to investigate the combustion characteristics of hydrogen injection. A parametric study has been carried out on high speed hydrogen injection, which shows the effect of grid resolution, Adaptive Mesh

Refinement (AMR), and injection timing on mixture formation inside the cylinder [8]. The characteristics of hydrogen under-expanded jets has been comprehensively investigated using Large Eddy Simulation (LES) with different nozzle pressure ratios [9].

Replacing the working fluid can drastically impact the combustion characteristics since it can increase the ideal thermal efficiency by using high specific heat ratio gases such as argon and xenon instead of nitrogen [10, 11]. A more recent experimental study has been carried out on the effect of noble gases on fuel propagation and thermal characteristics of gaseous direct injection; it has been shown that argon can be a potentially beneficial alternative working fluid [12]. In another study, penetration length, cone angle, mixedness, and flame properties have been extensively studied for hydrogen injection into different working fluids by numerical modeling, indicating that replacing nitrogen with argon and xenon can dramatically change the jet development in the chamber [13].

One of the main properties of turbulent gaseous injection is the mixing rate of the jet with the working fluid. The mixedness of the fuel with the in-cylinder gases can heavily influence the combustion process, since ignition occurs at the smallest length scales where turbulent mixing is important [14]. However, research on quantifying the mixedness is quite limited. In one study, mean spatial variation in combination with mean scalar dissipation were used to evaluate mixedness in the turbulent injection of gasoline into air [15]. The mean spatial variation (Equation 1) represents the inhomogeneity of the scalar population, and the mean scalar dissipation (Equation 2) indicates the rate of fine-scale mixing. ζ is the conserved scalar (e. g. mixture fraction), ζ_m is the average intensity over an N by P pixel region of interest, and λ_D is the strain-limited length scale.

$$\frac{\sigma}{\xi_m} = \frac{\left\{ [1/(NP)] \sum_{i,j}^{N,P} \left[\xi_{i,j} - \xi_m \right]^2 \right\}^{1/2}}{\xi_m} \tag{1}$$

$$\chi_m = \frac{1}{NP} \sum_{i,j}^{N,P} \frac{\left(\nabla \xi_{i,j} . \nabla \xi_{i,j} \right)}{\left(\xi_m / \lambda_D \right)^2} \tag{2}$$

This method has been implemented on the direct injection of hydrogen into different working fluids, and it has been shown that argon can provide a better mixing conditions for the injected gas when compared to nitrogen due to a higher diffusivity coefficient [16]. In the present study, a three dimensional transient simulation of hydrogen injection into a constant volume combustion chamber (CVCC) is used to investigate the effect of mixedness on ignition properties.

2. Methodology

2.1 Numerical setup

Numerical simulations of hydrogen injection into a CVCC are conducted using CONVERGE computational code [17]. The CVCC is a cylindrical type with a diameter of 20 mm and a length of 50 mm. Three nozzle diameters (0.8, 1.0, and 1.2 mm), five injection velocities (80, 90, 100, 110, and 120 m/s), and two working fluids (79% nitrogen, and argon, in combination with 21% oxygen) aided to generate a sufficient number of case studies. The injection duration is adjusted based on the related velocity and diameter in order to have an equal injected mass in all the cases. A varied time step size with a minimum of 1e-7 s, and a maximum of 3e-4 s is used. CFL number, which is a vital criterion for convergence, and Mach number automatically modify the time steps. A CHEMKIN format mechanism with 10 species and 21 reactions [18] including thermodynamic data is used for the hydrogen combustion. The initial conditions inside the chamber are kept constant for all the cases at 1000 K and 1 bar, and zero heat flux boundary condition is chosen for the chamber walls. For combustion modeling, SAGE detailed chemistry solver [19], which calculates the reaction rates for each elementary reaction, is used. A 1-equation dynamic structure type of Large Eddy Simulation (LES) is implemented for turbulence modeling. CONVERGE automatically generates the computational cells, with the required base grid size set to 1 mm in this simulation. Adaptive Mesh Refinement (AMR) is utilized for the velocity and temperature with maximum embedding levels of 4 and 3 and sub-grid criterion of 0.1 m/s and 2.5 K, respectively. This method of grid refinement is quite beneficial since it generates smaller cells in high gradient areas and leads to greater computational efficiency.

2.2 Data analysis

The the steepest slope of the OH radical concentration curve is used to measure the ignition delay [20]. The output image at the onset of combustion generated by CONVERGE is then exported to MATLAB for image processing. Equations 1 and 2 are then discretized and implemented on the region of interest in order to calculate the mixedness parameters.

3. Results and Discussion

Five injection velocities (V_{inj}), three injector diameters (d_{inj}), and two working fluids provide 30 different case studies which are shown in Figure 1. Blue and red dots represent nitrogen and argon, respectively, as the working fluid in combination with oxygen. A strong correlation exists between the mean spatial variation and the mean scalar dissipation. This result is in line with the previously generated plot by Probst and Ghandhi [15]. The trend line indicates that the two mixedness parameter are correlated with a quadratic function. This originates from the linear relation of spatial variation and the quadratic relation of spatial variation to the scalar magnitudes.

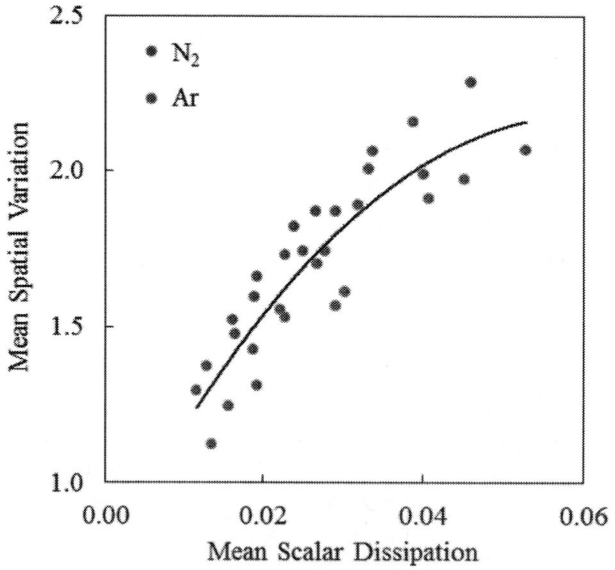

Figure 1. Mean spatial variation versus mean scalar dissipation for different working fluids, injection velocities, and injector diameters.

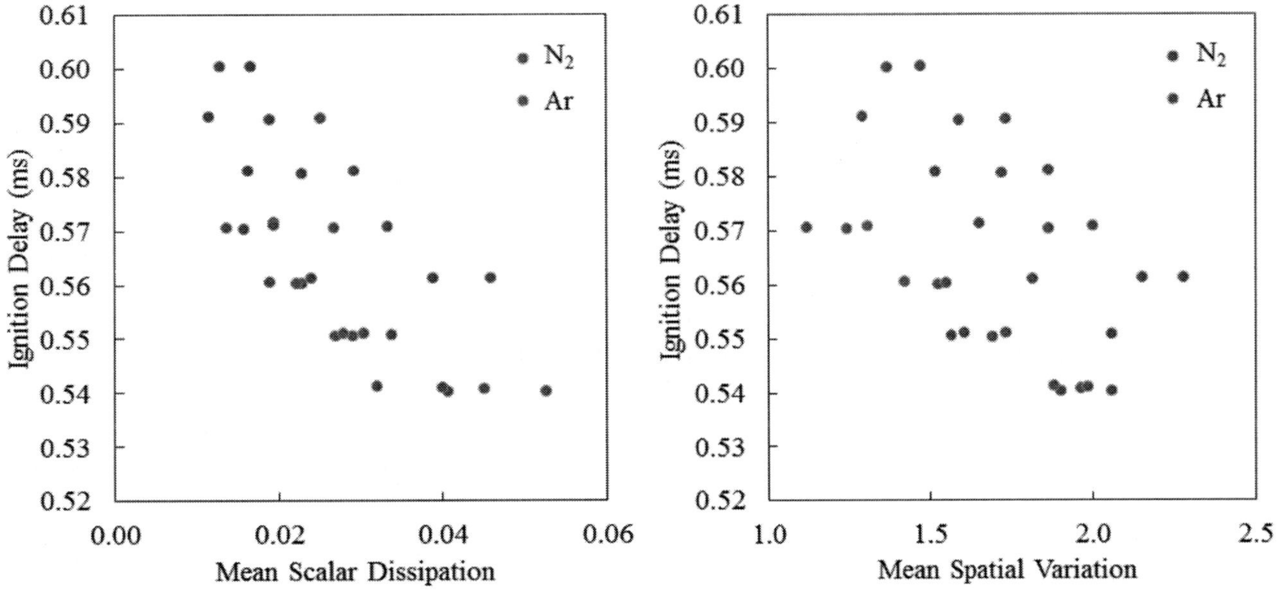

Figure 2. Ignition delay versus mixedness parameters

Figure 2 indicates the influence of the mixedness parameters on ignition delay. It can be interpreted that a higher mean scalar dissipation and a higher mean spatial variation lead to shorter ignition delay. In general, argon cases possess a shorter ignition delay when compared to nitrogen cases. In general, the ignition delay shows a more quadratic correlation with the mean scalar dissipation when compared to the mean spatial variation, which is more linear.

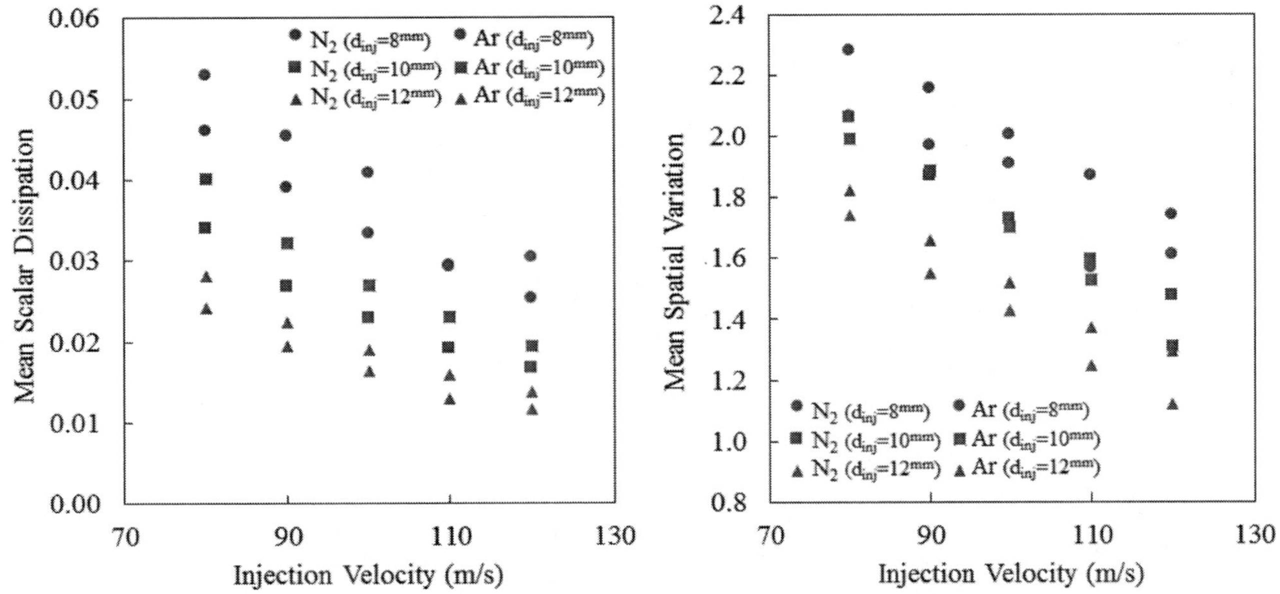

Figure 3. Effect of injection velocity and injector diameter on mixedness parameters.

The effects of injection strategies on mixedness can be found in Figure 3. As seen, increasing the injection velocity can strongly affect both mixedness parameters, which affects both the ignition timing and combustion properties. Increasing the nozzle diameter has a similar effect on both the mean scalar dissipation and spatial variation. This indicates that although argon provides a shorter ignition delay in general when compared to nitrogen, changing the injection parameters such as injection velocity and nozzle diameter can drastically impact the mixedness of the injected fuel into the working fluid; therefore, hydrogen shows a shorter ignition delay in nitrogen for some cases.

Investigating the effect of other parameters in non-premixed combustion of hydrogen such as various in-cylinder temperature and pressure, and a different injected fuel temperature can help to find a more detailed correlation of the combustion properties such as ignition delay with the determinant parameters. This procedure can be implemented on other gaseous fuels such as methane, or gas mixtures (e.g. syngas) in order to provide a larger number of case studies. Providing an experimental setup can effectively help to validate the numerical results and derive a general correlation between gaseous injection parameters and combustion properties such as ignition delay.

4. Conclusion

In this study, the effect of injection velocity, nozzle diameter, and working fluid on the mixedness parameters and ignition properties is investigated using a three dimensional transient simulation of hydrogen injection in a constant volume combustion chamber. The results indicate that increasing the injection velocity and nozzle diameter can heavily impact the mixing rate of the injected fuel and the working fluid, which dramatically change combustion properties, such as ignition delay. The relationship between two mixedness parameters, scalar dissipation and spatial variation, follow previous experimental results. The correlation of mixedness parameters to the ignition delay generated from this numerical modeling promises a potential general equation for the ignition delay calculation in all non-premixed gaseous combustion applications. An experimental study, which is currently in progress, can effectively aid to validate the modeling results and provide a comprehensive correlation for all gaseous injection cases.

5. Acknowledgements

This research was funded by the University of Massachusetts Lowell.

6. References

[1] S. Verhelst, "Recent progress in the use of hydrogen as a fuel for internal combustion engines," *international journal of hydrogen energy,* vol. 39, no. 2, pp. 1071-1085, 2014.

[2] S. Verhelst and T. Wallner, "Hydrogen-fueled internal combustion engines," *Progress in Energy and Combustion Science,* vol. 35, no. 6, pp. 490-527, 2009/12/01/ 2009.

[3] J. Duan, F. Liu, and B. Sun, "Backfire control and power enhancement of a hydrogen internal combustion engine," *International Journal of Hydrogen Energy,* vol. 39, no. 9, pp. 4581-4589, 2014.

[4] P. G. Aleiferis and M. F. Rosati, "Controlled autoignition of hydrogen in a direct-injection optical engine," *Combustion and Flame,* vol. 159, no. 7, pp. 2500-2515, 2012.

[5] A. Boretti, "Advantages of the direct injection of both diesel and hydrogen in dual fuel H 2 ICE," *international journal of hydrogen energy,* vol. 36, no. 15, pp. 9312-9317, 2011.

[6] P. Rahnama, A. Paykani, and R. D. Reitz, "A numerical study of the effects of using hydrogen, reformer gas and nitrogen on combustion, emissions and load limits of a heavy duty natural gas/diesel RCCI engine," *Applied Energy,* vol. 193, pp. 182-198, 2017.

[7] B. Afkhami, A. Kakaee, and K. Pouyan, "Studying engine cold start characteristics at low temperatures for CNG and HCNG by investigating low-temperature oxidation," *Energy conversion and management,* vol. 64, pp. 122-128, 2012.

[8] S. Sukumaran and S.-C. Kong, "Numerical study on mixture formation characteristics in a direct-injection hydrogen engine," *International Journal of Hydrogen Energy,* vol. 35, no. 15, pp. 7991-8007, 2010.

[9] A. Hamzehloo and P. Aleiferis, "Large eddy simulation of highly turbulent under-expanded hydrogen and methane jets for gaseous-fuelled internal combustion engines," *International Journal of Hydrogen Energy,* vol. 39, no. 36, pp. 21275-21296, 2014.

[10] A. Kato, S. Misawa, A. N. M. Yoshimatsu, and D. Sawada, "A Study of Rising Thermal Efficiency Using an Argon Circulated Hydrogen Engine (Third Report)," *JSAE paper,* vol. 20135721, 2013.

[11] N. J. Killingsworth, V. H. Rapp, D. L. Flowers, S. M. Aceves, J.-Y. Chen, and R. Dibble, "Increased efficiency in SI engine with air replaced by oxygen in argon mixture," *Proceedings of the Combustion Institute,* vol. 33, no. 2, pp. 3141-3149, 2011.

[12] M. Sierra-Aznar *et al.*, "Working fluid replacement in gaseous direct-injection internal combustion engines: A fundamental and applied experimental investigation," 2017.

[13] M. Shahsavan and J. Mack, "The effect of heavy working fluids on hydrogen combustion," in *10th US National Combustion Meeting*, College Park, MD, 2017.

[14] H. Pitsch and H. Steiner, "Scalar mixing and dissipation rate in large-eddy simulations of non-premixed turbulent combustion," *Proceedings of the Combustion Institute,* vol. 28, no. 1, pp. 41-49, 2000/01/01/ 2000.

[15] D. Probst and J. Ghandhi, "An experimental study of spray mixing in a direct injection engine," *International Journal of Engine Research,* vol. 4, no. 1, pp. 27-45, 2003.

[16] M. Shahsavan and J. H. Mack, "Mixedness Measurement in Gaseous Jet Injection," in *American Society for Engineering Education Northeast Section (ASEE-NE)*, Lowell, MA, 2017.

[17] K. Richards, P. Senecal, and E. Pomraning, "CONVERGE (Version 2.4), Convergent Science," *Inc., Madison, WI,* 2017.

[18] M. Ó Conaire, H. J. Curran, J. M. Simmie, W. J. Pitz, and C. K. Westbrook, "A comprehensive modeling study of hydrogen oxidation," *International Journal of Chemical Kinetics,* vol. 36, no. 11, pp. 603-622, 2004.

[19] P. Senecal *et al.*, "Multi-dimensional modeling of direct-injection diesel spray liquid length and flame lift-off length using CFD and parallel detailed chemistry," SAE Technical Paper0148-7191, 2003.

[20] S. S. Vasu, D. F. Davidson, and R. K. Hanson, "Jet fuel ignition delay times: Shock tube experiments over wide conditions and surrogate model predictions," *Combustion and Flame,* vol. 152, no. 1, pp. 125-143, 2008.

Spring Technical Meeting
Eastern States Section of the Combustion Institute
March 4-7, 2018
State College, Pennsylvania

The Effects of Piloting on Turbulent Flame Structure

*Ryan Shupp[1], Ankit Tyagi[1], Isaac Boxx[2], Stephen Peluso[1], and Jacqueline O'Connor[1]**

[1]Mechanical and Nuclear Engineering, Pennsylvania State University, University Park, PA
[2]DLR, German Aerospace Center, Stuttgart, Germany, 70596
**Corresponding Author Email: jxo22@psu.edu*

Pilot flames are commonly used to study turbulent flames as they present a means of preventing blow off at higher velocities. There are open questions regarding how pilot flames affect the main flame. This is true for both anchoring pilots, which are small pilot flames located at the base of the flame, and back-support pilots, which produce adiabatic or superadiabatic boundaries around the flame. The question of how the presence and size of pilot flames affect the properties of the main flames is important to understanding the behavior of the main flame. The purpose of this study is to explore how the presence of pilot flames affects the flame surface density, global consumption speed, and flame curvature statistics of a turbulent, premixed flame over a range of velocities. Natural gas was used as a fuel and the equivalence ratio was held constant at 1. Three different pilot flame configurations were used to explore the interaction between flame and pilot, including variations in the presence of both the anchoring pilots and the back-support pilots. The results of this study will affect our interpretation of flame sheet dynamics in a range of piloted flames.
Keywords: Turbulent Flames, Stabilization, Pilot Flames

1. Introduction

Piloting is used in many turbulent combustion experiments to help stabilize flames at high flow velocities and turbulence intensities. In this study, we focus on premixed flames only, as the structure and propagation of premixed flames is canonically different than that of non-premixed or even partially premixed flames. These piloting schemes vary between experiments, but typically can be put into two major categories: anchoring pilots, which are used to stabilize the base of the flame and prevent blow-off in higher velocity cases, and back-support pilots, which are used to produce an adiabatic or superadiabatic boundary layer to provide more favorable conditions for combustion. Table 1 provides examples of burners that have used pilot flames and describes the nature of the pilots used.

While the use of pilot flames is very common, it's unclear how these pilot flames change the structure and behavior of the turbulent flames they stabilized. The goal of this study is to begin to understand the impact of pilot flames on turbulent premixed flame behavior. In this particular study, we have found that the presence of pilot flames has a negligible effect on the turbulent global consumption speed and flame brush thickness of turbulent premised flames.

Table 1. List of experiments that utilize pilot flames to study turbulent premixed flames.

Authors	Type of Pilot	Experimental Goal	Pilot Fuel	Pilot Equivalence Ratio
Skiba *et al.* (Michigan)[1]	Back-Support	Study of flame structure responses to extreme levels of turbulence	Methane	0.98
Zhou *et al.* (Lund)[2]	Back-Support	Exploration of the relationship between reactive scalar and jet speed	Methane	0.4
Won *et al.* (Princeton)[3]	Anchoring	Research of turbulent flame propagation's response to low temperature fuel chemistry	Methane	1.00
Yuen *et al.* (Toronto) [4]	Anchoring	Characterization of flame structure in thin reactions regime	Methane, Ethylene	Not Provided
Dunn *et al.* (Sydney) [5]	Anchoring	Studying the finite-rate chemistry effects of a premixed jet	Natural Gas	1.00
Venkateswaran *et al.* (Georgia Tech) [6]	Anchoring	Understand the effects of syngas fuel blends on turbulent consumption speed	Methane	0.9
Steinberg *et al.* (Michigan) [7]	Back-Support	Studying the effect of turbulence-flame interactions on flame stretch	Methane	Varying

2. Experimental methods

2.1 Main Burner Setup

The experimental setup consists of a burner with an exit plane of dimensions 100mm x 10mm (Figure 1). The burner contains two stacked sections with heights of 178mm and 160mm. The first section contains the inlet for incoming flow of the air-fuel mixture and a ceramic honeycomb layer. The other section contains another honeycomb layer and two perforated plates. As a means of conditioning the flow of the air-fuel mixture (utilizing natural gas as the fuel and an equivalence ratio of 1), the flow is sent through both layers of honeycomb inside the burner prior to reaching the exit plane of the burner. The two perforated plates are placed at 30mm and 10mm upstream of the burner exit plane and are used to create turbulence in the flow field. The perforated plates, with a 3.175mm hole diameter and 40% open area, are able to generate a non-reacting average flow turbulence intensity of 18% at the exit of the burner. This turbulence intensity is normalized through the use of the bulk flow velocities.

2.2 Back Support Pilot Setup

The two identical back-support pilots each consist of rectangular exit planes with dimensions 90mm x 30mm. The exit plane of the back support pilots is 13mm upstream of the burner exit plane. The cavity of the back-support pilots is filled with ball bearings as a means of ensuring that that flow is relatively uniform throughout the entire rectangular pilot.

2.3 Anchoring Pilot Setup

The two identical anchoring pilots have a narrow rectangular exit plane with dimensions of 90mm x 4.8mm. The exit plane of the anchoring pilots is 6.4mm upstream of the burner exit plane. Similar to the back-support pilots, the anchoring pilots are filled with identical ball bearings and a layer of honey comb for flow conditioning.

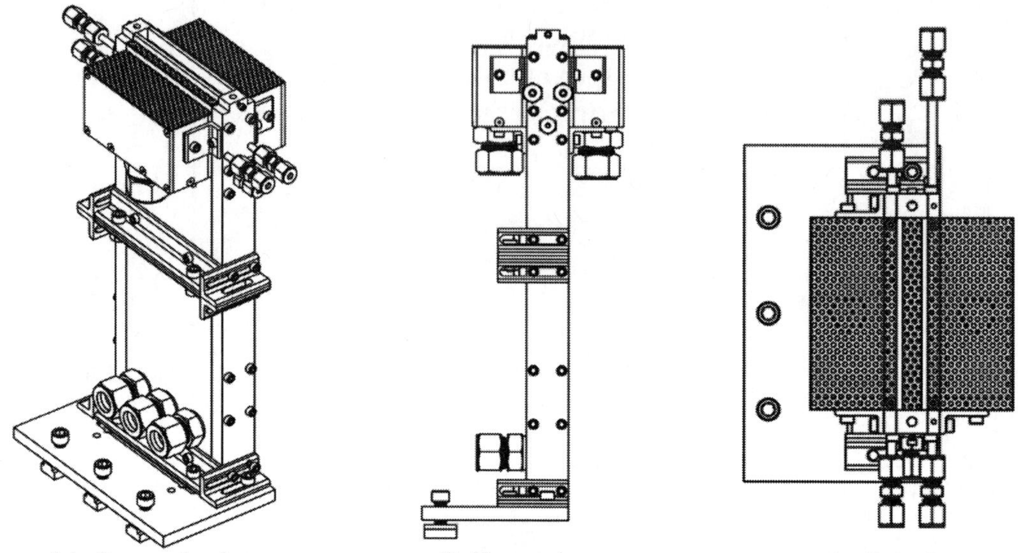

(a) Isometric view ***(b) Front view*** ***(c) Top view***

Figure 1: Sketches of the experiment's burner: *(a)* Isometric view of the burner configuration in its entirety, *(b)* Front view of the burner, *(c)* Top view of the burner.

2.4 OH-Planar Laser-Induced Fluorescence

OH-PLIF measurements are performed at a sampling rate of 10kHz. This high-speed diagnostic system consists of a 532nm Nd:YAG laser (Edgewave) pumping a dye laser (Sirah Credo). The maximum power output from the dye laser at 10 kHz repetition rate is 0.3mJ/pulse. The dye laser is calibrated to the $Q_1(6)$ line of the $A^2\Sigma^+ \leftarrow X2\Pi(1\text{-}0)$ band to excite the OH radicals at 282.94nm. A periscope and a set of three cylindrical lenses are used to obtain a collimated sheet with an approximate height of 21mm. A CMOS sensor camera (Photron FASTCAM SA1.1), coupled with an external intensifier (LaVision HS-IRO) and a 100 mm f/2.8 UV lens (Cerco) is used to acquire the images. A high transmissivity interference filter (LaVision 1108760 VZ) is used to reduce background noise. A more detailed description of this set up can be found in Ref. [8].

3. Results

3.1 Test Matrix

This study was run for three different pilot configurations, each over three different flow conditions with bulk flow velocities of 12m/s, 20m/s and 28m/s. The piloting configurations consisted of a condition where both anchoring and back-support pilots were used (denoted as "AB"), a condition where only anchoring pilots were used (denoted as "A"), and a condition where only back-support pilots were used (denoted as "B"). The test matrix in Table 2 provides the details of each of the nine cases examined in this study.

Table 2. Test matrix including variations in the bulk flow velocities, the Reynolds number based on the hydraulic diameter, and the pilot configuration for each case.

Case Number	Equivalence Ratio (Φ)	Bulk Flow Velocity U [m/s]	Reynolds Number (Hydraulic Diameter)	Pilot Configuration	u'[m/s]
12AB	1.0	12	15485	All	2.0
12A	1.0	12	15485	Anchoring	2.0
12B	1.0	12	15485	Back-support	2.0
20AB	1.0	20	25808	All	3.2
20A	1.0	20	25808	Anchoring	3.2
20B	1.0	20	25808	Back-support	3.2
20AB	1.0	28	36131	All	4.5
20A	1.0	28	36131	Anchoring	4.5
20B	1.0	28	36131	Back-support	4.5

3.2 Progress Variables

Figure 2 shows the stitched, time-averaged progress variables with red lines used to denote the three fields of view used for data collection, and Figure 3 shows an instantaneous LIF image of the flames in the first field of view. From the progress variable plots, there are no observable differences between the different pilot configurations as the time averaging diminishes any instantaneous topological differences that may exist in the flame structure. The instantaneous LIF views show the benefits that back-support pilots have on data collection. In the absence of back-support pilots, the absence of the additional OH around the flame makes edge detection and curvature calculations difficult and leads to a degree of ambiguity in regards to the behavior of the structure of the flame.

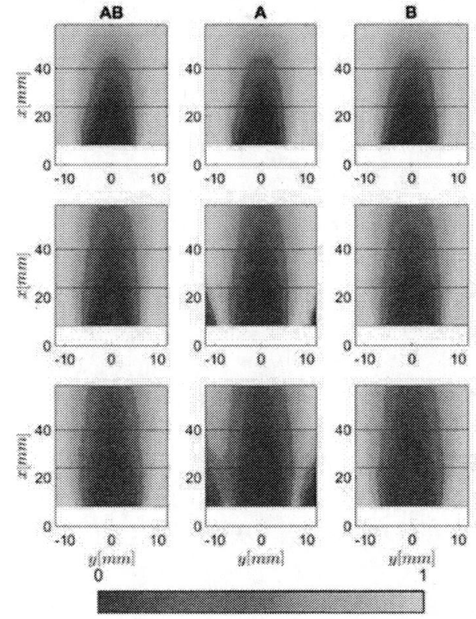

Figure 2: Stitched, time-averaged progress variable plots for each of the 9 cases.

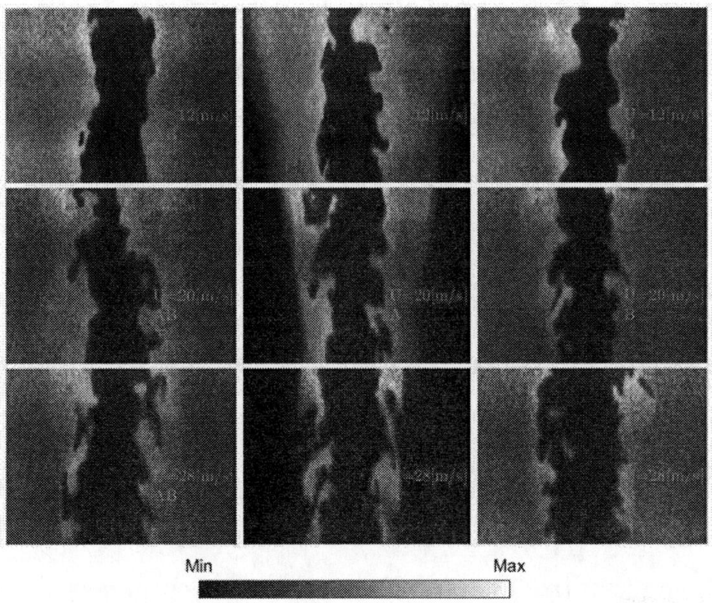

Figure 3: Instantaneous LIF images of the first field of view for each of the 9 cases.

3.3 Flame Brush Thickness

By taking an average of the instantaneous binarized images from the experiment, time-averaged progress variables ($\bar{c}$) were found and used as a means of calculating the flame brush thicknesses. According to studies conducted by Kheirkhah *et al.* [9] and Namazian *et al.* [10], the magnitude of the flame brush thickness can be approximated through the use of the maximum gradient of the progress variable in the cross-stream direction, as in Eq. 1.

$$\delta_t = \max\left(\frac{1}{\left|\frac{\partial \bar{c}}{\partial y}\right|}\right) \qquad (1)$$

Figure 4 provides the plots of the flame brush thicknesses for each of the velocities versus downstream position. The results show that at each of the velocities tested, the presence or absence of pilot flames had little to no effect on the slope of the flame brush thickness plot.

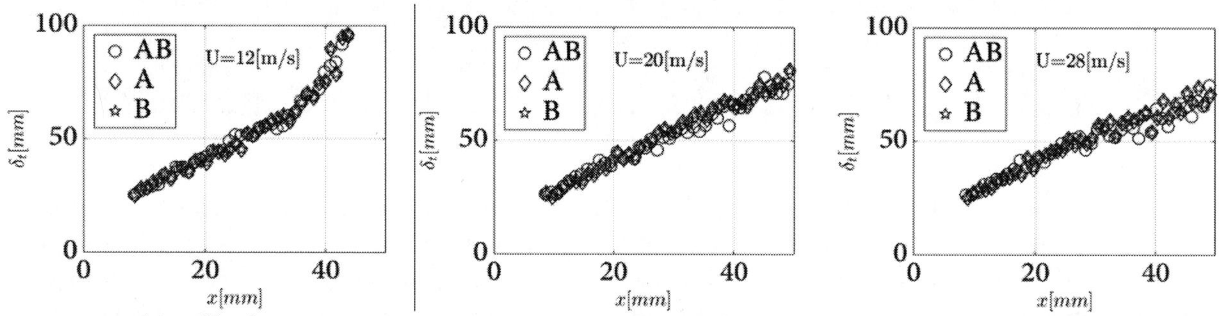

(a) 12m/s *(b) 20m/s* *(c) 28m/s*

Figure 4: Flame brush thickness vs. position for bulk flow velocities of (a) 12m/s, (b) 20m/s and (c) 28m/s.

3.4 Turbulent Global Consumption Speed

The turbulent global consumption speed can be calculated in the equation that follows, where the total reactant mass flow rate is divided by the product of the reactant gas density and the surface area of a contour for a given progress variable, as in Eq. 2.

$$S_{T,GC} = \dot{m}_R / \rho_R A_{\bar{c}} \qquad (2)$$

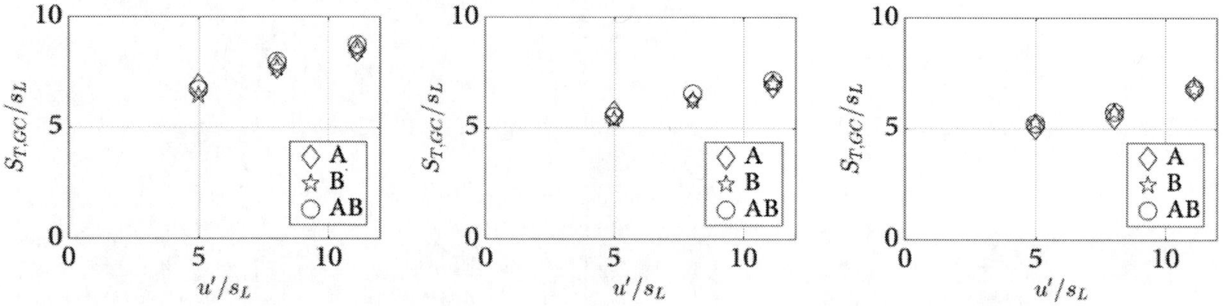

(a) $\bar{c} = 0.1$ *(b)* $\bar{c} = 0.2$ *(c)* $\bar{c} = 0.3$

Figure 5: Turbulent global consumption speeds for progress variables of (a) 0.1, (b) 0.2 and (c) 0.3.

We have utilized three different values of progress variables to calculate global consumption speed ($\bar{c} = 0.1, 0.2, 0.3$). The density of the reactants is taken at 300K. As we increase the bulk flow velocity, the turbulence intensity also increases and the global consumption speeds increase.

Conversely, as the progress variables increase, global consumption speeds decrease. This is due to the fact that larger progress variables result in a larger averaged surface area of the flame, as shown in Figure 5. As shown in the plots, there is virtually no difference in the global consumption speeds of flames when subjected to the various pilot flame configurations.

4. Conclusions

In summary, the overarching implication of this initial work is that any potential effects that pilot flames could have on the global behavior of turbulent flames can be regarded as negligible. The slope of the flame brush thickness data appeared to be consistent across each of the pilot conditions and was merely a function of bulk flow velocity. Furthermore, the global consumption speeds of the turbulent flames were likewise unaffected by the pilots, and were affected only by the bulk flow velocity and the progress variable. With pilot flames being a resource so frequently used in the combustion community, it is essential to ensure that data has not been skewed as a result. In the future, the authors will continue to explore potential effects that pilots may have on the local behavior of these flames. In particular, we will consider the impact of piloting on flame wrinkling and the frequency of flame interaction events and flame hole formation.

Acknowledgements

This work was supported by the Air Force Office of Scientific Research under Grant FA9550-16-1-0075 with program manager Dr. Chiping Li. The authors gratefully acknowledge Dr. Campbell Carter at the AFRL and LaVision GmbH for their advice and loan of equipment during these experiments.

References

1. Skiba, A.W., et al., *Premixed flames subjected to extreme levels of turbulence part I: Flame structure and a new measured regime diagram.* Combustion and Flame, 2017.
2. Zhou, B., et al., *Distributed reactions in highly turbulent premixed methane/air flames: Part I. Flame structure characterization.* Combustion and Flame, 2015. **162**(7): p. 2937-2953.
3. Won, S.H., et al., *The role of low temperature fuel chemistry on turbulent flame propagation.* Combustion and Flame, 2014. **161**(2): p. 475-483.
4. Yuen, F.T. and Ö.L. Gülder, *Premixed turbulent flame front structure investigation by Rayleigh scattering in the thin reaction zone regime.* Proceedings of the Combustion Institute, 2009. **32**(2): p. 1747-1754.
5. Dunn, M.J., A.R. Masri, and R.W. Bilger, *A new piloted premixed jet burner to study strong finite-rate chemistry effects.* Combustion and Flame, 2007. **151**(1): p. 46-60.
6. Venkateswaran, P., *Measurements and modeling of turbulent consumption speeds of syngas fuel blends.* 2013, Georgia Institute of Technology.
7. Steinberg, A.M., *The dynamics of turbulent premixed flames: Mechanisms and models for turbulence-flame interaction.* 2009, University of Michigan.
8. Tyagi, A., et al. *Structure of Flames in Flame Interaction Zones.* in *2018 AIAA Aerospace Sciences Meeting.* 2018.
9. Kheirkhah, S. and Ö. Gülder, *Topology and brush thickness of turbulent premixed V-shaped flames.* Flow, turbulence and combustion, 2014. **93**(3): p. 439-459.
10. Namazian, M., L. Talbot, and F. Robben. *Density fluctuations in premixed turbulent flames.* in *Symposium (International) on Combustion.* 1985. Elsevier.

Spring Technical Meeting
Eastern States Section of the Combustion Institute
March 4–7, 2018
State College, Pennsylvania

Challenges for Large Eddy Simulation of Partially Premixed Turbulent Combustion using Reduced-Order Manifold Flame Structure Models

Bruce A. Perry[1,*] *and Michael E. Mueller*[1]

[1]*Department of Mechanical and Aerospace Engineering, Princeton University, Princeton, NJ 08544, USA*
[*]*Corresponding author:* baperry@princeton.edu

Abstract: Partial premixing and the presence of multiple inlet streams are key features of many modern paradigms in combustion engine design but lead to systems that simultaneously involve premixed-like, nonpremixed-like, and intermediate combustion modes. In contrast, the classical reduced-order flame structure models used for Large Eddy Simulation (LES), such as "flamelet"-like models, are derived assuming a single dominant mode. A variety of reduced-order manifold approaches have been developed that attempt to model multi-modal combustion by switching between the classical single-mode models based on a parameter that indicates which mode is locally dominant. In this work, three approaches for including multi-modal effects in LES are compared for the case of a piloted jet burner with inhomogeneous inlets. These approaches are all based on an extended nonpremixed model that uses two mixture fractions to account for partial premixing of the fuel stream but use different methods of incorporating premixed combustion effects. Although the multi-modal models have shown improved predictions relative to single-mode models, they result in additional modeling challenges and still do not accurately predict the flame structure in all regions of the flame. Potential causes for the remaining inaccuracy in the predictions are assessed. In particular, the limitations of applying combustion models that rely on a 1D flame structure and are based on either premixed or nonpremixed combustion to a partially premixed system are discussed.
Keywords: ***Partially premixed combustion, Large Eddy Simulation, Reduced-order Manifolds***

1. Introduction

Many practical combustion systems involve inhomogeneous partial premixing of fuel and air, so combustion occurs outside of the classical premixed and nonpremixed combustion modes [1]. For example, flame lift off and the use of multiple fuel inlets in aviation gas turbines leads to a partially premixed environment. As reviewed by Masri [1], recent experimental and computational efforts have focused on moving beyond the traditional classification to understand partially premixed combustion. One particular area of need is to develop reduced-order models for partially premixed combustion that can be be used in Large Eddy Simulations (LES).

Reduced-order manifold models use the assumption of premixed or nonpremixed flame structures to project the high-dimensional thermochemical state space onto a low-dimensional manifold, typically parameterized by a mixture fraction (Z) that characterizes fuel-air mixing and/or a reaction progress variable (C). Examples of this type of model include the Flamelet/Progress Variable (FPV) approach [2] for nonpremixed combustion and the Flamelet Generated Manifolds (FGM) approach [3] for premixed combustion, which both use 1D flame calculations as component problems to generate the manifold. These models have been successfully applied in the corresponding regimes at much lower computational cost than transporting the full thermochemical state but are not applicable to partially premixed combustion without modification.

The FPV method can be extended to account for nonpremixed-like partially premixed combustion, where fuel and oxidizer may partially mix prior to the flame, but mixing between fuel and oxidizer remains the controlling factor in the flame. A second mixture fraction must be defined, which can be used to specify a variable boundary condition for 1D component flame calculations. This approach has enabled simulation of nonpremixed-like turbulent jet burners with inlet conditions that result in inhomogeneity in the fuel or air stream [4, 5]. The Partially Premixed Flamelet Tabulation (2PFT) approach of Franzelli et al. [6] extends this idea to include a treatment of premixed combustion but was developed and tested only for a laminar flame, and the scalar dissipation rate is used as a parameterizing variable instead of a second mixture fraction. The 1D component problem is a counterflow flame between air and a fuel stream, where the composition of the fuel stream is varied between pure fuel and a very lean fuel-air mixture. The premixed limit is recovered for flammable fuel stream mixtures when the strain rate becomes negligibly small.

Multi-modal combustion has also been modeled by determining the locally dominant combustion mode within a flame and switching between traditional nonpremixed or premixed models accordingly. The most rigorous approach for mode switching is based on a flamelet-like coordinate transformation to a 2D space, with separate parameterizing variables related to premixed and nonpremixed combustion [7]. The corresponding Knudsen regime index is defined based on the balance between premixed and nonpremixed terms in the resulting equation and is used to identify which limiting model should be used locally, due to complexity in the definition of the premixed coordinate that makes directly applying the 2D equations challenging [8]. The major limitation of the mode-switching approach is that it does not explicitly model intermediate combustion regimes.

To further study partially premixed combustion and provide a test case for model validation, a modified piloted jet burner has been developed at the University of Sydney [9] and subsequently been thoroughly characterized experimentally [10, 11]. The burner has a mixing tube upstream of the nozzle where concentric fuel and air streams are allowed to partially premix. The degree of premixing can be controlled by varying the length of the mixing tube. The intermediate mixing tube length (Lr75) case features inhomogeneous partially premixed conditions that result in multi-modal combustion and has been used to evaluate reduced-order manifold combustion models in these conditions. Wu and Ihme [12] showed that neither premixed nor nonpremixed models can provide accurate predictions for this configuration. Kleinheinz et al. [13] obtained improved results by applying the Knudsen mode switching model but did not explicitly account for intermediate combustion modes. Perry et al. [5] found that a nonpremixed-based two mixture fraction model accurately predicts flame structure except in a small region where premixed combustion dominates.

No single approach has been able to fully predict the partially premixed combustion that occurs for the Sydney jet burner with inhomogeneous inlets. The objective of this work is to determine whether a combination of existing approaches can do so, even though all assume an inherently 1D flame structure. In Section 2, the two mixture fraction model of Perry et al. [5] is extended to explicitly include premixed combustion using a similar methodology as 2PFT. Then, additional methods of incorporating premixed effects are considered, including both the Knudsen regime indicator and a phenomenological approach based on flammability limits. Predictions using all approaches are compared to the latest experimental data from the flame of interest in Section 3.

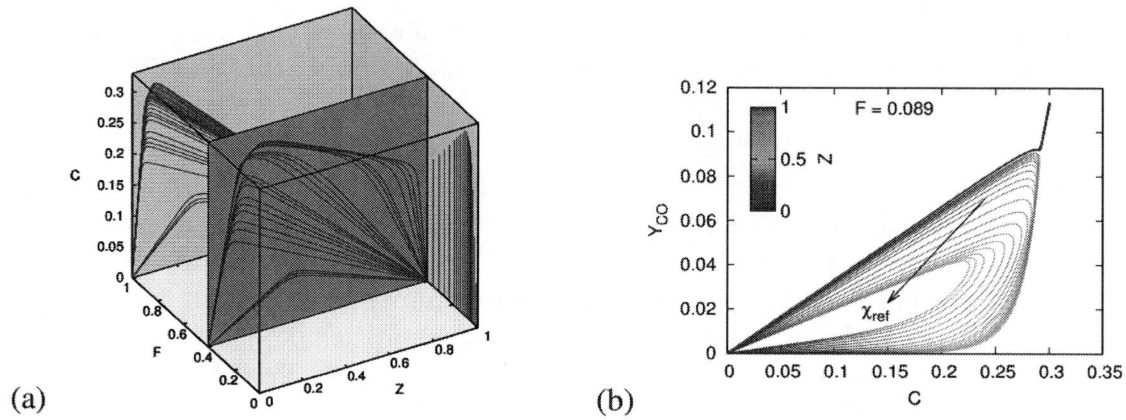

Figure 1: (a) Reduced-order manifold generation for the Nonpremixed+ approach. Each line within the blue and purple planes represents a solution to Eq. 1, with the boundary conditions $Y_F(Z=1)=1$ for the blue plane (recovering a standard FPV model), and $Y_F(Z=1)=0.4$ for the purple plane (allowing premixing in the fuel stream). A total of 35 conditions ranging from $F=0$ to $F=1$ were considered. The vertical lines in the red plane represent 1D premixed flame solutions, and as a whole this plane is equivalent to a standard FGM model. (b) Y_{CO} as a function of progress variable for nonpremixed flame solutions (colored by Z) with $F=0.089$ and the corresponding 1D premixed flame.

2. Modeling

The Sydney jet burner with inhomogeneous inlets [9] has three main inlet streams: the central fuel stream, the annular air stream that surrounds the fuel stream and partially premixes with it, and the air coflow. The pilot is a fourth inlet but will be modeled as a combination of the other inlets as is typically done for two-inlet systems. Following Perry et al. [5], two conserved-scalar mixture fractions (Z and Z^*) are defined and given boundary conditions such that Z corresponds to the mass fraction of material that originated from the central jet, including both the fuel stream and the annular air stream, and Z^* corresponds to the mass fraction of material that originated from the fuel stream. The fuel premixing fraction (F) is defined as $F \equiv Z^*/Z$. Mixing can be parameterized by Z and F, with the physical interpretation that Z again corresponds to jet/coflow mixing and F indicates the effective mixture originating from the nozzle.

The reduced-order mananfiold is generated using solutions to the steady 1D flame equation in mixture fraction space,

$$-\rho \frac{\chi_Z}{2} \frac{\partial^2 Y_i}{\partial Z^2} = \dot{m}_i , \tag{1}$$

where Y_i is the mass fraction of species i, $\dot{m}_i$ is the corresponding chemical source term, and $\chi_Z = 2D_Z \nabla Z \cdot \nabla Z$ is the scalar dissipation rate for Z. The value of F determines the fuel concentration at the $Z=1$ boundary when solving this equation. Equation 1 is solved with a range of boundary conditions corresponding to many values of F. For each boundary condition, the equation is also solved for several reference values of χ_Z. The resulting data is reparametrized using the progress variable, $C = Y_{H_2O} + Y_{H_2} + Y_{CO} + Y_{CO_2}$, instead of $\chi_{Z,\text{ref}}$, as shown in Fig. 1(a).

Three approaches are considered for expanding this manifold generation method to include premixed combustion:

Nonpremixed+ Approach: In Fig. 1(a), it can be seen that some of the 1D flame solutions with the partially premixed boundary condition in the purple plane take on increasingly vertical profiles near $Z = 1$. A perfectly premixed flame would be a vertical line (F is constant, Z is constant, but C increases across the flame), so this indicates that at low $\chi_{Z,\mathrm{ref}}$ the 1D flames contain a premixed-like component. This is verified in Fig. 1(b), which shows that the Y_{CO} vs. C profile for the 1D flames near $Z = 1$ becomes identical to a 1D premixed flame in the low $\chi_{Z,\mathrm{ref}}$ limit. Therefore, the nonpremixed-based two mixture fraction model may be extended to include premixed combustion by explicitly including 1D premixed flames at the $Z = 1$ boundary of the state space pictured in Fig 1(a), which forms the basis of the "Nonpremixed+" approach. Inclusion of premixed combustion in this manner is consistent, and it is conceptually similar to 2PFT, except the thermochemical manifold is parameterized entirely by the scalars Z, F, and C. The location within the resulting cubic state space determines whether the nonpremixed (blue), nonpremixed-like (purple), or premixed (red) model is accessed. The limitation of this approach is that it implicitly assumes an order of mixing and reaction: first, nonreactive mixing between fuel and air in the jet, then either premixed flame propagation or reactive mixing with the air coflow stream.

Flammability-based Mode Switching: This approach builds on the Nonpremixed+ approach by allowing for more general cases of premixed combustion. Underpinning the approach is the phenomenological argument that if the value of F indicates that the mixture of fuel and air in the central jet is flammable, premixed flame propogation should be expected. Therefore, a standard FGM premixed model is accessed if F is within the empirical flammability limits, taken to be $0.028 < F < 0.089$.

Knudsen Indicator Mode Switching: In this approach, the Knudsen regime indicator [7] is used to switch between the Nonpremixed+ model and a standard FGM premixed model. Past approaches used this regime indicator with standard FPV models (i.e. only the blue plane in Fig. 1(a)). This work explicitly includes the nonpremixed-like flames in the intermediate region (the purple plane), and the main purpose of the Knudsen switching is to be able to model premixed combustion of arbitrary mixtures, rather than limiting premixed combustion to the $Z = 1$ plane as in the Nonpremixed+ model alone.

The three mode switching approaches have been implemented in the low-Mach number reacting flow solver NGA [14, 15]. The same simulation approach (simulation of the upstream mixing tube prior to simulation of the flame) and computational grid for the flame were used as in [5]. The same subfilter closures for LES from that work are also utilized. The 1D calculations used to generate the thermochemical state manifold and the corresponding lookup table for LES were run using FlameMaster [16] with the GRI3.0 mechanism [17].

3. Results and Discussion

Predictions for temperature and the formation of CO from LES using the Nonpremixed+, Flammability, and Knudsen mode switching approaches are compared to experimental measurements from the burner with inhomogeneous inlets in Fig. 2. For reference, results using a standard premixed FGM approach and the two mixture fraction FPV approach of Perry et al. [5] are also plotted. For both temperature and CO predictions, excellent agreement is obtained using the premixed model for $x/D \leq 5$, but farther downstream both are overpredicted. In contrast, the nonpremixed model accurately predicts both quantities only for $x/D \geq 10$. This trend is consistent with the experimental observations, which indicate that premixed combustion near the nozzle transitions to

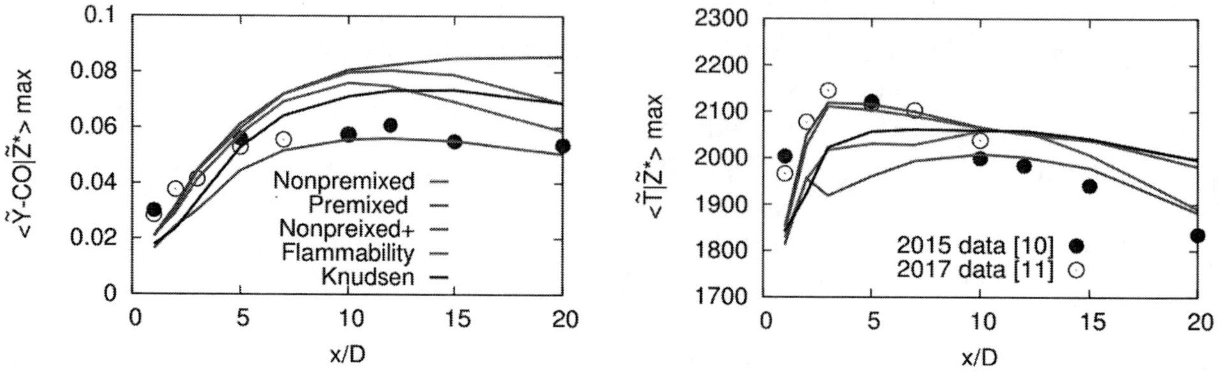

Figure 2: Peak conditional mean values of Y_{CO} (left) and Temperature (right) as a function of axial distance for the Lr75 flame. The three multi-modal approaches, a purely premixed model, and the nonpremixed model from [5] are compared to experimental data.

nonpremixed-like combustion downstream. It also suggests that it would be possible for a mode-switching approach that utilizes the two mixture fraction nonpremixed model to fully capture the behavior of this flame.

The qualitative trend that the premixed mode dominates upstream and the nonpremixed mode dominates downstream is reproduced by the mode switching models, and the Nonpremixed+ model in particular gives better predictions than either single mode model as a result. However, none of the approaches reproduce the transition at the correct location, instead predicting that premixed combustion persists too far downstream. The Nonpremixed+ approach is the closest to capturing the transition, with its predictions approaching the nonpremixed model at $x/D = 20$. The other two approaches use the Nonpremixed+ model as a basis and introduce additional ways for the mode to switch to premixed, so it follows that they predict premixed combustion in a larger region.

The mode-switching models used here fail to accurately capture the transition between modes because of the assumptions about the order of mixing and reaction that are presumed in the Nonpremixed+ model. In particular, it assumes that any and all mixing between the fuel and air in the jet occurs prior to nonpremixed combustion, which happens when the jet mixes with the coflow at constant F (in 1D along the Z-direction). Physically, nonpremixed-like behavior is also possible at $Z = 1$ (within the inhomogeneous jet, 1D nonpremixed flame in the F-direction), where the model presently assumes premixed combustion. This is a limitation of assuming combustion occurs in 1D in a specified direction within the reduced-order state space. Generating the manifold from solutions to a 2D equation, which is the subject of ongoing work [8], would bypass the issue.

4. Acknowledgements

The authors appreciate funding for this work provided by NASA under grant NNX16AP90A. B.A.P. is thankful to be supported by the National Science Foundation Graduate Research Fellowship Program under Grant No. DGE1148900. The authors also gratefully acknowledge valuable support in the form of computational time on the TIGRESS high performance computing center at Princeton University, which is jointly supported by the Princeton Institute for Computational Science and Engineering (PICSciE) and the Princeton University Office of Information Technology.

References

[1] A. R. Masri, Partial premixing and stratification in turbulent flames, Proc. Combust. Inst. 35 (2015) 1115–1136.

[2] C. D. Pierce and P. Moin, Progress-variable approach for large-eddy simulation of non-premixed turbulent combustion, J. Fluid Mech. 504 (2004) 73–97.

[3] J. A. Van Oijen and L. P. H. de Goey, Modelling of premixed laminar flames using Flamelet-Generated Manifolds, Combust. Sci. Tech 161 (2000) 113–137.

[4] M. Ihme and Y. C. See, LES flamelet modeling of a three-stream MILD combustor: Analysis of flame sensitivity to scalar inflow conditions, Proc. Combust. Inst. 33 (2011) 1309–1317.

[5] B. A. Perry, M. E. Mueller, and A. R. Masri, A two mixture fraction flamelet model for large eddy simulation of turbulent flames with inhomogeneous inlets, Proc. Combust. Inst. 36 (2017) 1767–1775.

[6] B. Franzelli, B. Fiorina, and N. Darabiha, A tabulated chemistry method for spray combustion, Proc. Combust. Inst. 34 (2013) 1659–1666.

[7] E. Knudsen and H. Pitsch, Capabilties and limitations of multi-regime flamelet combustion models, Combust. Flame 159 (2012) 242–264.

[8] M. E. Mueller, Physically-Derived Reduced-Order Manifolds for Multi-Modal Turbulent Combustion, 10th U.S. National Combustion Meeting (2017).

[9] S. Meares and A. R. Masri, A modified piloted burner for stabilizing turbulent flames of inhomogeneous mixtures, Combust. Flame 161 (2014) 484–495.

[10] R. S. Barlow, S. Meares, G. Magnotti, and A. R. Masri, Local extinction and near-field structure in piloted turbulent CH4/air jet flames with inhomogeneous inlets, Combust. Flame 162 (2015) 3516–3540.

[11] H. Cutcher, R. S. Barlow, G. Magnotti, and A. R. Masri, Turbulent flames with compositionally inhomogeneous inlets: Resolved measurements of scalar dissipation rates, Proc. Combust. Inst. 36 (2017) 1737–1745.

[12] H. Wu and M. Ihme, Compliance of combustion models for turbulent reacting flow simulations, Fuel 186 (2016) 853–863.

[13] K. Kleinheinz, T. Kubis, P. Trisjono, M. Bode, and H. Pitsch, Computational study of flame characteristics of a turbulent piloted jet burner with inhomogeneous inlets, Proc. Combust. Inst. 36 (2017) 1747–1757.

[14] O. Desjardins, G. Blanquart, G. Balarac, and H. Pitsch, High order conservative finite difference scheme for variable density low Mach number turbulent flows, J. Comp. Phys. 227 (2008) 7125–7159.

[15] J. F. MacArt and M. E. Mueller, Semi-implicit iterative methods for low Mach number turbulent reacting flows: Operator splitting versus approximate factorization, J. Comp. Phys. 326 (2016) 569–595.

[16] H. Pitsch, FlameMaster. A C++ program for 0D combustion and 1D flame calculations.

[17] G. P. Smith, D. M. Golden, and M. Frenklach, GRI-Mech 3.0, http://www.me.berkeley.edu/gri_mech.

Spring Technical Meeting
Eastern States Section of the Combustion Institute
March 4-7, 2018
State College, Pennsylvania

Topologically conditioned chemical pathways for turbulent lean premixed n-dodecane/air flames

Debolina Dasgupta[1], Wenting Sun[1], Marc Day[2], Andy J. Aspden[3], Tim Lieuwen[1]*

[1]*Department of Aerospace Engineering, Georgia Institute of Technology, 270 Ferst Drive, Atlanta, GA, 30318, U.S.A.*
[2]*Center for Computational Sciences and Engineering, Lawrence Berkeley National Laboratory, Berkeley, CA, 94720, U.S.A.*
[3]*School of Engineering, Newcastle University, NE1 7RU, UK*

[*]*Corresponding Author Email: ddasgupta3@gatech.edu*

Abstract: This paper analyzes turbulence-chemistry interactions in high Karlovitz number n-dodecane/air flames, specifically focusing on effects of turbulence on chemical pathways using the DNS Database by Aspden et al. (Proc. Combust. Institute, 36 (2017) 2005-2016). The metrics used to quantify the changes in chemical pathways are heat released by reactions and the rate of formation/consumption of species. These metrics are integrated along volumes defined by flame normal along an isotherm surface and then averaged along this surface. In our recent work with lean premixed turbulent n-dodecane/air flames (Dasgupta et al., AIAA SciTech 2018), we found a limited sensitivity of these integrated quantities with increasing turbulence intensities. However, increasing turbulence intensities leads to significant variation of scalar concentrations and their gradients along these surfaces which may not be captured by these averaged quantities. This variation in turn can affect the reactions rates and heat release of the reactions and potentially lead to different pathways along these surfaces. This work locally conditions the metrics on curvature and fuel consumption rate and look at their variation with increasing turbulence intensities.

Keywords: Turbulent Premixed Flames, Chemical Kinetics, Turbulence-chemistry interactions

1. Introduction

The manner in which turbulence interacts with chemistry in flames is an important question in turbulent combustion. This paper is part of a larger effort directed toward understanding how sensitive the chemical reaction pathways are to turbulence effects. A detailed discussion of the potential mechanisms is considered in our previous work[1] and briefly summarized here. First, unsteady stretching of the flame by turbulent disturbances can alter relative species and temperatures values at a given location, and hence reaction rates. Furthermore, the relative contributions of different reactions could be altered in turbulent flames due to unsteady kinetic and diffusive effects, leading to differences in relative values of concentrations of different species at a given location or isotherm. These can, in turn, change the reaction rates and disturb the chemical pathways locally. Additionally, increasing turbulence introduces increased energy in smaller fluid dynamical scales (and hence smaller eddies) that can cause convective stirring within the flame itself, and lead to spatially varying convective transport of species across the flame surface.

The effect of turbulence on the flame structure and chemical pathways has been explored for a number of fuels, including hydrogen, methane, propane, *n*-heptane and *n*-dodecane. For example, Carlsson et al. [2]and Aspden et al. [3, 4] described changes in the low temperature region with increasing turbulence for CH_4/Air and H_2/Air flames. They noted a large increase in heat release in these regions compared to 1D laminar flames. They reasoned that increasing turbulence intensities reduce the effective diffusive time scales leading to higher concentration of H radicals in the low temperature regions. This, in turn, promotes the reaction rates for $H+O_2(+M) \rightarrow HO_2(+M)$ and $H+CH_2O \rightarrow HCO+H_2$, leading to higher heat release at low temperatures. Aspden et al.[3] also observed a decorrelation between heat release and fuel consumption in regions of high curvature due to reactions $H+O_2(+M) \rightarrow HO_2(+M)$, $HO_2+H \rightarrow OH+OH$ and $HO_2+OH \rightarrow H_2O+O_2$. Aspden et al.[4, 5] noted a thickening of the flame's preheat zone in their DNS of CH_4/air flames and *n*-dodecane/air flames with increasing Karlovitz number (*Ka*). The profiles for products and stable intermediate species shifted from those similar to unstrained laminar flames towards "unity Lewis number" profiles for CH_4/air flames. This behavior is also observed in the profiles of primary fuel fragments for *n*-dodecane/air flames. However, concentrations of stable intermediates, such as HO_2, fall below unity Lewis number values. This response to turbulence is especially pronounced at lower temperatures (~1200K) where the *n*-dodecane molecules are consumed. The authors previously considered similar configurations for turbulent H_2/air[1] and CH_4/air flames[6], and found that the relative roles of the key reactions producing heat and/or key radicals were not substantially different between laminar and high Karlovitz flames, when the profiles were averaged across the flame surface. We noted, however, that large differences did occur near the leading edge of the flame brush when conditioned on fuel consumption based progress variable and curvature for both fuels. Both the CH_4 and H_2 fuels, however, have a relatively limited number of options for oxidation pathways. In the current study, we focus on turbulence-chemistry interactions in the combustion of *n*-dodecane, a much more complex fuel. This paper is a part of a larger effort to understand the effect of turbulence on the chemical pathways for *n*-dodecane. In our recent work for n-dodecane flames[7] we noted limited variation of the chemical pathways with increasing turbulence intensities even though there is a significant perturbation of the flame structure. We also noted a shift of the reaction rates towards higher temperature chemistry. This effect is particularly pronounced for reactions whose rates peak below 1200K. In this paper, we investigate the effects of the averaged metrics conditioned locally on surface curvature and fuel consumption rate.

2. Methodology

We begin with the DNS data set from Aspden et al.[5], which consists of a nominally one-dimensional, lean (ϕ=0.7), premixed turbulent *n*-dodecane/air flame. The DNS is based on a Low-Mach number reacting flow model with mixture-averaged transport for molecular diffusion, Soret and Dufour transport, gravity and radiative processes are neglected. The You et al.[8] model for reaction kinetics, thermodynamic properties and transport coefficients is used. This model consists of 56 chemical species and 289 reaction steps. The DNS utilizes reactant temperature T^u=298 K at a uniform pressure of 1 atm. A high aspect ratio domain (8:1:1) is used for the simulations with lateral periodic boundary conditions, free-slip fixed wall at the bottom and outflow at the top. A time-dependent, density-weighted forcing term in the momentum equations throughout the volume maintains the turbulence with an integral length scale of $l/l_O = 1$, where l and l_O denote the integral length scale and unstrained laminar flame length respectively. A flame is initialized near the top of the domain, and propagates downwards through the turbulent fuel mixture until it reaches the bottom of the domain. During this evolution, the flame reaches a quasi-stationary burning state from which flame statistics are extracted for analysis here. Five cases of increasing turbulence intensity (Karlovitz number = 1, 4, 12, 36 and 108) are analyzed where Karlovitz number is defined as:

$$Ka_F{}^2 = \frac{u'^3}{s_{L_O}{}^3}\frac{l_O}{l} \tag{1}$$

Here u' is the turbulent rms velocity and s_{L_O} is the unstretched laminar flame speed. For these cases, $s_{L_O} \approx 22.6$ cm/s and $l_O \approx$ 520 microns[5]. Table 1 summarizes these cases.

Table 1: Summary of cases [5]

Case	u' (m/s)	u'/s_{L_O}	Ka_F
A	0.226	1.00	1
B	0.570	2.52	4
C	1.18	5.25	12
D	2.47	17.3	36
E	5.12	22.7	108

Following closely the analysis procedures discussed in Dasgupta et al.[1] we quantify the "local" flame behavior via integrated flame sub-volumes, constructed to pass through localized sections of a triangulated isosurface that represents an instantaneous snapshot of the flame. Here, the reference isosurface is taken as an isotherm at the temperature of peak heat release in a steady, unstretched flame at the same equivalence ratio (ϕ=0.7), T_{ref} = 1460 K. Figure 1 shows the isotherm, T_{ref} = 1460 K(colored by local values of the heat release) for the cases considered. Since the volumetric statistics are integrated normal to the flame, they are insensitive to the precise isotherm used.

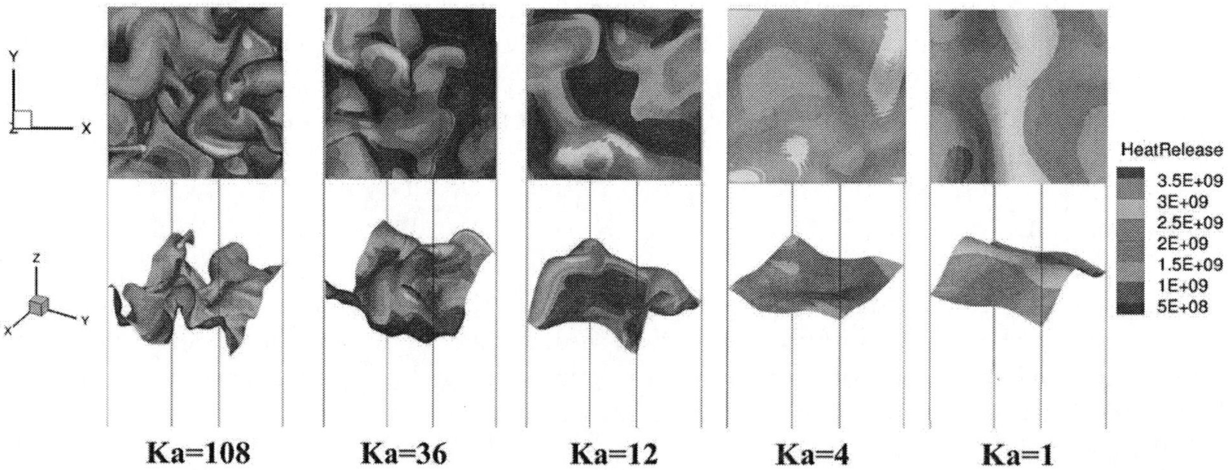

Figure 1. Isotherm, T_{ref} = 1460 K(colored by heat release). (Top) X-Y slices of the flame surface (bottom) 3D view of the flame surface. The rectangular domain is shown in black lines.

The reaction rate is integrated over the volume and normalized by the reference surface area to yield a line integrated average at each time instant:

$$\overline{RR_k}(\vec{x}, t) = \frac{\int RR_k(\vec{x}, t)\, d\Omega}{A_{ref}} \tag{2}$$

A_{ref} is the intersection area of the reference surface and the volume Ω. The average heat release rates and the radicals' consumption and production by each of the reactions can thus be calculated using these average reaction rates. The DNS data analyzed here is based upon a single time instant, extracted from the statistically stationary simulation results.
Different topological elements can be defined based on the principal curvatures of the iso-surface i.e. $\Bbbk_1$ and $\Bbbk_2$ ($\Bbbk_2 < \Bbbk_1$).
The heat release contribution of the individual reactions, $\overline{\dot{q}_k}$ within a certain element, e_i, on the reference surface is defined as:

$$\overline{\dot{q}_k}(x, y)|_{e_i} = \overline{RR_k}(x, y)|_{e_i} \left(\Delta h_f^o\big|_{reac} - \Delta h_f^o\big|_{prod}\right) \tag{3}$$

Similarly, the rate of formation/consumption of species, i in reaction k, within a certain element e_i, is defined as:

$$\overline{\dot{\omega}_{i,k}}(x, y)\big|_{e_i} = \overline{RR_k}(x, y)|_{e_i}(v''_{i,k} - v'_{i,k}) \tag{4}$$

where a generalized reaction is written as $v'_{i,k}M_i \rightarrow v''_{i,k}M_i$. The reaction rate, $\overline{RR_k}(x, y)$ can be <0 or >0. If $\overline{RR_k}(x, y) < 0$, then the reverse reaction from the mechanism is used for the analysis.

The area-weighted average heat release within an element can then be computed as:

$$\overline{\dot{q}_k}|_{e_i} = \frac{\sum_{j=1}^{j=N} \overline{\dot{q}_k}(x, y)|_{e_i} A_{ref,j}}{\sum_{j=1}^{j=N} A_{ref,j}} \tag{5}$$

and, the area weighted $\overline{\dot{\omega}_{i,k}}\big|_{e_i}$ as:

$$\overline{\dot{\omega}_{i,k}}\big|_{e_i} = \frac{\sum_{j=1}^{j=N} \overline{\dot{\omega}_{i,k}}(x, y)\big|_{e_i} A_{ref,j}}{\sum_{j=1}^{j=N} A_{ref,j}} \tag{6}$$

where N is the number of points on the reference isosurface within the given element configuration. $\overline{\dot{\omega}_{i,k}}\big|_{e_i} > 0$ implies that species, i is produced by reaction, k. We perform a separate accounting of the production and consumption of several key species below. The net heat release at e_i is:

$$\bar{\dot{q}}|_{e_i} = \sum_k \overline{\dot{q}_k}|_{e_i} \tag{7}$$

The net production of species, i at e_i is given as:

$$\overline{\dot{\omega}_{\iota,p}}\Big|_{e_i} = \sum_k \overline{\dot{\omega}_{\iota,k}}\Big|_{e_i} \tag{8}$$

Here if the reaction k, forms species, i then $\overline{\dot{\omega}_{\iota,k}} > 0$ else $\overline{\dot{\omega}_{\iota,k}} = 0$. Similarly, the net consumption is given as:

$$\overline{\dot{\omega}_{\iota,c}}\Big|_{e_i} = \sum_k \overline{\dot{\omega}_{\iota,k}}\Big|_{e_i} \tag{9}$$

Here if the reaction, k consumes species, i then $\overline{\dot{\omega}_{\iota,k}} < 0$ else $\overline{\dot{\omega}_{\iota,k}} = 0$. The subscript p and c denote production and consumption respectively. The values obtained in Eq. (7), (8) and (9) are used for normalizing the values obtained in Eq. (5) and (6).

3. Results and Discussion

This section contrasts and compares the metrics, (a) heat released by the reactions and (b) species production/consumption by different reactions, for the global analysis with the local conditioned analysis. We first condition the data on curvature followed by a progress variable based on the fuel consumption rate.

(a) Global characteristics

We briefly discuss the results obtained for the global analysis to compare with local conditioned analysis. The detailed results are presented in Dasgupta et al. [7]. Figure 2 plots the normalized heat release contribution of dominant reactions as a function of Karlovitz number.

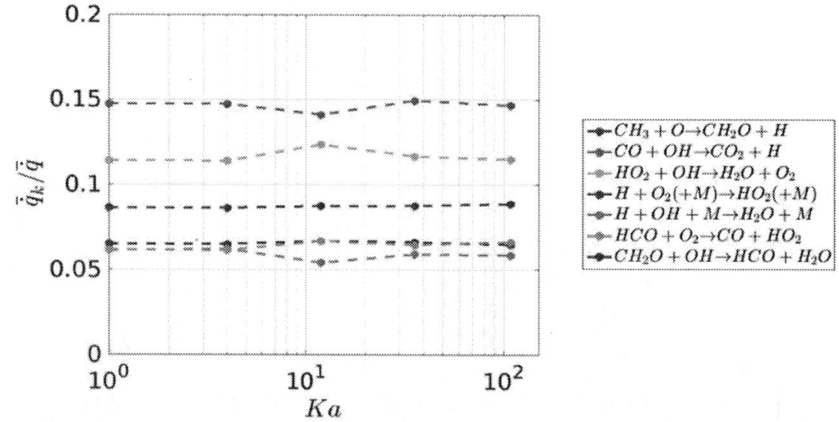

Figure 2. Variation of normalized heat release with turbulence intensities.

For the DNS, the variation in normalized heat release for the dominant heat release reaction, $CO+OH \rightarrow CO_2+H$, with increasing turbulence intensities is ~6% between Ka = 1 and Ka = 12. Similarly, the maximum variation for the second dominant heat release reaction, $HO_2+OH \rightarrow O_2+H_2O$, is ~5% between Ka=1 and Ka=12. Overall, all reactions show limited sensitivity to increasing turbulence intensities. A detailed discussion of the results can be found in our recent work Dasgupta et al.[7].

(b) Curvature conditioned results

This section discusses the chemical pathways conditioned on curvature. The metrics considered are heat release and species consumption/formation. Figure 3 below plots the fractional contribution of the different elements to the net heat release.

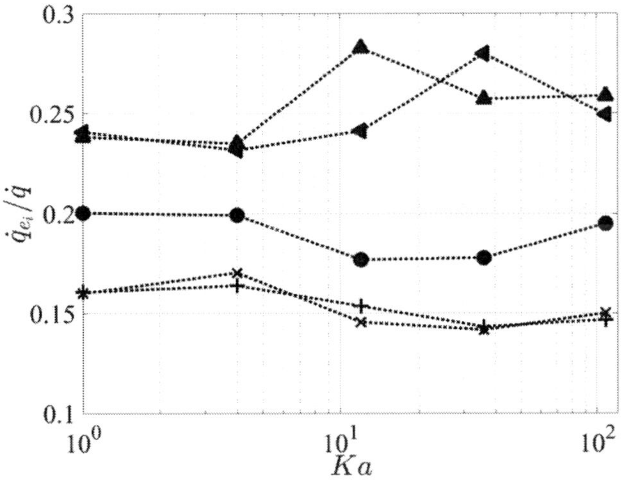

Figure 3. Variation of fractional heat release within each element with Karlovitz number. (.▲:Spherically convex towards products, .◀:Cylindrically convex towards products, .●:Saddle points, +:Spherically convex towards reactants, .x: Cylindrically convex towards reactants)

It can be observed that most of the heat release occurs in the negatively curved regions, as expected for this reactant mixture, where *Le*>1 and where burning is enhanced in negatively curved regions. Aspden et al.[5] showed a higher concentration of key species in these regions which in turn leads to increased reaction rates.

Figure 4 below plots the fraction heat release contribution of different reactions in each element. The heat release by a reaction is normalized by the total heat release within a certain element. It can be seen that the same reactions dominate the heat release in all 5 elements. The most dominant reaction changes from $CO+OH\rightarrow CO_2+H$ to $HO_2+OH\rightarrow O_2+H_2O$ in the positively curved elements with increasing turbulence intensities. Slices of reaction rates suggest a broadened region of reaction rate for $CO+OH\rightarrow CO_2+H$ which leads to lower overall values of the reaction rates and hence heat release whereas the reaction $HO_2+OH\rightarrow O_2+H_2O$ occurs in a thin region. This leads to a lowered contribution of the CO_2 formation reaction to the net heat release with increasing turbulence intensities. The most significant change is seen for the reaction $H+OH+M\rightarrow H_2O+M$ with a decrease in contribution of ~50% in the positively curved elements. $H+O_2(+M)\rightarrow HO_2(+M)$ and $HCO+O_2\rightarrow CO+HO_2$ change by ~40% with increasing turbulence intensities. Additionally, we note the growing contribution of the reaction $CH_2O+OH\rightarrow HCO+H_2O$ in these elements. This reaction has a higher heat release contribution than $H+O_2(+M)\rightarrow HO_2(+M)$ and $H+OH+M\rightarrow H_2O+M$ at higher turbulence intensities suggesting a slight alteration of the heat release pathway at higher Ka for the positively curved elements. One can also note a comparable heat release contribution of the reactions $CH_3+O\rightarrow CH_2O+H$ and

$HCO+O_2 \rightarrow CO+HO_2$ to the dominant reactions at a high Ka. These reactions have a relatively lower contribution at lower Ka.

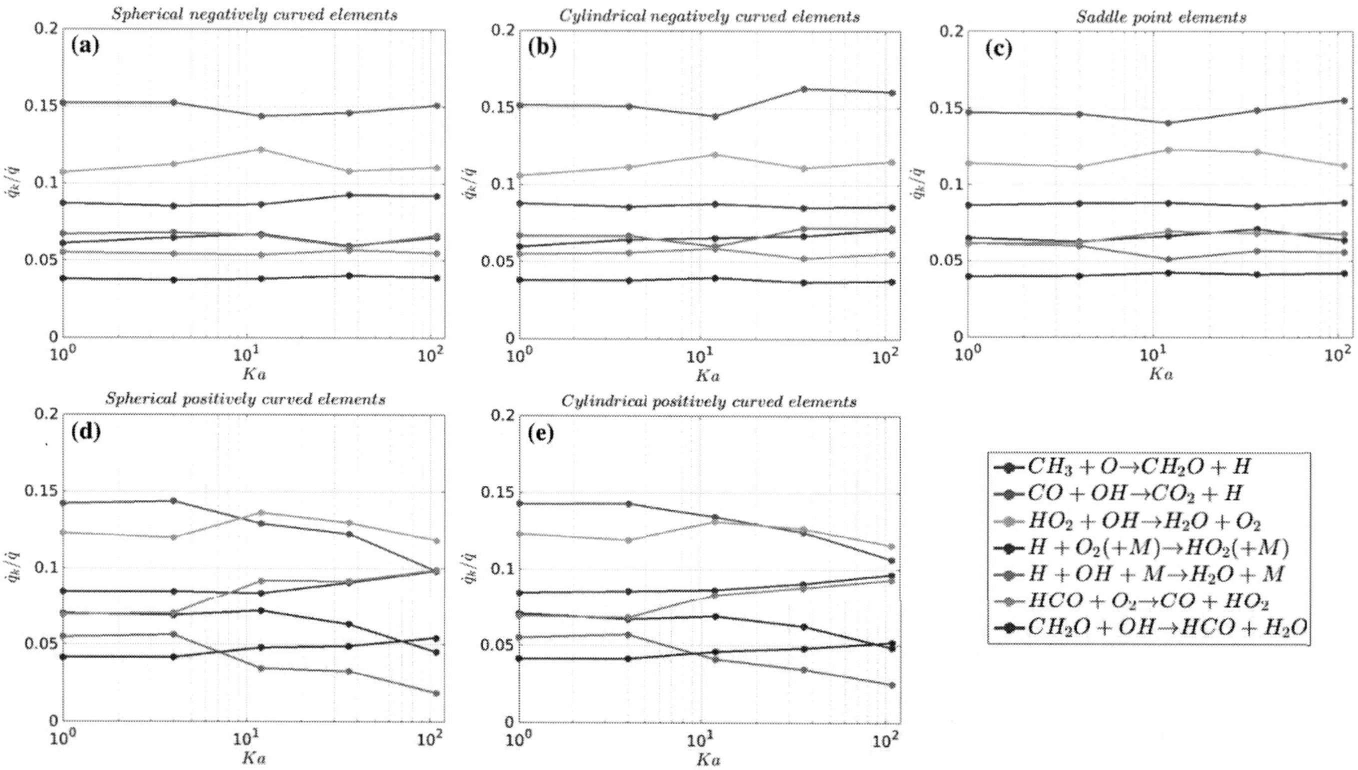

Figure 4. Normalized heat release by the dominant reactions in (a) Spherically convex towards reactants elements (b) Cylindrically convex towards reactants (c) Saddle point elements (d) Spherically convex towards products elements (e) Cylindrically convex towards products elements.

The contribution of the reactions in the negatively regions and saddle point elements behave similar to its global counter-part. For example, the contribution of the reaction $CO+OH \rightarrow CO_2+H$ changes by ~6% for the global characteristics and for these three elements.

Figure 5 below plots the normalized rate of consumption/production of certain key species. The same dominant reactions for HCO can be observed in Figure 5(a). For HCO consumption, it can be noted that $HCO+O_2 \rightarrow CO+HO_2$ is the dominant HCO consuming reaction in the positively curved elements (Figure 5(a, top)) followed by $HCO+M \rightarrow H+CO+M$. The contribution of $HCO+O_2 \rightarrow CO+HO_2$ increases by ~25% with increasing turbulence intensities whereas the contribution of $HCO+M \rightarrow H+CO+M$ decreased by ~25%. The same two reactions are observed for the negatively curved elements and saddle point elements (Figure 5(a, bottom)). The order of the dominant reactions is reversed for these elements with limited sensitivity to increasing turbulence intensities. For OH we can observe changes in the contributions of the secondary reactions with increasing turbulence intensities for the positively curved elements (Figure 5(b,top)). The consumption of OH by $CO+OH \rightarrow CO_2+H$ decreases by ~15% with increasing turbulence intensities. At higher turbulence intensities, the reactions $HO_2+OH \rightarrow H_2O+O_2$, $OH+H_2 \rightarrow H_2O+H$ and $CH_2O+OH \rightarrow HCO+H_2O$ have a higher consumption rate than $2OH \rightarrow H_2O+O$ which is the second dominant OH consumer at lower turbulence intensities. The contributions of these reactions do not change significantly for the negatively curved and saddle point elements (Figure 5(b, bottom)).

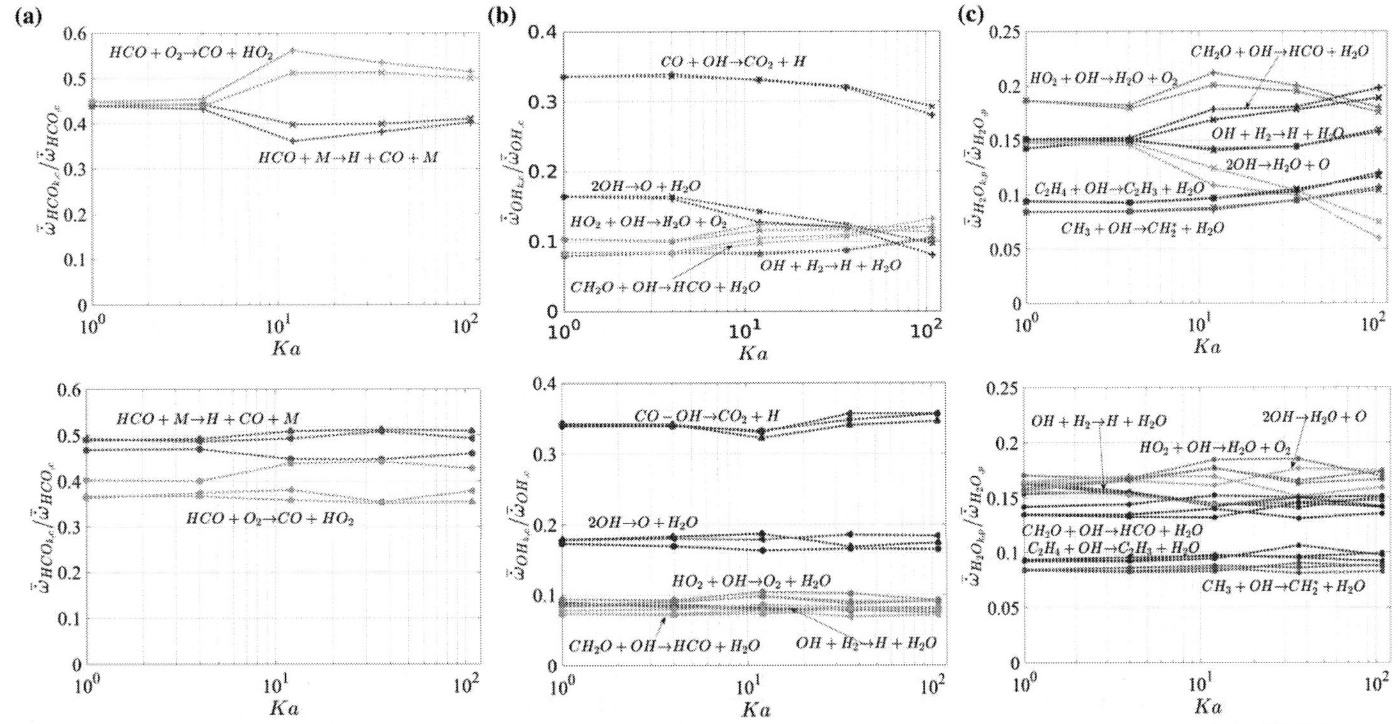

Figure 5. Normalized species consumption by the dominant reactions for (a) HCO consumption (b) OH consumption (c) H_2O productions. (.▲:Spherically convex towards products, .◄:Cylindrically convex towards products, .●:Saddle points, +:Spherically convex towards reactants, .x: Cylindrically convex towards reactants)

The pathways for water formation, one of the key products in hydrocarbon combustion, are strongly affected with increasing turbulence intensities for the positively curved elements (Figure 5(c, top)). For example, $CH_2O+OH \rightarrow HCO+H_2O$ takes over as the dominant H_2O producing reaction from $HO_2+OH \rightarrow H_2O+O_2$ at higher turbulence intensities. Also, the reactions $C_2H_4+OH \rightarrow C_2H_3+H_2O$ and $CH_3+OH \rightarrow CH_2^*+H_2O$ have a higher contribution to H_2O production than $2OH \rightarrow H_2O+O$ with increasing Ka. These contributions are increased by ~25% from Ka=1 to Ka=108. Other species such as HO_2, CH_3 show limited sensitivity to turbulence and curvature and their normalized consumption rates are plotted in the Appendix.

Overall, the reactions in the positively curved elements show a stronger sensitivity with increasing turbulence intensities compared to the negatively curved elements and saddle point elements.

(c) Progress variable conditioned results

The mean fuel consumption, $\bar{\omega}_{fuel}$ can be calculated as shown in Eq. (5) by integrating $\dot{\omega}_{fuel}$ over the volume Ω. The direction of propagation, Z is divided into 200 points between the lowest (Z_{min}) and the highest point (Z_{max}). The cumulative fuel consumption is then calculated along this direction and normalized by the total fuel consumption in the volume. A typical profile for the fuel consumption is shown in Figure 6. We define a progress variable, $\overline{c_i}$, at a location, Z_i, as:

$$\overline{c_i} = \frac{\sum_{Z_i<Z<Z_{max}} \bar{\dot{\omega}}_{fuel}}{\sum \bar{\dot{\omega}}_{fuel}} \quad (5)$$

The location, $\bar{c} = 0$, corresponds to unburnt reactants.

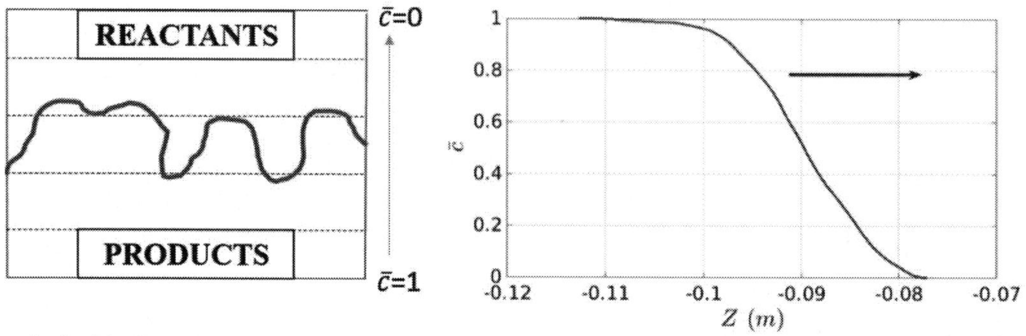

Figure 6. (left) Illustration of progress variable, $\bar{c}$. (right) Mapping of physical space to progress variable. The arrows denote the direction of propagation

Figure 7 below plots the fractional contribution of the dominant reactions to the total heat release at a given progress variable.

It is clear from Figure 7 that the same dominant heat release reactions show up at different progress variables. At the leading edge of the flame, i.e. $\bar{c} = 0.1$(in Figure 7(a)), the variations in the reactions with increasing Ka are the most significant. At higher Karlovitz numbers, the reaction $HO_2+OH \rightarrow O_2+H_2O$ takes over as the most dominant reaction from $CO+OH \rightarrow CO_2+H$. The heat release contributions of $H+O_2(+M) \rightarrow HO_2(+M)$ and $H+OH+M \rightarrow H_2O+M$ are attenuated by ~50% with increasing turbulence intensity. Additionally, the contribution of the reaction $HCO+O_2 \rightarrow CO+HO_2$ increases by ~35% with increasing turbulence intensities compared to a ~1% change for the global characteristics. The heat release variations for the reactions at the other progress variables (Figure 7(b), (c) and (d)), namely $\bar{c} = 0.3, 0.5, 0.7$, is similar to its global counter-part. For example, the reaction $CO+OH \rightarrow CO_2+H$ changes by ~6% for the global characteristics. This change is between ~1-6% for the progress variables away from the leading edge of the flame. Similarly, the change in the reaction $CH_3+O \rightarrow CH_2O+H$ is ~1-2% for the global behavior and these progress variables.

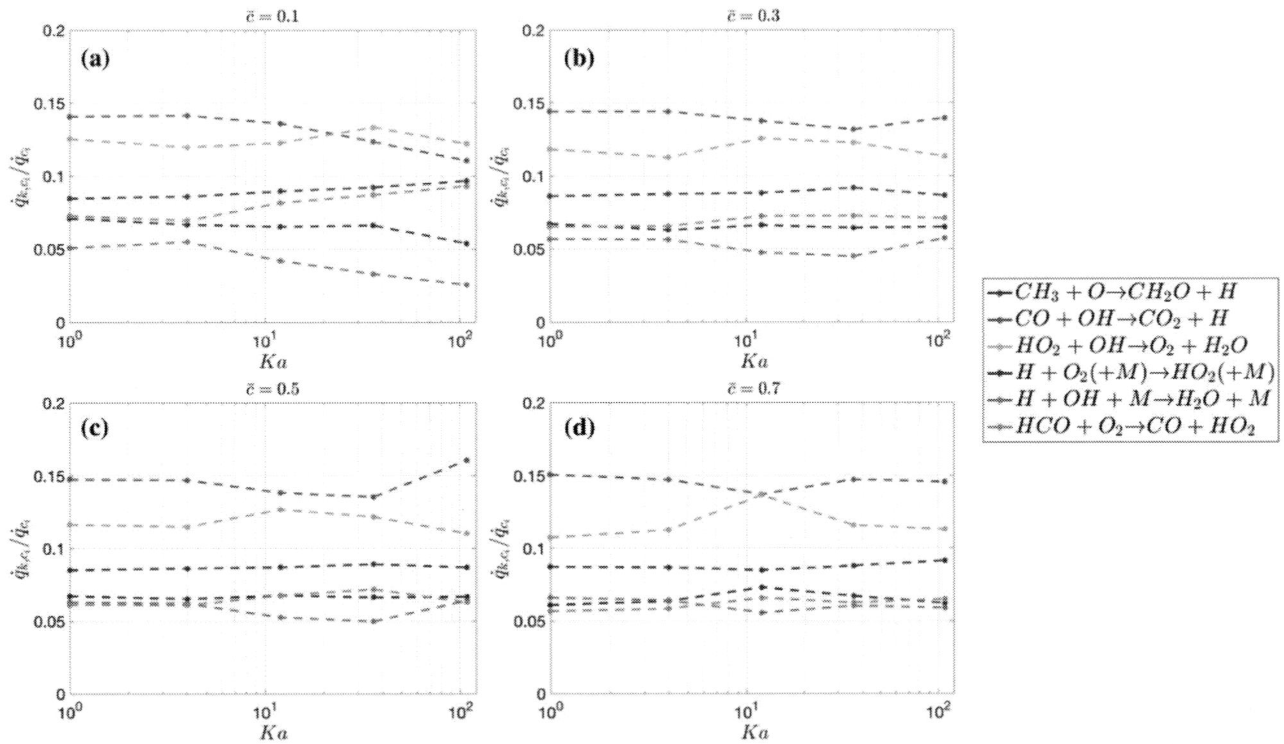

Figure 7. Normalized heat release by different reactions at (a) $\bar{c}=0.1$ (b) $\bar{c}=0.3$ (c) $\bar{c}=0.5$ (d) $\bar{c}=0.7$.

We next consider the individual behavior of the reactions at different progress variables and compare them with their laminar counterpart. Figure 8 shows the variation of the key reactions at different progress variables with increasing turbulence intensities.

The quantitative contributions of the reactions are close to the stretched flame calculations. However, certain reactions are relatively better represented than the others. For example, the contribution of the reactions $CO+OH\rightarrow CO_2+H$, $HO_2+OH\rightarrow O_2+H_2O$ and $H+O_2(+M)\rightarrow HO_2(+M)$ fall outside the bounds of the laminar calculations whereas $CH_3+O\rightarrow CH_2O+H$, $H+OH+M\rightarrow H_2O+M$ and $HCO+O_2\rightarrow CO+HO_2$ are closer to the laminar bounds. The y-axis for the plots covers a relatively small range and hence the changes in variations across the progress variables is of the order of ~10-15% with the maximum change of the order of ~30% for all the reactions. In general, similar to our previous observation, the reactions at $\bar{c}=0.1$ behave differently compared to other progress variables. For example, the reaction $CO+OH\rightarrow CO_2+H$ shows a ~20% decrease in contribution with increasing Ka at the leading edge whereas it shows a modest change of ~5% for the other progress variables and the global behavior. This stronger increase or decrease in contributions with increasing turbulence intensities at $\bar{c}=0.1$ can be observed for the others reactions as well such as $HCO+O_2\rightarrow CO+HO_2$ and $H+OH+M\rightarrow H_2O+M$.

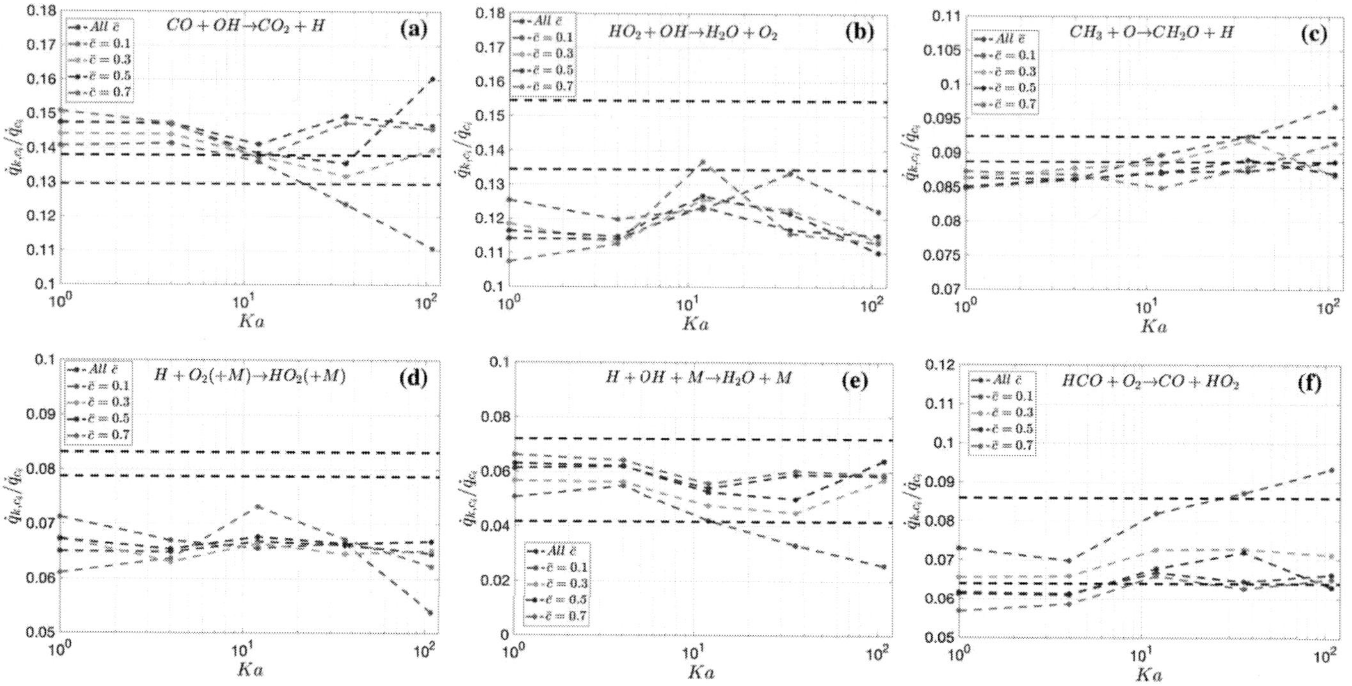

Figure 8. Progress variable conditioned heat release for (a)CO+OH→CO_2+H (b)HO_2+OH→O_2+H_2O (c)CH_3+O→CH_2O+H (d)H+O_2(+M)→HO_2(+M) (e)H+OH+M→H_2O+M (f)HCO+O_2→CO+HO_2 ; (Blue: All $\bar{c}$, Orange: $\bar{c}$ = 0.1, Yellow: $\bar{c}$ = 0.3, Purple: $\bar{c}$ = 0.5, Green: $\bar{c}$ = 0.7). Dashed horizontal lines denote range of values from OPPDIF calculations.

Figure 9 below plots the normalized rates for HCO consumption, OH consumption and H_2O production. At $\bar{c} = 0.1$, the dominant HCO consuming reaction is HCO+O_2→CO+HO_2 followed by HCO+M→H+CO+M. This observation is in line with our observation for the positively curved elements. The ordering of the reactions is flipped for $\bar{c} = 0.7$ similar to our observation for the global properties and the negatively curved conditions. The dominant OH consuming reaction does not change between the different progress variables. The reactions of secondary importance change with increasing turbulence intensities for $\bar{c}$=0.1. For example, the OH consumption by the reaction 2OH→O+H_2O is comparable to the other secondary reactions at Ka=108. However, this reaction consumes twice as much OH compared with the other reactions at lower turbulence intensities. Pathways for water formation have a stronger sensitivity to turbulence at progress variables closer to the leading edge of the flame compared to the other progress variables. For example, the H_2O production by the reaction 2OH→O+H_2O decreases by ~50% and by the reaction CH_2O+OH→HCO+H_2O increases by ~35% with increasing Ka. These numbers for the other $\bar{c}$ is ~5% for both the reactions. The progress variable conditioned results for other species such as CH_3 and HO_2 are plotted in the Appendix.

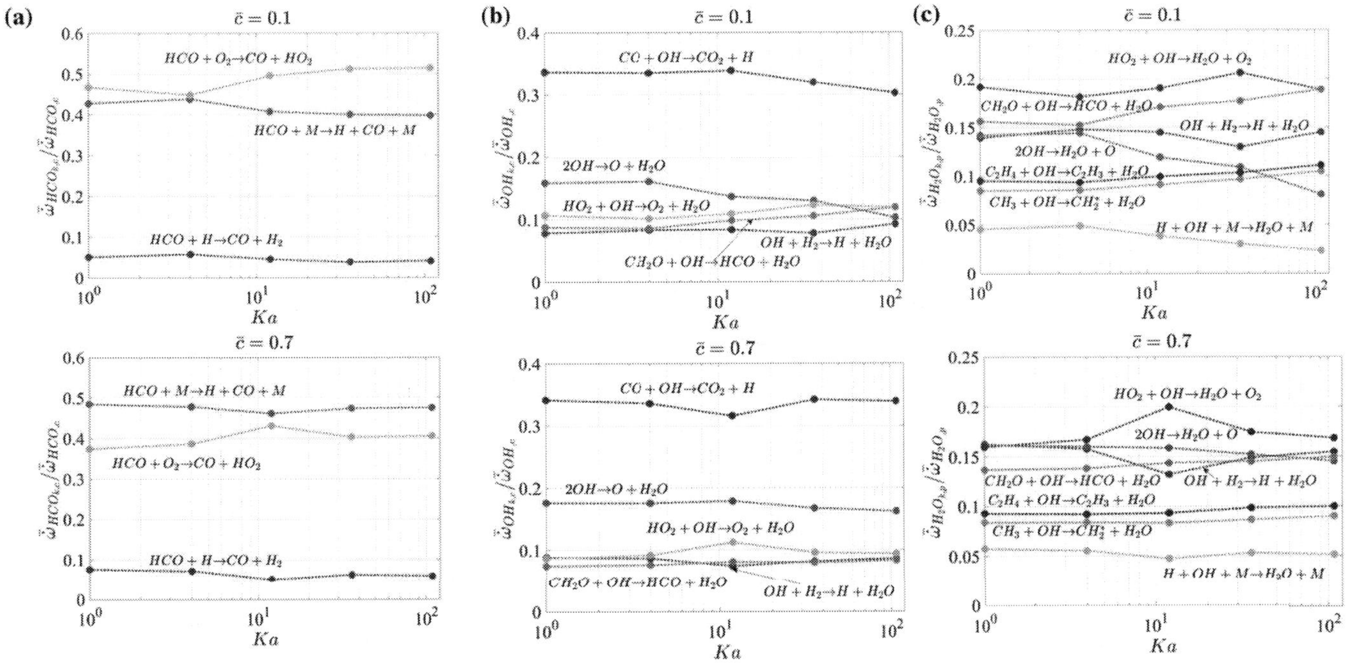

Figure 9. Normalized contribution of reactions for (a) HCO consumption (b) OH consumption (c) H_2O production for $\bar{c} = 0.1$ (top), 0.7 (bottom).

Overall, we see a stronger dependence of the reactions on Ka for progress variables closer to the leading edge of the flame compared to the rest of the flame. The behavior of the reactions at the other progress variables is closer to the global characteristics and are relatively insensitive to turbulence.

4. Conclusions

This work investigates the changes in chemical pathways for n-dodecane/air premixed turbulent flames when the metrics are conditioned on local quantities such as curvature and fuel consumption. We note that overall the same dominant heat release reactions are observed for these local conditioning compared to the global characteristics and limited sensitivity of these reactions to increasing turbulence intensities.

For the curvature conditioned results, it is observed that the negatively curved elements and the saddle point element behave similar to their global counterpart. For example, the contribution of the reaction $CO+OH \rightarrow CO_2+H$, which accounts for ~15% of the total heat release, changes by ~6% for the global characteristics and for these three elements. However, on the other hand more significant changes are observed for the positively curved elements. The most dominant reaction changes from $CO+OH \rightarrow CO_2+H$ to $HO_2+OH \rightarrow O_2+H_2O$ in the positively curved elements with increasing turbulence intensities. The most significant change is seen for the reaction $H+OH+M \rightarrow H_2O+M$ with a decrease in contribution of ~50% in the positively curved elements. A stronger sensitivity of the reaction path for different species is observed in these elements as well. For example, the contribution of the second dominant OH consuming reaction of $2OH \rightarrow H_2O+O$ decreases by ~50% with increasing turbulence intensities whereas for the negatively curved elements it changes by ~5%. Certain species show altered pathway between the different elements. For example, the dominant HCO consuming reaction for the positively curved elements is $HCO+O_2 \rightarrow CO+HO_2$. The dominant reaction changes to $HCO+M \rightarrow H+CO+M$ in the saddle point and negatively curved elements.

Progress variable based conditioning reveals a slightly different behavior at $\bar{c}$ closer to the leading edge of the flame compared to the rest of the flame. The reaction $HO_2+OH \rightarrow O_2+H_2O$ takes over as the most dominant reaction from $CO+OH \rightarrow CO_2+H$ with increasing turbulence intensities and a nearly 50% drop in the heat release contributions of the reactions $H+O_2(+M) \rightarrow HO_2(+M)$ and $H+OH+M \rightarrow H_2O+M$ at $\bar{c} = 0.1$. This difference in behavior at $\bar{c} = 0.1$ is also observed when we compare individual reactions at different progress variables. We also observe that some reactions, such as $H+OH+M \rightarrow H_2O+M$ and $CH_3+O \rightarrow CH_2O+H$ are bounded within their laminar maxima and minima. In contrast, reactions such as $CO+OH \rightarrow CO_2+H$, $H+O_2(+M) \rightarrow HO_2(+M)$ lie outside these laminar bounds. However, the difference in their contributions with respect to their laminar counter-part is of the order of ~10-30%. The reactions consuming or producing the species involved are sensitive to turbulence at progress variables closer to the leading edge of the flame compared to the other progress variables. For example, the contribution of the H_2O producing reaction, $2OH \rightarrow H_2O+O$ drops by ~50% from Ka=1 to Ka=108 for $\bar{c} = 0.1$ and by ~5% for $\bar{c} = 0.3, 0.7$.

Overall, the positively curved elements and the locations of the surface closer to the leading edge of the flame have altered pathways for heat release and key species consumption/production. The other locations behave similar to the global characteristics which has been discussed in our previous work[7].

Several questions remain for future work. First, in order to better understand the universality of this result, the reasons for the insensitivity need to be better understood, as well as situations where significant sensitivity could be observed. Some insights into this question might be gained by determining better what controls the fractional contributions of a given reaction. For example, why does the reaction $CO+OH \rightarrow CO_2+H$ account for ~15% of the heat release, and not 5% or 25%? Could this result be predicted from some simple calculation based upon certain global parameters (such as C/O ratios or equilibrium values), or is it a function of a myriad of details and individual reaction rates? Secondly, a comparison of the effects of turbulence on different fuels is warranted. Earlier studies indicate a higher low temperature activity for fuels such as H_2 and to some extent CH_4. Our recent study revealed a different effect on *n*-dodecane. It would be interesting to look at a comparative local analysis for all these fuels.

5. Appendix

Figure A1 plots the normalized contributions to certain key species consumption by different reactions.

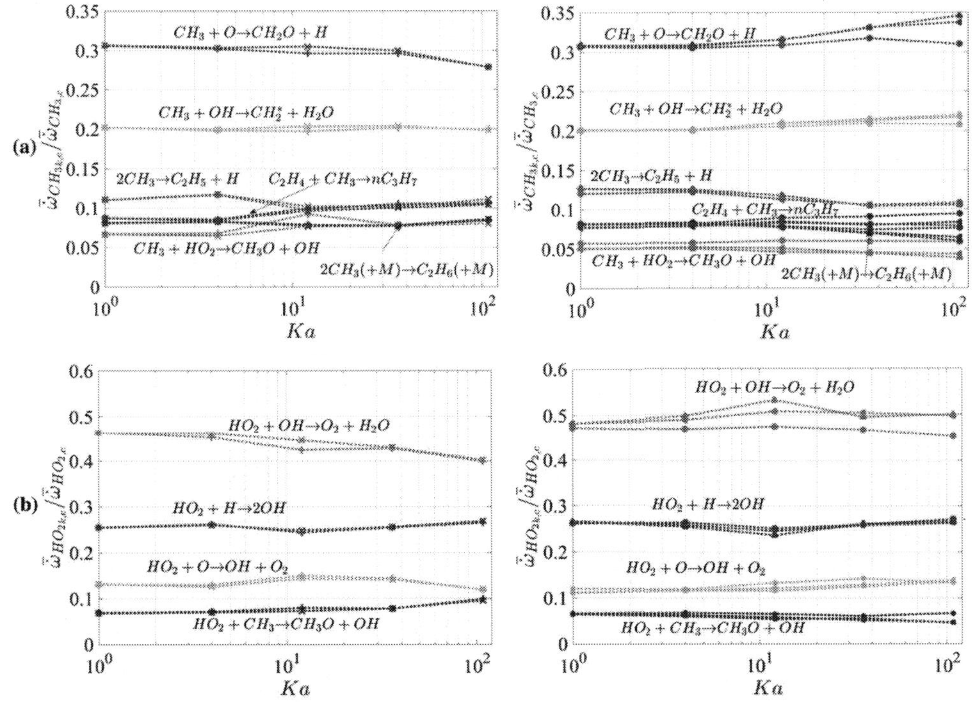

Figure A1. Normalized species consumption by the dominant reactions for (a) CH_3 consumption (b) HO_2 consumption. (.▲:Spherically convex towards products, .◄:Cylindrically convex towards products, .●:Saddle points, +:Spherically convex towards reactants, .x: Cylindrically convex towards reactants)

The dominant reactions for CH_3 consumption show limited sensitivity to increasing turbulence intensity (Figure A1(a)). The reactions of secondary importance for CH_3 consumption include recombination reactions. At higher turbulence intensities, $C_2H_4+CH_3\rightarrow nC_3H_7$ has a comparable consumption rate compared to $2CH_3\rightarrow C_2H_5+H$ for the positively curved elements. We do not observe these changes for the negatively curved elements (Figure A1(a, right)). The limited sensitivity of the HO_2 consumption reactions to increasing turbulence intensities and curvature is evident in Figure A1(b).

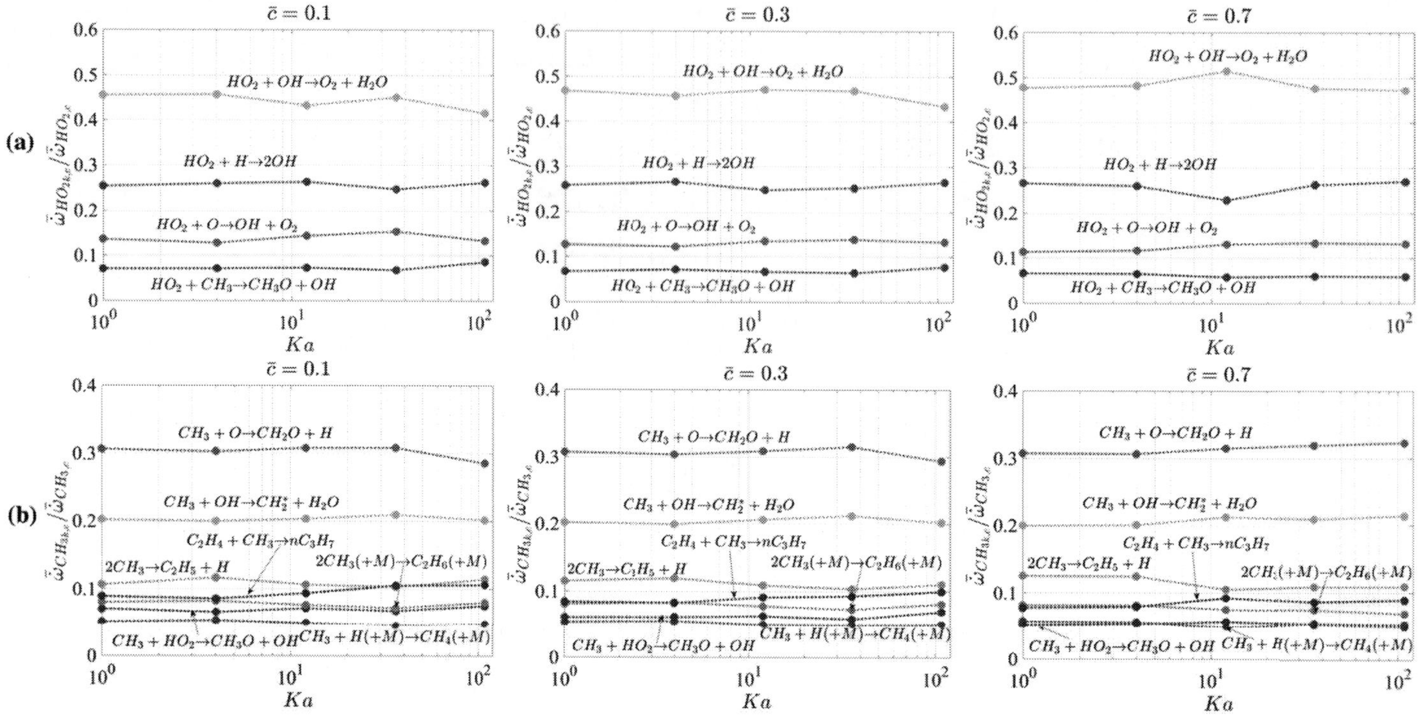

Figure A2. Normalized species consumption by different reactions for (a) HO_2 (b) CH_3 for $\bar{c} = 0.1$ (left), 0.3 (center), 0.7 (right).

The dominant reactions for CH_3 consumption do not change between the different progress variables. The reactions of secondary importance have varied contributions at higher turbulence intensities for $\bar{c} = 0.1$. For example, the reaction, C_2H_4+CH_3→nC_3H_7 has a higher rate of consumption at higher turbulence intensities compared to $2CH_3$→C_2H_5+H. Reactions involved in HO_2 consumption show limited sensitivity to increasing Ka and remain consistent across all the progress variables as seen in Figure A2(a).

6. Acknowledgement

The work was funded by the Air Force Office of Scientific Research (contract #FA9550-16-1-0442), contract monitor Dr. Chiping Li. This work used the Extreme Science and Engineering Discovery Environment(XSEDE), which is supported by National Science Foundation grant number ACI-1548562[9].

7. References

[1] D. Dasgupta, W. Sun, M. Day, T. Lieuwen, Effect of Turbulence-Chemistry interactions on chemical pathways for turbulent hydrogen-air premixed flames, Combustion and Flame 176 (2017) 191-201.

[2] H. Carlsson, R. Yu, X.-S. Bai, Direct numerical simulation of lean premixed CH4/air and H2/air flames at high Karlovitz numbers, International Journal of Hydrogen Energy 39 (2014) 20216-20232.

[3] A.J. Aspden, M.S. Day, J.B. Bell, Turbulence-chemistry interaction in lean premixed hydrogen combustion, Proceedings of the Combustion Institute 35 (2015) 1321-1329.
[4] A.J. Aspden, M.S. Day, J.B. Bell, Three- dimensional direct numerical simulation of turbulent lean premixed methane combustion with detailed kinetics, Combustion and Flame 166 (2016) 266–283.
[5] A.J. Aspden, J.B. Bell, M.S. Day, F.N. Egolfopoulos, Turbulence-Flame Interactions in Lean Premixed Dodecane Flames, Proceedings of the Combustion Institute 36 (2017) 2005-2016.
[6] D. Dasgupta, W. Sun, M. Day, T. Lieuwen, Turbulence effects on the chemical pathways for premixed methane/air flames, 55th Aerospace Sciences Meeting, Grapevine, Texas, 2017.
[7] D. Dasgupta, W. Sun, M. Day, A.J. Aspden, T. Lieuwen, Investigation of turbulence effects on chemical pathways for n-dodecane, 56th Aerospace Sciences Meeting, Kissimmee, Florida, 2018.
[8] X. You, F.N. Egolfopoulos, H. Wang, Detailed and simplified kinetic models of n-dodecane oxidation: The role of fuel cracking in aliphatic hydrocarbon combustion, Proceedings of the Combustion Institute 32 (2009) 403-410.
[9] J. Towns, T. Cockerill, M. Dahan, I. Foster, K. Gaither, A. Grimshaw, V. Hazlewood, S. Lathrop, D. Lifka, G.D. Peterson, R. Roskies, J. Ray Scott, N. Wilkins-Diehr, XSEDE: Accelerating Scientific Discovery, Computing in Science and Engineering 16 (2014) 62-74.

Spring Technical Meeting
Eastern States Section of the Combustion Institute
March 4–7, 2018
State College, Pennsylvania

On the accessed region of composition space in turbulent diffusion flames

A.S. Newale[1,], Y. Liang[1], S.B. Pope[1], and P. Pepiot[1]*

[1]*Sibley School of Mechanical and Aerospace Engineering, Cornell University, Ithaca, NY, USA*
[*]*Corresponding author:* asn79@cornell.edu

Abstract: The accessed region of composition space is defined as the set of all compositions that occur in a turbulent reacting flow. It is recognized that the accessed region of composition space is a small subset of the entire realizable region of the high dimensional compositional space (Pope, CTM 1997 [1]). In this work, we develop a technique to compare two datasets in the high dimensional composition space using in-situ adaptive tabulation (ISAT) (Pope, CTM 1997 [1]). The first step of the technique entails building two separate ISAT tables using the compositions in the two datasets. This is followed by a simple check to see whether compositions in the second database can be resolved using either a retrieve or a grow using the ISAT table built on the first database and vice-a-versa. We note that only compositions which are in close proximity in the high dimensional composition space and provide similar reaction mapping are selected by ISAT for either a retrieve or a grow. This technique is used to gauge the capability of a partially stirred reactor (PaSR) to emulate the accessed region of composition space in diffusion flames. Specifically, we compare the accessed region of composition space in Sandia flame D with a PaSR run meant to emulate the flame. The aforementioned technique is then used to gauge commonality in the two databases as well compositions that occur in one database but do not have an analog in the other database.
Keywords:* *ISAT, accessed, turbulent, diffusion

1. Introduction

The accessed region of composition space for a turbulent reactive flow is defined as the union of all compositions occurring at all positions, times, and realizations [2]. The realizable region of composition space can be defined as the union of all physically possible compositions in the high dimensional composition space [1]. We note that both the accessed and realizable regions are subsets of the high dimensional composition space. A crucial observation is that the accessed region of composition space is a small subset of the realizable region of composition space [1]. This observation is exploited to efficiently tabulate reaction mapping for compositions as they are encountered in-situ and forms the basis for the ISAT algorithm [1]. Pope [2] explored the geometry of the accessed region of composition space for several configurations. This work showed that simultaneous mixing and reaction can lead to a significantly higher dimensional accessed region of composition space compared to when mixing and reaction act sequentially or separately.

Chemical kinetics reduction techniques require samples in the high dimensional composition space for comparing the response of reduced models to that of the detailed model [3]. These samples are usually generated using simple $0-D$ configurations such as homogeneous ignition and

$1-D$ configurations such as freely propagating laminar flames. If the compositions used to reduce the detailed mechanism emulates the compositions belonging to the accessed region of the composition space in the turbulent reactive flow, a more specialized (in term of the number of retained species and reactions) and accurate mechanism can be obtained through the reduction process. This observation is irrespective of the specific technique used for reduction.

Partially stirred reactors (PaSR) have been used recently by Liang *et al.* [4] and Ren *et al.* [5] for assessing adaptive chemistry methodologies. These studies use PaSR as a surrogate for a cell in a transported PDF computation. Additionally, we note that the pre-partitioned adaptive chemistry methodology proposed by Liang *et al.* [4] relies on a database which mimics the compositions encountered at run time during the adaptive simulation. One possibility for generating such a database is simply to use a PaSR configuration intended to emulate the accessed region of composition space observed at runtime. However, the ability of PaSR to capture the accessed region of composition space in a turbulent combustion simulation is yet to be assessed quantitatively.

Both of the motivating applications listed above require a technique to compare two sets of compositions in the high dimensional composition space to gauge the commonality between them. The key contribution of this work is to develop a simple technique based on ISAT to do the same. This technique is then used to examine the capability of the PaSR to capture the accessed region of composition space in a turbulent combustion simulation. We also examine the capability of a reduced mechanism developed on a database obtained from PaSR to accurately integrate the compositions in the LES/PDF database.

2. Methods

The technique used for comparing compositions in high dimensional composition space is heavily dependent on ISAT and consequently before detailing it, the key elements of ISAT are reiterated for completeness sake. A more comprehensive description of ISAT can be found in Pope [1] and Lu *et al.* [6].

Given a specific input composition, ISAT returns to within a specified error tolerance (with a reasonable probability), the reaction mapping for that composition. A key concept utilized by ISAT is the region of accuracy, which is defined as the connected region in the high dimensional composition space such that for a given input composition, reaction mapping determined based on the linear approximation is within the user-specified error tolerance [6]. The region of accuracy is approximated in ISAT using ellipsoids. These are referred as ellipsoid of accuracy (EOA). When a composition is encountered by ISAT, it is resolved using one of the following operations:

- Retrieve: If the input composition is seen to be covered by a specific EOA, a linear approximation to the reaction mapping is returned.
- Grow: If the composition is not part of any existing EOA, it is directly integrated and EOAs in close proximity of the input compositions are examined for grow attempts. If the linear approximation to the reaction mapping obtained using one of the EOAs in close proximity, is within the user specified error tolerance of the exact reaction mapping, the EOA is grown.

- Add: If all attempts at a grow fail and the table is not full, then a new entry corresponding to the input composition is added to the ISAT table.
- Direct evaluation: If the specified limit for the size of the table is reached and a retrieve cannot be performed then the composition is integrated without saving the reaction mapping for subsequent use.

A crucial observation is that compositions that ISAT identifies for retrieves and grows are in close proximity of each other in the high dimensional composition space and provide a similar reaction mapping. This observation forms the crux of the technique explained below in detail. Given two sets of compositions, labeled D_1 and D_2, the following steps should be performed:

1. Build one ISAT table for each of set of the compositions.
2. Use the ISAT table built on D_1 and examine whether individual compositions in D_2 can be resolved using either a retrieve or a grow.
3. If the composition in D_2 can be resolved using an add or a retrieve, it is part of the region of the high dimensional composition space accessed by database 1. If not, it does not belong to the region of composition space accessed by D_1.
4. Repeat the operations 2 and 3 using the ISAT table built on D_2 and examine compositions in D_1.
5. Finally, identify compositions which belong to the intersection of accessed regions implied by compositions in both databases as those which can be resolved using either retrieve or grow using an ISAT table built on the other database.

3. Results and Discussion

The technique described in the previous section is used to the gauge the capability of PaSR to emulate the compositions encountered in a turbulent combustion simulation. Specifically, we compare a database obtained from a LES/PDF simulation of Sandia flame D containing 900,000 compositions with a database obtained from a PaSR meant to emulate Sandia flame D containing 30,000 compositions. The compositions in both databases are generated using the same 38 species C1-C4 skeletal mechanism [7]. The parameters used for PaSR are detailed in table 1 and the stream compositions are detailed in table 2. An error tolerance of 10^{-4} is used to build ISAT tables on both databases.

Figure 1 shows a scatter plot of particle temperatures versus mixture fraction for all the compositions in the LES/PDF database. The compositions are color coded to indicate whether they can be resolved using an ISAT table built on the PaSR database. The points shown by red can be resolved using the PaSR ISAT table using either an add or a grow, and points colored by green cannot be resolved by the PaSR ISAT table using the same operations. It is seen that 99.6% of the compositions in the LES/PDF database can be resolved using either a retrieve or an add using the PaSR ISAT table while only 0.4% of the compositions cannot be resolved using those operations.

Table 1: Parameter specification for PaSR runs

Parameter	Value
Number of streams	3
Stream type	Pilot/ Oxidizer/ Fuel
Stream relative flow rates	0.073/ 0.742/ 0.185
Pressure	1 bar
Number of particles	100
Pairing time	1 ms
Mixing time	1 ms
Residence time	1 ms
Time step	1 ms

Table 2: Stream composition for PaSR run

Species (mass fractions)	Pilot	Oxidizer	Fuel
CH4	0.0	0.0	1.0
N2	0.7368	0.77	0.647
O2	0.054	0.23	0.197
H2O	0.0942	0.0	0.0
CO	0.004	0.0	0.0
CO2	0.111	0.0	0.0
Temperature (K)	1880	291	294

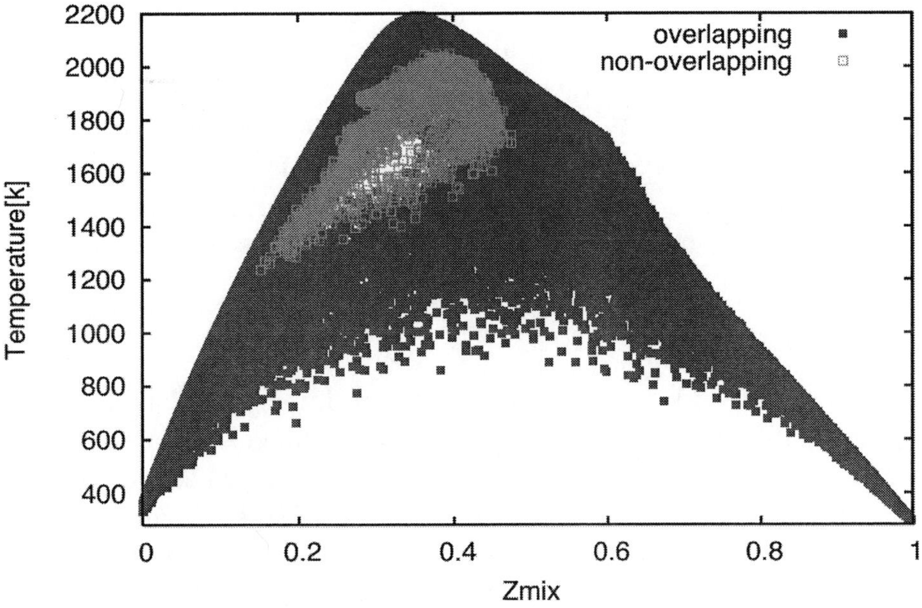

Figure 1: Scatter plot of particle temperature versus mixture fraction from LES/PDF computation. Compositions which are determined to in the accessed region of composition space of the PaSR are colored red, while those outside are labeled green.

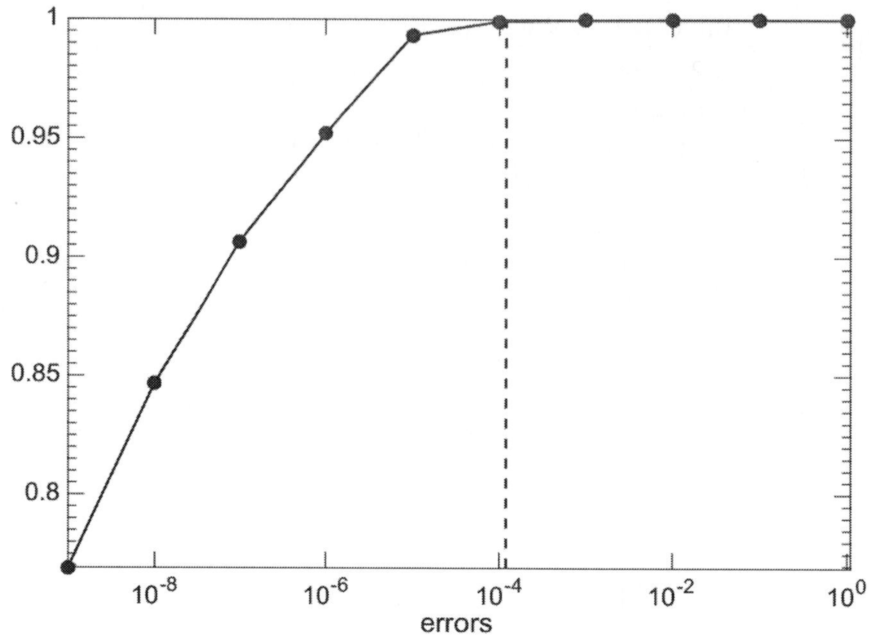

Figure 2: Cumulative distribution of incurred errors when integrating compositions in the LES/PDF database using a reduced mechanism derived on compositions from the PaSR database

Additionally, we examine the errors incurred when integrating compositions in the LES/PDF database using a reduced mechanism developed for the compositions in the PaSR database. Specifically, a reduced mechanism is developed using the directed relation graph with error propagation (DRGEP) [8] with an error tolerance of 10^{-4}. This reduced mechanism is then used to integrate the compositions in the LES/PDF database for one time step of size 10^{-4}s. The error is computed using the reference of the same integration performed with the detailed mechanism. The cumulative distribution of the error is shown in figure 2. We note that for more than 99.9% of the compositions the error is below the specified threshold. This shows that accurate and efficient reduced mechanism can derived a-*priori* from compositions representative of those encountered at runtime in a turbulent combustion simulation.

4. Conclusions

The key contribution of this work is the development of a simple technique based on ISAT which can be used to compare two sets of compositions in the high dimensional composition space. This technique is then used to show that for the case of Sandia flame D, a PaSR configuration can emulate the accessed region of composition space observed in a LES/PDF computation. Consequently, a reduced mechanism developed based on the compositions in the PaSR database is shown to be accurate and efficient for compositions in the LES/PDF database.

5. Acknowledgements

This material is based upon work supported by the U.S. Department of Energy Office of Science, Office of Basic Energy Sciences under Award Number DE-FG02-90ER14128.

References

[1] S. B. Pope, Computationally efficient implementation of combustion chemistry using in situ adaptive tabulation, (1997).

[2] S. B. Pope, Accessed compositions in turbulent reactive flows, Flow, turbulence and combustion 72 (2004) 219–243.

[3] T. Lu and C. K. Law, Toward accommodating realistic fuel chemistry in large-scale computations, Progress in Energy and Combustion Science 35 (2009) 192–215.

[4] Y. Liang, S. B. Pope, and P. Pepiot, A pre-partitioned adaptive chemistry methodology for the efficient implementation of combustion chemistry in particle PDF methods, Combustion and Flame 162 (2015) 3236–3253.

[5] Z. Ren, Y. Liu, T. Lu, L. Lu, O. O. Oluwole, and G. M. Goldin, The use of dynamic adaptive chemistry and tabulation in reactive flow simulations, Combustion and Flame 161 (2014) 127–137.

[6] L. Lu and S. B. Pope, An improved algorithm for in situ adaptive tabulation, Journal of Computational Physics 228 (2009) 361–386.

[7] G. Esposito and H. Chelliah, Skeletal reaction models based on principal component analysis: Application to ethylene–air ignition, propagation, and extinction phenomena, Combustion and Flame 158 (2011) 477–489.

[8] P. Pepiot-Desjardins and H. Pitsch, An efficient error-propagation-based reduction method for large chemical kinetic mechanisms, Combustion and Flame 154 (2008) 67–81.

Spring Technical Meeting
Eastern States Section of the Combustion Institute
March 4–7, 2018
State College, Pennsylvania

A Comprehensive Model for Non-Adiabatic Multi-Modal Combustion using Physically-Derived Reduced-Order Manifolds

A. Cody Nunno[1,*] *and Michael E. Mueller*[1]

[1]*Department of Mechanical and Aerospace Engineering, Princeton University, Engineering Quadrangle, Olden Street, Princeton, NJ, USA 08544*
[*]*Corresponding author:* anunno@princeton.edu

Abstract: Generally, turbulent combustion models fall into a hierarchy of model complexity based on the assumptions made about the underlying combustion processes. Models that make no assumption about the underlying combustion processes, such as Transported PDF (TPDF) methods, can model more complex combustion phenomena, specifically multi-modal combustion with heat losses, but this generality comes with increased computational cost. Manifold methods, including "flamelet"-like models, are more computationally efficient but at the cost of generality, usually restricting the combustion processes to a single asymptotic mode under adiabatic conditions and relying on one-dimensional component problems. Recent work from our group has attempted to increase the generality of manifold methods and has resulted in two-dimensional manifold equations for non-adiabatic premixed combustion, non-adiabatic nonpremixed combustion, and adiabatic multi-modal combustion. In this work, a new, more general manifold method is developed, capable of describing non-adiabatic multi-modal combustion. The approach relies on a three-dimensional manifold consisting of mixture fraction, a generalized progress variable, and a generalized enthalpy deficit parameter, which excludes the influence of mixture fraction and progress variable variations, defined on the unit cube. The governing equations for the thermochemical state on the manifold are derived, and the two-dimensional manifolds mentioned previously are shown to be recovered in appropriate limits.
***Keywords:* Low-dimensional manifolds, multi-modal turbulent combustion, heat loss, LES**

1. Introduction

In modeling turbulent combustion using Large Eddy Simulation (LES), there is a general hierarchy of model complexity regarding the combustion processes simulated. The highest model complexity, which is also the most general, makes no assumption about the underlying mode of combustion. While these models have a high degree of flexibility, these models are computationally expensive, scaling with the number of chemical species considered and often requiring a reduced-order description of chemical kinetics with fewer species. The Transported Probability Density Function (TPDF) [1, 2] approach and the Linear Eddy Model (LEM) [3] are such examples. Assumptions about the underlying combustion processes can be made *a priori* to reduce the scope of the combustion model while also to reducing the computational cost required. Typically, by restricting the combustion to a specific mode, the computational cost of the model can be decreased by projecting the thermochemical state onto a reduced-order manifold described by a small number of variables. Reduced-order manifolds can either be empirically-derived, using data from DNS and experiments to build the manifold [4, 5], or physically-derived, generating manifolds via solutions of the governing equations [6–8]. Attempts have been made to incorporate multi-modal combustion processes into manifold methods, primarily either using combustion mode-switching or mode-blending approaches [9, 10], which combine the results of premixed and nonpremixed manifolds. More recently, Nguyen et al. [11] and Mueller [12] have both derived two-dimensional manifold equations that describe truly multi-modal combustion. By considering the manifolds without the presumption of a specific combustion mode, these methods attempt to break the conventional model trade-off, by including more generality at a lower cost.

Heat losses further complicate the modeling of turbulent combustion within a reduced-order manifold context, requiring the addition of another dimension to any manifold method. In both premixed and nonpremixed contexts, reduced-order manifold methods have successfully integrated enthalpy loss using a number of methods [13–15]. However, no manifold-based model has been developed for multi-modal combustion with heat losses. To account for heat loss in multi-modal combustion, a new comprehensive approach must be developed.

Two goals are pursued in this work. First, a new framework for describing non-adiabatic, multi-modal turbulent combustion via a physically-derived reduced-order manifold is developed. The manifold will be directly constructed by a multi-dimensional set of governing equations using mixture fraction Z, a generalized progress variable Λ, and a generalized enthalpy deficit parameter H^*, the latter two of which will be explicitly defined within the work. Then, the model will be shown to recover the asymptotic limits of adiabatic multi-modal combustion, as well as both premixed and nonpremixed non-adiabatic combustion. The role of "unsteady" effects is also addressed.

2. Non-Adiabatic Multi-Modal Manifold Models

The evolution of species mass fractions, Y_k, is governed by the equation:

$$\frac{\partial \rho Y_k}{\partial t} + \frac{\partial \rho u_j Y_k}{\partial x_j} = \frac{\partial}{\partial x_j}\left(\rho D \frac{\partial Y_k}{\partial x_j}\right) + \dot{m}_k \,.$$

Here, D is the molecular diffusivity. For the purposes of this derivation, species Lewis numbers are presumed to be unity, often an appropriate assumption when turbulent mixing dominates at flame scales, but the model development is equally applicable to non-unity Lewis numbers. For non-adiabatic systems, a transport equation for a heat loss parameter H can also be constructed [16], where the heat loss parameter is governed by the same equation as gas-phase enthalpy, except defined such that H is zero at the adaiabtic limit and grows increasingly negative. For isobaric, low Mach number combustion, the governing equation is then given by

$$\frac{\partial \rho H}{\partial t} + \frac{\partial \rho u_j H}{\partial x_j} = \frac{\partial}{\partial x_j}\left(\rho D \frac{\partial H}{\partial x_j}\right) + \dot{q}_H \,.$$

Multi-modal combustion is inherently at minimum a two-stream problem, so two coordinates are required to describe the mixing and reaction within the system. The mixture fraction coordinate Z is a conserved scalar with unity Lewis number defined by its transport equation, such that Z is zero in the oxidizer stream and unity in the fuel stream. A second coordinate can be defined as a generalized progress variable Λ, specifically chosen such that it is independant of mixture fraction and defined to be zero for unburned gases and unity at thermodynamic equilibrium. Typical choices for a progress variable, such as linear combinations of product species mass fractions or temperature, cannot be used in this formulation, for these quantities are not independant of the mixture fraction. These two variables have been previously shown to adequately describe adiabatic multi-modal combustion [12].

For non-adiabatic multi-modal combustion, a new generalized enthalpy defecit parameter H^* is additionally defined to be independent of both mixture fraction and generalized progress variable, and the progress variable must also be redefined to ensure that all three variables are independent. The precise definition of both Λ and H^* will be left open, with the intent of defining it using the transport equation in a similar manner to how Λ was previously defined via the adiabatic multi-modal manifold equations [12]. With this approach, only the boundary conditions for H^* will be defined at this point, with zero corresponding to adiabatic and unity to "cold" isothermal conditions, that is, at the minimum environmental temperature to which heat is lost. By using this definition, all non-adiabatic (isobaric, two-stream) combustion processes can be described on a unit cube, between fuel and air, unburned and burned, and adiabatic and "cold".

To obtain a set of equations for manifold evolution, a coordinate transformation is performed on the physcial space governing equations: $(t, x_j) \to (Z, \Lambda, H^*)$. A detailed derivation of the governing equations under the coordinate transformation is not included here, but the species mass fraction manifold equation is now given by

$$\frac{\partial Y_k}{\partial \Lambda}\left[\rho\frac{\partial \Lambda}{\partial t}+\rho u_j\frac{\partial \Lambda}{\partial x_j}-\frac{\partial}{\partial x_j}\left(\rho D\frac{\partial \Lambda}{\partial x_j}\right)\right]+\frac{\partial Y_k}{\partial H^*}\left[\rho\frac{\partial H^*}{\partial t}+\rho u_j\frac{\partial H^*}{\partial x_j}-\frac{\partial}{\partial x_j}\left(\rho D\frac{\partial H^*}{\partial x_j}\right)\right]$$
$$=\frac{\rho\chi_{ZZ}}{2}\frac{\partial^2 Y_k}{\partial Z^2}+\rho\chi_{Z\Lambda}\frac{\partial^2 Y_k}{\partial Z\partial\Lambda}+\frac{\rho\chi_{\Lambda\Lambda}}{2}\frac{\partial^2 Y_k}{\partial \Lambda^2}+\rho\chi_{ZH^*}\frac{\partial^2 Y_k}{\partial Z\partial H^*}+\rho\chi_{\Lambda H^*}\frac{\partial^2 Y_k}{\partial \Lambda\partial H^*}+\frac{\rho\chi_{H^*H^*}}{2}\frac{\partial^2 Y_k}{\partial H^{*2}}+\dot{m}_k,$$

and the equation for heat loss parameter becomes

$$\frac{\partial H}{\partial \Lambda}\left[\rho\frac{\partial \Lambda}{\partial t}+\rho u_j\frac{\partial \Lambda}{\partial x_j}-\frac{\partial}{\partial x_j}\left(\rho D\frac{\partial \Lambda}{\partial x_j}\right)\right]+\frac{\partial H}{\partial H^*}\left[\rho\frac{\partial H^*}{\partial t}+\rho u_j\frac{\partial H^*}{\partial x_j}-\frac{\partial}{\partial x_j}\left(\rho D\frac{\partial H^*}{\partial x_j}\right)\right]$$
$$=\frac{\rho\chi_{ZZ}}{2}\frac{\partial^2 H}{\partial Z^2}+\rho\chi_{Z\Lambda}\frac{\partial^2 H}{\partial Z\partial\Lambda}+\frac{\rho\chi_{\Lambda\Lambda}}{2}\frac{\partial^2 H}{\partial \Lambda^2}+\rho\chi_{ZH^*}\frac{\partial^2 H}{\partial Z\partial H^*}+\rho\chi_{\Lambda H^*}\frac{\partial^2 H}{\partial \Lambda\partial H^*}+\frac{\rho\chi_{H^*H^*}}{2}\frac{\partial^2 H}{\partial H^{*2}}+\dot{q}_H.$$

The spatio-temporal evolution of both generalized progress variable and generalized enthalpy deficit parameter remain, leaving the equation unclosed. Therefore, transport equations are established for both Λ and H^* so as to replace the spatio-temporal terms by source terms in the manifold equations:

$$\rho\frac{\partial \Lambda}{\partial t}+\rho u_j\frac{\partial \Lambda}{\partial x_j}-\frac{\partial}{\partial x_j}\left(\rho D\frac{\partial \Lambda}{\partial x_j}\right)=\dot{m}_\Lambda,$$
$$\rho\frac{\partial H^*}{\partial t}+\rho u_j\frac{\partial H^*}{\partial x_j}-\frac{\partial}{\partial x_j}\left(\rho D\frac{\partial H^*}{\partial x_j}\right)=\dot{m}_{H^*}.$$

Substituting these source terms into the manifold equations for heat loss parameter and a reference species (chosen to be sufficiently progress variable-like, that is, monotonically increasing or decreasing from unburned to thermodynamic equilibrium at constant Z and constant H^*), a system of two equations with two unknowns is formed:

$$\frac{\partial Y_k}{\partial \Lambda}\dot{m}_\Lambda+\frac{\partial Y_k}{\partial H^*}\dot{m}_{H^*}=\mathscr{R}_R,$$

$$\frac{\partial H}{\partial \Lambda}\dot{m}_{\Lambda}+\frac{\partial H}{\partial H^*}\dot{m}_{H^*}=\mathscr{R}_H\,.$$

These are the same transport equations as above, with the right-hand side of the manifold equation for the reference species denoted by

$$\mathscr{R}_R=\frac{\rho\chi_{ZZ}}{2}\frac{\partial^2 Y_R}{\partial Z^2}+\rho\chi_{Z\Lambda}\frac{\partial^2 Y_R}{\partial Z\partial\Lambda}+\frac{\rho\chi_{\Lambda\Lambda}}{2}\frac{\partial^2 Y_R}{\partial\Lambda^2}+\rho\chi_{ZH^*}\frac{\partial^2 Y_R}{\partial Z\partial H^*}+\rho\chi_{\Lambda H^*}\frac{\partial^2 Y_R}{\partial\Lambda\partial H^*}+\frac{\rho\chi_{H^*H^*}}{2}\frac{\partial^2 Y_R}{\partial H^{*2}}+\dot{m}_R$$

and the right-hand side of the heat loss parameter equation denoted by

$$\mathscr{R}_H=\frac{\rho\chi_{ZZ}}{2}\frac{\partial^2 H}{\partial Z^2}+\rho\chi_{Z\Lambda}\frac{\partial^2 H}{\partial Z\partial\Lambda}+\frac{\rho\chi_{\Lambda\Lambda}}{2}\frac{\partial^2 H}{\partial\Lambda^2}+\rho\chi_{ZH^*}\frac{\partial^2 H}{\partial Z\partial H^*}+\rho\chi_{\Lambda H^*}\frac{\partial^2 H}{\partial\Lambda\partial H^*}+\frac{\rho\chi_{H^*H^*}}{2}\frac{\partial^2 H}{\partial H^{*2}}+\dot{q}_H\,.$$

Solving the two simultaneously leads to closure for the two source terms:

$$\dot{m}_{\Lambda}=\left[\frac{\frac{\partial H}{\partial H^*}\mathscr{R}_R-\frac{\partial Y_R}{\partial H^*}\mathscr{R}_H}{\frac{\partial Y_R}{\partial\Lambda}\frac{\partial H}{\partial H^*}-\frac{\partial Y_R}{\partial H^*}\frac{\partial H}{\partial\Lambda}}\right],$$

$$\dot{m}_{H^*}=\left[\frac{-\frac{\partial H}{\partial\Lambda}\mathscr{R}_R+\frac{\partial Y_R}{\partial\Lambda}\mathscr{R}_H}{\frac{\partial Y_R}{\partial\Lambda}\frac{\partial H}{\partial H^*}-\frac{\partial Y_R}{\partial H^*}\frac{\partial H}{\partial\Lambda}}\right].$$

With this solution, closure is achieved for the set of non-adiabatic multi-modal manifold equations.

3. Model Limits

In this section, the behavior of the model in three different sets of limits is explored. This is followed by a discussion of "unsteady" effects.

Adiabatic Combustion: The first test of the theoretical model is whether it recovers the adiabatic manifold equations in the absence of heat loss. In this limit, the heat loss source term is zero ($\nabla\dot{q}_H\to 0$), and gradients in heat loss are zero ($\nabla H^*\to 0$). In this limit, H is identically zero in accordance with its governing equation. The species manifold equation reduces to

$$\frac{\partial Y_k}{\partial\Lambda}\dot{m}_{\Lambda}=\frac{\rho\chi_{ZZ}}{2}\frac{\partial^2 Y_k}{\partial Z^2}+\rho\chi_{Z\Lambda}\frac{\partial^2 Y_k}{\partial Z\partial\Lambda}+\frac{\rho\chi_{\Lambda\Lambda}}{2}\frac{\partial^2 Y_k}{\partial\Lambda^2}+\dot{m}_k\,.$$

The source term $\dot{m}_{\Lambda}$ becomes

$$\dot{m}_{\Lambda}=\frac{1}{\partial Y_R/\partial\Lambda}\left(\frac{\rho\chi_{ZZ}}{2}\frac{\partial^2 Y_R}{\partial Z^2}+\rho\chi_{Z\Lambda}\frac{\partial^2 Y_R}{\partial Z\partial\Lambda}+\frac{\rho\chi_{\Lambda\Lambda}}{2}\frac{\partial^2 Y_R}{\partial\Lambda^2}+\dot{m}_k\right).$$

This equation is the same one established by the coordinate transformations in previous work for adiabatic multi-modal model [12].

Non-Adiabatic Premixed Combustion: In the premixed limit, gradients of mixture fraction become zero ($\nabla Z\to 0$). The species evolution under this simplification is

$$\frac{\partial Y_k}{\partial\Lambda}\dot{m}_{\Lambda}+\frac{\partial Y_k}{\partial H^*}\dot{m}_{H^*}=\frac{\rho\chi_{\Lambda\Lambda}}{2}\frac{\partial^2 Y_k}{\partial\Lambda^2}+\rho\chi_{\Lambda H^*}\frac{\partial^2 Y_k}{\partial\Lambda\partial H^*}+\frac{\rho\chi_{H^*H^*}}{2}\frac{\partial^2 Y_k}{\partial H^{*2}}+\dot{m}_k\,.$$

This is equivalent to the non-adiabatic premixed combustion manifold equations previously derived by the authors [17].

Non-Adiabatic Nonpremixed Combustion: In the nonpremixed limit, gradients of progress variable become zero ($\nabla\Lambda \to 0$). The species evolution under this simplification is

$$\frac{\partial Y_k}{\partial \Lambda}\dot{m}_\Lambda + \frac{\partial Y_k}{\partial H^*}\dot{m}_{H^*} = \frac{\rho\chi_{ZZ}}{2}\frac{\partial^2 Y_k}{\partial Z^2} + \rho\chi_{ZH^*}\frac{\partial^2 Y_k}{\partial Z \partial H^*} + \frac{\rho\chi_{H^*H^*}}{2}\frac{\partial^2 Y_k}{\partial H^{*2}} + \dot{m}_k\,.$$

Compared to equations derived by the authors for non-adiabatic nonpremixed combustion previously [18], the first term on the left-hand-side remains, and the source term for H^* is slightly different due to a different definition of this variable, which does not provide convenient limits of zero and unity independent of the mixture fraction and generalized progress variable. Retaining the first term on the left-hand-side does not make an *a priori* assumption of "steady" nonpremixed combustion as explained by Mueller [12].

"Unsteady" Effects: As explained by Mueller [12], the neglect of any time or pseudo-time term in the coordinate transformation implicitly assumes that the response of the manifold to changes in the scalar dissipation rates is fast compared to changes in the scalar dissipation rates. In other words, flow history effects are negligible. Other effects often attributed to "unsteadiness" including extinction and ignition, induced by either strain or heat losses, are accounted for in the model by the time-like terms on the left-hand-sides of the above manifold equations. If flow history effects are not negligible, due to, for example, slow chemical kinetics, then the manifold proposed above can be implemented within the context of a multiply-conditioned Conditional Moment Closure (CMC) approach.

4. Conclusions

In this work, a new physically-derived reduced-order manifold approach has been devised to model the effects of multi-modal combustion with heat loss. The manifold describes the thermochemical state space of all (isobaric, two-stream) combustion processes, both adiabatic and non-adiabatic, using mixture fraction Z, generalized progress variable Λ, and generalized enthalpy deficit parameter H^* as governing variables for the mapping. Under the transformation described in this work, the manifold equations were shown to simplify such that the limits of adiabatic multi-modal combustion and both premixed and nonpremixed non-adiabatic combustion are recovered when the appropriate spatial gradients and source terms vanish. Future research efforts will be devoted to developing a solver for these multi-dimensional manifold equations and implementation within a LES framework.

References

[1] S. B. Pope, PDF methods for turbulent reactive flows, Prog. Energy Combust. Sci. 11 (1985) 119–192.

[2] D. C. Haworth, Progress in probability density function methods for turbulent reacting flows, Prog. Energy Combust. Sci. 36 (2010) 168–259.

[3] P. A. McMurthy, S. Menon, and A. R. Kerstein, A linear eddy sub-grid model for turbulent reacting flows: Application to hydrogen-air combustion, Proc. Combust. Inst. 24 (1992) 271–278.

[4] J. C. Sutherland and A. Parente, Combustion modeling using principal component analysis, Proc. Combust. Inst. 32 (2009) 1563–1570.

[5] Y. Yang, S. B. Pope, and J. H. Chen, Empirical low-dimensional manifolds in composition space, Combust. Flame 160 (2013) 1967–1980.

[6] J. A. van Oijen and L. P. H. de Goey, Modelling of premixed laminar flames using flamelet-generated manifolds, Combust. Sci. Tech. 161 (2000) 113–137.

[7] C. D. Pierce and P. Moin, Progress variable approach for Large Eddy Simulation of non-premixed turbulent combustion, J. Fluid Mech. 504 (2004) 73–97.

[8] A. Y. Klimenko and R. W. Bilger, Conditional moment closure for turbulent combustion, Prog. Energy Combust. Sci. 25 (1999) 595–687.

[9] H. Yamashita, M. Shimada, and T. Takeno, A numerical study on flame stability at the transition point of jet diffusion flames, Proc. Combust. Inst. 26 (1996) 27–34.

[10] E. Knudsen and H. Pitsch, A general flamelet transformation useful for distinguishing between premixed and non-premixed modes of combustion, Combust. Flame 156 (2009) 678–696.

[11] P.-D. Nguyen, L. Vervisch, V. Subramanian, and P. Domingo, Multidimensional flamelet-generated manifolds for partially premixed combustion, Combust. Flame 157 (2010) 43–61.

[12] M. E. Mueller, Physically-Derived Reduced-Order Manifolds for Multi-Modal Turbulent Combustion, 10th U.S. National Combustion Meeting (2017).

[13] J. A. van Oijen, F. A. Lammer, and L. P. H. de Goey, Modeling of complex premixed burner systems by using flamelet-generated manifolds, Combust. Flame 127 (2001) 2124–2134.

[14] M. Ihme and H. Pitsch, Modeling of radiation and nitric oxide formation in turbulent non-premixed flames using a flamelet/progress variable formulation, Phys. Fluids 20 (2008) 055110.

[15] B. Fiorina, R. Baron, O. Gicquel, D. Thevenin, S. Carpentier, and N. Darabiha, Modelling non-adiabatic partially premixed flames using flame-prolongation of ILDM, Combust. Theory Model. 7 (2003) 449–470.

[16] M. E. Mueller and H. Pitsch, LES modeling of sooting turbulent nonpremixed flames, Combust. Flame 159 (2012) 2166–2180.

[17] A. C. Nunno, T. Grenga, and M. E. Mueller, Comparative analysis of methods for heat losses in turbulent premixed flames using Physically-Derived Reduced-Order Manifolds, Combust. Theory Model. (2018).

[18] A. C. Nunno and M. E. Mueller, Assumptions in Modeling Radiation Heat Losses in Turbulent Nonpremixed Combustion, Proc. Combust. Inst. (2019).

Spring Technical Meeting
Eastern States Section of the Combustion Institute
March 4-7, 2018
State College, Pennsylvania

DNS of Cavity Stabilized Premixed Turbulent Flame with a High-Order Immersed Boundary Method

A. H. Rauch[1,], K. Aditya[2], H. Kolla[2], J. H. Chen[2], H. K. Chelliah[1]*

[1]Dept. of Mechanical and Aerospace Engineering, University of Virginia, Thornton Hall, P.O. Box 400246, Charlottesville, VA 22904-426, USA
[2]Combustion Research Facility, Sandia National Laboratories, 7011 East Ave, Livermore, CA 94550, USA
[*]Corresponding Author Email: ahr6wa@virginia.edu

Abstract: A high-order immersed boundary method (IBM) is implemented into the DNS code S3D to analyze the interactions between combustion products generated in a cavity recirculation region and the high-speed flow stream. The S3D solver comprises of an 8th-order accurate central discretization in space with 10th-order accurate explicit filtering and 4th-order accurate explicit Runge-Kutta time integration. The immersed boundary is implemented via a ghost zone approach that represents the immersed boundary without the need to change the existing S3D solvers. Both a rectangular and linear close-up ramp cavity are investigated where the ramp is implemented by the IBM. Simulation conditions correspond to realistic operating conditions of a hypersonic vehicle with combustor inlet Mach number M=0.6. Due to limited computational resources turbulent, reacting flow simulations were performed at the lower combustor inlet Mach number M=0.3 for now. The ability of the IBM to deal with turbulence, compressibility and heat release is demonstrated. Additionally, qualitative analysis is performed revealing differences in flow and flame structure in the rectangular and ramp cavity.
Keywords: *DNS, Immersed Boundary Method, Cavity Flow, Turbulent Flame*

1. Introduction

To help stabilize the combustion in scram-/ramjets, cavities are frequently employed for their promising flame-holding properties and relatively low total pressure loss [1]. The idea is to create an optimum low-speed recirculation region in the cavity where a stable radical species generation region can exist which will act as an ignition source for the core flow.

In this investigation the stabilization of a premixed ethylene-air flame over a cavity is studied by Direct Numerical Simulation (DNS). Both a rectangular cavity [2] and a cavity with a linear-close up ramp are investigated where the close-up ramp is implemented via an Immersed Boundary Method (IBM) approach. The primary purpose of this paper is to demonstrate the capability of IBM to handle complex flow fields including turbulence, reacting flow, and flame-wall interactions. This is performed by qualitative comparison of the results from the rectangular and the ramp cavity geometries. Further, the interactions between the recirculation of partially-oxidized intermediate species from the cavity and the premixed main stream are investigated.

2. **Numerical Methods**

The DNS code adopted is S3D which solves the full compressible reacting Navier-Stokes equations coupled with detailed or reduced chemistry models [3]. It uses an 8th-order central scheme and a 10th-order explicit filter [4]. Time integration is performed with a six-stage, 4th-order explicit Runge-Kutta Method [5] and Navier-Stokes Characteristic Boundary conditions are applied [6, 7]. A 22 species ethylene-air reduced model [8] with a mixture averaged transport model is used for chemistry. S3D uses a regular cartesian mesh, but the IBM implementation allows for non-grid conformal bodies, such as the linear ramp, to be simulated as well.

2.1. Immersed Boundary Method

IBM allows for non-mesh conformal bodies to be simulated by adding a forcing term to the governing equations that mimic the presence of the boundary [9]. Its implementation in S3D will allow for more complex geometries to be simulated without the need to change the existing solvers. The approach taken in this work is similar to Minamoto et. al. [10].

A ghost zone approach is taken in which boundary conditions are imposed on 'ghost' cells that are just inside the solid region [9]. The ghost zone must be wide enough such that the fluid region is isolated from the solid region. It is thus max(O_f /2,O_d) where O_f is the filter order and O_d is the difference scheme order. Each ghost point has a corresponding image and wall point. The wall point is the point on the immersed boundary nearest to the ghost point. The image point is the reflection of the ghost point across the boundary. These three points lie along the normal to the immersed boundary and form a stencil which allows for the imposition of Dirichlet and Neumann conditions. Inverse distance weighting is used to interpolate from the fluid points surrounding the image point (neighbor points) onto it. This scheme is represented by,

$$\varphi_{IP} = \sum_{k=1}^{n} \omega_k \varphi_k \quad \text{, where } \omega_k = \frac{1}{d_k^2}\left(\sum_{i=1}^{n} \frac{1}{d_i^2}\right) , \tag{1}$$

where d_k is the distance of the image point from neighbor point 'k' and n is the number of neighbor points. This interpolation method is popularly used in IBM for its relative simplicity and stability. It is essentially a linear interpolation [11].

General Dirichlet (Eqn: 2) and Neumann (Eqn: 3) boundary conditions can be imposed as:

$$\phi_{GP} = 2\phi_{WP} - \phi_{IP} , \tag{2}$$
$$\phi_{GP} = \phi_{IP} , \tag{3}$$

on an arbitrary field variable Φ where the subscript 'GP' stands for ghost point, 'WP' for wall point and 'IP' for image point. In this study IBM is used to model an isothermal, no-slip wall. The physical boundary conditions and their numerical implementation are given in Table 1. Temperature is density weighted to minimize adverse effects due to high temperature gradients near the wall. Additionally a numerical boundary condition on the enthalpy is imposed to obtain the correct heat flux through the immersed boundary, it is analogous to the temperature condition:

$$h_{i,GP} = 2h_{i,WP} - h_{i,IP}, \qquad i = 1, N_{sc} , \tag{4}$$

where h_i is the specific enthalpy for species 'i' and N_{sc} is the number of species.

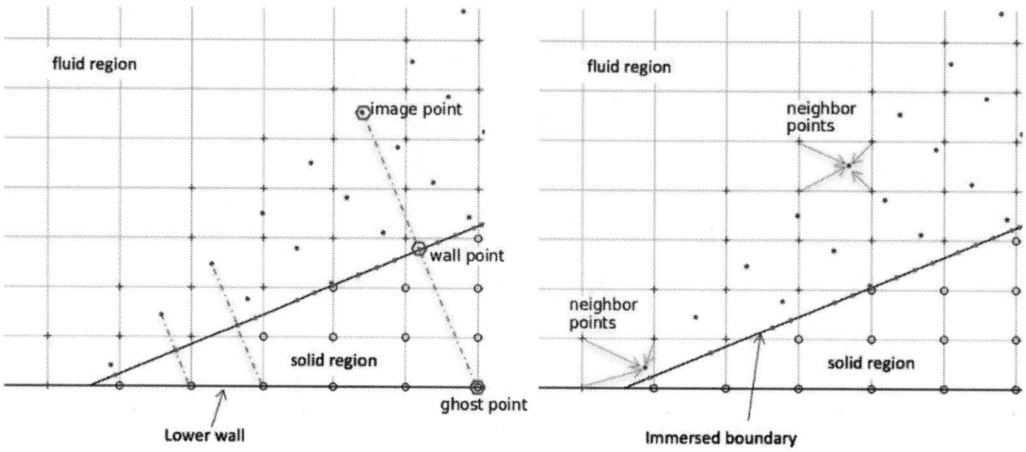

Figure 1: Schematic of Immersed Boundary implementation in 2D. The left figure shows the normal along which boundary conditions are imposed and image/wall points are determined. The right figure shows how neighbor points are used to interpolate onto image points. Black 'o' are ghost points, red '·' are image points, pink '·' are wall points and blue '+' are neighbor points.

Table 1: Physical boundary conditions for IBM and their numerical implementation. Subscript 'GP' stands for ghost point, 'WP' for wall point and 'IP' for image point.

Physical Boundary Condition	Numerical Implementation
$\partial p/\partial n = 0$	$p_{GP} = p_{IP}$
$T = T_{wall}$	$(\rho T)_{GP} = 2(\rho T)_{WP} - (\rho T)_{IP}$
$\vec{u}_{wall} = 0$	$\vec{u}_{GP} = -\vec{u}_{IP}$
$\partial \rho/\partial n = 0$	$\rho_{GP} = \rho_{IP}$
$\partial Y_i/\partial n = 0$	$Y_{i,GP} = Y_{i,IP}, \quad i = 1, N_{sc}$

2.2. Simulation Setup

Figure 2 illustrates the 2D computational domains considered in the present work. The imposed inflow conditions (Table 2) correspond to realistic operating conditions of a hypersonic vehicle near Mach 5 flight, with a fully premixed ethylene-air inflow. While these conditions correspond to a local combustor Mach number of Mach 0.6, simulations have been performed so far for Mach 0.3 due to limited computational resources. As this is a wall bounded flow, the defining turbulent length scale is the viscous length scale which is estimated to be 10μm at Mach=0.3. In order to adequately resolve the boundary layer, the grid spacing of half the viscous length scale is used off the wall. To initialize the chemical reaction, a kernel of hot products and radicals (determined from an adiabatic flame calculation) is placed near the leading edge of the cavity for a duration of 10μs.

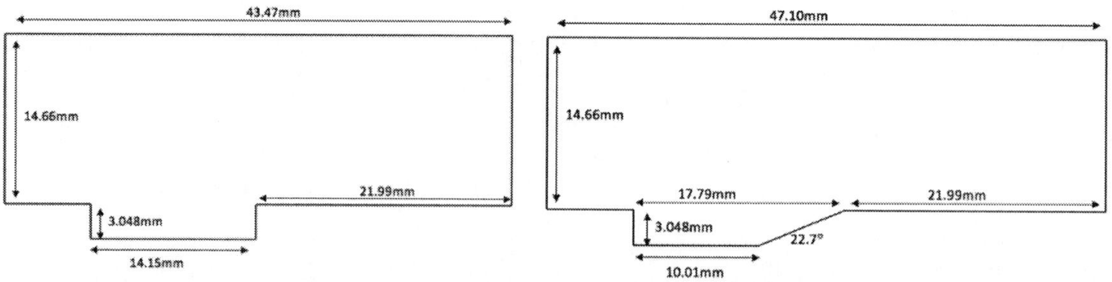

Figure 2: Schematic of the rectangular (left) and ramp (right) cavities simulated.

Table 2: Simulation conditions with a fully premixed ethylene-air mixture.

P [atm]	T [K]	T_{wall} [K]	ϕ	u'
1.72	1125	600	0.42	10%

3. Results and Discussion

To simulate turbulent flow it was necessary to feed in a turbulent velocity field that was generated in a 3D auxiliary channel simulation. The generated turbulent field was verified to be statistically stationary by analyzing the Reynolds stress profiles. 2D simulations were performed for both the ramp and the rectangular cavity. Additionally, a 3D simulation was performed for the rectangular cavity where the spanwise depth was sufficiently large to include a few integral length scales and the spanwise boundary conditions were periodic. Plots shown for the 3D cavity are an x-y plane slice taken along the centreline.

In Fig.'s 3, 5 contour lines of Y_{CH2O} are included as a rough indicator of the flame-front. From Fig. 3 noticeable entrainment of fresh reactants from the high-speed region into the ramp cavity is observed, this is not seen in the rectangular cavity. The ramp cavity simulations visually revealed increased spreading of the flame to the bulk region in addition to reduced unsteady vortex shedding (Fig. 4). In the 2D ramp and rectangular cavity simulations large scale convective structures are seen with no flame wrinkling (Fig.'s 4 a, b) whereas in the 3D rectangular cavity simulation siginificant flame wrinkling is observed (Fig. 4 c). This is due to the decay of turbulence in the 2D simulations. The absence of 3-dimensionality leads to an incorrect turbulent energy spectrum and a lack of forcing. This can be observed in Fig.'s 5 a, b where the inlet turbulence is seen to decay as near the outlet the velocity fluctuations are smaller and the velocity profile is nearing a turbulent mean velocity profile. Contrasting this to Fig. 5 c where the turbulence does not decay it is clear that this is due to the lack of dimensionality. Further, the absence of 3D structures inside the cavity reduces mixing. This is exemplified by the much more intense turbulence seen in Fig. 5 c than in Fig.'s 5 a, b.

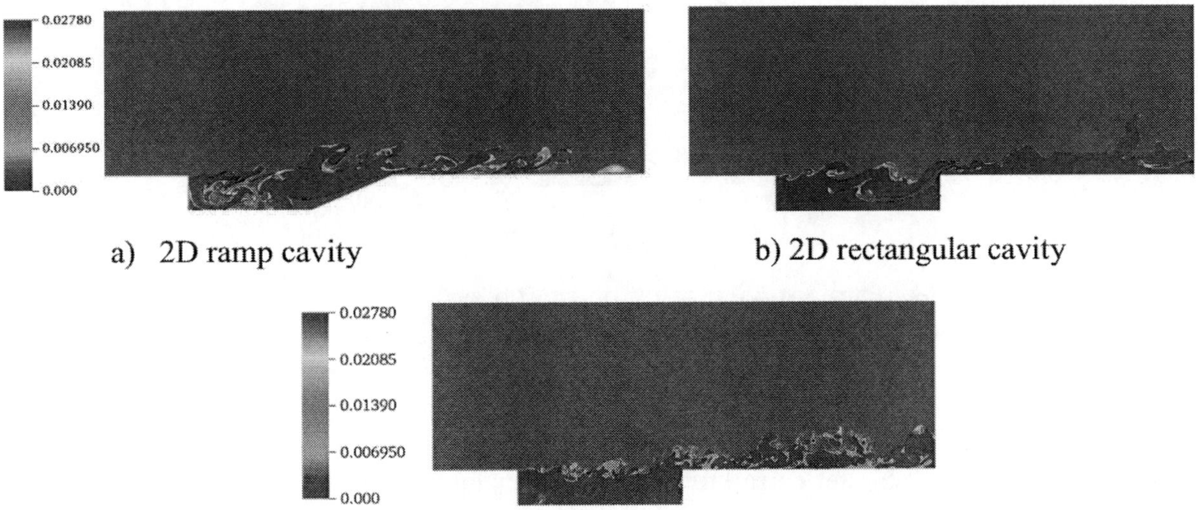

a) 2D ramp cavity

b) 2D rectangular cavity

c) 3D rectangular cavity

Figure 3: Contour plots of ethylene species mass fraction over both the ramp and rectangular cavity. The black line is a contour line of $Y_{CH2O} = 0.0006$.

a) 2D ramp cavity

b) 2D rectangular cavity

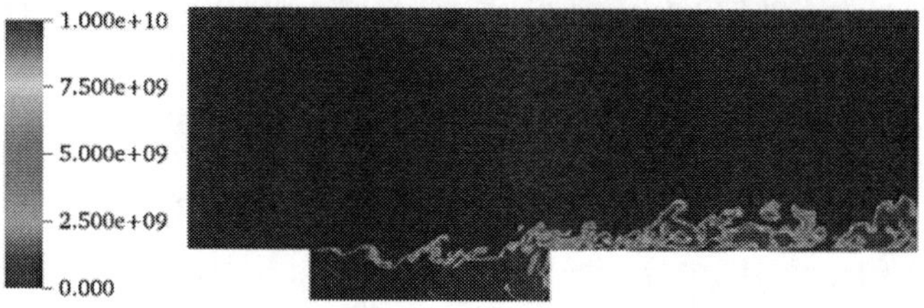

c) 3D rectangular cavity

Figure 4: Contour plots of heat release rate [J/m^3/s] over both the ramp and rectangular cavities.

a) 2D ramp cavity

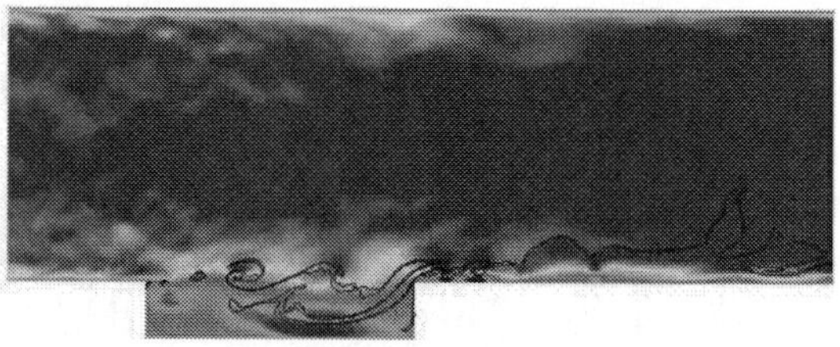

b) 2D rectangular cavity

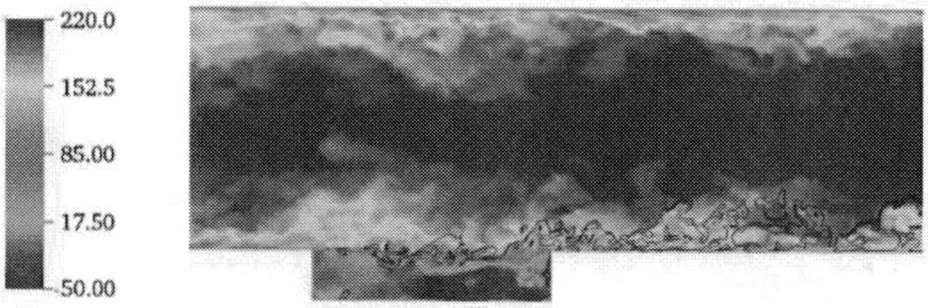

c) 3D rectangular cavity

Figure 5: Contour plots of stream-wise velocity [m/s] over both the ramp and rectangular cavities. The black line is a contour line of $Y_{CH2O} = 0.0006$.

4. Conclusions

The ramp cavity had a much higher entrainment of fresh reactants from the bulk high-speed region into the cavity than the rectangular cavity. Slightly higher spreading of the flame to the bulk high-speed region was observed for the ramp cavity. However, essential 3D turbulence effects were absent in the 2D cavity simulations resulting in a decay of turbulence. The turbulent mixing and flame wrinkling were much higher in the 3D rectangular cavity simulation than in the 2D ramp and rectangular cavity simulations which had large scale convective structures. In addition mixing in the 2D cavity simulations was inhibited by the lack of 3D structures.

The IBM implementation was able to resolve the velocity and thermal boundary layers for the modelled isothermal, no-slip wall with the use of the high-order accurate central scheme at the immersed boundary. Some instabilities were observed for flame-wall interactions that will be investigated further. Higher order interpolation and in particular Hermite interpolation will be investigated in future work to increase the order of accuracy of the IBM in S3D.

5. Acknowledgements

This work was supported by the National Science Foundation (grant number NSF-1511791) and the Air Force Office of Scientific Research (grant number FAA FA9550-15-1-0440), with Dr. Song-Charng Kong and Dr. Chiping Li as Technical Monitors.

6. References

[1] Tarun Mathur, Mark Gruber, Kevin Jackson, Jeff Donbar, Wayne Donaldson, Thomas Jackson, and Fred Billig. Supersonic combustion experiments with a cavity-based fuel injector. Journal of Propulsion and Power, 17(6):1305–1312, 2017/11/28 2001.

[2] J. Chen, A. Konduri, H. Kolla, A. Rauch, and H. Chelliah. Direct Numerical Simulation of a cavity-Stabilized Ethylene/Air Premixed Flame. In APS Meeting Abstracts, page 11, November 2016.

[3] Evatt R Hawkes, Ramanan Sankaran, James C Sutherland, and Jacqueline H Chen. Direct numerical simulation of turbulent combustion: fundamental insights towards predictive models. Journal of Physics: Conference Series, 16(1):65, 2005.

[4] Christopher A. Kennedy and Mark H. Carpenter. Several new numerical methods for compressible shear-layer simulations. Appl. Numer. Math., 14(4):397–433, June 1994.

[5] Christopher A. Kennedy, Mark H. Carpenter, and R.Michael Lewis. Low-storage, explicit runge-kutta schemes for the compressible navier stokes equations. Applied Numerical Mathematics, 35(3):177 – 219, 2000.

[6] Thierry Poinsot and Sanjiva Lele. Boundary conditions for direct simulations of compressible viscous flows. Journal of Computational Physics, 101, 08 1992.

[7] James C. Sutherland and Christopher A. Kennedy. Improved boundary conditions for viscous, reacting, compressible flows. Journal of Computational Physics, 191(2):502–524, 2003.

[8] Zhaoyu Luo, Chun Sang Yoo, Edward S. Richardson, Jacqueline H. Chen, Chung K. Law, and Tianfeng Lu. Chemical explosive mode analysis for a turbulent lifted ethylene jet flame in highly-heated coflow. Combustion and Flame, 159(1):265 – 274, 2012.

[9] Philippe Parnaudeau, Eric Lamballais, Dominique Heitz, and Jorge H. Silvestrini. Combination of the Immersed Boundary Method with Compact Schemes for DNS of Flows in Complex Geometry, chapter IB, pages 581–590. Springer Netherlands, Dordrecht, 2004.

[10] Y. Minamoto, K. Aoki, K. Osawa, T. Shi, A. Prodan, and M. Tanahashi. A high-order immersed boundary method for high-fidelity turbulent combustion simulations. In APS Meeting Abstracts, page 11, November 2016.

[11] Richard Franke. Scattered data interpolation: Tests of some methods. Mathematics of Computation, 38(157):181–200, 1982.

Spring Technical Meeting
Eastern States Section of the Combustion Institute
March 4-7, 2018
State College, Pennsylvania

Large Eddy Simulation and Probability Density Function Modelling of Swirling Cambridge/Sandia Turbulent Stratified Flame Series

Hasret Turkeri, Xinyu Zhao*

Department of Mechanical Engineering, University of Connecticut, Storrs, CT 06269, USA
**Corresponding Author Email: hasret.turkeri@uconn.edu*

Abstract: LES/PDF methodology is applied to Cambridge/Sandia turbulent stratified flame series. The finite-rate chemistry is represented by the 16-species reduced ARM1 mechanism. The *in situ* adaptation tabulation method is employed to accelerate the evaluation of chemical source terms. The simulations are performed for premixed flames under swirling condition. The results from LES/PDF simulations are compared with experimental measurements. The mean and r.m.s. profiles of velocity, temperature, and mass fractions of CH_4 are in very good agreement with measurements. The mean and r.m.s. of mass fraction profiles of CO are in agreement with the trend of experimental data but the peak values are underestimated mainly because of the less-than optimum mixing model parameter used in the simulations.
Keywords: ***LES/PDF modelling, Cambridge/Sandia Stratified Burner, Swirling Flow, Turbulent Premixed Flame***

1. Introduction

Swirling flow is widely used to stabilize lean premixed turbulent flames in gas turbines to satisfy the low-emission restrictions, and also is used to reduce the size of the devices by increasing the residence time [1]. Despite of its effectiveness, swirling introduces significant challenges in terms of combustion instabilities. Swirling flows are mainly characterized by having low frequency large-scale coherent flow structures that are inherently unstable. These unstable flow patterns increase the turbulent intensity and also introduce combustion oscillations, which makes modelling of the interaction between turbulence and reaction challenging [2-3].

The PDF method has proven to be very successful in non-premixed [4], premixed [5, 6] and partially premixed [7] combustion modes, because it has no intrinsic assumptions regarding the flame topologies. This is due to the exact treatment of the highly non-linear chemical reaction sources term without requiring any modeling. However, the conditional diffusion term remains in unclosed form, thus, needs to be modeled. Available LES/PDF studies have demonstrated that the interaction by exchange with the mean (IEM) mixing model with mean drift term [8] is an effective closure to the conditional diffusion term for the various jet flames investigated [9-12]. Therefore, it is of interest here to investigate the performance of the LES/PDF methodology under swirling conditions.

As a target flame, we select a turbulent methane/air stratified swirl burner (Cambridge/Sandia turbulent stratified swirl burner) developed by Sweeney et al. [13, 14] where extensive experimental measurements are available for model validation. The burner is designed to investigate the effect of stratification under swirl and non-swirl conditions. Here we only consider the swirling premixed case, SwB2, in which the two annular jets inject fuel/air mixtures at identical equivalence ratio. The flame is stabilized with a recirculation zone behind a bluff body at the centerline. The finite-rate chemistry is represented by an augmented reduced chemistry mechanism (ARM1) [15]. The *in situ* adaptive tabulation (ISAT) method [16, 17] is employed to accelerate the chemical calculations.

2. Methodology

In the LES method, the fields in a turbulent reacting flow are separated into large (filtered or resolved) and small (subfilter or unresolved) scales by applying a high band-pass filtering operation, for instance, the filtered density field $\bar{\rho}$ is defined as

$$\bar{\rho} \equiv \int_{-\infty}^{\infty} \rho(\mathrm{y},t) G(\mathrm{y}-\mathrm{x}) dy, \qquad [1]$$

where σ is density field and G is the LES filter. The transport equations for filtered mass and momentum equations are derived by applying the LES filter to the instantaneous transport equations. The effects of the small scale turbulent motions on the large scales are modelled by the dynamic Smagorinsky model [18]. For variable density flow, the filtered mass density function (FMDF) [19, 20] is defined as

$$F(\psi;\mathrm{x},t) \equiv \int_{-\infty}^{\infty} \rho(\mathrm{y},t)\delta(\psi-\phi(\mathrm{y},t))G(\mathrm{y}-\mathrm{x}) dy, \qquad [2]$$

where ϕ represents the $N=Ns+1$ compositional space consisting of the mass fractions of Ns species and sensible enthalpy. ψ is the sample space variable for the compositions and δ is the N-dimensional delta function. The Favre filtered PDF $\tilde{f}$ is related to the FMDF as $\tilde{f} = F / \bar{\rho}$. Then the modeled transport equation for $\tilde{f}$ can be derived from the transport equations of species and the definition of the Favre filtered PDF.

The numerical solution of filtered mass and momentum equations are performed using finite volume method. Because of its high dimensionality, Lagrangian based Monte Carlo method is employed to solve the modelled transport equation for $\tilde{f}$. In this method the flow is represented by large number of notional particles evolving by the following stochastic differential equations

$$dX_j^* = \left(\tilde{u}_j + \frac{1}{\bar{\rho}} \frac{\partial(\bar{\rho}\tilde{D}_T)}{\partial x_j} \right)^* dt + \left(2\tilde{D}_T^*\right)^{1/2} dW_j^*, \qquad [3]$$

$$d\phi_\alpha^* = -\Omega^*\left(\phi_\alpha^* - \tilde{\phi}_\alpha^*\right)dt + \frac{1}{\bar{\rho}^*} \frac{\partial}{\partial x_i}\left(\bar{\rho}\tilde{D}\frac{\partial \tilde{\phi}_\alpha}{\partial x_i} \right)^* dt + S_\alpha\left(\phi^*\right)dt, \qquad [4]$$

Equation (3) evolves particles in the physical space while Eq. (4) advances the particles in the composition space due to mixing and reaction. $\tilde{u}_j$ is the filtered velocity, $\tilde{D}_T$ and $\tilde{D}$ denotes turbulent and molecular diffusivity, respectively, and dW is the Wiener increment. In Eq. (4) the last term describes the chemical source terms, and the first two terms on the right-hand side account for molecular mixing and molecular diffusion, respectively. Ω is the mixing frequency and is modelled by $\Omega = C_m\left(\tilde{D} + \tilde{D}_T\right) / \Delta^2$, where Δ is LES filter size and C_m is a model parameter that is selected as $C_m=20$ in this study.

The density feedback from Monte Carlo solver to the finite volume solver is achieved by solving the transport equation for the specific volume [21]. In this method, a transport equation for resolved specific volume is solved in the finite volume based solver. The rate of change of specific volume due to molecular mixing and reaction is evaluated by the Monte Carlo solver and then is introduced into the transport equation as a source term. The *pdfFoam* solver [22] that is an unstructured LES/PDF solver developed entirely in OpenFOAM is used to solve the model equations of LES/PDF method.

3. Experimental configuration and simulation details

The Cambridge/Sandia stratified swirl burner was designed to investigate the effect of the stratification in swirling and non-swirling operation conditions. The burner consists of a central bluff body surrounded by two annular fuel jets, i.e., the inner and the outer jets. In the outermost annulus, air coflow is supplied to prevent entrainment of the ambient air. The bluff-body diameter is 12.7 mm, the inner jet annulus diameter is 23.6 mm with the wall thickness of 0.9 mm, and the outer jet annulus diameter is 34.8 mm with wall thickness of 1.65 mm. In the inner and outer jets, the inlet bulk velocities are 8.31 m/s and 18.7 m/s corresponding to Reynolds numbers of 5960 and 11500, respectively, and the air coflow has a laminar inlet velocity of 0.4 m/s. The flames are stabilized by a recirculation created behind the bluff body. Velocity measurements are provided by Zhou et al. [23], and the temperature and mass fraction measurements by Sweeney et al. [13, 14]. Here we only consider swirling case with purely premixed inlet conditions, and the moderately and highly stratified flames will be presented in a future study.

The cylindrical computational domain spans 200 mm in the axial z-direction and 200 mm in the radial r-direction. The domain is discretized using 2.3 million unstructured grids. The grid is concentrated near the nozzle along the axial direction and in the shear layer between the fuel jets in the radial directions.

For the fuel jets, the instantaneous inlet velocities are obtained through pipe simulations. First non-swirling annular pipe flow simulations are performed for each fuel jets and velocity field at a plane perpendicular to the axial direction are stored into feeding files. During the main LES/PDF simulations, the inlet velocities are interpolated to the fuel jets from the feeding files, and then the mean tangential velocity profiles measured at x=2 mm are added to the interpolated velocities. Finally, the interpolated axial velocities are adjusted according to the measured axial velocities at x=2 mm. The inlet velocity of the air coflow is set to be the uniform velocity profile of 0.4 m/s. At the bluff body and the wall-lips, no-slip boundary conditions are applied. Adiabatic boundary condition is assumed for temperature, and heat loss through the bluff body is neglected. The zero-gradient boundary conditions are imposed for all the fields (except pressure) at the outlet and at the far field. Approximately 20 particles per cell are used in the PDF solver. The methane/air combustion is represented by the augmented reduced mechanism (ARM1). ISAT is employed to accelerate the chemical kinetic calculations. The simulations are parallelized by domain decomposition technique using 192 cores. A constant time step of $\Delta t = 2\times10^{-6}$ s is used in the simulations corresponding to the Courant number of approximately 0.2. The simulations are performed for in total ten flow through times based on the inner stream velocity, and the statistics are collected after five flow through times. The computational cost of LES/PDF simulation is approximately 380 μs per grid cell per core per time step.

4. Results

In this section, numerical results from the LES/PDF simulation are presented and discussed. First, the velocity fields from the LES/PDF simulation are presented and compared with the experimental measurement. Then the temperature and mass fractions of CH_4 and CO profiles are presented at axial locations where experimental measurements are available.

Figure 1 shows the mean and r.m.s. of axial and circumferential velocity profiles with experimental measurement. Excellent agreements are demonstrated between measurement and numerical results. The swirl number, defined as the ratio of measured mean tangential velocity to axial velocity (i.e., $S = U_{tg}/U_z$) is 0.45, and the implementation that is described in Sec. 3 provides a sufficient description of both mean and r.m.s. velocities. It would be interesting to investigate the performance of the swirling treatment with higher swirl number, which will be part of the future research avenue.

Figures 2 and 3 present mean and r.m.s. profiles of temperature and mass fractions of CH_4 and CO obtained from the simulations, compared with experimental data. The temperature and mass fraction of CH_4 profiles are again in excellent agreement with measurements. Mean and r.m.s. of CO profiles are underestimated. This may mainly occur because the less-than optimal value of C_m is used in mixing frequency calculation. Used value of C_m=20 may underestimate the mixing frequency, thus, the reaction rate may be insufficient to reproduce the CO profiles. A sensitivity analysis on C_m model parameter will be performed to further analysis the discrepancy in CO profiles.

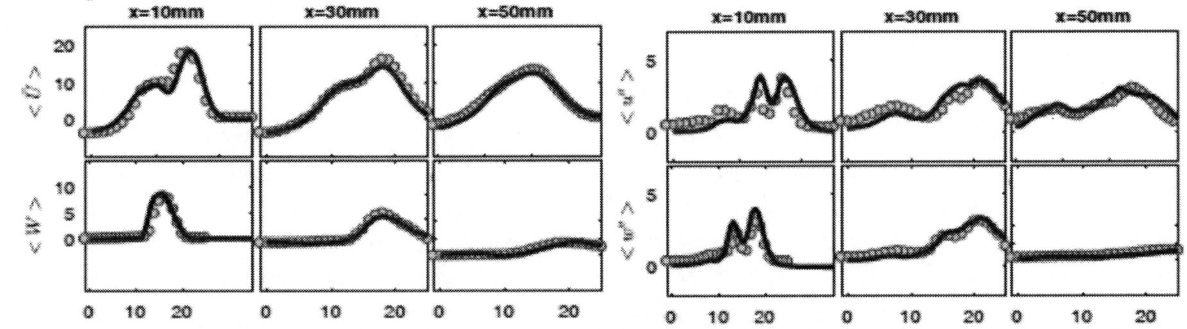

Figure 1: Mean axial and circumferential velocity profile (left), and r.m.s. axial and circumferential velocity profile (right). Symbols: experimental data; solid lines: results from LES/PDF simulations.

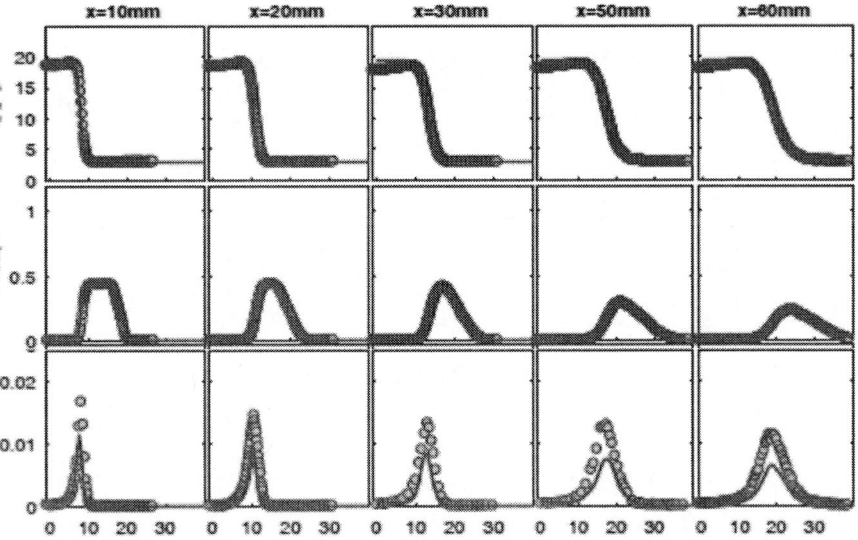

Figure 2: Mean temperature, and mass fraction of CH_4 and CO from top to bottom. Symbols: experimental data; solid lines: results from LES/PDF simulations.

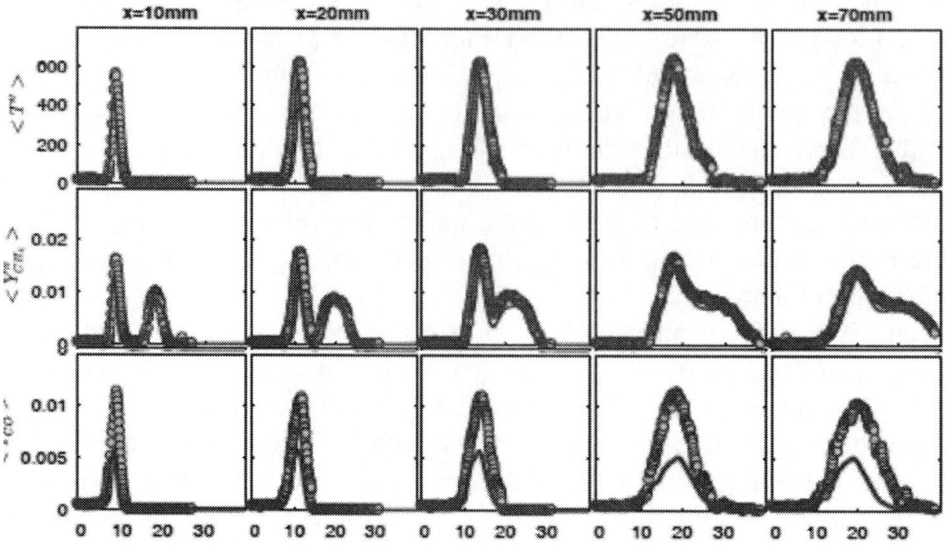

Figure 3: R.M.S. of temperature, and mass fraction of CH_4 and CO from top to bottom. Symbols: Experimental data; Solid lines: Results from LES/PDF simulations.

5. Conclusion

A LES/PDF methodology is applied to Cambridge/Sandia turbulent stratified flame series under swirling conditions. In this study, only the pure premixed case is studied. The mean and r.m.s. profiles of axial and circumferential velocities are found to be in very good agreement with experimental data. The temperature profiles and mass fraction of CH_4 demonstrate excellent consistency with experiment data. CO profiles obtained from the LES/PDF simulation can capture the trend in experimental data but the peak values in the profiles are underestimated. As a possible reason of this discrepancy, the effect of C_m model parameter in mixing frequency model will be examined in future studies. The moderately and highly stratified flames will also be studied using the same set of models in the future works.

6. Acknowledgment

Acknowledgement is made to the Donors of the American Chemical Society Petroleum Research Fund for partial support of this research.

7. References

[1] M. R. Johanson, D.Littlejohn, W. A. Nazeer, K. O. Smith, R. K. Cheng, Proc. Combust. Inst. 20 (2005) 2865-2874.
[2] Y. Huang, H. G. Sung, S. Y. Hsieh, V. Yang, J. Propul. Power 19 (2003) 782-794.
[3] T. Lieuwen, K. Mcmanus, J. Propul. Power 19 (2003) 721.
[4] M. Muradoglu, K. Liu, S. Pope, Pdf modeling of a bluff body stabilized turbulent flame, Combust. Flame 132 (1), (2003) 115–137.
[5] R. R. Tirunagari, S. B. Pope, Les/pdf for premixed combustion in the dns limit, Combust. Theor. Model.20 (5), (2016) 834–865.
[6] J. Kim, S. B. Pope, Effects of combined dimension reduction and tabulation on the simulations of a turbulent premixed flame using a large-eddy simulation/probability density function method, Combust. Theor. Model. 18 (3) (2014) 388–413.

[7] M. Juddoo, A. R. Masri, S. B. Pope, Turbulent piloted partially premixed flames with varying levels of o2/n2: Stability limits and pdf calculations, Combust. Theor. Model. 15 (6), (2011) 773–793.
[8] R. McDermott, S. B. Pope, A particle formulation for treating differential diffusion in filtered density function methods, J. Comput. Phys. 226 (2007) 947–993.
[9] Y. Yang, H. Wang, S. B. Pope, J. H. Chen, Large-eddy simulation/probability density function modeling of a non-premixed co/h2 temporally evolving jet flame, Proc. Combust. Inst. 34 (1) (2013) 1241–1249.
[10] D. H. Rowinski, S. B. Pope, Computational study of lean premixed turbulent flames using rans/pdf and les/pdf methods, Combust. Theor. Model.17 (4) (2013) 610–656.
[11] P. P. Popov, S. B. Pope, Large eddy simulation/probability density function simulations of bluff body stabilized flames, Combust. Flame 161 (12) (2014) 3100–3133.
[12] R. R. Tirunagari, S. B. Pope, An investigation of turbulent premixed counter flow flames using large-eddy simulations and probability density function methods, Combust. Flame 166 (2016) 229–242.
[13] M. S. Sweeney, S. Hochgreb, M. J. Dunn, R. S. Barlow, The structure of turbulent stratified and premixed methane/air flames I: Non-swirling flows, Combust. Flame 159 (9) (2012), 2896–2911.
[14] M. S. Sweeney, S. Hochgreb, M. J. Dunn, R. S. Barlow, The structure of turbulent stratified and premixed methane/air flames ii: Swirling flows, Combust. Flame 159 (9) (2012) 2912–2929.
[15] C. J. Sung, C. K. Law, J.-Y. Chen, Augmented reduced mechanisms for NO emission in methane oxidation, Combust. Flame 125 (2001) 906–919.
[16] S. B. Pope, Computationally efficient implementation of combustion chemistry using In Situ Adaptive Tabulation, Combust. Theor. Model.1 (1997) 41–63.
[17] L. Lu, S. R. Lantz, Z. Ren, S. B. Pope, Computationally efficient implementation of combustion chemistry in parallel PDF calculations, J. Comput. Phys. 28 (2009) 5490–5525.
[18] P. Moin, K. Squires, W. Cabot, S. Lee, A dynamic subgrid scale model for compressible turbulence and scalar transport, Phys. Fluids A 3 (1991) 2746–2757.
[19] P. J. Colucci, F. A. Jaberi, P. Givi, S. B. Pope, Filtered density function for large eddy simulation of turbulent reacting flows, Phys. Fluids 10 (1998) 499–515.
[20] F. A. Jaberi, P. J. Colucci, S. James, P. Givi, S. B. Pope, Filtered mass density function for large-eddy simulation of turbulent reacting flows, J. Comput. Phys. 401 (1999) 85–121.
[21] P. P. Popov, H. Wang, S. B. Pope, Specific volume coupling and convergence properties in hybrid particle/finite volume algorithms for turbulent reactive flows, J. Comput. Phys. 294 (2015) 110–126.
[22] H. Turkeri, S. B. Pope, M. Muradoglu, A general purpose les/pdf simulator on block structured meshes, submitted to Combust. Theor. Model..
[23] R. Zhou, S. Balusamy, M. S. Sweeney, R. S. Barlow, S. Hochgreb, Flow field measurements of a series of turbulent premixed and stratified methane/air flames, Combust. Flame 160 (2012) 2017–2028.

Spring Technical Meeting
Eastern States Section of the Combustion Institute
March 4–7, 2018
State College, Pennsylvania

Budgets of flame-conditioned second-order turbulence statistics in low and high Karlovitz number turbulent premixed jet flames

Jonathan F. MacArt[1,*], *Temistocle Grenga*[1], *and Michael E. Mueller*[1]

[1]*Department of Mechanical and Aerospace Engineering, Princeton University, Princeton, NJ*
[*]*Corresponding author:* `jmacart@princeton.edu`

Abstract: Small-scale heat release in low Karlovitz number turbulent premixed combustion significantly affects large-scale turbulence dynamics through nonlinear interactions occurring within the flame structure. These interactions result in flame-local changes to the alignment of the Reynolds stress and scalar fluxes with the mean gradients that lead to failure of commonly-used Boussinesq and gradient diffusion-type turbulence models. In this work, flame-conditioned second-order turbulence statistics, that is, velocity correlations and cross-correlations conditioned on a reaction progress variable, are computed following the concept of Conditional Moment Closure (CMC). By conditioning on a local flame coordinate, this statistical basis removes effects of flame motion that are responsible for the observed statistical misalignment at low Karlovitz number. The turbulent kinetic energy (TKE) is evaluated using data from Direct Numerical Simulation (DNS) of spatially-evolving turbulent premixed jet flames at low and high Karlovitz numbers. The evolution equations are derived, and budgets of these equations are computed in the two Karlovitz number cases. In analysis of these budgets, more precise mechanisms of interaction between combustion heat release and turbulence are suggested, and implications for modeling efforts are discussed.
***Keywords:* conditional velocity, turbulence statistics, turbulent premixed combustion, Direct Numerical Simulation**

1. Introduction

Turbulent combustion comprises numerous closely coupled interactions between turbulent flow and the combination of molecular transport and combustion chemistry that occur over a wide range of length and time scales. In turbulent premixed combustion, the relative strength of these competing physics are determined by the Karlovitz number [1], $Ka \equiv t_F/t_\eta$, where $t_F = \delta_F/s_L$ is the flame timescale, $t_\eta = (\nu/\varepsilon)^{1/2}$ is the Kolmogorov timescale, δ_F is the laminar flame thickness, s_L is the laminar flame speed, ε is the turbulent kinetic energy (TKE) dissipation rate, and ν is the kinematic viscosity. Scaling arguments [1] indicate that large-scale turbulence straining should be the dominant source of TKE at high Karlovitz number, but small-scale thermochemically-induced dilatation (volumetric expansion) should be the dominant source of TKE at low Karlovitz number [1]. A *critical Karlovitz number*, $Ka_{cr} \equiv \rho_u/\rho_b - 1$, where ρ_u and ρ_b are the densities of the unburned reactants and burned products, may be defined that delineates these two regimes [1, 2].

In Reynolds-averaged Navier-Stokes (RANS) simulations and Large Eddy Simulation (LES), models are required to close the "turbulence" terms, specifically, the Reynolds stress (subfilter stress) and the (subfilter) scalar flux, in addition to the "combustion" terms, principally, the mean (filtered) reaction source terms. Models for the "turbulence" terms are typically based on the

Boussinesq (Smagorinsky) or the gradient diffusion hypotheses and neglect any effects of small-scale heat release. Boussinesq-type (Smagorinsky-type) models for the velocity statistics assume statistical alignment of the Reynolds stress (subfilter stress) tensor with the mean (filtered) strain-rate tensor; however, this assumption has been shown to fail at Karlovitz numbers below Ka_{cr} in recent analysis of Direct Numerical Simulations (DNS) of turbulent premixed jet flames [2]. In particular, this failure at low Karlovitz numbers arises not only from mean dilatation effects but also from instantaneous flame-normal alignment of velocity and scalar statistics [2–4].

Fundamentally, traditional turbulence statistics cannot account for local affects due to flame-normal alignment and flame structure. A potentially promising alternative approach is to construct flame-conditioned turbulence statistics that are capable of capturing flame structure details directly. The "zone-conditioned" approach is the most common, in which velocity components are averaged in the unburned reactants and burned products [5, 6]; however, this approach assumes negligible probability of states between fully unburned and fully burned [7] and is therefore likely unsuitable for finite-thickness flames, particularly at moderate Karlovitz numbers. In this work, a more general framework is developed to condition the velocity on the local flame structure following the concept of Conditional Moment Closure (CMC) [8], in which the flame coordinate is a continuous function of the conditioning variable. Utilizing DNS databases of turbulent premixed jet flames at low and high Karlovitz numbers, the conditional TKE and its budgets are evaluated and compared in the two Karlovitz number regimes.

The geometric configuration comprises a central stoichiometric ($\phi = 1.0$) H_2/air turbulent jet separated by thin walls from symmetric laminar coflow jets of equilibrium products of the H_2/air premixture. Two geometrically similar configurations are simulated at in-flame Karlovitz numbers above and below Ka_{cr}. The Karlovitz number is varied by adjusting the bulk strain rate U_0/H, where U_0 is the central jet bulk velocity and H is the central jet height, while the bulk Reynolds number, $Re_0 = U_0 H/\nu_0 = 5,000$, is held constant. The two cases are denoted 5P-K1 and 5P-K2 and have in-flame Karlovitz numbers $Ka = 3.7$ and $Ka = 54.0$, respectively. The critical Karlovitz number for both cases is $Ka_{cr} = 6.7$. The DNS were computed using a semi-implicit, iterative algorithm [9, 10]. In the equations that follow, ρ is the density, u_i are velocity components, p is the pressure, and Y_k are species mass fractions. Further details may be found in MacArt *et al.* [2].

2. Conditional turbulent kinetic energy

In the present spatially-evolving configurations, data for the velocity field, density, pressure gradient, and scalars are analyzed on streamwise (y-z) planes at fixed locations x/H downstream of the nozzle exit plane. Conventional, *unconditional* means are obtained by averaging over the statistically homogeneous spanwise (z) direction and time. An unconditional, unweighted average is denoted by a plain overline, *i.e.*, $\overline{\phi}$. *Conditional* means are obtained by averaging into 50 bins of a sample space variable λ and time. A conditional, unweighted average is denoted by an overline with subscript, *i.e.*, $\overline{\phi}_\lambda$, while a conditional, density-weighted average is denoted by a tilde with subscript, $\widetilde{\phi}_\lambda \equiv \overline{\rho\phi}_\lambda/\overline{\rho}_\lambda$. Fluctuations about the density-weighted conditional mean are denoted by double primes, *i.e.*, $\phi'' \equiv \phi - \widetilde{\phi}_\lambda$.

The sample space coordinate λ is the sample space analog of the reaction progress variable C, which varies from $C = 0$ in the unburned reactants to $C = 1$ in the equilibrium burned products, based on the oxygen mass fraction. Density-weighted conditional probability density functions

(PDFs) of the progress variable are denoted $\widetilde{P}(\lambda;y)$. In order to facilitate comparison between the two Cases, for a single x-location, the cross-stream coordinate is remapped as the unconditional mean progress variable, $\widetilde{C}(y)$. Unconditional statistics are functions only of $\widetilde{C}(y)$ and are equivalent to integrating the corresponding conditional statistics, convoluted with $\widetilde{P}(\lambda;y)$, over λ, *i.e.*,

$$\widetilde{\phi}(y) = \int_0^1 \widetilde{\phi}_\lambda(\lambda;y)\widetilde{P}(\lambda;y)\,\mathrm{d}\lambda. \tag{1}$$

For simplicity in the following discussion, notation of the independent spatial variables is omitted.

The conditional TKE $\widetilde{k}_\lambda \equiv \widetilde{u_i''u_i''}_\lambda$ is analyzed in two steps. First, the conditional TKE equation is derived from conservation of momentum. Second, terms in the conditional TKE equation are analyzed in the two Karlovitz number cases.

2.1 Derivation of the conditional TKE equation

The derivation begins with the instantaneous momentum equation. Utilizing the PDF transport equation, an equation for the conditional velocity cross-correlation $\widetilde{u_iu_j}_\lambda$ is obtained. In this equation, the unclosed sample-space flux $\partial J_{u_iu_j}/\partial\lambda$ is closed using the analog of the primary closure proposed by Klimenko & Bilger [8]. Multiplying the equation for the conditional mean velocity $\widetilde{u}_{i,\lambda}$ [11] by $\widetilde{u}_{j,\lambda}$ and vice versa, summing, and following a similar procedure to the equation for $\widetilde{u_iu_j}_\lambda$, a transport equation for $\widetilde{u}_{i,\lambda}\widetilde{u}_{j,\lambda}$ is obtained. Subtracting this equation from the equation for $\widetilde{u_iu_j}_\lambda$ and taking the trace, the *conditional TKE equation* is obtained:

$$\begin{aligned}
&\frac{\partial \widetilde{u_i''u_i''}_\lambda}{\partial t} + \underbrace{\widetilde{u}_{k,\lambda}\frac{\partial \widetilde{u_i''u_i''}_\lambda}{\partial x_k}}_{\text{T1}} + \underbrace{2\widetilde{u_i''u_k''}_\lambda\frac{\partial \widetilde{u}_{i,\lambda}}{\partial x_k}}_{\text{T2}} + \frac{1}{\overline{\rho}_\lambda\widetilde{P}}\Bigg[\underbrace{\frac{\partial}{\partial x_k}\left(\overline{\rho}_\lambda\widetilde{u_i''u_i''u_k''}_\lambda\widetilde{P}\right)}_{\text{T3}} + \underbrace{\frac{\partial^2}{\partial x_k\partial\lambda}\left(\overline{\rho u_iu_iD\frac{\partial C}{\partial x_k}}_\lambda\widetilde{P}\right)}_{\text{T4}} \\
&\quad \underbrace{-2\widetilde{u}_{i,\lambda}\frac{\partial^2}{\partial x_k\partial\lambda}\left(\overline{\rho u_iD\frac{\partial C}{\partial x_k}}_\lambda\widetilde{P}\right)}_{\text{T5}} \underbrace{-\left(\widetilde{u_i''u_i''}_\lambda - \widetilde{u}_{i,\lambda}\widetilde{u}_{i,\lambda}\right)\frac{\partial^2}{\partial x_k\partial\lambda}\left(\overline{\rho D\frac{\partial C}{\partial x_k}}_\lambda\widetilde{P}\right)}_{\text{T6}}\Bigg] \\
&= \frac{1}{\overline{\rho}_\lambda\widetilde{P}}\Bigg[\underbrace{-\frac{\partial}{\partial\lambda}\left(\overline{\rho u_iu_i\dot{w}_C}_\lambda\widetilde{P}\right)}_{\text{T7}} + \underbrace{2\widetilde{u}_{i,\lambda}\frac{\partial}{\partial\lambda}\left(\overline{\rho u_i\dot{w}_C}_\lambda\widetilde{P}\right)}_{\text{T8}} + \underbrace{\left(\widetilde{u_i''u_i''}_\lambda - \widetilde{u}_{i,\lambda}\widetilde{u}_{i,\lambda}\right)\frac{\partial}{\partial\lambda}\left(\overline{\rho\dot{w}_C}_\lambda\widetilde{P}\right)}_{\text{T9}} \\
&\quad \underbrace{-\frac{1}{2}\frac{\partial}{\partial\lambda}\left(\overline{\rho\chi_{CC}}\widetilde{P}\right)\frac{\partial\widetilde{u_i''u_i''}_\lambda}{\partial\lambda}}_{\text{T10}}\Bigg] \underbrace{-\frac{2}{\overline{\rho}_\lambda}\overline{u_i''\frac{\partial p}{\partial x_i}}_\lambda}_{\text{T11}} + \underbrace{\frac{2}{\overline{\rho}_\lambda}\overline{u_i''\frac{\partial \tau_{ik}}{\partial x_k}}_\lambda}_{\text{T12}},
\end{aligned} \tag{2}$$

where D is the progress variable diffusivity, $\dot{w}_C$ is the progress variable source term, and $\chi_{CC} \equiv 2D(\partial C/\partial x_k)(\partial C/\partial x_k)$ is the progress variable dissipation rate. Terms in Eq. (2) correspond to physical sources and sinks of $\widetilde{u_i''u_i''}_\lambda$ as follows: convection (T1), production by the conditional mean shear (T2), turbulent transport (T3), cross-space terms (T4, T5, T6), chemical source-related terms (T7, T8, T9), flux in conditional sample space (T10), velocity-pressure gradient correlation (T11), and viscous transport and dissipation (T12).

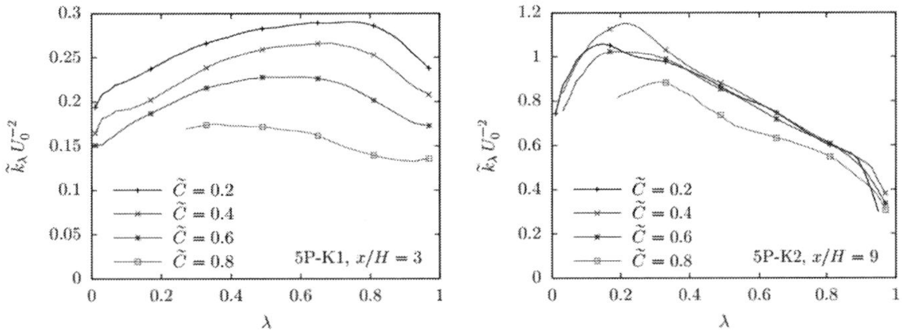

Figure 1: Normalized conditional TKE across the mean flame brush.

2.2 Results

The conditional TKE, normalized by U_0, is shown for Cases 5P-K1 and 5P-K2 in Fig. 1 as a function of λ at several locations $\widetilde{C}$ where $\widetilde{P} > 0.1$. In both Cases, $\widetilde{k}_\lambda$ decreases across the mean flame brush on moving from the turbulent central jet to the laminar coflow jet. In Case 5P-K2, $\widetilde{k}_\lambda$ also decreases with λ, since the strength of local turbulence-flame interactions is reduced at high Karlovitz number. However, in Case 5P-K1, $\widetilde{k}_\lambda$ increases with λ, highlighting the importance of instantaneous turbulence-flame interactions that contribute to the turbulent kinetic energy.

Budgets of terms in Eq. (2), normalized by H and U_0, are shown at $\widetilde{C} = 0.4$ in Fig. 2. In the top row of Fig. 2, the cross-space terms (T4, T5, T6) and chemistry terms (T7, T8, T9) are grouped respectively. In the bottom row of Fig. 2, terms with sample-space gradients are shown along with the residual of all terms in Eq. (2).

In Case 5P-K1, the velocity-pressure gradient correlation (T11) is the primary source of $\widetilde{k}_\lambda$ within the flame ($0.2 < \lambda < 0.9$) as is the case in unconditional TKE budgets in the same flame [2]. This source is balanced by the cross-space terms, viscous effects (T12), and chemical source terms. A negative spike in the velocity-pressure gradient correlation is observed on the unburned side of the flame and is balanced by the cross-space terms. This rapid change from the unburned turbulence ($\lambda = 0$) represents a "reconfiguring" of the turbulence to the flame structure, that is, the action of species transport on the conditional TKE through the pressure gradient.

Similar overall behavior is observed in Case 5P-K2; however, the relative magnitudes of the velocity-pressure gradient correlation, chemical source terms, and cross-space terms are reduced compared to the viscous terms. The reduction of these terms is evidence of the diminished influence of combustion heat release and instantaneous flame-normal alignment on the conditional TKE at high Karlovitz number. The nonzero velocity-pressure gradient source of conditional TKE highlights the slight influence of combustion heat release at non-infinite Karlovitz number.

The sample-space terms shown in the bottom row of Fig. 2 further demonstrate the relative influence of combustion heat release in the two Karlovitz number cases. T8, representing the source of TKE due to advection of heat release in conditional sample space, is large and positive on the unburned side of the 5P-K1 flame, becoming equivalently negative on the burned side. In contrast, in the 5P-K2 flame, T8 is positive on the unburned side of the flame but becomes much more significantly negative on the burned side, providing further evidence that turbulence-chemistry interactions are only important in the burned gases, where the Karlovitz and Reynolds numbers are significantly reduced from their in-flame values.

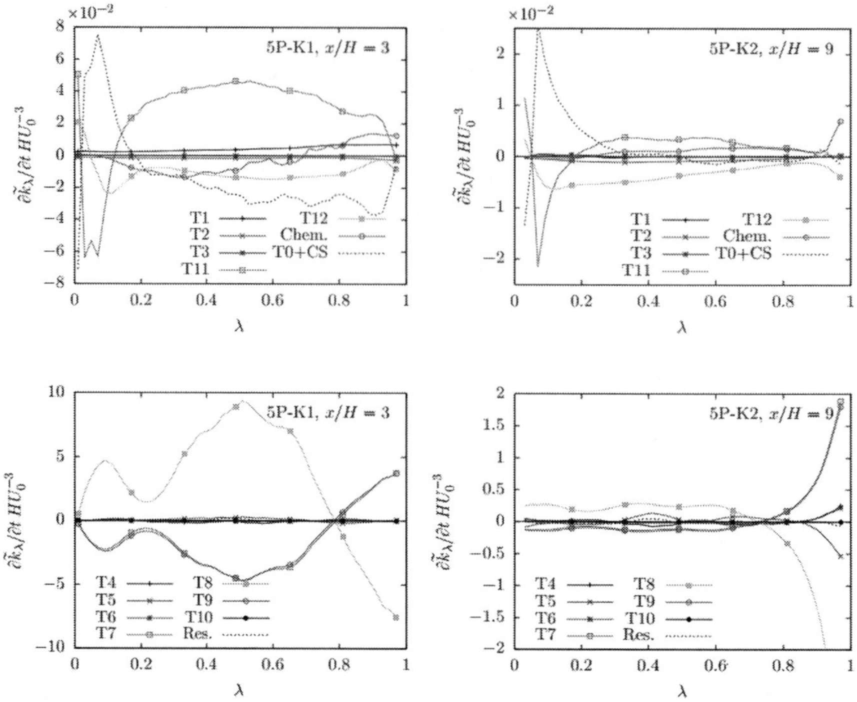

Figure 2: Normalized conditional TKE budgets at $\widetilde{C} = 0.4$. Top row: with combined terms (see text); bottom row: sample-space terms. Terms are labeled according to Eq. (2).

3. Conclusions

A conditional description of turbulence statistics for turbulent premixed combustion is employed based on the notion of CMC. Using data from DNS of spatially-evolving, turbulent premixed planar jet flames at Karlovitz numbers above and below the critical Karlovitz number, the conditional turbulent kinetic energy is evaluated and compared between the two Karlovitz number cases. The conditional statistical basis removes effects of flame motion from the turbulence statistics, thereby providing direct evidence for the flame-generated source of TKE at low Karlovitz number.

A balance equation for the conditional TKE is derived, and budgets of terms in this equation are evaluated and compared between the two Karlovitz number cases. Unlike traditional, unconditional TKE budgets, the conditional budgets explicitly include effects of the flame on the turbulence. In the low-Karlovitz number flame, the conditional velocity-pressure gradient correlation is the significant source of conditional TKE, and the same is true of the unconditional analog in traditional, unconditional TKE budgets. Interaction of the conditional velocity-pressure gradient correlation with cross-space terms (those with both physical- and conditional-space gradients) in the unburned gases provides evidence of the turbulence "reconfiguring" to the flame structure. Unlike in unconditional TKE budgets, pressure effects are evident in the high-Karlovitz number flame, but their magnitude is diminished compared to viscous effects. Finally, comparison of the various chemical source terms reveals significant action of the conditional advection of heat release throughout the low-Karlovitz number flame. This effect is greatly diminished in the high-Karlovitz number flame, where heat release and convective transport are relatively uncorrelated.

The conditional statistical description provides a fundamentally more complete foundation for turbulence modeling in reacting flows, for effects associated with flame motion and structure are distinguished from turbulence. Furthermore, the statistical basis on a flame coordinate allows for evaluation of the direct influence of combustion heat release on turbulence.

Acknowledgments

The authors gratefully acknowledge valuable support in the form of computational time on the TIGRESS high performance computer center at Princeton University, which is jointly supported by the Princeton Institute for Computational Science and Engineering (PICSciE) and the Princeton University Office of Information Technology's Research Computing department.

References

[1] R. W. Bilger, Some Aspects of Scalar Dissipation, Flow, Turbul. Combust. 72 (2004) 93–114.

[2] J. F. MacArt, T. Grenga, and M. E. Mueller, Effects of combustion heat release on velocity and scalar statistics in turbulent premixed jet flames at low and high Karlovitz numbers, Combust. Flame (2018) in press.

[3] P. E. Hamlington, A. Y. Poludnenko, and E. S. Oran, Interactions between turbulence and flames in premixed reacting flows, Phys. Fluids 23 (2011) 125111.

[4] H. Wang, E. R. Hawkes, and J. H. Chen, Turbulence-flame interactions in DNS of a laboratory high Karlovitz premixed turbulent jet flame, Phys. Fluids 28 (2016) 095107.

[5] J. Y. Chen, J. L. Lumley, and F. C. Gouldin, Modeling of wrinkled laminar flames with intermittency and conditional statistics, Symp. Combust. 21 (1988) 1483–1491.

[6] A. N. Lipatnikov, V. A. Sabelnikov, S. Nishiki, T. Hasegawa, and N. Chakraborty, DNS assessment of a simple model for evaluating velocity conditioned to unburned gas in premixed turbulent flames, Flow, Turbul. Combust. 94 (2015) 513–526.

[7] A. Lipatnikov, Conditioned moments in premixed turbulent reacting flows, Proc. Combust. Inst. 33 (2011) 1489–1496.

[8] A. Y. Klimenko and R. W. Bilger, Conditional moment closure for turbulent combustion, Prog. Energy Combust. Sci. 25 (1999) 595–687.

[9] O. Desjardins, G. Blanquart, G. Balarac, and H. Pitsch, High order conservative finite difference scheme for variable density low Mach number turbulent flows, J. Comput. Phys. 227 (2008) 7125–7159.

[10] J. F. MacArt and M. E. Mueller, Semi-implicit iterative methods for low Mach number turbulent reacting flows: Operator splitting versus approximate factorization, J. Comput. Phys. 326 (2016) 569–595.

[11] J. F. MacArt, T. Grenga, and M. E. Mueller, Conditional Turbulence Stats, Proc. Combust. Inst. (2018) submitted.

Spring Technical Meeting
Eastern States Section of the Combustion Institute
March 4-7, 2018
State College, Pennsylvania

Topology of Local Flame-Flame Interaction Events in Turbulent Flames

Ankit Tyagi[1], Isaac Boxx[2], Ryan Shupp[1], Stephen Peluso[1], and Jacqueline O'Connor[1,]*

[1]Mechanical and Nuclear Engineering, Pennsylvania State University, University Park, PA, USA
[2]DLR, German Aerospace Center, Stuttgart, Germany, 70596
[*]Corresponding Author Email: jxo22@psu.edu

Abstract: A large number of modern combustion devices have multiple, closely-spaced turbulent flames that interact with each other, resulting in interacting flowfields as well as scalar fields. The local structure and dynamics of adjacent flames is dependent on both these interacting fields. Local flame-flame interaction occurs frequently in turbulent flames, occasionally resulting in formation of unburned gas pockets, and in some cases, burned gas pockets. Unburned gas pockets can be a source of harmful pollutants, as they may result in toxic fuel emissions. The topology of interaction events can vary significantly in the presence of adjacent flames and it becomes crucial to understand the sensitivity of these flames to these local interaction events for improving the design and operability of multi-flame devices. In this study, we investigate two interacting premixed flames in a dual burner configuration and apply high-speed OH-planar laser-induced fluorescence (OH-PLIF) to obtain instantaneous flame-front locations of Bunsen flames. A non-rigid image registration technique is applied to flame images to track the topological changes occurring in small time steps. Flame-flame interaction events are identified using this technique and statistics conditioned on these events are compared. In particular, results are compared between single- and dual-burner configurations to illustrate the differences that interacting flames have on flame topology. Statistics of flame-flame interactions that result in flame pockets are also discussed.
Keywords: Turbulent flames, Flame-flame interactions

1. Introduction

Many modern combustion devices operate in multi-flame configurations and the interaction between adjacent flames in these devices can impact the flame shape, and the static and dynamic stability [1-5]. Flame-flame interactions affect the local structure and dynamics of turbulent flames and a detailed understanding of this local behavior is essential for predicting the global behavior of combustion devices. A number of studies have investigated flame-flame interactions [6-11] and results from these studies have shown that local flame-flame interactions can occur with either the consumption of reactant gases in between two flame-fronts (normal interactions), or the merging of the product sides of these flame-fronts (counter-normal interactions). Direct numerical simulations by Dunstan *et al.* [9, 10] and Griffiths *et al.* [6] have shown that the frequency of interactions vary with downstream distance and counter-normal interactions are more likely observed in shear layer regions with high strain rates and progress variables close to unity (towards products). Additionally, Chen *et al.* [12] have reported that flame wrinkling in turbulent flames leads to the beginning of interactions, such as a tunnel closure events, which lead to flame cusps and pocket formation. Results from their study have shown that these effects correspond to a sharp decrease in the local flame area, flame stretch, and the volumetric heat release. Global stretch rates calculated by Dunstan *et al.*[9, 10] also show bias to negative values and these rapid fluctuations in flame area and flame stretch from flame interactions also affect the flame speed. Chen *et al.*[12] have shown that the density-weighted displacement speed can increase by a factor of four during a pocket burnout event. These previous DNS studies have provided a detailed understanding of individual interaction events occurring in turbulent flames. However, analyzing flame-flame interactions for parametric variation of operating conditions becomes computationally expensive and experiments seem a viable approach towards obtaining statistical information for such dynamic events in turbulent flames. The goal of this study is to document the effect of adjacent flames on local flame-flame interaction events by providing statistical information of the characteristics of these events.

2. Methods / Experimental

2.1 Burner Configurations and Diagnostics

Measurements are performed in a dual slot-burner experimental facility, where the slot burners are mirror images along the experiment centerline (Figure 1). Stoichiometric premixed mixtures of fuel (natural gas) and air at room temperature enters the lower section of each burner and passes through two layers of ceramic honeycombs for flow conditioning. Two perforated plates placed at 30 mm and 10 mm upstream of the burner exit plane are used to create averaged turbulence intensities of 18% in the flowfield. The exit plane for each burner is 100 mm x 10 mm and this large-aspect ratio allows for obtaining time-averaged two-dimensional flowfields in our experiments. Each burner is equipped with two types of pilot flames, operated at stoichiometric conditions, along the 100 mm side of the slot: small 'anchoring' pilots, which are located close to the exit of the burners and help stabilize atmospheric pressure flames on the experiment, and larger 'back-support' pilots, which provide adiabatic or super-adiabatic combustion products around the flames. The burners are mounted to a small mill stage, which allows for changing fields-of-views (FOVs) for laser measurements. Conversion to single burner configuration involves attaching the back-support pilot flames of the right burner as a secondary pilot to the left burner to make the flow configuration symmetric on the left burner. High-speed flame-front measurements are performed using OH-planar laser-induced fluorescence (OH-PLIF) at a sampling rate of 10 kHz. The OH-PLIF system consists of a 532 nm Nd:YAG laser (Edgewave) pumping a dye laser (Sirah Credo). The output beam is calibrated to the $Q_1(6)$ line of the $A^2\Sigma^+ \leftarrow X^2\Pi$ (1-0) band to excite the OH radicals at 282.94 nm. The UV beam from the dye laser is passed through a periscope and a set of three cylindrical lenses to obtain a collimated sheet with an approximate height of 21 mm. The signal from the excited OH radicals is acquired using a CMOS sensor camera (Photron FASTCAM SA1.1), coupled with an external intensifier (LaVision HS-IRO) and a 100 mm f/2.8 UV lens (Cerco), resulting in a resolution of 0.1 mm/pixel. Measurements are performed in three FOVs with a 5mm overlap between two FOVs and FOV I begins at 5 mm downstream of the burner exit plane. Additional details of the experiment and laser diagnostics are provided in Ref. [13].

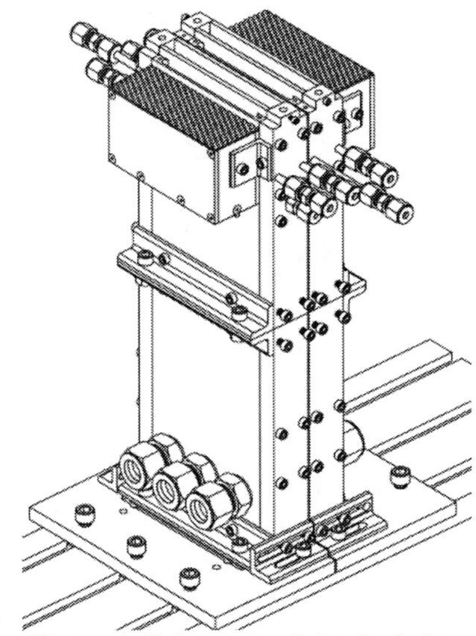

Figure 1: Schematic of the dual slot-burner experiment

2.2 Data Analysis Techniques

To obtain binarized images for edge-detection, raw OH-PLIF images are corrected for laser sheet intensity variations, and are median and bilateral filtered to reduce noise levels. Otsu's method is utilized for dynamic thresholding to obtain binarized images. Flame edges are obtained by tracing the edges of the binarized images using the '*bwboundaries*' function in MATLAB. Flame curvature of the traced flame edges is calculated by smoothing edges using the Savitzky-Golay filter, which uses a third order polynomial fit for sets of 15 pixels, along the entire flame edge. First and second partial derivatives of the smooth edge are calculated based on numerical approximations and are used to calculate the curvature of the flame edge. A non-rigid image registration technique (described as Automatic Feature Extraction in [9, 10]) is utilized to identify local flame-flame interaction events. In this technique, translation operations are performed on consecutive image pairs to account for local convection in the flowfield and resulting differences are recorded, which correspond to topological differences in these flames within 100 μs.

3. Results and Discussion

Figure 2 shows the stitched flame surface density images of the Bunsen-stabilized flames operating at a bulk flow velocity of 12 m/s and ϕ=1.0 in the dual- and single flame configurations. For the dual flame case, the inner side walls of the burners are adjacent, creating a center-to-center distance of 30 mm between the burner centerlines. Flame surface density is calculated by identifying instantaneous flame edges in a small interrogation region and averaging the results over 9999 frames. The single flame case is shown here as a reference for flame structure obtained at this flow velocity in our experiments. As shown in Figure 2, the global effect of interacting flames is particularly pronounced in FOVs II and III; the presence of adjacent flames in the dual-flame case changes the time-averaged flame structure by affecting the inner flame branches in these flames, as they are subjected to different boundary conditions. To compare the local structure of these flames, curvatures for each flame are calculated as outlined in Section 2.2. Figure 3 shows the probability density functions (PDFs) of flame curvatures in FOVs I, II, and III for the single- and dual flame cases. For the rest of the paper, 'Dual-left' corresponds to the left flame and 'Dual-right' corresponds to the right flame in the dual flame case. PDF trends of all flames in all FOVs show a good general agreement between each other; in general, the trends are distributed about zero, indicating similar amounts of wrinkling in these flames. While these curvature PDFs report the statistics of the local topology of these flames and provide us with information on the wrinkling of these flames, it is challenging to obtain any information on the dynamically changing topology of these flames.

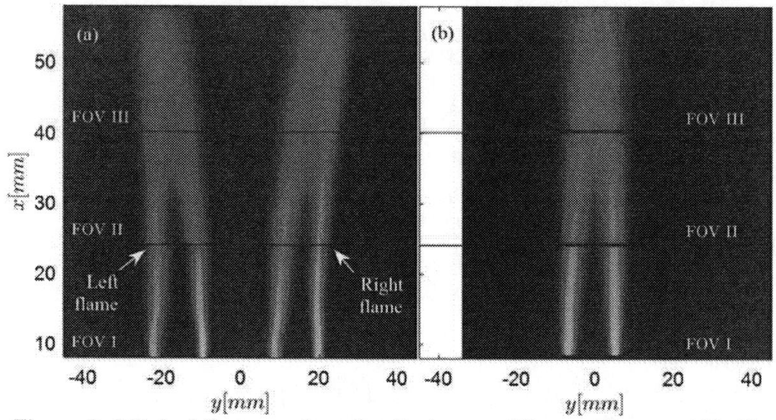

Figure 2: Stitched flame surface density image of Bunsen flames at U=12 m/s, ϕ=1.0 in the (a) dual flame, and (b) single flame configurations

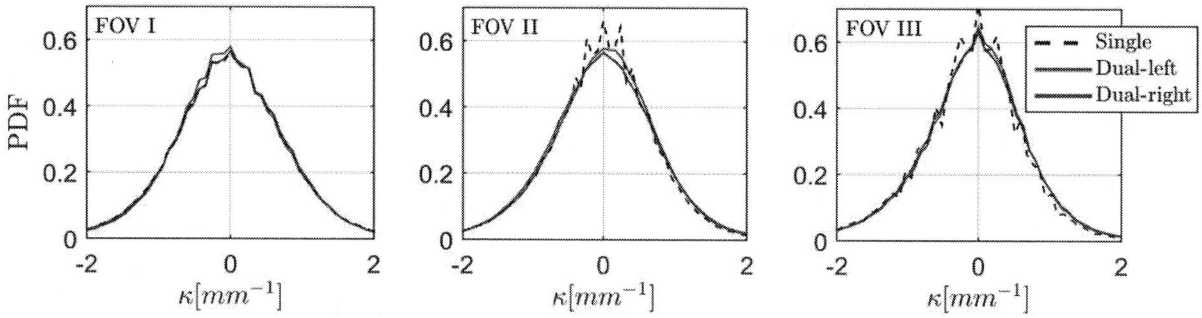

Figure 3: PDF of flame curvatures for the single flame and dual flame cases

The application of the non-rigid image registration technique permits identification and tracking of topological differences existing between two image pairs. Figure 4 shows a time-series of the flame-flame interaction events occurring in the left burner flame in FOV I where topological differences are highlighted in white closed contours. As shown with arrows in the figure, different types of flame-flame interactions can occur in turbulent flames: flame-front merging leading to pocket formations, merging of existing pockets in attached flame-fronts, breaking of pockets into several smaller pockets, etc. Flame tips are occasionally observed for these flames in FOV II (not shown here), where big flame pockets are formed that propagate into FOV III and either break into smaller pockets or get consumed further downstream.

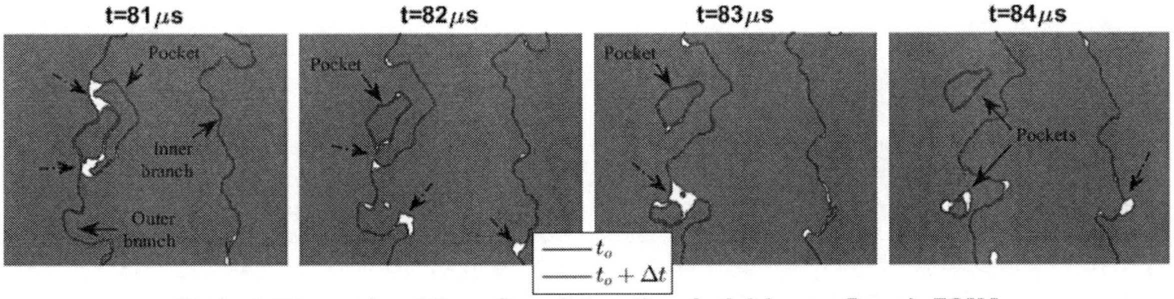

Figure 4: Time-series of flame-flame interactions for left burner flame in FOV I

All identified interactions over a range of 9998 images are reported in Figure 5(a). The total number of interaction events occurring in FOV I are very similar for both single and dual flames; however, these numbers vary significantly in FOV II and III. These trends indicate that near the flame attachment region, the dynamics of flames are similar for both single- and dual-burner cases. The presence of adjacent flames starts to affect the flames in FOVs II and III. To isolate the influence of flame pockets on attached flame-fronts and merging/breaking of flame pockets, pocket interactions are filtered out so that only interactions occurring along the main attached flame are counted. These filtered interactions are presented in Figure 5(b); the number of filtered interactions decrease for all flames as a function of FOV. FOVs II and III correspond to the flame tip region and a high frequency of formation of big flame pockets exist in these FOVs. Differences in the pocket-filtered interaction counts in FOVs II and III show that a large number of flame-flame interactions occur in the attached flame fronts in the dual flame case, which are lower for the single flame case. A high number of total flame interactions and a low number of filtered flame interactions in FOV II for the single flame case suggests that in this region, a high frequency of interactions is attributed to existing flame pockets that interact with the attached flame fronts or other existing pockets.

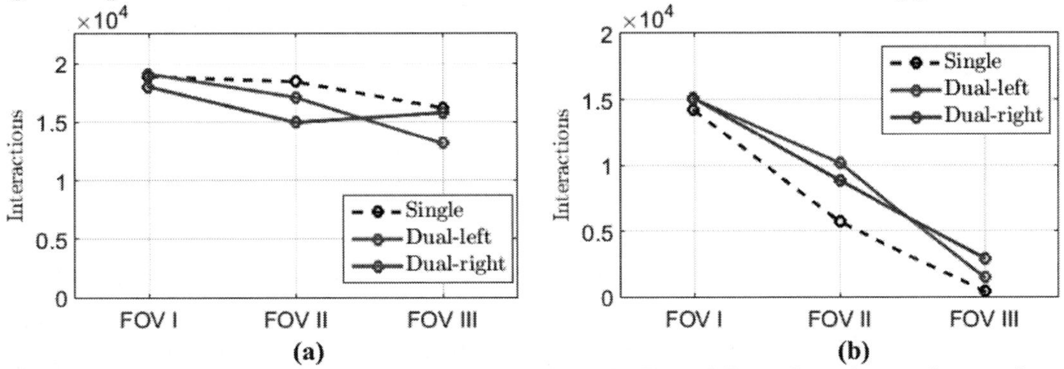

Figure 5: (a) Total number of flame interaction numbers, (b) filtered flame-flame interaction numbers

Centroids of the interaction zones, shown as white closed contours in Figure 4, correspond to a representative location of instantaneous, filtered flame-flame interactions and statistics conditioned on these centroids are presented in Figure 6. A comparison of PDF trends of filtered interaction conditioned averaged progress variables ($\bar{c}$) (shown in Figure 6(a)) show similarities between the left and right flames for the dual flame case, however, slight deviations can be seen for the single flame case. For all FOVs, a higher probability exists for interactions occurring at locations of $\bar{c} > 0.5$ of the flame brush. These results are congruent with results of Griffiths *et al.* [6] and Dunstan *et al.* [9, 10] that have reported high frequencies of interactions occurring along the leading and trailing edges of the flame brush. For FOV III, the PDF peaks close to $\bar{c} \approx 0.9$ for the single flame case, whereas, for the dual flame, this peaks is closer to $\bar{c} \approx 0.75$. PDFs of filtered interaction conditioned curvatures for all flames are shown in Figure 6(b). The PDF distributions for all FOVs shown in this figure are centered close to zero, which indicates that a normal-type distribution of interaction conditioned curvatures are present in all FOVs. Dunstan *et al.* [9, 10] have reported that for normal interactions, positive principal curvatures correspond to convex-normal interactions and negative principal curvatures correspond to tunnel-closure and pocket burn out interactions. The curvature distributions suggest that these types of interactions are present in all FOVs for the flames investigated here.

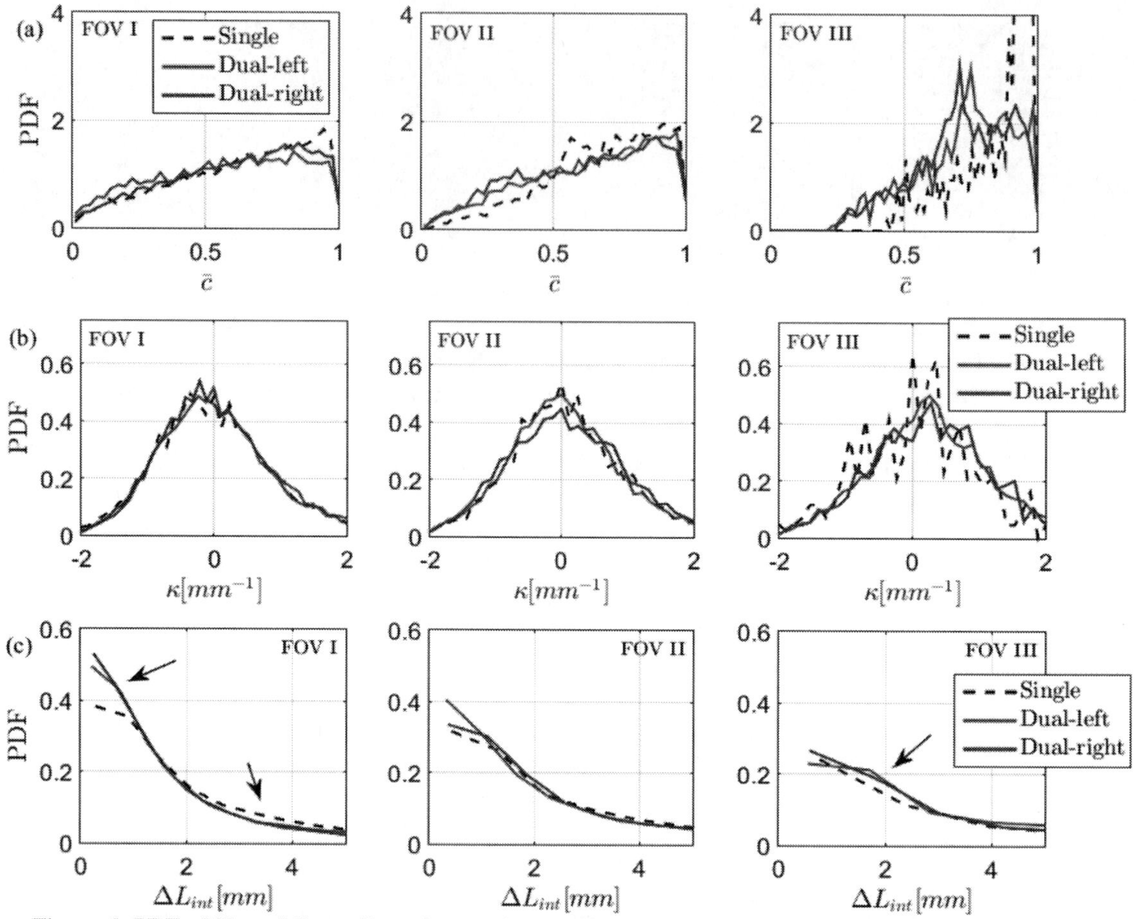

Figure 6: PDF of filtered flame-flame interaction conditioned (a) averaged progress variables ($\bar{c}$), (b) flame curvatures (κ), and (c) flame length scale differences before and after interactions (ΔL_{int})

Differences in the length scales due to flame-flame interactions are also extracted and probabilities of these length scales are presented in Figure 6(c). A close look at the PDFs of these length scales in FOV I shows higher probabilities of small length scales for the dual flame case, as compared with the single flame case. These differences are small for these flames in FOVs II and III. This is an important finding and will require further investigation of the velocity fields to understand the differences in the turbulence fields of these flames. Table 1 shows the number of pocket formations due to filtered flame-flame interactions in FOVs I and II for the single- and dual flame cases. Smaller differences exist for these flame in FOV I, however, these differences become large in FOV II. This is likely due to the differences in the averaged location of flame tip in the single- vs. dual flame case. These results indicate that the presence of adjacent flames can change the dynamics of flame-flame interactions in turbulent premixed flames.

Table 1:Number of filtered interactions leading to pocket formations for 9998 images

	FOV I	*FOV II*
Single	*1756*	*747*
Dual-left	*1596*	*1371*
Dual-right	*1479*	*1250*

4. Conclusions

High-speed OH-PLIF is implemented to study the effect of interacting flames on the local topology of flame fronts and statistics conditioned on local flame-front reduction events are presented. Instantaneous movies of these flames reveal that the local dynamics of flame-fronts can vary with the presence of adjacent flames and it is challenging to interpret this using flame-front curvature statistics. With the aid of image registration methodologies, dynamics of flame-flame interactions are tracked to capture these broad-range scale interactions. Higher probabilities of small length scales associated with flame-flame interactions in the branches of these flames indicate short times associated with consumption of these flame-front elements. Unburned gas pocket formation is a dominating phenomenon occurring further downstream of the attachment region and larger pockets are formed near the flame tip region. These larger flame pockets break into several smaller pockets and get consumed further downstream. Future work from this experiment will include dynamic tracking of flame pockets and capturing their consumption rates. These analysis techniques will be implemented to a large data set to study the effects of parametric variation in the current burner configuration. Instantaneous velocity fields will also be analyzed to support these findings.

5. Acknowledgements

This work was supported by the Air Force Office of Scientific Research under Grant FA9550-16-1-0075 with program manager Dr. Chiping Li. The authors would also like to thank the students of the Reacting Flow Dynamics Lab for their support.

6. References

1. Worth, N.A. and J.R. Dawson, *Tomographic reconstruction of OH* chemiluminescence in two interacting turbulent flames.* Measurement Science and Technology, 2012. **24**(2): p. 024013.
2. Samarasinghe, J., et al., *Three-Dimensional Chemiluminescence Imaging of Unforced and Forced Swirl-Stabilized Flames in a Lean Premixed Multi-Nozzle Can Combustor.* Journal of Engineering for Gas Turbines and Power, 2013. **135**(10): p. 101503.
3. Hegde, U., et al., *Flame driving of longitudinal instabilities in dump type ramjet combustors.* Combustion Science and Technology, 1987. **55**(4-6): p. 125-138.
4. Dawson, J.R. and N.A. Worth, *Flame dynamics and unsteady heat release rate of self-excited azimuthal modes in an annular combustor.* Combustion and Flame, 2014. **161**(10): p. 2565-2578.
5. Aguilar, M., et al., *Helical Flow Disturbances in a Multinozzle Combustor.* Journal of Engineering for Gas Turbines and Power, 2015. **137**(9): p. 091507.
6. Griffiths, R., et al., *Three-dimensional topology of turbulent premixed flame interaction.* Proceedings of the Combustion Institute, 2015. **35**(2): p. 1341-1348.
7. Sohrab, S., Z. Ye, and C. Law. *An experimental investigation on flame interaction and the existence of negative flame speeds.* in *Symposium (International) on Combustion*. 1985. Elsevier.
8. Chung, S., J. Kim, and C. Law. *Extinction of interacting premixed flames: theory and experimental comparisons.* in *Symposium (International) on Combustion*. 1988. Elsevier.
9. Dunstan, T., et al., *Flame interactions in turbulent premixed twin V-flames.* Combustion Science and Technology, 2013. **185**(1): p. 134-159.
10. Dunstan, T., et al., *The effects of non-unity Lewis numbers on turbulent premixed flame interactions in a twin V-flame configuration.* Combustion Science and Technology, 2013. **185**(6): p. 874-897.
11. Im, H.G. and J.H. Chen, *Preferential diffusion effects on the burning rate of interacting turbulent premixed hydrogen-air flames.* Combustion and flame, 2002. **131**(3): p. 246-258.
12. Chen, J.H., T. Echekki, and W. Kollmann, *The mechanism of two-dimensional pocket formation in lean premixed methane-air flames with implications to turbulent combustion.* Combustion and flame, 1999. **116**(1): p. 15-48.
13. Tyagi, A., et al. *Structure of Flames in Flame Interaction Zones.* in *2018 AIAA Aerospace Sciences Meeting*. 2018.

Spring Technical Meeting
Eastern States Section of the Combustion Institute
March 4–7, 2018
State College, Pennsylvania

Cellular Instability in Expanding Turbulent Flames

Zirui Liu[1], *Abhishek Saha*[1,*], *Sheng Yang*[1], *and Chung K. Law*[1]

[1]*Department of Mechanical and Aerospace Engineering, Princeton University, Princeton, New Jersey, 08544*
[*]*Corresponding author:* asaha@princeton.edu

Abstract: Cellular instability and turbulent eddies both can wrinkle an otherwise smooth laminar flamefront and thereby increase the surface area leading to flame acceleration, which is characteristically controlled by the length and time-scales of the individual processes. In this paper we report a study on turbulent flames over a large range of pressure conditions, where laminar flames are cellularly unstable. The effects of two mechanisms of cellular instabilities, namely Darrieus-Landau and diffusional-thermal instabilities on turbulent flames have been categorically investigated. First, a simple scaling analysis is presented to compare the strength of the instability and turbulence and to show that the classical premixed turbulent flame regime diagram can be subdivided into three regimes of: instability dominated, turbulence dominated and instability-turbulence interaction. Next, we present experimental results from a large range of conditions to show that irrespective of the mode of instability, flame propagation is indeed controlled by cellular instability, turbulence and both in these three regimes respectively. Furthermore, by modulating the mixture Lewis number, hence the strength of the diffusional-thermal instability, we would also demonstrate its role in turbulent flame propagation..
Keywords: ***Turbulent flame, Cellular instability***

1. Introduction

Turbulence facilitates the propagation of premixed flames in general by increasing the heat and mass transport rates as well as the surface area of the embedded flamelets by wrinkling them [1]. Since Damköhler laid the fundamental concept of scaling of the turbulent flame speed, S_T, extensive studies [1–8] have been conducted to quantify the extent of the augmentation, frequently through the normalized turbulent flame speed, $\overline{S_T} = S_T/S_{L0}$, where S_{L0} is the mixture's planar laminar flame speed.

The primary motivation of the present study is to extend these recent experimental and interpretive studies by incorporating the effects of Darrieus-Landau (DL) hydrodynamic and diffusional-thermal (DT) cellular flame-front instabilities on the structure and propagation of turbulent flames, especially in identifying the potential further increase in the turbulent flame speed through the corresponding increase of the surface area of the embedded flamelets due to the development of the instability cells. Previous experimental studies on turbulent flame propagation from our group [5, 9, 10] were purposefully designed to suppress the development of the DL cells so that the effects of "pure" turbulence can be unambiguously identified. The propagation of turbulent expanding flames without the presence of cellular flame-front instabilities was found to follow the scaling law [5], $\overline{S_T} = Re_{T,f}^{1/2} = [(U_{eff}R_f)/(S_{L0}\delta_{L0})]^{1/2}$, where U_{eff} is the effective fluctuating velocity of the tubrulent flow, R_f is mean flame radius and δ_{L0} is laminar flmae thiockness. A readily implementable approach to induce the development of DL cells is to reduce the thickness of the flamelets by elevating the chamber pressure. Furthermore, with a sufficiently high, constant chamber pressure, the propensity for the cell development will continuously increase, as the overall size of the expanding flame increases because of the corresponding decrease of the ratio of the flame thickness to the global flame size. In particular, our recent study [11] on expanding flames in quiescence and elevated pressures has identified three stages of flame propagation and cell development, namely an initial cell-free smooth propagation stage, followed by a transition stage with initiation of cell formation and rapid flame acceleration, which in turn leads to a final, saturated stage characterized by a power-law description of the flame propagation. These results also unambiguously demonstrate the phenomenon of self-acceleration of expanding laminar flames through the development of the cells of the flame-front instability.

In light of the fundamental interest discussed above, and further recognizing the relevance of high-pressure combustion within internal combustion engines, we have conducted a comprehensive study, phenomenologically and experimentally, on the role of the cellular instability on turbulent flame propagation. Leading-order scaling analysis is performed on the spectra of the DL cells and turbulent eddies, leading to the identification of three regimes of interaction, for which the acceleration exponents of the turbulent flames are determined. A modification of the classical regime diagram, which extends the previous experimental results that mostly fall within the flamelet regime, is also proposed to reflect the influence of DL instability on turbulent flames. We now present these investigations and their results

2. Methodology

In order to compare the behaviors of the flames in different regimes proposed in the following sections, and to validate the leading-order scaling analysis, experiments with constant-pressure expanding flames have been conducted in both laminar and turbulent environments. Experiments were conducted in our well-vetted, fan-stirred dual-chamber vessel. Four orthogonally located fans were used to generate well controlled nearly-isotropic turbulent flow field, whose intensity has been calibrated by high-speed Particle Image Velocimetry (PIV) reported in [12]. Since, as a first step, we are interested in interaction between DL instability and turbulence, the experimental conditions were so chosen such that the effective Lewis number (Le), defined as the ratio of the thermal diffusivity to the controlling mass diffusivity, remains close to unity in order to avoid the diffusional-thermal, stabilizing ($Le > 1$) or destabilizing ($Le < 1$) effects. Specifically, for hydrogen (H_2) flames, $Le \approx 1$ condition was achieved by using stoichiometric $H_2/O_2/N_2$ mixtures of unity equivalence ratio ($\phi = 1$). For hydrocarbon flames, a lean ($\phi = 0.8$) $C_3H_8/O_2/Ar$ mixture was selected as it is diffusionally neutral ($Le \approx 1$) because of the similarity in the molecular weights between the deficient species (C_3H_8) controlling mass diffusion and the abundant species (Ar) controlling thermal diffusion. Pressure and dilution ratio,$X_{inert}/(X_{inert} + X_{O_2})$, where X is mole fraction, were varied in order to modulate the planar laminar flame speed and flame thickness of the mixture.

Next, to identify the role of DT cells, we selected three $Le < 1$, $H_2/O_2/He/Ar$ mixtures at an equivalence ratio $\phi = 0.4$. The Lewis number was modulated by changing $(X_{Ar}/(X_{Ar} + X_{He}))$, where X_i is the mole fraction of species i. Since we are working with strongly fuel-lean mixtures, $Le \approx Le_{Fuel}$ can be assumed. For this part of the study, most of the experiments were conducted at

$1atm$, in order to suppress the DL instability.

3. Results and Discussion

3.1 Analysis for Interacting Hydrodynamic Instability and Turbulence

Recognizing that both cellular instability and turbulence involve a wide spectrum of length and time scales, we compare the inverse of the time scales associated with the common lengths scales. For DL instability, the inverse time scale is the growth rate of the instability and can be expressed as $\omega_{DL} = 1/\tau_{DL}(k)$, where τ_{DL} is the characteristic time at wave number k, the reciprocal of the corresponding length, i.e. $k = 1/l_k$. Similarly for turbulence we consider the simplest form of the interaction between turbulence and a planar flame, namely wrinkling of the flame by a turbulent eddy of wave number k. Since the only characteristic time here is the eddy turnover time, $\tau_T(k)$ and it is reasonable to assume that the flame surface growth rate due to this eddy is also on the order of the eddy turnover rate, $\omega_T(k)$, which is the reciprocal of the characteristic time for the wave number k, i.e. $\omega_T(k) = 1/\tau_T(k)$. Consequently, we aim to compare $\omega_{DL}(k)$ and $\omega_T(k)$ at the same wave number to assess the relative significance of DL instability and turbulence in increasing the flame surface area and consequently the propagation speed. The leading order of the growth rate for the DL instability at a wave number k can be expressed as [13], $\omega_{DL}(k) = S_L k X(\sigma)$, where S_L is the laminar flame speed in general, which more specifically can be represented by the unstretched, adiabatic, unburned, laminar flame speed, $S_{L0,u}$, and $X(\sigma)$ is the nondimensional growth rate depending on the thermal expansion, σ, and $X(\sigma) \approx 1$. So we can write, $\omega_{DL}(k) = S_{L0,u}k$.

Assuming Kolmogorov turbulence, we can write for the eddy turnover rate as, $\omega_T(k) = 1/\tau_T(k) = u'_k/l_k = u'_k k$, where u'_k is the eddy velocity at wave number k. Comparison of these two growth rates leads to

$$\frac{\omega_{DL}(k)}{\omega_T(k)} = \frac{S_{L0,u}}{u'_k} \tag{1}$$

It is noted that for Kolmogorov turbulence, the spectrum of k is bounded by the inverse of the Kolmogorov scale (λ_η) and the integral scale (λ_I). Consequently, u'_k is bounded by u'_η and U_{rms}, namely the eddy velocities of Kolmogorov and integral scales respectively, with the former being always weaker than the latter: $u'_\eta < U_{rms}$. Based on Eq. 1, three possible regimes can therefore be envisioned in terms of the relative values of $S_{L0,u}$, u'_η, and U_{rms}, which will now be discussed and further supported by experimental observations.

3.1.1 Regime I: Instability Dominated

In this regime, we have $S_{L0,u} > U_{rms}$, where U_{rms} is the turbulent fluctuating velocity at the largest length scale or smallest wave number. This regime of $U_{rms}/S_{L0,u} < 1$ coincides with the wrinkled flamelet regime in the turbulent flame regime diagram. Since $U_{rms} > u'_\eta$, we can expect the DL instability to be stronger than turbulence and hence this regime is termed as "*Instability Dominated Regime*". In Fig. 1(a), the normalized turbulent flame speeds, $\overline{S_T} = S_T/S_{L0,b}$ are plotted versus $Re_{T,f}$. The slope of the data shows the dependence of $\overline{S_T} = Re_{T,f}^{0.27}$ (black line) which is clearly weaker than the scaling $\overline{S_T} = Re_{T,f}^{0.5}$ (red line) derived for cellularly stable flames [5].

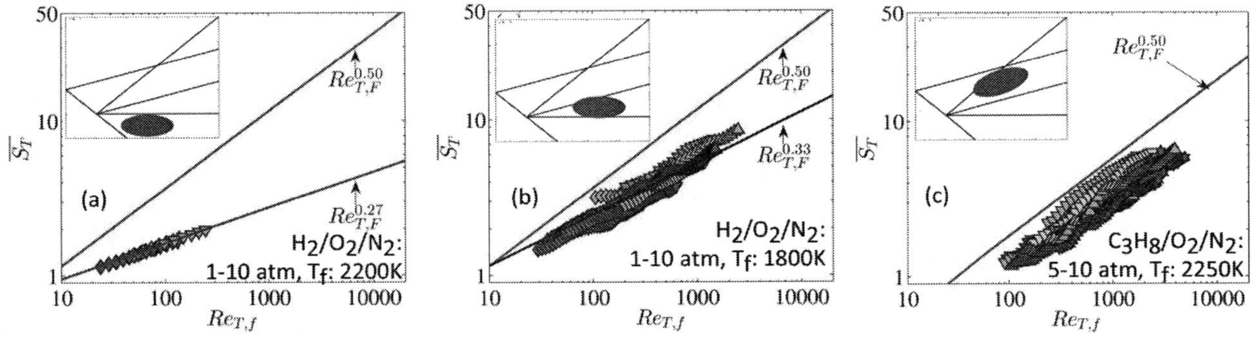

Figure 1: Experimental results: Normalized turbulent flame speeds, $\overline{S_T} = S_T/S_{L0,b}$ vs. $Re_{T,f}$ (=$(U_{eff}R_f)/(S_{L0}\delta_{L0})$) at different pressures and turbulence intensities (fan speeds in rpm). Mixture Lewis number ≈ 1. Inset: Experimental conditions on Borghi diagram (*a*): Regime I: Instability Dominated; (*b*): Regime II: Instability-Turbulence Interaction; (*c*): Regime III: Turbulence Dominated

3.1.2 Regime II: Instability-Turbulence Interaction

Next we focus on the intermediate regime with $u'_\eta < S_{L0,u} < U_{rms}$, where DL instability dominates at higher k with smaller u'_k, while turbulence dominates at lower k with larger u'_k; hence this regime is designated as the "*Instability-Turbulence Interaction Regime*". By definition, the upper and lower boundaries of this regime is defined by the two lines, $u'_\eta = S_{L0,u}$ and $S_{L0,u} = U_{rms}$, which are the upper, $Ka = (u'_\eta/S_{L0,u})^2 = 1$ limit, and the lower, $U_{rms}/S_{L0,u} = 1$, limit of the corrugated flamelet regime. The normalized turbulent flame speeds obtained for three different pressures and various fan speeds are shown in Fig. 1b, which clearly demonstrates a dependence of $\overline{S_T} = Re_{T,f}^{0.33}$, being weaker than $\overline{S_T} = Re_{T,f}^{0.5}$ for flames without cellular instabilities, but stronger than the exponent of $\overline{S_T} = Re_{T,f}^{0.27}$ observed in the "*Instability-Dominated Regime*".

3.1.3 Regime III: Turbulence Dominated

The third regime is characterized by $S_{L0,u} < u'_\eta$. Since $u'_\eta \leq u'_k$ for any wavenumber (k) in the turbulence spectrum, we expect turbulence to dominate in this regime. Since this regime satisfies the condition: $Ka = (u'_\eta/S_{L0,u})^2 > 1$, it coincides with the thickened flamelet and broken/distributed regimes of the turbulent flame regime diagram. Consequently, turbulence dominates in this regime, which is therefore referred to as "*Turbulence Dominated Regime*". The normalized turbulent flame speeds across all pressures and fan speeds (1b) in this regime show the $\overline{S_T} \approx Re_{T,f}^{1/2}$ dependence, similar to what is suggested for cellularly stable flames and as such confirming the minimal influence of DL instability on flame propagation.

3.2 Effect of DT instability

To evaluate the role of DT instability we also performed experiments with mixtures with $Le < 1$ using $H_2/O_2/Ar/He$ mixtures. Figure 2(a), (b) and (c) show the normalized turbulent flame speeds for three different Le, at *Instability Dominated*, *Instability-Turbulence Interaction* and *Turbulence*

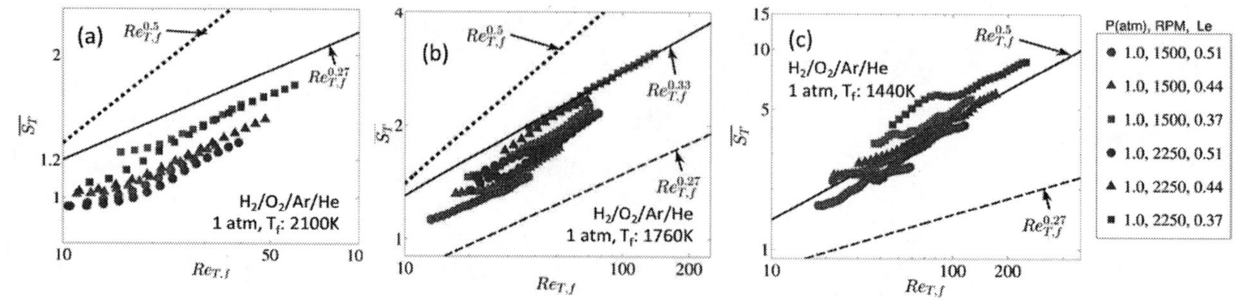

Figure 2: Experimental results: Normalized turbulent flame speeds, $\overline{S_T} = S_T/S_{L0,b}$ vs. $Re_{T,f}$ $(=(U_{eff}R_f)/(S_{L0}\delta_{L0}))$ at different Lewis number and turbulence intensities (fan speeds in rpm). (a): Regime I: Instability Dominated; (b): Regime II: Instability-Turbulence Interaction; (c): Regime III: Turbulence Dominated

Dominated regimes respectively. The scaling exponent (γ) of turbulent flames in all three regimes were found to be independent of Le and thus, DT instability, while the normalized flame speed, and thus the pre-factor B in $\overline{S_T} = B.Re^{\gamma}_{T,f}$, strongly increase with decreasing Le, implying the strong role of DT instability in the wrinkled flamelet and corrugated flamelet regimes, and that of weakened kinematic restoration in the thickened flamelet regime.

4. Conclusions

In summary, we have demonstrated the essential role of the cellular instability on the propagation of expanding turbulent flames with conditions representing various regimes of the turbulent flames. The leading-order scaling analysis of the growth rates over the wavenumber spectra suggests three distinct regimes based on the relative role of hydrodynamic instability cells and turbulence on flame propagation, which have been further validated experimentally with a wide range of conditions of diffusionally neutral ($Le \approx 1$) flames. Furthermore, we extended the experiments for $Le < 1$ mixtures to show that the diffusion-thermal instability does not alter the flame speed scaling with Reynolds number at a given regime, however, it enhances the total burning rate.

5. Acknowledgments

The authors acknowledge financial support from the US National Science Foundation (CBET, Grant No. 1510142).

References

[1] N. Peters, Turbulent combustion, Cambridge university press, 2000.

[2] S. B. Pope, PDF methods for turbulent reactive flows, Progress in Energy and Combustion Science 11 (1985) 119–192.

[3] R. G. Abdel-Gayed, D. Bradley, and M. Lawes, Turbulent burning velocities: a general correlation in terms of straining rates, Proceedings of the Royal Society A 414 (1987) 389–413.

[4] H. Kobayashi, K. Seyama, H. Hagiwara, and Y. Ogami, Burning velocity correlation of methane/air turbulent premixed flames at high pressure and high temperature, Proceedings of the Combustion Institute 30 (2005) 827–834.

[5] S. Chaudhuri, F. Wu, D. Zhu, and C. K. Law, Flame speed and self-similar propagation of expanding turbulent premixed flames, Physical Review Letters 108 (2012) 044503.

[6] J. B. Bell, M. S. Day, I. G. Shepherd, M. R. Johnson, R. K. Cheng, J. F. Grcar, V. E. Beckner, and M. J. Lijewski, Numerical simulation of a laboratory-scale turbulent V-flame, Proceedings of the National Academy of Sciences of the United States of America 102 (2005) 10006–10011.

[7] J. F. Driscoll, Turbulent premixed combustion: Flamelet structure and its effect on turbulent burning velocities, Progress in Energy and Combustion Science 34 (2008) 91–134.

[8] A. N. Lipatnikov and J. Chomiak, Molecular transport effects on turbulent flame propagation and structure, Progress in energy and combustion science 31 (2005) 1–73.

[9] S. Chaudhuri, F. Wu, and C. K. Law, Scaling of turbulent flame speed for expanding flames with Markstein diffusion considerations, Physical Review E 88 (2013) 033005.

[10] F. Wu, A. Saha, S. Chaudhuri, and C. K. Law, Propagation speeds of expanding turbulent flames of C 4 to C 8 n-alkanes at elevated pressures: Experimental determination, fuel similarity, and stretch-affected local extinction, Proceedings of the Combustion Institute 35 (2015) 1501–1508.

[11] S. Yang, A. Saha, F. Wu, and C. K. Law, Morphology and self-acceleration of expanding laminar flames with flame-front cellular instabilities, Combustion and Flame 171 (2016) 112–118.

[12] S. Chaudhuri, A. Saha, and C. K. Law, On flame–turbulence interaction in constant-pressure expanding flames, Proceedings of the Combustion Institute 35 (2015) 1331–1339.

[13] M. Matalon, Intrinsic flame instabilities in premixed and nonpremixed combustion, Annu. Rev. Fluid Mech. 39 (2007) 163–191.

Spring Technical Meeting
Eastern States Section of the Combustion Institute
March 4–7, 2018
State College, Pennsylvania

Numerical Simulations of Oxy-fuel Premixed Combustion in Supercritical CO_2-diluted Environment

Abdulafeez Adebiyi[1], *V'yacheslav Akkerman*[1,*], *and Konstantin Kemenov*[2]

[1]*Computational Fluid Dynamics and Applied Multi-Physics Center*
Center for Innovation in Gas Research and Utilization (CIGRU)
Center for Alternative Fuels, Engines and Emissions (CAFEE)
Department of Mechanical and Aerospace Engineering
West Virginia University, Morgantown, WV 26506, USA
[2]*Symplectic Research, Inc., Atlanta, GA 30366, USA*
[*]*Corresponding author:* `vyacheslav.akkerman@mail.wvu.edu`

Abstract: The design and development of advanced supercritical carbon dioxide (sCO_2) combustors for the next-generation gas turbines require deeper understanding of the flame dynamics and morphology at highly CO_2-diluted supercritical conditions. For this purpose, in the present work, we scrutinize the key parameters of supercritical oxy-methane premixed combustion, such as the velocity, thickness, and internal structure of the flame front. The analysis is performed by means of numerical simulations of the reacting flow equations with fully-compressible hydrodynamics, transport properties (heat conduction, diffusion and viscosity), and one-step Arrhenius chemical kinetics. The real-gas thermodynamics and molecular properties are incorporated into a reacting-flow Navier-Stokes solver along with the Peng-Robinson (PR) equation of state (EoS). It is shown that an increase in the CO_2-dilution rate makes the flame front thicker and reduces the unstretched laminar flame speed, as compared to non-diluted oxy-combustion. Starting with planar flames (attained by means of the slip boundary conditions at the computational domain), the analysis is then extended to corrugated ones (attained by non-slip walls of the computational domain). While the corrugated flame generally spreads faster than the planar one of similar thermal-chemical characteristics, the corrugated flame velocity not only reduces with a dilution rate, but also depends on a channel width.
***Keywords:* supercritical CO_2 combustion, laminar premixed flames, Peng-Robinson EoS.**

1. Introduction

The future generation of the fossil-fuel power systems is expected to be highly efficient to operate and exhibit extremely low (or zero) emission levels to address environmental concerns and comply with stricter government regulations. A promising technology to achieve such goals is to utilize supercritical CO_2 (sCO_2) as the working fluid in thermal power cycles. For example, the thermal cycle efficiency in the sCO_2-based Brayton power cycle can attain a value of 50%, and above [1], while the higher density of the working sCO_2 fluid can potentially decrease the overall size of turbomachinery in comparison with a conventional steam (Rankine) power cycle. Moreover, the sCO_2-based power cycles are suitable for implementation in zero-emission, fossil-fuel plants based on the oxy-combustion process which can use either pulverized coal or natural and synthesis (syngas) gases. In addition, the direct-fired sCO_2 cycle, where sCO_2 fluid resides in direct contact with combustion sources, offers some additional benefits including the facilitation of CO_2 capture and sequestration, higher cycle thermal efficiency, and potentially simpler designs. The design and development of the future sCO_2 combustors for direct-fired cycles requires deeper understanding of flame propagation under highly CO_2-diluted conditions in supercritical environment, where typical pressure and temperature are around 300 atm and 1100 K, respectively. Such a combination of conditions is unique, poses challenges for researchers, and remains largely unexplored.

While studies of the CO_2-dilution impacts on premixed flames at supercritical conditions are virtually non-existent, the topic has been extensively studied at the atmospheric conditions (or at the typical gas turbine pressure levels) experimentally for various fuel mixtures based on CH_4 [2–5], or H_2/CO (syngas) [6–9]. The main driving force behind this research is the attractiveness of oxy-combustion technology to revamp the existing air-powered industrial burners. In oxy-combustion, air is substituted with pure oxygen diluted with recycled flue gas consisting primarily of CO_2 and H_2O to reduce the flame temperature.

2. Governing Equations

The present direct numerical simulation (DNS) studies were conducted with an in-house research code, which solves a 2D system of the governing equations for a compressible reactive mixture. The code was proven to be robust and effective in numerous applications associated with aero-acoustics and combustion and was modified to include the real-gas thermodynamics, molecular properties, as well as the Peng-Robibson (PR) equation of state (EoS). In the general two-dimensional (2D) form, with respect to (r,z) coordinates, the governing equations include the continuity, momentum, and total energy equations. In addition, the reactive progress variable equation was employed and written as:

$$\frac{\partial(\rho Y_F)}{\partial t}+\frac{1}{r^\gamma}\frac{\partial}{\partial r}\left[r^\gamma\left(\rho u_i Y_F-\frac{\mu}{\mathrm{Sc}}\frac{\partial Y_F}{\partial r}\right)\right]+\frac{\partial}{\partial z}\left[\rho u_z Y_F-\frac{\mu}{\mathrm{Sc}}\frac{\partial Y_F}{\partial z}\right]=-\frac{\rho Y_F}{\tau_R}\exp\left(-E_a/R_u T\right), \quad (1)$$

where $\gamma=0$ for the 2D-planar Cartesian geometry and $\gamma=1$ for the cylindrical-axisymmetric geometry, respectively. A global one-step reaction mechanism representing methane CH_4 oxidation is assumed,

$$CH_4+2O_2+nCO_2\rightarrow 2H_2O+(1+n)CO_2, \quad (2)$$

with an Arrhenius-like reaction rate as shown in Eq. (1). Dilution is defined based on the molar fraction of CO_2 in the oxidizer stream as $\beta=n/(n+n_{O_2})=n/(n+2)$, where n and n_{O_2} denote the number of moles of diluted CO_2 and O_2, respectively. The reactive progress variable can be defined based on the mass fractions of the reactants and diluent as $Y=Y_F+Y_{CO_2,d}$, where $Y_F=Y_{CH_4}+Y_{O_2}$, and $Y_{CO_2,d}$ is the mass fraction of the diluted CO_2 in the reactants. Consequently, the reactive progress variable Y monotonically changes between the unity value in the fresh reactants and some value $Y^*=Y_{CO_2,d}>=0$ in the fully burnt mixture, respectively, where $Y^*=0$ corresponds to the pure undiluted case with $\beta=0$. It is further remarked that in the progress variable transport equation, Eq. (1), the reaction rate explicitly depends on Y_F only. In this manner, the diluted portion of CO_2 in the unburnt mixture, $Y_{CO_2,d}$, will play the role of inert gas such that it does not participate in the chemical reaction but affects all other thermal and physical processes. In this work, the PR EoS [10] is adopted to characterize the real gas effects at the supercritical conditions:

$$P(T,v_m)=\frac{R_u T}{v_m-b}-\frac{a(T)}{v_m(v_m+b)+b(v_m-b)}, \quad (3)$$

where v_m is the molar volume, R_u is the universal gas constant, and the cumulative quantities a and b for a mixture are given by the quadratic and linear mixing rules:

$$a(T) = \sum_{i=1}^{N}\sum_{j=1}^{N} X_i X_j (a_i a_j (1-k_{ij}))^{1/2}, \quad b = \sum_{i=1}^{N} X_i b_i. \tag{4}$$

Here, X_i is the mole fraction of species i, N is the total number of species, and k_{ij} characterize the binary interactions between species i and j:

$$a_i = 0.45724 \frac{R_u^2 T_{c,i}^2}{P_{c,i}} \left[1 + \left(0.37464 + 1.5422\omega_i - 0.26992\omega_i^2\right)\left(1 - T_{r,i}^{1/2}\right)\right]^2, \tag{5}$$

$$b_i = 0.0778 \frac{R_u T_{c,i}}{P_{c,i}}, \tag{6}$$

where for species i the respective critical pressure and temperature are $P_{c,i}$ and $T_{c,i}$, with the reduced temperature $T_{r,i} = T/T_{c,i}$, and the Pitzer's acentric factor ω_i. In particular, the following acentric factors are employed in this work: $\omega_{CH_4} = 0.01142$, $\omega_{O_2} = 0.0222$, $\omega_{CO_2} = 0.344$, $\omega_{H_2O} = 0.22394$. As for the binary interaction parameters, they are taken to be $k_{ij} = 0$ for $i = j$, while otherwise, $k_{CH_4,O_2} = 0.03$, $k_{CH_4,CO_2} = 0.0919$, $k_{CH_4,H_2O} = 0.5$, $k_{CO_2,O_2} = 0.0015$, $k_{H_2O,O_2} = 0.0046$, and $k_{H_2O,CO_2} = 0.1896$. The mole fractions for each species are calculated from the progress variable Y_F as $X_{CH_4} = Y_F/3$, $X_{O_2} = 2Y_F/3$, $X_{H_2O} = 2(1-Y_F)/3$, and $X_{CO_2} = (1+n-Y_F)/3$, with Y_F changing in the range $0 \leq Y_F \leq 3/(3+n)$. Obviously, in the case of no dilution ($n = \beta = 0$) we have $0 \leq Y_F \leq 1$ and $X_{CO_2} = (1-Y_F)/3$.

Molecular viscosity of the mixture is calculated employing the dense-gas Lucas model [11], which has often been used in the supercritical combustion DNS studies. The other transport properties such as thermal conduction and molecular diffusivity are computed based on the Prandtl, Pr, and Schmidt, Sc, numbers which are taken to be $Pr = Sc = 1$ such that the Lewis number is $Le = Sc/Pr = 1$ to avoid the diffusional-thermal instability. The specific heats are implemented as:

$$C_V = C_{V,id}(T) + C_{V,real}(T, v_m), \quad C_P = C_{P,id}(T) + C_{P,real}(T, v_m), \tag{7}$$

where the ideal gas components are computed using the NASA polynomial approximations [10], while the real gas correction is given by:

$$C_{V,real}(T, v_m) = \frac{Ta''}{b\sqrt{8}} \ln\left(\frac{v_m + b(1+\sqrt{2})}{v_m + b(1-\sqrt{2})}\right), \tag{8}$$

and $C_{P,real}$ is computed from the PR EoS as:

$$C_{P,real}(T, v_m) = C_{V,real}(T, v_m) - T\frac{(\partial P/\partial T)_{v_m}^2}{(\partial P/\partial v_m)_T}. \tag{9}$$

3. Computational Configuration

The governing equations are integrated employing a cell-centered, finite-volume computational scheme in the plane Cartesian geometry ($\gamma = 0$). The scheme is of the 2nd-order accuracy in time, and it features the 4th-order accurate discretization for the convective terms and the 2nd-order accurate discretization for the diffusive terms, respectively. Specifically, we deal with a semi-open 2D channel such that one (left) end of the channel is closed, and an initially planar flame front propagate from the closed end to the open (right) one. The non-reflecting boundary conditions are implemented at the open end. As for the upper and lower boundaries ("walls"), they are adiabatic ($\mathbf{n}\cdot\nabla T = 0$), and either slip ($\mathbf{n}\cdot\mathbf{u} = 0$), to keep the flame shape planar, or non-slip ($\mathbf{u} = 0$), which distorts the flame front. Here $\mathbf{n}$ is the unit normal vector at the walls. The channels of widths $D = 5$ μm and 10 μm are considered.

A structured rectangular mesh was employed with the cell sides being parallel to the radial and axial directions. The mesh is adaptive, to ensure that the flame is always kept within a fine grid zone, such that the axial size of the computational domain changes dynamically and follows the leading pressure wave. The 3rd-order splines are employed to re-interpolate the variables during a periodic grid reconstruction to preserve the 2nd-order accuracy of the numerical scheme. The grid size is chosen to ensure that the internal flame structure is resolved, with several grid points embedded in the flame front. In the majority of the present high-pressure simulations, the finest grid size was as small as 0.1 μm. The appropriateness of such a resolution is justified by the resolution test discussed below.

4. Results and Discussion

4.1 Planar Flames

In this subsection, we employ the slip boundary conditions at the walls to ensure planar flame propagation. To validate the computational platform, we first considered the conventional conditions (the initial temperature of the fuel mixture is $T_f = 300$ K and pressure is $P = 1$ atm), with the ideal-gas (ID) EoS, in order to reproduce the experimental measurements of the unstretched laminar flame speed for the stoichiometric oxy-methane mixture [12], which was found to be about $S_L = 2.9$ m/s. This experimental result was reproduced by the lowest curve in Fig. 1a. The computational thermal-chemical parameters chosen to obtain this particular value of S_L were chosen as follows: the activation energy $E_a = 126.885 kJ/kg$, and the energy release from the reaction $Q = 4896.577 kJ/kg$.

We next extended our analysis to the elevated initial temperature of $T_f = 800$ K, but keeping the atmospheric pressure, and compare the ID and PR EoS models. The results are shown by the two upper curves in Fig. 1a. We observe a substantial increase of the laminar flame speed as compared to the room temperature conditions with values of S_L being about 10 m/s, approximately. It is also seen that in both PR and ID cases, the laminar flame speed values are well correlated with the S_L for the PR case being little smaller in magnitude. Obviously, such a correlation is expected as long as the pressure is strongly subcritical.

However, the situation conceptually changes when we extend the investigation to the supercritical conditions. This consideration starts by Fig. 1b, which provides the resolution test for stoichiometric oxy-methane combustion (with no dilution) at $P = 300$ atm and $T_f = 800$ K (this is considered to be the "master-case" for the present work). Specifically, the time evolution of S_L is shown in Fig. 1b, for three various mesh sizes, namely: $\Delta x = 0.05$ μm, 0.1 μm and 0.2 μm. It is seen that all three meshes yield essentially the same value of S_L, about 1.11 m/s, approximately, thereby justifying that the resolution of 0.1 μm was be reasonable to be employed in the

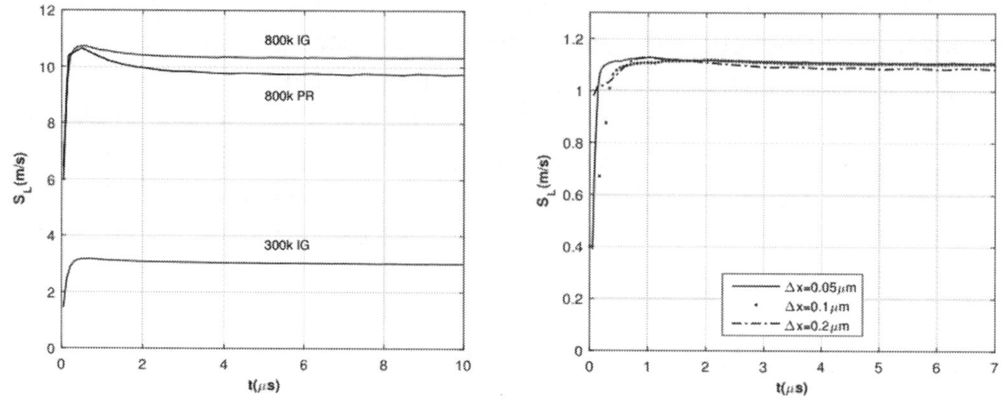

Figure 1: Time evolution of the unstretched laminar flame speed S_L: (a), (left): Comparison at the atmospheric pressure for the Peng-Rrobinson (PR, black) and ideal-gas (ID, red) EoS models; (b), (right): A resolution test: time evolution of the unstretched laminar flame speed S_L for $P = 300$ atm, $T_f = 800$ K and various mesh sizes: $\Delta x = 0.05$ μm (blue, solid), 0.1 μm (red, dotted) and 0.2 μm (black, dashed).

rest of simulations, for given $P = 300$ atm and $T_f = 800$ K. In fact, this grid resolution is in good agreement with that provided by the characteristic flame thickness δ_f. Indeed, the latter quantity is traditionally defined as

$$\delta_f = \mu_f / \mathrm{Pr} \rho_f S_L, \tag{10}$$

where μ_f and ρ_f are the dynamic viscosity and density in the unburnt matter. In our case, this formula yields $\delta_f \approx 0.3$ μm such that we have about three grid cells per δ_f with the grid size $\Delta x = 0.1$ μm, which is reasonable. It should be also noted, in this respect, that the quantity (10) is only a parameter of the length dimension (a characteristic transport length scale), while the real size of the burning zone may substantially exceed δ_f. For this reason, in the present study we used an alternative estimation for the flame thickness defined as:

$$L_F = (T_b - T_f)/|\nabla T|_{max}, \tag{11}$$

which better reproduces the width of the temperature gradient field. We next investigated the impact of the CO_2-dilution on the supercritical laminar flame propagation. For this purpose, we performed the simulation runs for the dilution rate in the range of $0 \leq \beta \leq 0.8$. The results are presented in Figs. 2-5. In particular, Fig. 2 exhibits the effect of dilution on the laminar flame speed. Here, the time evolution of S_L is shown in Fig. 2a for various dilution rates, $\beta = 0\%, 20\%, 40\%, 60\%$ and 80%, while Fig. 2b presents S_L versus β. It is seen that the dilution diminishes the laminar flame velocity significantly, namely, S_L reduces almost twice at $\beta = 40\%$, as compared to that at $\beta = 0$, and it drops to a small value, $S_L \approx 0.06$ m/s at $\beta = 80\%$. This effect is due to the fact that while an inert diluent does not participate in the combustion reaction, it nevertheless participates thermodynamically by absorbing the energy released from the reaction. It is also of interest to investigate the internal flame structure for various dilution rates. This question is addressed in Figs. 3 and 4. Specifically, Fig. 3 presents the instantaneous profiles of the progress variable Y, while the respective instantaneous temperature profiles are shown in Fig. 4a.

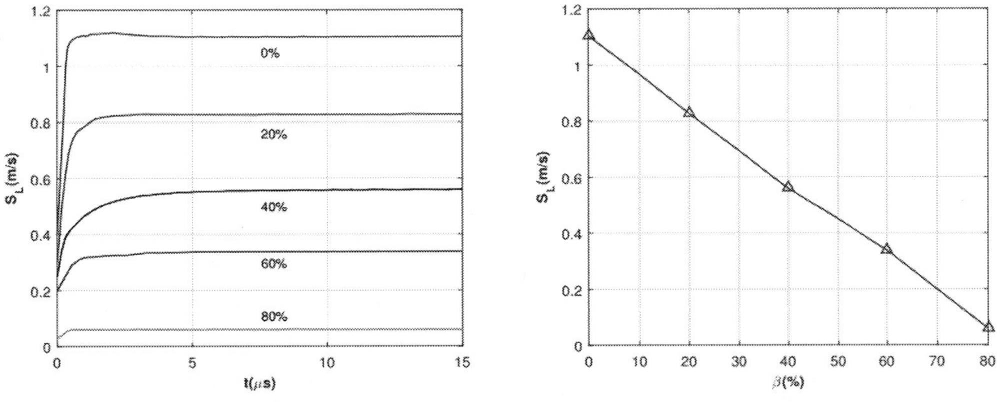

Figure 2: (a), (left): Time evolution of the unstretched laminar flame speed S_L for various dilution rates $\beta = 0\%, 20\%, 40\%, 60\%$ and 80%; and (b), (right): S_L vs β. $P = 300$ atm and $T_f = 800$ K in both cases.

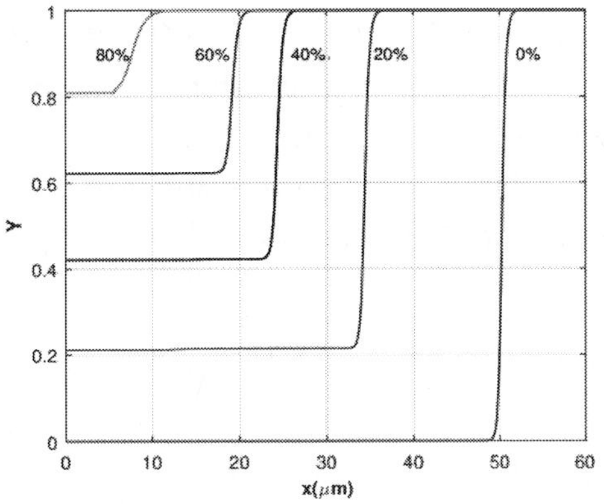

Figure 3: Instantaneous profiles of the progress variable Y for $P = 300$ atm, $T_f = 800$ K and various dilution rates $\beta = 0\%, 20\%, 40\%, 60\%$ and 80%.

In addition, based on the results of Fig. 4b, Fig. 4b, presents the adiabatic flame temperature T_{ad} versus the dilution rate β. The plots and outcomes of Figs. 4a and 4b, generally resemble that of Figs. 2(a,b): the presence of the inert diluent consumes the energy form the combustion reaction, and thereby diminishes the adiabatic flame temperature significantly, from $T_{ad} \approx 3000$ K at $\beta = 0$ till $T_{ad} \approx 1600$ K at $\beta = 80\%$. The dependence of T_{ad} on β is near-linear for $\beta \leq 60\%$, but it substantially drops thereafter.

One more result observed in Figs. 3 and 4 is that an increase in β makes the Y- and T-profiles smoother by reducing the gradients of temperature and progress variable fields which leads to thickening of the flame front. To address this point in more details, Fig. 5 shows the flame thickness L_f, defined based on the temperature gradient, Eq. (11), versus the dilution rate β. It is seen that the flame front gets thicker as β increases. Again, the dependence is practically linear up to $\beta \leq 60\%$, but it drastically rises thereafter. To be specific, being $L_f \approx 0.85$ μm in the case of no dilution, the

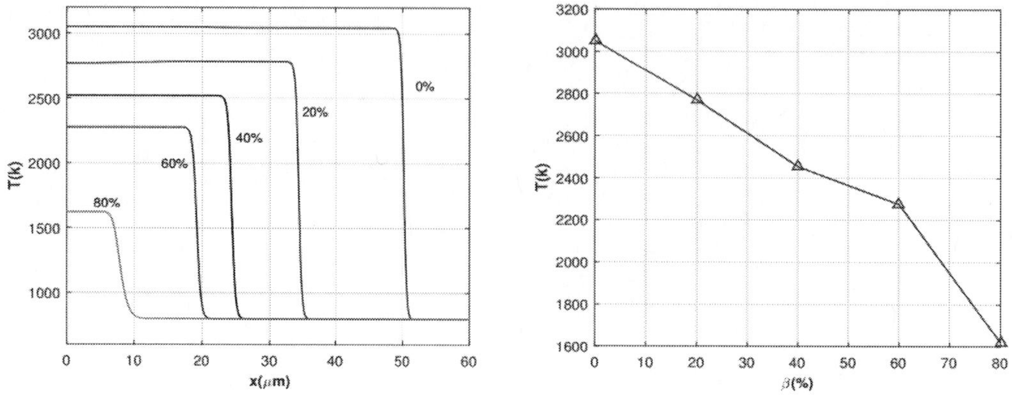

Figure 4: (a), (left): Instantaneous temperature profiles inside a flame front for various dilution rates $\beta = 0\%, 20\%, 40\%, 60\%$ and 80%; and (b), (right): the adiabatic flame temperature T_{ad} vs β. $P = 300$ atm and $T_f = 800$ K in both cases.

flame thickness attains the value of $L_f \approx 1.2$ μm at $\beta = 60\%$ and even $L_f \approx 2.2$ μm at $\beta = 80\%$. It is noted that all these values of L_f significantly exceed the value of $\delta_f \approx 0.3$ μm given by Eq. (10). This provides another justification for the grid resolution $\Delta x = 0.1$ μm employed in the present work. Indeed, we have from 8 (in the case of no dilution) till 22 (for $\beta = 80\%$) grid cells inside the burning zone, which is enough to resolve the flame front.

4.2 Corrugated Flames

In the previous subsection, we have demonstrated that the CO_2-dilution diminishes the velocity and temperature of a supercritical oxy-methane flame, and it makes burning zone thicker. In other words, the dilution mitigates oxy-methane combustion, with a risk of extinction. How to prevent this effect? The primary idea is to facilitate combustion and promote the flame propagation velocity by means of distortion of the flame shape. Indeed, it is well known that a corrugated flame has a larger surface area relative to a planar one involving the same mixture and under the same conditions. Thus, it is expected to consume more fuel per unit time and release more heat, thereby propagating more rapidly. In particular, flames in tubes are known to accelerate due to wall friction. However, the effect depends on a tube/channel width, the flame thickness and their ratio. On the one hand, a flame front is known to accelerate faster in a narrower channel than in a wider one. On the other hand, a finite (and noticeable) flame thickness may work against this trend because the wider the flame front is, the harder to corrugate it. The latter, however, will depend on the dilution rate which thickens the flame front. Consequently, a nontrivial interplay between the dilution rate and the channel width is expected, and this constitutes the focus of the present subsection.

The slip boundary conditions of the previous subsection are replaced now with the non-slip ones, in order to generate a corrugated flame front and scrutinize its dynamics and morphology. Two channels, of widths $D = 5\mu$m and 10μm are considered, with various dilution rates, $\beta = 0\%$, 20%, 40%, 60% and 80%, in both cases. The initial conditions are kept to be $P = 300$ atm and $T_f = 800$ K. We start the analysis of corrugated flames with the temperature snapshots of Fig. 6 which shows the evolution of a non-diluted flame ($\beta = 0$) in a channel of width $D = 5$ μm).

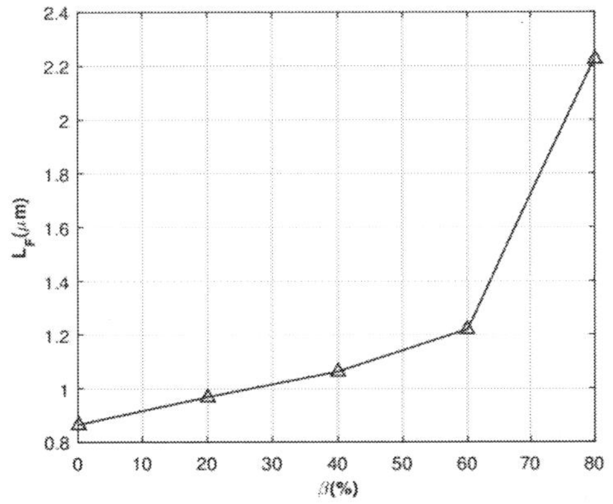

Figure 5: The flame thickness evaluated by Eq. (11) vs the dilution ratio β for $P = 300$ atm and $T_f = 800$ K.

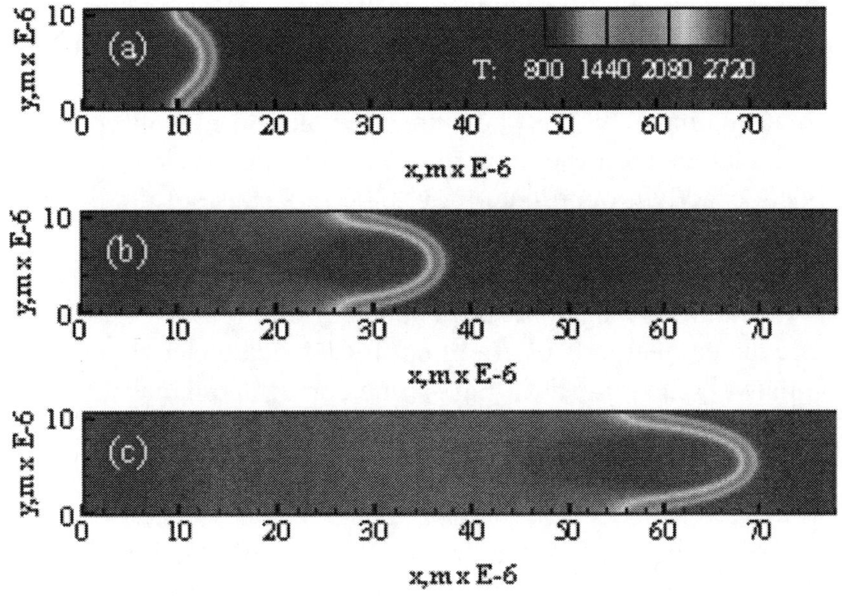

Figure 6: Consecutive color temperature snapshots describing a corrugated flame evolution in a channel of width $D = 5\mu m$ in the case of no dilution ($\beta = 0$) with $P = 300$ atm and $T_f = 800$ K.

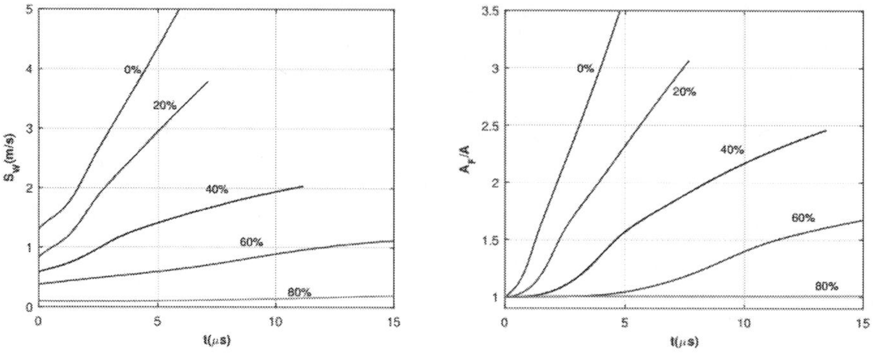

Figure 7: Time evolution of the instantaneous corrugated flame velocity S_w [(a), (left)] and scaled flame surface area (length) A_F/D [(b), (right)] in a channel of width $D = 5\mu m$ for various dilution rates $\beta = 0\%, 20\%, 40\%, 60\%$ and 80%, with $P = 300$ atm and $T_f = 800$ K in all cases.

Obviously, the flame surface area (and thereby propagation velocity) are promoted in the course of flame acceleration.

We next quantify this effect by means of the plots in Figs. 7 - 13. In particular, Fig. 7a compares the time evolution of the corrugated flame velocities S_w in a channel of width $D = 5\mu$m for various dilution rates in the range $0 \leq \beta \leq 0.8$. For simplicity, S_w is evaluated here by means of the flame tip velocity, which is a reasonable approximation accounting for a moderate and self-similar manner of flame propagation/acceleration. Figure 7a shows that for all dilution rates the S_w values eventually exceed the corresponding S_L values (for the same β) several times. On the other hand, the larger values of β are, the lower S_w, and the slower the acceleration trends are. Figure 7b shows the time evolution of the scaled flame surface area (taken to be the length of the flame front in 2D), A_f/D. It is seen that qualitatively, the flame speed evolution is well correlated with the flame length evolution, such that the corresponding plots in both figures show similar trends(at fixed β). Nevertheless, S_w is not exactly proportional to A_f/D: (i) because of the impact of the final flame thickness, and (ii) because the flame tip velocity still slightly differs from the total burning rate. Overall, both Fig. 7(a,b) supports the observation that the corrugated flames accelerate, but the dilution diminishes the flame propagation speed and acceleration.

Figures 8(a,b) are the counterparts of 7(a,b) but for a twice wider channel, $D = 10\mu m$. All the trends here are qualitatively the same as in Fig. 7: the corrugated flame accelerates, but the higher values of the dilution mitigates flame propagation, and moderates its acceleration. At the same time, it follows that an increase in the channel width also weakens the flame acceleration. Indeed, all the quantities and trends of Fig. 8 are a little lower than that of Fig. 7.

To scrutinize this effect further, we also compared the corrugated flame speed S_w and scaled flame length A_f/D time evolution in two channels with $D = 5$ μm and 10 μm, and with different dilution rate β. This is shown in Figs. 9 - 13 for $\beta = 0\%$, 20%, 40%, 60% and 80%. In all cases, the solid lines represent the narrower channel, $D = 5$ μm, while the wider channel, $D = 10$ μm is shown by the dashed lines. In line with our expectations, it is seen that an increase in the channel width moderates flame acceleration in all the cases considered. Nevertheless, the effect is minor in the case of high dilution such as $\beta = 80\%$.

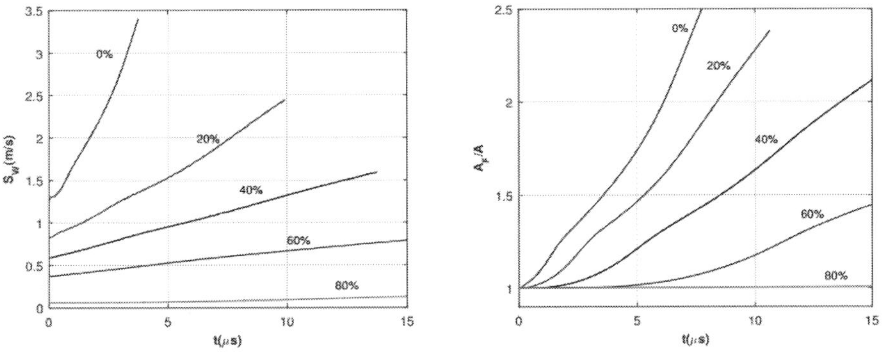

Figure 8: Time evolution of the instantaneous corrugated flame velocity S_w [(a), (left)] and scaled flame surface area (length) A_F/D [(b), (right)] in a channel of width $D = 10\mu m$ for various dilution rates $\beta = 0\%, 20\%, 40\%, 60\%$ and 80%, with $P = 300$ atm and $T_f = 800$ K in all cases.

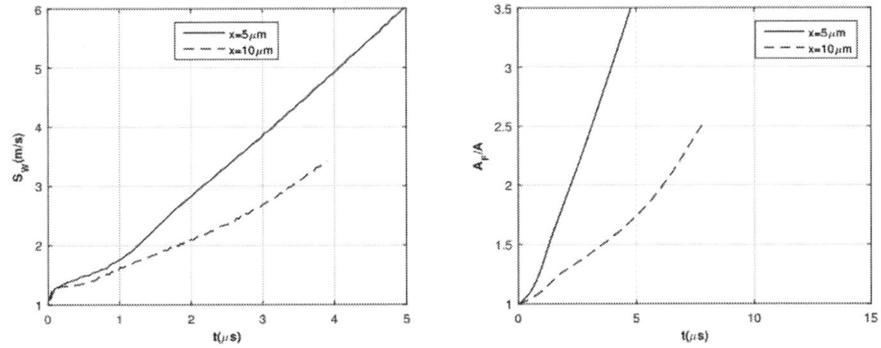

Figure 9: Time evolution of the instantaneous corrugated flame velocity S_w [(a), (left)] and scaled flame surface area (length) A_F/D [(b), (right)] in the channels of width $D = 5\mu m$ (solid, blue) and $10\mu m$ (dashed, red) in the case of no dilution ($\beta = 0$) with $P = 300$ atm and $T_f = 800$ K.

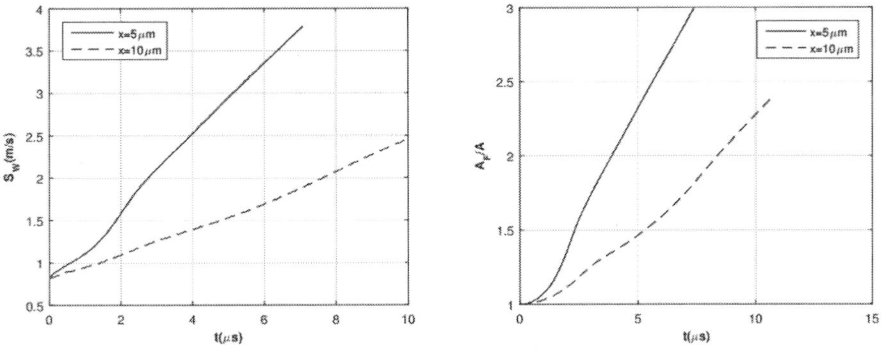

Figure 10: Time evolution of the instantaneous corrugated flame velocity S_w [(a), (left)] and scaled flame surface area (length) A_F/D [(b), (right)] in the channels of width $D = 5\mu m$ (solid, blue) and $10\mu m$ (dashed, red) for the dilution rate $\beta = 20\%$ with $P = 300$ atm and $T_f = 800$ K.

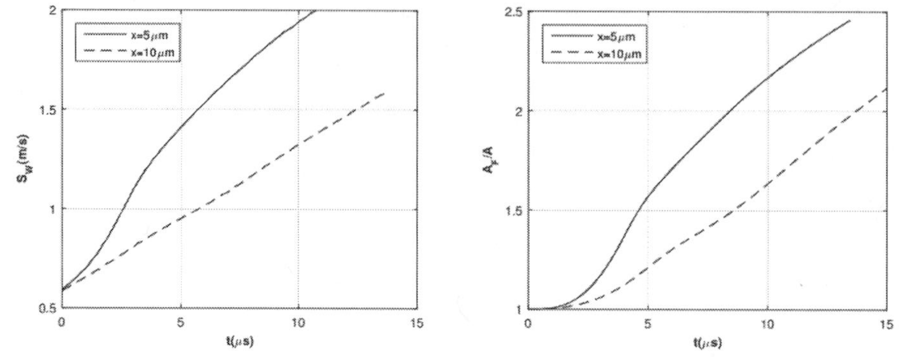

Figure 11: Time evolution of the instantaneous corrugated flame velocity S_w [(a), (left)] and scaled flame surface area (length) A_F/D [(b), (right)] in the channels of width $D = 5\mu m$ (solid, blue) and $10\mu m$ (dashed, red) for the dilution rate $\beta = 40\%$ with $P = 300$ atm and $T_f = 800$ K.

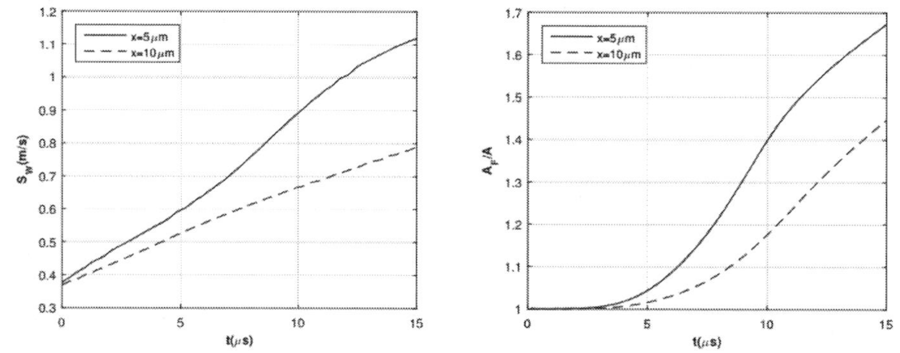

Figure 12: Time evolution of the instantaneous corrugated flame velocity S_w [(a), (left)] and scaled flame surface area (length) A_F/D [(b), (right)] in the channels of width $D = 5\mu m$ (solid, blue) and $10\mu m$ (dashed, red) for the dilution rate $\beta = 60\%$ with $P = 300$ atm and $T_f = 800$ K.

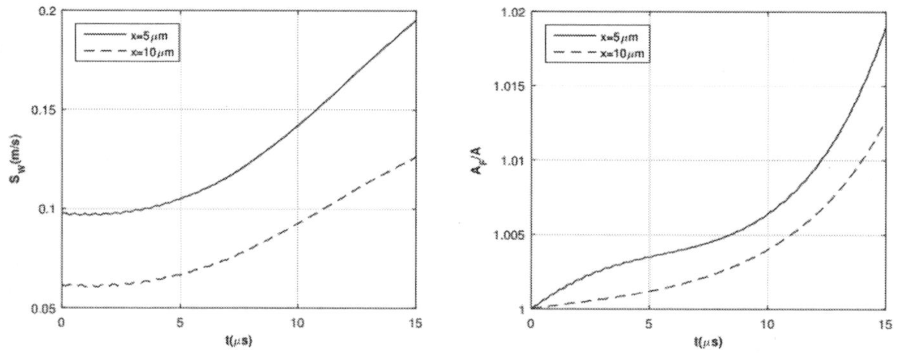

Figure 13: Time evolution of the instantaneous corrugated flame velocity S_w [(a), (left)]; and scaled flame surface area (length) A_F/D [(b), (right)] in the channels of width $D = 5\mu m$ (solid, blue) and $10\mu m$ (dashed, red) for the dilution rate $\beta = 80\%$ with $P = 300$ atm and $T_f = 800$ K.

5. Conclusions

In the present work, we explored the key parameters of supercritical oxy-methane premixed combustion, such as the flame speed, flame thickness, and internal structure of the flame front. The analysis is performed by means of the 2D numerical simulations of reacting mixtures with fully-compressible hydrodynamics, transport properties (heat conduction, diffusion and viscosity), and the one-step Arrhenius chemical kinetics. The real-gas thermodynamics and molecular properties are incorporated into a reacting-flow Navier-Stokes solver along with the Peng-Robinson EoS.

It is shown that an increase in a CO_2-dilution rate makes a flame front thicker and reduces the laminar flame speed significantly, as compared to non-diluted combustion. Starting with the planar flames, the analysis is then extended to the corrugated ones. While the corrugated flame generally propagates faster than the planar one of similar thermal-chemical characteristics, the corrugated flame velocity not only reduces with the dilution rate, but also depends on the channel width.

6. Acknowledgements

This research was funded by the Office of Science of the U.S. Department of Energy under SBIR Contract No. DE-SC0017241.

References

[1] Supercritical Carbon Dioxide Brayton Cycle, Quadrennial Technology Review 2015 Department of Energy, (2015).

[2] H. Kobayashi, H. Hagiwara, H. Kaneko, and Y. Ogami, Proc. Combust. Inst. 31 (2007) 1451–1458.

[3] X. Hu, Q. Yu, J. Lui, and N. Sun, Energy 70 (2014) 626–634.

[4] A. R. Khan, S. Anbusaravanan, L. Kalathi, R. Velamati, and C. Prathap, Fuel 196 (2017) 225–232.

[5] Z. Yang, X. Yu, J. Peng, L. Wang, Z. Dong, X. Li, S. Sun, S. Meng, and H. Xu, Exp. Therm. Fluid Sci. 81 (2017) 209–222.

[6] N. Ding, R. Arora, M. Norconk, and S.-Y. Lee, Int. J. Hydrogen Energy 36 (2011) 3222–3231.

[7] H. Kobayashi, Y. Otawara, J. Wang, F. Matsuno, Y. Ogami, M. Okuyama, T. Kudo, and S. Kadowaki, Proc. Combust. Inst. 34 (2013) 1437–1445.

[8] Y. Zhang, W. Shen, H. Zhang, Y. Wu, and J. Lu, Fuel 157 (2015) 115–121.

[9] R. Shang, Y. Zhang, M. Zhu, Z. Zhang, D. Zhang, and G. Li, Int. J. Hydrogen Energy 41 (2016) 15056–15067.

[10] D.-Y. Peng and D. Robinson, Ind. Eng. Chem. Fundam. 15 (1976) 59–64.

[11] B. E. Poling, J. M. Prausnitz, and J. P. O'Connel, "The Properties of Gases and Liquids", in: McGraw-Hill, 2001.

[12] J. Oh and D. Noh, Energy 45 (2012) 669–675.

Spring Technical Meeting
Eastern States Section of the Combustion Institute
March 4-7, 2018
State College, Pennsylvania

Thermoacoustic Linear Stability Model with Porous Media and Mean Flow Effects

Cody Dowd[1], Joseph Meadows[1]*

[1]Mechanical Engineering, Virginia Tech, Blacksburg VA, USA
[*]Corresponding Author Email: jwm84@vt.edu

Abstract: A 1-D numerical model was developed by Dowd and Meadows [1], for the linear stability analysis of the longitudinal modes for a series of constant cross-sectional area ducts without and with porous media. The model was developed to include the effects of a flame by using a n-Tau flame transfer function and is capable of including porous media by modeling the effects of its properties: porosity, flow resistivity, effective bulk modulus, and structure factor. The present study extends the capability of the model to include the effects of ring-shaped porous media and mean flow. The independent effects of each extension to the model are investigated and it is found that, due to reduction in porous material by using a ring shaped porous section, the acoustic damping is less than that of a solid cylindrical porous media section. The mean flow effects cause advection of the acoustic waves in the domain, and moving the acoustic nodes and anti-nodes can increase or decrease the acoustic damping performance when the anti-node is moved inside or outside of the porous section, respectively. The effects of neglecting mean flow are also demonstrated.
Keywords: Thermoacoustic Instability Modeling, Porous Media, Mean Flow

1. Introduction

In combustion processes where the acoustic pressure oscillation and unsteady heat release are in-phase, thermoacoustic instability may occur, according to the Rayleigh criterion [2]. For thermoacoustic instability to be prevalent in the system, the energy it produces must be greater than the energy dissipation of the system. Thermoacoustic instability is especially prevalent in lean premixed combustion (LPM). These instabilities occur near the natural frequencies of the combustor [3] and can have detrimental effects such as unwanted noise production, failure due to fatigue, and/or limiting the operational ranges. Passive methods have been used experimentally to mitigate these instabilities, such as changing shape and size of combustion chamber [4], adding Helmholtz resonators [5-6], and adding porous inert media (PIM) [7-9]. Numerical investigation with PIM was recently conducted using a 1-D model for the linear stability analysis of the longitudinal modes for a series of constant cross-sectional area ducts [1]. The current work expands this reduced order model to include ring shaped PIM and mean flow effects.

Reduced order modeling by Dowd and Meadows [1] was accomplished by utilizing a transfer matrix approach to solve for the frequency and mode shapes of a system with constant cross sectional area ducts. When including a flame, a linear assumption is made for the relationship between the flame and acoustic velocity. This allows for prediction of acoustic growth rates and subsequently the stability of the system. The model uses a $n-\tau$ flame transfer function (FTF) to describe the interaction between the flame and the acoustic velocity. The interaction index, n, and convective time delay, τ, have been shown to perform well for this type of analysis [10, 11]. Effects of porous media are modeled as a bulk fluid with the following properties: porosity, flow resistivity, structure factor, and effective bulk modulus. Physically, the structure factor is defined as the ratio of effective fluid density to its surrounding fluid density value. Effective bulk modulus is representative of the compressibility of the bulk fluid. As a solid cylindrical piece of PIM may be overly restrictive of the flow, a ring shaped PIM has been used experimentally [8]. Commonly, a low Mach number assumption is made and mean flow effects are neglected in reduced order modeling. This is not always the case in combustion systems and the mean flow effects can affect the thermoacoustic behavior of the system [12]. The present work expands on the previous reduced order model [1] to include the effects of changing from a solid cylindrical PIM section to a ring shaped section and including mean flow effects.

2. Numerical Methodology

The following discussion of the numerical model will present the independent details of the ring shape PIM and mean flow transfer matrix derivation. The process for deriving the final equations used in the numerical model will start from and follow methodology presented previously [1, 10, 12], details of the derivation are omitted for brevity. The transfer matrix approach uses the acoustic pressure and velocity to define jump conditions between duct interfaces. PIM and fluid properties are constant in each duct but change from duct to duct. Defining the jump conditions, acoustic pressure and velocity, a FTF, and acoustic boundary conditions, allow for the building of a set of equations that can be used to solve for the eigenfrequencies and mode shapes for a combustion system. The model was developed in systematic nature such that when nominal values are used for porous media, mean flow effects are neglected, and the inner radius is zero for ring shaped PIM, the equations presented reduce back to the acoustic governing equations commonly used in literature [10, 12]. Nominal values for PIM are as follow: porosity, $h = 1$, flow resistivity, $\sigma = 0$, structure factor, $s_t = 1$, and effective bulk modulus, $\kappa = \gamma P_0$, where γ is the ratio of specific heats and the subscript zero defines mean quantities. A porosity of 1 represents an entire volume represented by fluid, zero flow resistivity means there is no pressure drop, structure factor 1 means that the effective and surrounding densities are the same, and bulk modulus of γP_0 demonstrates the compressibility of surrounding fluid is not effected by PIM. The model neglects the effects of entropy waves and viscous effects.

Ring Shaped PIM Implementation

To change the PIM shape from a solid cylindrical section to a ring shaped section it is necessary to integrate the porous properties across the cross sectional area. By doing so the inner and outer radius (R_i and R_o, respectively) must be defined, where $0 < r < R_i$ is without porous material and $R_i < r < R_o$ is composed of porous material. The momentum and energy conservation equations presented in Dowd and Meadows [1] are represented below by equations 1 and 2, respectively.

$$\int_0^{R_o}\int_0^{2\pi}[\left(\frac{\partial p'}{\partial x}+\frac{s_t\rho_0}{h}\frac{\partial u'}{\partial t}+\sigma u'\right)r]drd\theta = 0 \tag{1}$$

$$\int_0^{R_o}\int_0^{2\pi}[\left(\frac{1}{\kappa}\frac{\partial p'}{\partial t}+\frac{1}{h}\frac{\partial u'}{\partial x}-\frac{\gamma-1}{\kappa}q'(x,t)\right)r]drd\theta = 0 \tag{2}$$

Where, ρ_0 is the mean density, q' is the unsteady heat release, and p' and u' are the acoustic pressure and velocity, respectively. Acoustic pressure is defined by equation 3 and used with Eulers equation to define acoustic velocity in equation 4.

$$p_j(x,t) = [A_j^+ exp(ik' x_j) + A_j^- exp(-ik'x_j)]exp(-i\omega t) \tag{3}$$

$$u_j'(x,t) = \left(\frac{R_i^2}{R_0^2\gamma P_0}+\frac{h(R_0^2-R_i^2)}{R_0^2\kappa}\right)\left(\frac{\omega}{k'}\right)[A_j^+ exp(ik'x) - A_j^- exp(-ik'x)]exp(-i\omega t) \tag{4}$$

Where, j ranges from 1 to the total number of ducts, A_j^+ and A_j^- are the forward and backward traveling wave amplitudes, respectively, ω is the frequency, and k' is the wave number. The jump conditions are defined for pressure and velocity by equations 5 and 6, respectively. Jump conditions are determined by integrating equations 1 and 2 across the boundary between ducts and taking the limit as x approaches the interface location.

$$[p_j']_-^+ = 0 \tag{5}$$

$$\left[u'\left(\pi R_i^2 + \frac{\pi(R_o^2-R_i^2)}{h}\right)\right]_-^+ = Q'(x,t) \tag{6}$$

By applying the $n-\tau$ FTF, the $Q'(x,t)$ term reduces to: $-n\pi R_o^2 u_j'(t-\tau)$, and the + and - signs represent the upstream and downstream duct, respectively. Physically, the jump conditions represent continuous pressure from one duct to another and a discontinuity in acoustic velocity when there is a change in porosity, change in area, or the presence of a flame between two interfaces. The boundary conditions are defined by reflection coefficients at the inlet and outlet of the domain. With the boundary conditions and jump conditions defined, it is now possible to establish a system of equations in a global transfer matrix and solve for the eigenfrequencies and mode shapes. An initial condition is defined in order to solve for the mode shapes. The transfer matrix approach is outlined in Dowd and Meadows [1] and process here is identical.

Mean Flow Implementation

Mean flow is incorporated into the numerical model by using the material derivative representation of the momentum and energy equations including PIM, similar to equations 1 and 2, respectively. The partial derivative with respect to time is replaced with the material derivative. The material derivative simplifies to $\frac{D(\ )}{Dt} = \frac{d(\ )}{dt} + \frac{d(\ )}{dx}u_0$, where u_0 is the mean velocity determined from the local Mach number (*Ma*). Following a similar process detailed in [1, 12], the resulting wave number is shown in equation 8. Previous analysis assumed waves move in the forward and backwards direction at the same speed since the speed of sound in each duct was constant. When incorporating mean flow effects there will be a biasing of the acoustic waves in the direction of the flow. For this reason there are two wave numbers defined, the forward and backward wavenumbers.

$$k_j'^{\pm} = \frac{\left[\mp\left(\frac{2s_t\rho_0\omega u_0}{\kappa} + \frac{i\sigma h}{\kappa}u_0\right) + \sqrt{\left(\frac{2s_t\rho_0\omega u_0}{\kappa} + \frac{i\sigma h}{\kappa}u_0\right)^2 + 4\left(1 - \frac{s_t\rho_0}{\kappa}u_0^2\right)\left(\frac{s_t\rho_0}{\kappa} + \frac{i\sigma h\omega}{\kappa}\right)}\right]}{2\left(1 - \frac{s_t\rho_0}{\kappa}u_0^2\right)} \tag{8}$$

The acoustic velocity can be found using the material derivative form of Eulers equation and acoustic pressure from equation 3. The resulting equation is given by equation 9.

$$u'(x,t) = \left[\left(\frac{\omega h}{\kappa k'^+} - \frac{u_0 h}{\kappa}\right)A_j^+ \exp(ik'^+x) - \left(\frac{\omega h}{\kappa k'^-} + \frac{u_0 h}{\kappa}\right)A_j^- \exp(-ik'^-x)\right]\exp(-i\omega t) \tag{9}$$

The jump conditions for acoustic pressure and velocity are defined by equations 10 and 11, respectively.

$$[p']_-^+ = \left[-\frac{\rho_0 s_t u_0}{h}u'\right]_-^+ \tag{10}$$

$$\left[\frac{S u_0 p'}{\kappa}\right]_-^+ + \left[\frac{u'S}{h}\right]_-^+ = \frac{\gamma-1}{\kappa}\Omega \tag{11}$$

Where, S is defined as the area of the cross section and the heat release Ω defined by: $\Omega = \int Sq'(x,t)dx$, where q' is the unsteady heat release. The jump conditions physically state that there is a discontinuity in pressure and velocity due to changing mean flow. The discontinuity introduced by mean flow into the acoustic pressure has been presented in other numerical work [13]. Velocity discontinuities due to area change, porosity change, and unsteady heat release are still present. The transfer matrix derivation follows similar methodology as Dowd and Meadows [1]. It is also important to note that for any of these equations, defining the inner radius as zero, no mean flow, and nominal PIM parameters simplify the equations to common acoustic governing equations found in literature.

3. Results and Discussion

The performance of the previous model was validated with analytical results [1], as well as compared to experimental work of Meadows and Agrawal [8] for cases with and without PIM. The simulation results show good agreement to experimental work. All results presented here are for the representative combustion system of the experimental work by Meadows and Agrawal [8]. Note that the inlet boundary condition is closed which approximates a choked inlet and the outlet is specified as an open boundary. Independent results are presented for the ring shaped PIM and then for the mean flow solver.

Ring Shape Results

Reducing a PIM section from a solid cylindrical section to a ring shaped is a reduction in PIM material, therefore the resulting acoustic damping will be less. This is demonstrated in Figure 1, where stability maps are compared for various inner PIM radii for a system with a flame, these are compared to a solid PIM and without PIM section. The PIM parameters used are 0.85, 1740, 9.9, 192360, for porosity, flow resistivity, structure factor, and effective bulk modulus, respectively. A stability map compares the normalized growth rate, $imag(\omega)/\omega_n$, and normalized convection time delay, $\tau\omega_n/2\pi$, where ω_n is the frequency response of a system without PIM. For the results presented, an interaction index of 0.1 is used and outer radii are given in Dowd and Meadows [1].

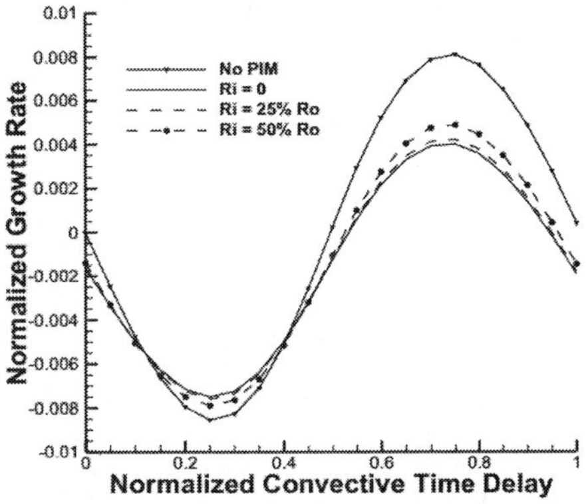

Figure 1: Stability diagrams for various inner radii.

From the results in Figure 1, it can be determined that the smaller volume the PIM section becomes, results in less damping. The stability is determined by the imaginary portion of the frequency response, when examining the temporal term $\exp(-i\omega t)$, with $\omega = a + bi$, a negative imaginary component results in a stable system advancing in time, while an unstable system is described by a positive value of b. Following this logic, a frequency response with a larger negative imaginary component is more stable and for the case with PIM, more damping is present.

To validate the solver, results are compared to the experimental work of Meadows and Agrawal [8], the experimental work did not have well defined acoustic boundary conditions and only present results for a convective time delay of 0.0236. The PIM radii were defined to match the converging PIM setup from Meadows and Agrawal [8] and the remaining conditions were maintained from previous investigation presented in figure 1. Experimental results showed peak sound pressure levels were observed around frequencies of 1/3 octave center frequency of 400 Hz, and when PIM was introduced the sound pressure levels were reduced drastically at the same frequencies. Results for the first mode from the numerical model give a frequency response of 424.1+0.2i and 416.6-2.2i for without PIM and PIM cases, respectively. The results are in agreement with the experimental results [8], despite not having experimentally well-defined acoustic boundary conditions.

Mean Flow Results

The inclusion of mean flow results is compared to results in the literature for verification purposes, as shown in Figure 2. Results are presented in comparison to other numerical formulation of Yu et al. [13]. The results of Yu et al. use linearized Euler equations and present independent mean flow results for a closed inlet and outlet boundary condition system. Figure 2 presents the normalized frequency, $real(\omega)/\omega_n$, with varying Mach number. No PIM or flame is used in this verification. Figure 2 shows the good agreement between the two numerical results, verifying the numerical implementation of mean flow.

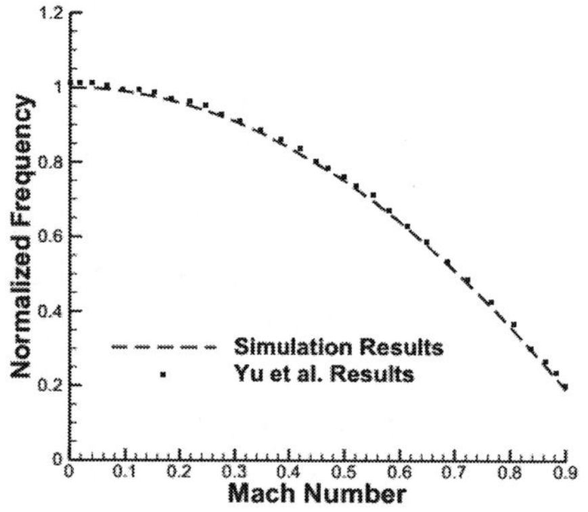

Figure 2: Comparison of normalized frequency variation with Mach number.

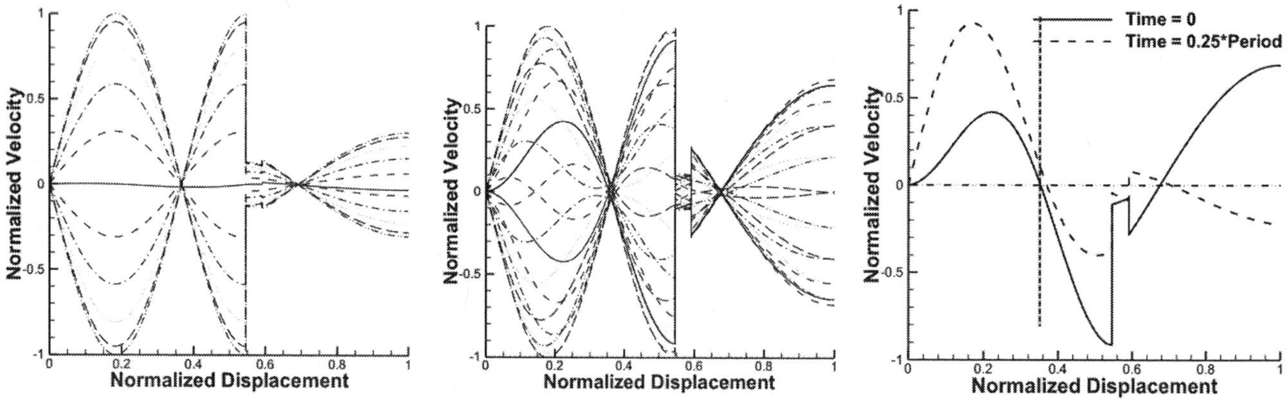

(A) (B) (C)

Figure 3: 5/4 mode wave (A) Ma = 0, (B) Mach = 0.25, (C) Ma = 0.25 at discrete times, horizontal and vertical lines represent reference locations of 0 and anti-node at t=0, respectively.

The effects of mean flow with PIM and flame are shown in Figure 3 where the 5/4 mode shapes of acoustic velocity are presented for Mach numbers 0 and 0.25. The mean flow in each duct is determined from the inlet Mach number, when an area or temperature change occurs, continuity is applied and the velocity is adjusted accordingly. Introduction of mean flow effects has been shown by [13], that results no longer represent standing waves; the resulting waves are advected with the flow and cause a shift of node and anti-node locations. Results are presented with flame for a convective time delay of 0.00343 and interaction index of 0.1, PIM parameters, geometry, and fluid properties in each duct follow set up discussed for Figure 1. The frequency response is 633.1-3.5i and 604.1-13.6i for the no mean flow and Mach number of 0.25 cases, respectively. The advection of waves is demonstrated in Figure 3C, where the velocity mode shape is presented for two time instances. In a system with no mean flow, the wave is moving at the speed of sound in both directions, which results in a standing wave. With mean flow, the differences in speed of the forward and backward traveling wave results in an oscillation of the node and anti-node locations. The displacement of the nodes and anti-nodes is shown in Figure 3C. The movement of nodes and anti-nodes has been shown to affect the stability of the system [1], the location of the velocity anti-node relative to PIM influences the effectiveness of PIM in mitigating thermoacoustic instabilities.

4. Conclusions

The numerical formulation for ring shaped PIM and mean flow effects are presented and results show good agreement with other experimental and numerical work. Ring shaped PIM shows a reduction in mitigation effects due the reduction in amount of PIM. Mean flow inclusion causes an oscillation of velocity and pressure node and anti-node locations, which can influences the stability of the system. As the numerical model is reduced order, the flow field is not resolved, fluid pressure and velocity are not evaluated, so the effect of PIM on the flow structure is not available to demonstrate the potential benefits of a ring shape vs solid cylindrical PIM section. This model could be coupled with CFD investigation to draw conclusions about the acoustic performance in relation to the flow performance (pressure drop, heat transfer, etc.). Future work will extend the model to include entropy waves and non-linear limit cycle prediction. Experimental investigation will also be conducted to validate the numerical models.

5. References

[1] C., Dowd, and J., Meadows, Thermoacoustic Instability Model with Porous Media: Linear Stability Analysis and the Impact of Porous Media, ASME Turbo Expo(2018), GT2018-75406 (Accepted)

[2] L., Rayleigh, The Theory of Sound - 2nd Edition, Macmillan, London, U.K., 1896, pp. 224-234.

[3] T., Lieuwen, Modeling premixed combustion-acoustic wave interactions: A review, Journal of Propulsion and Power 19 (2003) 765-781.

[4] J., Kornegay, D., Depperschimdt, A., Agrawal, Passive Control of Thermo-Acoustic Instability in Different Length Combustors Using a High-Strength Metallic Porous Insert, ASME Turbo Expo 2015, GT2105-43890.

[5] D., Gysling, G., Copeland, D., McCormick, V., Proscia, Combustion System Damping Augmentation with Helmholtz Resonators, Journal of Engineering for Gas Turbines and Power 122 (2000), 269-274.

[6] R., Kashani, J., Monfort, Low-Frequency Thermoacoustic Instability Mitigation Using Adaptive-Passive Acoustic Radiators, ASME Turbo Expo 2011, GT2011-46642.

[7] A., Agrawal, S., and Vijaykant, Passive Noise Attenuation System, Patent 8,109,362, 2012.

[8] J., Meadows, A., and Agrawal, Porous Inserts for Passive Control of Noise and Thermo-Acoustic Instabilities in LDI Combustion, Combustion Science and Technology 187 (2015), 1021-1035.

[9] J., Meadows, A., and Agrawal, Time-Resolved PIV of Lean Premixed Combustion Without and With Porous Inert Media for Acoustic Control, Combustion and Flame 162 (2015), 1063-1077.

[10] T., Poinsot, D., Veynante, Theoretical and Numerical Combustion, Edwards, Philadelphia USA, 2003.

[11] A., Dowling, S., Stow, Acoustic Analysis of Gas Turbine Combustors, J. of Propulsion and Power 19 (2003)

[12] T., Lieuwen, Unsteady Combustor Physics, Cambridge University Press, 2012.

[13] Y., Yu, J., Sisco, V., Sankaran, W., Anderson, Effects of Mean Flow, Entropy Waves, and Boundary Conditions of Longitudinal Combustion Instability, Combustion Science and Technology 182 (2010), 739-776.

Spring Technical Meeting
Eastern States Section of the Combustion Institute
March 4-7, 2018
State College, Pennsylvania

Thermoacoustic Analysis with Statistically Based Flame Transfer Function Extracted from Computational Fluid Dynamics

Shrihari Sampathkumar[1], Joseph W. Meadows[2,]*

[1,2]Department of Mechanical Engineering, Virginia Tech, Blacksburg, VA, USA
[*]Corresponding Author Email: jwm84@vt.edu

Abstract: Thermo-acoustic instabilities occur when the unsteady heat release rates and the acoustic pressure oscillations are in phase. For thermoacoustic modeling, the phase relationship is described using a flame transfer function (FTF). A common FTF used for thermoacoustic modeling defines the unsteady heat release term as a function of the velocity oscillations at a time delay, τ. Physically the time delay can represent the time for the fuel to convect from the injection point to the flame; however, in practical combustion systems there can exist a range of time delays due to multiple injection points, complex swirling flow, and/or recirculation zones. The present study describes a methodology of using a normally distributed probability density function (PDF) and a numerical PDF extracted from steady state CFD analysis of the convective fuel time delay. CFD analysis, with an FGM combustion model, is carried out to extract the numerical PDF of τ which is used as an input into a 1-D linear stability model based on the Transfer Matrix approach. Results are presented to compare the traditional FTF with the newly developed PDF based FTFs. The proposed methodology provides a framework for combined 1-D thermoacoustic analysis with a CFD analysis to improve prediction capabilities.
Keywords: Flame Transfer Function, CFD, Convective Time Delay, Thermoacoustics

1. Introduction

Lean Premixed (LPM) combustors have the advantage of reduced NO_x emissions, but are more susceptible to thermoacoustic instabilities. These instabilities occur when the energy added due to the in-phase relationship between unsteady heat release rates and pressure oscillations is greater than the energy lost due to damping. The Flame Transfer Function (FTF) which describes this phase relationship, often uses a singular value for fuel time delay, τ. Practical combustion systems however, might have a large range of τ values, making models with a single τ value ineffective. Therefore, the present study presents a model capable of predicting the stability of simplified combustion systems, considering a distributed convective fuel time delay, τ.

In order to include the effects of a distributed τ into the FTF, Armitage et al. [1] introduced a model which assumes a uniform spread in the convection time between $\tau-\Delta\tau$ and $\tau+\Delta\tau$. The frequencies predicted using the model with a uniform spread in τ was found to be closer to the experimentally determined values than analytical models that simply took τ as a single value. Polifke et al. [2] investigated the effect of elliptical nozzle shapes of varying eccentricity on the flame shape considering the distributed fuel transport time-lag using a τ-ζ model [3]. The fuel time-lag distribution was considered to be a Gaussian distribution with the standard deviation σ increasing with the eccentricity of the elliptical nozzle inlet. It was observed that a larger σ would reduce the instability and would eventually stabilize the investigated system.

Many FTFs are available in literature based on various physical mechanisms responsible for exciting instabilities. Though attempts have been made to include a distributed τ into the FTF, a comprehensive model that allows complicated distributions which do not follow a standard distribution or a multimodal distribution has not been studied in detail. The current study presents analytical and numerical PDFs of the convective fuel time delay, and incorporates the effects into a modified FTF for thermoacoustic stability analyses.

2. Numerical Methods

The transfer matrix approach presented by Poinsot and Veynante [4] is used to numerically solve for the stability of a combustor. The approach calculates the eigenfrequencies of a given system by solving a set of equations using acoustic pressure and velocity perturbations. The interaction between acoustic waves and the flame is included by using a FTF. Simple forms of the FTF are often used to predict stability, which affects the prediction accuracy.

The time taken by fuel to convect from the injection point to the flame is defined as the convective fuel time delay. Since the convective time delay is known to influence the stability of industrial gas turbines [5-6], the flame interaction is modeled using an n-τ FTF which is also able to capture the effects due to equivalence ratio (ϕ) oscillations convected from the fuel injection location to the flame. In n-τ, the τ stands for time delay and n is known as the interaction index. A 1-D solver, similar to the one developed by Dowd and Meadows [7] is used in this study. The solver is able to predict the stability of a system with a distributed convective time delay, rather than using a mean value of τ.

In constant cross-section ducts, the acoustic perturbation in pressure and velocity (p_1, u_1) is calculated by substituting the complex pressure and velocity (p_ω, u_ω) into the Helmholtz equation as:

$$p_1(x,t) = A^+ e^{i(kx-\omega t)} + A^- e^{i(-kx-\omega t)} \tag{1}$$

$$u_1(x,t) = \frac{1}{\rho_0 c_0}\left(A^+ e^{i(kx-\omega t)} - A^- e^{i(-kx-\omega t)}\right) \tag{2}$$

Where A is the amplitude and the +, - in the superscript is used to represent the forward and backward travelling waves respectively. k is the wave number, ω is the acoustic frequency and ρ_0, c_0 are the mean flow parameters. In a series of connected ducts, the jump conditions are obtained by integrating conservation equations across the interface and taking the limit from each side of the interface. The amplitudes of the wave between these sections can then be represented by equation (3).

$$\begin{pmatrix} A_{j+1}^+ \\ A_{j-1}^- \end{pmatrix} = T_j \begin{pmatrix} A_j^+ \\ A_j^- \end{pmatrix} + O_j \tag{3}$$

$$T_j = \frac{1}{2}\begin{bmatrix} e^{ikl_j}(1+\Gamma_j) & e^{-ikl_j}(1-\Gamma_j) \\ e^{ikl_j}(1-\Gamma_j) & e^{-ikl_j}(1+\Gamma_j) \end{bmatrix} \quad , where \quad \Gamma_j = \frac{\rho_{j+1} c_{j+1}}{\rho_j c_j} \frac{S_j}{S_{j+1}} \tag{4}$$

Where j is the duct index, l_j is the length and S_j is the cross sectional area of the duct. The O_j matrix is added in the case of reacting flows by assuming that the fluctuating heat release term $\dot{\Omega}_T^1$ is harmonic $(\Omega e^{i\omega t})$ and is hence defined as:

$$O_j = \frac{1}{2}\frac{\rho_{j+1}c_{j+1}}{S_{j+1}}\begin{pmatrix} \frac{\gamma-1}{\rho_j c_j^2}\Omega \\ -\frac{\gamma-1}{\rho_j c_j^2}\Omega \end{pmatrix} \tag{5}$$

The unsteady combustion term (Ω) has to be defined in order to completely define the system. The standard n-τ formulation defines the simplest scaling between $\dot{\Omega}_T^1$ and u_1 as

$$\frac{\gamma-1}{\rho_1 c_1^2}\dot{\Omega}_T^1 = S_j n e^{i\omega\tau} u_1(x) \tag{6}$$

In order to include the effects of a normally distributed τ into the FTF, the standard n-τ formulation which accounts for the effect of time delay τ is modified. The standard transfer function is multiplied by the probability density function of a normal distribution and integrated from -∞ to ∞, resulting in equation (6), where μ is the mean τ and σ^2 is the variance of the normal distribution. Equation (6) reduces to the standard n-τ FTF form when σ^2 tends to 0.

$$\frac{\gamma-1}{\rho_1 c_1^2}\dot{\Omega}_T^1 = S_1 n\, e^{i\omega\mu}\, e^{\frac{-\omega^2\sigma^2}{2}}\, u_1(x) \tag{7}$$

A simple swirl-stabilized premixed combustion system was modeled using Star-CCM+ 12.04. The geometry consists of a tube of diameter 0.051 m and a cylindrical combustion chamber of diameter 0.16 m, both of length 0.508 m. Figure 1 shows the isometric view of the computational geometry. Premixed fuel with an equivalence ratio of 0.5, a mass flow rate of 1.2 kg/s and swirl number of 0.6 is injected at the inlet.

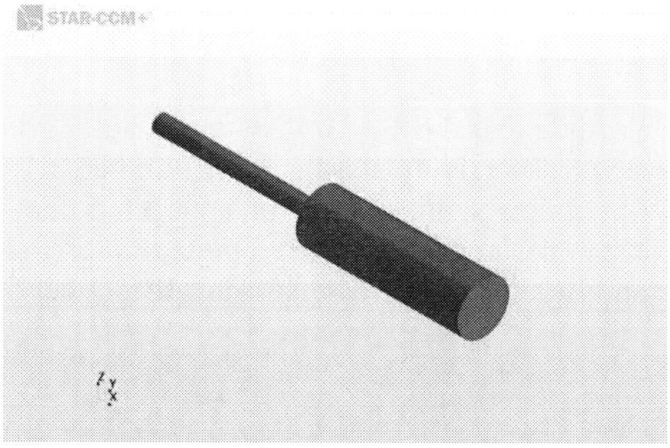

Figure 1: Isometric view of the computational geometry

Turbulence was modeled using the k-Ω model. Combustion is modeled using the Flamelet Generated Manifold (FGM) method, which reduces the computational expense of modeling combustion while taking detailed chemistry effects into account [8,9]. The controlling variable that is used to represent the combustion process is known as the progress variable.

The Lagrangian Multiphase Particle Tracking model in Star-CCM+ is used to inject 1571 massless particles at the inlet of the combustion chamber. The particles which reach the flame front defined by a progress variable of 0.5 are extracted along with their corresponding residence times in the computational domain. A histogram of the τ variation of the particles is generated. Iteration averaging is done over 250 iterations to nullify the slight variations observed in the histograms between different iterations. A numerical probability density function is then generated from the extracted histograms using equation (8)(6).

$$f_m = \begin{Bmatrix} p_1/w_1 \,, \tau_0 < \tau < \tau_1 \\ p_2/w_2 \,, \tau_1 < \tau < \tau_2 \\ . \\ . \\ p_m/w_m \,, \tau_{m-1} < \tau < \tau_m \end{Bmatrix} \qquad where \quad p_m = \frac{N_m}{N_{total}} \tag{8}$$

Where m is the bin index and $w_m = \tau_m - \tau_{m-1}$ is the bin width. p_m is the probability of the mth bin and N is the number of particles. The FTF extracted from the CFD data can therefore be calculated by integrating the numerical probability density function in a piecewise manner (equation (9), where τ_{m-1} and τ_m are the lower & upper limits of the τ values in a bin.

$$\frac{\gamma - 1}{\rho_1 c_1^2}\dot{\Omega}_T^1 = S_j n \left(\sum_{m=1}^{m_{max}} \int_{\tau_{m-1}}^{\tau_m} f_m \,.\, e^{i\omega\tau} d\tau \right) u_1(x) \tag{9}$$

3. Results and Discussion

The results of the numerical model is compared with the analytical model with a representative combustor, similar to the one used by Dowd and Meadows [7]. Results in **Error! Reference source not found.** compare the normalized growth rates, defined as imag(ω)/ω_n, by a normally distributed FTF with the FTF that utilizes the standard n-τ formulation, for the 3/4 wave mode. When a normal distribution is considered, a slight shift in the growth rates predicted can be observed. As the standard deviation σ decreases, the stability trend predicted by the normally distributed n-τ FTF converges to that of the standard n-τ FTF. Figure 3 shows this convergence at a fixed normalized τ value circled in Figure 2. The normalized σ is defined by σ/μ.

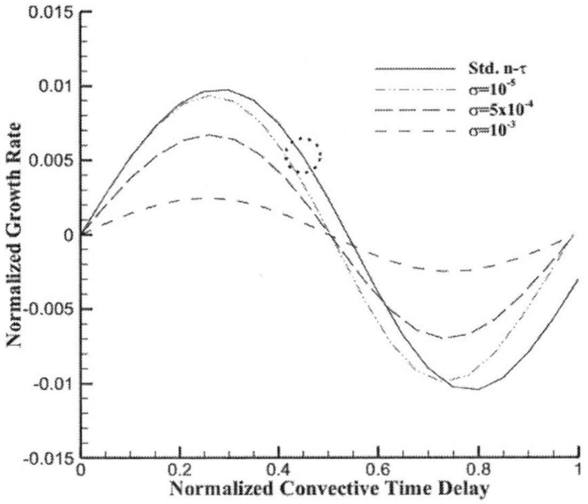

Figure 2 : Normalized growth rate for various normalized convective fuel time delays and standard deviation

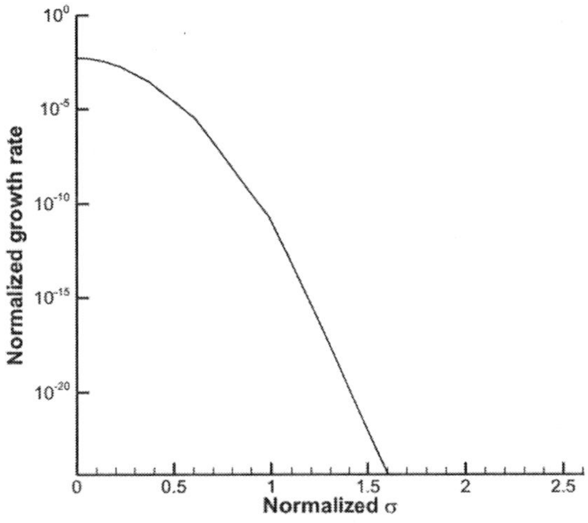

Figure 3 : Normalized growth rate for various σ values at a fixed mean fuel time delay (Circled in Figure 2)

Figure 4 shows the histogram of τ extracted from a single iteration of the CFD simulation. Statistical analyses performed on the histogram data show that the time delay does not follow a normal distribution or any other standard probability distribution. The iteration averaged histograms (Figure 5) also fail to follow any standard distributions. Therefore, a numerical probability density function is calculated using equation (9) and the numerical FTF (equation 10) is used to solve for the eigenfrequencies.

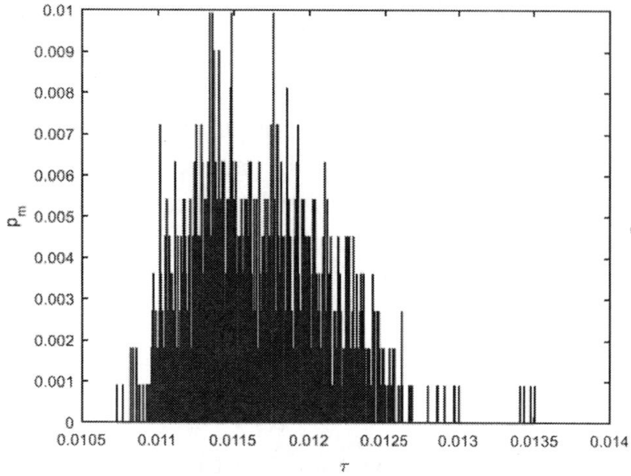

Figure 4 : Probability of bins extracted from single iteration

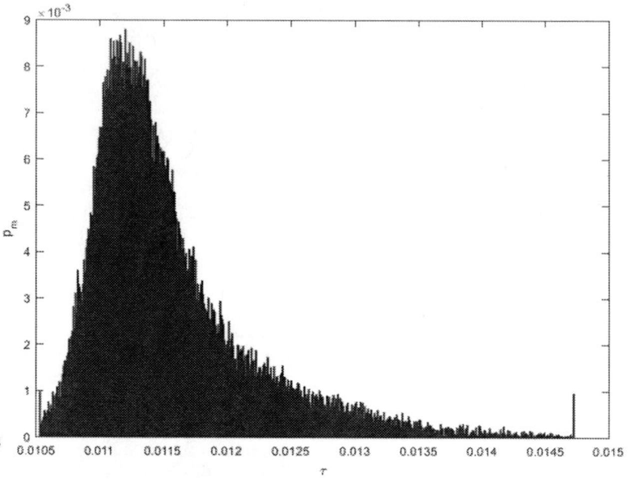

Figure 5 : Probability of bins – averaged over iterations

A representative combustor with dimensions like the geometry used for the CFD analysis is used to compare the effects of using a distributed τ on the eigenfrequencies of the system. The average x co-ordinate value of the flame front determined from CFD is taken as the flame location in the 1D solver (0.11 m from the combustor dump plane). The iteration averaged τ values, when approximately fitted to a normal distribution, have a μ of 0.0116 s and a σ of 6.52e-04 s. A comparison of the real and imaginary parts of the complex frequency (ω) computed from the different FTFs is shown in Table 1.

Table 1 : Comparison of eigenfrequencies obtained by using different FTFs

Mode	Standard n-τ		Normal Dist.		CFD	
	Real	Imag	Real	Imag	Real	Imag
1	31.31	-62.36	32.33	-64.15	32.25	-64.15
2	102.45	-44.13	107.00	-46.02	107.36	-46.33
3	138.49	1.94	139.04	1.24	139.12	1.20
4	200.30	-35.56	204.39	-0.31	204.36	-0.37
5	204.72	-0.10	207.87	-42.56	210.67	-43.13
6	287.37	-39.41	296.35	-51.59	303.41	-50.78
7	368.79	-43.44	378.79	-61.03	389.97	-56.80
8	442.29	-43.35	457.45	-65.06	472.34	-52.86
9	478.84	1.38	479.40	0.06	479.78	-0.37
10	524.45	-32.06	545.89	-0.21	545.10	-0.81

From the results in Table 1, the frequencies predicted by the numerical FTF show a slight shift from the standard FTF for every mode. However, the difference in growth rate between these FTFs is significant in several modes. In particular, for the 9th mode, the standard FTF predicts an unstable system while a stable system is predicted by the numerical FTF. Although the τ values from CFD do not exactly follow a normal distribution, it can be observed that the values are much closer to the numerical FTF than the standard FTF.

4. Conclusions

A 1-D thermoacoustic model with a normally distributed and a numerical PDF of the convective fuel time delay is used to modify the n-τ FTF for the prediction of acoustic pressure, velocity and growth rates in combustion systems. The numerical PDF of the convective fuel time delay is extracted from a steady-state CFD analysis of a swirl-stabilized combustor. An FTF with a normally distributed τ is also investigated and a shift is observed in the frequency and growth rates. On comparing the results, it is observed that the standard FTF predicts the growth rates much different compared to the one with a PDF of the convective fuel time delay, and, in some instances, predicts an unstable system while it is stable or vice versa. The accuracy of the numerical FTF has to be verified with experimental work in future investigations.

5. References

[1] C. a. Armitage, A. J. Riley, R. S. Cant, A. P. Dowling, and S. R. Stow, "Flame Transfer Function for Swirled LPP Combustion From Experiments and CFD," *Vol. 1 Turbo Expo 2004*, pp. 527–537, 2004.

[2] W. Polifke and A. Serbanovic, "Impact of the Fuel Time Lag Distribution in Elliptical Premix Nozzles on Combustion Stability," no. May, pp. 1–12, 2001.

[3] W. Polifke, A. Poncet, C. . Paschereit, and K. Dobbeling, "Reconstruction of acoustic transfer matrices by instationary Computational Fluid Dynamics," *J. Sound Vib.*, vol. 245, no. 3, pp. 483–510, 2001.

[4] T. Poinsot and D. Veynante, "Theoretical and Numerical Combustion," *Combust. Flame*, p. 534, 2005.

[5] H. C. Mongia, T. J. Held, G. C. Hsiao, and R. P. Pandalai, "Incorporation Of Combustion Instability Issues Into Design Process: Ge Aeroderivative And Aero Engines Experience," *Combust. Instab. Gas Turbine Engines*, vol. 210, pp. 43–63, 2006.

[6] T. Scarinci, "Combustion Instability and its Passive Control: Rolls Royce Aeroderivative Engine Experience," *Prog. Astronaut. Aeronaut.*, pp. 65–88, 2005.

[7] C. S. Dowd and J. W. Meadows, "Thermoacoustic Instability Model with Porous Media : Linear stability analysis and the Impact of Porous Media," pp. 1–13, 2018.

[8] S. Patil, J. Cooper, S. Orsino, J. Meadows, R. Valdes, and W. R. Laster, "Investigation of Single-Jet Combustor Near Lean Blowout Conditions Using Flamelet-Generated Manifold Combustion Model and Detailed Chemistry," *J. Eng. Gas Turbines Power*, vol. 138, no. 12, p. 121503, 2016.

[9] J. A. Van Oijen and L. P. H. De Goey, "Modelling Of Premixed Laminar Flames Using Flamelet-Generated Manifolds," vol. 137, pp. 113–137, 2000.

Spring Technical Meeting
Eastern States Section of the Combustion Institute
March 4-7, 2018
State College, Pennsylvania

Effects of Transverse Nozzle Location on High-Frequency Transverse Combustion Instabilities in Can Combustors

Vishal Acharya[1,*], *Timothy Lieuwen*

[1]*School of Aerospace Engineering, Georgia Institute of Technology, Atlanta, Georgia, USA*
[*]*Corresponding Author Email: vishal@aerospace.gatech.edu*

Abstract: During combustion instabilities, the flame is subjected to both "direct" acoustic and "indirect" vortical flow disturbances that are excited by acoustics. The vortical disturbances primarily stem from hydrodynamic instabilities in the flow, such as helical flow disturbances in a swirling flow. In this work, we consider the "direct" velocity coupled response of a premixed flame to the high frequency transverse acoustic modes in a cylindrical can combustor configuration. This work expands on prior work by the authors that analyzed the response of a premixed flame that is centered in the cylindrical combustor with the flame extending to the walls. Due to the non-compact nature of the flame, certain modes were always unstable in the range of control parameters investigated, others had region of stability. We extend this model in the current work to consider a flame that is not centered along the axis of the combustor; in this case, the azimuthal velocity component also influences the Rayleigh index. This is motivated by the fact that can combustors use multiple nozzles that are located off-axis and azimuthally distributed in the combustor. The results from this analysis help understand how different nozzle positions in the combustor head-end affect the Rayleigh Index and hence the overall stability of the combustor.
Keywords: Thermoacoustics, High-frequency instabilities, Can Combustors.

1. Introduction

This paper extends our work on the direct response of premixed flames to transverse acoustic oscillations in a cylindrical can combustor [1]. The work is motivated by the problem of combustion instabilities prevalent in lean premixed systems [2]. These instabilities occur as either: low-frequency longitudinal modes [2-5], low-frequency transverse modes in annular systems [6, 7] and high-frequency transverse modes in both can-type systems [8] as well as jet engine augmentors [9-14].

The connection between acoustic pressure fluctuations and the unsteady heat release stems from the acoustic energy equation. The acoustic energy density is given by:

$$e_1 = \frac{1}{2}\frac{p_1^2}{\rho_0 c_0^2} + \frac{1}{2}\rho_0 \left|\vec{u}_1\right|^2 \tag{1}$$

Using the linearized momentum and energy equations, the volume integrated acoustic energy equation is given by [15]:

$$\frac{\partial}{\partial t}\oiiint_V e_1 dV = \frac{\gamma - 1}{\gamma p_0}\oiiint_V \left(p_1 \dot{q}_1\right) dV - \oiint_S p_1\left(\vec{u}_1 \cdot d\vec{S}\right) \tag{2}$$

Here, p_1 is the acoustic pressure fluctuation, $\dot{q}_1$ is the unsteady heat release rate fluctuation, $\vec{u}_1$ is the acoustic velocity, Note that other source terms arise in the acoustic energy equation that are negligible in magnitude for high Reynolds number flows [15-17]. From the equation, the acoustic energy in the domain grows if the RHS is positive:

$$\frac{\gamma-1}{\gamma p_0}\oiiint_V (p_1\dot{q}_1)\,dV > \oiint_S p_1\left(\vec{u}_1\cdot d\vec{S}\right) \tag{3}$$

The integral on the LHS of this inequality is a gain term and corresponds to the Rayleigh Integral while that on the RHS corresponds to acoustic losses at the boundaries, specifically the inlet and outlet of the combustor volume. Thus, a necessary, but not sufficient condition for instability is that the Rayleigh Integral must be positive:

$$RI = \int_t \oint_V p_1\dot{q}_1 dV dt > 0 \tag{4}$$

Here, *RI* denotes the Rayleigh Index.

The response mechanisms of flames excited by longitudinal or transverse acoustic modes are qualitatively similar with the former having been treated extensively in literature [2, 18-21]. Among the response mechanisms identified were those due to velocity pressure and mixture composition oscillations [15, 19, 22, 23]. This paper focuses on the velocity-coupled flame response; i.e., the sensitivity of the heat release to flow disturbances. The acoustic velocity fluctuations arising from the natural modes of the combustor are the source fluctuations which also excite vortical velocity fluctuations that are due to the natural hydrodynamics in the flow [24-31]. These combine to result in the net flame response. However, experimental results from lower frequency axial disturbances generally show that the largest velocity fluctuations are vortical [32, 33] but the acoustic disturbances serve as the "clock" which sets the natural frequency of the system. An important difference between the two acoustic modes is that longitudinal acoustic modes are symmetric while transverse modes are non-axisymmetric about the flame. While longitudinal instabilities have received much attention, the direct flame response to transverse acoustic oscillations has relatively received less attention. For longitudinal oscillations, the phase between the acoustic pressure and velocity is approximately 90 degrees. Given that the heat release is most sensitive to the velocity, and that the sign of the Rayleigh criterion is controlled by the heat release –pressure phase, it turns out that the phase (or, equivalently, the time delay) between the velocity and heat release is critically important for controlling stability boundaries. For transverse oscillations, the phase need not be 90 degrees and, for traveling waves is closer to zero degrees. Thus, for a given velocity-heat release phase, the stability boundaries of a transverse and axial mode will be completely different.

For can combustor systems, the transverse modes are at high frequencies and a few recent studies have shed light on possible mechanisms for the flame response [34-37]. This also requires a detailed understanding of key mechanisms such as the high frequency excitation of vortical disturbances [36]. Prior modeling work has focused on phenomenological modeling of the flame dynamics [34, 35, 38, 39] where the velocity oscillations from the natural acoustic modes are the only displacement mechanism for the flame. While the fact that the oscillations are transverse does not introduce conceptually new challenges to the analysis, these modes are generally higher frequency and the flame is acoustically non-compact. For the compact flame case, it is only the spatially integrated unsteady heat release that controls system stability which leads to a considerable simplification to the analysis. In contrast, if the flame is non-compact, the same spatially integrated unsteady heat release can lead to substantially different *RI* values and, the spatially integrated flame transfer function is not a useful quantity. This is unfortunate, as it implies that a generic understanding of the flame response and the substantial body of theory and measurements on Flame Transfer Functions (*FTF*) cannot be applied. This also raises the question of how to generalize the spirit of that approach to developing some canonical problems for transverse mode instabilities interacting with non-compact flames. In the spirit of this, prior work by the authors presented a family of canonical problems that can be superposed to understand the response of a larger system (at least in the linear case) [1]. This work presented a theoretical framework for the response of a swirl-stabilized premixed flame to direct transverse acoustic velocity oscillations in a can-type circular combustor where the axis of the axisymmetric flame coincided with the axis of the combustor. The model presented accounted for the local flame dynamics due to local flame displacement, convection of flame disturbances and kinematic restoration processes. Using this local flame dynamics model, the Rayleigh Index was calculated as an indicator of the stability of that mode for the chosen control parameters of the flame and flow. The results indicated the relative stability of different transverse modes for a flame that is centered on the acoustic modes in the cylindrical can. A particularly significant result of this work was the demonstration that some modes (e.g., the radial $(l,0)$ modes) were always unstable, even as the dimensionless flame length changed. This is in significant contrast to typical results for compact flames where stability results are a non-monotonic function of parameters like flame length, and oscillates between positive and negative Rayleigh gain.

In this paper, we present an extension of our previous work to consider the effect of offsetting the nozzle from the center, on the stability of the different transverse acoustic modes. The nozzle offset has important effects on the local flame response and hence the *RI*. For example, for an axisymmetric flame that is centered in the combustor, the azimuthal flow fluctuations have no impact on the flame response. However, when the flame is offset from the center, the relative influence of the azimuthal fluctuations on the flame response increases depending on its azimuthal and radial position in the combustor. The analysis on nozzle offset is motivated by the use of multiple nozzles in can combustor with the azimuthally distributed nozzles being always offset from the combustor axis. In the next section, we briefly summarize the theoretical framework from our earlier work and then present the changes to the analytical representation of the velocity field relative to the co-ordinate system that is centered on the offset flame.

2. Mathematical Formulation

The flame schematic in as shown in Figure 1. The flame base is assumed to be anchored at the center-body implying:

$$\xi(r=R,\theta,t)=0 \tag{5}$$

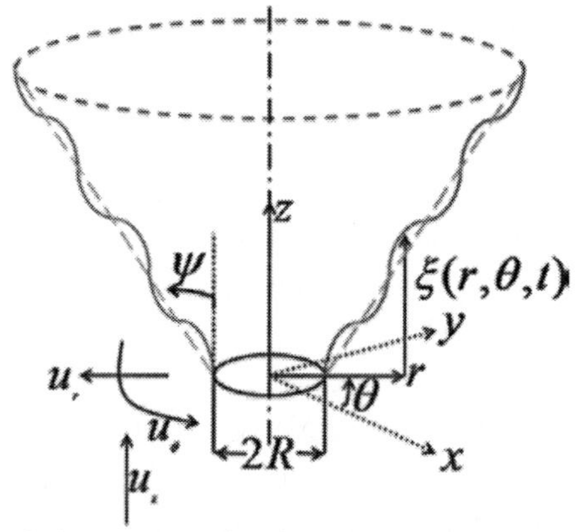

Figure 1 : Schematic of the center-body swirl-stabilized premixed flame [40].

The flame response model is based on the front tracking approach [41-45] for thin premixed flames which results in the following governing equation:

$$\frac{\partial \xi}{\partial t}+u_r\frac{\partial \xi}{\partial r}+\frac{u_\theta}{r}\frac{\partial \xi}{\partial \theta}+s_L\left[\left(\frac{\partial \xi}{\partial r}\right)^2+\frac{1}{r^2}\left(\frac{\partial \xi}{\partial \theta}\right)^2+1\right]^{1/2}=u_z \tag{6}$$

The assumptions for this equation have been discussed in prior work[40]. The velocity field can be decomposed into its unperturbed (subscript '0') and perturbed (subscript '1' for linear first-order) components as:

$$\begin{aligned} u_i(r,\theta,t)&=u_{i,0}(r,\theta)+u_{i,1}(r,\theta,t)\\ \xi(r,\theta,t)&=\xi_0(r,\theta)+\varepsilon\xi_1(r,\theta,t)+O(\varepsilon^2)\end{aligned} \tag{7}$$

Where the sub-script 'i' denotes the r, θ or z coordinate. The evolution equations for an unperturbed axisymmetric flame (ξ_0) and the local flame dynamics (ξ_1) may then be written as:

$$u_{r,0}\frac{d\xi_0}{dr}=u_{z,0}-s_L\left[1+\left(\frac{d\xi_0}{dr}\right)^2\right]^{1/2} \tag{8}$$

$$\frac{\partial \xi_1}{\partial t}+\vec{u}_{0,T}\cdot\vec{\nabla}\xi_1=\left(u_{z,1}-u_{r,1}\frac{d\xi_0}{dr}\right) \tag{9}$$

Here, the net tangential velocity vector along the flame surface, due to both the azimuthal and axial velocity components, is given by:

$$\vec{u}_{0,T}=\vec{u}_0-s_L\vec{e}_n \tag{10}$$

Where, $\vec{e}_n$ is the local unit normal vector, pointing from the nominal unperturbed flame surface into the products. The velocity disturbances used as source terms in Eq.(9) are obtained from the natural acoustic modes of a cylindrical combustor of radius R_C. The polar coordinate system used for the cylindrical combustor's axis and the relative position of the offset flame are shown in Figure 2.

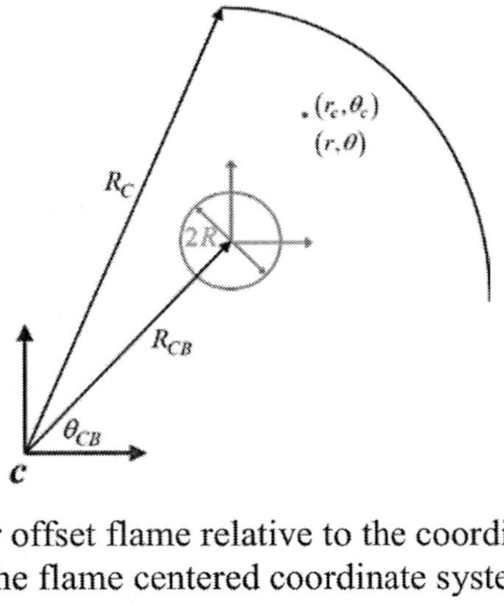

Figure 2 : Coordinate system for offset flame relative to the coordinate system for the cylindrical combustor (subscript 'c'). The flame centered coordinate system is as shown in Figure 1.

The pressure and acoustic velocity disturbances for the purely transverse modes are given by [15]:

$$\hat{p}_{c,1}(r_c,\theta_c)=J_{m_a}(\alpha_{l,m_a}\beta_f r_c)\cos(m_a\theta_c)$$
$$\hat{u}_{r_c,1}(r_c,\theta_c)=\frac{i}{2M_0}\left[J_{m_a+1}(\alpha_{l,m_a}\beta_f r_c)-J_{m_a-1}(\alpha_{l,m_a}\beta_f r_c)\right]\cos(m_a\theta_c) \tag{11}$$
$$\hat{u}_{\theta_c,1}(r_c,\theta_c)=\frac{im_a}{M_0\alpha_{l,m_a}\beta_f r_c}J_{m_a}(\alpha_{l,m_a}\beta_f r_c)\sin(m_a\theta_c)$$

Here,

$$\beta_f=\frac{L_f}{R_C}\qquad M_0=\frac{U_0}{c_0} \tag{12}$$

The overhat notation is used to denote complex frequency domain equivalents of the time-domain quantities. The coordinates have been normalized by the flame length, L_f and the velocities have been normalized by U_0. The acoustic azimuthal mode number is denoted by m_a and the radial mode number by l and α_{l,m_a} is determined as roots of:

$$J'_{m_a}(\alpha_{l,m_a})=0 \tag{13}$$

The first set of roots for different radial and azimuthal mode numbers are shown in Table 1.
The non-dimensional Eigen-frequency is given by:

$$\omega_C=\frac{\omega_0 R_C}{c_0}=\alpha_{l,m_a} \tag{14}$$

Table 1 : Roots of Eq.(13)

$\alpha_{l,m_a}/2\pi$	$m_a=0$	$m_a=1$	$m_a=2$	$m_a=3$	$m_a=4$	$m_a=5$
$l=1$	0	0.2930	0.4861	0.6686	0.8463	1.0211
$l=2$	0.6098	0.8485	1.0673	1.2757	1.4773	1.6743
$l=3$	1.1166	1.3586	1.5867	1.8058	2.0184	2.2261
$l=4$	1.6192	1.8631	2.0961	2.3214	2.5408	2.7554
$l=5$	2.1205	2.3656	2.6018	2.8312	3.0551	3.2747

The above radial and azimuthal velocity in the combustor reference frame are transformed to the reference frame of the flame's coordinate system, using the transformation:

$$\begin{matrix} r_c(r,\theta) \\ \theta_c(r,\theta) \end{matrix}\rightarrow\begin{cases} r_c^2=R_{CB}^2+r^2+2R_{CB}r\cos(\theta-\theta_{CB}) \\ \tan\theta_c=\dfrac{R_{CB}\sin\theta_{CB}+r\sin\theta}{R_{CB}\cos\theta_{CB}+r\cos\theta} \end{cases} \tag{15}$$

The velocity field in the cylinder's coordinate system are transformed to the flame's coordinate system using:

$$\hat{u}_{r,1}(r,\theta)=\begin{cases}\hat{u}_{r_c,1}(r_c(r,\theta),\theta_c(r,\theta))\cos(\theta-\theta_c(r,\theta)) \\ +\hat{u}_{\theta_c,1}(r_c(r,\theta),\theta_c(r,\theta))\sin(\theta-\theta_c(r,\theta))\end{cases}$$
$$\hat{u}_{\theta,1}(r,\theta)=\begin{cases}-\hat{u}_{r_c,1}(r_c(r,\theta),\theta_c(r,\theta))\sin(\theta-\theta_c(r,\theta)) \\ +\hat{u}_{\theta_c,1}(r_c(r,\theta),\theta_c(r,\theta))\cos(\theta-\theta_c(r,\theta))\end{cases} \tag{16}$$

Note that for the axisymmetric flame, the above azimuthal velocity component does not affect the flame response. However, note that the azimuthal velocity component in the acoustic field does affect the flame response through the radial velocity component. From Eq.(15), we can see that when the nozzle is not offset (i.e., $R_{CB} = 0$), we have: $r_c \equiv r$ and $\theta_c \equiv \theta$ and thus in Eq.(16) we have: $\hat{u}_{r,1} \equiv \hat{u}_{r_c,1}$ and the acoustic azimuthal velocity component does not affect the flame response. However, for any flame offset, this results in $\theta \neq \theta_c$ and thus, the acoustic azimuthal velocity component affects the flame response through the local radial velocity component of the disturbances at the flame. This has important implications on how the nozzle offset affects the local flame response and hence the stability of the mode.

3. Model Problem

For convenience of notation, the velocity field is decomposed into azimuthal components (mode number m) as:

$$\hat{u}_{i,1}(r,\theta) = \sum_{m=-\infty}^{+\infty} \hat{\Gamma}_{i,m}(r) e^{im\theta} \tag{17}$$

The following mean-flow and flame configuration is assumed similar to earlier work [46, 47]:

$$\left.\begin{aligned} u_{z,0} &= U_0 \\ u_{r,0} &= 0 \\ u_{\theta,0} &= \Omega r \end{aligned}\right\} \Rightarrow \xi_0(r) = (r - \frac{\beta_R}{\beta_f}) \cot\psi; s_L = \sin\psi \tag{18}$$

The non-dimensional center-body radius and radial flame extent are:

$$\beta_R = \frac{R}{R_C} \qquad \beta_{R_f} = \frac{R_f}{R_C} \tag{19}$$

Note that,

$$\beta_f = \left(\beta_{R_f} - \beta_R\right) \cot\psi \tag{20}$$

The frequency domain representation for the flame dynamics can be decomposed into its azimuthal components as:

$$\hat{\xi}_1(r,\theta) = \sum_{m=-\infty}^{\infty} \hat{\zeta}_{1,m}(r) e^{im\theta} \tag{21}$$

Applying this to Eq.(9) and solving, we have:

$$\hat{\zeta}_{1,m}(r) = -\frac{e^{-i\chi_m r}}{\sin^2\psi} \int_{\beta_R/\beta_f}^{r} e^{i\chi_m r} \hat{\Gamma}_{r,m}(r) dr \tag{22}$$

Here,

$$\begin{aligned} St &= \frac{\tilde{\omega}}{\cos\psi \sin\psi} \qquad St_\Omega = \frac{\Omega L_f}{u_0} \\ \tilde{\omega} &= \frac{\alpha_{l,m_a} \beta_f}{M_0} \qquad \sigma = \frac{St_\Omega}{\tilde{\omega}} \qquad \chi_m = St(m\sigma - 1) \end{aligned} \tag{23}$$

This local flame dynamics solution then feeds into the unsteady heat release rate calculation. Under the assumptions of the analysis, any oscillations in the local unsteady heat release rate are introduced purely by changes in the local flame area element given by:

$$dA = rdrd\theta\sqrt{1+\left(\frac{\partial\xi}{\partial r}\right)^2+\frac{1}{r^2}\left(\frac{\partial\xi}{\partial\theta}\right)^2} \tag{24}$$

Linearizing this equation, the unsteady heat release rate oscillations are given by:

$$\hat{q}_1 \sim \cos\psi\frac{\partial\hat{\xi}_1}{\partial r} = \cos\psi\sum_{m=-\infty}^{+\infty}\frac{d\hat{\zeta}_{1,m}(r)}{dr}e^{im\theta} \tag{25}$$

The Rayleigh Index in Eq. (4) is evaluated at the Eigen-frequencies:

$$RI \sim \cos\psi\int_t dt\int_r rdr\int_{\theta=0}^{2\pi}\mathrm{Re}\left\{\hat{p}_{c,1}\left(r_c(r,\theta),\theta_c(r,\theta)\right)e^{-i\bar{\omega}t}\right\}\mathrm{Re}\left\{\sum_{m=-\infty}^{+\infty}\frac{d\hat{\zeta}_{1,m}(r)}{dr}e^{im\theta}e^{-i\bar{\omega}t}\right\}d\theta \tag{26}$$

The integrated acoustic energy (from Eq.(2)) at the chosen Eigen-frequency is given by:

$$E = \int_t dt \oiiint_V e_1 dV \sim \int_t dt\int_{r_c} r_c dr_c \int_{\theta_c=0}^{2\pi}\left[\begin{array}{l}\mathrm{Re}\left\{\hat{p}_{c,1}(r_c,\theta_c)e^{-i\bar{\omega}t}\right\}\mathrm{Re}\left\{\hat{p}_{c,1}(r_c,\theta_c)e^{-i\bar{\omega}t}\right\}\\ +\frac{1}{2}\mathrm{Re}\left\{\hat{\vec{u}}_{c,1}(r_c,\theta_c)e^{-i\bar{\omega}t}\right\}\bullet\mathrm{Re}\left\{\hat{\vec{u}}_{c,1}(r_c,\theta_c)e^{-i\bar{\omega}t}\right\}\end{array}\right]d\theta_c \tag{27}$$

Note that each of these integrals (and Eq.(22)) can be computed numerically (for example, using quadrature). A normalized Rayleigh index is then defined as:

$$\Re = \frac{RI}{2E} \tag{28}$$

Physically, $\Re$ corresponds to the exponential growth or decay rate of the disturbances. This normalized Rayleigh Index can be calculated for different values of the control parameters and can then be used to determine if it is positive or negative depending on the offset of the flame.

4. Example Illustration

In this section, we present some illustrative results using the model problem to quantitatively depict the effects of nozzle positioning on the thermoacoustic stability of the flame. We consider only the first radial mode ($(l,m_a)=(2,0)$) and the first radial-azimuthal mode ($(l,m_a)=(1,1)$) from Table 1. The effect of nozzle positioning is determined using the radial offset parameter, R_{CB}, and the azimuthal offset parameter, θ_{CB}, both depicted in Figure 2.

First, consider the effect of nozzle position on the stability of the first radial mode as shown in Figure 3(a). Since the radial mode is axisymmetric, the azimuthal offset parameter has no effect and the plot shows the variation of $\Re$ with R_{CB} only. Note that for a nozzle that is centered in the combustor, $\Re > 0$ and as R_{CB} increases, the flame is relatively more stable. Beyond a certain value of R_{CB}, we have $\Re < 0$ indicating that this location of the flame in the radial mode shape becomes stable. This can be attributed to the fact that for no offset, the radial velocity of the mode is fully experienced by the flame resulting in maximum heat release oscillations. However, as the flame is offset, only the normal component of this radial velocity locally at the nozzle location results in heat release oscillations which are lower than that for a centered nozzle.

Next, consider the effect of nozzle position on the stability of the first radial-azimuthal mode as shown in Figure 3(b). Since this is an asymmetric mode, both the radial and azimuthal offset have effects on the stability of the mode. The figure shows the effect of azimuthal positioning of the nozzle for two different radial offsets, $R_{CB}=0.6$ corresponding to the red dots and

$R_{CB} = 0.3$ corresponding to the blue dots. The azimuthal position variation is shown only in the range $0 \le \theta_{CB} \le \pi/2$ since for the (1,1) mode, the results are symmetric about $\theta_{CB} = 0$ and $\theta_{CB} = \pi/2$ planes. As seen in the figure, the outer flames ($R_{CB} = 0.6$) are relatively more stable than the inner flames ($R_{CB} = 0.3$). Additionally, note that the inner flames are relatively more stable towards $\theta_{CB} = \pi/2$ but still have $\Re > 0$ over the entire range. In contrast, the outer flames which are relatively more stable, have $\Re > 0$ for lower azimuthal offsets and then result in $\Re < 0$ towards $\theta_{CB} = \pi/2$. This indicates that outer flames near the $\theta_{CB} = \pi/2, 3\pi/2$ azimuths are the most stable, for the chosen flame/flow parameters. These results have important implications on the design choice for nozzle positioning in can combustors with multi-nozzle configurations depending on the nature of the acoustic mode of interest.

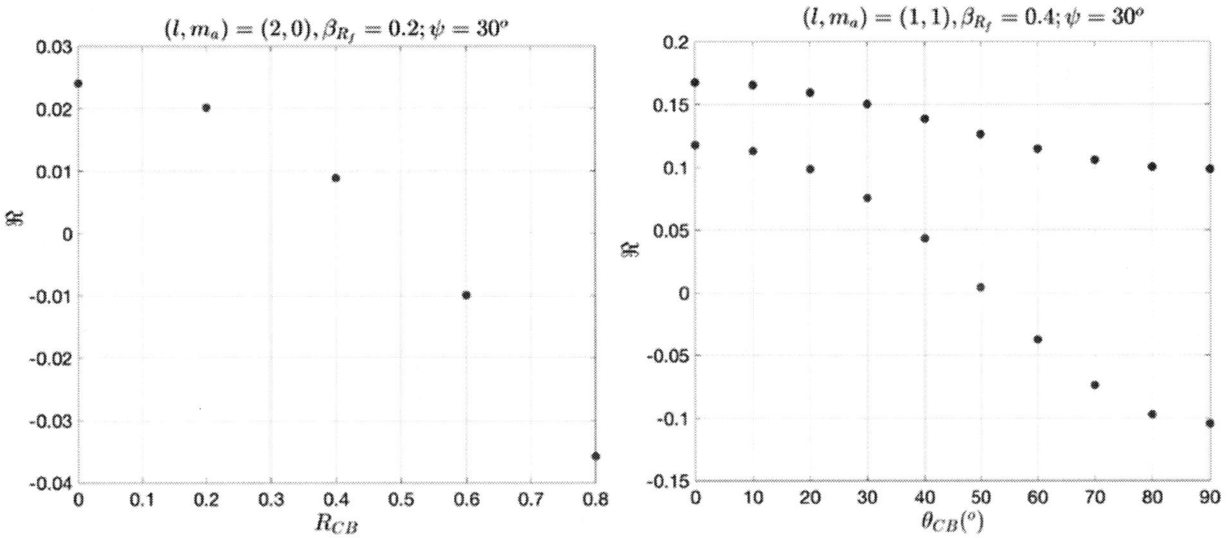

Figure 3 : (a) Effect of nozzle offset on the stability of the first radial mode. Nozzle offset is purely determined by the radial offset of the nozzle. (b) Effect of nozzle offset on the stability of the first radial-azimuthal mode. Red corresponds to $R_{CB} = 0.6$ and blue corresponds to $R_{CB} = 0.3$.

5. Conclusions

This paper presents the effects of nozzle positioning in a can combustor, on the stability of transverse modes. Earlier work by the authors developed a model using the G-equation that was applied to a nozzle centered in the cylindrical can. In the present work, the authors have extended this to consider the changes in the velocity mode shape relative to a flame whose axis is offset from the combustor axis.

The model considers the generation of local flame disturbances due to local flame displacement as well as flame disturbance convection and kinematic restoration, i.e. flame propagation. The local flame dynamics solution was determined in integral form using which, the normalized Rayleigh Index was calculated to determine mode stability. The example calculations depict the variations in stability of a given natural acoustic mode to different offset parameters. In earlier work, the radial mode was shown to be relatively less stable, i.e. $\Re > 0$ always, for a flame that is centered in the combustor. However, the results in this paper indicate that as the radial offset increased, the flame became relatively more stable resulting in $\Re < 0$. For the $(l, m_a) = (1,1)$ mode, results showed that outer flames were relatively more stable than inner flames. In addition, flames located closer to the $\theta_{CB} = \pi/2, 3\pi/2$ plane relative to the mode shape resulted in the most stable flames, specifically resulting in $\Re < 0$.

Collectively, these results give a guide to how for a given flame/flow configuration, the position of the nozzle in the combustor can be used to create a relatively more stable thermoacoustic system. An important point to note is that the analysis considers only velocity fluctuations due to the natural acoustic modes. The interaction of these acoustic modes with the natural fluid dynamic instabilities can result in high-frequency vortical velocity disturbances which also disturb the flame, which shall be a focus of future studies.

5. Acknowledgements

This work has been partially supported by the US Department of Energy through the UTSR program under contract DE-FE0031285 monitored by Mark Freeman.

6. References

[1] V. Acharya, T. Lieuwen, Modeling Premixed Flame Response to Transverse Acoustic Modes, AIAA Science and Technology Forum, AIAA, Kissimmee, FL, USA, 2018.
[2] T.C. Lieuwen, V. Yang, F.K. Lu, Combustion instabilities in gas turbine engines: operational experience, fundamental mechanisms and modeling, American Institute of Aeronautics and Astronautics2005.
[3] H.C. Mongia, T.J. Held, G.C. Hsiao, R.P. Pandalai, Incorporation of Combustion Instability Issues into Design Process: GE Aeroderivative and Aero Engines Experience, in: T.C. Lieuwen, V. Yang (Eds.), Combustion Instabilities in Gas Turbine Engines, AIAA, Washington D.C., 2005.
[4] J. Cohen, G. Hagen, A. Banaszuk, S. Becz, Attenuation of Combustor Pressure Oscillations Using Symmetry Breaking, 49th AIAA Aerospace Sciences Meeting, Orlando, FL, 2011.
[5] K. Smith, J. Blust, Combustion Instabilities in Industrial Gas Turbines: Solar Turbines' Experience, in: T.C. Lieuwen, V. Yang (Eds.), Combustion Instabilities in Gas Turbine Engines, AIAA, Washington D.C., 2005, pp. 29-42.
[6] W. Krebs, S. Bethke, J. Lepers, P. Flohr, B. Prade, Thermoacoustic Design Tools and Passive Control: Siemens Power Generation Approaches, in: T.C. Lieuwen, V. Yang (Eds.), Combustion Instabilities in Gas Turbine Engines, AIAA, Washington D.C., 2005, pp. 89-112.
[7] M. Zellhuber, C. Meraner, R. Kulkarni, W. Polifke, B. Schuermans, Large eddy simulation of flame response to transverse acoustic excitation in a model reheat combustor, Journal of engineering for gas turbines and power 135 (2013).
[8] J. Sewell, P. Sobieski, Monitoring of Combustion Instabilities: Calpine's Experience, in: T.C. Lieuwen, V. Yang (Eds.), Combustion Instabilities in Gas Turbine Engines, AIAA, Washington D.C., 2005, pp. 147-162.
[9] H.B. Ebrahimi. Overview of gas turbine augmentor design, operation, and combustion oscillation. In: editor^editors. 42nd AIAA/ASME/SAE/ASEE Joint Propulsion Conference; 2006; Sacramento, CA. p.
[10] D.E. Rogers, F.E. Marble, A mechanism for high-frequency oscillation in ramjet combustors and afterburners, Journal of Jet Propulsion 26 (1956) 456-462.
[11] I. Elias, Acoustical Resonances Produced by Combustion of a Fuel-Air Mixture in a Rectangular Duct, The Journal of the Acoustical Society of America 31 (1959) 296.

[12] W.E. Kaskan, A.E. Noreen, High-Frequency Oscillations of a Flame Held by a Bluff Body, Transactions of the ASME, (1955) 885-895.
[13] C.R. King, Experimental investigation of effects of combustion chamber length and inlet total temperature, total pressure, and velocity on afterburner performance, NACA, 1957.
[14] L.L. Staff, A Summary of Preliminary Investigations into the Characteristics of Combustion Screech in Ducted Burners, NACA, 1958.
[15] T.C. Lieuwen, Unsteady Combustor Physics, Cambridge University Press2012.
[16] M. Brear, F. Nicoud, M. Talei, A. Giauque, E. Hawkes, Disturbance energy transport and sound production in gaseous combustion, Journal of Fluid Mechanics 707 (2012) 53-73.
[17] A. Dowling, Thermoacoustic Sources and Instabilities, in: D. Crighton, A. Dowling, Ffowcs Williams JE, M. Heckl, F. Leppington (Eds.), Modern Methods in Analytical Acoustics: Lecture Notes, Springer-Verlag, New York, 1992, pp. 378-405.
[18] K. McManus, T. Poinsot, S. Candel, A review of active control of combustion instabilities, Progress in energy and combustion science 19 (1993) 1-29.
[19] S. Ducruix, T. Schuller, D. Durox, S. Candel, Combustion dynamics and instabilities: Elementary coupling and driving mechanisms, Journal of Propulsion and Power 19 (2003) 722-734.
[20] Y. Huang, V. Yang, Dynamics and stability of lean-premixed swirl-stabilized combustion, Progress in Energy and Combustion Science 35 (2009) 293-364.
[21] S. Candel, D. Durox, T. Schuller, J.-F. Bourgouin, J.P. Moeck, Dynamics of Swirling Flames, Annual Review of Fluid Mechanics 46 (2014) 147-173.
[22] K. Venkataraman, L. Preston, D. Simons, B. Lee, J. Lee, D. Santavicca, Mechanism of combustion instability in a lean premixed dump combustor, Journal of Propulsion and Power 15 (1999) 909-918.
[23] Shreekrishna, Response Mechanisms of Attached Premixed Flames Subjected to Harmonic Forcing, Aerospace Engineering, Georgia Institute of Technology, Atlanta, GA, 2011.
[24] T.J. Poinsot, A.C. Trouve, D.P. Veynante, S.M. Candel, E.J. Esposito, Vortex-driven acoustically coupled combustion instabilities, Journal of Fluid Mechanics 177 (1987) 265-292.
[25] K. Schadow, E. Gutmark, T. Parr, D. Parr, K. Wilson, J. Crump, Large-scale coherent structures as drivers of combustion instability, Combustion science and technology 64 (1989) 167-186.
[26] S.K. Thumuluru, T. Lieuwen, Characterization of acoustically forced swirl flame dynamics, Proceedings of the Combustion Institute 32 (2009) 2893-2900.
[27] P. Palies, D. Durox, T. Schuller, S. Candel, The combined dynamics of swirler and turbulent premixed swirling flames, Combustion and Flame 157 (2010) 1698-1717.
[28] Preetham, S. Hemchandra, T. Lieuwen, Dynamics of Laminar Premixed Flames Forced by Harmonic Velocity Disturbances, Journal of Propulsion and Power 24 (2008) 1390-1402.
[29] C.O. Paschereit, E. Gutmark, W. Weisenstein, Coherent structures in swirling flows and their role in acoustic combustion control, Physics of Fluids (1994-present) 11 (1999) 2667-2678.
[30] R. Balachandran, B. Ayoola, C. Kaminski, A. Dowling, E. Mastorakos, Experimental investigation of the nonlinear response of turbulent premixed flames to imposed inlet velocity oscillations, Combustion and Flame 143 (2005) 37-55.
[31] Y. Huang, V. Yang, Effect of swirl on combustion dynamics in a lean-premixed swirl-stabilized combustor, Proceedings of the Combustion Institute 30 (2005) 1775-1782.
[32] B. Emerson, U. Mondragon, V. Acharya, D.-H. Shin, C. Brown, V. McDonell, T. Lieuwen, Velocity and flame wrinkling characteristics of a transversely forced, bluff-body stabilized

flame, part I: experiments and data analysis, Combustion Science and Technology 185 (2013) 1056-1076.
[33] V. Acharya, B. Emerson, U. Mondragon, D.-H. Shin, C. Brown, V. McDonell, T. Lieuwen, Velocity and flame wrinkling characteristics of a transversely forced, bluff-body stabilized flame, Part II: Flame response modeling and comparison with measurements, Combustion Science and Technology 185 (2013) 1077-1097.
[34] J. Schwing, F. Grimm, T. Sattelmayer. A model for the thermo-acoustic feedback of transverse acoustic modes and periodic oscillations in flame position in cylindrical flame tubes. In: editor^editors. ASME Turbo Expo 2012: Turbine Technical Conference and Exposition; 2012: American Society of Mechanical Engineers. p. 553-566.
[35] J. Schwing, T. Sattelmayer, High-Frequency Instabilities in Cylindrical Flame Tubes: Feedback Mechanism and Damping, ASME Turbo Expo, San Antonio, TX, 2013.
[36] J. Schwing, T. Sattelmayer, N. Noiray, Interaction of Vortex Shedding and Transverse High-Frequency Pressure Oscillations in a Tubular Combustion Chamber, ASME Turbo Expo, Vancouver, Canada, 2011.
[37] D. Greenblatt, C. Paschereit, O. Lehmann, D. Luchtenburg, B. Noack, Spatiotemporal characterization of a conical swirler flow field under strong forcing, Journal of Engineering for Gas Turbines and Power 131 (2009) 031504-031501.
[38] T. Hummel, F. Berger, M. Hertweck, B. Schuermans, T. Sattelmayer, High-Frequency Thermoacoustic Modulation Mechanisms in Swirl-Stabilized Gas Turbine Combustors—Part II: Modeling and Analysis, Journal of Engineering for Gas Turbines and Power 139 (2017) 071502.
[39] T. Hummel, M. Schulze, B. Schuermans, T. Sattelmayer. Reduced Order Modeling of Transversal and Non-Compact Combustion Dynamics. In: editor^editors. 22nd International Congress on Sound and Vibration, July 12-16, Florence, Italy; 2015. p.
[40] V. Acharya, D.H. Shin, T. Lieuwen, Premixed Flames Excited by Helical Disturbances: Flame Wrinkling and Heat Release Oscillations, Journal of Propulsion and Power 29 (2013) 1282-1291.
[41] M. Fleifil, A.M. Annaswamy, Z.A. Ghoneim, A.F. Ghoniem, Response of a laminar premixed flame to flow oscillations: A kinematic model and thermoacoustic instability results, Combustion and Flame 106 (1996) 487-510.
[42] G.H. Markstein, Nonsteady flame propagation, Pergamon1964.
[43] M. Matalon, B. Matkowsky, Flames as gasdynamic discontinuities, Journal of Fluid Mechanics 124 (2006) 239-259.
[44] L. Boyer, J. Quinard, On the Dynamics of Anchored Flames, Combustion and Flame 82 (1990) 51-65.
[45] F. Williams, Turbulent combustion, in: J. Buckmaster (Ed.), The Mathematics of combustion1985, pp. 97-131.
[46] V. Acharya, Shreekrishna, D.H. Shin, T. Lieuwen, Swirl effects on harmonically excited, premixed flame kinematics, Combustion and Flame 159 (2012) 1139-1150.
[47] V. Acharya, T. Lieuwen, Premixed Flame Response to Helical Disturbances: Mean Flame Non-axisymmetry effects, Combustion and Flame 165 (2016) 188-197.

Spring Technical Meeting
Eastern States Section of the Combustion Institute
March 4-7, 2018
State College, Pennsylvania

Effects of the equivalence ratio transient durations on self-excited combustion instability time scales in a single nozzle combustor

*Xiaoling Chen, Wyatt Culler, Stephen Peluso, Domenic Santavicca, Jacqueline O'Connor**

Department of Mechanical and Nuclear Engineering, Pennsylvania State University, University Park, Pennsylvania, United States
**Corresponding Author Email: jxo22@engr.psu.edu*

Abstract: Fuel staging is an effective way to attenuate combustion instability in lean, premixed combustion systems. Combustion instability occurs due to the coupling between heat release rate oscillations and combustor acoustics. In our experiments, we seek to understand how transient fuel staging affects the structure and dynamics of a single flame to help inform our previous studies of multi-flame systems. In the current experiments, fuel staging is achieved by changing the equivalence ratio in the nozzle. The equivalence ratio in the nozzle is varied over two different timescales, or transient durations, which are 1 ms and 4 s. For each duration, we also vary the percentage change in the equivalence ratio, with levels of ±3.57% and ±7.14% of the mean equivalence ratio, ϕ=0.7. Flame images recorded using CH* chemiluminescence imaging are analyzed to show the variations of the phase difference between pressure and heat release oscillations during transients. Phase difference transition behaves similarly in the flame base region but differently in flame/wall interaction region. More similarities between short and long duration transients are observed in combustion instability onset direction than decay direction.
Keywords: Combustion instability, Flame transient, Time scale

1. Introduction

Lean-premixed combustion is used to reduce the NO_x emissions in gas turbine combustors. However, lean flames sometimes suffer from combustion instability, which is caused by the coupling between the combustor acoustic field and heat release rate oscillations [1]. Both passive and active control methods have been used to suppress the combustion instabilities [2]. Fuel splitting, or when fuel is unevenly distributed between nozzles in a multi-nozzle system, is an effective way to attenuate the combustion instabilities [3]. Samarasinghe *et al.* [4] recently studied the mechanism behind longitudinal combustion instability suppression via fuel staging in a multi-nozzle can combustor, showing that fuel staging suppresses instabilities through variations in the heat release rate fluctuation phase between the staged nozzle and the other nozzles.. The transient behavior of the combustion instability suppression by the fuel staging in the multi-nozzle combustor was also recently studied [5]. Time scales of the combustion instability transition were found to be dependent on the fuel staging amount, transient timescale, and transient direction.

In this paper, we consider the effects of the transient timescales on the combustion instability transition characteristics in a single nozzle combustor (SNC), to better understand the processes in each individual flame in the fuel staging transient timescale variation studies in the multi-nozzle combustor.

2. Methods / Experimental

2.1 Experimental Setup

A single nozzle, lean, premixed swirling combustor is operated in a transient matter by increasing and decreasing the equivalence ratio over different time durations. A quartz liner with a diameter of 110 mm and length of 305 mm is installed to provide optical access to the flame. The geometric swirl number of the nozzle is 0.7 and the diameter of the nozzle is 55 mm. An illustration of the combustor is displayed in **Figure 1**. Perfectly premixed fuel and air flow to the system after a choke plate, which prevents fluctuations in the equivalence ratio in the main flow. Additional fuel is injected through holes on the swirler blades 0.1 m upstream the dump plane, and is controlled by a solenoid valve that varies the staging fuel quantity and delivery timescale. Previous mixing measurements show that the staging fuel is completely mixed by the exit of the nozzle [6].

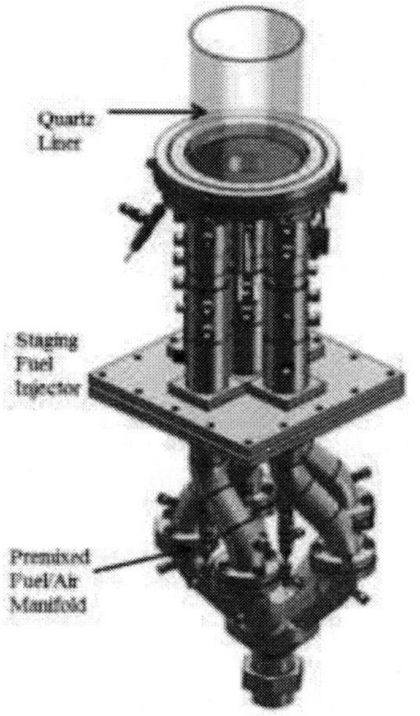

Figure 1. Geometry of the single nozzle combustor

2.2 Diagnostics Technique

An infrared (IR) absorption technique is applied to measure the temporal fuel concentration variation at the dump plane to obtain the fuel delivery characteristic timescale. Water-cooled, recess-mounted PCB dynamic pressure transducers mounted below the dump plate and on the center nozzle are used to record the pressure fluctuations within the combustor at a sampling rate of 16,000 Hz. A Photron SA4 high speed camera fitted with an Invisible Vision UVi 1850-10 intensifier, a Nikon AF Micro-Nikkor 60mm f/208 lens, and a 432±5nm bandpass filter is used to obtain high-speed CH* chemiluminescence images, a marker of heat release [7].

2.3 Data Processing Method

A logistic regression method is used to calculate the transient timescales, both the fuel delivery timescale as well as the instability growth and decay timescales. The transition median time, t_0, and timescale, τ, are calculated in the logistic fit, as described in detail in Ref. [5]. To visualize the instantaneous phase between pressure and heat release rate fluctuations from chemiluminescence imaging, we use a Hilbert transform on both the pressure and image time series to calculate the relative instantaneous phase between the two. This process is described in detail in Ref.[5].

2.4 Test Matrix

Two fuel variation durations and two directions are selected for the transient tests with three amplitudes for the non-reacting (IR measurements) and in the reacting conditions, two amplitudes. The detailed test matrix is displayed in **Table 1** and **Table 2**.

Table 1: Main fuel/air flow characteristic parameters

Fuel flow rate (*g/s*)	1.16
Air flow rate (*g/s*)	28.52
Inlet mixture temperature (*K*)	473.15
Nozzle flow velocity (*m/s*)	26
Inlet Reynolds number Re_d	17,000

Table 2: Test matrix for the SNC transient tests

Transient duration	Transient direction (Initial ϕ→ Final ϕ)				
	Non-reacting			Reacting	
1 ms	0.625→0.7	0.65→0.7	0.675→0.7	0.65→0.7	0.675→0.7
	0.7→0.625	0.7→0.65	0.7→0.675	0.7→0.65	0.7→0.675
4 s	0.625→0.7	0.65→0.7	0.675→0.7	0.65→0.7	0.675→0.7
	0.7→0.625	0.7→0.65	0.7→0.675	0.7→0.65	0.7→0.675

3. Results and Discussion

3.1 Fuel delivery characteristics

To better understand the transition from stable to unstable, or unstable to stable, operation, we first characterize the rate at which fuel is delivered to the combustor at different commanded transient durations.

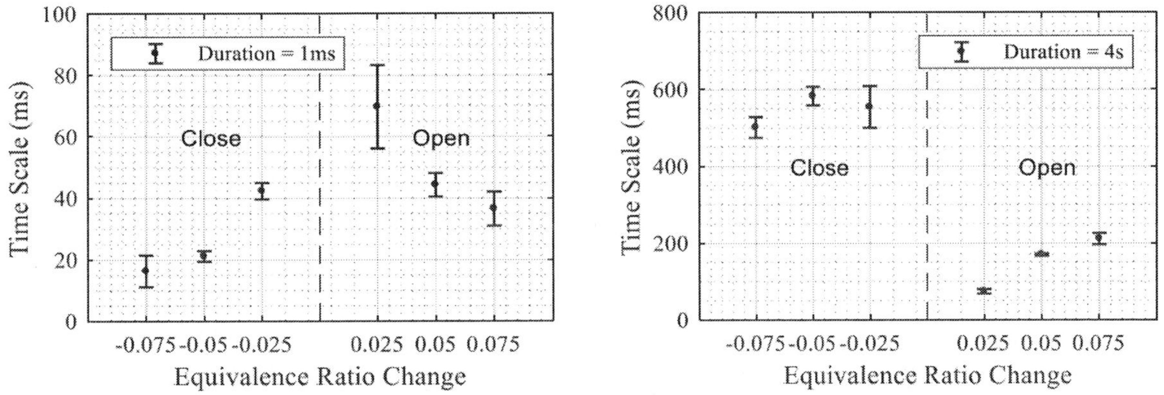

(a) Transient duration 1 ms (b) Transient duration 4 s

Figure 2: Dependence of fuel delivery time scales on ϕ change in two transient durations

In **Error! Reference source not found.**, the dependence of the fuel delivery timescales on equivalence ratio changes is illustrated, where the mean value is denoted by the red center and one standard deviation is denoted by the error bars. For the short-duration transients, the timescales associated with fuel delivery vary with equivalence ratio but not significantly with opening direction; the fuel delivery timescale is higher for larger equivalence ratio variations. For the long-duration transients, however, the equivalence ratio increase and decrease timescales are significantly different, where timescales associated with valve opening are much shorter than those of valve closing. These differences are likely due to both variations in the operation of the valve itself as well as the flow processes in the fuel circuit.

3.2 Pressure time scales

Similarly, logistic regression is also applied to combustor pressure fluctuation signals to obtain the timescales and median times of the pressure oscillation transition in both short and long duration transients. The dependence of the time scales/median times on the transient directions and amounts are plotted in boxplot [5] in Figure 3. In the boxplot, the red line indicates the median value of the ensemble and height of the box is determined by the first and third quartile of the data. Outliers of each case are denoted by the red crosses. The notch for the box represents the 95% confidence intervals, and the numbers above each box indicate the number of repeats in each ensemble at each condition.

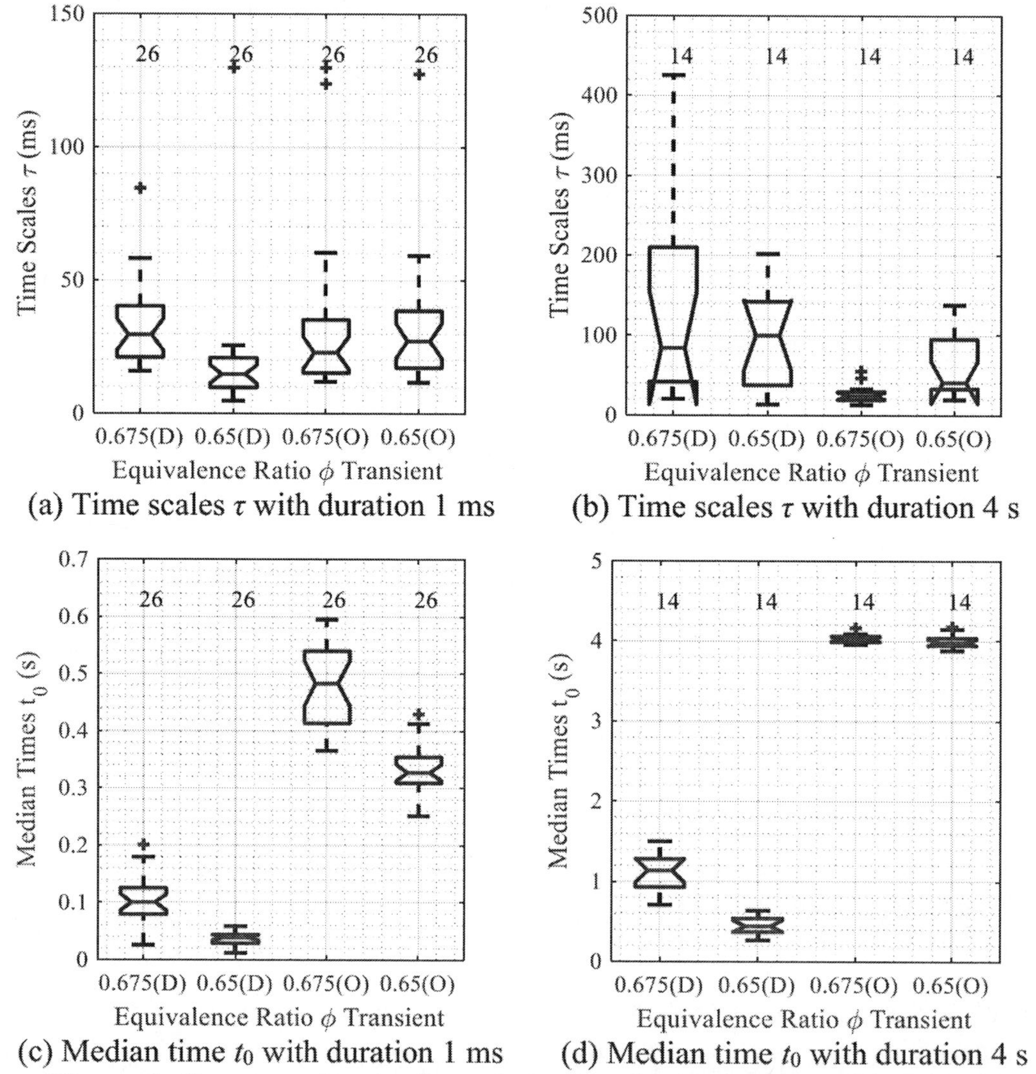

(a) Time scales τ with duration 1 ms (b) Time scales τ with duration 4 s

(c) Median time t_0 with duration 1 ms (d) Median time t_0 with duration 4 s

Figure 3: Pressure time scale and median time statistical distributions

The x-axis in the box plots shows the smaller equivalence ratio value during the transients and the capital letter inside the parenthesis indicates the direction of the transients. "D" represents combustion instability decay and "O" represents instability onset. For short duration transients, timescales of pressure oscillation transition are nearly independent of the ϕ change direction and amplitude, with the only exception that instability decay occurs more quickly with a smaller final ϕ. Differences between timescales for different ϕ change amplitudes for the long duration transients are not statistically significant. However, timescales for the long-duration transients are longer as compared with those of the short duration transients, which is congruent with the commanded transient duration. Instability onset occurs more repeatedly than instability decay, which is likely due to the differences in the fuel delivery characteristics for the long duration transient shown in **Error! Reference source not found.**. Dependence of the median times, t_0, on ϕ change directions and amplitudes are similar for the two transient durations; pressure onset occurs later than pressure decay and larger ϕ change amplitudes lead to an earlier transition of the pressure oscillation.

The pressure and ϕ time series with their logistic fits are plotted in one plot in Figure 4. Original data are plotted on top of the logistic curve in green (pressure) and blue (equivalence ratio). The median time, t_0, of the pressure transition occurs a little bit earlier than the equivalence ratio in direction of instability onset and opposite in direction of instability decay for short transient duration. For the long transient duration, the pressure transition always occurs earlier than the ϕ in the instability onset direction no matter the amplitude of ϕ change. In direction of the instability decay, t_0 of pressure is smaller that of ϕ for $\Delta\phi$=0.05 but larger for $\Delta\phi$=0.025. This is because the instability decays at a critical ϕ around 0.68 and it takes shorter time for the equivalence ratio to become this critical value in larger change of decreasing ϕ. A summary of the critical equivalence ratio in instability onset and decay with long transient duration is plotted in Figure 5.

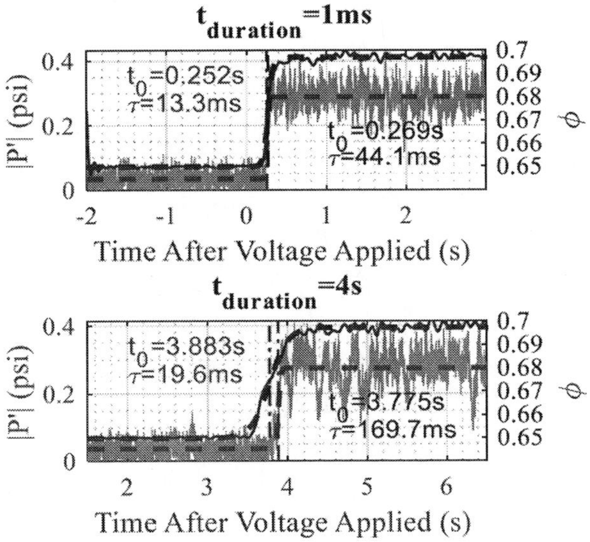

(a) 0.65 → 0.70 Direction

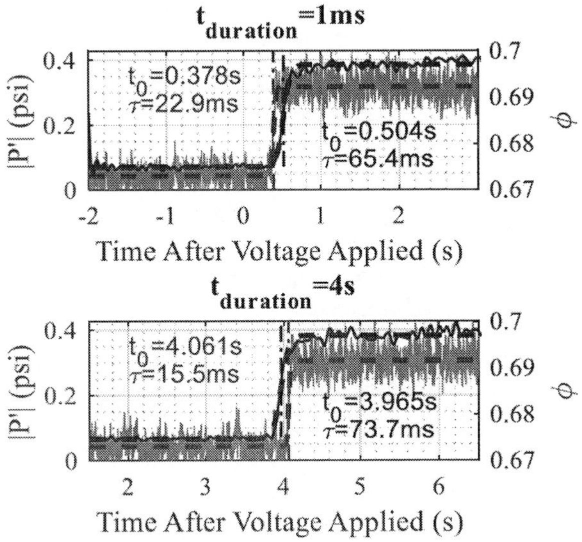

(b) 0.675 → 0.70 Direction

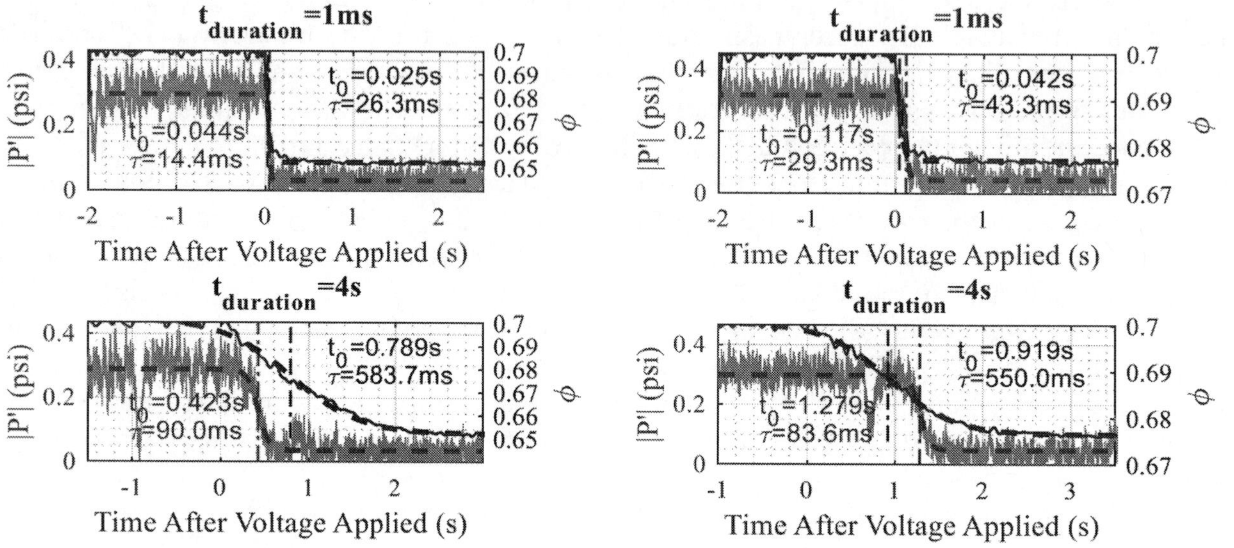

(c) 0.70 → 0.65 Direction (b) 0.70 → 0.675 Direction

Figure 4: Equivalence ratio and pressure oscillation variations

The critical equivalence ratio ϕ_{cr} is the equivalence ratio at the median time, t_0, of the pressure transition curve. The different ϕ_{cr} are plotted in Figure 5 for different transient directions and amplitudes, where the error bars show one standard deviation in the data from all the ensembles. The ϕ_{cr} is between 0.68 and 0.69 for all cases. No strong difference in the critical equivalence ratio is observed amongst all the cases, which supports the conclusion that a critical equivalence ratio indeed exists.

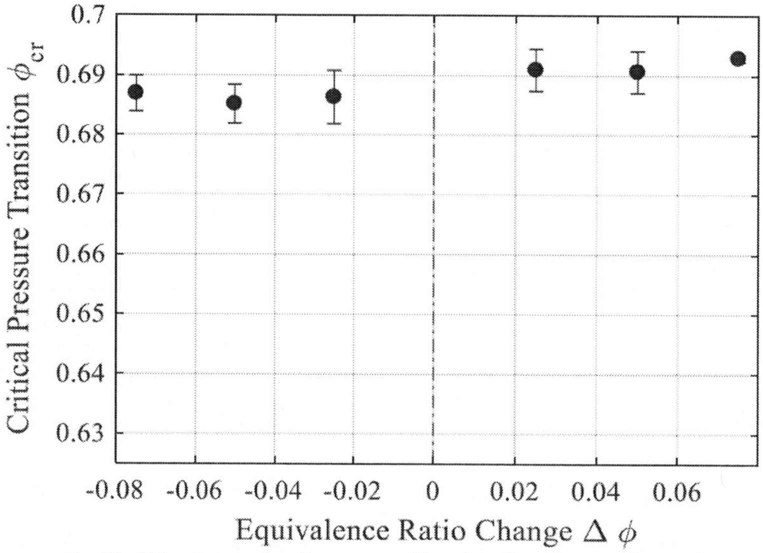

Figure 5: Critical equivalence ratio ϕ_{cr} for long duration transient

3.3 Phase image analysis

Instantaneous phase differences between pressure and heat release rate oscillations captured by CH* chemiluminescence intensity for both short and long duration transients are displayed in Figure 6. The white line is the $\pi/2$ iso-contour, which shows the delineation between regions that drive and those that damp the instability. The corresponding pressure time series is at the bottom of the phase images with left for short duration transients and right for long duration transients.

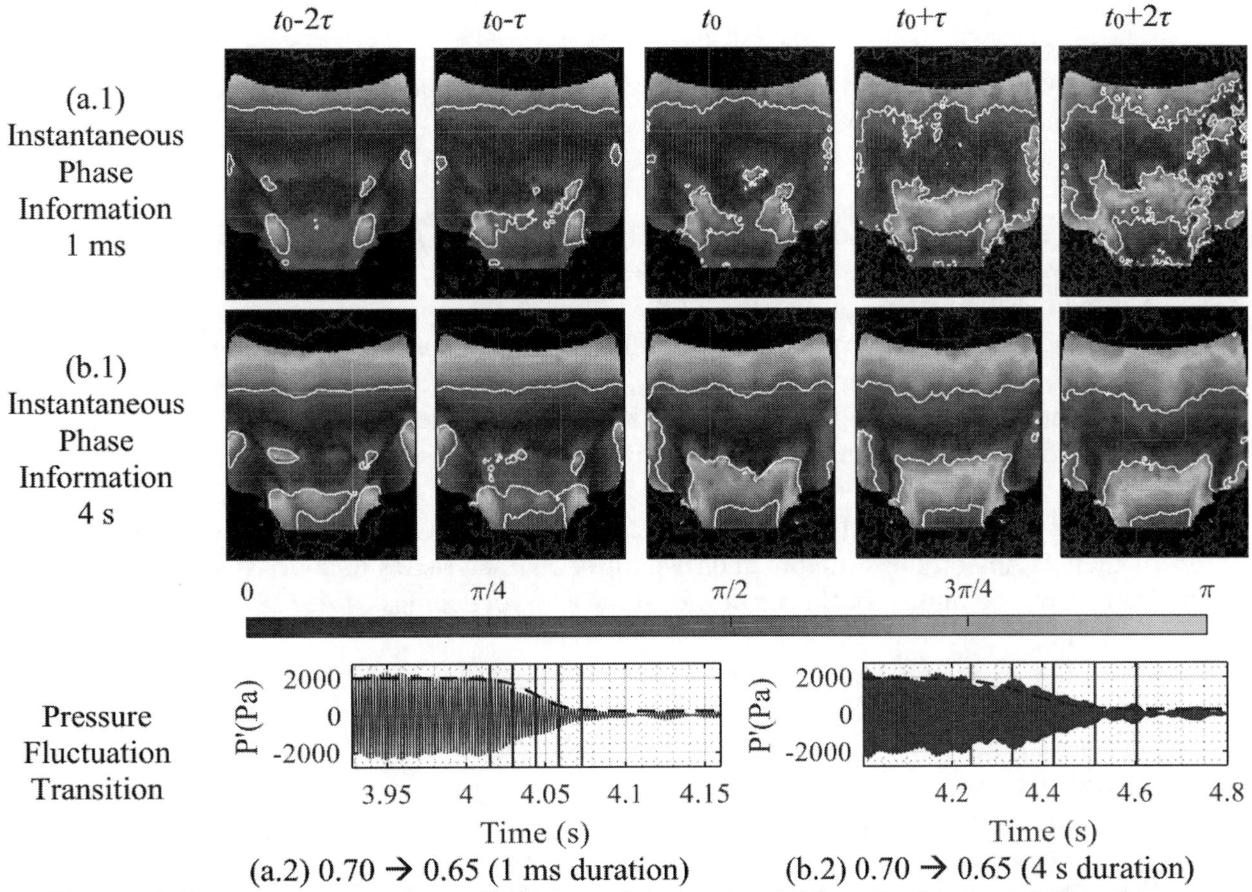

Figure 6: Instantaneous phase difference images transition during instability decay

There is little difference in the instantaneous phase distribution between the short duration and long duration transients in both the decay and onset directions, although only one (instability decay direction) is shown. This is different than the multi-nozzle combustor, where the onset and decay processes are quite different, especially for the intermediate steps during short duration transients. In the multi-nozzle combustor, the instability onset shows the entire center region of the combustor shifts in-phase very quickly, while the center region shifts first followed by the rest of the flame in instability decay direction. In long duration decay transient, out-of-phase appears mostly in both the flame/wall interaction and flame base regions. However, in short transient duration, the region slightly above the flame base transits to out-of-phase mostly and the originally in-phase region tends to have a more random distribution of phase difference. The short and long duration transients do show differences in the instantaneous phase transitions.

4. Conclusions

We compared the dependence of instability transition timescales on transient directions, timescales, and amplitudes in a lean, premixed single nozzle combustor. In both short and long duration transients, there is nearly no difference in timescales when the transient amplitudes doubles in both two directions. Comparison of the combustor pressure and equivalence ratio data shows that combustion instability transition strongly depends on the equivalence ratio and the critical equivalence ratio is approximately 0.68 for all cases regardless of instability transient duration or direction.

5. Acknowledgements

This research was funded by U.S. Department of Energy under award number DEFE0025495 and contract monitor Mark Freeman.

6. References

[1] T. C. Lieuwen and V. Yang, *Combustion Instabilities in Gas Turbine Engines: Operational Experience, Fundamental Mechanisms, and Modeling*. American Institute of Aeronautics and Astronautics, Inc., 2005.

[2] K. R. Mcmanus, T. Poinsoti, and S. Candel, "A review of active control of combustion instabilities," *Prog. Energy Combust. Sci.*, vol. 19, no. 1, pp. 1–29, 1993.

[3] G. Bulat, D. Skipper, R. McMillan, and K. Syed, "Active control of fuel splits in gas turbine DLE combustion systems," in *Proceedings of the ASME Turbo Expo*, 2007, pp. 1–10.

[4] J. Samarasinghe, W. Culler, B. D. Quay, D. A. Santavicca, and J. O'Connor, "The effect of fuel staging on the structure and instability characteristics of swirl-stabilized flames in a lean premixed multi-nozzle can combustor," in *Proceedings of ASME Turbo Expo*, 2017, pp. 1–11.

[5] W. Culler, J. Samarasinghe, B. D. Quay, D. A. Santavicca, and J. O'Connor, "The Effect of Transient Fuel Staging on Self-Excited Instabilities in a Multi-Nozzle Model Gas Turbine Combustor," in *Proceedings of ASME Turbo Expo*, 2017, pp. 1–11.

[6] P. Orawannukul, "An experimental study of flame response in a technically- premixed multi-nozzle gas turbine combustor," 2014.

[7] J. G. Lee and D. A. Santavicca, "Experimental Diagnostics for the Study of Combustion Instabilities in Lean Premixed Combustors," *J. Propuls. Power*, vol. 19, no. 5, pp. 735–750, 2003.

Spring Technical Meeting
Eastern States Section of the Combustion Institute
March 4-7, 2018
State College, Pennsylvania

Accounting for real-gas effects in high-density combustion chamber

*Chenwei Zheng and Ben Akih-Kumgeh**

Department of Mechanical and Aerospace Engineering
263 Link Hall, Syracuse University
Syracuse, NY 13244, USA
**Corresponding author:* bakihkum@syr.edu

Abstract: This paper investigates the high-density combustion in a GOX/kerosene combustion chamber using Large Eddy Simulations. Since the prevailing conditions are such that the ideal gas equation of state cannot be justifiably employed, a real gas equation of state is needed in conjunction with an appropriate mixing rule to establish the averaged state properties of the mixture. The chemical kinetics is incorporated in the simulations through the steady laminar flamelet. The flamelet model is based on a global reaction of n-dodecane. A new approach is also used to determine the parameters of a cubic equation of state from the transport data of the chemical kinetic model. The difference in the flow field resulting from the use of real gas model is established by comparing with simulations in idea gas model is used. The difference justified the use of the real gas model.
***Keywords:* CFD, equation of state, chemical kinetic model, turbulent combustion model**

1. Introduction

Some combustion systems, such as rocket engines, operate at high pressure over a range of temperatures. These extreme operating conditions lead to high densities of the fuel and air mixtures which invalidate the ideal gas modeling often used in analysis. The ideal gas model can cause errors in predicted combustion properties. For more physically realistic reacting flow simulations, real gas effects must therefore be taken into account.

The need to account for real gas effects in Computational Fluid Dynamics (CFD) simulations is being addressed. Sirignano [1] describes the use of cubic Equations of State (EOS) in spray combustion community. Reviews of work on real gas behavior in multiphase flows have been presented by Bellan [2] and Sazhin [3]. Also, vapor-liquid equilibrium (VLE) processes of $n-$alkanes from $C2$ to $C20$ were described by Zhu and Reitz [4]. Peng-Robinson (PR) and Soave-Redlich-Kwong (SRK) EOS were found to be superior to Redlich-Kwong (RK) EOS. Oefelein and Yang [5] employed Large Eddy Simulation (LES) to investigate two dimensional mixing and combustion of hydrogen and oxygen. Thermodynamic non-idealities was evaluated directly from fundamental theories over the fluid regime.

There are many models of real gas EOS, starting from the more theoretically Van der Waals equation to the very complex empirical EOS, such as the Benedict-Webb-Rubin (BWR) equation. Most practical real gas EOS are modifications of the Van der Waals equation of state. Currently, in order to simplify computational tasks associated with EOS evaluations, some research projects use cubic equation of state such as the SRK- and PR- EOS. While some internal combustion and gas turbine engines use ideal gas in combustion analysis, rocket engine operate very far from ideal gas conditions. These rockets require real gas models.

2. Real Gas Equation of State

According to Zhu and Reitz [4], the general form of cubic equation of states can be written as :

$$P = \frac{RT}{V-b} - \frac{a}{V^2+qbV+wb^2} \quad (1)$$

where V is the volume of the system and R is the universal gas constant, q and w are constants depending on the type of cubic EOS. For a pure species, $a = f(T, T_c, \omega)$ and $b = f(P_c, T_C)$ as can be seen for the EOS examples in table 1, where T_c and P_c are critical temperature and critical pressure, ω is the acentric factor, κ is a constant that depends on the acentric factor ω and T_r is defined as reduced temperature, $T_r = T/T_c$.

Table 1: Parameters for cubic EOS

Equation	b	a	q	w
Soave-Redlich-Kwong	$\frac{0.08664RT_c}{P_c}$	$\frac{0.42747R^2T_c^{2.5}}{P_c}[1+\kappa(1-T_r^{0.5})]^2$ $\kappa = (0.480+1.574\omega-0.176\omega^2)$	1	0
Peng-Robinson	$\frac{0.07780RT_c}{P_c}$	$\frac{0.45724R^2T_c^2}{P_c}[1+\kappa(1-T_r^{0.5})]^2$ $\kappa = 0.37464+1.54226\omega-0.26992\omega^2$	2	-1

For a multi-component system, the average parameters according to equation 1, a, b are assigned based on mixing rule. Therefore, in a combustion system, given each species T_c, P_c and ω, the EOS can be obtained.

The binary mixing rule with regards to a and b suggested by Peng and Robinson [6]:

$$a = \sum_i \sum_j x_i x_j a_{ij}; \qquad b = \sum_i x_i b_i; \qquad a_{ij} = (1-\delta_{ij}) a_i^{1/2} a_j^{1/2} \quad (2)$$

where δ_{ij} is an empirically determined binary interaction coefficient characterizing the binary formed by component i and component j. δ_{ij} is proposed at the range between -0.2 to 0.2 [7].

Another method to deal with the mixing rule is the pseudocritical method [8] based on corresponding state principle (CSP).

With the need to include more chemical kinetic effects in combustion simulations, there are many species whose critical properties are not well known for it would be too time consuming to implement manually.

This paper applies a framework to generate parameters of real gas EOS from transport and thermodynamics data already used in combustion chemical kinetic models. The study considers a simple case of jet impingement in a rocket engine. The reacting flow field is obtained with real and ideal gas models, revealing differences. A general method for real gas models is exhibited.

3. Large Eddy Simulation of GOX/Kerosene in a single element rocket combustor

The first part seeks to show that using LES of gaseous oxygen (GOX) and kerosene, the resulting flow fields can be dependent on the type of EOS used. The second part summarizes how RG EOS can be obtained from existing model references.

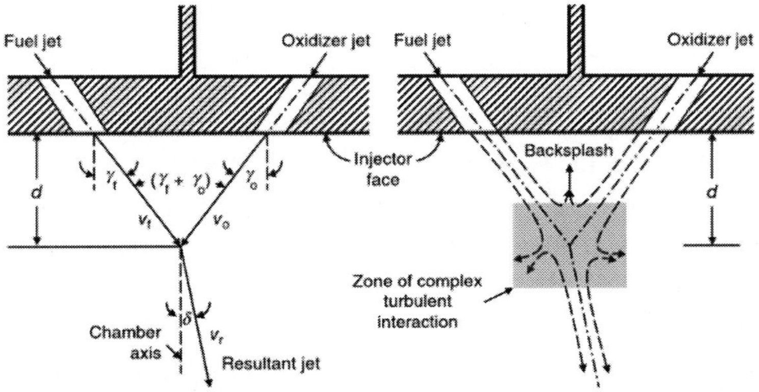

Fig. 1: Sketch of two impinging jets [9].

The problem we consider here consists of a double impinging stream as shown in Fig. 1, with GOX and kerosene (n-C12H26) introduced into a chamber of oxygen. The conditions are such that the chamber is at $T = 700$ K, and the subsonic GOX jet with inlet velocity at $v = 121.30$ m/s and the kerosene jet with a velocity at $v = 73.16$ m/s. Simulations are carried out at 200 atm, with constant injecting temperature at $T = 700$ K. The real gas EOS requires critical state parameters which for n-dodecane are $T_c = 658.20$ K, $P_c = 18.10$ bar, $\omega = 0.562$; for oxygen are $T_c = 154.58$ K, $P_c = 50.43$ bar, $\omega = 0.021$; for water are $T_c = 647.00$ K, $P_c = 220.64$ bar, $\omega = 0.344$ and for carbon dioxide are $T_c = 304.1$ K, $P_c = 73.8$ bar, $\omega = 0.239$, here ω is the acentric factor.

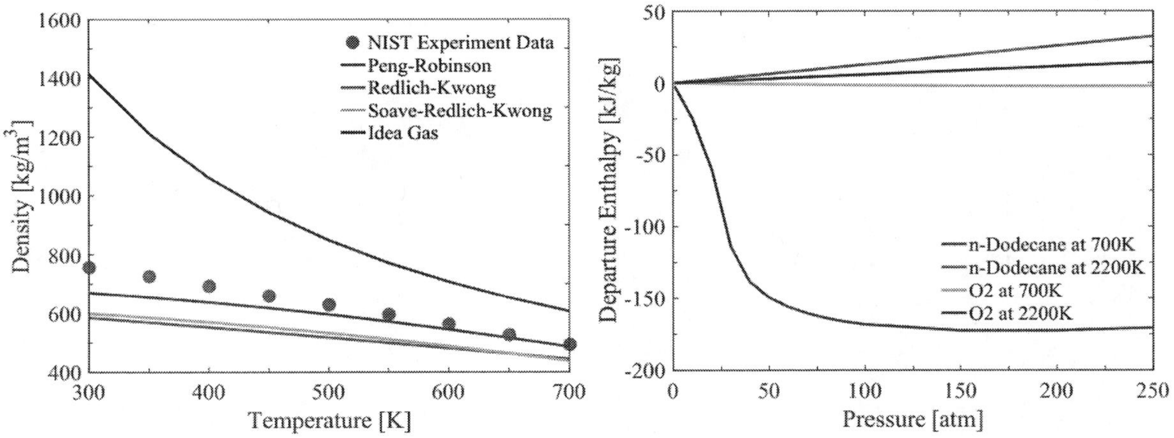

(a) n-c12h26, ρ-T at $P = 200$ atm. (b) GXO/Kerosene Enthalpy Departure.

Fig. 2: Variation of n-dodecane, oxygen density and enthalpy departure function with temperature and pressure, highlighting importance of real gas EOS.

The need for real gas effect is illustrated in Fig. 2 for the n-dodecane and oxygen. It can be seen that the use of real gas EOS greatly improve the prediction of experimental gas properties.

The enthalpy of these gases depart from the ideal gas at various temperature and pressures. The enthalpy differential is given by:

$$dh = C_p dT + \left[v - T \cdot \left(\frac{\partial v}{\partial T} \right) \Big|_P \right] dP \tag{3}$$

where the second terms are the departure terms. For an ideal gas the departure is zero but its significance in the real gas points to problems with energy conservation if ideal gas behavior is assumed.

In order to properly resolve the flow field, the turbulence model chosen here is LES with the sub-grid-scale (SGS) model, Wall-Adapting Local Eddy (WALE). To simulate combustion a one global reaction model with the steady laminar concept is employed. The mesh consists of $470,400$ cells in total with a minimum cell size $\Delta x_{min} = 0.01$ mm. The simulations are carried out using the CFD solver, STAR-CCM+. The time-step is $2.0E-8$ s with PISO unsteady scheme.

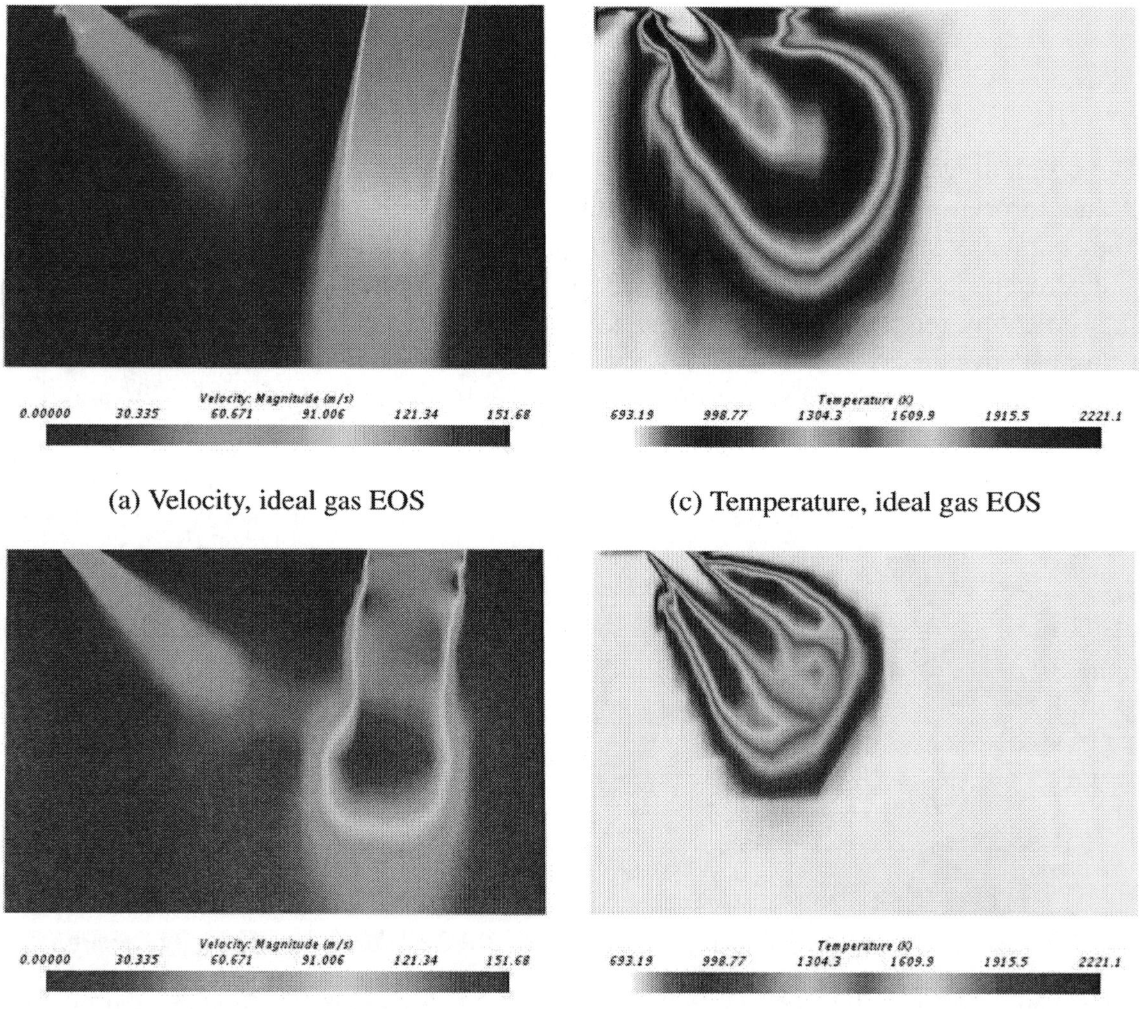

(a) Velocity, ideal gas EOS

(c) Temperature, ideal gas EOS

(b) Velocity, real gas EOS

(d) Temperature, real gas EOS

Fig. 3: The deviation of density prediction result in further departure in velocity and temperature

The resulting flow field can be compared using scalars such as the velocity and temperature. Fig. 3 show the velocity profile and temperature at 0.2 ms, obtained at a pressure of 200 atm. The simulations are still ongoing, but we see that flow fields are different, since the spread and penetration depth in the case of ideal gas EOS are greater than those of real gas EOS. These differences are also attached key differences in other dynamical behavior of the reacting flows. Extinction and combustion instability will be differently affected by the flow fields, with difference originally from the EOS.

4. Method of EOS parameter estimation

These simulations used a model with limited species. in the case of many species a method is needed to estimate the real gas EOS parameters. The goal would be to determine for each species in a given mechanism, its real gas EOS parameters, i.e. T_c, P_c, ω. Using an appropriate mixing rule, the EOS parameters for a given mixture combustion can be obtained from those of individual species.

There is a connection between the critical state properties and the parameters of intermolecular potentials (e.g. Lennard-Jones 12-6 potential). The latter are used in combustion flows to generate required transport properties (viscosity, conductivity and thermal diffusivity). One can reverse-engineer the available transport data to obtain the critical state parameters needed for the EOS.

Many empirical relations between parameters of intermolecular potentials and the critical state properties have been proposed [10]. Some require knowledge of the acentric factor while others do not. The most widely used of these equations are of the form:

$$\sigma\left(\frac{P_c}{T_c}\right)^{1/3} = a_\sigma - b_\sigma\omega; \text{ and } \frac{\varepsilon}{k_B T_c} = a_\varepsilon - b_\varepsilon\omega \tag{4}$$

Table 2: Parameters for estimation

Reference	Tee et al. [10]	Kee et al. [11]	Holley et al. [12]
a_σ	2.3551	2.393	2.3551
b_σ	0.0874	0	0.3955
a_ε	0.7915	0.75	0.8063
b_ε	0.1693	0	0.6802

where ε is the depth of the potential well and σ is the distance at which the potential is zero.

In chemical kinetic models, the estimation rules proposed in Kee et al. [11] are often used. It appears that the Tee et al. [10] performs better than the Kee et al. [11] when predicted transport variables are compared with observations. T_c, P_c, ω are observable quantities whereas the accuracy of σ, ε can only be inferred from their prediction of transport properties. This circularity can lead to a verification problem. Holley et al. [12] have suggested updated coefficients for the Tee correlation. It means is that with an appropriate correlation, one can obtain T_c and P_c from transport data.

There is therefore a method to obtain ral gas EOS from existing mechanism data and improve the simulation of high density combustion.

5. Conclusion

In this paper, the problem of real gas equations of state for high density flow simulations is considered. The importance of such equations of state is underlined by comparing the large eddy simulation results of GOX/kerosene into an high pressure vessel realized using large eddy simulations with ideal and real gas equations of state. It is then suggested that parameters of a cubic equation of state, namely, the critical state variables, can be determined from the transport data of chemical kinetic models. This makes it easier to account for high density behavior in flow simulation.

Acknowledgement

This research was supported in part through computational resources provided by Syracuse University.

References

[1] W. A. Sirignano, Fluid dynamics and transport of droplets and sprays, Cambridge university press, 1999.

[2] J. Bellan, Supercritical (and subcritical) fluid behavior and modeling: drops, streams, shear and mixing layers, jets and sprays, Progress in energy and combustion science (2000).

[3] S. Sazhin, Advanced models of fuel droplet heating and evaporation, Progress in energy and combustion science (2006).

[4] G. Zhu and R. Reitz, A model for high-pressure vaporization of droplets of complex liquid mixtures using continuous thermodynamics, International Journal of Heat and Mass Transfer (2002).

[5] J. Oefelein and V. Yang, Modeling high-pressure mixing and combustion processes in liquid rocket engines, Journal of Propulsion and Power (1998).

[6] D.-Y. Peng and D. B. Robinson, A new two-constant equation of state, Ind. Eng. Chem. Fundam 15 (1976) 59–64.

[7] A. M. Abudour, S. A. Mohammad, R. L. Robinson Jr, and K. A. Gasem, Generalized binary interaction parameters for the Peng–Robinson equation of state, Fluid Phase Equilibria 383 (2014) 156–173.

[8] B. E. Poling, J. M. Prausnitz, J. P. O'connell, et al., The properties of gases and liquids, vol. 5, Mcgraw-hill New York, 2001.

[9] S. Farokhi, Aircraft propulsion, John Wiley & Sons, 2014.

[10] L. S. Tee, S. Gotoh, and W. E. Stewart, Molecular parameters for normal fluids. Lennard-Jones 12-6 Potential, Industrial & Engineering Chemistry Fundamentals 5 (1966) 356–363.

[11] R. Kee, M. Coltrin, and P. Glarborg, Chemically reacting flow: theory and practice, 2005.

[12] A. Holley, X. You, E. Dames, H. Wang, and F. Egolfopoulos, Sensitivity of propagation and extinction of large hydrocarbon flames to fuel diffusion, Proceedings of the Combustion Institute (2009).

Spring Technical Meeting
Eastern States Section of the Combustion Institute
March 4-7, 2018
State College, Pennsylvania

Effect of the Presence of Water Molecules on Syngas Combustion: A ReaxFF Molecular Dynamics Study

Chowdhury Ashraf, Sharmin Shabnam and Adri C.T. van Duin *

Department of Mechanical and Nuclear Engineering, The Pennsylvania State University, University Park, PA 16802, USA

Corresponding Author Email: acv13@psu.edu

Abstract: Syngas (Synthesis Gas), a mixture of CO and H_2 gas, has a very high availability and great flexibility of usage. As such, syngas is expected to play a significant role in future energy generation. However, the varying amount of mixture components makes it difficult to develop a reliable reaction mechanism for syngas combustion. The main variable to consider for direct syngas combustion include the carbon monoxide/hydrogen ratio, fluctuating level of carbon dioxide and water and presence of other species that can significantly affect the combustion process. In this study, we investigate the effect of the presence of water molecules in syngas combustion. In our reactive molecular dynamics simulations, we consider two different carbon monoxide/hydrogen ratios and examine how water molecules, initially present in the system, affect carbon dioxide production at different temperatures. Our study indicates that a low amount of water molecules significantly increases carbon dioxide production, which is also observed experimentally. However, beyond that limit, water molecules have limited effect on CO oxidation. This latter effect is more prominent at higher temperatures. Overall, our simulations can be used as a computational inexpensive tool to predict optimal water levels for syngas combustion at any given condition.

Keywords: Syngas combustion, Water addition, Carbon dioxide production

1. Introduction

As the future energy demand continues to grow and energy sources like natural gas, coal and petroleum is expected to play a major role in upcoming energy production, a strong interest is generated towards the reformed fuels[1] which are both energy efficient and environment friendly. Due to the availability and flexibility of usage, the projected consumption of Syngas (Synthesis gas) for energy generation is likely to increase in future. Syngas is a mixture of CO and H_2, although fuel components may include CH_4 and inert gases like N_2, CO_2, and H_2O may also exist in the mixture gas[2]. Though syngas provides an opportunity for clean combustion, the varying amount of mixtures of its components difficult to develop a reliable reaction mechanism for syngas combustion. That is why a detailed study of chemical kinetics of syngas is required to completely understand the combustion process and to identify the reactions that are responsible for the quality of combustion. Previously, the effects of H_2O addition on CO oxidation[3,4], turbulent flame characteristics[5], laminar flame speed[6] of syngas have been studied experimentally and with kinetic modelling. In this paper, we applied the ReaxFF-MD method to study the effect of water addition on the production of CO_2 with two different syngas mixtures at varied temperatures. This paper is organized as follows, in Section 2, ReaxFF method is described while simulation setups for syngas combustion by ReaxFF method is described in section 3. We discussed the results in section 4 and conclusions are summarized in Section 5.

2. ReaxFF Reactive Force Field Method

ReaxFF is a general bond order[7] based empirical force field method. It uses the continuous bond-order/bond distance relation that enables it to properly describe bond formation and bond breaking. A more detail description of ReaxFF energy terms will be found in previous literature[8,9]. ReaxFF has already been successfully implemented to simulate different combustion phenomenon including hydrogen combustion[10], hydrocarbon oxidation[9], coal[11] and oxidation of Silicon Carbide[12].

3. Simulation Methodology

We studied two syngas mixtures containing different mixing ratio of CO and H_2. The combustion chemistry of both air-fuel mixtures has an equivalence ratio of $\phi<1$. The first system contains 40 molecules of both CO and H_2 in a cubic box with another 60 O_2 atoms. The second system contains 60 molecules of CO and 20 molecules of H_2 while the number of O_2 atoms remains same as before. The box dimensions were chosen such that the density of all the systems remain constant. The equivalence ratio (ϕ) of the two systems is 0.67. We used the CHO-2016 force field developed by Ashraf et al.[13] which is expanded from the ReaxFF CHO-2008[9] DFT-based training set by including reactions and transition state structures relevant to the syngas and oxidation initiation pathways. The study includes results obtained with CHO-2016 force field for syngas and methane oxidation simulation which showed significant improvement for C_1 chemistry predicted by ReaxFF.

Table 1: Simulation conditions for fuel mixtures

Mole fraction of water in fuel mixture (α)	0.012, 0.059, 0.111, 0.200, 0.273 and 0.333
Density, ρ (kg/dm^3)	0.32
Temperature, T (K)	2250K, 2500K, 2750K, 3000K

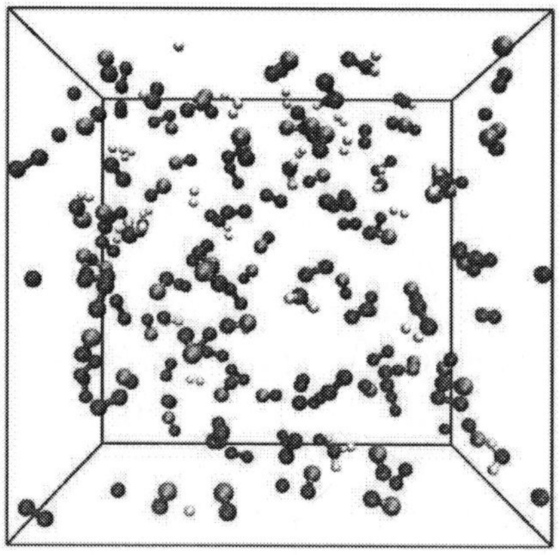

Figure 1: Snapshots of a representative system of syngas combustion containing 60 molecules of CO, 20 molecules of H_2, 60 molecules of O2 and 10 molecules of H_2O with a system density of 0.32 kg/dm^3. The red, cyan and white spheres indicate the oxygen, carbon and hydrogen atoms respectively.

After geometrically optimizing, we equilibrated the whole system at a temperature of 2000K for 25 ps with a time step of 0.25 fs. At this moderately low temperature, no reaction was observed. Next, we increased the temperature of the system at a rate of 0.005 K/time step from 2000K to four different temperatures from 2250 to 3000K with an interval of 250K. After increasing the temperature, we selected 10 different initial configurations to perform a series of NVT-MD simulations for 0.9ns with a time-step of 0.1 fs. Berendsen thermostat is used with a damping constant of 100fs in order to control the temperature. Figure 1 shows the initial structure of a representative MD simulation with this system.

4. Results and Discussion

4.1. Effect of water addition on CO and H_2 combustion

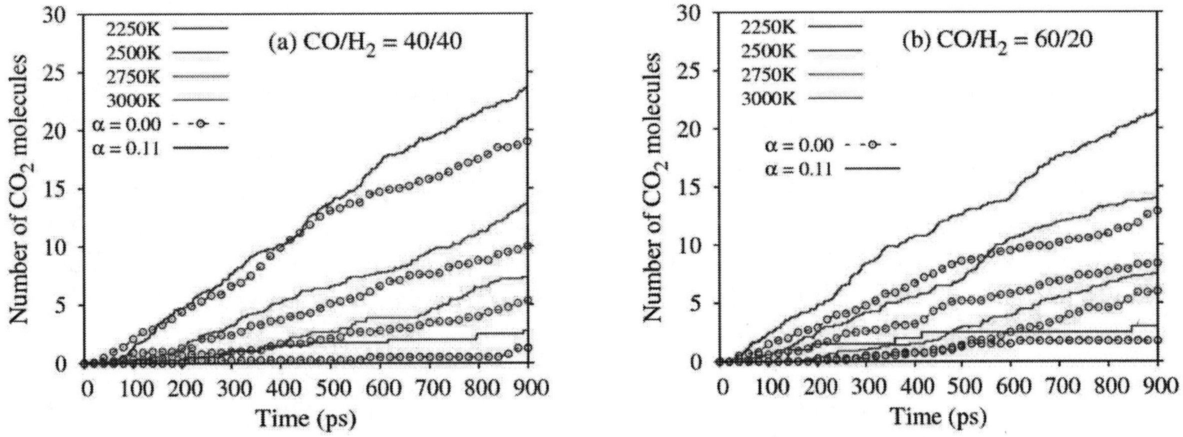

Figure 2: Time evolution of CO_2 molecules during ReaxFF-MD simulations of combustion of two syngas systems with (a) $CO/H_2/O_2$ =40/40/60 and (b) $CO/H_2/H_2O/O_2$ =60/20/10/60 (the numbers represent the number of molecules) with a system density of 0.32 kg/dm^3. The solid lines and the dashed lines with points indicate the dry and moist syngas systems respectively.

In figure 2, the number of CO_2 molecules generated as a function of time during the ReaxFF-MD simulations is presented for dry and moist syngas (α=0.11) combustion. The results are obtained for two different syngas systems with CO and H_2 ratios of 40/40 and 60/20. The overall effect of water addition is found to increase the CO_2 production, the effect being much more significant at higher temperatures. The first syngas system with equivalence ratio ϕ=0.67 demonstrates CO_2 production increase moderately for all the temperatures. However, for the second syngas system with higher CO/H_2 ratio but the same equivalence ratio, the effect of water addition varies at different temperatures.

At relatively lower temperatures (2250K and 2500K), the addition of water molecules shows moderate increase in the production of CO_2 molecules. However, at higher temperatures (2750K and 3000K), the results show a sharp increase in the number of CO_2 molecules produced. This is in accordance with the experiments where increasing water vapor facilitates OH and HO_2 radical production which in turn enhances CO oxidation to produce more CO_2[6].

Figure 3 shows the time evolution of water molecules produced during the ReaxFF-MD simulations of previously mentioned syngas combustion. For effective comparison we have excluded the number of initial water molecules present in case of moist syngas mixtures. The final amount of water produced by the combustion of H_2 is almost similar for both dry and moist syngas system. The results thus indicate that the water presence affects H_2 oxidation less significantly than CO oxidation. The reason can be attributed to the fact that addition of water consumes O radicals via $H_2O + O = OH + OH$ which would previously to oxidize hydrogen molecules.

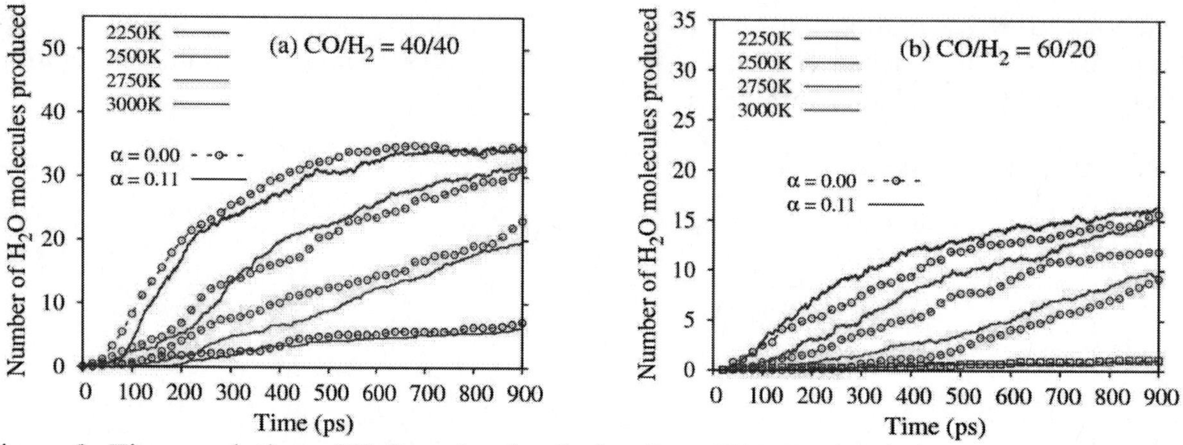

Figure 3: Time evolution of H_2O molecules during ReaxFF-MD simulations of combustion of two syngas systems with (a) $CO/H_2/O_2$ =40/40/60 and (b) $CO/H_2/H_2O/O_2$ =60/20/10/60 (the numbers represent the number of molecules) with a system density of 0.32 kg/dm^3. The solid lines and the dashed lines with points indicate the dry and moist syngas systems respectively.

4.2. Effect of water concentration on carbon dioxide production

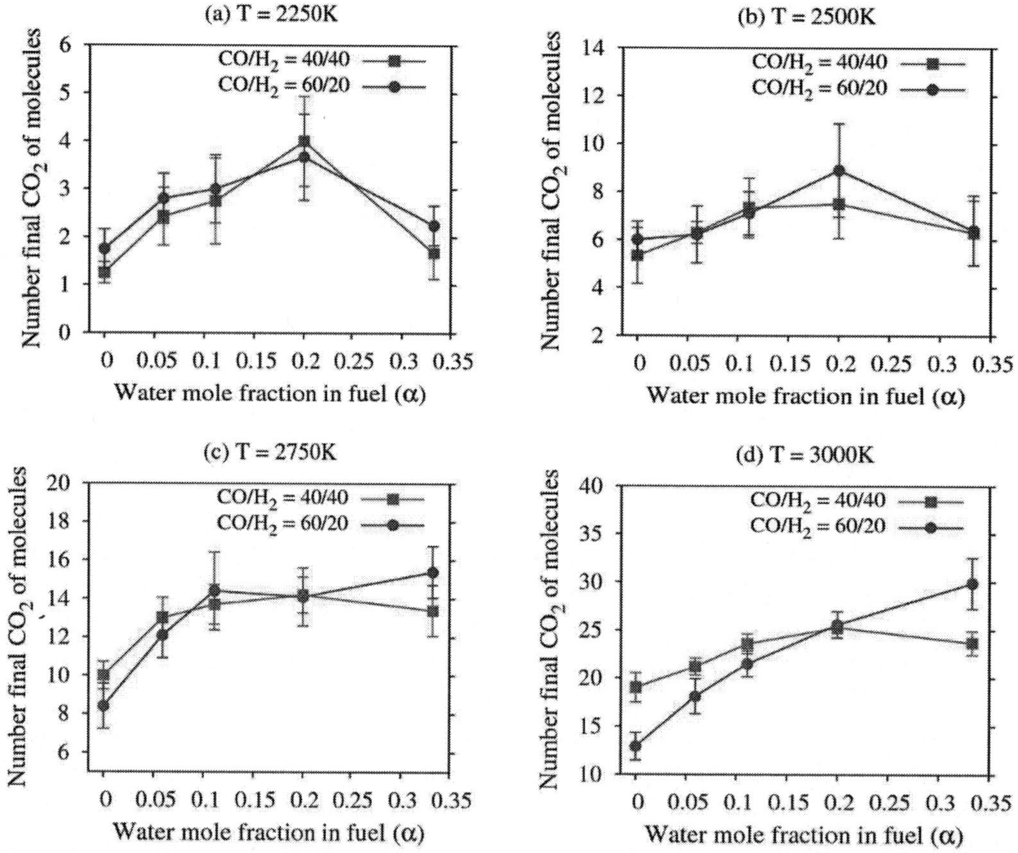

Figure 4: Variation of final CO_2 molecules with increasing mole fraction of water in unburnt fuel mixture at temperatures (a) T = 2250K, (a) T = 2500K, (a) T = 2750K and (d) T = 3000K during ReaxFF-MD simulations of combustion of two syngas systems with $CO/H_2/O_2$ =40/40/60 and (b) $CO/H_2/O_2$ =60/20/60 (the numbers represent the number of molecules) with a system density of 0.32 kg/dm^3.

To determine the effect of water concentration on the oxidation of CO during syngas combustion, we plotted the final number of CO_2 molecules at the end of ReaxFF-MD simulations at various temperatures and syngas mixing ratios in figure 4. Two general trends are observed here when the concentration of water is increases. When the temperature is relatively low (2250K and 2500K), we observed that the addition of water molecules initially increases the CO_2 production. But when the concentration of water reaches a certain threshold ($\alpha = 0.333$), the CO_2 production becomes close to the values found in combustion of dry syngas system. This trend is similar for both syngas mixtures we have considered.
However, at high temperatures (2750K and 3000K), with increasing water concentration, we observed the two mixtures behaved in a slightly different manner. In case of the relatively leaner mixture (CO/H_2 = 40/40), the CO_2 production does not show much variation when the water content increases. But our simulations indicate a sharp increase of CO_2 production for the mixture where CO/H_2 ratio is 60/20. With the inclusion of 40 water molecules ($\alpha = 0.333$) in the system, the number of CO_2 more than doubled compared to the system containing no water molecules. This is in accordance with the study by Xie et al. [14] that with the increase of CO/H_2 ratio, water molecules significantly increase the conversion of CO to CO_2 molecules since the OH radicals react with increased percentage of CO molecules in the system.

5. Conclusions

The effect of water addition on carbon dioxide production during syngas mixture combustion has been studied in this work by implementing the ReaxFF-MD simulations over a temperature range from 2250K to 3000K. Two different CO/H_2 mixture ratios with varied equivalence ratio have also been studied to observe the influence of H_2O addition with variable concentration on the amount of CO oxidation. The production of CO_2 for the two systems shows an overall increase due to the effect of water addition which is also in accordance with experimental literature. However, the water addition did not show any significant influence on the conversion of hydrogen to water molecules. The effect of varying concentration of water molecules have also been studied. The results indicate that the conversion of CO into CO_2 is enhanced by increasing water mole fraction in the fuel mixture. But for lower temperatures and lower CO/H_2 ratio, the effect diminishes with too much water molecules in the system. Overall, the results obtained from ReaxFF-MD simulations indicate the capability of this method to further study fuel oxidation.

6. Acknowledgements

This research was funded by the Air Force Office of Scientific Research (AFOSR) under grant no. FA9550-14-1-0355.

7. References

[1] R.D. Cortright, R.R. Davda, J.A. Dumesic, Hydrogen from catalytic reforming of biomass-derived hydrocarbons in liquid water, Nature. 418 (2002) 964. doi:10.1038/nature01009.
[2] Y. Xie, J. Wang, N. Xu, S. Yu, Z. Huang, Comparative study on the effect of CO2 and H2O dilution on laminar burning characteristics of CO/H2/air mixtures, Int. J. Hydrog. Energy. 39 (2014) 3450–3458. doi:10.1016/j.ijhydene.2013.12.037.
[3] T.J. Kim, R.A. Yetter, F.L. Dryer, New results on moist CO oxidation: high pressure, high temperature experiments and comprehensive kinetic modeling, Symp. Int. Combust. 25 (1994) 759–766. doi:10.1016/S0082-0784(06)80708-3.

[4] M. Abián, J. Giménez-López, R. Bilbao, M.U. Alzueta, Effect of different concentration levels of CO2 and H2O on the oxidation of CO: Experiments and modeling, Proc. Combust. Inst. 33 (2011) 317–323. doi:10.1016/j.proci.2010.05.078.
[5] J. Wang, F. Matsuno, M. Okuyama, Y. Ogami, H. Kobayashi, Z. Huang, Flame front characteristics of turbulent premixed flames diluted with CO2 and H2O at high pressure and high temperature, Proc. Combust. Inst. 34 (2013) 1429–1436. doi:10.1016/j.proci.2012.06.154.
[6] A.K. Das, K. Kumar, C.-J. Sung, Laminar flame speeds of moist syngas mixtures, Combust. Flame. 158 (2011) 345–353. doi:10.1016/j.combustflame.2010.09.004.
[7] D.W. Brenner, Empirical potential for hydrocarbons for use in simulating the chemical vapor deposition of diamond films, Phys. Rev. B. 42 (1990) 9458–9471. doi:10.1103/PhysRevB.42.9458.
[8] A.C.T. van Duin, S. Dasgupta, F. Lorant, W.A. Goddard, ReaxFF: A Reactive Force Field for Hydrocarbons, J. Phys. Chem. A. 105 (2001) 9396–9409. doi:10.1021/jp004368u.
[9] K. Chenoweth, A.C.T. van Duin, W.A. Goddard, ReaxFF Reactive Force Field for Molecular Dynamics Simulations of Hydrocarbon Oxidation, J. Phys. Chem. A. 112 (2008) 1040–1053. doi:10.1021/jp709896w.
[10] S. Agrawalla, A.C.T. van Duin, Development and Application of a ReaxFF Reactive Force Field for Hydrogen Combustion, J. Phys. Chem. A. 115 (2011) 960–972. doi:10.1021/jp108325e.
[11] F. Castro-Marcano, A.M. Kamat, M.F. Russo, A.C.T. van Duin, J.P. Mathews, Combustion of an Illinois No. 6 coal char simulated using an atomistic char representation and the ReaxFF reactive force field, Combust. Flame. 159 (2012) 1272–1285. doi:10.1016/j.combustflame.2011.10.022.
[12] D.A. Newsome, D. Sengupta, H. Foroutan, M.F. Russo, A.C.T. van Duin, Oxidation of Silicon Carbide by O2 and H2O: A ReaxFF Reactive Molecular Dynamics Study, Part I, J. Phys. Chem. C. 116 (2012) 16111–16121. doi:10.1021/jp306391p.
[13] C. Ashraf, A.C.T. van Duin, Extension of the ReaxFF Combustion Force Field toward Syngas Combustion and Initial Oxidation Kinetics, J. Phys. Chem. A. 121 (2017) 1051–1068. doi:10.1021/acs.jpca.6b12429.
[14] Y. Xie, J. Wang, N. Xu, S. Yu, M. Zhang, Z. Huang, Thermal and Chemical Effects of Water Addition on Laminar Burning Velocity of Syngas, Energy Fuels. 28 (2014) 3391–3398. doi:10.1021/ef4020586.

Spring Technical Meeting
Eastern States Section of the Combustion Institute
March 4–7, 2018
State College, Pennsylvania

High-Throughput Screening for Reactive and Energy-Transferring Collider Effects in Complex-Forming Reactions

Mark C. Barbet[1], *Kevin McCullough*[1], *and Michael P. Burke*[1,2,*]

[1]*Dept. of Mech Eng., Columbia University, New York, NY, USA*
[2]*Dept. of Chem Eng. and Data Sciences Institute, Columbia University, New York, NY, USA*
[*]*Corresponding author:* mpburke@columbia.edu

Abstract: The fate of ephemeral rovibrationally excited complexes, which mediate pressure- and mixture-dependent reactions, is influenced by their collisions with the surrounding molecules. In the mixtures commonly encountered in realistic combustion systems, these ephemeral complexes can undergo both (1) reactive collisions with radicals and other highly reactive molecules and (2) energy-transferring collisions with multiple molecules possessing different energy-transferring characteristics. Here, we present algorithms and associated results for high-throughput screening of both reactive and energy-transferring collider effects commonly neglected in combustion kinetic modeling. With regard to (1), existing kinetic models implicitly assume that reactive collisions are far less frequent than energy-transferring collisions and thus would not affect the fate of ephemeral complexes. However, recent work has shown that the high fractions of reactive colliders in combustion give rise to a commonly neglected type of reactions, "chemically termolecular reactions," that influence global combustion observables. This work presents a number of chemically termolecular reactions found to influence combustion predictions based on estimated values for their rate constants and thus warrant more quantitative investigation. With regard to (2), existing kinetic models often define rate constants for a common diluent (e.g. $M = N_2$, He, or Ar) along with "third-body efficiencies" for a small subset of other colliders, thus implicitly assuming collision efficiencies of unity for all colliders other than the select few listed explicitly. However, nearly all other species encountered in combustion (e.g. fuel or fuel intermediates), which possess more vibrational modes and may even be polar, are likely to be more effective in transferring energy than N_2, He, and Ar. Similarly, this work seeks to identify reaction-bath gas pairs involving bath gases with unspecified collision efficiencies that influence combustion predictions and thus warrant detailed characterization of their energy-transferring characteristics.
Keywords: ***Termolecular Reactions, Third Body Collider, Reactive Collision, Energy Transfer***

1. Introduction

The development of kinetic mechanisms has long relied on certain assumptions regarding the fate of rovibrationally excited complexes (formed from collision of two reactant species) – which have several possible pathways. In addition to simply dissociating back into two fragments, the energized complex may undergo a reactive collision with a third body to form new products or an energy-transferring collision with a third body that stabilizes the complex. This paper assesses the impact of processes involving (1) reactive and (2) energy-transferring third body collisions not typically considered in detailed kinetic modeling.

1.1 Reactive Collisions

Traditionally, pressure dependent reactions in kinetic models involving bath gases have assumed to only consist of termolecular association reactions. Recent research by Burke and Klippenstein [1] recently discovered a new class of reaction, termed "chemically termolecular" reactions, where the third species, rather than acting solely as a non-reactive means of energy transfer, reacts with the other species in the reaction. These reactions arise from the formation of rovibrationally excited intermediates–which are formed with non-Boltzmann energy distributions. Burke and Klippenstein found that the time required to thermalize such intermediates is on a similar order of magnitude as the time required for collisions with other species. In combustion mixtures, this leads to the distinct possibility of reactive collisions between the energetically activated intermediates and third-bodies, giving rise to reactions with entirely different rate laws than the reactions proceeding through thermalized intermediates.

Burke and Klippenstein found that reactions of this class, with the form $R+X+Y \rightarrow P_3$, can influence global observables. It is in principle possible, however, that any bimolecular reaction yielding an energetically excited complex could form the precursor complex for such a reaction–which could then proceed through a reactive collision with a third particle and form products. Given the enormous number of chemically termolecular reactions of potential importance, it is difficult to know *a priori* whether or not a given chemically termolecular reaction may be a significant new channel for a system to proceed through before performing a time consuming detailed analysis of the reaction. As such, we discuss here (briefly) a screening procedure developed for finding and analyzing these chemically termolecular reactions.

1.2 Energy Transferring Collisions

The other critical component of third-body collision reactions is the participation of third-bodies in energy-transferring collisions. These are typically association/dissociation reactions, and play a critical role in the evolution of a system. In the past decades, some work has been done to estimate and characterize relative collision efficiencies [2–8], usually for specific reactions. While it has been possible to find collision efficiencies for non-reactive collisions with bath gases, common combustion systems may involve large mole fractions of bath gases not typically given consideration when defining collision efficiencies in kinetic models. Although studies have previously estimated collision efficiencies and parameters for typical bath gases [4, 5, 8], this study attempts to quantify the impact of third-body efficiencies for species that are often not explicitly assigned a value in kinetic models. Here, we develop a procedure to manipulate existing kinetic models to assess the impact of these "missing" values and the degree to which additional efforts need to be undertaken to perform detailed analysis of collider efficiencies for any given combustion system.

2. Methods

2.1 Reactive Collisions

Based on the results found by Burke and Klippenstein [1], we developed an algorithm to rapidly identify and estimate the rate parameters for chemically termolecular reactions. By casting the problem in terms of a modified Lindemann mechanism, it is possible to derive estimates (under certain assumptions) for rate constants of chemically termolecular reactions from data present in existing kinetic models that describe the evolution of thermal species. Table 1 lists six reactions: reactions R1 through R3 describe the basic physical problem in terms of the rovibrationally excited complex RX^*. The reverse reaction of R1 is one of three possible pathways for this complex: simple dissociation to $R + X$. Reaction R2 then describes the stabilization of RX^* to a thermally distributed RX complex. Finally, Reaction R3 describes a reactive collision with a third body Y to yield products (a process not usually considered within kinetic models). However, given that most existing chemical mechanisms are cast in terms of reactions of thermal species alone, we can write this process in terms of the thermalized species in phenomenological reactions RI through RIII. RI and RIII are classes of reactions already contained in many mechanisms, while the the reaction RII belongs to the class of reaction discussed by Burke and Klippenstein [1].

Table 1: (Left)Elementary steps involving rovibrationally excited RX^* . (Right) Phenomenological reactions involving only thermal species, as may appear in typical mechanisms.

$R+X \leftrightarrow RX^*$	**R1**	$R+X \leftrightarrow RX$	**RI**
$RX^*+M \rightarrow RX+M$	**R2**	$RX+X+Y \rightarrow P_3$	**RII**
$RX^*+Y \rightarrow P_3$	**R3**	$RX+Y \rightarrow P_{III}$	**RIII**

Using methods analogous to standard analysis of Lindemann mechanisms as discussed in more detail in [9], simple estimates for rate constants for reactions of type RII have been developed. As such, we developed a script that screens existing mechanisms for reactions of type RI and RIII and uses these to find potential reactions RII. These new reactions are appended to existing input files, and can then be used to perform combustion simulations in Cantera software [10]. Sensitivity analysis may then be performed on the results of these simulations to find candidate reactions where energized complexes undergo reactive collisions with third bodies.

2.2 Energy-Transferring Collisions

Because the collider efficiencies typically included in a chemical mechanism only have a few major species listed, with other efficiencies taken to be one, it seems likely that the unlisted efficiencies may play a significant role in scenarios where significant fractions of additional species exist. For instance, we may expect fuels and fuel fragments to play a significant role under fuel rich conditions.

As such, we have developed a script to read mechanism files and manipulate the collider efficiencies in pressure dependent reactions. When the script encounters a pressure dependent reaction, it performs the following steps:

1. Check if the reaction lists any efficiencies
2. If efficiencies are present, find the maximum efficiency listed
3. Multiply the pre-exponential-factor(s) for the reaction by the maximum value
4. Normalize all the listed efficiencies listed by the maximum value

By doing this manipulation of efficiencies and running various combustion simulations, it is possible to assess to what degree third-body-efficiencies for unlisted species affect the influence of energy transferring collision reactions in a given scenario. If the models with the manupulated energy transferring reactions show little deviation from baseline models we may draw the conclusion that the current method of describing energy transferring reactions is sufficient for the given scenario. Otherwise, we may conclude that these energy transferring reactions need to undergo more in-depth analysis via master equation methods. Like the simulations for the analysis of reactive collisions, the combustion simulations are performed in Cantera software [10].

3. Results and Discussion

3.1 Reactive Collisions

Here, we summarize one of the results from [9] to illustrate the role of reactive collisions in some simple systems. The results in Figure 1 display flame speed predictions based on the FFCM-1 model [11]

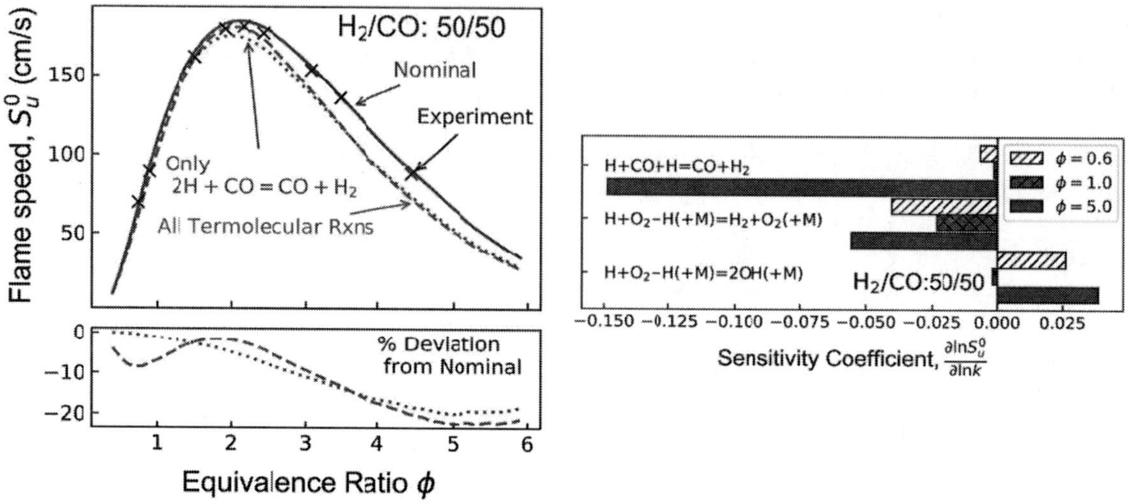

Figure 1: Top left: Laminar flame speeds for H_2/CO=50/50 mixtures in air at 1 atm and 295 K. Lines shows model predictions for the nominal model [11], the nominal model with all chemically termolecular reactions appended, the nominal model with only H + CO + H = CO + H_2 appended (as labeled in the figure). Symbols show experimental measurements [12]. Bottom left: Relative deviation from the nominal model for the two models with chemically termolecular reactions included. Right:Sensitivity coefficients for the five most sensitive chemically termolecular reactions for H_2/CO/air flame speeds.

using the nominal model and versions containing termolecular reactions found with our script. Experimental data from McLean et al.[12] for H_2/CO/air mixtures is also displayed. Via sensitivity analysis it was found that the influential chemically termolecular reactions primarily proceeded in the forward direction, acting as chain terminating reactions steps as shown in Figure 1. In particular, reactions such as $H + CO + H = CO + H_2$ primarily act to reduce flame speed predictions at fuel rich conditions, where excess CO is present.

3.2 Energy-Transferring Collisions

For this particular screening, we present results of flame speed simulations and ignition delay simulations for Methanol/Air mixtures. Results presented here make use of the kinetic mechanism for methanol developed by Li et al.[13]. The manipulations for the modified model are performed on this model according to the methods previously discussed.

The results shown in Figure 2 depict simulations of methanol/air free flames at atmospheric conditions. Given the nature of the changes to the model (manipulations of collision efficiencies), it was expected that if any effect was found, it would be in an regime with high fractions of colliders not accounted for in the development of the mechanism. In this case, it was expected that the primary effect on flame speeds would be in the fuel-rich region, where there is a higher likelihood of finding large concentrations of methanol and methanol fragments.

Indeed, referring to Figure 2, the differences between the two model formulations are most evident at equivalence ratios above 1.6. At very fuel rich conditions, relative differences between the two models was found to be as much as 40 percent. Sensitivity analysis suggests a number of possible reactions contributing, including HCO + M $\leftrightarrow$ CO + H + M and H + OH + M $\leftrightarrow$ H_2O + M. More in depth analysis is needed to properly understand these reactions' contributions.

The results shown in Figure 2 on the right show ignition delay times in shock tubes simulated using the

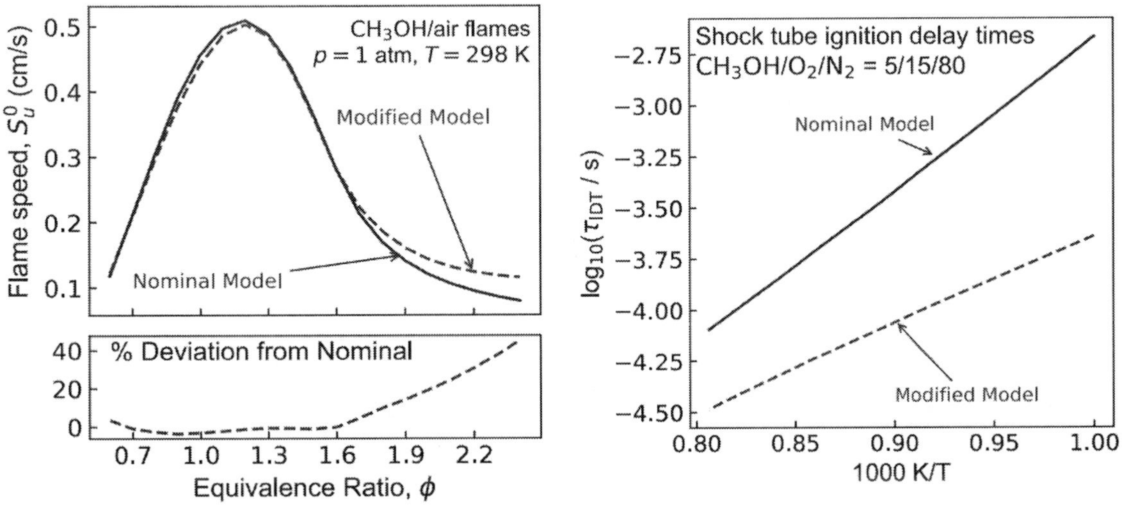

Figure 2: Top left: Laminar flame speeds for CH_3OH in air at 1 atm and 298 K. Lines shows model predictions for the nominal model [13] and the modified version of this model where the collider efficiencies have been manipulated(see text). Bottom Left: Relative percent deviation from the nominal model for the manipulated model. Right: Ignition delay times in a shock tube are given for the conditions in [14]. The model in [13] is used for the nominal simulation, and the modified model is manipulated based on this model as well.

same model as used in the previous flame speed simulations [13] at the conditions tested by Burke et al.[14] and simulated by [2]. Again, the manipulated mechanism was based on this kinetic model. Simulations using the modified kinetic mechanism show significantly shorter ignition delays throughout the temperature regime.

4. Conclusions

Here, we have briefly introduced procedures for the screening of kinetic mechanisms to find 1) chemically termolecular reactions, a class of phenomenological reactions resulting from reactive collision with third bodies and 2) scenarios where the current specifications of third-body collision efficiencies do not sufficiently describe the role of atypical bath gases in contributing to energy transfer in pressure dependent reactions. Both of these screening procedures utilize manipulations of existing kinetic models followed by simulations in common combustion software. In each case, we have briefly highlighted some initial results from these screening procedures, finding new chemically termolecular reactions that influence flame speed predictions, and finding cases where missing information about third-body colliders influences flame speeds and ignition delay times. Each of these scenarios may warrant further investigation via detailed analysis using master equation codes or other methods.

5. Acknowledgements

Acknowledgment is made to the Donors of the American Chemical Society Petroleum Research Fund for support of this research through a Doctoral New Investigator (DNI) award.

References

[1] M. P. Burke and S. J. Klippenstein, Ephemeral collision complexes mediate chemically termolecular transformations that affect system chemistry, Nature Chem. 9 (2017) 1078–1082.

[2] C. Olm, T. Varga, E. Valkó, H. Curran, and T. Turányi, Uncertainty quantification of a newly optimized methanol and formaldehyde combustion mechanism, Combust. Flame 186 (2017) 45–64.

[3] F. Fletcher, B. Rabinovitch, K. Watkins, and D. Locker, Energy Transfer in Thermal Methyl Isocyanide Isomerization, J. Phys. Chem. 70 (1966) 2823–2833.

[4] J. Michael, M.-C. Su, J. Sutherland, J. Carroll, and A. Wagner, Rate Constants For $H + O_2 + M \rightarrow HO_2 + M$ in Seven Bath Gases, J. Phys. Chem. A 106 (2002) 5297–5313.

[5] A. Jasper and J. Miller, Lennard-Jones parameters for combustion and chemical kinetics modeling from full-dimensional intermolecular potentials, Combust. Flame 161 (2014) 101–110.

[6] J. Troe, Theory of thermal unimolecular reactions at low pressures. I. Solutions of the master equation, J. Chem. Phys. 66 (1977) 4745–2822.

[7] J. Troe, Theory of thermal unimolecular reactions at low pressures. II. Strong collision rate constants. Applications, J. Chem. Phys. 66 (1977) 1426–2438.

[8] A. Jasper, C. Oana, and J. Miller, "Third-Body" collision efficiencies for combustion modeling: Hydrocarbons in atomic and diatomic baths, Proc. Combust. Inst. 35 (2015) 197–204.

[9] M. Barbet, K. McCullough, and M. Burke, A Framework for Automatic Discovery of Chemically Termolecular Reactions, (Awaiting Publication).

[10] D. G. Goodwin, H. K. Moffat, and R. L. Speth, *Cantera: An Object-oriented Software Toolkit for Chemical Kinetics, Thermodynamics, and Transport Processes*, Version 2.3.0, 2017.

[11] G. Smith, Y. Tao, and H. Wang, (2016), URL: `http://nanoenergy.stanford.edu/ffcm1`.

[12] I. McLean, D. Smith, and S. Taylor, Symp. (Int.) Combust. 25 (1994) 749–757.

[13] J. Li, Z. Zhao, A. Kzakov, M. Chaos, F. Dryer, and J. Scire, A Comprehensive KineticMechanism for CO, CH_2O,and CH_3OH Combustion, Int. J. Chem. Kinet. 39 (2007) 109–136.

[14] U. Burke, W. Metcalfe, S. Burke, and K. Heufer, A detailed chemical kinetic modeling, ignition delay time and jet-stirred reactor study of methanol oxidation, Combust. Flame 165 (2016) 125–136.

Spring Technical Meeting
Eastern States Section of the Combustion Institute
March 4–7, 2018
State College, Pennsylvania

Dynamic Evaluation of Multi-Component Pressure Dependence in Multi-Channel Reactions: A Case Study of CH_3+OH System

Lei Lei[1] *and Michael P. Burke*[2,*]

[1]*Department of Mechanical Engineering, Columbia University, New York, NY, 10027, United States*

[2]*Department of Mechanical Engineering, Department of Chemical Engineering, Data Science Institute, Columbia University, New York, NY, 10027, United States*

[*]*Corresponding author:* mpburke@columbia.edu

Abstract: Complex-forming reactions, whose rate constants depend on pressure and collisional energy transfer characteristics of the surrounding bath gas, play a major role the kinetics of combustion. In most realistic combustion environments, multiple species of distinct collisional energy transfer characteristics are present in significant quantities and thus contribute to collisional energy transfer involved in such reactions. Recent studies have indicated that certain representations of multi-component pressure dependence (i.e. "mixture rules") and/or a failure to implement a mixture rule can result in errors reaching an order of magnitude, whereas recently proposed mixture rules yield errors less than 3%. The present study compares the performance of various mixtures rules for representing multi-component pressure dependence of the multi-channel CH3 + OH reaction in flames, using a novel dynamic procedure for evaluating mixture effects as a function of reaction progress (viz. local temperature, pressure, and mixture composition). Results from this procedure, combining master equation simulations and kinetic-transport simulations, indicate that recently proposed mixture rules based on the reduced pressure provide a considerably more accurate representation of mixture effects for CH3 + OH than previous mixture rules based on the pressure. Furthermore, the present results demonstrate that mixture effects for the CH3 + OH reaction have a significant effect on predictions of the laminar flame speed.

***Keywords:* Mixture Rules, Dynamic Pressure Dependence, Multi-Channel Systems**

1. Introduction

In realistic chemical systems, collision energy transfer processes involved in complex-forming (pressure-dependent) reactions almost always involve contributions from multiple components in sufficient quantities, each of which often possesses distinct energy transferring characteristics. However, the majority of available fundamental experiments and theoretical studies of rate coefficients of these complex-forming reactions almost all focus on pure bath gas environments, based on which a "mixture rule" is often adopted to estimate the rate coefficients in multi-component environments. As a result, the accuracy of simulations in realistic combustion mixture environments depends on the validity and performance of the employed mixture rules – such that it is worthwhile to evaluate possible representations of multi-component pressure dependence, i.e. mixture rules, of important combustion reactions.

The effect of mixture composition on rate coefficients for complex-forming reactions has been extensively investigated for reactions with a single reaction channel [1–4], and more recently, multiple reaction channels (e.g. unimolecular dissociation to two sets of bimolecular fragments, chemically activated reactions) [5]. The results indicate a non-linear dependence of mixture composition on rate constants in the low-pressure and intermediate fall-off regime [1–3, 5] and significant deviations of the classic linear mixture rule (LMR,P) from master equation calculations (as high as a factor of 10 [5]). Newly proposed reduced-pressure-based mixture rules [3, 5] have been demonstrated to reduce the peak deviations of mixture rules from master equation calculations to within 5%, and hence provided a promising manner to incorporate the multi-component pressure dependence into combustion modeling.

Since many combustion systems involve spatially and/or temporally varying temperature, pressure, and mixture composition, here, we present a framework for dynamic evaluation of multi-component effects for the channel-specific rate constants of CH_3 + OH system (Figure 1) in laminar free flame. Multi-component effects are simulated using master equation calculations and estimated using various mixture rules, such that the performance of mixture rules can be evaluated in situations where temperature and species compositions undergo dramatic variations.

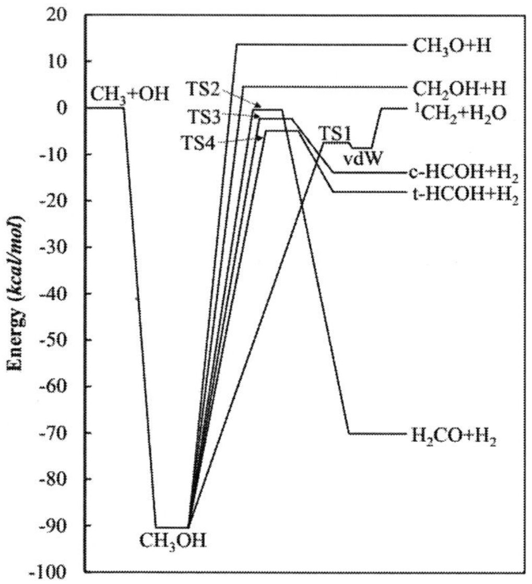

Figure 1: Potential Energy Surface of CH_3+OH System [6].

In the present study, we investigate the controlling features and suitability of mixture rules for representing multi-component pressure dependence for kinetic purposes. On the basis of our previous results [3, 5], the present study seeks to test the performance of our reduced-pressure-based mixture rules for a considerably more complicated reaction system and combustion system and more realistic mixture compositions. To achieve these, CH_3+OH is considered as a case study for a system of significant importance in gas phase combustion, particularly in hydrocarbon flames [7]; mixture compositions and multi-component effects are dynamically evaluated as the flame freely propagates across a fixed spatial location to best represent the species variations and their relative impacts on overall mixture rule performance.

To serve the above purpose, we first present various mixture rules for representing multi-component pressure dependence, master equation calculations and the dynamic evaluation method used to assess their performance. Second, we present the deviations of the various mixture rules against master equation calculations for CH_3 + OH system. Finally, we explore the suitability of various mixture rules for representing multi-component effects and show their implications on combustion kinetic modeling based on methane laminar free flame simulation. The results below suggest that newly proposed reduced-pressure-based mixture rules discussed below provide a reasonable means of incorporating and representing multi-component effects in combustion predictions.

2. Analytical and Computational Methods

2.1 Mixture Rules

Table 1: Mixture rules for multi-component pressure dependence[5]

LMR,P:	$k_{p,LMR,P}(T,P,\underline{X}) = \sum_i k_{i,p}(T,P)X_i$
LMR,R:	$k_{p,LMR,R}(T,P,\underline{X}) = \sum_i k_{i,p}(T,R_{i,LMR})\tilde{X}_{i,LMR}$ $R_{i,LMR}(T,P,\underline{X}) = \frac{\sum_i k_{0,i}(T)\ X_i[M]}{k_\infty(T)}$ $\tilde{X}_{i,LMR}(T,P,\underline{X}) = \frac{k_{0,i}(T)\ X_i}{\sum_j k_{0,j}(T)\ X_j}$
NMR,R:	$k_{p,NMR,R}(T,P,\underline{X}) = \sum_i k_{i,p}(T,R_{i,NMR})\tilde{X}_{i,NMR}$ $R_{i,NMR}(T,P,\underline{X}) = \frac{\sum_i f_i(T,\underline{X})\ k_{0,i}(T)\ X_i[M]}{k_\infty(T)}$ $\tilde{X}_{i,NMR}(T,P,\underline{X}) = \frac{f_i(T,\underline{X})\ k_{0,i}(T)\ X_i}{\sum_j f_j(T,\underline{X})\ k_{0,j}(T)\ X_j}$

Subscripts i and p refer to the $i-$th component and $p-$th reaction channel.

In the present study, three mixture rules are considered (as showed in Table 1). The classic linear mixture rule (LMR,P) estimates the rate constant of the $p-$th channel in the mixture, $k_p(T,P,\underline{X})$, by a linear sum of the rate constants of each consisting component at the same pressure, $k_{p,i}(T,P)$, weighted by the mole fraction of that component (which depicts the component-wise contribution to the overall reaction). Two mixture rules [3, 5] are newly developed based on the observation that rate coefficients for different components share much more similarities in the reduced-pressure scales, R, which actually measure the extent of falloff, than in the absolute pressure scales, P. Similarly, given that the quasi-steady state energy distributions in different components is more similar at the same R than at the same P, the reduced pressure provides a more suitable basis for estimating how the contribution of the $i-$th component is affected by the presence of other mixture components.

Table 2: Energy transfer parameters for colliders

	$\langle \Delta E_d \rangle\ (cm^{-1})$	$\mu_{i,0,A}(T)$
Collider A	$133.0 \times (T/298)^{0.8}$	1
Collider B	$184.1 \times (T/298)^{0.8}$	1.5
Collider C	$506.3 \times (T/298)^{0.8}$	4
Collider D	$992.3 \times (T/298)^{0.8}$	6
Collider E	$2296.9 \times (T/298)^{0.8}$	9

The linear mixture rule in reduced pressure, LMR,R, assumes that the rate constant of the $p-$th channel in the mixture, $k_p(T,P,\underline{X})$, is simply the linear sum of the contributions of the individual components evaluated at the same reduced pressure, $k_{p,i}(T,R)$, weighted by each component's fractional contribution to the reduced pressure. While LMR,R and LMR,P coincide into each other in the low-pressure limit, they yield substantially different estimates in the intermediate fall-off regime. Finally, the non-linear mixture rule in reduced pressure (NMR,R) differs itself from LMR,R by incorporating the component-wise activity coefficients, $f_i(T,\hat{\underline{X}})$, which account for nonlinearities in the low-pressure limit, into reduced pressure calculation (as described in [1, 5]). It is perhaps worth noting that the reduced pressure for multi-channel reactions is calculated according to the total unimolecular rate constants rather than the channel-specific rate constants, and the mixture rules presented in Table 1 for multi-channel reactions collapse to identical expressions to those presented for single-channel reactions [3] when only one channel is present.

2.2 Numerical Simulations

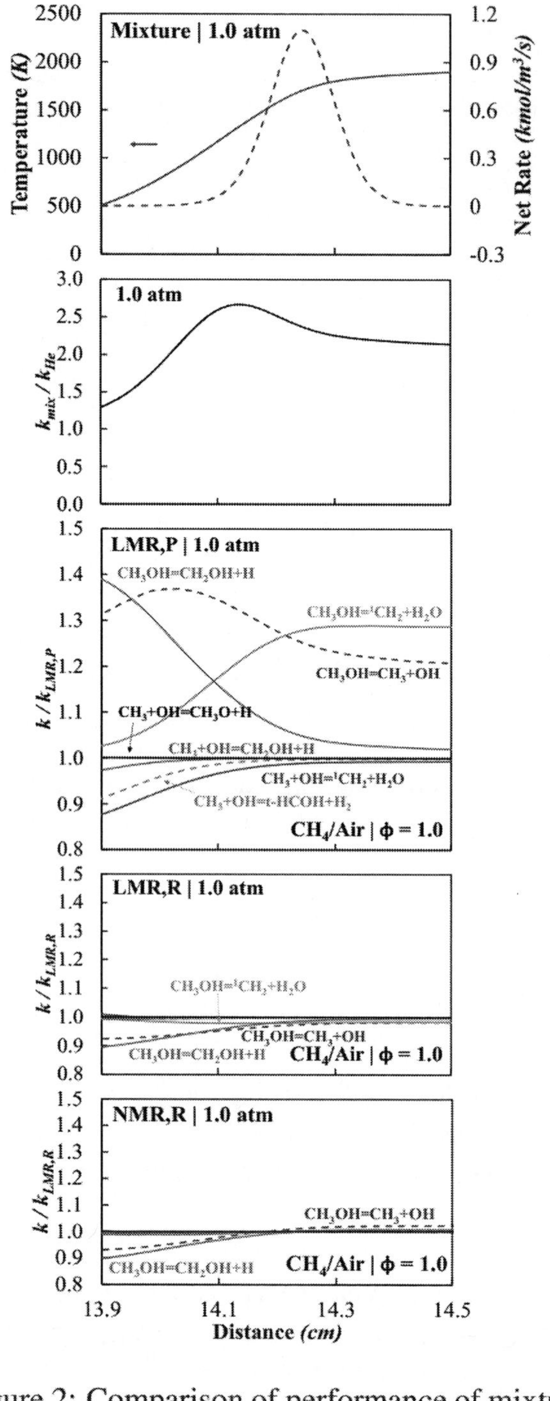

Figure 2: Comparison of performance of mixture rules in Table 1 on predicting rate coefficients of master equation calculations on various channels for five-component mixture as a function of distance from the flame front at 1 atm and equivalence ratio of 1.

Master equation calculations are conducted using the PAPER code [8, 9] on CH_3+OH system (single-well seven-channel) with the potential energy surface (PES, Figure 1) from Jasper et al. [6]. A single-exponential-down model is utilized for average energy transferred per down collision; and the collision frequency is estimated using a Lennard-Jones model. Multiple colliders with non-unity low-pressure "third-body efficiencies" with respect to He in [6] (collider A in Table 2), $\mu_{i,0,A}(T)$, are considered here, each of which has the same LJ parameters as He but different A values for exponential-down factors $\langle \Delta E_d \rangle(T) = A \times (T/298)^n$ such that each collider has a distinct $\mu_{i,0,A}(T)$ (Table 2). Qualitatively, a third-body efficiency of $\sim$1.5, $\sim$4, $\sim$6 and $\sim$10 relative to He have been respectively recommended for N_2, CO_2, small hydrocarbons and H_2O in various studies [10]. Though the enhancement on third-body efficiency is often due to both Z and $\langle \Delta E_d \rangle$, our previous studies [3, 5] and preliminary results indicated that sealing with $Z-$enhancement cases was more straightforward than $\langle \Delta E_d \rangle-$enhancement ones. Thus, we restrict our attention on the $\langle \Delta E_d \rangle-$enhancement mixtures in the present study.

Laminar flame speed simulations presented are performed in CH_4/air mixture (1 atm and T_i=298 K) and CH_4/O_2/He mixture (10 atm and T_i =298 K) at near-stoichiometric equivalence ratios, ϕ =0.6-1.4, using the FreeFlame module of Cantera [11] including multi-component transport and Soret effects using the AramcoMech 1.3 mechanism developed by Metcalfe et al. [7] for C1-C2 hydrocarbon oxidization. To evaluate the multi-component effects on flame speed, AramcoMech 1.3 is first solved at a given equivalence ratio, ϕ, from which the temperature, T, and mole fractions of major species, $\underline{X}$, (which sum up to accounts for over 96% of the entire mixture) are sampled every five grid points to ensure a thorough coverage of the domain. Master equation calculations are then conducted for each sampling point to compute channel-specific rate coefficients, $k_p(T,\underline{X})$ at the sampling condition. PLOG expressions are then fitted to these rate constants, with which the values recommended in AramcoMech 1.3 are replaced and flame speed simulations is re-performed. The above dynamic procedures are then repeated for all considered equivalence ratio and pressure combinations of interest.

3. Results and Discussion

Figure 2 presents the temperature and net rate of progress for CH_3 + OH consumption (top); comparison of total bimolecular rate constants in single-component He environment reported by [6] with those in mixture of Table 2 calculated using master equation (middle); and the deviations of three mixture rules in Table 1 from master equation calculations (bottom three) in a five-component mixture composed of B through E (Table 2, O_2 is assumed to have the same $\mu_{i,0,A}(T)$ as N_2) at 1 atm and equivalence ratio of 1 as a function of distance from the reactor inlet. In contrast to the single-component environment, the inclusion of mixture effect results in a non-monotonic increase in the total bimolecular rate constant of CH_3 + OH, where it first increases dramatically with temperature and peaks nearly a factor of 2.5 and then slightly drops and get steady at $\sim$2.0. As discussed before and demonstrated here, mixture effect is fairly substantial in the intermediate and higher temperature ranges, such that failing to incorporate multi-component effects may yield errors of nearly the same order as the single-component rate coefficients itself. With regard to general trends, deviations for all chemically activated channels peak at lowest temperatures and monotonically increase, while almost all collisionally activated channels (except for CH_3OH = $^1CH_2$ + H_2O) exhibits non-monotonic decreasing deviations with distance. At 1 atm, deviations from LMR,P are strictly positive for collisionally activated channels while negative for chemically activated channels, which is consistent with our previous results on other reactions [3, 5]. While the maximum deviations from LMR,P for collisionally and chemically activated channels reach $\sim$50% and $\sim$30%, both LMR,R and NMR,R yield significantly smaller deviations, reaching $\sim$10% for LMR,R and $\sim$2% for NMR,R for both collisionally and chemically activated reactions.

Acknowledgments

Acknowledgment is made to the Donors of the American Chemical Society Petroleum Research Fund for support of this research through a Doctoral New Investigator (DNI) award. The authors would also like to thank Ahren W. Jasper for sharing the theoretical data for the CH_3OH system [6].

References

[1] J. Tore, Ber. Bunsenges. Phys. Chem. 84 (1980) 829–834.

[2] J. Dove, S. Halperin, and S. Raynor, J. Chem. Phys. 81 (1984) 799–810.

[3] M. P. Burke and R. Song, Proc. Combust. Inst. 36 (2017) 245–253.

[4] H. Schranz, S. Nordholm, and L. Andersson, Chem. Phys. Lett. 161 (1989) 432–438.

[5] L. Lei and M. P. Burke, Solving Multi-Component Pressure Representation for Multi-Channel Unimolecular Reactions: $CH_2O(+M)$ System (Manuscript in Preparation), (2017).

[6] A.W.Jasper, S. Klippenstein, L. Harding, and B.Ruscic, J. Phys. Chem. A 111 (2007) 3932–3950.

[7] W. K. Metcalfe, S. M. Burke, S. S. Ahmed, and H. J. Curran, Int. J. Chem. Kinet. 45 (2013) 638–645.

[8] Y. Georgievskii, J. Miller, M. Burke, and S. Klippenstein, J. Phys. Chem. A 117 (2013) 12146–12154.

[9] Y. Georgievskii, A. Jasper, J. Zádor, J. Miller, M. Burke, C. Goldsmith, and S. Klippenstein, PAPER User Manual, v1, (2014).

[10] A. W. Jasper, C. M. Oana, and J. A. Miller, Proc. Combust. Inst. 35 (2015) 197–204.

[11] D. G. Goodwin, H. K. Moffat, and R. L. Speth, *Cantera: An Object-oriented Software Toolkie for Chemical Kinetics, Thermodynamics, and Transport Processes*, Version 2.3.0, 2017.

Spring Technical Meeting
Eastern States Section of the Combustion Institute
March 4-7, 2018
State College, Pennsylvania

Prompt Dissociations of Propyl Radicals from OH + C_3H_8 and their role in Combustion Simulations

Raghu Sivaramakrishnan[1,*], *C. Franklin Goldsmith*[2], *Sebastian L. Peukert*[1], *Joe V. Michael*[1]

[1] *Chemical Sciences & Engineering Division, Argonne National Laboratory, Argonne, IL, USA*
[2] *School of Engineering, Brown University, Providence, RI, USA*
[*] *Corresponding Author Email: raghu@anl.gov*

Abstract: OH + molecules are an important class of reactions that have limited direct measurements of channel specific rate constants, particularly at combustion relevant temperatures. In the present experiments, we have directly measured site-specific rate constants for abstraction of the secondary C-H bond in OH + C_3H_8 at high temperatures. Atomic resonance absorption spectroscopy (ARAS) was used to monitor the formation of H-atoms from shock-heated mixtures of tert-butylhydroperoxide and C_3H_8 at high temperatures. Simulations for the experimental H-atom profiles are sensitive only to abstraction of the secondary C-H bond leading to unambiguous measurements of the rate constants for this reaction. Simulations of the lower temperature data (T <1000 K) indicate that the H-atom profiles are also influenced to a minor extent by the thermal dissociation of iso-propyl, i-C_3H_7 → H + C_3H_6, at short time-scales. Direct dynamics calculations were performed to examine in greater detail the potential role of prompt dissociations of i-C_3H_7 and n-C_3H_7 (formed from the title reaction) in interpreting the lower temperature (< 1000 K) data from the present work. The role of propyl radical prompt dissociations in interpreting the present experiments and global combustion data will be discussed in the present work.
***Keywords:* Bimolecular Reactions, Alkyl Radicals, Direct Dynamics, Shock Tube**

1. Introduction

In practical combustion systems, the dominant fuel destruction process occurs most likely through hydrogen atom abstractions. Of course, these H-atom abstractions by atoms/radicals such as H, OH, CH_3, O, and HO_2, also serve as chain propagation steps within detailed combustion chemistry reaction mechanisms. The paramount nature of these abstraction reactions has resulted in an extensive literature on their kinetics [1], in particular for straight chain and branched alkanes. Among these processes, abstractions by OH are important under stoichiometric to fuel lean conditions in combustion. Total rate constants for OH + Alkanes [2, 3] are now fairly reasonably characterized by experiments at combustion relevant temperatures. However, channel specific rate constants have been more difficult to directly measure in either set of reactions. Laser Photolysis-Laser Induced Fluorescence (LP-LIF) studies by Tully and co-workers [4-6] utilized deuterium substitution in small alkanes (C_3H_8, n-C_4H_{10}, and iso-C_4H_{10}) and measured very precise kinetic isotope effects by monitoring the decay of OH. With high precision in the experimental data, Tully et al. [4-6] were able to conclude that reactivity at any abstracting site was independent to the isotopic content of the remaining alkyl groups in the molecules. This observation allowed the derivation of site-specific rate constants over the T-range 298-900 K. Recently, Badra et al. [7] have extended the Tully et al. [4-6] deuterium substitution strategy to measure site-specific branching ratios at higher temperatures, 840-1470 K. They used a shock tube monitoring OH by laser absorption, with the assumptions that Tully's branching ratio predictions extrapolated to these higher temperatures. While these recent measurements [7] provide useful benchmarks for assessing kinetic isotope effects at higher temperatures, they are not a direct probe for site-specific rate constants.

Consequently, we present here an alternative approach that relies on measurements of H-atoms as an eventual end-product from the radicals formed by hydrogen abstractions to obtain unambiguous site-specific rate constants. The caveat in the measurements is that in simple alkanes (like propane) at high temperatures, specific alkyl radicals (formed by abstraction) instantaneously dissociate to either an H-atom and an alkene or a smaller alkyl radical and an alkene. Measurements of H-atoms therefore lead to a channel specific rate constant measurement. The additional constraint of total rate constant measurements made in earlier studies [2] then allows unambiguous direct determinations of branching ratios in high temperature abstractions. This approach is amenable for probing not just OH + molecules but other classes of abstractions such as H/CH_3 + molecules. In these higher barrier abstractions, non-Arrhenius effects are more pronounced, and can thereby render Tully's indirect technique less accurate. In the present study, we have used measurements of H-atoms in reactions of OH + C_3H_8 and this is coupled with total rate constant measurements in prior studies [2] (under similar experimental conditions) to directly determine site-specific experimental abstraction branching ratios. The present branching ratio measurements are compared to previous theoretical predictions and also provide a basis for probing existing rate rules [3,7] for OH + Alkanes and to potentially aid in the benchmarking of high level theoretical calculations.

2. Methods / Experimental

The present experiments, in Kr diluent, were performed with the reflected shock tube technique using H-atom ARAS detection. The method and the apparatus currently used have been described earlier [8,9].In recent studies from this laboratory [10,11], it was shown that the highly sensitive H-ARAS technique can be used to directly determine H-abstraction rate constants by observing the formation of product H-atoms. Initially formed unstable radicals (from hydrogen abstraction reactions) instantaneously dissociate at high temperatures (T>1000 K) to give H-atoms. In the present example, abstraction at the two sites in simple alkanes such as C_3H_8 (primary and secondary sites) are known to produce unstable radicals that lead to H-atoms at only one unique site. For example, abstraction at the secondary site in C_3H_8 leads to i-C_3H_7 that instantaneously dissociates only to C_3H_6 and H-atoms, while abstraction at the primary site forms n-C_3H_7 that dissociates exclusively to CH_3 + C_2H_4 at typical combustion conditions [12].

Consequently, with prior knowledge of total abstraction rate constants, H-atom product measurements can be used to obtain site specific rate constants, and, therefore, site specific branching ratios. This technique has been applied in the present study for H-atom abstractions by OH from C_3H_8. With OH-multipass absorption as the diagnostic, total rate constants for the reaction,

$$OH + C_3H_8 \rightarrow Products \qquad (1)$$

was measured [2] at high temperatures (800-1300 K). As discussed earlier, there are two unique abstraction sites in propane forming,

$OH + C_3H_8 \rightarrow H_2O + i\text{-}C_3H_7 \rightarrow (H_2O + H + C_3H_6)$ (1a)
$OH + C_3H_8 \rightarrow H_2O + n\text{-}C_3H_7 \rightarrow (H_2O + CH_3 + C_2H_4)$ (1b)

Measuring H-atom formation in the present study, using H-atom ARAS as the diagnostic, leads to unambiguous determinations for the site-specific rate constants; k_{3a}. These site-specific rate constants in combination with the total rate constant constraints, under similar experimental conditions as the earlier experimental studies [2], provides direct determinations of the site-specific branching ratios in the title reaction.

3. Results and Discussion

Chemical Calibration Using OH + C_2H_6

Similar to our earlier studies [2,3], tert-butyl hydroperoxide (TBHP) was used as a thermal source for OH-radicals, and, at T > 900 K, TBH rapidly decomposes to form OH- and tert-butoxy-radicals that dissociate instantaneously to form acetone and CH_3. The rate constants for TBH dissociation were adopted from Pang et al. [13].In earlier experimental studies from this laboratory [14,15] that used OH-radical detection, simulations of $[OH]_t$ (from a calibrated curve-of-growth) were reproduced by adjusting $[TBH]_0$ in the gas phase. These values for $[TBH]_0$ (that ranged from $[TBH]/([TBH] + [H_2O]) = 0.25 - 0.34$) were in good agreement with calculations based on the assay from the supplier (Sigma-Aldrich) for the TBH/H_2O-solution (i.e., 32 mol% TBH and 68 mol% H_2O; $[TBH]/([TBH] + [H_2O]) = 0.32$).

However, in the present work, OH-radicals are not monitored. TBH/H_2O-solution is known to be "sticky", and its vapor-liquid equilibrium is not well established [16]. Consequently, an alternative procedure was devised to check for the integrity of the gas phase $[TBH]_0$ mole fractions in this work. The technique involves the use of a chemical calibrant; $OH + C_2H_6 \rightarrow C_2H_4 + H_2O + H$, is a reaction with well-established rate constants [17] that were measured by monitoring $[OH]_t$ over a wide temperature range. H-atom ARAS can now be used to monitor $[H]_t$ that evolves as a product from this calibrant reaction $OH + C_2H_6 \rightarrow C_2H_4 + H_2O + H$. Kinetics simulations confirm that $[H]_t$ profiles are only sensitive to the rate constant of this reaction over the experimental T-range (1000 -1350 K) with secondary reactions being negligible contributors. With the well determined OH + C_2H_6 rate constants, simulations of the $[H]_t$ profiles were then used to infer the fraction of $[TBH]_0$ in the gas phase mixture of TBH and H_2O. The average gas phase $[TBH]_0$ mole fraction from the calibration experiments performed immediately before and after the title reaction experiments was determined to be 22.24%. This $[TBH]_0$ mole fraction was used in the subsequent analyses of the OH + C_3H_8 experiment.

OH + C_3H_8

Using the same primary mix, a mixture of C_3H_8, ([TBH] + $[H_2O]$), and Kr was prepared and used in the branching ratio studies. A series of ethane + OH experiments before and after gave the ratio $[TBH]_0/([TBH]_0 + [H_2O]_0) = 0.2224$, thereby allowing $[TBH]_0$ and $[H_2O]_0$ to be specified in simulations for the OH + C_3H_8 experiments. In this case, H-atoms are formed only from abstraction from the secondary site. H-atom profiles were simulated using a in-house model with initial estimates for k_{1a} and k_{1b} from Sivaramakrishnan et al. [2]. Fig. 1 is the experimental profile for an OH + C_3H_8 experiment at 1059 K. The H-atom profile is simulated using a literature in-house detailed model (black solid line in Figure 1) to obtain the best-fit value for k_{1a}. Changes to this best fit k_{3a} by ± 30% (black dashed lines) lead to noticeable disagreement with the experimental profile. Sensitivity analyses reveal that the $[H]_t$ profile is influenced significantly only by k_{1a} indicating that reaction R1a is chemically isolated under the present experimental conditions.

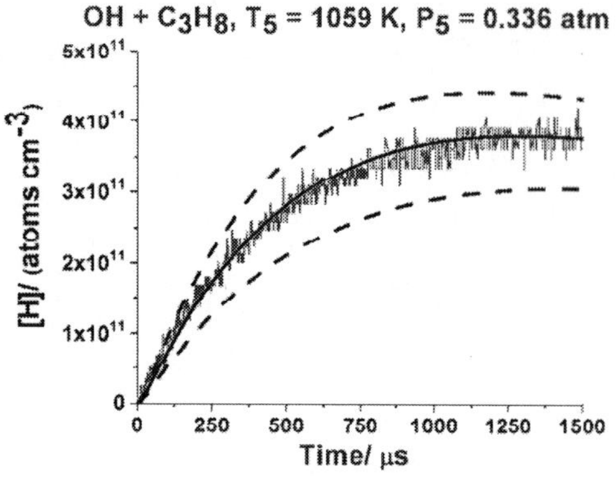

Figure 1: $[H]_t$ profile from OH + C_3H_8. Solid black line – Best fit k_{1a}, Dashed lines – 1.3 x k_{1a} and 0.7 k_{1a}.

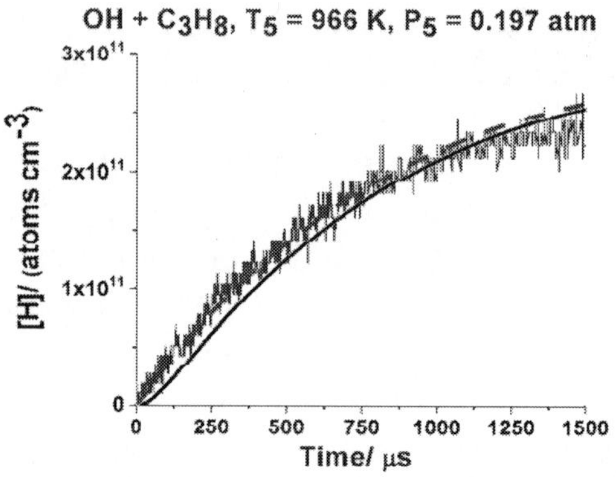

Figure 2: $[H]_t$ profile from a lower temperature OH + C_3H_8 experiment. Solid black line – Best fit k_{1a}, Dashed red line – Simulation with best fit k_{1a} assuming i-propyl dissociation is instantaneous.

Figure 2 represents another experimental $[H]_t$ profile, albeit at lower temperatures (T < 1000 K). Simulations using the detailed model were used to obtain the best fit k_{1a} for this experiment that is depicted by the solid black line. Changes to this best fit k_{1a} by ± 30% lead to significant under and over-predictions of the profile (not shown but similar to the dashed lines in Fig. 1). Interestingly, at early times (t < 500 μs) a noticeable delay in the rise of H-atoms is observed in the best fit simulation leading to a sigmoidal rise in contrast to the near linear rise of H-atoms in the experiment. Sensitivity analysis for this experiment reveals that apart from k_{1a}, rate constants for the thermal dissociation, i-C_3H_7 → H + C_3H_6, also influences $[H]_t$ at shorter time scales (t < 500 μs).

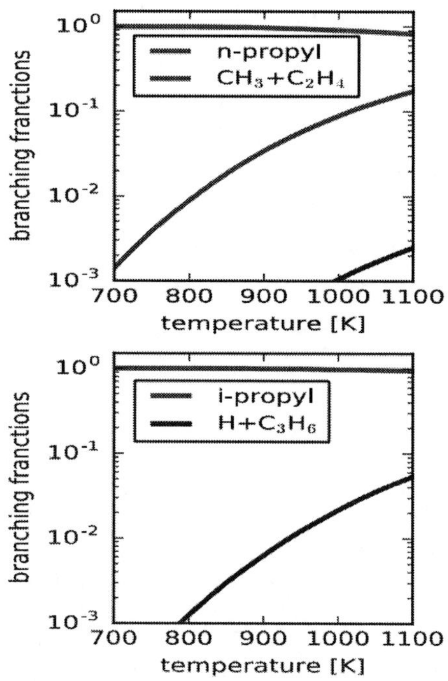

Figure 3: Branching fractions for C_3H_8+OH for abstraction of a primary hydrogen (top) and secondary hydrogen (bottom) at 0.2 atm.

An immediate resolution to resolve the early-time discrepancy in the simulation can be obtained by assuming that i-C_3H_7 instantaneously dissociates to H + C_3H_6, i. e. no i-C_3H_7 was assumed to exist in the model. This assumption with the same best fit k_{1a} is depicted by the dashed red line in Fig. 2. However at T < 1000 K, this assumption may not be appropriate. In recent work, Miller and Klippenstein [12] have characterized propyl decompositions, and their master equation analyses are in excellent agreement with available experimental data [18,19] for i-C_3H_7 and n-C_3H_7 decompositions. In particular, at T < 1000 K and pressures ~ 0.5 atm, the theoretical predictions for i-C_3H_7 decomposition are deep in the fall-off regime and slow enough that it may be time-resolvable on the time-scales of the present shock tube experiments. However, the experimental profiles depict that H-atoms are formed near-instantaneously. Increasing the rate constants from the theoretical predictions for i-C_3H_7 dissociation can resolve this discrepancy. Recent studies from our laboratory on HCO prompt dissociations [20,21] raise the question that prompt dissociations of propyl radicals may be active under the present experimental conditions. In particular, if i-C_3H_7 can be formed chemically activated from the abstraction process it can undergo prompt dissociation to explain the instantaneous H-atom rise. $\Delta H_R(0K)$ = -21.0 kcal/mol [22] for OH + $C_3H_8 \rightarrow$ i-C_3H_7 + H_2O, and if the additional thermal energy is accounted for at high temperatures, this may be sufficient to lead to prompt dissociation of the i-C_3H_7 radical.

Consequently, branching fractions for the prompt dissociation of *n*-propyl and *i*-propyl radicals were computed using a methodology similar to that in references [20,23,24]. Figure 3 summarizes the results from the direct dynamics calculations at P = 0.2 atm (conditions of relevance here). At T < 1000 K, the present calculations reveal that i-propyl prompt dissociation is < 3%. Therefore, inclusion of these prompt dissociation fractions in the model does not lead to noticeable changes in the simulated profiles and ultimately the best fit k_{1a} values.

Inclusion of the present direct dynamics results for OH + C_3H_8 in concert with thermal prompt dissociation of propyl radicals from H/O/CH_3/HO_2/O_2 + C_3H_8 reactions leads to a minor 2.5% change in laminar flame speed for 1atm, 600K, C_3H_8-air flame. The flame simulations were performed with an in-house propane model [25]. On the other hand, propane-air ignition delays at 1atm are retarded by up to 11% (see figure 4) relative to the baseline model (without propyl prompt dissociation) for fuel-rich mixtures when the present prompt dissociation trajectory results are used along with thermal prompt dissociations of propyl radicals from H/O/CH_3/HO_2/O_2 + C_3H_8 reactions using the same in-house propane model. In particular, the largest influence is due to prompt dissociation of propyl radicals from the H + C_3H_8 reaction that is known to compete with the H + O_2 chain branching reaction for consumption of H-atoms, thereby retarding ignition. Thermal prompt dissociation fractions utilized in the present simulations are a potential lower limit. Similar to our prior HCO prompt dissociation study [20], larger prompt dissociation fractions are expected for the H + C_3H_8 reaction in comparison to the OH + C_3H_8 reaction studied here. Consequently, ignition delays in fuel-rich C_3H_8-air mixtures are expected to be retarded further in simulations that include propyl prompt dissociations relative to the baseline simulations.

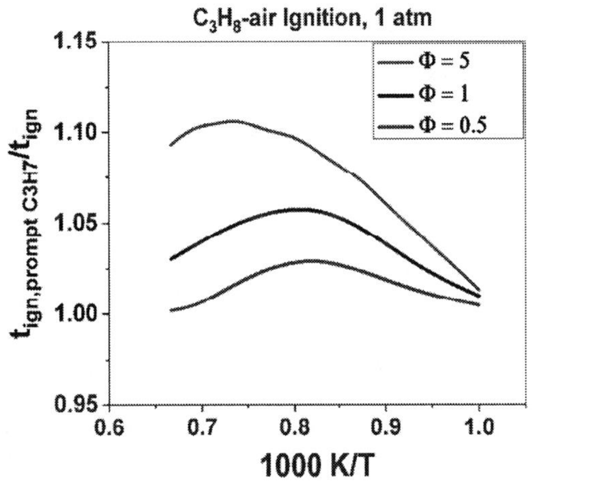

Figure 4: Ratio of ignition delays with and without propyl prompt dissociations.

4. Conclusions

OH + C_3H_8 has been studied in the present work at high temperatures using a shock-tube. Product H-atoms were measured behind reflected shock waves using atomic resonance absorption spectrometry (ARAS) in mixtures with tert-butylhydroperoxide as the OH precursors, and the alkane of interest in excess, at high temperatures (>900 K). Simulations using a kinetics model indicated that the H-atom profiles were sensitive to the abstraction rate constant at the secondary C-H bond, and therefore, best-fits to the H-atom profiles were used to determine channel-specific abstraction rate constants, k_{1a}. In combination with prior published total rate constants, the present results yield direct experimental determinations of abstraction branching ratios in the title reaction. Simulations of the experimental data at T<1000 K indicate the potential role for prompt dissociation of propyl radicals. Results from theoretical studies indicate that prompt dissociations of propyl radicals do not impact the present experimental observations. Inclusion of propyl prompt dissociation in simulations indicate that ignition delays may be retarded by > 10% in fuel-rich C_3H_8-air mixtures.

5. Acknowledgements

This material is based on work supported by the U.S. Department of Energy, Office of Basic Energy Sciences, Division of Chemical Sciences, Geosciences, and Biosciences through Argonne National Laboratory. Argonne is a U.S. Department of Energy laboratory managed by UChicago Argonne, LLC, under contract DE-AC02-06CH11357. Support for RS was provided as part of the Argonne-Sandia Consortium on High-Pressure Combustion Chemistry, FWP# 2009 ANL 59044. CFG gratefully acknowledges support from the U.S. National Science Foundation through Award Number CBET-1553366, with Dr. Song-Charng Kong as the program manager

6. References

[1] J.A. Manion, R.E. Huie, R.D. Levin, D.R. Burgess Jr., V.L. Orkin, W. Tsang, W.S. McGivern, J.W. Hudgens, V.D. Knyazev, D.B. Atkinson, E. Chai, A.M. Tereza, C.-Y. Lin, T.C. Allison, W.G. Mallard, F. Westley, J.T. Herron, R.F. Hampson, D.H. Frizzell, NIST Chemical Kinetics Database, NIST Standard Reference Database 17, Version 7.0 (Web Version), Release 1.4.3, Data version 2008.12, National Institute of Standards and Technology, Gaithersburg, Maryland, 20899-8320. Web address: http://kinetics.nist.gov/

[2] R. Sivaramakrishnan, N.K. Srinivasan, M.-C. Su, J.V. Michael, Proc. Combust. Inst. 32 (2009) 107-114.

[3] R. Sivaramakrishnan, J.V. Michael, J. Phys. Chem. A 113 (2009) 5047-5060.

[4] A.T. Droege, F.P. Tully, J. Phys. Chem. 90 (1986) 1949-1954.

[5] A.T. Droege, F.P. Tully, J. Phys. Chem. 90 (1986) 5937-5941.

[6] F.P. Tully, J.E.M. Goldsmith, A.T. Droege, J. Phys. Chem. 90 (1986) 5932-5937.

[7] J. Badra, E.F. Nasir, A. Farooq, J. Phys. Chem. A 118 (2014) 4652-4660.

[8] J.V. Michael, Prog. Energy Combust. Sci. 18 (1992) 327-347.

[9] J.V. Michael, in: J.R. Barker (Ed.), Advances in Chemical Kinetics and Dynamics, JAI, Greenwich, Vol. I, pp. 47-112, for original references, 1992.

[10] S.L. Peukert, J.V. Michael, J. Phys. Chem. A (117) 2013, 10186 – 10195.

[11] S.L. Peukert, N.J. Labbe, R. Sivaramakrishnan, J.V. Michael, J. Phys. Chem. A (117) 2013, 10228-10238.

[12] J.A. Miller, S.J. Klippenstein, J. Phys. Chem. A (117) 2013, 2718 – 2727.

[13] G.A. Pang, R.K. Hanson, D.M. Golden, C.T. Bowman, Z. Phys. Chem. 225 (2011) 1157-1178.

[14] N.K. Srinivasan, M.-C. Su, J.V. Michael, J. Phys. Chem. A 111 (2007) 3951-3958.

[15] R. Sivaramakrishnan, M.-C. Su, J. V. Michael, S. J. Klippenstein, L. B. Harding, Ruscic, B. J. Phys. Chem. A 114 (2010) 9425-9439.

[16] W. E. Vaughan, General Discussions, Discuss. Faraday Soc. 10 (1951) 330.

[17] L.N. Krasnoperov, J.V. Michael, J. Phys. Chem. A 108 (2004) 5643-5648.

[18] P.W. Seakins, S.H. Robertson, M.J. Pilling, I.R. Slagle, G.W. Gmurczyk, A. Bencsura, D. Gutman, W. Tsang, J. Phys. Chem. 97 (1993) 4450-4458.

[19] A. Bencsura, V.D. Knyazev, S.-B. Zing, I. R. Slagle, D. Gutman, Proc. Combust. Inst. 24 (1992) 629-635.

[20] N.J. Labbe, R. Sivaramakrishnan, C.F. Goldsmith, Y. Georgievskii, J.A. Miller, S.J. Klippenstein, J. Phys. Chem. Lett. 7 (2016) 85-89.

[21] N.J. Labbe, R. Sivaramakrishnan, C.F. Goldsmith, Y. Georgievskii, J.A. Miller, S.J. Klippenstein, Proc. Combust. Inst. 36 (2017) 525-532.

[22] B. Ruscic, ver. 1.112 of the Thermochemical Network (2013); available at https://ATcT.anl.gov

[23] M.P. Burke, C.F. Goldsmith, Y. Georgievskii, S.J. Klippenstein, Proc. Combust. Inst. 35 (2015) 205-213.

[24] C.F. Goldsmith, M.P. Burke, Y. Georgievskii, S.J. Klippenstein, Proc. Combust. Inst. 35 (2015) 283-290.

[25] J. A. Miller, S. J. Klippenstein, R. Sivaramakrishnan, C. F. Goldsmith, M. P. Burke, A. W. Jasper, J. Zador, N. Hansen, B. Ruscic, L. B. Harding, Y. Georgievskii, N. J. Labbe, P. Glarborg, Combustion Chemistry in the Twenty-First Century: Developing a Theory- Informed Chemical Kinetic Model for the Small-Hydrocarbon Fuels, In Preparation, Prog. Ener. Combust. Sci. 2018.

Spring Technical Meeting
Eastern States Section of the Combustion Institute
March 4-7, 2018
State College, Pennsylvania

Explosion Limits of $H_2/CH_4/O_2$ Mixtures: Analyticity and Dominant Kinetics

Wenkai Liang, Zirui Liu and Chung K. Law

Department of Mechanical and Aerospace Engineering, Princeton University, Princeton NJ 08544, USA

Abstract: The pressure-temperature explosion limits of $H_2/CH_4/O_2$ mixtures were analyzed computationally and analytically. It is shown that, with the addition of about 5% of H_2 to the mixture, the explosion limit evolves from being monotonic to the non-monotonic Z-shaped response characteristic of H_2/O_2 explosions. Such transition is much less sensitive compared with the $H_2/CO/O_2$ system, which is explained by the different response of the kinetic parameters corresponding to O and OH radical competitions by H_2 and CH_4. Further eigenvalue analysis with appropriate quasi-steady state simplifications leads to accurate analytic solutions for the explosion limits, with explicit expressions for the controlling pathways for different H_2 and CH_4 concentrations.
Keywords: methane; syngas; explosion limits; eigenvalue analysis.

1. Introduction

The explosion limits of homogeneous combustion mixtures, frequently expressed as functions of the system pressure and temperature, is a foundational phenomenon in combustion, indicating the mixtures' reactivity leading to chemical runaway. Among the explosion limits of various fuels, the H_2/O_2 system has commended extensive interest because not only it is a subset in the hierarchical structure of the oxidation models of hydrocarbon fuels [1], but it also has a sharp, non-monotonic, kinetic response to variations in the system parameters. Furthermore, recently there have also been considerable interests in hydrogen-enriched fuels of CH_4 and CO, commonly known as syngas and derived from coal and biomass [2], for its utilization in gas-turbines for propulsion and power generation [2-4].

The explosion limits of H_2/O_2 mixtures exhibits a characteristic Z-shaped curve in the pressure-vs-temperature plot, with the three segments respectively referred as the first, second and third limits. The phenomena have been studied extensively, with recent advances in accurate and comprehensive analytical studies [5, 6]. For the next level of fuel complexity, previous study [7] on $H_2/CO/O_2$ explosion limits show that while the oxidation of dry CO is an extremely slow process such that it is basically considered to be non-explosive, as is well known, the addition of a minute amount of H_2 greatly promotes its reactivity in the role of a catalyst, and the corresponding monotonic explosive curve of CO/O_2 will transition to the non-monotonic Z-shaped curve with 10^{-5} to 10^{-4} H_2 addition in the fuel mixture. Regarding the $H_2/CH_4/O_2$ mixtures, while several studies have been performed on the ignition delay times [8] and laminar flame speeds [9, 10], the corresponding explosion limits have not been studied. Recognizing that its chemistry represents the next level of complexity in fuels oxidation beyond the $H_2/CO/O_2$ chemistry, and that CH_4-based oxidation forms the foundational block of large hydrocarbon oxidation and is essential for natural gas applications, it is thus of interest to explore and quantify the extent of the kinetic interaction between CH_4 and H_2 and the influence of the H_2 chemistry on the overall reactivity of the mixture, particularly in terms of the Z-shaped H_2/O_2 explosion responses. Compared to other large hydrocarbons, the oxidation of $H_2/CH_4/O_2$ mixtures is governed by much smaller sizes, for which the skeletal and reduced mechanisms developed have 20-30 species [11, 12].

In view of the above considerations, we shall first simulate the explosion limits of $H_2/CH_4/O_2$ mixtures, with different kinetic models and various amounts of H_2 addition, and identify the roles and limits of H_2 addition and its subsequent dominance in terms of the Z-curve response compared with that of $H_2/CO/O_2$ mixtures. Based on the reduced kinetic models, analytical solution of the explosion limits can be derived. We shall then study the controlling chain branching and termination reactions as related to the separate and coupled chemistry of H_2 and CH_4 oxidation, and derive explicit expressions for the various $H_2/CH_4/O_2$ explosion limits.

2. Detailed Simulations

To begin with, the simulation results of the explosion limits with different kinetic mechanisms [13-15] are demonstrated in Fig. 1. The composition changes from pure CH_4 to pure H_2, which shows that the mixture becomes more explosive with H_2 addition. In general, the results show good agreements among different mechanisms when f_{H2} , which is defined as $[H_2]/([CH_4]+[H_2])$, is sufficiently large, *i.e.*, larger than 50%. However, for the low f_{H2} cases, the differences are larger as the controlling kinetics become dominated by CH_4 oxidation. Furthermore, as the fraction of H_2 is increased from 0 to around 10%, the explosion curve transitions from being monotonic to the non-monotonic Z-shaped, showing the three distinct explosion limits. Since this is an isothermal system, the larger amount of heat release for CH_4 is irrelevant. Also, by using the present isothermal system, we can isolate the effects purely due to the kinetic chain mechanisms.

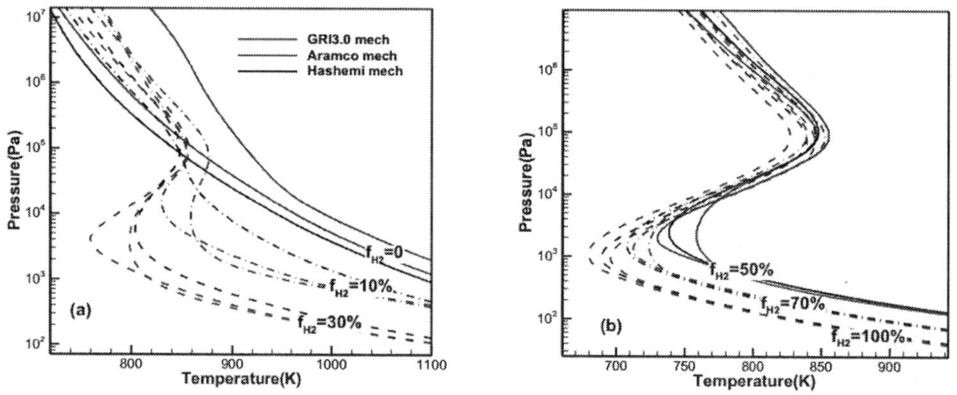

Figure 1: Explosion limits of $H_2/CH_4/O_2$ mixtures with different kinetic models.

Since the transition to non-monotonicity signifies the corresponding transition from CH_4- to H_2-dominated kinetics, we will focus on the transition with more detailed simulations. Figure 2(a) shows the monotonic-to-nonmonotonic transition for $H_2/CH_4/O_2$ mixtures. It is noted that the transition occurs at around f_{H2} = 4% to 5% by mole. As a comparison, we also show such transition of the $H_2/CO/O_2$ system using the NUIG mechanism [16]. The results show that the $H_2/CO/O_2$ system is much more sensitive to the H_2 addition, as the transition occurs at much lower f_{H2} values, around 10^{-5} to 10^{-4}. Such catalyzing effect is due to the highly sensitive competition of O radical between the branching channel with H_2 and the termination channel reacting with CO, as explained in previous work [7]. In the following sections, we shall identify the controlling kinetics for the corresponding monotonic-to-nonmonotonic transition for $H_2/CH_4/O_2$ mixtures through eigenvalue analysis.

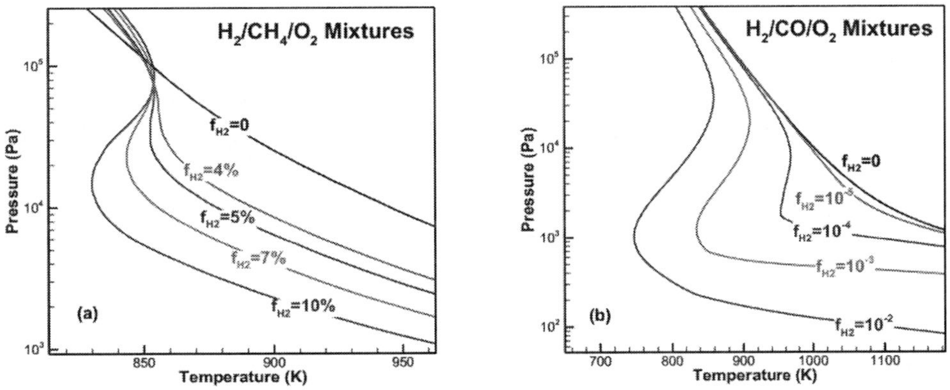

Figure 2: Transition from monotonic to nonmonotonic explosion limit curves of (a) $H_2/CH_4/O_2$ mixtures and (b) $H_2/CO/O_2$ mixtures.

3. Eigenvalue Analysis

Following the detailed simulations, we proceed to derive the analytic expressions of the explosion limit curves by applying eigenvalue analysis. As the first-step towards obtaining analytical solutions, we only consider reactions involving reactant-reactant and radical-reactant, such as $H_2 + O_2 \rightarrow H + HO_2$ and $H + O_2 \rightarrow O + OH$ respectively. The time-evolution of the intermediate species can then be written as a system of linear ordinary differential equations [7, 17]. Here, for the current reduced mechanism, nine intermediates species are identified, namely H, O, OH, HO_2, H_2O_2, CH_3, HCO, CH_2O and CH_3O, and this system is referred to as the 9-variable system. Furthermore, from previous studies of the H_2/O_2 [5, 6] and $CH_4/H_2/O_2$ [3] systems, the intermediate species O, OH, HCO, CH_2O and CH_3O can be assumed to be in quasi-steady state (QSS). Consequently, the time-evolving species can be reduced to only H, HO_2, H_2O_2, and CH_3, which is thus referred to as the 4-variable system.

To further facilitate interpretation, we have identified the following competing channels of the radicals with H_2 and CH_4 in the reduced system. For the O radical, the two competing reactions are $O + H_2 \rightarrow H + OH$ (R2) and $O + CH_4 \rightarrow CH_3 + OH$ (R9) respectively; for the OH radical, $OH + H_2 \rightarrow H + H_2O$ (R3) and $OH + CH_4 \rightarrow CH_3 + H_2O$ (R13); and for the HO_2 radical, $HO_2 + H_2 \rightarrow H + H_2O_2$ (R6) and $HO_2 + CH_4 \rightarrow CH_3 + H_2O_2$ (R18). Based on these three pairs of competing reactions, we can define three corresponding kinetic parameters as:

$$\alpha = \frac{k_2[H_2]}{k_2[H_2]+k_9[CH_4]},\ \beta = \frac{k_3[H_2]}{k_3[H_2]+k_{13}[CH_4]},\ \gamma = \frac{k_6[H_2]}{k_6[H_2]+k_{18}[CH_4]} \quad (1)$$

Specifically, α is the fraction of the O radical going through reaction R2 as compared to the total amount in going through R2 and R9. Similarly, β is the fraction of the OH radical going through reaction R3 as compared to the total amount in going through R3 and R13; γ is the fraction of the HO_2 radical going through reaction R6 as compared to the total amount in going through R6 and R18. As such, they respectively quantify the ratios of the O, OH and HO_2 radicals reacting with H_2 as compared to the channels in parallel reacting with CH_4, and degenerate to unity for the pure H_2/O_2 system. In addition, another pair of reactions for the CH_3O radical is found as $CH_3O + H_2 \rightarrow H + CH_3OH$ (R12) and $CH_3O + O_2 \rightarrow HO_2 + CH_2O$ (R22), leading to a kinetic parameter defined as

$$\eta = \frac{k_{12}[H_2]}{k_{12}[H_2]+k_{22}[O_2]} \tag{2}$$

Apart from all these kinetic parameters, several other gas-phase reactions are involved in the eigenvalue formulation. For the H_2/O_2 system, the two key reactions $H + O_2 \rightarrow O + OH$ (R1) and $H + O_2 + M \rightarrow HO_2 + M$ (R8) compete with each other forming the second limit of the Z-curve; and the reaction $H_2O_2 + M \rightarrow 2OH + M$ (R7) is important for high-pressure third limit. For the CH_4 oxidation, two reactions are important for the CH_3 radical, namely $CH_3 + O_2 \rightarrow O + CH_3O$ (R16) and $CH_3 + O_2 \rightarrow OH + CH_2O$ (R17). Furthermore, the shuffle reaction between the H and CH_3 radical, $H + CH_4 \leftrightarrow CH_3 + H_2$ (R11), is found to be important for the interaction of H_2 and CH_4 kinetics. Assuming an ideal gas, the total gas concentration [M] is proportional to the total pressure, that is $[M] = p/R^0T$. By using these kinetic parameters (α, β, γ and η) and the rate coefficients ($k_1, k_7, k_8, k_{16}, k_{17}, k_{11}$ and k_{11r}), the time-evolution equation of the system can be written as

$$\frac{dX}{dt} = LX + I \tag{3}$$

for the vector of the intermediate species concentrations

$$X = ([H], [HO_2], [H_2O_2], [CH_3])^T \tag{4}$$

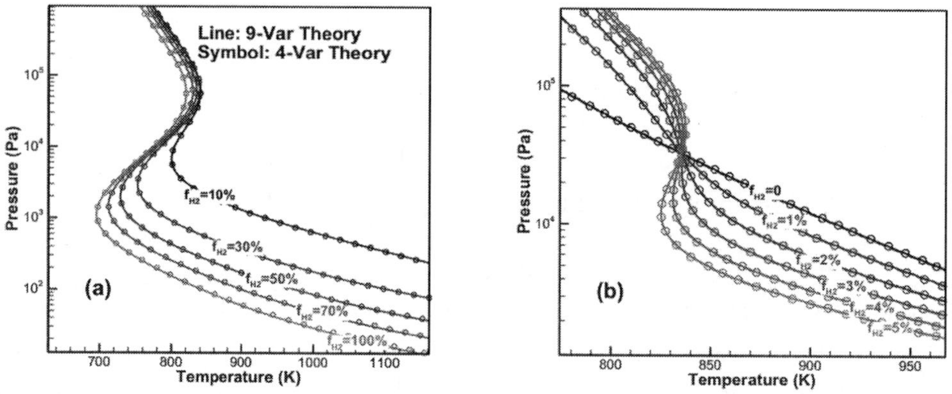

Figure 3: Theoretical predictions of explosion limit curves of $H_2/CH_4/O_2$ mixtures for (a) high- and (b) low-hydrogen levels.

The explosion limit can be expressed as the neutral condition, $\det(L) = 0$ [5, 6]. In Fig. 3 the prediction using this 4-variable theory with QSS species is compared with the prediction from the 9-variable theory without assuming QSS species. It is seen that these two theoretical eigen-value analyses yield almost identical predictions for all the fuel compositions, hence validating these QSS assumptions. Furthermore, we have found that the theoretical results in Fig.

3 qualitatively predict the effect of H_2 addition to the explosion limits of CH_4/O_2 mixtures, which has been shown in Fig. 1 and Fig. 2(a). Also, Fig. 3(b) demonstrates that the monotonic-to-nonmonotonic transition, which occurs around 5% H_2 in the fuel mixture, is accurately predicted by the linear analysis. Kinetically, the three parameters (α, β, γ) change from zero to one as H_2 concentration in fuel increases, for which they control such transition behavior. To understand the different effects of the H_2 concentration for the $H_2/CH_4/O_2$ and $H_2/CO/O_2$ mixtures shown in Fig. 2, we have also defined such kinetic parameters for the $H_2/CO/O_2$ system [7], namely for the O radical, $O + H_2 \rightarrow H + OH$ (R2) competing with $O + CO + M \rightarrow CO_2 + M$ (R-CO1), for the OH radical, $OH + H_2 \rightarrow H + H_2O$ (R3) competing with $OH + CO \rightarrow CO_2 + H$ (R-CO2) and for the HO_2 radical, $HO_2 + H_2 \rightarrow H + H_2O_2$ (R6) competing with $HO_2 + CO \rightarrow CO_2 + OH$ (R-CO3). Based on these three pairs of competing reactions, we can define three equivalent kinetic parameters for $H_2/CO/O_2$ as:

$$\alpha_{CO} = \frac{k_2[H_2]}{k_2[H_2]+k_{CO1}[CO][M]}, \quad \beta_{CO} = \frac{k_3[H_2]}{k_3[H_2]+k_{CO2}[CO]}, \quad \gamma_{CO} = \frac{k_6[H_2]}{k_6[H_2]+k_{CO3}[CO]} \tag{5}$$

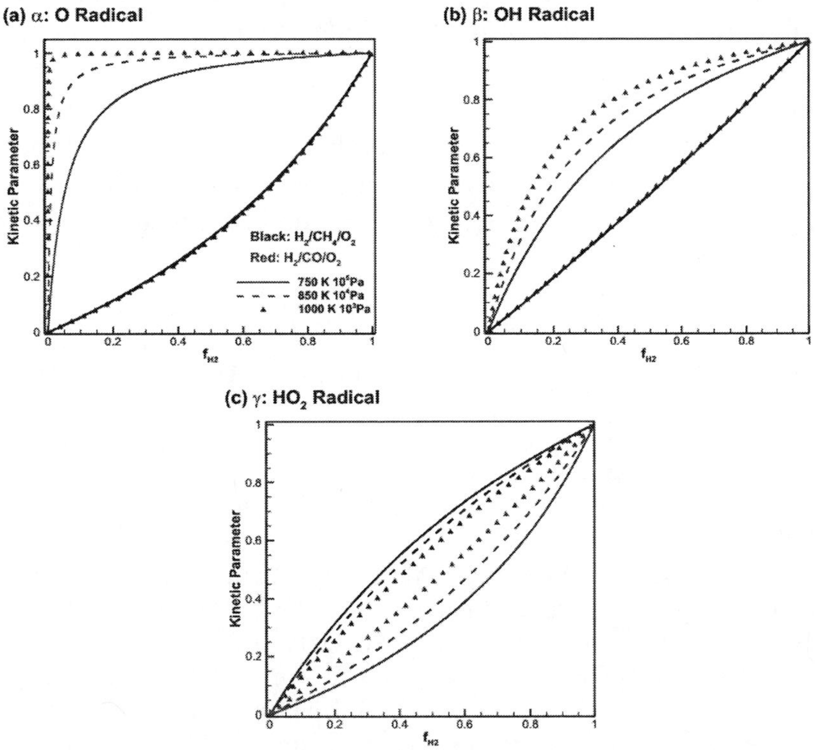

Figure 4: Kinetic parameters of $H_2/CH_4/O_2$ and $H_2/CO/O_2$ mixtures.

These three parameters for both $H_2/CH_4/O_2$ and $H_2/CO/O_2$ mixtures change with the H_2 percentage, f_{H2}, as demonstrated in Fig. 4. It is seen that for $H_2/CO/O_2$ mixtures α and β show a steep increase at small f_{H2} and saturate to unity very quickly, especially for the O radical at low pressures. However, comparing with the dramatic changes for $H_2/CO/O_2$ mixtures, they only have mild changes for the $H_2/CH_4/O_2$ mixtures. Such different responses explain the highly sensitive transition of $H_2/CO/O_2$ mixtures and relatively less sensitive transition for the $H_2/CH_4/O_2$ mixtures found in Fig. 2. This finding also indicates that the monotonic to nonmonotonic transition at low-pressure conditions is mainly controlled by the competition for the O radical by H_2 and CH_4 as well as their competition for the OH radical. Furthermore, for the γ parameter of the HO_2 radical, both systems show moderate transition from zero to one. Since HO_2 is essential at high-pressure conditions, this finding is also consistent with the slow change of the third limit shown in Fig. 1.

4. Conclusions

In this work, we have numerically and theoretically analyzed the explosion limits of $H_2/CH_4/O_2$ mixtures, with emphasis on the transition from the monotonic CH_4-controlled explosion boundary to the nonmonotonic H_2-controlled explosion boundary. Compared with the corresponding transition of $H_2/CO/O_2$ mixtures, the transition for the $H_2/CH_4/O_2$ mixtures is much less sensitive. Linear analysis is conducted based on the reduced model and the system can be further reduced to a 4-variable system with appropriate QSS assumptions. The less sensitive monotonic-to-nonmonotonic transition of the $H_2/CH_4/O_2$ system is explained based on the competing channels of key intermediate radicals consumed by H_2- and CH_4- related reactions. It is found that the kinetic parameters of CO show a much dramatic change compared to those of CH_4 at low H_2 levels. Furthermore, the essential competing radicals are identified as O and OH. The effects of chain carrier wall terminations are also identified.

5. Acknowledgements

This research was supported in part by the NSF under the technical monitoring of Dr. S. C. Kong. WL was in addition supported by a Maeder Fellowship and a Harari Fellowship of Princeton University, while ZL by a second year fellowship of Princeton University.

References

[1] A. L. Sánchez and F. A. Williams, Prog. Energy Combust. Sci., vol. 41, pp. 1–55, 2014.
[2] T. Lieuwen, V. Yang, and R. A. Yetter, eds. Synthesis gas combustion: fundamentals and applications. CRC Press, 2009.
[3] Z. M. Nikolaou, J. Y. Chen, and N. Swaminathan, Combust. Flame, vol. 160, no. 1, pp. 56–75, 2013.
[4] M. Fischer and X. Jiang, Combust. Flame, vol. 167, pp. 274–293, 2016.
[5] X. Wang and C. K. Law, Journal of Chemical Physics, 2013, 138: 134305.
[6] W. Liang and C. K. Law, An Analysis of the Explosion Limits of Hydrogen/Oxygen Mixtures with Nonlinear Chain Reactions. Physical Chemistry Chemical Physics. in press 2017.
[7] W. Liang, J. Liu, and C. K. Law, Combust. Flame, vol. 179, pp. 130–137, 2017.
[8] R. K. Cheng, A. K. Oppenheim, Combust. Flame. 1984, 58(2):125-39.
[9] E. Hu, Z. Huang, J. He, C. Jin, J. Zheng, Int. J. Hydrog. Energy. 2009 34(11):4876-88.
[10] F. Halter, C. Chauveau, N. Djebaili-Chaumeix, I. Gökalp, Proc. Combust. Inst., 2005 30(1):201-8.
[11] C. J. Sung, C. K. Law, and J.-Y. Chen, Proc. Combust. Inst., vol. 27, no. 1, pp. 295–304, 1998.
[12] T. Lu and C. K. Law, Combust. Flame, vol. 154, no. 4, pp. 761–774, 2008.
[13] G.P. Smith, D.M. Golden, M. Frenklach, N.W. Moriarty, B. Eiteneer, M. Goldenberg, C.T. Bowman, R.K. Hanson, S. Song, W.C. Gardiner, V.V. Lissianski. GRI 3.0 Mechanism. Gas Research Institute (http://www. me. berkeley. edu/gri_mech). 1999.
[14] W.K. Metcalfe, S.M. Burke, S.S. Ahmed, H.J. Curran, Int. J. Chem. Kinet. (2013) 45(10) 638–675.
[15] H. Hashemi, J. M. Christensen, S. Gersen, H. Levinsky, S. J. Klippenstein, and P. Glarborg, Combust. Flame, vol. 172, pp. 349–364, 2016.
[16] A. Kéromnès, W. K. Metcalfe, K. A. Heufer, N. Donohoe, A.K. Das, C. J. Sung, J. Herzler, C. Naumann, P. Griebel, O. Mathieu, M. C. Krejci, E. L. Petersen, W. J. Pitz, H. J. Curran, Combust. Flame 2013, 160:995-1011.
[17] W. Liang and C. K. Law, Combust. Flame, vol. 188, pp. 162–169, 2018.

Spring Technical Meeting
Eastern States Section of the Combustion Institute
March 4-7, 2018
State College, Pennsylvania

Measurement of Methane Autoignition Delays in a Shock Tube under Supercritical Carbon Dioxide Conditions

Miad Karimi[1], Bradley Ochs[1], Wenting Sun[1], Devesh Ranjan[1,2]*

[1]*School of Aerospace Engineering, Georgia Institute of Technology, 270 Ferst Drive, Atlanta, GA, 30332 U.S.A.*
[2]*School of Mechanical Engineering, Georgia Institute of Technology, 801 Ferst Drive, Atlanta, GA, 30332 U.S.A.*

Corresponding Author Email: miadkarimi@gatech.edu

Abstract: The directly fired supercritical carbon dioxide (sCO_2) power cycle has very high efficiency while allowing nearly complete carbon dioxide (CO_2) capture. The operating condition of sCO_2 power cycle (10 MPa to 30 MPa) combustors is dramatically different from conventional gas turbine combustors. However, combustion properties, e.g., autoigntion delays, have never been reported under sCO_2 conditions. In this study, autoignition delays of $CH_4/O_2/CO_2$ mixtures were measured above the critical pressure of CO_2 for the first time in a high-pressure shock tube. The experiments were performed at stoichiometric conditions with $CH_4/O_2/CO_2$=5:10:85 on mole fraction, averaged pressure 10 MPa, and temperature range of 1257 to 1383 K. Simulation results from Aramco 2.0, USC Mech II and HP-Mech were found to be close to each other (within 12%), while GRI 3.0 predicted approximately a factor of 3 shorter autoignition delay times. Experimental measurements fell between the prediction from USC Mech II and GRI 3.0 with a better agreement with USC Mech II. Sensitivity analysis and reaction flux analysis were performed at 10 MPa and 1300 K. It is observed that two reaction pathways exist for CH_3 at high pressure conditions: recombination to form C_2H_6 and oxidation to form CH_3O. The formation of C_2H_6 becomes dominant as pressure increases and temperature decreases. Further simulation shows that autoignition delay is not sensitive to the third body collision efficiency of CO_2.
Keywords: methane autoignition, supercritical CO2, reaction flux analysis

1. Introduction

In recent years, substantial progress has been made toward developing directly fired supercritical carbon-dioxide (sCO_2) power cycles [1, 2]. A pressurized oxy-combustion power cycle has the dual benefit of very high efficiency while simultaneously allowing nearly a complete capture of CO_2 for carbon sequestration [3]. In this technique, natural gas/syngas and pure oxygen are diluted with carbon dioxide (CO_2) before entering the combustor. As a result, the main products of combustion are water (H_2O) and CO_2 which act as the working fluid to drive the turbines. Unlike the operating conditions of conventional gas turbines for ground power generation (~ 2 MPa), the sCO_2 power cycle requires combustors to operate at much higher pressures (in the range of 10 to 30 MPa) and inlet temperature (approximately 900 K). One of the key requirements in designing such combustors is the comprehensive knowledge of chemical kinetics, in particular autoignition delay properties at those conditions.

In the past several decades, autoignition delays have been studied extensively for a variety of fuels [4-6]. However, the measurements of autoignition delays above the critical pressures of the mixtures are still limited [7, 8]. The autoignition of methane (CH_4) has been measured extensively in argon/ nitrogen baths over broad ranges of pressures. However, the measurement of CH_4 autoignition delay in CO_2 baths is very limited. To the authors' knowledge, only Vasu *et al.* [2], Pryor *et al.* [9], Koroglu *et al.* [10] and Hargis *et al.* [1] conducted measurements of $CH_4/O_2/CO_2/AR$ mixtures with different CO_2 concentrations up to 1.3 MPa (13 bar). Hargis *et al.* [1] and Koroglu *et al.* [10] studied autoignition delay times of CH_4 in shock tubes with high levels of CO_2 up to 1 MPa (10 bar) and 120 kPa (1.2 bar), respectively. Koroglu *et al* [10] reported that CO_2 dilution had a negligible effect on the autoignition delays. The experimental results were also compared with numerical predictions using GRI 3.0 [11] and Aramco 1.3 [12] and reasonable agreement was observed. Hargis *et al* also reported that the chemical effect of CO_2 dilution was minimal and within the uncertainty of the data, particularly for the experiments near 100 kPa (1 bar). The study showed a small effect of CO_2 addition for higher pressures (near 10 bar), with a general increase in autoignition delay time for the largest levels of CO_2 dilution (75%). Predictions from Aramco 1.3 [12] also showed a minimal effect of CO_2 addition on CH_4 ignition delay times in agreement with the shock tube data. In this work, autoignition delays of $CH_4/O_2/CO_2$ mixtures are measured above the critical pressure of CO_2 for the first time in a high-pressure shock tube. The results obtained from the experiments are also compared with predictions from different chemical kinetics models. Sensitivity and reaction flux analysis are conducted to understand the discrepancies among experiments and modeling.

2. Experimental Method

The shock tube used in this study is fabricated out of stainless steel 316 with 15.24 cm inner diameter and wall thickness of 5.08 cm. The total length of the shock tube is 20 m consisting of 10m long driver section and 10-m long driven section. The inner surface of the tube is electro polished with 0.2 μm surface finish. The driven section is equipped with twenty-five access port plugs which could be used for temperature/pressure measurements and optical diagnostics. Eight circumferential ports are located 1.24 cm away from the end wall enabling pressure and optical measurements.

The driven section is initially vacuumed down to 1.33 Pa using an Agilent IDP-15 Dry Scroll pump and finally down to 133 μPa by using an Agilent TwisTorr 304 FS AG turbo pump. The mixture is prepared in a high pressure mixing vessel according to Dalton's law of partial pressure using high purity gases (99.999%). The partial pressures are measured by three high accuracy (0.05%) MKS 960A absolute pressure sensors with ranges of 0.013 MPa, 0.133 MPa and 1.33 MPa. The prepared mixture is stirred inside the mixing vessel by a magnetic stir for two hours to ensure sufficient mixing prior to be fed into the driven section. A sample from the mixture is then taken and analyzed by micro gas chromatography (Inficon Fusion μGC) to estimate the uncertainty in the equivalence ratio (φ). Gas chromatography results revealed a maximum uncertainty of no more than 0.01 in equivalence ratio. Finally, the mixture settles inside the driven section for one hour to establish a homogenous condition. Helium is used as the driver gas for all experiments. The shock velocity is measured by four fast-response PCB 113B22 piezoelectric pressure transducers. The shock velocity obtained from these sensors are then extrapolated linearly to the end wall to account for shock attenuation. Using the obtained shock velocity, the temperature on the reflected shock is calculated by Gaseq [14]. The temperature uncertainty owing to the uncertainty in the shock velocity is found to be approximately ±10 K. The sidewall pressure histories are recorded by a Kistler 6045A PiezoStar dynamic pressure sensor located at 1.24 cm away from the end wall. OH* emission, used to monitor the ignition, is measured using a Hamamatsu H10723-113 photomultiplier (PMT) with a band pass filter centered at 307 nm ±10 nm, and collected from the sidewall. The uncertainty of PMT measurements is ±2 μs owing to the resolution of data acquisition system.

3. Results and Discussion

Experiments are conducted at stoichiometric conditions with $CH_4/O_2/CO_2$=5:10:85 on mole fraction, averaged pressure 10 MPa, and temperature range of 1257 to 1383 K. Simulation for these test conditions are performed using four different chemical kinetics models and compared with the experimental data. Davidson and Hanson [13] studied the effect of real gas on the calculation of temperature and pressure after the reflected shock. They found that at high temperature conditions (above 1000 K), ideal gas equation of state still performed well even though the pressure was above the critical point. For example, at 26.3 MPa (260 atm), 1300 K, the temperature and pressure predictions are only 3 K and 152 kPa (1.5 atm) lower than those obtained assuming an ideal gas. Therefore, standard normal shock relations still apply for the calculation of pressure and temperature after the reflected shock. Regarding the kinetic simulation, recent work from Kogekar *et al* [14] studied real gas effects on high pressure combustion by comparing simulated and experimentally measured shock tube ignition delay measurements for n-dodecane/O_2/N_2 mixtures. They found that real gas effects are negligible above 1100 K even though it is significant at low temperature conditions. Therefore, in this study, ideal gas is still assumed in all simulations. Another concern is the non-equilibrium vibrational excitation of CO_2 at elevated temperature conditions. Vibrational excitation of CO_2 was well studied in the past [15]. As reported in Ref. [16], the relaxation time of CO_2 at ~ 10 MPa (100 atm) and 1300 K is approximately 0.01 μs which is negligible in this study comparing to measured autoignition delays which are in the order of hundreds of microseconds. Therefore, the test gas after the reflected shock wave is still in thermodynamic equilibrium. Figure 1 summarizes the measured autoignition time of $CH_4/O_2/CO_2$ mixture (5:10:85).

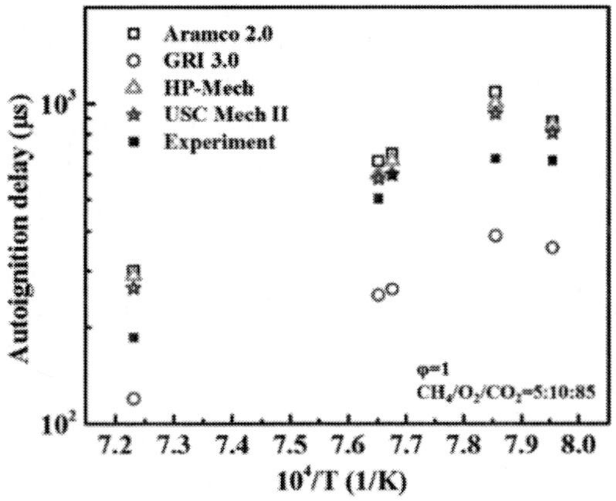

Figure 1: Comparison of measured autoignition delays of $CH_4/O_2/CO_2$ mixture (5:10:85) and simulation using different kinetic models

As shown in figure 1, simulation results from Aramco 2.0 [12], USC Mech II [17] and HP-Mech [18] are close to each other with a maximum deviation of 12%. GRI 3.0 [11] predicts a significantly shorter autoignition delay, having approximately a factor of 3 difference. Experimental measurements fall between the prediction of USC Mech II [17] and GRI 3.0 [11] and appear to have better agreement with predictions from USC Mech II [17]. The experimental results deviate approximately 40% for USC Mech II and 100% for GRI 3.0 [11]. To interpret the results shown in figure 1, a sensitivity analysis is conducted by perturbing the pre-exponential "A" factor of each reaction at 10 MPa (100 atm), 1300 K for $CH_4/O_2/CO_2$ mixture (5:10:85) using USC Mech II [17] and the results are shown in figure 2.

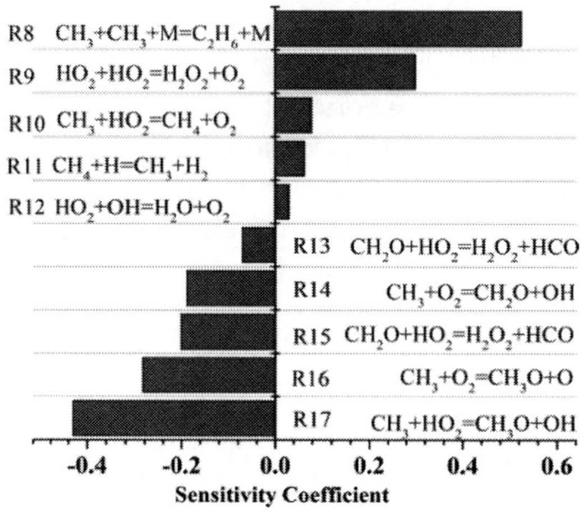

Figure 2. Sensitivity analysis of CH4/O2/CO2 mixture autoignition at 100 atm, 1300 K

The most dominant reactions on the determination of autoignition are identical from results using GRI 3.0 [11]. R8-R12 are the five most dominant reactions to hinder ignition. R13-R17 are the most dominant reactions to enhance ignition. Different from sensitivity analysis at low pressure conditions, $H+O_2=O+OH$ is the sixth most dominant reaction to enhance ignition. It can be seen that only R8 is related with the collision efficiency for CO_2. The third body efficiency of CO_2 was investigated by simply doubling its value, which has a negligible effect on autoignition. It should be noted that the change of the third body efficiency is different from the change of "A" factor because the later affects both forward and backward reactions while the former only affects the net reaction rate. To further understand the performance and differences among various kinetic models, a reaction flux analysis is conducted using a Global Pathway Selection (GPS) method [19] at 1300 K, 10 MPa (100 atm) for $CH_4/O_2/CO_2$ mixture (5:10:85). Even though the results from the sensitivity analysis are nearly identical, the reaction flux analysis reveals that different kinetic models have different reaction pathways. After the initiation reaction of CH_4 to form CH_3, a major reaction pathway of CH_3 oxidation is through

$$CH_3 \rightarrow CH_3O \rightarrow CH_2O \rightarrow HCO \rightarrow CO \qquad (1)$$

This is a well-known reaction pathway at low pressure, high temperature conditions with approximately 99% of carbon flux traveling through this pathway [20]. However, at the conditions of this study (100 MPa, 1300 K), a new reaction pathway becomes important:

$$CH_3 \rightarrow C_2H_6 \rightarrow C_2H_5 \rightarrow C_2H_4 \rightarrow C_2H_3 \rightarrow CH_2CHO \rightarrow CH_2CO \rightarrow HCCO \rightarrow CO \qquad (2)$$

The ratio between reaction pathways (1) and (2) are 0.95, 4, and 0.94 predicted by USC Mech II [18], GRI 3.0 [11], and Aramco 2.0 [12], respectively. Reaction pathway (2) is significantly longer/slower, therefore more carbon flux through reaction pathway (2) means longer autoignition delay. This explains why GRI 3.0 [11] predicts the shortest autoignition delay among the different kinetic models. Reaction pathway (2) is identical for USC Mech II [18] and Aramco 2.0 [12] but slightly different for GRI 3.0 [11]. It is interesting that the results of both the sensitivity analysis and reaction flux analysis are pressure and temperature dependent. At low pressure and high temperature (above 1400 K) conditions, the effect of reaction pathway (2) is negligible unless for a very fuel rich condition. That explains why previous sensitivity analyses [1, 10] did not capture reaction R8. However, this reaction pathway becomes dominant with decreasing temperature (below 1400 K) and increasing pressure.

4. Conclusions

In this study, methane autoignition delays at high carbon dioxide level are reported for the first time at supercritical carbon dioxide condition. The experiments were conducted for $CH_4/O_2/CO_2$ mixture with mole fraction 5:10:85 at averaged pressure 10.6 MPa (105 atm) and temperature of 1257 K to 1383 K. The measured autoignition delays are compared with simulation using existing kinetic models. It is found that GRI 3.0 [11] predicts the shortest autoignition delays by factor of 3 than the other three kinetic models which showed similar predictions with approximately 12% deviation among themselves. The experimental results show noticeable difference from simulations using different kinetic models and predictions from USC Mech II [18] showed a better agreement among others. Detailed analysis reveals that, unlike low pressure case, at high pressure conditions CH_3 recombination to form C_2H_6 through $CH_3+CH_3+M=C_2H_6+M$ becomes dominant instead of direct oxidation to form CH_3O. The branching ratio of these two reaction pathways determines the autoignition delays. With decrease in temperature, the effect of C_2H_6 formation increases at a given pressure.

5. Acknowledgements

This work is supported by U.S. Department of Energy (Award Number: DE-FE0025174, Program Manager: Seth Lawson)

6. References

1. Hargis, J.W. and E.L. Petersen, *Methane Ignition in a Shock Tube with High Levels of CO2Dilution: Consideration of the Reflected-Shock Bifurcation.* Energy & Fuels, 2015. **29**(11): p. 7712-7726.
2. Vasu, S.S., D.F. Davidson, and R.K. Hanson, *Shock Tube Study of Syngas Ignition in Rich CO2Mixtures and Determination of the Rate of H + O2+ CO2→ HO2+ CO2.* Energy & Fuels, 2011. **25**(3): p. 990-997.
3. *Technology Development for Supercritical Carbon Dioxide (SCO2) Based Power Cycles.* Available from: https://www.netl.doe.gov/research/coal/energy-systems/turbines/supercritical-co2-turbomachinery.

4. ERIC L. PETERSEN, M.R.H., DAVID F. DAVIDSON, RONALD K. HANSON and CRAIG T. BOWMAN, *<High pressure methane oxidation behind reflected shock waves.pdf>*. Twenty-Sixth Symposium (International) on Combustion/The Combustion Institute, 1996: p. pp. 799–806.
5. Gudiyella, S. and K. Brezinsky, *The high pressure study of n-propylbenzene pyrolysis.* Proceedings of the Combustion Institute, 2013. **34**(1): p. 1767-1774.
6. Wang, W., S. Gowdagiri, and M.A. Oehlschlaeger, *The high-temperature autoignition of biodiesels and biodiesel components.* Combustion and Flame, 2014. **161**(12): p. 3014-3021.
7. V. P. Zhukov, V.A.S., A. Yu. Starikovskii, *<spontanous.pdf>*. Combustion, Explosion, and Shock Waves, 2003. **39**: p. 487–495.
8. Petersen, E.L., D.F. Davidson, and R.K. Hanson, *Ignition Delay Times of Ram Accelerator CH/O/Diluent Mixtures.* Journal of Propulsion and Power, 1999. **15**(1): p. 82-91.
9. Pryor, O., et al., *Measurements and interpretation of shock tube ignition delay times in highly CO 2 diluted mixtures using multiple diagnostics.* Combustion and Flame, 2017. **180**: p. 63-76.
10. Koroglu, B., et al., *Shock tube ignition delay times and methane time-histories measurements during excess CO 2 diluted oxy-methane combustion.* Combustion and Flame, 2016. **164**: p. 152-163.
11. Gregory P. Smith, D.M.G., Michael Frenklach, Nigel W. Moriarty, Boris Eiteneer, Mikhail Goldenberg, C. Thomas Bowman, Ronald K. Hanson, Soonho Song, William C. Gardiner, Jr., Vitali V. Lissianski, and Zhiwei Qin
12. Zhou, C.-W., et al., *A comprehensive experimental and modeling study of isobutene oxidation.* Combustion and Flame, 2016. **167**: p. 353-379.
13. HANSON, D.F.D.A.R.K., *Real Gas Corrections in Shock Tube Studies at High Pressures.* IJC, 1996.
14. Kogekar, G., et al., *Impact of non-ideal behavior on ignition delay and chemical kinetics in high-pressure shock tube reactors.* Combustion and Flame, 2018. **189**: p. 1-11.
15. Kiefer, S.S.a.J.H., *Relaxation, Incubation, and Dissociation in CO2.* Physical Chemistry 2006. **111**: p. 3884-3890.
16. Simpson, C.J.S.M., T.R.D. Chandler, and A.C. Strawson, *Vibrational Relaxation in CO2 and CO2–Ar Mixtures Studied Using a Shock Tube and a Laser–Schlieren Technique.* The Journal of Chemical Physics, 1969. **51**(5): p. 2214-2219.
17. Hai Wang, X.Y., Ameya V. Joshi, Scott G. Davis, Alexander Laskin, Fokion Egolfopoulos & Chung K. Law, *High-Temperature Combustion Reaction Model of H2/CO/C1-C4 Compounds.* 2007.
18. Xueliang Yang, X.S., Jeffery Santer, Hao Zhao, and Yiguang Ju. *A high pressure mechanism for H2, CO, CH2O, CH4, CH3OH, C2H2, C2H4, C2H6 with EGR Effects (CO2 and H2O) and HCO prompt reactions* Available from: http://engine.princeton.edu/mechanism/HP-Mech.html.
19. Gao, X., S. Yang, and W. Sun, *A global pathway selection algorithm for the reduction of detailed chemical kinetic mechanisms.* Combustion and Flame, 2016. **167**: p. 238-247.
20. Sun, W., et al., *Kinetic effects of non-equilibrium plasma-assisted methane oxidation on diffusion flame extinction limits.* Combustion and Flame, 2012. **159**(1): p. 221-229.

Spring Technical Meeting
Eastern States Section of the Combustion Institute
March 4-7, 2018
State College, Pennsylvania

Color Camera Pyrometry and its Impact in Reaction Visualization of Energetic Materials

Dylan J. Kline[1], Rohit J. Jacob[2], Philip M. Guerieri[2], Michael R. Zachariah[1,3,]*

[1]Department of Chemical and Biomolecular Engineering, University of Maryland College Park, College Park, MD, 20742, USA
[2]Department of Mechanical Engineering, University of Maryland College Park, College Park, MD, 20742, USA
[3]Department of Chemistry and Biochemistry, University of Maryland College Park, College Park, MD, 20742, USA
[*]Corresponding Author Email: mrz@umd.edu

Abstract: Despite a continuously growing pool of information on novel energetic materials and mechanisms, the diagnostic tools used to evaluate temperature remains challenging. High-speed color camera videography can be used to visualize reactions dynamics in both space and time. Color camera pyrometry in particular has opened up new avenues to study the structure-function relationship of energetic materials and recent work has shed new light on the reaction mechanisms participating in the combustion of metal-laden hydrocarbon fuels and thermite mixtures. These case studies will demonstrate the ability for color camera pyrometry to map the temperature distribution of combustion events and assess the validity of previously proposed mechanisms based on macroscopic observations. Furthermore, a 32-channel VIS-NIR spectrometer will also be used to estimate flame temperature and emission at high sampling rates (>20kHz) as a means to cross-verify the phenomena observed by the color camera pyrometer.
Keywords: energetic materials, burn characterization, temperature measurements, pyrometry

1. Introduction

Energetic materials were developed for military use as early as 360 B.C. and have since led to advancements in propellants, pyrotechnics, and explosives through the efficient storage and release of the energy within their atomic bonds [1]. The primary energy release pathway in these compounds is chemical combustion, typically characterized by rapid heat release and light emission [2–4]. With an plateau inherent in the traditional CHNO systems due to their gaseous reaction products, recent research has been directed towards metal-based energetic composites that are prepared on the nanoscale to increase burn rates [5–7].

A large body of recent research has been directed at exploring the underlying mechanism responsible for the fast reaction rates observed in such Metastable Intermolecular Composites (MIC)/nanothermite combustion, however the experimental setups developed to quantify the reaction dynamics of these materials are limited since the results are sensitive to design variation. Interestingly, temperature measurements for these MIC's is not prevalent despite its direct relation to energy release and nature as a fundamental thermodynamic property. Emission measurements are favored as non-invasive methods to measure the temperature for MIC characterization. Weismiller et al. [8] and Kappagantula et al. [9] have both explored the reaction dynamics of these nanothermite systems through high-speed multi-wavelength and IR pyrometry, respectively. Although, due to the spatially dynamic nature of the flame front, point light sources could constantly move in and out of the field of view, affecting the temperature measurement. In either scenario, for probing of energetic nanomaterial combustion, the use of a spectrometer falls short in large volume studies for it is averaging the entire cross-sectional area that is being observed rather than having multiple point measurements to account for spatial resolution.

High speed color cameras can record videos over an entire area imaged by the sensor and post-processed with spectral response curves [10–14]. Numerous experiments have demonstrated the ability for color ratio pyrometry to effectively characterize thermally radiating particles. Color ratio pyrometry was first reported in 1994 using an infrared-sensitive pixel array to spatially resolve temperature profiles of thin filaments [15,16] and development of high-speed digital color cameras has since led to widespread research on the temperature characterization of flames, thin filaments, soot, and direct injection spark ignition (DISI) engines [12–14,17–19]. A paper by Densmore et al. detailed the setup for a color ratio pyrometer that was successfully used to characterize the temperature of an exploded C-4 charge at the U.S. Army Research Laboratory, which served as the basis for the experiments performed in this study [11,20–22].

The purpose of this work is to demonstrate how color camera pyrometry can be implemented as a tool to investigate the reaction mechanism of MIC's and other energetic materials. In one case study, we demonstrate and observed tunable reactivity of a nanothermite systems through systematic doping of the oxidizers and extend results of Sullivan et. al [23] by visualizing the impact of gas production on combustion. In the other, a free-falling, metal-laden hydrocarbon droplet undergoing combustion is analyzed with the same high-speed color camera pyrometer in parallel with a multi-wavelength pyrometer as a means to observe the impact microexplosions have on a reacting system.

A. Visualizing Reactivity of Thermites Tuned Through Oxidizer Variation[1]

A2. Experimental

Materials were prepared with commercially available aluminum nanoparticles (ALEX, Argonide Corp.) with 68.7% active content (TGA) and an average diameter of 50nm. The Fe_2O_3 and WO_3 nanopowders were produced by Sigma Aldrich with a diameter <100 nm. Samples were prepared by sonicating a known amount of oxide in 10mL of hexane for an hour, after which stoichiometric amount of aluminum was added to the slurry and resonicated to ensure adequate mixing. A set of 7 samples were made: Al/Fe_2O_3, Al/WO_3, and a set of 5 samples where we systematically adjusted the oxidizer by adding 20, 60, 70, 80, and 90% by mole of WO_3 to Fe_2O_3 while maintaining overall stoichiometry of the blend.

Wire ignition experiments were conducted in a stainless steel 6-way cross with windows to visualize the combustion with the high-speed camera. The chamber interior was painted black to minimize light reflections. The slurry was coated onto a 76 μm-diameter platinum wire and resistively heated in a 1atm argon environment for 4ms at a rate of $\sim 10^5$ K/s using the T-Jump apparatus detailed by Zhou et al. [24] Two videos were recorded per sample with a Vision Research Phantom Miro M110 and a macro lens at a framerate of 20,000 fps with the *f/#* and exposure times chosen to provide the best signal to noise ratio.

Video processing is done through an in-house code designed to extract raw pixels intensities, demosaic the color filter array, and estimate the temperature of the particles. Temperature is assigned by minimizing the error between theoretical and measured values for three channel ratios (green/red, blue/green, and blue/red) incorporating an inverse wavelength dependency. Further processing is done to determine pixels with an error in temperature >110 K.

A3. Results and Discussion

High speed color camera pyrometry videos enabled the identification of the different mechanisms of the Al/Fe_2O_3, the $Al/Fe_2O_3/xWO_3$ and Al/WO_3 reactions due to distinct features

---

[1] Adapted with permission from Rohit Jacob and Dylan Kline from paper titled "High Speed Temperature Measurements of Nanothermite Composites: Visualizing Tunable Reactivity" (under review).

that appeared throughout the reaction. For the Al/Fe_2O_3 mixture, the gas release from the thermite mixture produced a cloud of molten metal and little unreacted material was left as it propagated down the wire (Figure 1a). The Al/Fe_2O_3 reaction is believed to be limited by the oxidizer decomposition, and is characterized by the slow reaction of the aluminum with gas phase oxygen released from the oxidizer, as exemplified by its longer burn durations and slow rising pressure profile [25,23]. In comparison, the Al/WO_3 reaction is expected to occur in the condensed phase due to the lack of oxygen release from WO_3 [25]. Upon observing the videos of the Al/WO_3 reaction, the condensed phase nature of this reaction was evident by the absence of a reactive cloud, as the reaction seemingly occurred on the wire and continued on to the ends after melting the wire (Figure 1c). In line with the observations from the pressure cell, the $Al/Fe_2O_3/70\%WO_3$ sample (Figure 1b) shows both higher peak temperature, and a larger combustion zone which we attribute to higher gas release. The use of a color camera to measure temperature allows for the direct observation of the disaggregating role that the oxidizer gas release plays during reaction since the reaction cloud exhibits higher temperatures than the material remaining on the wire.

Figure 1: High speed pyrometry frames of a. Al/Fe_2O_3, b. $Al/Fe_2O_3/70\%$ WO_3 and c. Al/WO3 samples ignited on a hot wire at 1 atm. argon environment highlighting the gas production and enhanced reaction

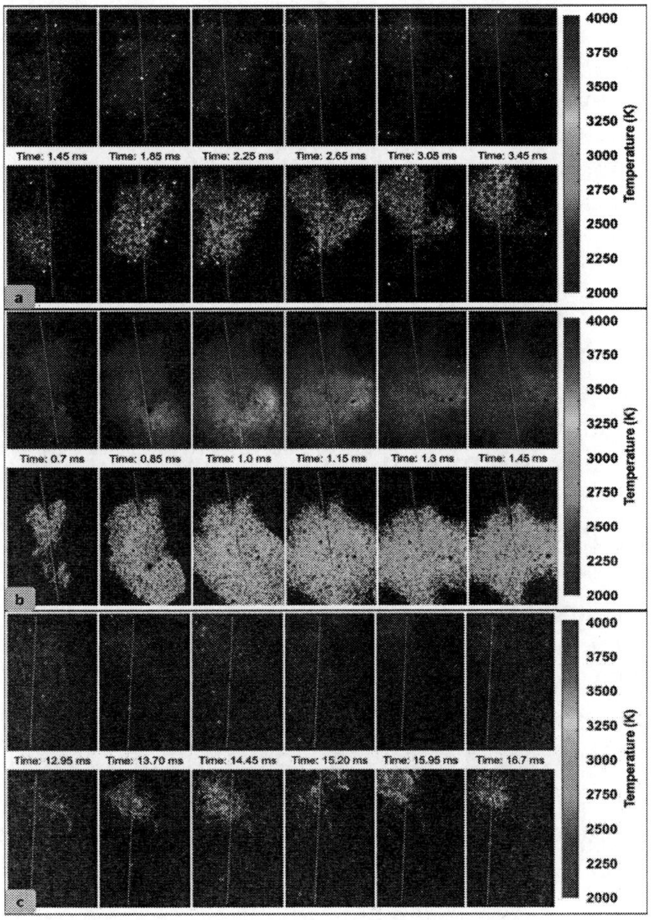

When data collected through high-speed color camera pyrometry was compared to the data presented by Sullivan et. al[23] and that collected through constant volume combustion, it was clear that gas generation due to mixture of the oxides was significantly impacting the peak pressure and burn time. Since the Al/Fe_2O_3 reaction is limited by the slow decomposition of the oxide [23], reactive cloud is small and burn time is in comparison to the suite of WO_3-doped thermites. In the case of Al/WO_3, the decomposition of the oxide occurs in the condensed phase [23], but the reactive cloud is still relatively small and the burn time is even longer than the Al/Fe_2O_3 case. Figure 1b demonstrated the most reactive mixture tested (Al/Fe_2O_3/70% WO_3) which has a large reactive cloud, short burn, time, and temperatures closer to the adiabatic flame temperature of Al/WO_3 rather than Al/Fe_2O_3 (~3447K vs ~2800K) [23]. Based on the low ignition temperature of the Al/WO_3 thermite in comparison to Al/Fe_2O_3 (1030K vs. 1410K) [25] and the reactive cloud temperature of the Al/Fe_2O_3/70% WO_3, it is proposed that, upon reaction, local hot spots of reacting Al/WO_3 leads to improved decomposition of the Fe_2O_3 (total decomposition at (~3300 K).[23] The released gas leads to deaggregation of the reacting material (leading to the large reactive cloud) and improves advection to propagate to other local initiation spots, ultimately resulting in a feedback loop that reduces the burn time within the composite.

B. Investigating the Impact of Metal Additives on Hydrocarbon Fuel Combustion[2]

B2. Methods / Experimental

Using a free-droplet combustion apparatus described in previous studies[26–28], 600-micron droplets were released from a capillary and ignited with a pilot methane flame at an approximate rate of 3 times per second. To gather magnified videos of entire combustion events with the high-speed color camera and a 32-channel spectrometer (475-500nm), the camera and collection fiber are mounted on a vertical stage that is allowed to free fall parallel to the tower.

While spatiotemporal maps are recorded for the videos and are presented here, the single-frame temperature measurements have allowed for a closer investigation on the impact of microexplosions on droplet combustion and the mechanisms at play. To calculate mean and median temperature of one frame, individual frames are binarized and areas of 3 or more contiguous pixels are included in the calculation.

In parallel with the high-speed camera, a 32-channel spectrometer was also mounted to the translational stage to measure the atomic emission of AlO species within the aluminized hydrocarbon droplet flames. The in-house assembled spectrometer is described by Jacob et al.[29] and consists of a 1 m collection optical fiber, a 0.5 m spectroscope (Acton SP 500i) with a 1800 lines/mm grating to disperse the light between 473-502 nm to detect an AlO emission peak at 484 nm, and a 32-channel PMT array interfaced with a high-speed data acquisition system (Vertilon IQSP 580). The sampling rate of the acquisition system was set at 5000 Hz, sufficient to resolve the sub-millisecond disruptions in the droplet flames.

Triisobutylaluminum (TiBAL) (Sigma Aldrich 257206 CAS 100-99-2), triisobutylamine (Sigma Aldrich 374989 CAS 1116-40-1), and Benzene (Sigma Aldrich 319953) were dissolved into purified Toluene solvent under within an Ar-atmosphere glovebox. Prior to experiments, ~0.5mL of material is loaded into a gastight syringe within the Ar glovebox and sealed in plastic bags. Furthermore, the syringe lead is purged with nitrogen before pumping into the capillary entering the tower. Solutions of TiBAl, TiBam, and benzene in toluene were each prepared at concentrations of 280 mM, 440 mM, and 810 mM.

---

[2] Adapted with permission by Phil Guerieri from paper titled "Toluene Free-Droplet Burning Rate Improvements upon Triisobutlyaluminum Dissolution" (in preparation).

B3. Results / Discussion

Although spatiotemporal maps have been collected and will be presented, only the single temperature measurements are discussed here. While analyzing the videos of the high-speed color camera pyrometer, it was evident that the multicomponent fuels led to microexplosions as expected due to different boiling points of each component [30–33]. Magnified color high-speed video of combusting droplets with accompanying pyrometric temperature estimation and concurrent emission spectroscopy between 474 nm and 502 nm wavelengths was collected for toluene, 810 mM TiBam in toluene, and 810 mM TiBAl in toluene to better observe droplet combustion behavior, estimate flame temperature effects, and detect the $\Delta v = 0$ emission band of AlO which indicates reaction of gas-phase aluminum with an oxidizer [34,35].

In the case of the TiBAl droplet combustion (shown in Figure 2), microexplosions at 35 ms, 55 ms, and 73 ms corresponded to a sharp decrease in temperature as the ambient oxygen mixed with the flame. However, prominent peaks in emission from AlO were evident as part of the combustion event, thus indicating gas-phase Al combustion. Through direct observation of the microexplosions upon TiBAl dissolution in toluene, it was suggested that gas-phase products formed within the droplet are liberated into the flame and allowed to combust, leading to catastrophic disassembly of the primary droplet. With smaller droplets formed as a result of the disruption, the burning surface area and burning rate increases for the composite. The observed AlO peaks occurring at the same time of disruption events serves as evidence for TiBAl breakdown attributable reactions with water (from toluene combustion) to release isobutene and hydrogen gas, thus leaving aluminum burning in air as a byproduct.[36,37]

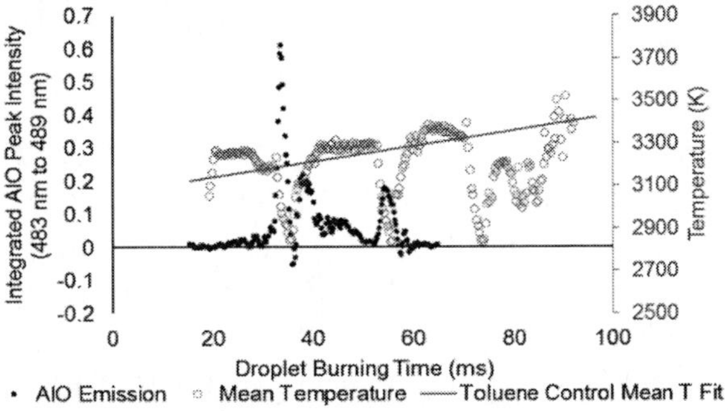

Figure 2: 810 mM TiBAl. Spatial mean of temperature and integrated size of AlO emission peak sensed in magnified video "TiBAl-2". Reference line corresponds to linear fit of Toluene Control temperature. Droplet leaves spectrometer view at 65 ms. Disruptions at 35 ms, 55 ms, and 73 ms.

4. Conclusion

Although high speed videography has allowed for the investigation of reaction dynamics in both space in time, color-camera pyrometry opens up new avenues to study the structure-function relationship of energetic composites. Herein, case studies were presented in which color-camera pyrometry was used to visualize the tunable reactivity of nanothermites and aluminized hydrocarbons. When combined with other diagnostics, such as a constant volume combustion cell or spectrometer, visualization of these mechanisms played a significant role in identifying the reaction mechanisms governing these composites.

5. Acknowledgements

This research was funded by the Army Research Office. Thank you to Rohit Jacob for assisting in the setup of the color camera pyrometer and constant volume combustion cell apparatus. Thank you to Phil Guerieri for optimization of MATLAB routines used as part of this project and for setting up the droplet experiments.

6. References

[1] United States, Military explosives., Headquarters, Dept. of the Army, Washington, D.C., 1989. file://catalog.hathitrust.org/Record/009782414.

[2] V.E. Zarko, A.A. Gromov, Energetic Nanomaterials: Synthesis, Characterization, and Application, Elsevier, 2016.

[3] S.R. Turns, An Introduction to Combustion: Concepts and Applications, 3rd ed., McGraw-Hill, New York, NY, 2012.

[4] M. Shukla, V.M. Boddu, J.A. Steevens, D. Reddy, J. Leszczynski, Energetic Materials From Cradle to Grave., Springer Verlag, 2017. http://public.eblib.com/choice/PublicFullRecord.aspx?p=5098667.

[5] D.D. Dlott, Thinking big (and small) about energetic materials, Mater. Sci. Technol. 22 (2006) 463–473. doi:10.1179/174328406X83987.

[6] M.R. Zachariah, NanoEnergetics: Hype, reality and future, Propellants, Explos. Pyrotech. 38 (2013) 7. doi:10.1002/prep.201380131.

[7] D. Sundaram, V. Yang, R.A. Yetter, Metal-based nanoenergetic materials: Synthesis, properties, and applications, Prog. Energy Combust. Sci. 61 (2017) 293–365. doi:10.1016/j.pecs.2017.02.002.

[8] M.R. Weismiller, J.G. Lee, R.A. Yetter, Temperature measurements of Al containing nano-thermite reactions using multi-wavelength pyrometry, Proc. Combust. Inst. 33 (2011) 1933–1940. doi:10.1016/j.proci.2010.06.094.

[9] K. Kappagantula, C. Crane, M. Pantoya, Factors influencing temperature fields during combustion reactions, Propellants, Explos. Pyrotech. 39 (2014) 434–443. doi:10.1002/prep.201300154.

[10] Vision Research, PHANTOM V21X / V31X / V41X / V61X / V12.1 / V71X / Miro310 / Miro110 Color & Spectral Response Curve, (n.d.). http://www.phantomhighspeed.com/Portals/0/Docs/SpectralResponse/spectral-response-V12-X1X-M310-M110.pdf.

[11] J.M. Densmore, M.M. Biss, K.L. McNesby, B.E. Homan, High-speed digital color imaging pyrometry, Appl. Opt. 50 (2011) 2659. doi:10.1364/AO.50.002659.

[12] H. Ma, R. Stevens, R. Stone, In-Cylinder Temperature Estimation from an Optical Spray-Guided DISI Engine with Color-Ratio Pyrometry (CRP), 2006. doi:10.4271/2006-01-1198.

[13] B.C. Connelly, S.A. Kaiser, M.D. Smooke, M.B. Long, Two-Dimensional Soot Pyrometry with a Color Digital Camera, Jt. Meet. U.S. Sect. Combust. Insitiute. 2 (2005).

[14] J.D. Maun, P.B. Sunderland, D.L. Urban, Thin-filament pyrometry with a digital still camera., Appl. Opt. 46 (2007) 483–488. doi:10.1364/AO.46.000483.

[15] K.L. Mcnesby, M.M. Biss, B.E. Homan, R.A. Benjamin, V.M.B. Sr, M. John, Imaging Detonations of Explosives, 2015.

[16] S. Pauzin, A. Giovannini, B. Bédat, Thin filament infrared pyrometry: instantaneous

temperature profile measurements in a weakly turbulent hydrocarbon premixed flame, Exp. Fluids. 17 (1994) 397–404. doi:10.1007/BF01877042.
[17] P.B. Kuhn, B. Ma, B.C. Connelly, M.D. Smooke, M.B. Long, Soot and thin-filament pyrometry using a color digital camera, Proc. Combust. Inst. 33 (2011) 743–750. doi:10.1016/j.proci.2010.05.006.
[18] T. Fu, X. Cheng, C. Shi, M. Zhong, T. Liu, X. Zheng, The set-up of a vision pyrometer, Meas. Sci. Technol. 17 (2006) 659–665. doi:10.1088/0957-0233/17/4/008.
[19] S. Simonini, S.J. Elston, C.R. Stone, Soot temperature and concentration measurements from colour charge coupled device camera images using a three-colour method, Proc. Inst. Mech. Eng. Part C J. Mech. Eng. Sci. 215 (2001) 1041–1052. doi:10.1177/095440620121500905.
[20] J.M. Densmore, B.E. Homan, M.M. Biss, K.L. McNesby, High-speed two-camera imaging pyrometer for mapping fireball temperatures, Appl. Opt. 50 (2011) 6267. doi:10.1364/AO.50.006267.
[21] J.M. Densmore, M.M. Biss, B.E. Homan, K.L. McNesby, Thermal imaging of nickel-aluminum and aluminum-polytetrafluoroethylene impact initiated combustion, J. Appl. Phys. 112 (2012). doi:10.1063/1.4762009.
[22] J.M. Densmore, K.T. Sullivan, A.E. Gash, J.D. Kuntz, Expansion behavior and temperature mapping of thermites in burn tubes as a function of fill length, Propellants, Explos. Pyrotech. 39 (2014) 416–422. doi:10.1002/prep.201400024.
[23] K. Sullivan, M.R. Zachariah, Simultaneous Pressure and Optical Measurements of Nanoaluminum Thermites: Investigating the Reaction Mechanism, J. Propuls. Power. 26 (2010) 467–472. doi:10.2514/1.45834.
[24] L. Zhou, N. Piekiel, S. Chowdhury, M.R. Zachariah, Time-resolved mass spectrometry of the exothermic reaction between nanoaluminum and metal oxides: The role of oxygen release, J. Phys. Chem. C. 114 (2010) 14269–14275. doi:10.1021/jp101146a.
[25] G.Q. Jian, S. Chowdhury, K. Sullivan, M.R. Zachariah, Nanothermite reactions: Is gas phase oxygen generation from the oxygen carrier an essential prerequisite to ignition?, Combust. Flame. 160 (2013) 432–437. doi:10.1016/j.combustflame.2012.09.009.
[26] P.M. Guerieri, R.J. Jacob, J.B. DeLisio, M.C. Rehwoldt, M.R. Zachariah, Stabilized microparticle aggregates of oxygen-containing nanoparticles in kerosene for enhanced droplet combustion, Combust. Flame. 187 (2018) 77–86. doi:10.1016/j.combustflame.2017.08.026.
[27] P.M. Guerieri, J.B. DeLisio, M.R. Zachariah, Nanoaluminum/Nitrocellulose microparticle additive for burn enhancement of liquid fuels, Combust. Flame. 176 (2017) 220–228. doi:10.1016/j.combustflame.2016.10.011.
[28] P.M. Guerieri, S. DeCarlo, B. Eichhorn, T. Connell, R.A. Yetter, X. Tang, Z. Hicks, K.H. Bowen, M.R. Zachariah, Molecular Aluminum Additive for Burn Enhancement of Hydrocarbon Fuels, J. Phys. Chem. A. 119 (2015) 11084–11093. doi:10.1021/acs.jpca.5b08580.
[29] R.J. Jacob, D.L. Ortiz-Montalvo, K.R. Overdeep, T.P. Weihs, M.R. Zachariah, Incomplete reactions in nanothermite composites, J. Appl. Phys. 121 (2017). doi:10.1063/1.4974963.
[30] C.K. Law, Internal boiling and superheating in vaporizing multicomponent droplets, AIChE J. 24 (1978) 626–632.
[31] C.K. Law, Recent advances in droplet vaporization and combustion, Prog. Energy Combust. Sci. 8 (1982) 171–201.

[32] A. Makino, C.K. Law, On the controlling parameter in the gasification behavior of multicomponent droplets, Combust. Flame. 73 (1988) 331–336.
[33] C.H. Wang, X.Q. Liu, C.K. Law, Combustion and microexplosion of freely falling multicomponent droplets, Combust. Flame. 56 (1984) 175–197.
[34] C.G. Parigger, J.O. Hornkohl, Computation of AlO B 2 Σ+→ X 2 Σ+ emission spectra, Spectrochim. Acta Part A Mol. Biomol. Spectrosc. 81 (2011) 404–411.
[35] C. Parigger, D.M. Surmick, A.C. Woods, A.B. Donaldson, J.L. Height, Measurement and Analysis of Aluminum Monoxide Flame Emission Spectra, Measurement. (2013).
[36] B.E. Bent, R.G. Nuzzo, L.H. Dubois, Surface Organometallic Chemistry in the Chemical Vapor Deposition of Aluminum Films Using Triisobutylaluminum: β-Hydride and β-Alkyl Elimination Reactions of Surface Alkyl Intermediates, J. Am. Chem. Soc. 111 (1989) 1634–1644. doi:10.1021/ja00187a016.
[37] D.A. Kaplowitz, R.J. Jouet, M.R. Zachariah, Aerosol synthesis and reactive behavior of faceted aluminum nanocrystals, J. Cryst. Growth. 312 (2010) 3625–3630. doi:10.1016/j.jcrysgro.2010.09.015.

Sub Topic: Fire

Spring Technical Meeting
Eastern States Section of the Combustion Institute
March 4-7, 2018
State College, Pennsylvania

Large Eddy Simulations of a Methane-Air Turbulent Line Fire Using Flamelet Combustion and Non-Gray Gas Radiation Models

Alexis Marchand[1], *Van Minh Le*[1,2], *Salman Verma*[1], *James White*[1], *Andre W. Marshall*[1], *Thomas Rogaume*[2], *Franck Richard*[2], *Jocelyn Luche*[2], *Arnaud Trouvé*[1,*]

[1]*Department of Fire Protection Engineering, University of Maryland, College Park, MD 20742, USA*
[2]*Institut Pprime (UPR 3346), Université de Poitiers, Isae-ENSMA, 88961 Futuroscope Chasseneuil Cedex, France*
[*]*Corresponding Author Email: atrouve@umd.edu*

Abstract: The general objective of this project is to develop an accurate combustion and radiation modeling framework for high-fidelity large eddy simulations (LES) of well-controlled turbulent laboratory-scale fires for which the fuel composition and fuel oxidation chemistry are known. This modeling framework is aimed at providing a solid basis for the development and validation of engineering-level models used in simulations of real-world fire problems for which the sources of fuel are diverse, complex, and in many cases, poorly characterized. The combustion model features a library of flamelet solutions corresponding to one-dimensional, steady, laminar, counterflow diffusion flames simulated with specialized software, a chemical kinetic mechanism and an equi-diffusive molecular transport model (*i.e.* unity Lewis numbers). Two different flamelet libraries are considered here: a first library generated with a solver called libOpenSMOKE and a detailed chemical kinetic mechanism developed for C_1-C_3 combustion chemistry and a second library generated with a solver called FlameMaster and the GRI-Mech v3.0 chemical kinetic mechanism developed for methane combustion chemistry. The radiation model features a banded Weighted-Sum-of-Gray-Gases model but (so far) no description of subgrid-scale turbulence-radiation interactions (TRI). This modeling framework is incorporated into a LES solver developed by FM Global and called FireFOAM, and is evaluated in simulations of a two-dimensional, plane, buoyancy-driven, methane-air, turbulent diffusion flame experimentally studied at the University of Maryland. The configuration corresponds to an intermediate validation step in our model development strategy without the complications of flame extinction. The flame structure is characterized by new micro-thermocouple measurements of the temporal mean and root-mean-square gas temperatures. Comparisons between simulated and measured temperatures show significant discrepancies that are explained by the large values of the width of the presumed probability density function (PDF) representing subgrid-scale variations of mixture fraction and by the absence of a model for subgrid-scale TRI.
Keywords: Buoyant turbulent diffusion flame, Large Eddy Simulation, Flamelet combustion, Weighted-Sum-of-Gray-Gases radiation

1. Introduction

Despite recent progress made in simulations of fires using Computational Fluid Dynamics (CFD), CFD models for fire applications face several technical challenges, in particular the need to provide accurate models for turbulent combustion [1–4]. The popular combustion model called the Eddy Dissipation Concept (EDC) model provides a cost-effective description of the location and intensity of heat release rate processes under a range of flow conditions but is known to have several limitations. First and foremost, EDC assumes that the rate of combustion is entirely determined by the rate of turbulent fuel-air mixing and therefore does not represent flame extinction/ignition phenomena, which also depend on combustion chemistry. In addition, EDC has a domain of validity restricted to turbulent flames and does not correctly represent laminar combustion. In the present study, we consider the flamelet modeling approach as an alternative approach that can overcome the limitations of EDC.

Another challenge in CFD descriptions of fires is the need to provide accurate models for thermal radiation [1–4]. This aspect will be included in our oral presentation but because of space restriction, we limit our discussion below to the case of adiabatic combustion in which radiant losses are simply neglected.

Our flamelet modeling framework is incorporated into a LES solver developed by FM Global and called FireFOAM [5]. In the present study, we consider two different flamelet libraries: a first library generated with a solver called libOpenSMOKE [6-8] and a detailed chemical kinetic mechanism developed for C_1-C_3 combustion chemistry [9]; and a second library generated with a solver called FlameMaster [10] and the GRI-Mech v3.0 chemical kinetic mechanism developed for methane combustion chemistry [11]. The change from our original choice of using libOpenSMOKE to the new choice of using FlameMaster is made in preparation for future simulations of fires with flame extinction/ignition phenomena that will require an unsteady flamelet treatment [12–14] (this treatment is available in FlameMaster but not in libOpenSMOKE). The performance of the two flamelet models is evaluated by comparisons to experimental data corresponding to an experimental configuration called the turbulent line burner [15].

2. Experiment and numerical configuration

a. Experimental configuration

The configuration adopted in the present study is the turbulent line burner studied at the University of Maryland [15]. The configuration corresponds to a buoyant, turbulent, methane-fueled, slot-burner diffusion flame with a controlled coflow (Fig. 1). Flames are stabilized above a 5-cm-wide and 50-cm-deep Wolfhard-Parker burner surrounded by a small horizontal plate and an oxidizer coflow. The coflow is 50-cm-wide and 75-cm-deep. The methane mass flow rate (velocity) is 1 g/s (6 cm/s) and the theoretical heat release rate of the flame is 50 kW. In the present study, the oxidizer coflow is made of pure air; the mass flow rate (velocity) of the air coflow is 85 g/s (25 cm/s). The flame structure is characterized through measurements of the temporal mean and root-mean-square (rms) gas temperature using 12.7-μm-diameter Type-S (Pt/10%Rh) thermocouples (response time ≈ 1 ms) with a sampling rate of 10,000 Hz. The radiation loss from the micro-thermocouples is assumed small and is simply neglected.

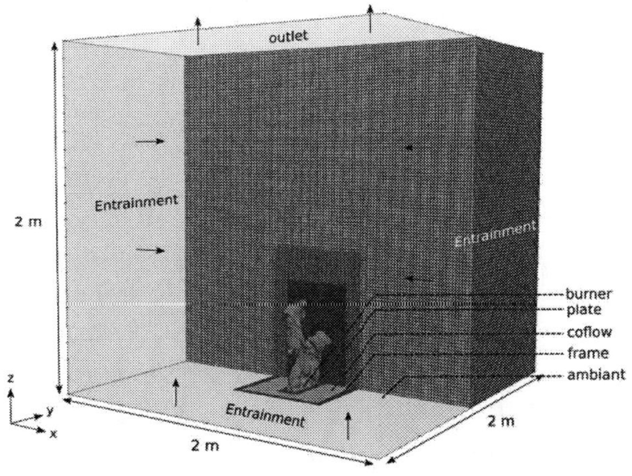

Figure 1: Computational domain and multi-level mesh refinement. The flame height is approximately 50 cm.

b. Numerical configuration

Numerical simulations are performed using FireFOAM [5], a fire modeling solver developed by FM Global and based on an open-source, general-purpose, CFD software package called OpenFOAM [16]. FireFOAM is a second-order accurate, finite volume solver with implicit time integration; the solver features advanced meshing capabilities (structured/unstructured polyhedral mesh); it also features a massively parallel computing capability. FireFOAM uses a Favre-filtered, compressible-flow, LES formulation and provides a choice between several modeling options for the treatment of turbulence, combustion and thermal radiation. In the current simulations: subgrid-scale (SGS) turbulence is described using the one-equation eddy viscosity model, which solves a transport equation for SGS turbulent kinetic energy.

The computational domain is presented in Fig. 1 and is 2-m-wide in the cross-stream *x*-direction, 2-m-long in the spanwise *y*-direction and 2-m-high in the vertical *z*-direction. The grid is block-structured and features three levels of resolution. The finest level of resolution used in the flame region corresponds to a uniform grid with cubic cells of size Δx_1 = 4.167 mm (*i.e.* 12 grid cells across the burner width). The total number of cells is 2.36 million. All FireFOAM simulations are performed for a duration of 30 s. Turbulent statistics are collected for the final 20 s of each simulation, after the flow and flame become statistically stationary and long enough for the statistics to be converged. The time step is controlled by a classical Courant-Friedrichs-Lewy (CFL) condition and is approximately equal to 0.25 ms. Each simulation is run using 192 processors on a large-scale Linux cluster with the baseline model requiring 14,000 CPU-hours.

c- Combustion Modeling

As mentioned in the introduction, we use libOpenSMOKE and FlameMaster to generate laminar flame solutions that are first parametrized in terms of a reduced set of variables: the mixture fraction Z, the stoichiometric value of the scalar dissipation rate χ_{st}, and an enthalpy deficit that is here ignored (we limit the present discussion to the case of adiabatic combustion).

These two flamelet libraries are then incorporated into FireFOAM using statistical averaging based on probability density functions (PDFs). The implementation of the flamelet library into FireFOAM is based on the following expression:

$$\tilde{q} = \int_{\Psi} q^{fl}(\Psi)\tilde{p}(\Psi)d(\Psi)$$

where $\tilde{q}$ designates the LES-filtered value of a quantity q (with q representing the gas temperature or a species mass fraction); where ψ represents the parametrization of the flamelet library, $\Psi = (Z, \chi_{st})$; where q^{fl} is the value of q found in the flamelet library for a given set of values for ψ; and where $\tilde{p}$ is the joint (mass-weighted) PDF representing SGS variations of Z and χ_{st}. Assuming statistical independence, we write: $\tilde{p}(Z, \chi_{st}) = \tilde{p}(Z) \times \tilde{p}(\chi_{st})$. In the present work, we adopt a classical presumed-PDF approach in which $\tilde{p}(Z)$ is a β-PDF and $\tilde{p}(\chi_{st})$ is a Dirac δ-function. $\tilde{p}(Z)$ is parameterized by the local LES-filtered mean and variance, $\tilde{Z}$ and $\widetilde{Z_v}$, with $Z_v = (Z - \tilde{Z})^2$; $\tilde{p}(\chi_{st})$ is parametrized by the local LES-filtered means $\widetilde{\chi_{st}}$. An efficient implementation of Eq. (1), consists of pre-tabulating $\tilde{q}$ in terms of the LES grid scale variables, $\tilde{Z}$, $\widetilde{Z_v}$ and $\widetilde{\chi_{st}}$. Following standard modeling choices, $\tilde{Z}$ is obtained through the solution of turbulent transport equations, whereas $\widetilde{Z_v}$ and $\widetilde{\chi_{st}}$ are obtained through algebraic closure models. In summary, the FireFOAM system of governing equations includes conservation equations for the LES-filtered mass density, flow velocity vector and mixture fraction; the flamelet model is informed by the values of $(\tilde{Z}, \widetilde{Z_v}, \widetilde{\chi_{st}})$ and provides in return the values of LES-filtered temperature and species mass fractions. In the following, we refer to the two steady laminar flamelet models as SLF/libOpenSMOKE and SLF/FlameMaster.

3. **Results and discussion**

Figure 2 presents typical laminar flame profiles stored in the flamelet libraries produced by SLF/libOpenSMOKE and SLF/FlameMaster. As explained in the previous section, the flamelet libraries store the variations of temperature and major species mass fractions as a function of mixture fraction Z and stretch rate χ_{st}. We find that SLF/libOpenSMOKE and SLF/FlameMaster provide similar results and conclude that the main features of the flame structure are not sensitive to the choice of the chemical kinetic mechanism.

Figure 3 presents representative results of FireFOAM simulations of the turbulent line burner obtained with the new laminar flamelet models. In Fig. 3, the simulated vertical variations of the temporal mean and *rms* values of temperature are plotted along the centerline of the burner and compared to experimental data. As expected from Fig. 2, we find that SLF/libOpenSMOKE and SLF/FlameMaster provide similar results. We also find that in the flame region ($0 \le z \le 0.5$ m), both models produce mean temperatures that are too low (by approximately 200 K) and rms temperatures that are too high (by approximately 200 K).

A sensitivity study suggests that these discrepancies are primarily due to overestimated values of the SGS Z-variance, $\widetilde{Z_v}$. For instance, in regions where $\tilde{Z} \approx Z_{st}$, where $Z_{st} = 0.055$ is the stoichiometric value of mixture fraction, we find high values of the fluctuation intensity, $((\widetilde{Z_v})^{1/2}/\widetilde{Z_v}) \ge 1$; high values of the fluctuation intensity correspond to a wide β-PDF in mixture fraction space, which in turn lead to reduced values of temperature, in particular for $\frac{Z_{st}}{2} \le \tilde{Z} \le (2 \times Z_{st})$ (the reduction may be as high as several 100s K). Thus, the low values of simulated temperatures can be explained by excessive values of $\widetilde{Z_v}$ which are responsible for increasing the effect of statistical filtering associated with the SGS *Z*-PDF. These results suggest that the SGS *Z*-PDF model should be reexamined in future work in order to improve the performance of the flamelet model.

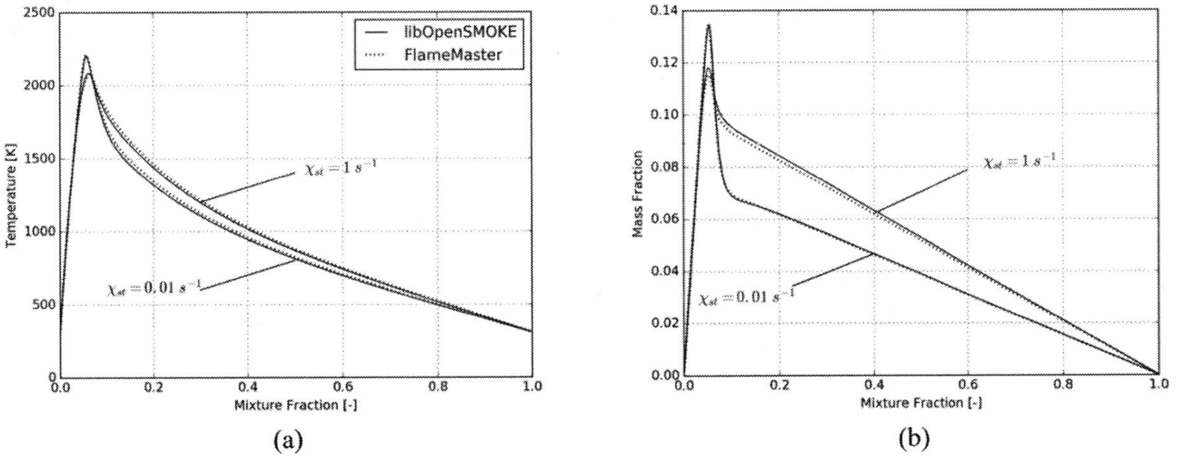

Figure 2: Flamelet library. Variations of (a) temperature and (b) CO_2 mass fraction across a stretched laminar flame, as obtained from SLF/libOpenSMOKE (solid line) and SLF/FlameMaster (dotted line) for two values of the rate of stretch, $\chi_{st} = 0.01$ and 1 s^{-1}.

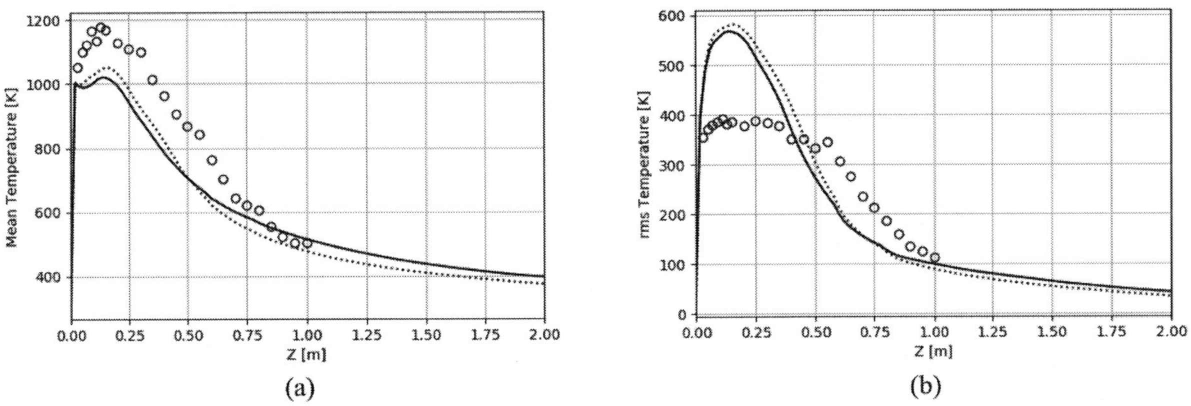

Figure 3: FireFOAM simulations of the turbulent line burner. Vertical variations of (a) mean temperature T and (b) rms temperature T_{rms} along the flame centerline. Comparison between experimental data (symbols) and LES simulations performed with SLF/libOpenSMOKE (solid line) and SLF/FlameMaster (dotted line).

4. Conclusion

This paper presents LES simulations of a buoyancy-driven methane-air turbulent diffusion flame with a new modeling framework that accounts for combustion with detailed chemistry. Two laminar flamelet combustion models have been implemented in FireFOAM and their performance evaluated against temperature measurements performed with micro-thermocouples. The flamelet models under-predict mean temperatures, a result that is explained by the large values of the modelled SGS variance of mixture fraction. These results suggest that the modeling of both the shape of the SGS Z-PDF (currently described as a β-PDF) and the SGS Z-variance requires additional investigation. Work is also in progress to develop a flamelet-based description of radiant emission and absorption effects that accounts for SGS turbulence-radiation interactions.

5. Acknowledgment

This project is financially supported at the University of Maryland (UMD) by the U.S. National Science Foundation (NSF, CBET Program, Award #1604907) and by FM Global. It is also financially supported at Université de Poitiers by the French Government program "Investissements d'Avenir" (LABEX INTERACTIFS, reference ANR-11-LABX-0017-01). The project is also supported by supercomputing resources made available by UMD (http://www.it.umd.edu/hpcc) and by NSF (XSEDE Program, Grant #TG-CTS140046). Dr. James White contributed to this work while a PhD student at UMD and is now employed at FM Global.

6. References

[1] V. Novozhilov, Prog. Energy Combust. Sci. 27 (2001) 611–666.
[2] A. Trouvé, Y. Wang, Intl. J. Computational Fluid Dynamics 24 (2010) 449–466.
[3] B. Merci, Fire Safety Science, Proc. Eleventh International Symposium, International Association for Fire Safety Science 24 (2014) 46–65.
[4] K. McGrattan, S. Miles, SFPE Handbook of Fire Protection Engineering (5th Ed.), Society of Fire Protection Engineering (2016) 1034–1065.
[5] FireFOAM, developed by FM Global, available at: https://github.com/fireFoam-dev, accessed: 1-20-2018.
[6] A. Cuoci, A. Frassoldati, T. Faravelli, E. Ranzi, Combust. Flame 156 (2009) 2010–2022.
[7] A. Cuoci, A. Frassoldati, T. Faravelli, E. Ranzi, Env. Eng. Sci. 25 (2008) 1407–1422.
[8] Flamelet model for OpenFOAM, developed by Holtzmann CFD, available at: https://holzmann-cfd.de/publications/flamelet-model, accessed: 11-16-2017.
[9] E. Ranzi, A. Frassoldati, R. Grana, A. Cuoci, T. Faravelli,A. Kelley, C. Law, Prog. Energy Combust. Sci. 38 (2012) 468–501.
[10] H. Pitsch, FlameMaster, A C++ Computer Program for 0D Combustion and 1D Laminar Flame Calculations, available at: https://www.itv.rwth-aachen.de/index.php?id=flamemaster&L=1, accessed: 1-20-2018.
[11] G.P. Smith, D.M. Golden, M. Frenklach, N.W. Moriarty, B. Eiteneer, M. Goldenberg, C.T. Bowman, R.K. Hanson, S. Song, W.C. Gardiner Jr., V.V. Lissianski, Z. Qin, GRI-Mech v3.0, available at: http://www.me.berkeley.edu/gri_mech/, accessed: 1-20-2018.
[12] M. Ihme, H. Pitsch, Phys. Fluids 20 (2008) 055110.
[13] M. Ihme, Y.C. See, Combust. Flame 157 (2010) 1850–1862.
[14] M.E. Mueller, H. Pitsch, Combust. Flame 159 (2012) 2166–2180.
[15] J.P. White, Measurement and Simulation of Suppression Effects in a Buoyant Turbulent Line Fire, Ph.D. thesis, University of Maryland, College Park, USA, 2016.
[16] OpenFOAM, developed by the OpenFOAM Foundation, available at: http://www.openfoam.org, accessed: 1-20-2018.

Spring Technical Meeting
Eastern States Section of the Combustion Institute
March 4-7, 2018
State College, Pennsylvania

Memory optimization of a radiative heat transfer solver for fire simulations

Andrew Caratenuto[1], Peiyu Zhang[1], Bifen Wu[1], Xinyu Zhao[1,]*

[1]Department of Mechanical Engineering, University of Connecticut, Storrs CT 06269, USA
[*]Corresponding Author Email: xinyu.zhao@uconn.edu

Abstract: The effects of thermal radiation in large-scale fire systems are extremely important to system behavior. Multiphysics combustion solvers that account for radiative heat transfer are necessary to accurately predict large scale fire behavior. High-fidelity radiation solvers tend to be computationally expensive and require excessive memory resources. However, recent developments in the high performance computing field have radically changed computing architectures and have reduced memory-to-core ratios, such as in new many-core systems. In this study, A MPI-parallelized Monte Carlo radiative heat transfer solver for a pool fire simulation is optimized on an Intel© Xeon Phi™ (code name "Knights Landing", or KNL) CPU for more efficient memory access. This CPU features 64 physical cores and can process with up to 256 logical threads, and therefore possesses a high capacity for processing multi-threaded applications. A line-by-line database of approximately 3 gigabytes is required to assess the optical properties. At any given instance, every entry in the database has the possibility of being accessed during the ray-tracing process. Before the optimization, a copy of the database must be loaded for each MPI process when running with MPI in parallel. The memory capacity of the Knights Landing CPU limits parallelization to below 64 MPI processes *per* node. After the optimization, 64-core concurrency is enabled. The addition of more MPI processes increases the overall efficiency of the pool fire simulation by reducing the total time-to-solution. A pre-vaporized n-Heptane pool fire is simulated, and the results are compared before and after the optimization to verify the implementation. Going forward, larger-scale simulations that are not limited to one computer node will be performed and the performance will be evaluated.
Keywords: Fire, Radiation, Many-core, Memory Optimization

1. Introduction

Thermal radiation has very significant effects on the behaviors of fire systems. During large-scale fires, the majority of the overall heat transfer to the ambient environment is accounted for by radiation [1]. Simulating these systems with increased accuracy provides vital information about fire propagation that can help prevent the consequences of large-scale fires, including damage to people, property, and the environment. Unfortunately, high-fidelity radiation solvers are extremely computationally expensive, and have very substantial memory requirements [2]. Many new high-performance CPUs offer computing environments with large amounts of CPU cores [3], allowing parallelism to be exploited on much greater scales. While parallelism typically decreases the expected computational time-to-solution, the characteristically small memory-to-core ratios of many-core CPUs challenge the feasibility of the solvers themselves. The immense memory requirements of high-fidelity radiation solvers present problems for parallel implementations, which often have insufficient memory per core to process a simulation. For this reason, memory optimization is often necessary in order to fully exploit the parallel capabilities of high-performance CPUs.

The objective of the study is to optimize a Fortran-based Monte Carlo radiative heat transfer solver for a pool fire simulation in order to increase its efficiency and feasibility. The target CPU is an Intel© Xeon Phi™ (code name "Knights Landing", or KNL) CPU featuring 64 physical cores and 256 logical processors, with a clock speed of 1.30 GHz. The strength of the CPU lies in its multi-threaded capabilities. As such, the efficiency of a solver is expected to increase as the simulation thread count approaches the maximum. However, the optical properties necessary for the simulation are described by a line-by-line database of roughly three gigabytes. When the solver is executed in parallel, each processor must individually load the database. Therefore, the total simulation memory requirement increases with the processor count. This trend causes the system memory to reach capacity before the maximum thread count can be exploited, limiting the usage of the KNL CPU's resources. It should be mentioned that such memory limitation can occur to any solvers that require the usage of extensive tabulation, such as the flamelet based turbulent combustion models [4]. Therefore, it is a worthwhile exercise to establish a paradigm to achieve such memory optimization.

This issue will be tackled using the shared memory features of MPI3 [5] in this study. Using MPI3, all processors can access the same copy of the line-by-line database, drastically reducing the memory requirements of the solver. This also keeps the required memory at a fixed value that no longer increases as additional processors are utilized. The memory optimization of the solver allows for the maximum amount of processors to be used, which is expected to reduce the computational time-to-solution and therefore increase the practicality of the radiation models.

2. Methods

2.1. The Monte Carlo Ray Tracing Solver

To adequately represent the radiative behaviors of fire, optical properties from all participating media should be considered, including spectral gas, wall, and soot. The Monte Carlo ray tracing (MCRT) solver used in this study can naturally account for these spectral effects while solving the radiative transfer equation for participating media. The radiative transfer equation is resolved by tracking and emitting a large number of energy rays. These rays are continuously tracked until their energies are absorbed by the participating media or until they exit the computational domain. The solver examined in this study can also evaluate radiation attenuation through scattering and absorption processes of discrete phases, such as water or spray droplets [6]. Selection algorithms of wavenumbers, propagation directions and emission points are based on random number relations, and these relations are pre-tabulated into the line-by-line database to account for the effects of nongray gases, soot, wall, and spray/water droplets if they are present. The line-by-line properties of CO_2, H_2O, and CO are tabulated, and such database for atmospheric flames amounts to approximately 2.6 GB. It has been identified that tracking is a key contributor to computational cost [6], and that increasing parallelism should in principle reduce the overall computational time. Therefore, it is significant to remove the limitations posed by the memory.

2.2. Shared Memory Techniques

The shared memory capabilities included with MPI3 are used to optimize the memory architecture of the solver. The target environment consists of only one node, as a first step towards simulations on a truly distributed system. A shared MPI communication network must be created first, including all processors that need to access the database. A window of shared memory large enough to fit the line-by-line database is then created by the master processor. The window exists only in the memory of this master processor, but is accessible by all processors in the shared communication network. The window is represented by a base pointer that points to the first memory location of the window. Through this base pointer, data can be placed inside of the shared window by associating a Fortran pointer with the target of the base pointer. All other processors use a query to find out the location through which this shared memory window can be accessed. This address is typically different for each processor. The database is then placed inside of the memory by modifying the target of the shared window base pointer. With one copy of the line-by-line database now accessible by all processors, the solver can be executed using the CPU's maximum processor count without being limited by the memory requirements of the database.

It should be noted that the optimization described in this preliminary study is not specific to the memory architecture of Intel Xeon Phi processor. Optimization that is particular to the Xeon Phi processor, namely those utilizing the different configurations of the unique MCDRAM, will be explored in the future.

2.3. Test Flame and numerical details

A 7.1-cm diameter n-heptane pool fire [7] is simulated in this study. The target pool fire is sufficiently large so that radiation loss is significant (the radiant fraction is approximately 23%). There are also relatively abundant well-documented experimental data, including radiative heat flux, temperature and soot distributions, flame heights, heat release rates, radiative feedback, etc. [7]. The flame is simulated using components of OpenFOAM 2.2.x [8] with in-house radiation libraries [9]. Navier-Stokes equations and transport equations for species mass fractions and sensible enthalpies are solved in the framework of large-eddy simulations. Turbulence is modeled using the WALE model [10], and combustion is closed using the perfect stirred reactor with constant $C_{mix} = 1$ [11]. A 40-species n-heptane chemical mechanism [12] is employed, to facilitate future studies of soot modeling under fire conditions. In addition, the usage of a skeletal chemical mechanism also provides a representative estimation of the required computational resources for such simulations. For the purpose of this study, validation with experimental measurements is not the primary goal; the optimization of the radiation solver is of priority here. Therefore, a simple algebraic soot model [13] is employed to provide all the necessary components for a fire mixture. A constant fuel mass flux is provided at the pool surface to avoid the complication of phase change [14] due to liquid fuel evaporation, which again is a simplification and cannot account for the feedback mechanisms between radiation or convection and the surface evaporation. Roughly 1.7 million grid cells are used, and the simulation was advanced in time for three seconds in physical time to attain the statistically steady state. Additional details about domain and environment properties can be found in Fig. 1 and Table 1. The flame structure and the baseline computational cost are discussed in the next section.

Table 1: Geometric and mesh information of the test pool fire flame.

Pool size (mm)	# of cells	(Δx_{min}, Δx_{max}) mm	mass flux (kg/m^{2}s)	Air velocity (m/s)
R0 = 35.5	1,692,000	(0.6, 3)	0.023	0.001

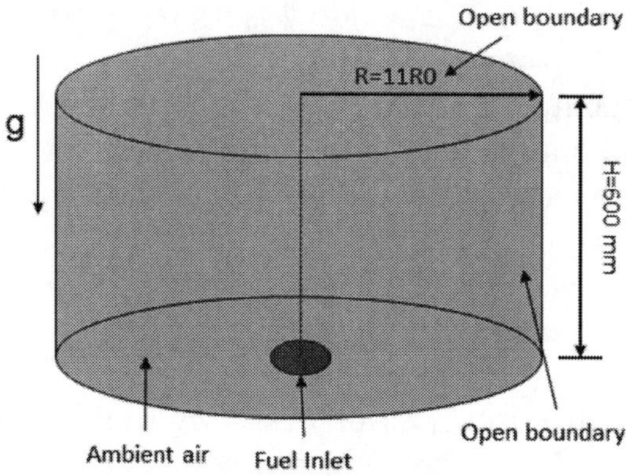

Figure 1: An illustration of the computational domain.

3. Preliminary results

3.1. Flame Structure

Preliminary simulation results using the test flame outlined in Sec. 2.3 are presented here. Figure 2 shows the comparison of the heat flux along the outer boundary of the computational domain obtained from the simulations compared with experimental measurements. Figure 2(a) shows a significant discrepancy between the simulation and the experiment. This is due to the fact that soot was not considered and soot radiation plays a significant role in such pool fires. Simulation in Fig. 2(b) employs an empirical soot model [15] instead to facilitate the creation of a more realistic fire radiation conditions. The improved level of heat flux collected from the simulation is observed in Fig. 2(b). Therefore, the simulation with the empirical soot model is chosen as the baseline to perform the radiation scaling study. It should be noted that the consideration of the soot radiation also adds to the size of the database in memory.

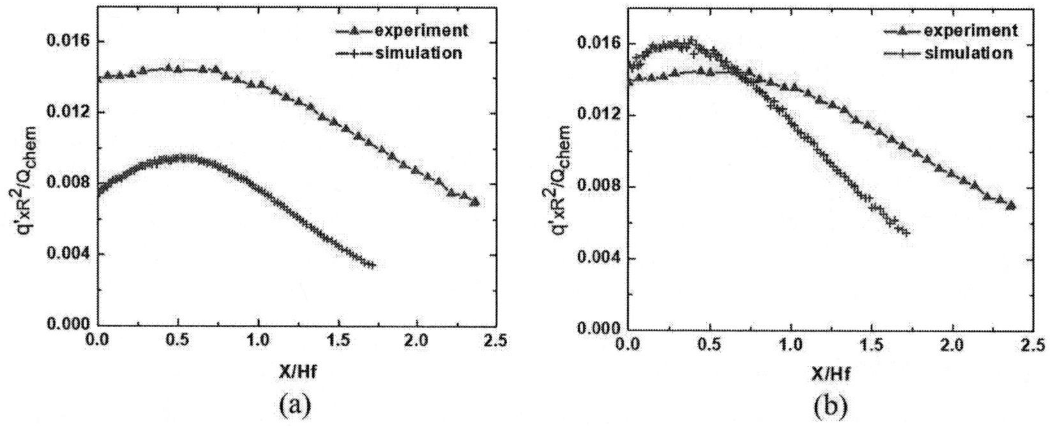

Figure 2: Comparison of heat flux obtained from experiment and baseline simulation: (a) no soot, radiant fraction 14.49% (b) with soot, radiant fraction 24.99%.

3.2. Scaling performance on the Xeon and Xeon-Phi processors

To profile the performance of the KNL CPU with a radiative heat transfer solver, a smaller coal fire simulation [16] with radiation was performed on KNL. A similar solver and the identical radiation library are employed to avoid the extensive computational cost associated with the pool fire test case described in Sec. 2.3. A smaller mesh (1% of the grid counts in the pool fire case) is associated with the coal fire simulation, and the difference of the scaling behavior can be studied preliminarily. The performance of the same code on a 28-core Intel Xeon workstation (Acadia) with a 2.60 GHz clock speed is used for comparison.

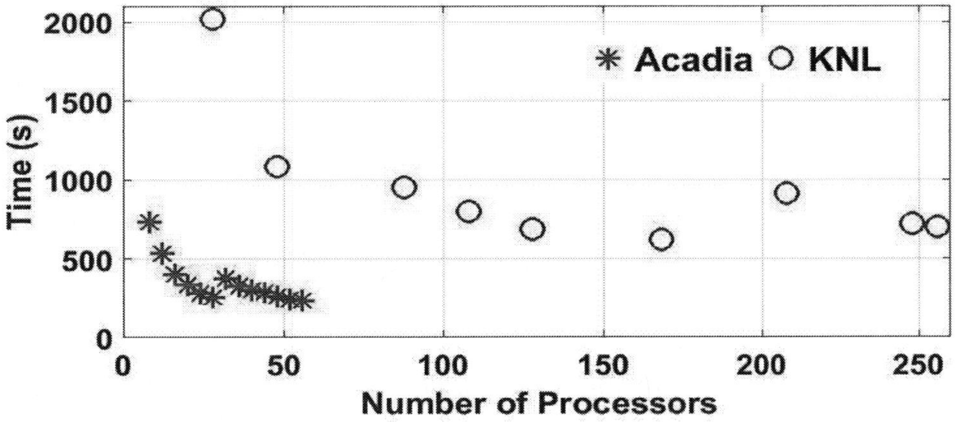

Figure 3: Execution time comparison of Acadia and KNL with coal fire simulation.

The low clock speed of the KNL CPU is a significant detriment to its performance, even as processor count is increased. However, there are differing scaling behaviors between the two CPUs that should be noted: Acadia, with 28 physical cores, has a loss in performance after the 28 processor mark, while KNL does not see this type of explicit performance loss until over 160 processes are invoked. This suggests that while the low clock speed of KNL is limiting, its parallelism can be harnessed to greatly increase efficiency at higher processor counts.

4. Conclusions

The memory optimization of a high-fidelity radiation solver is tackled in this study on a modern many-core Xeon Phi CPU. In particular, the shared memory function of MPI3 is explored to reduce the memory footprint of the MCRT solver when executing in parallel. A 7.1-cm heptane pool fire test flame is simulated and outlined in detail, and experimental results have been compared with baseline simulation results. The scaling performance of the radiation solver on the traditional CPU and the many-core system is compared. Going forward, the LES based pool fire solver will be optimized with shared memory as described to fully exploit the parallelism offered by the KNL CPU. The results will include more knowledge of shared memory implementation and potentially an improved radiative pool fire solver.

5. Acknowledgements

This material is based upon work supported by the National Science Foundation under Grant No. 1566259. The study is also partially funded by FM Global within the framework of the FM Global Strategic Research Program on Fire Modeling.

6. References

[1] J. de Ris, Fire radiation - A review, Symp. Int. Combust. Proc. 17 (1979) 1003-1016.

[2] G. Maragkos, T. Beji, B. Merci, Advances in modelling in CFD simulations of turbulent gaseous pool fires, Combust. Flame 181 (2017) 22-38.

[3] J. Slotnick, A. Khodadoust, J. Alonso, D. Darmofal, W. Gropp, E. Lurie, D. Mavriplis, CFD Vision 2030 Study: A Path to Revolutionary Computational Aerosciences, Langley Research Center, Hampton, Virginia, 2014.

[4] S. Weise, D. Messig, B. Meyer, C. Hasse, An abstraction layer for efficient memory management of tabulated chemistry and flamelet solutions, Combust. Theor. Model. 17 (2013), 411-430.

[5] MPI: A message-passing interface standard, Version 3.0, University of Tennessee, Knoxville, Tennessee, 2012. <http://mpi-forum.org/docs/mpi-3.0/mpi30-report.pdf>.

[6] B. Wu, X. Zhao, Radiation characteristics of water droplets in a fire-inspired environment: a Monte Carlo ray tracing study, Under review with J. Quant. Spectrosc. Radiat. Transfer (2017).

[7] M. Klassen, J. P. Gore, Structure and radiation properties of pool fires, Thermal Sciences and Propulsion Center, School of Mechanical Engineering, Purdue University, West Lafayette, IN, 1992.

[8] OpenFOAM, 2018. <https://www.openfoam.com>.

[9] B. Wu, S. P. Roy, X. Zhao, M. F. Modest, A computational study of radiative heat transfer in pulverized coal jet flame, East. States Section Combustion Inst. Spring Technical Meet. (2016).

[10] F. Nicoud, F. Ducros, Subgrid-scale stress modelling based on the square of the velocity gradient tensor, Flow Turbul. Combust. 62 (1999) 183-200.

[11] P. A. Niklas Nordin, Complex chemistry modeling of diesel spray combustion, Chalmers University of Technology, Gothenburg, Sweden, 2001.

[12] V. Raj Mohan, D. C. Haworth, Turbulence-chemistry interactions in a heavy-duty compression-ignition engine, Proc. Combust. Inst. 35 (2015) 3053-3060.

[13] T. Beji, J. P. Zhang, W. Yao, M. Delichatsios, A novel soot model for fires: Validation in a laminar non-premixed flame, Combust. Flame 158 (2011) 281-290.

[14] G. Krishnamoorthy, A comparison of angular discretization strategies for modeling radiative transfer in pool fire simulations, Heat Transf. Eng. 33 (2012) 1040-1051.

[15] V. P. Mutheramalingham, Implementation of soot model for aachenBomb tutorial, Chalmers University of Technology, Gothenburg, Sweden, 2016. <http://www.tfd.chalmers.se/~hani/kurser/OS_CFD_2015/VigneshPandian/Report_Vignesh_revised.pdf>.

[16] B. Wu, S. P. Roy, X. Zhao, M. Modest, Effect of multiphase radiation on coal combustion in a pulverized coal jet flame, J. Quant. Spectrosc. Radiat. Transfer 197 (2017) 154–165.

Sub Topic: Fire

Spring Technical Meeting
Eastern States Section of the Combustion Institute
March 4–7, 2018
State College, Pennsylvania

Development of 3D Pyrolysis in FDS

Morgan Bruns[1,*], *Randall McDermott*[1], *Salah Benkorichi*[2], *and Simo Hostikka*[3]

[1]*National Institute of Standards and Technology, Gaithersburg, MD, USA*
[2]*Omega Fire Engineering Ltd, Manchester, UK*
[3]*Aalto University, Espoo, Finland*
[*]*Corresponding author:* morgan.bruns@nist.gov

Abstract: Computational predictions of flame spread could support the design and development of safer products and buildings. Flame spread is a multiscale, multiphase, and multidimensional problem. In particular, downward flame spread is largely driven by multidimensional heat conduction within the condensed phase. The Fire Dynamics Simulator (FDS) currently utilizes a 1D pyrolysis model. A 3D heat transfer solver has recently been implemented in FDS. The 3D solver is verified against an analytical solution for heat conduction in a sphere with internal heat generation and a constant surface temperature boundary condition. In the present work, the pyrolysis source term from the 1D solid model in FDS has been ported to the 3D heat transfer model to enable multi-dimensional pyrolysis simulations. As with the 1D solid model, pyrolysis gas is assumed to be instantaneously transported to the nearest upward facing surface. The 3D pyrolysis model is verified against predictions of PMMA gasification from the 1D solid model in FDS.
Keywords:* *Pyrolysis, FDS, Burning Rate

1. Introduction

The development of computational fire models has primarily focused on the spread of heat and smoke generated by a prescribed fire. Consequently, fire models have proven useful for fire protection engineering of buildings. Such models are not currently well validated for predicting the spread and growth of a building fire involving a diverse range of fuel sources. This limitation arises from the fact that the physics governing the burning of condensed phase materials is relatively complicated. To begin with, there is a large diversity in the types of condensed phase fuels encountered in modern building fires, and these diverse fuels burn in significantly variable ways. For instance, many plastics used in consumer electronics melt and flow during burning, whereas wood materials used in furniture can char and crack. Modeling such diverse scenarios requires a certain level of generality if fire spread is ever to be adequately computationally predicted.

The project of reliably predicting flame spread for diverse fuel sources using computational models will require a number of improvements in existing pyrolysis models. Most current computational pyrolysis models solve a form of the heat equation that accounts for energy absorption (or generation) due to condensed phase chemical reactions as well as appropriate solid phase component conservation equations. Some models such as Gpyro [1] and ThermaKin [2] account for transport of pyrolysis gases, while other models such as the Fire Dynamics Simulator (FDS) [3] assume that pyrolysis gases are immediately transported to the surface of the material. Furthermore, some models account for multidimensional effects—Gpyro is three dimensional and ThermaKin is two dimensional, but the current version of the FDS solid model is one dimensional. For some problems such as the uniform burning of a thin slab, a one dimensional pyrolysis model is sufficient. However, in cases involving flame spread, multidimensional transport must be considered. For example, it has been observed that in downward flame spread on PMMA, approximately half of the energy transport ahead of the flame is due to conduction within the condensed phase [4].

The focus of this paper is on the development of an efficient 3D pyrolysis model. This model solves for heat conduction and chemical reactions on the Cartesian grid within solid obstructions in FDS. It is thus fully coupled to the gas phase. The chemical reactions are modeled using the same mathematical form of the existing 1D solid model within FDS. As with that model, it is assumed that generated pyrolysis gases are immediately transported to the nearest upward facing cell surface on the solid obstruction. In contrast to most existing pyrolysis models, the present methodology uses fully explicit time stepping. Additionally, material burn away is accounted for by the removal of solid cells with zero density. The goal is to create a robust and efficient pyrolysis solver that allows for fully coupled simulations on a relatively coarse grid.

In the next section, the governing equations for the 3D pyrolysis model are given, and the solution procedure is described in some detail. Conduction and internal heat generation in a sphere is used as a verification case. Finally, the complete 3D pyrolysis model is verified against gasification predictions for poly(methyl methacrylate) (PMMA) using the 1D solid model in FDS. The results are preliminary, and more work is needed before the goal of efficient flame spread simulations is achieved.

2. Model Development

The governing equation for 3D solid phase heat transfer solved in FDS is

$$\rho_{\rm s} c_{\rm s} \frac{\partial T_{\rm s}}{\partial t} = \frac{\partial}{\partial x_i}\left(k_{\rm s} \frac{\partial T_{\rm s}}{\partial x_i} \right) + \dot{q}_{\rm s}''' \tag{1}$$

where $\rho_{\rm s}$, $c_{\rm s}$, and $k_{\rm s}$ are the density, specific heat, and thermal conductivity of the solid, respectively. The source term, $\dot{q}_{\rm s}'''$, currently consists of chemical reactions or a user specified volumetric heating rate. Future work will require accounting for the contribution due to radiative absorption and emission. Thermal properties are treated the same as in the 1D heat conduction model as either volume or mass weighted sums of the component properties.

Equation (1) is discretized with a second-order finite-volume scheme for fluxes and time integrated with explicit Euler, with an update taking place during the second step of the two-step predictor-corrector time integration scheme. The "solid temperature" is stored in the cell center. The intercell fluxes are continuous. If the solid material properties of neighboring cells are the same, then thermal properties at face centers are taken as linear averages when computing the intercell flux. If the neighboring cells have different material properties (consider the x direction with cell indices i and $i+1$), then the cell interface temperature, $T_{s,i+\frac{1}{2}}$, is computed as follows:

$$T^n_{{\rm s},i+\frac{1}{2}} = \frac{T^n_{{\rm s},i} + \left[\frac{k_{{\rm s},i+1}}{k_{{\rm s},i}} \frac{\delta x_i}{\delta x_{i+1}} \right] T^n_{{\rm s},i+1}}{1 + \left[\frac{k_{{\rm s},i+1}}{k_{{\rm s},i}} \frac{\delta x_i}{\delta x_{i+1}} \right]} \tag{2}$$

The intercell flux at time level n may be obtained from either of the following (in practice the first relationship is used):

$$k_s \frac{\partial T_s}{\partial x}\bigg|_{i+\frac{1}{2}} = k_{s,i} \frac{T^n_{s,i+\frac{1}{2}} - T^n_{s,i}}{\frac{1}{2}\delta x_i} = k_{s,i+1} \frac{T^n_{s,i+1} - T^n_{s,i+\frac{1}{2}}}{\frac{1}{2}\delta x_{i+1}} \tag{3}$$

The time step is chosen to be the minimum of the LES time step, δt, and the time step required to satisfy the Von Neumann stability criterion.

For two-way coupling between the gas and the solid phase, the boundary condition ensures continuity of heat flux and temperature. On the gas phase side, the heat fluxes are due to convection, $\dot{q}''_c$, and radiation, $\dot{q}''_{r,in} - \dot{q}''_{r,out}$, and on the solid phase side the heat flux is due to conduction. The surface temperature links all these fluxes at an instant in time. Let T_w denote the interface temperature between the gas and solid, the face temperature of the "wall" cell. The flux continuity boundary condition implies

$$\dot{q}''_{r,in} - \dot{q}''_{r,out} + \dot{q}''_c = -k_s \frac{\partial T_s}{\partial n}\bigg|_w \tag{4}$$

where n is in the outward normal direction of the solid surface. To obtain the surface temperature, we take $\dot{q}''_{r,in}$ from the gas phase solution, linearize $\dot{q}''_{r,out}$, and use Newton's law of cooling for $\dot{q}''_c$. The flux condition may then be discretized as

$$(\dot{q}''_{r,in})^n + 3\varepsilon^n \sigma (T^n_w)^4 - 4\varepsilon^n \sigma (T^n_w)^3 T^{n+1}_w + h^n (T^n_g - T^{n+1}_w) = -2k^n_s \frac{T^n_s - T^{n+1}_w}{\delta n} \tag{5}$$

where the superscript n denotes the value at time step n and δn is the grid spacing in the direction across the interface. This leads to our final result for the boundary temperature

$$T^{n+1}_w = \frac{(\dot{q}''_{r,in})^n + 3\varepsilon^n \sigma (T^n_w)^4 + h^n T^n_g + 2k^n_s T^n_s/\delta n}{4\varepsilon^n \sigma (T^n_w)^3 + h^n + 2k^n_s/\delta n} \tag{6}$$

where ε is the solid surface emissivity, σ is the Stefan-Boltzmann constant, and h is the convection coefficient.

As the solid phase pyrolyzes, mass density within the Cartesian cells decreases. For non-charring materials, this density will eventually reach zero. This so-called burn away is handled by removing the zero density cells at each time step.

3. Verification and Validation

3.1 Heat Diffusion and Generation in a Sphere

In this example, a solid sphere of radius 0.1 m with internal heat generated at $\dot{q}'''_s$ = 200 kW/m^3 is studied. The initial temperature at the boundary is set to 20 °C. The objective of the test is to ensure that the heat conduction through the sphere and across the radius is computed properly. The heat conduction can be written as

$$\frac{\partial T}{\partial t} = \alpha \left[\frac{1}{r^2} \frac{\partial}{\partial r^2} \left(r^2 \frac{\partial T}{\partial r} \right) + \frac{\dot{q}'''}{k} \right] \tag{7}$$

where k is the thermal conductivity and α is the thermal diffusivity. The initial and boundary conditions are

$$T\,(r,t=0) = T_0 \quad ; \quad T\,(r=a,t) = T_0 \quad ; \quad \left.\frac{\partial T}{\partial r}\right|_{r=0} = 0 \tag{8}$$

where T_0 is the initial temperature and a is the radius of the sphere. The exact solution to Eqs. (7) and (8) is from [5] (Sec. 9.8, p. 243):

$$T = T_0 + \frac{\dot{q}'''}{6k}\left(a^2 - r^2\right) + 2\frac{\dot{q}'''a^3}{k\pi^3 r}\sum_{n=1}^{\infty}\frac{(-1)^n}{n^3}\sin\left(\frac{n\pi r}{a}\right)\exp\left(-\alpha t\frac{n^2\pi^2}{a^2}\right) \tag{9}$$

The calculation is run for 180 s. In Fig. 1 we show the temperature contours at $t = 165$ s for the medium resolution case with 50 cells across the sphere diameter (left) and a comparison between the analytical and numerical solutions at several points in time for the high resolution case with 100 cells across the sphere diameter (right). Fig. 2 shows the L_∞ error as a function of cell size.

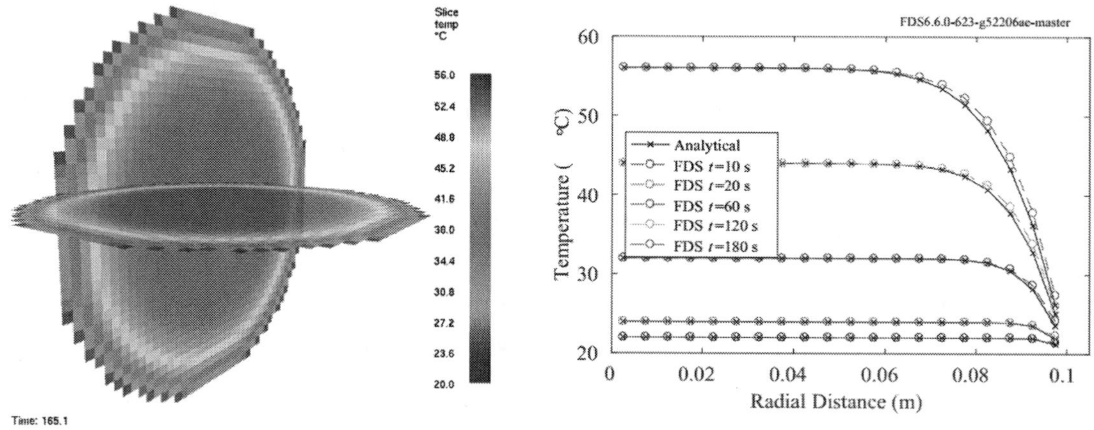

Figure 1: (Left) Temperature contours at $t = 165$ s for 50 cells across the diameter. (Right) A comparison between radial profiles of the analytical and numerical solution at various times in the simulation.

3.2 Gasification of PMMA

As a final test of the implementation of the methodology, the case of the gasification of a slab of poly(methyl methacrylate) (PMMA) is simulated. The predictions of the 3D pyrolysis model are compared to the mass loss rate per unit area predicted by the 1D solid model in FDS. The scenario involves the radiant heating of a horizontal slab of PMMA in an inert (nitrogen rich) gas environment. For the PMMA case considered, the sample was 2 mm by 2 mm square and 1 cm in thickness. An external radiant heat flux of 50 kW/m^2 was applied to the top surface of the slab. The sample is assumed to be perfectly insulated on all sides except for the upward facing irradiated surface. For the simulations, the material properties used are listed in Table 1.

The predicted results using the FDS 1D and 3D pyrolysis models are plotted in Fig. 3. Both simulation results are obtained with a single column of 35 cells across the thickness of the slab. Note that in-depth radiation is not considered as this has yet to be implemented in the 3D pyrolysis

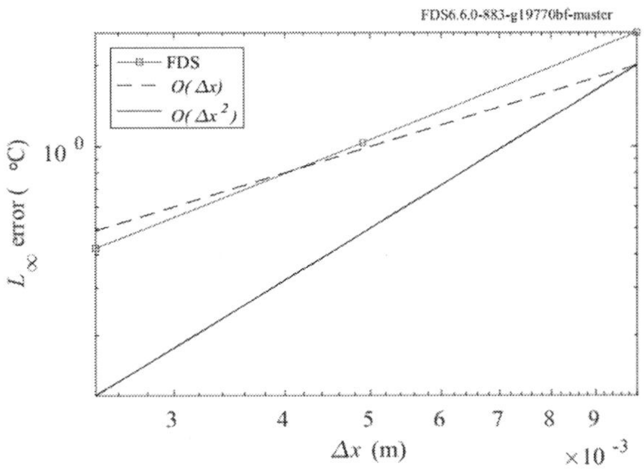

Figure 2: L_∞ error in sphere temperature at $t = 180$ s with varying cell widths.

Table 1: Input parameters for PMMA gasification. Courtesy S. Stoliarov, M. McKinnon and J. Li, University of Maryland.

Property	Units	PMMA	Unc. (%)	Ref.
Density	kg/m^3	1100	5	[6]
Conductivity	W/m/K	0.20	15	[6]
Specific Heat	kJ/kg/K	2.2	15	[7]
Emissivity		0.85	20	[8]
Pre-Exp. Factor	s^{-1}	8.5×10^{12}	50	[6]
Activation Energy	kJ/kmol	1.88×10^5	3	[6]
Heat of Reaction	kJ/kg	870	15	[7]

solver. The 1D solid model does not allow for burn away, but rather shrinks the 1D grid attached to the surface of the solid obstruction shown on the right hand side of the left part of Fig. 3. The right side of Fig. 3 shows a comparison between the 1D and 3D predictions of mass loss rate per unit area (MLRPUA). The 3D predictions track the average MLRPUA predicted by the well-validated 1D model fairly well—initially, the two curves overlap. The sharp peaks observed in the MLRPUA for the 3D solution correspond to the burn away of cells in the 3D Cartesian grid. This numerical artifact is due to the pyrolysis rate term predicting a loss of mass within a grid cell that exceeds the available mass at the time step preceding burn away. Current work is directed towards developing a procedure to properly limit this over-prediction of mass loss rate.

4. Conclusions

In an effort to create a methodology to efficiently predict fire growth and spread, a 3D pyrolysis model has been developed within FDS. The approach taken is somewhat unique compared to other pyrolysis models in that the time integration is fully explicit, and material burn away is modeled consistently by the removal of cells with solid density of zero. Furthermore, the fact that it has been developed within FDS means that issues such as gas phase coupling and parallelization are

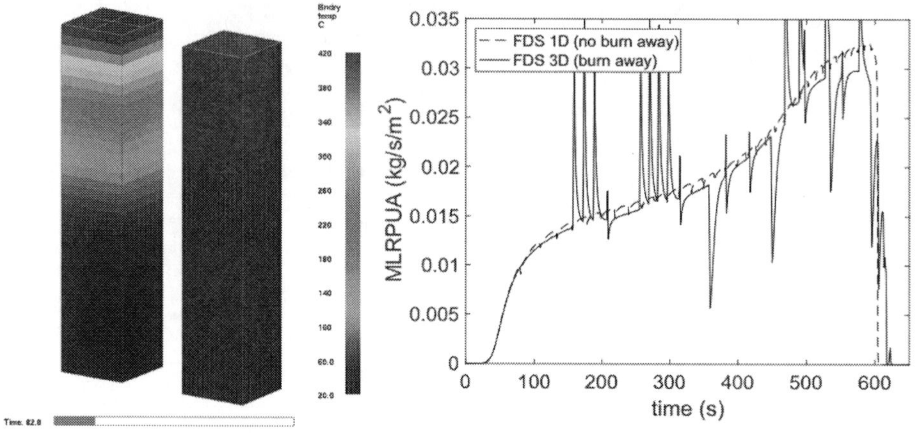

Figure 3: (Left) Visualization of 3D (on left) and 1D (on right) computational domains. (Right) Mass loss rate versus time for 1D and 3D FDS simulations.

necessarily accounted for. In this paper, a verification case (heat conduction in a sphere) and a code-to-code validation case (PMMA gasification) have been presented. More work is needed to make reliable computational flame spread predictions, but the presented methodology is a necessary step in achieving that ultimate goal. The next steps will involve accounting for in depth radiation, species transport, burn away, and material anisotropy.

References

[1] C. Lautenberger, Gpyro3d: A Three Dimensional Generalized Pyrolysis Model, Fire Safety Science 11 – Proceedings of the 11th International Symposium Christchurch, NZ, (2014), pp. 193–207.

[2] S. I. Stoliarov, I. T. Leventon, and R. E. Lyon, Two-dimensional model of burning for pyrolyzable solids, Fire and Materials 38 (2014) 391–408.

[3] K. McGrattan, S. Hostikka, R. McDermott, J. Floyd, C. Weinschenk, and K. Overholt, *Fire Dynamics Simulator, Technical Reference Guide*, sixth, 1018, NIST Special Publication, Vol. 1: Mathematical Model; Vol. 2: Verification Guide; Vol. 3: Validation Guide; Vol. 4: Configuration Management Plan, National Institute of Standards, Technology, Gaithersburg, Maryland, USA, and VTT Technical Research Centre of Finland, Espoo, Finland, 2013.

[4] A. Ito and T. Kashiwagi, Temperature Measurements in PMMA during Downward Flame Spread in Air using Holographic Interferometry, Proceedings of the Twenty-first Symposium (International) on Combustion The Combustion Institute, (1986), pp. 65–74.

[5] H. Carslaw and J. Jaegar, Conduction of Heat in Solids, 2nd, Oxford University Press, 1959.

[6] S. Stoliarov, S. Crowley, R. Lyon, and G. Linteris, Prediction of the burning rates of non-charring polymers, Combustion and Flame 156 (2009) 1068–1083.

[7] S. Stoliarov and R. Walters, Determination of the heats of gasification of polymers using differential scanning calorimetry, Polymer Degradation and Stability 93 (2008) 422–427.

[8] J. Hallman, Polymer surface reflectance-absorptance characteristics, Polymer Engineering and Science 14 (1974) 717–723.

Spring Technical Meeting
Eastern States Section of the Combustion Institute
March 4-7, 2018
State College, Pennsylvania

Critical Ignition Conditions of Structural Materials by Cylindrical Firebrands

*H. Salehizadeh, R.S.P. Hakes, M.J. Gollner**

University of Maryland, College Park, Department of Fire Protection Engineering, 3106 J.M. Patterson Bldg., College Park, MD 20742, United States
**Corresponding Author Email: mgollner@umd.edu*

Abstract: Smoldering firebrands (embers) are a major cause of ignition and eventual structural damage during wildfires at the Wildland Urban Interface (WUI). These small pieces of burning wood can loft several kilometers ahead of the main flame front and ignite spot fires in vegetation or directly ignite structural elements, such as decks. In this research, firebrands were modeled using small cylindrical wooden dowels which were ignited and placed in a small-scale wind tunnel as a pile. Two tests were performed under wind for each loading condition of firebrands, one studying ignition of a common building material, another measuring temperatures and heat fluxes. Both smoldering and flaming ignition of wood were found to be similarly dependent on wind speed. The methodology is being applied now to determine the critical conditions for ignition of different building materials during the firebrand attack.

1. Introduction:

Smoldering firebrands (embers) are a major cause of ignition and eventual structural damage during wildfires at the Wildland Urban Interface (WUI) [1, 2]. Despite their small size, several post-fire investigations have shown they are often responsible for at least half, if not more, of the ignitions and eventual home losses at the WUI [3-5]. While several studies have investigated the role that one or several firebrands play on the ignition of a structural component, many times whole piles of firebrands may land on an area prior to ignition, especially in crevices or corners [1, 6].

In order to design appropriate fire safety standards at the WUI and, someday, to model the propagation of these fires, the conditions leading to ignition of common WUI materials by piles of lofted firebrands must be quantified. The goal of this work is to apply a new methodology used to measure heat fluxes from piles of firebrands to various wind conditions and understand how these conditions relate to those which cause flaming and smoldering ignition of various WUI materials.

2. Experimental Setup

Experiments were performed in a small, enclosed wind tunnel which was able to create ambient wind conditions in the range of 0.5 m/s to 3.5 m/s, shown in Figure 1. A 30 cm x 30 cm inlet reduces into a 25 cm x 7.5 cm test section made of stainless steel and is exhausted through a high-temperature fan. The interior of the setup is painted with matte black to reduce re-radiation and a 20 cm x 6 cm window of borosilicate glass used to provide virtual access to the experiments. The bottom of the wind tunnel co a 15 cm x 12 cm hole which either a sensor array or fuel sample can be placed. A small hole, 1.5 cm in diameter just past the reducing section allows for a hot-wire anemometer to be inserted, measuring ambient wind speed. Wind speeds had a flat profile with height and no significant variation was seen when a blockage of firebrands was placed within the tunnel.

Figure 1: Small-scale combustion wind tunnel used for firebrand experiments.

Firebrands are simulated using cylindrical birch wooden dowels, 1.27 cm in diameter and 2.54 cm in length. All dowels are fully dried in an oven at 82°C before the test. Wooden dowels are placed flat in a mesh basket supported over a large propane burner, shown in Figure 2. Once all brands achieve flaming ignition (10-15 seconds depending on the pile size) the propane burner is turned off and the brands left to smolder until all flaming combustion ceases and only smoldering combustion remains. A funnel-like dumper with an outlet area of 10 cm x 10 cm is used to deposit the glowing firebrands over the sensors with a contact area almost identified to the sensor surface (10 cm x 10 cm).

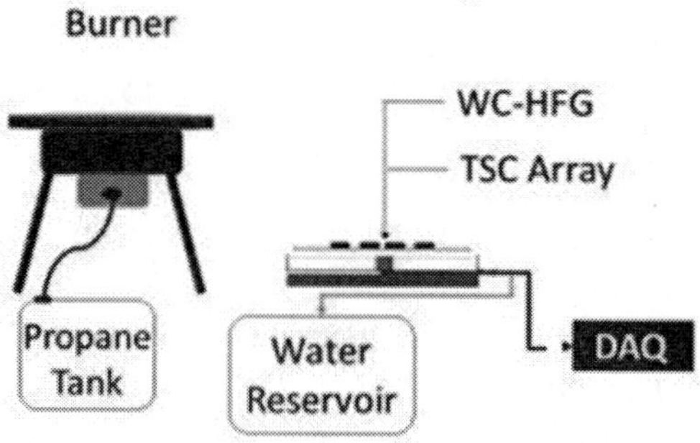

Figure 2: Schematic view of sensor array and burner. Figure from Hakes [7], not in scale.

Figure 3: Sensor array used to determine heat fluxes

Three sensors were used for thermal measurements over a 1.27 cm thick sheet of ceramic insulation board, simulating a relatively adiabatic surface. First, a 1.27 cm diameter water-cooled heat flux gauge (WC-HFG) was used to precisely measure the time-resolved heat flux at the center of the array. Unlike individual brands whose heating is dominated by conduction and surface contact, in piles it is thought that re-radiation plays a significant role in heating, justifying the use of radiation-based gauges. To provide a spatial representation of heat flux and temperature, sixteen thin-skin calorimeters (TSC), consisting of thin sheets of metal with a K-type thermocouple spot welded to their rear surface, were installed over the 10 cm x 10 cm region, shown in Figure 3. To measure temperatures along the surface, more accurately providing firebrand surface temperature, five fine-wire K-type thermocouples were placed adjacent to 5 TSCs, providing relative measurements of glowing firebrand temperatures along the surface during the tests.

For ignition tests, firebrands were placed under the same conditions but over WUI fuels. In this paper, experiments over marine-grade plywood, a common material used for decking and flooring systems, as well as outdoor patios and furnishing in residential houses, is used as a fuel. Samples were 15 cm x 15 cm x 0.95 cm. Ignition tests are done under the different wind conditions in the wind tunnel and firebrands are deposited over the fuel in the same way as with the heat flux tests. Plywood and the firebrands are fully dried under the same oven conditions mentioned above prior to starting the tests. To calculate the advancement of the smoldering fire front thorough the wood, five K-type thermocouples were pinned inside the fuel, 0.5 cm from the top surface. These measured both flame spread in depth under the pile and along the stream wise direction to measure the rate of smoldering front propagation within the fuel.

3. Results and discussion

3.1. Firebrand surface temperature

Measuring the surface temperatures of firebrands is valuable as it is a key component of the heating process and useful to numerically model smoldering ignition of fuels. Measurements are complicated, however, because firebrands naturally degrade into ash which makes continuous surface contact difficult. The focus here, however, is on temperatures at the recipient fuel surface, where heating would eventually cause ignition of the fuel. Temperatures measured both from the back surface of 16 TSC gauges and from 5 TCs placed between smoldering firebrands and TSCs were compared. Figures 5 and 5 show comparisons from 10 tests conducted with both TC and TSC gauges, with average temperatures shown as solid lines and the standard deviation between test results shown as a shaded area. While variation between the 16 gauges and between the 10 tests is high, expected due to the variability of wood and the random deposition pattern, averages are relatively uniform between tests and between TC and TSC measurements, indicating TSC measurements may provide a good average of spatial temperature profiles before ignition.

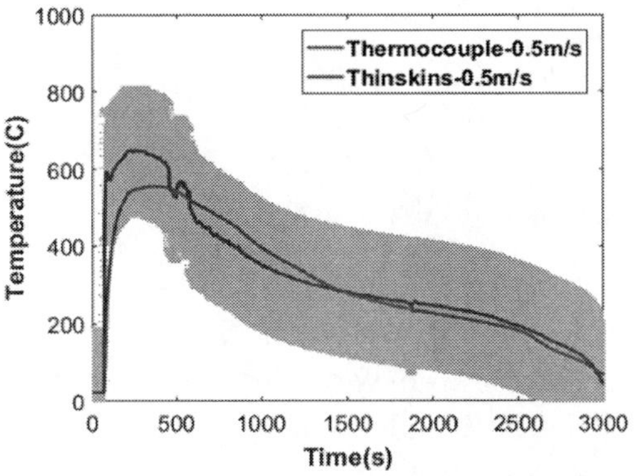

Figure 4: Average temperature of TC and TSC under a wind speed of 0.5 m/s.

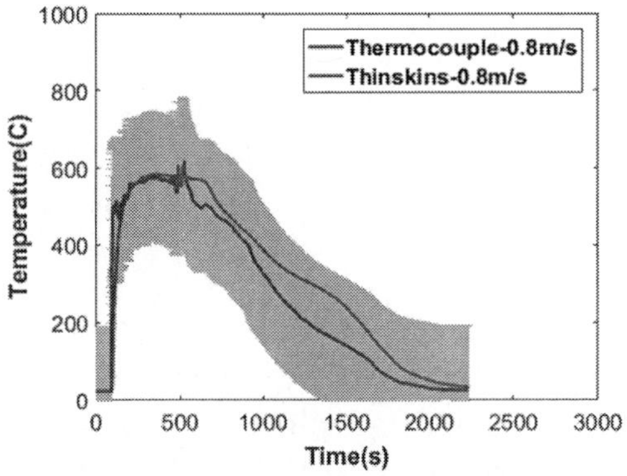

Figure 5: Average temperature of TC and TSC under a wind speed of 0.8 m/s.

Figures 6-8, averaged temperature measurements taken from TSCs are presented for two measurement locations, labeled TSC1 and TSC2. TSC1 measurements were taken in the center of the pile, shown in Figure 3, while TSC2 measurements were taken by TSCs at the edge of the pile. Figures 6 and 7 show results for tests in which 10 grams of smoldering firebrands were deposited over the sensor array with wind speeds of 0.8 and 1.7 m/s. Solid lines represent TSC1 and dashed TSC2. Somewhat higher temperatures are reached at the center of the pile, peaking at 600 – 800 C on average, where adjacent firebrands experience enhanced re-radiation and decreased convective cooling. At the edge of the pile temperatures tend to peak between 400-500 C. Figure 8 shows averaged results from tests at 0.8, 1.4 and 1.7 m/s with the same loading conditions. While increasing wind speeds should result in higher firebrand surface temperatures, to a point, at these speeds it appears that the location of the measurement within the pile is more important than the ambient wind speed, as long as wind is applied.

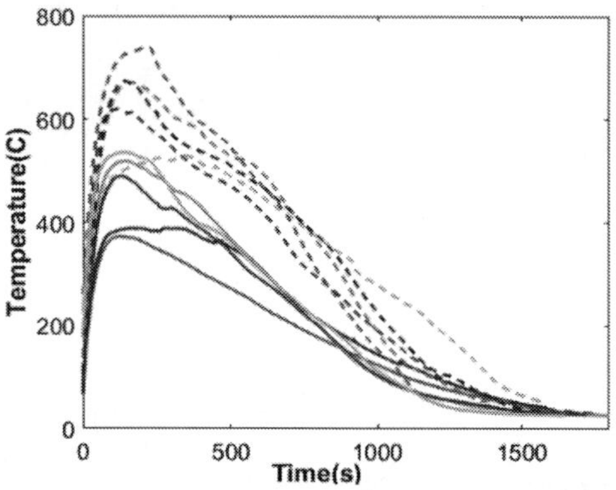

Figure 6: TSC1 (solid) and TSC2 (dashed) temperature measurements for a wind speed of 0.8 m/s.

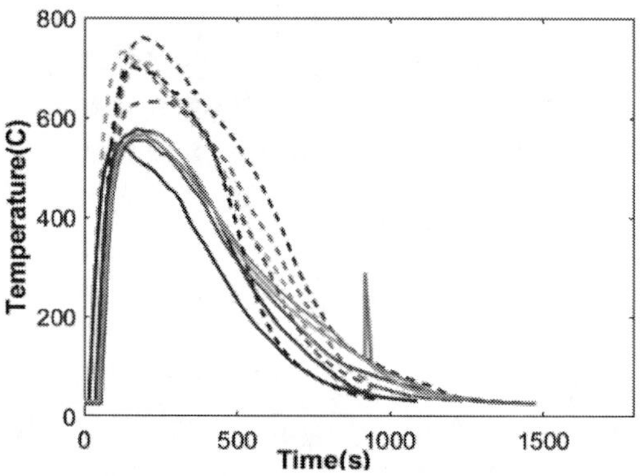

Figure 7: TSC1 and TSC2 temperature measurements for a wind speed of 1.7 m/s.

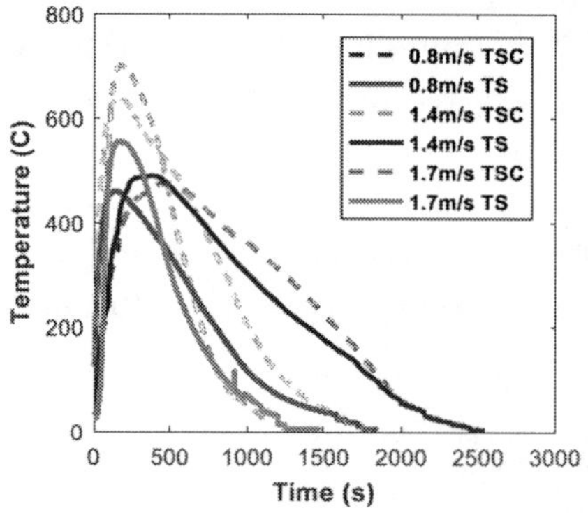

Figure 8: TSC1 and TSC2 temperatures for different wind speeds when 10 g of firebrands were deposited.

3.2. Ignition Tests

These experiments were conducted to determine the relationship between wind speed and the ignition of a solid wooden fuel when 10 grams of glowing firebrands were deposited over a 10 cm x 10 cm surface of a well-dried sheet of marine-grade plywood under wind speeds of 0.5-2 m/s. At 0.5 m/s, smoldering within the fuel is found to spread downward with an average speed of 0.0025 cm/s, an average from TC readings within the wood from five tests with 10 g of smoldering brands deposited. The horizontal spread rate was negligible. At 0.8 m/s flames appear above the dowels with a low intensity 10 s after deposition, with a downward smoldering rate of 0.0052 cm/s. The horizontal spread rate was also 0.002 cm/s. The flames in the fuel bed start to appear at 1m/s after ninety seconds from the beginning of the tests and the vertical and horizontal spread rate of the smoldering combustion in the fuel bed were 0.0038 and 0.0057 cm/s, where the horizontal rate was now larger than the downward rate for this case. When the wind speed reached 1.4 m/s, the transition to flaming occurred immediately above the fuel bed and high intensity flames appeared 5 seconds after deposition. The average vertical and horizontal smoldering spread rates were 0.004 and 0.0071 cm/s, respectively.

Figure 9: Small flame appeared over fuel bed under 1 m/s wind.

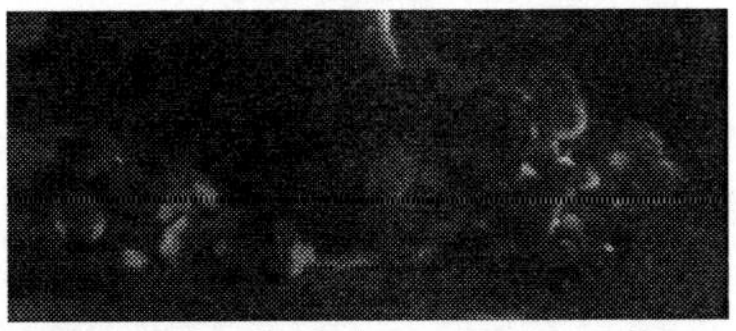

Figure 10: Low intensity flame appeared on firebrands under 0.8 m/s wind.

Figure 11: Intense flame appeared under 1.4 m/s wind.

3.3. Heat Flux

The wind speed was the main quantity of interest changed during the test to determine the effects of wind speed on the heat flux released from the glowing dowels to recipient fuels. First, the mass deposited remained constant (10 grams) in order to study the effects of wind alone. At least five tests were repeated for each wind speed of 0.5, 0.8, 1.4 and 1.7 m/s. The average heat fluxes measured by a WC-HFG in the center of an inert surface are presented in Figure 12. Tests were also conducted with half the deposited mass (5 g). Figure 13 shows the peak heat fluxes averaged between 5 repeated tests under different wind speeds for both loading conditions.

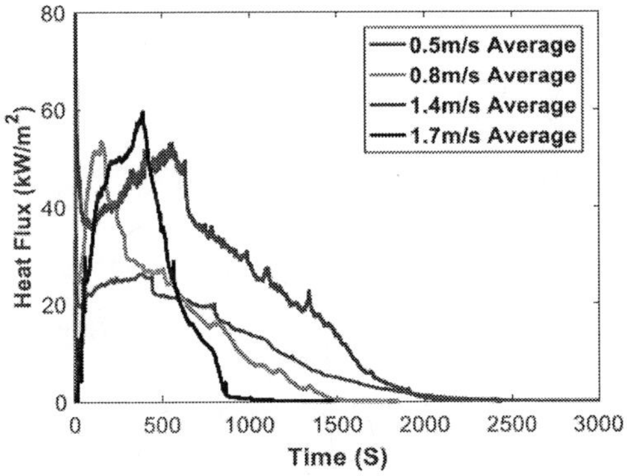

Figure 12: Heat fluxes measured by a WC-HFG averaged between 5 tests for each of 4 wind speeds.

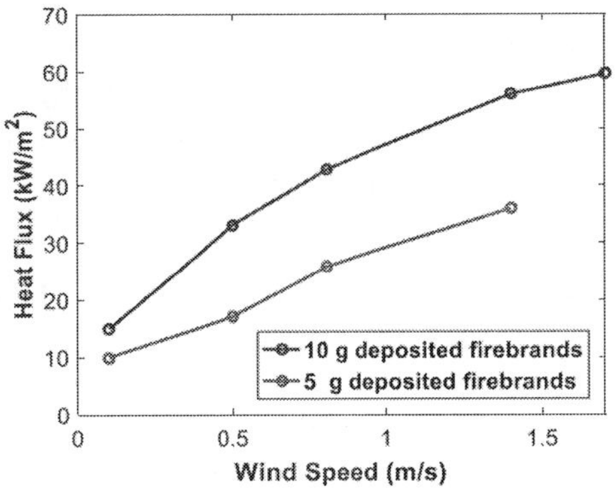

Figure 13: Peak heat fluxes as a function of wind speed.

4. Conclusions

A new methodology has been applied incorporating two styles of heat flux gauges and thermocouples to quantify temperatures and heat fluxes from firebrand piles to an inert surface. Corresponding experiments have been conducted over wood samples to understand the critical conditions present for ignition of fuels from piles of firebrands. Wind was found to play an important role for all measurements recorded. Even wind speeds as low as 1 m/s could produce ignition of thick sheets of plywood, depending on the pile size deposited. Higher temperatures were observed in the center of the pile, as expected, which highlights the need for spatial measurements over the contact area of the pile. Larger piles also produced higher peak heat fluxes at the center of the pile, highlighting the role of re-radiation within the pile influencing heat fluxes to recipient fuels. Future work will link the measured heat fluxes to ignition conditions over different materials, including changes in geometry such as L-shaped fuels and gaps which resemble deck boards.

5. Acknowledgements

The authors would like to thank Alicea Fitzpatrick and Seth Lattner for their assistance with laboratory experiments and the NIST WUI group, especially Nelson Bryner, Jiann Yang and Erica Kuligowski for their advice and support. This research was funded by National Institute of Standards and Technology Fire Research Grant Program under financial assistance award 70NANB16H284.

6. References

1. Caton, S.E., Hakes, R.S., Gorham, D.J., Zhou, A. and Gollner, M.J., 2017. Review of pathways for building fire spread in the wildland urban interface part I: exposure conditions. *Fire technology*, *53*(2), pp.429-473.
2. Hakes, R.S., Caton, S.E., Gorham, D.J. and Gollner, M.J., 2017. A review of pathways for building fire spread in the wildland urban interface Part II: Response of components and systems and mitigation strategies in the United States. *Fire technology*, *53*(2), pp.475-515.
3. Hakes, R.S.P., 2017. Thermal Characterization of Firebrand Piles, M.S. Thesis, University of Maryland, College Park. doi:10.13016/M2H70814D
4. Cohen J (2000) Examination of the home destruction in Los Alamos associated with the Cerro Grande Fire. http://www.treesearch.fs.fed.us/pubs/4686
5. Cohen J, Stratton R (2008) Home destruction examination grass valley fire. USDA R5-TP-026b. http://www.treesearch.fs.fed.us/pubs/31544
6. Maranghides A, McNamara D, Mell W, et al. (2013) A case study of a community affected by the Witch and Guejito fires: Report #2—evaluating the effects of hazard mitigation actions on structure ignitions. doi:10.6028/NIST.TN.1796
7. Manzello, S.L. and S. Suzuki. Exposing Decking Assemblies to Continuous Wind-Driven Firebrand Showers. In: Proceedings of the 11th International Fire Safety Science Symposium, pp. 1339-1352, 2014. doi: 10.3801/IAFSS.FSS.11- 1339.

Spring Technical Meeting
Eastern States Section of the Combustion Institute
March 4-7, 2018
State College, Pennsylvania

Fire Dynamics and Forensic Analysis of Compartment Fires

Shijin P. Kozhumal[1,*], *Gregory E. Gorbett*[1]
[1]*Department of Fire Protection and Paramedicine Sciences, Eastern Kentucky University, Richmond, KY, USA*
[*]*Corresponding Author Email: shijin.kozhumal@eku.edu*

Abstract: Identification of the origin and cause of ventilation controlled compartment fires can be challenging due to complex fuel characteristics and inadequate information on the time histories of flow, mixing, and fire spread. The present study proposes a numerical approach to investigate the fire spread in a single door compartment in the presence of burning sheets of wood. Three-dimensional, turbulent, unsteady, incompressible flow calculations have been carried out using the Fire Dynamics Simulator (FDS). The turbulent flow is modeled using Large Eddy Simulation with a time marching scheme for the momentum transport. The effect of combustion and radiation are accounted for using source terms in the energy transport equation. The unsteady fire spread has been predicted by considering the pyrolysis of the solid fuel and the resulting gas-phase combustion. Interactions between the flow dynamics and the burning behavior have been analyzed from the three-dimensional fields of flow, heat release, and temperature. The effect of ventilation on the burning behavior has been analyzed by simulating the burning of the wooden sheets. The burning rate rapidly increases as the fire spreads from its origin, and the mixing of fuel vapor with the entrained air dictates the extent of damage far more than the origin of the fire. Time histories of the temperature field, entrainment velocity, heat release rate, and burning rate have been analyzed to understand the heat and mass transfer mechanisms that dictate the fire spread. Analysis of the time-integrated wall heat flux suggests that the region of maximum damage depends on the behavior fire spread and it need not be close to the origin of the fire.
Keywords: Fire forensics; Fire Dynamics Simulator; Compartment fire; Fire modeling

1. Introduction

In most catastrophic fire incidents, a first fuel is ignited with an ignition source which starts a relatively small fire in the vicinity of the ignition source. The early stages of the fire damage are confined to the adjacent contents and lining materials within the compartment. However, as the burning rate increases, the air entrainment increases and the fuel vapors, formed through pyrolysis of the solid fuels or the evaporation of the liquid fuels, mix with the entrained air forming a flammable mixture detached from the true origin. At this stage, the bulk of the reaction happens in the region of mixing which need not necessarily be close to the true origin of the fire. The detached burning results in fire damage remote from the true origin, which poses significant challenges to fire investigators to identify origin and cause.

Several studies have been conducted that demonstrate the difficulty fire investigators may have when damage is created remotely from the true origin [1, 2]. Carman (2008) divided a residential bedroom into four quadrants and performed a survey of seminar attendees in an attempt to derive an error rate study of investigators. He reported a 5.7% success rate in determining the correct quadrant where the fire was started. Tinsley and Gorbett (2013) surveyed almost 600 professional investigators as to origin determination of a living room fire and reported a 77% success rate. Both studies illustrate the difficulty fire investigators may have in determining whether the damage they visualize was a result of the burning at the true origin or as a result of the detached burning.

Since the beginning of organized fire investigation in the late 1940's, fire investigators have relied on fire patterns as their basis for determining the fire origin [3]. Fire patterns are defined as the "visible or measurable physical changes, or identifiable shapes, formed by a fire effect or group of fire effects" [4, p. 12]. Absent the testimony of reliable eyewitnesses to the fire's inception, the investigator is required to determine the origin by observation and expert interpretation of the physical evidence (fire patterns) in an attempt to reconstruct the fire. As such, fire origin determination is largely a matter of fire pattern recognition and analysis [4].

Currently, fire investigators identify fire patterns by visible observation or through depth measurements of materials affected by fire. This analysis demands the coupling of the physical laws of fire dynamics with the investigator's inference regarding the damage. The pattern data is collected and then analyzed by the investigator and is assigned some weight or meaning comparative to all of the remaining damage within the compartment. The investigator attempts to identify the area(s) of damage that can best explain the collected data to arrive at a decision regarding the area of origin. In other words, the analysis of fire patterns involves identifying damage and then performing a comparative analysis to other materials and damage observed. The standard of care in the fire investigation profession is the 2017 edition of the National Fire Code NFPA 921 *Guide for Fire and Explosion Investigations*, as espoused by the National Fire Protection Association (NFPA).

Even though historic and current treatises espouse the use of fire patterns for fire investigations, only limited research has been conducted to study the scientific foundation of fire patterns. The National Institute of Standards and Technology (NIST), National Institute of Justice (NIJ), and the United States Fire Administration (USFA) have completed full-scale fire research specifically to address fire patterns [5, 6]. Due to the numerous parameters associated with full-scale fire tests and the limited number of studies conducted, there are still many questions unanswered.

The legal and science professions are currently scrutinizing forensic science, which is forcing the nation to question the scientific foundation of all forensics [7]. Recently, the National Academy of Sciences released a cautionary report regarding this type of analysis (2009). In this document, the authors outlined the need to improve the scientific foundations of the forensic disciplines, particularly those that are dependent on qualitative analyses and expert interpretation of observed patterns, including fire investigations [7].

A strong numerical study of this problem collaborated with experimental research will build a foundation for improving the understanding and use of fire patterns by fire investigators.

2. Numerical Methodology

The present study is carried out using Fire Dynamics Simulator (FDS), developed by the National Institute of Standards and Technology (NIST). The turbulent flow is simulated using Large Eddy Simulation (LES) with a time marching scheme for the momentum transport. The effect of combustion and radiation are accounted for using source terms in the energy transport equation. The species transport equations are solved using a predictor-corrector scheme. The energy conservation equation is defined implicitly via the divergence of the flow field, which contains the combustion and radiation source terms. The geometric mean of the local mesh spacing is used as the LES filter width in each direction. Eddy Dissipation Concept (EDC) [8] is used to calculate chemical source terms. The volumetric heat release rates are evaluated as products of the rate of production of species and their respective heats of formation. Both convective and radiative heat transfer between the gas phase and solid obstacles are considered. The species transport equations are solved using a predictor-corrector scheme. The radiation transport equation is solved using a finite volume based approach.

The compartment selected for the present study is a single room (3.66 m x 3.66 m x 2.44 m) with one doorway that serves as the ventilation opening. The doorway is 0.91 m wide and 2 m high. Wooden sheets of 0.5 m x 0.5 m x 0.01 m are considered with a density of 360 kg/m^3, a soot yield of 2%, and a heat of combustion of 17700 kJ/kg [9, 10]. A moisture content of 5% is considered. The sheets are arranged in blocks of five with a spacing of 0.1 m. Four such blocks are used with different spacing of 0.25 m, 0.4 m, and 0.5 m from one block to the other. The walls and ceiling are modeled as gypsum wallboard with 0.1 m thickness. An ignition source with a heat release rate of 1000 kW/m^2 is provided over an area of 0.3 m x 0.25 m, for 25 s, at the bottom of one of the fuel sheets as shown in Figure *1*. A computational domain of 6.67 m x 6.67 m x 3.94 m is used with open boundaries to closely predict the inflow of air to the compartment and outflow of combustion products to the ambient. An initial temperature of 20 ^{0}C is assumed. A non-uniform grid of 0.9 million cells is arrived at after a systematic grid refinement based on the gradients in the temperature and the heat release rate following [11]. The computational model has been validated by comparing with experimental measurements [12, 13].

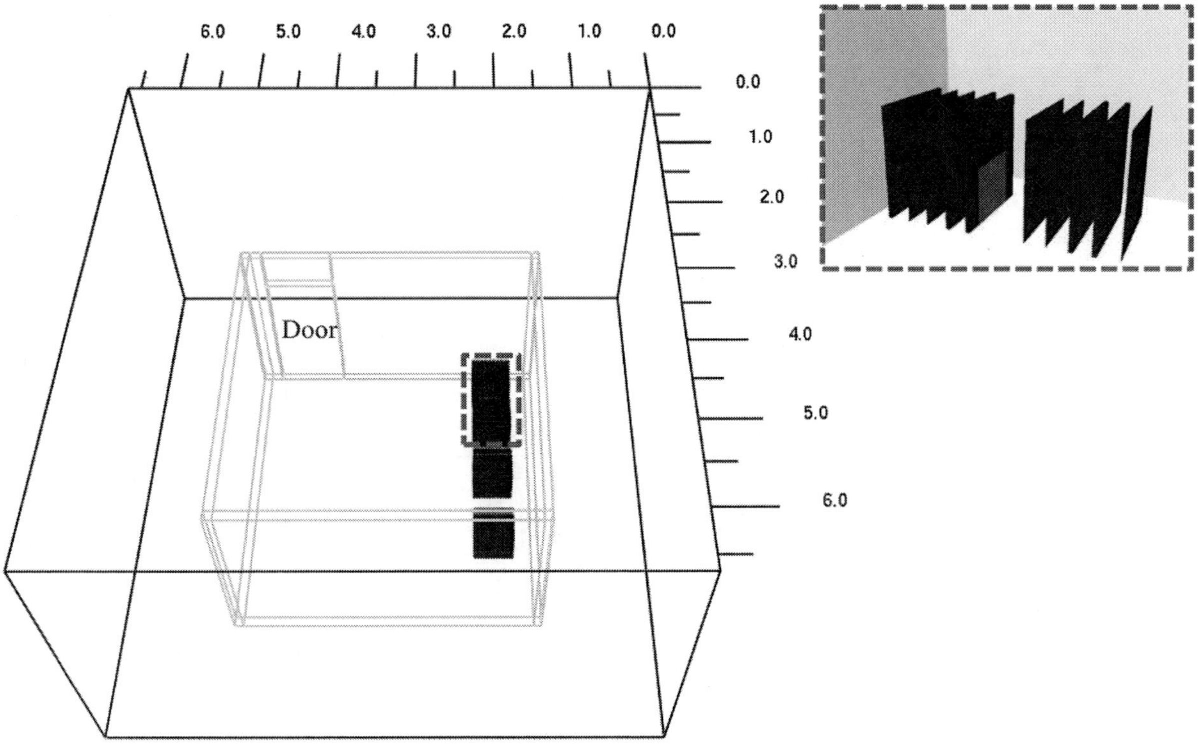

Figure 1: The computational domain. Boundaries of the computational domain as shown using dark lines and the boundaries of the compartment are shown using gray lines. The distances are shown in meters. A scaled view of two blocks of the wooden sheets with the ignition location (origin) is also shown.

3. Results and Discussion

Numerical simulations have been carried out to analyze the flow and fire dynamics of a single door compartment fire. The high heat flux acts as an ignition source and initiates pyrolysis of wood. Soon, the fire spreads upward due to buoyancy. The rates of pyrolysis and burning increase as more surface area gets exposed to sufficiently high heat flux. The increase in temperature leads to local expansion causing a net outflow of gases through the door (Figure 2, top).

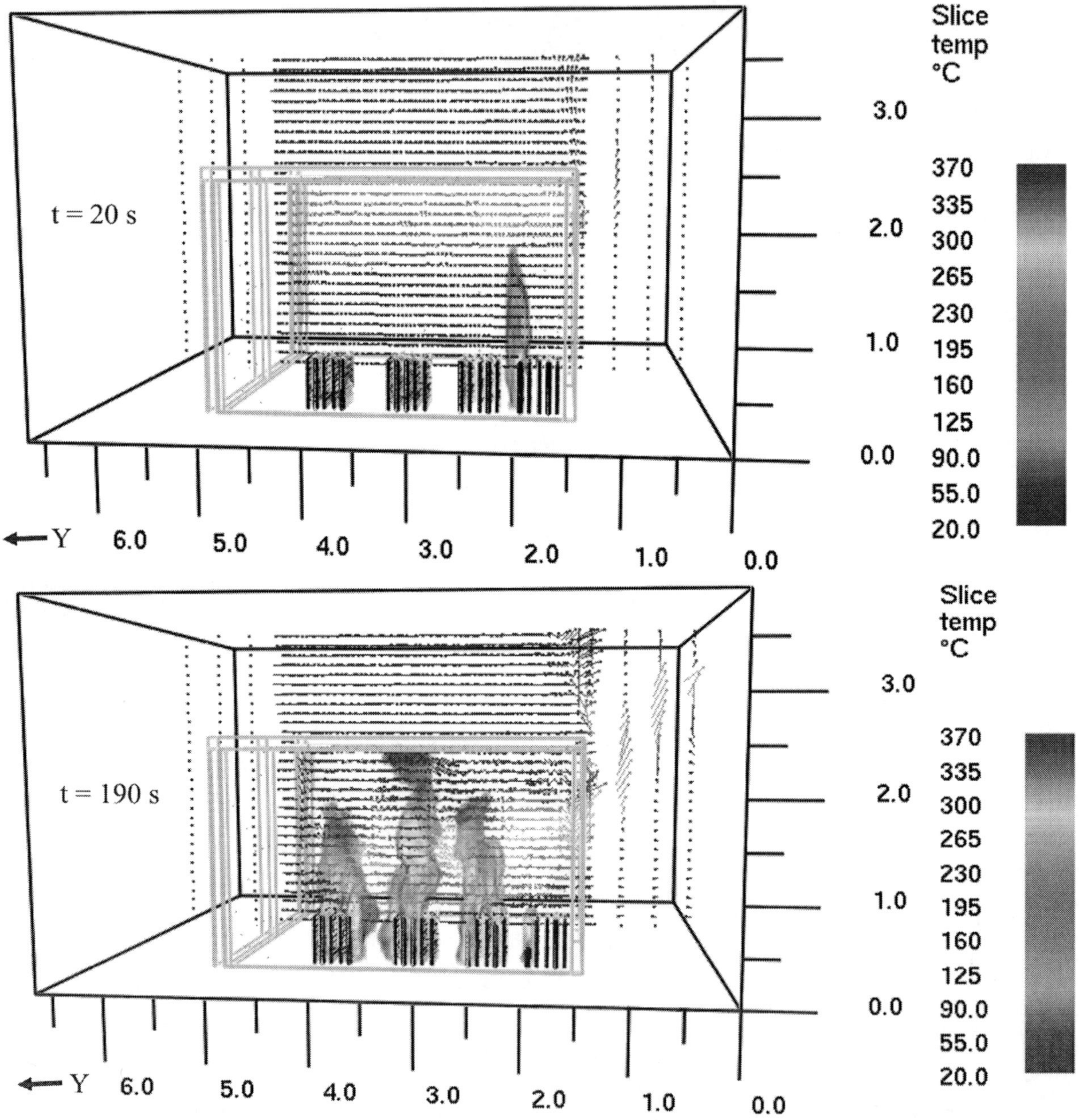

Figure 2: Velocity vectors colored by temperature, on a vertical plane normal to and passing through the middle of the door, at two instances, t = 20 s and t = 190 s after the ignition. The distances are shown in meters. The velocity vectors are superimposed with heat release rate visualizations. A series of partially transparent planes, which are colored when the heat release rate is above a cut-off value of 200 kW/m^3, is used for the three-dimensional representation.

As the burning rate increases the gas phase temperature increases and it leads to a rapid increase in radiative heat flux and pyrolysis rate. Thus, the fire sustains even after the removal of the ignition source after 25 s. The fire spreads to the adjacent sheet within 13 seconds and to the adjacent block of sheets at 53 seconds. All blocks of wooden sheets get ignited by 143 seconds from the ignition due to the increased radiative heat flux. The enhanced airflow and convective heat transfer also aid the process. An air circulation pattern is established as ambient air enters the compartment through the lower part of the door and hot gases leave the compartment through the top part of the door at a velocity as high as 3.05 m/s. The burning rate depends on both the heat flux, which dictates the rate of pyrolysis, and the mixing of fuel vapor with air. From 20 s to 190 s, the heat release rate grows from 150 kW to 1.15 MW. By 190 s from the ignition (Figure 2, bottom), the region of maximum burning rate shifts significantly from the location of initial ignition (i.e. true origin). In such scenarios, the region of maximum damage need not necessarily coincide with the ignition location. Hence the possible flow patterns and the behavior of fire spread should be considered in forensic fire analysis to identify the origin and cause of the fire.

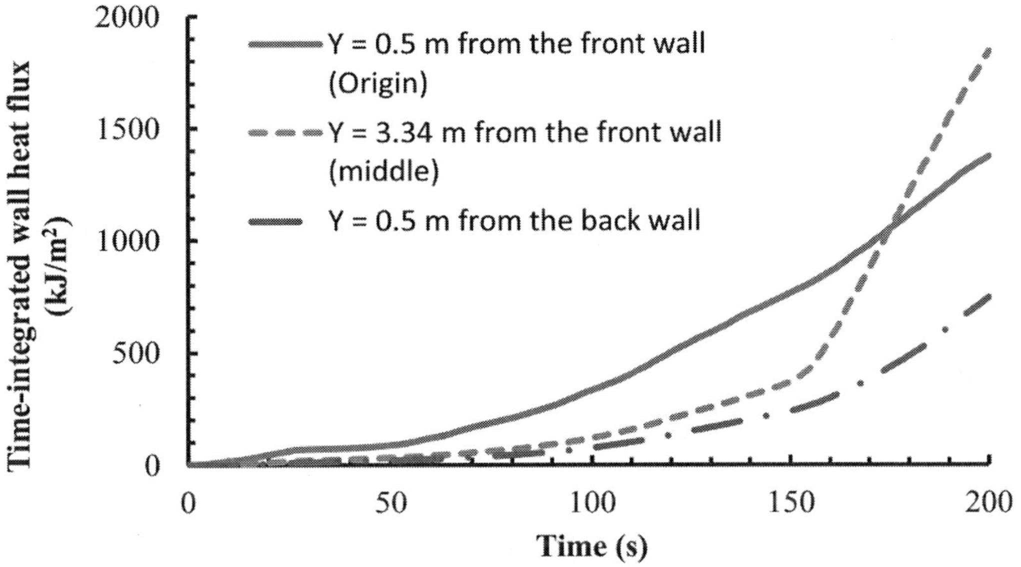

Figure 3: Time-integrated wall heat flux on the wall close to the wooden sheets at a height 1.8 m from the floor.

To predict the possible extent of damage and the fire patterns that the fire could leave, time history of time-integrated wall heat flux are extracted at three locations on the wall close to the burning wooden sheets at a height of 1.8 m (Figure 3). The location close to the origin (Y=0.5m from the front wall) receives more heat flux during the initial stage of fire spread. But later, as the fire spreads to more wooden sheets, the burning rate rapidly increases in the middle blocks of wooden sheets (Y=3.34 m from front wall) due to increased heat flux from both sides and enhanced mixing of the air and fuel vapor compared to the corners of the compartment. The time-integrated heat flux at the middle location exceeds that to the true origin after 175 seconds from ignition. The time-integrated heat flux for the first 100 seconds is an order of magnitude less than that for the second 100 seconds. This suggests that most of the damage in such compartment fires could happen during the stage of fire when the heat release rates are maximum, and that could possibly overshadow the initial fire patterns formed in the early stage of fire with low burning rates. In scenarios similar to the case considered, attempting to identify the origin of fire purely based on the extent of damage would be misleading.

4. Conclusions

Numerical investigations on the fire spread inside a single door compartment have been carried out considering, turbulence-combustion interactions, pyrolysis of solid fuels, fuel burn away, convective, conductive and radiative heat transfer, air entrainment, and gas phase combustion. At early stages of fire spread, the burning happens close to the origin in the abundance of oxygen. As the burning rate increases, the fire spreads to nearby fuel elements and the region of maximum heat release rate could shift based on the flow dynamics, heat transfer, and air-fuel mixing. Analysis of the time-integrated wall heat flux suggests that the region of maximum damage in the present scenario could be away from the origin. This calls for the necessity of an in-depth analysis of fire dynamics in fire forensics in addition to analyzing fire patterns. Systematic numerical experimental studies on different fire spread scenarios, with different fuel, fuel distribution, and ventilation are required to develop correlations between the flow, fire, and geometry characteristics with fire patterns to make forensic fire investigations more reliable.

References

[1] S. Carman, "Burn Pattern Development in Post-Flashover Fires," in *International Symposium on Fire Investigations*, Sarasota, 2008.

[2] A. Tinsley and G. Gorbett, "Fire Investigation Origin Determination Survey," *Fire and Arson Investigator-Journal of the International Association of Arson Investigators,* pp. 24-40, 2013.

[3] H. Rethoret, Fire Investigations, Toronto: Recording and Statistical Corp, Ltd, 1945.

[4] NFPA, Guide for Fire and Explosion Investigations, Quincy, MA: National Fire Protection Association, 2017.

[5] A. McGarry and J. Milke, "Full-Scale Room Fire Experiments Conducted at the University of Maryland," National Institute of Standards and Technology, Gaithersburg, 1997.

[6] J. Shanley, P. Kennedy and J. Ward, USFA Fire Burn Pattern Test, Vols. FA-178, Emmitsburg, MD: United States Fire Administration, 1997.

[7] National Academy of Science, Strengthening Forensic Science in the United States: A Path Forward, Washington D.C.: National Institute of Justice, 2009.

[8] B. Magnussen and B. Hjertage, "On mathematical modeling of turbulent combustion with special emphasis on soot formation and combustion," *Proc. Combust. Inst.,* vol. 16, p. 719–729, 1976.

[9] R. A. Susott, "Characterization of the Thermal Properties of Forest Fuels by Combustible Gas Analysis," *Forest Science,* vol. 28, pp. 404-420, 1982.

[10] S. P. Kozhumal, G. D. Cristina, N. S. Skowronski, A. Simeoni, A. S. Rangwala and S.-k. Im, "Numerical Investigation of Fire Dynamics in the Presence of Burning Obstacles under a Unidirectional Wind," in *10th US National Combustion Meeting*, College Park, MD, 2017.

[11] S. P. Kozhumal, V. Raghavan and V. Babu, "Numerical investigation of flame–vortex interactions in laminar cross-flow non-premixed flames in the presence of bluff bodies," *Combustion Theory and Modelling,* vol. 20, no. 4, p. 683–706, 2016.

[12] W. Mell, A. Maranghides, R. McDermott and S. L. Manzello, "Numerical simulation and experiments of burning douglas fir trees," *Combust. Flame,* vol. 156, p. 2023–2041, 2009.

[13] "Fire Dynamics Simulator Technical Reference Guide, Volume 3: Validation," NIST 1018-3.

Spring Technical Meeting
Eastern States Section of the Combustion Institute
March 4-7, 2018
State College, Pennsylvania

A study of cavity effect on flame spread over kerosene-soaked sand bed

Nadia Mofidi[1], Ali S. Rangwala[1]

[1]Department of Fire Protection Engineering, Worcester Polytechnic Institute, Worcester, MA, 01609

[*]Corresponding Author Email: nmofidi@wpi.edu

Abstract: *The flame spread behavior over a kerosene-soaked sand bed containing cavities was investigated experimentally. The experiments were conducted in a rectangular Pyrex tray with dimensions of 30 cm × 7.6 cm with depth of 3 cm containing kerosene-soaked sand with cavities of 2 cm depth and varying widths from 0.3 cm to 20 cm. The flame front movement and mass loss rate of kerosene were captured by a high-definition camera and a load cell, respectively. The flame spread velocity changed significantly over the cavities with different widths compared to that over planar surface. The different behavior of flame spread could be used to classify the cavities to three categories of thin, intermediate and wide. Each of these categories was studied in detail. A flame jump phenomenon was observed during the experiments with a specific behavior and velocity for each category. In case of thin cavities, flame did not penetrate inside and jumped over the cavity with the highest velocity among all cavities. In the intermediate cavities a partial downward movement of the flame was observed before the flame jumped to the trailing edge of the cavity. For the wide cavities, flame spread velocity was slower within the cavity and at a certain distance, it jumped to the trailing edge of the cavity. Moreover, the mass loss rate of kerosene during flame spread until extinction was studied in detail. The results showed that there is an optimum width of cavity among all experiments which leads to both maximum rate of flame spread and highest peak of mass loss rate. The results of this study can help to understand the fire behavior over deformed porous surfaces and it is pertinent in cleanup of shorelines contaminated by crude oil or studies related to flame spread behavior in wildland fires.*

Keywords: Flame spread rate, Cavity, Porous bed, Kerosene, Mass loss rate

1. Introduction

Transportation infrastructures of fossil fuels frequently deal with accidental spill of liquid hydrocarbons both in in-land and offshore routes. When a liquid hydrocarbon spills in-land, it can ignite and result in the flame spread over the substrate surface of the spilled fuel. This substrate could be soil, sand, water or snow depending on the location of the spill. The flame spread is a complicated phenomenon given a numerous underlying physical, chemical and environmental aspects involved. In particular, the medium that contains the spilled liquid hydrocarbon is important for determining the flame spread mechanisms. Although there have been numerous studies on investigating flame spread in pool fires [1-6], only few studies have been conducted on the flame spread over a liquid fuel soaked porous bed in the past [7]. Many parameters of the porous bed such as particle size of porous material, thickness and temperature of the bed had been investigated earlier [8-12]. Additionally, fuels with different thermal and physical properties and different concentrations had been tested [9, 13]. The effect of concurrent vs. countercurrent airflow on the flame spread in horizontal and inclined positions had been studied too [10, 12]. However, in the real world situations, the fuel-soaked ground has a variety of obstructions and cavities that deform the fuel surface. Yet, all the mentioned studies have investigated the flame spread over planar surfaces. Flame spread can be highly affected as a result of those deformations. These geometrical variations are addressed by conducting systematic lab-scale experiments in this study. The phenomenon of flame spread process under the influence of cavities in sand is explained with photographic visualization. Furthermore, the subsequent effect of the cavities on mass loss behavior of kerosene-soaked sand bed is analyzed. The knowledge obtained by this study can contribute to oil spill cleanup problems where crude oil is spilled on shorelines or in the snow and use of *in-situ* burning is recommended. Also, this knowledge will be helpful in understanding the fire behavior over deformed porous surfaces and flame spread behavior in wildfires.

2. Methods / Experimental

The setup was consisted of a Pyrex tray with outer dimensions of 30 cm × 7.6 cm with depth of 3 cm and wall thickness of 5 mm as shown in Fig. 1. The tray was filled with kerosene-soaked sand. Several numbers of cavities with different width, *w*, and constant depth, *u*, of 20 mm were created on the surface of the sand. All cavities were separated by a constant 40 mm spacing shown as *s*. The density of the sand used herein was 2,065 kg/m^3 with sand particle sizes in the range of 300 to 1000 microns. The flash point of Kerosene was around 65°C and the boiling point was in the range of 180°C to 230°C [14]. The sand used in this study was soaked with 16% of kerosene (by mass) which was close to (but slightly lower than) full saturation. Pouring additional kerosene resulted in a thin film of liquid fuel to be formed on the surface and cavity deformation. The entire tray was kept on a load cell to measure mass loss rate over the duration of experiments.

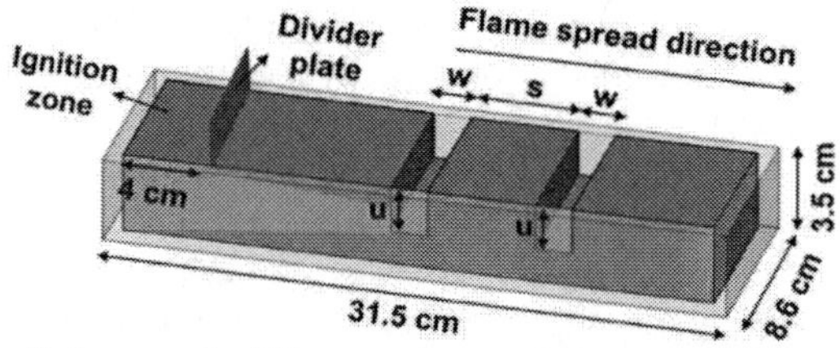

Figure 1. Schematic illustration of the experimental setup

The cavity widths were varied from 3 mm to 200 mm. The number of cavities created on the sand for each cavity width are presented in Table 1. In all experiments, the first 40 mm length of the tray was used as ignition zone initially separated from the rest of fuel surface by a divider plate. A pilot flame torch was used to ignite this zone. The divider plate was removed after 60 seconds from the onset of ignition for each experiment. All the experiments were allowed to burn until the fire naturally extinguished. A high-definition camera was used to record the videos of the flame spread to capture the flame movement through the planar surface as well as inside the cavities. The room temperature was maintained at 18 °C during the experiments.

Table 1. Experimental matrix

Cavity width (mm)	Baseline	3	5	10	20	30	40	80	160	200
Number of cavities	0	5	5	4	4	3	3	2	1	1

3. Results and Discussion

The observations of flame spread over the cavities with different widths were used to classify the cavities to three groups of thin, medium and wide as given in Table 2. In the group of thin cavities, when the flame reached the leading edge, it jumped immediately over the cavity with a remarkably higher velocity compared to baseline with no cavity (see Fig. 2.a). The measured average flame jump velocity (defined as the width of the cavity over the flame jump time) over the cavity with 3 mm width was around 3.5 mm/s, which is 5 times larger than the baseline with 0.7 mm/s. A similar flame jump phenomenon was observed for 5 mm wide cavity as well with the average velocity of 1.4 mm/s. Though, the higher rate of flame spread over the thin cavities could not reduce the total time of flame spread over the whole tray length compared to the baseline. The total flame spread duration over the tray in case of thin cavities and baseline were almost equal. The reason of this similarity was back to width of thin cavities which were much less than the total length of the tray that could be ignored.

Table 2. Classification of cavities based on duration of flame spread over them

Cavity width (mm)	3	5	10	20	30	40	80	160	200
Category	Thin		Medium			Wide			

Despite of thin cavities, the flame leading edge in medium cavities penetrated inside and moved downward slightly over the leading wall of the cavity (with the average rate of 0.2 mm/s) until the trailing edge of the cavity ignited as shown in Fig. 2b. The total vertical penetration length of flame inside the medium cavities till the initiation of a secondary flame on the trailing edge of the cavity were 8 mm, 10 mm, and 20 mm for cavity width of 10 mm, 20 mm, and 30 mm, respectively. In the latter case, the flame moved vertically all over the leading wall until ignition of trailing edge. Accounting for the downward movement in the medium cavities, the average flame spread velocity over the cavities for 10 mm case was around 2.7 mm/s. However, in case of 20 mm and 30 mm width, due to increased width and the reluctant downward penetration of flame over the leading wall of the cavity, the average spread rate across this cavity decreased to 0.8 mm/s. It was observed that the 30 mm cavity shows a critical width where the flame propagates over the entire leading wall until enough vapors were produced to achieve ignition of the top trailing edge of the cavity (Fig. 2b).

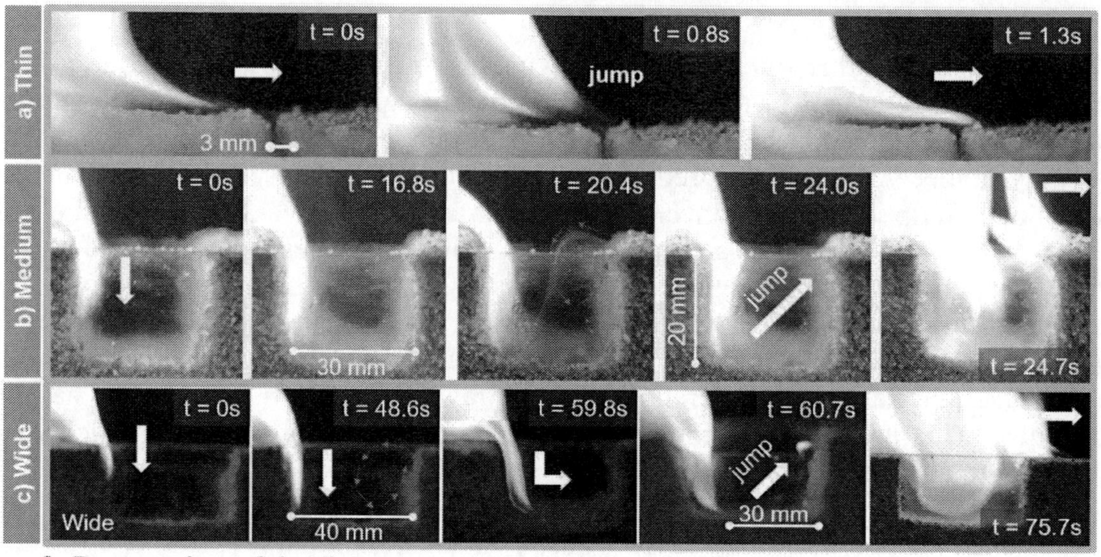

Figure 2. Progression of the flame over the thin, medium and wide cavities (3mm as thin, 30 mm as medium and 40 mm as wide cavities top to bottom)

In the case of wide cavities, the flame penetrates inside the cavity propagates downward on the leading wall and then further propagates horizontally on the cavity bottom and then jumps up to the trailing edge of the opposite wall as shown in Fig. 2c. Remarkably, the same critical distance of 30 mm was observed in the case of wide cavities as well. When the flame leading edge reached the distance of 30 mm from the trailing wall, the secondary flame initiated on the trailing edge of the wide cavities as shown in Fig. 2b and 2c. Moreover, as the flame moved inside the wide cavities, the rate of spread reduced up to 0.4 mm/s due to obstruction of air entrainment by trailing and side walls of the cavity. Additionally, the small arrows on Fig. 2 (b and c) shows the counter-clockwise circulation of vapor (visually observed) inside the largest medium cavity (with 30 mm width) and all wide cavities when the flame was getting close to the trailing wall of the cavity. During the experiment, the vapor was visually observed issuing out of the sand on the trailing wall and maybe a mixture of kerosene vapor mixing with entrained air coming in a counter-clockwise vortex.

Load cell outputs during the experiments presented different behavior of mass loss rate (MLR) in the presence of various cavity widths. The variation of mass loss rate in grams/second (g/s) is shown in Fig. 3 (left). Four cases, baseline, 10 mm wide cavity, 30 mm wide cavity, and 200 mm wide cavity are shown in this plot. MLR, for all cases, increased and reached a peak value. At these peaks, the flame had fully developed over the entire length of the tray. This trend was observed during the experiments with all cavity cases except the wide cavities. In the case of 200 mm wide cavity, a secondary peak (A) and a local minimum point (B) were observed. As discussed previously, the rate of the flame spread reduced during propagation inside the wide cavities. The declining MLR between point A and B is due to the slower flame propagation within the cavity. Point B was the critical point, where the secondary flame on the trailing edge of the cavity initiates. This abrupt increase in burning area increased the MLR until flame fully spread over the length of the tray to the point C. The time span of flame spread from ignition zone to the end of the tray is called phase I. The 10 mm wide cavity case had the highest MLR among other cases which relates to the flame spread rate characteristics in the presence of different cavities as discussed earlier. Meanwhile, the 30 mm cavity case had the minimum peak of MLR among all experiments.

After phase I, the burning of fully developed fire until natural extinguishment is called phase II of experiment. Here, the flame stood over the entire length of tray and consumed a finite amount of fuel in the porous bed. During phase II, MLR presented a continuously decreasing trend as shown in Fig. 3 (left). The total burning time was different for the various cases owing to the difference in the initial mass of the fuel present for each cavity case.

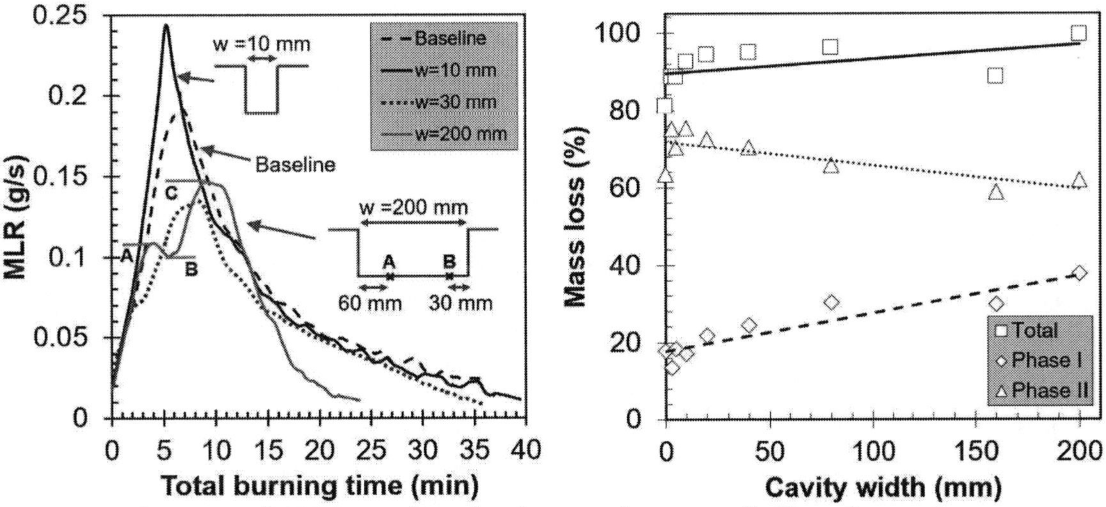

Figure 3: Mass loss rate (MLR) vs. time for four cavity cases (left) and mass loss percentage in entire duration of experiments for all cavity cases (right)

The mass loss percentage (mass loss over the total mass of the fuel) during phase I, phase II and the total burning time for all cavity cases are shown in Fig. 3 (right). The percentage of burnt fuel in phase I increased from around 13.5% to 38% along with the increase in the cavity width. While, this percentage decreased with increasing in cavity width in phase II from 75% to 62%. The total mass loss followed the trend of phase I. For baseline case, total mass loss was around 81% which is the lowest value among all experiments. Adding cavities to the kerosene-soaked bed increased the total mass loss by 8%. The 200 mm wide cavity case had the highest value of burnt mass percentage as 99.7%. This progression was predictable due to the persistent flaming inside the cavities before complete extinction.

4. Conclusions

A set of experiments were conducted over a rectangular Pyrex tray containing kerosene-soaked sand with a certain variety of cavities. The flame spread features over top surface of mixture as well as inside the cavities were investigated. Also, the mass loss of burning mixture in presence of different cavities was recorded and analyzed. The main conclusions are as follows:

(1) According to observed flame progression features, cavities could be classified to three groups; thin, intermediate and wide. In the thin cavity cases, the flame jumped over the cavity with a velocity of 2-5 times of the baseline. The flame did not penetrate inside the cavity despite the intermediate cavity cases where the flame moved partially downward over the first wall of the cavity and initiated the secondary flame on the trailing edge of the cavity. The 10 mm cavity (intermediate) had the largest value for flame jump velocity among all cavities. In the wide cavity cases, the flame moved all over the first wall as well as within the cavity. The propagating flame initiated the secondary flame on

the trailing edge of the cavity in a certain distance of 3 cm. The flame velocity within wide cavities reduced notably compared to baseline.

(2) Mass loss rate of kerosene had a general trend of increasing up to a peak point where the flame fully developed then decreased until extinction of fire. The 10 mm cavity and the 30 mm wide cavity owned the highest and lowest peaks respectively.

5. Acknowledgements

This research was supported by National Fire Protection Association (NFPA) graduate support grant.

6. References

[1] H. D. Ross, "Ignition of and flame spread over laboratory-scale pools of pure liquid fuels," *Progress in Energy and Combustion Science,* vol. 20, no. 1, pp. 17-63, 1994.

[2] K. Torrance, "Subsurface flows preceding flame spread over a liquid fuel," *Combustion Science and Technology,* vol. 3, no. 3, pp. 133-143, 1971.

[3] I. Glassman and F. L. Dryer, "Flame spreading across liquid fuels," *Fire Safety Journal,* vol. 3, no. 3, pp. 123-138, (1981).

[4] J. Ji, S. Lin, C. Zhao, K. Li, and Z. Gao, "Experimental study on initial temperature influence on flame spread characteristics of diesel and gasoline–diesel blends," *Fuel,* vol. 178, pp. 283-289, (2016).

[5] E. Degroote and P. Garcia Ybarra, "Flame propagation over liquid alcohols: Part I. Experimental results," *Journal of thermal analysis and calorimetry,* vol. 80, no. 3, pp. 541-548, (2005).

[6] D. White, C. L. Beyler, C. Fulper, and J. Leonard, "Flame spread on aviation fuels," *Fire Safety Journal,* vol. 28, no. 1, pp. 1-31, (1997).

[7] J. Zanganeh, B. Moghtaderi, and H. Ishida, "Combustion and flame spread on fuel-soaked porous solids," *Progress in Energy and Combustion Science,* vol. 39, no. 4, pp. 320-339, 2013/08/01/ (2013).

[8] K. Takeno and T. Hirano, "Flame spread over porous solids soaked with a combustible liquid," *Symposium (International) on Combustion,* vol. 21, no. 1, pp. 75-81, 1988/01/01/ (1988).

[9] J. Zanganeh and B. Moghtaderi, "Effect of fuel soaked time and fuel ratio on the flame spread rate over a porous bed wetted with liquid fuel," *Fire Safety Journal,* vol. 59, no. Supplement C, pp. 151-159, 2013/07/01/ (2013).

[10] H. Ishida, "Flame spread over fuel-soaked ground," *Fire Safety Journal,* vol. 10, no. 3, pp. 163-171, 1986/05/01/ (1986).

[11] W. Kong, C. Y. H. Chao, and J. Wang, "Behavior of non-spread diffusion flames of combustible liquid soaked in porous beds," *Proceedings of the Combustion Institute,* vol. 29, no. 1, pp. 251-257, 2002/01/01/ (2002).

[12] T. Suzuki, M. Kawamata, K. Matsumoto, and T. Hirano, "Behavior of the reverse flow in front of the leading flame edge spreading over fuel-soaked sand in an air stream," *Fire Safety Science,* vol. 3, pp. 227-236, (1991).

[13] k. Takeno and T. Hirano, "Behavior of combustible liquid soaked in porous beds during flame spread," *Symposium (International) on Combustion,* vol. 22, no. 1, pp. 1223-1230, 1989/01/01/ (1989).

[14] T. Asahara, "Yohzai Handbook (Solvent Handbook)," *Kohdansha,* p. 171, 1976.

Spring Technical Meeting
Eastern States Section of the Combustion Institute
March 4-7, 2018
State College, Pennsylvania

The Structure of a Turbulent Line Fire Subjected to Cross-Flow

Salman Verma[1], Arnaud Trouvé[1,]*

[1]*Department of Fire Protection Engineering, University of Maryland, College Park, MD, 20742, USA*
[*]*Corresponding Author Email: atrouve@umd.edu*

Abstract: The general objective of this project is to provide a basic understanding of the transition between the different flame regimes observed in fires with cross-flow and/or fires along inclined surfaces. We consider here a simplified configuration corresponding to a methane-air, buoyancy-driven, turbulent line flame stabilized on top of a horizontal floor surface and subjected to different air cross-flow velocities. At high values of the cross-flow velocity, the flame features a horizontal shape and develops as a boundary layer flame in the vicinity of the floor surface; the flow downstream of the flame is attached to the floor surface and air entrainment into the flame is one-sided. In contrast, at low values of the cross-flow velocity, the flame features a tilted vertical shape and develops as a pool-like flame; the flow downstream of the flame separates from the floor surface and air entrainment into the flame is two-sided. In the present study, we analyze the transition from an attached flame to a lifted flame using wall-resolved large eddy simulations (LES). Simulations are performed with an LES solver developed by FM Global and called FireFOAM. The simulated line burner is 50-cm wide and 5-cm long; the flame power is 50 kW; and the air cross-flow velocities range between 0.75 and 3 m/s. The LES simulations provide a detailed description of the different contributions to flow kinetic energy in the horizontal and vertical directions and thereby provide unique insights into the competing effects of the external momentum of the horizontal cross-flow and the internal momentum of the vertical buoyant motions produced by the combustion heat release and the resulting unstable thermal stratification. A new criterion is proposed to measure the relative strength of external/cross-flow-driven versus internal/buoyancy-driven motions and to thereby predict the transition from an attached to a lifted flame regime.
Keywords: Buoyant turbulent diffusion flame, Wind-driven flames, Boundary layer flame, Pool flame, Air entrainment, Large Eddy Simulation

1. Introduction

Developing a fundamental understanding of the effects of cross-flow on fires is of considerable interest in fire science; because of the ubiquity and importance of such effects in wildland, urban and industrial fires. One basic way in which cross-flow affects a fire is by changing the flame geometry: as the external momentum of the horizontal cross-flow increases, the flame transitions from a vertical pool-like flame to a horizontal flame attached to the downstream surface [1]. Among other things, attachment of the flame increases both the convective and radiative heat transfer (due to the improved view factor) to the downstream surface; this in turn increases the rate of flame spread (in cases for which the downstream surface contains fresh fuel).

The variations in flame geometry in fires subjected to cross-flow have been studied both experimentally [2] and numerically [3,4] over the past several decades. However, past experimental studies typically use a limited range of diagnostics, primarily video imaging and temperature measurements, while past numerical studies typically use low-order approaches, for instance Reynolds-Averaged Navier-Stokes (RANS) simulations [4,5], that do not resolve the relevant turbulent flow and flame scales. Thus, a basic understanding of the physics associated with variations in flame geometry resulting from changes in the magnitude of the cross-flow velocity is lacking and is the primary motivation of the present study.

In the present study, a methane-air, buoyancy-driven, turbulent line flame stabilized on top of a horizontal floor surface and subjected to different cross-flow velocities is simulated using a high-fidelity numerical approach, namely Large Eddy Simulations (LES). The primary objective is to gain fundamental insights into the transition from a vertical to a horizontal flame resulting from increases in the cross-flow velocity. This study features LES simulations that are wall-resolved, i.e., the simulations are performed with sufficient grid resolution to capture the wall gradients.

2. Numerical Solver and Configuration

Numerical simulations are performed using FireFOAM [6], a fire modeling solver developed by FM Global and based on an open-source, general-purpose, Computational Fluid Dynamics (CFD) software package called OpenFOAM [7]. FireFOAM is a second-order accurate, finite volume solver with implicit time integration; the solver features advanced meshing capabilities (structured/unstructured polyhedral mesh); it also features a massively parallel computing capability using Message Passing Interface protocols.

FireFOAM uses a Favre-filtered, compressible-flow, LES formulation and provides a choice between several modeling options for the treatment of turbulence, combustion and thermal radiation. In the present study: subgrid-scale (SGS) turbulence is described using the WALE (Wall-Adapting Local Eddy-viscosity) model; combustion is described using the classical concept of a global combustion equation combined with the Eddy Dissipation Concept (EDC) model; radiation is described by solving the radiative transfer equation (RTE) using a discrete-ordinates, finite-volume method and by assuming a non-scattering, non-absorbing, optically-thin, gray medium and using a prescribed global radiative loss fraction χ_{rad}.

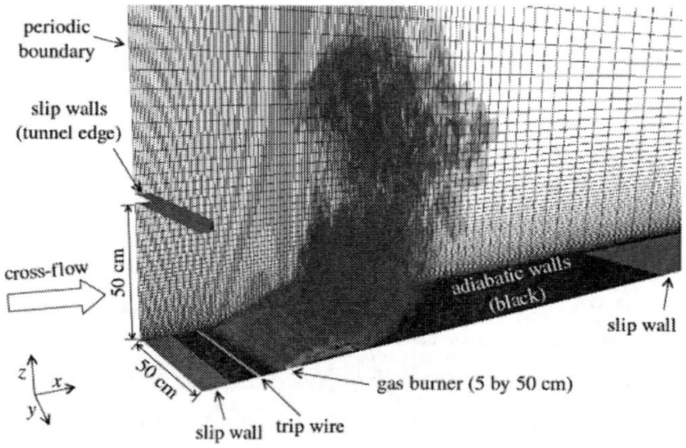

Figure 1: Illustration of the FireFOAM numerical configuration; the 50 kW methane diffusion flame is visualized using six isocontours of instantaneous volumetric heat release rate (0.1, 1, 10, 100, 1000, 10000 kW/m^3).

The numerical configuration is presented in Fig. 1. The computational domain is 780-cm-long in the streamwise x-direction, 50-cm-wide in the spanwise y-direction and 250-cm-high in the vertical z-direction. The line burner face is 5-cm-long in the x-direction and 50-cm-wide in the y-direction; and the spanwise center of its leading edge is the origin of the coordinate system $(x,y,z) = (0,0,0)$. The burner is flush-mounted on a 50-cm-wide horizontal solid plate that starts at $x = (-20)$ cm and ends at $x = 205$ cm. The cross-flow air is injected through a 50-cm-wide and 50-cm-high wind tunnel whose outlet is located on the inlet boundary of the computational domain, at $x = (-30)$ cm. A 5-mm-long, 50-cm-wide and 5-mm-high trip wire is placed at $x = (-10.5)$ cm in order to perturb the incoming flow and promote laminar-to-turbulent transition.

The computational grid is a rectangular Cartesian mesh. Grid spacing in the streamwise x-direction is uniform and is equal to 7 mm for $x \leq 80$ cm; beyond that location, the x-grid is stretched with a stretch factor equal to 1.09. Grid spacing in the spanwise y-direction is uniform and is equal to 7 mm. This streamwise (spanwise) resolution corresponds to 7 (70) grid cells across the burner length (width). Grid spacing in the vertical z-direction is non-uniform: the z-grid spacing is 2.1 mm at $z = 0$ (*i.e*, the first cell center is 1.05 mm above the bottom surface) and is 30 mm at $z = 50$ cm with a stretching factor of 1.06. For $z \geq 50$ cm, the z-grid is stretched with a stretch factor equal to 1.09. Note that with the present resolution, the trip wire is resolved by 1 (2) grid cells in the x- (z-) direction. The total number of cells is 1 million. Note also that these are preliminary coarse simulations; higher resolution simulations are currently in progress.

The methane mass flow rate is prescribed at the burner boundary and the air velocity is fixed at the tunnel outlet. The horizontal solid plate and the trip wire are both treated as no-slip adiabatic solid walls. The surface located at $z = 0$ between the tunnel outlet and the leading edge of the solid plate is treated as a slip wall. The surface located at $z = 0$ beyond the solid plate, at $x \geq 205$ cm, is also treated as a slip wall. The side boundaries at $y = (-25)$ and 25 cm correspond to periodic conditions. Other boundaries are treated as boundaries with open flow conditions.

In all cases, the methane mass flow rate is linearly increased from 0 to 1 g/s during the first five seconds and is then held constant for the remainder of the simulations. This is done to allow the cross-flow to establish itself over the line burner before the fire reaches its nominal value of the heat release rate equal to 50 kW. All simulations are performed for a duration of 30 s. Turbulent statistics are collected for the final 15 s of each simulation, after the flow and flame become statistically stationary and long enough for the statistics to be converged (to improve convergence, statistics are computed using both temporal- and spanwise-averaging). The time step is controlled by a classical Courant-Friedrichs-Lewy (CFL) condition and is approximately equal to 0.5 ms. Each simulation is run using 100 processors on a large-scale Linux cluster with a typical simulation requiring 5,000 CPU-hours.

3. Results and Discussion

We first consider the mean flame shape for a range of cross-flow velocities (Fig. 2). As U_∞ increases from 0.75 to 3 m/s, the flame transitions from a lifted (vertically tilted) flame to an attached (horizontal) flame. The vertical elevation of the flame decreases (from 50 to 10 cm); the flame length (loosely defined as the distance from the burner to the flame tip) increases (from 50 to 100 cm); and the flame drag (defined as the x-wall-distance downstream of the burner within the flame region) increases (from 30 to 90 cm). In the present configuration the transition from a vertical tilted flame to a horizontal flame is gradual and happens between $U_\infty = 1$ and 1.5 m/s.

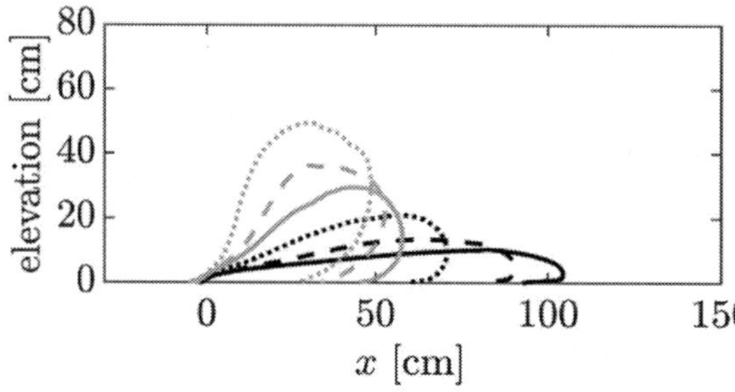

Figure 2: Mean flame shape marked using a particular isoline of the mean heat release rate per unit volume (50 kW/m^3). From left to right: U_∞ = 0.75, 1, 1.25, 1.5, 2, 3 m/s.

We now consider the spatial variations of the mean velocity vector ($\overline{U}$, $\overline{W}$) for two cases, U_∞ = 0.75 and 3 m/s (Fig. 3). In the 0.75-m/s case, the cross-flow is strongly deflected upwards; in addition, a reversed flow is induced by the flame on the downwind side leading to two-sided entrainment similar to pool fires. In contrast, in the 3-m/s case, the cross-flow is only weakly affected by the presence of the flame and entrainment is one-sided due to the flame being attached to the downstream surface. These differences in flow pattern are believed to be key ingredients in flame spread mechanisms.

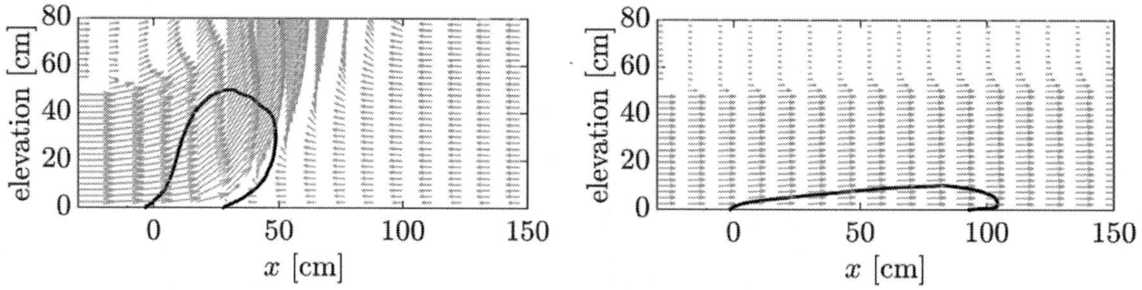

Figure 3: Spatial variations of the mean velocity vector ($\overline{U}$, $\overline{W}$): U_∞ = 0.75 m/s (left); U_∞ = 3 m/s (right). Solid lines are isolines of the mean heat release rate per unit volume (50 kW/m^3) and are used for marking the flame region.

Fig. 4 adopts a different perspective and presents z-profiles of different quantities of interest ($\overline{U}$, $\overline{W}$, U_{rms}, W_{rms}, B_z and S_{xz}) across the flame zone, at x = 50 cm and for U_∞ = 1.5 m/s. U_{rms} (W_{rms}) is defined as the mean amplitude of temporal fluctuations of the grid-resolved streamwise (vertical) flow velocity component; and B_z and S_{xz} are source terms responsible for producing turbulent fluctuations in the z- and x-directions, respectively, $B_z=(\overline{-w''(\partial \overline{p}/\partial z)})$ and $S_{xz}=(\overline{-\rho u'' w''(\partial \tilde{u}/\partial z)})$, where $\overline{q}$ ($\tilde{q}$) designates a straight (mass-weighted) temporal mean of a grid-resolved quantity q, and $q''= (q-\tilde{q})$. B_z represents production of (vertical) turbulence by buoyancy whereas S_{xz} represents production of (horizontal) turbulence by shear.

The case with U_∞=1.5-m/s is important because this flame is close to the point of transition between the horizontal and vertical flame regimes. Fig. 4 shows that: turbulence production in the flame zone is dominated by buoyancy (B_z is much larger than S_{xz}); the flow activity in the vertical direction is strongly turbulent ($W_{rms} > \overline{W}$); and the vertical turbulent velocities take values that have the same order of magnitude as those of the cross-flow velocity ($W_{rms} = O(U_\infty)$). These results suggest that flow deflection and transition to a vertically-oriented flame is a consequence of the vertical turbulent motions produced by buoyancy.

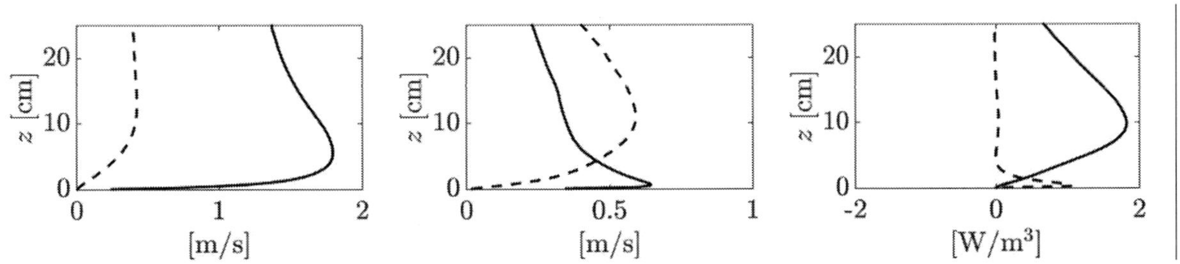

Figure 4: Vertical profiles for $U_\infty = 1.5$ m/s at $x = 50$ cm: $\overline{U}$ and $\overline{W}$ (left); U_{rms} and W_{rms} (middle); B_z and S_{xz} (right). $\overline{U}$, U_{rms}, B_z ($\overline{W}$, W_{rms}, S_{xz}) are plotted using solid (dashed) lines.

The magnitude of this effect can be measured globally by spatially-integrating the source term B_z and by comparing the resulting integrated value to the flow rate of kinetic energy (or equivalently the power) of the incoming cross-flow. We write:

$$R = \frac{\left(\int_0^{L_x}\left(\int_0^{L_z} B_z dz\right)dx\right)L_y}{\frac{1}{2}\rho_\infty U_\infty^3 L_z L_y} = \frac{\left(\int_0^{L_x}\left(\int_0^{L_z} B_z dz\right)dx\right)}{\frac{1}{2}\rho_\infty U_\infty^3 L_z} \tag{1}$$

where R is the ratio of the power of the production of z-turbulence due to buoyancy divided by the power of the incoming cross-flow. In Eq. 1, ρ_∞ is the mass density of ambient air, and L_x, L_y and L_z are the x-, y- and z sizes of the control volume under consideration (due to periodicity in the y-direction, L_y is simply dropped from the expression of R).

Fig. 5 presents the variations of the power ratio R with streamwise distance using $L_x = x$ and $L_z = H_t$ ($H_t = 50$ cm is the height of the wind tunnel). With these choices for L_x and L_z, R represents the cumulative effects of production of vertical turbulence by buoyancy up to a given location x compared to the total power of the flow delivered by the wind tunnel. We expect significant flow deflection and a transition to a vertical flame regime when R takes values close to 1 prior to the end of the flame zone. Fig. 5 shows that the 1.25-m/s-flame reaches a peak value of R equal to 0.4 and that the 1-m/s-flame reaches a peak value above 1. These estimates confirm that the 1.25-m/s-flame is in the transitional regime while the 1-m/s flame is in the vertical flame regime.

4. Conclusions

Fine-grained LES are performed to bring fundamental insight into the effects of cross-flow velocity on the structure of a methane-air, buoyancy-driven, turbulent line flame (50 kW) stabilized on top of a horizontal floor surface. As the cross-flow velocity is increased, the flame transitions from a pool-like flame characterized by a tilted vertical shape to a boundary layer flame characterized by a horizontal shape. The pool-like flame strongly deflects the incoming cross-flow upwards and features downwind flow separation and two-sided air entrainment into the flame. In contrast, the boundary layer flame does not significantly deflect the incoming cross-flow and features downwind flow attachment and one-sided air entrainment into the flame.

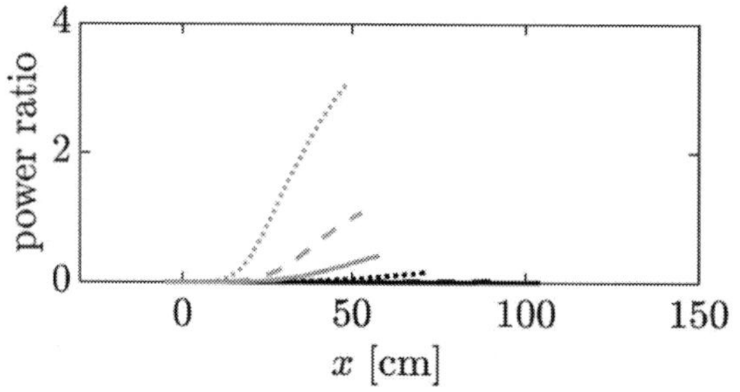

Figure 5: Streamwise evolution of the power ratio R (see Eq. 1), plotted from $x = 0$ to the downstream x-location of the flame tip. From left to right: $U_\infty = 0.75$, 1, 1.25, 1.5, 2, 3 m/s; note that the 2- and 3-m/s curves correspond to low values of R and are horizontal.

The present simulations are analyzed in terms of production of mean flow and turbulent flow kinetic energy. Results suggest that the transition between the (vertical) pool-like and (horizontal) boundary layer flame regimes is controlled by the strength of the vertical turbulent motions produced by buoyancy. A new non-dimensional number that measures the ratio of the power of the production of z-turbulence due to buoyancy divided by the power of the incoming cross-flow is introduced to explain the aforementioned transition. Work is in progress to explore the implications of this power ratio for scaling laws.

5. Acknowledgements

This project is financially supported by the U.S Forest Service, Rocky Mountain Research Station, with Dr. Mark Finney as Program Manager, and by FM Global. The project is also supported by supercomputing resources made available by the University of Maryland (http://hpcc.umd.edu) and by the National Science Foundation (XSEDE Program, Grant #ACI-1548562).

6. References

[1] W. Tang, C.H. Miller, M.J. Gollner, Local flame attachment and heat fluxes in wind-driven line fires, Proc. Combust. Inst. 36 (2017) 3253-3261.
[2] C.S. Lam, E.J. Weckman, Wind-blown pool fire, Part I: Experimental characterization of the thermal field, Fire Safety J. 75 (2015) 1-13.
[3] F.A. Albini, A model for the wind-blown flame from a line fire, Combust. Flame 43 (1981) 155-174.
[4] D. Morvan, B. Porterie, J.C. Loraud, M. Larini, A numerical investigation of cross wind effects on a turbulent buoyant diffusion flame, Combust. Sci. Technol. 164 (2001) 1-35.
[5] Y.L. Sinai, M.P. Owens, Validation of CFD modelling of unconfined pool fires with cross-wind: flame geometry, Fire Safety J. 24 (1995) 1-34.
[6] FireFOAM, developed by FM Global, available at: https://github.com/fireFoam-dev.
[7] OpenFOAM, developed by the OpenFOAM foundation, available at: http://www.openfoam.org.

Sub Topic: Fire

Spring Technical Meeting
Eastern States Section of the Combustion Institute
March 4-7, 2018
State College, Pennsylvania

Conditions for Formation of the Blue Whirl

Yu Hu[1,3], Sriram Bharath Hariharan[2], Michael J. Gollner[2], Elaine S. Oran[3]*

[1]*Key laboratory for Thermal Science and Power Engineering of Ministry of Education, Tsinghua University, Beijing 100084, P.R. China*
[2] *Department of Fire Protection Engineering, University of Maryland, College Park, USA*
[3]*Department of Aerospace Engineering, University of Maryland, College Park, USA*
[*]*Corresponding Author Email: mgollner@umd.edu*

Abstract: Recent experiments with fire whirls on water have led to the discovery of the blue whirl, a soot-free regime of the fire whirl. Upon ignition of fuel in a fixed-frame, self-entraining apparatus consisting of two half cylinders suspended over a water surface, a traditional yellow fire whirl formed, followed by a transition to a blue whirl. From previous observations of the shape and internal structure of the blue whirl, it is thought to be a bubble type vortex breakdown in reacting flow. While the surface boundary conditions required for this transition to occur are qualitatively known, the combination of swirl and heat release required have not yet been established. To maintain steady rates of burning, liquid n-heptane was continuously supplied to the surface of the water. By varying the gap between the two cylinders and the flow rate of fuel, several stable and transitional states of the fire whirl were observed, including the blue whirl and large yellow fire whirls. The heat-release rate ($\dot{Q}$) was calculated from the volumetric fuel flow rate (assuming complete combustion). The circulation (Γ) was calculated using velocity measurements at the inlet of the gap (using a hot-wire anemometer) and the radius of the enclosure. Combinations of the heat-release rate and the circulation leading to formation of fire whirls and blue whirls were recorded, and marked out on a graph of $\dot{Q}$ vs. Γ. The resulting graph shows that the stable blue whirl forms in a narrow range of fuel flow rate and circulation, close to the extinction limit found in existing literature. Implications of these findings and possible scaling approaches are discussed.
Keywords: Fire Whirl, Blue Whirl, Hot-wire Anemometer

1. Introduction

Reducing harmful emissions from combustion processes have long been a goal of the combustion research community. While there are many methods to reduce emissions from combustion, the blue whirl presents a unique opportunity as it appears to burn a variety of liquid hydrocarbon fuels with little visible soot (as indicated by its blue color), and is naturally formed without any mechanical forcing [1]. While the surface boundary conditions required for the blue whirl to occur are qualitatively known [2], the combination of heat release and swirl required, previously shown to govern formation of traditional fire whirls [3], have not yet been established for the blue whirl.

In this study, the conditions leading to formation of the blue whirl are further quantified. A fixed-frame, self-entraining apparatus consisting of two half cylinders suspended over a water surface is used to generate swirling flame patterns over a range of heat-release rates and ambient circulations.

The various stable and transitional states formed as a result of varying the gap between the two cylinders and the fuel flow rate, are distinguished using recorded videos. These corresponding ranges of heat-release rate ($\dot{Q}$) and ambient circulation (Γ), determined from inlet velocity measurements, then reveals the narrow range of blue whirl formation. Possible approaches for a scaling analysis, to better understand the physics of the blue whirl, are also presented.

2. Experimental Approach

Experiments were conducted using a setup similar to that of Xiao et al. [1], depicted in Figure 1. A 40 mm diameter steel pan was filled water, and liquid n-heptane was fed continuously using a syringe pump to the water surface through a small copper tube. Flow rates ranged from 0.5 to 10.0 ml/min. Two quartz half-cylinders with a diameter of 310 mm and a height of 610 mm were suspended directly over the water surface. By offsetting them with gaps of a particular width (15, 25, 35, 45 and 55 mm in this study), swirl was naturally induced in the test section following ignition of the fuel. High speed videos were captured using a SONY RX10II at *f*/5.6 with a frame rate of 60 fps at multiple ISO values. A DANTEC 55P16 hot-wire probe with a 54T42 Mini CTA anemometer were positioned at the center of the inlet gap to record tangential velocities. The vertical position of the probe was controlled using a robotic traverse mechanism.

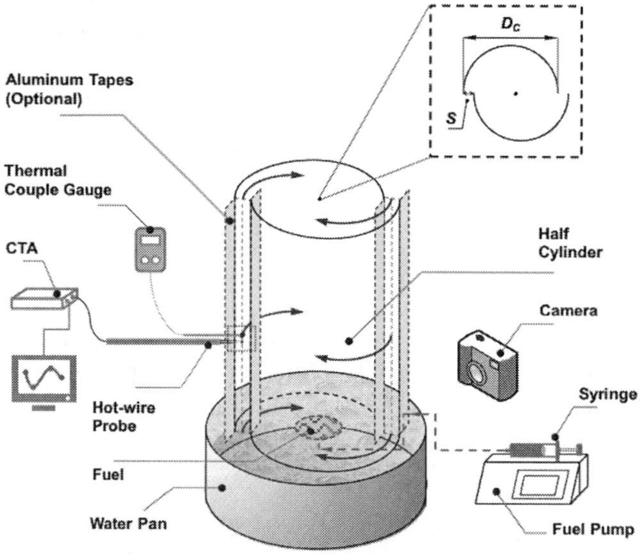

Figure 1. Schematic of the experimental setup.

Air entrainment into the enclosure for a given gap-size is controlled only by the buoyancy created by the flames within the enclosure. As a result of this naturally entraining configuration, the inlet velocity was observed to adjust itself for a given fuel flow rate. Although a steady-state process was not observed for all combinations, the fire whirl height was quite repeatable. Combinations which resulted in constantly varying fire whirl heights are termed "transition states," and are discussed later. The velocity at the inlet due to air entrainment, U, achieved a fixed value for most combinations of gap size (S) and fuel supply rate ($\dot{V}$). Results are presented in terms of inlet circulation, $\Gamma=\pi U D_c$, where D_c is the fixed diameter of the setup, and the heat-release rate of the fire $\dot{Q}$, estimated assuming complete combustion of the fuel on the surface, $\dot{Q} = \dot{V}\Delta H_c \rho_f$, where

ΔH_c is the heat of combustion, and ρ_f, the density of n-heptane. Therefore, by changing S and $\dot{V}$, various combination of Γ and $\dot{Q}$ are achieved, resulting in corresponding flame patterns. These are marked out as regions on graphs of both S vs. $\dot{V}$ and Γ vs. $\dot{Q}$, revealing the conditions for formation of the various swirling flames, including the blue whirl.

3. Results and Discussion

3.1 Flame Patterns

Figure 2 shows the typical flame patterns observed as $\dot{V}$ is increased for a fixed *S*. The blue whirl (Fig. 2 *A* and *B*) forms at small values of $\dot{V}$. It burns stably and completely blue, indicating soot-free burning. The blue whirl is small, usually ~4 cm in diameter and between 5 and 10 cm in height. The size of the blue whirl grows slightly as $\dot{V}$ is increased, however for all *S* studied in this apparatus, a blue whirl was not seen to form for $\dot{V}$ beyond 1.1 ml/min. Beyond this threshold (up to ~ 2-3 ml/min, dependent on *S*), transitional fire whirls, including the conical fire whirl, repeatedly transition between the fire whirl regime and blue whirl regime. This is depicted in Figs. 2 *C*, *D* and *E*, where the flame transitions back and forth between the blue whirl and a slender, tall yellow flame. In the transition from the tall slender flame to the blue whirl, movement of remnant soot within the conical region of the blue whirl resembles recirculating patterns that are characteristic of bubble type vortex breakdown [4]. As $\dot{V}$ approaches 1.1 mL/min the frequency of these transitions increases, until a stable blue whirl is sustained.

From Fig. 2, (*C*) to (*E*), the average flame diameter grows slightly while its height rapidly increases and decreases during the transitions. As the fuel supply rate is further increased, this transitional state evolves into a stable fire whirl (Figs. 2 *F*, *G* and *H*). The critical fuel supply rate for this transition is strongly dependent on *S*. These stable fire whirls have a much larger diameter than the blue and transitional blue whirl. As $\dot{V}$ increases, fire whirls as high as 60 cm are observed, roughly twice that of the transitional fire whirls. The dimensions of a stable fire whirl grow with $\dot{V}$, however, they are quite stable for a particular $\dot{V}$. When $\dot{V}$ increases high enough (~6 - 8 ml/min; dependent on *S*), a stable fire whirl transitions to, what we call in this paper, a large fire whirl (I). This process begins as a stable fire whirl and keeps growing in dimensions steadily, and upon reaching a limit, "bursts" into a large fire whirl. The flame here is much larger than the slender, stable fire whirl and fills most of the enclosure. This state is only maintained for few seconds, during which it consumes most of the fuel over the water surface, resulting in a transition back to a medium-size fire whirl, sometimes even a blue whirl depending on *S*. After that, a new cycle of lower consumption rate and high fuel supply rate starts over again, resulting in fuel accumulation over the water surface, slowly causing the size of the fire whirl to increase steadily. A stable large fire whirl was not observed, which may be attributed to the relative size of the fire whirl to the size of the enclosure. $\dot{V}$ was not increased beyond 10 ml/min in light of safety concerns. A larger apparatus would allow tests with a higher $\dot{V}$ to be conducted.

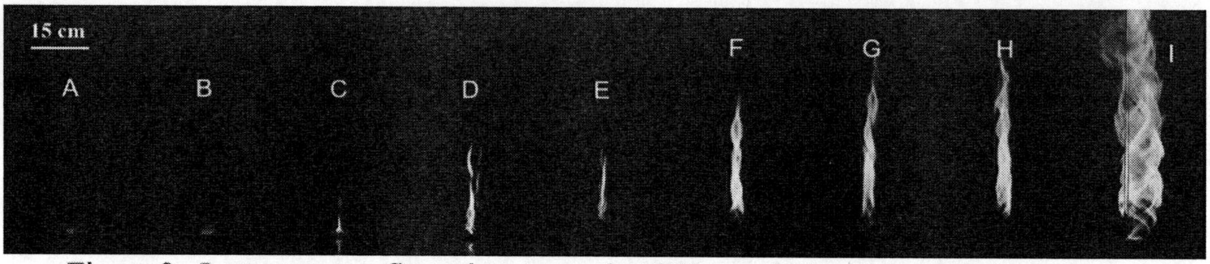

Figure 2 . Instantaneous flame images under different fuel supply rates for S=25mm.
(From (A) to (I), $\dot{V}$=0.5, 1.1, 1.5, 2.0, 3.0, 4.5, 6.0, 8.0 and 10.0 ml/min)

3.2 Specific Formation Conditions

Figure 3 distinguishes specific regions for the previously described fire whirl patterns, as regions on a graph of S vs. $\dot{V}$, thus indicating the conditions required to form the corresponding fire whirl patterns. A stable blue whirl is seen to form only in a narrow region close to the extinction limit. Additionally, it was found in experiments that the blue whirl was vulnerable to small external acoustic perturbations as it approached its extinction or transition limit.

In order to reduce the effect of any stray air movements outside the setup on the blue whirl, strips of aluminum tape were attached parallel to the cylinders. In addition, a small gap between the lower end of the quartz cylinders and the water surface was added to improve blue whirl stability.

Figure 3 shows the effect of adding the inlet tape extensions. Comparison of these graphs reveals that the range over which the blue whirl is formed can be extended merely by conditioning the inlet flow. The blue whirl extends further into the extinction region when the strips of tape are added, however, these measures have minimal effect on the other fire whirl regimes. These results provide quantitative information on the increased stability as a result of flow conditioning. However, the physical mechanisms are not yet fully apparent. It is thought that, by slightly lifting the cylinders up off the water surface, a small co-flow is introduced along the bottom edges. The additional strips of tape may lead to some flow conditioning at the inlet. As evident from Figure 3, these two effects widen the range of conditions that result in blue whirl formation.

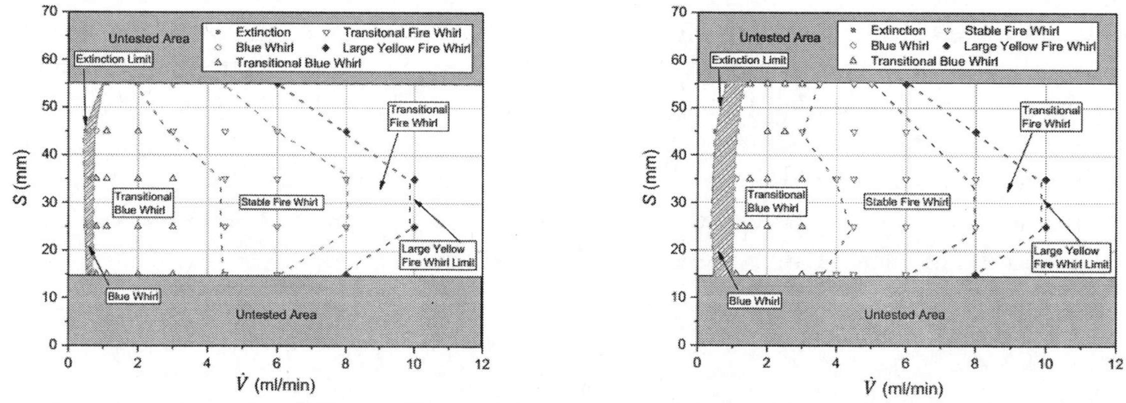

Figure 3. S-$\dot{V}$ regions for different fire whirl patterns, without (left) and with (right) inlet tapes.

The flame regions are also shown in a graph of Γ vs. $\dot{Q}$ in Figure 4, for experiments performed without tape. The heat-release rate, $\dot{Q}$ associated with the whirls was calculated from the volumetric fuel supply rate $\dot{V}$ (assuming complete combustion). The circulation (Γ) was calculated using velocity measurements at the inlet of the gap and the radius of the enclosure. The graph shows that the blue whirl forms in a narrow region near the extinction limit. This reflects that the required Γ vs. $\dot{Q}$ combinations are limited and, with an apparatus of a fixed radius, will only form within a specific regime.

A comparison similar to Figure 3 with strips of tape at the inlet are still in process; however, the authors speculate the difference in Γ vs. $\dot{Q}$ plots would not be as significant as in S *vs.* $\dot{V}$ plots because a relationship between Γ, S and $\dot{Q}$, based on the dimensional analysis as followed, indicates just that.

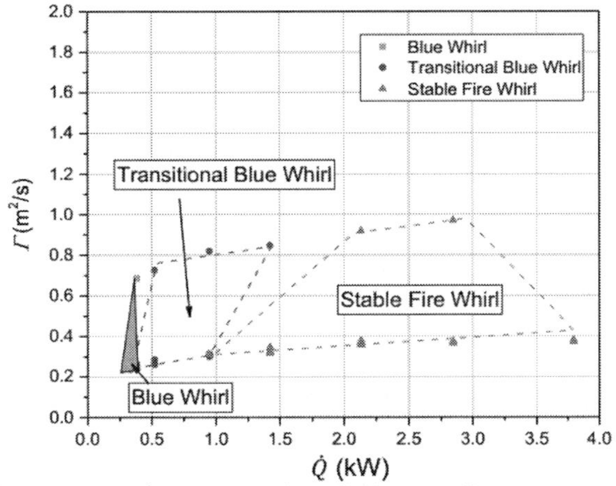

Figure 4. Γ-$\dot{Q}$ regions for different flame pattern.

3.3 Potential Scaling Approach

Following the work of Tohidi et al. [5] and Hartl and Smits [6], a dimensional analysis can be performed on the governing parameters of this fixed-frame fire whirl apparatus with a focus on the circulation,

$$\Gamma = f(\dot{Q}, S, D_c, T_0, \rho_0, \Delta T, \Delta \rho, C_{p,0}, g) \quad (1)$$

Eq. (1) includes a variety of assumptions including that (1) circulation is independent of height (as has been observed experimentally by Hartl and Smits [6]), (2) viscosity can be neglected compared to the inertial and buoyant forces, (3) heat conduction and mass diffusion terms are negligibly small and (4) combustion is infinitely fast and in steady state. By choosing the four parameters as basic physical quantities, including the dimeter of cylinders D_c, the acceleration to gravity g, the ambient air temperature T_0 and the density ρ_0, the dimensional analysis then yields

$$\Gamma^* = f'(\dot{Q}^*, S^*, \frac{\Delta T}{T_0}, \frac{\Delta \rho}{\rho_0}, \frac{g D_c}{C_{p,0} \Delta T}) \quad (2)$$

where $\Gamma^* = \Gamma/(D_c\sqrt{gD_c}) = \pi U_\theta/\sqrt{gD_c}$, which actually denotes the Froude number Fr_θ, based on the azimuthal velocity component. The authors use U_θ in Froude number instead of the axial velocity component (applied by Hartl and Smits [6]) because the ratio of azimuthal momentum to axial momentum has significant impact on vortex breakdown, which is thought to have a strong role in the formation of the blue whirl. $\dot{Q}^* = \dot{Q}/(C_{p,0}\Delta T \rho_0 {D_c}^2\sqrt{gD_c})$ represents the ratio of fire heat-release rate to the enthalpy rate and $S^* = S/D_c$ is defined as the non-dimensional gap size between cylinders. Following Kuwana et al. [7], $\Delta T/T_0$ and $\Delta\rho/\rho_0$ are assumed to be nearly constant in ordinary fires. Finally, $gD_c/C_{p,0}\Delta T$ indicates a ratio of potential energy to thermal energy which is small enough to be neglected here. Hence,

$$Fr_\theta = \Gamma^* = \Phi(\dot{Q}^*, S^*) \tag{3}$$

Based on Eq. (), and assuming complete combustion, ($\dot{V} \sim \dot{Q}$), it can be shown that

$$\Gamma^* = \psi(\dot{V}^*,\ S^*) \tag{4}$$

where $\dot{V}^* = \dot{V}/{D_c}^2\sqrt{gD_c}$, is the non-dimensional fuel supply rate. Obviously, Eq. (4) implies that a single Γ could correspond to multiple combinations of $\dot{V}$ and S and, as a result, a narrower range might be obtained as plots were projected from the graphic of S vs. $\dot{V}$ to that of Γ vs. $\dot{Q}$.

Further, consider a mass balance for a control volume that encircles the flame. By defining the mass of fuel entering as $\dot{m}_{fuel} = \dot{V} \cdot \rho_{fuel}$ and the mass of air required for stoichiometric combustion as $\dot{m}_{air} = U_\theta \rho_0 S L_C$ and $\dot{m}_{fuel} \sim \dot{m}_{air}$, where L_C is the height of the cylinders, and $U_\theta = \Gamma/D_c$ a linear dependence $\Gamma^* \sim \dot{V}^*/\ S^*$ is obtained. Similarly, if we define $\dot{m}_{fuel} = \dot{Q}/\Delta H_L$, where ΔH_L is the lower heating value of the fuel, a similar linear relationship $\Gamma^* \sim \dot{Q}^*/\ S^*$ is obtained. Both of these relationships could potentially serve as a basis for scaling the fire whirl regimes discussed here.

4. Conclusions

Experiments were conducted in a fix-framed, self-entraining fire whirl setup to investigate the conditions required for formation of the blue whirl. A number of stable and transitional fire whirl patterns were observed, including the blue whirl, transitional blue whirl, stable fire whirl, transitional fire whirl and large fire whirl. The flames varied widely in dimension, appearance and stability. These regimes were seen to occupy distinct regions on graphs of $\dot{V}$ vs. S, and Γ vs. $\dot{Q}$. The plots reveal that the blue whirl forms in a narrow range close to the extinction limit. Some measures, namely lifting the cylinders slightly and adding tape at the leading edge of the inlet were found to help stabilize the blue whirl and widen the limits of stable formation. Dimensional analysis indicates a non-dimensional relationship, $Fr_\theta = \Gamma^* = \Phi(\dot{Q}^*, S^*)$ where Γ^* is the Froude Number Fr_θ, based on the azimuthal velocity component. Two linear dependencies have also been found, $\Gamma^* \sim \dot{V}^*/\ S^*$ and $\Gamma^* \sim \dot{Q}^*/\ S^*$, both of which can potentially be applied for scaling of the blue whirl in the future.

5. Acknowledgements

This work was supported by NSF award CBET-1507623 and by the University of Maryland through the Minta Martin Endowment Funds in the Department of Aerospace Engineering and the Glenn L. Martin Institute Chaired Professorship at the A. James Clark School of Engineering. MJG acknowledges partial support from NSF award CBET-1554026. YH acknowledges support from the China Scholarship Council (CSC, No. 201606210356). Any opinions, findings and conclusions or recommendations expressed in this material are those of the authors and do not necessarily reflect the view of the NSF or CSC.

6. References

[1] H. Xiao, M.J. Gollner and E.S. Oran, From fire whirls to blue whirls and combustion with reduced pollution. Proceedings of the National Academy of Sciences, 113 (2016) 9457-9462.

[2] S.B. Hariharan, The Structure of the Blue Whirl, 70th Annual Meeting of the APS Division of Fluid Dynamics (2017), Abstract D35.00003.

[3] J. Lei, et al., Experimental investigation on flame patterns of buoyant diffusion flame in a large range of imposed circulations. Proceedings of the Combustion Institute, 36 (2017) 3149-3156.

[4] J.H. Faler and S. Leibovich, An experimental map of the internal structure of a vortex breakdown. Journal of Fluid Mechanics, 86(1978) 313-335.

[5] A. Tohidi, M.J. Gollner and H. Xiao, Fire Whirls. Annual Review of Fluid Mechanics, 50 (2018) 1, 187-213.

[6] K.A. Hartl and A.J. Smits, Scaling of a small scale burner fire whirl. Combustion and Flame, 163 (2016) 163 202-208.

[7] K. Kuwana, et al., Scaling fire whirls. Fire Safety Journal, 43 (2008) 252-257.

Spring Technical Meeting
Eastern States Section of the Combustion Institute
March 4-7, 2018
State College, Pennsylvania

Effects of Ullage on Combustion Efficiency and Plume Entrainment of Pit Fires

Veronica Kimmerly[1], Ali S. Rangwala[1]*

[1]Department of Fire Protection Engineering, Worcester Polytechnic Institute, Worcester, MA 01609, USA
**Corresponding Author Email: vmkimmerly@wpi.edu*

Abstract: Experiments have been conducted to study the effect of ullage height on air entrainment in buoyant diffusion flames in a cavity. The experimental apparatus is similar to that used by Zukoski et al. [1], however, with the capability of lowering the height of the gas burner to vary ullage. Ullage heights have been varied from 0 to 2D where D is the diameter of the burner. Experiments show that the ullage height alters the rate of air entrainment in the combusting part of the fire plume which alters the plume entrainment behavior in general. The entrainment rates are also influenced by the boundary condition of the wall of the cavity – an adiabatic wall is used in the current experiments. After a critical ullage height is reached, experimental data does not match classical correlations of air entrainment in turbulent buoyant plumes and a dimensional scaling analysis is presented to incorporate the influence of ullage in these cases. The implications of the results towards use of fire pits or burn pits as a means of waste disposal in remote areas is espoused.
Keywords: Entrainment, Pit Fire, Plume Correlation, Ullage

1. Introduction

In developing countries domestic waste incineration is often done in an open burn pit fire. Open burning of domestic waste has a significant impact on global emissions of air pollutants [1] and respiratory health of people living nearby [2]. It is unlikely that the use of open burn trash fires can be eliminated, but they can be made less hazardous to people and the environment with simple changes. Finally, developments in thermoelectric technology lead to the possibility of generating electricity from waste heat on a small scale [3-4]. Previous work has shown that increasing the ullage (freeboard height between top of the pit and top of the fuel) past a critical value leads to improved combustion, as measured by burning rate, for liquid pool fires [5]. It has been suggested that changes in air entrainment lead to this increase [6]. The effect of ullage on combustion efficiency offers a way to make these open pit trash fires less hazardous with minimal effort by simply optimizing the geometry of the pits. This paper seeks to further quantify the altered air entrainment in pit fires. Plume entrainment correlations have been experimentally developed for fires where the fuel surface is flush with the "ground" [7-8] and for fires located at the bottom of rack storage [9] but no work has been done on a pit geometry where the fuel is located below the "ground". In previous work an entrainment coefficient E based on total heat release rate has been identified in the theoretical equation for the mass of air entrained in strong plumes with self-preserving density profiles. This paper takes Zukoski's original setup for determining E but alters it so that the ullage of the pit can be varied.

2. Model and Experimental Setup

Fuel mass flow rate is controlled and the heat release rate ($\dot{Q}$) is determined from this via a known heat of combustion (ΔH_C) and an assumed combustion efficiency of $\chi = 0.7$.

$$\dot{Q} = \chi \dot{m}_{fuel} \Delta H_C \qquad (1)$$

Equations 2 and 3 show Heskestad's equation for plume entrainment ($\dot{m}_{ent}$) with a virtual origin (z_0) is shown for a convective heat release rate of $\dot{Q}_C = 0.8\,\dot{Q}$ [8], where z is height of the hot gas layer above the fuel. Assumed values are $\rho_\infty = 1.2\ kg/m^3$, $C_p = 1.0\ kJ/kgK$, $T_\infty = 293\ K$.

$$z_0 = -1.02D + 0.083\dot{Q}^{\frac{2}{5}} \qquad (2)$$

$$\dot{m}_{ent} = E\left(\frac{g\rho_\infty^2}{C_p T_\infty}\right)^{\frac{1}{3}} \dot{Q}_C^{\frac{1}{3}} (z - z_0)^{\frac{5}{3}} \qquad (3)$$

The actual mass of air entrained into the plume is found from the mass extraction rate ($\dot{m}_{ext}$) needed to maintain a steady state height of the hot gas layer. The equation for the extracted rate is shown in Equations 4 and 5 for a known extraction velocity (V_{ext}) and an extraction pipe radius $r_{ext} = 0.0565\ m$. The extracted hot gas density is determined with the ideal gas law assuming the gas is dry air at a known temperature (T_{ext}). Assumed values are $p = 1\ atm$ and $R = 287.058\ J/kgK$.

$$\rho_{ext} = {}^{p}/_{RT_{ext}} \qquad (4)$$

$$\dot{m}_{ext} = \rho_{ext} V_{ext} \pi r_{ext}^2 \qquad (5)$$

Then the parameter $\mathcal{M} = \dot{m}_{ext}/\dot{m}_{ent}$ is used and when it is approximately unity then the value of E is valid. $\mathcal{M}$ is plotted against the non-dimensionalized ullage (U/D) to demonstrate that the new E is valid across a range of ullages.

The burner (Figure 1) is square steel pan ($D = 10\ cm$) with a fuel supply pipe entering through the bottom of the pan. A gaseous fuel was chosen so that the ullage would remain constant throughout the burning period and lab grade methane gas was selected. A deflector plate and deflector mesh is located over the opening. Then glass beads (diameter 1.27 cm) are located above the mesh to finish dispersing the methane gas.

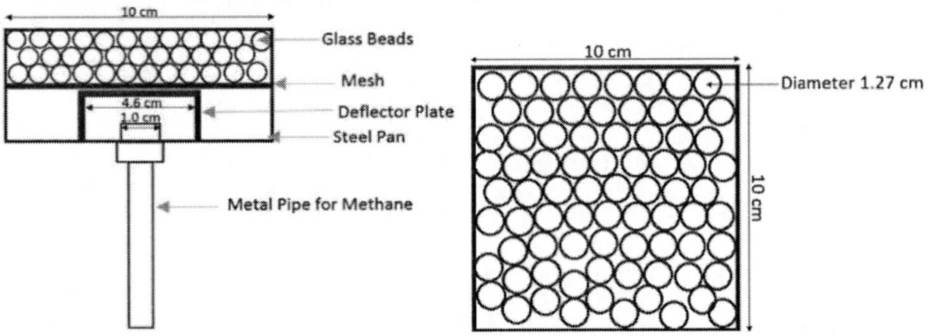

Figure 1: (Left) Cross Section of Square Glass Bead Burner and (Right) Top View

The "pit" (Figure 2) is constructed of 4 pieces of 1.5875 cm thick borosilicate glass and the corners are sealed with high temperature silicone sealant (RTV106 or FB136). Ullage is set by lifting the burner on a car jack at ullages of 0 to $2D$. The glass is insulated with 0.635 cm thick mineral wool on three sides; the front is left uninsulated for visualization purposes. The "pit" rests on 0.635 cm thick sheetrock with an artificial ground is constructed around the top of the pit of 0.635 cm thick sheetrock. The nearest edge is at least D distance from the pit entrance so that uniform air flow exists before entering the "pit". The "pit" sits below an extraction hood with an adjustable extraction rate. The exhaust pump controller is calibrated with a vane anemometer and ambient air to get V_{ext} for a given pump setting. Wire mesh surrounds the top of pit to reduce ambient air disturbances. Sixteen K-type 32 gauge chrome-aluminum thermocouples are arrayed vertically in the hood with a 1 cm spacing down from the angled section and are used to determine the location of the hot gas layer within the hood. There is also a K-type 32 gauge chrome-aluminum thermocouple located in the exhaust piping to determine T_{ext}.

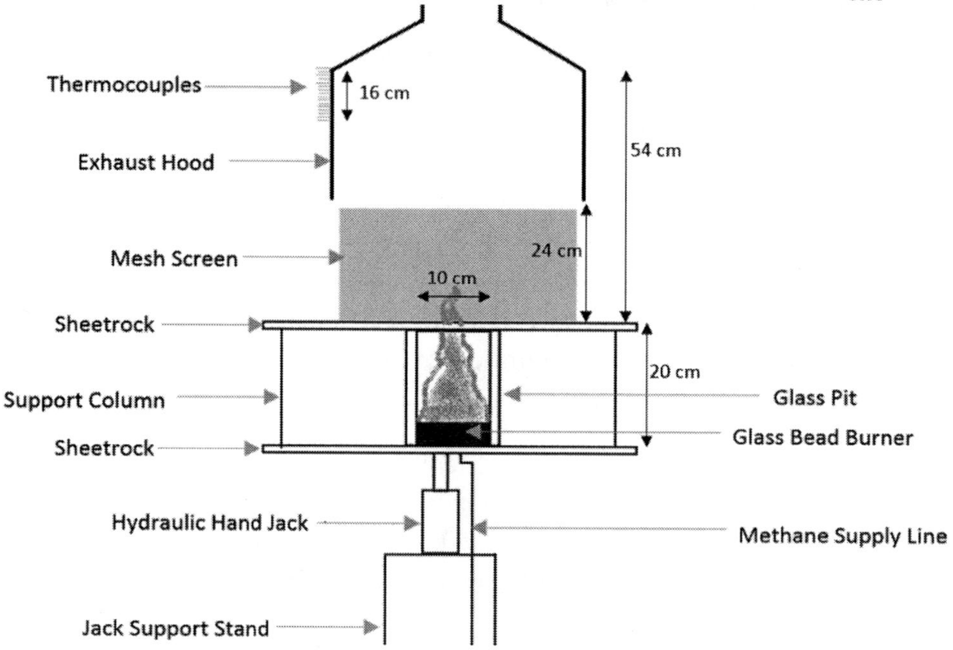

Figure 2: Pit Set Up

The exhaust gases are run through a gas analyzer to determine volume percent of oxygen, carbon dioxide, and carbon monoxide. The gas analyzer has a response delay of 47 seconds so a delay of at least 1 minute is used before data collection begins. $\dot{m}_{fuel}$ is controlled with a *SIERRA* mass flow controller and $\dot{m}_{fuel} = 4\ g/min$ methane was used for all runs. The experimental procedure is:

1. Ignite fuel with a pilot flame
2. Set ullage (U)
3. Wait one minute to allow the flame and gas analyzer to stabilize
4. Begin recording thermocouple and gas analyzer data for at least two minutes
5. Decrease U by 1 cm
6. Repeat steps 2 to 5 until each ullage is tested at least twice.

From the hood thermocouple data the location of the stable hot gas layer is determined for each run. The temperature of the exhaust gas is recorded and the average value is T_{ext}. The gas analyzer data is used to determine combustion efficiency by the carbon monoxide/carbon dioxide ratio (CO/CO_2).

3. Results and Discussion

In Zukoski's 1981 paper on determining the entrainment coefficient, they used a similar set up to the one discussed above but instead of a pit with changing ullage the whole "ground" level was lowered or raised to vary the height of the hot gas layer. Porous bed glass bead burners of 10 to 50 cm were used with methane gas measured through a laminar flow meter. The fire was located beneath a hood whose extraction rate could be controlled and measured. The location of the hot gas layer was found using vertical thermocouple arrays in the hood. Then Zukoski used the ratio of the measured mass flow rate ($\dot{m}_p$) to the theoretical mass entrainment rate ($\dot{m}_t$) for a chosen value of E ($M = \dot{m}_p/\dot{m}_t$) to determine if the entrainment coefficient was valid. The scatter in M across z values and burner diameters was $\pm 30\%$ difference from $M = 1$.

In this experimental setup, the location of the hot gas layer within the hood was found by temperature profile over the 16 thermocouples in the hood. The temperature results for $U = 8\ cm$ are shown in Figure 3. The temperature increases with increasing height until 49 cm where it jumps and then remains relatively constant at increasing heights. Therefore the bottom of the stable hot gas layer exists 49 cm above the top of the pit ($z = 49 + 8 = 57\ cm$). This procedure is repeated for each test run.

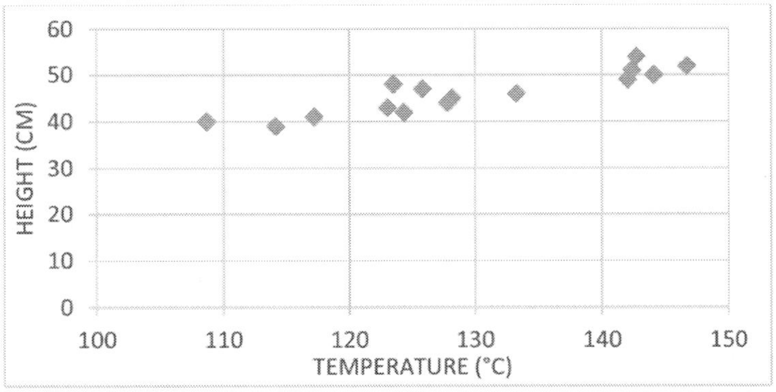

Figure 3: Temperature profile in the hood for $U = 8\ cm$

Heskestad showed that the virtual origin equation was valid for a fire in storage racks [9], and experimental results are compared with a virtual origin and without the virtual origin (done by setting $z_0 = 0$ in Equation 3). When Zukoski's entrainment coefficient of $E = 0.21$ is used it is seen in Figure 4 that $\mathcal{M}$ is less than unity for all the tested ullages with or without the virtual origin for a non-dimensionalized ullage U/D. Therefore $E = 0.21$ is not an appropriate assumption for a pit geometry.

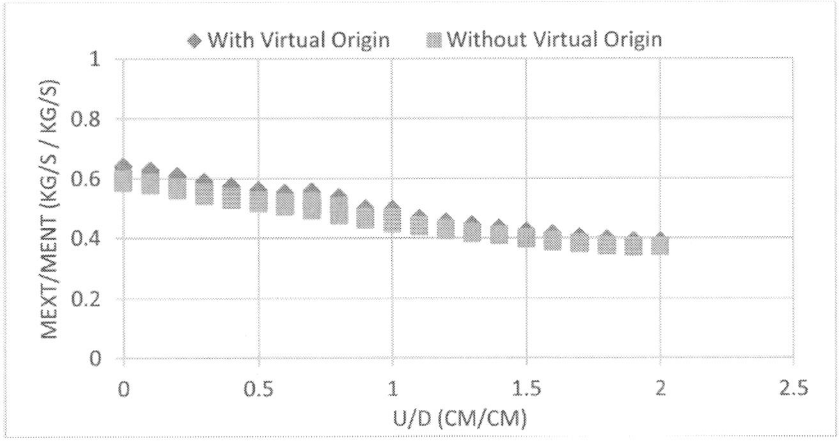

Figure 4: $\mathcal{M}$ with and without virtual origin for E = 0.21 versus U/D

The appropriate E for this geometry was found by varying E until smallest difference from unity in $\mathcal{M}$ was achieved. In this process it was found that at $U/D = 0.8$ there is a transition in the entrainment coefficient. Thus it was experimentally determined that $E = 0.12$ for $U/D \leq 0.8$ and $E = 0.092$ for $U/D > 0.8$. The scatter in $\mathcal{M}$ for the new E values is $\pm 15\%$ difference from unity and this is better than the scatter found by Zukoski as discussed previously.

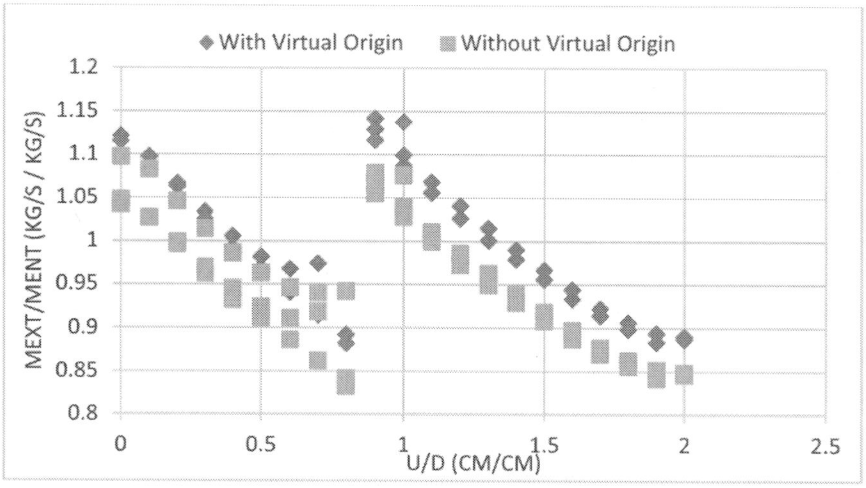

Figure 6: $\mathcal{M}$ with and without virtual origin and E = 0.092, 0.12 for U/D

The combustion efficiency was determined from the CO/CO_2 ratio and is shown for non-dimensionalized ullage. The results from these experiments shown that increasing ullage decreases the combustion efficiency as seen by the increasing CO/CO_2 ratio and that this trend is fairly linear as shown by the high R^2 value of 0.799. This contradicts the previous work of Shi et al. but it could be down to a difference in fuel type since a gaseous fuel was used here and the previous work used a liquid fuel.

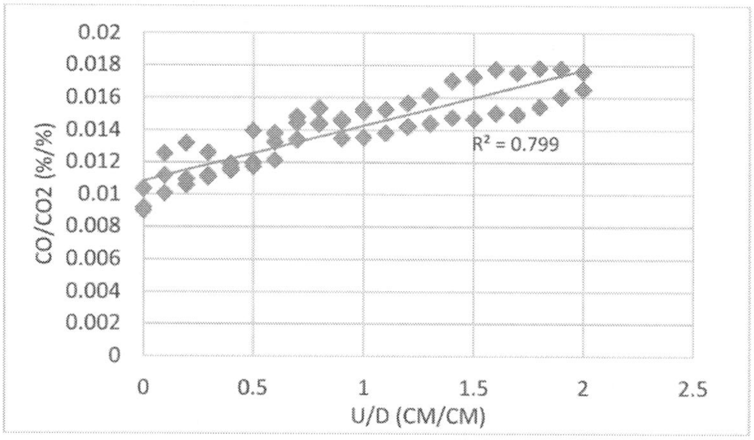

Figure 7: Combustion Efficiency for U/D

4. Conclusions and Future Work

A new entrainment coefficient has been found for a gas fire in a pit geometry and agrees within $\pm 15\%$ of experimental results proving the new E valid for predicting plume entrainment. The virtual origin shown to hold for this new geometry. Increasing ullage is shown to decrease the CO/CO_2 ratio which contradicts previous work. Further tests will be run with this experimental setup varying the burner diameter, type of fuel, and the boundary conditions of the pit. In particular testing liquid and solid fuels is important because the type of domestic waste fires that motivate this work are often composed of liquids and solids of varying chemical compositions. In addition, modelling will be developed, likely using Fire Dynamics Simulator, to visualize the altered air entrainment profiles.

5. Acknowledgements

This work was funded by the Presidential Fellowship from Worcester Polytechnic Institute.

6. References

[1] C. Wiedinmyer, R.J. Yokelson, B.K. Gullett, Global Emissions of Trace Gases, Particulate Matter, and Hazardous Air Pollutants from Open Burning of Domestic Waste, Environ. Sci. Technol. 48 (2014) 9523–9350.

[2] J. Aurell, B.K. Gullett, D. Yamamoto, Emissions from Open Burning of Simulated Military Waste from Forward Operating Bases, Environ. Sci. Technol. 46 (2012) 11004–11012.

[3] R.Y. Nuwayhid, D.M. Rowe, G. Min, Low cost stove-top thermoelectric generator for regions with unreliable electricity supply, Renew. Energy. 28 (2003) 205–222.

[4] D. Champier, J.P. Bedecarrats, M. Rivaletto, F. Strub, Thermoelectric power generation from biomass cook stoves, Energy. 35 (2010) 935–942.

[5] B.Z. Dlugogorski, M. Wilson, Effect of Ullage on Properties of Small-Scale Pool Fires, Dev. Chem. Eng. Miner. Process. 8 (2000) 149–166.

[6] X. Shi, A.K. Sahu, S. Nair, V. Raghavan, A.S. Rangwala, Effect of ullage on burning behavior of small-scale pool fires in a cavity, Proc. Combust. Inst. 36 (2017) 3113–3120.

[7] E.E. Zukoski, T. Kubota, B. Cetegen, Entrainment in fire plumes, Fire Saf. J. 3 (1981) 107–121.

[8] G. Heskestad, Fire Plumes, Flame Height, and Air Entrainment, in: SFPE Handb. Fire Prot. Eng., Springer New York, New York, NY, 2016: pp. 396–428.

[9] G. Heskestad, Flame Heights Of Fuel Arrays With Combustion In Depth, Fire Saf. Sci. 5 (1997) 427–438.

Spring Technical Meeting
Eastern States Section of the Combustion Institute
March 4-7, 2018
State College, Pennsylvania

Role of Low-Temperature Chemistry in Detonation of *n*-Heptane/Oxygen/Diluent Mixtures

Wenkai Liang[1], Rémy Mével[2, 3] and Chung K. Law [1, 2]

*[1]Department of Mechanical and Aerospace Engineering,
Princeton University, Princeton, NJ 08544
[2]Center for Combustion Energy,
Tsinghua University, Beijing 100084, China
[3]Department of Automotive Engineering,
Tsinghua University, Beijing 100084, China*

Abstract: Numerical simulation of the ZND detonation structure of mixtures of *n*-heptane, oxygen and diluent detonations was performed with detailed chemistry. Results show that low-temperature chemistry is activated behind high Mach number shock waves with increasing dilutions of CO_2, which leads to non-monotonous thermicity profiles with two stages of energy release caused by low- and high-temperature chemistries, respectively. For the evolution of the induction length and cell size, although the total length scales do not show negative response with increasing temperature, the length scale of the first-stage ignition demonstrates negative response when the post-shock temperature decreases within the NTC (negative temperature coefficient) regime.

Keywords: n-Heptane, Low Temperature Chemistry, Induction Length, Detonation Cell Size

1. Introduction

At low ($T<650$ K) to intermediate ($650 \text{ K}<T<1000$ K) temperatures, the oxidation of large hydrocarbon fuels is dominated by low-temperature chemistry (LTC) [1-3], which is frequently related to the phenomena of cool flames and the negative temperature coefficient (NTC), and is considered to be a key factor in engine knock [4]. Consequently, extensive efforts have been made to investigate LTC-affected phenomena in various combustion systems [3-8]. In particular, cool flames with oscillatory volumetric heat release and chemiluminescence have long been observed in homogeneous systems such as those in heated closed vessels [5] and jet-stirred reactors [3]. In non-homogeneous systems, the coupling between the LTC and the transport processes involved in either non-premixed or premixed systems have been investigated in counterflow flames [6], droplet burning [7] as well as flames in microchannels [8].

Studies on LTC-affected detonation waves, however, have been relatively less. The observations and the associated phenomena, while mostly global in nature, are nevertheless fundamentally significant and practically relevant. As notable examples, at the fundamental level, Romano *et al.* [9, 10] investigated the effects of the partial oxidation of cool flame on the *n*-pentane-oxygen detonation cell size [9] and deflagration-to-detonation transition (DDT) run-up distance [10], showing that when the mixture undergoes LTC reactions, there is a significant reduction of the cell size as well as the DDT run-up distance. Furthermore, it has also been shown that the detonation structure is highly sensitive to chemistry and double cellular structure has been correlated with non-monotonous energy release in a number of mixtures including $CH_3NO_2(-O_2)$ [11] and fuel-NO_2/N_2O_4 (fuel: H_2, CH_4,C_2H_6) [12, 13]. At the industrial safety level, the dynamic parameters of detonations (detonation cell size, direct initiation energy and critical tube diameter) [14], which are used for detonation risk assessment through empirical or semi-empirical models, require knowledge of the detailed structure of the chemical reaction rates, and hence can be strongly influenced by LTC. At the combustor level, super-knock [15] in spark-ignited (SI) engines is a challenge for low-speed and high-load operating conditions, and is attributed to the developing detonation resulting from pressure waves by end-gas autoignition. The LTC effect on the oxidation of large hydrocarbon fuels will fundamentally affect the autoignition kinetics and the subsequent detonation development. Hence, it is important to understand the LTC-affected detonation to help overcome the obstacle of super-knock.

Motivated by the above considerations, the present study is our first step towards attaining a comprehensive understanding of the dynamics and structure of LTC-affected detonations. In particular, we shall study herein the effects of LTC on the ZND structure of detonations using steady 1D numerical simulation, and to analyze the chemical kinetic effect on the energy release dynamics and on the detonation cell size.

2. ZND Structure

As a typical primary reference fuel (PRF) and a well-studied large hydrocarbon fuel with LTC, *n*-heptane is selected as the fuel and a skeletal mechanism consisting of 88 species and 387 reactions [16] is used in the simulation. To identify the conditions at which LTC would be activated after the leading shock, the ZND structure of detonation in *n*-heptane/oxygen mixtures with different dilutions of N_2 and CO_2 and at various initial pressures are simulated.

Figure 1 shows the pressure and temperature at the Neumann state as a function of the dilution mole fraction. It is seen that both pressure and temperature decrease as the dilution mole fraction increases, and hence the mixture at the Neumann state could enter the thermodynamic regime affected by the LTC. Moreover, as the initial pressure, P_1, increases from 10 kPa to 100 kPa, although the response of temperature does not change much for high dilution conditions, the pressure after the leading shock significantly increases. As the NTC regime shifts to higher temperature conditions [6] when the pressure is increased, it is expected that the LTC effect would be observed at lower dilution for higher initial pressures. Furthermore, by comparing the results for two different diluents, N_2 and CO_2, it is found that CO_2 is more effective as an inert to reduce the temperature at the Neumann state due to its higher heat capacity. For the case with N_2 dilution, even with X_{N2} = 90%, the temperature is still as high as 1200 K, whereas for the case of CO_2 dilution, the temperature can be reduced to lower than 1000 K and hence enters into the LTC-affected regime.

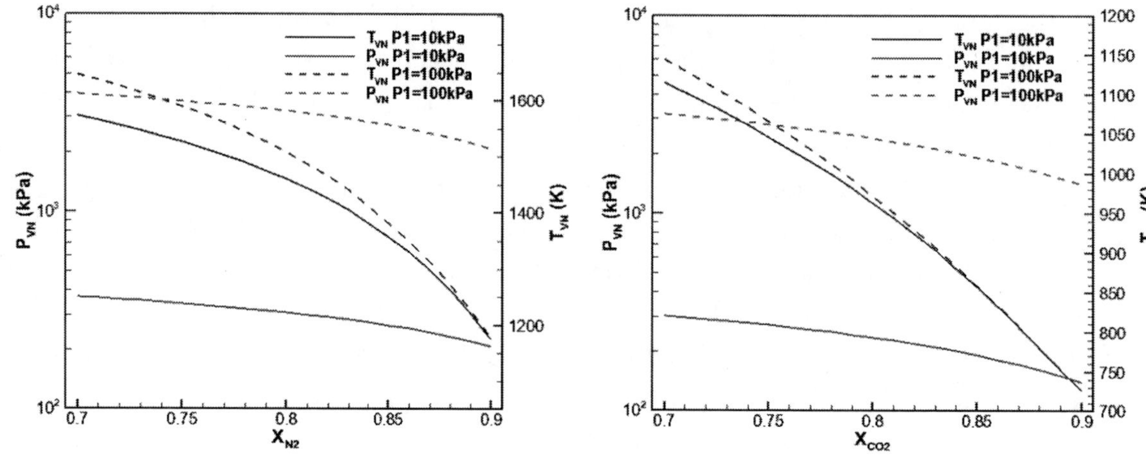

Figure 1: Pressure and temperature of *n*-heptane/oxygen mixtures with dilutions of N_2 and CO_2 at the Neumann state.

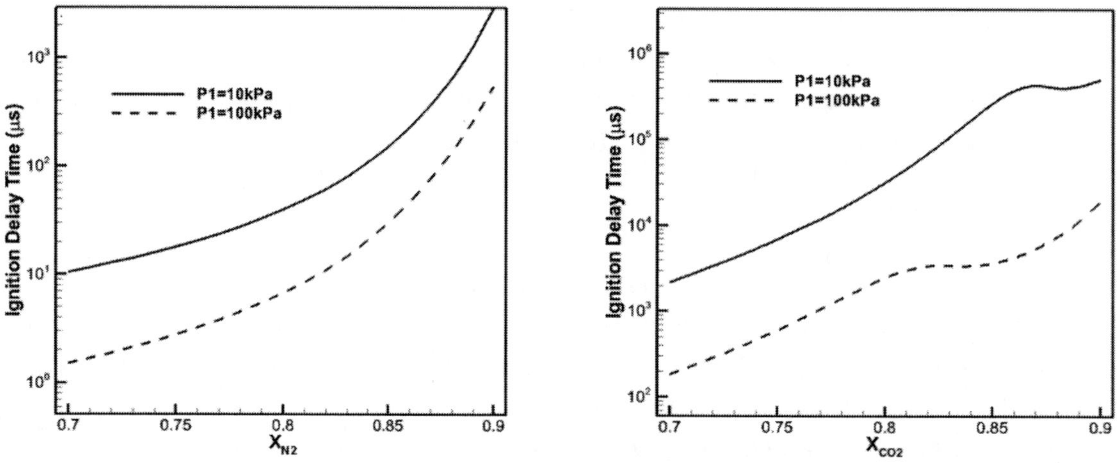

Figure 2: Adiabatic constant pressure ignition delay-time of *n*-heptane/oxygen mixtures with dilutions of N_2 and CO_2 at the Neumann state.

To further validate the conditions in which LTC would have significant effect on the detonation structure, the ignition delay times for the mixtures at the Neumann state were calculated and shown in Fig. 2. The results indicate that when the diluent is N_2, the ignition delay increases monotonically with dilution. However, for the CO_2 diluted cases, the ignition delay exhibits the negative dependence on the dilution mole fraction for $X_{CO2}>82\%$ at $P_1=100$ kPa and for $X_{CO2}>87\%$ at $P_1=10$ kPa. Moreover, as both the Neumann state pressure and temperature decrease with increasing dilution as shown in Fig. 1, the slope of the NTC regime is less steep compared with autoignition of constant pressure or volume.

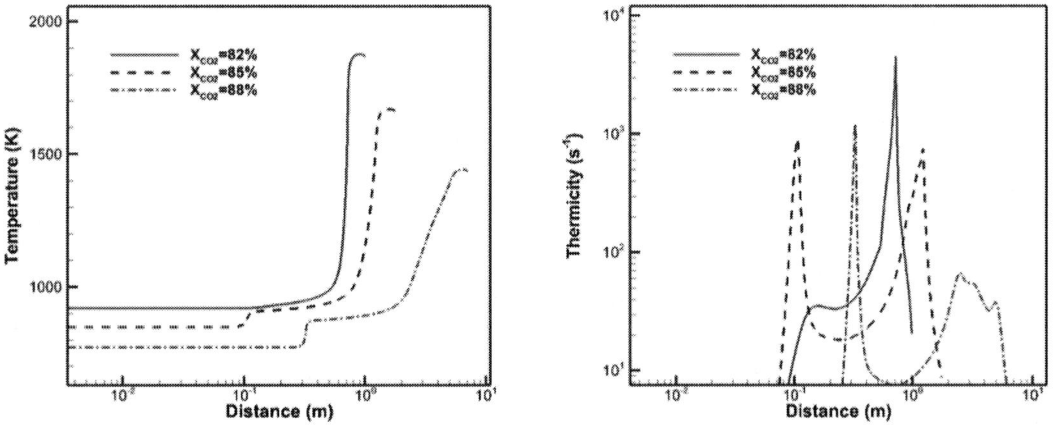

Figure 3: ZND temperature (left) and thermicity (right) profiles of stoichiometric *n*-heptane/oxygen mixtures with different CO_2 dilutions at P_1=100 kPa and T_1=298 K.

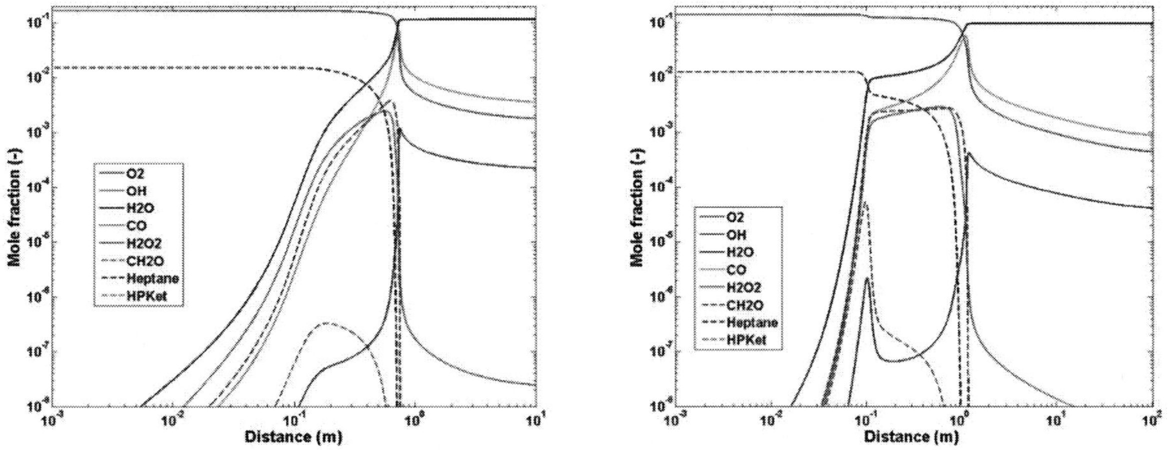

Figure 4: Species profiles of stoichiometric *n*-heptane/oxygen mixtures with 82% (left) and 85% (right) CO_2 dilutions at P_1=100 kPa and T_1=298 K.

Since the LTC is more effective at higher initial pressures and CO_2 dilutions, the ZND temperature and thermicity profiles of stoichiometric *n*-heptane/oxygen mixtures with different CO_2 dilutions at P_1=100 kPa and T_1=298 K are shown in Fig. 3. It is seen that as the dilution mole fraction increases from 82% to 85%, the temperature profile shows a two-stage increment and the thermicity profile changes from single-peak to double-peak. As the dilution increases to 88%, the second thermicity peak further splits into two peaks, leading to a three-peak profile and as such demonstrating the separation of intermediate- and high-temperature chemistries. Figure 4 plots the mole fractions of the key species for the single-peak and double-peak cases. Specifically, for the single-peak case with 82% dilution, the maximum concentrations for the intermediate species almost collapse at the same location with much less hydroperoxyketone as compared to the 85% case. However, for the case of double peaks with 85% dilution, there is an obvious separation of the maximal locations for different intermediate species. It is found that the end of the first-stage is indicated by the consumption of hydroperoxyketone, which is dominant for LTC, while the second stage finishes when there is significant consumption of hydrogen peroxide and formaldehyde, which are the signatures of intermediate- and high-temperature oxidation channels.

3. Induction Lengths and Cell Sizes

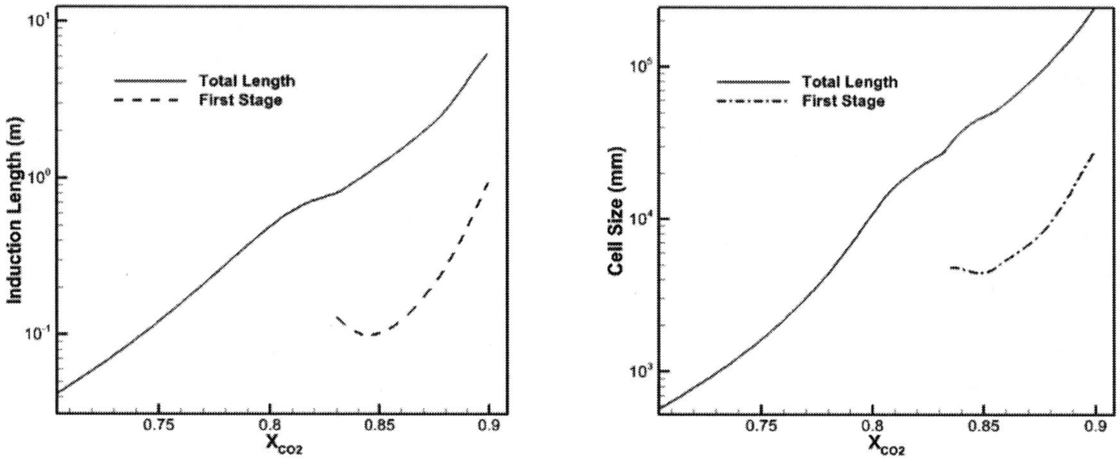

Figure 5: First stage and total induction lengths (left) and cell size (right) of stoichiometric *n*-heptane/oxygen mixtures as a function of CO_2 dilutions at P_1=100 kPa and T_1=298 K.

To further investigate the non-monotonous heat release profiles, the total and first-stage induction lengths corresponding to the first and second thermicity peaks are shown in Fig. 5. Furthermore, the cell sizes can be estimated by using Ng's model [17]:

$$\lambda = f(\chi)\Delta_i = \Delta_i \sum_{k=1}^{N}(a_k\chi^k + b_k\chi^{-k}) \tag{1}$$

where λ is the cell size, Δ_i the induction length and χ the stability parameter. The results are plotted in Fig. 5 as well. As there are two different induction lengths corresponding to first and second stage ignition, both the cell sizes associated with the first and second induction lengths have been evaluated. It is found that for the total induction length and cell size, the negative response of length scales disappears, although the slope of the length scales with dilution decreases within the NTC regime. However, for the induction length and cell size of the first stage heat release, the length scales still show the negative dependence on the dilution mole fraction. This result is mainly due to the fact that both temperature and pressure change as the dilution increases and the NTC regime would shift accordingly. In addition, we note that the estimated cell size for LTC-affected detonation is at least of several meters, which is much larger than those for typical fuels studied experimentally.

4. Conclusions

The structure of detonation propagating in n-heptane-oxygen mixtures of different CO_2 dilutions has been investigated through ZND numerical simulations. Results for mixtures with high CO_2 dilutions show a two-step energy release characteristic of the low-temperature chemistry, and a negative temperature dependence characteristic of the first-stage ignition delay.

5. Acknowledgements

WL was supported by a Maeder Fellowship and a Harari Fellowship of Princeton University, and RM by a start-up fund of the Center for Combustion Energy of Tsinghua University.

References

[1] J. F. Griffiths and S. K. Scott, Thermokinetic interactions: Fundamentals of spontaneous ignition and cool flames, Prog. Energy Combust. Sci., vol. 13, no. 3, pp. 161–197, 1987.
[2] F. Battin-Leclerc, Detailed chemical kinetic models for the low-temperature combustion of hydrocarbons with application to gasoline and diesel fuel surrogates, Prog. Energy Combust. Sci., vol. 34, no. 4, pp. 440–498, 2008.
[3] P. G. Lignola and E. Reverchon, Cool flames, Prog. Energy Combust. Sci., vol. 13, no. 1, pp. 75–96, 1987.
[4] J. E. Dec, Advanced compression-ignition engines—understanding the in-cylinder processes, Proc. Combust. Inst., vol. 32, no. 2, pp. 2727–2742, 2009.
[5] S. W. Benson, The kinetics and thermochemistry of chemical oxidation with application to combustion and flames, Prog. Energy Combust. Sci., vol. 7, no. 2, pp. 125–134, 1981.
[6] C. K. Law and P. Zhao, NTC-affected ignition in nonpremixed counterflow, Combust. Flame, 2011.
[7] V. Nayagam, D. L. Dietrich, M. C. Hicks, and F. A. Williams, Cool-flame, extinction during n-alkane droplet combustion in microgravity, Combust. Flame, vol. 162, no. 5, pp. 2140–2147, 2015.
[8] A. Yamamoto, H. Oshibe, H. Nakamura, T. Tezuka, S. Hasegawa, and K. Maruta, Stabilized three-stage oxidation of gaseous n-heptane/air mixture in a micro flow reactor with a controlled temperature profile, Proc. Combust. Inst., vol. 33, no. 2, pp. 3259–3266, 2011.
[9] M. P. Romano, M. I. Radulescu, A. J. Higgins, J. H. S. Lee, W. J. Pitz, and C. K. Westbrook, Sensitization of hydrocarbon-oxygen mixtures to detonation via cool-flame oxidation, Proc. Combust. Inst., vol. 29, no. 2, pp. 2833–2838, 2002.
[10] M. P. Romano, M. I. Radulescu, A. J. Higgins, and J. H. S. Lee, Sensitization of pentane-oxygen mixtures to DDT via cool flame oxidation, Combust. Flame, vol. 132, no. 3, pp. 387–394, 2003.
[11] M. O. Sturtzer, N. Lamoureux, C. Matignon, D. Desbordes, and H. N. Presles, On the origin of the double cellular structure of the detonation in gaseous nitromethane and its mixtures with oxygen, Shock Waves, vol. 14, no. 1–2, pp. 45–51, 2005.
[12] F. Joubert, D. Desbordes, and H. N. Presles, Detonation cellular structure in NO_2/N_2O_4-fuel gaseous mixtures, Combust. Flame, vol. 152, no. 4, pp. 482–495, 2008.
[13] D. Davidenko, R. Mével, and G. Dupré, Numerical study of the detonation structure in rich H_2-NO_2/N_2O_4 and very lean H_2-N_2O mixtures, Shock Waves, vol. 21, no. 2, pp. 85–99, 2011.
[14] J. H. S. Lee, Dynamic Parameters of Gaseous Detonations, Annu. Rev. Fluid Mech., vol. 16, no. 1, pp. 311–336, 1984.
[15] Z. Wang, Y. Qi, X. He, J. Wang, S. Shuai, and C. K. Law, Analysis of pre-ignition to super-knock: Hotspot-induced deflagration to detonation, Fuel, vol. 144, pp. 222–227, 2015.
[16] C. S. Yoo, T. Lu, J. H. Chen, and C. K. Law, Direct numerical simulations of ignition of a lean *n*-heptane/air mixture with temperature inhomogeneities at constant volume: Parametric study, Combust. Flame, vol. 158, no. 9, pp. 1727–1741, 2011.
[17] H. D. Ng, Y. Ju, and J. H. S. Lee, Assessment of detonation hazards in high-pressure hydrogen storage from chemical sensitivity analysis, Int. J. Hydrogen Energy, vol. 32, no. 1, pp. 93–99, 2007.

Spring Technical Meeting
Eastern States Section of the Combustion Institute
March 4-7, 2018
State College, Pennsylvania

Effects of Thermal and Fuel Stratifications and Turbulence Transport on Knocking Formation for Dimethyl Ether/Air Mixtures

Tianhan Zhang[1,], Weiqi Sun[1], Liang Wang[2], Yiguang Ju[1]*

[1]Department of Aerospace and Mechanical Engineering, Princeton University, Princeton, US
[2]School of Aerospace, Tsinghua University, Beijing, China
[*]Corresponding Author Email: tianhanz@princeton.edu

Abstract: The effects of thermal and fuel stratification and turbulent transport on knocking formation are numerically modeled in the negative temperature coefficient region using dimethyl ether/air mixtures with a detailed chemistry. The critical conditions of knocking formation with thermal and fuel concentration gradients are examined. The effects of turbulence timescales, length scale, and Reynolds number on knocking development and knocking strength are explored. The results show that either a thermal gradient, concentration gradient, or combine gradients can initiate knocking. A unified criterion and diagram for knocking formation including both thermal and fuel concentration gradients as well as the normalized length scale of gradient fields is demonstrated. The results show that turbulence transport can delay knocking/detonation transition and dramatically reduce detonation strength due to turbulent mixing. It is found that when turbulence time scale is much shorter than the ignition delay time of the gradient field, the knocking formation will be modified. With a given turbulence time scale, a larger the turbulence length scale will induce a broader mixing in the preheated zone in front of detonation wave, and thus reduces the knocking strength. Moreover, knocking formation in a smaller gradient kernel is more affected and weakened by turbulence transport. The present research provides important insights of knocking formation and control of knocking using stratifications and turbulence in the real engines.

Keywords: Knocking and Detonation; Turbulent transport; Concentration gradient; Temperature gradient

1. Introduction

Due to the increasing concerns on global warming and the air pollution by burning fossil fuels, development of advanced engines has regained unprecedented attention. Among these advanced engine concepts, the homogeneous charge compression ignition(HCCI) engines[1], the reactivity controlled compression ignition (RCCI)[2] engines and the spark assisted compression ignition(SACI)[3] engines have promising potentials for high fuel efficiency and low emissions. However, the increased compression ratio in compression ignition engines results in a big challenge of engine knocking[4] .

Recently, researches have shown the knockings are mainly caused by end-gas autoignition and the detonation. The latter is even more serious and results in a super knock. More recent studies have focused on detonation formation mechanism at the negative temperature coefficient (NTC) region. The hot spot detonation formation mechanism was proposed and investigated by Zel'dovich [5]. By using a temperature gradient and three different flame regimes: spontaneous ignition, detonation and deflagration, are identified. Bradley and coworkers[6], [7]performed numerical studies of the hot spot theory using with one-step and detailed chemistry, and obtained a detonation diagram by using normalized critical temperature gradient and chemical reaction time. Recently, low temperature chemistry and the corresponding cold spot model in detonation initiation have been studied by Z. Chen et al.[8]. More recently, the effect concentration gradient on pressure oscillation and knocking formation was investigated by Sun et al. [9], Zhang et al. [10]. The results showed that low temperature chemistry and concentration gradients have strong effects on the formation of supersonic ignition wave, pressure oscillation, and knocking formation.

In practical engines, turbulent mixing will strongly modify the temperature and concentration gradients and enhance the mixing between burned and unburned mixtures. However, very few studies have been carried out to understand the combined effect of temperature and fuel concentration gradients and the turbulent transport on detonation and knocking formation. Thus, the goal of this study is to answer the following questions: (1) What is the detonation diagram considering coupled temperature and fuel concentration gradients? (2) How does turbulence transport affect detonation and knocking formation in thermally and compositionally stratified flow? (3) What are the impact of turbulent length scales and time scales on detonation transition and detonation strength?

2. Numerical models

2.1 Chemistry mechanism and the ignition kernel model

Dimethyl ether(DME), one of the simplest fuels with low temperature chemistry and a high cetane number, is chosen for the present study. A comprehensively reduced DME mechanism (54 species and 418 reactions) from HP-Mech/DME [12] is obtained by using the multi-generation path flux analysis(PFA) [13]. The initial pressure is chosen as 40atm to model the working conditions of HCCI or spark assisted compression ignition engines.

Based on the observations in experimental knocking studies[14], knocking usually occurs close to the wall where there exists considerable temperature and concentration gradients due to wall heat losses and the impinged fuel droplets. To consider a combined effect of temperature and fuel concentration gradients, in both one-dimensional and two-dimensional modeling, the gradients are linearly distributed inside ignition kernel with a size of l_i at the end of the left boundary. At the right hand side boundary, an zero gradient boundary conditions are used. The resulting distribution functions for temperature (T) and equivalence ratio (ϕ) are: $\theta \equiv T \ or \ \phi$

$$\theta(x) = \begin{cases} \theta_0 + x * \nabla\theta & x < l_i \\ \theta_0 & x \geq l_i \end{cases} \tag{1}$$

Here l_i denotes for the ignition kernel size with temperature and concentration gradients, T_0 and Φ_0 the temperature and equivalence ratio out of ignition kernel, and ∇T and $\nabla\Phi$ the temperature and equivalence ratio gradients, which are constants in this study.
The critical gradients of temperature and fuel equivalence ratio are determined at the condition that the ignition wave speed is equal to the local acoustic wave speed, a. The ignition wave speed (v) in the ignition gradient kernel is obtained by gradient of ignition time (τ_{ig}), $v = \frac{d\tau_{ig}}{dx}$. As such, the critical temperature and fuel concentration gradients are given as: $\theta \equiv T \ or \ \phi$

$$(\nabla\theta)_{cri} = \left(\frac{d\theta}{dx}\right)_{cri} = a^{-1}\left(\frac{\partial\tau_{ig}}{\partial\theta}\right)^{-1}, \tag{2}$$

Based on the critical gradients, a non-dimensional gradient can be defined as: $\xi_\theta = \frac{\nabla\theta}{(\nabla\theta)_{cri}}$. This non-dimensional parameter describes the speed of ignition wave into three different categories: $\xi \to 0$, supersonic ignition wave; $\xi >> 1$, subsonic or transonic ignition wave; and $\xi \sim 1$ knocking or detonation. From Eq.4, the non-dimensional coupled temperature and concentration gradient can be given as:

$$\xi_{T\phi} = \xi_T + \xi_\phi \tag{3}$$

2.2 Numerical Method

For the 1D numerical simulation, the adaptive simulation of unsteady reactive(ASURF+) code[16] is used. The correlated dynamic adaptive chemistry and transport(CO-DACT) method [17] together with hybrid multi-timescale(HMTS)[18] method are adopted in order to improve computation efficiency. A 3rd-order weighted essentially non-oscillatory (WENO) scheme is used to capture the shock wave. The smallest grid cell is 12 μm and the smallest time step is 0.2 ns to better capture detonation time history. In the two-dimensional simulation, a newly developed multiscale adaptive reduced chemistry solver(MARCS) is used. The solver uses the 4th order Runge-Kutta scheme in time and 3rd order AUSMPW+ scheme in space. The uniform cell size of 10 μm and the smallest time step of 0.1ns are used in the simulation.

The calculation domain length is the same for 1D and 2D cases for better comparison. The domain length is 1 cm and the width varies from 1 mm to 2 mm due to the different length scales of the isentropic turbulence. The wall boundary conditions are applied on the left and the transmissive boundary conditions is applied on the right. The periodic boundary condition is used for the top and bottom boundary in 2D cases.

3. Results and Discussion

3.1 Critical gradients and the detonation peninsular for detonation formation with coupled temperature and equivalence ratio gradients

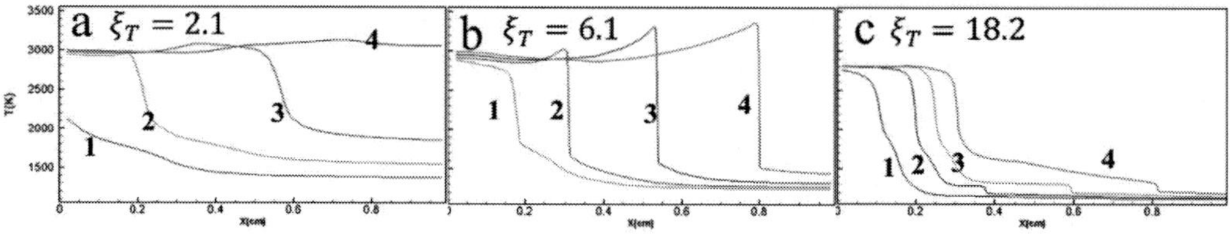

Figure 1 Laminar cases with temperature gradient:(a). ξ_T=2.1, supersonic autoignition. 1: 386μs, 2: 387μs, 3: 388μs, 4: 390μs. (b). ξ_T=6.1, detonation. 1:383μs, 2:384μs, 3: 385μs, 4:386μs (c) ξ_T=18.2, transonic propagation. 1:367μs, 2:370μs, 3:372μs, 4:375μs.

Figure 1 respectively show three typical results of ignition events with different temperature gradients in 1D. For the normalized temperature gradient is very small, for example, ξ_T=2.1, the mixture almost ignites simultaneously. Fig 1(a) shows the resulting ignition wave is a supersonic ignition wave. No detonation will occur. On the other hand, when the normalized temperature gradient is very large, for example, ξ_T=18.2, Fig.1(c) shows that the ignition wave is a subsonic to transonic propagation wave because the ignition wave is not fast enough to enable ignition and pressure wave coupling. However, if the temperature gradient is moderate, for example, ξ_T=6.1, the ignition wave speed is fast enough to couple with acoustic wave. As shown in Fig.1(b), a detonation wave is formed at 384μs. Therefore, Fig.1 clearly shows that there are three different ignition events, supersonic ignition wave, subsonic or transonic ignition wave, and detonation corresponding respectively to the low, high, and intermediate normalized temperature gradients.

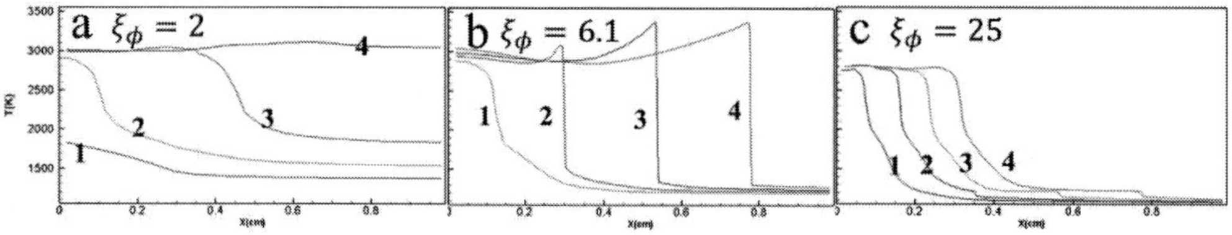

Figure 2 Laminar cases with equivalence ratio gradient:(a). ξ_ϕ=2, supersonic autoignition. 1: 386μs, 2: 387μs, 3: 388μs, 4: 390μs. (b). ξ_ϕ=6.1, detonation. 1:380μs, 2:381μs, 3: 382μs, 4:383μs (c) ξ_ϕ=25, transonic propagation. 1:353μs, 2:356μs, 3:358μs, 4:361μs.

Fig.2 shows the three ignition events, respectively, with low, high, and intermediate normalized concentration gradients ξ_ϕ. It is seen that similar to the temperature gradient case, there also are three different ignition regimes, supersonic ignition wave, subsonic or transonic ignition wave, and detonation corresponding respectively to the low, high, and intermediate normalized concentration gradients. Therefore, Figs.1 and 2 shows that both temperature and concentration gradients can cause strong knocking and detonation if their gradients are in the appropriate range.

To demonstrate the effect of coupled temperature and concentration gradients on detonation formation, Fig. 3(a) shows the boundary separating the detonation and supersonic ignition wave for different non-dimensional temperature and fuel concentration gradients. It is seen that the detonation limit linearly depends on the temperature and fuel concentration gradients. Therefore, Eq.(3) can be used appropriately to define the detonation boundary for coupled thermal and fuel stratifications.

Fig.3(b) shows the detonation diagram using the coupled temperature and concentration gradient, $\xi_{T\phi}$, and the normalized excitation time. A C-shaped detonation peninsular is bounded by a low limit, ξ_l, and an upper limit, ξ_u. For a high scalar gradient, $\xi>\xi_u$, a transonic deflagration ignition wave will be observed (Figs.1b and 2b). For a low scalar gradient, $\xi<\xi_l$, a supersonic ignition wave will be observed (Figs.1a and 2a). In addition, if $\varepsilon< \varepsilon_{cri}$, no matter how high the gradient is, no detonation can occur. This is the limit of chemical heat release time and the acoustic wave propagation time. In the following turbulent modeling, the initial gradient and the ignition kernel size will be chosen from the detonation diagram in Fig.3b.

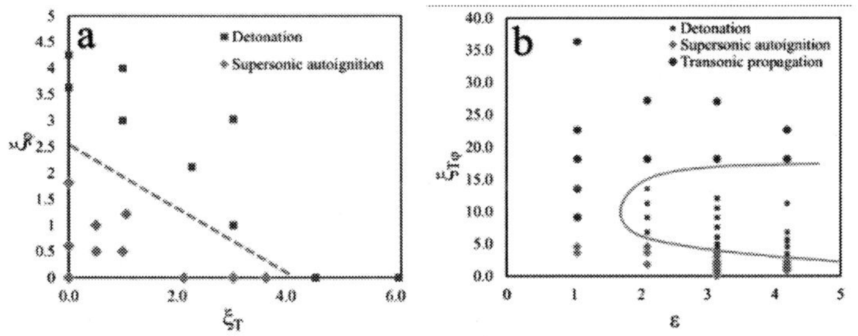

Figure 3 Detonation peninsular with coupled temperature and concentration gradients

3.2 Effect of turbulence transport on detonation formation and detonation strength

Table 1 shows the turbulence parameters for the 2D turbulence transport modeling in each case. Here l is the integral length scale and Re_λ is the Re number based on the Taylor microscale. The ignition delay time at P=40atm, T=920 K, Φ=1.0 is 389.8μs. Thus, Cases 1 and 2 are chosen to study the effect of turbulence timescales on detonation. The turbulence timescale for Cases 1is 83μs which is than the ignition time, while the timescale in Case 2 is 333μs, which comparable to ignition time.

Table 1 Turbulence parameters in 2D simulations

Case	l(mm)	τt(ms)	ε	ξT	ξϕ
1	1	83	2.09	0	5.4
2	1	333	2.09	0	5.4
3	1	83	3.14	4.5	0
4	1	83	2.09	6.8	0

Fig.4 shows the comparison of temperature and pressure distributions between Case 1 and Case 2 at two timesteps respectively for initial stage of ignition. At t=385μs, the ignition kernel of Case 2 is much larger than Case 1 because the turbulent mixing in Case 2 is slower and weaker. To better compare the difference in Case 1 and Case 2, the temperature and pressure distributions along the center line are plotted in Fig. 5.

It is more obvious to notice that the peak pressure rises due to detonation in Case 1 (line #4) is much less than in Case 2 (line #4), which means the detonation strength in Case 1 is reduced by turbulence mixing. From Fig.5, it is seen that the difference in pressure distribution between Case 1 and Case 2 starts from 386μs (line 2). The corresponding pressure and temperature contours at this time is shown in Fig. 4 at t=386μs. Comparing the pressure and temperature distributions at t=386μs in Fig. 4 of Case 2, the pressure wave and the ignition front have already coupled together, indicating a fully developed detonation front and leading to a sharp pressure rise over 280atm. However, in Case 1 at the same time, the ignition front is still uncoupled from the pressure wave. Therefore, this is not a fully developed detonation wave. Such turbulence mixing delays detonation formation.

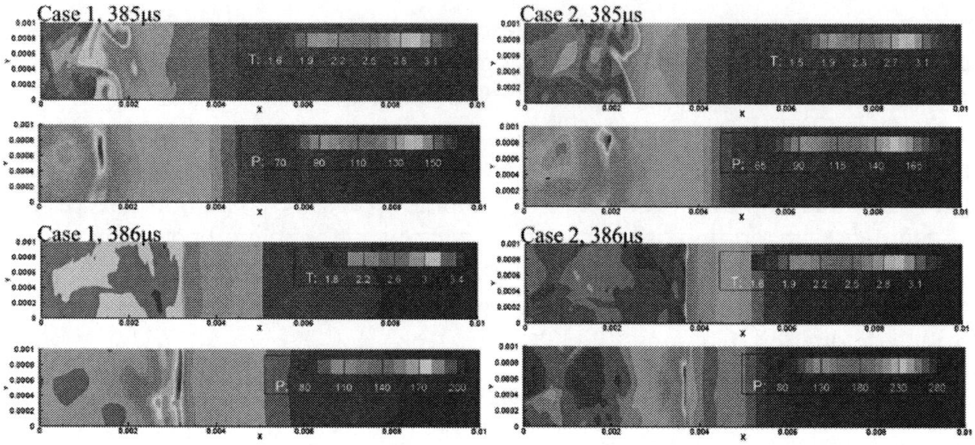

Figure 4 Comparison of temperature and pressure contours between Case 1 and Case 2.

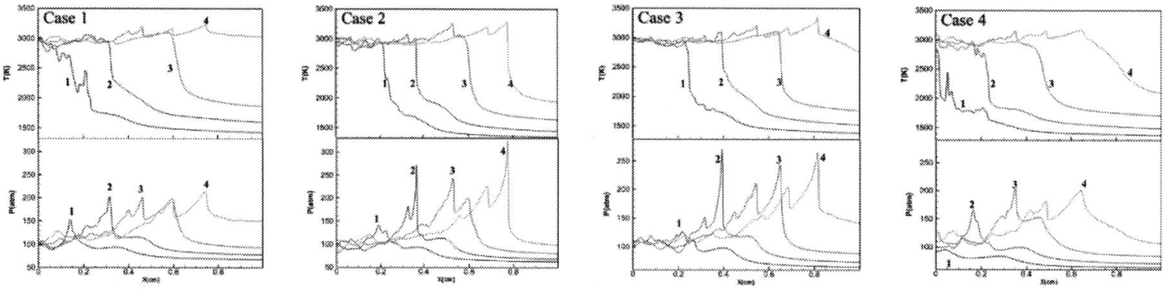

Figure 5 The pressure and temperature distributions along the centerline for Case 1, Case 2, Case 3 and Case 4. For Case 1 and Case 2: 1: 385μs, 2:386μs, 3: 387μs, 4: 388μs. For Case 3 and Case 4. 1:384.5μs, 2:385.5μs, 3: 386.5μs, 4:387.5μs.

Case 3 and Case 4 have the same turbulence length scales(1mm) and timescales(83μs), but with different ignition kernel sizes 3mm and 2mm, respectively. The temperature on the left boundary are 905K and the upstream temperature is 920K. The resulting ξ_T=4.5, 6.8 for Case 3 and Case 4 respectively. Case 3 which has a larger ignition kernel size results in a higher detonation pressure than Case 4, which has smaller ignition kernel size. Therefore, at the given turbulence length scale and timescale, the detonation with a smaller ignition kernel will be more weakened by turbulent mixing. This is because turbulence dissipation will reduce the temperature or concentration gradient quicker for a smaller ignition kernel.

4. Conclusion

The effects of thermal and fuel stratification and turbulent transport on knocking formation are numerically modeled in the negative temperature coefficient region using dimethyl ether/air mixtures with a detailed chemistry. The critical conditions of detonation formation with thermal and fuel concentration gradients are examined. The effects of turbulence timescales, length scale, and Reynolds number on knocking development and knocking strength are explored. The results show that either a thermal gradient, concentration gradient, or combine gradients can initiate knocking. A unified criterion and diagram for detonation formation including both thermal and fuel concentration gradients as well as the normalized length scale of gradient fields is demonstrated. The results show that turbulence transport can delay knocking/detonation transition and dramatically reduce detonation strength due to turbulent mixing. It is found that when turbulence time scale is much shorter than the ignition delay time of the gradient field, the knocking formation will be modified. With a given turbulence time scale, a larger the turbulence length scale will induce a broader mixing in the preheated zone in front of detonation wave, and thus reduces the knocking strength. Moreover, knocking formation in a smaller gradient kernel is more affected and weakened by turbulence transport. The present research provides important insights of knocking formation and control of knocking using stratifications and turbulence in the real engines.

5. Acknowledgement

The authors would like to thank the grant support from the Army Research Office with grant number W911NF-16-1-0076.

Reference

[1] R. H. Thring, "Homogeneous-Charge Compression-Ignition (HCCI) Engines," in *SAE Technical Paper*, 1989.

[2] S. L. Kokjohn, R. M. Hanson, D. a. Splitter, and R. D. Reitz, "Fuel reactivity controlled compression ignition (RCCI): a pathway to controlled high-efficiency clean combustion," *Int. J. Engine Res.*, vol. 12, no. 3, pp. 209–226, Jun. 2011.

[3] L. Manofsky, J. Vavra, D. N. Assanis, and A. Babajimopoulos, "Bridging the Gap between HCCI and SI: Spark-Assisted Compression Ignition," in *SAE Technical Paper*, 2011.

[4] G. T. Kalghatgi, "Developments in internal combustion engines and implications for combustion science and future transport fuels," *Proc. Combust. Inst.*, vol. 35, no. 1, pp. 101–115, 2015.

[5] Y. Zeldovich, "Regime classification of an exothermic reaction with nonuniform initial conditions," *Combust. Flame*, vol. 39, no. 2, pp. 211–214, Oct. 1980.

[6] X. J. J. Gu, D. R. R. Emerson, and D. Bradley, "Modes of reaction front propagation from hot spots," *Combust. Flame*, vol. 133, no. 1–2, pp. 63–74, Apr. 2003.

[7] L. Bates, D. Bradley, G. Paczko, and N. Peters, "Engine hot spots: Modes of auto-ignition and reaction propagation," *Combust. Flame*, vol. 166, pp. 80–85, 2015.

[8] P. Dai, Z. Chen, S. Chen, and Y. Ju, "Numerical experiments on reaction front propagation in n-heptane/air mixture with temperature gradient," *Proc. Combust. Inst.*, vol. 35, no. 3, pp. 3045–3052, 2015.

[9] W. Sun, S. H. Won, X. Gou, and Y. Ju, "Multi-scale modeling of dynamics and ignition to flame transitions of high pressure stratified n-heptane/toluene mixtures," *Proc. Combust. Inst.*, vol. 35, pp. 1049–1056, 2015.

[10] T. Zhang, W. Sun, and Y. Ju, "Multi-scale modeling of detonation formation with concentration and temperature gradients in n-heptane/air mixtures," *Proc. Combust. Inst.*, vol. 36, pp. 1539–1547, 2016.

[11] C. Qi and Z. Chen, "Modes of Reaction Front Propagation in n-Heptane / air Mixture with Concentration Non-uniformity," *Symp. Combust.*, pp. 1–6, 2015.

[12] H. Zhao, X. Yang, and Y. Ju, "Kinetic studies of ozone assisted low temperature oxidation of dimethyl ether in a flow reactor using molecular-beam mass spectrometry," *Combust. Flame*, vol. 173, pp. 187–194, 2016.

[13] W. Sun, Z. Chen, X. Gou, and Y. Ju, "A path flux analysis method for the reduction of detailed chemical kinetic mechanisms," *Combust. Flame*, vol. 157, no. 7, pp. 1298–1307, Jul. 2010.

[14] Z. Wang, Y. Qi, X. He, J. Wang, S. Shuai, and C. K. Law, "Analysis of pre-ignition to super-knock: Hotspot-induced deflagration to detonation," *Fuel*, vol. 144, pp. 222–227, 2015.

[15] S. Ishihara, K. Ishii, and H. Kataoka, "Mechanism of detonation transition from accelerating flames in a channel," *Proc. Combust. Inst.*, vol. 0, no. 2, p. , 2016.

[16] Z. Chen and Y. Ju, "Combustion Theory and Modelling Theoretical analysis of the evolution from ignition kernel to flame ball and planar flame Theoretical analysis of the evolution from ignition kernel to flame ball and planar flame," *Combust. Theory Model.*, vol. 11, no. 3, pp. 427–453, 2007.

[17] W. Sun, X. Gou, H. A. El-Asrag, Z. Chen, and Y. Ju, "Multi-timescale and correlated dynamic adaptive chemistry modeling of ignition and flame propagation using a real jet fuel surrogate model," 2015.

[18] X. Gou, W. Sun, Z. Chen, and Y. Ju, "A dynamic multi-timescale method for combustion modeling with detailed and reduced chemical kinetic mechanisms," *Combust. Flame*, vol. 157, no. 6, pp. 1111–1121, Jun. 2010.

Spring Technical Meeting
Eastern States Section of the Combustion Institute
March 4-7, 2018
State College, Pennsylvania

Large-Eddy Simulations of a Spark Ignition IC Engine

Samuel J. Kazmouz[1*], *Daniel C. Haworth*[1]

[1]*Department of Mechanical and Industrial Engineering, Penn State University, University Park, PA-16801 USA*
[*]*Corresponding Author Email: xuk13@psu.edu*

Abstract: Cycle-to-cycle variations of flow and combustion in internal combustion engines limit their fuel efficiency and emissions potential. Large-eddy simulation is an effective method to simulate fired engine operation, including accounting for cycle-to-cycle variations. In this research, large-eddy simulations of in-cylinder processes in a single-cylinder optically accessible spark-ignition piston engine are performed. The engine grid is created using ES-ICE, and is refined in the spark-plug vicinity, to better capture early flame development. STAR-CD is utilized to carry out the simulations, using a modified thickened flame model for turbulent premixed flame propagation and a Smagorinsky model for sub-grid-scale turbulence. Multiple engine cycles are simulated successively for steady-state motored and firing engine operation. Here the fired results are for homogeneous stoichiometric premixed reactants. The eventual goal is to develop a predictive tool for CFD simulations of direct-injection stratified-charge spark-ignition engines, which can be used to investigate cycle-to-cycle variations in flow and combustion.

Keywords: IC Engines, Cycle to Cycle Variation, Flame Modelling, LES

1. Introduction

Cycle-to-Cycle variations (CCV) in spark-ignition engines are observed in both flow and combustion parameters, and lead to losses in engine efficiency [1]. CCV have been extensively studied over the past few decades, through experimental (optical diagnostics) and computational (LES/RANS) research methods of in-cylinder flow and combustion [2]. Several factors have been identified as contributing sources of CCV, including variations in flow parameters, temperature, equivalence ratio, and spark discharge. In particular, early flame development has been found to influence, to a large extent, what happens throughout the subsequent cycle [3].

While previous CCV research focused on stoichiometric premixed homogeneous spark-ignition engines, other advanced combustion modes (e.g., fuel lean, stratified charge, high levels of EGR) are appealing because of their efficiency benefits, especially at low loads [4]. However, due to stability challenges, CCV becomes an even more critical issue for these modes of operations. This paper highlights the first step of an ongoing project aimed towards a better understanding of CCV in stratified-charge engine operation, through high fidelity simulations of in-cylinder flow and combustion. The work presented here is a necessary step to ensure that the models are performing as expected, before moving forward to stratified simulations.

In this paper, the LES results for a four-valve spark-ignition engine in motored and fired (stoichiometric premixed homogeneous) operation are presented. In Section 2, the engine configuration, operating conditions, and numerical/physical models are explained. In Section 3, the main results are shown and discussed. Section 4 recaps the main findings and gives the next steps of this research.

2. Method

2.1. Engine Configuration and Operating Conditions

The engine is a four-valve, four-stroke-cycle, single-cylinder, optical spark-ignition engine, which can also run globally lean with direct injection and a stratified charge. The exhaust ports merge into a single exhaust tube, and similarly the intake ports merge into a single intake tube. One intake port has a swirl control valve installed (a throttle) to control air flow through one of the intake valves, which is used to modify the large-scale in-cylinder flow structure.

LES is used to simulate engine operation in five modes. Four of them are motored, and one is fired. These cases are summarized in Table 1. The motored cases simulate 60 consecutive cycles, while the fired case simulates 20 consecutive cycles.

Table 1: Five modes of engine operation

Case Number	Mode	Mean Intake Manifold Pressure (kPa)	Swirl Control Valve Condition	Number of Consecutive Cycles
M1	Motored	40	open	60
M2	Motored	95	open	60
M3	Motored	40	closed	60
M4	Motored	95	closed	60
P1	Premixed Homogeneous Stoichiometric	40	open	20

2.2. Physical Models and Numerical Methods

STAR-CD v4, a Siemens PLM Software, was used to carry out the LES simulations. The grid contains approximately 4.7 million elements and includes the in-cylinder region, intake and exhaust ports. The grid size is roughly ~1 mm, and is refined in the spark-plug vicinity and valve-curtain regions. The temperature and pressure boundary conditions assigned to the intake and exhaust ports were developed using a GT-Power model, and are both temporally variable. A standard Smagorinsky turbulence model is used for subgrid-scale turbulence. Case P1 utilizes a simple energy deposition ignition model, where a sphere of a prescribed temperature and radius is artificially created at the spark plug tip for a specified duration. The ignition parameters are summarized in Table 2. In addition, Case P1 employs a two-step global propane-air mechanism and a modified thickened flame model (TFM) for turbulent flame propagation. The original TFM implementation [5] has been modified to account for wall interactions and heat transfer [6].

Table 2: Parameter for energy deposition ignition model

Source radius	1.37 (mm)
Source temperature	2500 (K)
Source duration	702-706 (4 CAD)

The premise of the TFM is to recover the correct burn rate even if the flame is "under-resolved." In a laminar flame, the laminar flame speed needs to remain the same while artificially thickening the flame to make it large enough to be resolved by the grid. This is done by multiplying the diffusivity and dividing the chemical source term in the species and enthalpy equations by a thickening factor *F*. Equations 1 and 2 show that the flame speed remains constant, and hence the burn rate is recovered, while the flame is thickened by a factor *F*.

$$s_l^0 \propto \sqrt{D\dot{\omega}} \tag{1}$$

$$\delta_l^0 \propto \sqrt{\frac{D}{\dot{\omega}}} \tag{2}$$

For turbulent flame propagation, and to account for wrinkling, another parameter Ξ (efficiency function) is required. The diffusivity and chemical source terms are multiplied by Ξ, to increase the burn rate due to turbulence. The calculation of Ξ is more involved than that of *F*. The resulting species equation is:

$$\frac{\partial \rho Y_m}{\partial t} + \frac{\partial \rho u_i Y_m}{\partial x_i} = \frac{\partial}{\partial x_i}\left(\rho \Xi F D_m \frac{\partial Y_m}{\partial x_i}\right) + \Xi \frac{\dot{\omega}_m}{F} \tag{3}$$

3. Results and Discussion

3.1 Motored Results

Computed in-cylinder pressure traces over 60 consecutive engine cycles are shown in Fig. 1 for four motored cases. Cases M1 and M3 have an average intake manifold pressure of 40 kPa, corresponding to part-load engine operation. This can be seen in the peak in-cylinder pressure, which is approximately 691 and 686 kPa, respectively. Alternatively, for cases M2 and M4, which have an average intake manifold pressure of 95 kPa (full load), the peak in-cylinder pressure is approximately 1670 and 1660 kPa, respectively.

Fig. 1 also shows the level of CCV in the pressure traces. For case M1, the range of observed peak pressures among the 60 cycles is approximately 690.75-692.25 kPa. This means that the CCV is of the order of 1.5 kPa (~0.22% of peak pressure). Similarly, for M2, M3, and M4, the levels of CCV are approximately 3 kPa (~0.18% of peak pressure), 2.5 kPa (~0.36% of peak pressure), and 5 kPa (~0.3% of peak pressure), respectively. The amount of CCV is expected to rise significantly when the engine switches from motored to fired operation.

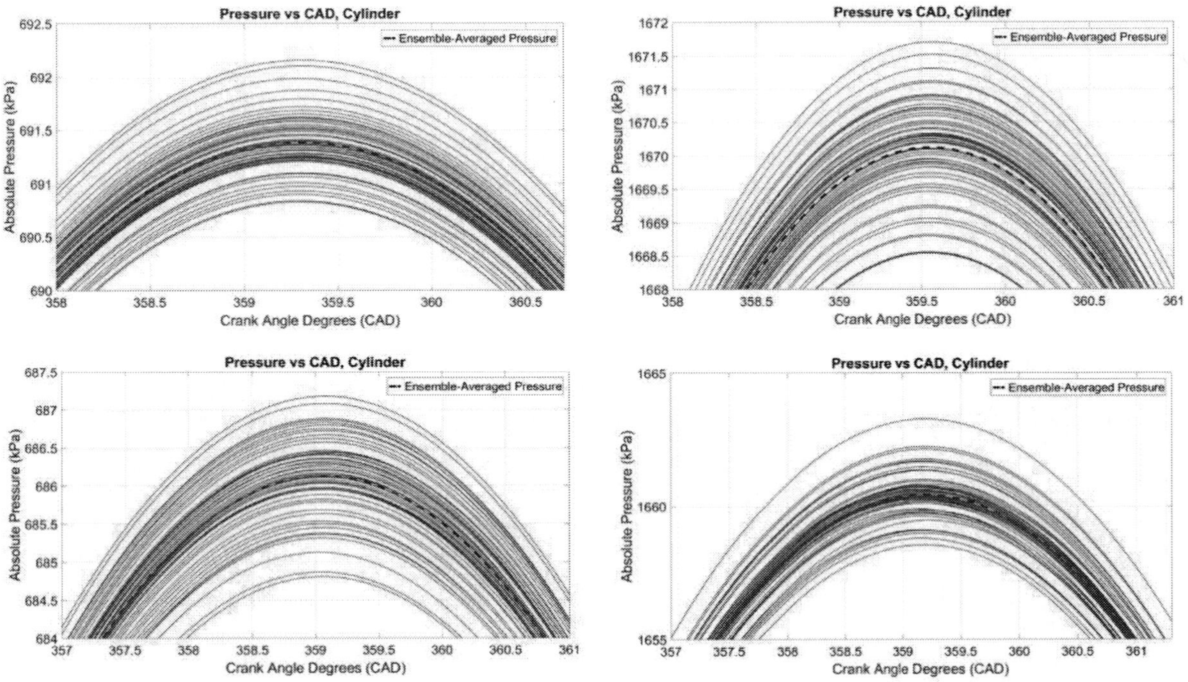

Figure 1: Peak in-cylinder pressure per cycle. From top left to bottom right: M1, M2, M3, M4

3.2 Homogeneous Premixed Results

A stoichiometric homogeneous premixed simulation was performed to ensure that the combustion and turbulence models implemented in the four-valve engine are working properly before venturing into stratified operation. Fig. 2 shows that the peak in-cylinder pressure is approximately 19 bar, which is typical for this operating condition. Furthermore, Fig. 2 shows that the CCV in the peak pressure is approximately 4 bars (~21% of peak pressure), which is significant. The engine also experiences misfires for this operating condition, as indicated by the pressure traces that are similar to those for motored operation.

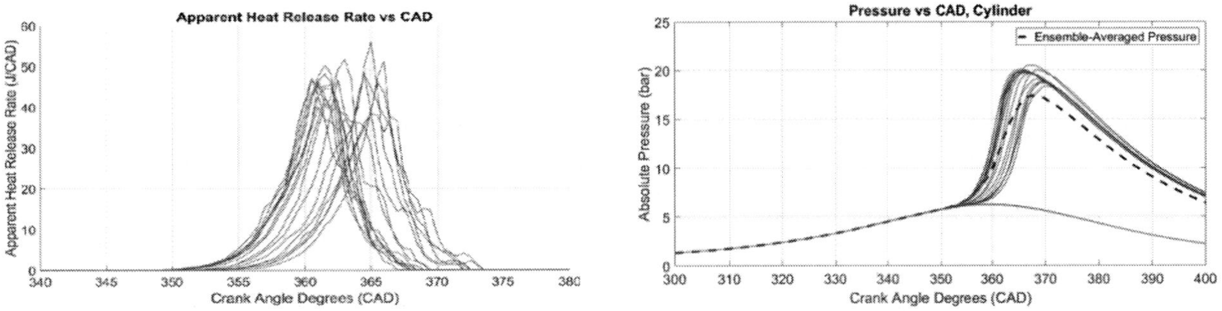

Figure 2: Left: Apparent Heat Release Rate. Right: Peak in-cylinder pressure for P1

Fig. 2 also shows the apparent heat release rate, which is calculated as shown in [3]. This also shows the level of CCV experienced in consecutive engine cycles. Given the level of CCV and misfires shown in Fig. 2, implementing a better ignition model is expected to be essential for success in simulating direct-injection stratified-charge combustion, in future work.

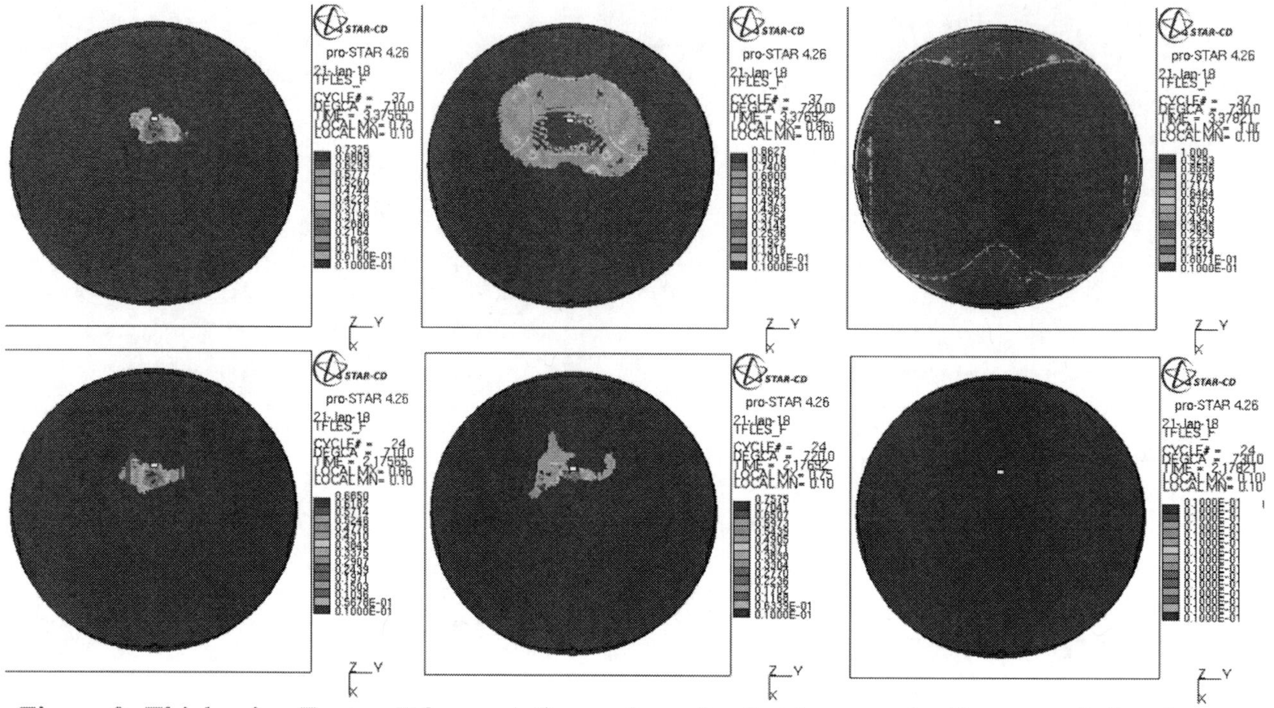

Figure 3: Thickening Factor *F* for a misfire cycle and a fast-burn cycle. From top left to bottom right: fast cycle at 710 CAD, fast cycle at 720 CAD, fast cycle at 730 CAD, misfire at 710 CAD, misfire at 720 CAD, misfire at 730 CAD.

To better visualize the flame propagation, Fig. 3 shows the computed thickening factor *F* for a fast cycle and a misfire, out of the 20 simulated cycles. The value of *F* is relevant only in the flame sheet. Hence, Fig. 3 gives a visual impression of the turbulent flame. The initial flame development (10 CAD before TDC) is similar for both cycles, while the flame at TDC is significantly different between the two cycles. For the misfire, the flame is still concentrated near the spark-plug tip. For the fast cycle, the flame is quickly making its way towards the piston and cylinder walls. Initial flame/kernel development is essential for full burn, and hence an improved ignition model is required. By 10 CAD after TDC, the misfire flame is extinguished, while the fast cycle flame has already reached the walls of the cylinder

4 Conclusions

The efficiency improvement offered by lean and stratified combustion in SI engines is an incentive to investigate CCV in such engines. This paper highlights initial work towards LES of a stratified-charge four-valve spark-ignition engine. Four motored cases have been simulated for 60 consecutive cycles, and a single stoichiometric homogeneous premixed fired case has been simulated for 20 consecutive cycles. A summary of the findings is as follows:

- LES of a realistic four-valve engine goes beyond previous simulations for an idealized engine [3,6], and offers more insight into real engines.
- The motored results are typical and as expected. CCV in the computed pressure traces is small.
- The premixed case shows typical in-cylinder pressure traces, and larger CCV than the motored results, as expected.
- Four misfires occurred out of 20 firing cycles, while some cycles were fast. This indicates a need for an improved ignition model, possibly ISSIM-LES [7].

Based on the work presented in this paper, the next steps are as follows:
- Spray modeling and validation.
- Extending the TFM to account for the presence of the spray fuel (Lagrangian droplets) and stratified combustion [8].
- Implement the ISSIM-LES ignition model for the TFM [9].
- Comparison and validation against experimental results.

5. Acknowledgements

The authors gratefully acknowledge funding from GM Global R&D, and Siemens PLM Software for providing access to, and support for, their STAR-CD CFD code.

6. References

Journal article:

[1] R. S. Davis, S. Keum, An Industry Perspective On Cycle-To-Cycle Variation. Presentation, Workshop on Understanding Cyclic Variability in Internal Combustion Engines-Advanced Photon Source (APS) Conference Center, Argonne National Laboratory, 2017.

[2] D. Haworth, In-Cylinder LES: A Historical Perspective. Presentation, Workshop on Understanding Cyclic Variability in Internal Combustion Engines-Advanced Photon Source (APS) Conference Center, Argonne National Laboratory, 2017.

[3] Y. Shekhawat, D.C. Haworth, A. d'Adamo, F. Berni, S. Fontanesi, P. Schiffmann, D.L. Reuss and V. Sick, An Experimental and Simulation Study of Early Flame Development in a Homogeneous-charge Spark-Ignition Engine, Oil & Gas Science and Technology - Rev IFP Energies nouvelles 72 (2017) 5-32

[4] M. Sjöberg, Mitigation Of Cycle-To-Cycle Combustion Variability To Enable Highly Efficient DISI Engines. Presentation, Workshop on Understanding Cyclic Variability in Internal Combustion Engines-Advanced Photon Source (APS) Conference Center, Argonne National Laboratory, 2017.

[5] T. Butler and P. O'Rourke, A numerical method for two- dimensional unsteady reacting flows, International Symposium on Combustion 16 (1977)1503-1515

[6] Y. Shekhawat, Large-Eddy Simulations of Motored Flow and Combustion in a Homogeneous-Charge Spark-Ignition Engine. Ph.D. The Pennsylvania State University, 2017.

[7] O. Colin and K. Truffin, A Spark Ignition Model For Large Eddy Simulation Based On An FSD Transport Equation (ISSIM-LES), Proc. Combust. Inst. 33 (2011)3097-3104

[8] F. Sacomano Filho, G. Kuenne, M. Chrigui, A. Sadiki, and J. Janicka, A consistent Artificially Thickened Flame approach for spray combustion using LES and the FGM chemistry reduction method: Validation in Lean Partially Pre-vaporized flames, Combustion and Flame 184 (2017)68-89.

[9] A. d'Adamo, S. Breda, S. Fontanesi, and G. Cantore, LES Modelling of Spark-Ignition Cycle-to-Cycle Variability on a Highly Downsized DISI Engine, SAE Int. J. Engines 8(5) (2015)

2018 Spring Technical Meeting
Eastern States Section of the Combustion Institute
Hosted by The Pennsylvania State University
March 04-07, 2018

Influence of Turbulence-Radiation Interactions in Engine Radiation Heat Transfer

C. Paul[1], D.C. Haworth[1], S. Roy[2], M.F. Modest[3]*

[1]Penn State, USA
[2]Marquette University, USA
[2]University of California Merced, USA

[*]Corresponding Author Email: cxp484@psu.edu

Abstract: Detailed radiation modelling in advanced high-efficiency piston engines is recently getting attention due to the trends toward higher operating pressures and higher levels of exhaust gas recirculation (EGR), which makes molecular gas radiation more prominent (absorption coefficient proportional to participating species concentration). Advanced high-efficiency engines also are expected to function closer to the limits of stable operation, where even small perturbations to the energy balance can have a large influence on system behavior. Hence, here line-by-line (LBL) spectral radiation property model and photon Monte Carlo (PMC) radiative transfer equation (RTE) solvers have been implemented in an OpenFOAM-based engine CFD code. The influence of turbulence-radiation interactions (TRI) is determined by comparing results obtained using local mean values of composition and temperature to compute radiative emission and absorption with those obtained using a particle-based transported probability density function (tPDF) method. Simulations have been performed for a full-load (peak pressure ~200 bar) heavy-duty diesel engine. Differences in computed NO and soot levels and wall heat transfer rates are shown for cases with and without TRI. Computed radiative emission and reabsorption with TRI are higher compared to those obtained from a no-TRI case for the same operating condition. With consideration of TRI, the net radiative heat loss is lower than for the no-TRI case for the same operating condition. With TRI, radiative emission from gases and soot both increase-. However, the increase in soot emission is more prominent than in gas emission.

Keywords: *Radiation modelling; Turbulence-radiation interactions; Transported probability density function method; Turbulence–chemistry interactions; Compression–ignition engine*

1. Introduction

The importance of accurately accounting for realistic chemical kinetics and turbulence-chemistry interactions in CFD simulations of compression-ignition engines is well established [1]. On the other hand, while it has long been known that radiation can account for a significant fraction of the total heat losses in very large bore, heavy-duty diesel engines [2], conventional wisdom had been that in-cylinder radiation was dominated by soot, and that radiation was of secondary importance in road-vehicle-scale engines (cars and trucks). Few in-cylinder CFD modeling studies have considered radiation heat transfer, and only a subset of those has considered radiatively participating medium effects by solving a radiative transfer equation (RTE) [3-6].

In the meantime, advanced high-efficiency engines that are currently being developed are expected to function close to the limits of stable operation, where even small perturbations to the energy balance can have a large influence on global system behavior. The trends toward higher operating pressures and higher levels of exhaust-gas recirculation (EGR) will make molecular gas radiation (absorption coefficient proportional to participating species concentration) more prominent. At the same time, increasing quantitative accuracy is being demanded from CFD-based models for in-cylinder processes, including accurate predictions of heat losses and pollutant emissions. For these reasons, radiative heat transfer under engine-relevant conditions has been revisited recently in [7]. There it has been shown that a detailed spectral model is required to account for correct reabsorption, and under engine-relevant conditions radiative emission by gas species is more important than the radiative emission by soot. While the main focus in [7] was on detailed radiation modeling and importance of overall radiation effects, in this work the main focus is on influence of turbulence-radiation interactions (TRI).

Here, the numerical simulations have been performed for the same Volvo truck diesel engine that was the subject of earlier experimental and modeling studies [1, 7]. The line-by-line (LBL) spectral radiation model and photon Monte Carlo (PMC) RTE solver have been used along with a transported probability density function (tPDF) method to perform coupled simulations (where the radiative source term feeds back to the CFD through the enthalpy equation). Preliminary results are presented emphasizing the influence of unresolved turbulent fluctuations on radiation (TRI) for a full-load operating condition with EGR, where experimental pressure and heat-release data are available.

2. Physical Models and Numerical Methods

An unsteady Reynolds-averaged formulation has been adopted, with a standard two-equation turbulence model with wall functions, and conventional stochastic-Lagrangian-parcel fuel injection and spray models (using n-heptane liquid fuel properties). A 42-species chemical mechanism (40-species n-heptane [8], plus two additional species for thermal NO) is used to represent the gas-phase chemistry, and a semi-empirical two-equation model is used for soot [9]. Gas-phase chemistry and soot are calculated at the notional particle level in a particle-based tPDF method [10] that accounts for the influences of turbulent fluctuations with respect to local mean values (turbulence-chemistry interactions – TCI). Radiative emission and reabsorption are calculated either at the finite-volume cell level using local mean values of composition and temperature (thereby neglecting the influence of turbulent fluctuations – no-TRI), or at the notional particle level that accounts for the influences of turbulent fluctuations (TRI). The models are essentially the same as those that were used in earlier modeling studies for a heavy-duty diesel engine [1, 7].

An up-to-date review of the theory and applications of radiative heat transfer in turbulent combustion systems can be found in [11]. Here the most recently available spectral radiation properties databases have been used. Radiative properties for molecular gases have been pre-calculated using the HITEMP2010 (CO_2, H_2O, CO) [12] and HITRAN2008 (CH_4, C_2H_4, etc.) [13] spectral databases for pressures from 0.1 bar to 80 bar, temperatures from 300 K to 3000 K, and for various mole fractions of the participating species. Simple pressure-based scaling is used to extrapolate to higher pressures, where needed. The spectral absorption coefficient for soot is evaluated using the small-particle limit (Rayleigh theory) [14] with the complex index of refraction from [15]; scattering is neglected. Models for participating liquid spray radiation have been developed recently [11], but spray radiation is negligible for the conditions that are of interest here, and the spray radiation models have not been used. From the spectral molecular-gas databases and the presumed soot radiation properties, a hierarchy of spectral models is constructed. These range from full LBL, to full-spectrum k-distributions, to gray-gas models with Planck-mean absorption coefficients [14].

Multiple RTE solvers have been implemented to calculate the local radiative intensity in situations where absorption is important. These include the stochastic photon Monte Carlo (PMC) method where no intrinsic assumptions are invoked regarding the directional distribution of radiative intensity, spherical-harmonics methods (SHM), and discrete-ordinates methods (DOM) [11, 14]. In the lowest-order SHM implementation (the P1 method), a single elliptic PDE must be solved. While DOM and variants have probably been used more widely than the others in combustion applications, recent work has shown that SHM methods provide a more favorable tradeoff between computational effort and accuracy as one goes to higher-order implementations [16].

PMC/LBL provides a benchmark against which the performance of simpler RTE solvers and/or spectral models can be compared. When combined with the stochastic Lagrangian particle methods that are used with transported PDF methods for turbulent reacting flows, PMC/LBL has proven to be a powerful approach for computing radiative transfer, including the effects of unresolved turbulent fluctuations in composition and temperature (turbulence-chemistry-soot-radiation interactions) in both Reynolds-averaged [17] and large-eddy simulations [18]. Results comparing different combination of spectral models and RTE solvers were presented in [7]. Here, to separate turbulence-chemistry interaction (TCI) effects from turbulence-radiation interaction (TRI) effects, tPDF simulations have been performed using particle-level chemistry and radiation (thereby accounting for both TCI and TRI) and with particle-level chemistry and cell-level radiation (thereby accounting for TCI, while ignoring TRI). Differences between results for PDF/PMC/LBL/noTRI and PDF/PMC/LBL/TRI illustrate effects of TRI.

3. Engine Configuration and Operating Conditions

A simplified model of Volvo 13L production six-cylinder heavy-duty truck engine has been built using OpenFOAM v2.3.x [19], and simulations have been performed for a baseline full-load diesel operating condition with EGR (15.8:1 compression ratio, 1213 r/min, 14.28 bar intake pressure, 561 K intake temperature, 282 mg fuel injected from 4.6° bTDC to 16.8° aTDC, initial O_2, N_2, CO_2 and H_2O mass fraction of 0.1884, 0.7575, 0.392 and 0.149 respectively). A relatively coarse (~85,000 cell) 60-degree sector mesh is centered on one of the six spray plumes, and includes the piston top-ring-land crevice. Simulations begin after intake-valve closure (IVC) with prescribed initial in-cylinder conditions, and end before exhaust-valve opening. A computational time step of 3.4 μs (0.025 CAD) is used, in all cases.

4. Results and Discussion

Results are presented for two different levels of in-cylinder soot. First, the "normal" engine operating conditions (Sec. 3) are explored. For this case, computed engine-out soot levels closely match the experimentally measured value. The computed peak total soot volume fraction for this full-load case is approximately 0.25 ppm, while maximum local soot volume fraction is approximately 8 ppm. Here radiative emission and reabsorption are calculated using radiation feedback to the CFD-solver.

Second, for the full-load no-radiation engine run the computed soot volume fraction in each computational particle and each computational cell is multiplied by a factor of 100. Then radiative emission and reabsorption are calculated in post-processing mode using PMC/LBL at the particle level (with TRI) and at the cell level (no-TRI). In this case the peak overall soot volume fractions become approximately 25 ppm. This high-soot case is designed to validate the effect of TRI for environments where the contribution of soot radiation is not negligible.

4.1 Normal Engine Operating Condition

Computed NO and soot mass for the full-load case are plotted in Fig. 1. With consideration of TRI, the engine-out NO (in ppm) decreases by approximately 1%, while the soot mass decreases by 5% (w.r.t no-TRI case). Computed total boundary layer heat loss through walls, radiative emission, radiative reabsorption and radiation reaching the walls are tabulated in Table 1. Instantaneous radiative emission, reabsorption and radiation reaching the walls comparing no-TRI and TRI cases are plotted in Fig. 2. Table 1 shows that total radiative emission and reabsorption are approximately 46% and 41% of total boundary layer heat loss, respectively. As a result, radiation reaching the walls is approximately 5% of total boundary layer heat loss. It is evident that 99% of the total radiative emission is contributed from gas emission. From Table 1 and Fig. 2, it is evident that with TRI total radiative emission and reabsorption increases by 1.34% and 2.84% (w.r.t to no-TRI case), respectively. As with TRI, increase in radiative reabsorption is more than increase in radiative emission. Hence, with consideration of TRI, net radiative heat loss is less than the no-TRI case under same operating condition.

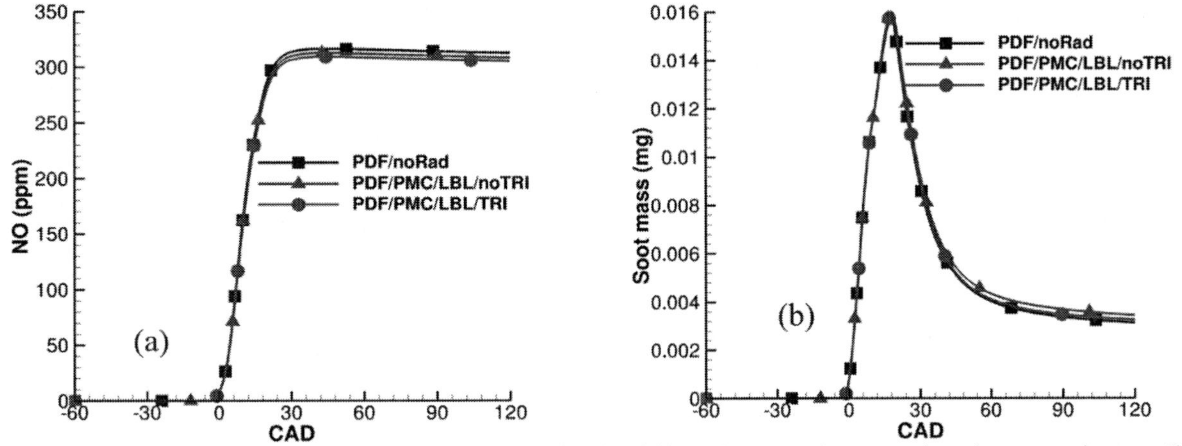

Figure 1: Computed NO (in ppm) and soot mass for the full-load case under normal engine operating condition.

Table 1: Radiative emission, reabsorption and radiation reaching walls for normal engine condition. BLHL- boundary layer heat loss through walls. All quantities are in joule.

Models	BLHL	Gas emis.	Soot emis.	Total emis.	Reabs.	Rad. Reaching walls
PDF/PMC/LBL/noTRI	231.96	106.891	0.235	107.126	94.206	12.919
PDF/PMC/LBL/TRI	232.27	108.341	0.272	108.613	96.869	11.743

4.2 Increased soot operating condition

As described earlier, here the soot volume fraction is increased 100 times (during radiative post-processing) for the no-radiation full-load case. Table 2 and Figure 3(a) shows that with TRI radiative emission and reabsorption increase by 4.84% and 5.37% (w.r.t no-TRI case), respectively. Due to increased soot, radiation reaching the walls becomes approximately 10% of the boundary layer heat loss. In this case 80% of total emission is due to gas emission and the rest is due to soot emission. Figure 3(b) shows the TRI effect on radiative emission from gas-species and soot. It is evident that with TRI, radiative emission from both gas and soot increases. However, the increase in radiative emission from gas-species are less than 3%, while the increase in soot emission is approximately 15% (w.r.t no-TRI run). This indicates that most of the soot belongs to higher temperature particles.

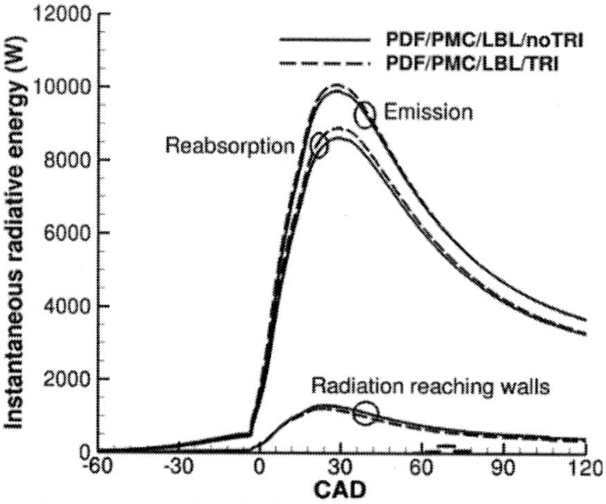

Figure 2: Radiative emission, reabsorption and radiation reaching walls for normal engine operating condition.

Table 2: Radiative emission, reabsorption and radiation reaching walls for increased soot engine condition. BLHL-boundary layer heat loss through walls. All quantities are in joule.

Models	BLHL	Gas emis.	Soot emis.	Total emis.	Reabs.	Rad. Reaching walls
PDF/PMC/LBL/noTRI	231.96	107.365	23.309	130.674	108.470	22.204
PDF/PMC/LBL/TRI	232.27	109.666	27.340	137.005	114.300	22.706

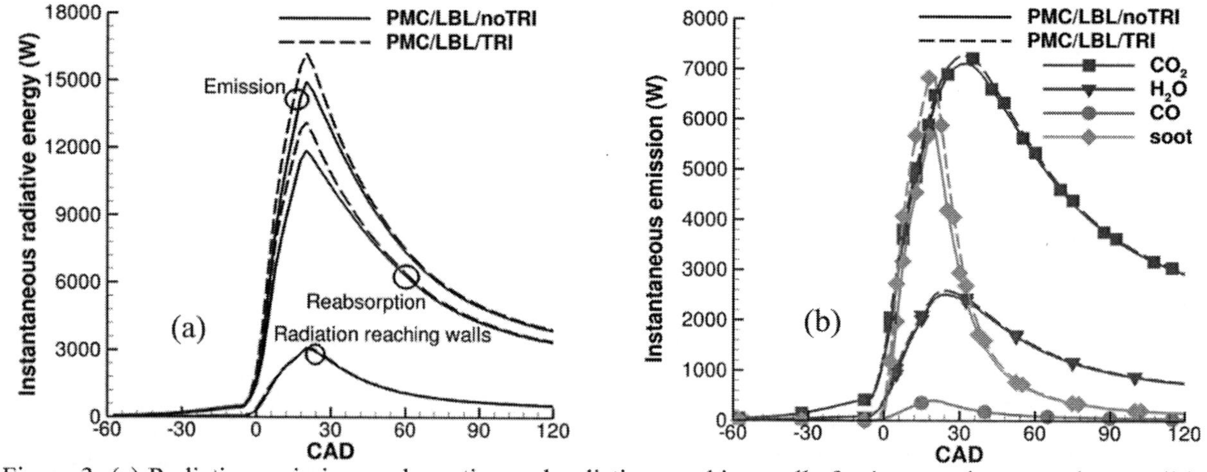

Figure 3: (a) Radiative emission, reabsorption and radiation reaching walls for increased soot engine condition. (b) Radiative emission by each of the gas species and soot comparing no-TRI and TRI case.

5. Conclusions

Preliminary coupled-radiation results have been presented for a PMC/LBL radiation model comparing no-TRI and TRI cases for a heavy-duty compression-ignition engine at a full-load operating condition. Overall, radiation effects are relatively small for this operating condition (~ 5%). With TRI, engine-out NO and soot emission decreases marginally, while radiative emission and reabsorption increase by a small amount (less than 3%). For the increased soot case the post-processing analysis shows that with TRI, the increase in soot emission is more prominent than the increase in gas-species emission.

6. Acknowledgements

This research has been funded by the U.S. National Science Foundation through grants CBET-258613 and CBET-1604446 (Haworth) and CBET-1258635 (Modest), and by the U.S. Department of Energy through Cooperative Agreement DE-EE0007278.

7. References

[1] Mohan, V.R. and Haworth, D.C., Turbulence–Chemistry Interactions in a Heavy-Duty Compression–Ignition Engine. Proc. Combust. Inst. 35 (2015) 3053-3060.

[2] Borman, G. and Nishiwaki, K., Internal-Combustion Engine Heat Transfer. Prog. Energy Combust. Sci. 13 (1987) 1-46.

[3] Abraham, J. and Magi, V., Modeling Radiant Heat Loss Characteristics in a Diesel Engine. SAE Technical Paper No. 970888 (1997), (1997).

[4] Mengüc, M.P., Viskanta, R., and Ferguson, C.R., Multidimensional Modeling of Radiative Heat Transfer in Diesel Engines. SAE Technical Paper No. 850503 (1985), (1985).

[5] Wiedenhoefer, J.F. and Reitz, R.D., A Multidimensional Radiation Model for Diesel Engine Simulation with Comparison to Experiment. Numer. Heat Transf. A. 44 (2003) 665-682.

[6] Yoshikawa, T. and Reitz, R., Effect of Radiation on Diesel Engine Combustion and Heat Transfer. J. Thermal Sci. Technol. 4 (2009) 86-97.

[7] Paul, C., Sircar, A., Ferreyro-Fernandez, S., Imren, A., Haworth, D.C., Roy, S., Ge, W., and Modest, M.F. Modeling Radiative Heat Transfer and Turbulence-Radiation Interactions in Engines. 2017. 10th U. S. National Combustion Meeting, College Park, Maryland, USA.

[8] Golovitchev, V.I., *Chalmers University of Technology, Gothenburg, Sweden*. 2000.

[9] Leung, K.M., Lindstedt, R.P., and Jones, W.P., A Simplified Reaction Mechanism for Soot Formation in Nonpremixed Flames. Combust. Flame. 87 (1991) 289-305.

[10] Haworth, D.C., Progress in Probability Density Function Methods for Turbulent Reacting Flows. Prog. Energy Combust. Sci. 36 (2010) 168-259.

[11] Modest, M.F. and Haworth, D.C., Radiative Heat Transfer in Turbulent Combustion Systems. 2016: Springer.

[12] Rothman, L.S., Gordon, I.E., Barber, R.J., Dothe, H., Gamache, R.R., Goldman, A., Perevalov, V.I., Tashkun, S.A., and Tennyson, J., Hitemp, the High-Temperature Molecular Spectroscopic Database. J. Quant. Spectrosc. Ra. 111 (2010) 2139-2150.

[13] Rothman, L.S., Gordon, I.E., Babikov, Y., Barbe, A., Benner, D.C., Bernath, P.F., Birk, M., Bizzocchi, L., Boudon, V., Brown, L.R., Campargue, A., Chance, K., Cohen, E.A., Coudert, L.H., Devi, V.M., Drouin, B.J., Fayt, A., Flaud, J.M., Gamache, R.R., Harrison, J.J., Hartmann, J.M., Hill, C., Hodges, J.T., Jacquemart, D., Jolly, A., Lamouroux, J., Roy, R.J.L., Li, G., Long, D.A., Lyulin, O.M., Mackie, C.J., Massie, S.T., Mikhailenko, S., MÃ¼ller, H.S.P., Naumenko, O.V., Nikitin, A.V., Orphal, J., Perevalov, V., Perrin, A., Polovtseva, E.R., Richard, C., Smith, M.A.H., Starikova, E., Sung, K., Tashkun, S., Tennyson, J., Toon, G.C., Tyuterev, V.G., and Wagner, G., The Hitran2012 Molecular Spectroscopic Database. J. Quant. Spectrosc. Ra. 130 (2013) 4-50.

[14] Modest, M.F., Radiative Heat Transfer. Third Ed. 2013: Academic Press, New York.

[15] Chang, H. and Charalampopoulos, T.T., Determination of the Wavelength Dependence of Refractive Indices of Flame Soot. Proceedings of the Royal Society of London A: Mathematical, Physical and Engineering Sciences. 430 (1990) 577-591.

[16] Pal, G., Gupta, A., Modest, M.F., and Haworth, D.C., Comparison of Accuracy and Computational Expense of Radiation Models in Simulation of Non-Premixed Turbulent Jet Flames. Combust. Flame. 162 (2015) 2487-2495.

[17] Mehta, R.S., Modest, M.F., and Haworth, D.C., Radiation Characteristics and Turbulence–Radiation Interactions in Sooting Turbulent Jet Flames. Combust. Theory Model. 14 (2010) 105-124.

[18] Gupta, A., Haworth, D.C., and Modest, M.F., Turbulence-Radiation Interactions in Large-Eddy Simulations of Luminous and Nonluminous Nonpremixed Flames. Proc. Combust. Inst. 34 (2013) 1281-1288.

[19] Openfoam, 2015 <Http://Www.Openfoam.Org/Download/Git.Php>.

2018 ESSCI Spring Technical Meeting
Eastern States Section of The Combustion Institute
March 04-07, 2018
State College, Pennsylvania

CFD-based non-equilibrium wall heat transfer models for engine-relevant conditions

A. Sircar[1] and D.C. Haworth[1]

[1]Department of Mechanical & Nuclear Engineering, The Pennsylvania State University, University Park, Pennsylvania, USA

Abstract: The goal of this research is to contribute towards the development of predictive, physics-based computational fluid dynamics (CFD) models for studying advanced internal combustion (IC) engines with high efficiency and low emissions. The lack of accurate submodels for in-cylinder heat transfer has been identified as a key shortcoming to this end. Recent measurements of wall layers in engines show discrepancies of up to 100% with respect to standard CFD boundary-layer models. Additionally, available experimental data are quite sparse (only a few data points on engine walls) and limited (available measurements are those of heat flux only). Predictive submodels are needed for medium-resolution engineering LES and for unsteady Reynolds-averaged simulations (URANS). Recently, some research groups have performed DNS studies on engine-relevant conditions using simple geometries which provide very useful data for benchmarking computational models. While some recent sophisticated (one-equation) models have been incorporated as a part of the current work, others of a more detailed nature are underway. These include solving additional partial differential equations (PDEs) within the thin boundary layer near the wall and fall under the category of non-equilibrium models. The main aim of this project is to develop such two-zone models for engine wall heat transfer and to test them against available DNS/experimental data in both SI (spark-ignition) and CI (compression-ignition) engines. Ultimately it is also intended to include the effect of radiation in these models.

***Keywords:* wall-models, non-equilibrium models, IC engines, URANS, LES**

1. Introduction

The Workshop on Clean and Efficient Combustion of 21st Century Transportation Fuels (organized by the DOE[2]) recognizes the need for accurate simulation of in-cylinder processes in IC engines in order to effectively utilize the limited natural reserve of fuels and reduce pollutant emissions at the same time. These in-cylinder processes are not only multi-scale in nature but they also entail a wide range of physical concepts. Hence the development of predictive models for these processes is of importance for their accurate simulation. One of the major hurdles in CFD simulations of IC engines (both CI and SI) is the use of reduced order modeling for wall heat transfer. There are a number of models available in the literature, but with great discrepancies (almost 100% [1]) amongst the predictions from them. These models have been extensively used for IC engine simulation in spite of the fact that they make use of approximations that are not valid in an engine environment especially with flame propagation and combustion. In reference to the multi-scale nature of these problems it should be noted that although LES can capture the large-scale phenomenon it is quite expensive in computational cost and complexity, and still requires wall models.

[2]https://science.energy.gov/~/media/bes/pdf/reports/files/Clean_and_Efficient_Combustion_of_21st_Century_Transportation_Fuels_rpt.pdf

For engineering computations, the URANS approach is often adopted although it doesn't resolve the flow-scales in the boundary layers, especially at high engine speeds. Typical boundary layer thicknesses in IC engines are of the order of 100 μm to 1000 μm under motored conditions and can vary depending on the level of swirl and firing [2]. The physical processes in the boundary layers of IC engines have great implications on heat exchange and combustion phasing [3]. These are modeled using wall-functions, which can be derived from the thin boundary-layer equations (reduced Navier Stokes equations in boundary layers) after making the following approximations [4]:

- A1: Flow is parallel to the chamber walls with negligible flow normal to the walls
- A2: Quasi-steady flow with negligible transients
- A3: Constant density and transport properties in the boundary layer
- A4: High Reynolds number flow, very thin laminar shear layer
- A5: Zero pressure gradients along the flow
- A6: No combustion and chemically inert mixture

Models which result from these assumptions are called equilibrium wall-function models for the flow; the corresponding models, by analogy, from a low-Mach number assumption, for temperature are known as the thermal wall-function models. It has been found that such models usually under-predict wall heat losses compared to experimental measurements [5]. This can be explained by the fact that approximations A1-A6 deviate greatly from the flow conditions in real engines. Ma et al. [6] have performed experiments using μ-PIV and LIF in a two-valve optical research engine (the TCC, or Transparent Combustion Chamber, engine) and showed that the boundary layer is driven by a pressure gradient which causes the flow to be 3D and unsteady with local flow reversal. Assumption A3 is strongly violated as well, since there is a large temperature gradient near the wall and transport properties are inherently temperature-dependent. These arguments have led researchers to believe that these assumptions need to be relaxed. Hence, a wide range of models is available in the literature that relax one or more of these assumptions [7]. Some of these are discussed in the next section. Since these equilibrium assumptions aren't sufficient to capture the dynamics of the boundary layers, non-equilibrium models have been proposed and have been shown to better reproduce experimental results [6].

Another major limiting factor in understanding which models best represent the physics of boundary layers in engines is the lack of good experimental data. Rakopoulos et al. [8] list a number of available experiments from the literature. However, these studies report only the heat flux values at discrete locations on the cylinder head, liner and piston. This is not sufficient for model validation since we need to be able to compare with fields of data. Also, measurements of other physical quantities, such as turbulence quantities, y+ values and turbulent Prandtl number, which go into models for wall heat losses, are required to make a comprehensive analysis of model performance.

2. Equilibrium wall-models and preliminary results

It is important to be able to validate the many models available in the literature. For this purpose, two main data sets are being targeted for the current study – experimental measurements of heat flux and velocity fields in the TCC engine [9] and DNS (Direct Numerical Simulations) of the Morse engine by Schmitt et al. [10]. DNS results are better for model validation since they provide complete fields of heat fluxes, turbulence quantities (integral length scales), velocity etc. Details of the DNS setup and corresponding data for this engine are provided in refs [10-13]. The results from the DNS study are used to validate some of the models incorporated in the URANS framework. OpenFOAM (OF) is used as the basic CFD code in which the wall-function models have been incorporated. The modelled energy equation (1) solved in OF2.3.x for reactive flows is solved in each finite volume cell while for wall cells the last term in the LHS represents the wall heat loss. Table 1 lists the models from the literature that have been incorporated into OF2.3.x.

$$\frac{\partial(\rho h)}{\partial t}+\nabla(\varphi h)+\frac{\partial\left(\rho\frac{1}{2}u^2\right)}{\partial t}+\nabla\left(\varphi\frac{1}{2}u^2\right)-\frac{dP}{dt}-\nabla^2\left(\left[\alpha+\alpha_t\right]h\right)=Q_{rxn} \tag{1}$$

where φ is the mass flux in each is cell and Q_{rxn} denotes the chemical energy term.

$$\alpha_{t,w}=\frac{\mu_{t,w}}{\mathrm{Pr}_t} \quad \text{where the subscript w denotes the value at the wall} \tag{2}$$

Table 1: Heat-loss models from the literature incorporated in OF2.3.x

Name of model	**Heat Loss model**	**Comments**
WernerWengle [14]	$q_W=\frac{\rho c_p u_\tau T\ln(T/T_W)}{C_{hw}\left\{7.843\tan^{-1}\left(0.0935y^+\right)\right\}}$	$y^+<40$ Modeled τ_W for computing u_τ
Angelberger [15]	$q_W=\frac{\rho c_p u_\tau T_W\ln(T/T_W)}{\theta^+}$	Viscous sub layer $\theta^+=\mathrm{Pr}\,y^+$ Turbulent layer $\theta^+=\mathrm{Pr}_t\left(u^++P\right)$
Jayatilleke [16]	$T^+=\frac{\rho c_p u_\tau\left(T_W-T\right)}{q_W}$ $T^+=\mathrm{Pr}\,y^++\frac{1}{2}\rho\,\mathrm{Pr}\frac{u_\tau}{q_W}U_P^2$ $T^+=\mathrm{Pr}_t\left[\frac{1}{\kappa}\ln\left(Ey^+\right)+P\right]+\frac{1}{2}\rho\frac{u_\tau}{q_W}Q$ $Q=\mathrm{Pr}_t U_P^2+\left(\mathrm{Pr}\quad\mathrm{Pr}_t\right)U_C^2$ $P=9.24\left[\phi^{3/4}-1\right]\left[1\quad 0.28e^{-0.007\phi}\right]$ $\phi=\mathrm{Pr}/\mathrm{Pr}_t$	 If $y^+<y^+$_thermal If $y^+>y^+$_thermal

The models discussed in Table 1 along with the OpenFOAM built-in model (referred to as constant Prt=0.85) are used to generate results for the compression stroke of the geometry of the configuration used in [13]. The LES and DNS of the same compression stroke by Mandanis et al. [13] are used for preliminary validation of these models. Figure 1 shows that these models predict the average pressure and temperature of the engine reasonably well. Figure 2 shows that there is a wide discrepancy in the total heat loss predictions. The average $y+$ values show that the first grid point lies within the log-law region.

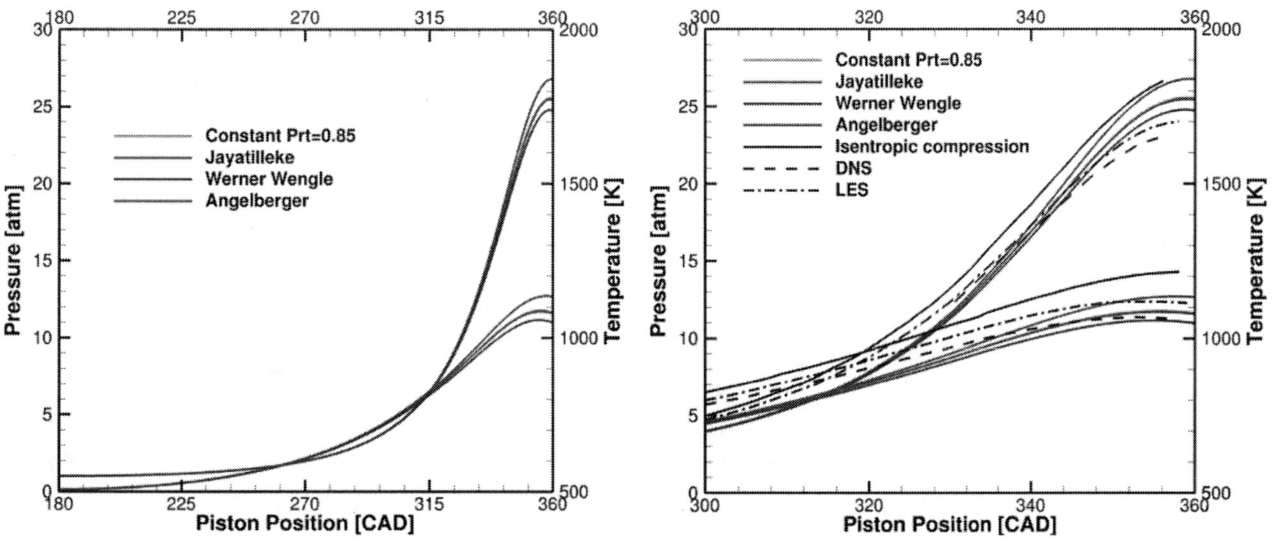

Figure 1: Predictions of average Pressure and Temperature from the models in Table 1

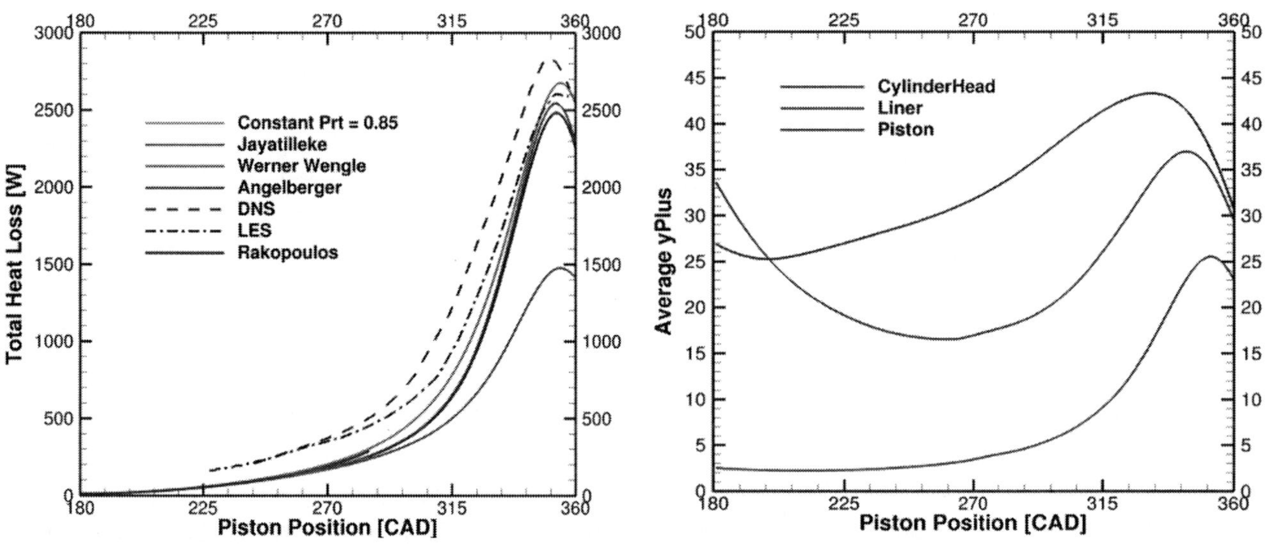

Figure 2: (left) Predictions of Total wall heat loss from the models in Table 1
(right) Variation of average $y+$ values of the OpenFOAM model

3. Non-equilibrium wall models

The next target is to incorporate non-equilibrium wall models as described in [6]. These models solve the thin boundary layer equations (1D/2D transient NS equations) on a grid which is laid between the last grid point (closest to the wall) in the main CFD mesh and the wall. This requires no additional modeling since it directly solves the relevant governing equation, hence inherently removing assumptions A1-A5. Ma et al. [6] discuss the implementation of this model using the zonal approach and different turbulence closures. It was shown that this model works better in predicting wall heat loss under motored conditions for the TCC engine.

The implementation of non-equilibrium wall heat transfer models is being done in OF-2.3.x. For initial testing of the implementation, a simple rectangular channel flow geometry is considered, as shown in Figure 3. The flow geometry and initial setup is taken from [17]. Figure

3 shows how the channel is divided into two zones (red and green). The red region is the main CFD mesh, while the green region is the near-wall region in which the thin boundary-layer equations are to be solved. Fully developed solutions with URANS and LES solvers have been validated with results from [17]. The task of coupling the three-dimensional solver (for the main mesh) and the locally one-dimensional solver (for the wall-zone) is underway. Such a hybrid solver is expected to yield more accurate results (high resolution of LES near the wall) at lower computational cost compared to solving the three-dimensional equations down to the wall.

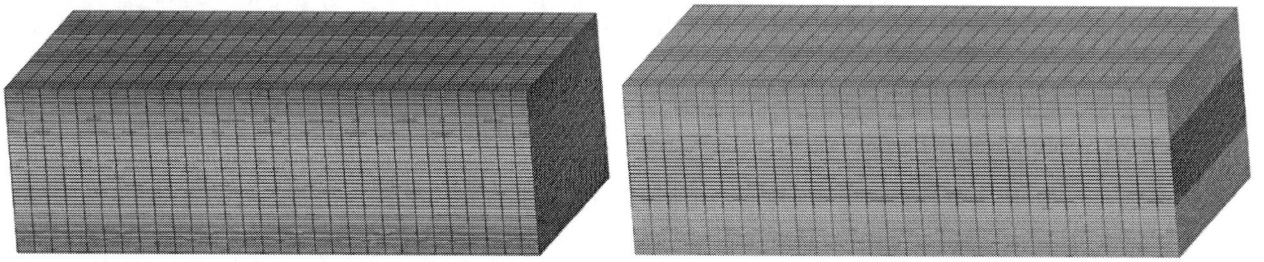

Figure 3: (left) One-zone mesh (right) Two-zone (wall-zone exaggerated) mesh – for a rectangular channel

The modeled thin boundary layer equations that are to be solved in the wall-zone are given below in 2D form.

$$\bar{\rho}\left(\frac{\partial\bar{u}}{\partial t}+\bar{u}\frac{\partial\bar{u}}{\partial x}+\bar{v}\frac{\partial\bar{u}}{\partial y}\right)=-\frac{\partial\bar{p}}{\partial x}+\frac{\partial}{\partial y}\left((\bar{\mu}+\mu_t)\frac{\partial\bar{u}}{\partial y}\right) \tag{3}$$

$$\bar{\rho}c_P\left(\frac{\partial\bar{T}}{\partial t}+\bar{u}\frac{\partial\bar{T}}{\partial x}+\bar{v}\frac{\partial\bar{T}}{\partial y}\right)=-\frac{\partial\bar{p}_0}{\partial t}+\frac{\partial}{\partial y}\left((\bar{\lambda}+\lambda_t)\frac{\partial\bar{T}}{\partial y}\right)+Q_{rad} \tag{4}$$

4. Conclusions and Future Work

These preliminary validation runs with equilibrium models show that the prediction is in relatively good agreement for a motored flow, except for the Angelberger model which predicts very low wall heat loss (similar to previous findings [5]). However, further comparisons are required of the turbulence quantities to be able to conclusively say anything about the actual performance of these models under fired conditions. This will be addressed further after the implementation and validation of the two-zone hybrid non-equilibrium model.

Next steps are

1. Implementation of the hybrid two-zone non-equilibrium model
2. Removing assumption A6, i.e. considering combustion
3. Including the effect of radiation in the boundary layer as well [18]

5. Acknowledgements

This material is based upon work supported by the Department of Energy, Office of Energy Efficiency and Renewable Energy (EERE) and the Department of Defense, Tank and Automotive Research, Development, and Engineering Center (TARDEC), under Award Number DE-EE0007278.

6. References

Journal articles:
[1] A.Y. Alharbi, V. Sick, Investigation of boundary layers in internal combustion engines using a hybrid algorithm of high speed micro-PIV and PTV, Experiments in Fluids, 49:949-959, 2010.
[2] P.H. Pierce, J.B. Ghandhi, J.K. Martin, Near-wall velocity characteristics in valved and ported motored engines, SAE Paper 1992; : 920152
[3] J.E. Dec, Advanced compression-ignition engines – understanding the in-cylinder processes, Proc Combust Inst 2009; 32: 2727-2742
[4] Z. Han, R.D. Reitz, A temperature wall-function formulation for variable-density turbulent flows with application to engine convective heat transfer modeling, Int. J. Heat Mass Transf, 40(3), 613-625, 1997.
[5] S. Keum, H. Park, et al., Modeling of heat transfer in internal combustion engines with variable density effect, Int J Engine Res 2011; 12(6): 513-526
[6] P.C. Ma, T. Ewan, C. Jainski, L. Lu, A. Dreizler, V. Sick, M. Ihme, Development and analysis of wall models for internal combustion engine simulations using high-speed micro-PIV measurements, Flow Turbulence Combustion, 2017 98:283-309
[7] K.Y. Huh, I.P. Chang, J.K. Martin, A comparison of boundary layer treatments for heat transfer in IC engines, SAE Paper 900252, 1990
[8] C.D. Rakopoulos, G.M. Kosmadakis, E.G. Pariotis, Critical evaluation of current heat transfer models used in CFD in-cylinder engine simulations and establishment of a comprehensive wall-function formulation, Applied Energy 87 2010, 1612-1630
[9] P. Schiffman, S. Gupta, D. Reuss et al., TCC-III engine benchmark for large-eddy simulation of IC engine flows, Oil & Gas Science and Technology 2017; 71(1)
[10] M. Schmitt, C.E. Frouzakis, A.G. Tomboulides, Y.M. Wright, K. Boulochos, Direct numerical simulation of multiple cycles in a valve/piston assembly, Physics of Fluids 26, 035105, 2014
[11] M. Schmitt, C.E. Frouzakis, A.G. Tomboulides, Y.M. Wright, K. Boulochos, Direct numerical simulation of the effect of compression on the flow, temperature and composition under engine-like conditions, Proceedings of the Combustion Institute, 35 3069-3077, 2015.
[12] M. Schmitt, C.E. Frouzakis, A.G. Tomboulides, Y.M. Wright, K. Boulochos, Direct numerical simulation of the compression under engine relevant conditions: Local wall heat flux distribution, Int J of Heat and Mass Transfer, 92 718-731, 2016.
[13] C. Mandanis, M. Schmitt, J. Koch, Y.M. Wright, K. Boulouchos, Wall heat flux and thermal stratification investigations during the compression stroke of an engine-like geometry: A comparison between LES and DNS, Flow Turbulence and Combustion, 2016
[14] H. Werner, H. Wengle, LES of turbulent flow over a square rib in a channel, Advances in Turbulence, Vol 2 418-423, 1989.
[15] C. Angelberger, T. Poinsot, B. Delhay, Improving near-wall combustion and wall heat transfer modeling in SI engine computations, SAE 972881 113-130, 1997
[16] C. Jayatilleke, The influence of Prandlt number and surface roughness on the resistance of the laminar sublayer to momentum and heat transfer, Progress in Heat and Mass Transfer, 1:193-321, 1969
[17] A. Gupta, Large-eddy simulation of turbulence/chemistry/radiation interactions in planar channel flow, Masters Thesis, Penn State 2007.
[18] Y. F. Zhang, R. Vicquelin, O. Gicquel, J. Taine, Practical indicators for assessing the magnitudes of wall radiative flux and of coupling effects between radiation and other heat transfer modes on the temperature law-of-the-wall in turbulent gaseous boundary layers, IJHMT, 120 76-85, 2018.

Spring Technical Meeting
Eastern States Section of the Combustion Institute
March 4-7, 2018
State College, Pennsylvania

Combustion Visualization in a Single-Cylinder Heavy-Duty CI Engine Converted to Natural Gas SI Operation

Jinlong Liu[1], *Cosmin E. Dumitrescu*[1*]

[1]*Mechanical and Aerospace Engineering, West Virginia University, Morgantown, WV, USA,*

[*]*Corresponding Author Email: cosmin.dumitrescu@mail.wvu.edu*

Abstract: This paper details an optical investigation of the combustion phenomena inside a spark-ignition internal combustion engine. The main objective was to provide an additional insight into the combustion characteristics of a conventional heavy-duty direct-injection engine converted to spark ignited natural gas (NG) operation. The compression-ignition-to-spark-ignition conversion was made through the addition of a NG injector in the intake manifold and by replacing the diesel injector with a NG spark plug. The engine was run at medium engine load under lean operating conditions, using pure methane as natural gas surrogate. The results suggested no knocking phenomena despite the increased compression ratio of the engine and a diesel-representative combustion geometry. Individual engine cycles showed strong evidence of turbulent flame wrinkling and counterclockwise rotation. The ignition point was usually different from the spark plug center. The moderate rate of pressure rise and lack of knocking showed promise for heavy-duty CI engines converted to NG SI operation.

Keywords: Optical IC Engine, Natural Gas, SI Combustion

1. Introduction

Internal combustion engines will continue to be the main power source in the transportation sector for the foreseeable future. While current and future exhaust aftertreatment technologies can meet increasingly stringent emission regulations, they increase engine cost and require advanced controlling strategies. At the same time, the use of alternative fuels can both reduce petroleum dependence and the cost of implementing aftertreatment devices. For example, natural gas (NG) has the potential to lower engine-out CO_2 emissions at similar power output compared to conventional fuels due to its higher hydrogen-to-carbon (H/C) ratio [1]. Additionally, NG will lower PM emissions by avoiding the creation of fuel-rich regions inside the cylinder [1]. More, NG's higher octane-number allows a higher compression ratio which increases engine efficiency

[1]. Furthermore, NG engines can run leaner than their gasoline counterparts, which would not only reduce carbon monoxide (CO) and unburned hydrocarbon (HC) emissions, but can also lower the engine-out nitrogen oxides (NOx) emissions due to lower combustion temperature at lean-mixture conditions [1]. Ideally, NG would replace conventional petroleum fuels (i.e., gasoline and diesel) in IC engines without the need of major modifications. However, NG has a higher auto-ignition temperature compared to diesel fuel, which can be solved by replacing the diesel injector with a spark plug and adding a port-fuel injector in the intake manifold for NG delivery [2]. Other solutions to overcome NG's higher auto-ignition temperature include dual-fuel combustion, partially-stratified injection strategy, and jet-ignition methods [3-6]. However, these solutions usually require major engine modifications which would increase cost and control complexity. Therefore, the former solution is preferred if appropriate calibrations are applied to help limiting high rate of pressure rise and avoiding knocking specially at medium and high load conditions. Albeit its abundance, low cost, and combustion advantages over conventional fuels, not many NG vehicles are registered in the United States (U.S.) [1]. As most NG combustion investigations in the literature were performed in setups not representative of engine geometry (e.g., constant-volume vessels, rapid-compression machines), the main objective of this study was to provide an additional insight into the combustion characteristics of a conventional heavy-duty direct-injection engine with optical access converted to spark ignited NG operation. Images of flame inception and propagation can be useful for the development and validation of combustion models that simulate the physics of turbulent flames in such environments.

2. Experimental Setup

Experiments were performed in an optically-accessible single-cylinder research engine (Ricardo/Cussons, U.K., Model Proteus), based on a commercial heavy-duty diesel engine (Volvo, Sweden, Model TD120). The original cylinder block was replaced with an extended version allowing the visualization of the combustion chamber from below through a "Bowditch" piston. As a result, the combustion bowl has a flat bottom and vertical walls compared to the original piston bowl. The original diesel injector was replaced by a spark plug (Stitt, USA, Model S-RSGN40XLBEX8.4-2) and NG was delivered using a port fuel injector. In-cylinder pressure was measured by a piezo-electric pressure transducer (Kistler, Model 6011), which was installed in the glow plug location and connected to a charge amplifier (Kistler, Model 5010). To protect the optical window from thermal and mechanical stresses, a skip-fired control strategy (one fuel injection event every 5 engine cycles) was applied in these experiments. At the same time, the cylinder head needed to be removed each time the window needed cleaning, which reduced the number of engine cycles that could be recorded before the image quality was affected. An aftermarket engine management system (Megasquirt, Model 3X) controlled engine operation conditions such as fuel injection and spark timing. An in-house-built data acquisition software (Scimitar) collected operating data such as engine speed, torque, air, coolant, and oil temperature, and air mass flow. The engine was operated at 900 rpm and at a lean equivalence ratio of 0.66, using 99.5% methane as NG surrogate, a fuel mass per cycle of 0.083g, and a spark timing of 20 CA BTDC. The intake pressure and temperature were 84 kPa and 300 K, respectively. A high-speed CMOS camera (Photron, USA, Model Fastcam SA5), fitted with a 50-mm Nikon lens, collected flame luminosity in the visible range during combustion cycles. Figure 1 shows the schematic of the research engine in the optical-access configuration. Additional details of the engine setup and operating conditions are provided in reference [1].

Figure 1: Optical research engine

The high-speed camera visualized the combustion chamber from below through the fused-silica piston window using a 45° mirror. The first acquisition image was synchronized with the engine's spark timing to capture combustion behavior. The camera used a longer exposure time to increase image quality during early flame propagation, which resulted in saturated pixels once the fully-turbulent flame encompassed the field of view. However, this did not affect the optical analysis, which was done in a MATLAB® environment.

3. Results

For the reasons mentioned in the *Experimental Setup* section, the results and discussion bellow were based on pressure traces and images from 21 engine cycles.

3.1 In-cylinder pressure analysis.

In-cylinder pressure traces shown in Figure 2a show cycle-to-cycle pressure variations, especially for the peak pressure, which is normal for NG SI combustion. Knocking is always a concern for CI engines retrofitted to NG SI operation. However, the low rates of pressure rise shown in Figure 2b suggest no knocking phenomena in the visualized cycles at this operating condition, despite the increased compression ratio of the engine and a diesel-representative combustion geometry. Specifically, the peak pressure rise rates for all the cycles were below 1.5 bar/deg CA, suggesting that flame propagation was dominant throughout the whole combustion process. The apparent heat release rate (AHRR) and cumulative heat release rate in Figures 2c and 2d suggest a start of combustion between 12 to 2 CA BTDC, even though the combustion images show an earlier flame propagation. As explained in reference [1], this was probably due to the location of the pressure transducer (further away from the spark plug), small pressure increases in the early flame stages, and the delay added by each component of the data acquisition system.

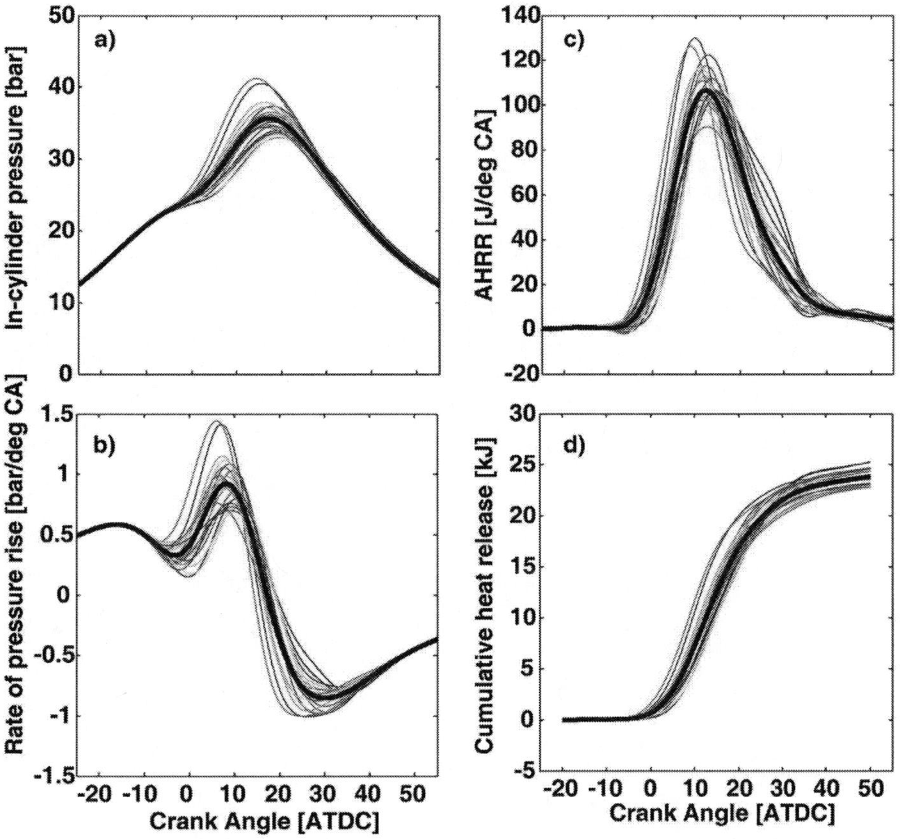

Figure 2. In-cylinder pressure (a), rate of pressure rise (b), apparent heat release rate (c), and cumulative heat release rate (d), for the 21 visualized engine cycles. Individual traces are shown with thin lines and their average with thick black line.

3.2 Flame luminosity analysis.

Flame luminosity recorded the flame inception and propagation for each fired cycle, until the flame completely encompassed the field of view. Figure 3 shows typical early flame development images from four different engine cycles. The inner circle in the images identifies the piston bowl edge. The spark plug, positioned in the middle of the bowl, is identified by the strong luminosity during ignition. The first row of images at 14.5 CA BTDC shows different spark intensities, which probably affected the flame inception. At the same time, the flame inception was different for each cycle. The second row of images at 11.7 CA BTDC images confirm that the different spark affected the early flame propagation. More, the flame rotated counterclockwise. Cycle 01 and Cycle 16 had similar 2D flame areas but with different shapes. Cycle 05 and Cycle 12 had late flame propagation but for different reasons. Cycle 05 had a slow flame development with a non-circular shape that spread in two opposite directions. Cycle 12 experienced the late flame inception mainly due to a low spark intensity. The rest of the recorded images (not shown here) show that the end of combustion was early for all cycles with an early flame developed. This suggests that the flame inception played the most important role in the flame propagation for this engine at the operating conditions investigated.

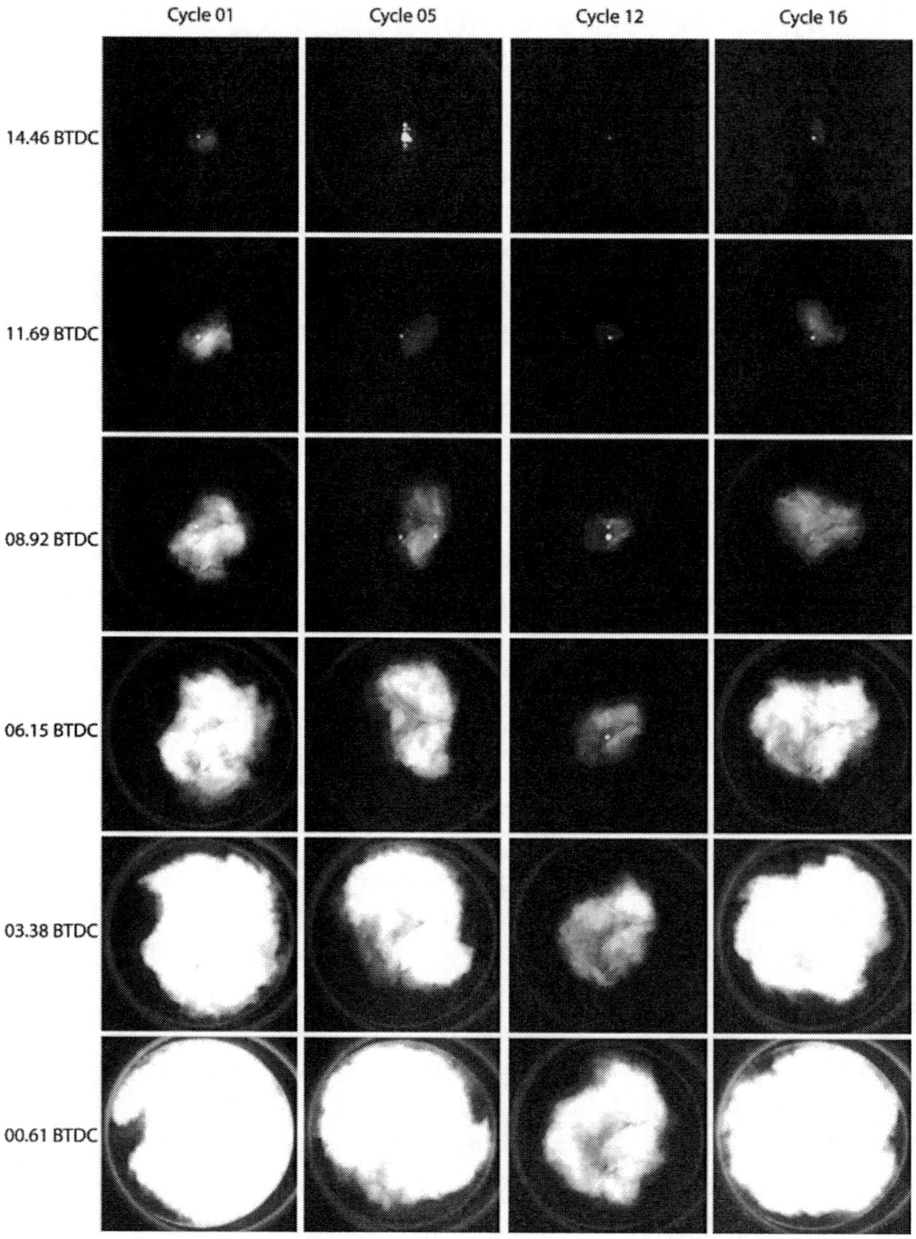

Figure 3: Early flame development for selected cycles.

4. Conclusions

This study investigated flame development in an optical CI engine that was converted to SI NG operation through the addition of a NG injector in the intake manifold and by replacing the diesel injector with a NG spark plug. The major conclusions were:

(1) Images and pressure data captured cycle-to-cycle variations characteristic of normal SI combustion but no knocking phenomena despite the increased compression ratio of the engine and a diesel-representative combustion geometry.

(2) Strong evidence of turbulent flame wrinkling was seen in the movies of individual cycles, accompanied with anti-clock rotation. The ignition point was usually different from the spark plug center.

(3) The moderate rate of pressure rise and lack of knocking showed promise for heavy-duty CI engines converted to NG SI operation.

5. Acknowledgement

This material is based upon work supported by the West Virginia University. The authors gratefully acknowledge CAFEE for its assistance with comprehensive/mechanical, electronic, and data-acquisition hardware/software systems, respectively. The research was conducted at the WVU's Advanced Combustion Laboratory in Morgantown, WV.

6. References

[1] Dumitrescu, Cosmin E., Vishnu Padmanaban, and Jinlong Liu, An Experimental Investigation of Early Flame Development in an Optical SI Engine Fueled with Natural Gas, ASME 2017 Internal Combustion Engine Division Fall Technical Conference, paper ICEF2017-3525.

[2] Ferguson, Colin R., and Allan T. Kirkpatrick. Internal combustion engines: Applied thermosciences. John Wiley & Sons, 2015.

[3] McTaggart-Cowan, G., K. Mann, N. Wu, and S. Munshi, An efficient direct-injection of natural gas engine for heavy duty vehicles, SAE 2014 World Congress & Exhibition, paper 2014-01-1332.

[4] Dronniou, Nicolas, Julian Kashdan, Bertrand Lecointe, Kyle Sauve, and Dominique Soleri, Optical Investigation of Dual-fuel CNG/Diesel Combustion Strategies to Reduce CO_2 Emissions, SAE 2014 World Congress & Exhibition, paper 2014-01-1313.

[5] Dahodwala, Mufaddel, Satyum Joshi, Erik Koehler, Michael Franke, and Dean Tomazic, Experimental and computational analysis of diesel-natural gas RCCI combustion in heavy-duty engines, SAE 2015 World Congress & Exhibition, paper 2015-01-0849.

[6] Andreassi, L., S. Cordiner, V. Mulone, C. Reynolds, and R. L. Evans, A mixed numerical-experimental analysis for the development of a partially stratified compressed natural gas engine, 7th International Conference on Engines for Automobile, paper 2005-24-029.

2018 Spring Technical Meeting
Eastern States Section of the Combustion Institute
Hosted by Pennsylvania State University
March 4-7, 2018

High Load Diesel Engine-Generator Power Improvement with Advanced Combustion Modes

Michael Walker[1], *Dianne Luning Prak*[2], *Len Hamilton*[1], *Jim Cowart*[1,*]

[1]*Mechanical Engineering Department, U.S. Naval Academy, Annapolis, MD, USA*
[2]*Chemistry Department, U.S. Naval Academy, Annapolis, MD, USA*
[*] *Corresponding Author Email: cowart@usna.edu*

Abstract: An advanced combustion mode, Partially Premixed Combustion (PPC), approach was applied in a Yanmar L100V6-L single cylinder diesel engine gen-set (electrical generator set with engine) under high operating loads in an effort to increase maximum power. Direct in-cylinder fueling was conventional Navy Jet Fuel (JP-5). Port injected fuel consisted also of JP-5, however, additional low reactivity fuels were also explored, neat isooctane (iC8) and iso-butanol. PPC (both single and dual fuel) modes are alternative advanced combustion approaches that generally result in lower engine-out soot. In this study, these modes were accomplished with port fuel injection through the intake manifold while the majority fraction of JP-5 fuel was delivered directly to the combustion chamber near the start of combustion (SOC). Fast in-cylinder pressure based combustion metrics were acquired along with exhaust soot opacity.

Results show that engine Indicated Gross Mean Effective Pressure (iGMEP) from 7.5 to 8.5 bar could be attained with PPC combustion as compared to only 7 bar with conventional operation. This led to power levels of 9.3 kW to 11.3 kW compared to 8.5 kW at conventional operation (gen-set rated for 6.7 kW) without a soot penalty. In general, JP-5/JP-5 (in-cylinder/port fuel) PPC led to the lowest power increase due to early combustion initiation. JP-5/iC8 provided for the highest power improvement due to the low reactivity port injected iC8 fuel, which resulted in later start and bulk combustion as compared to conventional operation and the other PPC modes. Iso-butanol PPC provided moderate power improvements.

Keywords: Internal Combustion Engine, PPC, Jet Fuel

1. Introduction

Portable and independent electric power generation is a core need of the military, particularly when operating in remote and isolated environments. Due to the power needs of modern technology and weapons systems, the demands of electric power continue to grow in these operational environments. A rapid and cost effective solution to meeting this need may be found in applying a retrofit approach to existing diesel generators in order to increase engine specific power. However, diesel engine power is limited by soot formation, an indicator of locally high fuel-air equivalence ratio, due to incomplete mixing of the air and fuel at high engine loads. More fuel cannot be delivered and reacted in the combustion chamber since the allotted diffusion time is very small at high engine RPM, and oxygen is limited at high engine loads. An alternative advanced combustion mode called partially premixed combustion (PPC, also known as Partially

Premixed Compression Ignition, PPCI, and Premixed Charge Compression Ignition, PCCI) delivers a portion of the fuel to the cylinder via the engine air intake system, which generally leads to lower engine-out particulate. This approach has the potential to lead to increased specific power at the same soot-opacity level.

Normal butanol and isobutanol have been studied as promising biofuels to use for advanced combustion modes due to prolonged ignition delay, low cetane number, and high volatility fuel characteristics. Cheng et al achieved PPC with butanol-diesel blends, showing soot emissions could be reduced by 70%, while nitrogen oxide concentration (NOx) increased at low loads [1]. Leermakers et al showed extremely high soot reduction with moderate butanol-diesel blends compared to diesel only operation using a PPC approach [2]. Zheng et al studied n-butanol, isobutanol, and octanol (2-ethylhexanol) under conventional diesel operation. They found that due to their oxygen content, these fuels produced much lower emissions of soot and carbon monoxide than were achieved with diesel fuel alone [3].

In a previous study by the authors [4] a single-fuel PPC approach was explored with a minority fraction of jet fuel injected into the intake manifold (retro-fit type application), while the majority fraction of jet fuel is delivered directly to the combustion chamber near the start of combustion (SOC). This afforded an opportunity for high levels of port-injected fuel leading to power increases from 5 to 23% as CR increased from 14 to 21.5. PPC allowed for these power increases due to smaller particles (and lower overall number of particles) in the exhaust that influence measured opacity less significantly than larger and more numerous conventional diesel exhaust particulates.

Manente et al used a single-fuel, PPC approach at 11.5 bar IMEP and found that high octane number fuels, like ethanol and gasoline, were able to achieve a reduction of soot and NOx with a lower amount of exhaust gas recirculation (EGR) as compared to diesel [5]. Noehre et al mapped out the operating range for PPC using standard diesel fuel under a range of engine operating conditions including CR, EGR, speed, and boost at loads up to 15 bar gross indicated mean effective pressure (gIMEP) [6]. They concluded that soot emission reduction was a result of good premixing, long ignition delay, and low local combustion temperature due to high EGR. Kalghatgi et al demonstrated the ability to use the soot reduction of PPC for high load extension [7]. Using PPC with gasoline, they reached mean IMEP of 15.95 bar and smoke opacity of 0.33%, while attaining the same level of smoke with diesel fuel, IMEP had to be below 6.5 bar.

While PPC has been studied quite extensively, there is a need for more research especially at high load conditions. This study explores both a single and dual-fuel PPC approach in order to modify the nature of diesel engine combustion via some port delivered fuel. The fuel chosen for direct injection in this study was Navy jet propulsion '5' (JP-5), a fuel commonly used in many Navy and Marine Corps platforms due to its low volatility for safety purposes. From a military standpoint, it is an attractive choice to reduce the significant logistical burden of bringing multiple fuels to the battlefield (Department of Defense (DoD) Directive 4140.43, the 'One Fuel Forward' initiative [8]). However, in this study, isobutanol and iso-octane along with JP-5 for reference, were studied in order to characterize the power gains that might be possible with PPC port injection. In essence, this study seeks to explore whether or not a port injection PPC approach is a viable option for increasing diesel generator power with various fuel combinations.

2. Experimental Setup

Engine Details: The engine used in this study was a Yanmar L100V6-L four-stroke single-cylinder diesel engine rated at 6.7 kW at 3600 RPM with specifications listed in Table 1. This engine is naturally aspirated and utilizes a Bosch Unit Pump System with a nominal mechanical injector opening pressure of 200 bar and nominally fixed start-of-injection (SOI) timing at 14 crank angle degrees (CAD) before top dead center (BTC). To attain the PPC alternative combustion mode, port injection was accomplished with a Cole-Parmer peristaltic pump continuously delivering fuel directly to the intake manifold immediately upstream of the cylinder head. The engine was coupled to a Central Maine Diesel S16W-150 generator controlled by an Inter-Lok control unit using constant speed control. The engine was outfitted with a fast Kistler in-cylinder (6125C) piezoelectric pressure sensor as well as a high-pressure injection line sensor (4067C2000) located near the inlet to the fuel injector. Engine position was measured using a BEI dual channel shaft encoder coupled directly to the crankshaft. An ETAS wide-band lambda meter provided overall air-fuel ratio information. Exhaust opacity was measured with a Wager 6500 opacity meter. Further equipment details can be found in previous papers [9, 10].

Fuels: Navy Jet Propulsion '5' (JP-5) fuel, isobutanol, and iso-octane (2,2,4-Trimethylpentane) were used in this investigation. For all of the fuels tested in this study, a range of physical properties were evaluated in previous work [4, 15]. The general categories of components that are present in JP-5 were determined using gas chromatograph/mass spectrometry (GC/MS) analysis in these same studies. This JP-5 fuel has a moderately higher flash point (less volatile) as compared to conventional commercial jet fuels (aromatic content is 17%). The analytical results of JP-5 chromatography correspond principally to the normal alkanes: decane, undecane, dodecane, tridecane, tetradecane, and pentadecane. Isobutanol and iso-octane were chosen due to their high volatilities and low reactivities so as to increase in-cylinder fuel-air equivalence ratios without a loss of control due to autoigniton before the main cylinder injection event. An oxygenated fuel, isobutanol was chosen in the hope of reducing soot by adding oxygen to the cylinder periphery. Additionally, the high latent heat of evaporation of isobutanol could further help delay autoignition due to evaporative cooling effects. A summary of key fuel properties is shown in Table 3.

Table 1: Engine specifications and details

Engine type	Yanmar L100V6-L
Bore × Stroke	86 mm × 75 mm
Displacement	435 cc (0.435 L)
Compression ratio	20.6
Combustion chamber	Re-entrant bowl, single cylinder
Net power rating	6.7 kW @ 3600 RPM

Table 2: Experimental Matrix

Port Fuel	DI Fuel	Power loads (kW)
(0%)	JP-5 (100%)	4, 5.5
JP-5 (22%)	JP-5 (78%)	5.5
iC8 (22% & 44%)	JP-5 (78% & 56%)	4, 5.5, 7
isoB (20% & 40%)	JP-5 (80% and 60%)	4, 5.5, 7

Table 3: Fuel Properties

Test	Units	JP-5 [4]	iC8 [12,13,14]	isoB [13]
Density 20°C	kg/L	0.8068	0.6839	0.801848
Flash Point	°C	66	-4	30
Viscosity 40°C	cSt	1.389	0.724	3.71
Surface Tension	dyne/cm	28.7	18.7	22.4
Speed of sound 20°C	m/s	1319	1103.2	1205.1
Bulk Modulus 20°C	MPa	1397	842	1165
Cetane number	DCN	46	17-19	8.5
T-10	°C	190	99	110
T-50	°C	208	99	110
T-90	°C	240	99	110

Test Matrix: The Yanmar engine was warmed up using conventional diesel combustion with Navy jet fuel JP-5. Engine load was set at 4 kW, then increased to 5.5 kW and 7 kW. This was accomplished by adding these electric loads to the generator with standard electrical plug and socket. First, a Conv. DI maximum-load operating point was determined so that engine speed was stable at rated RPM. Next, PPC operation was performed with JP-5 at a single port injection level, iso-octane (iC8) at two port injection levels, and isobutanol (isoB) at three injection levels. Nominal fuel fractions and loads for the different operating conditions are shown in Table 2 (previous work by the authors showed these intake port fueling levels to function well in a Waukesha diesel CFR engine without significant compression stroke fuel energy release [4]). For the baseline DI and JP-5 PPC (JP-5 DI and JP-5 port) engine speed could be seen to diminish before the higher load cases, thus these load cases are not shown in Fig. 1-8. DI JP-5 SOI was not adjusted in this study (stock configuration only) to maintain constant CAD50 despite the port injection event. At each stabilized operating point, twenty five seconds of pressure data were collected at a 50 kHz sampling frequency using MATLAB data acquisition software and National Instruments hardware. A single zone heat release analysis was then applied as outlined in references [16, 17].

3. Results and Discussion

In-cylinder instantaneous heat release rates (HRR) for each of the four combustion modes at the mid-load (5.5 kW) case are shown in Fig. 1. Stronger maximum rate of heat release is observed with iC8 PPC and isoB PPC compared to Conv. DI operation. It is believed that the addition of increased intake port fuel displaced air in the cylinder periphery, leading to a different flame behavior as compared to the base. The JP-5 PPC exhibits negitive temperature coefficient (NTC) behavior with early heat release followed by an endothermic period before the main phase heat release. JP-5 has a relatively high cetane number (CN), but coupled with the low temperature heat release, the main injection event occurs in a warmer combustion chamber, leading to shorter IGD, shown in Fig. 3, and less time for premixing of the main in-cylinder injection event, thus the main phase heat release peak (the premixed heat release phase) is lower than the Conv. DI. Because less premixing occurs coupled with lower initial heat release, JP-5 PPC shows primarily a diffusion controlled burn with more of the heat release after TDC. IsoB and iC8 PPC, due to their lower cetane numbers, allow for more time for mixing prior to SOC, thus there is a large initial phase heat release near TDC and less diffusion heat release. Fig. 2 shows the 50% heat release location (CAD50) as a function of engine load. Load is expressed as gross indicated mean effective pressure (gIMEP), which is the integrated work over the closed engine cycle period divided by the displaced volume of the cylinder. For all cases, CAD50 occurs within an acceptable window of 11-14 degrees after top dead center (ATC) for ideal combustion phasing. Notably CAD50 for the PPC cases is similar to that of Conv. DI, thus the comparison of power improvement is valid since combustion phasing is similar. Peak in-cylinder pressure is not shown but remained in an acceptable range of 65 to 70 bar across the range of power loads and combustion modes.

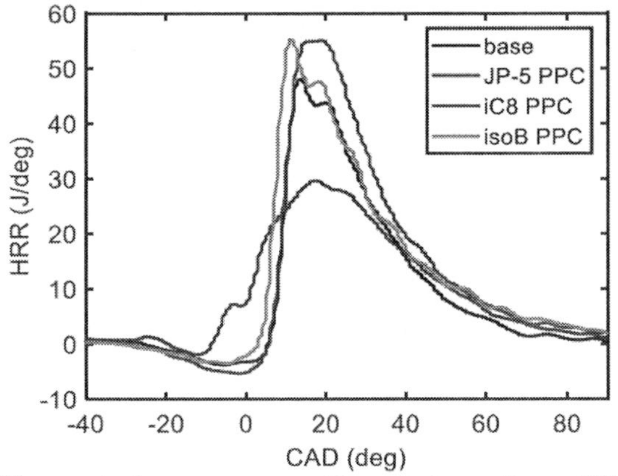

Figure 1: Instantaneous Heat Release Rate (HRR) as a function of piston position (0 CAD is TDC)

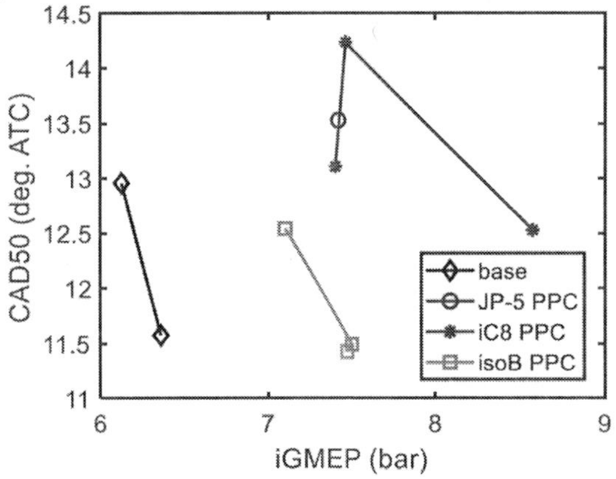

Figure 2: 50% Heat Release Location as a function of engine load and combustion mode

Ignition Delay (IGD), shown in Fig. 3, was defined as the difference between SOI and start of combustion (SOC). SOC was defined as the engine crank angle degrees where 10% of the heat release has occurred (CAD10). Burn Duration was defined as the difference between CAD10 and CAD90 (the engine crank angle degrees where 90% if the heat release has occurred). CAD90 was relatively constant near 28 degrees ATC, except for in the case of JP-5 PPC, where it was 17 degrees ATC. This was again due to the early autoignition nature of premixed JP-5, which is also the cause of the short IGD in Fig. 3. Another paper by the authors outlines this auto-ignition nature of JP-5 at various compression ratios [15]. Otherwise the Conv. DI, iC8 PPC, and isoB PPC have similar IGDs across the load range. At fixed SOI, burn duration strongly correlates with IGD, since CAD50 (shown in Fig. 2) is similar for all cases.

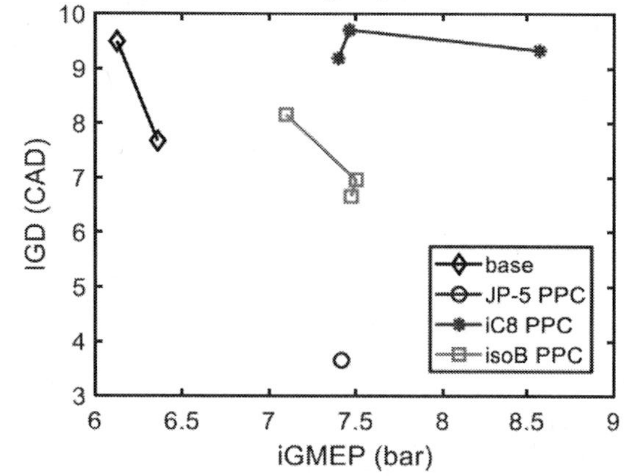

Figure 3: Ignition Delay as a function of engine load and combustion mode

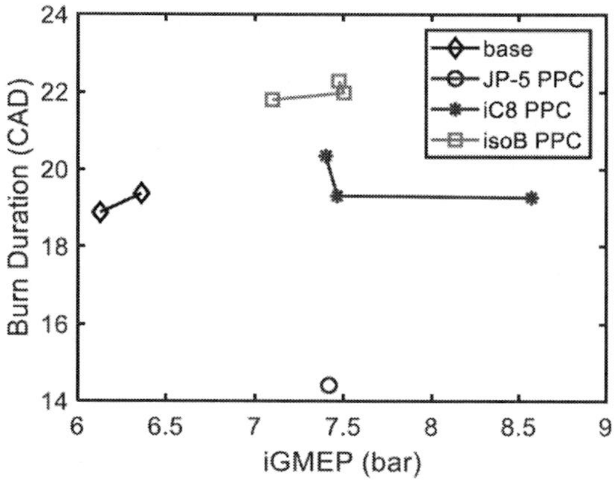

Figure 4: Burn Duration at varying engine loads

Fig. 5 shows the overall fuel-air equivalence ratio (phi, ϕ) as a function of load. Fig. 5 shows that phi increases linearly with load. The baseline case was fuel-load limited by a maximum fueling emission control device. With PPC, phi could be increased with a resulting increase in engine load. In Fig. 6, exhaust opacity is shown as a function of load. Notably, for iC8 PPC and isoB PPC, the exhaust opacity remains near 0.5%, similar to Conv. DI despite the increase in fuel mass and resulting engine load. Thus for these cases, combustion is not soot-limited. The ability of PPC to extend the high load limit of Conv. DI by maintaining low exhaust soot, as measured by opacity, is the critical factor in power improvement for military engines. Load could be extended considerably, up to 8.5 bar for iC8 PPC and 7.5 bar for isoB PPC, compared to 6.5 bar for Conv. DI. Thus, PPC shows the potential to delay the soot limit for this engine. At fixed engine speed, this torque improvement translated into a moderate power improvement over base conventional IDI operation. Likely due to the unique nature of JP-5 combustion shown previously, exhaust opacity was considerably higher at 2%. However, in an emissions context, 2% opacity is still relatively low compared to previous studies by the authors [4, 15].

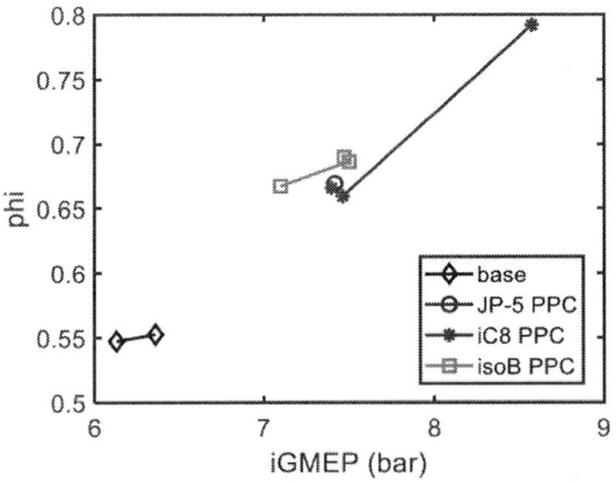

Figure 5: Fuel-Air Equivalence Ratio as a function of engine load

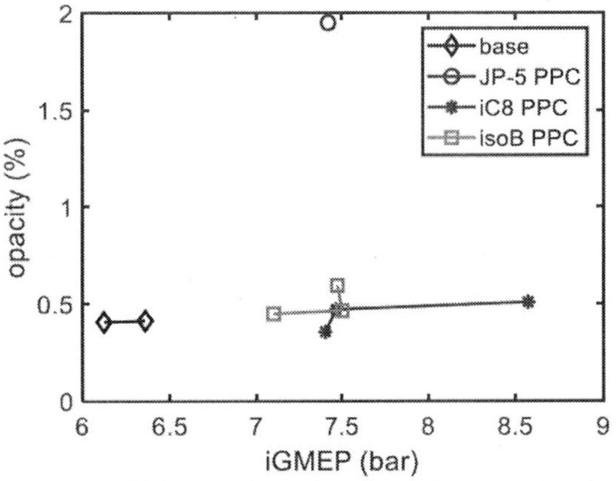

Figure 6: Exhaust Opacity at various engine loads and combustion modes

Fig. 7 and 8 illustrate the key finding of this study, relative torque/power improvements with JP-5 PPC, iC8 PPC, and isoB PPC. At constant engine speed of 3600 RPM, indicated power was directly proportional to iGMEP, shown in Fig. 7. More fuel could be added using PPC leading to increased torque levels without increased soot. Fig. 8 shows the comparison of the highest load case of the PPC mode compared to the highest load case of Conv. DI. The greatest improvements can be seen with iso-octane at nearly a 35% improvement. JP-5 and isobutanol show similar 15% improvements, although soot levels for JP-5 were significantly higher. Overall, iso-octane performed better in increasing power than JP-5, primarily due to the early heat release of JP-5.

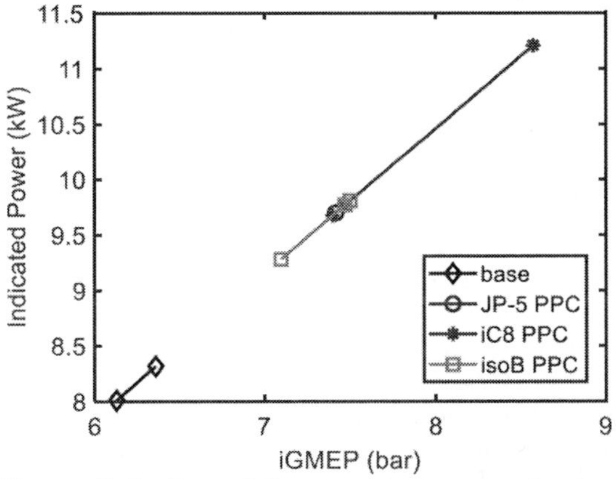

Figure 7: Indicated Power at increasing load, illustrating PPC high load extension at fixed engine speed

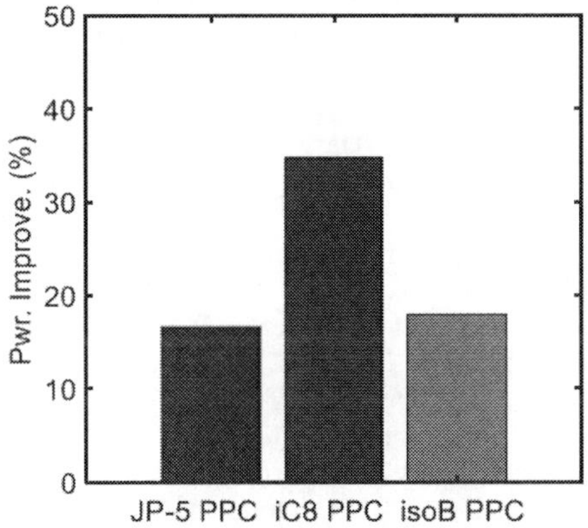

Figure 8: Power Improvement over baseline Conv. DI for each combustion mode at the high load limit

4. Conclusions

This study applied a PPC advanced combustion approach using port fuel injection in a Yanmar L100V6 single cylinder engine-generator in order to extend the diesel engine high load limit based on exhaust soot opacity evaluation. Isobutanol, iso-octane, and jet JP-5 fuel were used for port injection, while jet JP-5 fuel was the only fuel used for in-cylinder injection. It was seen that the Conv. DI soot limit occurred at leaner overall maximum fuel-air equivalence ratio, which afforded a greater opportunity to increase port injected fuel levels. Thus, power levels were then able to increase over conventional diesel combustion. Iso-octane provided the best improvements. Results show that engine Indicated Gross Mean Effective Pressure (iGMEP) from 7.5 to 8.5 bar could be attained with PPC combustion as compared to only 7 bar with conventional operation. This led to power levels of 9.3 kW to 11.3 kW compared to 8.5 kW at conventional operation without a soot penalty. In general, JP-5 PPC led to the lowest power increase due to early combustion initiation. JP-5/iC8 provided for the highest power improvement due to the low reactivity port injected iC8 fuel, which resulted in later start and bulk combustion as compared to conventional operation and the other PPC modes. Iso-butanol PPC provided moderate power improvements.

5. Acknowledgements

This research was funded by the Office of Naval Research with Dr. Maria Medeiros as Program Manager. The authors would also like to thank Mr. Steve Galindo for his assistance in the experimental setup.

6. References

1. Cheng, Xiaobei, Shuai LI, Jin Yang, Shijun Dong, and Zufeng Bao. "Effect of N-Butanol-Diesel Blends on Partially Premixed Combustion and Emission Characteristics in a Light-Duty Engine," SAE Technical Paper 2014-01-2675, 2014, doi: 10.4271/2014-01-2675.
2. Leermakers, C., Bakker, P., Somers, L., de Goey, L. et al., "Butanol-Diesel Blends for Partially Premixed Combustion," *SAE Int. J. Fuels Lubr.* 6(1):217-229, 2013, doi: 10.4271/2013-01-1683.
3. Zheng, Ming, Tie Li, and Xiaoye Han. "Direct Injection of Neat N-Butanol for Enabling Clean Low Temperature Combustion in a Modern Diesel Engine," *Fuel* 142: 28-37, 2013, doi: 10.1016/j.fuel.2014.10.075
4. Walker, M., Kelso, R., Bowes, K., Hamilton L., Luning Prak, D. and Cowart, J., "Partially Premixed Combustion Application for Diesel Power Improvement," ASME ICEF Technical Paper 2017-3544, 2017, Seattle, WA, October, 2017.
5. Manente, V., Johansson, B., and Tunestal, P., "Half load partially premixed combustion, PPC, with high octane number fuels. Gasoline and ethanol compared with diesel," Symposium on International Automotive Technology, SIAT2009-295, 2009.
6. Noehre, C., Andersson, M., Johansson, B., and Hultqvist, A., "Characterization of Partially Premixed Combustion," SAE Technical Paper 2006-01-3412, 2006, doi: 10.4271/2006-01-3412.
7. Kalghatgi, G. T., Risberg, P., and Ångström, H., "Partially Pre-Mixed Auto-Ignition of Gasoline to Attain Low Smoke and Low NOx at High Load in a Compression Ignition Engine and Comparison with a Diesel Fuel," SAE Technical Paper 2010-32-0090, 2010, doi: 10.4271/2010-32-0090.
8. Office of the Secretary of Defense, Directive 4140.43 *Fuel Standardization*, 1988; superseded by Directive 4140.25, *DOD Management Policy for Energy Commodities and Related Services*, 2004.
9. Arment, T., Cowart, J., Caton, P., and Hamilton, L., "The effect of ceramic thermal barrier combustion chamber coatings on the performance and efficiency of a small diesel engine," SAE Technical Paper 841359, 1984, doi: 10.4271/841359
10. Luning Prak, D., Cowart, J., Hamilton, J., Hoang, D., Brown, E., and Trulove, P., "Development of a Surrogate Mixture for Algal-Based Hydrotreated Renewable Diesel," *Energ. Fuel.* 2013. 27 (2) 954-961. doi: 10.1021/ef301879g
11. Luning Prak, D., Luning Prak, P. J., Trulove, P., and Cowart, J. S., "Formulation of Surrogate Mixtures based on physical and chemical analysis of hydrodepolymerized Cellulosic Diesel Fuel," *Energy and Fuels*, 30(9):7331-7341, 2016, doi: 10.1021/acs.energyfuels.6b01114.
12. Cowart, J., and Luning Prak, D., "Reference and Pure Component Fuel Characterization in an Ignition Quality Tester Derived Cetane Rating Unit," under review SAE Technical Paper 2017-01-0774, 2017, Detroit, MI, April 2017.
13. Yanowitz, J., Ratcliff, M. A., McCormick, R. L., Taylor, J. D., Murphy, M J. "Compendium of Experimental Cetane Numbers." National Renewable Energy Lab Technical Report NREL/TP-5400-67585, February 2017.
14. *CRC handbook of Chemistry and Physics*, 97th ed.; Haynes, W. M., Ed.; CRC Press: Boca Raton, FL, Internet version 2017. http://hbcponline.com (accessed Aug 15, 2017).
15. Cowart, J., Bowes, K., Walker, M., Hamilton, L, and Luning Prak, D., "Homogeneous Charge Compression Ignition Operation with Jet Fuel and Water Injection in a Single Cylinder Diesel CFR Engine," ASME ICEF Technical Pape 2017-3653, 2017, Seattle, WA, October, 2017.
16. Chun, K. M., and Heywood, J. B., "Estimating Heat Release and Mass of Mixture Burned from SI Engine Pressure Data," *Combustion Science and Technology* 54(1-6):133-143, 1987, doi: 10.1080/00102208708947049.
17. Gatowski, J. A., Balles, E. N., Chun, K. M., Nelson, F. E., Ekchian, J. A., and Heywood, J. B., "Heat Release Analysis of Engine Pressure Data," SAE Technical Paper 841359, 1984, doi: 10.4271/841359

Spring Technical Meeting
Eastern States Section of the Combustion Institute
March 4-7, 2018
State College, Pennsylvania

Compression-ignition of GDI Sprays in a Constant Volume Combustion Chamber

Libing Wang[1], Fujun Wang[1], William L. Roberts[2] and Tiegang Fang[1,]*

[1]Department of Mechanical and Aerospace Engineering,
North Carolina State University, Raleigh, NC 27695
[2]Clean Combustion Research Center,
King Abdullah University of Science and Technology, Thuwal, Saudi Arabia
[*]Corresponding Author Email: tfang2@ncsu.edu

Abstract: Gasoline compression-ignition combustion has the potential to produce high thermal efficiency and low emissions. Different ambient temperature and oxygen levels can have great effects on the auto-ignition. In this study, experiments were carried out in a constant volume combustion chamber using a hollow-cone gasoline direct injection (GDI) injector, under a given ambient gas density condition. Different oxygen levels were used to simulate different exhaust gas recirculation (EGR) levels. The effect of ambient temperature and EGR levels on the combustion development were investigated and analyzed. The heat release rate and cumulative heat release rate under each condition are compared. High speed videos of the flame development are also recorded and studied. With the increase of ambient temperature, the cumulative heat release rate of gasoline compression ignition decreases, while the peak luminosity and average luminosity increase by a large scale. The auto-ignition delay also decreases with the increase of ambient temperature. It is noted that the peak heat release rate first increases then decreases with the increase of ambient temperature. On the other hand, higher EGR level leads to lower heat release rate, longer flame duration, much longer auto-ignition delay and lower flame luminosity. The results also suggest that a high ambient temperature can help reduce the case-to-case variation of combustion process.

***Keywords:** gasoline compression ignition, heat release rate, flame luminosity, EGR*

1. Introduction

Facing the requirement of energy and environment, the need for higher efficiency and lower pollutant emissions of internal combustion engines has become more and more urgent. With the emission regulations becoming increasingly stringent, much effort has been made to improve the thermal efficiency and reduce pollutant emissions: new combustion concepts have been proposed and investigated, such as homogenous charge compression ignition (HCCI) [1], partially premixed compression ignition (PCCI) [2], reactivity controlled compression ignition (RCCI) [3], and gasoline compression ignition (GCI) [4]. Among these combustion concepts, gasoline compression ignition (GCI) is considered as the most promising [5,6], though HCCI and PCCI also offer good efficiency under uniformly lean condition. It is because the difficulty of auto-ignition controllability in HCCI and the complexity of fuel injection system in PCCI [7] make them not as user friendly as GCI. Furthermore, low temperature combustion (LTC) can be achieved more easily using GCI [8].

During the past decades, low temperature combustion (LTC) has drawn extensive attention of researchers worldwide due to its potential to achieving both high thermal efficiency and low emissions [9]. HCCI, which was initially demonstrated in a 4-stroke engine by Foster and Najt [10], is an important early concept to realize LTC. However, the multi-point spontaneous combustion and very short combustion duration of HCCI generate high pressure-rise rate, which causes serious engine vibration and noise. Therefore, the operating regions of HCCI are limited, and misfire and knock are likely to take place under light and heavy load conditions.

In GCI combustion mode, relatively high resistance to auto-ignition and good volatility of gasoline help increase ignition delay and air-fuel mixing time [11]. Furthermore, with high compression ratio and low heat transfer in GCI engines, thermal efficiency can be as high as diesel engines. Research groups in Tsinghua University, Delphi Powertrain and Oak Ridge National Laboratory achieved over 44% indicated thermal efficiency in GCI engines [12-14]. Meanwhile, most of the current GCI studies were done in a piston engine, and few experiments were conducted in a constant volume combustion chamber, where precise control over the ambient conditions can be realized before injection and combustion.

On the other hand, fuel volatility and chemical reactivity have a great impact on the in-cylinder processes with regard to LTC operability [15]. In order to better study the chemical kinetics, researchers proposed that gasoline surrogates, which is composed of several relatively simple hydrocarbons, could be used to represent real gasoline in scientific studies [16]. Iso-octane has been adopted as a single component gasoline surrogate for a long time. However, the RON (research octane number) of gasoline was lower than iso-octane, causing many ignition characteristic differences. Therefore, this single component surrogate cannot represent the combustion behavior of real gasoline. The primary reference fuel (PRF), which is a mixture of iso-octane and n-heptane, is usually used to match the RON or MON (motor octane number) of gasoline [17]. From the aspect of reproducing the combustion characteristics, PRF has advantage in determining the RON of fuels through adjusting the volume ratio between iso-octane and n-heptane. In this study, PRF95, which is a mixture of 95% isooctane (by volume) and 5% n-heptane (by volume), is used in the experiments to represent gasoline with RON of 95.

In the present study, compression ignition experiments of PRF95 were carried out in a constant volume combustion chamber (CVCC) under a given ambient gas density condition. Different oxygen levels were used to simulate different exhaust gas recirculation (EGR) levels. Both the pressure data and high-speed video of the combustion process were recorded and analyzed in order to study the effect of ambient temperature and EGR levels on the combustion development.

2. Experimental Setup

In this study, the constant volume combustion chamber (CVCC) used is a chamber with six ports and there are different accessory systems (gas inlet, outlet, fuel injector, spark plug, pressure sensor, etc.) connected to the chamber through the ports. The optical access was enabled by two quartz windows mounted on two ports opposite to each other. Other ports were sealed by solid stainless steel plugs with accessory components. The inner diameter of the quartz window is 100 mm, and the inner volume of the chamber is 0.95 L. More details of the chamber can be found in the previous publications [18-19].During the experiment, a premixed combustion of acetylene/air takes place first to generate a high temperature, high pressure environment which leads to the compression ignition condition. Three different gas components (acetylene, O2/N2 (50%/50%) and dry air) were used to formulate appropriate mixtures to obtain the desired ambient temperatures and remaining oxygen concentrations after the premix combustion. For all the experiments, gas density in chamber was kept at 3.5 kg/m^3, similar to the Spray-G density conditions. The fuel used is PRF95, which is a mixture of 95% isooctane (by volume) and 5% n-heptane (by volume). The injector is a gasoline direct injection hollow cone injector. Three ambient oxygen concentrations (10%, 15% and 21%) were selected to simulate different exhaust gas recirculation (EGR) levels. Seven different ambient temperatures (650K, 700K, 750K, 800K, 850K, 900K and 950K) were used to study the effect of ambient temperature on the auto-ignition of PRF95. Fuel injection pressure difference (from fuel rail pressure to the ambient) was kept at 10 MPa (100 bar) by a gasoline common rail fuel system for all cases. Injection duration was kept as 1.349 ms by an external pulse generator and injected fuel mass was 50 mg per injection. The thamber body, fuel injector body and gasoline common rail were all heated and the temperatures were all kept at 90 degrees Celsius. A Kistler 6041B transducer coupled with a Kistler 5004 charge amplifier was utilized to measure the transient pressure. A high-speed camera (Phantom V4.3 from Vision Research Inc.) was used to capture the combustion images. The F number (F#) was set at 5.6. The camera was set at 8113 frames per second (fps) with a resolution of 256×256 pixels. Different ND filters were used to ensure the luminosity signal received by the high-speed camera is within the dynamic range of the imager.

3. Results and Discussion

The effects of ambient temperature and EGR levels on the combustion development are investigated and analyzed in the experiments. The heat release rate and accumulative heat release under each condition are compared. High-speed videos of the flame development are also recorded and studied.

3.1 Heat release analysis

The pressure data inside the chamber was recorded at a sampling rate of 50000 samples per second before and during the injection and combustion process. By using the pressure data we are able to calculate the heat release rate and cumulative heat release inside the constant volume combustion chamber. Then the maximum cumulative heat release and maximum heat release rate can be found, as well as the timing for them after the trigger signal. Figure 1 shows the maximum cumulative heat release for all the three oxygen level and seven ambient temperature conditions. It is seen that in general the maximum cumulative heat release decrease with the increase of the ambient temperature and decrease of oxygen level. This shows under higher ambient temperature the combustion efficiency becomes lower and it is harder for the fuel to chemically burn completely due to soot formation and heat loss. The oxygen level can also play a similar role while the effect is minor. Figure 2 shows the peak heat release rate. It is seen that the peak heat release rate first increase and then decrease with the increase of ambient temperature for all three oxygen levels. Under a low ambient temperature, the oxygen level plays a major role affecting the peak heat release rate. However, when ambient temperature reaches 950 K, all three oxygen levels have quite close peak heat release rates. A similar phenomenon also appears for the peak heat release rate timing, as shown in Figure 3, which means under a higher ambient temperature, which causes the soot, the temperature plays the dominant role in affecting the combustion process, while the oxygen level only influences the maximum cumulative heat release by a small amount.

3.2 Flame luminosity analysis

The high-speed video of each combustion process was taken and analyzed. At each time step, the flame area and the spatially integrated flame luminosity (SIFL) data are calculated. Figure 4(a) shows the peak SIFL data for all the cases. It is seen that the peak SIFL increases strongly with the increase of the ambient temperature, and decreases with the decrease of the oxygen level. The small error bar shows that the case-to-case variation of flame luminosity is very small. Figure 4(b) shows the ignition delay based on the flame (also called flame delay), which is the time of the frame when flame first appears. It is seen that the oxygen level greatly affects the flame delay, and a higher ambient temperature contributes to a shorter flame delay. Flame delay under 10% oxygen is almost twice as other cases. On the other hand, a high ambient temperature can help reduce the case-to-case variation of the combustion development. Figure 5 are typical developments cases for the SIFL and flame area with time under 950 K for all three oxygen levels. It is seen that the peak flame luminosity for the 21% oxygen case appears the earliest and is the highest. From the flame area, it is seen that the 10% oxygen case has the longest flame delay and the slowest flame development, while its flame duration is the longest. Under a lower oxygen level, the flame is weaker and the combustion process takes more time to complete.

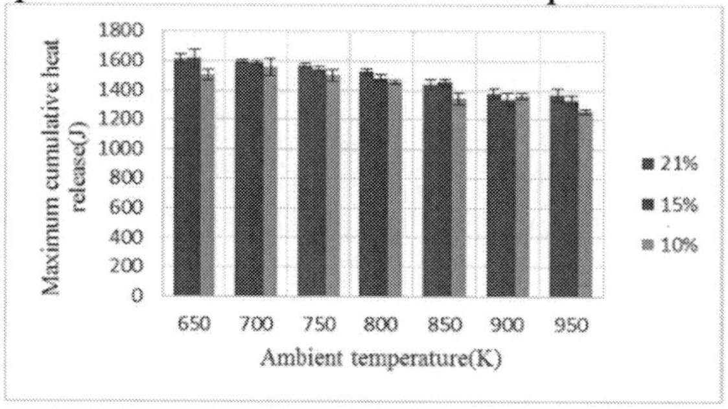

Figure 1: maximum cumulative heat release under different temperature and oxygen level

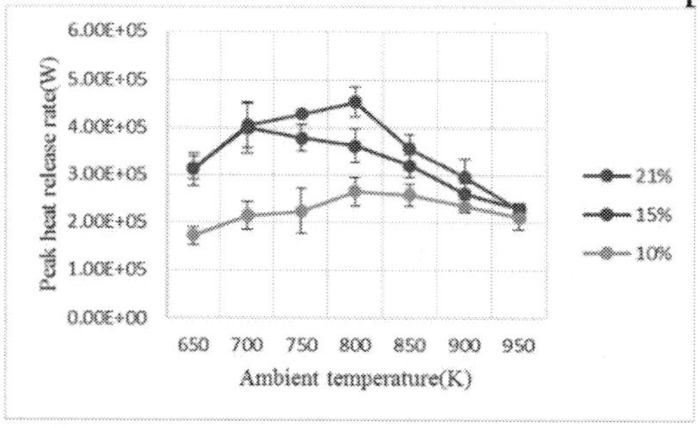

Figure 2: peak heat release rate under different temperature and oxygen level

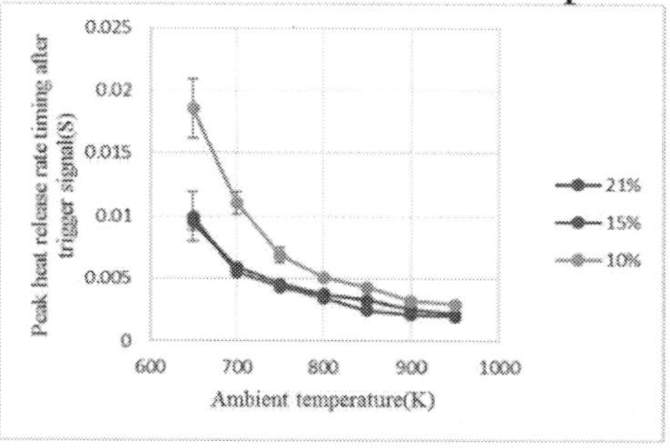

Figure 3: peak heat release rate timing under different temperature and oxygen level

4. Conclusions

In this study, experiments were carried out in a constant volume combustion chamber and the effects of ambient temperatures and oxygen levels on the combustion development were investigated and analyzed. Some key findings are summarized as follows:

1. The cumulative heat release decreases with the increase of ambient temperatures and oxygen levels. With the increase of the ambient temperature, the combustion efficiency decreases. It is noted that the peak heat release rate first increases then decreases with the increase of the ambient temperature for all three oxygen levels, while the peak heat release rate timing always decreases with the increase of the ambient temperature or oxygen level. Under a low ambient temperature,

the oxygen level plays a major role in affecting the peak heat release rate. However, under a high ambient temperature, the effect of oxygen levels on the peak heat release rate becomes minor and the ambient temperature plays a major role instead.

2. The peak SIFL increases strongly with the increase of the ambient temperature, and decreases with the increase of oxygen levels. Higher ambient temperatures contribute to a shorter auto-ignition delay. A lower oxygen level leads to longer flame duration and a much longer auto-ignition delay. Generally speaking, under lower oxygen levels, the flame is weaker and the combustion process takes more time to complete. The results also suggest that a high ambient temperature can help reduce the case-to-case variation of the combustion process.

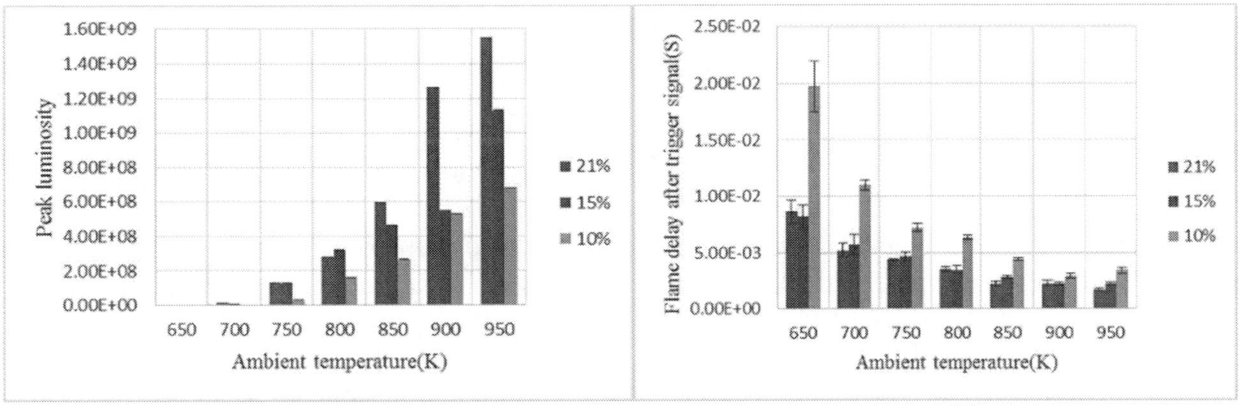

(a) (b)

Figure 4: peak integrated luminosity (a) and flame delay (b)under different temperature and oxygen level

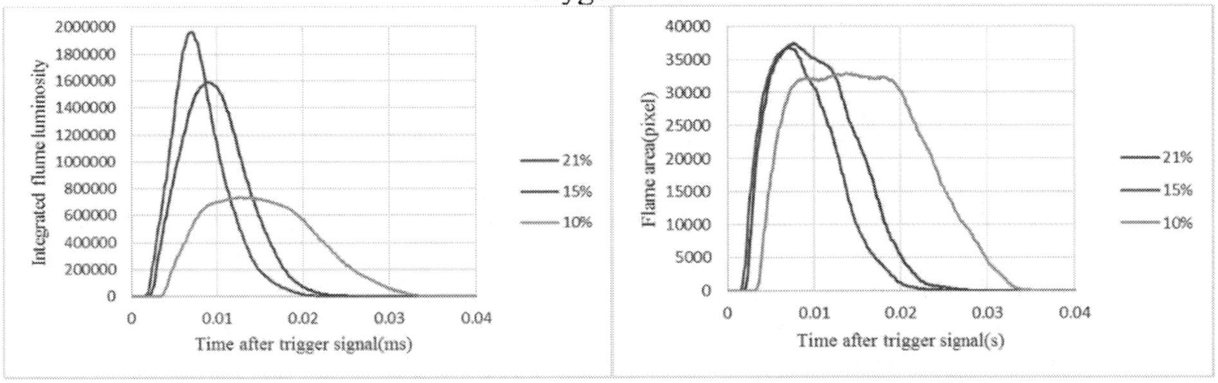

(a) (b)

Figure 5: integrated flame luminosity development (a) and flame area development (b) with time under 950K ambient temperature

5. Acknowledgements

This research was supported in part by the Saudi Aramco Public R&D Center through the Clean Combustion Research Center of the King Abdullah University of Science and Technology under the FUELCOM program.

6. References

[1] Yao M, Zheng Z, Liu H. Progress and recent trends in homogeneous charge compression ignition (HCCI) engines. Prog Energy Combust Sci 2009;35 (5):398–437.

[2] Rohani B, Park SS, Bae C. Effect of injection strategy on smoothness, emissions and soot characteristics of PCCI-conventional diesel mode transition. Appl Therm Eng 2016;93:1033–42.
[3] Reitz RD, Duraisamy G. Review of high efficiency and clean reactivity controlled compression ignition combustion in internal combustion engines. Prog Energy Combust Sci 2014;1:1–60.
[4] Jia G, Wang H, Tong L, Wang X, Zheng Z, Yao M.Experimental and numerical studies on three gasoline surrogates applied in gasoline compression ignition (GCI) mode. Applied Energy 192 (2017) 59–70.
[5] Cracknell R, Ariztegui Cortijo J, Dubois T, Engelen B, Manuelli P, Pellegrini L, et al. Modelling a Gasoline Compression Ignition (GCI) Engine Concept. SAE Int 2014;2014-01–13:1–54. doi:10.4271/2013-01-0911
[6] Rose K.D., Ariztegui J., Cracknell R.F., Dubois T., Hamje H.D.C., Rickeard LP, Heuser D.J. B., Schnorbus T. Exploring a gasoline compression ignition (GCI) engine concept. SAE Int J Fuels Lubr 2014:1–54. doi:10.4271/2013-01-0911
[7] Zhong S, Wyszynski ML, Megaritis A, Yap D, Xu H. Experimental Investigation into HCCI Combustion Using Gasoline and Diesel Blended Fuels. SAE Int J Engines 2005. doi:10.4271/2005-01-3733.
[8] Yanuandri Putrasari, Ock Taeck Lim. Performance and Emission of Gasoline Compression Ignition Engine Fueled with 5 and 20% Gasoline-Biodiesel Blends under Single Injection Strategy. / Energy Procedia 105 (2017) 1743 – 1750
[9] Wei H, Hua J, Pan M, Feng D, Zhou L, Pan J. Experimental investigation on knocking combustion characteristics of gasoline compression ignition engine. Energy 143 (2018) 624-633.
[10] Najt PM, Foster DE. Compression-ignited homogeneous charge combustion. 1983. SAE Technical Paper 0148-7191.
[11] Wang B, Wang Z, Shuai S, Yang H, Wang J. Combustion and emission characteristics of Multiple Premixed Compression Ignition (MPCI) fuelled with naphtha and gasoline in wide load range. Energy Convers Manag 2014;88: 79-87.
[12] Wang B, Wang Z, Shuai S, Xu H. Combustion and emission characteristics of multiple premixed compression ignition (MPCI) mode fuelled with different low octane gasolines. Appl Energy 2015;160:769-76.
[13] Dempsey AB, Curran S, Wagner R, Cannella W. Effect of premixed fuel preparation for partially premixed combustion with a low octane gasoline on a light-duty multicylinder compression ignition engine. J Eng Gas Turbines Power 2015;137:111506.
[14] Sellnau MC, Sinnamon J, Hoyer K, Kim J, Cavotta M, Husted H. Part-load operation of gasoline direct-injection compression ignition (GDCI) engine, vol. 1; 2013.
[15] Zheng M, Li T, Han X. Direct injection of neat n-butanol for enabling clean low temperature combustion in a modern diesel engine. Fuel 2015;142:28–37
[16] Dryer FL. Chemical kinetic and combustion characteristics of transportation fuels. Proc Combust Inst 2015;35(1):117–44.
[17] Wang H, Yao MF, Reitz RD. Development of a reduced primary reference fuel (PRF) mechanism for IC engine combustion simulations. Energy Fuels 2013;27 (12):7843–53.
[18] Jing, W., Z. Wu, W.L. Roberts, and T. Fang. "Spray combustion of biomass-based renewable diesel fuel using multiple injection strategy in a CVCC." Fuel 181 (2016): 718-728.
[19] Jing, W., Z. Wu, W.L. Roberts, and T. Fang. "Effects of Fuel Quantity on Soot Formation Process for Biomass-Based Renewable Diesel Fuel Combustion." J. Eng. Gas Turbines Power 139, no. 10 (2017): 102803.

Spring Technical Meeting
Eastern States Section of the Combustion Institute
March 4-7, 2018
State College, Pennsylvania

Ozonolysis activated autoignition in non-premixed coflow

Wenting Sun, Xiang Gao*

School of Aerospace Engineering, Georgia Institute of Technology, Atlanta, GA 30332
**Corresponding Author Email: wenting.sun@aerospace.gatech.edu*

Abstract: While the majority of fuel oxidation reactions require high-temperature conditions, those involving ozone (O3) and unsaturated hydrocarbons, e.g., ethylene (C_2H_4), occur very rapidly at low temperature, even at room temperature. The subsequent radical production and heat release could induce autoignition in a fuel/air mixture. The autoignition events activated by ozonolysis reactions were investigated in a non-premixed coflow burner using C_2H_4 as the fuel. Images of chemiluminescence showed that one of the major ozonolysis products, formaldehyde (CH_2O), accumulated before autoignition occurs, which confirmed that ozonolysis reactions initiated autoignition. As expected, the autoignition time, τign, decreased with the increase of O_3 concentration in the mixture. Overall, τign decreased as the jet-exit Reynolds number, Re, increased. At lower Re, this trend was explained by the inverse relation between the mixing timescale and flow velocity (based on a non-reacting 3D model). At higher Re, this trend was attributed to the mixing promoted via turbulence. Owing to the ozonolysis reactions, fuel can be "preprocessed" and therefore the reactivity of the fuel-oxidizer mixture increased. As a result, the flame propagation speed was increased significantly. This enhancement, owing to "in situ fuel preprocessing," is an important new flame stabilization and propagation mechanism.
Keywords: ozone assisted combustion, ozonolysis reaction

1. Introduction

Combustion plays a vital role in transportation and power generation. However, concerns of efficiency, emission, and operation at extreme conditions highlight the needs to enhance combustion processes. If the rate-limiting chemical pathways can be enhanced, the ignition and combustion process could be dramatically accelerated. Following this idea, addition of ozone (O_3) is proposed as a *combustion enhancer* [1]. O_3 is one of the strongest oxidizers. It can be efficiently and economically produced *in situ* at high pressure conditions and transported to the desired region from an injection location to modify fuel oxidization, therefore controlling the combustion process.

Previously, O_3 addition has shown its potential to accelerate ignition and control the ignition timing in engines [2-5], enhance the flame propagation [2, 6-9], and flame stability [10, 11]. The conventional understanding of these enhancements is that, due to its long lifetime, O_3 can be transported to the preheat zone and then decompose to release O atoms through the reaction $O_3+M \rightarrow O_2+O+M$ (where M represents a third-body collider). Fuel oxidation can be accelerated by reactions between O atoms and fuel [2, 6]. In previous studies, no direct O_3/fuel reactions were considered because the fuels used were saturated hydrocarbons or syngas, for which the reactions with O_3 are slow at room temperature. However, for unsaturated hydrocarbons, such as C_2H_4, reactions with O_3 are rapid even at room temperature. This is the so-called ozonolysis reaction. The ozonolysis reactions have been widely studied in the atmospheric chemistry community [12, 13] but typically are not considered in the combustion community.

Our previous work [9] showed that heat released during the mixing of O_3 and C_2H_4 may be lost in a non-adiabatic environment, resulting in decreased reactivity. A natural follow-up question is that, if the heat is preserved, could the heat released from ozonolysis reactions and radical/fuel fragment production be utilized to increase reactivity or even initiate autoignition? Such scenarios are relevant to practical applications such as hot-gas recirculation. Previously, this has been investigated using a jet in a hot coflow [19-28], wherein the temperature of the coflow is raised above the autoignition temperature to initiate combustion.

In the present work, ozonolysis reactions are studied as an alternative way to create an autoigniting environment, which avoids the difficulties associated with heating a large coflow to high temperature. Furthermore, the O_3 concentration can be easily and rapidly controlled without changing the equivalence ratio and/or adiabatic flame temperature. The autoignition events activated by ozonolysis reactions are investigated using a fuel-jet in an oxidizing coflow. This configuration serves as a platform for the investigation of flame stabilization in an autoigniting environment.

2. Methods / Experimental

The experiments were conducted at room conditions (298 K and 101 kPa). The burner is illustrated in Fig. 1(a). Here, fuel (C_2H_4) flows through a central jet and is surrounded by an oxidizing coflow (composed of $O_2/O_3/N_2$). The center tube has an outer diameter (OD) of 0.317 cm, and inner diameter (ID) of 0.106 cm. To remove the dilution effects from ambient air, an outer quartz tube is used, which has a length of 30 cm and an ID of 0.787 cm. N_2, O_2 (99.994% purity, Airgas), and C_2H_4 (99.999% purity, Airgas) were sent through a bank of calibrated flow controllers. O_2 was sent through three O_3 generators in parallel (Ozone Solutions, TG-40) to be partially ozonized. An O_3 monitor (Teledyne Instruments, Model 454) was connected downstream of the O_3 generator to measure O_3 concentration. The fluctuation of O_3 concentration estimated using the O_3 monitor was within 2.5% of the measured value. N_2, O_2, and O_3 were mixed first (as the oxidizer stream) and then directed to the coflow. A high-frame-rate camera (NAC, Memrecam GX-3) with an intensifier (Lambert Instruments, HiCATT) was used to capture the chemiluminescence from the flames. An optical narrow bandpass filter centered at 430 nm (±10 nm) was used to capture the chemiluminescence from CH* and CH_2O*.

To quantify the flow conditions, a jet-exit Reynolds number *Re* is defined

$$Re = \frac{\left(U_{\text{jet}} - U_{\text{coflow}}\right) D_{\text{nozzle exit}}}{\nu} \qquad (1)$$

where $D_{\text{nozzle exit}}$ is the average of ID and OD of the fuel tube, U_{jet} and U_{coflow} are velocities of the center fuel jet and coflow, respectively, and ν is kinematic viscosity of the mixture. In addition, a nominal flow velocity, U_0, is defined as the total flow rate (fuel and oxidizer) divided by the cross-sectional area of the quartz tube (based on ID).

The O_3 generator applies a high-voltage pulse (5-30 kHz) to convert O_2 to O_3. To minimize the fluctuation of O_3 concentration, a mixing buffer is added between the O_3 generators and the O_3 monitor. This stainless-steel buffer has a volume of approximately 0.3 L and is filled with small glass beads to allow O_2/O_3 to be well mixed before reaching the burner. During the experiment, initially the oxidizer and O_3 generators were turned on, but the fuel stream was turned off. When the generation of O_3 reached steady state, fuel was turned on, and autoignition may occur, which was recorded by the high-frame-rate camera. It was observed that certain products from ozonolysis reactions (and its subsequent reactions) are liquid at room temperature and collect on the wall of the quartz tube. Therefore, after each experiment, the quartz tube was replaced with a clean tube.

Please note that the full products and detailed reaction pathways of ozonolysis reactions are still uncertain. In the present work, the following ozonolysis reactions are based on previous work [29, 30].

$$C_2H_4 + O_3 = CH_2O + OH + HCO \quad (2a)$$
$$C_2H_4 + O_3 = CH_2O + 2H + CO_2 \quad (2b)$$

$$C_2H_4 + O_3 = CH_2O + H_2 + CO_2 \quad (2c)$$
$$C_2H_4 + O_3 = CH_2O + H_2O + CO \quad (2d)$$

Due to the lack of detailed kinetic information, the branching ratio is assumed to be 1 for each of these pathways. That is, they are assumed to have the same Arrhenius rates

$$k = Ae^{-E/(RT)} \quad (3)$$

where R is the universal gas constant, T is the temperature, $A = 1.81\times10^9$ mole/cm-s-K is the pre-exponential factor, and $E = 5227$ cal/mole is the activation energy, based on Ref. [16]. These reactions are combined with USC Mech II [31] and a sub-mechanism for other O_3 reactions [8, 9] for the kinetics model. Numerical simulations were conducted using CHEMKIN [32] and FLUENT [33] software packages. The geometry of the 3D model in FLUENT simulation is illustrated in Fig. 1(b), sharing the same dimensions with the experimental setup.

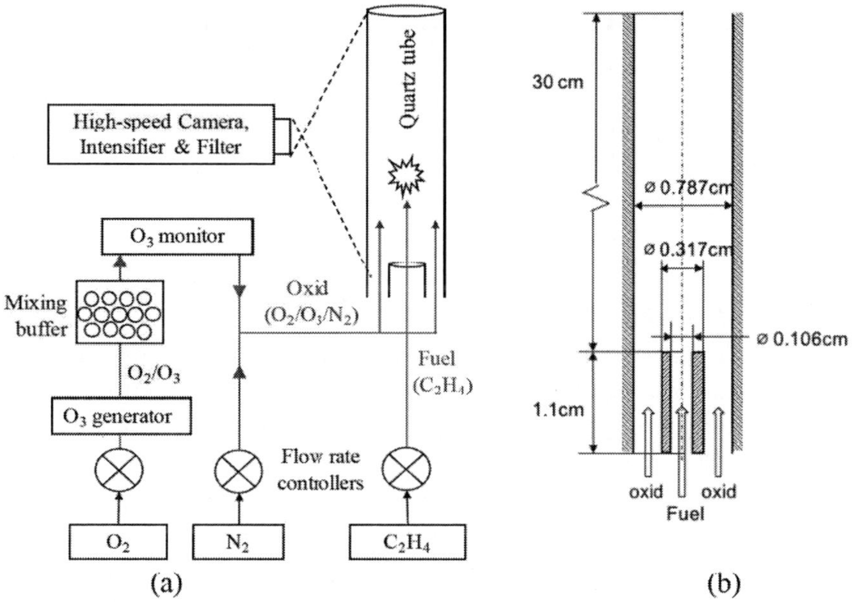

Fig. 1. (a) Schematic of the experimental setup, and (b) model geometry for 3D simulation

3. Results and Discussion

In the present work, the experimental conditions were set in such a way that autoignition can occur easily at room temperature and pressure. The ratio of N and O atoms (including these in O_2 and O_3) in the oxidizer was set as N/O=0.28. The overall equivalence ratio, ϕ, was set to 0.4.

For the study presented in this section, 7.7% of O_2 was converted to O_3, which corresponded to 3.7% of O_3 mole fraction in total reactants. U_0 was 2.37 m/s, which corresponded to average axial velocities of U_{jet} = 12.3 m/s and U_{coflow} = 2.6 m/s. The observed autoignition event from the intensified high-frame-rate camera is shown in Fig. 2. Autoignition was initiated by ozonolysis reaction between C_2H_4 and O_3. A simple demonstration is that if the fuel is CH_4, whose ozonolysis reactions are negligible at room temperature, no autoignition at such condition is observed regardless of O_3 concentration.

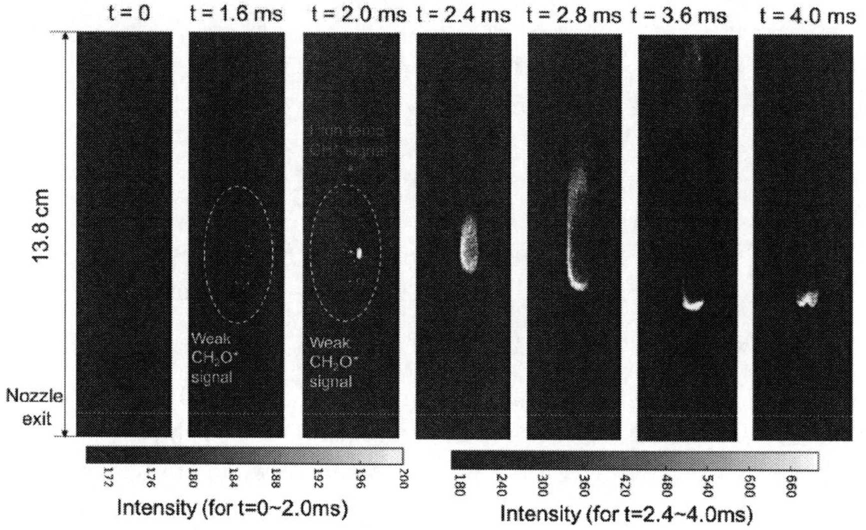

Fig. 2. Filtered chemiluminescence images at 430 nm of ozonolysis activated autoignition, *Re*=1317

To further investigate the role of ozonolysis reactions, simulation was conducted using the adiabatic steady laminar plug flow reactor model (PLUG) [34]. It is predicted that autoignition occurs at 5.5 cm downstream. This autoignition is a two-stage process. First stage ignition is activated by the ozonolysis reactions. Only when the temperature becomes sufficiently high (>800 K), the decomposition of O_3 (O_3+M=O_2+O+M) and other O_3 reactions become important. This confirmed that the autoignition was induced by ozonolysis reactions. In fact, if ozonolysis reactions are removed from the kinetic model, none of these reactions could occur. As a product of the ozonolysis reactions, CH_2O is produced prior to the significant temperature rise. The peak mole fraction of CH_2O is approximately 1%.

Chemiluminescence is used as a marker to indicate that the reactants were reacting upstream the location where the autoignition kernel was formed. As shown in Fig. 2, before the autoignition kernel was formed, a region with weak chemiluminescence was observed. Similar observations were made when the reactants were highly diluted using N_2. At such conditions, no autoignition could occur and no flame could exist. A temperature rise of approximately 100 K was measured at the exit of the quartz tube owing to ozonolysis reactions. Chemiluminescence with the 430-nm bandpass filter was observed at such condition using a long exposure time (2 ms). Besides the chemiluminescence and temperature rise, another evidence for the existence of ozonolysis at such conditions is that, CH_2O as a major ozonolysis product was also confirmed by Gas Chromatography (Inficon Fusion μGC) sampling C.

Further investigations are necessary to identify the emitters of the chemiluminescence observed at 430 nm, but one very possible emitter is CH_2O^*. CH_2O^* emission covers a wide band (centered at 420~430 nm [35]). Typically, CH_2O^* chemiluminescence is not used to study conventional high temperature (>1000 K) flame because at such condition CH* chemiluminescence (which is centered at 431.5 nm [36, 37]) from combustion is significantly stronger than CH_2O^* chemiluminescence. However, it is possible that, at low temperature, CH_2O^* becomes the major chemiluminescence emitter. In an early work, Sheinson and Williams [35] obtained a resolved chemiluminescence spectra from cool flames (473-673 K) and concluded that the primary emission source was CH_2O^*. Other possible emission sources such as CO^*, CH^*, C_2^*, OH^*, and HCO^* were shown to be not significant at such condition. Particularly, CH^* emissions only become comparable with CH_2O^* signal at higher temperature (~1073 K). Therefore, CH_2O^* can be used to identify cool flame owing to low temperature chemistry, such as demonstrated in Ref. [38]. Chemiluminescence from CH_2O^* was also observed at room temperature (no flame) from ozonolysis reactions of C_3H_4 (propadiene) [39]. A simulation conducted using a PLUG flow with a sub-mechanism of OH^*, CH^*, and CO_2^* [40, 41] shows that at the experimental interrogated condition the mole fraction of CH^* is less than 10^{-18} so its effect is negligible before hot ignition.

The above discussion implies that autoignition occurred in such a sequence: firstly, oxidizer and fuel were mixed in the shear layer; then ozonolysis started to release heat (as confirmed by temperature measurement) and produce species such as CH_2O (as confirmed using GC) and some chemiluminescence emitters (potentially CH_2O^* as discussed above); and finally autoignition was induced by the combined effect of reactive species and heat production from ozonolysis reactions.

4. Conclusions

In this work, ozonolysis between O_3 and C_2H_4, leading to autoignition, was studied in a non-premixed co-flow jet flame. Images of chemiluminescence and GC sampling showed that one major product of ozonolysis reactions, CH_2O, accumulated before autoignition occurred, and further reactions caused autoignition. The autoignition times, τ_{ign}, were investigated. As expected, τ_{ign} decreased with increasing O_3 concentration in the coflow. Overall, τ_{ign} decreased with the increase of jet exit Reynolds number *Re*. At low *Re*, mixing was enhanced as the flow velocity increased, therefore accelerating the ozonolysis process and subsequent autoignition because of faster mixing. At higher *Re*, this trend was attributed to turbulent mixing. Owing to ozonolysis reactions, fuel can be "preprocessed" with significantly increased reactivity. It was observed that the flame propagation speed could be increased by several times. This enhancement, due to "fuel preprocessing," could be an important enabling mechanism for flame/combustion stabilization for practical combustion systems.

5. Acknowledgements

This work is supported by the Air Force Office of Scientific Research (FA9550-16-1-0441).

6. References

[1] Y. Ju, W. Sun, Plasma assisted combustion: Dynamics and chemistry, Progress in Energy and Combustion Science, (2014).

[2] T. Nomaguchi, S. Koda. Spark ignition of methane and methanol in ozonized air. Symposium (International) on Combustion; 1989. p. 1677-1682.

[3] T. Tachibana, K. Hirata, H. Nishida, H. Osada, Effect of ozone on combustion of compression ignition engines, Combustion and Flame 85 (1991) 515-519.

[4] H. Nishida, T. Tachibana, Homogeneous charge compression ignition of natural gas/air mixture with ozone addition, Journal of propulsion and power 22 (2006) 151-157.

[5] F. Foucher, P. Higelin, C. Mounaïm-Rousselle, P. Dagaut, Influence of ozone on the combustion of n-heptane in a HCCI engine, Proceedings of the Combustion Institute 34 (2013) 3005-3012.

[6] T. Ombrello, S.H. Won, Y. Ju, S. Williams, Flame propagation enhancement by plasma excitation of oxygen. Part I: Effects of O3, Combustion and Flame 157 (2010) 1906-1915.

[7] F. Halter, P. Higelin, P. Dagaut, Experimental and detailed kinetic modeling study of the effect of ozone on the combustion of methane, Energy & Fuels 25 (2011) 2909-2916.

[8] Z.H. Wang, L. Yang, B. Li, Z.S. Li, Z.W. Sun, M. Aldén, K.F. Cen, A.A. Konnov, Investigation of combustion enhancement by ozone additive in CH4/air flames using direct laminar burning velocity measurements and kinetic simulations, Combustion and Flame 159 (2012) 120-129.
[9] X. Gao, Y. Zhang, S. Adusumilli, J. Seitzman, W. Sun, T. Ombrello, C. Carter, The effect of ozone addition on laminar flame speed, Combustion and Flame 162 (2015) 3914-3924.
[10] T.M. Vu, S.H. Won, T. Ombrello, M.S. Cha, Stability enhancement of ozone-assisted laminar premixed Bunsen flames in nitrogen co-flow, Combustion and Flame 161 (2014) 917–926.
[11] Y. Zhang, M. Zhu, Z. Zhang, R. Shang, D. Zhang, Ozone effect on the flammability limit and near-limit combustion of syngas/air flames with N 2, CO 2, and H 2 O dilutions, Fuel 186 (2016) 414-421.
[12] R. Atkinson, D. Baulch, R. Cox, J. Crowley, R. Hampson, R. Hynes, M. Jenkin, M. Rossi, J. Troe, Evaluated kinetic and photochemical data for atmospheric chemistry: Part 1? gas phase reactions of O x, HO x, NO x and SO x species, Atmospheric Chemistry and Physics Discussions 3 (2003) 6179-6699.
[13] R. Atkinson, D. Baulch, R. Cox, J. Crowley, R. Hampson Jr, J. Kerr, M. Rossi, J. Troe, Summary of evaluated kinetic and photochemical data for atmospheric chemistry, (2001) 1-56.
[14] C.C. Schubert, S. Schubert, R.N. Pease, The Oxidation of Lower Paraffin Hydrocarbons. I. Room Temperature Reaction of Methane, Propane, n-Butane and Isobutane with Ozonized Oxygen1, Journal of the American Chemical Society 78 (1956) 2044-2048.
[15] R.J. Morrissey, C. Schubert, The reactions of ozone with propane and ethane, Combustion and Flame 7 (1963) 263-268.
[16] W.B. DeMore, S.P. Sander, D.M. Golden, R.F. Hampson, M.J. Kurylo, C.J. Howard, A.R. Ravishankara, C.J. Kolb, M.J. Molina, Chemical Kinetic and Photochemical Data for Use in Stratospheric Modeling: Evaluation No. 11 of the NASA Panel for Data Evaluation, JPL Publication 94-26, (1994).
[17] R. Atkinson, D.L. Baulch, R.A. Cox, J.N. Crowley, R.F. Hampson, Jr., J.A. Kerr, M.J. Rossi, J. Troe, Summary of Evaluated Kinetic and Photochemical Data for Atmospheric Chemistry, IUPAC Subcommittee on Gas Kinetic Data Evaluation for Atmospheric Chemistry Web Version December 2001 (2001).
[18] Y. Shi, Y. Xu, L. Jia, Arrhenius parameters for the gas - phase reactions of O3 with two butenes and two methyl - substituted butenes over the temperature range of 295–351 K, International Journal of Chemical Kinetics 43 (2011) 238-246.
[19] B.B. Dally, A. Karpetis, R. Barlow, Structure of turbulent non-premixed jet flames in a diluted hot coflow, Proceedings of the combustion institute 29 (2002) 1147-1154.
[20] R. Cabra, T. Myhrvold, J. Chen, R. Dibble, A. Karpetis, R. Barlow, Simultaneous laser raman-rayleigh-lif measurements and numerical modeling results of a lifted turbulent H 2/N 2 jet flame in a vitiated coflow, Proceedings of the Combustion Institute 29 (2002) 1881-1888.
[21] C. Markides, E. Mastorakos, An experimental study of hydrogen autoignition in a turbulent co-flow of heated air, Proceedings of The Combustion Institute 30 (2005) 883-891.
[22] E. Oldenhof, M. Tummers, E. Van Veen, D. Roekaerts, Ignition kernel formation and lift-off behaviour of jet-in-hot-coflow flames, Combustion and Flame 157 (2010) 1167-1178.
[23] E. Oldenhof, M. Tummers, E. Van Veen, D. Roekaerts, Role of entrainment in the stabilisation of jet-in-hot-coflow flames, Combustion and Flame 158 (2011) 1553-1563.
[24] T. Lu, C.S. Yoo, J. Chen, C.K. Law, Three-dimensional direct numerical simulation of a turbulent lifted hydrogen jet flame in heated coflow: a chemical explosive mode analysis, Journal of Fluid Mechanics 652 (2010) 45-64.
[25] B. Choi, K. Kim, S. Chung, Autoignited laminar lifted flames of propane in coflow jets with tribrachial edge and mild combustion, Combustion and Flame 156 (2009) 396-404.
[26] B. Choi, S.-H. Chung, Autoignited laminar lifted flames of methane, ethylene, ethane, and n-butane jets in coflow air with elevated temperature, Combustion and Flame 157 (2010) 2348-2356.
[27] S. Deng, P. Zhao, M.E. Mueller, C.K. Law, Autoignition-affected stabilization of laminar nonpremixed DME/air coflow flames, Combustion and Flame 162 (2015) 3437-3445.
[28] S. Deng, P. Zhao, M.E. Mueller, C.K. Law, Stabilization of laminar nonpremixed DME/air coflow flames at elevated temperatures and pressures, Combustion and Flame 162 (2015) 4471-4478.

[29] J.D. Fenske, A.S. Hasson, S.E. Paulson, K.T. Kuwata, A. Ho, K. Houk, The pressure dependence of the OH radical yield from ozone-alkene reactions, The Journal of Physical Chemistry A 104 (2000) 7821-7833.

[30] S.E. Paulson, J.D. Fenske, A.D. Sen, T.W. Callahan, A novel small-ratio relative-rate technique for measuring OH formation yields from the reactions of O3 with alkenes in the gas phase, and its application to the reactions of ethene and propene, The Journal of Physical Chemistry A 103 (1999) 2050-2059.

[31] H. Wang, X. You, A.V. Joshi, S.G. Davis, A. Laskin, F. Egolfopoulos, C.K. Law, USC Mech Version II. High-Temperature Combustion Reaction Model of H2/CO/C1-C4 Compounds, (2007).

[32] CHEMKIN 10131, Reaction Design: San Diego, (2013).

[33] FLUENT release 17.1, ANSYS, Inc., (2016).

[34] R.S. Larson, PLUG: A Fortran program for the analysis of plug flow reactors with gas-phase and surface chemistry, Sandia Labs., Livermore, CA (United States), 1996.

[35] R.S. Sheinson, F.W. Williams, Chemiluminescence spectra from cool and blue flames: electronically excited formaldehyde, Combustion and Flame 21 (1973) 221-230.

[36] J. Kojima, Y. Ikeda, T. Nakajima, Spatially resolved measurement of OH*, CH*, and C2* chemiluminescence in the reaction zone of laminar methane/air premixed flames, Proceedings of the Combustion institute 28 (2000) 1757-1764.

[37] A.G. Gaydon, The Spectroscopy of Flames, Chapman and Hall Ltd1957.

[38] S. Deng, P. Zhao, D. Zhu, C.K. Law, NTC-affected ignition and low-temperature flames in nonpremixed DME/air counterflow, Combustion and Flame 161 (2014) 1993-1997.

[39] S. Toby, Chemiluminescence in the gas-phase reaction between ozone and allene, Journal of Luminescence 8 (1973) 94-96.

[40] T. Guiberti, D. Durox, T. Schuller, Flame chemiluminescence from CO2-and N2-diluted laminar CH4/air premixed flames, Combustion and Flame 181 (2017) 110-122.

[41] J. Luque, J. Jeffries, G. Smith, D. Crosley, K. Walsh, M. Long, M. Smooke, CH (AX) and OH (AX) optical emission in an axisymmetric laminar diffusion flame, Combustion and Flame 122 (2000) 172-175.

[42] A.E. Lutz, R.J. Kee, J.A. Miller, SENKIN: A FORTRAN program for predicting homogeneous gas phase chemical kinetics with sensitivity analysis, Sandia National Labs., Livermore, CA (USA), 1988.

AUTHOR INDEX

AUTHOR INDEX

AUTHOR INDEX

AUTHOR INDEX